DIGITAL SIGNAL PROCESSING

A Computer-Based Approach

McGraw-Hill Series in Electrical and Computer Engineering

Senior Consulting Editor

Stephen W. Director, *University of Michigan, Ann Arbor*

Circuits and Systems
Communications and Signal Processing
Computer Engineering
Control Theory and Robotics
Electromagnetics
Electronics and VLSI Circuits
Introductory
Power
Antennas, Microwaves, and Radar

Previous Consulting Editors

Ronald N. Bracewell, Colin Cherry, James F. Gibbons,
Willis W. Harman, Hubert Heffner, Edward W. Harold,
John C. Linvill, Simon Ramo, Ronald A. Rohrer,
Anthony E. Siegman, Charles Susskind, Frederick E. Terman,
John C. Truxal, Ernst Weber, and John R. Whinnery

DIGITAL SIGNAL PROCESSING
A Computer-Based Approach

Second Edition

Sanjit K. Mitra

Department of Electrical and Computer Engineering
University of California, Santa Barbara

McGraw-Hill
Irwin

Boston   Burr Ridge, IL   Dubuque, IA   Madison, WI
New York   San Francisco   St. Louis
Bangkok   Bogotá   Caracas   Lisbon   London   Madrid   Mexico City
Milan   New Delhi   Seoul   Singapore   Sydney   Taipei   Toronto

McGraw-Hill

A Division of The McGraw·Hill Companies

DIGITAL SIGNAL PROCESSING
Published by McGraw-Hill, an imprint of The McGraw-Hill Companies, Inc. 1221 Avenue of the Americas, New York, NY, 10020. Copyright © 2001, 1998, by The McGraw-Hill Companies, Inc. All rights reserved. No part of this publication may be reproduced or distributed in any form or by any means, or stored in a database or retrieval system, without the prior written consent of The McGraw-Hill Companies, Inc., including, but not limited to, in any network or other electronic storage or transmission, or broadcast for distance learning. Some ancillaries, including electronic and print components, may not be available to customers outside the United States.

This book is printed on acid-free paper.

2 3 4 5 6 7 8 9 0 QPF/QPF 0 9 8 7 6 5 4 3 2

ISBN   0-07-252261-5

Publisher: *Thomas Casson*
Executive editor: *Elizabeth A. Jones*
Senior sponsoring editor: *Catherine Fields Schultz*
Development editor: *Michelle L. Flomenhoft*
Senior marketing manager: *John Wannemacher*
Senior project manager: *Pat Frederickson*
Production supervisor: *Debra Sylvester*
Coordinator freelance design: *Pam Verros*
Supplement coordinator: *Rose Range*
Media technology producer: *Christopher Styles*
Cover illustration: *Goutam K. Mitra and Alicia Rodriguez*
Typeface: *10/12 Times Roman*
Printer: *Quebecor World Fairfield, PA*

Library of Congress Control Number: 2001092070

www.mhhe.com

About the Author

Sanjit K. Mitra received his M.S. and Ph.D. in electrical engineering from the University of California, Berkeley, and an Honorary Doctorate of Technology from Tampere University of Technology in Finland. After holding the position of assistant professor at Cornell University until 1965 and working at AT&T Bell Laboratories, Holmdel, New Jersey, until 1967, he joined the faculty of the University of California at Davis. Dr. Mitra then transferred to the Santa Barbara campus in 1977, where he served as department chairman from 1979 to 1982 and is now a Professor of Electrical and Computer Engineering. Dr. Mitra has published more than 500 journal and conference papers, and 11 books, and holds 5 patents. He served as President of the IEEE Circuits and Systems Society in 1986 and is currently a member of the editorial boards for four journals: *Multidimensional Systems and Signal Processing; Signal Processing; Journal of the Franklin Institute*; and *Automatika*. Dr. Mitra has received many distinguished industry and academic awards, including the 1973 F. E. Terman Award, the 1985 AT&T Foundation Award of the American Society of Engineering Education, the 1989 Education Award of the IEEE Circuits and Systems Society, the 1989 Distinguished Senior U.S. Scientist Award from the Alexander von Humboldt Foundation of Germany, the 1996 Technical Achievement Award of the IEEE Signal Processing Society, the 1999 Mac Van Valkenburg Society Award and the CAS Golden Jubilee Medal of the IEEE Circuits & System Society, and the IEEE Millennium Medal in 2000. He is an Academician of the Academy of Finland. Dr. Mitra is a Fellow of the IEEE, AAAS, and SPIE and is a member of EURASIP and the ASEE.

To my students

Contents

Preface

The field of digital signal processing (DSP) has seen explosive growth during the past three decades, as phenomenal advances both in research and application have been made. Fueling this growth have been the advances in digital computer technology and software development. Almost every electrical and computer engineering department in this country and abroad now offers one or more courses in digital signal processing, with the first course usually being offered at the senior level. This book is intended for a two-semester course on digital signal processing for seniors or first-year graduate students. It is also written at a level suitable for self-study by the practicing engineer or scientist.

Even though the first edition of this book was published barely two years ago, based on the feedback received from professors who adopted this book for their courses and many readers, it was clear that a new edition was needed to incorporate the suggested changes to the contents. A number of new topics have been included in the second edition. Likewise, a number of topics that are interesting but not practically useful have been removed because of size limitations. It was also felt that more worked-out examples were needed to explain new and difficult concepts.

The new topics included in the second edition are: calculation of total solution, zero-input response, zero-state response, and impulse response of finite-dimensional discrete-time systems (Sections 2.6.1–2.6.3), correlation of signals and its applications (Section 2.7), inverse systems (Section 4.9), system identification (Section 4.10), matched filter and its application (Section 4.14), sampling of bandpass signals (Section 5.3), design of highpass, bandpass, and bandstop analog filters (Section 5.5), effect of sample-and-hold operation (Section 5.11), design of highpass, bandpass, and bandstop IIR digital filters (Section 7.4), design of FIR digital filters with least-mean-square error (Section 7.8), constrained least-square design of FIR digital filters (Section 7.9), perfect reconstruction two-channel FIR filter banks (Section 10.9), cosine-modulated L-channel filter banks (Section 10.11), spectral analysis of random signals (Section 11.4), and sparse antenna array design (Section 11.14). The topics that have been removed from the first edition are as follows: state-space representation of LTI discrete-time systems from Chapter 2, signal flow-graph representation and state-space structures from Chapter 6, impulse invariance method of IIR filter design and FIR filter design based on the frequency-sampling approach from Chapter 7, reduction of product round-off errors from state-space structures from Chapter 9, and voice privacy system from Chapter 11. The fractional sampling rate conversion using the Lagrange interpolation has been moved to Chapter 10. Materials in each chapter are now organized more logically.

A key feature of this book is the extensive use of MATLAB® -based[1] examples that illustrate the program's powerful capability to solve signal processing problems. The book uses a three-stage pedagogical structure designed to take full advantage of MATLAB and to avoid the pitfalls of a "cookbook" approach to problem solving. First, each chapter begins by developing the essential theory and algorithms. Second, the material is illustrated with examples solved by hand calculation. And third, solutions are derived using MATLAB. From the beginning, MATLAB codes are provided with enough details to permit the students to repeat the examples on their computers. In addition to conventional theoretical problems requiring analytical solutions, each chapter also includes a large number of problems requiring solution via MATLAB. This book requires a minimal knowledge of MATLAB. I believe students learn the intricacies of problem solving with MATLAB faster by using tested, complete programs, and then writing simple programs to solve specific problems that are included at the ends of Chapters 2 to 11.

Because computer verification enhances the understanding of the underlying theories and, as in the first edition, a large library of worked-out MATLAB programs are included in the second edition. The original MATLAB programs of the first edition have been updated to run on the newer versions of MATLAB and the *Signal Processing Toolbox*. In addition, new MATLAB programs and code fragments have been added in this edition. The reader can run these programs to verify the results included in the book. Altogether there are 90 MATLAB programs in the text that have been tested under version 5.3 of MATLAB and version 4.2 of the *Signal Processing Toolbox*. Some of the programs listed in this book are not necessarily the fastest with regard to their execution speeds, nor are they the shortest. They have been written for maximum clarity without detailed explanations.

A second attractive feature of this book is the inclusion of 231 simple but practical examples that expose the reader to real-life signal processing problems which has been made possible by the use of computers in solving practical design problems. This book also covers many topics of current interest not normally found in an upper-division text. Additional topics are also introduced to the reader through problems at the end of each chapter. Finally, the book concludes with a chapter that focuses on several important, practical applications of digital signal processing. These applications are easy to follow and do not require knowledge of other advanced-level courses.

The prerequisite for this book is a junior-level course in linear continuous-time and discrete-time systems, which is usually required in most universities. A minimal review of linear systems and transforms is provided in the text, and basic materials from linear system theory are included, with important materials summarized in tables. This approach permits the inclusion of more advanced materials without significantly increasing the length of the book.

The book is divided into 11 chapters. Chapter 1 presents an introduction to the field of signal processing and provides an overview of signals and signal processing methods. Chapter 2 discusses the time-domain representations of discrete-time signals and discrete-time systems as sequences of numbers and describes classes of such signals and systems commonly encountered. Several basic discrete-time signals that play important roles in the time-domain characterization of arbitrary discrete-time signals and discrete-time systems are then introduced. Next, a number of basic operations to generate other sequences from one or more sequences are described. A combination of these operations is also used in developing a discrete-time system. The problem of representing a continuous-time signal by a discrete-time sequence is examined for a simple case. Finally, the time-domain characterization of discrete-time random signals is discussed.

Chapter 3 is devoted to the transform-domain representations of a discrete-time sequence. Specifically discussed are the discrete-time Fourier transform (DTFT), the discrete Fourier transform (DFT), and the z-transform. Properties of each of these transforms are reviewed and a few simple applications outlined. The chapter ends with a discussion of the transform-domain representation of a random signal.

This book concentrates almost exclusively on the linear time-invariant discrete-time systems, and

---

[1]MATLAB is a registered trademark of The MathWorks, Inc., 24 Prime Park Way, Natick, MA 01760-1500, Phone: 508-647-7000, **http://www.mathworks.com**.

Chapter 4 discusses their transform-domain representations. Specific properties of such transform-domain representations are investigated, and several simple applications are considered.

Chapter 5 is concerned primarily with the discrete-time processing of continuous-time signals. The conditions for discrete-time representation of a bandlimited continuous-time signal under ideal sampling and its exact recovery from the sampled version are first derived. Several interface circuits are used for the discrete-time processing of continuous-time signals. Two of these circuits are the anti-aliasing filter and the reconstruction filter, which are analog lowpass filters. As a result, a brief review of the basic theory behind some commonly used analog filter design methods is included, and their use is illustrated with MATLAB. Other interface circuits discussed in this chapter are the sample-and-hold circuit, the analog-to-digital converter, and the digital-to-analog converter.

A structural representation using interconnected basic building blocks is the first step in the hardware or software implementation of an LTI digital filter. The structural representation provides the relations between some pertinent internal variables with the input and the output, which in turn provides the keys to the implementation. There are various forms of the structural representation of a digital filter, and two such representations are reviewed in Chapter 6, followed by a discussion of some popular schemes for the realization of real causal IIR and FIR digital filters. In addition, it describes a method for the realization of IIR digital filter structures that can be used for the generation of a pair of orthogonal sinusoidal sequences.

Chapter 7 considers the digital filter design problem. First, it discusses the issues associated with the filter design problem. Then it describes the most popular approach to IIR filter design, based on the conversion of a prototype analog transfer function to a digital transfer function. The spectral transformation of one type of IIR transfer function into another type is discussed. Then a very simple approach to FIR filter design is described. Finally, the chapter reviews computer-aided design of both IIR and FIR digital filters. The use of MATLAB in digital filter design is illustrated.

Chapter 8 is concerned with the implementation aspects of DSP algorithms. Two major issues concerning implementation are discussed first. The software implementations of digital filtering and DFT algorithms on a computer using MATLAB are reviewed to illustrate the main points. This is followed by a discussion of various schemes for the representation of number and signal variables on digital machines, which is basic to the development of methods for the analysis of finite wordlength effects considered in Chapter 9. Algorithms used to implement addition and multiplication, the two key arithmetic operations in digital signal processing, are reviewed next, along with operations developed to handle overflow. Finally, the chapter outlines two general methods for the design and implementation of tunable digital filters, followed by a discussion of algorithms for the approximation of certain special functions.

Chapter 9 is devoted to analysis of the effects of the various sources of quantization errors; it describes structures that are less sensitive to these effects. Included here are discussions on the effect of coefficient quantization.

Chapter 10 discusses multirate discrete-time systems with unequal sampling rates at various parts. The chapter includes a review of the basic concepts and properties of sampling rate alteration, design of decimation and interpolation digital filters, and multirate filter bank design.

The final chapter, Chapter 11, reviews a few simple practical applications of digital signal processing to provide a glimpse of its potential.

The materials in this book have been used in a two-quarter course sequence on digital signal processing at the University of California, Santa Barbara, and have been extensively tested in the classroom for over 10 years. Basically, Chapters 2 through 6 form the basis of an upper-division course, while Chapters 7 through 10 form the basis of a graduate-level course.

Many topics included in this text can be omitted from class discussion, depending on the coverage of other courses in the curriculum. Because a senior-level course on random signals and systems is required of all electrical and computer engineering majors in most universities, materials in Sections 2.7, 3.10, and 4.9 can be excluded from an upper-division course on digital signal processing. However, these topics are important in the analysis of wordlength effects discussed in Chapter 9, and readers not familiar with this subject are encouraged to review these sections before reading Chapter 9. Likewise, Section 8.4 on number representation and Section 8.5 on arithmetic operations can similarly be omitted from discussion since most students taking a digital signal processing course usually take a course on digital hardware design.

This text contains 231 examples, 90 MATLAB programs, 684 problems, and 186 MATLAB exercises.

Every attempt has been made to ensure the accuracy of all materials in this book, including the MATLAB programs. I would, however, appreciate readers bringing to my attention any errors that may appear in the printed version for reasons beyond my control and that of the publisher. These errors and any other comments can be communicated to me by e-mail addressed to: **mitra@ece.ucsb.edu**.

Finally, I have been particularly fortunate to have had the opportunity to work with the outstanding students who were in my research group during my teaching career, which spans over 35 years. I have benefited immensely, and continue to do so, both professionally and personally, from my friendship and association with them, and to them I dedicate this book.

Sanjit K. Mitra

Acknowledgements

The preliminary versions of the complete manuscript for the first edition were reviewed by Dr. Hrvojc Babic of the University of Zagreb, Croatia; Dr. James F. Kaiser of Duke University; Dr. Wolfgang F. G. Mecklenbräuker of the Technical University of Vienna, Austria; and Dr. P. P. Vaidyanathan of the California Institute of Technology. A later version was reviewed by Dr. Roberto H. Bambmerger of Microsoft; Dr. Charles Boumann of Purdue University; Dr. Kevin Buckley of the University of Minnesota; Dr. John A. Flemming of the Texas A & M University; Dr. Jerry D. Gibson of the Southern Methodist University; Dr. John Gowdy of Clemson University; Drs. James Harris and Mahmood Nahvi of the California Polytechnic University, San Louis Obispo; Dr. Yih-Chyun Jenq of Portland State University; Dr. Troung Q. Ngyuen of Boston University; and Dr. Andreas Spanias of Arizona State University. Various parts of the manuscript were reviewed by Dr. C. Sidney Burrus of Rice University; Dr. Richard V. Cox of the AT&T Laboratories; Dr. Ian Galton of the University of California, San Diego; Dr. Nikil S. Jayant of the Georgia Institute of Technology; Dr. Tor Ramstad of the Norwegian University of Science and Technology, Trondheim, Norway; Dr. B. Ananth Shenoi of Wright State University; Dr. Hans W. Schüssler of the University of Erlangen-Nuremberg, Germany; Dr. Richard Schreier of Analog Devices and Dr. Gabor C. Temes of Oregon State University.

Reviews for the second edition were provided by Dr. Winser E. Alexander of North Carolina State University; Dr. Sohail A. Dianat of the Rochester Institute of Technology; Dr. Suhash Dutta Roy of the Indian Institute of Technology, New Delhi; Dr. David C. Farden of North Dakota State University; Dr. Abdulnasir Y. Hossein of Sultan Qaboos University, Sultanate of Omman; Dr. James F. Kaiser of Duke University; Dr. Ramakrishna Kakarala of the Agilent Laboratories; Dr. Wolfgang F. G. Mecklenbräuker of the Technical University of Vienna, Austria; Dr. Antonio Ortega of the University of Southern California; Dr. Stanley J. Reeves of Auburn University; Dr. George Symos of the University of Maryland, College Park; and Dr. Gregory A. Wornell of the Massachusetts Institute of Technology. Various parts of the manuscript for the second edition were reviewed by Dr. Dimitris Anastassiou of Columbia University; Dr. Rajendra K. Arora of the Florida State University; Dr. Ramdas Kumaresan of the University of Rhode Island; Dr. Upamanyu Madhow of the University of California, Santa Barbara; Dr. Urbashi Mitra of the University of Southern California; Dr. Randolph Moses of Ohio State University; Dr. Ivan Selesnick of Polytechnic University, Brooklyn, New York; and Dr. Gabor C. Temes of Oregon State University.

I thank all of them for their valuable comments, which have improved the book tremendously.

Many of my former and present research students reviewed various portions of the manuscript of both editions and tested a number of the MATLAB programs. In particular, I would like to thank Drs. Charles D. Creusere, Rajeev Gandhi, Michael Lightstone, Ing-Song Lin, Luca Lucchese, Debargha Mukherjee, Norbert Strobel, and Stefan Thurnhofer, and Messrs. Serkan Hatipoglu, Zhihai He, Eric Leipnik, Michael Moore, and Mylene Queiroz de Farias. I am also indebted to all former students in my ECE 158 and ECE 258A classes at the University of California, Santa Barbara, for their feedback over the years, which helped refine the book.

I thank Goutam K. Mitra and Alicia Rodriguez for the cover design of the book. Finally, I thank Patricia Monohon for her assistance in the preparation of the LaTeX files of the second edition.

Supplements

All MATLAB programs included in this book are available via anonymous file transfer protocol (FTP) from the Internet site **iplserv.ece.ucsb.edu** in the directory **/pub/mitra/Book_2e**.

A solutions manual prepared by Rajeev Gandhi, Serkan Hatipoglu, Zhihai He, Luca Lucchese, Michael Moore, and Mylene Queiroz de Farias and containing the solutions to all problems and MATLAB exercises is available to instructors from the publisher.

A companion book *Digital Signal Processing Laboratory Using MATLAB* by the author is also available from McGraw-Hill.

1 Signals and Signal Processing

Signals play an important role in our daily life. Examples of signals that we encounter frequently are speech, music, picture, and video signals. A signal is a function of independent variables such as time, distance, position, temperature, and pressure. For example, speech and music signals represent air pressure as a function of time at a point in space. A black-and-white picture is a representation of light intensity as a function of two spatial coordinates. The video signal in television consists of a sequence of images, called frames, and is a function of three variables: two spatial coordinates and time.

Most signals we encounter are generated by natural means. However, a signal can also be generated synthetically or by computer simulation. A signal carries information, and the objective of signal processing is to extract useful information carried by the signal. The method of information extraction depends on the type of signal and the nature of the information being carried by the signal. Thus, roughly speaking, signal processing is concerned with the mathematical representation of the signal and the algorithmic operation carried out on it to extract the information present. The representation of the signal can be in terms of basis functions in the domain of the original independent variable(s) or it can be in terms of basis functions in a transformed domain. Likewise, the information extraction process may be carried out in the original domain of the signal or in a transformed domain. This book is concerned with discrete-time representation of signals and their discrete-time processing.

This chapter provides an overview of signals and signal processing methods. The mathematical characterization of the signal is first discussed along with a classification of signals. Next, some typical signals are discussed in detail and the type of information carried by them is described. Then a review of some commonly used signal processing operations is provided and illustrated through examples. Advantages and disadvantages of digital processing of signals are then discussed. Finally, a brief review of some typical signal processing applications is included.

1.1 Characterization and Classification of Signals

Depending on the nature of the independent variables and the value of the function defining the signal, various types of signals can be defined. For example, independent variables can be continuous or discrete. Likewise, the signal can either be a continuous or a discrete function of the independent variables. Moreover, the signal can be either a real-valued function or a complex-valued function.

A signal can be generated by a single source or by multiple sources. In the former case, it is a scalar signal and in the latter case it is a vector signal, often called a multichannel signal.

A one-dimensional (1-D) signal is a function of a single independent variable. A two-dimensional (2-D) signal is a function of two independent variables. A multidimensional (M-D) signal is a function of more than one variable. The speech signal is an example of a 1-D signal where the independent variable is time. An image signal, such as a photograph, is an example of a 2-D signal where the two independent variables are the two spatial variables. Each frame of a black-and-white video signal is a 2-D image signal that is a function of two discrete spatial variables, with each frame occurring sequentially at discrete instants of time. Hence, the black-and-white video signal can be considered an example of a three-dimensional (3-D) signal where the three independent variables are the two spatial variables and time. A color video signal is a three-channel signal composed of three 3-D signals representing the three primary colors: red, green, and blue (RGB). For transmission purposes, the RGB television signal is transformed into another type of three-channel signal composed of a luminance component and two chrominance components.

The value of the signal at a specific value(s) of the independent variable(s) is called its *amplitude*. The variation of the amplitude as a function of the independent variable(s) is called its *waveform*.

For a 1-D signal, the independent variable is usually labeled as *time*. If the independent variable is continuous, the signal is called a *continuous-time signal*. If the independent variable is discrete, the signal is called a *discrete-time signal*. A continuous-time signal is defined at every instant of time. On the other hand, a discrete-time signal is defined at discrete instants of time, and hence, it is a sequence of numbers.

A continuous-time signal with a continuous amplitude is usually called an *analog signal*. A speech signal is an example of an analog signal. Analog signals are commonly encountered in our daily life and are usually generated by natural means. A discrete-time signal with discrete-valued amplitudes represented by a finite number of digits is referred to as a *digital signal*. An example of a digital signal is the digitized music signal stored in a CD-ROM disk. A discrete-time signal with continuous-valued amplitudes is called a *sampled-data signal*. This last type of signal occurs in switched-capacitor (SC) circuits. A digital signal is thus a quantized sampled-data signal. Finally, a continuous-time signal with discrete-valued amplitudes has been referred to as a *quantized boxcar signal*[Ste93]. Figure 1.1 illustrates the four types of signals.

The functional dependence of a signal in its mathematical representation is often explicitly shown. For a continuous-time 1-D signal, the continuous independent variable is usually denoted by t, whereas for a discrete-time 1-D signal, the discrete independent variable is usually denoted by n. For example, $u(t)$ represents a continuous-time 1-D signal and $\{v[n]\}$ represents a discrete-time 1-D signal. Each member, $v[n]$, of a discrete-time signal is called a *sample*. In many applications, a discrete-time signal is generated from a parent continuous-time signal by sampling the latter at uniform intervals of time. If the discrete instants of time at which a discrete-time signal is defined are uniformly spaced, the independent discrete variable n can be normalized to assume integer values.

In the case of a continuous-time 2-D signal, the two independent variables are the spatial coordinates, which are usually denoted by x and y. For example, the intensity of a black-and-white image can be expressed as $u(x, y)$. On the other hand, a digitized image is a 2-D discrete-time signal, and its two independent variables are discretized spatial variables often denoted by m and n. Hence, a digitized image can be represented as $v[m, n]$. Likewise, a black-and-white video sequence is a 3-D signal and can be represented as $u(x, y, t)$ where x and y denote the two spatial variables and t denotes the temporal variable time. A color video signal is a vector signal composed of three signals representing the three primary colors: red, green, and blue:

$$\mathbf{u}(x, y, t) = \begin{bmatrix} r(x, y, t) \\ g(x, y, t) \\ b(x, y, t) \end{bmatrix}.$$

There is another classification of signals that depends on the certainty by which the signal can be uniquely described. A signal that can be uniquely determined by a well-defined process such as a mathematical expression or rule, or table look-up, is called a *deterministic signal*. A signal that is generated in a random fashion and cannot be predicted ahead of time is called a *random signal*. In this text we are primarily concerned with the processing of discrete-time deterministic signals. However, since practical discrete-time systems employ finite wordlengths for the storing of signals and the implementation of the signal processing algorithms, it is necessary to develop tools for the analysis of finite wordlength effects on the performance of discrete-time systems. To this end, it has been found convenient to represent certain pertinent signals as random signals and employ statistical techniques for their analysis.

Some typical signal processing operations are reviewed in the following section.

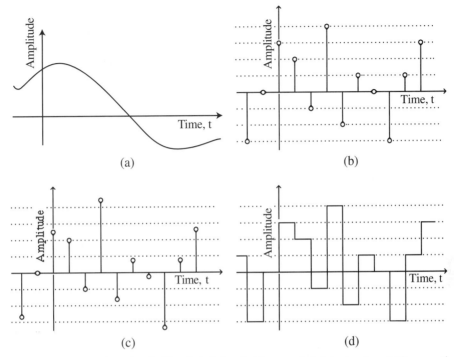

Figure 1.1: (a) A continuous-time signal, (b) a digital signal, (c) a sampled-data signal, and (d) a quantized boxcar signal.

1.2 Typical Signal Processing Operations

Various types of signal processing operations are employed in practice. In the case of analog signals, most signal processing operations are usually carried out in the time-domain, whereas, in the case of discrete-time signals, both time-domain and frequency-domain operations are employed. In either case, the desired operations are implemented by a combination of some elementary operations. These operations are also usually implemented in real-time or near real-time, even though, in certain applications, they may be implemented off-line.

1.2.1 Elementary Time-Domain Operations

The three most basic time-domain signal operations are scaling, delay, and addition. *Scaling* is simply the multiplication of the signal by a positive or a negative constant. In the case of analog signals, this operation is usually called *amplification* if the magnitude of the multiplying constant, called *gain*, is greater than one. If the magnitude of the multiplying constant is less than one, the operation is called *attenuation*. Thus, if $x(t)$ is an analog signal, the scaling operation generates a signal $y(t) = \alpha x(t)$ where α is the multiplying constant. Two other elementary operations are integration and differentiation. The *integration* of an analog signal $x(t)$ generates a signal $y(t) = \int_{-\infty}^{t} x(\tau)\, d\tau$, while its *differentiation* results in a signal $w(t) = dx(t)/dt$.

The *delay* operation generates a signal that is a delayed replica of the original signal. For an analog signal $x(t)$, $y(t) = x(t - t_0)$ is the signal obtained by delaying $x(t)$ by the amount t_0 which is assumed to be a positive number. If t_0 is negative, then it is an *advance* operation.

Many applications require operations involving two or more signals to generate a new signal. For example, $y(t) = x_1(t) + x_2(t) - x_3(t)$ is the signal generated by the *addition* of the three analog signals $x_1(t)$, $x_2(t)$, and $x_3(t)$. Another elementary operation is the *product* of two signals. Thus, the product of two signals $x_1(t)$ and $x_2(t)$ generates a signal $y(t) = x_1(t)x_2(t)$.

The elementary operations mentioned above are also carried out on discrete-time signals and are discussed in later parts of this text. Next we review some commonly used complex signal processing operations that are implemented by combining two or more of the elementary operations.

1.2.2   Filtering

One of the most widely used complex signal processing operations is *filtering*. Filtering is used to pass certain frequency components in a signal through the system without any distortion and to block other frequency components. The system implementing this operation is called a *filter*. The range of frequencies that is allowed to pass through the filter is called the *passband*, and the range of frequencies that is blocked by the filter is called the *stopband*. Various types of filters can be defined, depending on the nature of the filtering operation. In most cases, the filtering operation for analog signals is linear and is described by the convolution integral

$$y(t) = \int_{-\infty}^{\infty} h(t - \tau)x(\tau)\, d\tau, \tag{1.1}$$

where $x(t)$ is the input signal and $y(t)$ is the output of the filter characterized by an impulse response $h(t)$.

A *lowpass filter* passes all low-frequency components below a certain specified frequency f_c, called the *cutoff frequency*, and blocks all high-frequency components above f_c. A *highpass filter* passes all high-frequency components above a certain cutoff frequency f_c and blocks all low-frequency components below f_c. A *bandpass filter* passes all frequency components between two cutoff frequencies f_{c1} and f_{c2} where $f_{c1} < f_{c2}$, and blocks all frequency components below the frequency f_{c1} and above the frequency f_{c2}. A *bandstop filter* blocks all frequency components between two cutoff frequencies f_{c1} and f_{c2}, and passes all frequency components below the frequency f_{c1} and above the frequency f_{c2}. Figure 1.2(a) shows a signal composed of three sinusoidal components of frequencies 50 Hz, 110 Hz, and 210 Hz, respectively. Figure 1.2(b) to (e) shows the results of the above four types of filtering operations with appropriately chosen cutoff frequencies.

A bandstop filter designed to block a single frequency component is called a *notch filter*. A *multiband filter* has more than one passband and more than one stopband. A *comb filter* is designed to block frequencies that are integral multiples of a low frequency.

A signal may get corrupted unintentionally by an interfering signal called *interference* or *noise*. In many applications the desired signal occupies a low-frequency band from dc to some frequency f_L Hz, and it is corrupted by a high-frequency noise with frequency components above f_H Hz with $f_H > f_L$. In such cases, the desired signal can be recovered from the noise-corrupted signal by passing the latter through a lowpass filter with a cutoff frequency f_c where $f_L < f_c < f_H$. A common source of noise is power lines radiating electric and magnetic fields. The noise generated by power lines appears as a 60-Hz sinusoidal signal corrupting the desired signal and can be removed by passing the corrupted signal through a notch filter with a notch frequency at 60 Hz.[1]

1.2.3   Generation of Complex Signals

As indicated earlier, a signal can be real-valued or complex-valued. For convenience, the former is usually called a *real signal* while the latter is called a *complex signal*. All naturally generated signals are real-valued. In some applications, it is desirable to develop a complex signal from a real signal having more desirable

---

[1] In many countries, power lines generate 50-Hz noise.

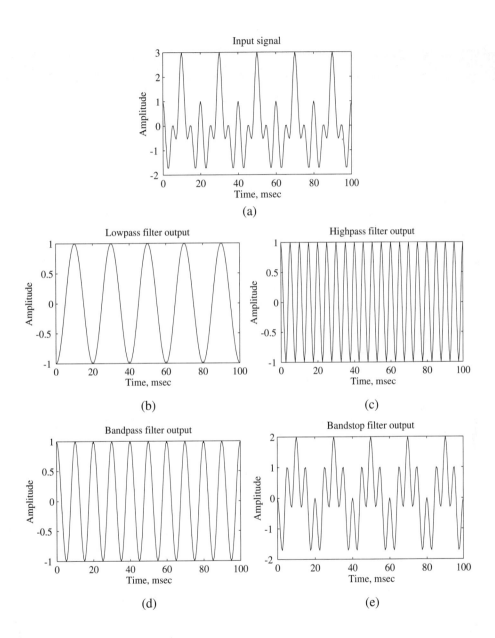

Figure 1.2: (a) Input signal, (b) output of a lowpass filter with a cutoff at 80 Hz, (c) output of a highpass filter with a cutoff at 150 Hz, (d) output of a bandpass filter with cutoffs at 80 Hz and 150 Hz, and (e) output of a bandstop filter with cutoffs at 80 Hz and 150 Hz.

properties. A complex signal can be generated from a real signal by employing a *Hilbert transformer* that is characterized by an impulse response $h_{HT}(t)$ given by [Fre94], [Opp83]

$$h_{HT}(t) = \frac{1}{\pi t}. \tag{1.2}$$

To illustrate the method, consider a real analog signal $x(t)$ with a continuous-time Fourier transform (CTFT) $X(j\Omega)$ given by

$$X(j\Omega) = \int_{-\infty}^{\infty} x(t)e^{-j\Omega t}\, dt. \tag{1.3}$$

$X(j\Omega)$ is called the spectrum of $x(t)$. The magnitude spectrum of a real signal exhibits even symmetry while the phase spectrum exhibits odd symmetry. Thus, the spectrum $X(j\Omega)$ of a real signal $x(t)$ contains both positive and negative frequencies and can therefore be expressed as

$$X(j\Omega) = X_p(j\Omega) + X_n(j\Omega), \tag{1.4}$$

where $X_p(j\Omega)$ is the portion of $X(j\Omega)$ occupying the positive frequency range and $X_n(j\Omega)$ is the portion of $X(j\Omega)$ occupying the negative frequency range. If $x(t)$ is passed through a Hilbert transformer, its output $\hat{x}(t)$ is given by the linear convolution of $x(t)$ with $h_{HT}(t)$:

$$\hat{x}(t) = \int_{-\infty}^{\infty} h_{HT}(t - \tau)x(\tau)\, d\tau. \tag{1.5}$$

The spectrum $\hat{X}(j\Omega)$ of $\hat{x}(t)$ is given by the product of the continuous-time Fourier transforms of $x(t)$ and $h_{HT}(t)$. Now the continuous-time Fourier transform $H_{HT}(j\Omega)$ of $h_{HT}(t)$ of Eq. (1.2) is given by

$$H_{HT}(j\Omega) = \begin{cases} -j, & \Omega > 0, \\ j, & \Omega < 0. \end{cases} \tag{1.6}$$

Hence,

$$\hat{X}(j\Omega) = H_{HT}(j\Omega)X(j\Omega) = -jX_p(j\Omega) + jX_n(j\Omega). \tag{1.7}$$

As the magnitude and the phase of $\hat{X}(j\Omega)$ are an even and odd function, respectively, it follows from Eq. (1.7) that $\hat{x}(t)$ is also a real signal. Consider the complex signal $y(t)$ formed by the sum of $x(t)$ and $\hat{x}(t)$:

$$y(t) = x(t) + j\hat{x}(t). \tag{1.8}$$

The signals $x(t)$ and $\hat{x}(t)$ are called, respectively, the *in-phase* and *quadrature components* of $y(t)$. By making use of Eqs. (1.4) and (1.7) in the continuous-time Fourier transform of $y(t)$, we obtain

$$Y(j\Omega) = X(j\Omega) + j\hat{X}(j\Omega) = 2X_p(j\Omega). \tag{1.9}$$

In other words, the complex signal $y(t)$, called an *analytic signal*, has only positive frequency components.

A block diagram representation of the scheme for the analytic signal generation from a real signal is sketched in Figure 1.3.

1.2.4 Modulation and Demodulation

For transmission of signals over long distances, a transmission media such as cable, optical fiber, or the atmosphere is employed. Each such medium has a bandwidth that is more suitable for the efficient transmission of signals in the high-frequency range. As a result, for the transmission of a low-frequency

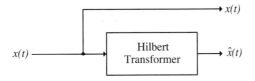

Figure 1.3: Generation of an analytic signal using a Hilbert transformer.

signal over a channel, it is necessary to transform the signal to a high-frequency signal by means of a modulation operation. At the receiving end, the modulated high-frequency signal is demodulated, and the desired low-frequency signal is then extracted by further processing. There are four major types of modulation of analog signals: amplitude modulation, frequency modulation, phase modulation, and pulse amplitude modulation. Of these schemes, amplitude modulation is conceptually simple and is discussed here [Fre94], [Opp83].

In the *amplitude modulation* scheme, the amplitude of a high-frequency sinusoidal signal $A \cos(\Omega_o t)$, called the *carrier signal*, is varied by the low-frequency bandlimited signal $x(t)$, called the *modulating signal*, generating a high-frequency signal, called the *modulated signal*, $y(t)$ according to

$$y(t) = Ax(t)\cos(\Omega_o t). \tag{1.10}$$

Thus, amplitude modulation can be implemented by forming the product of the modulating signal with the carrier signal. To demonstrate the frequency-translation property of the amplitude modulation process, let $x(t) = \cos(\Omega_1 t)$ where Ω_1 is much smaller than the carrier frequency Ω_o, i.e., $\Omega_1 \ll \Omega_o$. From Eq. (1.10) we therefore obtain

$$\begin{aligned} y(t) &= A\cos(\Omega_1 t) \cdot \cos(\Omega_o t) \\ &= \tfrac{A}{2}\cos\left((\Omega_o + \Omega_1)t\right) + \tfrac{A}{2}\cos\left((\Omega_o - \Omega_1)t\right). \end{aligned} \tag{1.11}$$

Thus, the modulated signal $y(t)$ is composed of two sinusoidal signals of frequencies $\Omega_o + \Omega_1$ and $\Omega_o - \Omega_1$ which are close to Ω_o as Ω_1 has been assumed to be much smaller than the carrier frequency Ω_o.

It is instructive to examine the spectrum of $y(t)$. From the properties of the continuous-time Fourier transform it follows that the spectrum $Y(j\Omega)$ of $y(t)$ is given by

$$Y(j\Omega) = \frac{A}{2}X\left(j(\Omega - \Omega_o)\right) + \frac{A}{2}X\left(j(\Omega + \Omega_o)\right), \tag{1.12}$$

where $X(j\Omega)$ is the spectrum of the modulating signal $x(t)$. Figure 1.4 shows the spectra of the modulating signal and that of the modulated signal under the assumption that the carrier frequency Ω_o is greater than Ω_m, the highest frequency contained in $x(t)$. As seen from this figure, $y(t)$ is now a bandlimited high-frequency signal with a bandwidth $2\Omega_m$ centered at Ω_o.

The portion of the amplitude-modulated signal between Ω_o and $\Omega_o + \Omega_m$ is called the *upper sideband* whereas the portion between Ω_o and $\Omega_o - \Omega_m$ is called the *lower sideband*. Because of the generation of two sidebands and the absence of a carrier component in the modulated signal, the process is called *double-sideband suppressed carrier* (DSB-SC) *modulation*.

The *demodulation* of $y(t)$, assuming $\Omega_o > \Omega_m$, is carried out in two stages. First, the product of $y(t)$ with a sinusoidal signal of the same frequency as the carrier is formed. This results in

$$r(t) = y(t)\cos\Omega_o t = Ax(t)\cos^2\Omega_o t \tag{1.13}$$

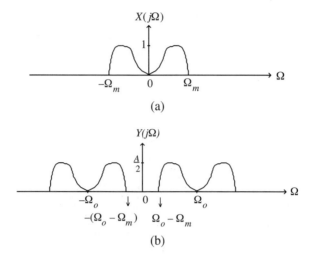

Figure 1.4: (a) Spectrum of the modulating signal $x(t)$, and (b) spectrum of the modulated signal $y(t)$. For convenience, both spectra are shown as real functions.

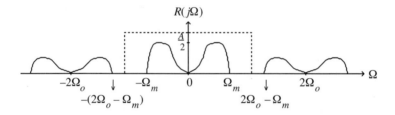

Figure 1.5: Spectrum of the product of the modulated signal and the carrier.

which can be rewritten as

$$r(t) = y(t) \cos \Omega_o t = \frac{A}{2} x(t) + \frac{A}{2} x(t) \cos(2\Omega_o t). \tag{1.14}$$

This result indicates that the product signal is composed of the original modulating signal scaled by a factor 1/2 and an amplitude-modulated signal with a carrier frequency $2\Omega_o$. The spectrum $R(j\Omega)$ of $r(t)$ is as indicated in Figure 1.5. The original modulating signal can now be recovered from $r(t)$ by passing it through a lowpass filter with a cutoff frequency Ω_c satisfying the relation $\Omega_m < \Omega_c < 2\Omega_o - \Omega_m$. The output of the filter is then a scaled replica of the modulating signal.

Figure 1.6 shows the block diagram representations of the amplitude modulation and demodulation schemes. The underlying assumption in the demodulation process outlined above is that a sinusoidal signal identical to the carrier signal can be generated at the receiving end. In general, it is difficult to ensure that the demodulating sinusoidal signal has a frequency identical to that of the carrier all the time. To get around this problem, in the transmission of amplitude-modulated radio signals, the modulation process is modified so that the transmitted signal includes the carrier signal. This is achieved by redefining the amplitude modulation operation as follows:

$$y(t) = A[1 + mx(t)] \cos(\Omega_o t), \tag{1.15}$$

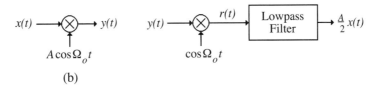

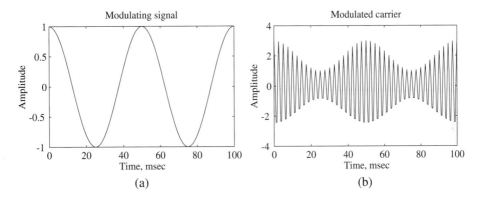

Figure 1.6: Schematic representations of the amplitude modulation and demodulation schemes: (a) modulator, and (b) demodulator.

Figure 1.7: (a) A sinusoidal modulating signal of frequency 20 Hz, and (b) modulated carrier with a carrier frequency of 400 Hz based on the DSB modulation.

where m is a number chosen to ensure that $[1 + mx(t)]$ is positive for all t. Figure 1.7 shows the waveforms of a modulating sinusoidal signal of frequency 20 Hz and the amplitude-modulated carrier obtained according to Eq. (1.15) for a carrier frequency of 400 Hz and $m = 0.5$. Note that the envelope of the modulated carrier is essentially the waveform of the modulating signal. As here the carrier is also present in the modulated signal, the process is called simply *double-sideband* (DSB) *modulation*. At the receiving end, the carrier signal is separated first and then used for demodulation.

1.2.5　Multiplexing and Demultiplexing

For an efficient utilization of a wideband transmission channel, many narrow-bandwidth low-frequency signals are combined to form a composite wideband signal that is transmitted as a single signal. The process of combining these signals is called *multiplexing* which is implemented to ensure that a replica of the original narrow-bandwidth low-frequency signals can be recovered at the receiving end. The recovery process is called *demultiplexing*.

　　One widely used method of combining different voice signals in a telephone communication system is the *frequency-division multiplexing* (FDM) scheme [Cou83], [Opp83]. Here, each voice signal, typically bandlimited to a low-frequency band of width $2\Omega_m$, is frequency-translated into a higher frequency band using the amplitude modulation method of Eq. (1.10). The carrier frequency of adjacent amplitude-modulated signals is separated by Ω_o, with $\Omega_o > 2\Omega_m$ to ensure that there is no overlap in the spectra of the individual modulated signals after they are added to form a baseband composite signal. This signal is then modulated onto the main carrier developing the FDM signal and transmitted. Figure 1.8 illustrates the frequency-division multiplexing scheme.

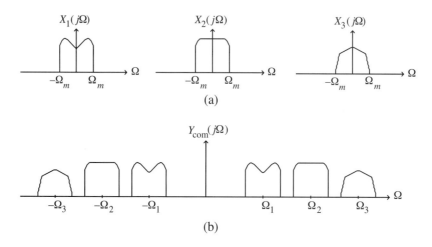

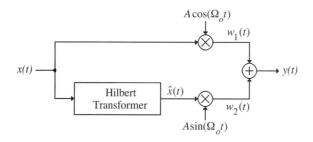

Figure 1.8: Illustration of the frequency-division multiplexing operation. (a) Spectra of three low-frequency signals, and (b) spectra of the modulated composite signal.

Figure 1.9: Single-sideband modulation scheme employing a Hilbert transformer.

At the receiving end, the composite baseband signal is first derived from the FDM signal by demodulation. Then each individual frequency-translated signal is first demultiplexed by passing the composite signal through a bandpass filter with a center frequency of identical value as that of the corresponding carrier frequency and a bandwidth slightly greater than $2\Omega_m$. The output of the bandpass filter is then demodulated using the method of Figure 1.6(b) to recover a scaled replica of the original voice signal.

In the case of the conventional amplitude modulation, as can be seen from Figure 1.4, the modulated signal has a bandwidth of $2\Omega_m$, whereas the bandwidth of the modulating signal is Ω_m. To increase the capacity of the transmission medium, a modified form of the amplitude modulation is often employed in which either the upper sideband or the lower sideband of the modulated signal is transmitted. The corresponding procedure is called *single-sideband (SSB) modulation* to distinguish it from the double-sideband modulation scheme of Figure 1.6(a).

One way to implement single-sideband amplitude modulation is indicated in Figure 1.9, where the Hilbert transformer is defined by Eq. (1.6). The spectra of pertinent signals in Figure 1.9 are shown in Figure 1.10.

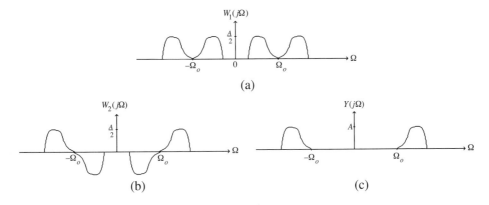

Figure 1.10: Spectra of pertinent signals in Figure 1.9.

1.2.6   Quadrature Amplitude Modulation

We observed earlier that DSB amplitude modulation is half as efficient as SSB amplitude modulation with regard to utilization of the spectrum. The *quadrature amplitude modulation* (QAM) method uses DSB modulation to modulate two different signals so that they both occupy the same bandwidth; thus QAM only takes up as much bandwidth as the SSB modulation method. To understand the basic idea behind the QAM approach, let $x_1(t)$ and $x_2(t)$ be two bandlimited low-frequency signals with a bandwidth of Ω_m as indicated in Figure 1.4(a). The two modulating signals are individually modulated by the two carrier signals $A\cos(\Omega_o t)$ and $A\sin(\Omega_o t)$, respectively, and are summed, resulting in a composite signal $y(t)$ given by

$$y(t) = Ax_1(t)\cos(\Omega_o t) + Ax_2(t)\sin(\Omega_o t). \tag{1.16}$$

Note that the two carrier signals have the same carrier frequency Ω_o but have a phase difference of $90°$. In general, the carrier $A\cos(\Omega_o t)$ is called the *in-phase component* and the carrier $A\sin(\Omega_o t)$ is called the *quadrature component*. The spectrum $Y(j\Omega)$ of the composite signal $y(t)$ is now given by

$$Y(j\Omega) = \tfrac{A}{2}\left\{X_1\left(j(\Omega - \Omega_o)\right) + X_1\left(j(\Omega + \Omega_o)\right)\right\}$$
$$+ \tfrac{A}{2j}\left\{X_2\left(j(\Omega - \Omega_o)\right) - X_2\left(j(\Omega + \Omega_o)\right)\right\}, \tag{1.17}$$

and is seen to occupy the same bandwidth as the modulated signal obtained by a DSB modulation.

   To recover the original modulating signals, the composite signal is multiplied by both the in-phase and the quadrature components of the carrier separately, resulting in two signals:

$$r_1(t) = y(t)\cos(\Omega_o t),$$
$$r_2(t) = y(t)\sin(\Omega_o t). \tag{1.18}$$

Substituting $y(t)$ from Eq. (1.16) in Eq. (1.18), we obtain after some algebra

$$r_1(t) = \tfrac{A}{2}x_1(t) + \tfrac{A}{2}x_1(t)\cos(2\Omega_o t) + \tfrac{A}{2}x_2(t)\sin(2\Omega_o t),$$
$$r_2(t) = \tfrac{A}{2}x_2(t) + \tfrac{A}{2}x_1(t)\sin(2\Omega_o t) - \tfrac{A}{2}x_2(t)\cos(2\Omega_o t). \tag{1.19}$$

Lowpass filtering of $r_1(t)$ and $r_2(t)$ by filters with a cutoff at Ω_m yields the two modulating signals. Figure 1.11 shows the block diagram representations of the quadrature amplitude modulation and demodulation schemes.

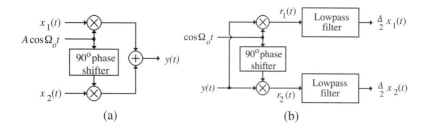

Figure 1.11: Schematic representations of the quadrature amplitude modulation and demodulation schemes: (a) modulator, and (b) demodulator.

As in the case of the DSB suppressed carrier modulation method, the QAM method also requires at the receiver an exact replica of the carrier signal employed in the transmitting end for accurate demodulation. It is therefore not employed in the direct transmission of analog signals, but finds applications in the transmission of discrete-time data sequences and in the transmission of analog signals converted into discrete-time sequences by sampling and analog-to-digital conversion.

1.2.7 Signal Generation

An equally important part of signal processing is synthetic signal generation. One of the simplest such signal generators is a device generating a sinusoidal signal, called an *oscillator*. Such a device is an integral part of the amplitude-modulation and demodulation system described in the previous two sections. It also has various other signal processing applications.

There are applications that require the generation of other types of periodic signals such as square waves and triangular waves. Certain types of random signals with a spectrum of constant amplitude for all frequencies, called *white noise*, often find applications in practice. One such application, is in the generation of discrete-time synthetic speech signals.

1.3 Examples of Typical Signals[2]

To better understand the breadth of the signal processing task, we now examine a number of examples of some typical signals and their subsequent processing in typical applications.

Electrocardiography (ECG) Signal

The electrical activity of the heart is represented by the ECG signal [Sha81]. A typical ECG signal trace is shown in Figure 1.12(a). The ECG trace is essentially a periodic waveform. One such period of the ECG waveform as depicted in Figure 1.12(b) represents one cycle of the blood transfer process from the heart to the arteries. This part of the waveform is generated by an electrical impulse originating at the sinoatrial node in the right atrium of the heart. The impulse causes contraction of the atria, which forces the blood in each atrium to squeeze into its corresponding ventricle. The resulting signal is called the P-wave. The atrioventricular node delays the excitation impulse until the blood transfer from the atria to the ventricles is completed, resulting in the P-R interval of the ECG waveform. The excitation impulse then causes contraction of the ventricles, which squeezes the blood into the arteries. This generates the

---

[2]This section has been adapted from *Handbook for Digital Signal Processing*, Sanjit K. Mitra and James F. Kaiser, Eds., ©1993, John Wiley & Sons. Adapted by permission of John Wiley & Sons.

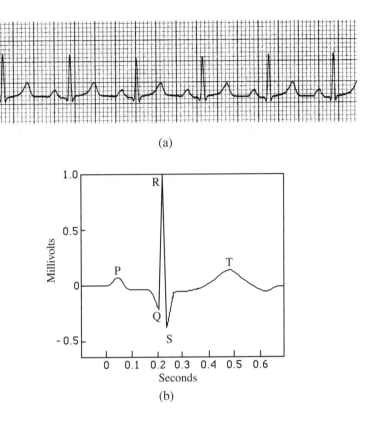

Figure 1.12: (a) A typical ECG trace, and (b) one cycle of an ECG waveform.

QRS part of the ECG waveform. During this phase the atria are relaxed and filled with blood. The T-wave of the waveform represents the relaxation of the ventricles. The complete process is repeated periodically, generating the ECG trace.

Each portion of the ECG waveform carries various types of information for the physician analyzing a patient's heart condition [Sha81]. For example, the amplitude and timing of the P and QRS portions indicate the condition of the cardiac muscle mass. Loss of amplitude indicates muscle damage, whereas increased amplitude indicates abnormal heart rates. Too long a delay in the atrioventricular node is indicated by a very long P-R interval. Likewise, blockage of some or all of the contraction impulses is reflected by intermittent synchronization between the P- and QRS-waves. Most of these abnormalities can be treated with various drugs, and the effectiveness of the drugs can again be monitored by observing the new ECG waveforms taken after the drug treatment.

In practice, there are various types of externally produced artifacts that appear in the ECG signal [Tom81]. Unless these interferences are removed, it is difficult for a physician to make a correct diagnosis. A common source of noise is the 60-Hz power lines whose radiated electric and magnetic fields are coupled to the ECG instrument through capacitive coupling and/or magnetic induction. Other sources of interference are the electromyographic signals that are the potentials developed by contracting muscles. These and other interferences can be removed with careful shielding and signal processing techniques.

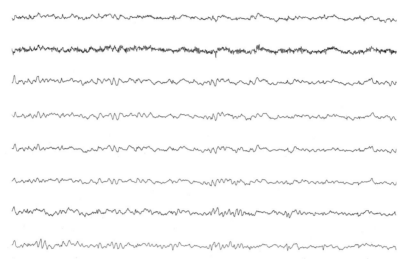

Figure 1.13: Multiple EEG signal traces.

Electroencephalogram (EEG) Signal

The summation of the electrical activity caused by the random firing of billions of individual neurons in the brain is represented by the EEG signal [Coh86], [Tom81]. In multiple EEG recordings, electrodes are placed at various positions on the scalp with two common electrodes placed on the earlobes, and potential differences between the various electrodes are recorded. A typical bandwidth of this type of EEG ranges from 0.5 to about 100 Hz, with the amplitudes ranging from 2 to 100 mV. An example of multiple EEG traces is shown in Figure 1.13.

Both frequency-domain and time-domain analyses of the EEG signal have been used for the diagnosis of epilepsy, sleep disorders, psychiatric malfunctions, etc. To this end, the EEG spectrum is subdivided into the following five bands: (1) the *delta* range, occupying the band from 0.5 to 4 Hz; (2) the *theta* range, occupying the band from 4 to 8 Hz; (3) the *alpha* range, occupying the band from 8 to 13 Hz; (4) the *beta* range, occupying the band from 13 to 22 Hz; and (5) the *gamma* range, occupying the band from 22 to 30 Hz.

The delta wave is normal in the EEG signals of children and sleeping adults. Since it is not common in alert adults, its presence indicates certain brain diseases. The theta wave is usually found in children even though it has been observed in alert adults. The alpha wave is common in all normal humans and is more pronounced in a relaxed and awake subject with closed eyes. Likewise, the beta activity is common in normal adults. The EEG exhibits rapid, low-voltage waves, called *rapid-eye-movement* (REM) waves, in a subject dreaming during sleep. Otherwise, in a sleeping subject, the EEG contains bursts of alpha-like waves, called *sleep spindles*. The EEG of an epileptic patient exhibits various types of abnormalities, depending on the type of epilepsy that is caused by uncontrolled neural discharges.

Seismic Signals

These types of signals are caused by the movement of rocks resulting from an earthquake, a volcanic erup-tion, or an underground explosion [Bol93]. The ground movement generates elastic waves that propagate through the body of the earth in all directions from the source of movement. Three basic types of elastic waves are generated by the earth movement. Two of these waves propagate through the body of the earth, one moving faster with respect to the other. The faster moving wave is called the *primary* or *P-wave*, while the slower moving one is called the *secondary* or *S-wave*. The third type of wave is known as the *surface wave*, which moves along the ground surface. These seismic waves are converted into electrical signals by a seismograph and are recorded on a strip chart recorder or a magnetic tape.

Because of the three-dimensional nature of ground movement, a seismograph usually consists of three separate recording instruments that provide information about the movements in the two horizontal directions and one vertical direction and develops three records as indicated in Figure 1.14. Each such record is a one-dimensional signal. From the recorded signals it is possible to determine the magnitude of the earthquake or nuclear explosion and the location of the source of the original earth movement.

Seismic signals also play an important role in the geophysical exploration for oil and gas [Rob80]. In this type of application linear arrays of seismic sources, such as high-energy explosives, are placed at regular intervals on the ground surface. The explosions cause seismic waves to propagate through the subsurface geological structures and reflect back to the surface from interfaces between geological strata. The reflected waves are converted into electrical signals by a composite array of geophones laid out in certain patterns and displayed as a two-dimensional signal that is a function of time and space, called a *trace gather*, as indicated in Figure 1.15. Before these signals are analyzed, some preliminary time and amplitude corrections are made on the data to compensate for different physical phenomena. From the corrected data, the time differences between reflected seismic signals are used to map structural deformations, whereas the amplitude changes usually indicate the presence of hydrocarbons.

Diesel Engine Signal

Signal processing is playing an important role in the precision adjustment of diesel engines during production [Jur81]. Efficient operation of the engine requires the accurate determination of the topmost point of piston travel (called the *top dead center*) inside the cylinder of the engine. Figure 1.16 shows the signals generated by a dual probe inserted into the combustion chamber of a diesel engine in place of the glow plug. The probe consists of a microwave antenna and a photodiode detector. The microwave probe captures signals reflected from the cylinder cavity caused by the up and down motion of the piston while the engine is running. Interestingly, the waveforms of these signals exhibit a symmetry around the top dead center independent of the engine speed, temperature, cylinder pressure, or air-fuel ratio. The point of symmetry is determined automatically by a microcomputer, and the fuel-injection pump position is then adjusted by the computer accurately to within ± 0.5 degree using the luminosity signal sensed by the photodiode detector.

Speech Signals

The linear acoustic theory of speech production has led to mathematical models for the representation of speech signals. A speech signal is formed by exciting the vocal tract and is composed of two types of sounds: *voiced* and *unvoiced* [Rab78], [Slu84]. The voiced sound, which includes the vowels and a number of consonants such as B, D, L, M, N, and R, is excited by the pulsatile airflow resulting from the vibration of the vocal folds. On the other hand, the unvoiced sound is produced downstream in the forward part of the oral cavity (mouth) with the vocal cords at rest and includes sounds like F, S, and SH.

Figure 1.17(a) depicts the speech waveform of a male utterance "every salt breeze comes from the sea" [Fla79]. The total duration of the speech waveform is 2.5 seconds. Magnified versions of the "A" and "S" segments in the word "salt" are sketched in Figure 1.17(b) and (c), respectively. The slowly varying low-frequency voiced waveform of "A" and the high-frequency unvoiced fricative waveform of "S" are evident from the magnified waveforms. The voiced waveform in Figure 1.17(b) is seen to be quasi-periodic and can be modeled by a sum of a finite number of sinusoids. The lowest frequency of oscillation in this representation is called the *fundamental frequency* or *pitch frequency*. The unvoiced waveform in Figure 1.17(c) has no regular fine structure and is more noise-like.

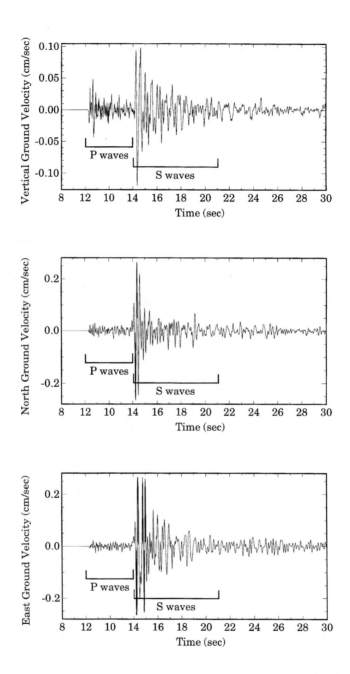

Figure 1.14: Seismograph record of the Northridge aftershock, January 29, 1994. Recorded at Stone Canyon Reservoir, Los Angeles, CA. (Courtesy of Institute for Crustal Research, University of California, Santa Barbara, CA.)

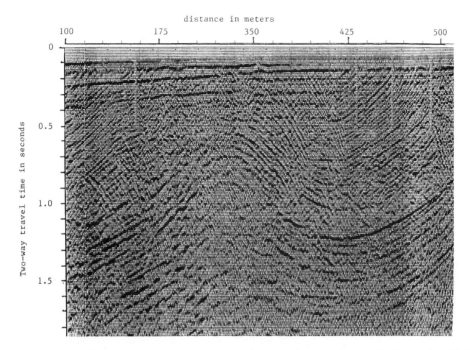

Figure 1.15: A typical seismic signal trace gather. (Courtesy of Institute for Crustal Research, University of California, Santa Barbara, CA.)

One of the major applications of digital signal processing techniques is in the general area of speech processing. Problems in this area are usually divided into three groups: (1) speech analysis, (2) speech synthesis, and (3) speech analysis and synthesis [Opp78]. Digital speech analysis methods are used in automatic speech recognition, speaker verification, and speaker identification. Applications of digital speech synthesis techniques include reading machines for the automatic conversion of written text into speech, and retrieval of data from computers in speech form by remote access through terminals or telephones. One example belonging to the third group is voice scrambling for secure transmission. Speech data compression for an efficient use of the transmission medium is another example of the use of speech analysis followed by synthesis. A typical speech signal after conversion into a digital form contains about 64,000 bits per second (bps). Depending on the desired quality of the synthesized speech, the original data can be compressed considerably, e.g., down to about 1000 bps.

Musical Sound Signal

The electronic synthesizer is an example of the use of modern signal processing techniques [Moo77], [Ler83]. The natural sound generated by most musical instruments is generally produced by mechanical vibrations caused by activating some form of oscillator that then causes other parts of the instrument to vibrate. All these vibrations together in a single instrument generate the musical sound. In a violin the primary oscillator is a stretched piece of string (cat gut). Its movement is caused by drawing a bow across it; this sets the wooden body of the violin vibrating, which in turn sets up vibrations of the air inside as well as outside the instrument. In a piano the primary oscillator is a stretched steel wire that is set into vibratory motion by the hitting of a hammer, which in turn causes vibrations in the wooden body (sounding board) of the piano. In wind or brass instruments the vibration occurs in a column of air, and a mechanical change in the length of the air column by means of valves or keys regulates the rate of vibration.

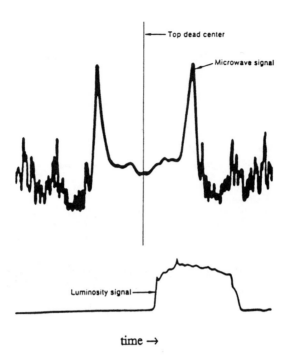

time →

Figure 1.16: Diesel engine signals. (Reproduced with permission from R. K. Jurgen, Detroit bets on electronics to stymie Japan, *IEEE Spectrum*, vol. 18, July 1981, pp. 29–32 ©1981 IEEE.)

The sound of orchestral instruments can be classified into two groups: *quasi-periodic* and *aperiodic*. Quasi-periodic sounds can be described by a sum of a finite number of sinusoids with independently varying amplitudes and frequencies. The sound waveforms of two different instruments, the cello and the bass drum, are indicated in Figure 1.18(a) and (b), respectively. In each figure, the top waveform is the plot of an entire isolated note, whereas the bottom plot shows an expanded version of a portion of the note: 10 ms for the cello and 80 ms for the bass drum. The waveform of the note from a cello is seen to be quasi-periodic. On the other hand, the bass drum waveform is clearly aperiodic. The tone of an orchestral instrument is commonly divided into three segments called the *attack* part, the *steady-state* part, and the *decay* part. Figure 1.18 illustrates this division for the two tones. Note that the bass drum tone of Figure 1.18(b) shows no steady-state part. A reasonable approximation of many tones is obtained by splicing together these parts. However, high-fidelity reproduction requires a more complex model.

Time Series

The signals described thus far are continuous functions with time as the independent variable. In many cases the signals of interest are naturally discrete functions of the independent variables. Often such signals are of finite duration. Examples of such signals are the yearly average number of sunspots, daily stock prices, the value of total monthly exports of a country, the yearly population of animal species in a certain geographical area, the annual yields per acre of crops in a country, and the monthly totals of international airline passengers over certain periods. This type of finite extent signal, usually called a *time series*, occurs in business, economics, physical sciences, social sciences, engineering, medicine, and many other fields. Plots of some typical time series are shown in Figures 1.19 to 1.21.

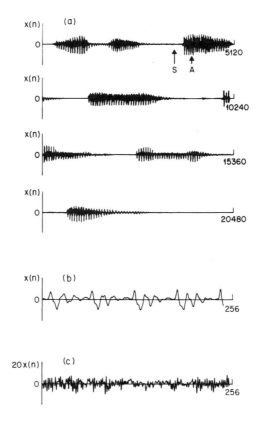

Figure 1.17: Speech waveform example: (a) sentence-length segment, (b) magnified version of the voiced segment (the letter A), and (c) magnified version of the unvoiced segment (the letter S). (Reproduced with permission from J. L. Flanagan et al., Speech coding, *IEEE Trans. on Communications*, vol. COM-27, April 1979, pp. 710–737 ©1979 IEEE.)

There are many reasons for analyzing a particular time series [Box70]. In some applications, there may be a need to develop a model to determine the nature of the dependence of the data on the independent variable and use it to forecast the future behavior of the series. As an example, in business planning, reasonably accurate sales forecasts are necessary. Some types of series possess seasonal or periodic components, and it is important to extract these components. The study of sunspot numbers is important for predicting climate variations. Invariably, the time series data are noisy, and their representations require models based on their statistical properties.

Images

As indicated earlier, an image is a two-dimensional signal whose intensity at any point is a function of two spatial variables. Common examples are photographs, still video images, radar and sonar images, and chest and dental x-rays. An image sequence, such as that seen in a television, is essentially a three-dimensional signal for which the image intensity at any point is a function of three variables: two spatial variables and time. Figure 1.22(a) shows the photograph of a digital image.

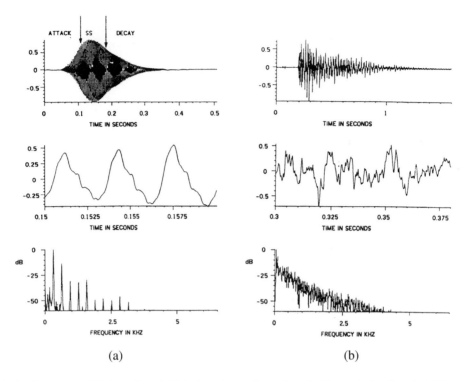

Figure 1.18: Waveforms of (a) the cello and (b) the bass drum. (Reproduced with permission from J. A. Moorer, Signal processing aspects of computer music: A survey, *Proceedings of the IEEE*, vol. 65, August 1977, pp. 1108–1137 ©1977 IEEE).

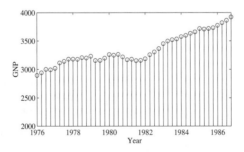

Figure 1.19: Seasonally adjusted quarterly Gross National Product of United States in 1982 dollars from 1976 to 1986. (Adapted from [Lüt91].)

The basic problems in image processing are image signal representation and modeling, enhancement, restoration, reconstruction from projections, analysis, and coding [Jai89].

Each picture element in a specific image represents a certain physical quantity; a characterization of the element is called the *image representation*. For example, a photograph represents the luminances of various objects as seen by the camera. An infrared image taken by a satellite or an airplane represents the temperature profile of a geographical area. Depending on the type of image and its applications, various

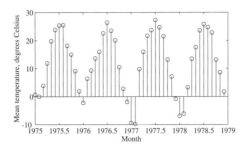

Figure 1.20: Monthly mean St. Louis, Missouri temperature in degrees Celsius for the years 1975 to 1978. (Adapted from [Mar87].)

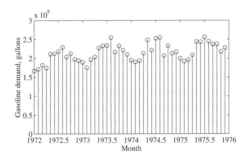

Figure 1.21: Monthly gasoline demand in Ontario, Canada (in millions of gallons) from January 1971 to December 1975. (Adapted from [Abr83].)

types of image models are usually defined. Such models are also based on perception, and on local or global characteristics. The nature and performance of the image processing algorithms depend on the image model being used.

Image enhancement algorithms are used to emphasize specific image features to improve the quality of the image for visual perception or to aid in the analysis of the image for feature extraction. These include methods for contrast enhancement, edge detection, sharpening, linear and nonlinear filtering, zooming, and noise removal. Figure 1.22(b) shows the contrast-enhanced version of the image of Figure 1.22(a) developed using a nonlinear filter [Thu00].

The algorithms used for elimination or reduction of degradations in an image, such as blurring and geometric distortion caused by the imaging system and/or its surroundings are known as *image restoration*. *Image reconstruction* from projections involves the development of a two-dimensional image slice of a three-dimensional object from a number of planar projections obtained from various angles. By creating a number of contiguous slices, a three-dimensional image giving an inside view of the object is developed.

Image analysis methods are employed to develop a quantitative description and classification of one or more desired objects in an image.

For digital processing, an image needs to be sampled and quantized using an analog-to-digital converter. A reasonable size digital image in its original form takes a considerable amount of memory space for storage. For example, an image of size 512×512 samples with 8-bit resolution per sample contains over 2 million bits. *Image coding* methods are used to reduce the total number of bits in an image without any degradation in visual perception quality as in speech coding, e.g., down to about 1 bit per sample on the average.

(a)   (b)

Figure 1.22: (a) A digital image, and (b) its contrast-enhanced version. (Reproduced with permission from *Nonlinear Image Processing*, S. K. Mitra and G. Sicuranza, Eds., Academic Press, New York NY, ©2000.)

1.4 Typical Signal Processing Applications[3]

There are numerous applications of signal processing that we often encounter in our daily life without being aware of them. Due to space limitations, it is not possible to discuss all of these applications. However, an overview of selected applications is presented.

1.4.1 Sound Recording Applications

The recording of most musical programs nowadays is usually made in an acoustically inert studio. The sound from each instrument is picked up by its own microphone closely placed to the instrument and is recorded on a single track in a multitrack tape recorder containing as many as 48 tracks. The signals from individual tracks in the master recording are then edited and combined by the sound engineer in a *mix-down* system to develop a two-track stereo recording. There are a number of reasons for following this approach. First, the closeness of each individual microphone to its assigned instrument provides a high degree of separation between the instruments and minimizes the background noise in the recording. Second, the sound part of one instrument can be rerecorded later if necessary. Third, during the mix-down process the sound engineer can manipulate individual signals by using a variety of signal processing devices to alter the musical balances between the sounds generated by the instruments, can change the timbre, and can add natural room acoustics effects and other special effects [Ble78], [Ear76].

Various types of signal processing techniques are utilized in the mix-down phase. Some are used to modify the spectral characteristics of the sound signal and to add special effects, whereas others are used to improve the quality of the transmission medium. The signal processing circuits most commonly used are: (1) compressors and limiters, (2) expanders and noise gates, (3) equalizers and filters, (4) noise reduction systems, (5) delay and reverberation systems, and (6) circuits for special effects [Ble78], [Ear76], [Hub89], [Wor89]. These operations are usually performed on the original analog audio signals and are implemented using analog circuit components. However, there is a growing trend toward all digital implementation and its use in the processing of the digitized versions of the analog audio signals [Ble78].

---

[3]This section has been adapted from *Handbook for Digital Signal Processing*, Sanjit K. Mitra and James F. Kaiser, Eds., ©1993, John Wiley & Sons. Adapted by permission of John Wiley & Sons.

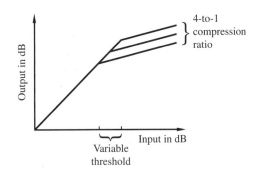

Figure 1.23: Transfer characteristic of a typical compressor.

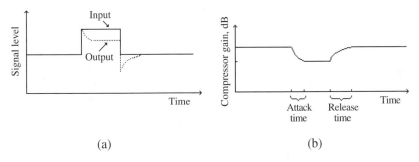

Figure 1.24: Parameters characterizing a typical compressor.

Compressors and limiters. These devices are used for the compression of the dynamic range of an audio signal. The compressor can be considered as an amplifier with two gain levels: the gain is unity for input signal levels below a certain threshold and less than unity for signals with levels above the threshold. The threshold level is adjustable over a wide range of the input signal. Figure 1.23 shows the transfer characteristic of a typical compressor.

The parameters characterizing a compressor are its compression ratio, threshold level, attack time, and release time, which are illustrated in Figure 1.24.

When the input signal level suddenly rises above a prescribed threshold, the time taken by the compressor to adjust its normal unity gain to the lower value is called the *attack time*. Because of this effect, the output signal exhibits a slight degree of overshoot before the desired output level is reached. A zero attack time is desirable to protect the system from sudden high-level transients. However, in this case, the impact of sharp musical attacks is eliminated, resulting in a dull "lifeless" sound [Wor89]. A longer attack time causes the output to sound more percussive than normal.

Similarly, the time taken by the compressor to reach its normal unity gain value when the input level suddenly drops below the threshold is called the *release time* or *recovery time*. If the input signal fluctuates rapidly around the threshold in a small region, the compressor gain also fluctuates up and down. In such a situation, the rise and fall of background noise results in an audible effect called *breathing* or *pumping*, which can be minimized with a longer release time for the compressor gain.

There are various applications of the compressor unit in musical recording [Ear76]. For example, it can be used to eliminate variations in the peaks of an electric bass output signal by clamping them to a constant level, thus providing an even and solid bass line. To maintain the original character of the instrument, it is

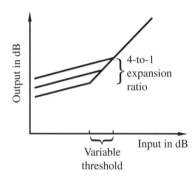

Figure 1.25: Transfer characteristic of a typical expander.

necessary to use a compressor with a long recovery time compared to the natural decay rate of the electric bass. The device is also useful to compensate for the wide variations in the signal level produced by a singer who moves frequently, changing the distance from the microphone.

A compressor with a compression ratio of 10-to-1 or greater is called a *limiter* since its output levels are essentially clamped to the threshold level. The limiter is used to prevent overloading of amplifiers and other devices caused by signal peaks exceeding certain levels.

Expanders and noise gates. The expander's function is opposite that of the compressor. It is also an amplifier with two gain levels: the gain is unity for input signal levels above a certain threshold and less than unity for signals with levels below the threshold. The threshold level is again adjustable over a wide range of the input signal. Figure 1.25 shows the transfer characteristic of a typical expander. The *expander* is used to expand the dynamic range of an audio signal by boosting the high-level signals and attenuating the low-level signals. The device can also be used to reduce noise below a threshold level.

The expander is characterized by its expansion ratio, threshold level, attack time, and release time. Here, the time taken by the device to reach the normal unity gain for a sudden change in the input signal to a level above the threshold is defined as the *attack time*. Likewise, the time required by the device to lower the gain from its normal value of one for a sudden decrease in the input signal level is called the *release time*.

The noise gate is a special type of expander that heavily attenuates signals with levels below the threshold. It is used, for example, to totally cut off a microphone during a musical pause so as not to pass the noise being picked up by the microphone.

Equalizers and filters. Various types of filters are used to modify the frequency response of a recording or the monitoring channel. One such filter, called the *shelving filter*, provides boost (rise) or cut (drop) in the frequency response at either the low or at the high end of the audio frequency range while not affecting the frequency response in the remaining range of the audio spectrum, as shown in Figure 1.26. *Peaking filters* are used for midband equalization and are designed to have either a bandpass response to provide a boost or a bandstop response to provide a cut, as indicated in Figure 1.27.

The parameters characterizing a low-frequency shelving filter are the two frequencies f_{1L} and f_{2L}, where the magnitude response begins tapering up or down from a constant level and the low-frequency gain levels in dB. Likewise, the parameters characterizing a high-frequency shelving filter are the two frequencies f_{1H} and f_{2H}, where the magnitude response begins tapering up or down from a constant level and the high-frequency gain levels in dB. In the case of a peaking filter, the parameters of interest are the center frequency f_0, the 3-dB bandwidth Δf of the bell-shaped curve, and the gain level at the center

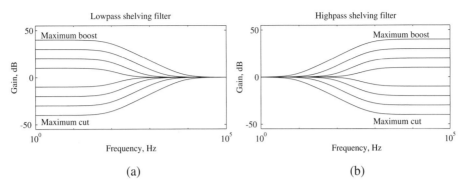

Figure 1.26: Frequency responses of (a) low-frequency shelving filter and (b) high-frequency shelving filter.

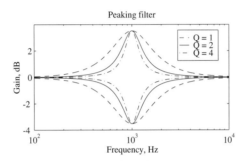

Figure 1.27: Peaking filter frequency response.

frequency. Most often, the quality factor $Q = f_0/\Delta f$ is used to characterize the shape of the frequency response instead of the bandwidth Δf.

A typical equalizer consists of a cascade of a low-frequency shelving filter, a high-frequency shelving filter, and three or more peaking filters with adjustable parameters to provide adjustment of the overall equalizer frequency response over a broad range of frequencies in the audio spectrum. In a *parametric equalizer*, each individual parameter of its constituent filter blocks can be varied independently without affecting the parameters of the other filters in the equalizer.

The *graphic equalizer* consists of a cascade of peaking filters with fixed center frequencies but adjustable gain levels that are controlled by vertical slides in the front panel. The physical position of the slides reasonably approximates the overall equalizer magnitude response, as shown schematically in Figure 1.28.

Other types of filters that also find applications in the musical recording and transfer processes are the *lowpass, highpass*, and *notch filters*. Their corresponding frequency responses are indicated in Figure 1.29. The notch filter is designed to attenuate a particular frequency component and has a narrow notch width so as not to affect the rest of the musical program.

Two major applications of equalizers and filters in recording are to correct certain types of problems that may have occurred during the recording or the transfer process and to alter the harmonic or timbral contents of a recorded sound purely for musical or creative purposes [Ear76]. For example, a direct transfer of a musical recording from old 78 rpm disks to a wideband playback system will be highly noisy due to the limited bandwidth of the old disks. To reduce this noise, a bandpass filter with a passband matching the bandwidth of the old records is utilized. Often, older recordings are made more pleasing by adding a

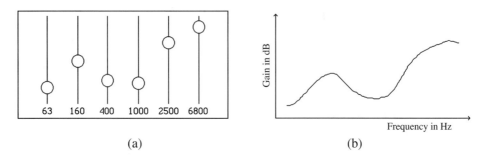

Figure 1.28: Graphic equalizer: (a) control panel settings, and (b) corresponding frequency response. (Adapted from [Ear76].)

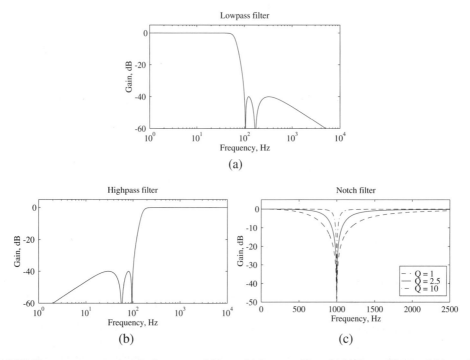

Figure 1.29: Frequency responses of other types of filters: (a) lowpass filter, (b) highpass filter, and (c) notch filter.

broad high-frequency peak in the 5- to 10-kHz range and by shelving out some of the lower frequencies. The notch filter is particularly useful in removing 60-Hz power supply hum.

In creating a program by mixing down a multichannel recording, the recording engineer usually employs equalization of individual tracks for creative reasons [Ear76]. For example, a "fullness" effect can be added to weak instruments such as the acoustical guitar, by boosting frequency components in the range 100 to 300 Hz. Similarly, by boosting the 2- to 4-kHz range, the transients caused by the fingers against the string of an acoustical guitar can be made more pronounced. A high-frequency shelving boost above the 1- to 2-kHz range increases the "crispness" in percussion instruments such as the bongo or snare drums.

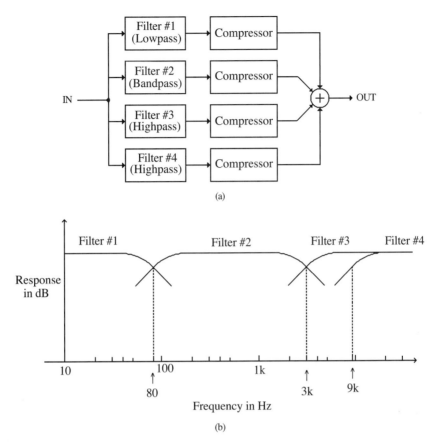

Figure 1.30: The Dolby A-type noise reduction scheme for the recording mode. (a) Block diagram, and (b) frequency responses of the four filters with cutoff frequencies as shown.

Noise reduction system. The overall dynamic range of human hearing is over 120 dB. However, most recording and transmission mediums have a much smaller dynamic range. The music to be recorded must be above the sound background or noise. If the background noise is around 30 dB, the dynamic range available for the music is only 90 dB, requiring dynamic range compression for noise reduction.

A noise reduction system consists of two parts. The first part provides the compression during the recording mode while the second part provides the complementary expansion during the playback mode. To this end, the most popular methods in musical recording are the Dolby noise reduction schemes, of which there are several types [Ear76], [Hub89], [Wor89].

In the Dolby A-type method used in professional recording, for the recording mode, the audio signal is split into four frequency bands by a bank of four filters; separate compression is provided in each band and the outputs of the compressors are combined, as indicated in Figure 1.30(a). Moreover, the compression in each band is restricted to a 20-dB input range from -40 to -20 dB. Below the lower threshold (-40 dB), very low level signals are boosted by 10 dB, and above the upper threshold (-20 dB), the system has unity gain, passing the high-level signals unaffected. The transfer characteristic for the record mode is thus as shown in Figure 1.31.

In the playback mode, the scheme is essentially the same as that in the recording mode, except here the compressors are replaced by expanders with complementary transfer characteristics, as indicated in Figure 1.31. Here, the expansion is limited to a 10-dB input range from -30 to -20 dB. Above the upper

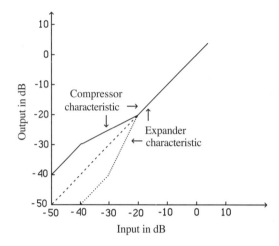

Figure 1.31: Compressor and expander transfer characteristic for the Dolby A-type noise reduction scheme.

threshold (-20 dB), very high level signals are cut by 10 dB, while below the lower threshold (-30 dB), the system has unity gain passing the low-level signals unaffected.

Note that for each band, a 2-to-1 compression is followed by a 1-to-2 complementary expansion such that the dynamic range of the signal at the input of the compressor is exactly equal to that at the expander output. This type of overall signal processing operation is often called *companding*. Moreover, the companding operation in one band has no effect on a signal in another band and may often be masked by other bands with no companding.

Delay and reverberation systems. Music generated in an inert studio does not sound natural compared to the music performed inside a room, such as a concert hall. In the latter case, the sound waves propagate in all directions and reach the listener from various directions and at various times, depending on the distance traveled by the sound waves from the source to the listener. The sound wave coming directly to the listener, called the *direct sound,* reaches first and determines the listener's perception of the location, size, and nature of the sound source. This is followed by a few closely spaced echoes, called *early reflections,* generated by reflections of sound waves from all sides of the room and reaching the listener at irregular times. These echoes provide the listener's subconscious cues as to the size of the room. After these early reflections, more and more densely packed echoes reach the listener due to multiple reflections. The latter group of echoes is referred to as the *reverberation.* The amplitude of the echoes decay exponentially with time as a result of attenuation at each reflection. Figure 1.32 illustrates this concept. The period of time in which the reverberation falls by 60 dB is called the *reverberation time.* Since the absorption characteristics of different materials are not the same at different frequencies, the reverberation time varies from frequency to frequency.

Delay systems with adjustable delay factors are employed to artificially create the early reflections. Electronically generated reverberation combined with artificial echo reflections are usually added to the recordings made in a studio. The block diagram representation of a typical delay-reverberation system in a monophonic system is depicted in Figure 1.33.

There are various other applications of electronic delay systems, some of which are described next [Ear76].

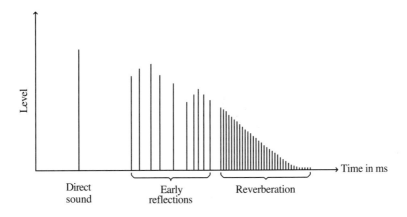

Figure 1.32: Various types of echoes generated by a single sound source in a room.

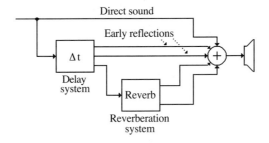

Figure 1.33: Block diagram of a complete delay-reverberation system in a monophonic system.

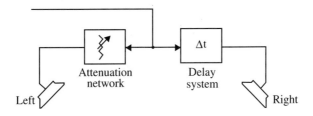

Figure 1.34: Localization of sound source using delay systems and attenuation network.

Special effects. By feeding in the same sound signal through an adjustable delay and gain control, as indicated in Figure 1.34, it is possible to vary the localization of the sound source from the left speaker to the right for a listener located on the plane of symmetry. For example, in Figure 1.34, a 0-dB loss in the left channel and a few milliseconds delay in the right channel give the impression of a localization of the sound source at the left. However, lowering of the left-channel signal level by a few-dB loss results in a phantom image of the sound source moving toward the center. This scheme can be further extended to provide a degree of *sound broadening* by phase shifting one channel with respect to the other through allpass networks[4] as shown in Figure 1.35.

---

[4]An allpass network is characterized by a magnitude spectrum which is equal to one for all frequencies.

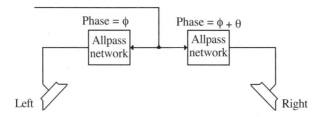

Figure 1.35: Sound broadening using allpass networks.

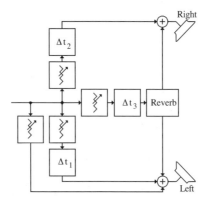

Figure 1.36: A possible application of delay systems and reverberation in a stereophonic system.

Another application of the delay-reverberation system is in the processing of a single track into a pseudo-stereo format while simulating a natural acoustical environment, as illustrated in Figure 1.36.

The delay system can also be used to generate a chorus effect from the sound of a soloist. The basic scheme used is illustrated in Figure 1.37. Each of the delay units has a variable delay controlled by a low-frequency pseudo-random noise source to provide a random pitch variation [Ble78].

It should be pointed out here that additional signal processing is employed to make the stereo submaster developed by the sound engineer more suitable for the record-cutting lathe or the cassette tape duplicator.

1.4.2 Telephone Dialing Applications

Signal processing plays a key role in the detection and generation of signaling tones for push-button telephone dialing [Dar76]. In telephones equipped with TOUCH-TONE® dialing, the pressing of each button generates a unique set of two-tone signals, called *dual-tone multifrequency (DTMF) signals*, that are processed at the telephone central office to identify the number pressed by determining the two associated tone frequencies. Seven frequencies are used to code the 10 decimal digits and the two special buttons marked " * " and " # ". The low-band frequencies are 697 Hz, 770 Hz, 852 Hz, and 941 Hz. The remaining three frequencies belonging to the highband are 1209 Hz, 1336 Hz, and 1477 Hz. The fourth high-band frequency of 1633 Hz is not presently in use and has been assigned for future applications to permit the use of four additional push-buttons for special services. The frequency assignments used in the TOUCH-TONE® dialing scheme are shown in Figure 1.38 [ITU84].

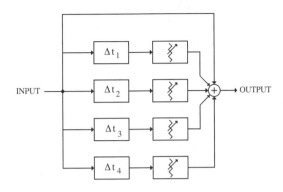

Figure 1.37: A scheme for implementing chorus effect.

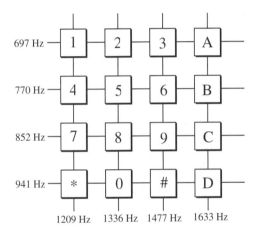

Figure 1.38: The tone frequency assignments for TOUCH-TONE® dialing.

The scheme used to identify the two frequencies associated with the button that has been pressed is shown in Figure 1.39. Here, the two tones are first separated by a lowpass and a highpass filter. The passband cutoff frequency of the lowpass filter is slightly above 1000 Hz, whereas that of the highpass filter is slightly below 1200 Hz. The output of each filter is next converted into a square wave by a limiter and then processed by a bank of bandpass filters with narrow passbands. The four bandpass filters in the low-frequency channel have center frequencies at 697 Hz, 770 Hz, 852 Hz, and 941 Hz. The four bandpass filters in the high-frequency channel have center frequencies at 1209 Hz, 1336 Hz, 1477 Hz, and 1633 Hz. The detector following each bandpass filter develops the necessary dc switching signal if its input voltage is above a certain threshold.

All the signal processing functions described above are usually implemented in practice in the analog domain. However, increasingly, these functions are being implemented using digital techniques.[5]

---

[5]See Section 11.1.

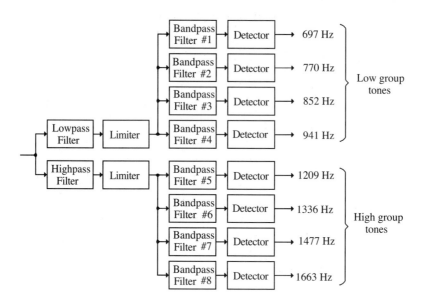

Figure 1.39: The tone detection scheme for TOUCH-TONE® dialing.

1.4.3 FM Stereo Applications

For wireless transmission of a signal occupying a low-frequency range, such as an audio signal, it is necessary to transform the signal to a high-frequency range by modulating it onto a high-frequency carrier. At the receiver, the modulated signal is demodulated to recover the low-frequency signal. The signal processing operations used for wireless transmission are modulation, demodulation, and filtering. Two commonly used modulation schemes for radio are amplitude modulation (AM) and frequency modulation (FM).

We next review the basic idea behind the FM stereo broadcasting and reception scheme as used in the United States [Cou83]. An important feature of this scheme is that at the receiving end, the signal can be heard over a standard monaural FM radio with a single speaker or over a stereo FM radio with two speakers. The system is based on the frequency-division multiplexing (FDM) method described earlier in Section 1.2.5.

The block diagram representations of the FM stereo transmitter and the receiver are shown in Figure 1.40(a) and (b), respectively. At the transmitting end, the sum and the difference of the left and right channel audio signals, $s_L(t)$ and $s_R(t)$, are first formed. Note that the summed signal $s_L(t) + s_R(t)$ is used in monaural FM radio. The difference signal $s_L(t) - s_R(t)$ is modulated using the double-sideband suppressed carrier (DSB-SC) scheme[6] using a subcarrier frequency f_{sc} of 38 kHz. The summed signal, the modulated difference signal, and a 19-kHz pilot tone signal are then added, developing the composite baseband signal $s_B(t)$. The spectrum of the composite signal is shown in Figure 1.40(c). The baseband signal is next modulated onto the main carrier frequency f_c using the frequency modulation method. At the receiving end, the FM signal is demodulated to derive the baseband signal $s_B(t)$ which is then separated into the low-frequency summed signal and the modulated difference signal using a lowpass filter and a bandpass filter. The cutoff frequency of the lowpass filter is around 15 kHz, whereas the center frequency of the bandpass filter is at 38-kHz. The 19-kHz pilot tone is used in the receiver to develop the 38-kHz

---

[6]See Section 1.2.4.

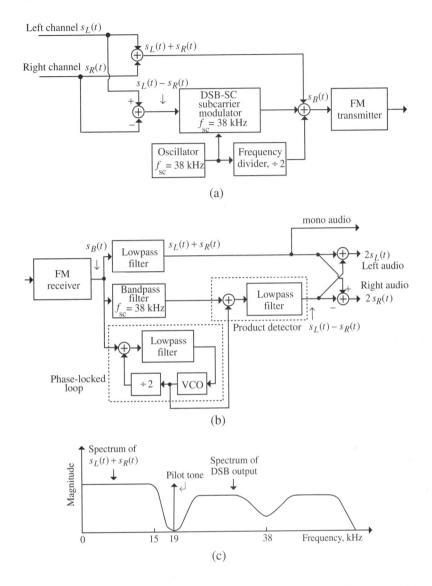

Figure 1.40: The FM stereo system: (a) transmitter, (b) receiver, and (c) spectrum of the composite baseband signal $s_B(t)$.

reference signal for a coherent subcarrier demodulation for recovering the audio difference signal. The sum and difference of the two audio signals create the desired left audio and right audio signals.

1.4.4 Electronic Music Synthesis

The generation of the sound of a musical instrument using electronic circuits is another example of the application of signal processing methods [All80], [Moo77]. The basis of such music synthesis is the

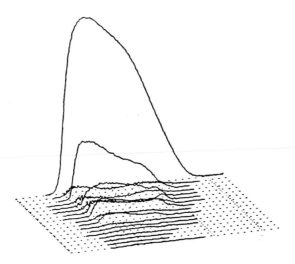

Figure 1.41: Perspective plot of the amplitude functions $A_k(t)$ for an actual note from a clarinet. (Reproduced with permission from J. A. Moorer, Signal processing aspects of computer music: A survey, *Proceedings of the IEEE*, vol. 65, August 1977, pp. 1108–1137 ©1977 IEEE.)

following representation of the sound signal $s(t)$:

$$s(t) = \sum_{k=1}^{N} A_k(t) \sin(2\pi f_k(t)t) \tag{1.20}$$

where $A_k(t)$ and $f_k(t)$ are the time-varying amplitude and frequency of the kth component of the signal. The frequency function $f_k(t)$ varies slowly with time. For an instrument playing an isolated tone, $f_k(t) = kf_o$, i.e.,

$$s(t) = A_k(t) \sin (2\pi k f_o t) \tag{1.21}$$

where f_o is called the *fundamental frequency*. In a musical sound with many tones, all other frequencies are usually integer multiples of the fundamental and are called *partial frequencies,* also called *harmonics*. Figure 1.41 shows, for example, the perspective plot of the amplitude functions as a function of time of 17 partial frequency components of an actual note from a clarinet. The aim of the synthesis is to produce electronically the $A_k(t)$ and $f_k(t)$ functions. To this end, the two most popular approaches followed are described next.

Subtractive synthesis. This approach, which nearly duplicates the sound generation mechanism of a musical instrument, is based on the generation of a periodic signal containing all required harmonics and the use of filters to selectively attenuate (i.e., *subtract*) unwanted partial frequency components. The frequency-dependent gain of the filters can also be used to boost certain frequencies. The desired variations in the amplitude functions are generated by an analog multiplier or a voltage-controlled amplifier. Additional variations in the amplitude functions can be provided by dynamically adjusting the frequency response characteristics of the filters.

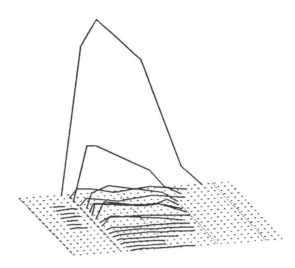

Figure 1.42: A piecewise-linear approximation to the amplitude functions of Figure 1.41. (Reproduced with permission from J. A. Moorer, Signal processing aspects of computer music: A survey, *Proceedings of the IEEE*, vol. 65, August 1977, pp. 1108–1137 ©1977 IEEE.)

Additive synthesis. Here, partial frequency components are generated independently by oscillators with time-varying oscillation frequencies. The amplitudes of the required signals are then individually modified, approximating the actual variations obtained by analysis and combined (i.e., *added*) to produce the desired sound signal. For example, a piecewise linear approximation of the clarinet note of Figure 1.41 is sketched in Figure 1.42 and can be used to generate a reasonable replica of the note. Usually, some alterations to the amplitude and frequency functions may be needed before the music that is generated sounds as close as possible to that of the original instrument.

1.4.5 Echo Cancellation in Telephone Networks

In a telephone network the central offices perform the necessary switching to connect two subscribers [Dut80], [Fre78], [Mes82].For economic reasons, a *two-wire* circuit is used to connect a subscriber to his/her central office, whereas the central offices are connected using *four-wire* circuits. A two-wire circuit is bidirectional and carries signals in both directions. A four-wire circuit uses two separate unidirectional paths for signal transmission in both directions. The latter is preferred for long-distance trunk connections since signals at intermediate points in the trunk can be equalized and amplified using repeaters and, if necessary, multiplexed easily. A hybrid coil in the central office provides the interface between a two-wire circuit and a four-wire circuit, as shown in Figure 1.43. The hybrid circuit ideally should provide a perfect impedance match to the two-wire circuit by impedance balancing so that the incoming four-wire receive signal is passed directly to the two-wire circuit connected to the hybrid with no portion appearing in the four-wire transmit path. However, to save cost, a hybrid coil is shared among several subscribers. Thus, it is not possible to provide a perfect impedance match in every case since the length of the subscriber lines vary. The resulting imbalance causes a large portion of the incoming receive signal from the distance talker to appear in the transmit path, and it is returned to the talker as an echo. Figure 1.44 illustrates the normal transmission between a talker and a listener as well as two possible major echo paths.

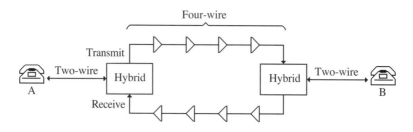

Figure 1.43: Basic 2/4-wire interconnection scheme.

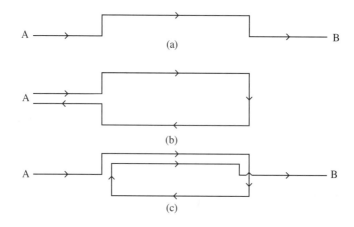

Figure 1.44: Various signal paths in a telephone network. (a) Transmission path from talker A to listener B, (b) echo path for talker A, and (c) echo path for listener B.

The effect of the echo can be annoying to the talker, depending on the amplitude and delay of the echo, i.e., on the length of the trunk circuit. The effect of the echo is worst for telephone networks involving geostationary satellite circuits, where the echo delay is about 540 ms.

Several methods are followed to reduce the effect of the echo. In trunk circuits up to 3000 km in length, adequate reduction of the echo is achieved by introducing additional signal loss in both directions of the four-wire circuit. In this scheme, an improvement in the signal-to-echo ratio is realized since the echo undergoes loss in both directions while the signals are attenuated only once.

For distances greater than 3000 km, echoes are controlled by means of an echo suppressor inserted in the trunk circuit, as indicated in Figure 1.45. The device is essentially a voice-activated switch implementing two functions. It first detects the direction of the conversation and then blocks the opposite path in the four-wire circuit. Even though it introduces distortion when both subscribers are talking by clipping parts of the speech signal, the echo suppressor has provided a reasonably acceptable solution for terrestrial transmission.

For telephone conversation involving satellite circuits, an elegant solution is based on the use of an echo canceler. The circuit generates a replica of the echo using the signal in the receive path and subtracts it from the signal in the transmit path, as indicated in Figure 1.46. Basically, it is an adaptive filter structure whose parameters are adjusted using certain adaptation algorithms until the residual signal is satisfactorily

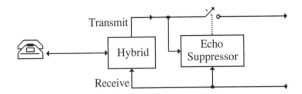

Figure 1.45: Echo suppression scheme.

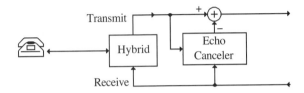

Figure 1.46: Echo cancellation scheme.

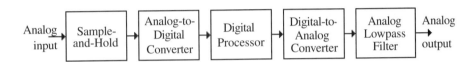

Figure 1.47: Scheme for the digital processing of an analog signal.

minimized.[7] Typically, an echo reduction of about 40 dB is considered satisfactory in practice. To eliminate the problem generated when both subscribers are talking, the adaptation algorithm is disabled when the signal in the transmit path contains both the echo and the signal generated by the speaker closer to the hybrid coil.

1.5 Why Digital Signal Processing?[8]

In some sense, the origin of digital signal processing techniques can be traced back to the seventeenth century when finite difference methods, numerical integration methods, and numerical interpolation methods were developed to solve physical problems involving continuous variables and functions. The more recent interest in digital signal processing arose in the 1950s with the availability of large digital computers. Initial applications were primarily concerned with the simulation of analog signal processing methods. Around the beginning of the 1960s, researchers began to consider digital signal processing as a separate field by itself. Since then, there have been significant and myriad developments and breakthroughs in both theory and applications of digital signal processing.

Digital processing of an analog signal consists basically of three steps: conversion of the analog signal into a digital form, processing of the digital version, and finally, conversion of the processed digital signal back into an analog form. Figure 1.47 shows the overall scheme in a block diagram form.

---

[7]For a review of adaptive filtering methods see [Cio93].

[8]This section has been adapted from *Handbook for Digital Signal Processing*, Sanjit K. Mitra and James F. Kaiser, Eds., ©1993, John Wiley & Sons. Adapted by permission of John Wiley & Sons.

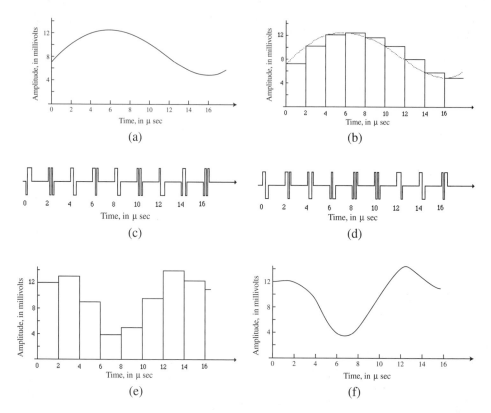

Figure 1.48: Typical waveforms of signals appearing at various stages in Figure 1.47. (a) Analog input signal, (b) output of the S/H circuit, (c) A/D converter output, (d) output of the digital processor, (e) D/A converter output, and (f) analog output signal. In (c) and (d), the digital HIGH and LOW levels are shown as positive and negative pulses for clarity.

Since the amplitude of the analog input signal varies with time, a *sample-and-hold* (S/H) circuit is used first to sample the analog input at periodic intervals and hold the sampled value constant at the input of the *analog-to-digital* (A/D) *converter* to permit accurate digital conversion. The input to the A/D converter is a staircase-type analog signal if the S/H circuit holds the sampled value until the next sampling instant. The output of the A/D converter is a binary data stream that is next processed by the digital processor implementing the desired signal processing algorithm. The output of the digital processor, another binary data stream, is then converted into a staircase-type analog signal by the *digital-to-analog* (D/A) *converter.* The lowpass filter at the output of the D/A converter then removes all undesired high-frequency components and delivers at its output the desired processed analog signal. Figure 1.48 illustrates the waveforms of the pertinent signals at various stages in the above process, where for clarity the two levels of the binary signals are shown as a positive and a negative pulse, respectively.

In contrast to the above, a direct analog processing of an analog signal is conceptually much simpler since it involves only a single processor, as illustrated in Figure 1.49. It is therefore natural to ask what the advantages are of digital processing of an analog signal.

There are of course many advantages in choosing digital signal processing. The most important ones are discussed next [Bel84], [Pro92].

Figure 1.49: Analog processing of analog signals.

Unlike analog circuits, the operation of digital circuits does not depend on precise values of the digital signals. As a result, a digital circuit is less sensitive to tolerances of component values and is fairly independent of temperature, aging, and most other external parameters. A digital circuit can be reproduced easily in volume quantities and does not require any adjustments either during construction or later while in use. Moreover, it is amenable to full integration, and with the recent advances in *very large scale integrated* (VLSI) circuits, it has been possible to integrate highly sophisticated and complex digital signal processing systems on a single chip.

In a digital processor, the signals and the coefficients describing the processing operation are represented as binary words. Thus, any desirable accuracy can be achieved by simply increasing the wordlength, subject to cost limitation. Moreover, the dynamic ranges for signals and coefficients can be increased still further by using floating-point arithmetic if necessary.

Digital processing allows the sharing of a given processor among a number of signals by timesharing, thus reducing the cost of processing per signal. Figure 1.50 illustrates the concept of timesharing where two digital signals are combined into one by time-division multiplexing. The multiplexed signal can then be fed into a single processor. By switching the processor coefficients prior to the arrival of each signal at the input of the processor, the processor can be made to look like two different systems. Finally, by demultiplexing the output of the processor, the processed signals can be separated.

Digital implementation permits easy adjustment of processor characteristics during processing, such as that needed in implementing adaptive filters. Such adjustments can be simply carried out by periodically changing the coefficients of the algorithm representing the processor characteristics. Another application of the changing of coefficients is in the realization of systems with programmable characteristics, such as frequency selective filters with adjustable cutoff frequencies. Filter banks with guaranteed complementary frequency response characteristics are easily implemented in digital form.

Digital implementation allows the realization of certain characteristics not possible with analog implementation, such as exact linear phase and multirate processing. Digital circuits can be cascaded without any loading problems unlike analog circuits. Digital signals can be stored almost indefinitely without any loss of information on various storage media such as magnetic tapes and disks, and optical disks. Such stored signals can later be processed off-line, such as in the compact disk player, the digital video disk player, the digital audio tape player, or simply by using a general purpose computer as in seismic data processing. On the other hand, stored analog signals deteriorate rapidly as time progresses and cannot be recovered in their original forms.

Another advantage is the applicability of digital processing to very low frequency signals, such as those occurring in seismic applications, where inductors and capacitors needed for analog processing would be physically very large in size.

Digital signal processing is also associated with some disadvantages. One obvious disadvantage is the increased system complexity in the digital processing of analog signals because of the need for additional pre- and postprocessing devices such as the A/D and D/A converters and their associated filters and complex digital circuitry.

A second disadvantage associated with digital signal processing is the limited range of frequencies available for processing. This property limits its application particularly in the digital processing of analog signals. As shown later, in general, an analog continuous-time signal must be sampled at a frequency that is

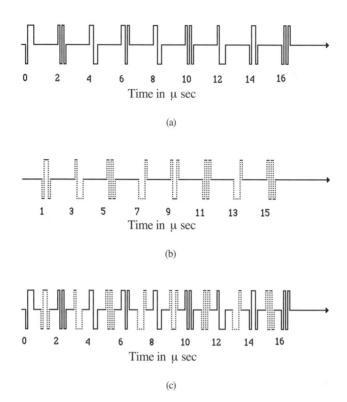

Figure 1.50: Illustration of the time-sharing concept. The signal shown in (c) has been obtained by time-multiplexing the signals shown in (a) and (b).

at least twice the highest frequency component present in the signal. If this condition is not satisfied, then signal components with frequencies above half the sampling frequency appear as signal components below this particular frequency, totally distorting the input analog signal waveform. The available frequency range of operation of a digital signal processor is primarily determined by the S/H circuit and the A/D converter, and as a result is limited by the state of the art of the technology. The highest sampling frequency reported in the literature presently is around 1 GHz [Pou87]. Such high sampling frequencies are not usually used in practice since the achievable resolution of the A/D converter, given by the wordlength of the digital equivalent of the analog sample, decreases with an increase in the speed of the converter. For example, the reported resolution of an A/D converter operating at 1 GHz is 6 bits [Pou87]. On the other hand, in most applications, the required resolution of an A/D converter is from 12 bits to around 16 bits. Consequently, a sampling frequency of at most 10 MHz is presently a practical upper limit. This upper limit, however, is getting larger and larger with advances in technology.

The third disadvantage stems from the fact that digital systems are constructed using active devices that consume electrical power. For example, the WE DSP32C Digital Signal Processor chip contains over 405,000 transistors and dissipates around 1 watt. On the other hand, a variety of analog processing algorithms can be implemented using passive circuits employing inductors, capacitors, and resistors that do not need power. Moreover, active devices are less reliable than passive components.

However, the advantages far outweigh the disadvantages in various applications, and with the continuing decrease in the cost of digital processor hardware, applications of digital signal processing are increasing rapidly.

2 Discrete-Time Signals and Systems in the Time-Domain

The signals arising in digital signal processing are basically discrete-time signals, and discrete-time systems are used to process these signals. As indicated in Figure 1.1(c), a discrete-time signal in its most basic form is defined at equally spaced discrete values of time, the independent variable, with the signal amplitude at these discrete times being continuous. Consequently, a discrete-time signal can be represented as a sequence of numbers, with the independent time variable represented as an integer in the range from $-\infty$ to $+\infty$. Discrete-time signal processing then involves the processing of a discrete-time signal by a discrete-time system to develop another discrete-time signal with more desirable properties or to extract certain information about the original discrete-time signal.

In many applications, it is increasingly becoming more attractive to process a continuous-time signal by discrete-time signal processing methods. To this end, the continuous-time signal is first converted into an "equivalent" discrete-time signal by periodic sampling; the discrete-time signal is then processed by a discrete-time system to generate another discrete-time signal, and the latter is converted into an equivalent continuous-time signal, if necessary. As we shall show later in this book, under certain (ideal) conditions, the conversion of a continuous-time signal can be carried out such that the discrete-time equivalent has all the information contained in the original continuous-time signal, and if necessary, can be converted back into the original continuous-time signal without any distortion.

Thus, to understand the theory of digital signal processing and the design of discrete-time systems, we need to know the characterization of discrete-time signals and systems in the time-domain, a subject we discuss in this chapter. It turns out that it is often convenient to characterize the discrete-time signals and systems in a transformed domain. This alternative representation is considered in the following two chapters.

In this chapter, we first discuss the time-domain representation of a discrete-time signal as a sequence of numbers and its various classifications. We then describe several basic discrete-time signals or sequences that play important roles in the time-domain characterization of arbitrary discrete-time signals and discrete-time systems. A number of basic operations that generate other sequences from one or more sequences are described next. As we show later, a discrete-time system is composed of a combination of these basic operations. The problem of representing a continuous-time signal by a discrete-time sequence is examined for a simple case. A more thorough mathematical treatment for the general case is deferred until Chapter 5 since it is based on a transform-domain representation of the discrete-time signal discussed in Chapter 3.

In the latter half of this chapter, we introduce the general concept of the processing of a discrete-time signal by a discrete-time system and the classification of such systems. Of these systems, the class of linear, time-invariant type is of exclusive interest in this book, and we describe its time-domain characterization in several different forms. We also introduce the concept of cross-correlation between a pair of discrete-time sequences which provides a measure of the degree of similarity between the pair.

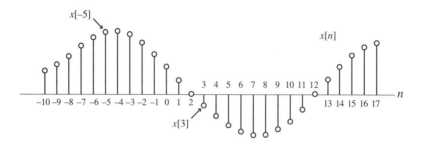

Figure 2.1: Graphical representation of a discrete-time sequence $\{x[n]\}$.

Most of the book deals with the processing of signals that are deterministic in nature. However, in some instances, the signals encountered could be random, and a discussion of the time-domain representation of discrete-time random signals is also included in this chapter.

Throughout this chapter and successive chapters, we make extensive use of MATLAB to illustrate through computer simulations the various concepts introduced.

2.1 Discrete-Time Signals

2.1.1 Time-Domain Representation

As indicated earlier, in digital signal processing, signals are represented as sequences of numbers called *samples*. A sample value of a typical discrete-time signal or sequence is denoted as $x[n]$ with the argument n being an integer in the range $-\infty$ and ∞. It should be noted that $x[n]$ is defined only for integer values of n and is undefined for noninteger values of n. The discrete-time signal is represented by $\{x[n]\}$. If a discrete-time signal is written as a sequence of numbers inside braces, the location of the sample value associated with the time index $n = 0$ is indicated by an arrow $\uparrow$ under it. The sample values to its right are for positive values of n, and the sample values to its left are for negative values of n. An example of a discrete-time signal with real-valued samples is given by

$$\{x[n]\} = \{\ldots, 0.95, -0.2, 2.17, 1.1, 0.2, -3.67, 2.9, -0.8, 4.1, \ldots\}. \tag{2.1}$$
$$\uparrow$$

For the above signal, $x[-1] = -0.2$, $x[0] = 2.17$, $x[1] = 1.1$, and so on. The graphical representation of a sequence $\{x[n]\}$ with real-valued samples is illustrated in Figure 2.1.

In some applications a discrete-time sequence $\{x[n]\}$ is generated by periodically sampling a continuous-time signal $x_a(t)$ at uniform time intervals:

$$x[n] = x_a(t)|_{t=nT} = x_a(nT), \qquad n = \ldots, -2, -1, 0, 1, 2, \ldots \tag{2.2}$$

as illustrated in Figure 2.2. The spacing T between two consecutive samples in Eq. (2.2) is called the *sampling interval* or *sampling period*. The reciprocal of the sampling interval T, denoted as F_T, is called the *sampling frequency*:

$$F_T = \frac{1}{T}. \tag{2.3}$$

The unit of sampling frequency is cycles per second, or hertz (Hz), if the sampling period is in seconds (sec).

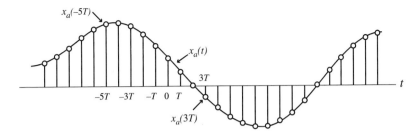

Figure 2.2: Sequence generated by sampling a continuous-time signal $x_a(t)$.

It should be noted that, whether or not a sequence $\{x[n]\}$ has been obtained by sampling, the quantity $x[n]$ is called the *nth sample* of the sequence. For a sequence $\{x[n]\}$, the nth sample value $x[n]$ can, in general, take any real or complex value. If $x[n]$ is real for all values of n, then $\{x[n]\}$ is a *real sequence*. On the other hand, if the nth sample value is complex for one or more values of n, then it is a *complex sequence*. By separating the real and imaginary parts of $x[n]$, we can write a complex sequence $\{x[n]\}$ as

$$\{x[n]\} = \{x_{re}[n]\} + j\,\{x_{im}[n]\}, \tag{2.4}$$

where $x_{re}[n]$ and $x_{im}[n]$ are the real part and the imaginary part of $x[n]$, respectively, and are thus real sequences. The complex conjugate sequence of $\{x[n]\}$ is usually denoted by $\{x^*[n]\}$ and written as $\{x^*[n]\} = \{x_{re}[n]\} - j\,\{x_{im}[n]\}$.[1] Often the braces are ignored to denote a sequence if there is no ambiguity.

As defined in the previous chapter, there are basically two types of discrete-time signals: sampled-data signals in which the samples are continuous-valued and digital signals in which the samples are discrete-valued. The pertinent signals in a practical digital signal processing system are digital signals obtained by quantizing the sample values either by *rounding* or by *truncation*. For example, the digital signal $\{\hat{x}[n]\}$ obtained by rounding the sample values of the discrete-time sequence $x[n]$ of Eq. (2.1) to the nearest integer values is given by

$$\{\hat{x}[n]\} = \{\ldots,\ 1,\ 0,\ 2,\ 1,\ 0,\ -4,\ 3,\ -1,\ 4,\ \ldots\}.$$
$$\uparrow$$

Figure 2.3 shows a digital signal with amplitudes taking discrete integer values in the range from -3 to 3.

For digital processing of a continuous-time signal, it is first converted into an equivalent digital signal by means of a sample-and-hold circuit followed by an analog-to-digital converter. The processed digital signal is then converted back into an equivalent continuous-time signal by a digital-to-analog converter followed by an analog reconstruction filter. Chapter 5 is concerned with the digital processing of continuous-time signals. It develops the mathematical foundation of the sampling process and describes the operations of various interface circuits between the continuous-time domain and the digital domain. Chapter 9 considers the effect of discretization of the amplitudes.

The discrete-time signal may be a *finite-length* or an *infinite-length sequence*. A finite-length (also called *finite-duration* or *finite-extent*) sequence is defined only for a finite time interval:

$$N_1 \leq n \leq N_2, \tag{2.5}$$

where $-\infty < N_1$ and $N_2 < \infty$ with $N_2 \geq N_1$. The *length or duration* N of the above finite-length sequence is

$$N = N_2 - N_1 + 1. \tag{2.6}$$

---

[1]The complex conjugation operation is denoted by the symbol *.

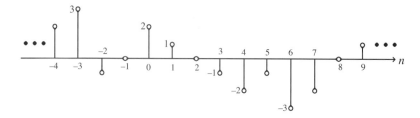

Figure 2.3: A digital signal.

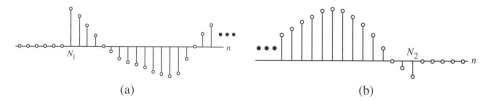

(a)   (b)

Figure 2.4: (a) A right-sided sequence, and (b) a left-sided sequence.

A length-N discrete-time sequence consists of N samples and is often referred to as an N-point sequence. A finite-length sequence can also be considered as an infinite-length sequence by assigning zero values to samples whose arguments are outside the above range. The process of lengthening a sequence by adding zero-valued samples is called *appending with zeros* or *zero-padding*.

There are three types of infinite-length sequences. A *right-sided sequence* $x[n]$ has zero-valued samples for $n < N_1$, i.e.,

$$x[n] = 0 \quad \text{for } n < N_1, \tag{2.7}$$

where N_1 is a finite integer that can be positive or negative. If $N_1 \geq 0$, a right-sided sequence is usually called a *causal sequence*. [2] Likewise, a *left-sided sequence* $x[n]$ has zero-valued samples for $n > N_2$, i.e.,

$$x[n] = 0 \quad \text{for } n > N_2, \tag{2.8}$$

where N_2 is a finite integer which can be positive or negative. If $N_2 \leq 0$, a left-sided sequence is usually called an *anticausal sequence*. A general *two-sided sequence* is defined for all values of n in the range $-\infty < n < \infty$. Figure 2.4 illustrates the above two types of one-sided sequences.

For simplicity, for finite-length sequences defined for positive values of the time index n beginning at $n = 0$, the first sample in the sequence will always be assumed to be the one associated with the time index $n = 0$ without the arrow being explicitly shown under it.

2.1.2 Operations on Sequences

A single-input, single-output discrete-time system operates on a sequence, called the *input sequence*, according to some prescribed rules and develops another sequence, called the *output sequence*, usually with more desirable properties. For example, the input may be a signal corrupted by an additive noise, and the discrete-time system is designed to remove the noise component from the input. In some applications, the discrete-time system can have more than one input and more than one output. An M-input, N-output discrete-time system operates on M input signals generating N output signals. The FM stereo

---

[2]The term *causal sequence* has its origin in the causality condition of a discrete-time system as discussed in Section 2.5.4.

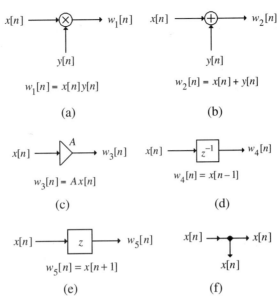

Figure 2.5: Schematic representations of basic operations on sequences: (a) modulator, (2) adder, (c) multiplier, (d) unit delay, (e) unit advance, and (f) pick-off node.

transmission system is a two-input, single-output system since here the left and right channel audio signals are combined into a high-frequency composite baseband signal. In most cases, the operation defining a particular discrete-time system is composed of some basic operations that we describe next.

Basic Operations

Let $x[n]$ and $y[n]$ be two known sequences. By forming the *product* of the sample values of these two sequences at each instant, we form a new sequence $w_1[n]$:

$$w_1[n] = x[n] \cdot y[n]. \tag{2.9}$$

In some applications, the product operation is also known as *modulation*. The device implementing the modulation operation is called a *modulator* and its schematic representation is shown in Figure 2.5(a).

An application of the product operation is in forming a finite-length sequence from an infinite-length sequence by multiplying the latter with a finite-length sequence called a *window sequence*. This process of forming the finite-length sequence is usually called *windowing*, which plays an important role in the design of certain types of digital filters (Section 7.6).

The second basic operation is the *addition* by which a new sequence $w_2[n]$ is obtained by adding the sample values of two sequences $x[n]$ and $y[n]$:

$$w_2[n] = x[n] + y[n]. \tag{2.10}$$

The device implementing the addition operation is called an *adder* and its schematic representation is shown in Figure 2.5(b).

The third basic operation is the scalar *multiplication*, whereby a new sequence is generated by multiplying each sample of a sequence $x[n]$ by a scalar A:

$$w_3[n] = Ax[n]. \tag{2.11}$$

The device implementing the multiplication operation is called a *multiplier* and its schematic representation is shown in Figure 2.5(c).

The *time-shifting* operation illustrated below in Eq. (2.12) shows the relation between $x[n]$ and its time-shifted version $w_4[n]$:

$$w_4[n] = x[n - N], \qquad (2.12)$$

where N is an integer. If $N > 0$, it is a *delaying* operation and if $N < 0$, it is an *advancing* operation. The device implementing the delay operation by one sample is called a *unit delay* and its schematic representation is shown in Figure 2.5(d). The reason for using the symbol z^{-1} will be clear after we have reviewed the z-transform of sequences in Chapter 3. The schematic representation of the *unit advance* operation is shown in Figure 2.5(e).

The *time-reversal* operation, also called the *folding operation*, is another useful scheme to develop a new sequence. An example is:

$$w_6[n] = x[-n], \qquad (2.13)$$

which is the time-reversed version of the sequence $x[n]$.

In Figure 2.5(f) we also show a *pick-off node* which is used to provide multiple copies of a sequence.

EXAMPLE 2.1   Consider the following two sequences of length 5 defined for $0 \le n \le 4$:

$$c[n] = \{3.2, \quad 41, \quad 36, \quad -9.5, \quad 0\},$$
$$d[n] = \{1.7, \quad -0.5, \quad 0, \quad 0.8, \quad 1\}.$$

Several new sequences of length 5 generated from the above sequences are given by

$$w_1[n] = c[n] \cdot d[n] = \{5.44, \quad -20.5, \quad 0, \quad -7.6, \quad 0\},$$
$$w_2[n] = c[n] + d[n] = \{4.9, \quad 40.5, \quad 36, \quad -8.7, \quad 1\},$$
$$w_3[n] = \frac{7}{2}c[n] = \{11.2, \quad 143.5, \quad 126, \quad -33.25, \quad 0\}.$$

As indicated by the above example, operations on two or more sequences to generate a new sequence can be carried out if all sequences are of the same length and defined for the same range of the time index n. However, in some situations, this problem can be circumvented by appending zero-valued samples to the sequence(s) of smaller lengths to make all sequences have the same range of the time index n. This process is illustrated in the following example.

EXAMPLE 2.2   Consider a sequence $\{g[n]\}$ of length 3 defined for $0 \le n \le 2$ given by

$$\{g[n]\} = \{-21, \quad 1.5, \quad 3\}.$$

It is clear that we cannot develop another sequence by operating on this sequence and any one of the length-5 sequences of Example 2.1. However, it is possible to treat $\{g[n]\}$ as a sequence of length 5 and defined for $0 \le n \le 4$ by appending it with two zero-valued samples:

$$\{g_e[n]\} = \{-21, \quad 1.5, \quad 3, \quad 0, \quad 0\}.$$

Examples of new sequences generated from $\{g_e[n]\}$ and $c[n]$ of the previous example are indicated below:

$$\{w_4[n]\} = \{c[n] \cdot g_e[n]\} = \{-67.2, \quad 61.5, \quad 108, \quad 0, \quad 0\},$$
$$\{w_5[n]\} = \{c[n] + g_e[n]\} = \{-17.8, \quad 42.5, \quad 39, \quad -9.5, \quad 0\}.$$

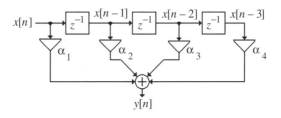

Figure 2.6: Discrete-time system of Example 2.3.

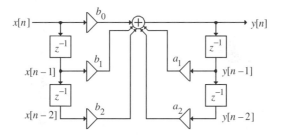

Figure 2.7: Discrete-time system of Example 2.4.

Combination of Basic Operations

In most applications, combinations of the above basic operations are used. Illustrations of such combinations are given in the following two examples.

EXAMPLE 2.3   Figure 2.6 shows the block diagram of a discrete-time system obtained by combining some of the basic operations. We analyze this figure to determine how the sequence $y[n]$ is generated from the sequence $x[n]$. From the definition of the unit delay block given in Figure 2.5(d), it follows that the leftmost delay block generates a sequence $x[n-1]$, the middle delay block develops $x[n-2]$ and finally, the rightmost delay block generates $x[n-3]$. The multipliers labeled $\alpha_1, \alpha_2, \alpha_3$, and α_4 generate sequences $\alpha_1 x[n], \alpha_2 x[n-1], \alpha_3 x[n-2]$, and $\alpha_4 x[n-3]$, respectively. The sequence $y[n]$ is then obtained by adding $\alpha_1 x[n], \alpha_2 x[n-1], \alpha_3 x[n-2]$, and $\alpha_4 x[n-3]$:

$$y[n] = \alpha_1 x[n] + \alpha_2 x[n-1] + \alpha_3 x[n-2] + \alpha_4 x[n-3]. \tag{2.14}$$

EXAMPLE 2.4   We next analyze the discrete-time system of Figure 2.7. Observe first that the two leftmost delay blocks generate the sequences $x[n-1]$ and $x[n-2]$, whereas the two rightmost delay blocks develop the sequences $y[n-1]$ and $y[n-2]$. These delayed sequences along with $x[n]$ are then applied to the five multipliers, labeled b_0, b_1, b_2, a_1, and a_2 developing the sequences $b_0 x[n], b_1 x[n-1], b_2 x[n-2], a_1 y[n-1]$, and $a_2 y[n-2]$ which are then added to yield $y[n]$:

$$y[n] = b_0 x[n] + b_1 x[n-1] + b_2 x[n-2] + a_1 y[n-1] + a_2 y[n-2]. \tag{2.15}$$

Sampling Rate Alteration

Another quite useful operation is the sampling rate alteration that is employed to generate a new sequence with a sampling rate higher or lower than that of a given sequence. Thus, if $x[n]$ is a sequence with a sampling rate of F_T Hz and it is used to generate another sequence $y[n]$ with a desired sampling rate of F_T' Hz, then the *sampling rate alteration ratio* is given by

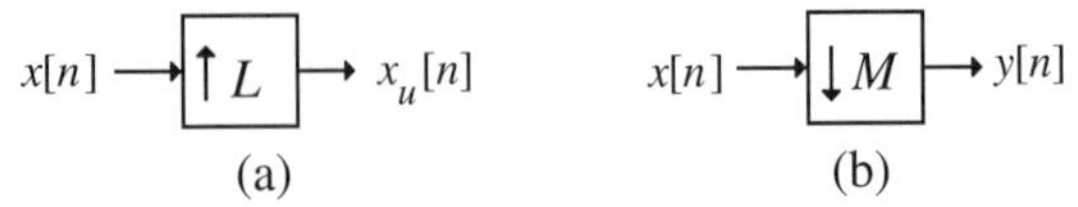

$$x[n] \longrightarrow \boxed{\uparrow L} \longrightarrow x_u[n] \qquad\qquad x[n] \longrightarrow \boxed{\downarrow M} \longrightarrow y[n]$$

(a) (b)

Figure 2.8: Representation of basic sampling rate alteration devices: (a) up-sampler, and (b) down-sampler

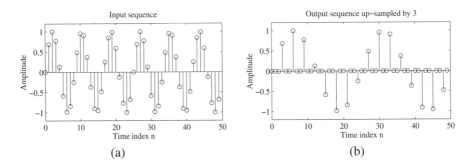

(a) (b)

Figure 2.9: Illustration of the up-sampling process.

$$\frac{F_T'}{F_T} = R. \tag{2.16}$$

If $R > 1$, the process is called *interpolation* and results in a sequence with a higher sampling rate. On the other hand, if $R < 1$, the sampling rate is decreased by a process called *decimation*.

The basic operations employed in the sampling rate alteration process are called *up-sampling* and *down-sampling*. These operations play important roles in multirate discrete-time systems and are considered in Chapter 10.

In up-sampling by an integer factor $L > 1$, $L - 1$ equidistant zero-valued samples are inserted by the up-sampler between each two consecutive samples of the input sequence $x[n]$ to develop an output sequence $y[n]$ according to the relation

$$x_u[n] = \begin{cases} x[n/L], & n = 0, \pm L, \pm 2L, \dots , \\ 0, & \text{otherwise.} \end{cases} \tag{2.17}$$

Note that the sampling rate of $y[n]$ is L times larger than that of the original sequence $x[n]$.

The block-diagram representation of the up-sampler, also called a *sampling rate expander*, is shown in Figure 2.8(a). Figure 2.9 illustrates the up-sampling operation for an up-sampling factor of $L = 3$.

Conversely, the down-sampling operation by an integer factor $M > 1$ on a sequence $x[n]$ consists of keeping every Mth sample of $x[n]$ and removing $M - 1$ in-between samples, generating an output sequence $y[n]$ according to the relation

$$y[n] = x[nM]. \tag{2.18}$$

This results in a sequence $y[n]$ whose sampling rate is $(1/M)$th that of $x[n]$. Basically, all input samples with indices equal to an integer multiple of M are retained at the output and all others are discarded.

The schematic representation of the *down-sampler* or *sampling rate compressor* is shown in Figure 2.8(b). Figure 2.10 illustrates the down-sampling operation for a down-sampling factor of $M = 3$.

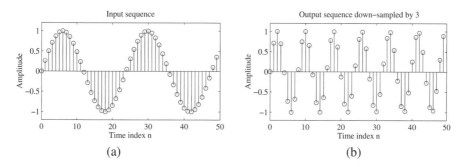

Figure 2.10: Illustration of the down-sampling process.

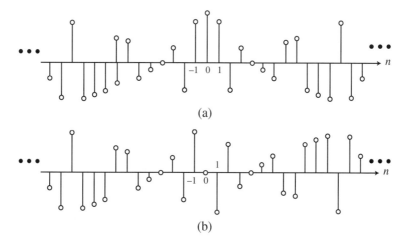

Figure 2.11: (a) An even sequence, and (b) an odd sequence.

2.1.3 Classification of Sequences

A discrete-time signal can be classified in various ways. One classification discussed earlier is in terms of the number of samples defining the sequence. Another classification is with respect to the symmetry exhibited by the samples with respect to the time index $n = 0$. A discrete-time signal can also be classified in terms of its other properties such as periodicity, summability, energy, and power.

Classification Based on Symmetry

A sequence $x[n]$ is called a *conjugate-symmetric sequence* if $x[n] = x^*[-n]$. A real conjugate-symmetric sequence is called an *even sequence*. A sequence $x[n]$ is called a *conjugate-antisymmetric sequence* if $x[n] = -x^*[-n]$. A real conjugate-antisymmetric sequence is called an *odd sequence*. For a conjugate-antisymmetric sequence $x[n]$, the sample value at $n = 0$ must be purely imaginary. Consequently, for an odd sequence $x[0] = 0$. Examples of even and odd sequences are shown in Figure 2.11.

Any complex sequence $x[n]$ can be expressed as a sum of its conjugate-symmetric part $x_{cs}[n]$ and its conjugate-antisymmetric part $x_{ca}[n]$:

$$x[n] = x_{cs}[n] + x_{ca}[n], \tag{2.19}$$

where

$$x_{cs}[n] = \frac{1}{2}\left(x[n] + x^*[-n]\right),\tag{2.20a}$$

$$x_{ca}[n] = \frac{1}{2}\left(x[n] - x^*[-n]\right).\tag{2.20b}$$

As indicated by Eqs. (2.20a) and (2.20b), the computation of the conjugate-symmetric and conjugate-antisymmetric parts of a sequence involves conjugation, time-reversal, addition, and multiplication operations. Because of the time-reversal operation, the decomposition of a finite-length sequence into a sum of a conjugate-symmetric sequence and a conjugate-antisymmetric sequence is possible, if the parent sequence is of odd length defined for a symmetric interval, $-M \leq 0 \leq M$.

EXAMPLE 2.5 Consider the finite-length sequence of length 7 defined for $-3 \leq n \leq 3$:

$$\{g[n]\} = \{0,\ 1+j4,\ -2+j3,\ 4-j2,\ -5-j6,\ -j2,\ 3\}.$$
$$\uparrow$$

To determine its conjugate-symmetric part $g_{cs}[n]$ and its conjugate-antisymmetric part $g_{ca}[n]$, we form

$$\{g^*[n]\} = \{0,\ 1-j4,\ -2-j3,\ 4+j2,\ -5+j6,\ j2,\ 3\},$$
$$\uparrow$$

whose time-reversed version is then given by

$$\{g^*[-n]\} = \{3,\ j2,\ -5+j6,\ 4+j2,\ -2-j3,\ 1-j4,\ 0\}.$$
$$\uparrow$$

Using Eq. (2.20a) we thus arrive at

$$\{g_{cs}[n]\} = \{1.5,\ 0.5+j3,\ -3.5+j4.5,\ 4,\ -3.5-j4.5,\ 0.5-j3,\ 1.5\}.$$
$$\uparrow$$

Likewise, using Eq. (2.20b) we get

$$\{g_{ca}[n]\} = \{-1.5,\ 0.5+j,\ 1.5-j1.5,\ -j2,\ -1.5-j1.5,\ -0.5-j,\ 1.5\}.$$
$$\uparrow$$

It can be easily verified that $g_{cs}[n] = g_{cs}^*[-n]$ and $g_{ca}[n] = -g_{ca}^*[-n]$.

Likewise, any real sequence $x[n]$ can be expressed as a sum of its even part $x_{ev}[n]$ and its odd part $x_{od}[n]$:

$$x[n] = x_{ev}[n] + x_{od}[n],\tag{2.21}$$

where

$$x_{ev}[n] = \frac{1}{2}\left(x[n] + x[-n]\right),\tag{2.22a}$$

$$x_{od}[n] = \frac{1}{2}\left(x[n] - x[-n]\right).\tag{2.22b}$$

For a length-N sequence defined for $0 \leq n \leq N-1$, the above definitions of symmetry are not applicable. The definitions of symmetry in the case of finite-length sequences are given instead using a modulo operation with all such symmetric and antisymmetric parts of a length-N sequence being also of length N and defined for the same range of values of the time index n. Thus, a length-N sequence $x[n]$ can be expressed as

$$x[n] = x_{pcs}[n] + x_{pca}[n], \quad 0 \leq n \leq N-1,\tag{2.23}$$

where $x_{pcs}[n]$ and $x_{pca}[n]$ denote, respectively, the *periodic conjugate-symmetric part* and the *periodic conjugate-antisymmetric part*, defined by [3]

$$x_{pcs}[n] = \frac{1}{2}\left(x[n] + x^*[\langle -n \rangle_N]\right), \quad 0 \le n \le N - 1, \tag{2.24a}$$

$$x_{pca}[n] = \frac{1}{2}\left(x[n] - x^*[\langle -n \rangle_N]\right), \quad 0 \le n \le N - 1. \tag{2.24b}$$

For a real sequence $x[n]$, the periodic conjugate-symmetric part is a real sequence, called the *periodic even part*, and denoted by $x_{pe}[n]$. Likewise, for a real sequence $x[n]$, the periodic conjugate-antisymmetric part is also a real sequence, called the *periodic odd part*, and denoted by $x_{po}[n]$.

A length-N sequence $x[n]$, defined for $0 \le n \le N - 1$ is said to be *periodic conjugate-symmetric* if $x[n] = x^*[\langle -n \rangle_N] = x^*[N - n]$, and is said to be *periodic conjugate-antisymmetric* if $x[n] = -x^*[\langle -n \rangle_N] = -x^*[N - n]$. A finite-length real periodic conjugate-symmetric sequence is called a *symmetric sequence* and a finite-length real periodic conjugate-antisymmetric sequence is called an *antisymmetric sequence*.

EXAMPLE 2.6 Consider the finite-length sequence of length 4 defined for $0 \le n \le 3$:

$$\{u[n]\} = \{1 + j4, \; -2 + j3, \; 4 - j2, \; -5 - j6\}.$$

To determine its periodic conjugate-symmetric part $u_{pcs}[n]$ and its periodic conjugate-antisymmetric part $u_{pca}[n]$ we form

$$\{u^*[n]\} = \{1 - j4, \; -2 - j3, \; 4 + j2, \; -5 + j6\}.$$

To compute the modulo-4 time-reversed version $\{u^*[\langle -n \rangle_4]\}$ we observe that $u^*[\langle -0 \rangle_4] = u^*[0] = 1 - j4$, $u^*[\langle -1 \rangle_4] = u^*[3] = -5 + j6$, $u^*[\langle -2 \rangle_4] = u^*[2] = 4 + j2$, and $u^*[\langle -3 \rangle_4] = u^*[1] = -2 - j3$. Hence,

$$\{u^*[\langle -n \rangle_4]\} = \{1 - j4, \; -5 + j6, \; 4 + j2, \; -2 - j3\}.$$

Using Eq. (2.24a) we thus arrive at

$$\{u_{pcs}[n]\} = \{1, \; -3.5 + j4.5, \; 4, \; -3.5 - j4.5\}.$$

Likewise, using Eq. (2.24b) we get

$$\{u_{pca}[n]\} = \{j4, \; 1.5 - j1.5, \; -j2, \; -1.5 - j1.5\}.$$

It can be easily verified that $u_{pcs}[n] = u^*_{pcs}[\langle -n \rangle_4]$ and $u_{pca}[n] = -u^*_{pca}[\langle -n \rangle_4]$.

The symmetric properties of sequences often simplify their respective frequency-domain representations and can be exploited in signal analysis. Implications of the symmetry conditions are considered in Chapter 3.

Periodic and Aperiodic Signals

A sequence $\tilde{x}[n]$ satisfying

$$\tilde{x}[n] = \tilde{x}[n + kN] \quad \text{for all } n \tag{2.25}$$

is called a *periodic* sequence with a *period* N where N is a positive integer and k is any integer. An example of a periodic sequence that has a period $N = 7$ samples is shown in Figure 2.12. A sequence is called an *aperiodic sequence* if it is not periodic. To distinguish a periodic sequence from an aperiodic

---

[3] $\langle k \rangle_N = k$ modulo N.

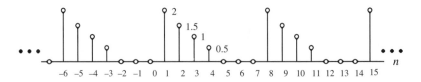

Figure 2.12: An example of a periodic sequence.

sequence, we shall denote the former with a "~" on top. The *fundamental period* N_f of a periodic signal is the smallest value of N for which Eq. (2.25) holds.

Energy and Power Signals

The total *energy* of a sequence $x[n]$ is defined by

$$\mathcal{E}_x = \sum_{n=-\infty}^{\infty} |x[n]|^2. \tag{2.26}$$

An infinite-length sequence with finite sample values may or may not have finite energy, as illustrated in the following example.

EXAMPLE 2.7   The infinite-length sequence $x_1[n]$ defined by

$$x_1[n] = \begin{cases} \frac{1}{n}, & n \geq 1, \\ 0, & n \leq 0, \end{cases} \tag{2.27}$$

has an energy equal to

$$\mathcal{E}_{x_1} = \sum_{n=1}^{\infty} \left(\frac{1}{n}\right)^2,$$

which converges to $\pi^2/6$, indicating that $x_1[n]$ has finite energy. However, the infinite-length sequence $x_2[n]$ defined by

$$x_2[n] = \begin{cases} \frac{1}{\sqrt{n}}, & n \geq 1, \\ 0, & n \leq 0, \end{cases} \tag{2.28}$$

has an energy equal to

$$\mathcal{E}_{x_2} = \sum_{n=1}^{\infty} \frac{1}{n},$$

which does not converge. As a result, the sequence $x_2[n]$ has infinite energy.

The *average power* of an aperiodic sequence $x[n]$ is defined by

$$\mathcal{P}_x = \lim_{K \to \infty} \frac{1}{2K+1} \sum_{n=-K}^{K} |x[n]|^2. \tag{2.29}$$

The average power of a sequence can be related to its energy by defining its energy over a finite interval $-K \leq n \leq K$ as

$$\mathcal{E}_{x,K} = \sum_{n=-K}^{K} |x[n]|^2. \tag{2.30}$$

Then

$$P_x = \lim_{K \to \infty} \frac{1}{2K+1} \mathcal{E}_{x,K}.$$ (2.31)

The average power of a periodic sequence $\tilde{x}[n]$ with a period N is given by

$$P_x = \frac{1}{N} \sum_{n=0}^{N-1} |\tilde{x}[n]|^2.$$ (2.32)

The average power of an infinite-length sequence may be finite or infinite.

EXAMPLE 2.8 Consider the causal sequence defined by

$$x[n] = \begin{cases} 3(-1)^n, & n \geq 0, \\ 0, & n < 0. \end{cases}$$

It follows from Eq. (2.26) that $x[n]$ has infinite energy. On the other hand, from Eq. (2.29) its average power is given by

$$P_x = \lim_{K \to \infty} \frac{1}{2K+1} \left(9 \sum_{n=0}^{K} 1 \right) = \lim_{K \to \infty} \frac{9(K+1)}{2K+1} = 4.5,$$

which is finite.

An infinite energy signal with finite average power is called a *power signal*. Likewise, a finite energy signal with zero average power is called an *energy signal*. An example of a power signal is a periodic sequence which has a finite average power but infinite energy. An example of an energy signal is a finite-length sequence which has finite energy but zero average power.

Other Types of Classification

A sequence $x[n]$ is said to be *bounded* if each of its samples is of magnitude less than or equal to a finite positive number B_x, i.e.,

$$|x[n]| \leq B_x < \infty.$$ (2.33)

The periodic sequence of Figure 2.12 is a bounded sequence with a bound $B_x = 2$.

A sequence $x[n]$ is said to be *absolutely summable* if

$$\sum_{n=-\infty}^{\infty} |x[n]| < \infty.$$ (2.34)

A sequence is said to be *square-summable* if

$$\sum_{n=-\infty}^{\infty} |x[n]|^2 < \infty.$$ (2.35)

A square-summable sequence therefore has finite energy and is an energy signal if it also has zero power.

2.2 Typical Sequences and Sequence Representation

We now consider several special sequences that play important roles in the analysis and design of discrete-time systems. For example, an arbitrary sequence can be expressed in terms of some of these basic

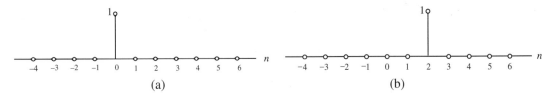

Figure 2.13: (a) The unit sample sequence $\{\delta[n]\}$, and (b) the shifted unit sample sequence $\{\delta[n-2]\}$.

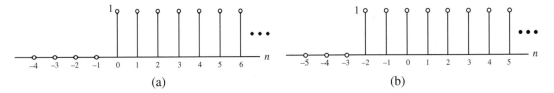

Figure 2.14: (a) The unit step sequence $\{\mu[n]\}$, and the shifted unit step sequence $\{\mu[n+2]\}$.

sequences. Another fundamental application that is the key behind discrete-time signal processing is the representation of a class of discrete-time systems in terms of the response of the system to certain basic sequences. This representation permits the computation of the response of the discrete-time system to arbitrary discrete-time signals if the latter can be expressed in terms of these basic sequences.

2.2.1 Some Basic Sequences

The most common basic sequences are the unit sample sequence, the unit step sequence, the sinusoidal sequence, and the exponential sequence. These sequences are defined next.

Unit Sample Sequence

The simplest and one of the most useful sequences is the *unit sample sequence*, often called the *discrete-time impulse* or the *unit impulse*, as shown in Figure 2.13(a). It is denoted by $\delta[n]$ and defined by

$$\delta[n] = \begin{cases} 1, & n = 0, \\ 0, & n \neq 0. \end{cases} \tag{2.36}$$

The unit sample sequence shifted by k samples is thus given by

$$\delta[n-k] = \begin{cases} 1, & n = k, \\ 0, & n \neq k. \end{cases}$$

Figure 2.13(b) shows $\delta[n-2]$. We shall show later in this section that any arbitrary sequence can be represented as a sum of weighted time-shifted unit sample sequences. In Section 2.6.1 we demonstrate that a certain class of discrete-time systems is completely characterized in the time-domain by its output response to a unit impulse input. Furthermore, knowing this particular response of the system, we can compute its response to any arbitrary input sequence.

Unit Step Sequence

A second basic sequence is the *unit step sequence* shown in Figure 2.14(a). It is denoted by $\mu[n]$ and is defined by

$$\mu[n] = \begin{cases} 1, & n \geq 0, \\ 0, & n < 0. \end{cases} \tag{2.37}$$

The unit step sequence shifted by k samples is thus given by

$$\mu[n-k] = \begin{cases} 1, & n \geq k, \\ 0, & n < k. \end{cases}$$

Figure 2.14(b) shows $\mu[n+2]$.

The unit sample and the unit step sequences are related as follows (Problem 2.19):

$$\mu[n] = \sum_{k=-\infty}^{n} \delta[k], \qquad \delta[n] = \mu[n] - \mu[n-1]. \tag{2.38}$$

Sinusoidal and Exponential Sequences

A commonly encountered sequence is the *real sinusoidal sequence* with constant amplitude of the form

$$x[n] = A\cos(\omega_o n + \phi), \qquad -\infty < n < \infty, \tag{2.39}$$

where A, ω_o, and ϕ are real numbers. The parameters A, ω_o, and ϕ are called, respectively, the *amplitude*, the *angular frequency*, and the *phase* of the sinusoidal sequence $x[n]$.

Figure 2.15 shows different types of sinusoidal sequences. The real sinusoidal sequence of Eq. (2.39) can be written alternatively as

$$x[n] = x_i[n] + x_q[n], \tag{2.40}$$

where $x_i[n]$ and $x_q[n]$ are, respectively, the *in-phase* and the *quadrature components* of $x[n]$, and are given by

$$x_i[n] = A\cos\phi\cos(\omega_o n), \qquad x_q[n] = -A\sin\phi\sin(\omega_o n). \tag{2.41}$$

Another set of basic sequences is formed by taking the nth sample value to be the nth power of a real or complex constant. Such sequences are termed *exponential sequences* and their most general form is given by

$$x[n] = A\alpha^n, \qquad -\infty < n < \infty, \tag{2.42}$$

where A and α are real or complex numbers. By expressing

$$\alpha = e^{(\sigma_o + j\omega_o)}, \qquad A = |A|e^{j\phi},$$

we can rewrite Eq. (2.42) as

$$x[n] = Ae^{(\sigma_o + j\omega_o)n} = |A|e^{\sigma_o n}e^{j(\omega_o n + \phi)} \tag{2.43a}$$

$$= |A|e^{\sigma_o n}\cos(\omega_o n + \phi) + j|A|e^{\sigma_o n}\sin(\omega_o n + \phi), \tag{2.43b}$$

to arrive at an alternative general form of a *complex exponential sequence* where σ_o, ϕ, and ω_o are now real numbers. If we write $x[n] = x_{re}[n] + jx_{im}[n]$, then from Eq. (2.43b):

$$x_{re}[n] = |A|e^{\sigma_o n}\cos(\omega_o n + \phi),$$

$$x_{im}[n] = |A|e^{\sigma_o n}\sin(\omega_o n + \phi).$$

Thus the real and imaginary parts of a complex exponential sequence are real sinusoidal sequences with constant ($\sigma_o = 0$), growing ($\sigma_o > 0$), or decaying ($\sigma_o < 0$) amplitudes for $n > 0$. Figure 2.16 depicts a

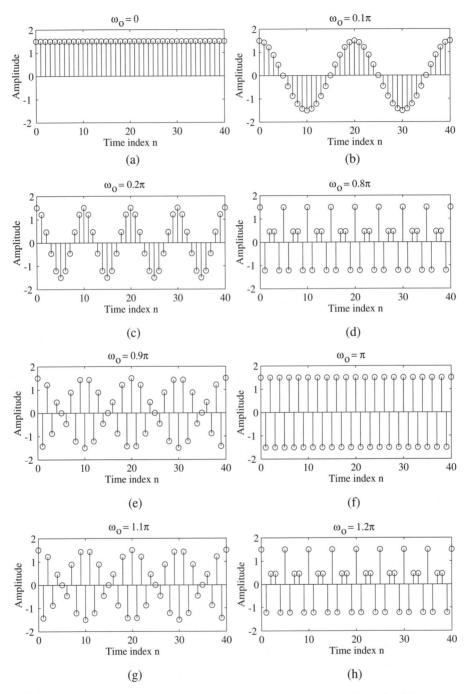

Figure 2.15: A family of sinusoidal sequences given by $x[n] = 1.5 \cos \omega_o n$: (a) $\omega_o = 0$, (b) $\omega_o = 0.1\pi$, (c) $\omega_o = 0.2\pi$, (d) $\omega_o = 0.8\pi$, (e) $\omega_o = 0.9\pi$, (f) $\omega_o = \pi$, (g) $\omega_o = 1.1\pi$, and (h) $\omega_o = 1.2\pi$.

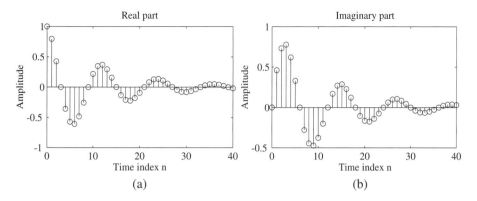

Figure 2.16: A complex exponential sequence $x[n] = e^{(-1/12+j\pi/6)n}$. (a) Real part and (b) imaginary part.

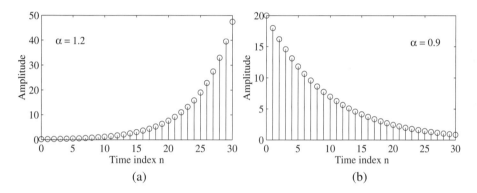

Figure 2.17: Examples of real exponential sequences: (a) $x[n] = 0.2(1.2)^n$, (b) $x[n] = 20(0.9)^n$.

complex exponential sequence with a decaying amplitude. Note that in the display of a complex exponential sequence, its real and imaginary parts are shown separately.

With both A and α real, the sequence of Eq. (2.42) reduces to a *real exponential sequence*. For $n \geq 0$ such a sequence with $|\alpha| < 1$ decays exponentially as n increases and with $|\alpha| > 1$ grows exponentially as n increases. Examples of real exponential sequences obtained for two values of α are shown in Figure 2.17.

We shall show later in Section 3.1 that a large class of sequences can be expressed in terms of complex exponential sequences of the form $e^{j\omega n}$.

Note that the sinusoidal sequence of Eq. (2.39) and the complex exponential sequence of Eq. (2.43a) with $\sigma_o = 0$ are periodic sequences of period N as long as $\omega_o N$ is an integer multiple of 2π, i.e., $\omega_o N = 2\pi r$ where N and r are positive integers. The smallest possible N satisfying this condition is the *fundamental period* of the sequence. To verify this, consider the two sinusoidal sequences $x_1[n] = \cos(\omega_o n + \phi)$ and $x_2[n] = \cos(\omega_o(n + N) + \phi)$. Now

$$x_2[n] = \cos(\omega_o(n + N) + \phi) = \cos(\omega_o n + \phi)\cos \omega_o N - \sin(\omega_o n + \phi)\sin \omega_o N$$

which will be equal to $\cos(\omega_o n + \phi) = x_1[n]$ only if $\sin \omega_o N = 0$ and $\cos \omega_o N = 1$. These two conditions are satisfied if and only if $\omega_o N$ is an integer multiple of 2π, i.e.,

$$\omega_o N = 2\pi r \tag{2.44a}$$

or

$$\frac{2\pi}{\omega_o} = \frac{N}{r}. \tag{2.44b}$$

If $2\pi/\omega_o$ is a noninteger rational number, then the period will be a multiple of $2\pi/\omega_o$. If $2\pi/\omega_o$ is not a rational number, then the sequence is aperiodic even though it has a sinusoidal envelope. For example, $x[n] = \cos(\sqrt{3}n + \phi)$ is an aperiodic sequence.

EXAMPLE 2.9   Let us determine the periods of the sinusoidal sequences of Figure 2.15. In Figure 2.15(a), $\omega_o = 0$, and Eq. (2.44a) is satisfied with $r = 0$ and any value of N. Here, the smallest value of N is equal to 1. Likewise, in Figure 2.15(b), $\omega_o = 0.1\pi$, and hence, applying Eq. (2.44a) we find that $N = 20$. Following similar lines we arrive at $N = 10$ for Figure 2.15(c), $N = 5$ for Figure 2.15(d), $N = 20$ for Figure 2.15(e), $N = 2$ for Figure 2.15(f), $N = 20$ for Figure 2.15(g), and $N = 5$ for Figure 2.15(h). These numbers are also evident from the figures.

The number ω_o in the two sequences of Eqs. (2.39) and (2.43a) is called the *angular frequency*. Since the time instant n is dimensionless, the unit of angular frequency ω_o and *phase* ϕ is simply radians. If the unit of n is designated as samples, then the unit of ω_o and ϕ is radians per sample. Often in practice, the angular frequency ω is expressed as

$$\omega = 2\pi f, \tag{2.45}$$

where f is *frequency* in cycles per sample.

Two interesting properties of these sequences are discussed next. Consider two complex exponential sequences $x_1[n] = e^{j\omega_1 n}$ and $x_2[n] = e^{j\omega_2 n}$ with $0 \le \omega_1 < 2\pi$ and $2\pi k \le \omega_2 < 2\pi(k + 1)$, where k is any positive or negative integer. If

$$\omega_2 = \omega_1 + 2\pi k, \tag{2.46}$$

then

$$x_2[n] = e^{j\omega_2 n} = e^{j(\omega_1 + 2\pi k)n} = e^{j\omega_1 n} = x_1[n].$$

Thus these two sequences are indistinguishable. Likewise, two sinusoidal sequences $x_1[n] = \cos(\omega_1 n + \phi)$ and $x_2[n] = \cos(\omega_2 n + \phi)$ with $0 \le \omega_1 < 2\pi$ and $2\pi k \le \omega_2 < 2\pi(k + 1)$, where k is any positive or negative integer, are indistinguishable from one another if $\omega_2 = \omega_1 + 2\pi k$.

The second interesting feature of discrete-time sinusoidal signals can be seen from Figure 2.15. The frequency of oscillation of the discrete-time sinusoidal sequence $x[n] = A\cos(\omega_o n)$ increases as ω_o increases from 0 to π, and then the frequency of oscillation decreases as ω_o increases from π to 2π. As a result of the first property, a frequency ω_o in the neighborhood of $\omega = 2\pi k$ is indistinguishable from a frequency $\omega_o - 2\pi k$ in the neighborhood of $\omega = 0$, and a frequency ω_o in the neighborhood of $\omega = \pi(2k + 1)$ is indistinguishable from a frequency $\omega_o - \pi(2k + 1)$ in the neighborhood of $\omega = \pi$ for any integer value of k. Therefore, frequencies in the neighborhood of $\omega_o = 2\pi k$ are usually called *low frequencies*, and frequencies in the neighborhood of $\omega_o = \pi(2k + 1)$ are called *high frequencies*. For example, $v_1[n] = \cos(0.1\pi n) = \cos(1.9\pi n)$ shown in Figure 2.15(b) is a low-frequency signal, whereas, $v_2[n] = \cos(0.8\pi n) = \cos(1.2\pi n)$ shown in Figure 2.15(d) and (h) is a high-frequency signal.

Another application of the modulation operation discussed earlier in Section 2.1.2 is to transform a sequence with low-frequency sinusoidal components to a sequence with high-frequency components by modulating the former with a sinusoidal sequence of very high frequency, as illustrated in the following example.

EXAMPLE 2.10   Let $x_1[n] = \cos(\omega_1 n)$ and $x_2[n] = 2\cos(\omega_2 n)$ with $\pi > \omega_2 \gg \omega_1 > 0$. The product sequence $y[n] = x_1[n]x_2[n]$ is then given by

$$y[n] = 2\cos(\omega_1 n) \cdot \cos(\omega_2 n).$$

Using a trigonometric identity we arrive at

$$y[n] = \cos((\omega_1 + \omega_2)n) + \cos((\omega_2 - \omega_1)n).$$

The new sequence $y[n]$ is thus composed of two sinusoidal sequences of frequencies $\omega_1 + \omega_2$ and $\omega_2 - \omega_1$. It should be noted that because of the property given by Eq. (2.46) if $\omega_1 + \omega_2 > \pi$, the high-frequency sinusoidal sequence $\cos((\omega_1 + \omega_2)n)$ appears as a low-frequency sinusoidal sequence $\cos((2\pi - \omega_1 - \omega_2)n)$ with an angular frequency of value smaller than π.[4]

2.2.2 Sequence Generation Using MATLAB

MATLAB includes a number of functions that can be used for signal generation. Some of these functions of interest are

          exp,   sin,   cos,   square,   sawtooth

For example, Program 2_1 given below can be employed to generate a complex exponential sequence of the form shown in Figure 2.16.

```
% Program 2_1
% Generation of complex exponential sequence
%
a = input('Type in real exponent = ');
b = input('Type in imaginary exponent = ');
c = a + b*i;
K = input('Type in the gain constant = ');
N = input ('Type in length of sequence = ');
n = 1:N;
x = K*exp(c*n);
stem(n,real(x));
xlabel('Time index n');ylabel('Amplitude');
title('Real part');
disp('PRESS RETURN for imaginary part');
pause
stem(n,imag(x));
xlabel('Time index n');ylabel('Amplitude');
title('Imaginary part');
```

Likewise, Program 2_2 listed below can be employed to generate a real exponential sequence of the form shown in Figure 2.17.

```
% Program 2_2
% Generation of real exponential sequence
%
a = input('Type in argument = ');
K = input('Type in the gain constant = ');
N = input ('Type in length of sequence = ');
```

_____

[4]The appearance of a high-frequency signal $\cos((\omega_1 + \omega_2)n)$ as a low-frequency signal $\cos((2\pi - \omega_1 - \omega_2)n)$ is called aliasing (see Section 2.3).

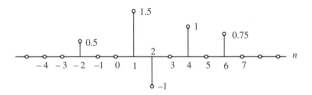

Figure 2.18: An arbitrary sequence $x[n]$.

```
n = 0:N;
x = K*a.^n;
stem(n,x);
xlabel('Time index n');ylabel('Amplitude');
title(['\alpha =  ',num2str(a)]);
```

Another type of sequence generation using MATLAB is given later in Example 2.14.

2.2.3 Representation of an Arbitrary Sequence

An arbitrary sequence can be represented in the time-domain as a weighted sum of some basic sequence and its delayed versions. A commonly used basic sequence in the representation is the unit sample sequence. For example, the sequence $x[n]$ of Figure 2.18 can be expressed as

$$x[n] = 0.5\delta[n+2] + 1.5\delta[n-1] - \delta[n-2] + \delta[n-4] + 0.75\delta[n-6]. \tag{2.47}$$

An implication of this type of representation is considered later in Section 2.5.1, where we develop the general expression for calculating the output sequence of certain types of discrete-time systems for an arbitrary input sequence.

Since the unit step sequence and the unit sample sequence are simply related through Eq. (2.38), it is also possible to represent an arbitrary sequence as a weighted combination of delayed unit step sequences (Problem 2.24).

2.3 The Sampling Process

We indicated earlier that often the discrete-time sequence is developed by uniformly sampling a continuous-time signal $x_a(t)$, as illustrated in Figure 2.2. The relation between the two signals is given by Eq. (2.2), where the time variable t of the continuous-time signal is related to the time variable n of the discrete-time signal only at discrete-time instants t_n given by

$$t_n = nT = \frac{n}{F_T} = \frac{2\pi n}{\Omega_T}, \tag{2.48}$$

with $F_T = 1/T$ denoting the sampling frequency and $\Omega_T = 2\pi F_T$ denoting the sampling angular frequency. For example, if the continuous-time signal is

$$x_a(t) = A\cos(2\pi f_o t + \phi) = A\cos(\Omega_o t + \phi), \tag{2.49}$$

the corresponding discrete-time signal is given by

$$\begin{aligned}
x[n] &= A\cos(\Omega_o nT + \phi) \\
&= A\cos\left(\frac{2\pi\Omega_o}{\Omega_T}n + \phi\right) = A\cos(\omega_o n + \phi),
\end{aligned} \tag{2.50}$$

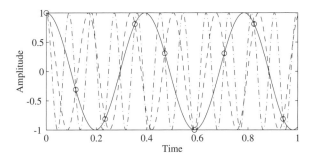

Figure 2.19: Ambiguity in the discrete-time representation of continuous-time signals. $g_1(t)$ is shown with the solid line, $g_2(t)$ is shown with the dashed line, $g_3(t)$ is shown with the dashed-dot line, and the sequence obtained by sampling is shown with circles.

where

$$\omega_o = \frac{2\pi \Omega_o}{\Omega_T} = \Omega_o T \qquad (2.51)$$

is the normalized digital angular frequency of the discrete-time signal $x[n]$. The unit of the normalized digital angular frequency ω_o is radians per sample, while the unit of the normalized analog angular frequency Ω_o is radians per second and the unit of the analog frequency f_o is hertz if the unit of the sampling period T is in seconds.

EXAMPLE 2.11 Consider the three sequences generated by uniformly sampling the three cosine functions of frequencies 3 Hz, 7 Hz, and 13 Hz, respectively: $g_1(t) = \cos(6\pi t)$, $g_2(t) = \cos(14\pi t)$, and $g_3(t) = \cos(26\pi t)$ with a sampling rate of 10 Hz, i.e., with $T = 0.1$ sec. The derived sequences are therefore:

$$g_1[n] = \cos(0.6\pi n), \qquad g_2[n] = \cos(1.4\pi n), \qquad g_3[n] = \cos(2.6\pi n).$$

Plots of these sequences (shown with circles) and their parent time functions are given in Figure 2.19. Note from these plots that each sequence has exactly the same sample value for any given n. This also can be verified by observing that

$$g_2[n] = \cos(1.4\pi n) = \cos((2\pi - 0.6\pi)n) = \cos(0.6\pi n)$$

and

$$g_3[n] = \cos(2.6\pi n) = \cos((2\pi + 0.6\pi)n) = \cos(0.6\pi n).$$

As a result, all three sequences above are identical and it is difficult to associate a unique continuous-time function with this sequence. In fact, all cosine waveforms of frequencies given by $(10k \pm 3)$ Hz, with k being any nonnegative integer, lead to the sequence $g_1[n] = \cos(0.6\pi n)$ when sampled at a 10-Hz rate.

In the general case, the family of continuous-time sinusoids

$$x_{a,k}(t) = A\cos(\pm(\Omega_o t + \phi) + k\Omega_T t), \qquad k = 0, \pm 1, \pm 2, \ldots \qquad (2.52)$$

leads to identical sampled signals:

$$x_{a,k}(nT) = A\cos\left((\Omega_o + k\Omega_T)nT + \phi\right) = A\cos\left(\frac{2\pi(\Omega_o + k\Omega_T)n}{\Omega_T} + \phi\right)$$

$$= A\cos\left(\frac{2\pi \Omega_o n}{\Omega_T} + \phi\right) = A\cos(\omega_o n + \phi) = x[n]. \qquad (2.53)$$

The above phenomenon of a continuous-time sinusoidal signal of higher frequency acquiring the identity of a sinusoidal sequence of lower frequency after sampling is called *aliasing*. Since there are an infinite number of continuous-time functions that can lead to a given sequence when sampled periodically, additional conditions need to be imposed so that the sequence $\{x[n]\} = \{x_a(nT)\}$ can uniquely represent the parent continuous-time function $x_a(t)$. In which case, $x_a(t)$ can be fully recovered from a knowledge of $\{x[n]\}$.

EXAMPLE 2.12   Determine the discrete-time signal $v[n]$ obtained by uniformly sampling at a sampling rate of 200 Hz a continuous-time signal $v_a(t)$ composed of a weighted sum of five sinusoidal signals of frequencies 30 Hz, 150 Hz, 170 Hz, 250 Hz, and 330 Hz, as given below:

$$v_a(t) = 6\cos(60\pi t) + 3\sin(300\pi t) + 2\cos(340\pi t)$$
$$+ 4\cos(500\pi t) + 10\sin(660\pi t).$$

The sampling period $T = 1/200 = 0.005$ sec. Hence, the generated discrete-time signal $v[n]$ is given by

$$\begin{aligned}
v[n] &= 6\cos(0.3\pi n) + 3\sin(1.5\pi n) + 2\cos(1.7\pi n) + 4\cos(2.5\pi n) \\
&\quad + 10\sin(3.3\pi n) \\
&= 6\cos(0.3\pi n) + 3\sin\left((2\pi - 0.5\pi)n\right) + 2\cos\left((2\pi - 0.3\pi)n\right) \\
&\quad + 4\cos\left((2\pi + 0.5\pi)n\right) + 10\sin\left((4\pi - 0.7\pi)n\right) \\
&= 6\cos(0.3\pi n) - 3\sin(0.5\pi n) + 2\cos(0.3\pi n) + 4\cos(0.5\pi n) \\
&\quad - 10\sin(0.7\pi n) \\
&= 8\cos(0.3\pi n) + 5\cos(0.5\pi n + 0.6435) - 10\sin(0.7\pi n).
\end{aligned}$$

The discrete-time signal $v[n]$ is composed of a weighted sum of three discrete-time sinusoidal signals of normalized angular frequencies: $0.3\pi, 0.5\pi$, and 0.7π.

It should be noted that an identical discrete-time signal is also generated by uniformly sampling at a 200-Hz sampling rate the following continuous-time signal composed of a weighted sum of three sinusoidal continuous-time signals of frequencies 30 Hz, 50 Hz, and 70 Hz:

$$w_a(t) = 8\cos(60\pi t) + 5\cos(100\pi t + 0.6435) - 10\sin(140\pi t).$$

Another example of a continuous-time signal generating the same discrete-time sequence is

$$u_a(t) = 2\cos(60\pi t) + 4\cos(100\pi t) + 10\sin(260\pi t) + 6\cos(460\pi t) + 3\sin(700\pi t),$$

which is composed of a weighted sum of five sinusoidal continuous-time signals of frequencies 30 Hz, 50 Hz, 130 Hz, 230 Hz, and 350 Hz, respectively.

It follows from Eq. (2.51) that if $\Omega_T > 2|\Omega_o|$, then the corresponding normalized digital frequency ω_o of the discrete-time signal $x[n]$ obtained by sampling the parent continuous-time signal $x_a(t)$ will be in the range $-\pi < \omega < \pi$, implying no aliasing. On the other hand, if $\Omega_T < 2\Omega_o$, the normalized digital frequency will fold into a lower digital frequency $\omega_o = \langle 2\pi\Omega_o/\Omega_T\rangle_{2\pi}$ in the range $-\pi < \omega < \pi$ because of aliasing. Hence, to prevent aliasing, the sampling frequency Ω_T should be greater than 2 times the frequency Ω_o of the sinusoidal signal being sampled. Generalizing the above result, we observe that if we have an arbitrary continuous-time signal $x_a(t)$ that can be represented as a weighted sum of a number of sinusoidal signals, then $x_a(t)$ can also be represented uniquely by its sampled version $\{x[n]\}$ if the sampling frequency Ω_T is chosen to be greater than 2 times the highest frequency contained in $x_a(t)$. The condition to be satisfied by the sampling frequency to prevent aliasing is called the *sampling theorem*, which is formally derived later in Section 5.2.

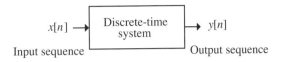

Figure 2.20: Schematic representation of a discrete-time system.

The discrete-time signal obtained by sampling a continuous-time signal $x_a(t)$ may be represented as a sequence $\{x_a[nT]\}$. However, we shall use the more common notation $\{x[n]\}$ for simplicity (with T assumed to be normalized to 1 sec). It should be noted that when dealing with the sampled version of a continuous-time function, it is essential to know the numerical value of the sampling period T.

2.4 Discrete-Time Systems

The function of a discrete-time system is to process a given *input sequence* to generate an *output sequence*. In most applications, the discrete-time system used is a single-input, single-output system, as shown schematically in Figure 2.20. The output sequence is generated sequentially, beginning with a certain value of the time index n, and thereafter progressively increasing the value of n. If the beginning time index is n_o, the output $y[n_o]$ is first computed, then $y[n_o + 1]$ is computed, and so on. We restrict our attention in this text to this class of discrete-time systems with certain specific properties as described later in this section.

In a practical discrete-time system, all signals are digital signals, and operations on such signals also lead to digital signals. Such a discrete-time system is usually called a digital filter. However, if there is no ambiguity, we shall refer to a discrete-time system also as a digital filter whether or not it has been implemented using finite precision arithmetic.

Simple Discrete-Time Systems

The devices implementing the basic operations shown in Figures 2.5 and 2.8 can be considered as elementary discrete-time systems. The modulator and the adder are examples of two-input, single-output discrete-time systems. The remaining devices are examples of single-input, single-output discrete-time systems. More complex discrete-time systems are obtained by combining two or more of these elementary discrete-time systems as illustrated in Figures 2.6 and 2.7, respectively. Some additional examples of discrete-time systems are given below.

EXAMPLE 2.13 A very simple example of a discrete-time system is the *accumulator* defined by the input-output relation

$$y[n] = \sum_{\ell=-\infty}^{n} x[\ell]$$

$$= \sum_{\ell=-\infty}^{n-1} x[\ell] + x[n] = y[n-1] + x[n]. \tag{2.54}$$

The output $y[n]$ at time instant n is the sum of the input sample value $x[n]$ at time instant n and the previous output $y[n-1]$ at time instant $n-1$, which is the sum of all previous input sample values from $-\infty$ to the time instant $n-1$. The system therefore cumulatively adds, i.e., it accumulates all input sample values from $-\infty$ to n.

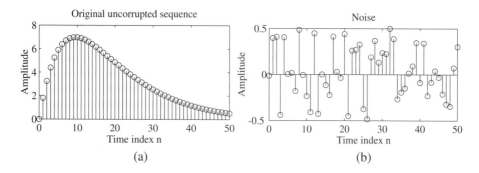

Figure 2.21: (a) The original uncorrupted sequence $s[n]$, and (b) the noise sequence $d[n]$.

The above equation can also be written in the form

$$y[n] = \sum_{\ell=-\infty}^{-1} x[\ell] + \sum_{\ell=0}^{n} x[\ell] = y[-1] + \sum_{\ell=0}^{n} x[\ell], \qquad n \geq 0. \tag{2.55}$$

The second form of the accumulator is used for a causal input sequence, in which case $y[-1]$ is called the *initial condition*.

We next illustrate the operation of a discrete-time system by two examples.

EXAMPLE 2.14   Another simple example of a discrete-time system is the M-point *moving-average filter* defined by

$$y[n] = \frac{1}{M} \sum_{k=0}^{M-1} x[n-k]. \tag{2.56}$$

Such a system is often used in smoothing random variations in data. Consider for example a signal $s[n]$ corrupted by a noise $d[n]$ for $n \geq 0$, resulting in a measured data given by

$$x[n] = s[n] + d[n].$$

We would like to reduce the effect of the noise $d[n]$ and get a better estimate of $s[n]$ from $x[n]$. To this end, the moving-average filter of Eq. (2.56) often gives reasonably good results [Sch75]. We illustrate this approach using MATLAB and assume for simplicity the original uncorrupted signal is given by

$$s[n] = 2[n(0.9)^n]. \tag{2.57}$$

The MATLAB program to generate $s[n]$ and $d[n]$ is given below where the noise $d[n]$ is computed using the function rand. Both sequences are shown in Figure 2.21.

```
% Program 2_3
%  Illustration of Signal Generation
%
% Random noise generation
R = 50;
d = rand(R,1)-0.5;
% Generate the uncorrupted signal
m = 0:1:R-1;
```

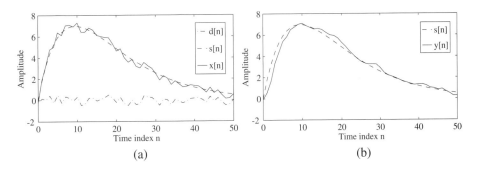

Figure 2.22: Pertinent signals of Example 2.14: $s[n]$ is the original uncorrupted sequence, $d[n]$ is the noise sequence, $x[n] = s[n] + d[n]$, and $y[n]$ is the output of the moving-average filter.

```
s = 2*m.*(0.9.^m);
subplot(2,1,1)
stem(m,s)
xlabel('Time index n'); ylabel('Amplitude')
title('Original uncorrupted sequence')
subplot(2,1,2)
stem(m,d)
xlabel('Time index n'); ylabel('Amplitude')
title('Noise')
```

Program 2_4 has been used to generate the smoothed output $y[n]$ from the noise corrupted signal $x[n]$ using the moving-average system of Eq. (2.56). Figure 2.22 shows the plots of pertinent signals generated by this program for $M = 3$. During execution, the program requests the input data, which is the desired number M of input samples to be added. It should be noted that to illustrate the effect of noise smoothing more clearly, the discrete-time signals have been plotted as continuous curves using the function plot.

```
% Program 2_4
% Signal Smoothing by a Moving-Average Filter
%
R = 50;
d = rand(R,1)-0.5;
m = 0:1:R-1;
s = 2*m.*(0.9.^m);
x = s + d';
plot(m,d,'r-',m,s,'b--',m,x,'g:')
xlabel('Time index n'); ylabel('Amplitude')
legend('r-','d[n]','b--','s[n]','g:','x[n]');
pause
M = input('Number of input samples = ');
b = ones(M,1)/M;
y = filter(b,1,x);
plot(m,s,'r-',m,y,'b--')
legend('r-','s[n]','b--','y[n]');
xlabel ('Time index n');ylabel('Amplitude')
```

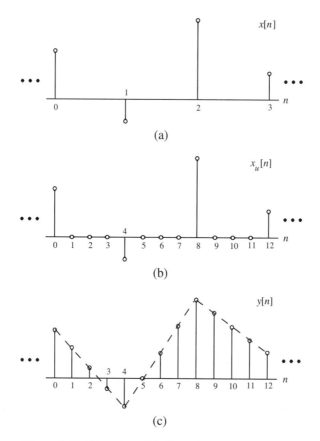

Figure 2.23: Illustration of the linear interpolation method.

Note that in Figure 2.22(b), the output $y[n]$ of the 3-point moving-average filter is nearly equal to the desired uncorrupted input $s[n]$, except that it is delayed by one sample. We shall show later in Section 4.2.6 that a delay of $(M-1)/2$ samples is inherent in an M-point moving-average filter.

EXAMPLE 2.15 Another example of a discrete-time system is the linear interpolation method often employed to estimate sample values between pairs of adjacent sample values of a discrete-time sequence. The linear interpolation is implemented by first passing the input sequence $x[n]$ to be interpolated through an up-sampler whose output $x_u[n]$ is then passed through a second discrete-time system that "fills in" the zero-valued samples inserted by the up-sampler with values obtained by a linear interpolation of the pair of input samples surrounding the zero-valued samples. The interpolated samples thus lie on a straight line joining the pair of input samples, as illustrated in Figure 2.23 for a factor-of-4 interpolation.

If we denote the output of the factor-of-2 interpolator as $y[n]$, then it is computed using the relation

$$y[n] = x_u[n] + \frac{1}{2}\left(x_u[n-1] + x_u[n+1]\right). \tag{2.58}$$

Likewise, for a factor-of-3 interpolator, the output is obtained using the relation

$$y[n] = x_u[n] + \frac{1}{3}\left(x_u[n-1] + x_u[n+2]\right) + \frac{2}{3}\left(x_u[n-2] + x_u[n+1]\right). \tag{2.59}$$

2.4.1 Classification of Discrete-Time Systems

There are various types of classification of discrete-time systems that are described next. These classifications are based on the input-output relation of the system.

Linear System

The most widely used discrete-time system, and the one that we shall be exclusively concerned with in this text, is a *linear system* for which the superposition principle always holds. More precisely, for a linear discrete-time system, if $y_1[n]$ and $y_2[n]$ are the responses to the input sequences $x_1[n]$ and $x_2[n]$, respectively, then for an input

$$x[n] = \alpha x_1[n] + \beta x_2[n],$$

the response is given by

$$y[n] = \alpha y_1[n] + \beta y_2[n].$$

The superposition property must hold for any arbitrary constants, α and β, and for all possible inputs, $x_1[n]$ and $x_2[n]$. The above property makes it very easy to compute the response to a complicated sequence that can be decomposed as a weighted combination of some simple sequences, such as the unit sample sequences or the complex exponential sequences. In this case, the desired output is given by a similarly weighted combination of the outputs to the simple sequences.

EXAMPLE 2.16 Consider the discrete-time accumulator of Example 2.13. From its input-output relation given by Eq. (2.54) we observe that the outputs $y_1[n]$ and $y_2[n]$ for inputs $x_1[n]$ and $x_2[n]$ are given by

$$y_1[n] = \sum_{\ell=-\infty}^{n} x_1[\ell],$$

$$y_2[n] = \sum_{\ell=-\infty}^{n} x_2[\ell].$$

The output $y[n]$ due to an input $\alpha x_1[n] + \beta x_2[n]$ is then given by

$$y[n] = \sum_{\ell=-\infty}^{n} (\alpha x_1[\ell] + \beta x_2[\ell]) = \alpha \sum_{\ell=-\infty}^{n} x_1[\ell] + \beta \sum_{\ell=-\infty}^{n} x_2[\ell] = \alpha y_1[n] + \beta y_2[n].$$

Hence, the discrete-time system of Eq. (2.54) is a linear system.

Consider the modified form of the accumulator defined by Eq. (2.55) with a causal input applied at $n = 0$. Now, the outputs $y_1[n]$ and $y_2[n]$ for inputs $x_1[n]$ and $x_2[n]$ are given by

$$y_1[n] = y_1[-1] + \sum_{\ell=0}^{n} x_1[\ell],$$

$$y_2[n] = y_2[-1] + \sum_{\ell=0}^{n} x_2[\ell].$$

Hence, the output $y[n]$ for an input $\alpha x_1[n] + \beta x_2[n]$ is of the form

$$y[n] = y[-1] + \sum_{\ell=0}^{n} (\alpha x_1[\ell] + \beta x_2[\ell]) = y[-1] + \alpha \sum_{\ell=0}^{n} x_1[\ell] + \beta \sum_{\ell=0}^{n} x_2[\ell]. \tag{2.60}$$

On the other hand,

$$\alpha y_1[n] + \beta y_2[n] = \alpha y_1[-1] + \alpha \sum_{\ell=0}^{n} x_1[\ell] + \beta y_2[-1] + \beta \sum_{\ell=0}^{n} x_2[\ell], \tag{2.61}$$

which is equal to $y[n]$ of Eq. (2.60) if

$$y[-1] = \alpha y_1[-1] + \beta y_2[-1].$$

For the system of Eq. (2.55) to be linear, the above condition must hold for all initial conditions $y[-1]$, $y_1[-1]$, and $y_2[-1]$ and all constants α and β. This condition cannot be satisfied unless the system of Eq. (2.55) is initially at rest with a zero initial condition. For nonzero initial conditions the discrete-time system of Eq. (2.55) is nonlinear.

It can be easily verified that the discrete-time systems of Eqs. (2.14), (2.15), (2.17), (2.18), (2.56), and (2.58) are linear systems (Problem 2.25). However, the linearity of the discrete-time system of Eq. (2.15) depends on the type of input being applied.

EXAMPLE 2.17   Consider the discrete-time system given by [Kai80]

$$y[n] = x^2[n] - x[n-1]x[n+1].$$

The outputs $y_1[n]$ and $y_2[n]$ for inputs $x_1[n]$ and $x_2[n]$ are given by

$$y_1[n] = x_1^2[n] - x_1[n-1]x_1[n+1],$$

$$y_2[n] = x_2^2[n] - x_2[n-1]x_2[n+1].$$

The output $y[n]$ due to an input $\alpha x_1[n] + \beta x_2[n]$ is given by

$$
\begin{aligned}
y[n] = {} & \{\alpha x_1[n] + \beta x_2[n]\}^2 \\
& - \{\alpha x_1[n-1] + \beta x_2[n-1]\}\{\alpha x_1[n+1] + \beta x_2[n+1]\} \\
= {} & \alpha^2 \left\{ x_1^2[n] - x_1[n-1]x_1[n+1] \right\} + \beta^2 \left\{ x_2^2[n] - x_2[n-1]x_2[n+1] \right\} \\
& + \alpha\beta \left\{ 2x_1[n]x_2[n] - x_1[n-1]x_2[n+1] - x_1[n+1]x_2[n-1] \right\}.
\end{aligned}
$$

On the other hand,

$$
\begin{aligned}
\alpha y_1[n] + \beta y_2[n] = {} & \alpha \left\{ x_1^2[n] - x_1[n-1]x_1[n+1] \right\} \\
& + \beta \left\{ x_2^2[n] - x_2[n-1]x_2[n+1] \right\} \neq y[n].
\end{aligned}
$$

Therefore, the system is *nonlinear*.

Shift-Invariant System

The *shift-invariance* property is the second condition imposed on most digital filters in practice. For a shift-invariant discrete-time system, if $y_1[n]$ is the response to an input $x_1[n]$, then the response to an input

$$x[n] = x_1[n - n_0]$$

is simply

$$y[n] = y_1[n - n_0],$$

where n_0 is any positive or negative integer. This relation between the input and output must hold for any arbitrary input sequence and its corresponding output. In the case of sequences and systems with indices n related to discrete instants of time, the above restriction is more commonly called the *time-invariance* property. The time-invariance property ensures that for a specified input, the output of the system is independent of the time the input is being applied.

A *linear time-invariant* (LTI) discrete-time system satisfies both the linearity and the time-invariance properties. Such systems are mathematically easy to analyze, and characterize, and as a consequence, easy to design. In addition, highly useful signal processing algorithms have been developed utilizing this class of systems over the last several decades. In this text, we consider almost entirely this type of discrete-time system.

EXAMPLE 2.18 The up-sampler of Eq. (2.17) is a time-varying system. To show this, we observe from Eq. (2.17) that its output $y_1[n]$ for an input $x_1[n] = x[n - n_o]$ is given by

$$y_1[n] = \begin{cases} x_1[n/L], & n = 0, \pm L, \pm 2L, \ldots, \\ 0, & \text{otherwise}, \end{cases}$$

$$= \begin{cases} x[(n - Ln_o)/L], & n = 0, \pm L, \pm 2L, \ldots, \\ 0, & \text{otherwise}. \end{cases}$$

But from Eq. (2.17)

$$y[n - n_o] = \begin{cases} x[(n - n_o)/L], & n = n_o, n_o \pm L, n_o \pm 2L, \\ 0, & \text{otherwise}, \end{cases}$$

$$\neq y_1[n].$$

Likewise, it can be shown that the down-sampler of Eq. (2.18) is a time-varying system.

Causal System

In addition to the above two properties, we impose, for practicality, additional restrictions of causality and stability on the class of discrete-time systems we deal with in this text. In a *causal* discrete-time system, the n_oth output sample $y[n_o]$ depends only on input samples $x[n]$ for $n \leq n_o$ and does not depend on input samples for $n > n_o$. Thus, if $y_1[n]$ and $y_2[n]$ are the responses of a causal discrete-time system to the inputs $u_1[n]$ and $u_2[n]$, respectively, then

$$u_1[n] = u_2[n] \quad \text{for } n < N$$

implies also that

$$y_1[n] = y_2[n] \quad \text{for } n < N.$$

Simply speaking, for a causal system, changes in output samples do not precede changes in the input samples. It should be pointed out here that the definition of causality given above can be applied only to discrete-time systems with the same sampling rate for the input and the output.[5]

It can be easily shown that the discrete-time systems of Eqs. (2.14), (2.15), (2.54), (2.55), and (2.56) are causal systems. However, the discrete-time systems defined by Eqs. (2.58) and (2.59) are noncausal systems. It should be noted that these two noncausal systems can be implemented as causal systems by simply delaying the output by one and two samples, respectively.

---

[5]If the input and output sampling rates are not the same, the definition of causality has to be modified.

Stable System

There are various definitions of stability. We define a discrete-time system to be *stable* if and only if, for every bounded input, the output is also bounded. This implies that, if the response to $x[n]$ is the sequence $y[n]$ and if

$$|x[n]| < B_x$$

for all values of n, then

$$|y[n]| < B_y$$

for all values of n where B_x and B_y are finite constants. This type of stability is usually referred to as *bounded-input, bounded-output* (BIBO) *stability*.

EXAMPLE 2.19　Consider the M-point moving-average filter of Example 2.14. It follows from Eq. (2.56) that for a bounded input $|x[n]| < B_x$, the magnitude of the nth sample of output is given by

$$|y[n]| = \left| \frac{1}{M} \sum_{k=0}^{M-1} x[n-k] \right| \le \frac{1}{M} \sum_{k=0}^{M-1} |x[n-k]| \le \frac{1}{M} (M) B_x \le B_x,$$

indicating that the system is BIBO stable.

Passive and Lossless Systems

A discrete-time system is said to be *passive* if, for every finite energy input sequence $x[n]$, the output sequence $y[n]$ has, at most, the same energy, i.e.,

$$\sum_{n=-\infty}^{\infty} |y[n]|^2 \le \sum_{n=-\infty}^{\infty} |x[n]|^2 < \infty. \tag{2.62}$$

If the above inequality is satisfied with an equal sign for every input sequence, the discrete-time system is said to be *lossless*.

EXAMPLE 2.20　Consider the discrete-time system defined by $y[n] = \alpha x[n-N]$ with N a positive integer. Its output energy is given by

$$\sum_{n=-\infty}^{\infty} |y[n]|^2 = |\alpha|^2 \sum_{n=-\infty}^{\infty} |x[n]|^2.$$

Hence, it is a passive system if $|\alpha| \le 1$ and is a lossless system if $|\alpha| = 1$.

As we shall see later, in Section 9.9, the passivity and the losslessness properties are crucial to the design of discrete-time systems with very low sensitivity to changes in the filter coefficients.

2.4.2　Impulse and Step Responses

The response of a digital filter to a unit sample sequence $\{\delta[n]\}$ is called the *unit sample response*, or simply, the *impulse response*, and is denoted as $\{h[n]\}$. Correspondingly, the response of a discrete-time system to a unit step sequence $\{\mu[n]\}$, denoted as $\{s[n]\}$, is its *unit step response* or simply, the *step response*. As we show next, a linear time-invariant digital filter is completely characterized in the time-domain by its impulse response or its step response.

EXAMPLE 2.21 The impulse response $\{h[n]\}$ of the discrete-time system of Eq. (2.14) is obtained by setting $x[n] = \delta[n]$ resulting in

$$h[n] = \alpha_1 \delta[n] + \alpha_2 \delta[n-1] + \alpha_3 \delta[n-2] + \alpha_4 \delta[n-3].$$

The impulse response is thus a finite-length sequence of length 4 given by

$$\{h[n]\} = \{\alpha_1, \quad \alpha_2, \quad \alpha_3, \quad \alpha_4\}.$$
$$\uparrow$$

EXAMPLE 2.22 The impulse response $\{h[n]\}$ of the discrete-time accumulator of Eq. (2.54) is obtained by setting $x[n] = \delta[n]$ resulting in

$$h[n] = \sum_{\ell=-\infty}^{n} \delta[\ell],$$

which from Eq. (2.38) is precisely the unit step sequence $\mu[n]$.

EXAMPLE 2.23 The impulse response $\{h[n]\}$ of the factor-of-2 interpolator of Eq. (2.58) is obtained by setting $x_u[n] = \delta[n]$ and is given by

$$h[n] = \delta[n] + \frac{1}{2}\left(\delta[n-1] + \delta[n+1]\right).$$

The impulse response is seen to be a finite-length sequence of length 3 and can be written alternately as

$$\{h[n]\} = \{0.5, \ 1, \ 0.5\}.$$
$$\uparrow$$

2.5 Time-Domain Characterization of LTI Discrete-Time Systems

In most cases an LTI discrete-time system is designed as an interconnection of simple subsystems. Each subsystem in turn is implemented with the aid of the basic building blocks discussed earlier in Section 2.1.2. In order to be able to analyze such systems in the time-domain, we need to develop the pertinent relationships between the input and the output of an LTI discrete-time system, and the characterization of the interconnected system.

2.5.1 Input-Output Relationship

A consequence of the linear, time-invariance property is that an LTI discrete-time system is completely specified by its impulse response; i.e., knowing the impulse response, we can compute the output of the system to any arbitrary input. We develop this relationship now.

Let $h[n]$ denote the impulse response of the LTI discrete-time system of interest, i.e., the response to an input $\delta[n]$. We first compute the response of this filter to the input $x[n]$ of Eq. (2.47). Since the discrete-time system is time-invariant, its response to $\delta[n-1]$ will be $h[n-1]$. Likewise, the responses to $\delta[n+2]$, $\delta[n-4]$, and $\delta[n-6]$ will be, respectively, $h[n+2]$, $h[n-4]$, and $h[n-6]$. Because of linearity, the response of the LTI discrete-time system to the input

$$x[n] = 0.5\delta[n+2] + 1.5\delta[n-1] - \delta[n-2] + \delta[n-4] + 0.75\delta[n-6]$$

will be simply

$$y[n] = 0.5h[n + 2] + 1.5h[n - 1] - h[n - 2] + h[n - 4] + 0.75h[n - 6].$$

It follows from the above result that an arbitrary input sequence $x[n]$ can be expressed as a weighted linear combination of delayed and advanced unit sample sequences in the form

$$x[n] = \sum_{k=-\infty}^{\infty} x[k]\delta[n - k], \tag{2.63}$$

where the weight $x[k]$ on the right-hand side denotes specifically the kth sample value of the sequence $\{x[n]\}$. The response of the LTI discrete-time system to the sequence $x[k]\delta[n - k]$ will be $x[k]h[n - k]$. As a result, the response $y[n]$ of the discrete-time system to $x[n]$ will be given by

$$y[n] = \sum_{k=-\infty}^{\infty} x[k]h[n - k], \tag{2.64a}$$

which can be alternately written as

$$y[n] = \sum_{k=-\infty}^{\infty} x[n - k]h[k] \tag{2.64b}$$

by a simple change of variables. The above sum in Eqs. (2.64a) and (2.64b) is called the *convolution sum* of the sequences $x[n]$ and $h[n]$, and represented compactly as:

$$y[n] = x[n] \circledast h[n] \tag{2.65}$$

where the notation $\circledast$ denotes the convolution sum.[6]

The convolution sum operation satisfies several useful properties. First, the operation is *commutative*, i.e.,

$$x_1[n] \circledast x_2[n] = x_2[n] \circledast x_1[n]. \tag{2.66}$$

Second, the convolution operation, for stable and single-sided sequences, is *associative*, i.e.,

$$(x_1[n] \circledast x_2[n]) \circledast x_3[n] = x_1[n] \circledast (x_2[n] \circledast x_3[n]), \tag{2.67}$$

and last, the operation is *distributive*, i.e.,

$$x_1[n] \circledast (x_2[n] + x_3[n]) = x_1[n] \circledast x_2[n] + x_1[n] \circledast x_3[n]. \tag{2.68}$$

Proof of these properties is left as an exercise (Problems 2.37 to 2.39).

The convolution sum operation of Eq. (2.64a) can be interpreted as follows. We first time-reverse the sequence $h[k]$ arriving at $h[-k]$. We then shift $h[-k]$ to the right by n sampling periods if $n > 0$, or to the left by n sampling periods if $n < 0$, to form the sequence $h[n - k]$. Next we form the product sequence $v[k] = x[k]h[n - k]$. Summing all samples of $v[k]$ then yields the nth sample of $y[n]$ of the convolution sum. The process of generating $v[k]$ is illustrated in Figure 2.24. This process is implemented for each value of n in the range $-\infty < n < \infty$. The representation of the alternate form of the convolution sum operation given by Eq. (2.64b) is obtained by interchanging the sequences $x[k]$ and $h[k]$ in Figure 2.24.

---

[6]In the literature, the symbol for the convolution sum is $*$ without the circle. However, as the superscript $*$ is always used for denoting the complex conjugation operation, in this text we have adopted the symbol $\circledast$ to denote the convolution sum operation.

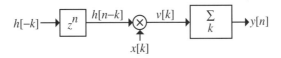

Figure 2.24: Schematic representation of the convolution sum operation.

It is clear from the above discussion that the impulse response $\{h[n]\}$ completely characterizes an LTI discrete-time system in the time-domain because, knowing the impulse response, we can compute, in principle, the output sequence $y[n]$ for any given input sequence $x[n]$ using the convolution sum of Eq. (2.64a) or (2.64b). The computation of an output sample is simply a sum of products involving fairly simple arithmetic operations such as additions, multiplications, and delays. However, in practice, the convolution sum can be employed to compute the output sample at any instant only if either the impulse response sequence and/or the input sequence is of finite length, resulting in a finite sum of products. Note that if both the input and the impulse response sequences are of finite length, the output sequence is also of finite length. In the case of a discrete-time system with an infinite-length impulse response, it is obviously not possible to compute the output using the convolution sum if the input is also of infinite length. We shall therefore consider alternative time-domain descriptions of such systems that involve only finite sums of products.

EXAMPLE 2.24   We systematically develop the sequence $y[n]$ generated by the convolution of the two finite-length sequences $x[n]$ and $h[n]$ shown in Figure 2.25(a). As can be seen from the plot of the shifted time-reversed version $\{h[n-k]\}$ for $n < 0$ sketched in Figure 2.25(b) for $n = -3$, for any value of the sample index k, the kth sample of either $\{x[k]\}$ or $\{h[n-k]\}$ is zero. As a result the product of the kth samples is always zero for any k, and consequently, the convolution sum of Eq. (2.64a) leads to

$$y[n] = 0 \quad \text{for } n < 0.$$

Consider now the calculation of $y[0]$. We form $\{h[-k]\}$ as shown in Figure 2.25(c). The product sequence $\{x[k]h[-k]\}$ is plotted in Figure 2.25(d), which has a single nonzero sample for $k = 0$, $x[0]h[0]$. Thus,

$$y[0] = x[0]h[0] = -2.$$

For the computation of $y[1]$, we shift $\{h[-k]\}$ to the right by one sample period to form $\{h[1-k]\}$ as sketched in Figure 2.25(e). The product sequence $\{x[k]h[1-k]\}$ shown in Figure 2.25(f) has one nonzero sample for $k = 0$, $x[0]h[1]$. As a result,

$$y[1] = x[0]h[1] + x[1]h[0] = -4 + 0 = -4.$$

To calculate $y[2]$, we form $\{h[2-k]\}$, as indicated in Figure 2.25(g). The resulting product sequence $\{x[k]h[2-k]\}$ is plotted in Figure 2.25(h) which leads to

$$y[2] = x[0]h[2] + x[1]h[1] + x[2]h[0] = 0 + 0 + 1 = 1.$$

This process is continued, resulting in

$$y[3] = x[0]h[3] + x[1]h[2] + x[2]h[1] + x[3]h[0] = 2 + 0 + 0 + 1 = 3,$$
$$y[4] = x[1]h[3] + x[2]h[2] + x[3]h[1] + x[4]h[0] = 0 + 0 - 2 + 3 = 1,$$
$$y[5] = x[2]h[3] + x[3]h[2] + x[4]h[1] = -1 + 0 + 6 = 5,$$
$$y[6] = x[3]h[3] + x[4]h[2] = 1 + 0 = 1,$$
$$y[7] = x[4]h[3] = -3.$$

From the plot of $\{h[n-k]\}$ for $n > 7$ as shown in Figure 2.25(i) and the plot of $\{x[k]\}$ in Figure 2.25(a), it can be seen that there is no overlap between these two sequences. As a result,

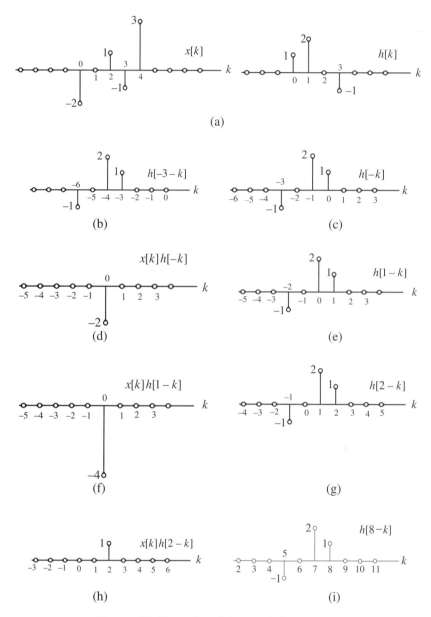

Figure 2.25: Illustration of the convolution process.

$$y[n] = 0 \quad \text{for } n > 7.$$

The sequence $\{y[n]\}$ as obtained above is sketched in Figure 2.26.

It should be noted that the sum of the indices of each sample product inside the summation of either Eq. (2.64a) or Eq. (2.64b) is equal to the index of the sample being generated by the convolution sum operation. For example, the sum in the computation of $y[3]$ in the above example involves the products

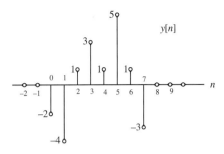

Figure 2.26: Sequence generated by the convolution.

$x[0]h[3]$, $x[1]h[2]$, $x[2]h[1]$, and $x[3]h[0]$. The sum of the indices in each of these four products is equal to 3 which is the index of the sample $y[3]$.

As can be seen from Example 2.24, the convolution of two finite-length sequences results in a finite-length sequence. In this example, the convolution of a sequence $\{x[n]\}$ of length 5 with a sequence $\{h[n]\}$ of length 4 resulted in a sequence $\{y[n]\}$ of length 8. In general, if the lengths of the two sequences being convolved are M and N, then the resulting sequence after convolution is of length $M + N - 1$ (Problem 2.40).

In MATLAB, the statement $c = \texttt{conv(a,b)}$ implements the convolution of two finite-length sequences a and b, generating the finite-length sequence c. The process is illustrated in the following example.

EXAMPLE 2.25   We repeat the problem of the previous example using Program 2_5 given below.

```
% Program 2_5
% Illustration of Convolution
%
a = input('Type in the first sequence = ');
b = input('Type in the second sequence = ');
c = conv(a, b);
M = length(c)-1;
n = 0:1:M;
disp('output sequence =');disp(c)
stem(n,c)
xlabel('Time index n'); ylabel('Amplitude');
```

During execution, the input data requested are the two sequences to be convolved that are entered in a vector format inside square brackets as indicated below:

```
a = [-2    0    1    -1    3]
b = [ 1    2    0    -1]
```

The program then computes the convolution and displays the resulting sequence shown below:

```
output sequence =
  -2    -4    1    3    1    5    1    -3
```

and provides a plot, as indicated in Figure 2.27. As expected, the result is identical to that derived in the previous example.

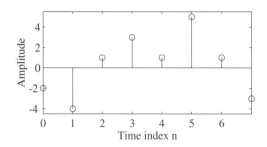

Figure 2.27: Sequence generated by convolution using MATLAB.

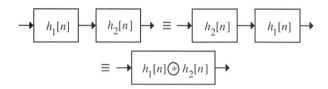

Figure 2.28: The cascade connection.

2.5.2 Simple Interconnection Schemes

Two widely used schemes for developing complex LTI discrete-time systems from simple LTI discrete-time system sections are described next.

Cascade Connection

In Figure 2.28, the output of one filter is connected to the input of a second filter and the two filters are said to be connected in *cascade*. The overall impulse response $h[n]$ of the cascade of two filters of impulse responses $h_1[n]$ and $h_2[n]$ is given by

$$h[n] = h_1[n] \circledast h_2[n]. \tag{2.69}$$

Note that, in general, the ordering of the filters in the cascade has no effect on the overall impulse response because of the commutative property of convolution.

It can be shown that the cascade connection of two stable systems is stable. Likewise, the cascade connection of two passive (lossless) systems is passive (lossless).

The cascade connection scheme is employed in the development of an *inverse system*. If the two LTI systems in the cascade connection of Figure 2.28 are such that

$$h_1[n] \circledast h_2[n] = \delta[n], \tag{2.70}$$

then the LTI system $h_2[n]$ is said to be the inverse of the LTI system $h_1[n]$, and vice versa. As a result of the above relation, if the input to the cascaded system is $x[n]$, its output is also $x[n]$. An application of this concept is in the recovery of a signal from its distorted version appearing at the output of a transmission channel. This is accomplished by designing an inverse system if the impulse response of the channel is known.

The following example illustrates the development of an inverse system.

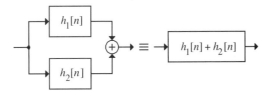

Figure 2.29: The parallel connection.

EXAMPLE 2.26   From Example 2.22 the impulse response of the discrete-time accumulator is the unit step sequence $\mu[n]$. Therefore, from Eq. (2.70) the inverse system must satisfy the condition

$$\mu[n] \circledast h_2[n] = \delta[n]. \qquad (2.71)$$

It follows from Eq. (2.71) that $h_2[n] = 0$ for $n < 0$ and

$$h_2[0] = 1,$$

$$\sum_{\ell=0}^{n} h_2[\ell] = 0, \qquad n \geq 1.$$

As a result,

$$h_2[1] = -1$$

and

$$h_2[n] = 0, \qquad \text{for } n \geq 2.$$

Thus the impulse response of the inverse system is given by

$$h_2[n] = \delta[n] - \delta[n-1],$$

which is called a *backward difference system*.

Parallel Connection

The connection scheme of Figure 2.29 is called the *parallel* connection, and here the outputs of the two filters are added to form the new output while the same input is fed to both filters. The impulse response of the overall filter is given here by

$$h[n] = h_1[n] + h_2[n]. \qquad (2.72)$$

It is a simple exercise to show that the parallel connection of two stable systems is stable. However, the parallel connection of two passive (lossless) systems may or may not be passive (lossless).

EXAMPLE 2.27   Consider the discrete-time system of Figure 2.30 composed of an interconnection of four simple discrete-time systems with impulse responses given by

$$h_1[n] = \delta[n] + \tfrac{1}{2}\delta[n-1], \qquad h_2[n] = \tfrac{1}{2}\delta[n] - \tfrac{1}{4}\delta[n-1],$$

$$h_3[n] = 2\delta[n], \qquad h_4[n] = -2\left(\tfrac{1}{2}\right)^n \mu[n].$$

The overall impulse response $h[n]$ is given by

$$h[n] = h_1[n] + h_2[n] \circledast (h_3[n] + h_4[n]) = h_1[n] + h_2[n] \circledast h_3[n] + h_2[n] \circledast h_4[n].$$

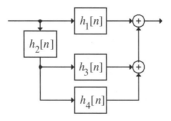

Figure 2.30: The discrete-time system of Example 2.27.

Now

$$h_2[n] \circledast h_3[n] = (\tfrac{1}{2}\delta[n] - \tfrac{1}{4}\delta[n-1]) \circledast 2\delta[n] = \delta[n] - \tfrac{1}{2}\delta[n-1]$$

and

$$h_2[n] \circledast h_4[n] = \left(\tfrac{1}{2}\delta[n] - \tfrac{1}{4}\delta[n-1]\right) \circledast \left(-2\left(\tfrac{1}{2}\right)^n \mu[n]\right)$$

$$= -\left(\tfrac{1}{2}\right)^n \mu[n] + \tfrac{1}{2}\left(\tfrac{1}{2}\right)^{n-1} \mu[n-1] = -\left(\tfrac{1}{2}\right)^n \mu[n] + \left(\tfrac{1}{2}\right)^n \mu[n-1]$$

$$= -\left(\tfrac{1}{2}\right)^n \delta[n] = -\delta[n].$$

Therefore

$$h[n] = \delta[n] + \tfrac{1}{2}\delta[n-1] + \delta[n] - \tfrac{1}{2}\delta[n-1] - \delta[n] = \delta[n].$$

2.5.3 Stability Condition in Terms of the Impulse Response

Recall from Section 2.4.1 that a discrete-time system is defined to be stable, or precisely, bounded-input, bounded-output (BIBO) stable, if the output sequence $y[n]$ of the system remains bounded for all bounded input sequences $x[n]$. We now develop the stability condition for an LTI discrete-time system. We show that an LTI digital filter is BIBO stable if and only if its impulse response sequence $\{h[n]\}$ is absolutely summable, i.e.,

$$S = \sum_{n=-\infty}^{\infty} |h[n]| < \infty. \tag{2.73}$$

We prove the above statement for a real impulse response $h[n]$. The extension of the proof for a complex impulse response sequence is left as an exercise (Problem 2.59). Now, if the input sequence $x[n]$ is bounded, i.e., $|x[n]| \le B_x < \infty$, then the output amplitude, from Eq. (2.64b), is

$$|y[n]| = \left| \sum_{k=-\infty}^{\infty} h[k]x[n-k] \right| \le \sum_{k=-\infty}^{\infty} |h[k]|\,|x[n-k]|$$

$$\le B_x \sum_{k=-\infty}^{\infty} |h[k]| = B_x S < \infty. \tag{2.74}$$

Thus, $S < \infty$ implies $|y[n]| \le B_y < \infty$, indicating that $y[n]$ is also bounded. To prove the converse, assume $y[n]$ is bounded, i.e., $|y[n]| \le B_y$. Now, consider the input given by

$$x[n] = \begin{cases} \operatorname{sgn}\left(h[-n]\right), & \text{if } h[-n] \ne 0, \\ K, & \text{if } h[-n] = 0, \end{cases} \tag{2.75}$$

where $\text{sgn}(c) = +1$ if $c > 0$ and $\text{sgn}(c) = -1$ if $c < 0$, and $|K| \le 1$. Note that since $|x[n]| \le 1$, $\{x[n]\}$ is obviously bounded. For this input, $y[n]$ at $n = 0$ is

$$y[0] = \sum_{k=-\infty}^{\infty} \text{sgn}(h[k])h[k] = S \le B_y < \infty. \qquad (2.76)$$

Therefore $|y[n]| \le B_y$ implies $S < \infty$.

EXAMPLE 2.28 Consider a causal LTI discrete-time system with an impulse response given by

$$h[n] = (\alpha)^n \mu[n].$$

For this system

$$S = \sum_{n=-\infty}^{\infty} |\alpha^n| \mu[n] = \sum_{n=0}^{\infty} |\alpha|^n = \frac{1}{1 - |\alpha|} \qquad \text{if } |\alpha| < 1.$$

Therefore $S < \infty$ if $|\alpha| < 1$ for which the above system is BIBO stable. On the other hand, if $|\alpha| = 1$, the above system is not BIBO stable.

EXAMPLE 2.29 Now an LTI discrete-time system with an impulse response given by

$$h[n] = \begin{cases} \alpha^n, & N_1 \le n \le N_2, \\ 0, & \text{otherwise}, \end{cases} \qquad (2.77)$$

has only a finite number of nonzero impulse response samples for finite values of N_1 and N_2. Hence, the impulse response sequence is absolutely summable independent of the value of α as long as it is not infinite. Hence, the system of Eq. (2.77) is BIBO stable.

2.5.4 Causality Condition in Terms of the Impulse Response

We now develop the condition for an LTI discrete-time system to be causal. Let $x_1[n]$ and $x_2[n]$ be two input sequences with

$$x_1[n] = x_2[n] \qquad \text{for } n \le n_o. \qquad (2.78)$$

From Eq. (2.64b) the corresponding output samples at $n = n_o$ of an LTI discrete-time system with an impulse response $\{h[n]\}$ are then given by

$$y_1[n_o] = \sum_{k=-\infty}^{\infty} h[k]x_1[n_o - k] = \sum_{k=0}^{\infty} h[k]x_1[n_o - k] + \sum_{k=-\infty}^{-1} h[k]x_1[n_o - k], \qquad (2.79a)$$

$$y_2[n_o] = \sum_{k=-\infty}^{\infty} h[k]x_2[n_o - k] = \sum_{k=0}^{\infty} h[k]x_2[n_o - k] + \sum_{k=-\infty}^{-1} h[k]x_2[n_o - k]. \qquad (2.79b)$$

If the LTI discrete-time system is also causal, then $y_1[n_o]$ must be equal to $y_2[n_o]$. Now, because of Eq. (2.78), the first sum on the right-hand side of Eq. (2.79a) is equal to the first sum on the right-hand side of Eq. (2.79b). This implies that the second sums on the right-hand side of the above two equations must be equal. As $x_1[n]$ may not be equal to $x_2[n]$ for $n > n_o$, the only way these two sums will be equal is if they are each equal to zero, which is satisfied if

$$h[k] = 0 \qquad \text{for } k < 0. \qquad (2.80)$$

As a result, an LTI discrete-time system is *causal* if and only if its impulse response sequence $\{h[n]\}$ is a causal sequence satisfying the condition of Eq. (2.80) .

It follows from Example 2.21 that the discrete-time system of Eq. (2.14) is a causal system since its impulse response satisfies the causality condition of Eq. (2.80). Likewise, from Example 2.22 we observe that the discrete-time accumulator of Eq. (2.54) is also a causal system. On the other hand, from Example 2.23 it can be seen that the factor-of-2 linear interpolator defined by Eq. (2.58) is a noncausal system because its impulse response does not satisfy the causality condition of Eq. (2.80). However, a noncausal discrete-time system with a finite-length impulse response can often be realized as a causal system by inserting a delay of an appropriate amount. For example, a causal version of the discrete-time factor-of-2 linear interpolator is obtained by delaying the output by one sample period with an input-output relation given by

$$y[n] = x_u[n-1] + \frac{1}{2}(x_u[n-2] + x_u[n]).$$

2.6   Finite-Dimensional LTI Discrete-Time Systems

An important subclass of LTI discrete-time systems is characterized by a linear constant coefficient difference equation of the form

$$\sum_{k=0}^{N} d_k y[n-k] = \sum_{k=0}^{M} p_k x[n-k], \qquad (2.81)$$

where $x[n]$ and $y[n]$ are, respectively, the input and the output of the system, and $\{d_k\}$ and $\{p_k\}$ are constants. The *order* of the discrete-time system is given by $\max(N, M)$, which is the order of the difference equation characterizing the system. It is possible to implement an LTI system characterized by Eq. (2.81) since the computation here involves two finite sum of products even though such a system, in general, has an impulse response of infinite length.

The output $y[n]$ can then be computed recursively from Eq. (2.81). If we assume the system to be causal, then we can rewrite Eq. (2.81) to express $y[n]$ explicitly as a function of $x[n]$:

$$y[n] = -\sum_{k=1}^{N} \frac{d_k}{d_0} y[n-k] + \sum_{k=0}^{M} \frac{p_k}{d_0} x[n-k], \qquad (2.82)$$

provided $d_0 \neq 0$. The output $y[n]$ can be computed for all $n \geq n_o$ knowing $x[n]$ and the initial conditions $y[n_o - 1], y[n_o - 2], \dots, y[n_o - N]$.

2.6.1   Total Solution Calculation

The procedure for computing the solution of the constant coefficient difference equation of Eq. (2.81) is very similar to that employed in solving the constant coefficient differential equation in the case of an LTI continuous-time system. In the case of the discrete-time system of Eq. (2.81) the output response $y[n]$ also consists of two components which are computed independently and then added to yield the total solution:

$$y[n] = y_c[n] + y_p[n]. \qquad (2.83)$$

In the above equation the component $y_c[n]$ is the solution of Eq. (2.81) with the input $x[n] = 0$; i.e., it is the solution of the homogeneous difference equation:

$$\sum_{k=0}^{N} d_k y[n-k] = 0, \qquad (2.84)$$

and the component $y_p[n]$ is a solution of Eq. (2.81) with $x[n] \neq 0$. $y_c[n]$ is called the *complementary solution*, while $y_p[n]$ is called the *particular solution* resulting from the specified input $x[n]$, often called the *forcing function*. The sum of the complementary and the particular solutions as given by Eq. (2.83) is called the *total solution*.

We first describe the method of computing the complementary solution $y_c[n]$. To this end we assume that it is of the form

$$y_c[n] = \lambda^n. \tag{2.85}$$

Substituting the above in Eq. (2.84) we arrive at

$$\sum_{k=0}^{N} d_k y[n-k] = \sum_{k=0}^{N} d_k \lambda^{n-k}$$

$$= \lambda^{n-N}(d_0 \lambda^N + d_1 \lambda^{N-1} + \cdots + d_{N-1}\lambda + d_N) = 0. \tag{2.86}$$

The polynomial $\sum_{k=0}^{N} d_k \lambda^{N-k}$ is called the *characteristic polynomial* of the discrete-time system of Eq. (2.81). Let $\lambda_1, \lambda_2, \ldots, \lambda_N$ denote its N roots. If these roots are all distinct, then the general form of the complementary solution is given by

$$y_c[n] = \alpha_1 \lambda_1^n + \alpha_2 \lambda_2^n + \cdots + \alpha_N \lambda_N^n, \tag{2.87}$$

where $\alpha_1, \alpha_2, \ldots, \alpha_N$ are constants determined from the specified initial conditions of the discrete-time system. The complementary solution takes a different form in the case of multiple roots. For example, if λ_1 is of multiplicity L and the remaining $N-L$ roots, $\lambda_2, \lambda_3, \ldots, \lambda_{N-L}$, are distinct, then Eq. (2.87) takes the form

$$y_c[n] = \alpha_1 \lambda_1^n + \alpha_2 n \lambda_1^n + \alpha_3 n^2 \lambda_1^n + \cdots + \alpha_L n^{L-1} \lambda_1^n + \alpha_{L+1} \lambda_2^n + \cdots + \alpha_N \lambda_{N-L}^n. \tag{2.88}$$

Next, we consider the determination of the particular solution $y_p[n]$ of the difference equation of Eq. (2.81). Here the procedure is to assume that the particular solution is also of the same form as the specified input $x[n]$ if $x[n]$ has the form $\lambda_0^n (\lambda_0 \neq \lambda_i, i = 1, 2, \ldots, N)$ for all n. Thus, if $x[n]$ is a constant, then $y_p[n]$ is also assumed to be constant. Likewise, if $x[n]$ is a sinusoidal sequence, then $y_p[n]$ is also assumed to be a sinusoidal sequence, and so on.

We illustrate below the determination of the total solution by means of an example.

EXAMPLE 2.30   Let us determine the total solution for $n \geq 0$ of a discrete-time system characterized by the following difference equation:

$$y[n] + y[n-1] - 6y[n-2] = x[n], \tag{2.89}$$

for a step input $x[n] = 8\mu[n]$ and with initial conditions $y[-1] = 1$ and $y[-2] = -1$.

We first determine the form of the complementary solution. Setting $x[n] = 0$ and $y[n] = \lambda^n$ in Eq. (2.89) we arrive at

$$\lambda^n + \lambda^{n-1} - 6\lambda^{n-2} = \lambda^{n-2}(\lambda^2 + \lambda - 6)$$

$$= \lambda^{n-2}(\lambda + 3)(\lambda - 2) = 0,$$

and hence the roots of the characteristic polynomial $\lambda^2 + \lambda - 6$ are: $\lambda_1 = -3, \lambda_2 = 2$. Therefore the complementary solution is of the form

$$y_c[n] = \alpha_1(-3)^n + \alpha_2(2)^n. \tag{2.90}$$

For the particular solution we assume

$$y_p[n] = \beta.$$

Substituting the above in Eq. (2.89) we get

$$\beta + \beta - 6\beta = 8\mu[n],$$

which for $n \geq 0$ yields $\beta = -2$.

The total solution is therefore of the form

$$y[n] = \alpha_1(-3)^n + \alpha_2(2)^n - 2, \qquad n \geq 0. \tag{2.91}$$

The constants α_1 and α_2 are chosen to satisfy the specified initial conditions. From Eqs. (2.89) and (2.91) we get

$$y[-2] = \alpha_1(-3)^{-2} + \alpha_2(2)^{-2} - 2 = -1,$$
$$y[-1] = \alpha_1(-3)^{-1} + \alpha_2(2)^{-1} - 2 = 1.$$

Solving these two equations we arrive at

$$\alpha_1 = -1.8, \ \alpha_2 = 4.8.$$

Thus, the total solution is given by

$$y[n] = -1.8(-3)^n + 4.8(2)^n - 2, \qquad n \geq 0.$$

If the input excitation is of the same form as one of the terms in the complementary solution, then it is necessary to modify the form of the particular solution as illustrated in the following example.

EXAMPLE 2.31   We determine the total solution for $n \geq 0$ of the difference equation of Eq. (2.89) for an input $x[n] = (2)^n \mu[n]$ with the same initial conditions as in the previous example.

As indicated in the previous example, the complementary solution contains a term $\alpha_2(2)^n$, which is of the same form as the specified input. Hence we need to select a form for the particular solution which is distinct and does not contain any terms similar to those contained in the complementary solution. We assume

$$y_p[n] = \beta n(2)^n.$$

Substituting the above in Eq. (2.89) we get

$$\beta n(2)^n + \beta(n-1)(2)^{n-1} - 6\beta(n-2)(2)^{n-2} = (2)^n \mu[n].$$

For $n \geq 0$ we obtain from the above equation $\beta = 0.4$. The total solution is now of the form

$$y[n] = \alpha_1(-3)^n + \alpha_2(2)^n + 0.4n(2)^n, \qquad n \geq 0. \tag{2.92}$$

To determine the values of α_1 and α_2, we make use of the specified initial conditions. From Eqs. (2.89) and (2.92) we arrive at

$$y[-2] = \alpha_1(-3)^{-2} + \alpha_2(2)^{-2} + 0.4(-2)(2)^{-2} = -1,$$
$$y[-1] = \alpha_1(-3)^{-1} + \alpha_2(2)^{-1} + 0.4(-1)(2)^{-1} = 1,$$

which when solved yields $\alpha_1 = -5.04$, $\alpha_2 = -0.96$. Therefore the total solution is given by

$$y[n] = -5.04(-3)^n - 0.96(2)^n + 0.4n(2)^n, \qquad n \geq 0.$$

2.6.2 Zero-Input Response and Zero-State Response

An alternate approach to determining the total solution $y[n]$ of the difference equation of Eq. (2.81) is by computing its *zero-input response* $y_{zi}[n]$ and *zero-state response* $y_{zs}[n]$. The component $y_{zi}[n]$ is obtained by solving Eq. (2.81) by setting the input $x[n] = 0$, and the component $y_{zs}[n]$ is obtained by solving Eq. (2.81) by applying the specified input with all initial conditions set to zero. The total solution is then given by $y_{zi}[n] + y_{zs}[n]$.

This approach is illustrated by the following example.

EXAMPLE 2.32   We determine the total solution of the discrete-time system of Example 2.30 by computing the zero-input response and the zero-state response.

The zero-input response $y_{zi}[n]$ of Eq. (2.89) is given by the complementary solution of Eq. (2.90) where the constants α_1 and α_2 are chosen to satisfy the specified initial conditions. Now from Eq. (2.89) we get

$$y[0] = -y[-1] + 6y[-2] = -1 - 6 = -7,$$
$$y[1] = -y[0] + 6y[-1] = 7 + 6 = 13.$$

Next from Eq. (2.90) we get

$$y[0] = \alpha_1 + \alpha_2,$$
$$y[1] = -3\alpha_1 + 2\alpha_2.$$

Solving these two sets of equations we arrive at $\alpha_1 = -5.4$, $\alpha_2 = -1.6$. Therefore

$$y_{zi}[n] = -5.4(-3)^n - 1.6(2)^n, \quad n \geq 0.$$

The zero-state response is determined from Eq. (2.91) by evaluating the constants α_1 and α_2 to satisfy the zero initial conditions. From Eq. (2.89) we get

$$y[0] = x[0] = 8,$$
$$y[1] = x[1] - y[0] = 0.$$

Next, from Eq. (2.91) and the above set of equations we arrive at $\alpha_1 = 3.6$, $\alpha_2 = 6.4$. Thus, the zero-state response for $n \geq 0$ with initial conditions $y_{zs}[-2] = y_{zs}[-1] = 0$ is given by

$$y_{zs}[n] = 3.6(-3)^n + 6.4(2)^n - 2.$$

Hence, the total solution $y[n]$ is given by the sum $y_{zi}[n] + y_{zs}[n]$ resulting in

$$y[n] = -1.8(-3)^n + 4.8(2)^n - 2, \quad n \geq 0,$$

which is identical to that derived in Example 2.30 as expected.

2.6.3 Impulse Response Calculation

The impulse response $h[n]$ of a causal LTI discrete-time system is the output observed with input $x[n] = \delta[n]$. Thus, it is simply the zero-state response with $x[n] = \delta[n]$. Now for such an input, $x[n] = 0$ for $n > 0$, and thus, the particular solution is zero, i.e., $y_p[n] = 0$. Hence the impulse response can be computed from the complementary solution of Eq. (2.87) in the case of simple roots of the characteristic equation by determining the constants α_i to satisfy the zero initial conditions. A similar procedure can be followed in the case of multiple roots of the characteristic equation. A system with all zero initial conditions is often called a *relaxed* system.

EXAMPLE 2.33   In this example we determine the impulse response $h[n]$ of the causal discrete-time system of Example 2.30. From Eq. (2.90) we get

$$h[n] = \alpha_1(-3)^n + \alpha_2(2)^n, \qquad n \geq 0.$$

From the above, we arrive at

$$h[0] = \alpha_1 + \alpha_2,$$
$$h[1] = -3\alpha_1 + 2\alpha_2.$$

Next, from Eq. (2.89) with $x[n] = \delta[n]$ we get

$$h[0] = 1,$$
$$h[1] + h[0] = 0.$$

Solution of the above two sets of equations yields $\alpha_1 = 0.6$, $\alpha_2 = 0.4$.
   Thus, the impulse response is given by

$$h[n] = 0.6(-3)^n + 0.4(2)^n, \qquad n \geq 0.$$

It follows from the form of the complementary solution given by Eq. (2.88) that the impulse response of a finite-dimensional LTI system characterized by a difference equation of the form of Eq. (2.81) is of infinite length. However, as illustrated by the following example, there exist infinite impulse response LTI discrete-time systems that cannot be characterized by the difference equation form of Eq. (2.81).

EXAMPLE 2.34   The system defined by the impulse response

$$h[n] = \frac{1}{n^2}\mu[n-1]$$

does not have a representation in the form of a linear constant coefficient difference equation. It should be noted that the above system is causal and also BIBO stable.

Since the impulse response $h[n]$ of a causal discrete-time system is a causal sequence, Eq. (2.82) can also be used to calculate recursively the impulse response for $n \geq 0$ by setting initial conditions to zero values, i.e., by setting $y[-1] = y[-2] = \cdots = y[-N] = 0$, and using a unit sample sequence $\delta[n]$ as the input $x[n]$. The step response of a causal LTI system can similarly be computed recursively by setting zero initial conditions and applying a unit step sequence as the input. It should be noted that the causal discrete-time system of Eq. (2.82) is linear only for zero initial conditions (Problem 2.45).

2.6.4 Output Computation Using MATLAB

The causal LTI system of the form of Eq. (2.82) can be simulated in MATLAB using the function `filter` already made use of in Program 2.4. In one of its forms, the function

```
y = filter(p,d,x)
```

processes the input data vector x using the system characterized by the coefficient vectors p and d to generate the output vector y assuming zero initial conditions. The length of y is the same as the length of x. Since the function implements Eq. (2.82), the coefficient d_0 must be nonzero.
   The following example illustrates the computation of the impulse and step responses of an LTI system described by Eq. (2.82).

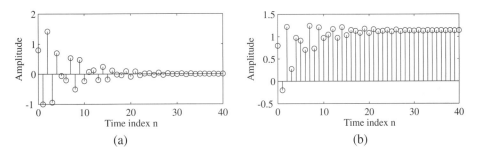

Figure 2.31: (a) Impulse response and (b) step response of the system of Eq. (2.93).

EXAMPLE 2.35 Program 2_6 given below can be employed to compute the impulse response of a causal finite-dimensional LTI discrete-time system of the form of Eq. (2.82). The program calls for the following input data: desired length of the impulse response and the filter coefficient vectors p and d that must be entered inside square brackets. The program then plots the impulse response sequence. To illustrate its application we determine the first 41 samples of the impulse response of the causal LTI system defined by

$$y[n] + 0.7y[n-1] - 0.45y[n-2] - 0.6y[n-3]$$
$$= 0.8x[n] - 0.44x[n-1] + 0.36x[n-2] + 0.02x[n-3]. \qquad (2.93)$$

```
% Program 2_6
% Illustration of Impulse Response Computation
%
N = input('Desired impulse response length = ');
p = input('Type in the vector p =');
d = input('Type in the vector d =');
x = [1 zeros(1,N-1)];
y = filter(p,d,x);
k = 0:1:N-1;
stem(k,y)
xlabel('Time index n'); ylabel('Amplitude')
```

The input data entered are:

```
N = 41
p = [0.8  -0.44  0.36  0.02]
d = [1  0.7  -0.45 -0.6]
```

The impulse response sequence generated by the program is then plotted as indicated in Figure 2.31(a). Note that the impulse response of a discrete-time finite-dimensional system can also be computed in MATLAB using the function impz (Problem M2.11).

To determine the step response we replace in the above program the statement x = [1 zeros(1,N-1)] with the statement x = [ones(1,N)]. The computed first 41 samples of the step response are indicated in Figure 2.31(b).

2.6.5 Location of Roots of Characteristic Equation for BIBO Stability

It should be noted that the impulse response samples of a stable LTI system decay to zero values as the time index n becomes very large. Likewise, the step response samples of a stable LTI system approach a constant value as n becomes very large. From the plots of Figure 2.31(a) and (b) we can conclude that most likely the LTI system of Eq. (2.93) is BIBO stable. However, it is impossible to check the stability of a system just by examining only a finite segment of its impulse or step response as in these figures.

The BIBO stability of a causal LTI system characterized by a constant coefficient difference equation of the form of Eq. (2.81) can be inferred from the values of the roots λ_i of its characteristic polynomial. To establish the stability conditions, recall that the form of the impulse response is the same as that of the complementary solution. From Eq. (2.87), assuming all the roots to be distinct, we have

$$h[n] = \sum_{i=1}^{N} \alpha_i \lambda_i^n \mu[n]. \tag{2.94}$$

The constants α_i in the above expression are determined to satisfy zero initial conditions. From Eq. (2.94) we get

$$\sum_{n=0}^{\infty} |h[n]| = \sum_{n=0}^{\infty} \left| \sum_{i=1}^{N} \alpha_i (\lambda_i)^n \right| \leq \sum_{i=1}^{N} |\alpha_i| \sum_{n=0}^{\infty} |\lambda_i|^n. \tag{2.95}$$

It follows from the above equation that if $|\lambda_i| < 1$ for all values of i, then $\sum_{n=0}^{\infty} |\lambda_i|^n < \infty$ and as a result $\sum_{n=0}^{\infty} |h[n]| < \infty$, i.e., the impulse response is absolutely summable implying BIBO stability of the causal LTI discrete-time system. However, the impulse response sequence is not absolutely summable if one or more of the roots λ_i has a magnitude greater than or equal to one. It should be noted that the discrete-time system of Example 2.30 described in Eq. (2.89) is clearly an unstable system as both roots of the characteristic equation have magnitudes greater than one.

In the case of multiple roots of the characteristic equation, the impulse response will contain terms of the form $n^K \lambda_i^n$. As a result, the expression for $\sum_{n=0}^{\infty} |h[n]|$ will contain the term

$$\sum_{n=0}^{\infty} |n^K (\lambda_i)^n|,$$

which converges if $|\lambda_i| < 1$ (Problem 2.73) and as a result, here also the impulse response is absolutely summable.

Summarizing, a causal LTI system characterized by a linear constant coefficient difference equation of the form of Eq. (2.81) is BIBO stable, if the magnitude of each of the roots of its characteristic equation is less than one. This condition is both necessary and sufficient.

2.6.6 Classification of LTI Discrete-Time Systems

Linear time-invariant (LTI) discrete-time systems are usually classified either according to the length of their impulse response sequences or according to the method of calculation employed to determine the output samples.

Classification Based on Impulse Response Length

If $h[n]$ is of finite length, i.e.,

$$h[n] = 0 \quad \text{for } n < N_1 \quad \text{and } n > N_2 \quad \text{with } N_1 < N_2, \tag{2.96}$$

then it is known as a *finite impulse response* (FIR) discrete-time system, for which the convolution sum reduces to

$$y[n] = \sum_{k=N_1}^{N_2} h[k]x[n - k]. \tag{2.97}$$

Note that the above convolution sum, being a finite sum, can be used to calculate $y[n]$ directly. The basic operations involved are simply multiplication and addition. Note that the calculation of the present value of the output sequence involves the value of the input sample at $n = N_1$ and $N_2 - N_1$ previous values of the input sequence along with the $N_2 - N_1 + 1$ impulse response samples describing the FIR discrete-time system.

Examples of FIR discrete-time systems are the moving-average system of Eq. (2.56) and the linear interpolators of Eqs. (2.58) and (2.59).

If $h[n]$ is of infinite length, then it is known as an *infinite impulse response* (IIR)) discrete-time system. For a causal IIR discrete-time system with a causal input $x[n]$, the convolution sum can be expressed in the form:

$$y[n] = \sum_{k=0}^{n} x[k]h[n-k],$$

which can be used to compute the output samples. However, for increasing n, the computational complexity increases caused by the growing number of terms in the sum.

The class of IIR filters we are concerned with in this text is the causal system characterized by the linear constant coefficient difference equation of Eq. (2.82). Note that here also the basic operations needed in the output calculations are multiplication and addition, and involve a finite sum of terms for all values of n. An example of such an IIR system is the accumulator of Eqs. (2.54) and (2.55). Another example is described next.

EXAMPLE 2.36   The familiar numerical integration formulas that are used to numerically solve integrals of the form

$$y(t) = \int_0^t x(\tau)\, d\tau$$

can be shown to be characterized by linear constant coefficient difference equations and, hence, are examples of IIR systems. If we divide the interval of integration into n equal parts of length T, then the above integral can be rewritten as

$$y(nT) = y((n-1)T) + \int_{(n-1)T}^{nT} x(\tau)\, d\tau, \qquad (2.98)$$

where we have set $t = nT$ and used the notation

$$y(nT) = \int_0^{nT} x(\tau)\, d\tau.$$

Using the trapezoidal method we integrate the integral on the right-hand side of Eq. (2.98) and arrive at (with $y(0) = 0$)

$$y(nT) = y((n-1)T) + \frac{T}{2}\{x((n-1)T) + x(nT)\}.$$

Denoting $y(nT) = y[n]$ and $x(nT) = x[n]$ we rewrite the above as

$$y[n] = y[n-1] + \frac{T}{2}\{x[n] + x[n-1]\}, \qquad (2.99)$$

which is recognized as the difference equation representation of a first-order IIR discrete-time system.

Classification Based on the Output Calculation Process

If the output sample can be calculated sequentially, knowing only the present and past input samples, the filter is said to be a *nonrecursive* discrete-time system. If, on the other hand, the computation of the output involves past output samples in addition to the present and past input samples, it is known as a *recursive* discrete-time system. An example of a nonrecursive system is the FIR discrete-time system implemented using Eq. (2.97). The IIR discrete-time system implemented using the difference equation of Eq. (2.82) is an example of a recursive system. This equation permits the recursive computation of the output response beginning at some instant $n = n_o$ and for progressively higher values of n provided the initial conditions $y[n_o - 1]$ through $y[n_o - N]$ are known. However, it is possible to implement an FIR system using a recursive computational scheme and an IIR system using a nonrecursive computational scheme [Gol68]. The former case is illustrated in the example below.

EXAMPLE 2.37   Consider the FIR discrete-time system given by the impulse response

$$h[n] = \begin{cases} 1, & 0 \leq n \leq M - 1, \\ 0, & \text{otherwise,} \end{cases} \tag{2.100}$$

The above discrete-time system is commonly known as the *running sum filter*. Comparing the above with Eq. (2.56) we observe that the above running sum filter is essentially an M-point moving-average filter with a gain of M. Substituting the above in Eq. (2.97) we get the input-output relation

$$y[n] = \sum_{k=0}^{M-1} x[n - k]. \tag{2.101}$$

To develop the recursive version of this filter, we note from the above

$$y[n - 1] = \sum_{k=1}^{M} x[n - k],$$

which can be rewritten as

$$y[n - 1] - x[n - M] = \sum_{k=1}^{M-1} x[n - k] = \sum_{k=0}^{M-1} x[n - k] - x[n],$$

or equivalently,

$$y[n] = x[n] - x[n - M] + y[n - 1], \tag{2.102}$$

which is a difference equation of the form of Eq. (2.82). The representation of Eq. (2.102) has been called a *recursive running sum* (RRS) digital filter [Ada83].

Classification Based on the Coefficients

A third classification scheme is based on the real or complex nature of the impulse response sequence. Thus, a discrete-time system with a real-valued impulse response is defined as a *real discrete-time system*. Likewise, for a *complex discrete-time system*, the impulse response is a complex-valued sequence.

2.7   Correlation of Signals

There are applications where it is necessary to compare one reference signal with one or more signals to determine the similarity between the pair and to determine additional information based on the similarity. For example, in digital communications, a set of data symbols are represented by a set of unique discrete-time sequences. If one of these sequences is transmitted, the receiver has to determine which particular sequence has been received by comparing the received signal with every member of possible sequences from the set. Similarly, in radar and sonar applications, the received signal reflected from the target is the delayed version of the transmitted signal and by measuring the delay, one can determine the location of the target. The detection problem gets more complicated in practice, as often the received signal is corrupted by additive random noise.

2.7.1 Definitions

A measure of similarity between a pair of energy signals, $x[n]$ and $y[n]$, is given by the *cross-correlation sequence* $r_{xy}[\ell]$ defined by

$$r_{xy}[\ell] = \sum_{n=-\infty}^{\infty} x[n]y[n-\ell], \quad \ell = 0, \pm 1, \pm 2, \ldots \tag{2.103}$$

The parameter ℓ called *lag*, indicates the time-shift between the pair. The time sequence $y[n]$ is said to be shifted by ℓ samples with respect to the reference sequence $x[n]$ to the right for positive values of ℓ, and shifted by ℓ samples to the left for negative values of ℓ.

The ordering of the subscripts xy in Eq. (2.103) specifies that $x[n]$ is the reference sequence which remains fixed in time whereas the sequence $y[n]$ is being shifted with respect to $x[n]$. If we wish to make $y[n]$ the reference sequence and shift the sequence $x[n]$ with respect to $y[n]$, then the corresponding cross-correlation sequence is given by

$$r_{yx}[\ell] = \sum_{n=-\infty}^{\infty} y[n]x[n-\ell]$$

$$= \sum_{m=-\infty}^{\infty} y[m+\ell]x[m] = r_{xy}[-\ell]. \tag{2.104}$$

Thus, $r_{yx}[\ell]$ is obtained by time-reversing the sequence $r_{xy}[\ell]$.

The *autocorrelation* sequence of $x[n]$ is given by

$$r_{xx}[\ell] = \sum_{n=-\infty}^{\infty} x[n]x[n-\ell] \tag{2.105}$$

obtained by setting $y[n] = x[n]$ in Eq. (2.103). Note from Eq. (2.105) that $r_{xx}[0] = \sum_{n=-\infty}^{\infty} x^2[n] = \mathcal{E}_x$, the energy of the signal $x[n]$. From Eq. (2.104) it follows that $r_{xx}[\ell] = r_{xx}[-\ell]$ implying that $r_{xx}[\ell]$ is an even function for real $x[n]$.

An examination of Eq. (2.103) reveals that the expression for the cross-correlation looks quite similar to that of the convolution given by Eq. (2.64a). This similarity is much clearer if we rewrite Eq. (2.103) as

$$r_{xy}[\ell] = \sum_{n=-\infty}^{\infty} y[n]x[-(\ell-n)] = y[\ell] \circledast x[-\ell]. \tag{2.106}$$

The above result implies that the cross-correlation of the sequence $y[n]$ with the reference sequence $x[n]$ can be computed by processing $y[n]$ with an LTI discrete-time system of impulse response $x[-n]$. Likewise, the autocorrelation of $x[n]$ can be determined by passing it through an LTI discrete-time system of impulse response $x[-n]$.

2.7.2 Properties of Autocorrelation and Cross-correlation Sequences

We next derive some basic properties of the autocorrelation and cross-correlation sequences [Pro92]. Consider two finite-energy sequences $x[n]$ and $y[n]$. Now, the energy of the combined sequence $ax[n] + y[n - \ell]$ is also finite and nonnegative. That is

$$\sum_{n=-\infty}^{\infty} (ax[n] + y[n - \ell])^2 = a^2 \sum_{n=-\infty}^{\infty} x^2[n] + 2a \sum_{n=-\infty}^{\infty} x[n]y[n - \ell] + \sum_{n=-\infty}^{\infty} y^2[n - \ell]$$

$$= a^2 r_{xx}[0] + 2a r_{xy}[\ell] + r_{yy}[0] \geq 0 \qquad (2.107)$$

where $r_{xx}[0] = \mathcal{E}_x > 0$ and $r_{yy}[0] = \mathcal{E}_y > 0$ are energies of the sequences $x[n]$ and $y[n]$, respectively. We can rewrite Eq. (2.107) as

$$\begin{bmatrix} a & 1 \end{bmatrix} \begin{bmatrix} r_{xx}[0] & r_{xy}[\ell] \\ r_{xy}[\ell] & r_{yy}[0] \end{bmatrix} \begin{bmatrix} a \\ 1 \end{bmatrix} \geq 0$$

for any finite value of a. Or in other words, the matrix

$$\begin{bmatrix} r_{xx}[0] & r_{xy}[\ell] \\ r_{xy}[\ell] & r_{yy}[0] \end{bmatrix}$$

is positive semidefinite. This implies

$$r_{xx}[0] r_{yy}[0] - r_{xy}^2[\ell] \geq 0$$

or equivalently

$$|r_{xy}[\ell]| \leq \sqrt{r_{xx}[0] r_{yy}[0]} = \sqrt{\mathcal{E}_x \mathcal{E}_y}. \qquad (2.108)$$

The above inequality provides an upper bound for the cross-correlation sequence samples. If we set $y[n] = x[n]$, the above reduces to

$$|r_{xx}[\ell]| \leq r_{xx}[0] = \mathcal{E}_x. \qquad (2.109)$$

This is a significant result as it states that at zero lag ($\ell = 0$), the sample value of the autocorrelation sequence has its maximum value.

To derive an additional property of the cross-correlation sequence consider the case

$$y[n] = \pm b x[n - N]$$

where N is an integer and $b > 0$ is an arbitrary number. In this case $\mathcal{E}_y = b^2 \mathcal{E}_x$, and therefore

$$\sqrt{\mathcal{E}_x \mathcal{E}_y} = \sqrt{b^2 \mathcal{E}_x^2} = b \mathcal{E}_x.$$

Using the above result in Eq. (2.108) we get

$$-b \, r_{xx}[0] \leq r_{xy}[\ell] \leq b \, r_{xx}[0].$$

2.7.3 Correlation Computation Using MATLAB

The cross-correlation and the autocorrelation sequences can be easily computed using MATLAB as illustrated in the following two examples.

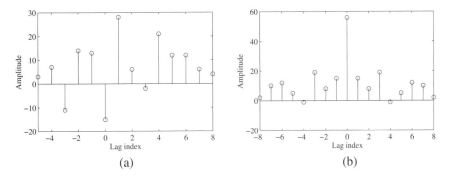

Figure 2.32: (a) Cross-correlation sequence and (b) autocorrelation sequence.

EXAMPLE 2.38   Consider the two finite-length sequences

$$x[n] = [1 \quad 3 \quad -2 \quad 1 \quad 2 \quad -1 \quad 4 \quad 4 \quad 2],$$
$$y[n] = [2 \quad -1 \quad 4 \quad 1 \quad -2 \quad 3].$$

Using MATLAB we determine and plot the cross-correlation sequence $r_{xy}[\ell]$.

We use Program 2.7 to compute the cross-correlation sequence of two finite-length sequences.

```
% Program 2_7
% Computation of Cross-correlation Sequence
%
x = input('Type in the reference sequence = ');
y = input('Type in the second sequence = ');
% Compute the correlation sequence
n1 = length(y)-1; n2 = length(x)-1;
r = conv(x,fliplr(y));
k = (-n1):n2';
stem(k,r);
xlabel('Lag index'); ylabel('Amplitude');
v = axis;
axis([-n1 n2 v(3:end)]);
```

As the program is run, it requests the input data consisting of the x and y vectors, which are entered inside square brackets. The program then computes the cross-correlation sequence using the function conv with the vector y time-reversed, and plots it as shown in Figure 2.32(a).

EXAMPLE 2.39   In this example, we evaluate and plot the autocorrelation of the sequence $x[n]$ of the previous example. Next, by adding a random noise $d[n]$ to $x[n]$, we compute and plot the autocorrelation of the noise-corrupted sequence.

Program 2.7 can also be used to compute the autocorrelation sequence of a finite-length sequence. To this end, the vector x is entered twice as the program requests the input data. The plot of the autocorrelation sequence $r_{xx}[\ell]$ generated by the program for the sequence $x[n]$ of the previous example is shown in Figure 2.32(b). As expected at zero lag, $r_{xx}[0]$ is the maximum.

Program 2.7 is run again by computing the cross-correlation of $x[n]$ and $y[n] = x[n - N]$ for $N = 4$. As can be seen from the plots in Figure 2.33(a), the peak of the cross-correlation is precisely the value of N used thus demonstrating the fact that the cross-correlation can be employed to compute the exact value of the delay N.

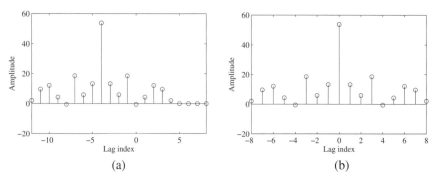

Figure 2.33: (a) Delay estimation from cross-correlation sequence and (b) autocorrelation sequence of a noise-corrupted aperiodic sequence.

Next, Program 2.7 is modified to generate a sequence formed by adding a random noise computed using the function rand to $x[n]$, and the autocorrelation of the noise-corrupted sequence is plotted in Figure 2.33(b). As expected, the autocorrelation still exhibits a pronounced peak at zero lag.

It should be noted that the autocorrelation and cross-correlation sequences can also be computed using the MATLAB function xcorr. However, the correlation sequences generated using this function are the time-reversed version of those generated using Programs 2.7 and 2.8. The cross-correlation $r_{xy}[\ell]$ of two sequences $x[n]$ and $y[n]$ can be computed using the statement r = xcorr(x,y), while the autocorrelation $r_{xx}[\ell]$ of the sequence $x[n]$ is determined using the statement r = xcorr(x).

2.7.4 Normalized Forms of Correlation

For convenience in comparing and displaying, normalized forms of autocorrelation and cross-correlation given by

$$\rho_{xx}[\ell] = \frac{r_{xx}[\ell]}{r_{xx}[0]}, \tag{2.110}$$

$$\rho_{xy}[\ell] = \frac{r_{xy}[\ell]}{\sqrt{r_{xx}[0]\, r_{yy}[0]}} \tag{2.111}$$

are often used. It follows from Eqs. (2.108) and (2.109) that $|\rho_{xx}[\ell]| \leq 1$ and $|\rho_{xy}[\ell]| \leq 1$ independent of the range of values of $x[n]$ and $y[n]$.

2.7.5 Correlation Computation for Power and Periodic Signals

In the case of power and periodic signals, the autocorrelation and cross-correlation sequences are defined slightly differently.

For a pair of power signals, $x[n]$ and $y[n]$, the cross-correlation sequence is defined as

$$r_{xy}[\ell] = \lim_{K \to \infty} \frac{1}{2K+1} \sum_{n=-K}^{K} x[n]y[n-\ell], \tag{2.112}$$

and the autocorrelation sequence of $x[n]$ is given by

$$r_{xx}[\ell] = \lim_{K \to \infty} \frac{1}{2K+1} \sum_{n=-K}^{K} x[n]x[n-\ell]. \tag{2.113}$$

Likewise, if $\tilde{x}[n]$ and $\tilde{y}[n]$ are two periodic signals with period N, then their cross-correlation sequence is given by

$$r_{\tilde{x}\tilde{y}}[\ell] = \frac{1}{N} \sum_{n=0}^{N-1} \tilde{x}[n]\,\tilde{y}[n-\ell] \qquad (2.114)$$

and the autocorrelation sequence of $x[n]$ is given by

$$r_{\tilde{x}\tilde{x}}[\ell] = \frac{1}{N} \sum_{n=0}^{N-1} \tilde{x}[n]\,\tilde{x}[n-\ell]. \qquad (2.115)$$

It follows from the above definitions that both $r_{\tilde{x}\tilde{y}}[\ell]$ and $r_{\tilde{x}\tilde{x}}[\ell]$ are also periodic sequences with a period N.

The periodicity properties of the autocorrelation sequence can be exploited to determine the period N of a periodic signal that may have been corrupted by an additive random disturbance. Let $\tilde{x}[n]$ be a positive periodic signal corrupted by the random noise $d[n]$ resulting in the signal

$$w[n] = \tilde{x}[n] + d[n]$$

which is observed for $0 \le n \le M-1$ where $M \gg N$. The autocorrelation of $w[n]$ is given by

$$\begin{aligned}
r_{ww}[\ell] &= \frac{1}{M} \sum_{n=0}^{M-1} w[n]\,w[n-\ell] \\
&= \frac{1}{M} \sum_{n=0}^{M-1} (\tilde{x}[n]+d[n])\,(\tilde{x}[n-\ell]+d[n-\ell]) \\
&= \frac{1}{M} \sum_{n=0}^{M-1} \tilde{x}[n]\,\tilde{x}[n-\ell] + \frac{1}{M} \sum_{n=0}^{M-1} d[n]\,d[n-\ell] \\
&\quad + \frac{1}{M} \sum_{n=0}^{M-1} \tilde{x}[n]\,d[n-\ell] + \frac{1}{M} \sum_{n=0}^{M-1} d[n]\,\tilde{x}[n-\ell] \\
&= r_{\tilde{x}\tilde{x}}[\ell] + r_{dd}[\ell] + r_{\tilde{x}d}[\ell] + r_{d\tilde{x}}[\ell]. \qquad (2.116)
\end{aligned}$$

Now in the above equation, $r_{xx}[\ell]$ is a periodic sequence with a period N and hence it will have peaks at $\ell = 0, N, 2N, \ldots$, with the same amplitudes as ℓ approaches M. As $\tilde{x}[n]$ and $d[n]$ are not correlated, samples of cross-correlation sequences $r_{xd}[\ell]$ and $r_{dx}[\ell]$ are likely to be very small relative to the amplitudes of $r_{xx}[\ell]$. The autocorrelation of the disturbance signal $d[n]$ shows a peak at $\ell = 0$ with other samples having rapidly decreasing amplitudes with increasing values of $|\ell|$. Hence the peaks of $r_{ww}[\ell]$ for $\ell > 0$ are essentially due to the peaks of $r_{\tilde{x}\tilde{x}}[\ell]$ and can be used to determine whether $\tilde{x}[n]$ is a periodic sequence and its period if the peaks occur at periodic intervals.

2.7.6 Correlation Computation of a Periodic Sequence Using MATLAB

We illustrate next the determination of the period of a noise-corrupted periodic sequence using MATLAB.

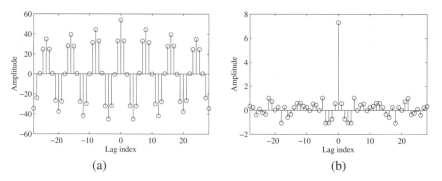

Figure 2.34: (a) Autocorrelation sequence of the noise-corrupted sinusoid, and (b) autocorrelation sequence of the noise.

EXAMPLE 2.40   We determine the period of the sinusoidal sequence $x[n] = \cos(0.25\pi n)$, $0 \le n \le 95$, corrupted by an additive uniformly distributed random noise of amplitude in the range $[-0.5, 0.5]$.

To this end, we use Program 2_8 to compute the autocorrelation sequence of the noise-corrupted sinusoidal sequence.

```
% Program 2_8
% Computation of Autocorrelation of a
% Noise Corrupted Sinusoidal Sequence
%
N = 96;
n = 1:N;
x = cos(pi*0.25*n); % Generate the sinusoidal sequence
d = rand(1,N) - 0.5; % Generate the noise sequence
y = x + d; % Generate the noise-corrupted sinusoidal sequence
r = conv(y, fliplr(y)); % Compute the correlation sequence
k = -28:28;
stem(k, r(68:124));
xlabel('Lag index'); ylabel('Amplitude');
```

The plot generated by running this program is shown in Figure 2.34(a). As can be seen from this plot there is a strong peak at zero lag. However, there are distinct peaks at lags that are multiples of 8 indicating the period of the sinusoidal sequence to be 8 as expected.[7] Figure 2.34(b) shows the plot of the autocorrelation sequence $r_{dd}[n]$ of the noise component. As can be seen from this plot, $r_{dd}[n]$ shows a very strong peak only at zero lag. The amplitudes are considerably smaller at other values of the lag as sample values of the noise sequence are uncorrelated with each other.

2.8 Random Signals

The underlying assumption on the discrete-time signals we have considered so far is that they can be uniquely determined by well-defined processes such as a mathematical expression or a rule or a lookup table. Such a signal is usually called a *deterministic signal* since all sample values of the sequence are

---

[7]The decaying amplitudes of the peaks as the lag index ℓ increases are due to the finite lengths of the periodic sequences which result in reducing the number of nonzero products in the computation of the convolution sum.

well defined for all values of the time index. For example, the sinusoidal sequence of Eq. (2.39) and the exponential sequence of Eq. (2.42) are deterministic sequences.

Signals for which each sample value is generated in a random fashion and cannot be predicted ahead of time comprise another class of signals. Such a signal, called a *random signal* or a *stochastic signal*, cannot be reproduced at will, not even using the process generating the signal, and therefore needs to be modeled using statistical information about the signal. Some common examples of random signals are speech, music, and seismic signals. The error signal generated by forming the difference between the ideal sampled version of a continuous-time signal and its quantized version generated by a practical analog-to-digital converter is usually modeled as a random signal for analysis purposes.[8] The noise sequence $d[n]$ of Figure 2.21(b) generated using the rand function of MATLAB is also an example of a random signal.

The discrete-time *random signal* or process consists of a typically infinite collection or *ensemble* of discrete-time sequences $\{X[n]\}$. One particular sequence in this collection $\{x[n]\}$ is called a *realization* of the random process. At a given time index n, the observed sample value $x[n]$ is the value taken by the *random variable* $X[n]$. Thus, a random process is a family of random variables $\{X[n]\}$. In general, the range of sample values is a continuum. We review in this section the important statistical properties of the random variable and the random process.

2.8.1 Statistical Properties of a Random Variable

The statistical properties of a random variable depend on its probability distribution function or, equivalently, on its probability density function, which are defined next. The probability that the random variable X takes a value in a specified range from $-\infty$ to α is given by its *probability distribution function*

$$P_X(\alpha) = \text{Probability}[X \leq \alpha]. \qquad (2.117)$$

The *probability density function* of X is defined by

$$p_X(\alpha) = \frac{\partial P_X(\alpha)}{\partial \alpha}, \qquad (2.118)$$

if X can assume a continuous range of values. From Eq. (2.118) the probability distribution function is therefore given by

$$P_X(\alpha) = \int_{-\infty}^{\alpha} p_X(u)\, du. \qquad (2.119)$$

The probability density function satisfies the following two properties:

$$p_X(\alpha) \geq 0, \qquad (2.120a)$$

$$\int_{-\infty}^{\infty} p_X(\alpha)\, d\alpha = 1. \qquad (2.120b)$$

Likewise, the probability distribution function satisfies the following properties, which follow from Eqs. (2.119), (2.120a), and (2.120b):

$$0 \leq P_X(\alpha) \leq 1, \qquad (2.121a)$$

$$P_X(\alpha_1) \leq P_X(\alpha_2), \qquad \text{for all } \alpha_2 \geq \alpha_1, \qquad (2.121b)$$

$$P_X(-\infty) = 0, \qquad P_X(+\infty) = 1, \qquad (2.121c)$$

$$\text{Probability}[\alpha_1 < X \leq \alpha_2] = P_X(\alpha_2) - P_X(\alpha_1). \qquad (2.121d)$$

---

[8]See Section 9.5.1.

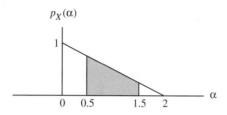

Figure 2.35: Probability density function of Eq. (2.125).

A random variable is characterized by a number of statistical properties. For example, the rth *moments* are defined by

$$\mu_r = E(X^r) = \int_{-\infty}^{\infty} \alpha^r p_X(\alpha)\, d\alpha, \tag{2.122}$$

where r is any nonnegative integer and $E(\cdot)$ denotes the expectation operator. A random variable is completely characterized by all its moments. In most cases, all such moments are not known a priori or are difficult to evaluate. Three more commonly used statistical properties characterizing a random variable are the *mean* or *expected value* m_X, the *mean-square value* $E(X^2)$, and the *variance* σ_X^2 as defined below:

$$m_X = E(X) = \int_{-\infty}^{\infty} \alpha p_X(\alpha)\, d\alpha, \tag{2.123a}$$

$$E(X^2) = \int_{-\infty}^{\infty} \alpha^2 p_X(\alpha)\, d\alpha, \tag{2.123b}$$

$$\sigma_X^2 = E\left([X - m_X]^2\right) = \int_{-\infty}^{\infty} (\alpha - m_X)^2 p_X(\alpha)\, d\alpha. \tag{2.123c}$$

These three properties provide adequate information about a random variable in most practical cases. It can be easily shown that

$$\sigma_X^2 = E(X^2) - (m_X)^2. \tag{2.124}$$

The square root of the variance, σ_X, is called the *standard deviation* of the random variable X. It follows from Eq. (2.124) that the variance and the mean-square value are equal for a random variable with zero mean.

It can be shown that the mean value m_X is the best constant representing a random variable X in a minimum mean-squared error sense, i.e., $E([X - \kappa]^2)$ is a minimum for $\kappa = m_X$ and the minimum mean-square error is given by its variance σ_X^2 (Problem 2.78). This implies that if the variance is small, then the value assumed by X is likely to be close to m_X and if the variance is large, the value assumed by X is likely to be far from m_X.

We illustrate the concepts introduced so far by means of an example.

EXAMPLE 2.41　　Let a random variable X be characterized by a probability density function

$$p_X(\alpha) = \begin{cases} 1 - \frac{\alpha}{2}, & \text{for } 0 \le \alpha \le 2, \\ 0, & \text{otherwise,} \end{cases} \tag{2.125}$$

as shown in Figure 2.35. Its probability distribution function is thus given by

$$P_X(\alpha) = \int_{-\infty}^{\alpha} p_X(u)\, du = \int_0^{\alpha} \left(1 - \frac{u}{2}\right) du = \alpha - \frac{\alpha^2}{4}, \qquad \text{for } 0 < \alpha \le 2. \tag{2.126}$$

From Eqs. (2.121d) and (2.126) we can compute the probability that X is in a specified range. For example, the probability that X is in the range $0.5 < X \le 1.5$ is given by

$$\text{Probability}\,[0.5 < X \le 1.5] = P_X(1.5) - P_X(0.5) = \frac{15}{16} - \frac{7}{16} = \frac{1}{2},$$

which is the area of the shaded portion in Figure 2.35. From Eq. (2.123a) the mean value is obtained as

$$m_X = \int_0^2 \alpha \left(1 - \frac{\alpha}{2}\right) d\alpha = \frac{2}{3},$$

and from Eq. (2.123b) the mean-square value is computed as

$$E(X^2) = \int_0^2 \alpha^2 \left(1 - \frac{\alpha}{2}\right) d\alpha = \frac{2}{3}.$$

Substituting the above results in Eq. (2.124) we arrive at the variance given by

$$\sigma_X^2 = E(X^2) - (m_X)^2 = \frac{2}{9}.$$

Two probability density functions, commonly encountered in digital signal processing applications, are the *uniform density function* defined by

$$p_X(\alpha) = \begin{cases} \frac{1}{b-a}, & \text{for } a \le \alpha \le b, \\ 0, & \text{otherwise,} \end{cases} \tag{2.127}$$

and the *Gaussian density function*, also called the *normal density function*, defined by

$$p_X(\alpha) = \frac{1}{\sigma_X \sqrt{2\pi}} e^{-(\alpha - m_X)^2 / 2\sigma_X^2}, \tag{2.128}$$

where the parameters m_X and σ_X are, respectively, the mean value and the standard deviation of X and lie in the range $-\infty < m_X < \infty$ and $\sigma_X > 0$. These density functions are plotted in Figure 2.36. Various other density functions are defined in the literature (Problem 2.79).

EXAMPLE 2.42 Determine the mean and the variance of a uniformly distributed random variable X defined by Eq. (2.127).
From Eqs. (2.123a) and (2.123b), we arrive at

$$m_X = \frac{1}{b-a} \int_a^b \alpha \, d\alpha = \frac{b+a}{2}, \tag{2.129}$$

and

$$E\left(X^2\right) = \frac{1}{b-a} \int_a^b \alpha^2 \, d\alpha = \frac{b^2 + a^2 + ab}{3}.$$

Substituting the above values in Eq. (2.124), we obtain

$$\sigma_X^2 = \frac{(b-a)^2}{12}. \tag{2.130}$$

In the case of two random variables X and Y their joint statistical properties as well as their individual statistical properties are of practical interest. The probability that X takes a value in a specified range from

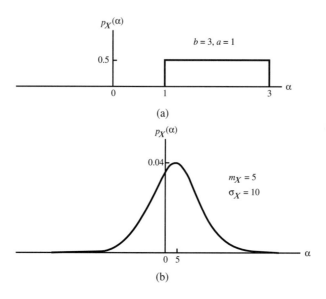

Figure 2.36: (a) Uniform and (b) Gaussian probability density functions.

$-\infty$ to α and that Y takes a value in a specified range from $-\infty$ to β is given by their *joint probability distribution function*

$$P_{XY}(\alpha, \beta) = \text{Probability}\,[X \le \alpha, Y \le \beta] \tag{2.131}$$

or, equivalently, by their *joint probability density function*

$$p_{XY}(\alpha, \beta) = \frac{\partial^2 P_{XY}(\alpha, \beta)}{\partial\alpha\,\partial\beta}. \tag{2.132}$$

The joint probability distribution function is thus given by

$$P_{XY}(\alpha, \beta) = \int_{-\infty}^{\alpha} \int_{-\infty}^{\beta} p_{XY}(u, v)\, du\, dv. \tag{2.133}$$

The joint probability density function satisfies the following two properties:

$$p_{XY}(\alpha, \beta) \ge 0, \tag{2.134a}$$

$$\int_{-\infty}^{\infty} \int_{-\infty}^{\infty} p_{XY}(\alpha, \beta)\, d\alpha\, d\beta = 1. \tag{2.134b}$$

The joint probability distribution function satisfies the following properties, which are a direct consequence of Eqs. (2.133), (2.134a), and (2.134b):

$$0 \le P_{XY}(\alpha, \beta) \le 1, \tag{2.135a}$$

$$P_{XY}(\alpha_1, \beta_1) \le P_{XY}(\alpha_2, \beta_2) \quad \text{for } \alpha_2 \ge \alpha_1 \text{ and } \beta_2 \ge \beta_1, \tag{2.135b}$$

$$P_{XY}(-\infty, -\infty) = 0, \quad P_{XY}(+\infty, +\infty) = 1, \tag{2.135c}$$

The joint statistical properties of two random variables X and Y are described by their cross-correlation and cross-covariance, as defined by

$$\phi_{XY} = E\,(XY) = \int_{-\infty}^{\infty} \int_{-\infty}^{\infty} \alpha\beta p_{X,Y}(\alpha, \beta)\, d\alpha\, d\beta, \tag{2.136}$$

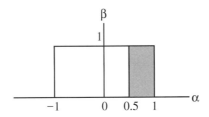

Figure 2.37: Range of the random variables with the joint probability density function of Eq. (2.139).

$$\gamma_{XY} = E\left([X - m_X][Y - m_Y]\right)$$
$$= \int_{-\infty}^{\infty} \int_{-\infty}^{\infty} (\alpha - m_X)(\beta - m_Y) p_{X,Y}(\alpha, \beta) \, d\alpha d\beta$$
$$= \phi_{XY} - m_X m_Y, \tag{2.137}$$

where m_X and m_Y are, respectively, the mean of the random variables X and Y. The two random variables X and Y are said to be *linearly independent or uncorrelated* if

$$E(XY) = E(X)E(Y), \tag{2.138a}$$

and *statistically independent* if

$$P_{X,Y}(\alpha, \beta) = P_X(\alpha) P_Y(\beta). \tag{2.138b}$$

It can be shown that if the random variables X and Y are statistically independent, then they are also linearly independent (Problem 2.80). However, if X and Y are linearly independent, they may not be statistically independent.

The statistical independence property makes it easier to compute the statistical properties of a random variable that is a function of several independent random variables. For example, if X and Y are statistically independent random variables with means m_X and m_Y, respectively, then it can be shown that the mean of the random variable $V = aX + bY$, where a and b are constants, is given by $m_V = am_X + bm_Y$. Likewise, if the variances of X and Y are σ_X^2 and σ_Y^2, respectively, the variance of V is given by $\sigma_V^2 = a^2\sigma_X^2 + b^2\sigma_Y^2$ (Problem 2.82).

EXAMPLE 2.43   Consider the two random variables X and Y described by a uniformly distributed joint probability density function given by

$$p_{X,Y}(\alpha, \beta) = \begin{cases} A, & -1 \le \alpha \le 1, 0 \le \beta \le 1, \\ 0, & \text{otherwise.} \end{cases} \tag{2.139}$$

Determine the value of the constant A, and then compute the probability that X and Y lie in the range ($0.5 \le \alpha \le 1, 0 \le \beta \le 1$) shown by the shaded region in Figure 2.37.

Invoking the property of Eq. (2.134b), we obtain

$$A \int_{-1}^{1} \int_{0}^{1} d\alpha \, d\beta = A \left[\int_{-1}^{1} d\alpha \right] \left[\int_{0}^{1} d\beta \right] = 2A = 1,$$

and hence, $A = 1/2$. Next, applying Eq. (2.133), we get

$$\text{Probability}\,[0.5 \le X \le 1, 0 \le Y \le 1] = \frac{1}{2} \int_{0}^{1} \int_{0.5}^{1} d\alpha \, d\beta = \frac{1}{4}.$$

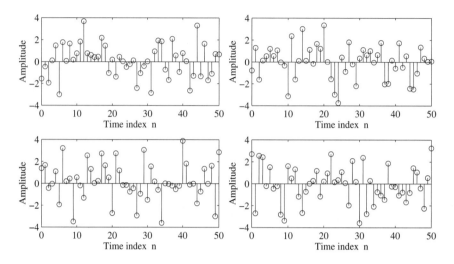

Figure 2.38: Sample realizations of the random sinusoidal signal of Eq. (2.140) for $\omega_o = 0.06\pi$.

2.8.2   Statistical Properties of a Random Signal

As indicated earlier, the random discrete-time signal is a sequence of random variables and consists of a typically infinite collection or ensemble of discrete-time sequences. Figure 2.38 shows four possible realizations of a random sinusoidal signal

$$\{X[n]\} = \{A \cos(\omega_o n + \Phi)\},\tag{2.140}$$

with $\omega_o = 0.06\pi$, where the amplitude A and the phase Φ are statistically independent random variables with uniform probability distribution in the range $0 \leq \alpha \leq 4$ for the amplitude and in the range $0 \leq \phi \leq 2\pi$ for the phase.

The statistical properties of the random signal $\{X[n]\}$ at time index n are given by the statistical properties of the random variable $X[n]$. Thus, the *mean* or *expected value* of $\{X[n]\}$ at time index n is given by:

$$m_{X[n]} = E(X[n]) = \int_{-\infty}^{\infty} \alpha p_{X[n]}(\alpha; n)\, d\alpha.\tag{2.141}$$

The *mean-square value* of $\{X[n]\}$ at time index n is given by

$$E\left(X[n]^2\right) = \int_{-\infty}^{\infty} \alpha^2 p_{X[n]}(\alpha; n)\, d\alpha.\tag{2.142}$$

The *variance* $\sigma_{X[n]}^2$ of $\{X[n]\}$ at time index n is defined by

$$\sigma_{X[n]}^2 = E\left(\{X[n] - m_{X[n]}\}^2\right) = E\left(X[n]^2\right) - \left(m_{X[n]}\right)^2.\tag{2.143}$$

In general, the mean, mean-square value, and variance of a random discrete-time signal are functions of the time index n, and can be considered as sequences.

So far we have assumed the random variables and the random signals to be real-valued. It is straight-forward to generalize the treatment to complex-valued random variables and random signals. For example, the nth sample of a complex-valued random signal $\{X[n]\}$ is of the form

$$X[n] = X_{re}[n] + jX_{im}[n],\tag{2.144}$$

where $\{X_{re}[n]\}$ and $\{X_{im}[n]\}$ are real-valued sequences called the real and imaginary parts of $\{X[n]\}$, respectively. The mean value of a complex sequence at time index n is thus given by

$$m_{X[n]} = E\left(X[n]\right) = E\left(X_{re}[n]\right) + j E\left(X_{im}[n]\right) = m_{X_{re}[n]} + j m_{X_{im}[n]}. \qquad (2.145)$$

Likewise, the variance $\sigma^2_{X[n]}$ of $\{X[n]\}$ at time index n is given by

$$\sigma^2_{X[n]} = E\left(\left|X[n] - m_{X[n]}\right|^2\right) = E\left(|X[n]|^2\right) - \left(\left|m_{X[n]}\right|\right)^2. \qquad (2.146)$$

Often, the statistical relation of the samples of a random discrete-time signal at two different time indices m and n is of interest. One such relation is the *autocorrelation*, which for a complex random discrete-time signal $\{X[n]\}$ is defined by

$$\phi_{XX}[m, n] = E\left(X[m]X^*[n]\right), \qquad (2.147)$$

where $*$ denotes complex conjugation. Substituting Eq. (2.144) in Eq. (2.147) we obtain the expression for the autocorrelation of $X[n]$:

$$\begin{aligned}
\phi_{XX}[m, n] = {} & \phi_{X_{re}X_{re}}[m, n] + \phi_{X_{im}X_{im}}[m, n] \\
& - j\phi_{X_{re}X_{im}}[m, n] + j\phi_{X_{im}X_{re}}[m, n],
\end{aligned} \qquad (2.148)$$

where

$$\phi_{X_{re}X_{re}}[m, n] = E\left(X_{re}[m]X_{re}[n]\right), \qquad (2.149a)$$

$$\phi_{X_{im}X_{im}}[m, n] = E\left(X_{im}[m]X_{im}[n]\right), \qquad (2.149b)$$

$$\phi_{X_{re}X_{im}}[m, n] = E\left(X_{re}[m]X_{im}[n]\right), \qquad (2.149c)$$

$$\phi_{X_{im}X_{re}}[m, n] = E\left(X_{im}[m]X_{re}[n]\right). \qquad (2.149d)$$

Another relation is the *autocovariance* of $\{X[n]\}$, defined by

$$\begin{aligned}
\gamma_{XX}[m, n] &= E\left((X[m] - m_{X[m]})(X[n] - m_{X[n]})^*\right) \\
&= \phi_{XX}[m, n] - m_{X[m]}(m_{X[n]})^*.
\end{aligned} \qquad (2.150)$$

As can be seen from the above, both the autocorrelation and the autocovariance are functions of two time indices m and n and can be considered as two-dimensional sequences.

EXAMPLE 2.44   We compute in this example the statistical properties of the random signal of Eq. (2.140). Now the probability density functions of the amplitude A and the phase Φ are given by

$$p_A(\alpha) = \begin{cases} \frac{1}{4}, & 0 \le \alpha \le 4, \\ 0, & \text{otherwise} \end{cases} \qquad (2.151)$$

and

$$p_\Phi(\phi) = \begin{cases} \frac{1}{2\pi}, & 0 \le \phi \le 2\pi, \\ 0, & \text{otherwise}, \end{cases} \qquad (2.152)$$

respectively. Since the two random variables are statistically independent, their joint probability density function is

$$p_{A\Phi}(\alpha, \phi) = \begin{cases} \frac{1}{8\pi}, & 0 \le \alpha \le 4, 0 \le \phi \le 2\pi, \\ 0, & \text{otherwise}. \end{cases} \qquad (2.153)$$

The mean value of the random process $\{X[n]\}$ is thus

$$
\begin{aligned}
m_{X[n]} &= \frac{1}{8\pi} \int_0^4 \int_0^{2\pi} \alpha \cos(\omega_o n + \phi) \, d\alpha \, d\phi \\
&= \frac{1}{8\pi} \left(\int_0^4 \alpha \, d\alpha \right) \left(\int_0^{2\pi} \cos(\omega_o n + \phi) \, d\phi \right) \\
&= \frac{1}{\pi} \left(\sin(\omega_o n + 2\pi) - \sin(\omega_o n) \right) = 0.
\end{aligned}
\tag{2.154}
$$

The mean-square value is given by

$$
\begin{aligned}
E\left(X^2[n]\right) &= \frac{1}{8\pi} \int_0^4 \int_0^{2\pi} \alpha^2 \cos^2(\omega_o n + \phi) \, d\alpha \, d\phi \\
&= \frac{1}{8\pi} \left(\int_0^4 \alpha^2 \, d\alpha \right) \left(\int_0^{2\pi} \cos^2(\omega_o n + \phi) \, d\phi \right) = \frac{8}{3},
\end{aligned}
\tag{2.155}
$$

which is also the variance since the random process is of zero mean.

The autocorrelation function is given by

$$
\begin{aligned}
\phi_{XX}[m, n] &= E\left(X[m]X[n]\right) \\
&= \frac{1}{8\pi} \int_0^4 \alpha^2 \, d\alpha \int_0^{2\pi} \cos(\omega_o m + \phi) \cos(\omega_o n + \phi) \, d\phi \\
&= \frac{8}{3} \cos\left(\omega_o(m - n)\right).
\end{aligned}
\tag{2.156}
$$

The correlation between two different random discrete-time signals $\{X[n]\}$ and $\{Y[n]\}$ is described by the *cross-correlation function*

$$
\phi_{XY}[m, n] = E\left(X[m]Y^*[n]\right) = \int_{-\infty}^{\infty} \int_{-\infty}^{\infty} \alpha\beta^* \, p_{X[m], Y[n]}(\alpha, m, \beta, n) \, d\alpha \, d\beta,
\tag{2.157}
$$

and the *cross-covariance function*

$$
\begin{aligned}
\gamma_{XY}[m, n] &= E\left((X[m] - m_{X[m]})(Y[n] - m_{Y[n]})^*\right) \\
&= \phi_{XY}[m, n] - m_{X[m]}(m_{Y[n]})^*,
\end{aligned}
\tag{2.158}
$$

where $p_{X[m], Y[n]}(m, \alpha, n, \beta)$ is the joint probability density function of $X[n]$ and $Y[n]$. Both the cross-correlation and the cross-covariance functions can also be considered as two-dimensional sequences. The two random discrete-time signals $\{X[n]\}$ and $\{Y[n]\}$ are uncorrelated if $\gamma_{XY}[m, n] = 0$ for all values of the time indices m and n.

2.8.3 Wide-Sense Stationary Random Signal

In general, the statistical properties of a random discrete-time signal $\{X[n]\}$, such as the mean and variance of the random variable $X[n]$, and the autocorrelation and the autocovariance functions, are time-varying functions. The class of random signals often encountered in digital signal processing applications are the so-called *wide-sense stationary* (WSS) random processes for which some of the key statistical properties are either independent of time or of the time origin. More specifically, for a wide-sense stationary random process $\{X[n]\}$, the mean $E(X[n])$ has the same constant value m_X for all values of the time index n, and the autocorrelation and the autocovariance functions depend only on the difference of the time indices m and n, i.e.,

$$m_X = E(X[n]), \text{ for all } n, \tag{2.159}$$

$$\phi_{XX}[\ell] = \phi_{XX}[n+\ell, n] = E(X[n+\ell]X^*[n]), \text{ for all } n \text{ and } \ell, \tag{2.160}$$

$$\gamma_{XX}[\ell] = \gamma_{XX}[n+\ell, n] = E\left((X[n+\ell]-m_X)(X[n]-m_X)^*\right)$$

$$= \phi_{XX}[\ell] - |m_X|^2, \text{ for all } n \text{ and } \ell. \tag{2.161}$$

Note that in the case of a WSS random process, the autocorrelation and the autocovariance functions are one-dimensional sequences.

The mean-square value of a WSS random process $\{X[n]\}$ is given by

$$E\left(|X[n]|^2\right) = \phi_{XX}[0], \tag{2.162}$$

and the variance is given by

$$\sigma_X^2 = \gamma_{XX}[0] = \phi_{XX}[0] - |m_X|^2. \tag{2.163}$$

It follows from Eqs. (2.154) and (2.156) that the random process of Eq. (2.140) is a wide-sense stationary signal.

The cross-correlation and cross-covariance functions between two WSS random processes $\{X[n]\}$ and $\{Y[n]\}$ are given by

$$\phi_{XY}[\ell] = E\left(X[n+\ell]Y^*[n]\right), \tag{2.164}$$

$$\gamma_{XY}[\ell] = E\left((X[n+\ell]-m_X)(Y[n]-m_Y)^*\right)$$

$$= \phi_{XY}[\ell] - m_X(m_Y)^*. \tag{2.165}$$

The symmetry properties satisfied by the autocorrelation, autocovariance, cross-correlation, and cross-covariance functions are:

$$\phi_{XX}[-\ell] = \phi_{XX}^*[\ell], \tag{2.166a}$$

$$\gamma_{XX}[-\ell] = \gamma_{XX}^*[\ell], \tag{2.166b}$$

$$\phi_{XY}[-\ell] = \phi_{YX}^*[\ell], \tag{2.166c}$$

$$\gamma_{XY}[-\ell] = \gamma_{YX}^*[\ell]. \tag{2.166d}$$

From the above symmetry properties it can be seen that sequences $\phi_{XX}[\ell]$, $\gamma_{XX}[\ell]$, $\phi_{XY}[\ell]$, and $\gamma_{XY}[\ell]$ are always two-sided sequences.

Some additional useful properties concerning these functions are:

$$\phi_{XX}[0]\phi_{YY}[0] \geq |\phi_{XY}[\ell]|^2, \tag{2.167a}$$

$$\gamma_{XX}[0]\gamma_{YY}[0] \geq |\gamma_{XY}[\ell]|^2, \tag{2.167b}$$

$$\phi_{XX}[0] \geq |\phi_{XX}[\ell]|, \tag{2.167c}$$

$$\gamma_{XX}[0] \geq |\gamma_{XX}[\ell]|. \tag{2.167d}$$

A consequence of the above properties is that the autocorrelation and autocovariance functions of a WSS random process assume their maximum values at $\ell = 0$. In addition, it can be shown that, for a WSS signal with nonzero mean, i.e., $m_{X[n]} \neq 0$, and with no periodic components,

$$\lim_{|\ell|\to\infty} \phi_{XX}[\ell] = |m_{X[n]}|^2. \tag{2.168}$$

If $X[n]$ has a periodic component, then $\phi_{XX}[\ell]$ will contain the same periodic component as illustrated in Example 2.40.

EXAMPLE 2.45   Determine the mean and variance of a WSS real signal with an autocorrelation function given by

$$\phi_{XX}[\ell] = \frac{10 + 21\ell^2}{1 + 3\ell^2}. \tag{2.169}$$

Applying Eq. (2.168), we obtain $\left|m_{X[n]}\right|^2 = 7$, and hence, the mean value is $m_{X[n]} = \pm\sqrt{7}$. Next, from Eqs. (2.162) and (2.169) we arrive at the mean-square value as 10 and as a result, the variance is equal to $\sigma^2_{X[n]} = 10 - 7 = 3$.

2.8.4 Concept of Power in a Random Signal

The average power of a deterministic sequence $x[n]$ was defined earlier and is given by Eq. (2.29). To compute the power associated with a random signal $\{X[n]\}$ we use instead the following definition:

$$\mathcal{P}_X = E\left(\lim_{N\to\infty} \frac{1}{2N+1} \sum_{n=-N}^{N} |X[n]|^2\right). \tag{2.170}$$

In most practical cases, the expectation and summation operators in Eq. (2.170) can be interchanged, resulting in a more simple expression given by

$$\mathcal{P}_X = \lim_{N\to\infty} \frac{1}{2N+1} \sum_{n=-N}^{N} E\left(|X[n]|^2\right). \tag{2.171}$$

In addition, if the random signal has a constant mean-square value for all values of n, as in the case of a WSS signal, then Eq. (2.171) reduces to

$$\mathcal{P}_X = E\left(|X[n]|^2\right) \tag{2.172}$$

From Eqs. (2.162) and (2.163) it follows that for a WSS signal, the average power is given by

$$\mathcal{P}_X = \phi_{XX}[0] = \sigma_X^2 + |m_X|^2. \tag{2.173}$$

2.8.5 Ergodic Signal

In many practical situations, the random signal of interest cannot be described in terms of a simple analytical expression, as in Eq. (2.140), to permit computations of its statistical properties, which invariably involves the evaluation of definite integrals or summations. Often a finite portion of a single realization of the random signal is available, from which some estimation of the statistical properties of the ensemble must be made. Such an approach can lead to meaningful results if the ergodicity condition is satisfied. More precisely, a stationary random signal is defined to be an *ergodic signal* if all its statistical properties can be estimated from a single realization of sufficiently large finite length.

For an ergodic signal, time averages equal ensemble averages derived via the expectation operator in the limit as the length of the realization goes to infinity. For example, for a real ergodic signal we can compute the mean value, variance, and autocovariance as:

$$m_X = \lim_{M\to\infty} \frac{1}{2M+1} \sum_{n=-M}^{M} x[n], \tag{2.174a}$$

$$\sigma_X^2 = \lim_{M \to \infty} \frac{1}{2M + 1} \sum_{n=-M}^{M} (x[n] - m_X)^2, \tag{2.174b}$$

$$\gamma_{XX}[\ell] = \lim_{M \to \infty} \frac{1}{2M + 1} \sum_{n=-M}^{M} (x[n] - m_X)(x[n + \ell] - m_X). \tag{2.174c}$$

The limiting operation required to compute the ensemble averages by means of time averages is still not practical in most situations and therefore replaced with a finite sum to provide an estimate of the desired statistical properties. For example, approximations to Eq. (2.174a)–(2.174c) that are often used are:

$$\hat{m}_X = \frac{1}{M + 1} \sum_{n=0}^{M} x[n], \tag{2.175a}$$

$$\hat{\sigma}_X^2 = \frac{1}{M + 1} \sum_{n=0}^{M} (x[n] - m_X)^2, \tag{2.175b}$$

$$\hat{\gamma}_{XX}[\ell] = \frac{1}{M + 1} \sum_{n=0}^{M} (x[n] - m_X)(x[n + \ell] - m_X). \tag{2.175c}$$

2.9 Summary

In this chapter we introduced some important and fundamental concepts regarding the characterization of discrete-time signals and systems in the time-domain. Certain basic discrete-time signals that play important roles in discrete-time signal processing have been defined, along with basic mathematical operations used for generating more complex signals and systems. The relation between a continuous-time signal and the discrete-time signal generated by sampling the former at uniform time intervals has been examined.

This text deals almost exclusively with linear, time-invariant (LTI) discrete-time systems that find numerous applications in practice. These systems are defined and their convolution sum representation in the time-domain is derived. The concepts of causality and stability of LTI systems are introduced. Also discussed is an important class of LTI systems described by an input-output relation composed of a linear constant coefficient difference equation and the procedure for computing its output for a given input and initial conditions. The LTI discrete-time system is usually classified in terms of the length of its impulse response. The concepts of the autocorrelation of a sequence and the cross-correlation between a pair of sequences are introduced. Finally, the chapter concludes with a review of the time-domain characterization of a discrete-time random signal in terms of some of its statistical properties.

For further details on discrete-time signals and systems, we refer the reader to the texts by Cadzow [Cad73], Gabel and Roberts [Gab87], Haykin and Van Veen [Hay99], Jackson [Jac91], Lathi [Lat98], Oppenheim and Willsky [Opp83], Strum and Kirk [Stu88], and Ziemer et al. [Zie83]. Additional materials on probability theory and statistical properties of random discrete-time signals can be found in Cadzow [Cad87], Papoulis [Pap65], Peebles [Pee87], Stark and Woods [Sta94], and Therrien [The92].

Further insights can often be obtained by considering the frequency-domain representations of discrete-time signals and LTI discrete-time systems. These are discussed in the following two chapters.

2.10 Problems

2.1 Consider the following length-7 sequences defined for $-3 \le n \le 3$:

$$x[n] = \{3 \quad -2 \quad 0 \quad 1 \quad 4 \quad 5 \quad 2\},$$
$$y[n] = \{0 \quad 7 \quad 1 \quad -3 \quad 4 \quad 9 \quad -2\},$$
$$w[n] = \{-5 \quad 4 \quad 3 \quad 6 \quad -5 \quad 0 \quad 1\}.$$

Generate the following sequences: (a) $u[n] = x[n] + y[n]$, (b) $v[n] = x[n] \cdot w[n]$, (c) $s[n] = y[n] - w[n]$, and (d) $r[n] = 4.5y[n]$.

2.2 Analyze the block diagrams of Figure P2.1 and develop the relation between $y[n]$ and $x[n]$.

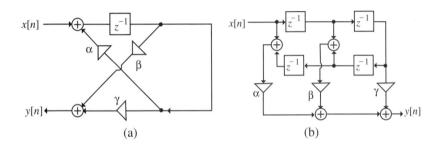

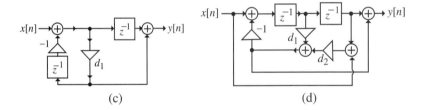

Figure P2.1

2.3 Determine the even and odd parts of the sequences $x[n]$, $y[n]$, and $w[n]$ of Problem 2.1.

2.4 Let $g[n]$ and $h[n]$ be even and odd real sequences, respectively. For each of the following sequences, determine if it is even or odd.
   (a) $x[n] = g[n]g[n]$, (b) $u[n] = g[n]h[n]$, (c) $v[n] = h[n]h[n]$.

2.5 Let $\tilde{x}_1[n]$, $\tilde{x}_2[n]$, and $\tilde{x}_3[n]$ be three periodic sequences with fundamental periods, N_1, N_2, and N_3, respectively. Is a linear combination of these three periodic sequences a periodic sequence? If it is, what is its fundamental period?

2.6 Determine the periodic conjugate symmetric and periodic conjugate antisymmetric parts of the following sequences:
   (a) $\{x[n]\} = \{A\alpha^n\}$, $-N \le n \le N$, where A and α are complex numbers,
   (b) $\{h[n]\} = \{-2 + j5 \quad 4 - j3 \quad 5 + j6 \quad 3 + j \quad -7 + j2\}$.
      $\uparrow$

2.7 Which ones of the following sequences are bounded sequences?

(a) $\{x[n]\} = \{A\alpha^n\}$, where A and α are complex numbers, and $|\alpha| < 1$,

(b) $\{y[n]\} = A\alpha^n \mu[n]$, where A and α are complex numbers, and $|\alpha| < 1$,

(c) $\{h[n]\} = C\beta^n \mu[n]$, where C and β are complex numbers, and $|\beta| > 1$,

(d) $\{g[n]\} = 4 \sin(\omega_a n)$,

(e) $\{v[n]\} = 3 \cos^3(\omega_b n^2)$.

2.8 (a) Show that a causal real sequence $x[n]$ can be fully recovered from its even part $x_{\text{ev}}[n]$ for all $n \geq 0$, whereas it can be recovered from its odd part $x_{\text{od}}[n]$ for all $n > 0$.

(b) Is it possible to fully recover a causal complex sequence $y[n]$ from its conjugate antisymmetric part $y_{\text{ca}}[n]$? Can $y[n]$ be fully recovered from its conjugate symmetric part $y_{\text{cs}}[n]$? Justify your answers.

2.9 Show that the even and odd parts of a real sequence are, respectively, even and odd sequences.

2.10 Show that the periodic conjugate symmetric part $x_{\text{pcs}}[n]$ and the periodic conjugate antisymmetric part $x_{\text{pca}}[n]$ of a length-N sequence $x[n]$, $0 \leq n \leq N - 1$, as defined in Eqs. (2.24a) and (2.24b) can be alternately expressed as

$$x_{\text{pcs}}[n] = x_{\text{cs}}[n] + x_{\text{cs}}[n - N], \quad 0 \leq n \leq N - 1 \tag{2.176a}$$

$$x_{\text{pca}}[n] = x_{\text{ca}}[n] + x_{\text{ca}}[n - N], \quad 0 \leq n \leq N - 1 \tag{2.176b}$$

where $x_{\text{cs}}[n]$ and $x_{\text{ca}}[n]$ are, respectively, the conjugate symmetric and conjugate antisymmetric parts of $x[n]$.

2.11 Show that the periodic conjugate symmetric part $x_{\text{pcs}}[n]$ and the periodic conjugate antisymmetric part $x_{\text{pca}}[n]$ of a length-N sequence $x[n]$, $1 \leq n \leq N - 1$ as defined in Eqs. (2.24a) and (2.24b) can also be expressed as

$$x_{\text{pcs}}[n] = \tfrac{1}{2}(x[n] + x^*[N - n]), \quad 1 \leq n \leq N - 1, \tag{2.177a}$$

$$x_{\text{pcs}}[0] = \text{Re}(x[0]), \tag{2.177b}$$

$$x_{\text{pca}}[n] = \tfrac{1}{2}(x[n] - x^*[N - n]), \quad 1 \leq n \leq N - 1 \tag{2.177c}$$

$$x_{\text{pca}}[0] = j \, \text{Im}(x[0]). \tag{2.177d}$$

2.12 Show that an absolutely summable sequence has finite energy, but a finite energy sequence may not be absolutely summable.

2.13 Show that the square-summable sequence $x_1[n] = \frac{1}{n}$ of Eq. (2.27) is not absolutely summable.

2.14 Show that the sequence $x_2[n] = \frac{\cos \omega_c n}{\pi n}$, $1 \leq n \leq \infty$, is square-summable but not absolutely summable.

2.15 Let $x_{\text{ev}}[n]$ and $x_{\text{od}}[n]$ denote, respectively, the even and odd parts of a square-summable sequence $x[n]$. Prove the following result:

$$\sum_{n=-\infty}^{\infty} x^2[n] = \sum_{n=-\infty}^{\infty} x_{\text{ev}}^2[n] + \sum_{n=-\infty}^{\infty} x_{\text{od}}^2[n].$$

2.16 Compute the energy of the length-N sequence

$$x[n] = \cos\left(\frac{2\pi k n}{N}\right), \quad 0 \leq n \leq N - 1.$$

2.17 Determine the average power and the energy of the following sequences:

(a) $x_1[n] = \mu[n]$,

(b) $x_2[n] = n\mu[n]$,

(c) $x_3[n] = A_o e^{j\omega_o n}$,

(d) $x_4[n] = A\sin((2\pi n/M) + \phi)$.

2.18 Express the sequence $x[n] = 1$, $-\infty < n < \infty$, in terms of the unit step sequence $\mu[n]$.

2.19 Verify the relation between the unit sample sequence $\delta[n]$ and the unit step sequence $\mu[n]$ given in Eq. (2.38).

2.20 The following sequences represent one period of a sinusoidal sequence of the form $x[n] = A\cos(\omega_o n + \phi)$:

(a) $\{0 \ -\sqrt{2} \ -2 \ -\sqrt{2} \ 0 \ \sqrt{2} \ 2 \ \sqrt{2}\}$,

(b) $\{\sqrt{2} \ \sqrt{2} \ -\sqrt{2} \ -\sqrt{2}\}$,

(c) $\{3 \ -3\}$,

(d) $\{0 \ 1.5 \ 0 \ -1.5\}$.

Determine the values of the parameters A, ω_o, and ϕ for each case.

2.21 Determine the fundamental period of the following periodic sequences:

(a) $\tilde{x}_1[n] = e^{-j0.4\pi n}$,

(b) $\tilde{x}_2[n] = \sin(0.6\pi n + 0.6\pi)$,

(c) $\tilde{x}_3[n] = 2\cos(1.1\pi n - 0.5\pi) + 2\sin(0.7\pi n)$,

(d) $\tilde{x}_4[n] = 3\sin(1.3\pi n) - 4\cos(0.3\pi n + 0.45\pi)$,

(e) $\tilde{x}_5[n] = 5\sin(1.2\pi n + 0.65\pi) + 4\sin(0.8\pi n) - \cos(0.8\pi n)$,

(f) $\tilde{x}_6[n] = n$ modulo 6.

2.22 Determine the fundamental period of the sinusoidal sequence $x[n] = A\cos(\omega_o n)$ for the following values of the angular frequency ω_o:

(a) 0.14π, (b) 0.24π, (c) 0.34π, (d) 0.68π, (e) 0.75π.

2.23 A continuous-time sinusoidal signal $x_a(t) = \cos\Omega_o t$ is sampled at $t = nT$, $-\infty \le n \le \infty$, generating the discrete-time sequence $x[n] = x_a(nT) = \cos(\Omega_o nT)$. For what values of T is $x[n]$ a periodic sequence? What is the fundamental period of $x[n]$ if $\Omega_o = 18$ radians and $T = \pi/6$ seconds?

2.24　(a) Express the sequences $x[n]$, $y[n]$, and $w[n]$ of Problem 2.1 as a linear combination of delayed unit sample sequences.

(b) Express the sequences $x[n]$, $y[n]$, and $w[n]$ of Problem 2.1 as a linear combination of delayed unit step sequences.

2.25 Show that the discrete-time systems described by the following equations are linear systems:

(a) Eq. (2.14), (b) Eq. (2.15), (c) Eq. (2.17), (d) Eq. (2.18), (e) Eq. (2.56), (f) Eq. (2.58), and (g) Eq. (2.59).

2.26 For each of the following discrete-time systems, where $y[n]$ and $x[n]$ are, respectively, the output and the input sequences, determine whether or not the system is (1) linear, (2) causal, (3) stable, and (4) shift-invariant:

(a) $y[n] = n^2 x[n]$,

(b) $y[n] = x^4[n]$,

(c) $y[n] = \beta + \sum_{\ell=0}^{3} x[n - \ell]$, β is a nonzero constant,

(d) $y[n] = \beta + \sum_{\ell=-3}^{3} x[n - \ell]$, β is a nonzero constant,

(e) $y[n] = \alpha x[-n]$, α is a nonzero constant,

(f) $y[n] = x[n - 5]$.

2.27 The second derivative $y[n]$ of a sequence $x[n]$ at time instant n is usually approximated by

$$y[n] = x[n + 1] - 2x[n] + x[n - 1].$$

If $y[n]$ and $x[n]$ denote the output and input of a discrete-time system, is the system linear? Is it time-invariant? Is it causal?

2.28 The *median filter* is often used for the smoothing of signals corrupted by impulse noise [Reg93]. It is implemented by sliding a window of odd length over the input sequence $x[n]$ one sample at a time. At the nth instant, the input samples inside the window are rank ordered from the largest to the smallest in values, and the sample at the middle is the median value. The output $y[n]$ of the median filter is then given

$$y[n] = \text{med}\{x[n - K], \ldots, x[n - 1], x[n], x[n + 1], \ldots, x[n + K]\}.$$

For example, med$\{2, -3, 10, 5, -1\} = 2$. Is the median filter a linear or nonlinear discrete-time system? Is it time-invariant? Justify your answer.

2.29 Consider the discrete-time system characterized by the input-output relation [Cad87]

$$y[n] = \frac{1}{2}\left(y[n - 1] + \frac{x[n]}{y[n - 1]}\right), \tag{2.178}$$

where $x[n]$ and $y[n]$ are, respectively, the input and output sequences. Show that the output $y[n]$ of the above system for an input $x[n] = \alpha \mu[n]$ with $y[-1] = 1$ converges to $\sqrt{\alpha}$ as $n \to \infty$ when α is a positive number. Is the above system linear or nonlinear? Is it time-invariant? Justify your answer.

2.30 An algorithm for the calculation of the square root of a number α is given by [Mik92]

$$y[n] = x[n] - y^2[n - 1] + y[n - 1], \tag{2.179}$$

where $x[n] = \alpha \mu[n]$ with $0 < \alpha < 1$. If $x[n]$ and $y[n]$ are considered as the input and output of a discrete-time system, is the system linear or nonlinear? Is it time-invariant? As $n \to \infty$, show that $y[n] \to \sqrt{\alpha}$. Note that $y[-1]$ is a suitable initial approximation to $\sqrt{\alpha}$.

2.31 Develop a general expression for the output $y[n]$ of an LTI discrete-time system in terms of its input $x[n]$ and the unit step response $s[n]$ of the system.

2.32 A periodic sequence $\tilde{x}[n]$ with a period N is applied as an input to an LTI discrete-time system characterized by an impulse response $h[n]$ generating an output $y[n]$. Is $y[n]$ a periodic sequence? If it is, what is its period?

2.33 Consider the following sequences: (i) $x_1[n] = 2\delta[n - 1] - 0.5\delta[n - 3]$, (ii) $x_2[n] = -3\delta[n - 1] + \delta[n + 2]$, (iii) $h_1[n] = 2\delta[n] + \delta[n - 1] - 3\delta[n - 3]$, and (iv) $h_2[n] = -\delta[n - 2] - 0.5\delta[n - 1] + 3\delta[n - 3]$. Determine the following sequences obtained by a linear convolution of a pair of the above sequences: (a) $y_1[n] = x_1[n] \circledast h_1[n]$, (b) $y_2[n] = x_2[n] \circledast h_2[n]$, (c) $y_3[n] = x_1[n] \circledast h_2[n]$, and (d) $y_4[n] = x_2[n] \circledast h_1[n]$.

2.34 Let $g[n]$ be a finite-length sequence defined for $N_1 \le n \le N_2$, with $N_2 > N_1$. Likewise, let $h[n]$ be a finite-length sequence defined for $M_1 \le n \le M_2$, with $M_2 > M_1$. Define $y[n] = g[n] \circledast h[n]$. (a) What is the length of $y[n]$? (b) What is the range of the index n for which $y[n]$ is defined?

2.35 Let $y[n] = x_1[n] \circledast x_2[n]$ and $v[n] = x_1[n - N_1] \circledast x_2[n - N_2]$. Express $v[n]$ in terms of $y[n]$.

2.36 Let $g[n] = x_1[n] \circledast x_2[n] \circledast x_3[n]$ and $h[n] = x_1[n - N_1] \circledast x_2[n - N_2] \circledast x_3[n - N_3]$. Express $h[n]$ in terms of $g[n]$.

2.37 Prove that the convolution sum operation is commutative and distributive.

2.38 Consider the following three sequences:

$$x_1[n] = A \text{ (a constant)}, \qquad x_2[n] = \mu[n], \qquad x_3[n] = \begin{cases} 1, & \text{for } n = 0, \\ -1, & \text{for } n = 1, \\ 0, & \text{otherwise.} \end{cases}$$

Show that $x_3[n] \circledast x_2[n] \circledast x_1[n] \neq x_2[n] \circledast x_3[n] \circledast x_1[n]$.

2.39 Prove that the convolution operation is associative for stable and single-sided sequences.

2.40 Show that the convolution of a length-M sequence with a length-N sequence leads to a sequence of length $(M + N - 1)$.

2.41 Let $x[n]$ be a length-N sequence given by

$$x[n] = \begin{cases} 1, & 0 \leq n \leq N - 1, \\ 0, & \text{otherwise.} \end{cases}$$

Determine $y[n] = x[n] \circledast x[n]$ and show that it is a triangular sequence with a maximum sample value of N. Determine the locations of the samples with the following values: $N/4$, $N/2$, and N.

2.42 Let $x[n]$ and $h[n]$ be two length-N sequences given by

$$x[n] = \begin{cases} 1, & 0 \leq n \leq N - 1, \\ 0, & \text{otherwise,} \end{cases}$$

$$h[n] = \begin{cases} n + 1, & 0 \leq n \leq N - 1, \\ 0, & \text{otherwise.} \end{cases}$$

Determine the location and the value of the largest positive sample of $y[n] = x[n] \circledast h[n]$ without performing the convolution operation.

2.43 Consider two real sequences $h[n]$ and $g[n]$ expressed as a sum of their respective even and odd parts, i.e., $h[n] = h_{ev}[n] + h_{od}[n]$, and $g[n] = g_{ev}[n] + g_{od}[n]$. For each of the following sequences, determine if it is even or odd.
(a) $h_{ev}[n] \circledast g_{ev}[n]$,   (b) $h_{od}[n] \circledast g_{ev}[n]$,   (c) $h_{od}[n] \circledast g_{od}[n]$.

2.44 Let $y[n]$ be the sequence obtained by a linear convolution of two causal finite-length sequences $h[n]$ and $x[n]$. For each pair of $y[n]$ and $h[n]$ listed below, determine $x[n]$. The first sample in each sequence is its value at $n = 0$.
(a) $\{y[n]\} = \{-1, -1, 11, -3, 30, 28, 48\}$, $\{h[n]\} = \{-1, 2, 3, 4\}$,
(b) $\{y[n]\} = \{1, 3, 6, 10, 15, 14, 12, 9, 5\}$, $\{h[n]\} = \{1, 2, 3, 4, 5\}$,
(c) $\{y[n]\} = \{-14 - j5, -3 - j17, -2 + j5, -9.73 + j12.5, 5.8 + j5.67\}$, $\{h[n]\} = \{3 + j2, -1 + j4, 2 + j\}$.

2.45 Consider a causal discrete-time system characterized by a first-order linear, constant-coefficient difference equation given by

$$y[n] = ay[n - 1] + bx[n], \quad n \geq 0,$$

where $y[n]$ and $x[n]$ are, respectively, the output and input sequences. Compute the expression for the output sample $y[n]$ in terms of the initial condition $y[-1]$ and the input samples.

(a) Is the system time-invariant if $y[-1] = 1$? Is the system linear if $y[-1] = 1$?

(b) Repeat part (a) if $y[-1] = 0$.

(c) Generalize the results of parts (a) and (b) to the case of an Nth-order causal discrete-time system given by Eq. (2.93).

2.46 A causal LTI discrete-time system is said to have an *overshoot* in its step response if the response exhibits an oscillatory behavior with decaying amplitudes around a final constant value. Show that the system has no overshoot in its step response if the impulse response $h[n]$ of the system is nonnegative for all $n \geq 0$.

2.47 The sequence of Fibonacci numbers $f[n]$ is a causal sequence defined by

$$f[n] = f[n-1] + f[n-2], \quad n \geq 2$$

with $f[0] = 0$ and $f[1] = 1$.

(a) Develop an exact formula to calculate $f[n]$ directly for any n.

(b) Show that $f[n]$ is the impulse response of a causal LTI system described by the difference equation [Joh89]

$$y[n] = y[n-1] + y[n-2] + x[n-1].$$

2.48 Consider a first-order complex digital filter characterized by a difference equation

$$y[n] = \alpha y[n-1] + x[n],$$

where $x[n]$ is the real input sequence, $y[n] = y_{re}[n] + j y_{im}[n]$ is the complex output sequence with $y_{re}[n]$ and $y_{im}[n]$ denoting its real and imaginary parts, and $\alpha = a + jb$ is a complex constant. Develop an equivalent two-output, single-input real difference equation representation of the above complex digital filter. Show that the single-input, single-output digital filter relating $y_{re}[n]$ to $x[n]$ is described by a second-order difference equation.

2.49 Determine the expression for the impulse response of the factor-of-3 linear interpolator of Eq. (2.59).

2.50 Determine the expression for the impulse response of the factor-of-L linear interpolator.

2.51 Let $h[0]$, $h[1]$, and $h[2]$ denote the first three impulse response samples of the first-order causal LTI system of Problem 2.56. Show that the coefficients of the difference equation characterizing this system can be uniquely determined from these impulse response samples.

2.52 Let a causal IIR digital filter be described by the difference equation

$$\sum_{k=0}^{N} d_k y[n-k] = \sum_{k=0}^{M} p_k x[n-k], \tag{2.180}$$

where $y[n]$ and $x[n]$ denote the output and the input sequences, respectively. If $h[n]$ denotes its impulse response, show that

$$p_k = \sum_{n=0}^{k} h[n] d_{k-n}, \quad k = 0, 1, \ldots, M.$$

From the above result, show that $p_n = h[n] \circledast d_n$.

2.53 Consider a cascade of two causal stable LTI systems characterized by impulse responses $\alpha^n \mu[n]$ and $\beta^n \mu[n]$, where $0 < \alpha < 1$ and $0 < \beta < 1$. Determine the expression for the impulse response $h[n]$ of the cascade.

2.54 Determine the impulse response $g[n]$ of the inverse system of the LTI discrete-time system of Example 2.28.

2.55 Determine the impulse response $g[n]$ characterizing the inverse system of the LTI discrete-time system of Problem 2.45.

2.56 Consider the causal LTI system described by the difference equation

$$y[n] = p_0 x[n] + p_1 x[n-1] - d_1 y[n-1],$$

where $x[n]$ and $y[n]$ denote, respectively, its input and output. Determine the difference equation representation of its inverse system.

2.57 Determine the expression for the impulse response of each of the LTI systems shown in Figure P2.2.

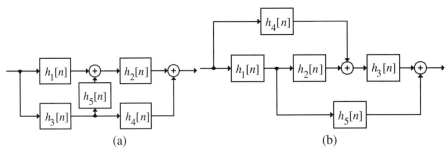

(a)   (b)

Figure P2.2

2.58 Determine the overall impulse response of the system of Figure P2.3, where

$$h_1[n] = 2\delta[n-2] - 3\delta[n+1], \qquad h_2[n] = \delta[n-1] + 2\delta[n+2],$$

$$h_3[n] = 5\delta[n-5] + 7\delta[n-3] + 2\delta[n-1] - \delta[n] + 3\delta[n+1].$$

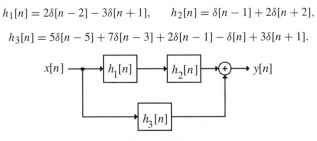

Figure P2.3

2.59 Prove that the BIBO stability condition of Eq. (2.73) also holds for an LTI digital filter with a complex impulse response.

2.60 Is the cascade connection of two stable LTI systems also stable? Justify your answer.

2.61 Is the parallel connection of two stable LTI systems also stable? Justify your answer.

2.62 Prove that the cascade connection of two passive (lossless) LTI systems is also passive (lossless).

2.63 Is the parallel connection of two passive (lossless) LTI systems also passive (lossless)? Justify your answer.

2.64 Consider a causal FIR filter of length $L+1$ with an impulse response given by $\{g[n]\}$, $n = 0, 1, \ldots, L$. Develop the difference equation representation of the form of Eq. (2.81) where $M + N = L$ of a causal finite-dimensional IIR digital filter with an impulse response $\{h[n]\}$ such that $h[n] = g[n]$ for $n = 0, 1, \ldots, L$.

2.65 Compute the output of the accumulator of Eq. (2.55) for an input $x[n] = n\mu[n]$ and the following initial conditions: (a) $y[-1] = 0$, and (b) $y[-1] = -2$.

2.66 In the rectangular method of numerical integration, the integral on the right-hand side of Eq. (2.98) is expressed as

$$\int_{(n-1)T}^{nT} x(\tau)d\tau = T \cdot x\left((n-1)T\right). \tag{2.181}$$

Develop the difference equation representation of the rectangular method of numerical integration.

2.67 Develop a recursive implementation of the time-varying linear discrete-time system characterized by

$$y[n] = \begin{cases} \frac{1}{n}\sum_{\ell=1}^{n} x[\ell], & n > 0, \\ 0, & n \leq 0. \end{cases}$$

2.68 Determine the total solution for $n \geq 0$ of the difference equation

$$y[n] + 0.5y[n-1] = 2\mu[n],$$

with the initial condition $y[-1] = 2$.

2.69 Determine the total solution for $n \geq 0$ of the difference equation

$$y[n] + 0.1y[n-1] - 0.06y[n-2] = 2^n\mu[n],$$

with the initial condition $y[-1] = 1$, and $y[-2] = 0$.

2.70 Determine the total solution for $n \geq 0$ of the difference equation

$$y[n] + 0.1y[n-1] - 0.06y[n-2] = x[n] - 2x[n-1],$$

with the initial condition $y[-1] = 1$, and $y[-2] = 0$, when the forcing function is $x[n] = 2^n\mu[n]$.

2.71 Determine the impulse response $h[n]$ of the LTI system described by the difference equation

$$y[n] + 0.5y[n-1] = x[n].$$

2.72 Determine the impulse response $h[n]$ of the LTI system described by the difference equation

$$y[n] + 0.1y[n-1] - 0.06y[n-2] = x[n] - 2x[n-1].$$

2.73 Show that the sum $\sum_{n=0}^{\infty} |n^K(\lambda_i)^n|$ converges if $|\lambda_i| < 1$.

2.74   (a) Evaluate the autocorrelation sequence of each of the sequences of Problem 2.1.

   (b) Evaluate the cross-correlation sequence $r_{xy}[\ell]$ between the sequences $x[n]$ and $y[n]$, and the cross-correlation sequence $r_{xw}[\ell]$ between the sequences $x[n]$ and $w[n]$ of Problem 2.1.

2.75 Determine the autocorrelation sequence of each of the following sequences and show that it is an even sequence in each case. What is the location of the maximum value of the autocorrelation sequence in each case?

   (a) $x_1[n] = \alpha^n\mu[n]$,

   (b) $x_2[n] = \begin{cases} 1, & 0 \leq n \leq N-1, \\ 0, & \text{otherwise.} \end{cases}$

2.76 Determine the autocorrelation sequence and its period of each of the following periodic sequences.

(a) $\tilde{x}_1[n] = \cos(\pi n/M)$, where M is a positive integer,

(b) $\tilde{x}_2[n] = n$ modulo 6,

(c) $\tilde{x}_3[n] = (-1)^n$.

2.77 Let X and Y be two random variables. Show that $E(X + Y) = E(X) + E(Y)$ and $E(cX) = cE(X)$, where c is a constant.

2.78 Determine the value of the constant κ that minimizes the mean-square error $E([X - \kappa]^2)$, and then find the minimum value of the mean-square error.

2.79 Compute the mean value and the variance of the random variables with the probability density functions listed below [Par60].

(a) *Cauchy distribution:* $p_X(x) = \frac{\alpha/\pi}{x^2+\alpha^2}$,

(b) *Laplacian distribution:* $p_X(x) = \frac{\alpha}{2}e^{-\alpha|x|}$,

(c) *Binomial distribution:* $p_X(x) = \sum_{\ell=0}^{n} \binom{n}{\ell} p^\ell (1-p)^{n-\ell} \delta(x - \ell)$,

(d) *Poisson distribution:* $p_X(x) = \sum_{\ell=0}^{\infty} \frac{e^{-\alpha}\alpha^\ell}{\ell!} \delta(x - \ell)$.

(e) *Rayleigh distribution:* $p_X(x) = \frac{x}{\alpha^2}e^{-x^2/2\alpha^2} \mu(x)$.

In the above equations, $\delta(x)$ is the Dirac delta function and $\mu(x)$ is the unit step function.

2.80 Show that if the two random variables X and Y are statistically independent, then they are also linearly independent.

2.81 Prove Eq. (2.124).

2.82 Let $x[n]$ and $y[n]$ be two statistically independent stationary random signals with means m_x and m_y, and variances σ_x^2 and σ_y^2, respectively. Consider the random signal $v[n]$ obtained by a linear combination of $x[n]$ and $y[n]$, i.e., $v[n] = ax[n] + by[n]$, where a and b are constants. Show that the mean m_v and the variance σ_v^2 of $v[n]$ are given by $m_v = am_x + bm_y$ and $\sigma_v^2 = a^2\sigma_x^2 + b^2\sigma_y^2$, respectively.

2.83 Let $x[n]$ and $y[n]$ be two independent zero-mean WSS random signals with autocorrelations $\phi_{xx}[n]$ and $\phi_{yy}[n]$, respectively. Consider the random signal $v[n]$ obtained by a linear combination of $x[n]$ and $y[n]$, i.e., $v[n] = ax[n] + by[n]$, where a and b are constants. Express the autocorrelation and cross-correlations, $\phi_{vv}[n]$, $\phi_{vx}[n]$, and $\phi_{vy}[n]$, in terms of $\phi_{xx}[n]$ and $\phi_{yy}[n]$. What would be the results if either $x[n]$ or $y[n]$ was zero-mean?

2.84 Prove the symmetry properties of Eqs. (2.166a) through (2.166d).

2.85 Verify the inequalities of Eqs. (2.167a) through (2.167d).

2.86 Prove Eq. (2.168).

2.87 Determine the mean and variance of a WSS real signal with an autocorrelation function given by

$$\phi_{XX}[\ell] = \frac{9 + 11\ell^2 + 14\ell^4}{1 + 3\ell^2 + 2\ell^4}.$$

2.11 MATLAB Exercises

M 2.1 Write a MATLAB program to generate the following sequences and plot them using the function stem: (a) unit sample sequence $\delta[n]$, (b) unit step sequence $\mu[n]$, and (c) ramp sequence $n\mu[n]$. The input parameters specified by the user are the desired length L of the sequence and the sampling frequency F_T in Hz. Using this program generate the first 100 samples of each of the above sequences with a sampling rate of 20 kHz.

M 2.2 The square wave and the sawtooth wave are two periodic sequences as sketched in Figure P2.4. Using the functions sawtooth and square write a MATLAB program to generate the above two sequences and plot them using the function stem. The input data specified by the user are: desired length L of the sequence, peak value A, and the period N. For the square wave sequence an additional user-specified parameter is the duty cycle, which is the percent of the period for which the signal is positive. Using this program generate the first 100 samples of each of the above sequences with a sampling rate of 20 kHz, a peak value of 7, a period of 13, and a duty cycle of 60% for the square wave.

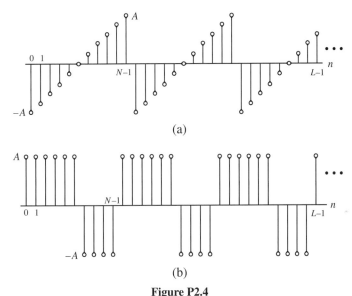

(a)

(b)

Figure P2.4

M 2.3   (a) Using Program 2_1 generate the sequences shown in Figures 2.16 and 2.17.

   (b) Generate and plot the complex exponential sequence $2.5e^{(-0.4+j\pi/5)n}$ for $0 \le n \le 100$ using Program 2_1.

M 2.4   (a) Write a MATLAB program to generate a sinusoidal sequence $x[n] = A\cos(\omega_o n + \phi)$ and plot the sequence using the stem function. The input data specified by the user are the desired length L, amplitude A, the angular frequency ω_o, and the phase ϕ where $0 < \omega_o < \pi$ and $0 \le \phi \le 2\pi$. Using this program generate the sinusoidal sequences shown in Figure 2.15.

   (b) Generate sinsusoidal sequences with the angular frequencies given in Problem 2.22. Determine the period of each sequence from the plot and verify the result theoretically.

M 2.5 Generate the sequences of Problem 2.21(b) to 2.21(e) using MATLAB.

M 2.6 Write a MATLAB program to plot a continuous-time sinusoidal signal and its sampled version and verify Figure 2.19. You need to use the hold function to keep both plots.

M 2.7 Using the program developed in the previous problem, verify experimentally that the family of continuous-time sinusoids given by Eq. (2.53) lead to identical sampled signals.

M 2.8 Using Program 2_4 investigate the effect of signal smoothing by a moving average filter of lengths 5, 7, and 9. Does the signal smoothing improve with an increase in the length? What is the effect of the length on the delay between the smoothed output and the noisy input?

M 2.9 Write a MATLAB program implementing the discrete-time system of Eq. (2.178) in Problem 2.29 and show that the output $y[n]$ of this system for an input $x[n] = \alpha\mu[n]$ with $y[-1] = 1$ converges to $\sqrt{\alpha}$ as $n \to \infty$.

M 2.10 Write a MATLAB program to compute the square root using the algorithm of Eq. (2.179) in Problem 2.30 and show that the output $y[n]$ of this system for an input $x[n] = \alpha\mu[n]$ with $y[-1] = 1$ converges to $\sqrt{\alpha}$ as $n \to \infty$. Plot the error as a function of n for several different values of α. How would you compute the square-root of a number α with a value greater than one?

M 2.11 Using the function `impz` write a MATLAB program to compute and plot the impulse response of a causal finite-dimensional discrete-time system characterized by a difference equation of the form of Eq. (2.81). The input data to the program are the desired length of the impulse response, and the constants $\{p_k\}$ and $\{d_k\}$ of the difference equation. Generate and plot the first 41 samples of the impulse response of the system of Eq. (2.93).

M 2.12 Using Program 2_7, determine the autocorrelation and the cross-correlation sequences of Problem 2.74. Are your results the same as those determined in Problem 2.74?

M 2.13 Modify Program 2_7 to determine the autocorrelation sequence of a sequence corrupted with a uniformly distributed random signal generated using the M-function `randn`. Using the modified program demonstrate that the autocorrelation sequence of a noise-corrupted signal exhibits a peak at zero lag.

M 2.14   (a) Write a MATLAB program to generate the random sinusoidal signal of Eq. (2.140) and plot four possible realizations of the random signal. Comment on your results.

   (b) Compute the mean and variance of a single realization of the above random signal using Eqs. (2.174a) and (2.174b). How close are your answers to those given in Example 2.44?

M 2.15 Using the M-function `rand` generate a uniformly distributed length-1000 random sequence in the range $(-1, 1)$. Using Eqs. (2.174a) and (2.174b), compute the mean and variance of the random signal.

3 Discrete-Time Signals in the Transform Domain

In Section 2.2.3 we pointed out that any arbitrary sequence can be represented in the time-domain as a weighted linear combination of delayed unit sample sequences $\{\delta[n - k]\}$. An important consequence of this representation, derived in Section 2.5.1, is the input-output characterization of an LTI digital filter in the time-domain by means of the convolution sum describing the output sequence in terms of a weighted linear combination of its delayed impulse responses. We consider in this chapter an alternate description of a sequence in terms of complex exponential sequences of the form $\{e^{-j\omega n}\}$ and $\{z^{-n}\}$ where z is a complex variable. This leads to three particularly useful representations of discrete-time sequences and LTI discrete-time systems in a transform domain.[1] These transform-domain representations are reviewed here along with the conditions for their existence and their properties. MATLAB has been used extensively to illustrate various concepts and implement a number of useful algorithms. Applications of these concepts are discussed in the following chapters.

The first transform-domain representation of a discrete-time sequence we discuss is the discrete-time Fourier transform by which a time-domain sequence is mapped into a continuous function of a frequency variable. Because of the periodicity of the discrete-time Fourier transform, the parent discrete-time sequence can be simply obtained by computing its Fourier series representation. We then show that for a length-N sequence, N equally spaced samples of its discrete-time Fourier transform are sufficient to describe the frequency-domain representation of the sequence and from these N frequency samples, the original N samples of the discrete-time sequence can be obtained by a simple inverse operation. These N frequency samples constitute the discrete Fourier transform of a length-N sequence, a second transform-domain representation. We next consider a generalization of the discrete-time Fourier transform, called the z-transform, the third type of transform-domain representation of a sequence. Finally, the transform-domain representation of a random signal is discussed. Each of these representations is an important tool in signal processing and is used often in practice. A thorough understanding of these three transforms is therefore very important to make best use of the signal processing algorithms discussed in this book.

3.1 The Discrete-Time Fourier Transform

The *discrete-time Fourier transform* (DTFT) or, simply, the Fourier transform of a discrete-time sequence $x[n]$ is a representation of the sequence in terms of the complex exponential sequence $\{e^{-j\omega n}\}$ where ω is the real frequency variable. The DTFT representation of a sequence, if it exists, is unique and the original sequence can be computed from its DTFT by an inverse transform operation. We first define the forward transform and derive its inverse transform. We then describe the condition for its existence and summarize its important properties.

---

[1] Periodic sequences can be represented in the frequency domain by means of a discrete Fourier series (see Problem 3.34)

3.1.1 Definition

The discrete-time Fourier transform $X(e^{j\omega})$ of a sequence $x[n]$ is defined by

$$X(e^{j\omega}) = \sum_{n=-\infty}^{\infty} x[n]e^{-j\omega n}. \tag{3.1}$$

In general $X(e^{j\omega})$ is a complex function of the real variable ω and can be written in rectangular form as

$$X(e^{j\omega}) = X_{re}(e^{j\omega}) + jX_{im}(e^{j\omega}), \tag{3.2}$$

where $X_{re}(e^{j\omega})$ and $X_{im}(e^{j\omega})$ are, respectively, the real and imaginary parts of $X(e^{j\omega})$, and are real functions of ω. $X(e^{j\omega})$ can alternately be expressed in the polar form as

$$X(e^{j\omega}) = |X(e^{j\omega})|e^{j\theta(\omega)}, \tag{3.3}$$

where

$$\theta(\omega) = \arg\{X(e^{j\omega})\}. \tag{3.4}$$

The quantity $|X(e^{j\omega})|$ is called the *magnitude function* and the quantity $\theta(\omega)$ is called the *phase function* with both functions again being real functions of ω. In many applications, the Fourier transform is called the *Fourier spectrum* and, likewise, $|X(e^{j\omega})|$ and $\theta(\omega)$ are referred to as the *magnitude spectrum* and *phase spectrum,* respectively. The complex conjugate of $X(e^{j\omega})$ is denoted as $X^*(e^{j\omega})$. The relations between the rectangular and polar forms of $X(e^{j\omega})$ follow from Eqs. (3.2) and (3.3), and are given by

$$X_{re}(e^{j\omega}) = |X(e^{j\omega})|\cos\ \theta(\omega),$$

$$X_{im}(e^{j\omega}) = |X(e^{j\omega})|\sin\ \theta(\omega),$$

$$|X(e^{j\omega})|^2 = X_{re}^2(e^{j\omega}) + X_{im}^2(e^{j\omega}),$$

$$\tan\theta(\omega) = \frac{X_{im}(e^{j\omega})}{X_{re}(e^{j\omega})}.$$

It can be easily shown that for a real sequence $x[n]$, $|X(e^{j\omega})|$ and $X_{re}(e^{j\omega})$ are even functions of ω, whereas $\theta(\omega)$ and $X_{im}(e^{j\omega})$ are odd functions of ω (Problem 3.1).

Note from Eq. (3.3) that if we replace $\theta(\omega)$ with $\theta(\omega) + 2\pi k$, where k is any integer, $X(e^{j\omega})$ remains unchanged implying that the phase function cannot be uniquely specified for any Fourier transform. Unless otherwise stated, we will assume that the phase function $\theta(\omega)$ is restricted to the following range of values,

$$-\pi \leq \theta(\omega) < \pi,$$

called the *principal value*. As illustrated in Example 3.8, the discrete-time Fourier transforms of some sequences exhibit discontinuities of 2π in their phase responses. In such cases, it is often useful to consider an alternate type of phase function that is a continuous function of ω derived from the original phase function by removing the discontinuities of 2π. The process of removing the discontinuities is called "*unwrapping the phase*" and the new phase function will be denoted as $\theta_c(\omega)$ with the subscript "*c*" indicating that it is a continuous function of ω.[2]

We illustrate the DTFT computation in the following two examples.

---

[2]In some cases, discontinuities of π may still be present after phase unwrapping (see Table 3.1 for an example).

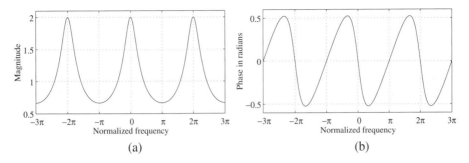

Figure 3.1: Magnitude and phase of $X(e^{j\omega}) = 1/(1 - 0.5e^{-j\omega})$.

EXAMPLE 3.1   Consider the unit sample sequence $\delta[n]$. Its Fourier transform obtained using Eq. (3.1) is given by

$$\Delta(e^{j\omega}) = \sum_{n=-\infty}^{\infty} \delta[n] e^{-j\omega n} = 1, \tag{3.5}$$

where we have used the sampling property of the impulse function.

EXAMPLE 3.2   Consider a causal sequence

$$x[n] = \alpha^n \mu[n], \quad |\alpha| < 1.$$

Its Fourier transform $X(e^{j\omega})$ obtained using Eq. (3.1) is given by

$$X(e^{j\omega}) = \sum_{n=-\infty}^{\infty} \alpha^n \mu[n] e^{-j\omega n} = \sum_{n=0}^{\infty} \alpha^n e^{-j\omega n}$$

$$= \sum_{n=0}^{\infty} (\alpha\, e^{-j\omega})^n = \frac{1}{1 - \alpha\, e^{-j\omega}}, \tag{3.6}$$

as $|\alpha e^{-j\omega}| = |\alpha| < 1$. The magnitude and phase of the above Fourier transform have been plotted in Figure 3.1 for $\alpha = 0.5$.

It should be noted here that for most practical discrete-time sequences, their DTFTs can be expressed in terms of a sum of a convergent geometric series which can be summed in a simple closed form as illustrated by the above example. We take up the issue of the convergence of a general DTFT later in this section.

As can be seen from the definition and also from Figure 3.1, the Fourier transform $X(e^{j\omega})$ of a sequence $x[n]$ is a continuous function of ω. It is also a periodic function in ω with a period 2π. To verify this latter property observe that

$$X(e^{j(\omega_1 + 2\pi k)}) = \sum_{n=-\infty}^{\infty} x[n] e^{-j(\omega_1 + 2\pi k)n} = \sum_{n=-\infty}^{\infty} x[n] e^{-j\omega_1 n} e^{-j2\pi kn}$$

$$= \sum_{n=-\infty}^{\infty} x[n] e^{-j\omega_1 n} = X(e^{j\omega_1}).$$

It therefore follows that Eq. (3.1) represents the Fourier series representation of the periodic function $X(e^{j\omega})$. As a result, the Fourier coefficients $x[n]$ can be computed from $X(e^{j\omega})$ using the Fourier integral given by

$$x[n] = \frac{1}{2\pi} \int_{-\pi}^{\pi} X(e^{j\omega}) e^{j\omega n} \, d\omega, \tag{3.7}$$

called the *inverse discrete-time Fourier transform*. Equations (3.1) and (3.7) constitute a discrete-time Fourier transform pair for the sequence $x[n]$.

To verify that Eq. (3.7) is indeed the inverse of Eq. (3.1) we substitute the expression for $X(e^{j\omega})$ from Eq. (3.1) in Eq. (3.7) arriving at

$$x[n] = \frac{1}{2\pi} \int_{-\pi}^{\pi} \left(\sum_{\ell=-\infty}^{\infty} x[\ell] e^{-j\omega\ell} \right) e^{j\omega n} d\omega.$$

The order of integration and the summation on the right-hand side of the above equation can be interchanged if the summation inside the brackets converges uniformly, i.e., if $X(e^{j\omega})$ exists. Under this condition we get from the above

$$\sum_{\ell=-\infty}^{\infty} x[\ell] \left(\frac{1}{2\pi} \int_{-\pi}^{\pi} e^{j\omega(n-\ell)} d\omega \right) = \sum_{\ell=-\infty}^{\infty} x[\ell] \frac{\sin \pi(n-\ell)}{\pi(n-\ell)}.$$

Now,

$$\frac{\sin \pi(n-\ell)}{\pi(n-\ell)} = \begin{cases} 1, & n = \ell, \\ 0, & n \neq \ell, \end{cases}$$
$$= \delta[n-\ell].$$

Hence,

$$\sum_{\ell=-\infty}^{\infty} x[\ell] \frac{\sin \pi(n-\ell)}{\pi(n-\ell)} = \sum_{\ell=-\infty}^{\infty} x[\ell] \delta[n-\ell] = x[n],$$

using the sampling property of the impulse function.

3.1.2 Convergence Condition

Now, an infinite series of the form of Eq. (3.1) may or may not converge. The Fourier transform $X(e^{j\omega})$ of $x[n]$ is said to exist if the series in Eq. (3.1) converges in some sense. If we denote

$$X_K(e^{j\omega}) = \sum_{n=-K}^{K} x[n] e^{-j\omega n}, \tag{3.8}$$

then for *uniform convergence* of $X(e^{j\omega})$, the absolute value of the error $\left(X(e^{j\omega}) - X_K(e^{j\omega}) \right)$ must approach zero for each value of ω as K approaches ∞, i.e.,

$$\lim_{K \to \infty} \left| X(e^{j\omega}) - X_K(e^{j\omega}) \right| = 0.$$

Now, if $x[n]$ is an *absolutely summable sequence*, i.e., if

$$\sum_{n=-\infty}^{\infty} |x[n]| < \infty, \tag{3.9}$$

then

$$\left| X(e^{j\omega}) \right| = \left| \sum_{n=-\infty}^{\infty} x[n] e^{-j\omega n} \right| \le \sum_{n=-\infty}^{\infty} |x[n]| < \infty$$

for all values of ω. Thus, Eq. (3.9) is a sufficient condition for the existence of the DTFT $X(e^{j\omega})$ of the sequence $x[n]$. Note the sequence $x[n] = \alpha^n \mu[n]$ of Example 3.2 is absolutely summable as

$$\sum_{n=-\infty}^{\infty} |\alpha^n| \mu[n] = \sum_{n=0}^{\infty} |\alpha|^n = \frac{1}{1 - |\alpha|} < \infty,$$

and its Fourier transform $X(e^{j\omega})$ therefore converges to $1/(1 - \alpha e^{-j\omega})$ uniformly.

Since

$$\sum_{n=-\infty}^{\infty} |x[n]|^2 \le \left(\sum_{n=-\infty}^{\infty} |x[n]| \right)^2,$$

an absolutely summable sequence has always a finite energy. However, a finite-energy sequence is not necessarily absolutely summable. The sequence $x_1[n]$ of Example 2.7 is such a sequence. To represent such sequences by a discrete-time Fourier transform, it is necessary to consider a *mean-square convergence* of $X(e^{j\omega})$, in which case the total energy of the error $\left(X(e^{j\omega}) - X_K(e^{j\omega}) \right)$ must approach zero at each value of ω as K goes to ∞, i.e.,

$$\lim_{K \to \infty} \int_{-\pi}^{\pi} \left| X(e^{j\omega}) - X_K(e^{j\omega}) \right|^2 d\omega = 0. \tag{3.10}$$

In such a case, the error $\left| X(e^{j\omega}) - X_K(e^{j\omega}) \right|$ may not go to zero as K goes to ∞, and the DTFT is no longer bounded, i.e., the absolute summability condition of Eq. (3.9) does not hold. The following example considers such a sequence.

EXAMPLE 3.3 Consider the DTFT

$$H_{LP}(e^{j\omega}) = \begin{cases} 1, & 0 \le |\omega| \le \omega_c, \\ 0, & \omega_c < |\omega| \le \pi, \end{cases} \tag{3.11}$$

shown in Figure 3.2. This DTFT finds application in digital filtering and we shall consider it again in Section 4.4.1. The inverse DTFT of $H_{LP}(e^{j\omega})$ is given by

$$h_{LP}[n] = \frac{1}{2\pi} \int_{-\omega_c}^{\omega_c} e^{j\omega n} d\omega$$

$$= \frac{1}{2\pi} \left(\frac{e^{j\omega_c n}}{jn} - \frac{e^{-j\omega_c n}}{jn} \right) = \frac{\sin \omega_c n}{\pi n}, \qquad -\infty < n < \infty. \tag{3.12}$$

We shall show later in Example 3.7 that the energy of the above sequence is given by $\frac{\omega_c}{\pi}$, and therefore, $h_{LP}[n]$ is a finite-energy sequence. However, it is not absolutely summable. As a result

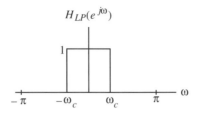

Figure 3.2: Frequency response plot of Eq. (3.11).

$$\sum_{n=-\infty}^{\infty} h_{LP}[n]e^{-j\omega n} = \sum_{n=-\infty}^{\infty} \frac{\sin \omega_c n}{\pi n}e^{-j\omega n}$$

does not uniformly converge to $H_{LP}(e^{j\omega})$ of Eq. (3.11) for all values of ω, but converges to $H_{LP}(e^{j\omega})$ in the mean-square sense.

The mean-square convergence property of the sequence $h_{LP}[n]$ discussed in the previous example can be further illustrated by examining the plot of the function

$$H_{LP,K}(e^{j\omega}) = \sum_{n=-K}^{K} \frac{\sin \omega_c n}{\pi n}e^{-j\omega n}, \tag{3.13}$$

for various values of K as shown in Figure 3.3. It can be seen from this figure that, independent of the number of terms in the above sum, there are ripples in the plot of $H_{LP}(e^{j\omega})$ around both sides of the point $\omega = \omega_c$. The number of ripples increases as K increases with the height of the largest ripple remaining the same for all values of K. As K goes to infinity, the condition of Eq. (3.10) holds indicating the convergence of $H_{LP,K}(e^{j\omega})$ to $H_{LP}(e^{j\omega})$. The oscillatory behavior in the plot of $H_{LP,K}(e^{j\omega})$ approximating a DTFT $H_{LP}(e^{j\omega})$ in the mean-square sense at a point of discontinuity, as indicated in Figure 3.3, is commonly known as the *Gibbs phenomenon*. We shall return to this phenomenon in the design of FIR filters based on the windowed Fourier series discussed in Section 7.6.3.

The DTFT can be defined for a certain class of sequences which are neither absolutely summable nor square-summable. Examples of such sequences are the unit step sequence of Eq. (2.37), the sinusoidal sequence of Eq. (2.39), and the complex exponential sequence of Eq. (2.42) which are neither absolutely summable nor square-summable. For this type of sequence a discrete-time Fourier transform representation is possible by using Dirac delta functions. A Dirac delta function $\delta(\omega)$ is a function of ω with infinite height, zero width, and unit area. It is the limiting form of a unit area pulse function $p_\Delta(\omega)$ shown in Figure 3.4 as Δ goes to 0 satisfying

$$\lim_{\Delta \to 0} \int_{-\infty}^{\infty} p_\Delta(\omega)\, d\omega = \int_{-\infty}^{\infty} \delta(\omega)\, d\omega. \tag{3.14}$$

The discrete-time Fourier transforms resulting from the use of Dirac delta functions are not continuous functions of ω.

EXAMPLE 3.4   Consider the complex exponential sequence

$$x[n] = e^{j\omega_o n}.$$

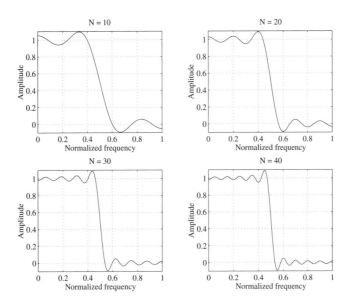

Figure 3.3: Frequency response plots of Eq. (3.13) for various values of $N = 2K$.

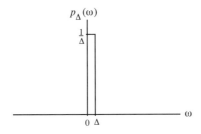

Figure 3.4: Unit area pulse function.

Its DTFT is given by

$$X(e^{j\omega}) = \sum_{k=-\infty}^{\infty} 2\pi\delta(\omega - \omega_o + 2\pi k), \qquad (3.15)$$

where $\delta(\omega)$ is an impulse function of ω and $-\pi \leq \omega_o \leq \pi$. The function on the right-hand side of the above equation is a periodic function of ω with a period 2π and is called a *periodic impulse train*.

To show the above result, we compute the inverse DTFT of Eq. (3.15):

$$x[n] = \frac{1}{2\pi} \int_{-\pi}^{\pi} \sum_{k=-\infty}^{\infty} 2\pi\delta(\omega - \omega_o + 2\pi k)\, e^{j\omega n}\, d\omega$$

$$= \int_{-\pi}^{\pi} \delta(\omega - \omega_o)\, e^{j\omega n}\, d\omega = e^{j\omega_o n},$$

where we have used the sampling property of the impulse function $\delta(\omega)$.

Table 3.1: Commonly used discrete-time Fourier transform pairs.

Sequence	Discrete-Time Fourier Transform
$\delta[n]$	1
1	$\displaystyle\sum_{k=-\infty}^{\infty} 2\pi\,\delta(\omega + 2\pi k)$
$\mu[n]$	$\displaystyle\frac{1}{1 - e^{-j\omega}} + \sum_{k=-\infty}^{\infty} \pi\,\delta(\omega + 2\pi k)$
$e^{j\omega_o n}$	$\displaystyle\sum_{k=-\infty}^{\infty} 2\pi\,\delta(\omega - \omega_o + 2\pi k)$
$\alpha^n \mu[n], \quad (|\alpha| < 1)$	$\displaystyle\frac{1}{1 - \alpha e^{-j\omega}}$

Table 3.1 lists the discrete-time Fourier transforms of some commonly encountered sequences.

3.1.3 Bandlimited Signals

A full-band discrete-time signal has a spectrum occupying the whole frequency range $0 \leq |\omega| \leq \pi$. If the spectrum is limited to a portion of the frequency range $0 \leq |\omega| \leq \pi$, it is called a *bandlimited* signal. A *lowpass* discrete-time signal has a spectrum occupying the frequency range $0 \leq |\omega| \leq \omega_p < \pi$, where ω_p is called the *bandwidth* of the signal. A *bandpass* discrete-time signal has a spectrum occupying the frequency range $0 < \omega_L \leq |\omega| \leq \omega_H < \pi$, where $\omega_H - \omega_L$ is its bandwidth.

3.1.4 Discrete-Time Fourier Transform Properties

There are a number of important properties of the discrete-time Fourier transform which are useful in digital signal processing applications. These are listed here without proof. However, their proofs are quite straightforward and have been left as exercises. We list the general properties in Table 3.2, and the symmetry properties in Tables 3.3 and 3.4.

The following examples illustrate some applications of a few of the properties of the DTFT.

EXAMPLE 3.5   Determine the DTFT of the sequence

$$y[n] = (n + 1)\alpha^n \mu[n], \qquad |\alpha| < 1.$$

Let $x[n] = \alpha^n \mu[n], |\alpha| < 1$. We can therefore write

$$y[n] = n\,x[n] + x[n].$$

From Table 3.1, the DTFT of $x[n]$ is given by

$$X(e^{j\omega}) = \frac{1}{1 - \alpha e^{-j\omega}}.$$

Using the differentiation property of the DTFT given in Table 3.2, we observe that the DTFT of $n\,x[n]$ is given by

$$j\frac{dX(e^{j\omega})}{d\omega} = j\frac{d}{d\omega}\left(\frac{1}{1-\alpha e^{-j\omega}}\right) = \frac{\alpha e^{-j\omega}}{(1-\alpha e^{-j\omega})^2}.$$

Next using the linearity property of the DTFT given in Table 3.2 we arrive at the DTFT of $y[n]$ as

$$Y(e^{j\omega}) = \frac{\alpha e^{-j\omega}}{(1-\alpha e^{-j\omega})^2} + \frac{1}{1-\alpha e^{-j\omega}} = \frac{1}{(1-\alpha e^{-j\omega})^2}.$$

EXAMPLE 3.6 Determine the DTFT $V(e^{j\omega})$ of the sequence $v[n]$ given by

$$d_0 v[n] + d_1 v[n-1] = p_0\delta[n] + p_1\delta[n-1], \qquad |d_1/d_0| < 1. \tag{3.16}$$

From Example 3.1 and also Table 3.1, the DTFT of $\delta[n]$ is simply 1. Next, from Table 3.2, using the time-shifting property of the DTFT we observe that the DTFT of $\delta[n-1]$ is $e^{-j\omega}$ and the DTFT of $v[n-1]$ is $e^{-j\omega}V(e^{j\omega})$. Using the linearity property of Table 3.2 we then obtain from Eq. (3.16) the following equation

$$d_0 V(e^{j\omega}) + d_1 e^{-j\omega}V(e^{j\omega}) = p_0 + p_1 e^{-j\omega}.$$

Solving the above equation we arrive at

$$V(e^{j\omega}) = \frac{p_0 + p_1 e^{-j\omega}}{d_0 + d_1 e^{-j\omega}}.$$

The expression for the DTFT given above is a rational function in $e^{j\omega}$, i.e., a ratio of polynomials in $e^{j\omega}$. The two polynomials are each of first order. In the general case, the DTFTs we shall encounter in this book are ratios of polynomials of higher order and are of the form

$$X(e^{j\omega}) = \frac{P(e^{j\omega})}{D(e^{j\omega})} = \frac{p_0 + p_1 e^{-j\omega} + \cdots + p_M e^{-j\omega M}}{d_0 + d_1 e^{-j\omega} + \cdots + d_N e^{-j\omega N}}. \tag{3.17}$$

3.1.5 Energy Density Spectrum

One important application of Parseval's relation given in Table 3.2 is in the computation of the energy of a finite-energy sequence. Recall from Eq. (2.26) that the total energy of a finite-energy sequence $g[n]$ is given by

$$\mathcal{E}_g = \sum_{n=-\infty}^{\infty} |g[n]|^2.$$

If $h[n] = g[n]$, then from Parseval's relation we observe

$$\mathcal{E}_g = \sum_{n=-\infty}^{\infty} |g[n]|^2 = \frac{1}{2\pi}\int_{-\pi}^{\pi} |G(e^{j\omega})|^2\, d\omega. \tag{3.18}$$

Thus the energy of the sequence $g[n]$ can be computed by evaluating the integral on the right. The quantity

$$S_{gg}(e^{j\omega}) = |G(e^{j\omega})|^2 \tag{3.19}$$

is called the *energy density spectrum* of the sequence $g[n]$. The area under this curve in the range $-\pi \le \omega \le \pi$ divided by 2π is the energy of the sequence.

Table 3.2: General properties of the discrete-time Fourier transform of sequences.

Type of Property	Sequence	Discrete-Time Fourier Transform
	$g[n]$	$G(e^{j\omega})$
	$h[n]$	$H(e^{j\omega})$
Linearity	$\alpha g[n] + \beta h[n]$	$\alpha G(e^{j\omega}) + \beta H(e^{j\omega})$
Time-shifting	$g[n - n_o]$	$e^{-j\omega n_o} G(e^{j\omega})$
Frequency-shifting	$e^{j\omega_o n} g[n]$	$G\left(e^{j(\omega - \omega_o)}\right)$
Differentiation in frequency	$ng[n]$	$j\dfrac{dG(e^{j\omega})}{d\omega}$
Convolution	$g[n] \circledast h[n]$	$G(e^{j\omega})H(e^{j\omega})$
Modulation	$g[n]h[n]$	$\frac{1}{2\pi}\int_{-\pi}^{\pi} G(e^{j\theta})H(e^{j(\omega-\theta)})\, d\theta$
Parseval's relation	$\displaystyle\sum_{n=-\infty}^{\infty} g[n]h^*[n] = \frac{1}{2\pi}\int_{-\pi}^{\pi} G(e^{j\omega})H^*(e^{j\omega})\, d\omega$	

Table 3.3: Symmetry relations of the discrete-time Fourier transform of a complex sequence.

Sequence	Discrete-Time Fourier Transform
$x[n]$	$X(e^{j\omega})$
$x[-n]$	$X(e^{-j\omega})$
$x^*[-n]$	$X^*(e^{j\omega})$
$\mathrm{Re}\{x[n]\}$	$X_{\mathrm{cs}}(e^{j\omega}) = \frac{1}{2}\{X(e^{j\omega}) + X^*(e^{-j\omega})\}$
$j\mathrm{Im}\{x[n]\}$	$X_{\mathrm{ca}}(e^{j\omega}) = \frac{1}{2}\{X(e^{j\omega}) - X^*(e^{-j\omega})\}$
$x_{\mathrm{cs}}[n]$	$X_{\mathrm{re}}(e^{j\omega})$
$x_{\mathrm{ca}}[n]$	$jX_{\mathrm{im}}(e^{j\omega})$

Note: $X_{\mathrm{cs}}(e^{j\omega})$ and $X_{\mathrm{ca}}(e^{j\omega})$ are the conjugate-symmetric and conjugate-antisymmetric parts of $X(e^{j\omega})$, respectively. Likewise, $x_{\mathrm{cs}}[n]$ and $x_{\mathrm{ca}}[n]$ are the conjugate-symmetric and conjugate-antisymmetric parts of $x[n]$, respectively.

Table 3.4: Symmetry relations of the discrete-time Fourier transform of a real sequence.

Sequence	Discrete-Time Fourier Transform
$x[n]$	$X(e^{j\omega}) = X_{\text{re}}(e^{j\omega}) + jX_{\text{im}}(e^{j\omega})$
$x_{\text{ev}}[n]$ $x_{\text{od}}[n]$	$X_{\text{re}}(e^{j\omega})$ $jX_{\text{im}}(e^{j\omega})$
Symmetry relations	$X(e^{j\omega}) = X^*(e^{-j\omega})$ $X_{\text{re}}(e^{j\omega}) = X_{\text{re}}(e^{-j\omega})$ $X_{\text{im}}(e^{j\omega}) = -X_{\text{im}}(e^{-j\omega})$ $|X(e^{j\omega})| = |X(e^{-j\omega})|$ $\arg\{X(e^{j\omega})\} = -\arg\{X(e^{-j\omega})\}$

Note: $x_{\text{ev}}[n]$ and $x_{\text{od}}[n]$ denote the even and odd parts of $x[n]$, respectively.

EXAMPLE 3.7   We compute the energy of the sequence $h_{LP}[n]$ of Eq. (3.12). From Eq. (3.18) we get

$$\sum_{n=-\infty}^{\infty} |h_{LP}[n]|^2 = \frac{1}{2\pi} \int_{-\pi}^{\pi} |H_{LP}(e^{j\omega})|^2 \, d\omega$$

$$= \frac{1}{2\pi} \int_{-\omega_c}^{\omega_c} d\omega = \frac{\omega_c}{\pi} < \infty.$$

Hence, $h_{LP}[n]$ is a finite-energy sequence.

Recall from Eq. (2.105) that the autocorrelation sequence $r_{gg}[\ell]$ of $g[n]$ can be expressed as

$$r_{gg}[\ell] = \sum_{n=-\infty}^{\infty} g[n]g[-(\ell-n)] = g[\ell] \circledast g[-\ell]. \tag{3.20}$$

Now from Table 3.3, the DTFT of $g[-\ell]$ is $G(e^{-j\omega})$. Therefore, using the convolution property of the DTFT given in Table 3.2, we observe that the DTFT of $g[\ell] \circledast g[-\ell]$ is given by $G(e^{j\omega})G(e^{-j\omega}) = |G(e^{j\omega})|^2$, where we have used the fact that for a real sequence $g[n]$, $G(e^{-j\omega}) = G^*(e^{j\omega})$. As a result, the energy density spectrum $S_{gg}(e^{j\omega})$ of a real sequence $g[n]$ can be computed by taking the DTFT of its autocorrelation sequence $r_{gg}[\ell]$, i.e.,

$$S_{gg}(e^{j\omega}) = \sum_{\ell=-\infty}^{\infty} r_{gg}[\ell]e^{j\omega\ell}. \tag{3.21}$$

Analogously, the DTFT $S_{gh}(e^{j\omega})$ of the cross-correlation sequence $r_{gh}[\ell]$ of two sequences $g[n]$ and $h[n]$ is called the *cross-energy density spectrum*:

$$S_{gh}(e^{j\omega}) = \sum_{\ell=-\infty}^{\infty} r_{gh}[\ell]e^{j\omega\ell}. \tag{3.22}$$

3.1.6 DTFT Computation Using MATLAB

The *Signal Processing Toolbox* in MATLAB includes a number of M-files to aid in the DTFT-based analysis of discrete-time signals. Specifically, the functions that can be used are freqz, abs, angle, and unwrap. In addition, the built-in MATLAB functions real and imag are also useful in some applications.

The function freqz can be used to compute the values of the DTFT of a sequence, described as a rational function in $e^{j\omega}$ in the form of Eq. (3.17) at a prescribed set of discrete frequency points $\omega = \omega_\ell$. For a reasonably accurate plot, a fairly large number of frequency points should be selected. There are various forms of this function:

```
    H   = freqz(num,den,w),        H   = freqz(num,den,f,FT),
[H,w]   = freqz(num,den,k),    [H,f]   = freqz(num,den,k,FT),
[H,w]   = freqz(num,den,k,'whole'),
[H,f]   = freqz(num,den,k,'whole',FT),  freqz(num,den).
```

The function freqz returns the frequency response values as a vector H of a DTFT defined in terms of the vectors num and den containing the coefficients $\{p_i\}$ and $\{d_i\}$, respectively, at a prescribed set of frequency points. In H = freqz(num,den,w), the prescribed set of frequencies between 0 and 2π are given by the vector w. In H = freqz(num,den,f,FT) the vector f is used to provide the prescribed frequency points whose values must be in the range 0 to FT/2 with FT being the sampling frequency. The total number of frequency points can be specified by k in the argument of freqz. In this case, the DTFT values H are computed at k equally spaced points between 0 and π, and returned as the output data vector w or computed at k equally spaced points between 0 and FT/2, and returned as the output data vector f. For faster computation, it is recommended that the number k be chosen as a power of 2, such as 256 or 512. By including 'whole' in the argument of freqz, the range of frequencies becomes 0 to 2π or 0 to FT, as the case may be. After the DTFT values have been determined, they can be plotted either showing their real and imaginary parts using the functions real and imag or in terms of their magnitude and phase components using the functions abs and angle. The function angle computes the phase angle in radians. If desired, the phase can be unwrapped using the function unwrap. freqz(num,den) with no output arguments computes and plots the magnitude and phase response values as a function of frequency in the current figure window.

We illustrate the DTFT computation using MATLAB in the following example.

EXAMPLE 3.8   Program 3_1 can be employed to determine the values of the DTFT of a real sequence described as a rational function in $e^{-j\omega}$.

```
% Program 3_1
% Discrete-Time Fourier Transform Computation
%
% Read in the desired length of DFT
k = input('Number of frequency points = ');
% Read in the numerator and denominator coefficients
num = input('Numerator coefficients = ');
```

```
den = input('Denominator coefficients = ');
% Compute the frequency response
w = 0:pi/k:pi;
h = freqz(num, den, w);
% Plot the frequency response
subplot(2,2,1)
plot(w/pi,real(h));grid
title('Real part')
xlabel('\omega/\pi'); ylabel('Amplitude')
subplot(2,2,2)
plot(w/pi,imag(h));grid
title('Imaginary part')
xlabel('\omega/\pi'); ylabel('Amplitude')
subplot(2,2,3)
plot(w/pi,abs(h));grid
title('Magnitude Spectrum')
xlabel('\omega/\pi'); ylabel('Magnitude')
subplot(2,2,4)
plot(w/pi,angle(h));grid
title('Phase Spectrum')
xlabel('\omega/\pi'); ylabel('Phase, radians')
```

The input data requested by the program are the number of frequency points k at which the DTFT is to be computed, the vectors num and den containing the coefficients of the numerator and the denominator of the DTFT, respectively, given in ascending powers of $e^{-j\omega}$. These vectors must be entered inside square brackets. The program computes the DTFT values at the prescribed frequency points and plots the real and imaginary parts, and the magnitude and phase spectrums. It should be noted that because of the symmetry relations of the DTFT of a real sequence as indicated in Table 3.3, the DTFT is evaluated only at k equally spaced values of ω between 0 and π.

As an example we consider the plotting of the following DTFT:

$$X(e^{j\omega}) = \frac{0.008 - 0.033e^{-j\omega} + 0.05e^{-j2\omega} - 0.033e^{-j3\omega} + 0.008e^{-j4\omega}}{1 + 2.37e^{-j\omega} + 2.7e^{-j2\omega} + 1.6e^{-j3\omega} + 0.41e^{-j4\omega}}.$$

The input data used for the DTFT given above are as follows:

```
k = 256
num = [0.008    -0.033    0.05    -0.033    0.008]
den = [1    2.37    2.7    1.6    0.41].
```

The program first computes the DTFT at the specified 256 discrete frequency points equally spaced between 0 and π. It then computes the real and imaginary parts, and the magnitude and phase of the DTFT at these frequencies, which are plotted as shown in Figure 3.5. As can be seen from this figure, the phase spectrum displays a discontinuity of 2π at around $\omega = 0.72$. This discontinuity can be removed using the function unwrap and the unwrapped phase plotted as shown in Figure 3.6.

3.1.7 Linear Convolution Using DTFT

An important property of the DTFT is given by the convolution theorem in Table 3.2, which states that the DTFT $Y(e^{j\omega})$ of a sequence $y[n]$ generated by the linear convolution of two sequences, $g[n]$ and $h[n]$, is simply given by the product of their respective DTFTs, $G(e^{j\omega})$ and $H(e^{j\omega})$. This implies that the linear

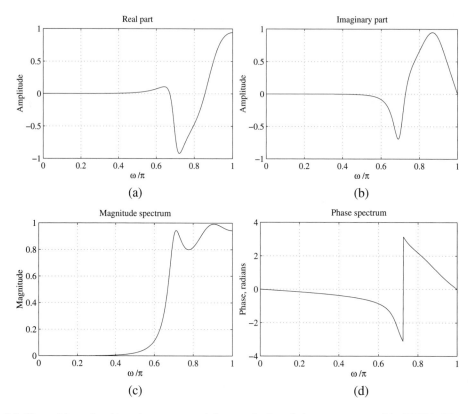

Figure 3.5: Plots of the real and imaginary parts, and the magnitude and phase spectrums of the DTFT of Example 3.8.

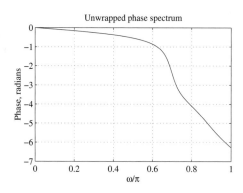

Figure 3.6: Unwrapped phase spectrum of the DTFT of Example 3.8.

convolution $y[n]$ of two sequences, $g[n]$ and $h[n]$, can be implemented by computing first their DTFTs, $G(e^{j\omega})$ and $H(e^{j\omega})$, forming the product $Y(e^{j\omega}) = G(e^{j\omega})H(e^{j\omega})$, and then computing the inverse DTFT of the product. In some applications, particularly in the case of infinite-length sequences, this DTFT-based approach may be more convenient to carry out than the direct convolution.

3.2 The Discrete Fourier Transform

In the case of a finite-length sequence $x[n]$, $0 \leq n \leq N - 1$, there is a simpler relation between the sequence and its discrete-time Fourier transform $X(e^{j\omega})$. In fact, for a length-N sequence, only N values of $X(e^{j\omega})$, called the *frequency samples,* at N distinct frequency points, $\omega = \omega_k$, $0 \leq k \leq N - 1$, are sufficient to determine $x[n]$, and hence, $X(e^{j\omega})$, uniquely. This leads to the concept of the discrete Fourier transform, a second transform-domain representation that is applicable only to a finite-length sequence. In this section we define the discrete Fourier transform, usually known as the DFT, and develop the inverse transformation, often abbreviated as IDFT. We then summarize its major properties and study especially two of its unique properties. Several important applications of the DFT, such as the numerical computation of the DTFT and implementation of linear convolution, are also discussed here.

3.2.1 Definition

The simplest relation between a finite-length sequence $x[n]$, defined for $0 \leq n \leq N - 1$, and its DTFT $X(e^{j\omega})$ is obtained by uniformly sampling $X(e^{j\omega})$ on the ω-axis between $0 \leq \omega \leq 2\pi$ at $\omega_k = 2\pi k/N$, $0 \leq k \leq N - 1$. From Eq. (3.1),

$$X[k] = X(e^{j\omega})\big|_{\omega=2\pi k/N} = \sum_{n=0}^{N-1} x[n]e^{-j2\pi kn/N}, \qquad 0 \leq k \leq N - 1. \tag{3.23}$$

Note that $X[k]$ is also a finite-length sequence in the frequency domain and is of length N. The sequence $X[k]$ is called the *discrete Fourier transform* (DFT) of the sequence $x[n]$.[3] Using the commonly used notation

$$W_N = e^{-j2\pi/N}, \tag{3.24}$$

we can rewrite Eq. (3.23) as

$$X[k] = \sum_{n=0}^{N-1} x[n]W_N^{kn}, \qquad 0 \leq k \leq N - 1. \tag{3.25}$$

The *inverse discrete Fourier transform* (IDFT) is given by

$$x[n] = \frac{1}{N} \sum_{k=0}^{N-1} X[k]W_N^{-kn}, \qquad 0 \leq n \leq N - 1. \tag{3.26}$$

To verify the above relation we multiply both sides of Eq. (3.26) by $W_N^{\ell n}$ and sum the result from $n = 0$ to $n = N - 1$, resulting in

$$\sum_{n=0}^{N-1} x[n]W_N^{\ell n} = \sum_{n=0}^{N-1} \left(\frac{1}{N} \sum_{k=0}^{N-1} X[k]W_N^{-kn} \right) W_N^{\ell n}$$

$$= \frac{1}{N} \sum_{n=0}^{N-1}\sum_{k=0}^{N-1} X[k]W_N^{-(k-\ell)n}. \tag{3.27}$$

---

[3]A generalization of the DFT concept is the *nonuniform discrete Fourier transform* (NDFT) obtained by sampling the DTFT at nonuniformly spaced frequency points [Bag98]. The NDFT is investigated in Problem 3.109.

An interchange of the order of summation on the right-hand side of Eq. (3.27) yields

$$\sum_{n=0}^{N-1} x[n]W_N^{\ell n} = \frac{1}{N} \sum_{k=0}^{N-1}\sum_{n=0}^{N-1} X[k]W_N^{-(k-\ell)n}$$

$$= \frac{1}{N} \sum_{k=0}^{N-1} X[k]\left[\sum_{n=0}^{N-1} W_N^{-(k-\ell)n}\right].$$

The right-hand side of the above equation reduces to $X[\ell]$ by virtue of the following identity (Problem 3.30):

$$\sum_{n=0}^{N-1} W_N^{-(k-\ell)n} = \begin{cases} N, & \text{for } k - \ell = rN, r \text{ an integer,} \\ 0, & \text{otherwise.} \end{cases} \tag{3.28}$$

thus verifying Eq. (3.26) is indeed the IDFT of $X[k]$.

The DFT computation is illustrated in the following two examples.

EXAMPLE 3.9   Consider the length-N sequence defined for $0 \leq n \leq N - 1$:

$$x[n] = \begin{cases} 1, & n = 0, \\ 0, & \text{otherwise.} \end{cases} \tag{3.29}$$

Its N-point DFT is obtained by applying Eq. (3.25) resulting in

$$X[k] = 1. \tag{3.30}$$

Now consider the length-N sequence defined for $n = 0, 1, \ldots, N - 1$

$$y[n] = \begin{cases} 1, & n = m, \\ 0, & \text{otherwise.} \end{cases} \tag{3.31}$$

Its DFT is given by

$$Y[k] = W_N^{km}. \tag{3.32}$$

EXAMPLE 3.10   Compute the N-point DFT of the length-N sequence

$$x[n] = \cos(2\pi rn/N), \qquad 0 \leq n \leq N - 1, 0 \leq r \leq N - 1. \tag{3.33}$$

Note that using a trigonometric identity and the notation of Eq. (3.24), we can rewrite $x[n]$,

$$x[n] = \tfrac{1}{2}\left(e^{j2\pi rn/N} + e^{-j2\pi rn/N}\right) = \tfrac{1}{2}\left(W_N^{-rn} + W_N^{rn}\right). \tag{3.34}$$

Substituting the above in Eq. (3.25), we arrive at its DFT,

$$X[k] = \frac{1}{2}\left[\sum_{n=0}^{N-1} W_N^{-(r-k)n} + \sum_{n=0}^{N-1} W_N^{(r+k)n}\right]. \tag{3.35}$$

Making use of the identity of Eq. (3.28) in Eq. (3.35) we obtain the DFT of the length-N sequence $x[n]$ of Eq. (3.33):

$$X[k] = \begin{cases} N/2, & \text{for } k = r, \\ N/2, & \text{for } k = N - r, \\ 0, & \text{otherwise.} \end{cases} \tag{3.36}$$

As can be seen from Eqs. (3.25) and (3.26), the computation of the DFT and the IDFT requires, respectively, approximately N^2 complex multiplications and $N(N-1)$ complex additions. However, elegant methods have been developed to reduce the computational complexity to about $N(\log_2 N)$ operations. These techniques are usually called fast Fourier transform (FFT) algorithms and are discussed in Section 8.3.2. As a result of the availability of these fast algorithms, the DFT and the IDFT, and their variations, are often used in digital signal processing applications for various purposes.

3.2.2 Matrix Relations

The DFT samples defined in Eq. (3.25) can be expressed in matrix form as

$$\mathbf{X} = \mathbf{D}_N \mathbf{x}, \tag{3.37}$$

where $\mathbf{X}$ is the vector composed of the N DFT samples,

$$\mathbf{X} = [\, X[0] \quad X[1] \quad \cdots \quad X[N-1]\,]^T, \tag{3.38}$$

$\mathbf{x}$ is the vector of N input samples,

$$\mathbf{x} = [\, x[0] \quad x[1] \quad \cdots \quad x[N-1]\,]^T, \tag{3.39}$$

and $\mathbf{D}_N$ is the $N \times N$ *DFT matrix* given by

$$\mathbf{D}_N = \begin{bmatrix} 1 & 1 & 1 & \cdots & 1 \\ 1 & W_N^1 & W_N^2 & \cdots & W_N^{N-1} \\ 1 & W_N^2 & W_N^4 & \cdots & W_N^{2(N-1)} \\ \vdots & \vdots & \vdots & \ddots & \vdots \\ 1 & W_N^{N-1} & W_N^{2(N-1)} & \cdots & W_N^{(N-1)\times(N-1)} \end{bmatrix}. \tag{3.40}$$

Likewise, the IDFT relations can be expressed in matrix form as

$$\begin{bmatrix} x[0] \\ x[1] \\ \vdots \\ x[N-1] \end{bmatrix} = \mathbf{D}_N^{-1} \begin{bmatrix} X[0] \\ X[1] \\ \vdots \\ X[N-1] \end{bmatrix}, \tag{3.41}$$

where $\mathbf{D}_N^{-1}$ is the $N \times N$ *IDFT matrix* given by

$$\mathbf{D}_N^{-1} = \frac{1}{N} \begin{bmatrix} 1 & 1 & 1 & \cdots & 1 \\ 1 & W_N^{-1} & W_N^{-2} & \cdots & W_N^{-(N-1)} \\ 1 & W_N^{-2} & W_N^{-4} & \cdots & W_N^{-2(N-1)} \\ \vdots & \vdots & \vdots & \ddots & \vdots \\ 1 & W_N^{-(N-1)} & W_N^{-2(N-1)} & \cdots & W_N^{-(N-1)\times(N-1)} \end{bmatrix}. \tag{3.42}$$

It follows from Eqs. (3.40) and (3.42) that

$$\mathbf{D}_N^{-1} = \frac{1}{N} \mathbf{D}_N^*. \tag{3.43}$$

3.2.3   DFT Computation Using MATLAB

There are four built-in functions in MATLAB for the computation of the DFT and the IDFT:

```
fft(x),        fft(x,N),        ifft(X),        ifft(X,N)
```

All of these functions make use of FFT algorithms which are computationally highly efficient compared to the direct computation of DFT and the inverse DFT.

The function fft(x) computes the R-point DFT of a vector x, with R being the length of x. For computing the DFT of a specific length N, the function fft(x,N) is used. Here, if $R > N$, it is truncated to the first N samples, whereas, if $R < N$, the vector x is zero-padded at the end to make it into a length-N sequence. Likewise, the function ifft(X) computes the R-point IDFT of a vector X, where R is the length of X, while ifft(X,N) computes the IDFT of X, with the size N of the IDFT being specified by the user. As before, if $R > N$, it is automatically truncated to the first N samples, whereas, if $R < N$, the DFT vector X is zero-padded at the end by the program to make it into a length-N DFT sequence.

In addition, the function dftmtx(N) in the *Signal Processing Toolbox* of MATLAB can be used to compute the $N \times N$ DFT matrix $\mathbf{D}_N$ defined in Eq. (3.40). To compute the inverse of the $N \times N$ DFT matrix, one can use the function conj(dftmt(N))/N.

We illustrate the application of the above M-files in the following three examples.

EXAMPLE 3.11   Using MATLAB we determine the M-point DFT $U[k]$ of the following N-point sequence

$$u[n] = \begin{cases} 1, & 0 \le n \le N-1, \\ 0, & \text{otherwise.} \end{cases} \tag{3.44}$$

To this end we can use Program 3_2 given below. During execution the program requests the input data N and M. To ensure correct DFT values M must be greater than or equal to N. After they are entered, it computes the M-point DFT and plots the original N-point sequence, and the magnitude and phase of the DFT sequence as indicated in Figure 3.7 for $N = 8$ and $M = 16$.

```
% Program 3_2
% Illustration of DFT Computation
%
% Read in the length N of sequence and the desired
% length M of the DFT
N = input('Type in the length of the sequence = ');
M = input('Type in the length of the DFT = ');
%   Generate the length-N time-domain sequence
u = [ones(1,N)];
% Compute its M-point DFT
U = fft(u,M);
% Plot the time-domain sequence and its DFT
t = 0:1:N-1;
stem(t,u)
title('Original time-domain sequence')
xlabel('Time index n'); ylabel('Amplitude')
pause
subplot(2,1,1)
k = 0:1:M-1;
stem(k,abs(U))
title('Magnitude of the DFT samples')
xlabel('Frequency index k'); ylabel('Magnitude')
```

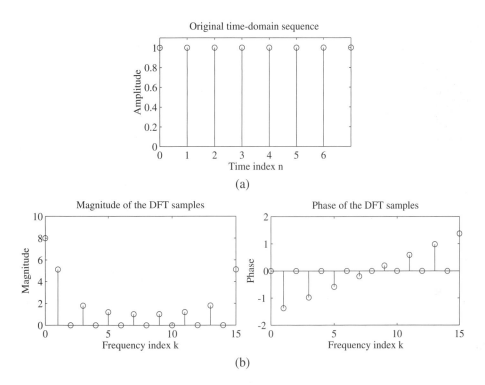

Figure 3.7: (a) Original length-N sequence of Eq. (3.44) and (b) its M-point DFT for $N = 8$ and $M = 16$.

```
subplot(2,1,2)
stem(k,angle(U))
title('Phase of the DFT samples')
xlabel('Frequency index k'); ylabel('Phase')
```

EXAMPLE 3.12   We next illustrate the computation of the IDFT using MATLAB. The K-point DFT sequence is given by

$$V[k] = \begin{cases} k/K, & 0 \le k \le K - 1, \\ 0, & \text{otherwise.} \end{cases} \tag{3.45}$$

We make use of Program 3_3 to determine its IDFT sequence $v[n]$.

```
% Program 3_3
% Illustration of IDFT Computation
%
% Read in the length K of the DFT and the desired
% length N of the IDFT
K = input('Type in the length of the DFT = ');
N = input('Type in the length of the IDFT = ');
%   Generate the length-K DFT sequence
k = 1:K;
U = (k-1)/K;
```

```
%  Compute its N-point IDFT
u = ifft(U,N);
% Plot the DFT and its IDFT
k=1:K;
stem(k-1,U)
xlabel('Frequency index k'); ylabel('Amplitude')
title('Original DFT samples')
pause
subplot(2,1,1)
n = 0:1:N-1;
stem(n,real(u))
title('Real part of the time-domain samples')
xlabel('Time index n'); ylabel('Amplitude')
subplot(2,1,2)
stem(n,imag(u))
title('Imaginary part of the time-domain samples')
xlabel('Time index n'); ylabel('Amplitude')
```

As the program is run, it calls for the input data consisting of the length of the DFT and the length of the IDFT. It then computes the IDFT of the ramp DFT sequence of Eq. (3.45) and plots the original DFT sequence and its IDFT as indicated in Figure 3.8. Note that even though the DFT sequence is real, its IDFT is a complex time-domain sequence as expected.

EXAMPLE 3.13   Let $r = 3$ and $N = 16$ for the finite-length sequence $x[n]$ of Eq. (3.33). From Eq. (3.36) its 16-point DFT is therefore given by

$$X[k] = \begin{cases} 8, & \text{for } k = 3, \\ 8, & \text{for } k = 13, \\ 0, & \text{otherwise.} \end{cases}$$

We determine the DTFT $X(e^{j\omega})$ of the length-16 $x[n]$ by computing a 512-point DFT using the MATLAB program indicated below.

```
% Program 3_4
% Numerical Computation of DTFT Using DFT
%
% Generate the length-16 sinusoidal sequence
k = 0:15;
x = cos(2*pi*k*3/16);
% Compute its 512-point DFT
X = fft(x);
XE = fft(x,512);
% Plot the frequency response
L = 0:511;
plot(L/512,abs(XE))
hold
plot(k/16,abs(X),'o')
xlabel('Normalized angular frequency')
ylabel('Magnitude')
```

Figure 3.9 shows the plot of DTFT $X(e^{j\omega})$ along with the DFT samples $X[k]$. As indicated in this figure the DFT values $X[k]$, indicated by circles, are precisely the frequency samples of the DTFT $X(e^{j\omega})$ at $\omega = \pi k/8$, $0 \le k \le 15$.

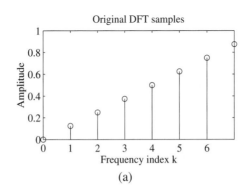

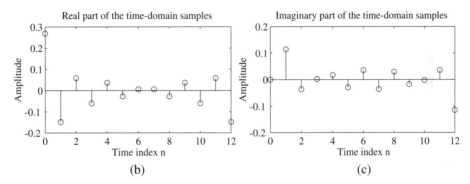

Figure 3.8: (a) Original DFT sequence of length $K = 8$, and (b) its 13-point IDFT.

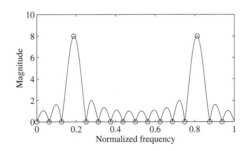

Figure 3.9: The magnitudes of the DTFT $X(e^{j\omega})$ and the DFT $X[k]$ of the sequence $x[n]$ of Eq. (3.33) with $r = 3$ and $N = 16$. The DTFT is plotted as a solid line and the DFT samples are shown by circles.

3.3 Relation between the DTFT and the DFT, and Their Inverses

We now examine the explicit relation between the DTFT and the N-point DFT of a length-N sequence, and the relation between the DTFT of a length-M sequence and the N-point DFT obtained by sampling the DTFT.

3.3.1 DTFT from DFT by Interpolation

As indicated by Eq. (3.23), the N-point DFT $X[k]$ of a length-N sequence $x[n]$ is simply the frequency samples of its DTFT $X(e^{j\omega})$ evaluated at N uniformly spaced frequency points, $\omega = \omega_k = 2\pi k/N$ radians, $0 \leq k \leq N-1$. Given the N-point DFT $X[k]$ of a length-N sequence, it is also possible to determine its DTFT $X(e^{j\omega})$ uniquely. To this end, we first determine $x[n]$ using the IDFT relation of Eq. (3.26) and then compute its DTFT using Eq. (3.1), resulting in

$$X(e^{j\omega}) = \sum_{n=0}^{N-1} x[n]e^{-j\omega n} = \frac{1}{N}\sum_{n=0}^{N-1}\left[\sum_{k=0}^{N-1} X[k]W_N^{-kn}\right]e^{-j\omega n}$$

$$= \frac{1}{N}\sum_{k=0}^{N-1} X[k]\sum_{n=0}^{N-1} e^{j2\pi kn/N}e^{-j\omega n}, \tag{3.46}$$

where we have used Eq. (3.24). Now the right-hand summation in the above expression can be rewritten as

$$\sum_{n=0}^{N-1} e^{-j(\omega - 2\pi k/N)n} = \frac{1 - e^{-j(\omega N - 2\pi k)}}{1 - e^{-j[\omega - (2\pi k/N)]}}$$

$$= \frac{e^{-j[(\omega N - 2\pi k)/2]}}{e^{-j[(\omega N - 2\pi k)/2N]}} \cdot \frac{\sin\left(\frac{\omega N - 2\pi k}{2}\right)}{\sin\left(\frac{\omega N - 2\pi k}{2N}\right)}$$

$$= \frac{\sin\left(\frac{\omega N - 2\pi k}{2}\right)}{\sin\left(\frac{\omega N - 2\pi k}{2N}\right)} \cdot e^{-j[\omega - (2\pi k/N)][(N-1)/2]} \tag{3.47}$$

Substituting Eq. (3.47) in Eq. (3.46) we obtain the desired relation expressing $X(e^{j\omega})$ in terms of $X[k]$:

$$X(e^{j\omega}) = \frac{1}{N}\sum_{k=0}^{N-1} X[k]\frac{\sin\left(\frac{\omega N - 2\pi k}{2}\right)}{\sin\left(\frac{\omega N - 2\pi k}{2N}\right)} \cdot e^{-j[\omega - (2\pi k/N)][(N-1)/2]} \tag{3.48}$$

3.3.2 Sampling the DTFT

Consider a sequence $\{x[n]\}$ with a discrete-time Fourier transform (DTFT) $X(e^{j\omega})$. We sample $X(e^{j\omega})$ at N equally spaced points $\omega_k = 2\pi k/N$, $0 \leq k \leq N-1$, developing the N *frequency samples* $\{X(e^{j\omega_k})\}$. These N frequency samples can be considered as an N-point DFT $Y[k]$ whose N-point inverse DFT is a length-N sequence $\{y[n]\}$, $0 \leq n \leq N-1$.

Now, $X(e^{j\omega})$ is a periodic function of ω with a Fourier series representation given by Eq. (3.1). Its Fourier coefficients $x[n]$ are given by Eq. (3.7). It is instructive to develop the relation between $x[n]$ and $y[n]$.

From Eq. (3.1),

$$Y[k] = X(e^{j\omega_k}) = X(e^{j(2\pi k/N)}) = \sum_{\ell=-\infty}^{\infty} x[\ell]W_N^{k\ell}, \tag{3.49}$$

where $W_N = e^{-j(2\pi/N)}$. An inverse DFT of $Y[k]$ yields

$$y[n] = \frac{1}{N}\sum_{k=0}^{N-1} Y[k]W_N^{-kn}. \tag{3.50}$$

Substituting Eq. (3.49) in Eq. (3.50), we get

$$y[n] = \frac{1}{N} \sum_{k=0}^{N-1} \sum_{\ell=-\infty}^{\infty} x[\ell] W_N^{k\ell} W_N^{-kn}$$

$$= \sum_{\ell=-\infty}^{\infty} x[\ell] \left[\frac{1}{N} \sum_{k=0}^{N-1} W_N^{-k(n-\ell)} \right]. \quad (3.51)$$

Recall from Eq. (3.28) that

$$\frac{1}{N} \sum_{k=0}^{N-1} W_N^{-k(n-r)} = \begin{cases} 1, & \text{for } r = n + mN, \\ 0, & \text{otherwise.} \end{cases} \quad (3.52)$$

Making use of the above identity in Eq. (3.51), we finally arrive at the desired relation

$$y[n] = \sum_{m=-\infty}^{\infty} x[n + mN], \quad 0 \le n \le N - 1. \quad (3.53)$$

The above relation indicates that $y[n]$ is obtained from $x[n]$ by adding an infinite number of shifted replicas of $x[n]$ to $x[n]$, with each replica shifted by an integer multiple of N sampling instants, and observing the sum only for the interval $0 \le n \le N - 1$. To apply Eq. (3.53) to finite-length sequences we assume the samples outside the specified range are zeros. Thus, if $x[n]$ is a finite-length sequence of length M less than or equal to N, then $y[n] = x[n]$ for $0 \le n \le N - 1$, otherwise there is a time-domain aliasing of samples of $x[n]$ in generating $y[n]$, and $x[n]$ cannot be recovered from $y[n]$, as illustrated in the following example.

EXAMPLE 3.14   Let $x[n]$ be a length-6 sequence given by

$$\{x[n]\} = \{0 \quad 1 \quad 2 \quad 3 \quad 4 \quad 5\}.$$
$$\uparrow$$

By sampling the discrete-time Fourier transform $X(e^{j\omega})$ of $\{x[n]\}$ at 4 equally spaced points given by $\omega_k = 2\pi k/4$, $0 \le k \le 3$, and then applying a 4-point inverse-DFT to these samples, we arrive at a sequence $y[n]$ which from Eq. (3.53) is given by

$$y[n] = x[n] + x[n + 4] + x[n - 4], \quad 0 \le n \le 3,$$

i.e.,

$$\{y[n]\} = \{4 \quad 6 \quad 2 \quad 3\}.$$
$$\uparrow$$

3.3.3 Numerical Computation of the DTFT Using the DFT

The DFT provides a practical approach to the numerical computation of the DTFT of a finite-length sequence, particularly if fast algorithms are available for the computation of the DFT. Let $X(e^{j\omega})$ be the DTFT of a length-N sequence $x[n]$. We wish to evaluate $X(e^{j\omega})$ at a dense grid of frequencies $\omega_k = 2\pi k/M, 0 \le k \le M - 1$, where $M \gg N$:

$$X(e^{j\omega_k}) = \sum_{n=0}^{N-1} x[n]e^{-j\omega_k n} = \sum_{n=0}^{N-1} x[n]e^{-j2\pi kn/M}. \tag{3.54}$$

Define a new sequence $x_e[n]$ obtained from $x[n]$ by augmenting with $M - N$ zero-valued samples:

$$x_e[n] = \begin{cases} x[n], & 0 \le n \le N-1, \\ 0, & N \le n \le M-1. \end{cases} \tag{3.55}$$

Making use of $x_e[n]$ in Eq. (3.54) we arrive at

$$X(e^{j\omega_k}) = \sum_{n=0}^{M-1} x_e[n]e^{-j2\pi kn/M}, \tag{3.56}$$

which is seen to be an M-point DFT $X_e[k]$ of the length-M sequence $x_e[n]$. The DFT $X_e[k]$ can be computed very efficiently using the FFT algorithm if M is an integer power of 2.

The MATLAB function freqz, described in Section 3.1.6, employs the above approach to evaluate the frequency response of a rational DTFT expressed as a rational function in $e^{-j\omega}$ at a prescribed set of discrete frequencies. It computes the DFTs of the numerator and the denominator separately by considering each as finite-length sequences, and then expresses the ratio of the DFT samples at each frequency point to evaluate the DTFT.

3.4 Discrete Fourier Transform Properties

Like the DTFT, the DFT also satisfies a number of properties that are useful in signal processing applications. Some of these properties are essentially identical to those of the DTFT, while some others are somewhat different. A summary of the DFT properties are included in Tables 3.5, 3.6, and 3.7. Their proofs are again quite straightforward and have been left as exercises. Most of these properties can also be verified using MATLAB. We discuss next those properties that are different from their counterparts for the DTFT.

3.4.1 Circular Shift of a Sequence

This property is analogous to the time-shifting property of the DTFT as given in Table 3.2, but with a subtle difference. Let us consider length-N sequences defined for $0 \le n \le N-1$. Such sequences have sample values equal to zero for $n < 0$ and $n \ge N$. If $x[n]$ is such a sequence, then, for any arbitrary integer n_o, the shifted sequence $x_1[n] = x[n - n_o]$ is no longer defined for the range $0 \le n \le N-1$. We therefore need to define another type of a shift that will always keep the shifted sequence in the range $0 \le n \le N-1$. This is achieved by defining a new type of shift, called the *circular shift*, using a modulo operation according to[4]

$$x_c[n] = x\left[\langle n - n_o\rangle_N\right]. \tag{3.57}$$

For $n_o > 0$ (right circular shift), the above equation implies

$$x_c[n] = \begin{cases} x[n - n_o], & \text{for } n_o \le n \le N-1, \\ x[N - n_o + n], & \text{for } 0 \le n < n_o. \end{cases} \tag{3.58}$$

The concept of a circular shift of a finite-length sequence is illustrated in Figure 3.10. Figure 3.10(a) shows a length-6 sequence $x[n]$. Figure 3.10(b) shows its circularly shifted version shifted to the right by

---

[4] $\langle m \rangle_N = m$ modulo N.

Table 3.5: General properties of the DFT.

Type of Property	Length-N Sequence	N-point DFT
	$g[n]$	$G[k]$
	$h[n]$	$H[k]$
Linearity	$\alpha g[n] + \beta h[n]$	$\alpha G[k] + \beta H[k]$
Circular time-shifting	$g[\langle n - n_o \rangle_N]$	$W_N^{kn_o} G[k]$
Circular frequency-shifting	$W_N^{-k_o n} g[n]$	$G[\langle k - k_o \rangle_N]$
Duality	$G[n]$	$N g[\langle -k \rangle_N]$
N-point circular convolution	$\displaystyle\sum_{m=0}^{N-1} g[m] h[\langle n - m \rangle_N]$	$G[k]H[k]$
Modulation	$g[n]h[n]$	$\displaystyle\frac{1}{N} \sum_{m=0}^{N-1} G[m] H[\langle k - m \rangle_N]$
Parseval's relation	$\displaystyle\sum_{n=0}^{N-1} |x[n]|^2 = \frac{1}{N} \sum_{k=0}^{N-1} |X[k]|^2$	

Table 3.6: Symmetry properties of the DFT of a complex sequence.

Length-N Sequence	N-point DFT
$x[n]$	$X[k]$
$x^*[n]$	$X^*[\langle -k \rangle_N]$
$x^*[\langle -n \rangle_N]$	$X^*[k]$
$\mathrm{Re}\{x[n]\}$	$X_{\mathrm{pcs}}[k] = \frac{1}{2}\{X[\langle k \rangle_N] + X^*[\langle -k \rangle_N]\}$
$j\,\mathrm{Im}\{x[n]\}$	$X_{\mathrm{pca}}[k] = \frac{1}{2}\{X[\langle k \rangle_N] - X^*[\langle -k \rangle_N]\}$
$x_{\mathrm{pcs}}[n]$	$\mathrm{Re}\{X[k]\}$
$x_{\mathrm{pca}}[n]$	$j\,\mathrm{Im}\{X[k]\}$

Note: $x_{\mathrm{pcs}}[n]$ and $x_{\mathrm{pca}}[n]$ are the periodic conjugate-symmetric and periodic conjugate-antisymmetric parts of $x[n]$, respectively. Likewise, $X_{\mathrm{pcs}}[k]$ and $X_{\mathrm{pca}}[k]$ are the periodic conjugate-symmetric and periodic conjugate-antisymmetric parts of $X[k]$, respectively.

Table 3.7: Symmetry properties of the DFT of a real sequence.

Length-N Sequence	N-point DFT
$x[n]$	$X[k] = \text{Re}\{X[k]\} + j\,\text{Im}\{X[k]\}$
$x_{\text{pe}}[n]$	$\text{Re}\{X[k]\}$
$x_{\text{po}}[n]$	$j\,\text{Im}\{X[k]\}$
Symmetry relations	$X[k] = X^*[\langle -k\rangle_N]$
	$\text{Re}\,X[k] = \text{Re}\,X[\langle -k\rangle_N]$
	$\text{Im}\,X[k] = -\,\text{Im}\,X[\langle -k\rangle_N]$
	$\lvert X[k]\rvert = \lvert X[\langle -k\rangle_N]\rvert$
	$\arg X[k] = -\arg X[\langle -k\rangle_N]$

Note: $x_{\text{pe}}[n]$ and $x_{\text{po}}[n]$ are the periodic even and periodic odd parts of $x[n]$, respectively.

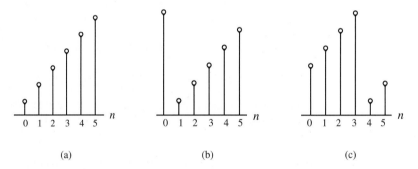

(a)   (b)   (c)

Figure 3.10: Illustration of a circular shift of a finite-length sequence. (a) $x[n]$, (b) $x[\langle n-1\rangle_6] = x[\langle n+5\rangle_6]$, and (c) $x[\langle n-4\rangle_6] = x[\langle n+2\rangle_6]$.

1 sample period or, equivalently, shifted to the left by 5 sample periods. Likewise, Figure 3.10(c) depicts its circularly shifted version shifted to the right by 4 sample periods or, equivalently, shifted to the left by 2 sample periods.

As can be seen from Figure 3.10(b) and (c), a right circular shift by n_o is equivalent to a left circular shift by $N - n_o$ sample periods. It should be noted that a circular shift by an integer number n_o greater than N is equivalent to a circular shift by $\langle n_o\rangle_N$.

If we view the length-N sequence displayed on the circumference of a cylinder at N equally spaced points, then the circular shift operation can be considered as a clockwise or anticlockwise rotation of the sequence by n_o sample spacings on the cylinder.

Figure 3.11: Two length-4 sequences.

3.4.2 Circular Convolution

This property is analogous to the linear convolution of Eq. (2.64), but with a subtle difference. Consider two length-N sequences, $g[n]$ and $h[n]$, respectively. Their linear convolution results in a length-$(2N-1)$ sequence $y_L[n]$ given by

$$y_L[n] = \sum_{m=0}^{N-1} g[m]h[n-m], \qquad 0 \leq n \leq 2N-2, \tag{3.59}$$

where we have assumed that both N-length sequences have been zero-padded to extend their lengths to $2N-1$.[5] The longer length of $y_L[n]$ results from the time-reversal of the sequence $h[n]$ and its linear shifting to the right. The first nonzero value of $y_L[n]$ is $y_L[0] = g[0]h[0]$, and the last nonzero value of $y_L[n]$ is $y_L[2N-2] = g[N-1]h[N-1]$.

To develop a convolution-like operation resulting in a length-N sequence $y_C[n]$, we need to define a circular time-reversal and then apply a circular time-shift. The resulting operation, called a *circular convolution*, is defined below:[6]

$$y_C[n] = \sum_{m=0}^{N-1} g[m]h\left[\langle n-m \rangle_N\right]. \tag{3.60}$$

Since the above operation involves two length-N sequences, it is often referred to as an *N-point circular convolution*, denoted as

$$y_C[n] = g[n] \circledN h[n]. \tag{3.61}$$

Like the linear convolution, the circular convolution is commutative (Problem 3.65), i.e.,

$$g[n] \circledN h[n] = h[n] \circledN g[n]. \tag{3.62}$$

We illustrate the concept of circular convolution through several examples.

EXAMPLE 3.15   Determine the 4-point circular convolution of the two length-4 sequences $g[n]$ and $h[n]$ given by

$$g[n] = \{1 \quad 2 \quad 0 \quad 1\}, \qquad h[n] = \{2 \quad 2 \quad 1 \quad 1\}, \tag{3.63}$$

and sketched in Figure 3.11. The result is a length-4 sequence $y_C[n]$ given by

$$y_C[n] = g[n] \circled{4} h[n] = \sum_{m=0}^{3} g[m]h[\langle n-m \rangle_4], \qquad 0 \leq n \leq 3. \tag{3.64}$$

From Eq. (3.64), $y_C[0]$ is given by

---

[5]As indicated in Section 2.5.1, the sum of the indices of each sample product inside the summation is equal to the index of the sample being generated by the linear convolution operation.

[6]Note that here the sum of the indices of each sample product inside the summation modulo N is equal to the index of the sample being generated by the circular convolution operation.

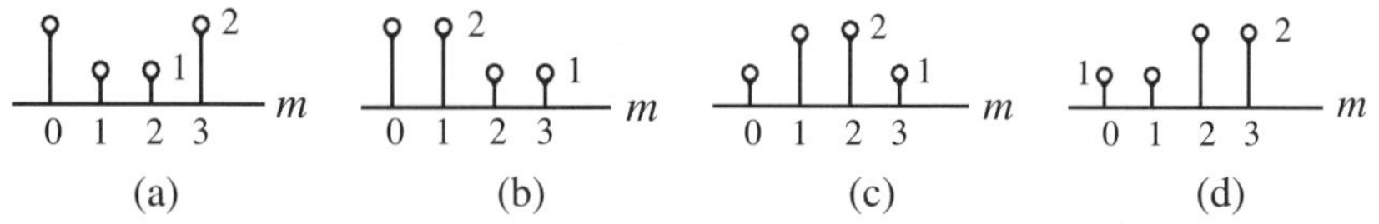

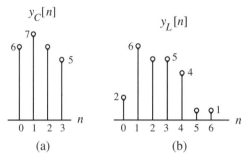

Figure 3.12: The circularly time reversed sequence and its circularly shifted versions: (a) $h[\langle -m\rangle_4]$, (b) $h[\langle 1 - m\rangle_4]$, (c) $h[\langle 2 - m\rangle_4]$, and (d) $h[\langle 3 - m\rangle_4]$.

Figure 3.13: Results of convolution of the two sequences of Figure 3.11. (a) Circular convolution, and (b) linear convolution.

$$y_C[0] = \sum_{m=0}^{3} g[m]h[\langle -m\rangle_4]. \qquad (3.65)$$

The circularly time reversed sequence $h[\langle -m\rangle_4]$ is shown in Figure 3.12(a). By forming the products of $g[m]$ with that of $h[\langle -m\rangle_4]$ for each value of m in the range $0 \le m < 3$ and summing the products, we arrive at

$$y_C[0] = g[0]h[0] + g[1]h[3] + g[2]h[2] + g[3]h[1]$$
$$= (1 \times 2) + (2 \times 1) + (0 \times 1) + (1 \times 2) = 6. \qquad (3.66)$$

Next, from Eq. (3.64) we compute $y_C[1]$ as

$$y_C[1] = \sum_{m=0}^{3} g[m]h[\langle 1 - m\rangle_4]. \qquad (3.67)$$

The sequence $h[\langle 1 - m\rangle_4]$ obtained by circularly time shifting $h[\langle -m\rangle_4]$ to the right by one sample period is shown in Figure 3.12(b). Summing the products $g[m]h[\langle 1 - m\rangle_4]$ for each value of m, we arrive at

$$y_C[1] = g[0]h[1] + g[1]h[0] + g[2]h[3] + g[3]h[2]$$
$$= (1 \times 2) + (2 \times 2) + (0 \times 1) + (1 \times 1) = 7. \qquad (3.68)$$

Continuing this process, we determine the remaining two samples of $y_C[n]$ as

$$y_C[2] = g[0]h[2] + g[1]h[1] + g[2]h[0] + g[3]h[3]$$
$$= (1 \times 1) + (2 \times 2) + (0 \times 2) + (1 \times 1) = 6. \qquad (3.69)$$
$$y_C[3] = g[0]h[3] + g[1]h[2] + g[2]h[1] + g[3]h[0]$$
$$= (1 \times 1) + (2 \times 1) + (0 \times 2) + (1 \times 2) = 5. \qquad (3.70)$$

The length-4 sequence $y_C[n]$ obtained by a 4-point circular convolution of the two length-4 sequences $g[n]$ and $h[n]$ is sketched in Figure 3.13(a).

The circular convolution of the two length-N sequences can also be computed by forming the product of their N-point DFTs and then applying an N-point IDFT, as indicated in Table 3.5. The process is illustrated in the following example.

EXAMPLE 3.16 We compute the circular convolution of the two length-4 sequences of Eq. (3.63) via a DFT-based approach.

Now the 4-point DFT $G[k]$ of the length-4 sequence $g[n]$ of Eq. (3.63) is given by

$$G[k] = g[0] + g[1]e^{-j2\pi k/4} + g[2]e^{-j4\pi k/4} + g[3]e^{-j6\pi k/4}$$
$$= 1 + 2e^{-j\pi k/2} + e^{-j3\pi k/2}, \qquad k = 0, 1, 2, 3. \tag{3.71}$$

Therefore,

$$G[0] = 1 + 2 + 1 = 4,$$
$$G[1] = 1 - j2 + j = 1 - j,$$
$$G[2] = 1 - 2 - 1 = -2,$$
$$G[3] = 1 + j2 - j = 1 + j. \tag{3.72}$$

The above DFT samples can also be computed using the matrix relation of Eq. (3.37):

$$\begin{bmatrix} G[0] \\ G[1] \\ G[2] \\ G[3] \end{bmatrix} = \mathbf{D}_4 \begin{bmatrix} g[0] \\ g[1] \\ g[2] \\ g[3] \end{bmatrix} = \begin{bmatrix} 1 & 1 & 1 & 1 \\ 1 & -j & -1 & j \\ 1 & -1 & 1 & -1 \\ 1 & j & -1 & -j \end{bmatrix} \begin{bmatrix} 1 \\ 2 \\ 0 \\ 1 \end{bmatrix} = \begin{bmatrix} 4 \\ 1-j \\ -2 \\ 1+j \end{bmatrix}, \tag{3.73}$$

where $\mathbf{D}_4$ is the 4×4 DFT matrix. Likewise the 4-point DFT of the sequence $h[n]$ of Eq. (3.63) is given by

$$\begin{bmatrix} H[0] \\ H[1] \\ H[2] \\ H[3] \end{bmatrix} = \mathbf{D}_4 \begin{bmatrix} h[0] \\ h[1] \\ h[2] \\ h[3] \end{bmatrix} = \begin{bmatrix} 1 & 1 & 1 & 1 \\ 1 & -j & -1 & j \\ 1 & -1 & 1 & -1 \\ 1 & j & -1 & -j \end{bmatrix} \begin{bmatrix} 2 \\ 2 \\ 1 \\ 1 \end{bmatrix} = \begin{bmatrix} 6 \\ 1-j \\ 0 \\ 1+j \end{bmatrix}. \tag{3.74}$$

Applying the 4-point IDFT to the product $Y_C[k] = G[k]H[k]$,

$$\begin{bmatrix} Y_C[0] \\ Y_C[1] \\ Y_C[2] \\ Y_C[3] \end{bmatrix} = \begin{bmatrix} G[0]H[0] \\ G[1]H[1] \\ G[2]H[2] \\ G[3]H[3] \end{bmatrix} = \begin{bmatrix} 24 \\ -j2 \\ 0 \\ j2 \end{bmatrix}, \tag{3.75}$$

we arrive at the desired circular convolution result:

$$\begin{bmatrix} y_C[0] \\ y_C[1] \\ y_C[2] \\ y_C[3] \end{bmatrix} = \frac{1}{4}\mathbf{D}_4^* \begin{bmatrix} 24 \\ -j2 \\ 0 \\ j2 \end{bmatrix} = \frac{1}{4} \begin{bmatrix} 1 & 1 & 1 & 1 \\ 1 & j & -1 & -j \\ 1 & -1 & 1 & -1 \\ 1 & -j & -1 & j \end{bmatrix} \begin{bmatrix} 24 \\ -j2 \\ 0 \\ j2 \end{bmatrix} = \begin{bmatrix} 6 \\ 7 \\ 6 \\ 5 \end{bmatrix}, \tag{3.76}$$

which is identical to the results given in Eqs. (3.66), (3.68) to (3.70).

EXAMPLE 3.17 Now let us extend the two length-4 sequences of Eq. (3.63) to length 7 by appending each with three zero-valued samples, i.e., by forming

$$g_e[n] = \begin{cases} g[n], & 0 \le n \le 3, \\ 0, & 4 \le n \le 6, \end{cases} \tag{3.77}$$

$$h_e[n] = \begin{cases} h[n], & 0 \le n \le 3, \\ 0, & 4 \le n \le 6. \end{cases} \tag{3.78}$$

We next determine the 7-point circular convolution of $g_e[n]$ and $h_e[n]$:

$$y[n] = \sum_{m=0}^{6} g_e[m]h_e[\langle n - m \rangle_7]. \tag{3.79}$$

From the above,

$$\begin{aligned} y[0] = g_e[0]h_e[0] + g_e[1]h_e[6] + g_e[2]h_e[5] + g_e[3]h_e[4] \\ + g_e[4]h_e[3] + g_e[5]h_e[2] + g_e[6]h_e[1]. \end{aligned} \tag{3.80}$$

Substituting the values from Eqs. (3.77) and (3.78) in Eq. (3.80), we arrive at

$$y[0] = g[0]h[0] = (1 \times 2) = 2.$$

Next we compute $y[1]$. From Eq. (3.79),

$$\begin{aligned} y[1] = g_e[0]h_e[1] + g_e[1]h_e[0] + g_e[2]h_e[6] + g_e[3]h_e[5] \\ + g_e[4]h_e[4] + g_e[5]h_e[3] + g_e[6]h_e[2], \end{aligned} \tag{3.81}$$

which leads to

$$y[1] = g[0]h[1] + g[1]h[0] = (1 \times 2) + (2 \times 2) = 6$$

using Eqs. (3.77) and (3.78). Continuing the above procedure, we arrive at the remaining samples of $y[n]$:

$$y[2] = g[0]h[2] + g[1]h[1] + g[2]h[0] = (1 \times 1) + (2 \times 2) + (0 \times 2) = 5,$$
$$y[3] = g[0]h[3] + g[1]h[2] + g[2]h[1] + g[3]h[0]$$
$$= (1 \times 1) + (2 \times 1) + (0 \times 2) + (1 \times 2) = 5,$$
$$y[4] = g[1]h[3] + g[2]h[2] + g[3]h[0] = (2 \times 1) + (0 \times 1) + (1 \times 2) = 4,$$
$$y[5] = g[2]h[3] + g[3]h[2] = (0 \times 1) + (1 \times 1) = 1,$$
$$y[6] = g[3]h[3] = (1 \times 1) = 1.$$

It is evident from the above that $y[n]$ is precisely the result $y_L[n]$, shown in Figure 3.13(b), obtained by a linear convolution of $g[n]$ and $h[n]$.

The N-point circular convolution operation of Eq. (3.61) can be written in matrix form as

$$\begin{bmatrix} y_C[0] \\ y_C[1] \\ y_C[2] \\ \vdots \\ y_C[N-1] \end{bmatrix} = \begin{bmatrix} h[0] & h[N-1] & h[N-2] & \cdots & h[1] \\ h[1] & h[0] & h[N-1] & \cdots & h[2] \\ h[2] & h[1] & h[0] & \cdots & h[3] \\ \vdots & \vdots & \vdots & \ddots & \vdots \\ h[N-1] & h[N-2] & h[N-3] & \cdots & h[0] \end{bmatrix} \begin{bmatrix} g[0] \\ g[1] \\ g[2] \\ \vdots \\ g[N-1] \end{bmatrix}. \tag{3.82}$$

The elements in each diagonal of the $N \times N$ matrix of Eq. (3.82) are equal. Such a matrix is called a *circulant matrix*.

3.5   Computation of the DFT of Real Sequences

In most practical applications, sequences of interest are real. In such cases, the symmetry properties of the DFT given in Table 3.7 can be exploited to make the DFT computations more efficient.

3.5.1 *N*-Point DFTs of Two Real Sequences Using a Single *N*-Point DFT

Let $g[n]$ and $h[n]$ be two real sequences of length N each with $G[k]$ and $H[k]$ denoting their respective N-point DFTs. These two N-point DFTs can be computed efficiently using a single N-point DFT $X[k]$ of a complex length-N sequence $x[n]$ defined by

$$x[n] = g[n] + jh[n]. \tag{3.83}$$

From the above, $g[n] = \text{Re}\{x[n]\}$ and $h[n] = \text{Im}\{x[n]\}$.

From Table 3.6 we arrive at

$$G[k] = \tfrac{1}{2}\left\{X[k] + X^*[\langle -k \rangle_N]\right\}, \tag{3.84}$$

$$H[k] = \tfrac{1}{2j}\left\{X[k] - X^*[\langle -k \rangle_N]\right\}. \tag{3.85}$$

Note that $X^*[\langle -k \rangle_N] = X^*[\langle N - k \rangle_N]$.

EXAMPLE 3.18   In this example we illustrate the computation of the 4-point DFTs of the two real sequences of Example 3.15 using a single 4-point DFT.

From Eq. (3.63) the complex sequence $x[n] = g[n] + jh[n]$ is given by

$$x[n] = \{1 + j2 \quad 2 + j2 \quad j \quad 1 + j\}.$$
$$\uparrow$$

Its DFT is then

$$\begin{bmatrix} X[0] \\ X[1] \\ X[2] \\ X[3] \end{bmatrix} = \begin{bmatrix} 1 & 1 & 1 & 1 \\ 1 & -j & -1 & j \\ 1 & -1 & 1 & -1 \\ 1 & j & -1 & -j \end{bmatrix} \begin{bmatrix} 1 + j2 \\ 2 + j2 \\ j \\ 1 + j \end{bmatrix} = \begin{bmatrix} 4 + j6 \\ 2 \\ -2 \\ j2 \end{bmatrix}. \tag{3.86}$$

From the above,

$$X^*[k] = \{4 - j6 \quad 2 \quad -2 \quad -j2\}.$$

Therefore

$$X^*[\langle 4 - k \rangle_4] = \{4 - j6 \quad -j2 \quad -2 \quad 2\}. \tag{3.87}$$

Substituting Eqs. (3.86) and (3.87) in Eqs. (3.84) and (3.85), we get

$$G[k] = \{4 \quad 1 - j \quad -2 \quad 1 + j\},$$
$$H[k] = \{6 \quad 1 - j \quad 0 \quad 1 + j\}, \tag{3.88}$$

verifying the results obtained in Example 3.16.

3.5.2 2*N*-Point DFT of a Real Sequence Using a Single *N*-Point DFT

Let $v[n]$ be a real sequence of length $2N$ with $V[k]$ denoting its $2N$-point DFT. Define two real sequences $g[n]$ and $h[n]$ of length N each as

$$g[n] = v[2n], \qquad h[n] = v[2n + 1], \qquad 0 \le n < N, \tag{3.89}$$

with $G[k]$ and $H[k]$ denoting their N-point DFTs. Now define a complex length-N sequence $x[n]$ according to Eq. (3.83). The DFTs $G[k]$ and $H[k]$ can be computed from the N-point DFT $X[k]$ of the sequence $x[n]$ by means of Eqs. (3.84) and (3.85).

Now,

$$V[k] = \sum_{n=0}^{2N-1} v[n]W_{2N}^{nk} = \sum_{n=0}^{N-1} v[2n]W_{2N}^{2nk} + \sum_{n=0}^{N-1} v[2n+1]W_{2N}^{(2n+1)k}$$

$$= \sum_{n=0}^{N-1} g[n]W_N^{nk} + \sum_{n=0}^{N-1} h[n]W_N^{nk} W_{2N}^{k}$$

$$= \sum_{n=0}^{N-1} g[n]W_N^{nk} + W_{2N}^{k} \sum_{n=0}^{N-1} h[n]W_N^{nk}.$$

Note that the first sum on the last expression is simply an N-point DFT $G[k]$ of the length-N sequence $g[n]$, whereas the second sum is an N-point DFT $H[k]$ of the length-N sequence $h[n]$. Therefore, we can express the $2N$-point DFT $V[k]$ as

$$V[k] = G[\langle k \rangle_N] + W_{2N}^{k} H[\langle k \rangle_N], \qquad 0 \le k \le 2N - 1, \tag{3.90}$$

where we have used the identity $W_{2N}^{2k} = W_N^k$ and used the modulo sign for the argument of the two N-point DFTs on the right-hand side since k here is in the range $0 \le k \le 2N - 1$.

EXAMPLE 3.19   Let us determine the 8-point DFT $V[k]$ of the length-8 real sequence $v[n]$ given below:

$$v[n] = \{1 \quad 2 \quad 2 \quad 2 \quad 0 \quad 1 \quad 1 \quad 1\}.$$
$$\uparrow$$

We form two length-4 real sequences

$$g[n] = v[2n] = \{1 \quad 2 \quad 0 \quad 1\}, \quad h[n] = v[2n+1] = \{2 \quad 2 \quad 1 \quad 1\}, \qquad 0 \le n \le 3.$$

From Eq. (3.90),

$$V[k] = G[\langle k \rangle_4] + W_8^k H[\langle k \rangle_4], \qquad 0 \le k \le 7.$$

The 4-point DFTs, $G[k]$ and $H[k]$ of the two length-4 sequences $g[n]$ and $h[n]$ were computed in the previous example and are given in Eq. (3.88). Substituting these DFT values in the above equation we finally arrive at

$$V[0] = G[0] + H[0] = 4 + 6 = 10,$$

$$V[1] = G[1] + W_8^1 H[1] = (1 - j) + (1 - j) \cdot e^{-j\pi/4} = 1.0 - j2.4142,$$

$$V[2] = G[2] + W_8^2 H[2] = -2 + 0 \cdot e^{-j\pi/2} = -2,$$

$$V[3] = G[3] + W_8^3 H[3] = (1 + j) + (1 + j) \cdot e^{-j3\pi/4} = 1.0 - j0.4142,$$

$$V[4] = G[0] + W_8^4 H[0] = 4 + 4 \cdot e^{-j\pi} = -2,$$

$$V[5] = G[1] + W_8^5 H[1] = (1 - j) + (1 - j) \cdot e^{-j5\pi/4} = 1.0 + j0.4142,$$

$$V[6] = G[2] + W_8^6 H[2] = -2 + 0 \cdot e^{-j3\pi/2} = -2,$$

$$V[7] = G[3] + W_8^7 H[3] = (1 + j) + (1 + j) \cdot e^{-j7\pi/4} = 1.0 + j2.4142.$$

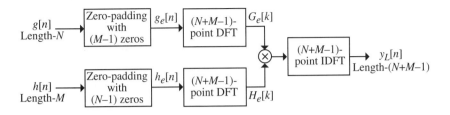

Figure 3.14: DFT-based implementation of the linear convolution of two finite-length sequences.

3.6 Linear Convolution Using the DFT

Linear convolution is a key operation in most signal processing applications. Since an N-point DFT can be implemented very efficiently using approximately $N(\log_2 N)$ arithmetic operations, it is of interest to investigate methods for the implementation of the linear convolution using the DFT. Earlier in Example 3.17, we have already illustrated for a very specific case how to implement a linear convolution using a circular convolution. We first generalize this example for the linear convolution of two finite-length sequences of unequal lengths. Later we consider the implementation of the linear convolution of a finite-length sequence with an infinite-length sequence.

3.6.1 Linear Convolution of Two Finite-Length Sequences

Let $g[n]$ and $h[n]$ be finite-length sequences of lengths N and M, respectively. Denote $L = M + N - 1$. Define two length-L sequences,

$$g_e[n] = \begin{cases} g[n], & 0 \le n \le N - 1, \\ 0, & N \le n \le L - 1, \end{cases} \tag{3.91}$$

$$h_e[n] = \begin{cases} h[n], & 0 \le n \le M - 1, \\ 0, & M \le n \le L - 1, \end{cases} \tag{3.92}$$

obtained by appending $g[n]$ and $h[n]$ with zero-valued samples. Then

$$y_L[n] = g[n] \circledast h[n] = y_C[n] = g_e[n] \textcircled{L} h_e[n]. \tag{3.93}$$

To implement Eq. (3.93) using the DFT, we first zero-pad $g[n]$ with $(M - 1)$ zeros to obtain $g_e[n]$, and zero-pad $h[n]$ with $(N - 1)$ zeros to obtain $h_e[n]$. Then we compute the $(N + M - 1)$-point DFTs of $g_e[n]$ and $h_e[n]$, respectively, resulting in $G_e[k]$ and $H_e[k]$. An $(N + M - 1)$-point IDFT of the product $G_e[k]H_e[k]$ results in $y_L[n]$. The process involved is sketched in Figure 3.14.

The following example uses MATLAB to illustrate the above approach.

EXAMPLE 3.20   Program 3_5 can be used to determine the linear convolution of two finite-length sequences via the DFT-based approach and to compare the result using a direct linear convolution. The input data to be entered in vector format inside square brackets are the two sequences to be convolved. The program plots the result of the linear convolution obtained using the DFT-based approach, and the difference between this result and the sequence obtained using a direct linear convolution using the M-file `conv`. Using this program we verify the result of Example 3.17 as demonstrated in Figure 3.15.

```
% Program 3_5
% Linear Convolution Via the DFT
```

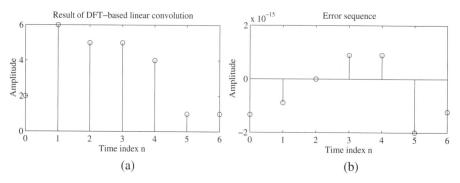

Figure 3.15: Plots of the output and the error sequences.

```
%
% Read in the two sequences
x = input('Type in the first sequence = ');
h = input('Type in the second sequence = ');
% Determine the length of the result of convolution
L = length(x)+length(h)-1;
% Compute the DFTs by zero-padding
XE = fft(x,L);
HE = fft(h,L);
% Determine the IDFT of the product
y1 = ifft(XE.*HE);
% Plot the sequence generated by DFT-based
% convolution and the error from direct linear
% convolution
k = 0:1:L-1;
subplot(2,1,1)
stem(k,y1)
xlabel('Time index n');ylabel('Amplitude');
title('Result of DFT-based linear convolution')
y2 = conv(x,h);
error = y1-y2;
subplot(2,1,2)
stem(k,abs(error))
xlabel('Time index n');ylabel('Amplitude')
title('Magnitude of error sequence')
```

3.6.2 Linear Convolution of a Finite-Length Sequence with an Infinite-Length Sequence

We consider now the DFT-based implementation of

$$y[n] = \sum_{\ell=0}^{M-1} h[\ell]x[n-\ell] = h[n] \circledast x[n], \tag{3.94}$$

where $h[n]$ is a finite-length sequence of length M and $x[n]$ is of infinite length (or a finite-length sequence of length much greater than M). There are two different approaches to solving this problem, as described below [Sto66].

Overlap-Add Method

In this method, we first segment $x[n]$, assumed to be a causal sequence here without any loss of generality, into a set of contiguous finite-length subsequences $x_m[n]$ of length N each:

$$x[n] = \sum_{m=0}^{\infty} x_m[n - mN], \qquad (3.95)$$

where

$$x_m[n] = \begin{cases} x[n + mN], & 0 \le n \le N - 1, \\ 0, & \text{otherwise.} \end{cases} \qquad (3.96)$$

Substituting Eq. (3.95) in Eq. (3.94) we get

$$y[n] = \sum_{m=0}^{\infty} y_m[n - mN], \qquad (3.97)$$

where $y_m[n] = h[n] \circledast x_m[n]$. Since $h[n]$ is of length M and $x_m[n]$ is of length N, the linear convolution $h[n] \circledast x_m[n]$ is of length $(N + M - 1)$. As a result, the desired linear convolution of Eq. (3.94) has been broken up into a sum of an infinite number of short-length linear convolutions of length $(N + M - 1)$ each. Each of these short convolutions can be implemented using the method outlined in Figure 3.14, where now the DFTs (and the IDFT) are computed on the basis of $(N + M - 1)$ points. There is one more subtlety to take care of before we can implement Eq. (3.97) using the DFT-based method.

Now the first short convolution in Eq. (3.97) given by $h[n] \circledast x_0[n]$, which is of length $(N + M - 1)$, is defined for $0 \le n \le N + M - 2$. The second short convolution in Eq. (3.97), given by $h[n] \circledast x_1[n]$, is also of length $(N + M - 1)$ but is defined for $N \le n \le 2N + M - 2$. This implies that there is an overlap of $M - 1$ samples between these two short linear convolutions in the range $N \le n \le N + M - 2$. Likewise, the third convolution in Eq. (3.97), given by $h[n] \circledast x_2[n]$, is defined for $2N \le n \le 3N + M - 2$, causing an overlap between the samples of $h[n] \circledast x_1[n]$ and $h[n] \circledast x_2[n]$ for $2N \le n \le 2N + M - 2$. In general, there will be an overlap of $M - 1$ samples between the samples of the short convolutions $h[n] \circledast x_{r-1}[n]$ and $h[n] \circledast x_r[n]$ for $rN \le n \le rN + M - 2$.

This process is illustrated in Figure 3.16. Figure 3.16(b) shows the first three length-7 ($N = 7$) segments $x_m[n]$ of the sequence $x[n]$ of Figure 3.16(a). Each of these segments is convolved with a length-5 ($M = 5$) sequence $h[n]$, resulting in length-11 ($N + M - 1 = 11$) short linear convolutions $y_m[n]$ shown in Figure 3.16(c). As can be seen from Figure 3.16(c), the last $M - 1 = 4$ samples of $y_0[n]$ overlap with the first 4 samples of $y_1[n]$. Likewise, the last $M - 1 = 4$ samples of $y_1[n]$ overlap with the first 4 samples of $y_2[n]$, and so on. Therefore, the desired sequence $y[n]$ obtained by a linear convolution of $x[n]$ and $h[n]$ is given by

$$
\begin{aligned}
y[n] &= y_0[n], & 0 \le n \le 6, \\
y[n] &= y_0[n] + y_1[n - 7], & 7 \le n \le 10, \\
y[n] &= y_1[n - 7], & 11 \le n \le 13, \\
y[n] &= y_1[n - 7] + y_2[n - 14], & 14 \le n \le 17, \\
y[n] &= y_2[n - 14], & 18 \le n \le 20, \\
&\ \ \vdots
\end{aligned}
$$

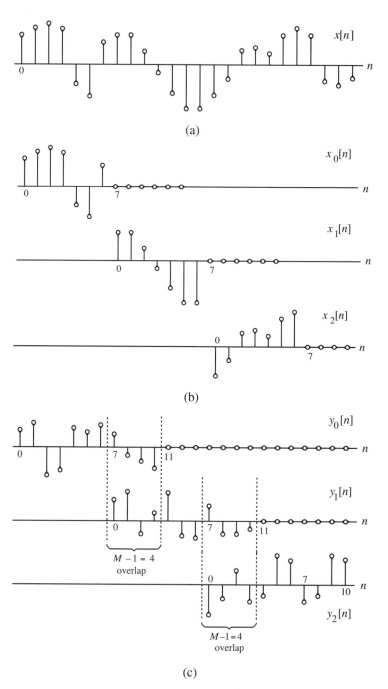

Figure 3.16: (a) Original $x[n]$, (b) segments $x_m[n]$ of $x[n]$, and (c) linear convolution of $x_m[n]$ with $h[n]$.

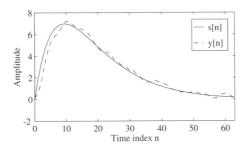

Figure 3.17: Uncorrupted input signal $s[n]$ (shown with solid line) and the filtered noisy signal $y[n]$ (shown with dashed line).

The above procedure is called the *overlap-add method* since the results of the short linear convolutions overlap and the overlapped portions are added to get the correct final result.

The M-file fftfilt can be used to implement the above method. There are two different forms of this function:

$$y = \texttt{fftfilt(h,x)}, \quad y = \texttt{fftfilt(h,x,r)}$$

where h is the impulse response vector of the FIR filter, x is the input vector segmented into successive sections, and y is the filtered output. In the first form the input data x is segmented into successive sections of length 512 each, whereas in the second form the input is segmented into sections of length r specified by the user. We illustrate its use in the following example.

EXAMPLE 3.21   We consider the filtering of the noise-corrupted signal of Example 2.14 using a length-3 moving average filter. To this end, we use Program 3_6 given below. The computer-generated results are indicated in Figure 3.17.

```
% Program 3_6
% Illustration of Overlap-Add Method
%
% Generate the noise sequence
R = 64;
d = rand(R,1)-0.5;
% Generate the uncorrupted sequence and add noise
for m = 1:1:R;
        s(m)  = 2*(m-1)*((0.9)^(m-1));
        x(m)  = s(m) + d(m);
end
k = 0:1:R-1;
% Read in the length of the moving average filter
M = input('Length of moving average filter = ');
% Generate the moving average filter coefficients
h = ones(1,M)/M;
% Perform the overlap-add filtering operation
y = fftfilt(h,x,4);
% Plot the results
plot(k,s,'r-',k,y,'b--')
xlabel('Time index n');ylabel('Amplitude')
legend( 'r-','s[n]','b--','y[n]')
```

Overlap-Save Method

In implementing the previous method using the DFT, we need to compute two $(N + M - 1)$-point DFTs and one $(N + M - 1)$-point IDFT since the overall linear convolution of Eq. (3.94) was expressed as a sum of short-length linear convolutions of length $(N + M - 1)$ each. It is possible to implement the linear convolution of Eq. (3.94) by performing instead circular convolutions of length shorter than $(N + M - 1)$. To this end, it is necessary to segment $x[n]$ into overlapping blocks $x_m[n]$, keep the terms of the circular convolution of $h[n]$ with $x_m[n]$ that corresponds to the terms obtained by a linear convolution of $h[n]$ and $x_m[n]$, and throw away the other parts of the circular convolution.

To understand the correspondence between the linear and circular convolutions, consider a length-4 sequence $x[n]$ and a length-3 sequence $h[n]$. Let $y_L[n]$ denote the result of a linear convolution of $x[n]$ with $h[n]$. The six samples of $y_L[n]$ are given by

$$y_L[0] = h[0]x[0],$$
$$y_L[1] = h[0]x[1] + h[1]x[0],$$
$$y_L[2] = h[0]x[2] + h[1]x[1] + h[2]x[0],$$
$$y_L[3] = h[0]x[3] + h[1]x[2] + h[2]x[1],$$
$$y_L[4] = h[1]x[3] + h[2]x[2],$$
$$y_L[5] = h[2]x[3]. \tag{3.98}$$

If we append $h[n]$ with a single zero sample and convert it into a length-4 sequence $h_e[n]$, the 4-point circular convolution $y_C[n]$ of $h_e[n]$ and $x[n]$ is given by

$$y_C[0] = h[0]x[0] + h[1]x[3] + h[2]x[2],$$
$$y_C[1] = h[0]x[1] + h[1]x[0] + h[2]x[3],$$
$$y_C[2] = h[0]x[2] + h[1]x[1] + h[2]x[0],$$
$$y_C[3] = h[0]x[3] + h[1]x[2] + h[2]x[1]. \tag{3.99}$$

Comparing Eqs. (3.98) and (3.99), we observe that the first two terms of the circular convolution do not correspond to the first two terms of the linear convolution, whereas the last two terms of the circular convolution are precisely the same as the third and fourth terms of the linear convolution, i.e.,

$$y_L[0] \neq y_C[0], \qquad y_L[1] \neq y_C[1],$$
$$y_L[2] = y_C[2], \qquad y_L[3] = y_C[3].$$

In the general case of an N-point circular convolution of a length-M sequence $h[n]$ with a length-N sequence $x[n]$ with $N > M$, the first $M - 1$ samples of the circular convolution are incorrect and are rejected, while the remaining $N - M + 1$ samples correspond to the correct samples of the linear convolution of $h[n]$ and $x[n]$.

Now consider an infinitely long or a very long sequence $x[n]$. We break it up as a collection of smaller length (length-4) sequences $x_m[n]$ as indicated below:

$$x_m[n] = x[n + 2m], \qquad 0 \leq n \leq 3, \quad 0 \leq m \leq \infty. \tag{3.100}$$

Next, we form

$$w_m[n] = h[n] \textcircled{4} x_m[n],$$

or equivalently,

$$w_m[0] = h[0]x_m[0] + h[1]x_m[3] + h[2]x_m[2],$$
$$w_m[1] = h[0]x_m[1] + h[1]x_m[0] + h[2]x_m[3],$$
$$w_m[2] = h[0]x_m[2] + h[1]x_m[1] + h[2]x_m[0],$$
$$w_m[3] = h[0]x_m[3] + h[1]x_m[2] + h[2]x_m[1]. \tag{3.101}$$

Computing the above for $m = 0, 1, 2, 3, \ldots$, and substituting the values of $x_m[n]$ from Eq. (3.100), we arrive at

$$
\begin{aligned}
w_0[0] &= h[0]x[0] + h[1]x[3] + h[2]x[2], &&\leftarrow \quad \textit{Reject}\\
w_0[1] &= h[0]x[1] + h[1]x[0] + h[2]x[3], &&\leftarrow \quad \textit{Reject}\\
w_0[2] &= h[0]x[2] + h[1]x[1] + h[2]x[0] = y[2], &&\leftarrow \quad \textit{Save}\\
w_0[3] &= h[0]x[3] + h[1]x[2] + h[2]x[1] = y[3], &&\leftarrow \quad \textit{Save}
\end{aligned}
$$

$$
\begin{aligned}
w_1[0] &= h[0]x[2] + h[1]x[5] + h[2]x[4], &&\leftarrow \quad \textit{Reject}\\
w_1[1] &= h[0]x[3] + h[1]x[2] + h[2]x[5], &&\leftarrow \quad \textit{Reject}\\
w_1[2] &= h[0]x[4] + h[1]x[3] + h[2]x[2] = y[4], &&\leftarrow \quad \textit{Save}\\
w_1[3] &= h[0]x[5] + h[1]x[4] + h[2]x[3] = y[5], &&\leftarrow \quad \textit{Save}
\end{aligned}
$$

$$
\begin{aligned}
w_2[0] &= h[0]x[4] + h[1]x[5] + h[2]x[6], &&\leftarrow \quad \textit{Reject}\\
w_2[1] &= h[0]x[5] + h[1]x[4] + h[2]x[7], &&\leftarrow \quad \textit{Reject}\\
w_2[2] &= h[0]x[6] + h[1]x[5] + h[2]x[4] = y[6], &&\leftarrow \quad \textit{Save}\\
w_2[3] &= h[0]x[7] + h[1]x[6] + h[2]x[5] = y[7], &&\leftarrow \quad \textit{Save}
\end{aligned}
$$

It should be noted that to determine $y[0]$ and $y[1]$, we need to form $x_{-1}[n]$:

$$
x_{-1}[0] = 0, \quad x_{-1}[1] = 0, \quad x_{-1}[2] = x[0], \quad x_{-1}[3] = x[1],
$$

and compute $w_{-1}[n] = h[n] \circledfour x_{-1}[n]$ for $0 \leq n \leq 3$, reject $w_{-1}[0]$ and $w_{-1}[1]$, and save $w_{-1}[2] = y[0]$ and $w_{-1}[3] = y[1]$.

Generalizing above, let $h[n]$ be a sequence of length M, and $x_m[n]$, the mth section of an infinitely long sequence $x[n]$ defined by

$$
x_m[n] = x[n + m(N - M + 1)], \quad 0 \leq n \leq N - 1, \tag{3.102}
$$

be of length N with $M \leq N$. If $w_m[n]$ denotes the N-point circular convolution of $h[n]$ and $x_m[n]$, i.e., $w_m[n] = h[n] \circledN x_m[n]$, then we reject the first $M - 1$ samples of $w_m[n]$ and "abut" the remaining $N - M + 1$ saved samples of $w_m[n]$ to form $y_L[n]$, the linear convolution of $h[n]$ and $x[n]$. If we denote the saved portion of $w_m[n]$ as $y_m[n]$, i.e.,

$$
y_m[n] = \begin{cases} 0, & 0 \leq n \leq M - 2, \\ w_m[n], & M - 1 \leq n \leq N - 1, \end{cases} \tag{3.103}
$$

then,

$$
y_L[n + m(N - M + 1)] = y_m[n], \quad M - 1 \leq n \leq N - 1. \tag{3.104}
$$

The above process is illustrated in Figure 3.18. The approach is called the *overlap-save method* since the input is segmented into overlapping sections and part of the results of the circular convolution are saved and abutted to determine the linear convolution result.

3.7 The z-Transform

The discrete-time Fourier transform provides a frequency-domain representation of discrete-time signals and LTI systems. Because of the convergence condition, in many cases, the discrete-time Fourier transform of a sequence may not exist, and as a result, it is not possible to make use of such frequency-domain characterization in these cases. A generalization of the discrete-time Fourier transform defined in Eq. (3.1) leads to the z-transform, which may exist for many sequences for which the discrete-time Fourier transform

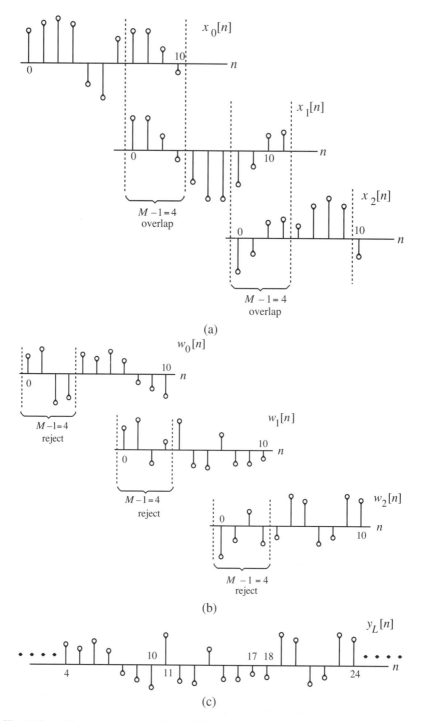

Figure 3.18: Illustration of the overlap-save method. (a) Overlapped segments of the sequence $x[n]$ of Figure 3.16(a), (b) sequences generated by an 11-point circular convolution, and (c) sequence obtained by rejecting the first four samples of $w_i[n]$ and abutting the remaining samples.

does not exist. Moreover, the use of z-transform techniques permits simple algebraic manipulations. Consequently, the z-transform has become an important tool in the analysis and design of digital filters.

We first define the z-transform of a sequence and study its properties by treating it as a generalization of the discrete-time Fourier transform. This leads to the concept of the region of convergence of a z-transform that is investigated in detail. We then describe the inverse transform operation and point out two straightforward approaches for the computation of the inverse of a real rational z-transform. Next, the properties of the z-transform are reviewed.

3.7.1 Definition

For a given sequence $g[n]$, its z-transform $G(z)$ is defined as

$$G(z) = \mathcal{Z}\{g[n]\} = \sum_{n=-\infty}^{\infty} g[n]z^{-n}, \tag{3.105}$$

where $z = \text{Re}(z) + j\text{Im}(z)$ is a complex variable. If we let $z = re^{j\omega}$, then the right-hand side of the above expression reduces to

$$G(re^{j\omega}) = \sum_{n=-\infty}^{\infty} g[n]r^{-n}e^{-j\omega n}, \tag{3.106}$$

which can be interpreted as the discrete-time Fourier transform of the modified sequence $\{g[n]r^{-n}\}$. For $r = 1$ (i.e., $|z| = 1$), the z-transform of $g[n]$ reduces to its discrete-time Fourier transform, provided the latter exists. The contour $|z| = 1$ is a circle in the z-plane of unity radius and is called the *unit circle*.

Like the discrete-time Fourier transform, there are conditions on the convergence of the infinite series of Eq. (3.105). For a given sequence, the set $\mathcal{R}$ of values of z for which its z-transform converges is called the *region of convergence* (ROC). It follows from our earlier discussion on the uniform convergence of the discrete-time Fourier transform that the series of Eq. (3.106) converges if $g[n]r^{-n}$ is absolutely summable, i.e., if

$$\sum_{n=-\infty}^{\infty} \left| g[n]r^{-n} \right| < \infty. \tag{3.107}$$

In general, the region of convergence $\mathcal{R}$ of a z-transform of a sequence $g[n]$ is an annular region of the z-plane:

$$R_{g-} < |z| < R_{g+} \tag{3.108}$$

where $0 \le R_{g-} < R_{g+} \le \infty$. It should be noted that the z-transform as defined by Eq. (3.105) is a form of a Laurent series and is an analytic function at every point in the ROC. This in turn implies that the z-transform and all its derivatives are continuous functions of the complex variable z in the ROC.

EXAMPLE 3.22   Let us determine the z-transform $X(z)$ of the causal sequence $\alpha^n \mu[n]$ and its region of convergence. Applying the definition of Eq. (3.105), we obtain

$$X(z) = \sum_{n=-\infty}^{\infty} \alpha^n \mu[n]z^{-n} = \sum_{n=0}^{\infty} \alpha^n z^{-n}. \tag{3.109}$$

The above power series converges to

$$X(z) = \frac{1}{1 - \alpha z^{-1}}, \qquad \text{for } |\alpha z^{-1}| < 1, \tag{3.110}$$

indicating that the ROC is the annular region $|z| > |\alpha|$.

Table 3.8: Some commonly used z-transform pairs.

Sequence	z-Transform	ROC
$\delta[n]$	1	All values of z
$\mu[n]$	$\dfrac{1}{1-z^{-1}}$	$\|z\| > 1$
$\alpha^n \mu[n]$	$\dfrac{1}{1-\alpha z^{-1}}$	$\|z\| > \|\alpha\|$
$(r^n \cos \omega_o n)\mu[n]$	$\dfrac{1-(r\cos\omega_o)z^{-1}}{1-(2r\cos\omega_o)z^{-1}+r^2 z^{-2}}$	$\|z\| > r$
$(r^n \sin \omega_o n)\mu[n]$	$\dfrac{(r\sin\omega_o)z^{-1}}{1-(2r\cos\omega_o)z^{-1}+r^2 z^{-2}}$	$\|z\| > r$

The z-transform $\mu(z)$ of the unit step sequence $\mu[n]$ can be obtained from Eq. (3.110) by setting $\alpha = 1$:

$$\mu(z) = \frac{1}{1-z^{-1}}, \qquad \text{for } |z^{-1}| < 1. \tag{3.111}$$

The ROC of $\mu(z)$ is thus the annular region $1 < |z| \le \infty$. Note that the unit step sequence is not absolutely summable, and, as a result, its Fourier transform does not converge uniformly.

EXAMPLE 3.23   Consider the anticausal sequence $x[n] = -\alpha^n \mu[-n-1]$. Using Eq. (3.105) we arrive at the expression for its z-transform:

$$X(z) = -\sum_{n=-\infty}^{-1} \alpha^n z^{-n} = -\sum_{m=1}^{\infty} \alpha^{-m} z^m$$

$$= -\alpha^{-1} z \sum_{n=0}^{\infty} \alpha^{-m} z^m = -\frac{\alpha^{-1} z}{1-\alpha^{-1} z} = \frac{1}{1-\alpha z^{-1}}, \qquad \text{for } |\alpha^{-1} z| < 1, \tag{3.112}$$

where now the ROC is the annular region $|z| < |\alpha|$.

It should be noted that in both of the above examples, the z-transforms are identical even though their parent sequences are different. The only way a unique sequence can be associated with a z-transform is by specifying its ROC. We shall discuss further the importance of the ROC in the following section.

It follows from the above that the Fourier transform $G(e^{j\omega})$ of a sequence $g[n]$ converges uniformly if and only if the ROC of the z-transform $G(z)$ of the sequence includes the unit circle. On the other hand, the existence of the Fourier transform does not always imply the existence of the z-transform. For example, the finite-energy sequence $h_{LP}[n]$ of Eq. (3.12) has a Fourier transform $h_{LP}(e^{j\omega})$ given by Eq. (3.11) which converges in the mean-square sense. However, this sequence does not have a z-transform as $h_{LP}[n] r^{-n}$ is not absolutely summable for any value of r.

Some commonly used z-transform pairs are listed in Table 3.8.

3.7.2 Rational z-Transforms

In the case of LTI discrete-time systems that we are concerned with in this text, all pertinent z-transforms are rational functions of z^{-1}, i.e., are ratios of two polynomials in z^{-1}:

$$G(z) = \frac{P(z)}{D(z)} = \frac{p_0 + p_1 z^{-1} + \cdots + p_{M-1} z^{-(M-1)} + p_M z^{-M}}{d_0 + d_1 z^{-1} + \cdots + d_{N-1} z^{-(N-1)} + d_N z^{-N}}, \qquad (3.113)$$

where the *degree* of the numerator polynomial $P(z)$ is M and that of the denominator polynomial $D(z)$ is N. An alternate representation of a rational z-transform is as a ratio of two polynomials in z:

$$G(z) = z^{(N-M)} \frac{p_0 z^M + p_1 z^{M-1} + \cdots + p_{M-1} z + p_M}{d_0 z^N + d_1 z^{N-1} + \cdots + d_{N-1} z + d_N}. \qquad (3.114)$$

The above equation can be alternately written in factored form as

$$G(z) = \frac{p_0}{d_0} \frac{\prod_{\ell=1}^{M} (1 - \xi_\ell z^{-1})}{\prod_{\ell=1}^{N} (1 - \lambda_\ell z^{-1})} = z^{(N-M)} \frac{p_0}{d_0} \frac{\prod_{\ell=1}^{M} (z - \xi_\ell)}{\prod_{\ell=1}^{N} (z - \lambda_\ell)}. \qquad (3.115)$$

At a root $z = \xi_\ell$ of the numerator polynomial, $G(\xi_\ell) = 0$, and as a result, these values of z are known as the *zeros* of $G(z)$. Likewise, at a root $z = \lambda_\ell$ of the denominator polynomial, $G(\lambda_\ell) \to \infty$, and these points in the z-plane are called the *poles* of $G(z)$. Observe from the expression in Eq. (3.115) that there are M finite zeros and N finite poles of $G(z)$. It also follows from the above expression that there are additional $(N - M)$ zeros at $z = 0$ (the origin in the z-plane) if $N > M$ or additional $(M - N)$ poles at $z = 0$ if $N < M$. For example, the z-transform $\mu(z)$ of Eq. (3.111) can be rewritten as

$$\mu(z) = \frac{z}{z - 1}, \qquad \text{for } |z| > 1, \qquad (3.116)$$

which has a zero at $z = 0$ and a pole at $z = 1$.

A physical interpretation of the concepts of poles and zeros can be given by plotting the log-magnitude $20 \log_{10} |G(z)|$. Now $20 \log_{10} |G(z)|$ is a two-dimensional function of $\text{Re}(z)$ and $\text{Im}(z)$. Hence its plot will describe a surface in the complex z-plane as illustrated in Figure 3.19 for the rational z-transform

$$G(z) = \frac{1 - 2.4 z^{-1} + 2.88 z^{-2}}{1 - 0.8 z^{-1} + 0.64 z^{-2}}.$$

It can be seen from this figure that the magnitude plot exhibits very large peaks around the points $z = 0.4 \pm j0.6928$ which are the poles of $G(z)$, and very narrow and deep wells around the location of the zeros at $z = 1.2 \pm j1.2$.

3.8 Region of Convergence of a Rational z-Transform

The ROC of a z-transform is an important concept for a variety of reasons. As we shall show later, without the knowledge of the ROC, there is no unique relationship between a sequence and its z-transform. Hence, the z-transform must always be specified with its ROC. Moreover, if the ROC of a z-transform of sequence includes the unit circle, the Fourier transform of the sequence is obtained simply by evaluating the z-transform on the unit circle. In the following chapter, we shall point out the relationship between the ROC of the z-transform of the impulse response of a causal LTI system and its BIBO stability. It is thus of interest to investigate the ROC more thoroughly.

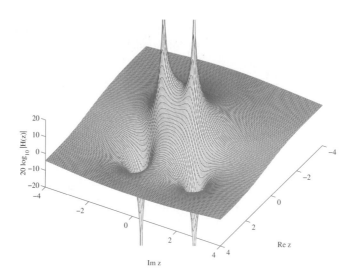

Figure 3.19: The 3-D plot of $20 \log_{10} |G(z)|$ as a function of $\text{Re}(z)$ and $\text{Im}(z)$.

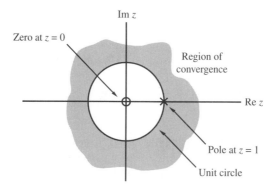

Figure 3.20: The pole-zero plot and the region of convergence of $\mathcal{Z}\{\mu[n]\}$.

Now, the ROC of a rational z-transform is bounded by the locations of its poles. To understand this relationship between the poles and the ROC, it is instructive to examine the plot of the poles and the zeros of a z-transform. Figure 3.20 shows the pole-zero plot of the z-transform $\mu(z)$ of Eq. (3.116), where the location of the pole is indicated by a cross "×" and the location of the zero is indicated by a circle "○". In this figure the ROC, shown as the shaded area, is the region of the z-plane just outside the circle centered at the origin and going through the pole at $z = 1$, and extending all the way to $|z| = \infty$.

EXAMPLE 3.24   Determine the ROC of the z-transform $H(z)$ of the sequence $h[n] = (-0.6)^n \mu[n]$. From Eq. (3.110) we arrive at

$$H(z) = \frac{1}{1 + 0.6 z^{-1}} = \frac{z}{z + 0.6}, \tag{3.117}$$

provided $|z| > 0.6$. This implies that the ROC is just outside the circle going through the point $z = -0.6$ and extending all the way to $z = \infty$, as shown in Figure 3.21. Note that $H(z)$ has a zero at $z = 0$ and a pole at $z = -0.6$.

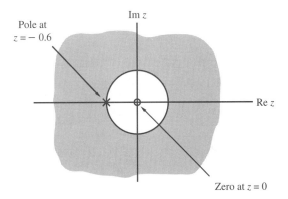

Pole at
$z = -0.6$

Im z

Re z

Zero at $z = 0$

Figure 3.21: Pole-zero plot of $\mathcal{Z}\{(-0.6)^n \mu[n]\}$.

In general, the ROC depends on the type of the sequence of interest as defined earlier in Section 2.1.1. We examine in the next four examples the ROCs of the *z*-transforms of several different types of sequences.

EXAMPLE 3.25   Consider a finite-length sequence $g[n]$ defined for $-M \leq n \leq N$, where M and N are nonnegative integers and $|g[n]| < \infty$. Its *z*-transform is given by

$$G(z) = \sum_{n=-M}^{N} g[n]z^{-n} = \frac{\sum_{n=0}^{N+M} g[n-M]z^{N+M-n}}{z^N}. \tag{3.118}$$

Note from Eq. (3.118) that $G(z)$ has M poles at $z = \infty$ and N poles at $z = 0$. As can be seen from the above, in general, the *z*-transform of a finite-length bounded sequence converges everywhere in the *z*-plane except possibly at $z = 0$ and/or at $z = \infty$.

EXAMPLE 3.26   A right-sided sequence $u_1[n]$ with nonzero sample values only for $n \geq 0$ is sometimes called a causal sequence. Its *z*-transform is given by

$$U_1(z) = \sum_{n=0}^{\infty} u_1[n]z^{-n}. \tag{3.119}$$

It can be shown that $U_1(z)$ converges exterior to a circle $|z| = R_1$, including the point $z = \infty$. On the other hand, a right-sided sequence $u_2[n]$ with nonzero sample values only for $n \geq -M$ with M nonnegative has a *z*-transform $U_2(z)$ with M poles at $z = \infty$, and therefore its ROC is exterior to a circle $|z| = R_2$, excluding the point $z = \infty$.

EXAMPLE 3.27   A left-sided sequence $v_1[n]$ with nonzero sample values only for $n \leq 0$, often called an anticausal sequence, has a *z*-transform given by

$$V_1(z) = \sum_{n=-\infty}^{0} v_1[n]z^{-n}, \tag{3.120}$$

and converges interior to a circle $|z| = R_3$, including the point $z = 0$. However, a left-sided sequence $v_2[n]$ with nonzero sample values only for $n \leq N$ with N nonnegative has a *z*-transform $V_2(z)$ with N poles at $z = 0$. As a result, the ROC of $V_2(z)$ is interior to a circle $|z| = R_4$ excluding the point $z = 0$.

The z-transform of a two-sided sequence $w[n]$ can be expressed as

$$W(z) = \sum_{n=-\infty}^{\infty} w[n]z^{-n} = \sum_{n=0}^{\infty} w[n]z^{-n} + \sum_{n=-\infty}^{-1} w[n]z^{-n}. \tag{3.121}$$

The first term on the right-hand side of Eq. (3.121) can be interpreted as the z-transform of a right-sided sequence, and it converges exterior to the circle $|z| = R_5$. The second term, on the other hand, is the z-transform of a left-sided sequence, and it converges interior to the circle $|z| = R_6$. As a result, if $R_5 < R_6$, there is an overlapping ROC given by $R_5 < |z| < R_6$. If, however, $R_6 < R_5$, the z-transform does not exist, as illustrated by the following example.

EXAMPLE 3.28   The two-sided sequence defined by

$$u[n] = \alpha^n,$$

where α can be a real or complex number, does not have a z-transform, regardless of the absolute value $|\alpha|$. This follows by noting that the z-transform expression can be rewritten as

$$U(z) = \sum_{n=0}^{\infty} \alpha^n z^{-n} + \sum_{n=-\infty}^{-1} \alpha^n z^{-n}. \tag{3.122}$$

The first term on the right-hand side of Eq. (3.122) converges for $|z| > |\alpha|$, whereas the second term converges for $|z| < |\alpha|$, and hence, there is no overlap of the two ROCs.

For a sequence with a rational z-transform, the ROC of the z-transform cannot contain any poles and is bounded by the poles. This property of such z-transforms can be seen from the z-transform of a unit step sequence given by Eq. (3.116) and illustrated by the pole-zero plot of Figure 3.20. Another example is the sequence defined in Example 3.24 and its z-transform given by Eq. (3.117), with the corresponding pole-zero plot given in Figure 3.21.

To show that it is bounded by the poles, assume that the z-transform $X(z)$ has simple poles at α and β, with $|\alpha| < |\beta|$. If the sequence is also assumed to be a right-sided sequence, then it is of the form

$$x[n] = \left(r_1(\alpha)^n + r_2(\beta)^n \right) \mu[n - N_o], \tag{3.123}$$

where N_o is a positive or negative integer. Now the z-transform of a right-sided sequence $(\gamma)^n \mu[n - N_o]$ exists if

$$\sum_{n=N_o}^{\infty} \left| (\gamma)^n z^{-n} \right| < \infty$$

for some z. It can be seen that the above holds for $|z| > |\gamma|$ but not for $|z| \leq |\gamma|$. The right-sided sequence of Eq. (3.123) has thus an ROC defined by $|\beta| < |z| \leq \infty$. A similar argument shows that if $X(z)$ is the z-transform of a left-sided sequence of the form of Eq. (3.123), with $\mu[n - N_o]$ replaced by $\mu[-n - N_o]$, then its ROC is defined by $0 \leq |z| < |\alpha|$. Finally, for a two-sided sequence, some of the poles contribute to terms for $n < 0$ and the others to terms for $n \geq 0$. The ROC is thus bounded on the outside by the pole with the smallest magnitude that contributes for $n < 0$ and on the inside by the pole with the largest magnitude that contributes for $n \geq 0$.

Figure 3.22 shows the three possible ROCs of a rational z-transform with poles at $z = \alpha$ and $z = \beta$ and with each ROC associated with a unique sequence. In general, if the rational z-transform has N poles with R distinct magnitudes, then it has $R + 1$ ROCs and, as a result, $R + 1$ distinct sequences having the same

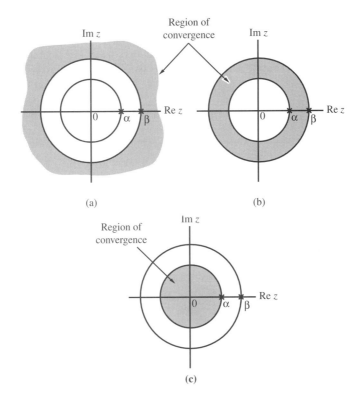

Figure 3.22: The pole-zero plot of a rational z-transform with three possible ROCs corresponding to three different sequences. (a) Right-sided sequence, (b) two-sided sequence, and (c) left-sided sequence.

rational z-transform. Consequently, a rational z-transform with a specified ROC has a unique sequence as its inverse z-transform. A rational z-transform without a specified ROC is thus not meaningful.

MATLAB can be used to determine the ROCs of a rational z-transform. To this end, several functions need to be used. The statement $[z,p,k] = \mathtt{tf2zp(num,den)}$ determines the zeros, poles, and the gain constant of a rational z-transform expressed as a ratio of polynomials in descending powers of z as in Eq. (3.113). The input arguments are the row vectors num and den containing the coefficients of the numerator and the denominator polynomials in descending powers of z.[7] The output files are the column vectors z and p containing the zeros and poles of the rational z-transform, and the gain constant k. The statement $[\mathtt{num,den}] = \mathtt{zp2tf(z,p,k)}$ is used to implement the reverse process.

From the zero-pole description, the factored form of the transfer function can be obtained using the function $\mathtt{sos} = \mathtt{zp2sos(z,p,k)}$. The statement computes the coefficients of each second-order factor given as an $L \times 6$ matrix sos, where

$$\mathtt{sos} = \begin{bmatrix} b_{01} & b_{11} & b_{21} & a_{01} & a_{11} & a_{21} \\ b_{02} & b_{12} & b_{22} & a_{02} & a_{12} & a_{22} \\ \vdots & \vdots & \vdots & \vdots & \vdots & \vdots \\ b_{0L} & b_{1L} & b_{2L} & a_{0L} & a_{1L} & a_{2L} \end{bmatrix},$$

where the kth row contains the coefficients of the numerator and the denominator of the kth second-order

[7]Or, equivalently, in ascending powers of z^{-1}.

factor of the z-transform $G(z)$:

$$G(z) = \prod_{k=1}^{L} \frac{b_{0k} + b_{1k}z^{-1} + b_{2k}z^{-2}}{a_{0k} + a_{1k}z^{-1} + a_{2k}z^{-2}}.$$

The pole-zero plot of a rational z-transform can also be plotted by using the M-file zplane. The z-transform can be described either in terms of its zeros and poles given as vectors zeros and poles or in terms of the numerator and the denominator polynomials entered as vectors num and den containing coefficients in descending powers of z:

$$\text{zplane(zeros,poles),}\qquad \text{zplane(num,den)}$$

It should be noted that the argument zeros and poles must be entered as column vectors, whereas the arguments num and den need to be entered as row vectors. The function zplane plots the zeros and poles in the current figure window, with the zero indicated by the symbol "o" and the pole indicated by the symbol "×." The unit circle is also included in the plot for reference. The automatic scaling included in the function can be overwritten if necessary.[8]

The following two examples illustrate the application of the above functions.

EXAMPLE 3.29   Express the following z-transform in factored form, plot its poles and zeros, and then determine its ROCs.

$$G(z) = \frac{2z^4 + 16z^3 + 44z^2 + 56z + 32}{3z^4 + 3z^3 - 15z^2 + 18z - 12}. \qquad (3.124)$$

To carry out the factorization we use the following MATLAB program:

```
% Program 3_7
% Determination of the Factored Form
% of a Rational z-Transform
%
num = input('Type in the numerator coefficients = ');
den = input('Type in the denominator coefficients = ');
[z,p,k] = tf2zp(num,den);
m = abs(p);
disp('Zeros are at');disp(z);
disp('Poles are at');disp(p);
disp('Gain constant');disp(k);
disp('Radius of poles');disp(m);
sos = zp2sos(z,p,k);
disp('Second-order sections');disp(real(sos));
zplane(num,den)
```

During execution the program asks for the coefficients of the numerator and denominator polynomials. These are entered as row vectors inside square brackets in descending powers of z. The program then computes and prints the locations of the poles and zeros, the value of the gain constant, the radius of the poles, and the coefficients of the second-order sections as indicated below:[9]

---

[8]See the manual for the *Signal Processing Toolbox* for details [Kra94].
[9]Note that in MATLAB, the usual notation for $\sqrt{-1}$ is "i" instead of "j."

```
    Zeros are at
     -4.0000
     -2.0000
     -1.0000 + 1.0000i
     -1.0000 - 1.0000i

    Poles are at
     -3.2361
      1.2361
      0.5000 + 0.8660i
      0.5000 - 0.8660i

  Gain constant
      0.6667

  Radius of poles
      3.2361
      1.2361
      1.0000
      1.0000

  Second-order sections
    0.6667    0.4000    0.5333    1.0000    2.0000   -4.0000
    1.0000    2.0000    2.0000    1.0000   -1.0000    1.0000
```

Therefore the factored form of the *z*-transform of Eq. (3.124) is given by

$$G(z) = \frac{(0.6667 + 0.4z^{-1} + 0.5333z^{-2})(1.0 + 2.0z^{-1} + 2.0z^{-2})}{(1.0 + 2.0z^{-1} - 4.0z^{-2})(1.0 - 1.0z^{-1} + 1.0z^{-2})}$$

$$= 0.6667 \frac{(1 + 6z^{-1} + 8z^{-2})(1 + 2z^{-1} + 2z^{-2})}{(1 + 2z^{-1} - 4z^{-2})(1 - z^{-1} + z^{-2})}. \tag{3.125}$$

The pole-zero plot developed by the program is shown in Figure 3.23. From Eq. (3.125) the four regions of convergence are thus seen to be

$$\mathcal{R}_1 : \infty \geq |z| > 3.2361,$$
$$\mathcal{R}_2 : 3.2361 > |z| > 1.2361,$$
$$\mathcal{R}_3 : 1.2361 > |z| > 1,$$
$$\mathcal{R}_4 : 1 > |z| \geq 0.$$

EXAMPLE 3.30 We now consider the determination of the rational *z*-transform from its zero and pole locations. The zeros are at $\xi_1 = 0.21$, $\xi_2 = 3.14$, $\xi_3 = -0.3 + j0.5$, $\xi_4 = -0.3 - j0.5$; the poles are at $\lambda_1 = -0.45$, $\lambda_2 = 0.67$, $\lambda_3 = 0.81 + j0.72$, $\lambda_4 = 0.81 - j0.72$; and the gain constant k is 2.2. The MATLAB program that is employed to compute the corresponding rational *z*-transform is as follows:

```
% Program 3_8
% Determination of the Rational z-Transform
% from its Poles and Zeros
%
```

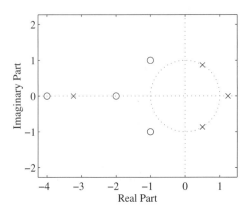

Figure 3.23: Pole-zero plot of the z-transform of Eq. (3.124).

```
format long
zr = input('Type in the zeros as a row vector = ');
pr = input('Type in the poles  as a row vector = ');
% Transpose zero and pole row vectors
z = zr'; p = pr';
k = input('Type in the gain constant = ');
[num, den] = zp2tf(z, p, k);
disp('Numerator polynomial coefficients'); disp(num);
disp('Denominator polynomial coefficients'); disp(den);
```

The program during execution asks for the input data, the zero and pole locations, and the gain constant. The program then displays the coefficients of the numerator and the denominator polynomials in descending powers of z, as indicated below:

```
Numerator polynomial coefficients
  Columns 1 through 4
   2.2   -6.05   -2.22332   -1.63592

  Column 5
  -0.4932312


Denominator polynomial coefficients
  Columns 1 through 4
   1.0   -1.840   1.22940    0.23004

  Column 5
  -0.35411175
```

From the above we thus arrive at the desired expression:

$$G(z) = \frac{2.2z^4 - 6.05z^3 - 2.22332z^2 - 1.63592z - 0.4932312}{z^4 - 1.84z^3 + 1.2294z^2 + 0.23004z - 0.35411175}.$$

3.9 Inverse *z*-Transform

We now derive the expression for the inverse *z*-transform and outline two methods for its computation.

3.9.1 General Expression

Recall that, for $z = re^{j\omega}$, the *z*-transform $G(z)$ given by Eq. (3.105) is merely the Fourier transform of the modified sequence $g[n]r^{-n}$. Accordingly, by the inverse Fourier transform relation of Eq. (3.7), we have

$$g[n]r^{-n} = \frac{1}{2\pi} \int_{-\pi}^{\pi} G(re^{j\omega})e^{j\omega n}\, d\omega. \tag{3.126}$$

By making the change of variable $z = re^{j\omega}$, the above equation can be converted into a contour integral given by

$$g[n] = \frac{1}{2\pi j} \oint_{C'} G(z)z^{n-1}\, dz, \tag{3.127}$$

where C' is a counterclockwise contour of integration defined by $|z| = r$. But the integral remains unchanged when C' is replaced with any contour C encircling the point $z = 0$ in the ROC of $G(z)$. The contour integral in Eq. (3.127) can be evaluated using the Cauchy's residue theorem [Chu90] resulting in

$$g[n] = \sum \left[\text{residues of } G(z)z^{n-1} \text{ at the poles inside } C \right]. \tag{3.128}$$

Note that Eq. (3.128) needs to be evaluated at all values of n and is not pursued here. Two simple methods for the inverse transform computation are reviewed next.

3.9.2 Inverse Transform by Partial-Fraction Expansion

The expression of Eq. (3.127) can be computed in a number of ways. A rational *z*-transform $G(z)$ with a causal inverse transform $g[n]$ has an ROC that is exterior to a circle. In this case it is more convenient to express $G(z)$ in a partial-fraction expansion form and then determine $g[n]$ by summing the inverse transforms of the individual simpler terms in the expansion. A rational $G(z)$ can be expressed as

$$G(z) = \frac{P(z)}{D(z)}, \tag{3.129}$$

where $P(z)$ and $D(z)$ are polynomials in z^{-1} as indicated in Eq. (3.113). If the degree M of the numerator polynomial $P(z)$ is greater than or equal to the degree N of the denominator polynomial $D(z)$, we can divide $P(z)$ by $D(z)$ and re-express $G(z)$ as

$$G(z) = \sum_{\ell=0}^{M-N} \eta_\ell z^{-\ell} + \frac{P_1(z)}{D(z)}, \tag{3.130}$$

where the degree of the polynomial $P_1(z)$ is less than that of $D(z)$. The rational function $P_1(z)/D(z)$ is called a *proper fraction*.

EXAMPLE 3.31　Consider the rational z-transform

$$\frac{2 + 0.8z^{-1} + 0.5z^{-2} + 0.3z^{-3}}{1 + 0.8z^{-1} + 0.2z^{-2}}.$$

It can be seen that the degree of its numerator polynomial is 3, whereas the degree of its denominator polynomial is 2. Since the numerator degree is greater than the denominator degree, the above rational z-transform is not a proper fraction. We can express it as a sum of a polynomial in z^{-1} and a proper fraction in the form of Eq. (3.130) by dividing the numerator by the denominator so that the numerator of the remainder is a first-degree polynomial. To this end we apply the long division with both polynomials expressed in a reverse order which results in

$$-3.5 + 1.5z^{-1} + \frac{5.5 + 2.1z^{-1}}{1 + 0.8z^{-1} + 0.2z^{-2}}.$$

Simple Poles

In most practical cases, $G(z)$ is a proper fraction with simple poles. Let the poles of $G(z)$ be at $z = \lambda_k$, $0 \le k \le N$, where λ_k are distinct. A partial-fraction expansion of $G(z)$ then is of the form

$$G(z) = \sum_{\ell=1}^{N} \frac{\rho_\ell}{1 - \lambda_\ell z^{-1}}, \tag{3.131}$$

where the constants ρ_ℓ in the above expression, called the *residues*, are given by

$$\rho_\ell = (1 - \lambda_\ell z^{-1}) G(z) \Big|_{z=\lambda_\ell}. \tag{3.132}$$

Each term of the sum on the right-hand side of Eq. (3.131) has an ROC given by $|z| > |\lambda_\ell|$ and, thus, an inverse transform of the form $\rho_\ell (\lambda_\ell)^n \mu[n]$. Therefore, the inverse transform $g[n]$ of $G(z)$ is given by

$$g[n] = \sum_{\ell=1}^{N} \rho_\ell (\lambda_\ell)^n \mu[n]. \tag{3.133}$$

Note that the above approach with slight modifications can also be used to determine the inverse z-transform of a noncausal sequence with a rational z-transform.

The inverse transform computation via partial-fraction expansion of a rational z-transform with simple poles is considered in the following example.

EXAMPLE 3.32　Let the z-transform of a causal sequence $h[n]$ be given by

$$H(z) = \frac{z(z + 2.0)}{(z - 0.2)(z + 0.6)} = \frac{1 + 2.0z^{-1}}{(1 - 0.2z^{-1})(1 + 0.6z^{-1})}. \tag{3.134}$$

A partial-fraction expansion of $H(z)$ is then of the form

$$H(z) = \frac{\rho_1}{1 - 0.2z^{-1}} + \frac{\rho_2}{1 + 0.6z^{-1}}. \tag{3.135}$$

Using Eq. (3.132), we obtain

$$\rho_1 = (1 - 0.2z^{-1}) H(z) \Big|_{z=0.2} = \frac{1 + 2.0z^{-1}}{1 + 0.6z^{-1}} \Big|_{z=0.2} = 2.75,$$

and

$$\rho_2 = (1 + 0.6z^{-1})H(z)\Big|_{z=-0.6} = \frac{1 + 2.0z^{-1}}{1 - 0.2z^{-1}}\Big|_{z=-0.6} = -1.75.$$

Substituting the above in Eq. (3.135), we arrive at

$$H(z) = \frac{2.75}{1 - 0.2z^{-1}} - \frac{1.75}{1 + 0.6z^{-1}}.$$

The inverse transform of the above is given by

$$h[n] = 2.75(0.2)^n \mu[n] - 1.75(-0.6)^n \mu[n]. \tag{3.136}$$

Multiple Poles

If $G(z)$ has multiple poles, the partial-fraction expansion is slightly of different form. For example, if the pole at $z = v$ is of multiplicity L and the remaining $N - L$ poles are simple and at $z = \lambda_\ell, 1 \le \ell \le N - L$, then the general partial-fraction expansion of $G(z)$ takes the form

$$G(z) = \sum_{\ell=0}^{M-N} \eta_\ell z^{-\ell} + \sum_{\ell=1}^{N-L} \frac{\rho_\ell}{1 - \lambda_\ell z^{-1}} + \sum_{i=1}^{L} \frac{\gamma_i}{(1 - vz^{-1})^i}, \tag{3.137}$$

where the constants γ_i (no longer called the residues for $i \ne 1$) are computed using the formula

$$\gamma_i = \frac{1}{(L-i)!(-v)^{L-i}} \frac{d^{L-i}}{d(z^{-1})^{L-i}} \left[(1 - vz^{-1})^L G(z) \right]\Big|_{z=v}, \quad 1 \le i \le L, \tag{3.138}$$

and the residues ρ_ℓ are calculated using Eq. (3.132). Techniques for determining the inverse z-transform of terms like $\gamma_i/(1 - vz^{-1})^i$ are described later in Example 3.41.

3.9.3 Partial-Fraction Expansion Using MATLAB

The M-file `residuez` can be used to develop the partial-fraction expansion of a rational z-transform and to convert a z-transform expressed in a partial-fraction form to its rational form. For the former case, the statement is `[r,p,k] = residuez(num,den)`, where the input data are the vectors `num` and `den` containing the coefficients of the numerator and the denominator polynomials, respectively, expressed in descending powers of z, and the output files are the vector `r` of the residues and the numerator constants, the vector `p` of corresponding poles, and the vector `k` containing the constants η_ℓ. The statement `[num,den] = residuez(r,p,k)` is employed to carry out the reverse operation. The applications of these functions are considered in the following two examples.

EXAMPLE 3.33 Using MATLAB we determine the partial-fraction expansion of the z-transform $G(z)$ given by

$$G(z) = \frac{18z^3}{18z^3 + 3z^2 - 4z - 1}. \tag{3.139}$$

To this end we use the following MATLAB program:

```
Program 3_9
% Partial-Fraction Expansion of Rational z-Transform
%
num = input('Type in numerator coefficients = ');
```

```
den = input('Type in denominator coefficients = ');
[r,p,k] = residuez(num,den);
disp('Residues');disp(r')
disp('Poles');disp(p')
disp('Constants');disp(k)
```

During execution the program calls for the input data that are the vectors num and den of the coefficients of the numerator and the denominator polynomials, respectively, in descending powers of z. These data are entered inside square brackets as follows:

```
num = [18]
den = [18   3   -4   -1]
```

The output data are the residues and the constants, the poles of the expansion of $G(z)$ in the form of Eq. (3.137). For our above example, these are as given below.

```
Residues
    0.3600     0.2400     0.4000


Poles
    0.5000    -0.3333    -0.3333


Constants
```

Note also that the z-transform of Eq. (3.139) has double poles at $z = -1/3 = -0.3333$. The first entry in both the residues and poles given above corresponds to the simple pole factor $(1 - 0.5z^{-1})$, the second entry corresponds to the simple pole factor $(1 + 0.3333z^{-1})$, and the third entry corresponds to the factor $(1 + 0.3333z^{-1})^2$ in the partial-fraction expansion. Thus the desired expansion is given by

$$G(z) = \frac{0.36}{1 - 0.5z^{-1}} + \frac{0.24}{1 + 0.3333z^{-1}} + \frac{0.4}{(1 + 0.3333z^{-1})^2}. \tag{3.140}$$

EXAMPLE 3.34   We now consider the determination of the rational form of a z-transform from its partial-fraction expansion representation as given by Eq. (3.140). The MATLAB program that can be used to find the rational form is given below.

```
% Program 3_10
% Partial-Fraction Expansion to Rational z-Transform
%
r = input('Type in the residues = ');
p = input('Type in the poles = ');
k = input('Type in the constants = ');
[num, den] = residuez(r,p,k);
disp('Numerator polynomial coefficients'); disp(num)
disp('Denominator polynomial coefficients'); disp(den)
```

During execution, the program requests the input data vector r of residues, the vector p of pole locations, and the vector k of constants, with each entered using square brackets as follows:

```
r = [0.4 0.24 0.36]
p = [-0.3333  -0.3333  0.5]
```

```
        k = [0]
```

It then prints the coefficients of the numerator and the denominator polynomials, respectively, in descending powers of z as indicated below. It can be seen that the coefficients are exactly the same as in Eq. (3.139) if we multiply each coefficient by 18.

```
    Numerator polynomial coefficients
        1.0000    -0.0000         0            0

    Denominator polynomial coefficients
        1.0000     0.1667    -0.2222    -0.0556
```

3.9.4 Inverse z-Transform via Long Division

For causal sequences, the z-transform $G(z)$ can be expanded into a power series in z^{-1}. In the series expansion, the coefficient multiplying the term z^{-n} is then the nth sample $g[n]$. For a rational $G(z)$, a convenient way to determine the power series is to express the numerator and the denominator as polynomials in z^{-1}, and then obtain the power series expansion by long division.

EXAMPLE 3.35   Consider the causal $H(z)$ of Example 3.32 as given in Eq. (3.134). Multiplying the numerator and the denominator by z^{-2}, we get

$$H(z) = \frac{1 + 2.0z^{-1}}{1 + 0.4z^{-1} - 0.12z^{-2}}. \tag{3.141}$$

Long division of the numerator by the denominator yields

$$
\begin{array}{r}
1 + 1.6z^{-1} - 0.52z^{-2} + 0.4z^{-3} - 0.2224z^{-4} + \cdots \\
1 + 0.4z^{-1} - 0.12z^{-2} \enclose{longdiv}{1 + 2.0z^{-1}} \\
1 + 0.4z^{-1} - 0.12z^{-2} \\
\hline
1.6z^{-1} + 0.12z^{-2} \\
1.6z^{-1} + 0.64z^{-2} - 0.192z^{-3} \\
\hline
-0.52z^{-2} + 0.192z^{-3} \\
-0.52z^{-2} - 0.208z^{-3} + 0.0624z^{-4} \\
\hline
0.400z^{-3} - 0.0624z^{-4} \\
0.400z^{-3} + 0.1600z^{-4} - 0.0480z^{-5} \\
\hline
-0.2224z^{-4} + 0.0480z^{-5} \\
\cdots \cdots \\
\cdots \cdots
\end{array}
$$

As a result, $H(z)$ can be expressed as

$$H(z) = 1.0 + 1.6z^{-1} - 0.52z^{-2} + 0.4z^{-3} - 0.2224z^{-4} + \cdots$$

which leads to

$$\{h[n]\} = \{\underset{\uparrow}{1.0}, \quad 1.6, \quad -0.52, \quad 0.4, \quad -0.2224, \ldots\}, \qquad \text{for } n \geq 0.$$

By computing the values of $h[n]$ from Eq. (3.136) for $n = 0, 1, 2, 3$, and 4, the correctness of the sample values obtained via the long-division method can be easily verified.

3.9.5 Inverse *z*-Transform Using MATLAB

The inverse of a rational *z*-transform can also be readily calculated using MATLAB. The function `impz` can be utilized for this purpose. Three versions of this function are as follows:

```
[h,t] = impz(num,den),   [h,t] = impz(num,den,L),
[h,t] = impz(num,den,L,FT)
```

where the input data consist of the vectors `num` and `den` containing the coefficients of the numerator and the denominator polynomials of the *z*-transform given in descending powers of *z*, the output impulse response vector `h`, and the time index vector `t`. In the first form, the length `L` of `h` is determined automatically by the computer with `t = 0:L-1`, whereas in the remaining two forms it is supplied by the user through the input datum `L`. In the last form, the time interval is scaled so that the sampling interval is equal to the reciprocal of `FT`. The default value of `FT` is 1.

Another way of arriving at this result using MATLAB is by making use of the M-file `filter` as indicated below:

```
y = filter(num,den,x)
```

where `y` is the output vector containing the coefficients of the power series representation of $H(z)$ in increasing powers of z^{-1}. The numerator and denominator coefficients of $H(z)$ expressed in ascending powers of z^{-1} are two of the input data vectors `num` and `den`. The length of `x` is the same as that of `y`, and its elements are all zeros except for the first one which is a 1.

We present next two examples to illustrate the use of both functions.

EXAMPLE 3.36   We first illustrate the use of the function `impz` to determine the first 11 coefficients of the inverse *z*-transform of Eq. (3.134). To this end, we use the following MATLAB program:

```
% Program 3_11
% Power Series Expansion of a Rational z-Transform
%
% Read in the number of inverse z-transform coefficients
% to be computed
L = input('Type in the length of output vector = ');
% Read in the numerator and denominator coefficients of
% the z-transform
num = input('Type in the numerator coefficients = ');
den = input('Type in the denominator coefficients = ');
% Compute the desired number of inverse transform
% coefficients
[y,t] = impz(num,den,L);
disp('Coefficients of the power series expansion');
disp(y)
```

The input data called by the program are the desired number of coefficients and the vectors of the numerator and denominator coefficients of the rational *z*-transform, with these last entered inside square brackets. In our example, these are as follows:

```
L = 11
num = [1   2]
den = [1   0.4   -0.12]
```

The output data are the desired inverse *z*-transform coefficients as indicated below:

```
Coefficients of the power series expansion
  Columns 1 through 7
    1.0000   1.6000   -0.5200   0.4000   -0.2224   0.1370
   -0.0815

  Columns 8 through 11
    0.0490   -0.0294   0.0176   -0.0106
```

whose first five coefficients are seen to be identical to that derived in Example 3.35 using the long-division approach.

EXAMPLE 3.37   We determine the inverse *z*-transform of Eq. (3.134). The MATLAB program that can be used to determine the inverse *z*-transform of a rational *z*-transform is given below.

```
% Program 3_12
% Power Series Expansion of a Rational z-Transform
%
% Read in the number of inverse z-transform coefficients
% to be computed
N = input('Type in the length of output vector = ');
% Read in the numerator and denominator coefficients of
% the z-transform
num = input('Type in the numerator coefficients = ');
den = input('Type in the denominator coefficients = ');
% Compute the desired number of inverse transform
% coefficients
x = [1 zeros(1, N-1)];
y = filter(num, den, x);
disp('Coefficients of the power series expansion');
disp(y)
```

As the program is run, it asks for the desired length of the output vector y followed by the vectors containing the numerator and denominator coefficients entered inside square brackets and in descending powers of z. After computing the output vector, it prints its elements that are the coefficients of the power series expansion of the rational *z*-transform. The program assumes that the *z*-transform is a proper fraction with the degree of the denominator being greater than or equal to that of the numerator.

An application of the above program to the *z*-transform of Eq. (3.134) yields the following output:

```
Coefficients of the power series expansion
    1.0000   1.6000   -0.5200   0.4000   -0.2224
```

which is seen to be identical to that derived in Example 3.35 using the long-division approach.

3.10   *z*-Transform Properties

We summarize in Table 3.9 some specific properties of the *z*-transform. An understanding of these properties makes the application of the *z*-transform techniques to the analysis and design of digital filters often easier. The reader is therefore encouraged to verify these properties. We consider the applications of some of these properties in the next several examples.

Table 3.9: Some useful properties of the z-transform.

Property	Sequence	z-Transform	ROC
	$g[n]$	$G(z)$	$\mathcal{R}_g$
	$h[n]$	$H(z)$	$\mathcal{R}_h$
Conjugation	$g^*[n]$	$G^*(z^*)$	$\mathcal{R}_g$
Time-reversal	$g[-n]$	$G(1/z)$	$1/\mathcal{R}_g$
Linearity	$\alpha g[n] + \beta h[n]$	$\alpha G(z) + \beta H(z)$	Includes $\mathcal{R}_g \cap \mathcal{R}_h$
Time-shifting	$g[n - n_o]$	$z^{-n_o} G(z)$	$\mathcal{R}_g$, except possibly the point $z = 0$ or ∞
Multiplication by an exponential sequence	$\alpha^n g[n]$	$G(z/\alpha)$	$\lvert\alpha\rvert\mathcal{R}_g$
Differentiation of $G(z)$	$ng[n]$	$-z\dfrac{dG(z)}{dz}$	$\mathcal{R}_g$, except possibly the point $z = 0$ or ∞
Convolution	$g[n] \circledast h[n]$	$G(z)H(z)$	Includes $\mathcal{R}_g \cap \mathcal{R}_h$
Modulation	$g[n]h[n]$	$\frac{1}{2\pi j} \oint_C G(v)H(z/v)v^{-1}\, dv$	Includes $\mathcal{R}_g \mathcal{R}_h$
Parseval's relation	$\displaystyle\sum_{n=-\infty}^{\infty} g[n]h^*[n] = \frac{1}{2\pi j} \oint_C G(v)H^*(1/v^*)v^{-1}\, dv$		

Note: If $\mathcal{R}_g$ denotes the region $R_{g^-} < \lvert z \rvert < R_{g^+}$ and $\mathcal{R}_h$ denotes the region $R_{h^-} < \lvert z \rvert < R_{h^+}$, then $1/\mathcal{R}_g$ denotes the region $1/R_{g^+} < \lvert z \rvert < 1/R_{g^-}$ and $\mathcal{R}_g\mathcal{R}_h$ denotes the region $R_{g^-}R_{h^-} < \lvert z \rvert < R_{g^+}R_{h^+}$.

EXAMPLE 3.38   Consider the two-sided sequence

$$v[n] = \alpha^n \mu[n] - \beta^n \mu[-n - 1].$$

Denote $x_1[n] = \alpha^n \mu[n]$ and $x_2[n] = -\beta^n \mu[-n - 1]$. Therefore, due to the linearity property of the z-transform, the z-transform $V(z)$ of the sequence $v[n]$ is given by the sum $X_1(z) + X_2(z)$, where $X_1(z)$ and $X_2(z)$ are the z-transforms of $x_1[n]$ and $x_2[n]$, respectively.

Now, from Eq. (3.110)

$$X_1(z) = \frac{1}{1 - \alpha z^{-1}}, \qquad \text{for } \lvert z \rvert > \lvert \alpha \rvert,$$

and from Eq. (3.112)

$$X_2(z) = \frac{1}{1 - \beta z^{-1}}, \qquad \text{for } \lvert z \rvert < \lvert \beta \rvert.$$

Therefore, using the linearity property of Table 3.9, we arrive at

$$V(z) = \frac{1}{1 - \alpha z^{-1}} + \frac{1}{1 - \beta z^{-1}}.$$

The ROC of $V(z)$ is given by the overlap regions of the ROCs of $X_1(z)$ and $X_2(z)$, respectively. If $\lvert\beta\rvert > \lvert\alpha\rvert$, then there is an overlap between the ROCs of $X_1(z)$ and $X_2(z)$, and the ROC of $V(z)$ is an annular region $\lvert\alpha\rvert < \lvert z \rvert < \lvert\beta\rvert$. However, if $\lvert\beta\rvert < \lvert\alpha\rvert$, then there is no overlap, and as a result $V(z)$ does not exist.

EXAMPLE 3.39   Determine the *z*-transform $X(z)$ of the causal real sequence $x[n] = r^n(\cos\omega_o n)\mu[n]$ and its ROC. Now we can express $x[n]$ as a sum of two causal exponential sequences:

$$x[n] = \tfrac{1}{2}r^n e^{j\omega_o n}\mu[n] + \tfrac{1}{2}r^n e^{-j\omega_o n}\mu[n].$$

We can write the above expression as $x[n] = v[n] + v^*[n]$, where

$$v[n] = \tfrac{1}{2}\alpha^n \mu[n],$$

with $\alpha = re^{j\omega_o}$. The *z*-transform $V(z)$ of $v[n]$ from Table 3.8 is given by

$$V(z) = \frac{1}{2} \cdot \frac{1}{1 - \alpha z^{-1}} = \frac{1}{2} \cdot \frac{1}{1 - re^{j\omega_o} z^{-1}}, \qquad |z| > |\alpha| = r.$$

Using Table 3.9, we obtain the *z*-transform of $v^*[n]$ as $V^*(z^*)$:

$$V^*(z^*) = \frac{1}{2} \cdot \frac{1}{1 - \alpha^* z^{-1}} = \frac{1}{2} \cdot \frac{1}{1 - re^{-j\omega_o} z^{-1}}, \qquad |z| > |\alpha| = r.$$

Therefore, by the linearity property of the *z*-transform, we obtain

$$X(z) = V(z) + V^*(z^*) = \frac{1}{2}\left(\frac{1}{1 - re^{j\omega_o} z^{-1}} + \frac{1}{1 - re^{-j\omega_o} z^{-1}}\right)$$

$$= \frac{1 - (r\cos\omega_o)z^{-1}}{1 - (2r\cos\omega_o)z^{-1} + r^2 z^{-2}}, \qquad |z| > r.$$

EXAMPLE 3.40   Determine the *z*-transform $Y(z)$ and the ROC of sequence $y[n] = (n+1)\alpha^n\mu[n]$. Let $x[n] = \alpha^n\mu[n]$. Then we can write $y[n] = n\,x[n] + x[n]$. From Eq. (3.110), the *z*-transform of $x[n]$ is given by

$$X(z) = \frac{1}{1 - \alpha z^{-1}}, \qquad |z| > |\alpha|.$$

Next, using the differentiation property of Table 3.9, we arrive at the *z*-transform of $n\,x[n]$ as

$$-z\frac{dX(z)}{dz} = \frac{\alpha z^{-1}}{(1 - \alpha z^{-1})^2}, \qquad |z| > |\alpha|.$$

Using the linearity property of Table 3.9 we obtain

$$Y(z) = \frac{1}{1 - \alpha z^{-1}} + \frac{\alpha z^{-1}}{(1 - \alpha z^{-1})^2} = \frac{1}{(1 - \alpha z^{-1})^2}, \qquad |z| > |\alpha|.$$

EXAMPLE 3.41   Determine the inverse *z*-transform of

$$G(z) = \frac{z^3}{\left(z - \frac{1}{2}\right)\left(z + \frac{1}{3}\right)^2}, \qquad |z| > \tfrac{1}{2}. \tag{3.142}$$

As the ROC is exterior to a circle of radius $1/2$, the inverse transform is a right-sided sequence. The partial-fraction expansion of $G(z)$ has been determined in Example 3.33 and is given by Eq. (3.140). From Table 3.8, the inverse *z*-transform of $1/(1 - 0.5z^{-1})$ is $(0.5)^n\mu[n]$, and the inverse *z*-transform of $1/(1 + (1/3)z^{-1})$ is $(-1/3)^n\mu[n]$.

To determine the inverse z-transform of $1/(1 + (1/3)z^{-1})^2$ we observe from Table 3.9 that the z-transform of $n(-1/3)^n \mu[n]$ is given by

$$-z(d/dz)\left(1/\left(1 + \frac{1}{3}z^{-1}\right)\right) = -\frac{1}{3}z^{-1}/\left(1 + \frac{1}{3}z^{-1}\right)^2.$$

As a result the inverse z-transform of $1/(1 + (1/3)z^{-1})^2$ is given by $-3(n-1)(-1/3)^{n-1}\mu[n-1]$. Therefore the inverse z-transform of $G(z)$ of Eq. (3.142) is

$$g[n] = \left[0.24\left(-\frac{1}{3}\right)^n + 0.36\left(\frac{1}{2}\right)^n\right]\mu[n] + 0.36(n-1)\left(-\frac{1}{3}\right)^n \mu[n-1].$$

The time-reversal property and the convolution theorem of Table 3.9 can be employed to develop the expression for the z-transform of the cross-correlation sequence $r_{gh}[\ell]$ of two sequences $g[n]$ and $h[n]$ in terms of their z-transforms. Let the z-transform of $g[n]$ be denoted by $G(z)$ with an ROC $\mathcal{R}_g$, and let the z-transform of $hg[n]$ be denoted by $H(z)$ with an ROC $\mathcal{R}_h$. Now, recall from Eq. (2.106) that the cross-correlation sequence $r_{gh}[\ell]$ can be expressed in terms of convolution as

$$r_{gh}[\ell] = g[\ell] \circledast h[-\ell].$$

Using the time-reversal property, we observe that the z-transform of $h[-\ell]$ is simply $H(z^{-1})$. Therefore, using the convolution theorem we obtain from the above equation that

$$\mathcal{Z}\left\{r_{gh}[\ell]\right\} = G(z)H(z^{-1}), \qquad (3.143)$$

with the ROC given by at least $\mathcal{R}_g \cap (1/\mathcal{R}_h)$.

As in the case of the discrete-time Fourier transform, the Parseval's relation for the z-transform given in Table 3.9 can also be used to compute the energy of a sequence. To establish the required formula, we let $g[n] = h[n]$, where $g[n]$ is a real sequence, in the expression given in Table 3.9. This leads to

$$\sum_{n=-\infty}^{\infty} g^2[n] = \frac{1}{2\pi j}\oint_C G(z)G(z^{-1})z^{-1}\,dz. \qquad (3.144)$$

where C is a closed contour in the ROC of $G(z)G(z^{-1})$. Note that if the ROC of $G(z)$ includes the unit circle, then that of $G(z^{-1})$ will also include the unit circle. In fact, for an absolutely summable sequence $g[n]$, the ROC of its z-transform $G(z)$ must include the unit circle. In this case, we can let $z = e^{j\omega}$ in Eq. (3.144), which then reduces to Eq. (3.18).

3.11 Transform-Domain Representations of Random Signals

The notion of random discrete-time signals was introduced in Section 2.8 along with their statistical characterizations in the time-domain. These infinite-length signals have infinite energy and do not have transform-domain characterizations like the deterministic signals. However, the autocorrelation and the autocovariance sequences of stationary random signals, defined by Eqs. (2.147) and (2.150), are of finite energy and, in most practical cases, their transform-domain representations do exist. We discuss these representations in this section.

3.11.1 Discrete-Time Fourier Transform Representation

The discrete-time Fourier transform of the autocorrelation sequence $\phi_{XX}[\ell]$ of a WSS sequence $X[n]$, defined in Eq. (2.147), is given by

$$\Phi_{XX}(e^{j\omega}) = \sum_{\ell=-\infty}^{\infty} \phi_{XX}[\ell]e^{-j\omega\ell}, \qquad |\omega| < \pi, \tag{3.145}$$

and is usually referred to as the *power density spectrum* or simply, the *power spectrum* [10] of $X[n]$. It is denoted by $\mathcal{P}_{XX}(\omega)$. The above relation between the autocorrelation sequence and the power spectrum is more commonly known as the *Wiener-Khintchine theorem*. A sufficient condition for the existence of the power spectrum $\mathcal{P}_{XX}(\omega)$ is that the autocorrelation sequence $\phi_{XX}[\ell]$ be absolutely summable. Likewise, the discrete-time Fourier transform of the autocovariance sequence $\gamma_{XX}[\ell]$ of a WSS sequence $x[n]$, defined in Eq. (2.150), is given by

$$\Gamma_{XX}(e^{j\omega}) = \sum_{\ell=-\infty}^{\infty} \gamma_{XX}[\ell]e^{-j\omega\ell}, \qquad |\omega| < \pi. \tag{3.146}$$

A sufficient condition for the existence of $\Gamma_{XX}(e^{j\omega})$ is that the autocovariance sequence $\gamma_{XX}[\ell]$ be absolutely summable. Applying the inverse discrete-time Fourier transform to Eq. (3.145) and using the notation $\mathcal{P}_{XX}(\omega) = \Phi_{XX}(e^{j\omega})$, we arrive at

$$\phi_{XX}[\ell] = \frac{1}{2\pi} \int_{-\pi}^{\pi} \mathcal{P}_{XX}(\omega)e^{j\omega\ell}\, d\omega. \tag{3.147}$$

It follows from Eqs. (3.147) and (2.162) that

$$E\left(|X[n]|^2\right) = \phi_{XX}[0] = \frac{1}{2\pi} \int_{-\pi}^{\pi} \mathcal{P}_{XX}(\omega)\, d\omega. \tag{3.148}$$

Thus, $\phi_{XX}[0]$ represents the *average power* in the random signal $X[n]$. Similarly, the inverse transform of Eq. (3.146) yields

$$\gamma_{XX}[\ell] = \frac{1}{2\pi} \int_{-\pi}^{\pi} \Gamma_{XX}(e^{j\omega})e^{j\omega\ell}\, d\omega. \tag{3.149}$$

From Eqs. (3.149) and (2.163) we get

$$\sigma_X^2 = \gamma_{XX}[0] = \frac{1}{2\pi} \int_{-\pi}^{\pi} \Gamma_{XX}(e^{j\omega})\, d\omega$$

$$= \frac{1}{2\pi} \int_{-\pi}^{\pi} \mathcal{P}_{XX}(\omega)\, d\omega - |m_X|^2. \tag{3.150}$$

Applying the discrete-time Fourier transform to both sides of Eq. (2.166a), we can show that the power spectrum $\mathcal{P}_{XX}(\omega)$ of a WSS random discrete-time signal $\{X[n]\}$ is a real-valued function of ω. In addition, if $\{X[n]\}$ is a real random signal, $\mathcal{P}_{XX}(\omega)$ is an even function of ω, i.e., $\mathcal{P}_{XX}(\omega) = \mathcal{P}_{XX}(-\omega)$. We shall demonstrate later in Section 4.13.2 that for a real-valued WSS random signal, $\mathcal{P}_{XX}(\omega) \geq 0$.

---

[10]Also called the *power spectral density*.

Likewise, the discrete-time Fourier transform of the cross-correlation sequence $\phi_{XY}[\ell]$ of two jointly stationary random signals $\{X[n]\}$ and $\{Y[n]\}$, given by

$$\Phi_{XY}(e^{j\omega}) = \sum_{\ell=-\infty}^{\infty} \phi_{XY}[\ell] e^{-j\omega\ell}, \qquad |\omega| < \pi, \tag{3.151}$$

is usually referred to as the *cross-power spectral density* or *cross-power spectrum*. It is denoted by $\mathcal{P}_{XY}(\omega)$ and, in general, it is a complex function of ω. A sufficient condition for the existence of $\mathcal{P}_{XY}(\omega)$ is that the cross-correlation sequence $\phi_{XY}[\ell]$ be absolutely summable. Applying the discrete-time Fourier transform to both sides of Eq. (2.166c), we can show that $\mathcal{P}_{XY}(\omega) = \mathcal{P}_{YX}^*(\omega)$. Similarly, we can define the discrete-time Fourier transform of the cross-covariance sequence $\gamma_{XY}[\ell]$ as

$$\Gamma_{XY}(e^{j\omega}) = \sum_{\ell=-\infty}^{\infty} \gamma_{XY}[\ell] e^{-j\omega\ell}, \qquad |\omega| < \pi. \tag{3.152}$$

A sufficient condition for the existence of $\Gamma_{XY}(e^{j\omega})$ is that the cross-covariance sequence $\gamma_{XY}[\ell]$ be absolutely summable.

The relation between the discrete-time Fourier transforms of the autocorrelation sequence and the autocovariance sequence can be derived from Eq. (2.161) and is given by

$$\Gamma_{XX}(e^{j\omega}) = \mathcal{P}_{XX}(\omega) - 2\pi |m_X|^2 \delta(\omega), \qquad |\omega| < \pi, \tag{3.153}$$

where we have used the notation $\mathcal{P}_{XX}(\omega) = \Phi_{XX}(e^{j\omega})$. Likewise, the relation between the discrete-time Fourier transforms of the cross-correlation sequence and the cross-covariance sequence follows from Eq. (2.165) and is given by

$$\Gamma_{XY}(e^{j\omega}) = \mathcal{P}_{XY}(\omega) - 2\pi m_X m_Y^* \delta(\omega), \qquad |\omega| < \pi, \tag{3.154}$$

where we have used the notation $\mathcal{P}_{XY}(\omega) = \Phi_{XY}(e^{j\omega})$.

3.11.2 *z*-Transform Representation

As can be seen from Eqs. (3.153) and (3.154), the Fourier transforms of the sequences $\gamma_{XX}[\ell]$ and $\gamma_{XY}[\ell]$ contain impulse functions. As a result, their z-transforms do not exist in general. However, for zero-mean stationary random signals, the z-transform of the autocorrelation sequence, $\Phi_{XX}(z)$, and that of the cross-correlation sequence, $\Phi_{XY}(z)$, may exist under certain conditions. Since the autocorrelation and cross-correlation sequences are two-sided sequences, their region of convergence must be an annular region of the form

$$R_1 < |z| < \frac{1}{R_1}. \tag{3.155}$$

We can generalize some of the results of the previous section if the z-transforms exist. For example, from the symmetry properties of the power spectrum $\mathcal{P}_{XX}(\omega)$ and cross-power spectrum $\mathcal{P}_{XY}(\omega)$, it follows that $\Phi_{XX}(z) = \Phi_{XX}^*(1/z^*)$ and $\Phi_{XY}(z) = \Phi_{YX}^*(1/z^*)$. It also follows from Eqs. (3.150) and (3.153) that

$$\sigma_X^2 = \frac{1}{2\pi j} \oint_C \Phi_{XX}(z) z^{-1} \, dz, \tag{3.156}$$

where C is a closed counterclockwise contour in the ROC of $\Phi_{XX}(z)$.

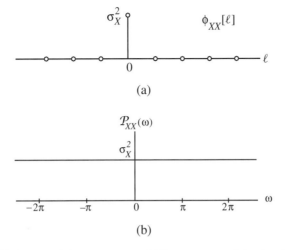

Figure 3.24: (a) Autocorrelation sequence and (b) power spectral density of a white noise.

3.11.3 White Noise

A random process $\{X[n]\}$ for which any pair of samples, $X[m]$ and $X[n]$ are uncorrelated with $m \neq n$, i.e., $E(X[m]X[n]) = E(X[m])E(X[n])$, is called a *white random process*. For a WSS white random process, the autocorrelation sequence is given by

$$\phi_{XX}[\ell] = \sigma_X^2 \delta[\ell] + m_X^2, \tag{3.157}$$

and the corresponding power spectrum is given by

$$\mathcal{P}_{XX}(\omega) = \sigma_X^2 + 2\pi m_X^2 \delta(\omega), \qquad |\omega| < \pi. \tag{3.158}$$

A zero-mean white WSS random process has an autocorrelation sequence $\phi_{XX}[\ell]$ that is an impulse sequence of area σ_x^2 and a power spectrum $\mathcal{P}_{XX}(\omega)$ that is of constant value σ_x^2 for all values of ω, as indicated in Figure 3.24. Such a random process is more commonly called *white noise* and plays an important role in digital signal processing.

3.12 Summary

Three different frequency-domain representations of an aperiodic discrete-time sequence have been introduced and their properties reviewed. Two of these representations, the discrete-time Fourier transform (DTFT) and the z-transform, are applicable to any arbitrary sequence, whereas the third one, the discrete Fourier transform (DFT), can be applied only to finite-length sequences. Each of these representations consists of a pair of expressions: the analysis equation and the synthesis equation. The analysis equation is used to convert from the time-domain representation to the frequency-domain representation, while the synthesis equation is used for the reverse process.

Relations between these three transforms have been established. The chapter ends with a discussion on the transform-domain representation of a random discrete-time sequence.

For future convenience we summarize below these three frequency-domain representations.

Discrete-Time Fourier Transform (DTFT)

Analysis equation:

$$X(e^{j\omega}) = \sum_{n=-\infty}^{\infty} x[n]e^{-j\omega n}. \tag{3.159}$$

Synthesis equation:

$$x[n] = \frac{1}{2\pi} \int_{-\pi}^{\pi} X(e^{j\omega})e^{j\omega n} \, d\omega. \tag{3.160}$$

Discrete Fourier Transform (DFT)

Analysis equation:

$$X[k] = \sum_{n=0}^{N-1} x[n]W_N^{kn}, \qquad 0 \le k \le N-1, \tag{3.161}$$

Synthesis equation:

$$x[n] = \frac{1}{N} \sum_{k=0}^{N-1} X[k]W_N^{-kn}, \qquad 0 \le n \le N-1, \tag{3.162}$$

where

$$W_N = e^{-j2\pi/N}. \tag{3.163}$$

z-Transform

Analysis equation:

$$X(z) = \sum_{n=-\infty}^{\infty} x[n]z^{-n}. \tag{3.164}$$

Synthesis equation:

$$x[n] = \frac{1}{2\pi j} \oint_C X(z)z^{n-1} \, dz, \quad C \text{ in ROC of } X(z). \tag{3.165}$$

We apply the concepts developed in this chapter to the representation, analysis, and design of the so-called linear, time-invariant (LTI) discrete-time systems in the following chapter.

3.13 Problems

3.1 Let $X(e^{j\omega})$ denote the DTFT of a real sequence $x[n]$. Show that the real part $X_{re}(e^{j\omega})$ and the magnitude function $|X(e^{j\omega})|$ of $X(e^{j\omega})$ are even functions of ω, and the imaginary part $X_{im}(e^{j\omega})$ and the phase funtion $\arg\{X(e^{j\omega})\}$ are odd functions of ω.

3.2 Determine $X_{re}(e^{j\omega})$, $X_{im}(e^{j\omega})$, $|X(e^{j\omega})|$, and $\arg\{X(e^{j\omega})\}$ of the DTFT $X(e^{j\omega})$ evaluated in Example 3.2.

3.3 Derive the DTFTs of the following sequences given in Table 3.1: (a) $\mu[n]$, and (b) $e^{j\omega_o n}$.

3.4 Show that the DTFT of $x[n] = 1$, $-\infty < n < \infty$, is given by $X(e^{j\omega}) = \sum_{k=-\infty}^{\infty} 2\pi\delta(\omega + 2\pi k)$.

3.5 Prove the following properties of the DTFT listed in Table 3.2: (a) Time-shifting, (b) frequency-shifting, (c) differentiation in frequency, (d) convolution, (e) modulation, and (f) Parseval's relation.

3.6 Let $X(e^{j\omega})$ denote the DTFT of a complex sequence $x[n]$. Express the DTFTs of the following sequences in terms of $X(e^{j\omega})$: (a) $x[-n]$, (b) $x^*[-n]$, (c) $\text{Re}\{x[n]\}$, (d) $j\text{Im}\{x[n]\}$, (e) $x_{cs}[n]$, and (f) $x_{ca}[n]$.

3.7 Let $X(e^{j\omega})$ denote the DTFT of a real sequence $x[n]$. Determine the inverse of the following DTFTs in terms of $x[n]$: (a) $X_{re}(e^{j\omega})$, (b) $jX_{im}(e^{j\omega})$.

3.8 Let $X(e^{j\omega})$ denote the DTFT of a real sequence $x[n]$. Prove the following symmetry relations: (a) $X(e^{j\omega}) = X^*(e^{-j\omega})$, (b) $X_{re}(e^{j\omega}) = X_{re}(e^{-j\omega})$, (c) $X_{im}(e^{j\omega}) = -X_{im}(e^{-j\omega})$, (d) $|X(e^{j\omega})| = |X(e^{-j\omega})|$, and (e) $\arg\{X(e^{j\omega})\} = -\arg\{X(e^{-j\omega})\}$.

3.9 Let $X(e^{j\omega})$ denote the DTFT of a real sequence $x[n]$.

(a) Show that if $x[n]$ is even, then it can be computed from $X(e^{j\omega})$ using: $x[n] = \frac{1}{\pi} \int_0^\pi X(e^{j\omega}) \cos \omega n \, d\omega$.

(b) Show that if $x[n]$ is odd, then it can be computed from $X(e^{j\omega})$ using: $x[n] = \frac{j}{\pi} \int_0^\pi X(e^{j\omega}) \sin \omega n \, d\omega$.

3.10 Determine the DTFT of the causal sequence $x[n] = A\alpha^n \cos(\omega_o n + \phi)\mu[n]$, where A, α, ω_o, and ϕ are real.

3.11 Determine the DTFT of each of the following sequences.

(a) $x_1[n] = \alpha^n \mu[n+1]$, $|\alpha| < 1$,   (b) $x_2[n] = n\alpha^n \mu[n]$, $|\alpha| < 1$,

(c) $x_3[n] = \begin{cases} \alpha^{|n|}, & |n| \le M, \\ 0, & \text{otherwise,} \end{cases}$ (d) $x_4[n] = \alpha^n \mu[n-3]$, $|\alpha| < 1$,

(e) $x_5[n] = n\alpha^n \mu[n+2]$, $|\alpha| < 1$,   (f) $x_6[n] = \alpha^n \mu[-n-1]$, $|\alpha| > 1$.

3.12 Determine the DTFT of each of the following sequences:

(a) $y_1[n] = \begin{cases} 1, & -N \le n \le N, \\ 0, & \text{otherwise,} \end{cases}$

(b) $y_2[n] = \begin{cases} 1 - \frac{|n|}{N}, & -N \le n \le N, \\ 0, & \text{otherwise,} \end{cases}$

(c) $y_3[n] = \begin{cases} \cos(\pi n/2N), & -N \le n \le N, \\ 0, & \text{otherwise.} \end{cases}$

3.13 Show that the inverse DTFT of

$$X(e^{j\omega}) = \frac{1}{(1 - \alpha e^{-j\omega})^m}, \quad |\alpha| < 1,$$

is given by

$$x[n] = \frac{(n+m-1)!}{n!(m-1)!} \alpha^n \mu[n].$$

3.14 Evaluate the inverse DTFT of each of the following DTFTs:

(a) $X_a(e^{j\omega}) = \sum_{k=-\infty}^{\infty} \delta(\omega + 2\pi k)$,   (b) $X_b(e^{j\omega}) = \frac{1 - e^{j\omega(N+1)}}{1 - e^{j\omega}}$,

(c) $X_c(e^{j\omega}) = 1 + 2\sum_{\ell=0}^{N} \cos \omega \ell$,   (d) $X_d(e^{j\omega}) = \frac{j\alpha e^{j\omega}}{(1 - \alpha e^{j\omega})^2}$, $|\alpha| < 1$.

3.15 Determine the inverse DTFT of each of the following DTFTs:

(a) $H_1(e^{j\omega}) = 1 + 2\cos \omega + 3\cos 2\omega$,   (b) $H_2(e^{j\omega}) = (3 + 2\cos \omega + 4\cos 2\omega)\cos(\omega/2)e^{-j\omega/2}$,

(c) $H_3(e^{j\omega}) = j(3 + 4\cos \omega + 2\cos 2\omega)\sin \omega$,   (d) $H_4(e^{j\omega}) = j(4 + 2\cos \omega + 3\cos 2\omega)\sin(\omega/2)e^{j\omega/2}$.

3.16 Determine the inverse DTFT of each of the following DTFTs:

(a) $H_1(e^{j\omega}) = 1 + 2\cos\omega + 3\cos^2\omega,$ (b) $H_2(e^{j\omega}) = (2 + 3\cos\omega + 4\cos^2\omega)\cos(\omega/2)e^{j\omega/2},$

(c) $H_3(e^{j\omega}) = j(3 + 4\cos\omega + 2\cos^2\omega)\sin\omega,$ (d) $H_4(e^{j\omega}) = j(4 + 2\cos\omega + 3\cos^2\omega)\sin(\omega/2)e^{-j\omega/2}.$

3.17 Let $X(e^{j\omega})$ denote the DTFT of a real sequence $x[n]$. Express the inverse DTFT $y[n]$ of $Y(e^{j\omega}) = X(e^{j3\omega})$ in terms of $x[n]$.

3.18 Let $X(e^{j\omega})$ denote the DTFT of a real sequence $x[n]$. Define $Y(e^{j\omega}) = \frac{1}{2}\left\{X(e^{j\omega/2}) + X(-e^{j\omega/2})\right\}$. Determine the inverse DTFT $y[n]$ of $Y(e^{j\omega})$.

3.19 Without computing the DTFT, determine which ones of the following sequences have real-valued DTFTs, and which ones have imaginary-valued DTFTs:

(a) $x_1[n] = \begin{cases} |n|, & -N \le n \le N, \\ 0, & \text{otherwise}, \end{cases}$ (b) $x_2[n] = \begin{cases} n^3, & -N \le n \le N, \\ 0, & \text{otherwise}, \end{cases}$

(c) $x_3[n] = \frac{\sin\omega_c n}{\pi n},$ (d) $x_4[n] = \begin{cases} 0, & \text{for } n \text{ even}, \\ \frac{2}{\pi n}, & \text{for } n \text{ odd}, \end{cases}$

(e) $x_5[n] = \begin{cases} 0, & n = 0, \\ \frac{\cos\pi n}{n}, & |n| > 0. \end{cases}$

3.20 Without computing the inverse DTFT, determine which ones of the following DTFTs have an inverse that is an even sequence, and which ones have an inverse that is an odd sequence:

(a) $Y_1(e^{j\omega}) = \begin{cases} |\omega|, & 0 \le |\omega| \le \omega_c, \\ 0, & \omega_c < |\omega| \le \pi, \end{cases}$ (b) $Y_2(e^{j\omega}) = j\omega, \quad 0 \le |\omega| \le \pi,$

(c) $Y_3(e^{j\omega}) = \begin{cases} j, & -\pi < \omega < 0, \\ -j & 0 < \omega < \pi. \end{cases}$

3.21 Without computing the inverse DTFT, determine which one of the DTFTs of Figure P3.1 has an inverse that is an even sequence, and which one has an inverse that is an odd sequence.

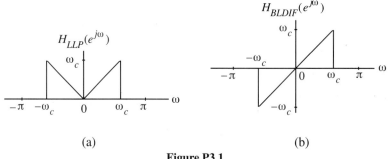

(a)   (b)

Figure P3.1

3.22 Let $X(e^{j\omega})$ denote the DTFT of a real sequence $x[n]$. Determine the DTFT $Y(e^{j\omega})$ of the sequence $y[n] = x[n] \circledast x[-n]$ in terms of $X(e^{j\omega})$ and show that it is a real-valued function of ω.

3.23 A sequence $x[n]$ has a zero-phase DTFT $X(e^{j\omega})$ as sketched in Figure P3.2. Sketch the DTFT of the sequence $x[n]e^{-j\pi n/3}$.

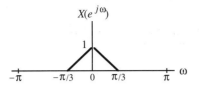

Figure P3.2

3.24 Using Parseval's relation evaluate the following integrals: (a) $\int_0^\pi \frac{4}{5+4\cos\omega} d\omega$, (b) $\int_0^\pi \frac{1}{3.25-3\cos\omega} d\omega$, and (c) $\int_0^\pi \frac{4}{(5-4\cos\omega)^2} d\omega$.

3.25 Let $x[n]$ be a length-9 sequence given by

$$\{x[n]\} = \{3 \quad 0 \quad 1 \quad -2 \quad -3 \quad 4 \quad 1 \quad 0 \quad -1\}$$
$$\uparrow$$

with a DTFT $X(e^{j\omega})$. Evaluate the following functions of $X(e^{j\omega})$ without computing the transform itself:

(a) $X(e^{j0})$,   (b) $X(e^{j\pi})$,

(c) $\int_{-\pi}^{\pi} X(e^{j\omega}) d\omega$,   (d) $\int_{-\pi}^{\pi} \left| X(e^{j\omega}) \right|^2 d\omega$,

(e) $\int_{-\pi}^{\pi} \left| \frac{dX(e^{j\omega})}{d\omega} \right|^2 d\omega$.

3.26 Repeat Problem 3.25 for the length-9 sequence

$$\{x[n]\} = \{-2 \quad 4 \quad -1 \quad 5 \quad -3 \quad -2 \quad 0 \quad 4 \quad 3\}.$$
$$\uparrow$$

3.27 Let $G_1(e^{j\omega})$ denote the discrete-time Fourier transform of the sequence $g_1[n]$ shown in Figure P3.3(a). Express the DTFTs of the remaining sequences in Figure P3.3 in terms of $G_1(e^{j\omega})$. Do not evaluate $G_1(e^{j\omega})$.

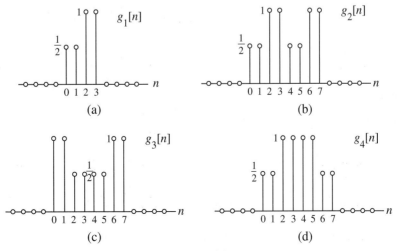

Figure P3.3

3.28 Let $y[n]$ denote the linear convolution of three sequences, $x_1[n]$, $x_2[n]$, and $x_3[n]$, i.e., $y[n] = x_1[n] \circledast x_2[n] \circledast x_3[n]$. Show that

(a) $\displaystyle \sum_{n=-\infty}^{\infty} y[n] = \left(\sum_{n=-\infty}^{\infty} x_1[n] \right) \left(\sum_{n=-\infty}^{\infty} x_2[n] \right) \left(\sum_{n=-\infty}^{\infty} x_3[n] \right),$

(b) $\displaystyle \sum_{n=-\infty}^{\infty} (-1)^n y[n] = \left(\sum_{n=-\infty}^{\infty} (-1)^n x_1[n] \right) \left(\sum_{n=-\infty}^{\infty} (-1)^n x_2[n] \right) \left(\sum_{n=-\infty}^{\infty} (-1)^n x_3[n] \right).$

3.29 Let $x[n]$ be a causal and absolutely summable real sequence with a DTFT $X(e^{j\omega})$. If $X_{\mathrm{re}}(e^{j\omega})$ and $X_{\mathrm{im}}(e^{j\omega})$ denote the real and imaginary parts of $X(e^{j\omega})$, show that they are related as

$$X_{\mathrm{im}}(e^{j\omega}) = -\frac{1}{2\pi} \int_{-\pi}^{\pi} X_{\mathrm{re}}(e^{j\phi}) \cot\left(\frac{\omega - \phi}{2} \right) d\phi, \qquad (3.166a)$$

$$X_{\mathrm{re}}(e^{j\omega}) = \frac{1}{2\pi} \int_{-\pi}^{\pi} X_{\mathrm{im}}(e^{j\phi}) \cot\left(\frac{\omega - \phi}{2} \right) d\phi + x[0]. \qquad (3.166b)$$

The above equations are called the *discrete Hilbert transform relations*.

3.30 Verify the identity of Eq. (3.28).

3.31 The *periodic convolution* of two periodic sequences, $\tilde{x}[n]$ and $\tilde{h}[n]$, of period N each, is defined by

$$\tilde{y}[n] = \sum_{r=0}^{N-1} \tilde{x}[r]\tilde{h}[n - r]. \qquad (3.167)$$

Show that $\tilde{y}[n]$ is also a periodic sequence of period N.

3.32 Determine the periodic sequence $\tilde{y}[n]$ obtained by a periodic convolution of the following two periodic sequences of period 5 each:

$$\tilde{x}[n] = \begin{cases} 0, & \text{for } n = 0, 2, \\ 1, & \text{for } n = 1, \\ -2, & \text{for } n = 3, \\ 3, & \text{for } n = 4, \end{cases} \qquad \tilde{h}[n] = \begin{cases} 2, & \text{for } n = 0, \\ 0, & \text{for } n = 1, 3, \\ 1, & \text{for } n = 2, \\ -2, & \text{for } n = 4. \end{cases}$$

3.33 Determine the periodic sequence $\tilde{y}[n]$ obtained by a periodic convolution of the following two periodic sequences of period 5 each:

$$\tilde{x}[n] = \begin{cases} 2, & \text{for } n = 0, 2, \\ -1, & \text{for } n = 1, \\ 3, & \text{for } n = 3, \\ -2, & \text{for } n = 4, \end{cases} \qquad \tilde{h}[n] = \begin{cases} 1, & \text{for } n = 0, \\ 2, & \text{for } n = 1, \\ -3, & \text{for } n = 2, 4, \\ 0, & \text{for } n = 3. \end{cases}$$

3.34 Let $\tilde{x}[n]$ be a periodic sequence with period N, i.e., $\tilde{x}[n] = \tilde{x}[n + \ell N]$, where ℓ is any integer. The sequence $\tilde{x}[n]$ can be represented by a Fourier series given by a weighted sum of periodic complex exponential sequences $\tilde{\Psi}_k[n] = e^{j2\pi kn/N}$. Show that, unlike the Fourier series representation of a periodic continuous-time signal, the Fourier series representation of a periodic discrete-time sequence requires only N of the periodic complex exponential sequences $\tilde{\Psi}_k[n]$, $k = 0, 1, \ldots, N - 1$, and is of the form

$$\tilde{x}[n] = \frac{1}{N} \sum_{k=0}^{N-1} \tilde{X}[k] e^{j2\pi kn/N}, \qquad (3.168a)$$

where the Fourier coefficients $\tilde{X}[k]$ are given by

$$\tilde{X}[k] = \sum_{k=0}^{N-1} \tilde{x}[n] e^{-j2\pi kn/N}. \tag{3.168b}$$

Show that $\tilde{X}[k]$ is also a periodic sequence in k with a period N. The set of equations in Eq. (3.168) represent the *discrete Fourier series* pair.

3.35 Determine the discrete Fourier series coefficients, defined in Eq. (3.168b), of the following periodic sequences:

(a) $\tilde{x}_1[n] = \cos(\pi n/4)$,

(b) $\tilde{x}_2[n] = \sin(\pi n/3) + 3\cos(\pi n/4)$.

3.36 Show using Eqs. (3.168a) and (3.168b) that the periodic impulse train

$$\tilde{p}[n] = \sum_{\ell=-\infty}^{\infty} \delta[n + \ell N]$$

can be expressed in the form

$$\tilde{p}[n] = \frac{1}{N} \sum_{\ell=0}^{N-1} e^{j2\pi kn/N}.$$

3.37 Let $x[n]$ be an aperiodic sequence with a DTFT $X(e^{j\omega})$. Define

$$\tilde{X}[k] = X(e^{j\omega})\Big|_{\omega=2\pi k/N} = X(e^{j2\pi k/N}), \qquad -\infty < k < \infty \ .$$

Show that $\tilde{X}[k]$ is a periodic sequence in k with a period N. Let $\tilde{X}[k]$ be the discrete Fourier series coefficients of the periodic sequence $\tilde{x}[n]$. Show using Eqs. (3.168a) and (3.168b) that

$$\tilde{x}[n] = \sum_{r=-\infty}^{\infty} x[n + rN].$$

3.38 Let $\tilde{x}[n]$ and $\tilde{y}[n]$ be two periodic sequences with period N. Denote their discrete Fourier series coefficients, defined in Eq. (3.168b), as $\tilde{X}[k]$ and $\tilde{Y}[k]$, respectively.

(a) Let $\tilde{g}[n] = \tilde{x}[n]\tilde{y}[n]$ with $\tilde{G}[k]$ denoting its discrete Fourier series coefficients. Show using Eqs. (3.168a) and (3.168b) that $\tilde{G}[k]$ can be expressed in terms of $\tilde{X}[k]$ and $\tilde{Y}[k]$ as

$$\tilde{G}[k] = \frac{1}{N} \sum_{\ell=0}^{N-1} \tilde{X}[\ell]\tilde{Y}[k - \ell]. \tag{3.169}$$

(b) Let $\tilde{H}[k] = \tilde{X}[k]\tilde{Y}[k]$ denote the discrete Fourier series coefficients of a periodic sequence $\tilde{h}[n]$. Show using Eqs. (3.168a) and (3.168b) that $\tilde{h}[n]$ can be expressed in terms of $\tilde{x}[n]$ and $\tilde{y}[n]$ as

$$\tilde{h}[n] = \sum_{r=0}^{N-1} \tilde{x}[r]\tilde{y}[n - r]. \tag{3.170}$$

3.39 Prove the following general properties of the DFT listed in Table 3.5: (a) linearity, (b) circular time-shifting, (c) circular frequency-shifting, (d) duality, (e) N-point circular convolution, (f) modulation, and (g) Parseval's relation.

3.40 Let $x[n]$ be a length-N complex sequence with an N-point DFT $X[k]$. Determine the N-point DFTs of the following length-N sequences in terms of $X[k]$: (a) $x^*[n]$, (b) $x^*[\langle -n \rangle_N]$, (c) $\text{Re}\{x[n]\}$, (d) $j\text{Im}\{x[n]\}$, (e) $x_{pcs}[n]$, and (f) $x_{pca}[n]$.

3.41 Let $x[n]$ be a length-N real sequence with an N-point DFT $X[k]$. Determine the N-point DFTs of the following length-N sequences in terms of $X[k]$: (a) $x_{pe}[n]$, and (b) $x_{po}[n]$.

3.42 Let $x[n]$ be a length-N real sequence with an N-point DFT $X[k]$. Prove the following symmetry properties of $X[k]$: (a) $X[k] = X^*[\langle -k \rangle_N]$, (b) $\text{Re}X[k] = \text{Re}X[\langle -k \rangle_N]$, (c) $\text{Im}X[k] = -\text{Im}X[\langle -k \rangle_N]$, (d) $|X[k]| = |X[\langle -k \rangle_N]|$, and (e) $\arg X[k] = -\arg X[\langle -k \rangle_N]$.

3.43 Consider the following length-8 sequences defined for $0 \le n \le 7$:

(a) $\{x_1[n]\} = \{1 \quad 1 \quad 1 \quad 0 \quad 0 \quad 0 \quad 1 \quad 1\}$,

(b) $\{x_2[n]\} = \{1 \quad 1 \quad 0 \quad 0 \quad 0 \quad 0 \quad -1 \quad -1\}$,

(c) $\{x_3[n]\} = \{0 \quad 1 \quad 1 \quad 0 \quad 0 \quad 0 \quad -1 \quad -1\}$,

(d) $\{x_4[n]\} = \{0 \quad 1 \quad 1 \quad 0 \quad 0 \quad 0 \quad 1 \quad 1\}$.

Which sequences have a real-valued 8-point DFT? Which sequences have an imaginary-valued 8-point DFT?

3.44 Let $x[n]$, $0 \le n \le N-1$, be a length-N sequence with an N-point DFT $X[k]$, $0 \le k \le N-1$.

(a) If $x[n]$ is a symmetric sequence satisfying the condition $x[n] = x[N-1-n]$, show that $X[N/2] = 0$ for N even.

(b) If $x[n]$ is a antisymmetric sequence satisfying the condition $x[n] = -x[N-1-n]$, show that $X[0] = 0$.

(c) If $x[n]$ is a sequence satisfying the condition $x[n] = -x[n+M]$ with $N = 2M$, show that $X[2\ell] = 0$ for $\ell = 0, 1, \ldots, M-1$.

3.45 Let $x[n]$, $0 \le n \le N-1$, be an even-length sequence with an N-point DFT $X[k]$, $0 \le k \le N-1$. If $X[2m] = 0$ for $0 \le m \le \frac{N}{2} - 1$, show that $x[n] = -x[n + \frac{N}{2}]$.

3.46 Let $x[n]$, $0 \le n \le N-1$, be a length-N sequence with an N-point DFT $X[k]$, $0 \le k \le N-1$. Determine the N-point DFTs of the following length-N sequences in terms of $X[k]$:

(a) $w[n] = \alpha x[\langle n - m_1 \rangle_N] + \beta x[\langle n - m_2 \rangle_N]$, where m_1 and m_2 are positive integers less than N,

(b) $g[n] = \begin{cases} x[n], & \text{for } n \text{ even,} \\ 0, & \text{for } n \text{ odd,} \end{cases}$

(c) $y[n] = x[n] \textcircled{N} x[n]$.

3.47 Let $x[n]$, $0 \le n \le N-1$, be an even-length sequence with an N-point DFT $X[k]$, $0 \le k \le N-1$. Determine the N-point DFTs of the following length-N sequences in terms of $X[k]$:

$u[n] = x[n] - x[n - \frac{N}{2}]$,

$v[n] = x[n] + x[n - \frac{N}{2}]$,

$y[n] = (-1)^n x[n]$.

3.48 Let $x[n]$, $0 \le n \le N-1$, be a length-N sequence with an N-point DFT $X[k]$, $0 \le k \le N-1$. Determine the N-point inverse DFTs of the following length-N DFTs in terms of $x[n]$:

(a) $W[k] = \alpha X[\langle k - m_1 \rangle_N] + \beta X[\langle k - m_2 \rangle_N]$, where m_1 and m_2 are positive integers less than N,

(b) $G[k] = \begin{cases} X[k], & \text{for } k \text{ even,} \\ 0, & \text{for } k \text{ odd,} \end{cases}$

(c) $Y[k] = X[k] \textcircled{N} X[k]$.

3.49 Let $x[n]$, $0 \le n \le N - 1$, be a length-N sequence with an N-point DFT $X[k]$, $0 \le k \le N - 1$.

(a) Show that if N is even and if $x[n] = -x[n + \frac{N}{2}]$ for all n, then $X[k] = 0$ for k even.

(b) Show that if N is an integer multiple of 4 and if $x[n] = -x[n + \frac{N}{4}]$ for all n, then $X[k] = 0$ for $k = 4\ell$, $0 \le \ell \le \frac{N}{4} - 1$.

3.50 Let $x[n]$, $0 \le n \le N - 1$, be a length-N real sequence with an N-point DFT $X[k]$, $0 \le k \le N - 1$.

(a) Show that $X[N - k] = X^*[k]$.

(b) Show that $X[0]$ is real.

(c) If N is even, show that $X[N/2]4$ is real.

3.51 Let $G[k]$ and $H[k]$ denote the 7-point DFTs of two length-7 sequences, $g[n]$ and $h[n]$, respectively.

(a) If $G[k] = \{1 + j2 \quad -2 + j3 \quad -1 - j2 \quad 0 \quad 8 + j4 \quad -3 + j \quad 2 + j5\}$ and $h[n] = g[\langle n - 3 \rangle_7]$, determine $H[k]$ without computing the DFT.

(b) If $g[n] = \{-3.1 \quad 2.4 \quad 4.5 \quad -6 \quad 1 \quad -3 \quad 7\}$ and $H[k] = G[\langle k - 4 \rangle_7]$, determine $h[n]$ without computing the DFT.

3.52 Let $Y[k]$ denote the MN-point DFT of a length-N sequence $x[n]$ appended with $(M - 1)N$ zeros. Show that the N-point DFT $X[k]$ can be simply obtained from $Y[k]$ as follows:

$$X[k] = Y[kM], \quad 0 \le k \le N - 1.$$

3.53 Let $x[n]$, $0 \le n \le N - 1$, be a length-N sequence with an MN-point DFT $X[k]$, $0 \le k \le MN - 1$. Define

$$y[n] = x[\langle n \rangle_N], \quad 0 \le n \le MN - 1.$$

How would you compute the MN-point DFT $Y[k]$ of $y[n]$ knowing only $X[k]$?

3.54 Consider the length-12 sequence, defined for $0 \le n \le 11$,

$$\{x[n]\} = \{3 \quad -1 \quad 2 \quad 4 \quad -3 \quad -2 \quad 0 \quad 1 \quad -4 \quad 6 \quad 2 \quad 5\},$$

with a 12-point DFT given by $X[k]$, $0 \le k \le 11$. Evaluate the following functions of $X[k]$ without computing the DFT:

(a) $X[0]$,   (b) $X[6]$,   (c) $\sum_{k=0}^{11} X[k]$,   (d) $\sum_{k=0}^{11} e^{-j(4\pi k/6)} X[k]$,   (e) $\sum_{k=0}^{11} |X[k]|^2$.

3.55 Let $X[k]$ be a 14-point DFT of a length-14 real sequence $x[n]$. The first 8 samples of $X[k]$ are given by

$$X[0] = 12, \qquad X[1] = -1 + j3, \quad X[2] = 3 + j4, \qquad X[3] = 1 - j5,$$
$$X[4] = -2 + j2, \quad X[5] = 6 + j3, \qquad X[6] = -2 - j3, \quad X[7] = 10.$$

Determine the remaining samples of $X[k]$. Evaluate the following functions of $x[n]$ without computing the IDFT of $X[k]$:

(a) $x[0]$,   (b) $x[7]$,   (c) $\sum_{n=0}^{13} x[n]$,   (d) $\sum_{n=0}^{13} e^{j(4\pi n/7)} x[n]$,   (e) $\sum_{n=0}^{13} |x[n]|^2$.

3.56 Let $g[n]$ and $h[n]$ be two finite-length sequences of length 7 each. If $y_L[n]$ and $y_C[n]$ denote the linear and 7-point circular convolutions of $g[n]$ and $h[n]$, respectively, express $y_C[n]$ in terms of $y_L[n]$.

3.57 The even samples of the 11-point DFT of a length-11 real sequence are given by $X[0] = 4$, $X[2] = -1 + j3$, $X[4] = 2 + j5$, $X[6] = 9 - j6$, $X[8] = -5 - j8$, and $X[10] = \sqrt{3} - j2$. Determine the missing odd samples of the DFT.

3.58 The following six samples of the 11-point DFT $X[k]$ of a real length-11 sequence are given: $X[0] = 12$, $X[2] = -3.2 - j2$, $X[3] = 5.3 - j4.1$, $X[5] = 6.5 + j9$, $X[7] = -4.1 + j0.2$, and $X[10] = -3.1 + j5.2$. Determine the remaining five samples.

3.59 A length-10 sequence $x[n]$ has a real-valued 10-point DFT $X[k]$. The first six samples of $x[n]$ are given by: $x[0] = 2.5$, $x[1] = 0.7 - j0.08$, $x[2] = -3.25 + j1.12$, $x[3] = -2.1 + j4.6$, $x[4] = 2.87 + j2$, and $x[5] = 5$. Find the remaining four samples of $x[n]$.

3.60 A 498-point DFT $X[k]$ of a real-valued sequence $x[n]$ has the following DFT samples: $X[0] = 2$, $X[11] = 7 + j3.1$, $X[k_1] = -2.2 - j1.5$, $X[112] = 3 - j0.7$, $X[k_2] = -4.7 + j1.9$, $X[249] = 2.9$, $X[309] = -4.7 - j1.9$, $X[k_3] = 3 + j0.7$, $X[412] = -2.2 + j1.5$, and $X[k_4] = 7 - j3.1$. Remaining DFT samples are assumed to be of zero value.

(a) Determine the values of the indices k_1, k_2, k_3, and k_4.

(b) What is the dc value of $\{x[n]\}$?

(c) Determine the expression for $\{x[n]\}$ without computing the IDFT.

(d) What is the energy of $\{x[n]\}$?

3.61 A 316-point DFT $X[k]$ of a real-valued sequence $x[n]$ has the following DFT samples: $X[0] = 3 + j\alpha$, $X[17] = 1.5$, $X[k_1] = j2.3$, $X[k_2] = 4.2$, $X[110] = -j1.7$, $X[158] = 13 + j\beta$, $X[k_3] = \gamma + j1.7$, $X[179] = 4.2 + j\delta$, $X[210] = \epsilon - j2.3$, and $X[k_4] = 1.5$. Remaining DFT samples are assumed to be of zero value.

(a) Determine the values of the indices k_1, k_2, k_3, and k_4.

(b) Determine the values of α, β, δ, and ϵ.

(c) What is the dc value of $\{x[n]\}$?

(d) Determine the expression for $\{x[n]\}$ without computing the IDFT.

(e) What is the energy of $\{x[n]\}$?

3.62 A length-8 sequence is given by $\{x[n]\} = \{-4, 5, 2, -3, 0, -2, 3, 4\}$, $0 \le n \le 7$, with an 8-point DFT given by $X[k]$. Without computing the IDFT, determine the sequence $y[n]$ whose 8-point DFT is given by $Y[k] = W_4^{3k} X[k]$.

3.63 Let $X[k]$ denote the 6-point DFT of the length-6 real sequence $x[n]$ shown in Figure P3.4. Without computing the IDFT, determine the length-6 sequence $g[n]$ whose 6-point DFT is given by $G[k] = W_3^{2k} X[k]$.

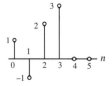

Figure P3.4

3.64 Let $g[n]$ and $h[n]$ be two finite-length sequences as given below:

$$\{g[n]\} \quad = \quad \{-3 \quad 2 \quad 4\}, \qquad \{h[n]\} \quad = \quad \{2 \quad -4 \quad 0 \quad 1\}.$$
$$\qquad\qquad\qquad\quad \uparrow \qquad\qquad\qquad\qquad\qquad\qquad \uparrow$$

(a) Determine $y_L[n] = g[n] \circledast h[n]$.

(b) Extend $g[n]$ to a length-4 sequence $g_e[n]$ by zero-padding and compute $y_C[n] = g_e[n] \textcircled{4} h[n]$.

(c) Determine $y_C[n]$ using the DFT-based approach.

3.65 Show that the circular convolution is commutative.

3.66 Let $y[n] = x_1[n] \circledN x_2[n] \circledN x_3[n]$ where the sequences $x_i[n]$, $1 \leq i \leq 3$, are defined for $0 \leq n \leq N - 1$. Prove the following equalities:

(a) $\displaystyle \sum_{n=0}^{N-1} y[n] = \left(\sum_{n=0}^{N-1} x_1[n] \right) \left(\sum_{n=0}^{N-1} x_2[n] \right) \left(\sum_{n=0}^{N-1} x_3[n] \right)$,

(b) $\displaystyle \sum_{n=0}^{N-1} (-1)^n y[n] = \left(\sum_{n=0}^{N-1} (-1)^n x_1[n] \right) \left(\sum_{n=0}^{N-1} (-1)^n x_2[n] \right) \left(\sum_{n=0}^{N-1} (-1)^n x_3[n] \right)$ for N even.

3.67 Let $X[k]$ denote the N-point DFT of a length-N sequence $x[n]$. Determine the N-point DFT $Y[k]$ of the N-point sequence $y[n] = \cos((2\pi \ell n)/N)x[n]$.

3.68 Let $x[n]$ be a length-N sequence with an N-point DFT given by $X[k]$. Assume N is divisible by 4. Define a sequence

$$y[n] = x[4n], \qquad 0 \leq n \leq \tfrac{N}{4} - 1.$$

Express the $(N/4)$-point DFT $Y[k]$ of $y[n]$ in terms of $X[k]$.

3.69 The 8-point DFT of a length-8 complex sequence $v[n] = x[n] + jy[n]$ is given by

$$V[0] = -2 + j3, \quad V[1] = 1 + j5, \quad V[2] = -4 + j7, \quad V[3] = 2 + j6,$$

$$V[4] = -1 - j3, \quad V[5] = 4 - j, \quad V[6] = 3 + j8, \quad V[7] = j6.$$

Without computing the IDFT of $V[k]$, determine the 8-point DFTs $X[k]$ and $Y[k]$ of the real sequences $x[n]$ and $y[n]$, respectively.

3.70 Compute the 4-point DFTs of $g_e[n]$ and $h[n]$ of Problem 3.64 using a single 4-point DFT.

3.71 Determine the 4-point DFTs of the following pair of length-4 sequences by computing a single DFT:

$$\{g[n]\} \quad = \quad \{-2 \quad 1 \quad -3 \quad 4\}, \qquad \{h[n]\} \quad = \quad \{1 \quad 2 \quad -3 \quad 2\}.$$

$$\quad\quad\quad \uparrow \quad\quad\quad\quad\quad\quad\quad\quad\quad\quad\quad\quad\quad\quad\quad \uparrow$$

3.72 Let $P[k]$ denote the $(M+1)$-point DFT of the numerator coefficients and $D[k]$ denote the $(N+1)$-point DFT of the denominator coefficients of a rational discrete-time Fourier transform $X(e^{j\omega})$ of the form of Eq. (3.17). Determine the exact expressions of the DTFT $X(e^{j\omega})$ for $M = N = 3$ if the 4-point DFTs of its numerator and denominator coefficients are as given below:

(a) $\{P[k]\} = \{4, \ 1 + j7, \ 2, \ 1 - j7\}$, $\{D[k]\} = \{4.5, \ 1.5 + j, \ -5.5, \ 1.5 - j\}$,

(b) $\{P[k]\} = \{7, \ 7 + j2, \ -9, \ 7 - j2\}$, $\{D[k]\} = \{0, \ 4 + j6, \ -4, \ 4 - j6\}$.

3.73 Consider a length-N sequence $x[n]$ with a DTFT $X(e^{j\omega})$. Define an M-point DFT $\hat{X}[k] = X(e^{j\omega_k})$, where $\omega_k = 2\pi k/M, k = 0, 1, \ldots, M - 1$. Denote the inverse DFT of $\hat{X}[k]$ as $\hat{x}[n]$, which is a length-M sequence. Express $x[n]$ in terms of $\hat{x}[n]$ and show that $x[n]$ can be fully recovered from $\hat{x}[n]$ only if $M \geq N$.

3.74 Let $X(e^{j\omega})$ denote the DTFT of the sequence

$$\{x[n]\} \quad = \quad \{1 \quad 1 \quad 1 \quad 1 \quad 1 \quad 1 \quad 1 \quad 1\}.$$
$$\uparrow$$

(a) For the DFT sequence $X_1[k]$ obtained by sampling $X(e^{j\omega})$ at uniform intervals of $\pi/5$ starting from $\omega = 0$, determine the IDFT $x_1[n]$ of $X_1[k]$ without computing $X(e^{j\omega})$ and $X_1[k]$. Can you recover $x[n]$ from $x_1[n]$?

(b) For the DFT sequence $X_2[k]$ obtained by sampling $X(e^{j\omega})$ at uniform intervals of $\pi/3$ starting from $\omega = 0$, determine the IDFT $x_2[n]$ of $X_2[k]$ without computing $X(e^{j\omega})$ and $X_2[k]$. Can you recover $x[n]$ from $x_2[n]$?

3.75 Let $x[n]$ be a length-N sequence with $X[k]$ denoting its N-point DFT. We represent the DFT operation as $X[k] = \mathcal{F}\{x[n]\}$. Determine the sequence $y[n]$ obtained by applying the DFT operation 6 times to $x[n]$, i.e,

$$y[n] = \mathcal{F}\{\mathcal{F}\{\mathcal{F}\{\mathcal{F}\{\mathcal{F}\{\mathcal{F}\{x[n]\}\}\}\}\}\}.$$

3.76 Let $x[n]$ and $h[n]$ be two length-40 sequences defined for $0 \le n \le 39$. It is known that $h[n] = 0$ for $0 \le n \le 11$ and $28 \le n \le 39$. Denote the 40-point circular convolution of these two sequences as $u[n]$ and their linear convolution as $y[n]$. Determine the range of n for which $y[n] = u[n]$.

3.77 The linear convolution of a length-55 sequence with a length-1100 sequence is to be computed using 64-point DFTs and IDFTs.

(a) Determine the smallest number of DFTs and IDFTs needed to compute the above linear convolution using the overlap-add approach.

(b) Determine the smallest number of DFTs and IDFTs needed to compute the above linear convolution using the overlap-save approach.

3.78   (a) Consider a length-N sequence $x[n]$, $0 \le n \le N - 1$, with an N-point DFT $X[k]$, $0 \le k \le N - 1$. Define a sequence $y[n]$ of length LN, $0 \le n \le NL - 1$, given by

$$y[n] = \begin{cases} x[n/L], & n = 0, L, 2L, \ldots, (N-1)L, \\ 0, & \text{otherwise,} \end{cases} \tag{3.171}$$

where L is a positive integer. Express the NL-point DFT $Y[k]$ of $y[n]$ in terms of $X[k]$.

(b) The 7-point DFT $X[k]$ of a length-7 sequence $x[n]$ is shown in Figure P3.5. Sketch the 21-point DFT $Y[k]$ of a length-21 sequence $y[n]$ generated using Eq. (3.171).

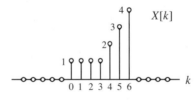

Figure P3.5

3.79 Consider two real, symmetric length-N sequences, $x[n]$ and $y[n]$, $0 \le n \le N - 1$ with N even. Define the length-$(N/2)$ sequences:

$$x_0[n] = x[2n + 1] + x[2n], \quad x_1[n] = x[2n + 1] - x[2n],$$
$$y_0[n] = y[2n + 1] + y[2n], \quad y_1[n] = y[2n + 1] - y[2n],$$

where $0 \le n \le \frac{N}{2} - 1$. It can be easily shown that $x_0[n]$ and $y_0[n]$ are real, symmetric sequences of length-$(N/2)$ each. Likewise, the sequences $x_1[n]$ and $y_1[n]$ are real and antisymmetric sequences. Denote the $(N/2)$-point DFTs of $x_0[n]$, $x_1[n]$, $y_0[n]$, and $y_1[n]$ by $X_0[k]$, $X_1[k]$, $Y_0[k]$, and $Y_1[k]$, respectively. Define a length-$(N/2)$ sequence $u[n]$:

$$u[n] = x_0[n] + y_1[n] + j(x_1[n] + y_0[n]).$$

Determine $X_0[k]$, $X_1[k]$, $Y_0[k]$, and $Y_1[k]$ in terms of the $(N/2)$-point DFT $U[k]$ of $u[n]$.

3.80 Let $X[k]$ denote the N-point DFT of a length-N sequence $x[n]$ with N even. Define two length-$(N/2)$ sequences given by

$$g[n] = \tfrac{1}{2}(x[2n] + x[2n+1]), \quad h[n] = \tfrac{1}{2}(x[2n] - x[2n+1]), \quad 0 \le n \le \tfrac{N}{2} - 1.$$

If $G[k]$ and $H[k]$ denote $(N/2)$-point DFTs of $g[n]$ and $h[n]$, respectively, determine the N-point DFT $X[k]$ from these two $(N/2)$-point DFTs.

3.81 Let $X[k]$ denote the N-point DFT of a length-N sequence $x[n]$ with N even. Define two length-$(N/2)$ sequences given by

$$g[n] = a_1 x[2n] + a_2 x[2n+1], \quad h[n] = a_3 x[2n] + a_4 x[2n+1], \quad 0 \le n \le \tfrac{N}{2} - 1,$$

where $a_1 a_4 \ne a_2 a_3$. If $G[k]$ and $H[k]$ denote $(N/2)$-point DFTs of $g[n]$ and $h[n]$, respectively, determine the N-point DFT $X[k]$ from these two $(N/2)$-point DFTs.

3.82 The *generalized discrete Fourier transform* (GDFT) is a generalization of the conventional DFT to allow shifts in either or both indices of the transform kernel [Bon76]. The N-point generalized discrete Fourier transform $X_{\mathrm{GDFT}}[k, a, b]$ of a length-N sequence $x[n]$ is defined by

$$X_{\mathrm{GDFT}}[k, a, b] = \sum_{n=0}^{N-1} x[n] \exp\left(-j\frac{2\pi(n+a)(k+b)}{N}\right). \tag{3.172}$$

Show that the inverse GDFT is given by

$$x[n] = \frac{1}{N} \sum_{k=0}^{N-1} X_{\mathrm{GDFT}}[k, a, b] \exp\left(j\frac{2\pi(n+a)(k+b)}{N}\right). \tag{3.173}$$

3.83 Show that for a causal sequence $x[n]$ defined for $n \ge 0$, and with a z-transform $X(z)$,

$$x[0] = \lim_{z \to \infty} X(z).$$

The above result is known as the *initial value theorem*.

3.84 Consider the z-transform

$$G(z) = \frac{(z+0.4)(z-0.91)(z^2+0.3z+0.4)}{(z^2-0.6z+0.6)(z^2+3z+5)}. \tag{3.174}$$

There are three possible nonoverlapping regions of convergence (ROCs) of this z-transform. Discuss the type of inverse z-transform (left-sided, right-sided, or two-sided sequences) associated with each of the three ROCs. It is not necessary to compute the exact inverse transform.

3.85 Consider the following sequences:

(i) $x_1[n] = (0.4)^n \mu[n]$,   (ii) $x_2[n] = (-0.6)^n \mu[n]$,

(iii) $x_3[n] = (0.3)^n \mu[n-4]$,   (iv) $x_4[n] = (-0.3)^n \mu[-n-2]$.

(a) Determine the ROCs of the z-transform of each of the above sequences.

(b) From the ROCs determined in part (a), determine the ROCs of the following sequences:

$$\begin{array}{lll}
\text{(i)} & y_1[n] = x_1[n] + x_2[n], & \text{(ii)} \quad y_2[n] = x_1[n] + x_3[n], \\
\text{(iii)} & y_3[n] = x_1[n] + x_4[n], & \text{(iv)} \quad y_4[n] = x_2[n] + x_3[n], \\
\text{(v)} & y_5[n] = x_2[n] + x_4[n], & \text{(vi)} \quad y_6[n] = x_3[n] + x_4[n].
\end{array}$$

3.86 Derive the z-transforms and the ROCs given in Table 3.8 of the following sequences: (a) $\delta[n]$, (b) $\alpha^n \mu[n]$, (c) $(r^n \cos \omega_o n)\mu[n]$, and (d) $(r^n \sin \omega_o n)\mu[n]$.

3.87 Show that the following three sequences have the same z-transform: (a) $x_1[n] = 6\left((0.5)^n - (0.3)^n\right)\mu[n]$, (b) $x_2[n] = -6(0.3)^n \mu[n] - 6(0.5)^n \mu[-n-1]$, and (c) $x_3[n] = 6\left((0.3)^n - (0.5)^n\right)\mu[-n-1]$.

3.88 Determine the z-transform of each of the following sequences and their respective ROCs. Assume $|\beta| > |\alpha|$. Show their pole-zero plots and indicate clearly the ROC in these plots.

(a) $x_1[n] = (\alpha^n + \beta^n)\mu[n]$,

(b) $x_2[n] = \alpha^n \mu[-n-1] + \beta^n \mu[n]$,

(c) $x_3[n] = \alpha^n \mu[n] + \beta^n \mu[-n-1]$.

3.89 Prove the following general properties of the z-transform listed in Table 3.9: (a) linearity, (b) time-reversal, (c) time-shifting, (d) multiplication by an exponential sequence, (e) differentiation of the z-transform, (f) convolution, (g) modulation, and (h) Parseval's relation.

3.90 Let the z-transform of a sequence $x[n]$ be $X(z)$ with $\mathcal{R}_x$ denoting its ROC. Express the z-transforms of the real and imaginary parts of $x[n]$ in terms of $X(z)$. Show also their respective ROCs.

3.91 The z-transform $X(z)$ of the length-9 sequence of Problem 3.25 is sampled at six points $\omega_k = \pi k/3, 0 \le k \le 5$, on the unit circle yielding the frequency samples

$$\tilde{X}[k] = X(z)|_{z=e^{j\pi k/3}}, \qquad 0 \le k \le 5.$$

Determine, without evaluating $\tilde{X}[k]$, the periodic sequence $\tilde{x}[n]$ whose discrete Fourier series coefficients are given by $\tilde{X}[k]$. What is the period of $\tilde{x}[n]$?

3.92 Let $X(z)$ denote the z-transform of the length-12 sequence $x[n]$ of Problem 3.54. Let $X_0[k]$ represent the samples of $X(z)$ evaluated on the unit circle at 9 equally spaced points given by $z = e^{j(2\pi k/9)}, 0 \le k \le 8$, i.e.,

$$X_0[k] = X(z)|_{z=e^{j(2\pi k/9)}}, \qquad 0 \le k \le 8.$$

Determine the 9-point IDFT $x_0[n]$ of $X_0[k]$ without computing the latter function.

3.93 Let $X(z)$ denote the z-transform of $x[n] = (0.4)^n \mu[n]$.

(a) Determine the inverse z-transform of $X(z^2)$ without computing $X(z)$.

(b) Determine the inverse z-transform of $(1 + z^{-1})X(z^2)$ without computing $X(z)$.

3.94 Determine the z-transforms of the sequences of Problem 3.11 and their ROCs. Show that the ROC includes the unit circle for each z-transform. Evaluate the z-transform evaluated on the unit circle for each sequence and show that it is precisely the DTFT of the respective sequence computed in Problem 3.11.

3.95 Determine the z-transforms of the sequences of Problem 3.12 and their ROCs. Show that the ROC includes the unit circle for each z-transform. Evaluate the z-transform evaluated on the unit circle for each sequence and show that it is precisely the DTFT of the respective sequence computed in Problem 3.12.

3.96 Determine the z-transforms of the following sequences and their respective ROCs: (a) $x_1[n] = -\alpha^n \mu[-n-1]$, (b) $x_2[n] = \alpha^n \mu[n+1]$, and (c) $x_3[n] = \alpha^n \mu[-n]$.

3.97 Determine the z-transform of the two-sided sequence $v[n] = \alpha^{|n|}$. What is its ROC?

3.98 Evaluate the inverse z-transforms of the following z-transforms:

(a) $Y_1(z) = \dfrac{z(z-1)}{(z+1)(z+\frac{1}{3})}$, $\quad |z| > 1$.

(b) $Y_2(z) = \dfrac{z(z-1)}{(z+1)(z+\frac{1}{3})}$, $\quad |z| < \dfrac{1}{3}$.

(c) $Y_3(z) = \dfrac{z(z-1)}{(z+1)(z+\frac{1}{3})}$, $\quad \dfrac{1}{3} < |z| < 1$.

3.99 Determine the inverse z-transform of the following z-transforms:

(a) $X_a(z) = \dfrac{4 - 3z^{-1} + 3z^{-2}}{(z+2)(z-3)^2}$, $\quad |z| > 3$.

(b) $X_b(z) = \dfrac{4 - 3z^{-1} + 3z^{-2}}{(z+2)(z-3)^2}$, $\quad |z| < 2$.

(c) $X_c(z) = \dfrac{4 - 3z^{-1} + 3z^{-2}}{(z+2)(z-3)^2}$, $\quad 2 < |z| < 3$.

3.100 Consider a rational z-transform $G(z) = P(z)/D(z)$ where $P(z)$ and $D(z)$ are polynomials in z^{-1}. Let ρ_ℓ denote the residue of $G(z)$ at a simple pole at $z = \lambda_\ell$. Show that

$$\rho_\ell = -\lambda_\ell \left.\frac{P(z)}{D'(z)}\right|_{z=\lambda_\ell},$$

where $D'(z) = \dfrac{dD(z)}{dz^{-1}}$.

3.101 Consider the z-transform $G(z)$ of Eq. (3.113) with $M < N$. If $G(z)$ has only simple poles, show that p_0/d_0 is equal to the sum of the residues in the partial-fraction expansion of $G(z)$ [Mit98b].

3.102 Show that the inverse z-transform $h[n]$ of the following rational z-transform

$$H(z) = \frac{1}{1 - 2r(\cos\theta)z^{-1} + r^2 z^{-2}}, \quad |z| > r > 0$$

is given by

$$h[n] = \frac{r^n \sin(n+1)\theta}{\sin\theta} \cdot \mu[n].$$

3.103 Prove the following properties of the z-transform listed in Table 3.9: (a) Conjugation, (b) time-reversal, (c) time-shifting, (d) multiplication by an exponential sequence, (e) differentiation, (f) convolution, (g) modulation, and (h) Parseval's relation.

3.104 Let the z-transform of sequence $x[n]$ be $X(z)$ with an ROC $\mathcal{R}_x$. Show that

$$\mathcal{Z}\{\text{Re}(x[n])\} = \frac{1}{2}\{X(z) + X^*(z^*)\},$$

$$\mathcal{Z}\{\text{Im}(x[n])\} = \frac{1}{2}\{X(z) - X^*(z^*)\}.$$

3.105 Determine the inverse z-transforms, $x_1[n]$ and $x_2[n]$, of the following rational z-transforms:

(a) $X_1(z) = \dfrac{1}{1 - z^{-3}}, \quad |z| > 1$,

(b) $X_2(z) = \dfrac{1}{1 - z^{-2}}, \quad |z| > 1$,

by expanding each in a power series and computing the inverse z-transform of the individual terms in the power series. Compare the results with that obtained using a partial-fraction approach.

3.106 Determine the inverse z-transforms of the following z-transforms:

(a) $X_1(z) = \log(1 - \alpha z^{-1}), \quad |z| > |\alpha|$,

(b) $X_2(z) = \log\left(\dfrac{\alpha - z^{-1}}{\alpha}\right), \quad |z| > 1/|\alpha|$,

(c) $X_3(z) = \log\left(\dfrac{1}{1 - \alpha z^{-1}}\right), \quad |z| > |\alpha|$,

(d) $X_4(z) = \log\left(\dfrac{\alpha}{\alpha - z^{-1}}\right), \quad |z| > 1/|\alpha|$.

3.107 The z-transform of a causal sequence $x[n]$ is given by $\alpha z^{-1}/(1 - \alpha z^{-1})^2$. Using Tables 3.8 and 3.9, determine $x[n]$.

3.108 The z-transform of a right-sided sequence $h[n]$ is given by

$$H(z) = \frac{z - 2}{(z + 0.4)(z - 0.2)}.$$

Find its inverse z-transform $h[n]$ via the partial-fraction approach. Verify the partial fraction using MATLAB.

3.109 A generalization of the DFT concept leads to the *nonuniform discrete Fourier transform* (NDFT) $X_{\text{NDFT}}[k]$ defined by [Bag98]

$$X_{\text{NDFT}}[k] = X(z_k) = \sum_{n=0}^{N-1} x[n] z_k^{-n}, \quad 0 \le k \le N - 1, \tag{3.175}$$

where $z_0, z_1, \ldots, z_{N-1}$, are N distinct points located arbitrarily in the z-plane. The NDFT has been applied to the efficient design of digital filters, antenna array design, and dual-tone multifrequency detection [Bag98]. The NDFT can be expressed in a matrix form as

$$\begin{bmatrix} X_{\text{NDFT}}[0] \\ X_{\text{NDFT}}[1] \\ \vdots \\ X_{\text{NDFT}}[N-1] \end{bmatrix} = \mathbf{D}_N \begin{bmatrix} x[0] \\ x[1] \\ \vdots \\ x[N-1] \end{bmatrix}, \tag{3.176}$$

where

$$\mathbf{D}_N = \begin{bmatrix} 1 & z_0^{-1} & z_0^{-2} & \cdots & z_0^{-(N-1)} \\ 1 & z_1^{-1} & z_1^{-2} & \cdots & z_1^{-(N-1)} \\ 1 & z_2^{-1} & z_2^{-2} & \cdots & z_2^{-(N-1)} \\ \vdots & \vdots & \vdots & \ddots & \vdots \\ 1 & z_{N-1}^{-1} & z_{N-1}^{-2} & \cdots & z_{N-1}^{-(N-1)} \end{bmatrix} \tag{3.177}$$

is the $N \times N$ NDFT matrix. The matrix $\mathbf{D}_N$ is known as the *Vandermonde matrix*. Show that it is nonsingular provided the N sampling points z_k are distinct. In which case, the inverse NDFT is given by

$$
\begin{bmatrix} x[0] \\ x[1] \\ \vdots \\ x[N-1] \end{bmatrix} = \mathbf{D}_N^{-1} \begin{bmatrix} X_{\mathrm{NDFT}}[0] \\ X_{\mathrm{NDFT}}[1] \\ \vdots \\ X_{\mathrm{NDFT}}[N-1] \end{bmatrix}. \tag{3.178}
$$

3.110 In general, for large N, the Vandermonde matrix is usually ill-conditioned (except for the case when the NDFT reduces to the conventional DFT), and a direct inverse computation is not advisable. A more efficient way is to directly determine the z-transform $X(z)$,

$$
X(z) = \sum_{n=0}^{N-1} x[n] z^{-n}, \tag{3.179}
$$

and hence, the sequence $x[n]$, from the given N-point NDFT $X_{\mathrm{NDFT}}[k]$ by using some type of polynomial interpolation method [Bag98]. One popular method is the Lagrange interpolation formula, which expresses $X(z)$ as

$$
X(z) = \sum_{k=0}^{N-1} \frac{I_k(z)}{I_k(z_k)} X_{\mathrm{NDFT}}[k], \tag{3.180}
$$

where

$$
I_k(z) = \prod_{\substack{i=0 \\ i \neq k}}^{N-1} (1 - z_i z^{-1}). \tag{3.181}
$$

Consider the z-transform $X(z) = 4 - 2z^{-1} + 3z^{-2} + z^{-3}$ of a length-4 sequence $x[n]$. By evaluating $X(z)$ at $z_0 = -1/2$, $z_1 = 1$, $z_2 = 1/2$, and $z_3 = 1/3$, determine the 4-point NDFT of $x[n]$ and then use the Lagrange interpolation method to show that $X(z)$ can be uniquely determined from these NDFT samples.

3.111 The *discrete cosine transform* (DCT) is used frequently in image-coding applications based on the compression of transform coefficients [Lim90]. We develop here the DCT computation algorithm via the DFT. Let $x[n]$ be a length-N sequence defined for $0 \leq n \leq N - 1$. First $x[n]$ is extended to a length $2N$ by zero-padding:

$$
x_e[n] = \begin{cases} x[n], & 0 \leq n \leq N-1, \\ 0, & N \leq n \leq 2N-1. \end{cases}
$$

Then a length-$2N$ sequence $y[n]$ is formed from $x_e[n]$ according to

$$
y[n] = x_e[n] + x_e[2N - 1 - n], \qquad 0 \leq n \leq 2N - 1,
$$

and its $2N$-point DFT $Y[k]$ is computed as

$$
Y[k] = \sum_{n=0}^{2N-1} y[n] W_{2N}^{nk}, \qquad 0 \leq k \leq 2N - 1.
$$

The N-point DCT $C_x[k]$ of $x[n]$ is then defined by

$$
C_x[k] = \begin{cases} W_{2N}^{k/2} Y[k], & 0 \leq k \leq N-1, \\ 0, & \text{otherwise.} \end{cases}
$$

Show that

$$
C_x[k] = \sum_{n=0}^{N-1} 2x[n] \cos\left(\frac{(2n+1)k\pi}{2N} \right), \qquad 0 \leq k \leq N - 1. \tag{3.182}
$$

Note from the above definition that the DCT coefficients of a real sequence are real, whereas in general the DFT of a real sequence is always complex. The DCT defined by Eq. (3.182) is often referred to as the *even symmetrical DCT*.

3.112 To form the inverse discrete cosine transform (IDCT) of an N-point DCT $C_x[k]$, first a $2N$-point DFT $Y[k]$ is formed according to

$$Y[k] = \begin{cases} W_{2N}^{-k/2}C_x[k], & 0 \le k \le N - 1, \\ 0, & k = N, \\ -W_{2N}^{-k/2}C_x[2N - k], & N + 1 \le k \le 2N - 1, \end{cases}$$

and its $2N$-point IDFT $y[n]$ is computed as

$$y[n] = \frac{1}{2N} \sum_{k=0}^{2N-1} Y[k]W_{2N}^{-kn}, \qquad 0 \le n \le 2N - 1.$$

The N-point IDCT of $C_x[k]$ is then given by

$$x[n] = \begin{cases} y[n], & 0 \le n \le N - 1, \\ 0, & \text{otherwise.} \end{cases}$$

(a) Show that

$$x[n] = \begin{cases} \frac{1}{N} \sum_{k=0}^{N-1} \alpha(k)C_x[k] \cos\left(\frac{(2n+1)\pi k}{2N}\right), & 0 \le n \le N - 1, \\ 0, & \text{otherwise,} \end{cases} \tag{3.183}$$

where

$$\alpha(k) = \begin{cases} 1/2, & k = 0, \\ 1, & 1 \le k \le N - 1. \end{cases} \tag{3.184}$$

(b) Prove that $x[n]$ given by Eq. (3.183) is indeed the inverse DCT of the DCT coefficients $C_x[k]$ given by Eq. (3.182).

3.113 Let $g[n]$ and $h[n]$ be two length-N sequences with N-point DCTs given by $C_g[k]$ and $C_h[k]$, respectively. Show that the N-point DCT $C_y[k]$ of the sequence $y[n] = \alpha g[n] + \beta h[n]$ is given by $C_y[k] = \alpha C_g[k] + \beta C_h[k]$, where α and β are arbitrary constants.

3.114 If the N-point DCT of a length-N sequence $x[n]$ is given by $C_x[k]$, show that the N-point DCT of $x^*[n]$ is given by $C_x^*[k]$.

3.115 If the N-point DCT of a length-N sequence $x[n]$ is given by $C_x[k]$, show that

$$\sum_{n=0}^{N-1} |x[n]|^2 = \frac{1}{2N} \sum_{k=0}^{n-1} \alpha(k)|C_x[k]|^2,$$

where $\alpha(k)$ is given by Eq. (3.184).

3.116 The N-point *discrete Hartley transform* (DHT) $X_{\text{DHT}}[k]$ of a length-N sequence $x[n]$ is defined by [Bra83]

$$X_{\text{DHT}}[k] = \sum_{n=0}^{N-1} x[n]\left(\cos\left(\frac{2\pi nk}{N}\right) + \sin\left(\frac{2\pi nk}{N}\right)\right), \qquad k = 0, 1, \ldots, N - 1. \tag{3.185}$$

As can be seen from the above, the DHT of a real sequence is also a real sequence. Show that the inverse discrete Hartley transform (DHT) is given by

$$x[n] = \frac{1}{N} \sum_{k=0}^{N-1} X_{\text{DHT}}[k]\left(\cos\left(\frac{2\pi nk}{N}\right) + \sin\left(\frac{2\pi nk}{N}\right)\right), \qquad n = 0, 1, \ldots, N - 1. \tag{3.186}$$

3.117 Let $X_{DHT}[k]$ denote the N-point DHT of a length-N sequence $x[n]$.

(a) Show that the DHT of $x[\langle n - n_0\rangle_N]$ is given by

$$X_{DHT}[k] \cos\left(\frac{2\pi n_0 k}{N}\right) + X_{DHT}[-k] \sin\left(\frac{2\pi n_0 k}{N}\right),$$

(b) Determine the N-point DHT of $x[\langle -n\rangle_N]$,

(c) Prove the Parseval's relation:

$$\sum_{n=0}^{N-1} x^2[n] = \frac{1}{N} \sum_{k=0}^{N-1} X_{DHT}^2[k]. \tag{3.187}$$

3.118 Develop the relation between the N-point DHT $X_{DHT}[k]$ and the N-point DFT $X[k]$ of a length-N sequence $x[n]$.

3.119 Let the N-point DHTs of the three length-N sequences $x[n]$, $g[n]$, and $y[n]$ be denoted by $X_{DHT}[k]$, $G_{DHT}[k]$, and $Y_{DHT}[k]$, respectively. If $y[n] = x[n] \circledN g[n]$, show that

$$Y_{DHT}[k] = \tfrac{1}{2} X_{DHT}[k](G_{DHT}[k] + G_{DHT}[\langle -k\rangle_N])$$
$$+ \tfrac{1}{2} X_{DHT}[\langle -k\rangle_N](G_{DHT}[k] - G_{DHT}[\langle -k\rangle_N]). \tag{3.188}$$

3.120 The Hadamard transform $X_{HT}[k]$ of a length-N sequence $x[n]$, $n = 0, 1, \ldots, N-1$, is given by [Gon87]

$$X_{HT}[k] = \frac{1}{N} \sum_{n=0}^{N-1} x[n](-1)^{\sum_{i=0}^{\ell-1} b_i(n) b_i(k)}, \qquad k = 0, 1, \ldots, N-1, \tag{3.189}$$

where $b_i(r)$ is the ith bit in the binary representation of r, and $N = 2^\ell$. In matrix form, the Hadamard transform can be represented as

$$\mathbf{X}_{HT} = \mathbf{H}_N \mathbf{x},$$

where

$$\mathbf{X}_{HT} = [X_{HT}[0] \quad X_{HT}[1] \quad \cdots \quad X_{HT}[N-1]]^T,$$
$$\mathbf{x} = [x[0] \quad x[1] \quad \cdots \quad x[N-1]]^T.$$

(a) Determine the form of the Hadamard matrix $\mathbf{H}_N$ for $N = 2, 4$, and 8.

(b) Show that

$$\mathbf{H}_4 = \begin{bmatrix} \mathbf{H}_2 & \mathbf{H}_2 \\ \mathbf{H}_2 & -\mathbf{H}_2 \end{bmatrix}, \qquad \mathbf{H}_8 = \begin{bmatrix} \mathbf{H}_4 & \mathbf{H}_4 \\ \mathbf{H}_4 & -\mathbf{H}_4 \end{bmatrix}.$$

(c) Determine the expression for the inverse Hadamard transform.

3.121 In this problem we consider the computation of a limited number of NDFT samples $X(z_\ell)$ of a length-N sequence $x[n]$ at $z_\ell = A V^{-\ell}$, $0 \le \ell \le L - 1$ where $A = A_o e^{j\theta_o}$ and $V = V_o e^{j\phi_o}$ with A_o and V_o being positive real numbers, and in general $L < N$. The contour spirals toward the origin as ℓ increases if $V_o > 1$ and it spirals outward if $V_o < 1$. The *chirp-z transform* (CZT) is then defined by [Rab69]

$$X(z_\ell) = \sum_{n=0}^{N-1} x[n] A^{-n} V^{\ell n}, \qquad 0 \le \ell \le L - 1. \tag{3.190}$$

Convert the above expression into a convolution sum using the identity

$$\ell n = \tfrac{1}{2}(\ell^2 + n^2 - (\ell - n)^2),$$

and show that the discrete-time system of Figure P3.6 can be employed to compute the CZT.

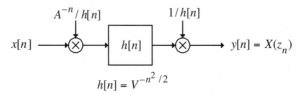

$$h[n] = V^{-n^2/2}$$

Figure P3.6

3.122 We wish to compute the L-point chirp-z transform (CZT) samples $X(z_\ell)$, $\ell = 0, 1, 2, \ldots, L-1$, of a length-$N$ sequence $x[n]$ according to Eq. (3.190), where $z_\ell = AV^{-\ell}$ with $A = A_o e^{j\theta_o}$ and $V = V_o e^{j\phi_o}$. What are the values of A_o, θ_o, V_o, and ϕ_o if the CZT needs to be calculated at points $\{z_\ell\}$ on the real axis in the z-plane such that $z_\ell = \alpha^\ell$, $0 \le \ell \le L-1$, for α real and $\alpha \ne \pm 1$?

3.123 Let $y_L[n]$ denote the length-$(2N-1)$ sequence obtained by a linear convolution of two length-N sequences $h[n]$ and $x[n]$, i.e., $y_L[n] = x[n] \circledast h[n]$. From the convolution theorem of z-transform (see Table 3.9) it is known that $Y_L(z) = X(z)H(z)$, where $Y_L(z)$, $X(z)$, and $H(z)$ denote the z-transforms of the sequences $y_L[n]$, $x[n]$, and $h[n]$, respectively. This implies that the samples of the sequence $y_L[n]$ are simply given by the coefficients of the product of the two polynomials $X(z)$ and $H(z)$. Now, let $y_C[n]$ denote the N-point circular convolution of $x[n]$ and $h[n]$, i.e., $y_C[n] = x[n] \circledN h[n]$. It can be shown that the samples of the sequence $y_C[n]$ can be obtained from the coefficients of the polynomial $Y_C(z) = Y_L(z) \bmod (1 - z^{-N})$. Verify the above result for $N = 3, 4,$ and 5.

Note: $(y[N+m]z^{-N-m}) \bmod (1 - z^{-N}) = y[N+m]z^{-m}$ where $0 \le m < N$.

3.124 Consider a sequence $x[n]$ with a z-transform $X(z)$. Define a new z-transform $\hat{X}(z)$ given by the complex natural logarithm of $X(z)$, i.e., $\hat{X}(z) = \ln X(z)$. The inverse z-transform of $\hat{X}(z)$ to be denoted by $\hat{x}[n]$ is called the *complex cepstrum* of $x[n]$ [Tri79]. Assume that the ROCs of both $X(z)$ and $\hat{X}(z)$ include the unit circle.

(a) Relate the DTFT $X(e^{j\omega})$ of $x[n]$ to the DTFT $\hat{X}(e^{j\omega})$ of its complex cepstrum $\hat{x}[n]$.

(b) Show that the complex cepstrum of a real sequence is a real-valued sequence.

(c) Let $\hat{x}_{ev}[n]$ and $\hat{x}_{od}[n]$ denote, respectively, the even and odd parts of a real-valued complex cepstrum $\hat{x}[n]$. Express $\hat{x}_{ev}[n]$ and $\hat{x}_{od}[n]$ in terms of $X(e^{j\omega})$, the DTFT of $x[n]$.

3.125 Determine the complex cepstrum $\hat{x}[n]$ of a sequence $x[n] = a\delta[n] + b\delta[n-1]$, where $|b/a| < 1$. Comment on your results.

3.126 Let $x[n]$ be a sequence with a rational z-transform $X(z)$ given by

$$X(z) = K \frac{\prod_{k=1}^{N_\alpha}(1 - \alpha_k z^{-1}) \prod_{k=1}^{N_\gamma}(1 - \gamma_k z)}{\prod_{k=1}^{N_\beta}(1 - \beta_k z^{-1}) \prod_{k=1}^{N_\delta}(1 - \delta_k z)}$$

where α_k and β_k are the zeros and poles of $X(z)$ that are strictly inside the unit circle, and $1/\gamma_k$ and $1/\delta_k$ are the zeros and poles of $X(z)$ that are strictly outside the unit circle [Rab78].

(a) Determine the exact expression for the complex cepstrum $\hat{x}[n]$ of $x[n]$.

(b) Show that $\hat{x}[n]$ is a decaying bounded sequence as $|n| \to \infty$.

(c) If $\alpha_k = \beta_k = 0$, show that $\hat{x}[n]$ is an anticausal sequence.

(d) If $\gamma_k = \delta_k = 0$, show that $\hat{x}[n]$ is a causal sequence.

3.127 Let $x[n]$ be a sequence with a rational z-transform $X(z)$ with poles and zeros strictly inside the unit circle. Show that the complex cepstrum $\hat{x}[n]$ of $x[n]$ can be computed using the recursion relation [Rab78]:

$$\hat{x}[n] = \begin{cases} 0, & n < 0, \\ \log(x[0]), & n = 0, \\ \frac{x[n]}{x[0]} - \sum_{k=0}^{n-1} \frac{k}{n} \cdot \hat{x}[k] \cdot \frac{x[n-k]}{x[0]}, & n > 0. \end{cases}$$

3.128 A zero-mean, white noise sequence $x[n]$ with a variance σ_x^2 is fed into an LTI system with an impulse response $h[n] = (0.6)^n \mu[n]$ generating the output $v[n]$. The signal $v[n]$ is then fed into a second LTI system with an impulse response $g[n] = (0.8)^n \mu[n]$ producing the output $y[n]$. Determine the variances σ_v^2 and σ_y^2 of the signals $v[n]$ and $y[n]$.

3.14 Matlab Exercises

M 3.1 Using Program 3_1 determine and plot the real and imaginary parts and the magnitude and phase spectra of the following DTFT for various values of r and θ:

$$G(e^{j\omega}) = \frac{1}{1 - 2r(\cos\theta)e^{-j\omega} + r^2 e^{-j2\omega}}, \qquad 0 < r < 1.$$

M 3.2 Using Program 3_1 determine and plot the real and imaginary parts and the magnitude and phase spectra of the DTFTs of the sequences of Problem 3.12 for $N = 10$.

M 3.3 Using Program 3_1 determine and plot the real and imaginary parts, and the magnitude and phase spectra of the following DTFTs:

(a) $X(e^{j\omega}) = \dfrac{0.0761(1 - 0.7631e^{-j2\omega} + e^{-j4\omega})}{1 + 1.355e^{-j2\omega} + 0.6196e^{-j4\omega}}$,

(b) $X(e^{j\omega}) = \dfrac{0.0518 - 0.1553e^{-j\omega} + 0.1553e^{-j2\omega} + 0.0518e^{-j3\omega}}{1 + 1.2828e^{-j\omega} + 1.0388e^{-j2\omega} + 0.3418e^{-j3\omega}}$.

M 3.4 Using Matlab verify the following general properties of the DTFT as listed in Table 3.2: (a) Linearity, (b) time-shifting, (c) frequency-shifting, (d) differentiation-in-frequency, (e) convolution, (f) modulation, and (g) Parseval's relation. Since all data in Matlab have to be finite-length vectors, the sequences to be used to verify the properties are thus restricted to be of finite length.

M 3.5 Using Matlab verify the symmetry relations of the DTFT of a complex sequence as listed in Table 3.3.

M 3.6 Using Matlab verify the symmetry relations of the DTFT of a real sequence as listed in Table 3.4.

M 3.7 Write a Matlab program to compute the $N \times N$ DFT matrix $\mathbf{D}_N$ of Eq. (3.40) and then form its inverse. Using this program verify the relation given by Eq. (3.43) for $N = 3, 4, 5,$ and 6.

M 3.8 Using Matlab compute the N-point DFTs of the length-N sequences of Problem 3.12 for $N = 3, 5, 7,$ and 10. Compare your results with that obtained by evaluating the DTFTs computed in Problem 3.12 at $\omega = 2\pi k/N$, $k = 0, 1, \ldots, N - 1$.

M 3.9 Write a Matlab program to compute the circular convolution of two length-N sequences via the DFT-based approach. Using this program determine the circular convolution of the following pairs of sequences:

(a) $g[n] = \{3,\ 4,\ -2,\ 0,\ 1,\ -4\}$, $h[n] = \{1,\ -3,\ 0,\ 4,\ -2,\ 3\}$,

(b) $x[n] = \{2 + j3,\ 3 - j,\ -1 + j2,\ j3,\ 2 + j4\}$,
    $v[n] = \{-3 - j2,\ 1 + j4,\ 1 + j2,\ 5 + j3,\ 1 + j2\}$,

(c) $x[n] = \sin(\pi n/2)$, $y[n] = 2^n$, $0 \le n \le 4$.

M 3.10 Using MATLAB prove the following general properties of the DFT listed in Table 3.5: (a) linearity, (b) circular time-shifting, (c) circular frequency-shifting, (d) duality, (e) N-point circular convolution, (f) modulation, and (g) Parseval's relation.

M 3.11 Using MATLAB verify the symmetry relations of the DFT of a complex sequence as listed in Table 3.6.

M 3.12 Using MATLAB verify the symmetry relations of the DFT of a real sequence as listed in Table 3.7.

M 3.13 Verify the results of Problem 3.54 by computing the DFT $X[k]$ of the sequence $x[n]$ given using MATLAB and then evaluate the functions of $X[k]$ listed.

M 3.14 Verify the results of Problem 3.55 by computing the IDFT $x[n]$ of the DFT $X[k]$ given using MATLAB and then evaluate the functions of $x[n]$ listed.

M 3.15 Write a MATLAB function to implement the overlap-save method. Using this function, demonstrate the filtering of the noise-corrupted signal of Example 2.14 using a length-3 moving average filter by modifying Program 3_6.

M 3.16 Using MATLAB determine the factored form of the following z-transforms:

(a) $G_1(z) = \dfrac{4z^4 + 15.6z^3 + 6z^2 + 2.4z - 6.4}{3z^4 + 2.4z^3 + 6.3z^2 - 11.4z + 6}$,

(b) $G_2(z) = \dfrac{2z^4 + 6.4z^3 + 4.9z^2 - 0.1z - 0.6}{5z^4 + 15.5z^3 + 31.7z^2 + 22.52z + 4.8}$,

and show their pole-zero plots. Determine all possible ROCs of each of the above z-transforms, and describe the type of their inverse z-transforms (left-sided, right-sided, two-sided sequences) associated with each of the ROCs.

M 3.17 Using Program 3_9 determine the partial-fraction expansions of the z-transforms listed in Problem 3.98 and then determine their inverse z-transforms.

M 3.18 Using Program 3_9 determine the partial-fraction expansions of the z-transforms listed in Problem 3.99 and then determine their inverse z-transforms.

M 3.19 Using Program 3_10 determine the z-transform as a ratio of two polynomials in z^{-1} from each of the partial-fraction expansions listed below:

(a) $X_1(z) = -2 + \dfrac{10}{4 + z^{-1}} - \dfrac{8}{2 + z^{-1}}$, $|z| > 0.5$,

(b) $X_2(z) = 3.5 - \dfrac{2}{1 - 0.5z^{-1}} - \dfrac{3 + z^{-1}}{1 - 0.25z^{-2}}$, $|z| > 0.5$,

(c) $X_3(z) = \dfrac{5}{(3 + 2z^{-1})^2} - \dfrac{4}{3 + 2z^{-1}} + \dfrac{3}{1 + 0.81z^{-2}}$, $|z| > 0.9$,

(d) $X_4(z) = 4 + \dfrac{10}{5 + 2z^{-1}} + \dfrac{z^{-1}}{6 + 5z^{-1} + z^{-2}}$, $|z| > 0.5$.

M 3.20 Using Program 3_11 determine the first 30 samples of the inverse z-transforms of the rational z-transforms determined in Problem M3.19. Show that these samples are identical to those obtained by explicitly evaluating the exact inverse z-transforms.

M 3.21 Write a MATLAB program to compute the NDFT and the inverse NDFT using the Lagrange interpolation method. Verify your program by computing the NDFT of a length-25 sequence and reconstructing the sequence from its computed NDFT.

4 LTI Discrete-Time Systems in the Transform Domain

We showed in Section 2.5 that a linear, time-invariant (LTI) discrete-time system is completely characterized in the time-domain by its impulse response sequence $\{h[n]\}$. As a result, the transform-domain representation of a discrete-time signal can equally be applied to the transform-domain representation of an LTI discrete-time system. Such transform-domain representations provide additional insight into the behavior of such systems, and also make it easier to design and implement them for specific applications.

In this chapter we discuss the use of the DTFT and the z-transform in transforming the time-domain representations of an LTI discrete-time system to alternative characterizations. Specific properties of such transform-domain representations are investigated and several simple applications are considered. As in the earlier chapters, MATLAB has been used extensively to illustrate various concepts and applications.

4.1 Finite-Dimensional LTI Discrete-Time Systems

The LTI discrete-time systems we shall be concerned with in this book are characterized by linear constant coefficient difference equations of the form of Eq. (2.81). Applying the discrete-time Fourier transform (DTFT) to this equation and making use of the linearity and the time-shifting properties of Table 3.2 we arrive at the input-output relation of the LTI system in the transform-domain given by

$$\sum_{k=0}^{N} d_k e^{-j\omega k} Y(e^{j\omega}) = \sum_{k=0}^{M} p_k e^{-j\omega k} X(e^{j\omega}), \tag{4.1}$$

where $Y(e^{j\omega})$ and $X(e^{j\omega})$ are the DTFTs of the sequences $y[n]$ and $x[n]$, respectively. In developing Eq. (4.1) it has been tacitly assumed that $Y(e^{j\omega})$ and $X(e^{j\omega})$ exist. The above equation can be alternately written as

$$\left(\sum_{k=0}^{N} d_k e^{-j\omega k}\right) Y(e^{j\omega}) = \left(\sum_{k=0}^{M} p_k e^{-j\omega k}\right) X(e^{j\omega}). \tag{4.2}$$

The input-output relation of the LTI system in the z-domain is obtained by applying the z-transform to both sides of Eq. (2.81) and making use of the linearity and time-shifting properties of Table 3.9 resulting in

$$\sum_{k=0}^{N} d_k z^{-k} Y(z) = \sum_{k=0}^{M} p_k z^{-k} X(z), \tag{4.3}$$

where $Y(z)$ and $X(z)$ denote the z-transforms of $y[n]$ and $x[n]$ with associated ROCs, respectively. A more convenient form of Eq. (4.3) is given by

$$\left(\sum_{k=0}^{N} d_k z^{-k} \right) Y(z) = \left(\sum_{k=0}^{M} p_k z^{-k} \right) X(z). \tag{4.4}$$

4.2 The Frequency Response

Most discrete-time signals encountered in practice can be represented as a linear combination of a very large, maybe infinite, number of sinusoidal discrete-time signals of different angular frequencies. Thus, knowing the response of the LTI system to a single sinusoidal signal, we can determine its response to more complicated signals by making use of the superposition property of the system. Since a sinusoidal signal can be expressed in terms of an exponential signal, the response of the LTI system to an exponential input is of practical interest. This leads to the concept of frequency response, a transform-domain representation of the LTI discrete-time system. We first define the frequency response, investigate its properties, and describe some of its applications. The computation of the time-domain representation of the LTI system from its frequency response is outlined.

4.2.1 Definition

An important property of an LTI system is that for certain types of input signals, called *eigen functions,* the output signal is the input signal multiplied by a complex constant. We consider here one such eigen function as the input. Recall from Section 2.5.1, the input-output relationship of an LTI discrete-time system as shown in Figure 4.1, with an impulse response $h[n]$, is given by the convolution sum of Eq. (2.64b) and is of the form

$$y[n] = \sum_{k=-\infty}^{\infty} h[k]x[n-k] \tag{4.5}$$

where $y[n]$ and $x[n]$ are, respectively, the output and the input sequences. Now if the input $x[n]$ is a complex exponential sequence of the form

$$x[n] = e^{j\omega n}, \quad -\infty < n < \infty, \tag{4.6}$$

then from Eq. (4.5) the output is given by

$$y[n] = \sum_{k=-\infty}^{\infty} h[k]e^{j\omega(n-k)} = \left(\sum_{k=-\infty}^{\infty} h[k]e^{-j\omega k} \right) e^{j\omega n}, \tag{4.7}$$

which can be rewritten as

$$y[n] = H(e^{j\omega})e^{j\omega n}, \tag{4.8}$$

where we have used the notation

$$H(e^{j\omega}) = \sum_{n=-\infty}^{\infty} h[n]e^{-j\omega n}. \tag{4.9}$$

It can be seen from Eq. (4.8) that for a complex exponential input signal $e^{j\omega n}$, the output of an LTI discrete-time system is also a complex exponential signal of the same frequency multiplied by a complex constant $H(e^{j\omega})$. Thus, $e^{j\omega n}$ is an eigen function of the system. Another example of an eigen function is given in Problem 4.1.

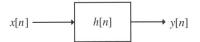

Figure 4.1: An LTI discrete-time system.

The quantity $H(e^{j\omega})$ defined above is called the *frequency response* of the LTI discrete-time system, and it provides a frequency-domain description of the system. Note from Eq. (4.9) that $H(e^{j\omega})$ is precisely the discrete-time Fourier transform (DTFT) of the impulse response $h[n]$ of the system.

Equation (4.8) implies that for a complex sinusoidal input sequence $x[n]$ of angular frequency ω as in Eq. (4.6), the output $y[n]$ is also a complex sinusoidal sequence of the same angular frequency but weighted by a complex amplitude $H(e^{j\omega})$ that is a function of the input frequency ω and the system's impulse response coefficients $h[n]$. We shall show later in Section 4.2.7 that $H(e^{j\omega})$ completely characterizes the LTI discrete-time system in the frequency domain.

Just like any other discrete-time Fourier transform, in general, $H(e^{j\omega})$ is also a complex function of ω with a period 2π and can be expressed in terms of its real and imaginary parts or its magnitude and phase. Thus,

$$H(e^{j\omega}) = H_{\text{re}}(e^{j\omega}) + jH_{\text{im}}(e^{j\omega}) = |H(e^{j\omega})|e^{j\theta(\omega)}, \tag{4.10}$$

where $H_{\text{re}}(e^{j\omega})$ and $H_{\text{im}}(e^{j\omega})$ are, respectively, the real and imaginary parts of $H(e^{j\omega})$, and

$$\theta(\omega) = \arg\{H(e^{j\omega})\}. \tag{4.11}$$

The quantity $|H(e^{j\omega})|$ is called the *magnitude response* and the quantity $\theta(\omega)$ is called the *phase response* of the LTI discrete-time system. Design specifications for the discrete-time systems, in many applications, are given in terms of the magnitude response or the phase response or both. In some cases, the magnitude function is specified in decibels as defined below:

$$\mathcal{G}(\omega) = 20\log_{10}|H(e^{j\omega})| \quad \text{dB} \tag{4.12}$$

where $\mathcal{G}(\omega)$ is called the *gain function*. The negative of the gain function, $a(\omega) = -\mathcal{G}(\omega)$, is called the *attenuation* or *loss function*.

It should be noted that the magnitude and phase functions are real functions of ω, whereas the frequency response is a complex function of ω. For a discrete-time system characterized by a real impulse response $h[n]$ it follows from Table 3.4 that the magnitude function is an even function of ω, i.e., $|H(e^{j\omega})| = |H(e^{-j\omega})|$, and the phase function is an odd function of ω, i.e., $\theta(\omega) = -\theta(-\omega)$. Likewise, $H_{\text{re}}(e^{j\omega})$ is even, and $H_{\text{im}}(e^{j\omega})$ is odd.

4.2.2 Frequency Response Computation Using MATLAB

The M-file function freqz(h,w) in MATLAB can be used to determine the values of the frequency response of a prescribed impulse response vector h at a set of given frequency points w. From these frequency response values, one can then compute the real and imaginary parts using the functions real and imag, and the magnitude and phase using the functions abs and angle as illustrated in the following example.

EXAMPLE 4.1 Consider the moving-average filter of Eq. (2.56). Comparing Eq. (2.56) with Eq. (2.97) we note that the impulse response of the moving-average filter is given by

$$h[n] = \begin{cases} \frac{1}{M}, & 0 \le n \le M-1, \\ 0, & \text{otherwise.} \end{cases} \tag{4.13}$$

From Eq. (4.9) its frequency response is thus given by

$$H(e^{j\omega}) = \frac{1}{M} \sum_{n=0}^{M-1} e^{-j\omega n} = \frac{1}{M} \left(\sum_{n=0}^{\infty} e^{-j\omega n} - \sum_{n=M}^{\infty} e^{-j\omega n} \right)$$

$$= \frac{1}{M} \left(\sum_{n=0}^{\infty} e^{-j\omega n} \right) \left(1 - e^{-jM\omega} \right) = \frac{1}{M} \frac{1 - e^{-jM\omega}}{1 - e^{-j\omega}}$$

$$= \frac{1}{M} \frac{\sin(M\omega/2)}{\sin(\omega/2)} e^{-j(M-1)\omega/2}. \tag{4.14}$$

From the above the magnitude and phase responses of the moving-average filter of Eq. (4.13) are obtained as

$$|H(e^{j\omega})| = \left| \frac{1}{M} \frac{\sin(M\omega/2)}{\sin(\omega/2)} \right|, \tag{4.15}$$

$$\theta(\omega) = -\frac{(M-1)\omega}{2} + \pi \sum_{k=0}^{\lfloor M/2 \rfloor} \mu \left(\omega - \frac{2\pi k}{M} \right), \tag{4.16}$$

where $\mu(\omega)$ is a step function in ω. Figure 4.2 shows the magnitude and phase responses of the moving-average filters for $M = 5$ and $M = 14$. These plots have been obtained using the following MATLAB program.

```
% Program 4_1
% Generate the filter coefficients
h1 = ones(1,5)/5; h2 = ones(1,14)/14;
w = 0:pi/255:pi;
% Compute the frequency responses
H1 = freqz(h1, 1, w);
H2 = freqz(h2, 1, w);
% Compute and plot the magnitude responses
m1 = abs(H1); m2 = abs(H2);
plot(w/pi,m1,'r-',w/pi,m2,'b--');
ylabel('Magnitude'); xlabel('\omega/\pi');
legend('r-','M=5','b--','M=14');
pause
% Compute and plot the phase responses
ph1 = angle(H1)*180/pi; ph2 = angle(H2)*180/pi;
plot(w/pi,ph1,w/pi,ph2);
ylabel('Phase, degrees');xlabel('\omega/\pi');
legend('r-','M=5','b--','M=14');
```

It can be seen from Figure 4.2 that in the range from $\omega = 0$ to $\omega = \pi$, the magnitude has a maximum value of unity at $\omega = 0$, and has zeros at $\omega = 2\pi k/M$ with $k = 1, 2, \ldots, \lfloor M/2 \rfloor$.[1] The phase function exhibits discontinuities of π at each zero of $H(e^{j\omega})$ and is linear elsewhere with a slope of $-(M-1)/2$. It should be noted that both the magnitude and phase functions are periodic in ω with a period 2π.

The phase responses of discrete-time systems when determined by a computer may also exhibit jumps by an amount of 2π caused by the way the arctangent function is computed, for example, in the function angle in MATLAB. The phase response can be made a continuous function of ω by unwrapping the phase response across the jumps by adding multiples of $\pm 2\pi$. The MATLAB function unwrap can be employed

---

[1] $\lfloor x \rfloor$ denotes the integer part of x.

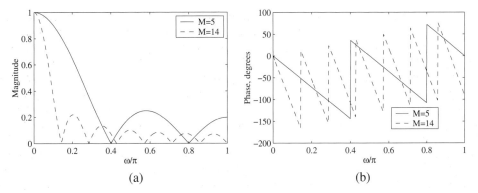

Figure 4.2: Magnitude and phase responses of the moving-average filters of length 5 and 14.

to this end, provided the computed phase is in radians.[2] The application of `unwrap` is illustrated later in this chapter in Figure 4.28. These jumps should not be confused with the jumps caused by the zeros of the frequency response as shown, for example, in Figure 4.2(b).

4.2.3 Steady-State Response

Note that the frequency response function also determines the *steady-state response* of the LTI discrete-time system to a sinusoidal input as shown in the following example.

EXAMPLE 4.2   Determine the steady-state output $y[n]$ of a real coefficient LTI discrete-time system with a frequency response $H(e^{j\omega})$ for an input

$$x[n] = A\cos(\omega_o n + \phi), \qquad -\infty < n < \infty, \tag{4.17}$$

where A is real.

Now using trigonometric identities, we can express the input $x[n]$ as a sum of two complex exponential sequences:

$$x[n] = g[n] + g^*[n],$$

where $g[n] = (1/2)Ae^{j\phi}e^{j\omega_o n}$. From Eq. (4.8), the output of the LTI discrete-time system for the input $e^{j\omega_o n}$ is simply $H(e^{j\omega_o})e^{j\omega_o n}$. Because of linearity, the response $v[n]$ to the input $g[n]$ is then given by

$$v[n] = \tfrac{1}{2}Ae^{j\phi}H(e^{j\omega_o})e^{j\omega_o n}.$$

Likewise, the output $v^*[n]$ to the input $g^*[n]$ is the complex conjugate of $v[n]$, i.e.,

$$v^*[n] = \tfrac{1}{2}Ae^{-j\phi}H(e^{-j\omega_o})e^{-j\omega_o n}.$$

Combining these two expressions, we arrive at the expression for the desired output $y[n]$ as

$$y[n] = v[n] + v^*[n] = \tfrac{1}{2}Ae^{j\phi}H(e^{j\omega_o})e^{j\omega_o n} + \tfrac{1}{2}Ae^{-j\phi}H(e^{-j\omega_o})e^{-j\omega_o n}$$

$$= \tfrac{1}{2}A|H(e^{j\omega_o})|\left\{e^{j\theta(\omega_o)}e^{j\phi}e^{j\omega_o n} + e^{-j\theta(\omega_o)}e^{-j\phi}e^{-j\omega_o n}\right\}$$

$$= A|H(e^{j\omega_o})|\cos(\omega_o n + \theta(\omega_o) + \phi). \tag{4.18}$$

---

[2]Care should be taken in using `unwrap` as it can give wrong answers sometimes if the computed phase response is sparse with rapidly changing values.

Thus, the output signal $y[n]$ has the same sinusoidal waveform as the input signal $x[n]$ of Eq. (4.17) with two differences: (1) the amplitude is multiplied by $|H(e^{j\omega_o})|$, the value of the magnitude function of the discrete-time system at $\omega = \omega_o$; and (2) the output signal has a phase *lag* relative to the input by an amount $\theta(\omega_o)$, the value of the phase of the discrete-time system at $\omega = \omega_o$.

4.2.4 Response to a Causal Exponential Sequence

An underlying assumption in developing Eq. (4.18) is that the system is initially relaxed before the application of the input of Eq. (4.17). However, in practice, the excitation to a discrete-time system is usually a causal sequence and applied at some sample index $n = n_o$. Hence, the output for such an input will be different from the one shown in Eq. (4.18) which we investigate here. Without any loss of generality, we develop next the expression for the output response when the input is a causal exponential sequence applied at $n = 0$.

Now, from Eq. (4.5), the output response $y[n]$ for an input

$$x[n] = e^{j\omega n}\mu[n]$$

is given by

$$y[n] = \left(\sum_{k=0}^{n} h[k]e^{j\omega(n-k)}\right)\mu[n] = \left(\sum_{k=0}^{n} h[k]e^{-j\omega k}\right)e^{j\omega n}\mu[n].$$

Thus, the output $y[n] = 0$ for $n < 0$, and for $n \geq 0$ it is given by

$$y[n] = \left(\sum_{k=0}^{n} h[k]e^{-j\omega k}\right)e^{j\omega n}$$

$$= \left(\sum_{k=0}^{\infty} h[k]e^{-j\omega k}\right)e^{j\omega n} - \left(\sum_{k=n+1}^{\infty} h[k]e^{-j\omega k}\right)e^{j\omega n}$$

$$= H(e^{j\omega})e^{j\omega n} - \left(\sum_{k=n+1}^{\infty} h[k]e^{-j\omega k}\right)e^{j\omega n}. \tag{4.19}$$

The first term in the output in Eq. (4.19) is the same as that given by Eq. (4.8) and is called the *steady-state response*:

$$y_{sr}[n] = H(e^{j\omega})e^{j\omega n}.$$

The second term in Eq. (4.19) is called the *transient response*:

$$y_{tr}[n] = -\left(\sum_{k=n+1}^{\infty} h[k]e^{-j\omega k}\right)e^{j\omega n}.$$

To determine the effect of the second term on the output response, we observe that

$$|y_{tr}[n]| = \left|\sum_{k=n+1}^{\infty} h[k]e^{-j\omega(k-n)}\right| \leq \sum_{k=n+1}^{\infty} |h[k]| \leq \sum_{k=0}^{\infty} |h[k]|. \tag{4.20}$$

Now, for a causal and stable LTI discrete-time system, the impulse response is absolutely summable, and as a result, the transient response $y_{tr}[n]$ is a bounded sequence. Moreover, as $n \to \infty$, $\sum_{k=n+1}^{\infty} |h[k]| \to 0$, and hence, the transient response decays to zero as n gets very large. On the other hand, for a causal FIR LTI discrete-time system with an impulse response of length $N + 1$, $h[n] = 0$ for $n > N$. Hence, $y_{tr}[n] = 0$ for $n > N - 1$. Thus, here the output $y[n]$ reaches the steady-state value $y_{sr}[n] = H(e^{j\omega})e^{j\omega n}$ for $n = N$.

4.2.5 The Concept of Filtering

One application of an LTI discrete-time system is to pass certain frequency components in an input sequence without any distortion (if possible) and to block other frequency components. Such systems are called *digital filters* and are one of the main subjects of discussion in this text. The key to the filtering process is the inverse discrete-time Fourier transform given in Eq. (3.7) which expresses an arbitrary input sequence as a linear weighted sum of an infinite number of exponential sequences, or equivalently, as a linear weighted sum of sinusoidal sequences. As a result, by appropriately choosing the values of magnitude function of the LTI digital filter at frequencies corresponding to the frequencies of the sinusoidal components of the input, some of these sinusoidal sequences can be selectively heavily attenuated or filtered with respect to the others.

We now explain the concept of filtering and then define the most commonly desired filter characteristics. To understand the mechanism behind the design of such a system, consider a real coefficient LTI discrete-time system characterized by a magnitude function

$$|H(e^{j\omega})| \cong \begin{cases} 1, & |\omega| \leq \omega_c, \\ 0, & \omega_c < |\omega| \leq \pi. \end{cases} \tag{4.21}$$

We apply an input $x[n] = A \cos \omega_1 n + B \cos \omega_2 n$ to this system, where $0 < \omega_1 < \omega_c < \omega_2 < \pi$. Because of linearity, it follows from Eq. (4.18) that the output $y[n]$ of this system is of the form

$$y[n] = A|H(e^{j\omega_1})| \cos(\omega_1 n + \theta(\omega_1)) + B|H(e^{j\omega_2})| \cos(\omega_2 n + \theta(\omega_2)). \tag{4.22}$$

Making use of Eq. (4.21) in Eq. (4.22), we get

$$y[n] \cong A|H(e^{j\omega_1})| \cos(\omega_1 n + \theta(\omega_1)),$$

indicating the LTI discrete-time system acts like a lowpass filter.

EXAMPLE 4.3   In this example, we consider the design of a very simple digital filter [Ham89]. The input signal consists of a sum of two cosine sequences of angular frequencies 0.1 rad/samples and 0.4 rad/samples, respectively. We need to design a highpass filter that will pass the high-frequency component of the input but block the low-frequency part.

For simplicity we assume the filter to be an FIR filter of length 3 with an impulse response

$$h[0] = h[2] = \alpha, \quad h[1] = \beta.$$

Hence, from Eq. (4.5), the digital filtering is performed using the difference equation

$$\begin{aligned} y[n] &= h[0]x[n] + h[1]x[n-1] + h[2]x[n-2] \\ &= \alpha x[n] + \beta x[n-1] + \alpha x[n-2] \end{aligned} \tag{4.23}$$

where $y[n]$ and $x[n]$ represent, respectively, the output and the input sequences. Thus, our design objective is to choose suitable values for the filter parameters, α and β, so that the output of the filter is a cosine sequence with a frequency 0.4 rad/samples.

Now, from Eq. (4.9) the frequency response of the above FIR filter is given by

$$H(e^{j\omega}) = h[0] + h[1]e^{-j\omega} + h[2]e^{-j2\omega}$$

$$= \alpha(1 + e^{-j2\omega}) + \beta e^{-j\omega} = 2\alpha \left(\frac{e^{j\omega} + e^{-j\omega}}{2} \right) e^{-j\omega} + \beta e^{-j\omega}$$

$$= (2\alpha \cos\omega + \beta)e^{-j\omega}. \tag{4.24}$$

The magnitude and phase functions of this filter are

$$|H(e^{j\omega})| = |2\alpha \cos\omega + \beta|, \tag{4.25}$$

$$\theta(\omega) = -\omega. \tag{4.26}$$

In order to stop the low-frequency component from appearing at the output of the filter, the magnitude function at $\omega = 0.1$ should be equal to zero. Similarly, to pass the high-frequency component without any attenuation, we need to ensure that the magnitude function at $\omega = 0.4$ is equal to 1. Thus, the two conditions that must be satisfied are

$$2\alpha \cos(0.1) + \beta = 0,$$

$$2\alpha \cos(0.4) + \beta = 1.$$

Solving the above two equations, we arrive at

$$\alpha = -6.76195, \qquad \beta = 13.456335. \tag{4.27}$$

Note that the above solution guarantees the positiveness of $(2\alpha \cos\omega + \beta)$. Substituting Eq. (4.27) in Eq. (4.23) we obtain the input-output relation of the desired FIR filter as

$$y[n] = -6.76195(x[n] + x[n-2]) + 13.456335x[n-1], \tag{4.28}$$

with

$$x[n] = \{ \cos(0.1n) + \cos(0.4n) \}\mu[n]. \tag{4.29}$$

To verify the filtering action, we implement the filter of Eq. (4.28) on MATLAB and calculate the first 100 output samples beginning from $n = 0$. Note that the input has been assumed to be a causal sequence with the first nonzero sample occurring at $n = 0$, and for calculating $y[0]$ and $y[1]$ we set $x[-1] = x[-2] = 0$. The MATLAB program used to calculate the output is given below.

```
% Program 4_2
% Set up the filter coefficients
b = [-6.76195 13.456335 -6.76195];
%  Set initial conditions to zero values
zi = [0 0];
% Generate the two sinusoidal sequences
n = 0:99;
x1 = cos(0.1*n);
x2 = cos(0.4*n);
%  Generate the filter output sequence
y = filter(b, 1, x1+x2, zi);
% Plot the input and the output sequences
plot(n,y,'r-',n,x2,'b--',n,x1,'g-.');grid
axis([0 100 -1.2 4]);
ylabel('Amplitude'); xlabel('Time index n');
legend('r-','y[n]','b--','x2[n]','g-.','x1[n]')
```

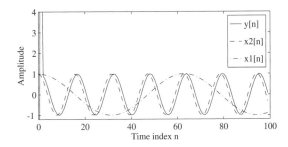

Figure 4.3: Output $y[n]$ (solid line), low-frequency input $x_1[n]$ (dash-dotted line), and high-frequency input $x_2[n]$ (dashed line) signals of the FIR filter of Eq. (4.28).

Table 4.1: Input and output sequences of the filter of Example 4.3.

n	$\cos(0.1n)$	$\cos(0.4n)$	$x[n]$	$y[n]$
0	1.0	1.0	2.0	-13.52390
1	0.9950041	0.9210609	1.9160652	13.956333
2	0.9800665	0.6967067	1.6767733	0.9210616
3	0.9553364	0.3623577	1.3176942	0.6967064
4	0.9210609	-0.0291995	0.8918614	0.3623572
5	0.8775825	-0.4161468	0.4614357	-0.0292002
6	0.8253356	-0.7373937	0.0879419	-0.4161467

Figure 4.3 shows the plots of the output $y[n]$ and the two sinusoids of the input $x[n]$ where the output values have been clipped to the range $(-1.2, 4)$ to show the steady state output more clearly. The first seven samples of the output are shown in Table 4.1 along with that of the two sinusoids. From this table and also from Figure 4.2, we observe that, neglecting the least significant digit,

$$y[n] = \cos(0.4(n-1)), \quad \text{for } n = 2, 3, 4, 5 \text{ and } 6.$$

Several comments are in order here. First, computation of the present value of output requires the knowledge of the present and two previous input samples. Hence, the first two output samples are the result of the assumed zero input sample values at $n = -1$ and $n = -2$, and are therefore the transient part of the output. Since the impulse response is of length $N + 1 = 3$, the steady-state is reached at $n = N = 2$. Second, the output is a delayed version of the high-frequency component $\cos(0.4n)$ of the input, and the delay is one sample period.

4.2.6 Phase and Group Delays

The output signal $y[n]$ of a frequency-selective LTI discrete-time system with a frequency response $H(e^{j\omega})$ exhibits some delay relative to the input signal $x[n]$ caused by the nonzero phase response $\theta(\omega) = \arg\{H(e^{j\omega})\}$ of the system. If the input is a sinusoidal signal of frequency ω_o as given by Eq. (4.17), the output is also a sinusoidal signal of the same frequency ω_o but lagging in phase by $\theta(\omega_o)$

radians as demonstrated in Eq. (4.18). We can rewrite Eq. (4.18) as

$$y[n] = A|H(e^{j\omega_o})| \cos\left(\omega_o\left(n + \frac{\theta(\omega_o)}{\omega_o}\right) + \phi\right), \tag{4.30}$$

indicating a time delay, more commonly known as *phase delay* at $\omega = \omega_o$ given by [3]

$$\tau_p(\omega_o) = -\frac{\theta(\omega_o)}{\omega_o}. \tag{4.31}$$

When the input signal contains many sinusoidal components with different frequencies that are not harmonically related, each component will go through different phase delays when processed by a frequency-selective LTI discrete-time system, and the output signal, in general, will not look like the input signal. In such cases, the signal delay is defined using a different parameter. To develop the necessary expression, we consider a discrete-time signal $x[n]$ obtained by a double-sideband suppressed carrier (DSB-SC) modulation with a carrier frequency ω_c of a low-frequency sinusoidal signal of frequency ω_o [Hay99]:[4]

$$x[n] = A \cos(\omega_o n) \cos(\omega_c n). \tag{4.32}$$

As indicated in Example 2.10, $x[n]$ can be rewritten as

$$x[n] = \frac{A}{2} \cos(\omega_\ell n) + \frac{A}{2} \cos(\omega_u n), \tag{4.33}$$

where $\omega_\ell = \omega_c - \omega_o$ and $\omega_u = \omega_c + \omega_o$.

If the above signal is processed by an LTI discrete-time system with a frequency response $H(e^{j\omega})$, it follows from Eq. (4.22) that the output signal is of the form

$$y[n] = \frac{A}{2} \cos(\omega_\ell n + \theta(\omega_\ell)) + \frac{A}{2} \cos(\omega_u n + \theta(\omega_u))$$

$$= A \cos\left(\omega_c n + \frac{\theta(\omega_u) + \theta(\omega_\ell)}{2}\right) \cos\left(\omega_o n + \frac{\theta(\omega_u) - \theta(\omega_\ell)}{2}\right), \tag{4.34}$$

assuming $|H(e^{j\omega})| \cong 1$ in the frequency range $\omega_\ell \leq \omega \leq \omega_u$. Thus the output is also in the form of a modulated carrier signal with the same carrier frequency ω_c and the modulation frequency ω_o as the input, but the two components in the output have different phase lags relative to their corresponding components in the input.

Consider the case when the modulated input given by Eq. (4.33) is a *narrowband signal* with the frequencies ω_ℓ and ω_u very close to the carrier frequency ω_c, i.e., ω_o is very small. In the neighborhood of ω_c we can express the unwrapped phase response $\theta_c(\omega)$ of the LTI discrete-time system approximately as

$$\theta_c(\omega) \cong \theta_c(\omega_c) + \frac{d\theta_c(\omega)}{d\omega}\bigg|_{\omega=\omega_c} (\omega - \omega_c), \tag{4.35}$$

by making a Taylor's series expansion and keeping only the first two terms. Using the above formula we evaluate the time delays of the carrier and the modulating components. In the former case it is given by

$$-\frac{\theta_c(\omega_u) + \theta_c(\omega_\ell)}{2\omega_c} \cong -\frac{\theta_c(\omega_c)}{\omega_c}, \tag{4.36}$$

---

[3]The minus sign indicates phase lag.
[4]See Section 1.2.4 for a review of the DSB-SC modulation scheme for analog signals.

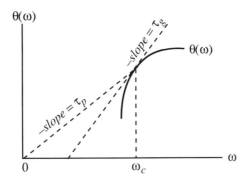

Figure 4.4: Evaluation of the phase delay and the group delay.

which is seen to be the same as the phase delay if only the carrier signal is passed through the system. On the other hand, in the latter case, it is given by

$$-\frac{\theta_c(\omega_u) - \theta_c(\omega_\ell)}{\omega_u - \omega_\ell} \cong -\left.\frac{d\theta_c(\omega)}{d\omega}\right|_{\omega=\omega_c}. \tag{4.37}$$

The parameter

$$\tau_g(\omega_c) = -\left.\frac{d\theta_c(\omega)}{d\omega}\right|_{\omega=\omega_c}$$

is called the *group delay* or *envelope delay* caused by the system at $\omega = \omega_c$. In the general case, the group delay is defined by

$$\tau_g(\omega) = -\frac{d\theta_c(\omega)}{d\omega}. \tag{4.38}$$

The group delay is a measure of the linearity of the phase function as a function of the frequency and is the time delay between the waveforms of underlying continuous-time signals whose sampled versions, sampled at $t = nT$, are precisely the input and the output discrete-time signals. If the phase function is in radians and the angular frequency ω is in radians per second, then the group delay is in seconds. Figure 4.4 illustrates the evaluation of the phase delay and the group delay of a typical phase function. Figure 4.5 shows the waveform of an amplitude-modulated input signal and that of the output generated by an LTI system. As can be seen from this figure, the carrier component at the output is delayed by the phase delay and the envelope of the output signal is delayed by the group delay relative to the waveform of the underlying continuous-time input signal.

The waveform of the underlying continuous-time output shows distortion when the group delay of the LTI system is not constant over the bandwidth of the modulated signal. If the distortion is unacceptable, a delay equalizer is usually cascaded with the LTI system so that the overall group delay of the cascade is approximately linear over the band of interest. However, to keep the magnitude response of the parent LTI system unchanged the equalizer must have a constant magnitude response at all frequencies.[5] It should be noted that the group delay is equal to the phase delay up to the first phase discontinuity.

For the filter of Example 4.3, the phase function $\theta(\omega) = -\omega$. Hence, the group delay is given by $\tau_g(\omega) = 1$ which is also evident from Figure 4.3, and pointed out earlier. Likewise, for the moving-

---

[5]See Section 4.6.3.

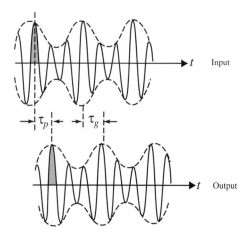

Figure 4.5: Illustration of the concept of the phase delay and the group delay. (Adapted with permission from A. Williams and F. J. Taylor, *Electronic Filter Designer's Handbook*, 3rd edition, McGraw-Hill, New York NY, 1995.)

average filter of Eq. (4.13) the group delay is given by

$$\tau_g(\omega) = \frac{M-1}{2},\tag{4.39}$$

or in other words, the moving-average filter exhibits a constant group delay for all frequencies.

The group delay can be determined using the M-file `grpdelay` in MATLAB. We shall illustrate its use later in this chapter.

4.2.7 Frequency-Domain Characterization of the LTI Discrete-Time System

We now derive the frequency-domain representation of an LTI discrete-time system. If $Y(e^{j\omega})$ and $X(e^{j\omega})$ denote the DTFTs of the output and input sequences, $y[n]$ and $x[n]$, respectively, then from Eq. (4.5) by taking the DTFT of both sides we obtain

$$Y(e^{j\omega}) = \sum_{n=-\infty}^{\infty} y[n]e^{-j\omega n}$$

$$= \sum_{n=-\infty}^{\infty} \left(\sum_{k=-\infty}^{\infty} h[k]x[n-k] \right) e^{-j\omega n}.\tag{4.40}$$

Interchanging the summation signs on the right-hand side of Eq. (4.40) and rearranging we arrive at

$$Y(e^{j\omega}) = \sum_{k=-\infty}^{\infty} h[k] \left(\sum_{n=-\infty}^{\infty} x[n-k]e^{-j\omega n} \right)$$

$$= \sum_{k=-\infty}^{\infty} h[k] \left(\sum_{\ell=-\infty}^{\infty} x[\ell]e^{-j\omega(\ell+k)} \right)$$

$$= \sum_{k=-\infty}^{\infty} h[k] \left(\sum_{\ell=-\infty}^{\infty} x[\ell]e^{-j\omega\ell} \right) e^{-j\omega k}.\tag{4.41}$$

The quantity inside the parentheses on the right-hand side of Eq. (4.41) is recognized as $X(e^{j\omega})$, the DTFT of the input sequence $x[n]$. Substituting this notation and rearranging, we finally obtain

$$Y(e^{j\omega}) = \left(\sum_{k=-\infty}^{\infty} h[k]e^{-j\omega k} \right) X(e^{j\omega}) = H(e^{j\omega})X(e^{j\omega}) \tag{4.42}$$

where $H(e^{j\omega})$ is the frequency response of the LTI system as defined in Eq. (4.9). Equation (4.42) thus relates the input and the output of an LTI system in the frequency domain. It should be noted that the development of Eq. (4.42) provides a proof of the convolution property of the DTFT given in Table 3.2.

From Eq. (4.42) we obtain

$$H(e^{j\omega}) = \frac{Y(e^{j\omega})}{X(e^{j\omega})}. \tag{4.43}$$

Thus, the frequency response of an LTI discrete-time system is given by the ratio of the DTFT $Y(e^{j\omega})$ of the output sequence $y[n]$ to the DTFT $X(e^{j\omega})$ of the input sequence $x[n]$. For example, for an LTI system characterized by a linear constant coefficient difference equation of the form of Eq. (2.81), the expression for its frequency response $H(e^{j\omega})$ can be simply derived from Eq. (4.2) resulting in

$$H(e^{j\omega}) = \frac{\sum_{k=0}^{M} p_k e^{-j\omega k}}{\sum_{k=0}^{N} d_k e^{-j\omega k}}. \tag{4.44}$$

4.3 The Transfer Function

A generalization of the frequency response function $H(e^{j\omega})$ leads to the concept of transfer function, which is defined next. As we have seen, the frequency response function does provide valuable information on the behavior of an LTI digital filter in the frequency-domain. However, being a complex function of the frequency variable ω, it is difficult to manipulate it for the realization of a digital filter. On the other hand, the z-transform of the impulse response of an LTI system, called the transfer function, is a polynomial in z^{-1}, and for a system with a real impulse response, it is a polynomial with real coefficients. Moreover, in most practical cases, the LTI digital filter of interest is characterized by a linear difference equation with constant and real coefficients. The transfer function of such a filter is a real rational function of the variable z^{-1}, i.e., a ratio of two polynomials in z^{-1} with real coefficients, and is thus more amenable for the synthesis.

We first develop the input-output relation of an LTI system in the z-domain from its various time-domain descriptions and arrive at different forms of the transfer function representation of the system. We then study its properties and in particular develop the conditions for the BIBO stability of a causal LTI system.

4.3.1 Definition

Consider the LTI digital discrete-time system of Figure 4.1 with an impulse response $h[n]$. The input-output relation of this system is given by Eq. (4.5) where $y[n]$ and $x[n]$ are, respectively, the output and the input sequences. If $Y(z)$, $X(z)$, and $H(z)$ denote the z-transforms of $y[n]$, $x[n]$, and $h[n]$, respectively, then by taking the z-transforms of both sides of Eq. (4.5) and following the steps similar to that used in the development of Eq. (4.42) we arrive at the input-output relation of the filter in the z-domain given by

$$Y(z) = H(z)X(z). \tag{4.45}$$

From the above we get

$$H(z) = \frac{Y(z)}{X(z)}. \tag{4.46}$$

The quantity $H(z)$, which is the z-transform of the impulse response sequence $h[n]$ of the filter, is more commonly called the *transfer function* or the *system function*. Thus, the transfer function $H(z)$ of an LTI discrete-time system is given by the ratio of the z-transform $Y(z)$ of the output sequence $y[n]$ to the z-transform $X(z)$ of the input sequence $x[n]$. It should be noted that Eq. (4.45) also follows from the convolution property of the z-transform given in Table 3.9.

The inverse z-transform of the transfer function $H(z)$ yields the impulse response $h[n]$. For a causal rational transfer function, the methods outlined in Section 3.9 can be used to compute its impulse response. For example, an analytical form of the impulse response can be determined via a partial-fraction expansion using MATLAB Program 3_9. On the other hand, a fixed number of impulse response samples starting at $n = 0$ can be computed using MATLAB Programs 3_11 or 3_12.

4.3.2   Derivation of the Transfer Function Expression

In the case of an FIR digital filter, the input-output relation in the time-domain is given by Eq. (2.97), and by taking the z-transform of both sides of this equation, we arrive at

$$Y(z) = \left(\sum_{n=N_1}^{N_2} h[n]z^{-n} \right) X(z),$$

from which we obtain

$$H(z) = \sum_{n=N_1}^{N_2} h[n]z^{-n}. \tag{4.47}$$

For a causal FIR filter, $0 \leq N_1 \leq N_2$. Note that all poles of $H(z)$ of a causal FIR filter are at the origin in the z-plane, and as a result, the ROC of $H(z)$ is the entire z-plane excluding the point $z = 0$.

In the case of an IIR digital filter, the transfer function expression in general is an infinite series. However, for the finite-dimensional IIR filter characterized by the difference equation of Eq. (2.81), the transfer function expression follows directly from Eq. (4.4) and is given by

$$H(z) = \frac{Y(z)}{X(z)} = \frac{p_0 + p_1 z^{-1} + p_2 z^{-2} + \cdots + p_M z^{-M}}{d_0 + d_1 z^{-1} + d_2 z^{-2} + \cdots + d_N z^{-N}}. \tag{4.48}$$

This is seen to be a rational function in z^{-1}, i.e., it is a ratio of two polynomials in z^{-1}. By multiplying the numerator and the denominator of the right-hand side by z^M and z^N, respectively, the transfer function can be expressed as a rational function in z:

$$H(z) = z^{(N-M)} \frac{p_0 z^M + p_1 z^{M-1} + p_2 z^{M-2} + \cdots + p_M}{d_0 z^N + d_1 z^{N-1} + d_2 z^{N-2} + \cdots + d_N}. \tag{4.49}$$

An alternate way to express the transfer function of Eq. (4.48) is to factor out the numerator and denominator polynomials leading to

$$H(z) = \frac{p_0}{d_0} \frac{\prod_{k=1}^{M}(1 - \xi_k z^{-1})}{\prod_{k=1}^{N}(1 - \lambda_k z^{-1})}, \tag{4.50a}$$

or to represent the transfer function of Eq. (4.49) in factored form, as

$$H(z) = \frac{p_0}{d_0} z^{(N-M)} \frac{\prod_{k=1}^{M}(z - \xi_k)}{\prod_{k=1}^{N}(z - \lambda_k)}, \tag{4.50b}$$

where $\xi_1, \xi_2, \ldots, \xi_M$ are the finite zeros, and $\lambda_1, \lambda_2, \ldots, \lambda_N$ are the finite poles of $H(z)$. If $N > M$, there are additional $(N - M)$ zeros at $z = 0$, and if $N < M$, there are additional $(M - N)$ poles at $z = 0$. For a causal IIR filter, the impulse response is a causal sequence. The ROC of the causal IIR transfer function $H(z)$ of Eq. (4.50b) is thus exterior to the circle going through the pole furthest from the origin, i.e., the ROC is given by

$$|z| > \max_k |\lambda_k|.$$

EXAMPLE 4.4   Consider the moving-average filter of Example 4.1 with an impulse response $h[n]$ given by Eq. (4.13). The transfer function of the moving-average FIR filter is then given by

$$H(z) = \frac{1}{M} \sum_{n=0}^{M-1} z^{-n} \tag{4.51}$$

$$= \frac{1 - z^{-M}}{M(1 - z^{-1})} = \frac{z^M - 1}{M[z^{M-1}(z - 1)]}. \tag{4.52}$$

From Eq. (4.52) it is seen that the transfer function has M zeros on the unit circle at $z = e^{j2\pi k/M}$, $k = 0, 1, 2, \ldots, M - 1$. There is an $(M - 1)$th-order pole at the origin $(z = 0)$ and a single pole at $z = 1$. But the pole at $z = 1$ exactly cancels a zero at the same place resulting in a transfer function with all poles at the origin. This is always the characteristic of a causal FIR filter for which the ROC is the entire z-plane except the origin.

EXAMPLE 4.5   A causal LTI IIR digital filter is characterized by a constant coefficient difference equation given by

$$y[n] = x[n - 1] - 1.2x[n - 2] + x[n - 3] + 1.3y[n - 1] - 1.04y[n - 2] + 0.222y[n - 3]. \tag{4.53}$$

From Eqs. (4.53) and (4.48) we arrive directly at the expression for the transfer function $H(z)$ as a rational function in z^{-1}:

$$H(z) = \frac{Y(z)}{X(z)} = \frac{z^{-1} - 1.2z^{-2} + z^{-3}}{1 - 1.3z^{-1} + 1.04z^{-2} - 0.222z^{-3}}. \tag{4.54}$$

Multiplying the numerator and the denominator of the right-hand-side term in the above equation by z^3 we obtain the expression for the transfer function as a rational function in z:

$$H(z) = \frac{z^2 - 1.2z + 1}{z^3 - 1.3z^2 + 1.04z - 0.222}. \tag{4.55}$$

The above transfer function can also be expressed in a factored form. To this end we make use of MATLAB Program 3_7 of Example 3.29 whose output are the zeros, poles, and gain constant as indicated below:

```
Zeros are at
    0.6000 + 0.8000i
    0.6000 - 0.8000i

Poles are at
    0.5000 + 0.7000i
    0.5000 - 0.7000i
    0.3000

Gain constant
    1
```

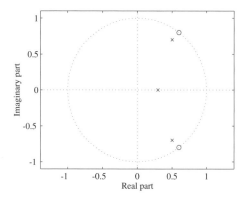

Figure 4.6: Pole-zero plot of the IIR transfer function of Eq. (4.54).

The transfer function of Eq. (4.55) has three zeros: one at $z = 0.6 - j0.8$, the second at $z = 0.6 + j0.8$, and the third at $z = \infty$. Likewise, it has three poles: one at $z = 0.5 - j0.7$, the second at $z = 0.5 + j0.7$, and the third at $z = 0.3$. Thus the factored form of the transfer function is given by

$$H(z) = \frac{(z - 0.6 + j0.8)(z - 0.6 - j0.8)}{(z - 0.3)(z - 0.5 + j0.7)(z - 0.5 - j0.7)}. \tag{4.56}$$

A pole-zero plot of the above transfer function obtained using the function `zplane` in MATLAB is sketched in Figure 4.6. Note from Eq. (4.56) and also from Figure 4.6 that the two poles furthest from the origin are of magnitude $\sqrt{0.74}$. Hence, the ROC of the transfer function of Eq. (4.55) is given by $|z| > \sqrt{0.74}$.

4.3.3 Frequency Response from Transfer Function

If the ROC of $H(z)$ includes the unit circle, then the frequency response $H(e^{j\omega})$ of the LTI digital filter can be obtained from its transfer function $H(z)$ by simply evaluating it on the unit circle, i.e.,

$$H(e^{j\omega}) = H(z)|_{z=e^{j\omega}}. \tag{4.57}$$

As indicated in Eq. (4.10), the frequency response $H(e^{j\omega})$ can be written in terms of its real and imaginary parts, $H_{re}(e^{j\omega})$ and $H_{im}(e^{j\omega})$, or in terms of its magnitude and phase functions, $|H(e^{j\omega})|$ and $\arg[H(e^{j\omega})]$, respectively. For a real-coefficient transfer function $H(z)$, it can be shown that

$$|H(e^{j\omega})|^2 = H(e^{j\omega})H^*(e^{j\omega}) = H(e^{j\omega})H(e^{-j\omega}) = H(z)H(z^{-1})|_{z=e^{j\omega}}. \tag{4.58}$$

For a stable rational transfer function $H(z)$ in the form of Eq. (4.50b), the factored form of the frequency response $H(e^{j\omega})$ is obtained by substituting $z = e^{j\omega}$ resulting in

$$H(e^{j\omega}) = \frac{p_0}{d_0} e^{j\omega(N-M)} \frac{\prod_{k=1}^{M}(e^{j\omega} - \xi_k)}{\prod_{k=1}^{N}(e^{j\omega} - \lambda_k)}. \tag{4.59}$$

The above form is convenient to visualize the contributions of the zero factor $(z - \xi_k)$ and the pole factor $(z - \lambda_k)$ of the transfer function $H(z)$ to the overall frequency response. From Eq. (4.59), the expression for the magnitude function is thus given by

$$|H(e^{j\omega})| = \left|\frac{p_0}{d_0}\right| |e^{j\omega}|^{(N-M)} \frac{\prod_{k=1}^{M} |e^{j\omega} - \xi_k|}{\prod_{k=1}^{N} |e^{j\omega} - \lambda_k|}$$

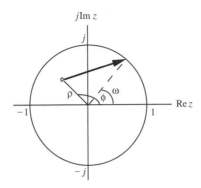

Figure 4.7: Geometric interpretation of frequency response computation of a rational transfer function.

$$= \left| \frac{p_0}{d_0} \right| \frac{\prod_{k=1}^{M} |e^{j\omega} - \xi_k|}{\prod_{k=1}^{N} |e^{j\omega} - \lambda_k|}. \tag{4.60}$$

Likewise, from Eq. (4.59), the phase response for a rational transfer function is of the form

$$\arg H(e^{j\omega}) = \arg(p_0/d_0) + \omega(N - M)$$

$$+ \sum_{k=1}^{M} \arg(e^{j\omega} - \xi_k) - \sum_{k=1}^{N} \arg(e^{j\omega} - \lambda_k). \tag{4.61}$$

The magnitude-squared function, for a real-coefficient rational transfer function can be computed using Eq. (4.59) which leads to

$$|H(e^{j\omega})|^2 = \left| \frac{p_0}{d_0} \right|^2 \frac{\prod_{k=1}^{M} (e^{j\omega} - \xi_k)(e^{-j\omega} - \xi_k^*)}{\prod_{k=1}^{N} (e^{j\omega} - \lambda_k)(e^{-j\omega} - \lambda_k^*)}. \tag{4.62}$$

4.3.4 Geometric Interpretation of Frequency Response Computation

For an LTI digital filter with a rational transfer function $H(z)$, the factored form of the frequency response expression given by Eq. (4.59) is convenient to develop a geometric interpretation of the frequency response computation from the pole-zero plot of the transfer function as ω is varied from 0 to 2π on the unit circle in the z-plane. The geometric interpretation can be used to obtain a sketch of the response as a function of the frequency.

If we examine the expression for the frequency response given in Eq. (4.59), we observe that a typical factor is of the form

$$(e^{j\omega} - \rho e^{j\phi}),$$

where $\rho e^{j\phi}$ is a zero if the factor is from the numerator, i.e., a zero factor, or is a pole if it is from the denominator, i.e., a pole factor. In the z-plane the factor $(e^{j\omega} - \rho e^{j\phi})$ represents a vector starting from the point $z = \rho e^{j\phi}$ and ending on the unit circle at $z = e^{j\omega}$ as shown in Figure 4.7. As ω is varied from 0 to 2π, the tip of the vector moves counterclockwise from the point $z = 1$ tracing the unit circle and back to the point $z = 1$.

As indicated by Eq. (4.60), the magnitude response $|H(e^{j\omega})|$ at a specific value of ω is given by the product of the magnitudes of all zero vectors divided by the product of the magnitudes of all pole vectors. Likewise, from Eq. (4.61) we observe that the phase response $\arg H(e^{j\omega})$ at a specific value of ω is obtained by adding the phase of the term p_0/d_0 and the linear-phase term $\omega(N - M)$ to the sum of the angles of all zero vectors minus the sum of the angles of all pole vectors. Thus, an approximate plot of the magnitude and phase responses of the transfer function of an LTI digital filter can be developed by examining its pole and zero locations.

Now, a zero vector has the smallest magnitude when $\omega = \phi$, and the pole vector has the largest magnitude when $\omega = \phi$. If the digital filter is to be designed to highly attenuate signal components in a specified range of frequencies, we need to place zeros of the transfer function very close to or on the unit circle in this range. Similarly, to highly emphasize signal components in a specified range of frequencies, we need to place poles of the transfer function very close to the unit circle in this range.

4.3.5 Stability Condition in Terms of Pole Locations

We describe in this text several methods for determining the transfer function of an FIR or IIR filter meeting the prescribed frequency-domain specifications. However, before the digital filter is implemented we need to ensure that the transfer function derived will lead to a stable structure. We now establish the condition to be satisfied by a causal LTI digital filter described by a rational transfer function $H(z)$ in order that it be BIBO stable.

Recall from our earlier discussion in Section 2.5.3 that an LTI digital filter is BIBO stable if and only if its impulse response sequence $h[n]$ is absolutely summable, i.e.,

$$S = \sum_{n=-\infty}^{\infty} |h[n]| < \infty. \tag{4.63}$$

The above stability condition in terms of the impulse response samples is difficult to test for a system with an impulse response of infinite length. We now develop a stability condition in terms of the pole locations of the transfer function $H(z)$ which is much easier to test if the pole locations are known.

Recall from Section 3.7.1, the ROC of the z-transform of $h[n]$, $H(z)$, is defined by values of $|z| = r$ for which $h[n]r^{-n}$ is absolutely summable. Thus if the ROC includes the unit circle $|z| = 1$, then the digital filter is stable, and vice versa. In addition, for a stable and causal digital filter for which $h[n]$ is a right-sided sequence, the ROC will include the unit circle and the entire z-plane outside the unit circle including the point $z = \infty$.

As indicated earlier, an FIR digital filter with bounded impulse response coefficients is always stable. On the other hand, an IIR filter may be unstable if not designed properly. In addition, an originally stable IIR filter characterized by infinite precision coefficients may also become unstable after implementation due to the unavoidable quantization of all coefficients, as illustrated by the following example.

EXAMPLE 4.6   Consider the causal second-order IIR transfer function given by

$$H(z) = \frac{1}{1 - 1.845z^{-1} + 0.850586z^{-2}}. \tag{4.64}$$

The plot of the impulse response coefficients of the above transfer function using a slightly modified Program 3_11 is shown in Figure 4.8(a). As seen from this figure, the impulse response coefficient $h[n]$ decays rapidly to zero value as n increases indicating that the absolute summability condition of Eq. (4.63) is satisfied. Hence, $H(z)$ is a stable transfer function.

Now consider the case when the transfer function coefficients are rounded to values with two digits after the decimal point. Then the transfer function of Eq. (4.64) takes the form

$$\hat{H}(z) = \frac{1}{1 - 1.85z^{-1} + 0.85z^{-2}}. \tag{4.65}$$

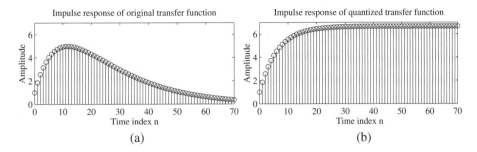

Figure 4.8: (a) Impulse response of original transfer function of Eq. (4.64), and (b) impulse response of transfer function of Eq. (4.65).

A plot of its impulse response coefficients is shown in Figure 4.8(b). In this case, the impulse response coefficient $h[n]$ increases rapidly to a constant value as n increases. As a result the absolute summability condition of Eq. (4.63) is violated, implying that $\hat{H}(z)$ is an unstable transfer function.

The stability testing of an IIR transfer function is therefore an important problem. However, it is difficult to compute the sum S of Eq. (4.63) analytically in most cases. For a causal IIR transfer function, it can be computed approximately on a computer by replacing the right-hand side of Eq. (4.63) with the following finite sum

$$S_K = \sum_{n=0}^{K-1} |h[n]|, \tag{4.66}$$

and iteratively computing Eq. (4.66) until the difference between a series of consecutive values of S_K is smaller than some arbitrarily chosen small number, which is typically 10^{-6}. For a transfer function of very high order this approach may not be satisfactory. We now develop an alternate stability condition based on the location of the poles of the transfer function. For practical reasons, we restrict our attention to causal transfer functions.

Consider the IIR digital filter described by the rational transfer function $H(z)$ of Eq. (4.48). If the digital filter is assumed to be causal, i.e., the impulse response sequence $\{h[n]\}$ is a right-sided sequence, the ROC of $H(z)$ is exterior to the circle going through the pole that is farthest from the origin. But stability requires that $\{h[n]\}$ be absolutely summable, which in turn implies that the discrete-time Fourier transform $H(e^{j\omega})$ of $\{h[n]\}$ exists. Now, the z-transform $H(z)$ of a sequence $\{h[n]\}$ reduces to the Fourier transform $H(e^{j\omega})$ by letting $z = e^{j\omega}$ if the unit circle lies within the ROC of the transfer function. Therefore we conclude that all poles of a stable causal transfer function $H(z)$ must be strictly inside the unit circle, as indicated in Figure 4.9.

EXAMPLE 4.7 The factored form of the transfer function $H(z)$ of the causal LTI digital filter of Eq. (4.64) is given by

$$H(z) = \frac{1}{(1 - 0.902z^{-1})(1 - 0.943z^{-1})},$$

which has a real pole at $z = 0.902$ and a real pole at $z = 0.943$. Since the magnitude of each pole of $H(z)$ is less than 1, all poles are inside the unit circle in the z-plane. Hence, the transfer function is stable.

On the other hand, the factored form of the transfer function $\hat{H}(z)$ of Eq. (4.65) is given by

$$\hat{H}(z) = \frac{1}{(1 - z^{-1})(1 - 0.85z^{-1})},$$

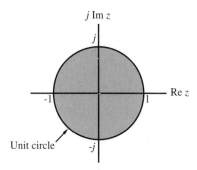

Figure 4.9: Stability region (shown shaded) in the z-plane for pole locations of a stable casual transfer function.

which now has one pole on the unit circle at $z = 1$ and the other pole inside the unit circle at $z = 0.85$, indicating that it is unstable.

4.4 Types of Transfer Functions

The time-domain classification of a digital transfer function based on the length of its impulse response sequence leads to the *finite impulse response* (FIR) and the *infinite impulse response* (IIR) transfer functions. We describe here several other types of classifications. In the case of digital transfer functions with frequency-selective frequency responses, one classification is based on the shape of the magnitude function $|H(e^{j\omega})|$ or the form of the phase function $\theta(\omega)$. Based on this, four types of ideal filters are usually defined. These ideal filters have doubly infinite impulse responses and are unrealizable. We describe here very simple realizable FIR and IIR digital filter approximations. In a number of applications, these simple filters are quite adequate and provide satisfactory performances. The importance of transfer functions with linear phase is then pointed out and the possible realizations of these transfer functions with FIR filters are discussed.

4.4.1 Ideal Filters

As pointed out in Example 4.3, a digital filter designed to pass signal components of certain frequencies without any distortion should have a frequency response of value equal to one at these frequencies, and should have a frequency response of value equal to zero at all other frequencies to totally block signal components with those frequencies. The range of frequencies where the frequency response takes the value of one is called the *passband*, and the range of frequencies where the frequency response is equal to zero is called the *stopband* of the filter.

The frequency responses of the four popular types of ideal digital filters with real impulse response coefficients are shown in Figure 4.10. For the *lowpass filter* of Figure 4.10(a), the passband and the stopband are given by $0 \leq \omega \leq \omega_c$ and $\omega_c < \omega \leq \pi$, respectively. For the *highpass filter* of Figure 4.10(b), the stopband is given by $0 \leq \omega < \omega_c$, while the passband is given by $\omega_c \leq \omega \leq \pi$. The passband region of the *bandpass filter* of Figure 4.10(c) is $\omega_{c1} \leq \omega \leq \omega_{c2}$ and the stopband regions are given by $0 \leq \omega < \omega_{c1}$ and $\omega_{c2} < \omega < \pi$. Finally, for the *bandstop filter* of Figure 4.10(d), the passband regions are $0 \leq \omega \leq \omega_{c1}$ and $\omega_{c2} \leq \omega \leq \pi$, while the stopband is from $\omega_{c1} < \omega < \omega_{c2}$. The frequencies ω_c, ω_{c1}, and ω_{c2} are called the *cutoff frequencies* of their respective filters. Note from this figure that an ideal

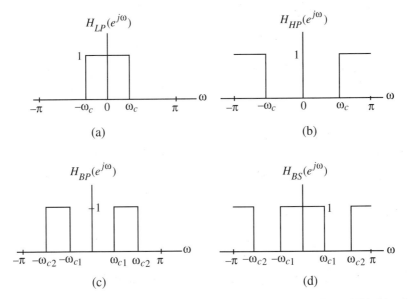

Figure 4.10: Four types of ideal filters: (a) ideal lowpass filter, (b) ideal highpass filter, (c) ideal bandpass filter, and (d) ideal bandstop filter.

filter thus has a magnitude response equal to unity in the passband and zero in the stopband, and has a zero phase everywhere.

We have already encountered the frequency response $H_{LP}(e^{j\omega})$ of the ideal lowpass filter of Figure 4.10(a) in Example 3.3 where we computed its impulse response given in Eq. (3.12). We repeat it here for convenience:

$$h_{LP}[n] = \frac{\sin \omega_c n}{\pi n}, \qquad -\infty < n < \infty. \tag{4.67}$$

In this example, we have shown that the above impulse response is not absolutely summable, and hence, the corresponding transfer function is not BIBO stable. Note also that the above impulse response is not causal and is of doubly infinite length. The remaining three frequency responses of Figure 4.10 also are characterized by doubly infinite, noncausal impulse responses and are not absolutely summable. As a result, the ideal filters with the ideal "brick wall" characteristics of Figure 4.10 cannot be realized by a finite dimensional LTI filter.

In order to develop stable and realizable transfer functions, the ideal frequency response specifications of Figure 4.10 are relaxed by including a *transition band* between the passband and the stopband to permit the magnitude response to decay more gradually from its maximum value in the passband to the zero value in the stopband. Moreover, the magnitude response is allowed to vary by a small amount both in the passband and the stopband. Typical magnitude specifications used for the design of a lowpass filter are shown in Figure 7.1. Chapter 7 is devoted to a discussion of various filter design methods that lead to stable and realizable transfer functions meeting such relaxed specifications. In the following two sections we describe several very simple low-order FIR and IIR digital filters that exhibit selective frequency response characteristics providing a first-order approximation to the ideal characteristics of Figure 4.10. Frequency responses with sharper characteristics can often be obtained by cascading one or more of these simple filters, which in many applications are quite satisfactory.

$$u[n] = v[-n], \quad y[n] = w[-n]$$

Figure 4.11: Implementation of a zero-phase filtering scheme.

4.4.2 Zero-Phase and Linear-Phase Transfer Functions

A second classification of a transfer function is with respect to its phase characteristics. In many applications, it is necessary to ensure that the digital filter designed does not distort the phase of the input signal components with frequencies in the passband. One way to avoid any phase distortions is to make the frequency response of the filter real and nonnegative, i.e., to design the filter with a *zero phase* characteristic. However, it is impossible to design a causal digital filter with a zero phase. For non-real-time processing of real-valued input signals of finite length, zero-phase filtering can be very simply implemented if the causality requirement is relaxed. To this end, one of two feasible schemes can be followed. In one scheme, the finite-length input data is processed through a causal real-coefficient filter $H(z)$ whose output is then time-reversed and processed by the same filter, once again as indicated in Figure 4.11.

To verify the above scheme, let $v[-n] = u[n]$. Also, let $X(e^{j\omega})$, $V(e^{j\omega})$, $U(e^{j\omega})$, $W(e^{j\omega})$, and $Y(e^{j\omega})$ denote the discrete-time Fourier transforms of $x[n]$, $v[n]$, $u[n]$, $w[n]$, and $y[n]$, respectively. Now, from Figure 4.11 and making use of the symmetry relations given in Tables 3.3 and 3.4, we arrive at the relations between the various Fourier transforms as

$$V(e^{j\omega}) = H(e^{j\omega})X(e^{j\omega}), \quad W(e^{j\omega}) = H(e^{j\omega})U(e^{j\omega}),$$
$$U(e^{j\omega}) = V^*(e^{j\omega}), \quad Y(e^{j\omega}) = W^*(e^{j\omega}).$$

Combining the above equations we obtain

$$Y(e^{j\omega}) = W^*(e^{j\omega}) = H^*(e^{j\omega})U^*(e^{j\omega}) = H^*(e^{j\omega})V(e^{j\omega})$$
$$= H^*(e^{j\omega})H(e^{j\omega})X(e^{j\omega}) = |H(e^{j\omega})|^2 X(e^{j\omega}).$$

Therefore the overall arrangement of Figure 4.11 implements a zero-phase filter with a frequency response $|H(e^{j\omega})|^2$.

The function `filtfilt` in MATLAB implements the above scheme.

A second scheme to achieve zero-phase filtering is outlined in Problem 4.67.

In the case of a causal transfer function with a nonzero phase response, the phase distortion can be avoided by ensuring that the transfer function has a unity magnitude and a *linear-phase* characteristic in the frequency band of interest. The most general type of such a filter has a frequency response given by

$$H(e^{j\omega}) = e^{-j\omega D}, \tag{4.68}$$

which has a linear-phase response from $\omega = 0$ to $\omega = 2\pi$. Note that the above filter has a unity magnitude response and a linear phase with a group delay of amount D at all frequencies, i.e.,

$$|H(e^{j\omega})| = 1, \quad \tau(\omega) = D. \tag{4.69}$$

The output of this filter to an input $x[n] = Ae^{j\omega n}$ is then given by

$$y[n] = Ae^{-j\omega D}e^{j\omega n} = Ae^{j\omega(n-D)}.$$

If $x_a(t)$ and $y_a(t)$ represent the continuous-time signals whose sampled versions, sampled at $t = nT$, are $x[n]$ and $y[n]$ given above, then the delay between $x_a(t)$ and $y_a(t)$ is precisely the group delay of amount

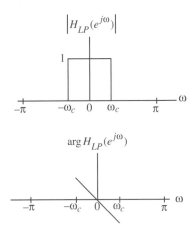

Figure 4.12: Frequency response of an ideal lowpass filter with a linear-phase response in the passband.

D. Note that if D is an integer, then the output sequence $y[n]$ is identical to the input sequence $x[n]$, but delayed by D samples. If D is not an integer, $y[n]$, being delayed by a fractional part, is not identical to $x[n]$. But in this latter case, the waveform of the underlying continuous-time output is identical to the waveform of the underlying continuous-time input and delayed by D units of time.

If we desire to pass input signal components in a certain frequency range undistorted in both magnitude and phase, then the transfer function should exhibit a unity magnitude response and linear-phase response in the band of interest. Figure 4.12 shows the frequency response of a lowpass transfer function with a linear-phase characteristic in the passband. Since the signal components in the stopband are blocked, the phase response in the stopband can be of any shape.

EXAMPLE 4.8   Determine the impulse response of an ideal lowpass filter with a linear-phase response. The desired frequency response is now given by

$$H_{LP}(e^{j\omega}) = \begin{cases} e^{-j\omega n_o}, & 0 < |\omega| < \omega_c, \\ 0 & \omega_c \leq |\omega| \leq \pi. \end{cases} \qquad (4.70)$$

Applying the frequency-shifting property of the Fourier transform given in Table 3.2 to Eq. (4.67) we arrive at the impulse response of the linear-phase lowpass filter of Eq. (4.70) as

$$h_{LP}[n] = \frac{\sin \omega_c(n - n_o)}{\pi(n - n_o)}, \qquad -\infty < n < \infty. \qquad (4.71)$$

As before, the above filter is noncausal and of doubly infinite length and, hence, is unrealizable. However, by truncating the impulse response to a finite number of terms, we can develop a realizable FIR filter approximation to the ideal linear-phase lowpass filter. It should be noted that the truncated approximation may or may not exhibit linear phase, depending on the value of n_0 chosen. If we choose $n_0 = N/2$ with N a positive integer, the truncated and shifted approximation

$$\hat{h}_{LP}[n] = \frac{\sin \omega_c(n - N/2)}{\pi(n - N/2)}, \qquad 0 \leq n \leq N, \qquad (4.72)$$

will be a length $N + 1$ causal linear-phase FIR filter. Figure 4.13 shows the filter coefficients obtained using the M-file `sinc` in MATLAB for two different values of N.

Because of the symmetry of the impulse response coefficients as indicated in this figure, the frequency response of the truncated approximation can be expressed as

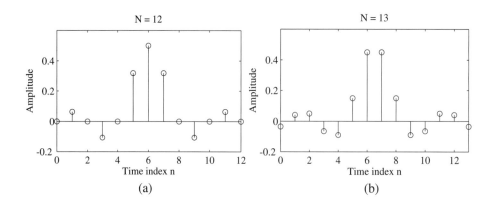

Figure 4.13: FIR approximation to the ideal linear-phase lowpass filter.

$$\hat{H}_{LP}(e^{j\omega}) = \sum_{n=0}^{N} \hat{h}_{LP}[n]e^{-j\omega n} = e^{-j\omega N/2}\breve{H}_{LP}(\omega), \qquad (4.73)$$

where $\breve{H}_{LP}(\omega)$, called the *zero-phase response* or *amplitude response*, is a real function of ω.

4.4.3 Types of Linear-Phase FIR Transfer Functions

In the previous section we pointed out why it is important to have a transfer function with a linear-phase property. It turns out it is always possible to design an FIR transfer function with an exact linear-phase response, while it is nearly impossible to design a linear-phase IIR transfer function. Recall that the length-3 FIR transfer function of Example 4.3 has a linear-phase response, as indicated in Eq. (4.26). This filter was characterized by a symmetric impulse response of the form $h[0] = h[2]$. Note also the symmetry of the impulse responses of the truncated approximations to the ideal linear-phase lowpass filter shown in Figure 4.13. We show now that a causal FIR transfer function of length $N + 1$,

$$H(z) = \sum_{n=0}^{N} h[n]\, z^{-n},$$

has a linear phase if its impulse response $h[n]$ is either symmetric, i.e.,

$$h[n] = h[N - n], \qquad 0 \le n \le N, \qquad (4.74)$$

or is antisymmetric, i.e.,

$$h[n] = -h[N - n], \qquad 0 \le n \le N. \qquad (4.75)$$

Since the length of the impulse response can be either even or odd, we can define four types of symmetry for the impulse response as demonstrated in Figure 4.14. It follows from Eq. (4.75) that for an antisymmetric FIR filter of odd length, i.e., N even, $h[N/2] = 0$. We next examine each of these four cases [Rab75].

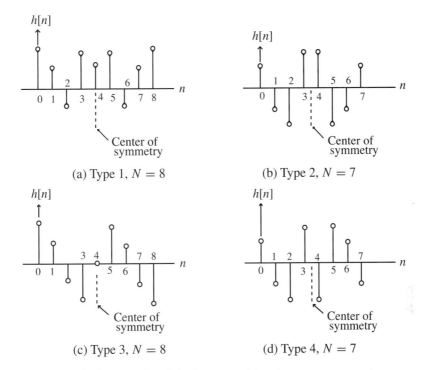

(a) Type 1, $N = 8$

(b) Type 2, $N = 7$

(c) Type 3, $N = 8$

(d) Type 4, $N = 7$

Figure 4.14: Illustration of the four types of impulse response symmetry.

Type 1: Symmetric Impulse Response with Odd Length

In this case, the degree N is even. Assume $N = 8$ for simplicity. The transfer function of the corresponding filter is given by

$$H(z) = h[0] + h[1]z^{-1} + h[2]z^{-2} + h[3]z^{-3} + h[4]z^{-4}$$
$$+ h[5]z^{-5} + h[6]z^{-6} + h[7]z^{-7} + h[8]z^{-8}. \tag{4.76}$$

But from Eq. (4.74) for $N = 8$, $h[0] = h[8]$, $h[1] = h[7]$, $h[2] = h[6]$, and $h[3] = h[5]$. Then, Eq. (4.76) reduces to

$$H(z) = h[0](1 + z^{-8}) + h[1](z^{-1} + z^{-7}) + h[2](z^{-2} + z^{-6})$$
$$+ h[3](z^{-3} + z^{-5}) + h[4]z^{-4}$$
$$= h[0]z^{-4}(z^4 + z^{-4}) + h[1]z^{-4}(z^3 + z^{-3}) + h[2]z^{-4}(z^2 + z^{-2})$$
$$+ h[3]z^{-4}(z + z^{-1}) + h[4]z^{-4}$$
$$= z^{-4}\{h[0](z^4 + z^{-4}) + h[1](z^3 + z^{-3}) + h[2](z^2 + z^{-2})$$
$$+ h[3](z + z^{-1}) + h[4]\}. \tag{4.77}$$

As a result, the corresponding frequency response is given by

$$H(e^{j\omega}) = e^{-j4\omega}\{2h[0]\cos(4\omega) + 2h[1]\cos(3\omega) + 2h[2]\cos(2\omega)$$
$$+ 2h[3]\cos(\omega) + h[4]\}, \tag{4.78}$$

obtained using the fact that $(z^m + z^{-m})/2|_{z=e^{j\omega}} = \cos(m\omega)$. Note that the quantity inside the braces in the above expression is a real function of ω and can assume positive or negative values in the range $0 \leq |\omega| \leq \pi$. Here the phase is given by

$$\theta(\omega) = -4\omega + \beta,$$

where β is either 0 or π, and hence, it is a linear function of ω in the generalized sense. The group delay is given by

$$\tau_g(\omega) = -\frac{d\theta(\omega)}{d\omega} = 4, \tag{4.79}$$

indicating a constant group delay of four samples.

In the general case for Type 1 FIR filters, the frequency response is of the form

$$H(e^{j\omega}) = e^{-jN\omega/2}\breve{H}(\omega), \tag{4.80}$$

where the *amplitude response* $\breve{H}(\omega)$, also called the *zero-phase response*, is given by

$$\breve{H}(\omega) = h\left[\tfrac{N}{2}\right] + 2\sum_{n=1}^{N/2} h\left[\tfrac{N}{2} - n\right]\cos(\omega n). \tag{4.81}$$

EXAMPLE 4.9   Consider the transfer function

$$H_0(z) = \tfrac{1}{6}\left[\tfrac{1}{2} + z^{-1} + z^{-2} + z^{-3} + z^{-4} + z^{-5} + \tfrac{1}{2}z^{-6}\right], \tag{4.82}$$

which is seen to be a slightly modified version of a length-7 moving-average FIR filter. The above transfer function has a symmetric impulse response and therefore a linear-phase characteristic. A plot of its magnitude response is shown in Figure 4.15 along with that of a length-7 conventional moving-average filter. Note the improved magnitude response obtained by simply changing the first and last impulse response coefficients of a moving-average FIR filter [Ham89].

It should be noted that the transfer function $H_0(z)$ of Eq. (4.82) can be expressed as

$$H_0(z) = \frac{1}{2}(1 + z^{-1}) \cdot \frac{1}{6}(1 + z^{-1} + z^{-2} + z^{-3} + z^{-4} + z^{-5}),$$

which is seen to be a cascade of a length-2 moving-average filter with a length-6 moving-average filter, thus giving a double zero at $z = -1$, i.e., $\omega = \pi$.

Type 2: Symmetric Impulse Response with Even Length

Here the degree N is odd. Let $N = 7$. By making use of the symmetry of the impulse response coefficients given by Eq. (4.74), the transfer function of the FIR filter can be written as:

$$\begin{aligned} H(z) &= h[0](1 + z^{-7}) + h[1](z^{-1} + z^{-6}) + h[2](z^{-2} + z^{-5}) \\ &\quad + h[3](z^{-3} + z^{-4}) \\ &= z^{-7/2}\{h[0](z^{7/2} + z^{-7/2}) + h[1](z^{5/2} + z^{-5/2}) \\ &\quad + h[2](z^{3/2} + z^{-3/2}) + h[3](z^{1/2} + z^{-1/2})\}. \end{aligned}$$

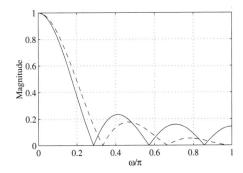

Figure 4.15: Magnitude responses of the length-7 moving-average lowpass filter (solid line) and the modified lowpass filter of Eq. (4.82) (dashed line).

The frequency response function is thus given by

$$H(e^{j\omega}) = e^{-j7\omega/2}\{2h[0]\cos(\tfrac{7\omega}{2}) + 2h[1]\cos(\tfrac{5\omega}{2})$$
$$+ 2h[2]\cos(\tfrac{3\omega}{2}) + 2h[3]\cos(\tfrac{\omega}{2})\}. \tag{4.83}$$

As before, here, the quantity inside the braces in the above expression is a real function of ω, and as in the previous case, it can assume positive or negative values in the range $0 \le |\omega| \le \pi$. Here the phase is given by

$$\theta(\omega) = -\tfrac{7}{2}\omega + \beta,$$

where β is again either 0 or π. As a result, the phase is also a linear function of ω in the generalized sense. The corresponding group delay is

$$\tau_g(\omega) = -\frac{d\theta(\omega)}{d\omega} = \frac{7}{2}, \tag{4.84}$$

indicating a group delay of $(7/2)$ samples.

The expression for the frequency response in the general case for Type 2 FIR filters is of the form

$$H(e^{j\omega}) = e^{-jN\omega/2}\breve{H}(\omega), \tag{4.85}$$

where the amplitude response is given by

$$\breve{H}(\omega) = 2\sum_{n=1}^{(N+1)/2} h\left[\tfrac{N+1}{2} - n\right]\cos\left(\omega(n - \tfrac{1}{2})\right). \tag{4.86}$$

Type 3: Antisymmetric Impulse Response with Odd Length

Here the degree N is even. Consider $N = 8$. Then applying the symmetry condition of Eq. (4.75) on the expression for the transfer function, we arrive at

$$H(z) = z^{-4}\{h[0](z^4 - z^{-4}) + h[1](z^3 - z^{-3})$$
$$+ h[2](z^2 - z^{-2}) + h[3](z - z^{-1})\}, \tag{4.87}$$

where we have used the fact that $h[4] = 0$. As a result, using the fact that $(z^m - z^{-m})|_{z=e^{j\omega}} = 2e^{j\pi/2}\sin(m\omega)$, the frequency response is given by

$$H(e^{j\omega}) = e^{-j4\omega}e^{j\pi/2}\{2h[0]\sin 4\omega + 2h[1]\sin 3\omega + 2h[2]\sin 2\omega$$
$$+ 2h[3]\sin\omega\}. \tag{4.88}$$

It also exhibits a generalized linear-phase response given by

$$\theta(\omega) = -4\omega + \tfrac{\pi}{2} + \beta,$$

where β is either 0 or π. The group delay here is given by

$$\tau_g(\omega) = 4, \tag{4.89}$$

implying a constant group delay of four samples.

The expression for the frequency response in the general case of Type 3 FIR filters is given by

$$H(e^{j\omega}) = je^{-jN\omega/2}\breve{H}(\omega), \tag{4.90}$$

where the amplitude response is of the form

$$\breve{H}(\omega) = 2\sum_{n=1}^{N/2} h\left[\tfrac{N}{2} - n\right]\sin(\omega n). \tag{4.91}$$

Type 4: Antisymmetric Impulse Response with Even Length

In this case the degree N is odd. For $N = 7$, the transfer function can be expressed as

$$H(z) = z^{-7/2}\{h[0](z^{7/2} - z^{-7/2}) + h[1](z^{5/2} - z^{-5/2})$$
$$+ h[2](z^{3/2} - z^{-3/2}) + h[3](z^{1/2} - z^{-1/2})\}. \tag{4.92}$$

The corresponding frequency response is thus given by

$$H(e^{j\omega}) = e^{-j7\omega/2}e^{j\pi/2}\{2h[0]\sin(7\omega/2) + 2h[1]\sin(5\omega/2)$$
$$+ 2h[2]\sin(3\omega/2) + 2h[3]\sin(\omega/2)\}. \tag{4.93}$$

It has a generalized linear-phase response

$$\theta(\omega) = -\tfrac{7}{2}\omega + \tfrac{\pi}{2} + \beta,$$

where β is either 0 or π. The group delay is constant and is given by

$$\tau_g(\omega) = \tfrac{7}{2}. \tag{4.94}$$

Here, the frequency response in the general case for Type 4 FIR filters is given by

$$H(e^{j\omega}) = je^{-jN\omega/2}\breve{H}(\omega), \tag{4.95}$$

where now the amplitude response is given by

$$\breve{H}(\omega) = 2\sum_{n=1}^{(N+1)/2} h\left[\tfrac{N+1}{2} - n\right]\sin\left(\omega(n - \tfrac{1}{2})\right). \tag{4.96}$$

General Form of Frequency Response

In each of the four types of causal linear-phase FIR filters, the frequency response $H(e^{j\omega})$ is of the form

$$H(e^{j\omega}) = e^{-jN\omega/2}e^{j\beta}\breve{H}(\omega).$$

It should be noted that the amplitude response $\breve{H}(\omega)$ for each of the four types of linear-phase FIR filters can become negative over certain frequency ranges, typically in the stopband, as indicated in Figure 3.3. The magnitude and phase responses of the linear-phase FIR filter are given by

$$|H(e^{j\omega})| = |\breve{H}(\omega)|,$$

$$\theta(\omega) = \begin{cases} -\frac{N\omega}{2} + \beta, & \text{for } \breve{H}(\omega) \geq 0, \\ -\frac{N\omega}{2} + \beta - \pi, & \text{for } \breve{H}(\omega) < 0. \end{cases}$$

The group delay in each case is

$$\tau(\omega) = \frac{N}{2}.$$

Note that, even though the group delay is constant, since in general $|H(e^{j\omega})|$ is not a constant, the output waveform is not a replica of the input waveform.

An FIR filter with a frequency response that is a real function of ω is often called a *zero-phase filter*. Such a filter must have a noncausal impulse response.

4.4.4 Zero Locations of Linear-Phase FIR Transfer Functions

Let us now study the zero locations of a linear-phase FIR transfer function. Consider first an FIR filter with a symmetric impulse response. Its transfer function $H(z)$ can be written as

$$H(z) = \sum_{n=0}^{N} h[n]z^{-n} = \sum_{n=0}^{N} h[N-n]z^{-n}, \tag{4.97}$$

using the symmetry condition of Eq. (4.74). By making a change of variable $m = N - n$, we can rewrite the rightmost expression in Eq. (4.97) as

$$H(z) = \sum_{m=0}^{N} h[m]z^{-N+m} = z^{-N}\sum_{n=0}^{N} h[m]z^{m} = z^{-N}H(z^{-1}). \tag{4.98}$$

Similarly, the transfer function $H(z)$ of an FIR filter with an antisymmetric response satisfying the condition of Eq. (4.75) can be expressed as

$$H(z) = \sum_{n=0}^{N} h[n]z^{-n} = -\sum_{n=0}^{N} h[N-n]z^{-n} = -z^{-N}H(z^{-1}). \tag{4.99}$$

A real-coefficient polynomial $H(z)$ satisfying the condition of Eq. (4.98) is called a *mirror-image polynomial* (MIP). Likewise, a real-coefficient polynomial $H(z)$ satisfying the condition of Eq. (4.99) is called an *antimirror-image polynomial* (AIP).

In either case, it follows from Eqs. (4.98) and (4.99) that if $z = \xi_0$ is a zero of $H(z)$, so is $z = 1/\xi_0$. Moreover, for an FIR filter with a real impulse response, the zeros occur in complex conjugate pairs.

Hence, a zero at $z = \xi_0$ is associated with a zero at $z = \xi_0^*$. Therefore, a complex zero that is not on the unit circle is associated with a set of four zeros given by

$$z = re^{\pm j\phi}, \qquad z = \frac{1}{r}e^{\pm j\phi}.$$

For a zero on the unit circle, its reciprocal is also its complex conjugate. Hence, in this case the zeros appear as a pair

$$z = e^{\pm j\phi}.$$

A real zero is paired with its reciprocal zero appearing at

$$z = \rho, \qquad z = \frac{1}{\rho}.$$

Note that a zero at $z = \pm 1$ is its own reciprocal, implying it can appear only singly. However, a Type 2 FIR filter must have a zero at $z = -1$ since from Eq. (4.98) we note that

$$H(-1) = (-1)^N H(-1) = -H(-1),$$

implying $H(-1) = 0$. In the case of a Type 3 or 4 FIR filter, Eq. (4.99) implies that $H(1) = -H(1)$, indicating that the filter must have a zero at $z = 1$. On the other hand, only a Type 3 FIR filter is restricted to have a zero at $z = -1$ since here

$$H(-1) = -(-1)^N H(-1) = -H(-1),$$

forcing $H(-1) = 0$. Figure 4.16 shows some examples of zero locations of all four types of FIR filters.

As can be seen from the above discussion, the principal difference between the four types of linear-phase FIR filters is with respect to the number of zeros at $z = 1$ and $z = -1$. Summarizing, we conclude that:

(a) Type 1 FIR Filter: Either an even number or no zeros at $z = 1$ and $z = -1$.

(b) Type 2 FIR Filter: Either an even number or no zeros at $z = 1$, and an odd number of zeros at $z = -1$.

(c) Type 3 FIR Filter: An odd number of zeros at $z = 1$ and $z = -1$.

(d) Type 4 FIR Filter: An odd number of zeros at $z = 1$, and either an even number or no zeros at $z = -1$.

The presence of zeros at $z = \pm 1$ leads to the following limitations on the use of these linear-phase FIR filters for designing filters. For example, since the Type 2 FIR filter always has a zero at $z = -1$, it cannot be used to design a highpass filter. Likewise, the Type 3 FIR filter has zeros at both $z = 1$ and $z = -1$ and, as a result, cannot be used to design either a lowpass or a highpass or a bandstop filter. Similarly, the Type 4 FIR filter is not appropriate to design a lowpass filter due to the presence of a zero at $z = 1$. Finally, the Type 1 FIR filter has no such restrictions and can be used to design almost any type of filter.

4.4.5 Bounded Real Transfer Functions

A causal stable real-coefficient transfer function $H(z)$ is defined as a *bounded real (BR) transfer function* [Vai84] if

$$|H(e^{j\omega})| \leq 1, \qquad \text{for all values of } \omega. \qquad (4.100)$$

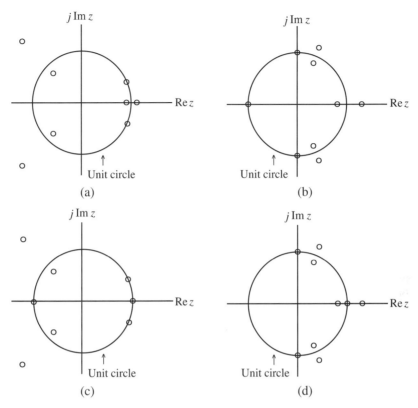

Figure 4.16: Examples of zero locations of linear-phase FIR transfer functions: (a) Type 1, (b) Type 2, (c) Type 3, and (d) Type 4 filters.

If the input and output of a digital filter characterized by a BR transfer function $H(z)$ are given by $x[n]$ and $y[n]$, respectively, with $X(e^{j\omega})$ and $Y(e^{j\omega})$ denoting their respective discrete-time Fourier transforms, then Eq. (4.100) implies that

$$|Y(e^{j\omega})|^2 \leq |X(e^{j\omega})|^2. \tag{4.101}$$

Integrating Eq. (4.101) from $-\pi$ to π, and applying Parseval's relation (see Table 3.2), we arrive at

$$\sum_{n=-\infty}^{\infty} y^2[n] \leq \sum_{n=-\infty}^{\infty} x^2[n]. \tag{4.102}$$

Or in other words, for all finite-energy inputs, the output energy is less than or equal to the input energy implying that a digital filter characterized by a BR transfer function can be viewed as a *passive* structure.

If Eq. (4.100) is satisfied with an equal sign, then from Eq. (4.102), the output energy is equal to the input energy, and such a digital filter is therefore a *lossless* system. A causal stable real-coefficient transfer function $H(z)$ with a frequency response $H(e^{j\omega})$ of unity magnitude is thus called a *lossless bounded real (LBR) transfer function* [Vai84].

The BR and LBR transfer functions are the keys to the realization of digital filters with low coefficient sensitivity (see Section 9.9).

4.5   Simple Digital Filters

In Chapter 7 we outline various methods of designing frequency-selective filters satisfying prescribed specifications. In this section we describe several low-order FIR and IIR digital filters with reasonable selective frequency responses that often are satisfactory in a number of applications.

4.5.1   Simple FIR Digital Filters

FIR digital filters considered here have integer-valued impulse response coefficients. These filters are employed in a number of practical applications, primarily because of their simplicity, which makes them amenable for inexpensive hardware implementation.

Lowpass FIR Digital Filters

The simplest lowpass FIR filter is the moving-average filter of Eq. (4.13) with $M = 2$ which has a transfer function

$$H_0(z) = \tfrac{1}{2}(1 + z^{-1}) = \frac{z+1}{2z}. \tag{4.103}$$

The above transfer function has a zero at $z = -1$ and a pole at $z = 0$. It follows from our discussion in Section 4.3.4 that the pole vector has a magnitude of unity, the radius of the unit circle, for all values of ω. On the other hand, as ω increases from 0 to π, the magnitude of the zero vector decreases from a value of 2, the diameter of the unit circle, to zero. Hence, the magnitude response $|H_0(e^{j\omega})|$ is a monotonically decreasing function of ω from $\omega = 0$ to $\omega = \pi$. The maximum value of the magnitude function is unity at $\omega = 0$, and the minimum value is zero at $\omega = \pi$, i.e.,

$$|H_0(e^{j0})| = 1, \qquad |H_0(e^{j\pi})| = 0.$$

From Eq. (4.103), it follows that the frequency response of the above filter is given by

$$H_0(e^{j\omega}) = e^{-j\omega/2} \cos\left(\tfrac{\omega}{2}\right), \tag{4.104}$$

whose magnitude response is given by $\cos(\omega/2)$ which is seen to be a monotonically decreasing function of ω (see Figure 4.17(a)). The frequency $\omega = \omega_c$ at which $|H_0(e^{j\omega_c})| = \frac{1}{\sqrt{2}}|H_0(e^{j0})|$ is of practical interest since here the gain $\mathcal{G}(\omega_c)$ in dB is given by

$$
\begin{aligned}
\mathcal{G}(\omega_c) &= 20\log_{10}|H_0(e^{j\omega_c})| = 20\log_{10}|H_0(e^{j0})| - 20\log_{10}\sqrt{2} \\
&= 0 - 3.0103 \cong -3.0\,\text{dB},
\end{aligned}
$$

since the dc gain $\mathcal{G}(0) = 20\log_{10}|H_0(e^{j0})| = 0$. Thus, the gain $\mathcal{G}(\omega)$ at $\omega = \omega_c$ is approximately 3 dB less than that at zero frequency. As a result, ω_c is called the *3-dB cutoff frequency*. To determine the expression for ω_c we set $|H_0(e^{j\omega_c})|^2 = \cos^2(\omega_c/2) = 1/2$, which yields $\omega_c = \pi/2$. This result checks with that given in the plot of Figure 4.17(a). The 3-dB cutoff frequency ω_c can be considered as the passband edge frequency, and as a result, for this filter the passband width is approximately $\pi/2$. The stopband is here from $\pi/2$ to π. Note from Eq. (4.103), the transfer function $H_0(z)$ has a zero at $z = -1$ or $\omega = \pi$, which is in the stopband of the filter.

A cascade of the simple FIR filters of Eq. (4.103) results in an improved lowpass frequency response, as illustrated in Figure 4.17(b) for a cascade of three sections. The 3-dB cutoff frequency of a cascade of M sections of the lowpass filter of Eq. (4.103) can be shown to be given by (Problem 4.49)

$$\omega_c = 2\cos^{-1}(2^{-1/2M}). \tag{4.105}$$

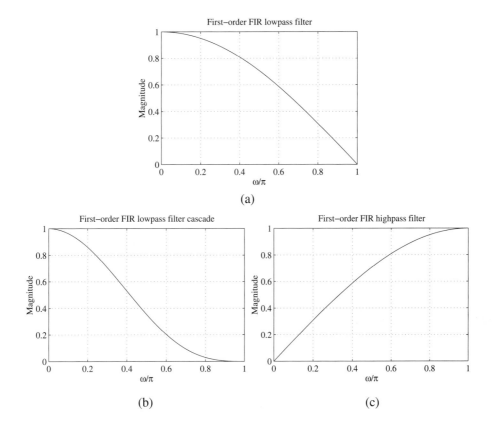

Figure 4.17: Magnitude responses of FIR filters: (a) first-order lowpass FIR filter $H_0(z)$ of Eq. (4.103), (b) cascade of first-order lowpass FIR filters, and (c) first-order highpass filter $H_1(z)$ of Eq. (4.106).

For $M = 3$, the above expression yields $\omega_c = 0.3002\pi$, which also checks with the plot in Figure 4.17(b). Thus, the cascade of first-order sections yields a sharper magnitude response but at the expense of a decrease in the passband width.

A better approximation to the ideal lowpass filter is given by the higher-order moving-average filter of Eq. (4.13). Signals with rapid fluctuations in sample values are generally associated with high-frequency components that are essentially eliminated by a moving-average filter of the type of Eq. (4.13), resulting in a much smoother output waveform as illustrated earlier in Example 2.14. Example 4.9 discusses a simple approach to improve the magnitude response of a moving-average lowpass filter.

The moving-average filter is often employed as the basic building block in the design of lowpass filters used in sampling rate alteration and is considered in Section 11.12.

Highpass FIR Digital Filters

The simplest highpass FIR filter is obtained by replacing z with $-z$ in Eq. (4.103), resulting in a transfer function

$$H_1(z) = \tfrac{1}{2}(1 - z^{-1}).$$ \hfill (4.106)

The corresponding frequency response is given by

$$H_1(e^{j\omega}) = je^{-j\omega/2}\sin(\tfrac{\omega}{2}), \tag{4.107}$$

whose magnitude response is shown in Figure 4.17(c). The monotonically increasing behavior of the magnitude function can again be demonstrated by examining the pole-zero pattern of the transfer function. The 3-dB cutoff frequency of this highpass filter is also at $\pi/2$. The transfer function $H_1(z)$ has a zero at $z = 1$ or $\omega = 0$, which is in the stopband of the filter.

Improved highpass frequency response can be obtained by cascading several sections of the simple highpass filter of Eq. (4.106). Alternatively, a higher-order highpass filter of the form

$$H_1(z) = \frac{1}{M}\sum_{n=0}^{M-1}(-1)^n z^{-n}, \tag{4.108}$$

obtained by replacing z with $-z$ in the expression for the transfer function of a moving-average lowpass filter yields a sharper magnitude response.

An application of the FIR highpass filters is in moving-target-indicator (MTI) radars. In these radars, interfering signals, called *clutters*, are generated from fixed objects in the path of the radar beam [Sko62]. The clutter, generated mainly from ground echoes and weather returns, has frequency components primarily near zero frequency (dc) and can be removed by filtering the radar return signal through a *two-pulse canceler*, which is the first-order highpass filter of Eq. (4.106). Often, the frequency components of the clutter occupy a small band near dc, and for a more effective removal it is necessary to use a highpass filter with a sharper magnitude response and a slightly broader stopband. To this end, a cascade of two two-pulse cancelers, called a *three-pulse canceler*, provides an improved performance. Problem 4.56 and Exercise M4.4 describe two simple IIR highpass filters proposed for clutter rejection.

4.5.2 Simple IIR Digital Filters

We now describe several simple IIR digital filters with first-order and second-order transfer functions and sketch their corresponding frequency responses. In many applications, use of such filters provides satisfactory results. Often, more complex frequency responses can be achieved by cascading these simple transfer functions.

Lowpass IIR Digital Filters

A first-order lowpass IIR digital filter has a transfer function given by

$$H_{LP}(z) = \frac{1-\alpha}{2}\frac{1+z^{-1}}{1-\alpha z^{-1}}, \tag{4.109}$$

where $|\alpha| < 1$ for stability. The above transfer function has a zero at $z = -1$, i.e., $\omega = \pi$, which is in the stopband of the filter. It has a real pole at $z = \alpha$. As ω increases from 0 to π, the magnitude of the zero vector decreases from a value of 2 to 0, whereas, for a positive value of α, the magnitude of the pole vector increases from a value of $1 - \alpha$ to $1 + \alpha$. The maximum value of the magnitude function is unity at $\omega = 0$, and the minimum value is zero at $\omega = \pi$, i.e.,

$$|H_{LP}(e^{j0})| = 1, \qquad |H_{LP}(e^{j\pi})| = 0.$$

Therefore, $|H_{LP}(e^{j\omega})|$ is a monotonically decreasing function of ω from $\omega = 0$ to $\omega = \pi$ (see Figure 4.18).

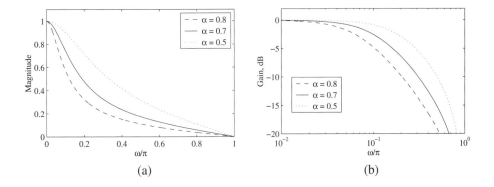

Figure 4.18: Magnitude and gain responses of the first-order lowpass filter of Eq. (4.109) for three values of α.

From Eq. (4.109), the squared magnitude function can be easily derived:

$$|H_{LP}(e^{j\omega})|^2 = \frac{(1-\alpha)^2(1+\cos\omega)}{2(1+\alpha^2-2\alpha\cos\omega)} \tag{4.110}$$

The derivative of $|H_{LP}(e^{j\omega})|^2$ with respect to ω is given by

$$\frac{d|H_{LP}(e^{j\omega})|^2}{d\omega} = \frac{-(1-\alpha)^2(1+2\alpha+\alpha^2)\sin\omega}{2(1+\alpha^2-2\alpha\cos\omega)^2}$$

which is always nonpositive in the range $0 \leq \omega \leq \pi$ verifying again the monotonically decreasing behavior of the magnitude function. To determine the 3-dB cutoff frequency ω_c we set $|H_{LP}(e^{j\omega_c})|^2 = 1/2$ in Eq. (4.110) and arrive at the equation

$$\frac{(1-\alpha)^2(1+\cos\omega_c)}{2(1+\alpha^2-2\alpha\cos\omega_c)} = \frac{1}{2} \quad \text{or} \quad (1-\alpha)^2(1+\cos\omega_c) = 1+\alpha^2-2\alpha\cos\omega_c,$$

which when solved yields

$$\cos\omega_c = \frac{2\alpha}{1+\alpha^2}. \tag{4.111a}$$

The above quadratic equation can be solved for α yielding two solutions. The solution resulting in a stable transfer function $H_{LP}(z)$ is given by

$$\alpha = \frac{1-\sin\omega_c}{\cos\omega_c}. \tag{4.111b}$$

Plots of the magnitude and the gain responses of the above lowpass transfer function are sketched in Figure 4.18 for several values of α.

It follows from Eq. (4.110) that the first-order lowpass transfer function $H_{LP}(z)$ of Eq. (4.109) is a bounded real (BR) function if $|\alpha| < 1$.

Highpass IIR Digital Filters

A first-order highpass transfer function $H_{HP}(z)$ is given by

$$H_{HP}(z) = \frac{1+\alpha}{2}\frac{1-z^{-1}}{1-\alpha z^{-1}}, \tag{4.112}$$

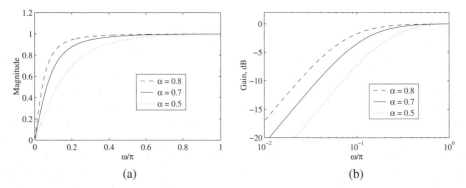

Figure 4.19: Magnitude and gain responses of the first-order highpass filter of Eq. (4.112) for three values of α.

where $|\alpha| < 1$ for stability. Its 3-dB cutoff frequency ω_c is also given by Eqs. (4.111a) and (4.111b). Plots of the magnitude and the gain responses of the above highpass transfer function for several values of α are shown in Figure 4.19.

It can be shown that the first-order highpass transfer function of Eq. (4.112) is a BR function if $|\alpha| < 1$.

EXAMPLE 4.10   Design a first-order highpass digital filter with a 3-dB cutoff frequency of 0.8π. Now, $\sin(\omega_c) = \sin(0.8\pi) = 0.58778525$ and $\cos(\omega_c) = \cos(0.8\pi) = -0.80902$, which when substituted in Eq. (4.111b) yields $\alpha = -0.50952449$. Therefore, from Eq. (4.112), the transfer function of the desired first-order highpass filter is given by

$$H_{HP}(z) = 0.245237755 \left(\frac{1 - z^{-1}}{1 + 0.50952449 z^{-1}} \right).$$

Bandpass IIR Digital Filters

A second-order bandpass digital filter is described by the transfer function

$$H_{BP}(z) = \frac{1 - \alpha}{2} \frac{1 - z^{-2}}{1 - \beta(1 + \alpha)z^{-1} + \alpha z^{-2}}. \tag{4.113}$$

Its squared magnitude function is given by

$$|H_{BP}(e^{j\omega})|^2 = \frac{(1 - \alpha)^2(1 - \cos 2\omega)}{2[1 + \beta^2(1 + \alpha)^2 + \alpha^2 - 2\beta(1 + \alpha)^2 \cos \omega + 2\alpha \cos 2\omega]}, \tag{4.114}$$

which goes to zero at $\omega = 0$ and at $\omega = \pi$. It assumes a maximum value of unity at $\omega = \omega_o$, called the *center frequency* of the bandpass filter, where

$$\omega_o = \cos^{-1}(\beta). \tag{4.115}$$

The frequencies ω_{c1} and ω_{c2} where the squared magnitude response goes to 1/2 are called the *3-dB cutoff frequencies*, and their difference, B_w, assuming $\omega_{c2} > \omega_{c1}$, called the *3-dB bandwidth*, is given by

$$B_w = \omega_{c2} - \omega_{c1} = \cos^{-1}\left(\frac{2\alpha}{1 + \alpha^2} \right). \tag{4.116}$$

Plots of the magnitude response of the bandpass filter of Eq. (4.113) are given in Figure 4.20 for several values of α and β.

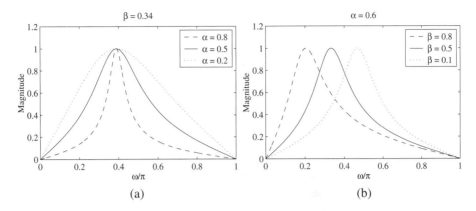

Figure 4.20: Magnitude response of the second-order bandpass filter of Eq. (4.113): (a) three specific values of α with $\beta = 0.34$, and (b) three specific values of β with $\alpha = 0.6$.

It can be shown that the bandpass transfer function of Eq. (4.113) is a BR function if $|\alpha| < 1$ and $|\beta| < 1$.

EXAMPLE 4.11 Design a second-order bandpass digital filter with center frequency at 0.4π and a 3-dB bandwidth of 0.1π.

From Eq. (4.115),
$$\beta = \cos(\omega_o) = \cos(0.4\pi) = 0.309016994,$$

and from Eq. (4.116),
$$\frac{2\alpha}{1 + \alpha^2} = \cos(B_w) = \cos(0.1\pi) = 0.951056516.$$

The solution of the above quadratic equation yields two values for α: 1.37638192 and 0.726542528. The corresponding second-order bandpass transfer functions are given by

$$H'_{BP}(z) = -0.18819096 \frac{1 - z^{-2}}{1 - 0.7343423986z^{-1} + 1.37638192z^{-2}}, \tag{4.117a}$$

and

$$H''_{BP}(z) = 0.136728736 \frac{1 - z^{-2}}{1 - 0.53353098z^{-1} + 0.726542528z^{-2}}. \tag{4.117b}$$

The poles of the transfer function $H'_{BP}(z)$ of Eq. (4.117a) are at $z = 0.367171199 \pm j1.114256358$ and have a magnitude of 1.173193, indicating that they are outside the unit circle. Therefore this transfer function is unstable. On the other hand, the poles of the transfer function $H''_{BP}(z)$ of Eq. (4.117b) are at $z = 0.2667655 \pm j0.8095546$ and have a magnitude of 0.85237464. As a result, $H''_{BP}(z)$ is BIBO stable. In Example 4.19 we outline a simpler approach to the testing of the stability of a second-order IIR transfer function.

Figure 4.21 shows the plots of the magnitude function and the group delay of $H''_{BP}(z)$. The group delay has been computed using the M-file `grpdelay` in MATLAB.

Bandstop IIR Digital Filters

Finally, a second-order bandstop digital filter has a transfer function of the form

$$H_{BS}(z) = \frac{1 + \alpha}{2} \frac{1 - 2\beta z^{-1} + z^{-2}}{1 - \beta(1 + \alpha)z^{-1} + \alpha z^{-2}}. \tag{4.118}$$

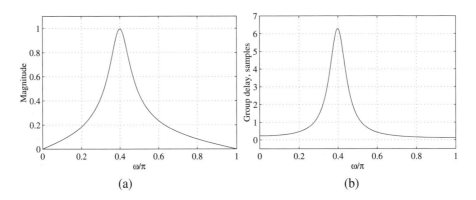

Figure 4.21: Magnitude response and the group delay of the bandpass transfer function of Eq. (4.117b).

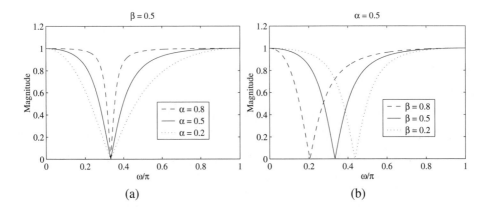

Figure 4.22: Frequency response of the second-order bandstop filter of Eq. (4.118). (a) Three specific values of α with $\beta = 0.5$, and (b) three specific values of β with $\alpha = 0.5$.

Its magnitude response is plotted in Figure 4.22 for various values of α and β. Here, the magnitude response takes the maximum value of unity at $\omega = 0$ and $\omega = \pi$, and goes to zero at $\omega = \omega_o$, where ω_o is given by Eq. (4.115). Since the magnitude response goes to zero at ω_o, ω_o is called the *notch frequency*, and the digital filter of Eq. (4.118) is more commonly called a *notch filter*. The *3-dB notch bandwidth* B_w is given by Eq. (4.116).

The bandstop transfer function of Eq. (4.118) is again a BR function if $|\alpha| < 1$ and $|\beta| < 1$.

Higher-Order IIR Digital Filters

By cascading the simple digital filters described above, we can implement digital filters with sharper magnitude responses. For example, for a cascade of K first-order lowpass sections characterized by the transfer function of Eq. (4.109), the overall structure has a transfer function $G_{LP}(z)$ given by

$$G_{LP}(z) = \left(\frac{1 - \alpha}{2} \frac{1 + z^{-1}}{1 - \alpha z^{-1}} \right)^K . \tag{4.119}$$

Using Eq. (4.110) we obtain the corresponding squared-magnitude function as

$$|G_{LP}(e^{j\omega})|^2 = \left[\frac{(1-\alpha)^2(1+\cos\omega)}{2(1+\alpha^2 - 2\alpha\cos\omega)} \right]^K. \tag{4.120}$$

To determine the relation between its 3-dB cutoff frequency ω_c and the parameter α, we set

$$\left[\frac{(1-\alpha)^2(1+\cos\omega_c)}{2(1+\alpha^2 - 2\alpha\cos\omega_c)} \right]^K = \frac{1}{2}, \tag{4.121}$$

which when solved for α yields, for a stable $G_{LP}(z)$,

$$\alpha = \frac{1 + (1-C)\cos\omega_c - \sin\omega_c\sqrt{2C - C^2}}{1 - C + \cos\omega_c} \tag{4.122}$$

where

$$C = 2^{(K-1)/K}. \tag{4.123}$$

It should be noted that Eq. (4.122) reduces to Eq. (4.111b) for $K = 1$.

> **EXAMPLE 4.12**   Design a lowpass filter with a 3-dB cutoff frequency at $\omega_c = 0.4\pi$ using a single-stage realization and as a cascade of four first-order lowpass sections, and compare their gain responses.
>   From Eq. (4.111b) or, equivalently from Eq. (4.122) with $K = 1$, i.e., $C = 1$, and $\omega_c = 0.4\pi$, we arrive at $\alpha = 0.1584$ for a single-stage design. Likewise, from Eq. (4.123), $C = 1.6818$ for $K = 4$, which when substituted in Eq. (4.122) along with $\omega_c = 0.4\pi$, yields $\alpha = -0.251$. Figure 4.23 shows the gain responses of a single first-order lowpass filter (plot marked $K = 1$) and a cascade of four identical first-order lowpass filters (plot marked $K = 4$). As can be seen from this figure, cascading has resulted in a sharper roll-off in the gain response, as expected.

Likewise, a cascade of first-order highpass sections results in a highpass filter with a sharper roll-off in the gain response.

Figure 4.24 shows the magnitude responses of a single second-order bandpass filter (plot marked 1), a cascade of two identical second-order bandpass filters (plot marked 2), and a cascade of three identical second-order bandpass filters (plot marked 3). All bandpass sections are characterized by $\alpha = 0.2$ and $\beta = 0.34$. Since the parameter β of all second-order sections is identical, the center frequency of the cascade structure is the same as that of the single section. However, the 3-dB bandwidth decreases with an increase in the number of sections. Similarly, a cascade of identical second-order bandstop filters results in a higher-order bandstop filter with an identical center frequency but with a narrower 3-dB bandwidth.

4.5.3 Comb Filters

The simple filters of the previous two sections are characterized either by a single passband and/or a single stopband. There are a number of applications where filters with multiple passbands and stopbands are required. The *comb filter* is an example of this latter type of filter. In its most general form, a comb filter has a frequency response that is a periodic function of ω with a period $2\pi/L$, where L is a positive integer. If $H(z)$ is a filter with a single passband and/or a single stopband, a comb filter can easily be generated from it by replacing each delay in its realization with L delays, resulting in a structure with a transfer function given by $G(z) = H(z^L)$. If the magnitude response $|H(e^{j\omega})|$ exhibits a peak at ω_p, then the magnitude response of $|G(e^{j\omega})|$ will exhibit L peaks at $\omega_p k/L$, $0 \le k \le L - 1$. Likewise, if the magnitude response $|H(e^{j\omega})|$ has a notch at ω_o, then the magnitude response of $|G(e^{j\omega})|$ will have L

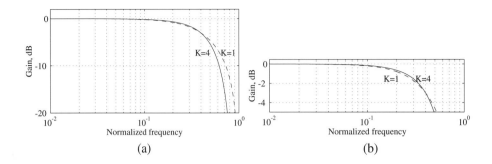

Figure 4.23: (a) Gain responses of a single first-order lowpass filter ($K = 1$) and a cascade of four identical first-order lowpass filters ($K = 4$) with a 3-dB cutoff frequency of $\omega_c = 0.4\pi$. (b) Passband details.

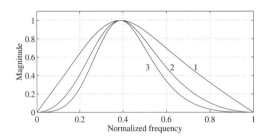

Figure 4.24: Gain responses of a single second-order bandpass filter, a cascade of two identical second-order bandpass filters, and a cascade of three identical second-order bandpass filters. All sections characterized by $\alpha = 0.2$ and $\beta = 0.34$.

notches at $\omega_o k/L$, $0 \le k \le L - 1$. It should be noted that a comb filter can be generated from either an FIR or an IIR prototype filter.

To illustrate the generation of the comb filter consider the prototype lowpass FIR filter of Eq. (4.103), which has a lowpass magnitude response as indicated in Figure 4.17(a). The comb filter generated from $H_0(z)$ has a transfer function

$$G_0(z) = H_0(z^L) = \tfrac{1}{2}(1 + z^{-L}), \tag{4.124}$$

with a corresponding magnitude response as sketched in Figure 4.25(a). The new filter is essentially a notch filter with L notch frequencies in the range $0 \le \omega < 2\pi$ located at $\omega = (2k + 1)\pi/L$ and has L peaks in its magnitude response located at $\omega = 2\pi k/L$, $0 \le k \le L - 1$.

The comb filter $G_1(z)$ generated from the prototype highpass filter of Eq. (4.106) has a transfer function given by

$$G_1(z) = H_1(z^L) = \tfrac{1}{2}(1 - z^{-L}), \tag{4.125}$$

with a magnitude response as indicated in Figure 4.25(b). This comb filter again has L notch frequencies in the range $0 \le \omega < 2\pi$ located at $\omega = 2\pi k/L$, $0 \le k \le L - 1$, which are exactly at the locations of the peaks of the comb filter $G_0(z)$ of Eq. (4.124). Likewise its magnitude response has L peaks at $\omega = (2k + 1)\pi/L$, $0 \le k \le L - 1$, which are precisely the locations of the notch frequencies of $G_0(z)$.

Depending on the application, comb filters with other types of periodic magnitude responses can be easily generated by appropriately choosing the prototype filter. For example, the M-point moving-average filter of Eq. (4.52):

$$H(z) = \frac{1 - z^{-M}}{M(1 - z^{-1})}$$

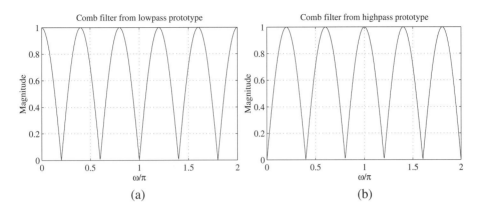

Figure 4.25: Magnitude responses of FIR comb filters: (a) generated from a prototype lowpass filter of Eq. (4.103) with $L = 5$, and (b) generated from a prototype highpass filter of Eq. (4.106) with $L = 5$.

has been used as the prototype. This filter has a peak magnitude at $\omega = 0$, and $M - 1$ notches at $\omega = 2\pi\ell/M$, $1 \leq \ell \leq M - 1$. The comb filter generated from this prototype has a transfer function

$$G(z) = H(z^L) = \frac{1 - z^{-LM}}{M(1 - z^{-L})},$$

whose magnitude response has L peaks at $\omega = 2\pi k/L$, $0 \leq k \leq L - 1$, and $L(M - 1)$ notches at $\omega = 2\pi k/LM$, $1 \leq k \leq L(M - 1)$. By choosing L and M appropriately, peaks and notches can be created at desired locations. In ionospheric measurements of electron concentration, the weak lunar spectral components are usually masked by the strong solar spectral components. These two spectral components have been separated by using two such comb filters [Ber76].

One of the applications of a comb filter considered in Section 11.5.1 is in creating special audio effects in musical sound processing, where both FIR and IIR prototypes are employed. Comb filters with multiple notch frequencies find applications in the cancellation of periodic interferences. The comb filter is also employed in LORAN navigation systems for the suppression of cross-rate interferences [Jac96].

An interesting application of the comb filters $G_0(z)$ and $G_1(z)$ of Eqs. (4.124) and (4.125), respectively, is in digital color television receivers for separating the luminance component containing the intensity information and the chrominance components containing the color information from the composite video signal [Orf96]. The basic structure for this purpose is as shown in Figure 4.26, where the delay chain is chosen to provide a line delay, i.e., the time to scan one horizontal line. However, a complete separation of the two components is not possible with this structure. Moreover, the filter $G_0(z)$ acts as a lowpass filter by averaging two successive horizontal lines of the video signal and blurs the luminance component. On the other hand, improved separation of the two components can be achieved by the structure of Figure 4.26 when the delay chain is chosen to provide a frame delay. Unfortunately, the separation fails if there is some motion from frame to frame.

4.6 Allpass Transfer Function

We now turn our attention to a very special type of IIR transfer function that is characterized by unity magnitude for all frequencies. Such a transfer function, called an *allpass transfer function*, has many useful applications in digital signal processing [Reg88]. We define the allpass transfer function, examine

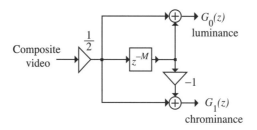

Figure 4.26: Filter structure for the separation of the luminance and chrominance components of a composite video signal.

some of its key properties, and outline one of its common applications. Later in this chapter and elsewhere in the text we discuss various other applications. One important application considered in this chapter is the development of an algebraic test for the BIBO stability of a causal IIR transfer function.

4.6.1 Definition

An IIR transfer function $A(z)$ with unity magnitude response for all frequencies, i.e.,

$$|A(e^{j\omega})|^2 = 1, \qquad \text{for all } \omega \tag{4.126}$$

is called an *allpass transfer function*. Now an Mth-order causal real-coefficient allpass transfer function is of the form

$$A_M(z) = \pm\frac{d_M + d_{M-1}z^{-1} + \cdots + d_1z^{-M+1} + z^{-M}}{1 + d_1z^{-1} + \cdots + d_{M-1}z^{-M+1} + d_Mz^{-M}}. \tag{4.127}$$

If we denote the denominator polynomial of the allpass function $A_M(z)$ as $D_M(z)$,

$$D_M(z) = 1 + d_1z^{-1} + \cdots + d_{M-1}z^{-M+1} + d_Mz^{-M}, \tag{4.128}$$

then it follows that $A_M(z)$ can be written as

$$A_M(z) = \pm\frac{z^{-M}D_M(z^{-1})}{D_M(z)}. \tag{4.129}$$

Note from above that if $z = re^{j\phi}$ is a pole of a real-coefficient allpass transfer function, then it has a zero at $z = (1/r)e^{-j\phi}$. The numerator of an allpass transfer function is said to be the *mirror-image polynomial* of the denominator, and vice versa. We shall use the notation $\tilde{D}_M(z)$ to denote the mirror-image polynomial of a degree-M polynomial, i.e., $\tilde{D}_M(z) = z^{-M}D_M(z^{-1})$. Equation (4.129) implies that the poles and the zeros of a real-coefficient allpass function exhibit *mirror-image symmetry* in the z-plane, as shown in Figure 4.27 for the following third-order allpass function:

$$A_3(z) = \frac{-0.2 + 0.18z^{-1} + 0.4z^{-2} + z^{-3}}{1 + 0.4z^{-1} + 0.18z^{-2} - 0.2z^{-3}}. \tag{4.130}$$

To show that the magnitude of $A_M(e^{j\omega})$ is indeed equal to one for all ω, it follows from Eq. (4.129) that

$$A_M(z^{-1}) = \pm\frac{z^M D_M(z)}{D_M(z^{-1})}.$$

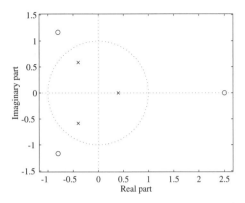

Figure 4.27: Pole-zero plot of the real coefficient allpass transfer function of Eq. (4.130).

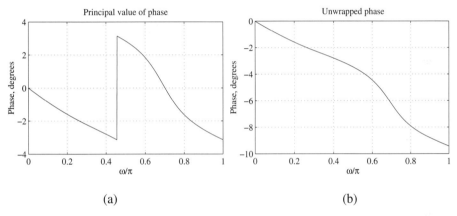

(a) (b)

Figure 4.28: (a) The principal value of the phase function, and (b) the unwrapped phase function of the allpass transfer function of Eq. (4.130).

Therefore,

$$A_M(z)A_M(z^{-1}) = \frac{z^{-M}D_M(z^{-1})}{D_M(z)}\frac{z^M D_M(z)}{D_M(z^{-1})} = 1.$$

Hence,

$$|A_M(e^{j\omega})|^2 = A_M(z)A_M(z^{-1})|_{z=e^{j\omega}} = 1. \tag{4.131}$$

As shown earlier in Section 4.3.5, the poles of a causal stable transfer function must lie inside the unit circle. As a result, all zeros of a causal stable allpass transfer function lie outside the unit circle in a mirror-image symmetry with its poles situated inside the unit circle.

It is interesting to examine the behavior of the phase function of an allpass transfer function. Figure 4.28(a) shows the principal value of the phase of the stable third-order allpass transfer function of Eq. (4.130). Note the discontinuity by the amount of 2π in the phase $\theta(\omega)$. If we unwrap the phase by removing the discontinuity, we arrive at the unwrapped phase function $\theta_c(\omega)$ indicated in Figure 4.28(b). As can be seen from this figure, the unwrapped phase function is a nonpositive continuous function of ω in the range $0 < \omega < \pi$. This property of the unwrapped phase function holds for any arbitrary causal stable allpass function.

Figure 4.29: Use of an allpass filter as a delay equalizer.

4.6.2 Properties

We now state three very useful and important properties of a causal stable allpass function without proof [Reg88].

Property 1. It follows from Section 4.3.5 that a causal stable real coefficient allpass transfer function is a lossless bounded real (LBR) transfer function or, equivalently, a causal stable allpass filter is a lossless structure.

Property 2. The second property is concerned with the magnitude of a stable allpass function $A(z)$. It can be shown very simply that (Problem 4.83)

$$|A(z)| \begin{cases} < 1 & \text{for } |z| > 1, \\ = 1 & \text{for } |z| = 1, \\ > 1 & \text{for } |z| < 1. \end{cases} \tag{4.132}$$

Property 3. The last property of interest is with regard to the change in phase for a real stable allpass function over the frequency range $\omega = 0$ to $\omega = \pi$. Let $\tau(\omega)$ denote the group delay function of the allpass filter $A(z)$, i.e.,

$$\tau(\omega) = -\frac{d}{d\omega}[\theta_c(\omega)],$$

where $\theta_c(\omega)$ is the unwrapped form of the phase function $\theta(\omega) = \arg\{A(e^{j\omega})\}$ in order that the group delay $\tau(\omega)$ be well behaved. Now the unwrapped phase function $\theta_c(\omega)$ of a stable allpass function $A(z)$ is a monotonically decreasing function of ω so that $\tau(\omega)$ is everywhere positive in the range $0 < \omega < \pi$. Therefore, an Mth-order stable real allpass transfer function satisfies the property (Problem 4.84)

$$\int_0^\pi \tau(\omega)\, d\omega = M\pi. \tag{4.133}$$

Or in other words, the change in the phase of an Mth-order allpass function as ω goes from 0 to π is $M\pi$ radians.

4.6.3 A Simple Application

A simple but often used application of an allpass filter is as a *delay equalizer*. Let $G(z)$ be the transfer function of a digital filter that has been designed to meet a prescribed magnitude response. The nonlinear phase response of this filter can be corrected by cascading it with an allpass filter section $A(z)$ so that the overall cascade with a transfer function $G(z)A(z)$ has a constant group delay over the frequency domain of interest (see Figure 4.29). Since the allpass filter has a unity magnitude response, the magnitude response of the cascade is still equal to $|G(e^{j\omega})|$, while the overall delay is given by the sum of the group delays of $G(z)$ and $A(z)$. The allpass is designed so that the overall group delay is approximately a constant in the frequency region of interest.

Various other applications of the allpass filter are described in the latter parts of this book.

4.7 Minimum-Phase and Maximum-Phase Transfer Functions

Another useful classification of a transfer function is in terms of the behavior of its phase response. Consider the two first-order transfer functions $H_1(z)$ and $H_2(z)$:

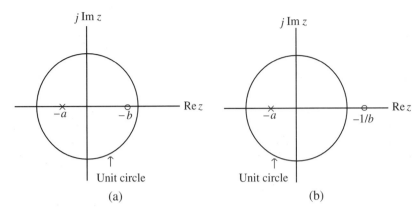

Figure 4.30: Pole-zero plots of the transfer functions of Eq. (4.134): (a) $H_1(z)$ and (b) $H_2(z)$.

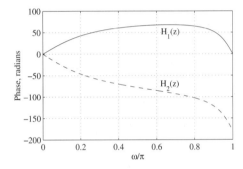

Figure 4.31: Unwrapped phase response of Eq. (4.134) for $a = 0.8$, $b = -0.5$.

$$H_1(z) = \frac{z+b}{z+a}, \quad H_2(z) = \frac{bz+1}{z+a}, \quad |a| < 1, \quad |b| < 1. \tag{4.134}$$

As can be seen from the pole-zero plots given in Figure 4.30, both transfer functions have a pole inside the unit circle at the same location at $z = -a$ and are therefore stable. On the other hand, the zero of $H_1(z)$ is inside the unit circle at $z = -b$, whereas the zero of $H_2(z)$ is outside the unit circle at $z = -1/b$ situated in a mirror-image symmetry with respect to the zero of $H_1(z)$. However, the two transfer functions have an identical magnitude function, since $H_1(z)H_1(z^{-1}) = H_2(z)H_2(z^{-1})$.

From Eq. (4.134),

$$\arg[H_1(e^{j\omega})] = \theta_1(\omega) = \tan^{-1}\frac{\sin\omega}{b+\cos\omega} - \tan^{-1}\frac{\sin\omega}{a+\cos\omega}, \tag{4.135a}$$

$$\arg[H_2(e^{j\omega})] = \theta_2(\omega) = \tan^{-1}\frac{b\sin\omega}{1+b\cos\omega} - \tan^{-1}\frac{\sin\omega}{a+\cos\omega}. \tag{4.135b}$$

Figure 4.31 shows the unwrapped phase responses of the two transfer functions. From this figure it can be seen that $H_2(z)$ has an excess phase lag with respect to $H_1(z)$.

Generalizing the above result, a causal stable transfer function with all zeros outside the unit circle has an excess phase compared to a causal stable transfer function with identical magnitude but having all zeros

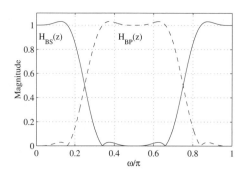

Figure 4.32: Delay-complementary linear-phase bandstop and bandpass FIR filters of Eqs. (4.139a) and (4.139b).

inside the unit circle. As a result, a causal stable transfer function with all zeros inside the unit circle is called a *minimum-phase transfer function*, whereas a causal stable transfer function with all zeros outside the unit circle is called a *maximum-phase transfer function*.

It can be easily shown that any nonminimum-phase transfer function can be expressed as the product of a minimum-phase transfer function and a stable allpass transfer function (Problem 4.87).

4.8 Complementary Transfer Functions

A set of digital transfer functions with complementary characteristics often finds useful applications in practice, such as efficient realizations of the transfer functions, low sensitivity realizations, and filter bank design and implementation. We describe next four useful complementary relations and indicate some of their applications.

4.8.1 Delay-Complementary Transfer Functions

A set of L transfer functions $\{H_0(z), H_1(z), \ldots, H_{L-1}(z)\}$ is defined to be *delay-complementary* of each other if the sum of their transfer functions is equal to some integer multiple of the unit delay [Vai93], i.e.,

$$\sum_{k=0}^{L-1} H_k(z) = \beta z^{-n_0}, \quad \beta \neq 0, \tag{4.136}$$

where n_0 is a nonnegative integer.

A delay-complementary pair $\{H_0(z), H_1(z)\}$ can be readily designed if one of the pair is a known Type 1 linear-phase FIR transfer function of odd length. Let $H_0(z)$ be a Type 1 linear-phase frequency-selective FIR transfer function of length $M = 2K + 1$ with a magnitude response equal to $1 \pm \delta_p$ in the passband and less than or equal to δ_s in the stopband where δ_p and δ_s are very small numbers. From Eq. (4.80) its frequency response is of the form

$$H_0(e^{j\omega}) = e^{-jK\omega} \breve{H}_0(\omega), \tag{4.137}$$

where $\breve{H}_0(\omega)$ is the *amplitude response* of the FIR transfer function $H_0(z)$. Its delay-complementary transfer function $H_1(z)$ defined for $\beta = 1$ and $n_0 = K$ has a frequency response given by

$$H_1(e^{j\omega}) = e^{-jK\omega} \breve{H}_1(\omega) = e^{-jK\omega}[1 - \breve{H}_0(\omega)]. \tag{4.138}$$

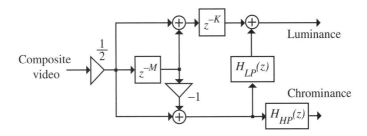

Figure 4.33: Luminance and chrominance components separation filter structure with improved vertical details.

Now, in the passband, $1 - \delta_p \leq \check{H}_0(\omega) \leq 1 + \delta_p$, and in the stopband, $-\delta_s \leq \check{H}_0(\omega) \leq \delta_s$. It follows therefore from Eq. (4.138), in the stopband of $H_1(z)$, $-\delta_p \leq \check{H}_0(\omega) \leq \delta_p$, and in the passband of $H_1(z)$, $1 - \delta_s \leq \check{H}_0(\omega) \leq 1 + \delta_s$. As a result, $H_1(z)$ has a complementary magnitude response characteristic to that of $H_0(z)$ with a stopband exactly identical to the passband of $H_0(z)$ and a passband exactly equal to the stopband of $H_0(z)$. For example, if $H_0(z)$ is a lowpass filter, $H_1(z)$ will be a highpass filter, and vice versa. The frequency ω_o at which $\check{H}_0(\omega_o) = \check{H}_1(\omega_o) = 0.5$ the gain responses of both filters are 6 dB below their maximum values. As a result, ω_o is called the *6-dB crossover frequency*.

EXAMPLE 4.13   Consider the Type 1 bandstop transfer function $H_{BS}(z)$ given below [Aca83]:

$$H_{BS}(z) = \tfrac{1}{64}(1 + z^{-2})^4(1 - 4z^{-2} + 5z^{-4} + 5z^{-8} - 4z^{-10} + z^{-12}). \qquad (4.139a)$$

Its delay-complementary bandpass transfer function is given by

$$H_{BP}(z) = z^{-10} - H_{BS}(z) = -\tfrac{1}{64}(1 - z^{-2})^4(1 + 4z^{-2} + 5z^{-4} + 5z^{-8} + 4z^{-10} + z^{-12}). \qquad (4.139b)$$

Figure 4.32 shows the plots of the magnitude responses of $H_{BS}(z)$ and $H_{BP}(z)$.

An interesting application of the delay-complementary FIR transfer function pair is in digital television receivers [Orf96]. It has been pointed out earlier that the structure of Figure 4.26 for the separation of the luminance and chrominance components of the composite video signal tends to blur the luminance output, resulting in the loss of vertical details. The low-frequency vertical details can be recovered from the output of the comb filter $G_1(z)$ of Figure 4.26 by a lowpass filter and added to the output of the comb filter $G_0(z)$. The vertical details can be removed from the output of $G_1(z)$ by filtering it through a filter that is delay-complementary to the lowpass filter. In practice the lowpass filter $H_{LP}(z)$ employed is a bandstop filter whose low-frequency passband coincides with the frequency range of the desired vertical details while its delay-complementary filter $H_{HP}(z)$ is a bandpass transfer function. The overall structure is thus as shown in Figure 4.33, where the delay-chain of length K in the top path is chosen to equalize the total delay in both the top and bottom paths.

One set of delay-complementary linear-phase bandpass/bandstop filters proposed for use in Figure 4.33 is the one given in Eqs. (4.139a) and (4.139b) for which $K = 10$. Other linear-phase bandstop FIR transfer functions suitable for vertical details recovery are given in Problem 4.91.

Delay-complementary filter sets can also be used as crossover filters for separating the digital audio input signals into two or three subsignals occupying different frequency bands which are then used to drive the appropriate speakers of a loudspeaker system [Orf96]. Design of such delay-complementary crossover filters is considered in Exercises M7.30 and M7.31. Another important application of the delay-complementary property is in the realization of low-sensitivity FIR digital filters discussed in Section 9.7.3.

4.8.2   Allpass-Complementary Transfer Functions

A set of M digital transfer functions $\{H_i(z)\}$, $0 \le i \le M - 1$, is defined to be *allpass-complementary* of each other, if the sum of their transfer functions is equal to an allpass function $A(z)$ [Gar80], [Neu84a], i.e.,

$$\sum_{i=0}^{M-1} H_i(z) = A(z). \qquad (4.140)$$

4.8.3   Power-Complementary Transfer Functions

A set of M digital transfer functions $\{H_i(z)\}$, $0 \le i \le M - 1$, is defined to be *power-complementary* of each other, if the sum of the squares of their magnitude responses is equal to a constant K for all values of ω [Neu84a], [Vai93], i.e.,

$$\sum_{i=0}^{M-1} |H_i(e^{j\omega})|^2 = K, \qquad \text{for all } \omega, \qquad (4.141a)$$

where $K > 0$ is a constant. By analytic continuation, the above property is equivalent to

$$\sum_{i=0}^{M-1} H_i(z^{-1}) H_i(z) = K, \qquad \text{for all } z, \qquad (4.141b)$$

for a real-coefficient $H_i(z)$. Usually, by scaling the transfer functions, the power-complementary property is defined with respect to $K = 1$.

For a pair of power-complementary transfer functions, $H_0(z)$ and $H_1(z)$, the frequency ω_o where $|H_0(e^{j\omega_o})|^2 = |H_1(e^{j\omega_o})|^2 = \frac{1}{2}$, is called the *crossover frequency*. At this frequency the gain responses of both filters are 3 dB below their maximum values. As a result, ω_o is also called the *3-dB cutoff frequency* of both filters.

Consider two transfer functions $H_0(z)$ and $H_1(z)$ described by

$$H_0(z) = \tfrac{1}{2}[A_0(z) + A_1(z)], \qquad (4.142a)$$

$$H_1(z) = \tfrac{1}{2}[A_0(z) - A_1(z)], \qquad (4.142b)$$

where $A_0(z)$ and $A_1(z)$ are stable allpass transfer functions. It follows from the above that the sum of the two transfer functions is an allpass function $A_0(z)$, and hence, $H_0(z)$ and $H_1(z)$ of Eqs. (4.142a) and (4.142b) are an *allpass-complementary* pair. It can be easily shown that the two filters described by Eq. (4.142a) and (4.142b) also form a *power-complementary* pair (Problem 4.93). It can also be easily shown that the transfer functions $H_0(z)$ and $H_1(z)$ of Eqs. (4.142a) and (4.142b) are also bounded real transfer functions (Problem 4.94).

4.8.4   Doubly-Complementary Transfer Functions

A set of M transfer functions satisfying both the allpass-complementary property of Eq. (4.140) and the power-complementary property of Eq. (4.141a) is known as a *doubly-complementary* set [Neu84a].

A pair of doubly-complementary IIR transfer functions, $H_0(z)$ and $H_1(z)$, with a sum of allpass decomposition in the form of Eqs. (4.142a) and (4.142b) can be simply realized by a parallel connection of the constituent allpass filters, as indicated in Figure 4.34. We shall demonstrate in Section 9.9.2 that such realizations also ensure low sensitivity in the passband with respect to the multiplier coefficients.

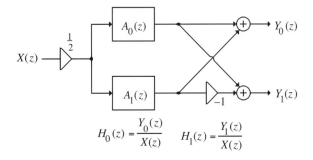

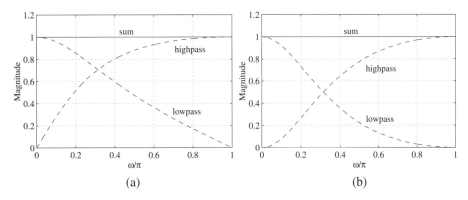

Figure 4.34: Parallel allpass realization of doubly-complementary IIR transfer functions.

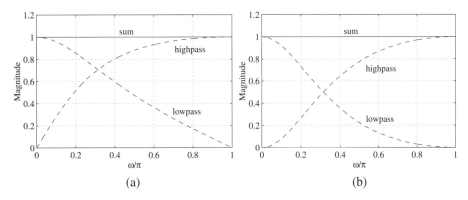

Figure 4.35: Illustration of the complementary properties of the first-order lowpass and highpass transfer functions with $\alpha = 0.3$. (a) Allpass complementary and (b) power-complementary.

EXAMPLE 4.14   The first-order lowpass transfer function of Eq. (4.109) can be expressed as:

$$H_{LP}(z) = \frac{1}{2}\left[\frac{1 - \alpha + z^{-1} - \alpha z^{-1}}{1 - \alpha z^{-1}}\right]$$

$$= \frac{1}{2}\left[1 + \frac{-\alpha + z^{-1}}{1 - \alpha z^{-1}}\right] = \frac{1}{2}[A_0(z) + A_1(z)], \tag{4.143}$$

where

$$A_0(z) = 1, \qquad A_1(z) = \frac{-\alpha + z^{-1}}{1 - \alpha z^{-1}}. \tag{4.144}$$

From Eq. (4.142b) its power-complementary highpass transfer function is thus given by

$$H_{HP}(z) = \frac{1}{2}[A_0(z) - A_1(z)]$$

$$= \frac{1}{2}\left[1 - \frac{-\alpha + z^{-1}}{1 - \alpha z^{-1}}\right] = \frac{1 + \alpha}{2} \cdot \frac{1 - z^{-1}}{1 - \alpha z^{-1}}, \tag{4.145}$$

which is precisely the first-order highpass transfer function of Eq. (4.112).

Figure 4.35 demonstrates the allpass complementary and the power-complementary properties of the first-order lowpass and highpass transfer functions of Eq. (4.143) and Eq. (4.145), respectively, for $\alpha = 0.3$.

It can be easily shown that the bandpass transfer function $H_{BP}(z)$ of Eq. (4.113) and the bandstop transfer function $H_{BS}(z)$ of Eq. (4.118) form a doubly-complementary pair (Problem 4.96).

4.8.5 Power-Symmetric Filters and Conjugate Quadrature Filters

A real-coefficient causal digital filter with a transfer function $H(z)$ is said to be a *power-symmetric filter* if it satisfies the condition [Vai93]:

$$H(z)H(z^{-1}) + H(-z)H(-z^{-1}) = K, \tag{4.146}$$

where $K > 0$ is a constant. It can be shown that the gain function $\mathcal{G}(\omega)$ of a power-symmetric transfer function at $\omega = \pi/2$ is given by $10 \log_{10} K - 3$ dB (Problem 4.97).

If we define $G(z) = H(-z)$, then it follows from the above equation that $H(z)$ and $G(z)$ are power-complementary as

$$H(z)H(z^{-1}) + G(z)G(z^{-1}) = \text{a constant}. \tag{4.147}$$

If $H(z)$ of Eq. (4.146) is the transfer function of an FIR digital filter of order N, then the FIR digital filter with a transfer function

$$G(z) = z^{-N}H(-z^{-1}) \tag{4.148}$$

is called the *conjugate quadratic filter* of $H(z)$ and vice versa [Vai88a]. Note that by definition, $G(z)$ is also a power-symmetric causal filter. It follows from Eqs. (4.146) and (4.148) that a pair of conjugate quadratic FIR filters $H(z)$ and $G(z)$ are also power-complementary as they satisfy Eq. (4.147).

> **EXAMPLE 4.15**   Let $H(z) = 1 - 2z^{-1} + 6z^{-2} + 3z^{-3}$. We form
>
> $$\begin{aligned} H(z)H(z^{-1}) + H(-z)H(-z^{-1}) &= (1 - 2z^{-1} + 6z^{-2} + 3z^{-3})(1 - 2z + 6z^2 + 3z^3) \\ &\quad + (1 + 2z^{-1} + 6z^{-2} - 3z^{-3})(1 + 2z + 6z^2 - 3z^3) \\ &= (3z^3 + 4z + 50 + 4z^{-1} + 3z^{-3}) \\ &\quad + (-3z^3 - 4z + 50 - 4z^{-1} - 3z^{-3}) = 100. \end{aligned}$$
>
> Hence, $H(z)$ is a power-symmetric transfer function.

4.8.6 Magnitude-Complementary Filters

A set of M digital filters $\{G_i(z)\}$, $i = 0, 1, \ldots, M - 1$, is defined to be *magnitude-complementary* of each other if the sum of their magnitude responses is equal to a constant [Reg87c], i.e.,

$$\sum_{i=0}^{M-1} |G_i(e^{j\omega})| = \beta \qquad \text{for all } \omega, \tag{4.149}$$

where β is a positive nonzero constant.

Consider two real-coefficient doubly-complementary transfer functions $H_0(z)$ and $H_1(z)$ that are related according to Eqs. (4.142a) and (4.142b). Define

$$G_0(z) = H_0^2(z) = \tfrac{1}{4}[A_0(z) + A_1(z)]^2, \tag{4.150a}$$

$$G_1(z) = -H_1^2(z) = -\tfrac{1}{4}[A_0(z) - A_1(z)]^2. \tag{4.150b}$$

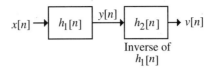

Figure 4.36: Cascade of a discrete-time system $h_1[n]$ with its inverse discrete-time system $h_2[n]$.

It follows from above that $G_0(z) + G_1(z) = A_0(z)A_1(z)$, i.e., $G_0(z)$ and $G_1(z)$ are an allpass-complementary pair. It can be easily shown that (Problem 4.110)

$$|G_0(e^{j\omega})| + |G_1(e^{j\omega})| = |H_0(e^{j\omega})|^2 + |H_1(e^{j\omega})|^2 = 1, \qquad (4.151)$$

i.e., $G_0(z)$ and $G_1(z)$ are a magnitude-complementary pair of transfer functions.

4.9 Inverse Systems

As indicated in Section 2.5.2, two LTI causal discrete-time systems with impulse responses $h_1[n]$ and $h_2[n]$ are inverses of each other if

$$h_1[n] \circledast h_2[n] = \delta[n]. \qquad (4.152)$$

An application of the inverse system design, as pointed out earlier, is in the recovery of a signal $x[n]$ that has been transmitted through an imperfect transmission channel. The received signal $y[n]$, in general, will be different from $x[n]$ as it will be distorted by the impulse response $h_1[n]$ of the channel. To recover the original signal $x[n]$ we need to pass $y[n]$ through a system with an impulse response $h_2[n]$ which is the inverse of the channel's impulse response (see Figure 4.36). The output $v[n]$ of the inverse system will be identical to the desired input $x[n]$.

4.9.1 Representation in the z-Domain

It is easy to characterize the inverse system in the z-domain. Taking the z-transform of both sides of Eq. (4.152) we get

$$H_1(z)\, H_2(z) = 1, \qquad (4.153)$$

where $H_1(z)$ and $H_2(z)$ are the z-transforms of $h_1[n]$ and $h_2[n]$, respectively. From Eq. (4.153) it follows that the transfer function $H_2(z)$ of the inverse system is simply the reciprocal of $H_1(z)$, i.e.,

$$H_2(z) = \frac{1}{H_1(z)}. \qquad (4.154)$$

For a rational transfer function $H_1(z)$

$$H_1(z) = \frac{P(z)}{D(z)}, \qquad (4.155)$$

the transfer function $H_2(z)$ of the inverse system is then given by

$$H_2(z) = \frac{D(z)}{P(z)}. \qquad (4.156)$$

It follows from Eqs. (4.155) and (4.156) that the poles (zeros) of the inverse system $H_2(z)$ are the zeros (poles) of the system $H_1(z)$.

EXAMPLE 4.16   Determine the inverse of the causal LTI system characterized by the transfer function:

$$H_1(z) = \frac{\left(z - \frac{1}{4}\right)\left(z + \frac{1}{5}\right)}{\left(z + \frac{1}{2}\right)\left(z - \frac{1}{3}\right)}, \qquad |z| > \frac{1}{2}.$$

Note that $H_1(z)$ is a stable system. Using Eq. (4.154) we arrive at the transfer function of the inverse system:

$$H_2(z) = \frac{\left(z + \frac{1}{2}\right)\left(z - \frac{1}{3}\right)}{\left(z - \frac{1}{4}\right)\left(z + \frac{1}{5}\right)}.$$

There are three possible ROCs of the above inverse system:

$$\mathcal{R}_1: \ |z| > \frac{1}{4},$$

$$\mathcal{R}_2: \ \frac{1}{5} < |z| < \frac{1}{4},$$

$$\mathcal{R}_3: \ |z| < \frac{1}{5}.$$

If we take $\mathcal{R}_1$ as the ROC, then the inverse system is stable and has a causal impulse response. If we instead choose $\mathcal{R}_3$ as the ROC, then the inverse system is unstable and has an anticausal impulse response. If, on the other hand, $\mathcal{R}_2$ is selected as the ROC, the inverse system is unstable and has a two-sided impulse response.

It follows from the above example that to obtain a unique inverse, the ROC needs to be known a priori. To this end, it is a usual practice to look for a causal inverse of a causal system. The causal inverse of the parent causal system with a minimum-phase transfer function is always stable. However, the inverse of a nonminimum-phase system is unstable if causality is imposed.

4.9.2   Recursive Computation of the Input Signal

If the parent causal system has a known impulse response $h_1[n]$ and is excited by a causal input signal $x[n]$, then knowing the output signal $y[n]$ for $n \geq 0$, we can determine the samples of the input signal using a recursive relation without determining the inverse system. To develop this relation we recall that the input-output relation in the time-domain is given by

$$y[n] = x[n] \circledast h_1[n] = \sum_{k=0}^{n} x[k]\, h_1[n - k], \qquad n \geq 0. \tag{4.157}$$

From Eq. (4.157) for $n = 0$ we have

$$y[0] = x[0]\, h_1[0],$$

$$x[0] = \frac{y[0]}{h_1[0]}. \tag{4.158}$$

To determine $x[n]$ for $n \geq 1$, we rewrite Eq. (4.157) as

$$y[n] = x[n]\, h_1[0] + \sum_{k=0}^{n-1} x[k]\, h_1[n - k],$$

which yields

$$x[n] = \frac{y[n] - \sum_{k=0}^{n-1} x[k] h_1[n-k]}{h_1[0]}, \qquad n \geq 1, \tag{4.159}$$

provided $h_1[0] \neq 0$.

EXAMPLE 4.17   The first 11 samples of the impulse response of a causal second-order IIR discrete-time system are given by

$$h_1[n] = \{2 \quad 4 \quad -5 \quad -3 \quad 13 \quad -7 \quad -19 \quad 33 \quad 5 \quad -71 \quad 61 \quad \cdots\}.$$
$$\uparrow$$

Using the above method we determine the first 11 samples of the input sequence $x[n]$ which produces the output sequence given below:

$$y[n] = \{2 \quad 8 \quad 9 \quad 17 \quad 18 \quad 20 \quad -3 \quad 8 \quad 32 \quad -25 \quad -27 \quad \cdots\}.$$
$$\uparrow$$

The MATLAB program given below can be used to determine the samples of the desired input sequence.

```
% Program 4_3
% Determination of the Input Signal from
% Given Impulse Response and Output Signal
% of same length N
h1 = input('Type in the impulse response samples = ');
y = input('Type in the output response samples = ');
N = length(y);
x = [y(1)/h1(1) zeros(1,N-1)];
for k = 2:N
    x(k) = (y(k) - fliplr(h1(2:k)*x(1:k-1)'))/h1(1);
end
disp('Input Samples');disp(x);
```

The output data generated by running the above program for the specified input is given by

```
Input Samples
1   2   3   4   5   5   4   3   2   1   0
```

To verify that the above result is correct we convolve the input sequence generated with the given impulse response using conv which results in the output sequence:

```
2   8   9   17   18   20   -3   8   32   -25   -27
```

The above result is seen to be identical to the output sequence specified earlier.

The process of determining the sequence $x[n]$ from the convolution sum given in Eq. (4.157) using Eqs. (4.158) and (4.159) is called *deconvolution*. An alternate interpretation of the deconvolution algorithm given by these two equations can be established by treating the convolution sum as a polynomial multiplication in the z-domain. If $Y(z)$, $X(z)$, and $H_1(z)$ denote the z-transforms of the sequences $y[n]$, $x[n]$, and $h_1[n]$, respectively, then in the z-domain, the convolution sum of Eq. (4.157) can be written as

$$Y(z) = X(z)H_1(z). \tag{4.160}$$

Hence $X(z)$ can be found by dividing the polynomial $Y(z)$ by the polynomial $H_1(z)$, i.e.,

$$X(z) = \frac{Y(z)}{H_1(z)}. \tag{4.161}$$

Now, the z-transforms $Y(z)$, $X(z)$, and $H_1(z)$ are polynomials in z^{-1}. Hence, $X(z)$ can be found by a long division of $Y(z)$ by $H_1(z)$ as indicated in Example 3.35 for the inverse z-transform computation.

4.10 System Identification

There are applications where the objective is to determine either the impulse response $h[n]$ or the transfer function $H(z)$ of an unknown initially relaxed causal LTI system by exciting it with a known input sequence $x[n]$ and observing the corresponding output $y[n]$. The system identification is thus dual to the problem of determining the input $x[n]$ knowing the impulse response $h[n]$ and the output $y[n]$ described in the previous section. In the time-domain, the system identification problem can be solved by interchanging the roles of the input and the impulse response in the method outlined above.

The recursive relation for computing the impulse response samples $h[n]$ of a causal LTI system from the specified causal input sequence $x[n]$ and the observed output sequence $y[n]$ is therefore as follows:

$$h[0] = \frac{y[0]}{x[0]}, \qquad h[n] = \frac{y[n] - \sum_{k=0}^{n-1} h[k]\,x[n-k]}{x[0]}, \qquad n \geq 1,$$

provided that $x[0] \neq 0$. The above process can be implemented in MATLAB by a simple modification of Program 4_3.

Alternately, $h[n]$ can be determined in the z-domain by dividing the z-transform $Y(z)$ of $y[n]$ by the z-transform $X(z)$ of $x[n]$. To this end again the function deconv can be employed.

If the causal LTI system has a rational transfer function $H(z)$ of known order M, the numerator and the denominator coefficients of the transfer function can be determined from the first $2M + 1$ impulse response coefficients. The method to determine $H(z)$ is described in Section 8.1.3.

A second method of system identification is based on computing the energy density spectrum $S_{xx}(e^{j\omega})$ of the input signal $x[n]$, and the cross-energy density spectrum $S_{yx}(e^{j\omega})$ of the input signal $x[n]$ and the output signal $y[n]$. These spectrums can be evaluated by taking the DTFTs of the autocorrelation sequence $r_{xx}[\ell]$ of $x[n]$, and the cross-correlation sequence $r_{yx}[\ell]$ of $y[n]$ and $x[n]$.

Consider a stable LTI discrete-time system with an impulse response $h[n]$. The input-output relation of this system is given by

$$y[n] = \sum_{k=-\infty}^{\infty} h[k]x[n-k]. \tag{4.162}$$

We assume that the autocorrelation sequence $r_{xx}[\ell]$ of the input is known. The cross-correlation sequence $r_{yx}[\ell]$ is defined by

$$r_{yx}[\ell] = \sum_{n=-\infty}^{\infty} y[n]x[n-\ell]. \tag{4.163}$$

Substituting Eq. (4.162) in the above equation we get

$$r_{yx}[\ell] = \sum_{n=-\infty}^{\infty} \left(\sum_{k=-\infty}^{\infty} h[k]x[n-k] \right) x[n-\ell]$$

$$= \sum_{k=-\infty}^{\infty} h[k] \left(\sum_{n=-\infty}^{\infty} x[n-k]x[n-\ell] \right)$$

$$= \sum_{k=-\infty}^{\infty} h[k]r_{xx}[\ell-k]. \tag{4.164}$$

If the system is assumed to have a causal finite-length impulse response of length N, then Eq. (4.164) reduces to

$$r_{yx}[\ell] = \sum_{k=0}^{N-1} h[k]r_{xx}[\ell-k]. \tag{4.165}$$

In the z-domain, Eq. (4.164) is equivalent to

$$S_{yx}(z) = H(z)S_{xx}(z),$$

where $S_{xx}(z)$ and $S_{yx}(z)$ are the z-transforms of $r_{xx}[\ell]$ and $r_{yx}[\ell]$, respectively, and $H(z)$ is the transfer function of the LTI system. On the unit circle, the above equation reduces to

$$S_{yx}(e^{j\omega}) = H(e^{j\omega})S_{xx}(e^{j\omega}) = H(e^{j\omega})|X(e^{j\omega})|^2, \tag{4.166}$$

using Eq. (3.19). From Eq. (4.166) it follows that the frequency response of the LTI system can be expressed as

$$H(e^{j\omega}) = \frac{S_{yx}(e^{j\omega})}{S_{xx}(e^{j\omega})}. \tag{4.167}$$

If $x[n]$ is selected to have a constant energy spectrum for all values of ω, i.e., $S_{xx}(e^{j\omega}) = 1/K, 0 \leq |\omega| \leq \pi$, then the above equation reduces to

$$H(e^{j\omega}) = K S_{yx}(e^{j\omega}).$$

It follows from Eq. (4.167) that the frequency response $H(e^{j\omega})$ of an LTI discrete-time system can be determined by taking the ratio of the cross-energy spectrum of the output and the input sequences, and the energy spectrum of the input. The frequency response is proportional to the cross-energy spectrum if the input has a constant energy spectrum.

In some applications, where the input signal $x[n]$ is not known, the system can be identified by computing the autocorrelation of the output signal $y[n]$ which is defined by

$$r_{yy}[\ell] = \sum_{n=-\infty}^{\infty} y[n]y[n-\ell]. \tag{4.168}$$

Substituting Eq. (4.162) in the above equation we arrive at

$$r_{yy}[\ell] = \sum_{n=-\infty}^{\infty} \left(\sum_{k=-\infty}^{\infty} h[k]x[n-k] \right) \left(\sum_{k=-\infty}^{\infty} h[m-\ell]x[n-m+\ell] \right)$$

$$= \sum_{k=-\infty}^{\infty} h[k] \sum_{m=-\infty}^{\infty} h[m-\ell] \left(\sum_{n=-\infty}^{\infty} x[n-k]x[n-m+\ell] \right)$$

$$= \sum_{k=-\infty}^{\infty} h[k] \sum_{m=-\infty}^{\infty} h[m-\ell] r_{xx}[m-\ell-k]$$

$$= h[\ell] \circledast h[-\ell] \circledast r_{xx}[\ell]. \tag{4.169}$$

In the z-domain, Eq. (4.169) becomes

$$S_{yy}(z) = H(z)H(z^{-1})S_{xx}(z),$$

where $S_{yy}(z)$ is the z-transform of $r_{yy}[\ell]$. On the unit circle, the above equation reduces to

$$S_{yy}(e^{j\omega}) = |H(e^{j\omega})|^2 S_{xx}(e^{j\omega}).$$

For an input signal with a flat energy density spectrum, we then have

$$S_{yy}(e^{j\omega}) = K|H(e^{j\omega})|^2,$$

or equivalently in the z-domain, with $K = 1$, this becomes

$$S_{yy}(z) = H(z)H(z^{-1}). \tag{4.170}$$

If the system is characterized by a rational transfer function $H(z) = P(z)/D(z)$, then

$$S_{yy}(z) = \frac{A(z)}{B(z)} = \frac{P(z)P(z^{-1})}{D(z)D(z^{-1})},$$

indicating that the numerator and the denominator polynomials of $S_{yy}(z)$ exhibit mirror-image symmetry. To determine $H(z)$ one can determine the roots of the polynomials $A(z)$ and $B(z)$ and associate appropriate factors of these polynomials with the numerator and the denominator of $H(z)$.

It should be noted from the above discussion that the autocorrelation of the output signal can provide only the magnitude response of the system but not the phase response. A solution of Eq. (4.170) thus leads to many possible answers. A single solution can, however, be obtained by imposing some additional constraints on the phase properties of the system.

EXAMPLE 4.18　Let the output energy density spectrum $S_{yy}(e^{j\omega})$ of a causal stable LTI discrete-time system excited by an input sequence with a unity energy spectrum be given by

$$S_{yy}(e^{j\omega}) = \frac{1.04 + 0.4 \cos\omega}{1.25 - \cos\omega}.$$

Using trigonometric identities, we rewrite the above as

$$S_{yy}(e^{j\omega}) = \frac{1.04 + 0.2(e^{j\omega} + e^{-j\omega})}{1.25 - 0.5(e^{j\omega} + e^{-j\omega})}.$$

Substituting $z = e^{j\omega}$ in the above and making use of Eq. (4.170) we arrive at

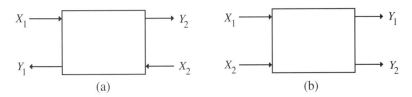

Figure 4.37: A digital two-pair.

$$H(z)H(z^{-1}) = \frac{1.04 + 0.2(z + z^{-1})}{1.25 - 0.5(z + z^{-1})} = -0.4\left(\frac{z^2 + 5.2z + 1}{z^2 - 2.5z + 1}\right)$$

$$= -0.4\frac{(z + 5)(z + 0.2)}{(z - 2)(z - 0.5)} = \frac{(z^{-1} + 0.2)(1 + 0.2z^{-1})}{(z^{-1} - 0.5)(1 - 0.5z^{-1})}.$$

Therefore, for a minimum-phase system the transfer function is given by

$$H(z) = \frac{1 + 0.2z^{-1}}{1 - 0.5z^{-1}},$$

whereas, for a non-minimum-phase system the transfer function is

$$H(z) = \frac{z^{-1} + 0.2}{1 - 0.5z^{-1}}.$$

4.11 Digital Two-Pairs

The LTI discrete-time systems considered so far are single-input, single-output structures characterized by a transfer function. Often, such a system can be efficiently realized by interconnecting two-input, two-output structures, more commonly called *two-pairs* [Mit73b]. Figure 4.37 shows two commonly used block diagram representations of a two-pair, with Y_1 and Y_2 denoting the two outputs and X_1 and X_2 denoting the two inputs, where the dependence on the variable z has been omitted for simplicity. We consider here the transform-domain characterizations of such digital filter structures and discuss several two-pair interconnection schemes for the development of more complex structures. Later, we outline minimum-multiplier realizations of allpass transfer functions based on the two-pair representation.

4.11.1 Characterization

The input-output relation of a digital two-pair is given by

$$\begin{bmatrix} Y_1 \\ Y_2 \end{bmatrix} = \begin{bmatrix} t_{11} & t_{12} \\ t_{21} & t_{22} \end{bmatrix} \begin{bmatrix} X_1 \\ X_2 \end{bmatrix}, \tag{4.171}$$

In the above relation the matrix τ given by

$$\tau = \begin{bmatrix} t_{11} & t_{12} \\ t_{21} & t_{22} \end{bmatrix} \tag{4.172}$$

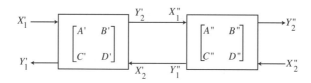

Figure 4.38: Γ-cascade connection of two-pairs.

is called the *transfer matrix* of the two-pair. From Eq. (4.171) it follows that the transfer parameters can be found as follows:

$$t_{11} = \frac{Y_1}{X_1}\Big|_{X_2=0}, \quad t_{12} = \frac{Y_1}{X_2}\Big|_{X_1=0}, \quad t_{21} = \frac{Y_2}{X_1}\Big|_{X_2=0}, \quad t_{22} = \frac{Y_2}{X_2}\Big|_{X_1=0}. \tag{4.173}$$

An alternative characterization of the two-pair is in terms of its chain parameters as

$$\begin{bmatrix} X_1 \\ Y_1 \end{bmatrix} = \begin{bmatrix} A & B \\ C & D \end{bmatrix} \begin{bmatrix} Y_2 \\ X_2 \end{bmatrix}, \tag{4.174}$$

where the matrix **Γ** given by

$$\Gamma = \begin{bmatrix} A & B \\ C & D \end{bmatrix} \tag{4.175}$$

is called the *chain matrix* of the two-pair.

The relations between the transfer parameters and the chain parameters can be easily derived and are given as

$$t_{11} = \frac{C}{A}, \quad t_{12} = \frac{AD - BC}{A}, \quad t_{21} = \frac{1}{A}, \quad t_{22} = -\frac{B}{A}, \tag{4.176a}$$

$$A = \frac{1}{t_{21}}, \quad B = -\frac{t_{22}}{t_{21}}, \quad C = \frac{t_{11}}{t_{21}}, \quad D = \frac{t_{12}t_{21} - t_{11}t_{22}}{t_{21}}. \tag{4.176b}$$

4.11.2 Two-Pair Interconnection Schemes

Two or more two-pairs can be connected in cascade in two different ways. The cascade connection of two-pairs shown in Figure 4.38 is called a Γ-*cascade*. We next determine the characterization of the overall two-pair. Let the individual two-pairs be characterized by their chain parameters as

$$\begin{bmatrix} X_1' \\ Y_1' \end{bmatrix} = \begin{bmatrix} A' & B' \\ C' & D' \end{bmatrix} \begin{bmatrix} Y_2' \\ X_2' \end{bmatrix}, \quad \begin{bmatrix} X_1'' \\ Y_1'' \end{bmatrix} = \begin{bmatrix} A'' & B'' \\ C'' & D'' \end{bmatrix} \begin{bmatrix} Y_2'' \\ X_2'' \end{bmatrix}. \tag{4.177}$$

But from Figure 4.38, $X_1'' = Y_2'$, and $Y_1'' = X_2'$. Substituting these relations in the first equation of Eq. (4.177) and combining the two equations, we arrive at

$$\begin{bmatrix} X_1' \\ Y_1' \end{bmatrix} = \begin{bmatrix} A' & B' \\ C' & D' \end{bmatrix} \begin{bmatrix} A'' & B'' \\ C'' & D'' \end{bmatrix} \begin{bmatrix} Y_2'' \\ X_2'' \end{bmatrix} \tag{4.178}$$

Therefore, the chain matrix of the overall cascade is given by the product of the individual chain matrices, as indicated in Eq. (4.178).

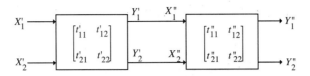

Figure 4.39: τ-cascade connection of two-pairs.

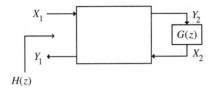

Figure 4.40: A constrained two-pair.

A second type of cascade connection, called the τ-*cascade*, is shown in Figure 4.39. If the individual two-pairs are described by their transfer matrices

$$\begin{bmatrix} Y_1' \\ Y_2' \end{bmatrix} = \begin{bmatrix} t_{11}' & t_{12}' \\ t_{21}' & t_{22}' \end{bmatrix} \begin{bmatrix} X_1' \\ X_2' \end{bmatrix}, \qquad \begin{bmatrix} Y_1'' \\ Y_2'' \end{bmatrix} = \begin{bmatrix} t_{11}'' & t_{12}'' \\ t_{21}'' & t_{22}'' \end{bmatrix} \begin{bmatrix} X_1'' \\ X_2'' \end{bmatrix}, \tag{4.179}$$

then it follows that the overall cascade is characterized by a transfer matrix that is the product of the transfer matrix of the constituent two-pairs, i.e.,

$$\begin{bmatrix} Y_1'' \\ Y_2'' \end{bmatrix} = \begin{bmatrix} t_{11}'' & t_{12}'' \\ t_{21}'' & t_{22}'' \end{bmatrix} \begin{bmatrix} t_{11}' & t_{12}' \\ t_{21}' & t_{22}' \end{bmatrix} \begin{bmatrix} X_1' \\ X_2' \end{bmatrix} \tag{4.180}$$

Another useful interconnection is the constrained two-pair indicated in Figure 4.40. If we denote the transfer function Y_1/X_1 as $H(z)$, then it can be shown that $H(z)$ can be expressed in terms of the two parameters and the constraining transfer function $G(z)$ as

$$H(z) = \frac{Y_1}{X_1} = \frac{C + D \cdot G(z)}{A + B \cdot G(z)} \tag{4.181a}$$

$$= t_{11} + \frac{t_{12}t_{21}G(z)}{1 - t_{22}G(z)}. \tag{4.181b}$$

4.12 Algebraic Stability Test

We have shown in Section 4.3.5 that the BIBO stability of a causal rational transfer function requires that all its poles be inside the unit circle. For very-high-order transfer functions it is difficult to determine the pole locations analytically, and the use of some type of root finding computer program is necessary. We outline here a simple algebraic stability test procedure that does not require the determination of the pole locations. The algorithm is based on the realization of an allpass transfer function with a denominator that is the same as that of the transfer function of interest.

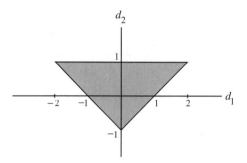

Figure 4.41: Stability triangle for a second-order digital transfer function. Stability region is the shaded area.

4.12.1 The Stability Triangle

For a second-order transfer function the stability can be checked easily by examining its denominator coefficients. Let

$$D(z) = 1 + d_1 z^{-1} + d_2 z^{-2} \tag{4.182}$$

denote the denominator of the transfer function. In terms of its poles, $D(z)$ can be expressed as

$$D(z) = (1 - \lambda_1 z^{-1})(1 - \lambda_2 z^{-1}) = 1 - (\lambda_1 + \lambda_2)z^{-1} + \lambda_1 \lambda_2 z^{-2}. \tag{4.183}$$

Comparing Eqs. (4.182) and (4.183) we obtain

$$d_1 = -(\lambda_1 + \lambda_2), \quad d_2 = \lambda_1 \lambda_2. \tag{4.184}$$

Now for stability of the transfer function, its poles must be inside the unit circle, i.e.,

$$|\lambda_1| < 1, \quad |\lambda_2| < 1.$$

Since from Eq. (4.182) the coefficient d_2 is given by the product of the poles, we must have

$$|d_2| < 1. \tag{4.185}$$

Now the roots of the polynomial $D(z)$ are given by

$$\lambda_1 = -\frac{d_1 + \sqrt{d_1^2 - 4d_2}}{2}, \quad \lambda_2 = -\frac{d_1 - \sqrt{d_1^2 - 4d_2}}{2}. \tag{4.186}$$

It can be shown (Problem 4.120) that a second coefficient condition is obtained using Eq. (4.186) and is given by

$$|d_1| < 1 + d_2. \tag{4.187}$$

The region in the (d_1, d_2)-plane where the two coefficient conditions of Eqs. (4.185) and (4.187) are satisfied is a triangle, as sketched in Figure 4.41, and is known as the *stability triangle* for a second-order digital transfer function [Jac96].

4.12.2 A Stability Test Procedure

It is impossible to develop simple conditions on the coefficients of the denominator polynomial of a higher-order transfer function similar to that developed for a second-order function as given above. However, a number of methods have been proposed to determine the stability of an Mth-order transfer function $H(z)$ without factoring out the roots of its denominator polynomial $D_M(z)$. We outline below one such method [Vai87e].

Let

$$D_M(z) = \sum_{i=0}^{M} d_i z^{-i}, \tag{4.188}$$

where we assume for simplicity $d_0 = 1$. We first form an Mth-order allpass transfer function from $D_M(z)$ as

$$
\begin{aligned}
A_M(z) &= \frac{\tilde{D}_M(z)}{D_M(z)} \\
&= \frac{d_M + d_{M-1}z^{-1} + d_{M-2}z^{-2} + \cdots + d_1 z^{-(M-1)} + z^{-M}}{1 + d_1 z^{-1} + d_2 z^{-2} + \cdots + d_{M-1} z^{-(M-1)} + d_M z^{-M}}.
\end{aligned}
\tag{4.189}
$$

If we express

$$D_M(z) = \prod_{i=1}^{M} (1 - \lambda_i z^{-1}),$$

it then follows that the coefficient d_M is the product of all roots, i.e.,

$$d_M = (-1)^M \prod_{i=1}^{M} \lambda_i.$$

Now for stability we must have $|\lambda_i| < 1$, which implies the condition $|d_M| < 1$. If we define

$$k_M = A_M(\infty) = d_M,$$

then a necessary condition for stability of $A_M(z)$, and hence, the original transfer function $H(z)$, is given by

$$k_M^2 < 1. \tag{4.190}$$

Assume the above condition holds. Next, we form a new function $A_{M-1}(z)$ according to

$$A_{M-1}(z) = z \left[\frac{A_M(z) - k_M}{1 - k_M A_M(z)} \right] = z \left[\frac{A_M(z) - d_M}{1 - d_M A_M(z)} \right]. \tag{4.191}$$

Substituting in the above equation the expression for $A_M(z)$ from Eq. (4.189), we arrive at

$$A_{M-1}(z) = \frac{\begin{array}{l}(d_{M-1} - d_M d_1) + (d_{M-2} - d_M d_2)z^{-1} + \cdots \\ + (d_1 - d_M d_{M-1})z^{-(M-2)} + (1 - d_M^2)z^{-(M-1)}\end{array}}{\begin{array}{l}(1 - d_M^2) + (d_1 - d_M d_{M-1})z^{-1} + \cdots \\ + (d_{M-2} - d_M d_2)z^{-(M-2)} + (d_{M-1} - d_M d_1)z^{-(M-1)}\end{array}}, \tag{4.192}$$

which is seen to be an $(M-1)$th-order allpass function. $A_{M-1}(z)$ can be rewritten in the form

$$A_{M-1}(z) = \frac{d'_{M-1} + d'_{M-2}z^{-1} + \cdots + d'_1 z^{-(M-2)} + z^{-(M-1)}}{1 + d'_1 z^{-1} + \cdots + d'_{M-2}z^{-(M-2)} + d'_{M-1}z^{-(M-1)}}, \tag{4.193}$$

where

$$d'_i = \frac{d_i - d_M d_{M-i}}{1 - d_M^2}, \qquad i = 1, 2, \ldots, M-1. \tag{4.194}$$

Now from Eq. (4.191) the poles λ_0 of $A_{M-1}(z)$ are given by the roots of the equation

$$A_M(\lambda_0) = \frac{1}{k_M}. \tag{4.195}$$

By assumption Eq. (4.190) holds. Hence the above equation implies that

$$|A_M(\lambda_0)| > 1. \tag{4.196}$$

If $A_M(z)$ is a stable allpass function, then according to Eq. (4.132), $|A_M(z)| < 1$ for $|z| > 1$, $|A_M(z)| = 1$ for $|z| = 1$, and $|A_M(z)| > 1$ for $|z| < 1$. Therefore, if $A_M(z)$ is a stable allpass function, then $|\lambda_0| < 1$, or in other words, $A_{M-1}(z)$ is a stable allpass function. Thus, if $A_M(z)$ is a stable allpass function and $k_M^2 < 1$, then $A_{M-1}(z)$ is a stable allpass function.

We now prove the converse; that is, if $A_{M-1}(z)$ is a stable allpass function and $k_M^2 < 1$, then $A_M(z)$ is a stable allpass function. To this end, we invert the relation of Eq. (4.192) to arrive at

$$A_M(z) = \frac{k_M + z^{-1}A_{M-1}(z)}{1 + k_M z^{-1}A_{M-1}(z)}. \tag{4.197}$$

If ζ_0 is a pole of $A_M(z)$, then

$$\zeta_0^{-1}A_{M-1}(\zeta_0) = -\frac{1}{k_M}. \tag{4.198}$$

By assumption Eq. (4.190) holds. Therefore, $|\zeta_0^{-1}A_{M-1}(\zeta_0)| > 1$, i.e.,

$$|A_{M-1}(\zeta_0)| > |\zeta_0|. \tag{4.199}$$

Assume $A_{M-1}(z)$ is a stable allpass function; then according to Eq. (4.132), $|A_{M-1}(z)| \leq 1$ for $|z| \geq 1$. Now, if $|\zeta_0| \geq 1$, then from Eq. (4.132), $|A_{M-1}(\zeta_0)| \leq 1$, which contradicts Eq. (4.199). On the other hand, if $|\zeta_0| < 1$, then from Eq. (4.132), $|A_{M-1}(\zeta_0)| > 1$, satisfying the condition of Eq. (4.199). Thus, if Eq. (4.190) holds and if $A_{M-1}(z)$ is a stable allpass function, then $A_M(z)$ is also a stable allpass function.

Summarizing, a necessary and sufficient set of conditions for the allpass function $A_M(z)$ to be stable is therefore:

(a) $k_M^2 < 1$, and

(b) The allpass function $A_{M-1}(z)$ is stable.

Thus, once we have checked the condition $k_M^2 < 1$, we merely test for the stability of the lower-order allpass function $A_{M-1}(z)$. The process can now be repeated, generating a set of coefficients

$$k_M, k_{M-1}, \ldots, k_2, k_1$$

and a set of allpass functions of decreasing orders,

$$A_M(z), A_{M-1}(z), \ldots, A_2(z), A_1(z), A_0(z) = 1.$$

The allpass function $A_M(z)$ is stable if and only if $k_i^2 < 1$ for all i.

EXAMPLE 4.20   Let us test the stability of the transfer function

$$H(z) = \frac{1}{4z^4 + 3z^3 + 2z^2 + z + 1}. \tag{4.200}$$

From the above, we arrive at a fourth-order allpass function whose denominator is the same as that of $H(z)$ given above with the numerator being its mirror-image polynomial:

$$A_4(z) = \frac{\frac{1}{4}z^4 + \frac{1}{4}z^3 + \frac{1}{2}z^2 + \frac{3}{4}z + 1}{z^4 + \frac{3}{4}z^3 + \frac{1}{2}z^2 + \frac{1}{4}z + \frac{1}{4}} = \frac{d_4 z^4 + d_3 z^3 + d_2 z^2 + d_1 z + 1}{z^4 + d_1 z^3 + d_2 z^2 + d_3 z + d_4}. \tag{4.201}$$

Note that $k_4 = A_4(\infty) = d_4 = 1/4$. Hence $|k_4| < 1$.

Using Eq. (4.194), we next determine the coefficients $\{d_i'\}$ of the third-order allpass transfer function $A_3(z)$ derived from $A_4(z)$:

$$d_i' = \frac{d_i - d_4 d_{4-i}}{1 - d_4^2}, \qquad i = 1, 2, 3. \tag{4.202}$$

Substituting the appropriate values of d_i from Eq. (4.201) in the above expression, we arrive at

$$A_3(z) = \frac{d_3' z^3 + d_2' z^2 + d_1' z + 1}{z^3 + d_1' z^2 + d_2' z + d_3'} = \frac{\frac{1}{15}z^3 + \frac{2}{5}z^2 + \frac{11}{15}z + 1}{z^3 + \frac{11}{15}z^2 + \frac{2}{5}z + \frac{1}{15}}.$$

It can be seen that $k_3 = A_3(\infty) = d_3' = 1/15$. Thus, $|k_3| < 1$.

Following the above procedure, we derive the next two lower-order allpass functions:

$$A_2(z) = \frac{\frac{79}{224}z^2 + \frac{159}{224}z + 1}{z^2 + \frac{159}{224}z + \frac{79}{224}},$$

$$A_1(z) = \frac{\frac{53}{101}z + 1}{z + \frac{53}{101}}.$$

Note from the above that $k_2 = A_2(\infty) = 79/224$, $k_1 = A_1(\infty) = 53/101$ implying that $|k_2| < 1$ and $|k_1| < 1$.

It should be noted that it is not necessary to derive $A_1(z)$ since $A_2(z)$ can be tested for stability using conditions of Eqs. (4.185) and (4.187). These two conditions are obviously satisfied for the $A_2(z)$ given above.

Since all of the stability test conditions are satisfied, $A_4(z)$ and, hence, $H(z)$ of Eq. (4.200) are stable transfer functions.

EXAMPLE 4.21   Determine whether all zeros of the denominator polynomial of a causal transfer function

$$D_4(z) = z^4 + 3.25z^3 + 3.75z^2 + 2.75z + 0.5$$

are inside the unit circle.

We form the allpass function

$$A_4(z) = \frac{\tilde{D}_4(z)}{D_4(z)} = \frac{0.5z^4 + 2.75z^3 + 3.75z^2 + 3.25z + 1}{z^4 + 3.25z^3 + 3.75z^2 + 2.75z + 0.5}.$$

Note from the above that $k_4 = A_4(\infty) = 0.5$, implying $|k_3| < 1$. Next, using Eq. (4.196) we derive the third-order allpass function $A_3(z)$, which can be shown to be equal to

$$A_3(z) = \frac{1.5z^3 + 2.5z^2 + 2.5z + 1}{z^3 + 2.5z^2 + 2.5z + 1.5}.$$

It can be seen from the above that $k_3 = A_3(\infty) = 1.5$, implying $|k_3| > 1$. Therefore, the allpass function $A_4(z)$ is unstable or, equivalently, not all zeros of $D_4(z)$ are inside the unit circle.

The M-file `poly2rc` in MATLAB can be utilized to determine the stability test parameters $\{k_i\}$ very efficiently on a computer. We illustrate its application below.

EXAMPLE 4.22   We test the stability of the transfer function of Eq. (4.200) using MATLAB. To this end we make use of the following program:

```
% Program 4_4
% Stability Test of a Rational Transfer Function
%
% Read in the polynomial coefficients
den = input('Type in the denominator coefficients =');
% Generate the stability test parameters
k = poly2rc(den);
knew = fliplr(k);
disp('The stability test parameters are');disp(knew);
stable = all(abs(k) < 1)
```

The input data is the vector `den` of the coefficients of the denominator polynomial entered inside a square bracket in descending powers of z as indicated below:

```
den = [4    3    2    1    1]
```

The output data are the stability test parameters $\{k_i\}$. The program also has a logical output which is `stable = 1` if the transfer function is stable, otherwise, `stable = 0`.

The output data generated for the transfer function of Eq. (4.200) are as follows:

```
The stability test parameters are
    0.2500    0.0667    0.3527    0.5248

stable =
    1
```

Note that the stability test parameters are identical to those computed in Example 4.20.

4.13 Discrete-Time Processing of Random Signals

As we shall observe later in the text, there are occasions when we need to study the effect of processing a random discrete-time signal by a linear time-invariant discrete-time system. More precisely, we need to determine the statistical properties of the output signal $\{y[n]\}$ generated by a stable LTI system with an impulse response $\{h[n]\}$ when its input $x[n]$ is a particular realization of a wide-sense stationary (WSS) random process $\{X[n]\}$. For simplicity we assume the input sequence and the impulse response to be real-valued. Now, the output of an LTI system is given by the linear convolution of the input and the impulse response of the system, i.e.,

$$y[n] = \sum_{k=-\infty}^{\infty} h[k]x[n-k]. \tag{4.203}$$

Since the function of a random variable is also a random variable, it follows from the above that the output $y[n]$ is also a sample sequence of an output random process $\{Y[n]\}$.

4.13.1 Statistical Properties of the Output Signal

Since the input $x[n]$ is a sample sequence of a stationary random process, its mean m_x is a constant independent of the time index n.[6] The mean $E(y[n])$ of the output random process $y[n]$ is then given by

$$m_y = E(y[n]) = E\left(\sum_{k=-\infty}^{\infty} h[k]x[n-k]\right)$$

$$= \sum_{k=-\infty}^{\infty} h[k]E(x[n-k]) = m_x \sum_{k=-\infty}^{\infty} h[k] = m_x H(e^{j0}), \tag{4.204}$$

which is a constant independent of the time index n.

The autocorrelation function of the output of the LTI discrete-time system for a real-valued input is given by

$$\phi_{yy}[n+\ell, n] = E(y[n+\ell]y[n])$$

$$= E\left(\left\{\sum_{i=-\infty}^{\infty} h[i]x[n+\ell-i]\right\}\left\{\sum_{k=-\infty}^{\infty} h[k]x[n-k]\right\}\right)$$

$$= \sum_{i=-\infty}^{\infty} h[i] \sum_{k=-\infty}^{\infty} h[k]E(x[n+\ell-i]x[n-k])$$

$$= \sum_{i=-\infty}^{\infty} h[i] \sum_{k=-\infty}^{\infty} h[k]\phi_{xx}[n+\ell-i, n-k]. \tag{4.205}$$

Since the input is a sample sequence of a WSS random process, its autocorrelation sequence depends only on the difference $\ell + k - i$ of the time indices $n + \ell - i$ and $n - k$, i.e.,

$$\phi_{xx}[n+\ell-i, n-k] = \phi_{xx}[\ell+k-i]. \tag{4.206}$$

---

[6]For convenience, we drop the notational difference between the random process $\{X[n]\}$ and its particular realization $\{x[n]\}$.

Substituting the above in Eq. (4.205), we arrive at

$$\phi_{yy}[n + \ell, n] = \sum_{i=-\infty}^{\infty} h[i] \sum_{k=-\infty}^{\infty} h[k]\phi_{xx}[\ell + k - i] = \phi_{yy}[\ell] \qquad (4.207)$$

indicating that the output autocorrelation sequence depends on the difference ℓ of the time indices $n + \ell$ and n. As a result of Eqs. (4.204) and (4.207), it follows that the output $y[n]$ is also a sample sequence of a WSS random process.

Substituting $m = i - k$ in Eq. (4.207) we arrive at

$$\phi_{yy}[\ell] = \sum_{m=-\infty}^{\infty} \phi_{xx}[\ell - m] \sum_{k=-\infty}^{\infty} h[k]h[m + k]$$

$$= \sum_{m=-\infty}^{\infty} \phi_{xx}[\ell - m]r_{hh}[m], \qquad (4.208)$$

where

$$r_{hh}[m] = \sum_{k=-\infty}^{\infty} h[k]h[m + k] = h[m] \circledast h[-m] \qquad (4.209)$$

is called the *aperiodic autocorrelation sequence* of the impulse response sequence $\{h[n]\}$. It should be noted that $r_{hh}[m]$ is the autocorrelation of a finite-energy deterministic sequence and is not the same as the autocorrelation of an infinite-energy WSS random signal.

The cross-correlation function between the output and the input sequences of the LTI system for a real-valued input is given by

$$\phi_{yx}[n + \ell, n] = E\left(y[n + \ell]x[n]\right)$$

$$= E\left(\sum_{i=-\infty}^{\infty} h[i]x[n + \ell - i]x[n]\right)$$

$$= \sum_{i=-\infty}^{\infty} h[i]E\left(x[n + \ell - i]x[n]\right)$$

$$= \sum_{i=-\infty}^{\infty} h[i]\phi_{xx}[\ell - i] = \phi_{yx}[\ell]. \qquad (4.210)$$

The above result indicates that the cross-correlation sequence depends on ℓ, the difference of the time indices $n + \ell$ and n.

4.13.2 Transform-Domain Representation

We now consider the z-transform representation of Eq. (4.207). As indicated in Section 3.11.2, the z-transform of $\phi_{xx}[\ell]$ may exist if the input random signal is of zero mean. From Eq. (4.204), for a zero-mean random input, the output of an LTI system is also a zero-mean random signal. In this case we obtain from Eq. (4.207) by taking the z-transform of both sides

$$\Phi_{yy}(z) = \Psi(z)\Phi_{xx}(z), \qquad (4.211)$$

where $\Phi_{xx}(z)$, $\Phi_{yy}(z)$, and $\Psi(z)$ are, respectively, the z-transforms of $\phi_{xx}[\ell]$, $\phi_{yy}[\ell]$, and $\psi[r]$. But, from Eq. (4.209) $\Psi(z) = H(z)H(z^{-1})$, which when substituted in Eq. (4.211) yields

$$\Phi_{yy}(z) = H(z)H(z^{-1})\Phi_{xx}(z). \tag{4.212}$$

On the unit circle, Eq. (4.212) reduces to

$$\Phi_{yy}(e^{j\omega}) = |H(e^{j\omega})|^2\Phi_{xx}(e^{j\omega}) \tag{4.213}$$

Using the notations $\mathcal{P}_{xx}(\omega)$ and $\mathcal{P}_{yy}(\omega)$ to denote the input and output power spectral densities, $\Phi_{xx}(e^{j\omega})$ and $\Phi_{yy}(e^{j\omega})$, respectively, we can rewrite Eq. (4.213) as

$$\mathcal{P}_{yy}(\omega) = |H(e^{j\omega})|^2\mathcal{P}_{xx}(\omega) \tag{4.214}$$

Now, from Eq. (2.162), for a zero-mean WSS process $y[n]$, the total average power is given by the mean-square value $E(y^2[n]) = \phi_{yy}[0]$. But, $\phi_{yy}[\ell]$ is given by the inverse Fourier transform of $\Phi_{yy}(e^{j\omega})$:

$$\phi_{yy}[\ell] = \frac{1}{2\pi}\int_{-\pi}^{\pi}\Phi_{yy}(e^{j\omega})e^{j\omega\ell}\,d\omega. \tag{4.215}$$

Therefore, from Eqs. (4.213) and (4.214) the total average power in the output signal $y[n]$ is given by

$$\begin{aligned}
E(y^2[n]) &= \phi_{yy}[0] \\
&= \frac{1}{2\pi}\int_{-\pi}^{\pi}\Phi_{yy}(e^{j\omega})\,d\omega \\
&= \frac{1}{2\pi}\int_{-\pi}^{\pi}|H(e^{j\omega})|^2\mathcal{P}_{xx}(\omega)\,d\omega. \tag{4.216}
\end{aligned}$$

For a real-valued WSS random signal $x[n]$ the autocorrelation sequence $\phi_{xx}[\ell]$ is an even sequence, and hence, $\Phi_{xx}(e^{j\omega})$ is an even function of ω. Assume the LTI system $h[n]$ to be an ideal filter with a square magnitude response

$$|H(e^{j\omega})|^2 = \begin{cases} 1, & \omega_{c1} \leq |\omega| \leq \omega_{c2}, \\ 0, & 0 \leq |\omega| < \omega_{c1}, \omega_{c2} < |\omega| < \pi. \end{cases} \tag{4.217}$$

In this case, Eq. (4.216) reduces to

$$\begin{aligned}
\phi_{yy}[0] &= \frac{1}{\pi}\int_{-\omega_{c2}}^{-\omega_{c1}}\mathcal{P}_{xx}(\omega)\,d\omega + \frac{1}{\pi}\int_{\omega_{c1}}^{\omega_{c2}}\mathcal{P}_{xx}(\omega)\,d\omega \\
&= \frac{2}{\pi}\int_{\omega_{c1}}^{\omega_{c2}}\mathcal{P}_{xx}(\omega)\,d\omega. \tag{4.218}
\end{aligned}$$

Since the total average output power $\phi_{yy}[0]$ is always nonnegative independent of the bandwidth of the linear filter, it follows from the above that

$$\mathcal{P}_{xx}(\omega) \geq 0, \tag{4.219}$$

proving that the power spectral density function of a real WSS random signal is also nonnegative in addition to being a real and even function.

Likewise, from the z-transforms of both sides of Eq. (4.210) we obtain

$$\Phi_{yx}(z) = H(z)\,\Phi_{xx}(z), \tag{4.220}$$

where $\Phi_{yx}(z)$ is the z-transform of $\phi[\ell]$. On the unit circle, the above equation reduces to

$$\Phi_{yx}(e^{j\omega}) = H(e^{j\omega})\, \Phi_{xx}(e^{j\omega}). \tag{4.221}$$

The function $\Phi_{yx}(e^{j\omega})$ is the *cross-spectral density* or *cross-power spectrum*, denoted by $\mathcal{P}_{yx}(\omega)$. Note that if $x[n]$ is a WSS white noise sequence, its power spectrum is a constant K at all frequencies. In this case, the above equation reduces to

$$\Phi_{yx}(e^{j\omega}) = K \cdot H(e^{j\omega}).$$

An application of Eq. (4.221) is in determining the frequency response of an unknown system by exciting it with a WSS random signal, and then computing the cross-power spectrum and the input-power spectrum. A ratio of these two power spectrums then yields an estimate of the frequency response of the system. Since both power spectrums are real functions of ω, the ratio provides only the magnitude response. To this end, the function tfe in MATLAB can be used. Some of the forms of this function are

```
Txy = tfe(x,y)
Txy = tfe(x,y,nfft)
[Txy, f] = tfe(x,y,nfft,FT)
Txy = tfe(x,y,nfft,FT,window)
```

EXAMPLE 4.23   To illustrate the application of tfe, we first design a causal FIR discrete-time system using the function fir1 which is based on the windowed Fourier series approach.[7] The white noise input sequence with a zero-mean Gaussian distribution is then generated using the function randn. Next, the output of the FIR filter is computed using the function filter. Finally, the magnitude response of the FIR filter is evaluated using tfe. The program used is given below.

```
num = fir1(30,0.2);
x = randn(20000,1);
y = filter(num,1,x);
tfe(x,y,1024,[],[],512);
pause
[h1,w] = freqz(num,1,512);
plot(w/pi,20*log10(abs(h1)));grid
xlabel('Normalized Frequency');ylabel('Gain, dB');
```

Figure 4.42 shows the plot of the estimated gain response along with the actual gain response.

Variance of the Output Signal

Now we develop the expression for the variance of the output random signal when the input to the LTI system is a real-valued white random process. From Eq. (3.150) we get

$$\sigma_y^2 = \gamma_{yy}[0] = \frac{1}{2\pi} \int_{-\pi}^{\pi} \mathcal{P}_{yy}(\omega)\, d\omega - m_y^2. \tag{4.222}$$

Substituting Eq. (4.214) in the above equation and making use of Eq. (3.158), we arrive at

$$\sigma_y^2 = \frac{\sigma_x^2}{2\pi} \int_{-\pi}^{\pi} |H(e^{j\omega})|^2\, d\omega, \tag{4.223}$$

---

[7]See Section 7.10.4

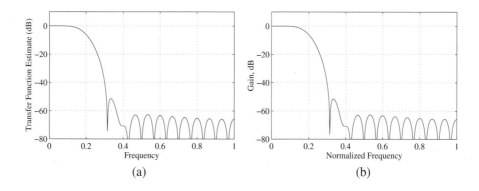

Figure 4.42: (a) Estimated gain response, and (b) actual gain response.

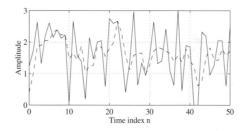

Figure 4.43: A typical uniformly distributed random sequence (solid line) and the output of a 3-point running average filter (dashed line).

which can be alternatively written as

$$\sigma_y^2 = \frac{\sigma_x^2}{2\pi j} \oint_C H(z)H(z^{-1})z^{-1} \, dz, \tag{4.224}$$

where C is a counterclockwise closed contour in the ROC of $H(z)H(z^{-1})$.

Using Parseval's relation in Eq. (4.223), we can also express the output variance as

$$\sigma_y^2 = \sigma_x^2 \sum_{n=-\infty}^{\infty} |h[n]|^2. \tag{4.225}$$

EXAMPLE 4.24   Let $\{X[n]\}$ be a uniformly distributed random signal with a probability density function $p_{X[n]}(\alpha)$, as indicated in Figure 2.36(a). From Eqs. (2.129) and (2.130) the mean and the variance of $X[n]$ are given by

$$m_x = 2, \qquad \sigma_x^2 = \tfrac{1}{3}, \qquad \text{for all values of } n. \tag{4.226}$$

Figure 4.43 shows a typical realization (solid line) of the above random signal.

We process the above random signal by a causal 3-point moving-average filter with an impulse response

$$\{h[n]\} = \left\{ \tfrac{1}{3}, \quad \tfrac{1}{3}, \quad \tfrac{1}{3} \right\}.$$

The output $y[n]$ obtained for the input of Figure 4.43 is also shown in the same figure by the dashed line. The first two samples of $y[n]$ are transients. Applying Eq. (4.204) we arrive at the mean value of the output of the 3-point moving-average filter given by

$$m_y = m_x(h[0] + h[1] + h[2]) = 2\left(\tfrac{1}{3} + \tfrac{1}{3} + \tfrac{1}{3}\right) = 2,$$

and from Eq. (4.225) the variance of the output is obtained as

$$\sigma_y^2 = \sigma_x^2(h^2[0] + h^2[1] + h^2[2]) = \tfrac{1}{3}\left(\tfrac{1}{9} + \tfrac{1}{9} + \tfrac{1}{9}\right) = \tfrac{1}{9}.$$

EXAMPLE 4.25   A zero-mean WSS white noise sequence $x[n]$ with an autocorrelation sequence

$$\phi_{xx}[n] = \sigma_x^2 \delta[n]$$

is being filtered by a causal stable LTI discrete system with a transfer function

$$H(z) = \frac{p_0}{1 + d_1 z^{-1}}, \qquad |d_1| < 1. \tag{4.227}$$

We determine the autocorrelation sequence $\phi_{yy}[n]$ of the output $y[n]$ of the linear system and the cross-correlation sequence $\phi_{yx}[n]$ between the output and the input.

Now the z-transform $\Phi_{yy}(z)$ of $\phi_{yy}[n]$ is related to the z-transform $\Phi_{xx}(z)$ of $\phi_{xx}[n]$ through Eq. (4.212). Substituting $H(z)$ from Eq. (4.227) and $\Phi_{xx}(z) = \sigma_x^2$ in Eq. (4.212) we arrive at

$$\Phi_{yy}(z) = \sigma_x^2 \left(\frac{p_0}{1 + d_1 z^{-1}}\right)\left(\frac{p_0}{1 + d_1 z}\right), \qquad |d_1| < |z| < \frac{1}{|d_1|}. \tag{4.228}$$

An inverse z-transform of the above is readily obtained by using the partial-fraction approach as described in Section 3.9.2 and is found to be

$$\phi_{yy}[n] = \frac{(p_0 \sigma_x)^2}{1 - (d_1)^2}\left(1 - (d_1)^{|n|}\right).$$

4.14 Matched Filter

In practice, a deterministic signal $x[n]$ transmitted through a channel is usually corrupted by a random signal $e[n]$. The noise-corrupted signal when processed by an LTI system with an impulse response $h[n]$, generates an output signal that contains the desired noise-free output $y[n]$ and an interference signal $d[n]$, where

$$y[n] = h[n] \circledast x[n],$$
$$d[n] = h[n] \circledast e[n].$$

The LTI system which maximizes the signal-to-noise ratio at its output is called the *matched filter*. We develop next the characterization of such a system.

4.14.1 Characterization of the Matched Filter

Now the output at time instant n_o can be expressed as

$$y[n_o] = \frac{1}{2\pi}\int_{-\pi}^{\pi} H(e^{j\omega})X(e^{j\omega})e^{j\omega n_o}\, d\omega, \tag{4.229}$$

where $X(e^{j\omega})$ and $H(e^{j\omega})$ are the DTFTs of $x[n]$ and $h[n]$, respectively. If the random signal $e[n]$ is assumed to be a zero-mean WSS process, from Eq. (4.216) the total average power of $d[n]$ is given by

$$E(d^2[n]) = \frac{1}{2\pi} \int_{-\pi}^{\pi} \left| H(e^{j\omega}) \right|^2 \mathcal{P}_{ee}(\omega)\, d\omega, \qquad (4.230)$$

where $\mathcal{P}_{ee}(\omega)$ is the power spectrum of $e[n]$.

Hence, the output signal-to-noise ratio can be expressed as

$$\left(\frac{S}{N} \right)_{\text{out}} = \frac{y^2[n_o]}{E(d^2[n])} = \frac{\left| \frac{1}{2\pi} \int_{-\pi}^{\pi} H(e^{j\omega}) X(e^{j\omega}) e^{j\omega n_o}\, d\omega \right|^2}{\frac{1}{2\pi} \int_{-\pi}^{\pi} \left| H(e^{j\omega}) \right|^2 \mathcal{P}_{ee}(\omega)\, d\omega}. \qquad (4.231)$$

To simplify the above expression, we make use of the *Schwartz inequality* which states

$$\left| \int_{-\pi}^{\pi} A(e^{j\omega}) B(e^{j\omega})\, d\omega \right|^2 \leq \int_{-\pi}^{\pi} \left| A(e^{j\omega}) \right|^2 d\omega \int_{-\pi}^{\pi} \left| B(e^{j\omega}) \right|^2 d\omega,$$

where the equality is obtained when $A(e^{j\omega}) = K \cdot B^*(e^{j\omega})$ with K a real constant. To this end, we rewrite

$$H(e^{j\omega}) X(e^{j\omega}) e^{j\omega n_o} = \left(H(e^{j\omega}) \sqrt{\mathcal{P}_{ee}(\omega)} \right) \left(\frac{X(e^{j\omega}) e^{j\omega n_o}}{\sqrt{\mathcal{P}_{ee}(\omega)}} \right).$$

Substituting the above in Eq. (4.231) we arrive at

$$\left(\frac{S}{N} \right)_{\text{out}} \leq \frac{1}{2\pi} \frac{\int_{-\pi}^{\pi} \left| H(e^{j\omega}) \right|^2 \mathcal{P}_{ee}(\omega)\, d\omega \int_{-\pi}^{\pi} \frac{|X(e^{j\omega})|^2}{\mathcal{P}_{ee}(\omega)}\, d\omega}{\int_{-\pi}^{\pi} \left| H(e^{j\omega}) \right|^2 \mathcal{P}_{ee}(\omega)\, d\omega}$$

$$\leq \frac{1}{2\pi} \int_{-\pi}^{\pi} \frac{\left| X(e^{j\omega}) \right|^2}{\mathcal{P}_{ee}(\omega)}\, d\omega. \qquad (4.232)$$

The maximum signal-to-noise ratio is obtained when the equality is achieved at which

$$\sqrt{\mathcal{P}_{ee}(\omega)} H(e^{j\omega}) = K \frac{X^*(e^{j\omega}) e^{-j\omega n_o}}{\sqrt{\mathcal{P}_{ee}(\omega)}}$$

or

$$H(e^{j\omega}) = K \frac{X^*(e^{j\omega}) e^{-j\omega n_o}}{\mathcal{P}_{ee}(\omega)} \qquad (4.233)$$

For a white noise process $e[n]$, $\mathcal{P}_{ee}(\omega) = \sigma_e^2$; then

$$H(e^{j\omega}) = \frac{K}{\sigma_e^2} X^*(e^{j\omega}) e^{-j\omega n_o}. \qquad (4.234)$$

Taking the inverse DTFT of the above we obtain

$$h[n] = \frac{K}{2\pi \sigma_e^2} \int_{-\pi}^{\pi} X^*(e^{j\omega}) e^{-j\omega n_o} e^{j\omega n}\, d\omega$$

$$= \frac{K}{\sigma_e^2} \left(\frac{1}{2\pi} \int_{-\pi}^{\pi} X^*(e^{j\omega}) e^{j\omega(n - n_o)}\, d\omega \right)$$

$$= \frac{K}{\sigma_e^2} x^*[n_o - n] = C \cdot x[n_o - n], \qquad (4.235)$$

for a real $x[n]$, where n_o is the time instant of peak signal output. The LTI discrete-time system developing the maximum output signal-to-noise ratio is said to be "matched" to the input and, hence, is called a *matched filter*.

4.14.2 Spread Spectrum Communication Systems

Matched filters are used for detection of signals in radar and digital communication systems. We describe here the basic idea behind the matched filter–based detection in the latter application [Vit95], [Mad98].

Signal processing for digital communication is typically performed on the (in general, complex-valued) baseband representation of (real-valued) passband signals. Consider the example of multiuser spread spectrum communication, in which the transmitted sequence is obtained by modulating a low-rate sequence of data *symbols* by a high-rate *spreading sequence,* typically chosen to be pseudo-random and periodic. The elements of the spreading sequence are termed *chips.* The number of chips per symbol is called the *processing gain,* denoted here by N. While the symbols and the chips can take complex values in general, we restrict attention here to binary ± 1 sequences. In this case, the symbols are called *bits.*

Let $b_0, b_1, b_2, \ldots$ denote the sequence of bits to be transmitted. Then the narrowband sequence to be transmitted consists of groups of bits with the first group containing N bits with each bit being b_0, the second group containing N bits with each bit being b_1, and so on. The resulting sequence is then multiplied element by element by the spreading sequence, and the product sequence is transmitted.

For example, a typical data sequence for a single user, its length-5 spreading code, and the resulting encoded product sequence could be as indicated below:

Data:	$+1$	$+1$	$+1$	$+1$	$+1$	-1	-1	-1	-1	-1	-1	-1	-1	-1	-1
Spreading Code:	$+1$	-1	$+1$	-1	$+1$	$+1$	-1	$+1$	-1	$+1$	$+1$	-1	$+1$	-1	$+1$
Product Sequence:	$+1$	-1	$+1$	-1	$+1$	-1	$+1$	-1	$+1$	-1	-1	$+1$	-1	$+1$	-1

In the above example, the spreading sequence has a period equal to the processing gain, which is termed a *short* spreading sequence. In many applications, the spreading sequence may be aperiodic, or have a period much larger than the processing gain, which is termed a *long* spreading sequence. For long spreading sequences, a different matched filter is needed for every bit, because the spreading sequence changes from bit to bit. In this case, a preferable implementation may be correlation of the received sequence with the spreading sequence over the N chips corresponding to a given bit.

In order to support several simultaneous users, different low-rate narrowband data streams are modulated by different high-rate spreading sequences. All product sequences are then summed into a single composite signal and transmitted over the same channel.

At the receiving end, to decode a particular data sequence, the composite signal is passed through a matched filter whose impulse response is the time-reversed version of one period of the spreading sequence used to generate its corresponding product sequence. The output of the matched filter then shows peaks at the Nth time instants. The signs of the matched filter output at these instants are precisely the bit of the data being transmitted. However, the outputs of the matched filters whose coefficients are the time-reversed version of the periods of the other spreading sequences will not show string peaks at the Nth time instants.

The above concepts are illustrated in the following example.

EXAMPLE 4.26   Let the data to be transmitted be given by $\{-1 \ -1 \ +1 \ +1 \ -1\}$. Using the function rand we generate a series of spreading sequences s_i of processing gain 5 for four different users :

$$s_1 : \{-1 \ +1 \ -1 \ +1 \ +1\}, \quad s_2 : \{-1 \ -1 \ -1 \ -1 \ +1\},$$
$$s_3 : \{-1 \ +1 \ -1 \ -1 \ +1\}, \quad s_4 : \{+1 \ -1 \ -1 \ -1 \ +1\}.$$

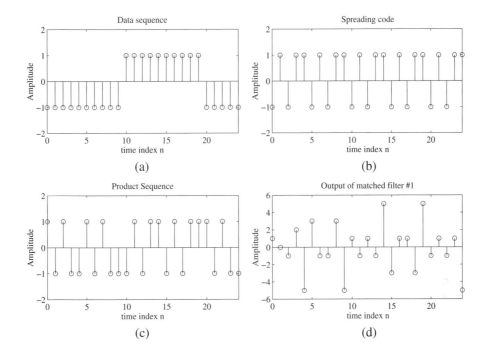

Figure 4.44: (a) The low-rate data sequence, (b) the spreading sequence s_1, (c) the product sequence, and (d) the output of matched filter #1.

We assume that user 1 is the user of interest and thus the data of interest is transmitted via the spreading sequence s_1. The low-rate data sequence, the corresponding spreading code, and the product sequence are indicated in Figure 4.44(a) to (c), respectively. The output of matched filter #1 whose coefficients are the time-reversed version of the period of the sequence s_1 is shown in Figure 4.44(d). As expected, the output shows strong peaks of magnitude 5 at time instants $n = 4, 9, 14, 19, 24$. The signs of these peaks are given by the sequence $\{-1 \ -1 \ +1 \ +1 \ -1\}$, which is precisely the data being transmitted.

Figure 4.45 (a) to (d) shows the outputs of the matched filters whose coefficients are matched to the other four spreading sequences. As can be seen, the sample values at time instants $n = 4, 9, 14, 19, 24$ have magnitudes much smaller than 5.

4.15 Summary

The frequency-domain representations of discrete-time sequences, considered in the previous chapter, are applied in this chapter to develop the frequency-domain representations of an LTI discrete-time system. One such representation is the frequency response, obtained by applying the discrete-time Fourier transform (DTFT) to the impulse response sequence, which characterizes an LTI discrete-time system uniquely in the frequency-domain. The DTFT of the output sequence of an LTI discrete-time system is simply the product of its frequency response and the DTFT of the input sequence, which provides the input-output relation of the system in the frequency domain.

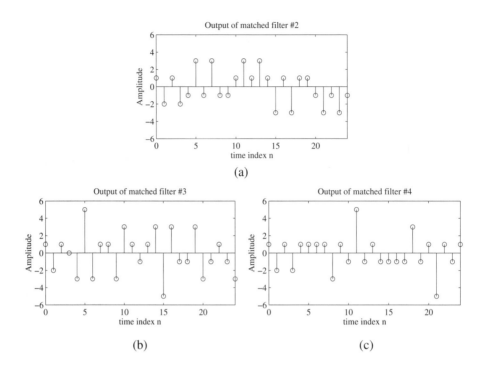

Figure 4.45: (a) The output of matched filter #2, (b) the output of matched filter #3, and (c) the output of matched filter #4.

A generalization of the frequency response concept is the transfer function defined by the z-transform of the impulse response sequence. An alternative input-output relation is thus given by the product of the transfer function and the z-transform of the input sequence that results in the z-transform of the output sequence. For a stable causal LTI discrete-time system, all poles of its transfer function must be strictly inside the unit circle. For a stable causal transfer function, the frequency response is given simply by its values on the unit circle in the z-plane.

The concept of filtering is introduced and several ideal filters are defined. Several simple approximations to the ideal filters are next introduced. In addition, various special types of transfer functions that are often encountered in practice are reviewed. The concept of complementary transfer functions relating a set of transfer functions is discussed, and several types of complementary conditions are introduced.

The inverse system design is encountered in estimating the unknown input of a discrete-time system from its known output. The determination of the transfer function of the inverse of a causal LTI discrete-time system with a rational transfer function is outlined. The recursive computation of the unknown causal input signal from the impulse response of a causal LTI system and its known output is outlined. Next, two methods are outlined for the system identification problem. In one approach, a recursive algorithm is described for determining the impulse response of a causal initially relaxed system from its known input and output sequences. In the second method, the frequency response of the system is determined from the cross-energy spectrum of the output and the input signal, and the energy spectrum of the input. Alternately, the square magnitude function of the system can be determined from the energy spectrum of the output and the input signals.

An important building block in the design of a single-input, single-output LTI discrete-time system is the digital two-pair, which is a two-input, two-output LTI discrete-time system. Characterizations of the digital two-pairs and their interconnections are discussed. A very simple algebraic procedure for testing the stability of a causal LTI transfer function is then introduced. Finally, the chapter concludes with a discussion on the statistical properties and transform-domain representation of the output signal of an LTI discrete-time system generated by a random input signal.

4.16 Problems

4.1 Show that the function $u[n] = z^n$, where z is a complex constant, is an eigenfunction of an LTI discrete-time system. Is $v[n] = z^n \mu[n]$ with z a complex constant also an eigenfunction of an LTI discrete-time system?

4.2 Determine a closed-form expression for the frequency response $H(e^{j\omega})$ of the LTI discrete-time system characterized by an impulse response

$$h[n] = \delta[n] - \alpha\delta[n - R], \tag{4.236}$$

where $|\alpha| < 1$. What are the maximum and the minimum values of its magnitude response? How many peaks and dips of the magnitude response occur in the range $0 \le \omega < 2\pi$? What are the locations of the peaks and the dips? Sketch the magnitude and the phase responses for $R = 5$.

4.3 Determine a closed-form expression for the frequency response $G(e^{j\omega})$ of the LTI discrete-time system characterized by an impulse response

$$g[n] = h[n] \circledast h[n] \circledast h[n], \tag{4.237}$$

where $h[n]$ is given by Eq. (4.236).

4.4 Determine a closed-form expression for the frequency response $G(e^{j\omega})$ of an LTI discrete-time system with an impulse response given by

$$g[n] = \begin{cases} \alpha^n, & 0 \le n \le M - 1, \\ 0 & \text{otherwise,} \end{cases}$$

where $|\alpha| < 1$. What is the relation of $G(e^{j\omega})$ to $H(e^{j\omega})$ of Eq. (4.14)? Scale the impulse response by multiplying it with a suitable constant so that the dc value of the magnitude response is unity.

4.5 Show that the group delay $\tau(\omega)$ of an LTI discrete-time system characterized by a frequency response $H(e^{j\omega})$ can be expressed as

$$\tau(\omega) = \text{Re}\left\{ \frac{j\frac{dH(e^{j\omega})}{d\omega}}{H(e^{j\omega})} \right\}. \tag{4.238}$$

4.6 A noncausal LTI FIR discrete-time system is characterized by an impulse response $h[n] = a_1\delta[n - 2] + a_2\delta[n - 1] + a_3\delta[n] + a_4\delta[n + 1] + a_5\delta[n + 2]$. For what values of the impulse response samples will its frequency response $H(e^{j\omega})$ have a zero phase?

4.7 A causal LTI FIR discrete-time system is characterized by an impulse response $h[n] = a_1\delta[n] + a_2\delta[n - 1] + a_3\delta[n - 2] + a_4\delta[n - 3] + a_5\delta[n - 4] + a_6\delta[n - 5] + a_7\delta[n - 6]$. For what values of the impulse response samples will its frequency response $H(e^{j\omega})$ have a linear phase?

4.8 An FIR LTI discrete-time system is described by the difference equation

$$y[n] = a_1x[n + k] + a_2x[n + k - 1] + a_3x[n + k - 2]$$
$$+ a_2x[n + k - 3] + a_1x[n + k - 4],$$

where $y[n]$ and $x[n]$ denote, respectively, the output and the input sequences. Determine the expression for its frequency response $H(e^{j\omega})$. For what values of the constant k will the system have a frequency response $H(e^{j\omega})$ that is a real function of ω?

4.9 Consider the cascade of two causal LTI systems: $h_1[n] = \alpha\delta[n] + \delta[n-1]$, and $h_2[n] = \beta^n \mu[n]$, $|\beta| < 1$. Determine the frequency response $H(e^{j\omega})$ of the overall system. For what values of α and β will $|H(e^{j\omega})| = 1$?

4.10 The input-output relation of a nonlinear discrete-time system in the frequency domain is given by

$$Y(e^{j\omega}) = |X(e^{j\omega})|^\alpha e^{j\arg X(e^{j\omega})} \qquad\qquad (4.239)$$

where $0 < \alpha \le 1$, and $X(e^{j\omega})$ and $Y(e^{j\omega})$ denote the DTFTs of the input and output sequences. Determine the expression for its frequency response $H(e^{j\omega}) = Y(e^{j\omega})/X(e^{j\omega})$ and show that it has zero phase. The nonlinear algorithm described by Eq. (4.239) is known as the *alpha-rooting method* and has been used in image enhancement [Jai89].

4.11 Determine the expression for the frequency response $H(e^{j\omega})$ of a causal IIR LTI discrete-time system characterized by the input-output relation

$$y[n] = x[n] + \alpha y[n-R], \qquad |\alpha| < 1,$$

where $y[n]$ and $x[n]$ denote, respectively, the output and the input sequences. Determine the maximum and the minimum values of its magnitude response. How many peaks and dips of the magnitude response occur in the range $0 \le \omega < 2\pi$? What are the locations of the peaks and the dips? Sketch the magnitude and the phase responses for $R = 5$.

4.12 An IIR LTI discrete-time system is described by the difference equation

$$y[n] + a_1 y[n-1] + a_2 y[n-2] = b_0 x[n] + b_1 x[n-1] + b_2 x[n-2],$$

where $y[n]$ and $x[n]$ denote, respectively, the output and the input sequences. Determine the expression for its frequency response. For what values of the constants b_i will the magnitude response be a constant for all values of ω?

4.13 Determine the input-output relation of a factor-of-2 up-sampler in the frequency domain.

4.14 The magnitude response of a digital filter with a real-coefficient transfer function $H(z)$ is shown in Figure P4.1. Plot the magnitude response of the filter $H(z^4)$.

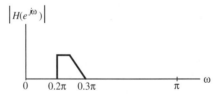

Figure P4.1

4.15 Consider an LTI discrete-time system with an impulse response $h[n] = (0.4)^n \mu[n]$. Determine the frequency response $H(e^{j\omega})$ of the system and evaluate its value at $\omega = \pm\pi/4$. What is the steady-state output $y[n]$ of the system for an input $x[n] = \sin(\pi n/4)\mu[n]$?

4.16 An FIR filter of length 3 is defined by a symmetric impulse response, i.e., $h[0] = h[2]$. Let the input to this filter be a sum of two cosine sequences of angular frequencies 0.2 rad/samples and 0.5 rad/samples, respectively. Determine the impulse response coefficients so that the filter passes only the high-frequency component of the input.

4.17 (a) Design a length-5 FIR bandpass filter with an antisymmetric impulse response $h[n]$, i.e., $h[n] = -h[4-n]$, $0 \le n \le 4$, satisfying the following magnitude response values: $|H(e^{j\pi/4})| = 0.5$ and $|H(e^{j\pi/2})| = 1$.

(b) Determine the exact expression for the frequency response of the filter designed, and plot its magnitude and phase responses.

4.18 (a) Design a length-4 FIR bandpass filter with an antisymmetric impulse response $h[n]$, i.e., $h[n] = -h[4-n]$, $0 \le n \le 4$, satisfying the following magnitude response values: $|H(e^{j\pi/4})| = 1$ and $|H(e^{j\pi/2})| = 0.5$.

(b) Determine the exact expression for the frequency response of the filter designed, and plot its magnitude and phase responses.

4.19 An FIR filter of length 5 is defined by a symmetric impulse response, i.e., $h[n] = h[4 - n]$, $0 \le n \le 4$. Let the input to this filter be a sum of three cosine sequences of angular frequencies: 0.2 rad/samples, 0.5 rad/samples, and 0.8 rad/samples, respectively. Determine the impulse response coefficients so that the filter passes only the midfrequency component of the input.

4.20 The frequency response $H(e^{j\omega})$ of a length-4 FIR filter with real impulse response has the following specific values: $H(e^{j0}) = 2$, $H(e^{j\pi/2}) = 7 - j3$, and $H(e^{j\pi}) = 0$. Determine $H(z)$.

4.21 The frequency response $H(e^{j\omega})$ of a length-4 FIR filter with a real and antisymmetric impulse response has the following specific values: $H(e^{j\pi}) = 8$, and $H(e^{j\pi/2}) = -2 + j2$. Determine $H(z)$.

4.22 Consider the two LTI causal digital filters with impulse responses given by

$$h_A[n] = 0.3\delta[n] - \delta[n-1] + 0.3\delta[n-2],$$
$$h_B[n] = 0.3\delta[n] + \delta[n-1] + 0.3\delta[n-2].$$

(a) Sketch the magnitude responses of the two filters and compare their characteristics.

(b) Let $h_A[n]$ be the impulse response of a causal digital filter with a frequency response $H_A(e^{j\omega})$. Define another digital filter whose impulse response $h_C[n]$ is given by

$$h_C[n] = (-1)^n h_A[n], \qquad \text{for all } n.$$

What is the relation between the frequency response $H_C(e^{j\omega})$ of this new filter and the frequency response $H_A(e^{j\omega})$ of the parent filter?

4.23 As indicated in Example 2.36, the trapezoidal integration formula can be represented as an IIR digital filter represented by a difference equation given by

$$y[n] = y[n-1] + \tfrac{1}{2}\{x[n] + x[n-1]\},$$

with $y[-1] = 0$. Determine the frequency response of the above filter.

4.24 A recursive difference equation representation of the Simpson's numerical integration formula is given by [Ham89]

$$y[n] = y[n-2] + \tfrac{1}{3}\{x[n] + 4x[n-1] + x[n-2]\}.$$

Evaluate the frequency response of the above filter and compare it with that of the trapezoidal method of Problem 4.23.

4.25 This problem develops expressions for the even and odd parts of a real-coefficient transfer function in terms of the mirror-image and antimirror-image parts of its numerator and denominator polynomials [Ram89].

(a) Show that any real-coefficient polynomial $B(z)$ in z^{-1} of degree N can be expressed as: $B(z) = B_{mi}(z) + B_{ai}(z)$, where $B_{mi}(z) = \tfrac{1}{2}\{B(z) + z^{-N}B(z^{-1})\}$ and $B_{ai}(z) = \tfrac{1}{2}\{B(z) - z^{-N}B(z^{-1})\}$ are, respectively, mirror-image and antimirror-image polynomials.

(b) Let $H(z) = N(z)/D(z)$ be a real rational function of z^{-1}. Show that the even and odd parts of $H(z)$ can be expressed as

$$H_{ev}(z) = \frac{1}{2}\left\{H(z) + H(z^{-1})\right\} = \frac{D_{mi}(z)N_{mi}(z) - D_{ai}(z)N_{ai}(z)}{D_{mi}^2(z) - D_{ai}^2(z)},$$

$$H_{od}(z) = \frac{1}{2}\left\{H(z) - H(z^{-1})\right\} = \frac{D_{mi}(z)N_{ai}(z) - D_{ai}(z)N_{mi}(z)}{D_{mi}^2(z) - D_{ai}^2(z)}.$$

In the above equations, $D_{mi}(z)$ and $D_{ai}(z)$ are, respectively, the mirror-image and antimirror-image parts of $D(z)$. Likewise, $N_{mi}(z)$ and $N_{ai}(z)$ are, respectively, the mirror-image and antimirror-image parts of $N(z)$.

(c) Show that $H_{re}(e^{j\omega}) = H_{ev}(z)\big|_{z=e^{j\omega}}$, and $H_{im}(e^{j\omega}) = -jH_{od}(z)\big|_{z=e^{j\omega}}$.

(d) Using the above approach, determine the real and imaginary parts of

$$H(z) = \frac{3z^2 + 4z + 6}{3z^3 + 2z^2 + z + 1}.$$

4.26 In this problem we consider the determination of a real rational, causal, stable discrete-time transfer function

$$H(z) = \frac{P(z)}{D(z)} = \frac{\sum_{i=0}^{N} p_i z^{-i}}{\sum_{i=0}^{N} d_i z^{-i}}$$

from the specified real part of its frequency response [Dut83]:

$$H_{re}(e^{j\omega}) = \frac{\sum_{i=0}^{N} a_i \cos(i\omega)}{\sum_{i=0}^{N} b_i \cos(i\omega)} = \frac{A(e^{j\omega})}{B(e^{j\omega})}. \tag{4.240}$$

(a) Show that

$$H_{re}(e^{j\omega}) = \frac{1}{2}\left[H(z) + H(z^{-1})\right]\Big|_{z=e^{j\omega}}$$

$$= \frac{1}{2}\left[\frac{P(z)D(z^{-1}) + P(z^{-1})D(z)}{D(z)D(z^{-1})}\right]\Big|_{z=e^{j\omega}}. \tag{4.241}$$

(b) Comparing Eqs. (4.240) and (4.241) we get

$$B(e^{j\omega}) = D(z)D(z^{-1})\big|_{z=e^{j\omega}}$$

$$A(e^{j\omega}) = \frac{1}{2}\left[P(z)D(z^{-1}) + P(z^{-1})D(z)\right]\big|_{z=e^{j\omega}}. \tag{4.242}$$

The spectral factor $D(z)$ can be determined, except for the scale factor K, from the roots of $B(z) = B(e^{j\omega})\big|_{z=e^{j\omega}}$ inside the unit circle. Show that

$$K = \sqrt{B(1)}/\prod_{i=1}^{N}(1 - z_i).$$

(c) To determine $P(z)$, Eq. (4.242) can be rewritten through analytic continuation as

$$A(z) = \frac{1}{2}\left[P(z)D(z^{-1}) + P(z^{-1})D(z)\right].$$

Substituting the polynomial forms of $P(z)$ and $D(z)$, and equating coefficients of $(z^i + z^{-i})/2$ on both sides of the above equation we arrive at a set of $N + 1$ equations which can be solved for the numerator coefficients $\{p_i\}$. Using the above approach, determine $H(z)$ for which

$$H_{re}(e^{j\omega}) = \frac{1 + \cos\omega + \cos 2\omega}{17 - 8\cos 2\omega}.$$

4.27 Let $H(z)$ be the transfer function of a causal stable LTI discrete-time system. Let $G(z)$ be the transfer function obtained by replacing z^{-1} in $H(z)$ with $\frac{\alpha+z^{-1}}{1+\alpha z^{-1}}$, i.e., $G(z) = H(z)|_{z^{-1}=\frac{\alpha+z^{-1}}{1+\alpha z^{-1}}}$. Show that $G(1) = H(1)$ and $G(-1) = H(-1)$.

4.28 Consider the digital filter structure of Figure P4.2, where $H_1(z)$, $H_2(z)$, and $H_3(z)$ are FIR digital filters with transfer functions given by

$$H_1(z) = \tfrac{2}{3} + \tfrac{2}{5}z^{-1} + \tfrac{4}{7}z^{-2}, \quad H_2(z) = \tfrac{4}{3} + \tfrac{8}{5}z^{-1} + \tfrac{3}{7}z^{-2}, \quad H_3(z) = 3 + 2z^{-1} + 4z^{-2}.$$

Determine the transfer function $H(z)$ of the composite filter.

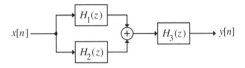

Figure P4.2

4.29 Determine the transfer function of a causal LTI discrete-time system described by the following difference equation:

$$y[n] = 5x[n] - 5x[n-1] + 0.4x[n-2] + 0.32x[n-3]$$
$$- 0.5y[n-1] + 0.34y[n-2] + 0.08y[n-3].$$

Express the transfer function in a factored form and sketch its pole-zero plot. Is the system BIBO stable?

4.30 Determine the transfer function of a causal LTI discrete-time system described by the following difference equation:

$$y[n] = 5x[n] - 10.5x[n-1] + 11.7x[n-2] + 0.3x[n-3] - 4.4x[n-4]$$
$$- 0.9y[n-1] + 0.76y[n-2] + 0.016y[n-3] + 0.096y[n-4].$$

Express the transfer function in a factored form and sketch its pole-zero plot. Is the system BIBO stable?

4.31 Determine the expression for the impulse response $\{h[n]\}$ of the following causal IIR transfer function:

$$H(z) = \frac{-0.1z^{-1} + 2.19z^{-2}}{(1 - 0.8z^{-1} + 0.41z^{-2})(1 + 0.3z^{-1})}.$$

4.32 The transfer function of a causal LTI discrete-time system is given by

$$H(z) = \frac{6 - z^{-1}}{1 + 0.5z^{-1}} + \frac{2}{1 - 0.4z^{-1}}.$$

(a) Determine the impulse response $h[n]$ of the above system.

(b) Determine the output $y[n]$ of the above system for all values of n for an input

$$x[n] = 1.2(-0.2)^n \mu[n] - 0.2(0.3)^n \mu[n].$$

4.33 Using z-transform methods, determine the explicit expression for the output $y[n]$ of each of the following causal LTI discrete-time systems with impulse responses and inputs as indicated:

(a) $h[n] = (0.8)^n \mu[n]$, $x[n] = (0.5)^n \mu[n]$,

(b) $h[n] = (0.5)^n \mu[n]$, $x[n] = (0.5)^n \mu[n]$.

4.34 Using z-transform methods, determine the explicit expression for the impulse response $h[n]$ of a causal LTI discrete-time system which develops an output $y[n] = 4(0.75)^n \mu[n]$ for an input $x[n] = 3(0.25)^n \mu[n]$.

4.35 A causal LTI discrete-time system is described by the difference equation

$$y[n] - 0.8y[n-1] + 0.15y[n-2] = x[n],$$

where $x[n]$ and $y[n]$ are, respectively, the input and the output sequences of the system.

(a) Determine the transfer function $H(z)$ of the system.

(b) Determine the impulse response $h[n]$ of the system.

(c) Determine the step response $s[n]$ of the system.

4.36 Consider the discrete-time system of Figure P4.3. For $H_0(z) = 1 + \alpha z^{-1}$, find a suitable $F_0(z)$ so that the output $y[n]$ is a delayed and scaled replica of the input.

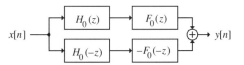

Figure P4.3

4.37 Let the transfer functions of Figure P4.3 be expressed in a polyphase form as

$$H_0(z) = E_0(z^2) + z^{-1}E_1(z^2), \quad F_0(z) = R_0(z^2) + z^{-1}R_1(z^2).$$

(a) Determine the condition so that the output $y[n]$ is a scaled and delayed replica of the input $x[n]$.

(b) Show that the transfer functions of Problem 4.36 satisfy this condition.

(c) Find at least another set of nontrivial realizable transfer functions $H_0(z)$ and $F_0(z)$ that satisfy this condition.

4.38 We have shown that a real-coefficient FIR transfer function $H(z)$ with a symmetric impulse response has a linear-phase response. As a result, the all-pole IIR transfer function $G(z) = 1/H(z)$ will also have a linear-phase response. What are the practical difficulties in implementing $G(z)$? Justify your answer mathematically.

4.39 Consider the following causal IIR transfer function:

$$H(z) = \frac{3z^3 + 2z^2 + 5}{(0.5z + 1)(z^2 + z + 0.6)}.$$

Is $H(z)$ a stable transfer function? If it is not stable, find a stable transfer function $G(z)$ such that $\left|G(e^{j\omega})\right| = \left|H(e^{j\omega})\right|$. Is there any other transfer function having the same magnitude response as $H(z)$?

4.40 Check the stability of the following causal IIR transfer function:

$$H(z) = \frac{(z^2 + 2z - 3)(z^2 - 3z + 5)}{(z^2 + 3.7z + 1.8)(z^2 - 0.4z + 0.35)}.$$

If it is unstable, find a stable transfer function $G(z)$ such that $\left|G(e^{j\omega})\right| = \left|H(e^{j\omega})\right|$. How many other transfer functions have the same magnitude response as $H(z)$?

4.41 The *notch filter* is used to suppress a particular sinusoidal component of frequency ω_o of an input signal $x[n]$ and has a transfer function with zeros at $z = e^{\pm j\omega_o}$. For each filter given below, (i) determine the notch frequency ω_o, (ii) show the form of the corresponding sinusoidal sequence to be suppressed, and (iii) verify by computing the output $y[n]$ by convolution that in the steady state, $y[n] = 0$ when the sinusoidal sequence is applied at the input of the filter.

(a) $H_1(z) = 1 + z^{-2}$, (b) $H_2(z) = \left(1 - \frac{\sqrt{3}}{2}z^{-1}\right) + z^{-2}$, (c) $H_3(z) = \left(1 + \sqrt{2}z^{-1}\right) + z^{-2}$.

4.42 Let $G_L(z)$ and $G_H(z)$ represent ideal lowpass and highpass filters with magnitude responses as sketched in Figure P4.4(a). Determine the transfer functions $H_k(z) = Y_k(z)/X(z)$ of the discrete-time system of Figure P4.4(b), $k = 0, 1, 2, 3$, and sketch their magnitude responses.

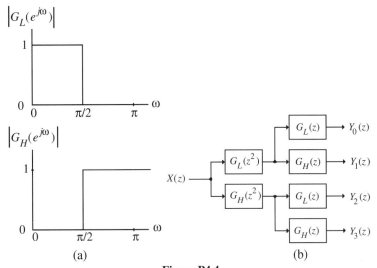

Figure P4.4

4.43 Let $H_{LP}(z)$ denote the transfer function of an ideal real coefficient lowpass filter with a cutoff frequency of ω_p. Sketch the magnitude response of $H_{LP}(-z)$ and show that it is a highpass filter. Determine the relation between the cutoff frequency of this highpass filter in terms of ω_p and its impulse response in terms of the impulse response $h_{LP}[n]$ of the parent lowpass filter.

4.44 Let $H_{LP}(z)$ denote the transfer function of an ideal real coefficient lowpass filter having a cutoff frequency of ω_p with $\omega_p < \pi/2$. Consider the complex coefficient transfer function $H_{LP}(e^{j\omega_o}z)$, where $\omega_p < \omega_o < \pi - \omega_p$. Sketch its magnitude response for $-\pi \le \omega \le \pi$. What type of filter does it represent? Now consider the transfer function $G(z) = H_{LP}(e^{j\omega_o}z) + H_{LP}(e^{-j\omega_o}z)$. Sketch its magnitude response for $-\pi \le \omega \le \pi$. Show that $G(z)$ is a real-coefficient bandpass filter with a passband centered at ω_o. Determine the width of its passband in terms of ω_p and its impulse response $g[n]$ in terms of the impulse response $h_{LP}[n]$ of the parent lowpass filter.

4.45 Let $H_{LP}(z)$ denote the transfer function of an ideal real coefficient lowpass filter with a cutoff frequency of ω_p with $0 < \omega_p < \pi/3$. Show that the transfer function $F(z) = H_{LP}(e^{j\omega_o}z) + H_{LP}(e^{-j\omega_o}z) + H_{LP}(z)$ where $\omega_o = \pi - \omega_p$ is a real-coefficient bandstop filter with the stopband centered at $\omega_o/2$. Determine the width of its stopband in terms of ω_p and its impulse response $f[n]$ in terms of the impulse response $h_{LP}[n]$ of the parent lowpass filter.

4.46 Show that the structure shown in Figure P4.5 implements the highpass filter of Problem 4.43.

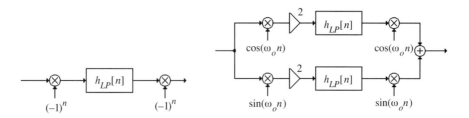

Figure P4.5   **Figure P4.6**

4.47 Show that the structure shown in Figure P4.6 implements the bandpass filter of Problem 4.44.

4.48 Let $H(z)$ be an ideal real-coefficient lowpass filter with a cutoff at ω_c, where $\omega_c = \pi/M$. Figure P4.7 shows a single-input, M-output filter structure, called an M-band analysis filter bank, where $H_k(z) = H(ze^{-j2\pi k/M})$, $k = 0, 1, \ldots, M - 1$. Sketch the magnitude response of each filter and describe the operation of the filter bank.

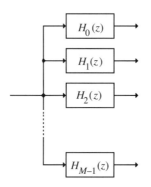

Figure P4.7

4.49 Consider a cascade of M sections of the first-order FIR lowpass filter of Eq. (4.103). Show that its 3-dB cutoff frequency is given by Eq. (4.105).

4.50 Consider a cascade of M sections of the first-order FIR highpass filter of Eq. (4.106). Develop the expression for its 3-dB cutoff frequency.

4.51 Verify that the value of α given by Eq. (4.111b) ensures that the transfer function $H_{LP}(z)$ of Eq. (4.109) is stable.

4.52 Show by trigonometric manipulation that Eq. (4.111a) can be alternately expressed as

$$\tan\left(\frac{\omega_c}{2}\right) = \frac{1 - \alpha}{1 + \alpha}. \tag{4.243}$$

Next show that the transfer function $H_{LP}(z)$ of Eq. (4.109) is stable for a value of α given by

$$\alpha = \frac{1 - \tan(\omega_c/2)}{1 + \tan(\omega_c/2)}. \tag{4.244}$$

4.53 Design a first-order lowpass IIR digital filter for each of the following normalized 3-dB cutoff frequencies: (a) 0.4 rad/samples, (b) 0.3π.

4.54 Show that the 3-dB cutoff frequency ω_c of the first-order highpass IIR digital filter of Eq. (4.112) is given by Eq. (4.111a).

4.55 Design a first-order highpass IIR digital filter for each of the following normalized 3-dB cutoff frequencies: (a) 0.25 rad/samples, (b) 0.4π.

4.56 The following first-order IIR transfer function has been proposed for clutter removal in MTI radars [Urk58]:

$$H(z) = \frac{1 - z^{-1}}{1 - kz^{-1}}.$$

Determine the magnitude response of the above transfer function and show that it has a highpass response. Scale the transfer function so that it has a 0-dB gain at $\omega = \pi$. Sketch the magnitude responses for $k = 0.95, 0.9$, and -0.5, respectively.

4.57 Show that the center frequency ω_o and the 3-dB bandwidth B_w of the second-order IIR bandpass filter of Eq. (4.113) are given by Eqs. (4.115) and (4.116), respectively.

4.58 Design a second-order bandpass IIR digital filter for each of the following specifications: (a) $\omega_o = 0.45\pi$, $B_w = 0.2\pi$, (b) $\omega_o = 0.6\pi$, $B_w = 0.15\pi$.

4.59 Show that the notch frequency ω_o and the 3-dB notch bandwidth B_ω of the second-order IIR bandstop filter of Eq. (4.118) are given by Eqs. (4.115) and (4.116), respectively.

4.60 Design a second-order bandstop IIR digital filter with for each of the following specifications: (a) $\omega_o = 0.4\pi$, $B_w = 0.15\pi$, (b) $\omega_o = 0.55\pi$, $B_w = 0.25\pi$.

4.61 Consider a cascade of K identical first-order lowpass digital filters with a transfer function given by Eq. (4.109). Show that the coefficient α of the first-order section is related to the 3-dB cutoff frequency ω_c of the cascade according to Eq. (4.122) with the parameter C given by Eq. (4.123).

4.62 Consider a cascade of K identical first-order highpass digital filters with a transfer function given by Eq. (4.112). Express the coefficient α of the first-order section in terms of the 3-dB cutoff frequency ω_c of the cascade.

4.63 If $H(z)$ is a lowpass filter, show that $H(-z)$ is a highpass filter. If ω_p and ω_s represent the passband and stopband edge frequencies of $H(z)$, determine the locations of the passband and stopband edge frequencies of $H(-z)$. Determine the relations between the impulse response coefficients of the two filters.

4.64 Using the method of Problem 4.63, develop the transfer function $G_{HP}(z)$ of a first-order IIR highpass filter from the transfer function $H_{LP}(z)$ of the first-order IIR lowpass filter given by Eq. (4.109). Is it the same as that of the highpass transfer function of Eq. (4.112)? If not, determine the location of its 3-dB cutoff frequency as a function of the parameter α.

4.65 Let $H(z)$ be an ideal lowpass filter with a cutoff frequency at $\pi/2$. Sketch the magnitude responses of the following systems: (a) $H(z^2)$, (b) $H(z)H(z^2)$, (c) $H(-z)H(z^2)$, and (d) $H(z)H(-z^2)$.

4.66 Let $H(z)$ be a lowpass filter with unity passband magnitude, a passband edge at ω_p, and a stopband edge at ω_s, as shown in Figure P4.8.

(a) Sketch the magnitude response of the digital filter $G_1(z) = H(z^M)F_1(z)$, where $F_1(z)$ is a lowpass filter with unity passband magnitude, a passband edge at ω_p/M, and a stopband edge at $(2\pi - \omega_s)/M$. What are the bandedges of $G_1(z)$?

(b) Sketch the magnitude response of the digital filter $G_2(z) = H(z^M)F_2(z)$ where $F_2(z)$ is a bandpass filter with unity passband magnitude, and with passband edges at $(2\pi - \omega_p)/M$ and $(2\pi + \omega_p)/M$, and stopband edges at $(2\pi - \omega_s)/M$ and $(2\pi + \omega_s)/M$, respectively. What are the bandedges of $G_2(z)$?

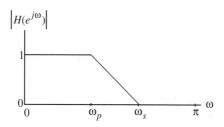

Figure P4.8

4.67 Let a causal LTI discrete-time system be characterized by a real impulse response $h[n]$ with a DTFT $H(e^{j\omega})$. Consider the system of Figure P4.9, where $x[n]$ is a finite-length sequence. Determine the frequency response of the overall system $G(e^{j\omega})$ in terms of $H(e^{j\omega})$, and show that it has a zero-phase response.

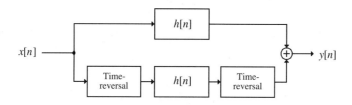

Figure P4.9

4.68 Show that the amplitude response $\breve{H}(\omega)$ of Type 1 and Type 3 linear-phase FIR transfer functions is a periodic function of ω with a period 2π, and the amplitude response $\breve{H}(\omega)$ of Type 2 and Type 4 linear-phase FIR transfer functions is a periodic function of ω with a period 4π.

4.69 A length-9 Type 1 real-coefficient FIR filter has the following zeros: $z_1 = -0.5$, $z_2 = 0.3 + j0.5$, $z_3 = -\frac{1}{2} + j\frac{\sqrt{3}}{2}$. (a) Determine the locations of the remaining zeros. (b) What is the transfer function $H_1(z)$ of the filter?

4.70 A length-10 Type 2 real-coefficient FIR filter has the following zeros: $z_1 = 3$, $z_2 = j0.8$, $z_3 = j$. (a) Determine the locations of the remaining zeros. (b) What is the transfer function $H_2(z)$ of the filter?

4.71 A length-13 Type 3 real-coefficient FIR filter has the following zeros: $z_1 = -0.3 + j0.5$, $z_2 = j0.8$, $z_3 = -0.3$. (a) Determine the locations of the remaining zeros. (b) What is the transfer function $H_3(z)$ of the filter?

4.72 A length-10 Type 4 real-coefficient FIR filter has the following zeros: $z_1 = -1.2 + j1.4$, $z_2 = \frac{1}{2} + j\frac{\sqrt{3}}{2}$, $z_3 = \frac{1}{4} + j\frac{\sqrt{15}}{4}$. (a) Determine the locations of the remaining zeros. (b) What is the transfer function $H_4(z)$ of the filter?

4.73 Show analytically that an FIR filter with a constant group delay must have either a symmetric or an antisymmetric impulse response.

4.74 Consider the following five FIR transfer functions:

$$\text{(i) } H_1(z) = -0.5 + 0.45z^{-1} + 0.58z^{-2} + 1.02z^{-3} + 0.1z^{-4} - 0.03z^{-5} - 0.18z^{-6},$$
$$\text{(ii) } H_2(z) = -0.3 + 0.11z^{-1} + 0.3z^{-2} + 1.22z^{-3} + 0.3z^{-4} + 0.11z^{-5} - 0.3z^{-6},$$
$$\text{(iii) } H_3(z) = 1 + 0.6z^{-1} + 0.49z^{-2} - 0.48z^{-3} - 0.14z^{-4} - 0.12z^{-5} + 0.09z^{-6},$$
$$\text{(iv) } H_4(z) = 0.25 - 0.6z^{-1} - 0.14z^{-2} + 0.97z^{-4} + 0.06z^{-5} + 0.36z^{-6},$$
$$\text{(v) } H_5(z) = 0.09 - 0.12z^{-1} - 0.14z^{-2} - 0.48z^{-3} + 0.49z^{-4} + 0.06z^{-5} + z^{-6}.$$

Using the M-file `zplane` determine the zero locations of each and then answer the following questions:

(a) Does any one of the FIR filters have a linear-phase response? If so, which one?

(b) Does any one of the FIR filters have a minimum-phase response? If so, which one?

(c) Does any one of the FIR filters have a maximum-phase response? If so, which one?

4.75 A third-order FIR filter has a transfer function $G_1(z)$ given by

$$G_1(z) = (6 - z^{-1} - 12z^{-2})(2 + 5z^{-1}).$$

(a) Determine the transfer functions of all other FIR filters whose magnitude responses are identical to that of $G_1(z)$.

(b) Which one of these filters has a minimum-phase transfer function and which one has a maximum-phase transfer function?

(c) If $g_k[n]$ denotes the impulse response of the kth FIR filter determined in part (a), compute the partial energy of the impulse response given by

$$\mathcal{E}_k[n] = \sum_{m=0}^{n} g_k[m]^2, \quad 0 \leq n \leq 3,$$

for all values of k, and show that

$$\sum_{m=0}^{n} |g_k[m]|^2 \leq \sum_{m=0}^{n} |g_{\min}[m]|^2,$$

and

$$\sum_{m=0}^{\infty} |g_k[m]|^2 = \sum_{m=0}^{\infty} |g_{\min}[m]|^2,$$

for all values of k, and where $g_{\min}[n]$ is the impulse response of the minimum-phase FIR filter determined in part (a).

4.76 The z-transforms of five sequences of length 7 are given below:

$$H_1(z) = 0.00836529 - 0.001782726z^{-1} - 0.075506z^{-2} + 0.3413956532z^{-3}$$
$$+ 0.13529123z^{-4} - 1.50627293z^{-5} + 3.3207295z^{-6},$$
$$H_2(z) = 3.3207295 - 1.50627293z^{-1} + 0.13529123z^{-2} + 0.34139565z^{-3}$$
$$- 0.07550603z^{-4} - 0.001782726z^{-5} + 0.00836592z^{-6},$$
$$H_3(z) = 0.0269311 - 0.0840756z^{-1} + 0.02580603z^{-2} + 0.940498549z^{-3}$$
$$- 2.2508765z^{-4} + 2.53245711z^{-5} + 1.03147943z^{-6},$$
$$H_4(z) = 1.03147943 + 2.5324571z^{-1} - 2.2508765z^{-2} + 0.9404985z^{-3}$$

$$+0.02580602z^{-4} - 0.0840756z^{-5} + 0.02693111z^{-6},$$
$$H_5(z) = 0.16667 - 0.05556z^{-1} - 0.75z^{-2} + 3.5z^{-3} - 0.75z^{-4}$$
$$- 0.05556z^{-5} + 0.16667z^{-6}.$$

The magnitude of the DFT for each of the above sequences is the same. Which one of the above z-transforms has all its zeros outside the unit circle? Which one has all its zeros inside the unit circle? How many other real sequences of length 7 exist that have the same DFT magnitude as those given above?

4.77 Let the first four impulse response samples of a causal linear-phase FIR transfer function be given by $h[0] = a$, $h[1] = b$, $h[2] = c$, and $h[3] = d$. Determine the remaining impulse response samples of $H(z)$ of lowest order for each type of linear-phase filter.

4.78 The first six samples of the impulse response of an FIR filter $H(z)$ are given by $h[0] = -1$, $h[1] = 2$, $h[2] = -3$, $h[3] = -4$, $h[4] = 5$, and $h[5] = 6$. Determine the remaining impulse response samples of $H(z)$ of lowest order for each type of linear-phase filter. Using zplane determine the zero locations for $H(z)$ for each type of linear-phase filter. Does $H(z)$ have a zero at $z = 1$ and/or $z = -1$? Do the zeros on the unit circle appear in complex conjugate pairs? Do the zeros not on the unit circle appear in mirror-image symmetry? Justify your answers.

4.79 Let $H_1(z)$, $H_2(z)$, $H_3(z)$, and $H_4(z)$ be, respectively, Type 1, Type 2, Type 3, and Type 4 linear-phase FIR filters. Are the following filters composed of a cascade of the above filters' linear phase? If they are, what are their types?

(a) $G_a(z) = H_1(z)H_1(z)$, (b) $G_b(z) = H_1(z)H_2(z)$, (c) $G_c(z) = H_1(z)H_3(z)$,
(d) $G_d(z) = H_1(z)H_4(z)$, (e) $G_e(z) = H_2(z)H_2(z)$, (f) $G_f(z) = H_3(z)H_3(z)$,
(g) $G_g(z) = H_4(z)H_4(z)$, (h) $G_h(z) = H_2(z)H_3(z)$, (i) $G_i(z) = H_3(z)H_4(z)$.

4.80 Let $h[n], 0 \le n \le N$, denote the impulse response of a Type 1 FIR filter of length $N+1$. The frequency response is of the form $H(e^{j\omega}) = \breve{H}(\omega)e^{-j\omega N/2}$, where the amplitude response $\breve{H}(\omega)$ is as indicated in Figure P4.10(a).

(a) Construct another FIR filter of length $N + 1$ with an impulse response defined by

$$g[n] = \begin{cases} h[n], & \text{for all } n \text{ except } n = m, \\ \alpha, & n = m, \end{cases} \qquad 0 \le n \le N.$$

Determine a suitable m and α such that the frequency response of the new filter is of the form $G(e^{j\omega}) = \breve{G}(\omega)e^{-j\omega N/2}$, with its amplitude response $\breve{G}(\omega)$ as indicated in Figure P4.10(b) [Her70].

(b) Show that $\breve{H}(\omega)$ can be expressed as the square magnitude response of a minimum-phase FIR filter $F(z)$ of order $N/2$ and develop a method to construct $F(z)$.

(c) Can $\breve{G}(\omega)$ also be expressed as the square magnitude response of a minimum-phase FIR filter? If not, why not?

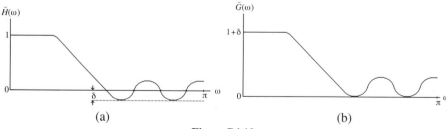

(a)   (b)

Figure P4.10

4.81 The *time constant K* of an LTI stable causal discrete-time system with an impulse response $h[n]$ is given by the value of the total time interval n at which the partial energy of the impulse response is within 95% of the total energy, i.e.,

$$\sum_{n=0}^{K} |h[n]|^2 = 0.95 \sum_{n=0}^{\infty} |h[n]|^2.$$

Determine the time constant K of the first-order causal transfer function $H(z) = 1/(1 + \alpha z^{-1})$, $|\alpha| < 1$.

4.82 Let $F_1(z)$ denote one of the factors of a linear-phase FIR transfer function $H(z)$. Determine at least one other factor $F_2(z)$ of $H(z)$ for the following choices of $F_1(z)$:

(a) $F_1(z) = 1 + 2z^{-1} + 3z^{-2}$, (b) $F_1(z) = 3 + 5z^{-1} - 4z^{-2} - 2z^{-3}$.

4.83 Consider the first-order causal and stable allpass transfer function given by

$$A_1(z) = \frac{1 - d_1^* z}{z - d_1}.$$

Determine the expression for $\left(1 - |A_1(z)|^2\right)$ and then show that

$$\left(1 - |A_1(z)|^2\right) \begin{cases} < 0, & \text{for } |z|^2 < 1, \\ = 0, & \text{for } |z|^2 = 1, \\ > 0, & \text{for } |z|^2 > 1. \end{cases}$$

Now, using the above approach, show that Property 2 given by Eq. (4.132) holds for any arbitrary causal stable allpass transfer function.

4.84 Derive Property 3 of a stable allpass transfer function given by Eq. (4.133).

4.85   (a) Show that the phase delay $\tau_p(\omega) = -\theta(\omega)/\omega$ of the first-order allpass transfer function

$$A_1(z) = \frac{d_1 + z^{-1}}{1 + d_1 z^{-1}},$$

is given by $\tau_p(\omega) \cong (1 - d_1)/(1 + d_1) = \delta$ [Ste96].

(b) Design a first-order allpass filter with a phase delay of $\delta = 0.5$ samples and operating at a sampling rate of 20 kHz. Determine the error in samples at 1 kHz in the phase delay from its design value of 0.5 samples.

4.86 Consider the second-order allpass transfer function

$$A_2(z) = \frac{d_2 + d_1 z^{-1} + z^{-2}}{1 + d_1 z^{-1} + d_2 z^{-2}}.$$

If δ denotes the desired low-frequency approximate value of the phase delay $\tau_p(\omega) = -\theta(\omega)/\omega$, show that [Fet72]

$$d_1 = 2\left(\frac{2 - \delta}{1 + \delta}\right), \qquad d_2 = \frac{(2 - \delta)(1 - \delta)}{(2 + \delta)(1 + \delta)}.$$

4.87 Let $G(z)$ be a causal stable nonminimum-phase transfer function, and let $H(z)$ denote another causal stable transfer function that is minimum-phase with $\left|G(e^{j\omega})\right| = \left|H(e^{j\omega})\right|$. Show that $G(z) = H(z)A(z)$, where $A(z)$ is a stable causal allpass transfer function.

4.88 The transfer function of a typical transmission channel is given by

$$H(z) = \frac{(z + 1.4)(z^2 + 2z + 4)}{(z + 0.8)(z - 0.6)}.$$

In order to correct for the magnitude distortion introduced by the channel on a signal passing through it, we wish to connect a stable digital filter characterized by a transfer function $G(z)$ at the receiving end. Determine $G(z)$.

4.89 Let $H(z)$ be a causal stable minimum-phase transfer function, and let $G(z)$ denote another causal stable transfer function which is nonminimum-phase with $\left|G(e^{j\omega})\right| = \left|H(e^{j\omega})\right|$. If $h[n]$ and $g[n]$ denote their respective impulse responses, show that

(a) $|g[0]| \leq |h[0]|$,

(b) $\sum_{\ell=0}^{n} |g[\ell]|^2 \leq \sum_{\ell=0}^{n} |h[\ell]|^2$.

4.90 Is the transfer function

$$H(z) = \frac{(z + 3)(z - 2)}{(z - 0.25)(z + 0.5)}$$

minimum-phase? If it is not minimum-phase, then construct a minimum-phase transfer function $G(z)$ such that $\left|G(e^{j\omega})\right| = \left|H(e^{j\omega})\right|$. Determine their corresponding unit sample responses, $g[n]$ and $h[n]$, for $n = 0, 1, 2, 3, 4$. For what values of m is $\sum_{n=0}^{m} |g[n]|^2$ bigger than $\sum_{n=0}^{m} |h[n]|^2$?

4.91 The following bandstop FIR transfer functions $H_{BS}(z)$ have also been proposed for the recovery of vertical details in the structure of Figure 4.33 employed for the separation of the luminance and the chrominance components [Aca83], [Pri80], [Ros75]:

(a) $H_{BS}(z) = \frac{1}{4}(1 + z^{-2})^2$,

(b) $H_{BS}(z) = \frac{1}{16}(1 + z^{-2})^2(-1 + 6z^{-2} - z^{-4})$,

(c) $H_{BS}(z) = \frac{1}{32}(1 + z^{-2})^2(-3 + 14z^{-2} - 3z^{-4})$.

Develop their delay-complementary transfer functions $H_{BP}(z)$.

4.92 Let $A_0(z)$ and $A_1(z)$ be two causal stable allpass transfer functions. Define two causal stable IIR transfer functions as follows:

$$H_0(z) = A_0(z) + A_1(z), \qquad H_1(z) = A_0(z) - A_1(z).$$

Show that the numerators of $H_0(z)$ and $H_1(z)$ are, respectively, a symmetric and an antisymmetric polynomial.

4.93 Show that the two transfer functions of Eqs. (4.142a) and (4.142b) are a power-complementary pair.

4.94 Show that the two transfer functions of Eqs. (4.142a) and (4.142b) are each a BR function.

4.95 Consider the transfer function $H(z)$ given by

$$H(z) = \frac{1}{M} \sum_{k=0}^{M-1} A_k(z),$$

where $A_k(z)$ are stable real-coefficient allpass functions. Show that $H(z)$ is a BR function.

4.96 Show that the bandpass transfer function $H_{BP}(z)$ of Eq. (4.113) and the bandstop transfer function $H_{BS}(z)$ of Eq. (4.118) form a doubly-complementary pair.

4.97 Show that the value of the gain function $\mathcal{G}(\omega)$ of a power-symmetric transfer function defined by Eq. (4.146) at $\omega = \pi/2$ is given by $10\log_{10} K - 3$ dB.

4.98 Consider the real-coefficient stable IIR transfer function $H(z) = A_0(z^2) + z^{-1}A_1(z^2)$, where $A_0(z)$ and $A_1(z)$ are stable allpass transfer functions. Show that $H(z)$ is a power-symmetric transfer function.

4.99 Show that the following FIR transfer functions satisfy the power-symmetric condition:

(a) $H_a(z) = \frac{1}{2} - z^{-1} + \frac{21}{2}z^{-2} - \frac{27}{2}z^{-3} - 5z^{-4} - \frac{5}{2}z^{-5}$,

(b) $H_b(z) = 1 + 3z^{-1} + 14z^{-2} + 22z^{-3} - 12z^{-4} + 4z^{-5}$.

4.100 Let $H(z) = a(1 + bz^{-1})$, where a and b are constants. Then $H(z)H(z^{-1})$ is of the form $cz + d + cz^{-1}$. Determine the condition on c and d so that $H(z)$ is a power-symmetric FIR transfer function with $K = 1$. Show that $a = 1/2$ and $b = 1$ satisfy the power-symmetric condition. Determine two other possible sets of values for a and b to ensure the power-symmetric condition. Using MATLAB show that $H(z)$ and $G(z) = -z^{-1}H(-z^{-1})$ are power-complementary for the above values of the constants a and b.

4.101 Let $H(z) = a(1 + bz^{-1})(1 + d_1 z^{-1} + d_2 z^{-2})$, where a, b, d_1, and d_2 are constants. Then $H(z)H(z^{-1})$ is of the form $(cz + d + cz^{-1})[d_2 z^2 + d_1(1 + d_2)z + (1 + d_1^2 + d_2^2) + d_1(1 + d_2)z^{-1} + d_2 z^{-2}]$. Determine the condition on c and d in terms of d_1 and d_2 so that $H(z)$ is a power-symmetric FIR transfer function with $K = 1$. For $d_1 = d_2 = 1$, evaluate the constraint on c and d, and using it determine one realizable set of values for a and b. Using MATLAB show that $H(z)$ and $G(z) = -z^{-3}H(-z^{-1})$ are power-complementary for these values of the constants a and b.

4.102 Show that

$$H(z) = \frac{0.1 + 0.5z^{-1} + 0.45z^{-2} + 0.45z^{-3} + 0.5z^{-4} + 0.1z^{-5}}{1 + 0.9z^{-2} + 0.2z^{-4}}$$

is a power-symmetric IIR transfer function.

4.103 Show that the following FIR transfer functions are BR functions:

(a) $H_1(z) = \frac{1}{1+\alpha}\left(1 + \alpha z^{-1}\right), \quad \alpha > 0$,

(b) $H_2(z) = \frac{1}{1+\beta}\left(1 - \beta z^{-1}\right), \quad \beta > 0$,

(c) $H_3(z) = \frac{1}{(1+\alpha)(1+\beta)}\left(1 + \alpha z^{-1}\right)\left(1 - \beta z^{-1}\right), \quad \alpha > 0, \quad \beta > 0$,

(d) $H_4(z) = \frac{1}{3.36}(1 + 0.4z^{-1})(1 + 0.5z^{-1})(1 + 0.6z^{-1})$.

4.104 Show that the following IIR transfer functions are BR functions:

(a) $H_1(z) = \frac{2 + 2z^{-1}}{3 + z^{-1}}$,

(b) $H_2(z) = \frac{1 - z^{-1}}{4 + 2z^{-1}}$,

(c) $H_3(z) = \frac{1 - z^{-2}}{4 + 2z^{-1} + 2z^{-2}}$,

(d) $H_4(z) = \frac{3 + 6z^{-1} + 3z^{-2}}{6 + 5z^{-1} + z^{-2}}$,

(e) $H_5(z) = \frac{3 + 2z^{-1} + 3z^{-2}}{4 + 2z^{-1} + 2z^{-2}}$,

(f) $H_6(z) = \dfrac{3 + 9z^{-1} + 9z^{-2} + 3z^{-3}}{12 + 10z^{-1} + 2z^{-2}}.$

4.105 If $A_1(z)$ and $A_2(z)$ are two LBR functions, show that $A_1(1/A_2(z))$ is also an LBR function.

4.106 Let $G(z)$ be an LBR function of order N. Define

$$F(z) = z\left(\frac{G(z) + \alpha}{1 + \alpha G(z)}\right),$$

where $|\alpha| < 1$. Show that $F(z)$ is also LBR. What is the order of $F(z)$? Develop a realization of $G(z)$ in terms of $F(z)$.

4.107 If $G(z)$ is a BR function, show that $G(1/A(z))$ is a BR function, where $A(z)$ is an LBR function.

4.108 Show that each of the following pairs of transfer functions are doubly complementary:

(a) $H(z) = \dfrac{2 + 2z^{-1}}{3 + z^{-1}},\ G(z) = \dfrac{1 - z^{-1}}{3 + z^{-1}},$

(b) $H(z) = \dfrac{-1 + z^{-2}}{4 + 2z^{-1} + 2z^{-2}},\ G(z) = \dfrac{3 + 2z^{-1} + 3z^{-2}}{4 + 2z^{-1} + 2z^{-2}}.$

4.109 Determine the power-complementary transfer function of each of the following BR transfer functions:

(a) $H_a(z) = \dfrac{2(1 + z^{-1} + z^{-2})}{3 + 2z^{-1} + z^{-2}},$

(b) $H_b(z) = \dfrac{3(1.5 + 6.5z^{-1} + 6.5z^{-2} + 1.5z^{-3})}{18 + 21z^{-1} + 8z^{-2} + z^{-3}}.$

4.110 Verify Eq. (4.151).

4.111 Figures P4.11(a) and P4.11(b) show, respectively, the DPCM (*differential pulse-code modulation*) coder and decoder often employed for the compression of digital signals [Jay84]. The linear predictor $P(z)$ in the encoder develops a prediction $\hat{x}[n]$ of the input signal $x[n]$ and the difference signal $d[n] = x[n] - \hat{x}[n]$ is quantized by the quantizer $\mathcal{Q}$ developing the quantized output $u[n]$ which is represented with fewer bits than that of $x[n]$. The output of the encoder is transmitted over a channel to the decoder. In the absence of any errors due to transmission and quantization, the input $v[n]$ to the decoder is equal to $u[n]$ and the decoder generates the output $t[n]$ which is equal to the input $x[n]$. Determine the transfer function $H(z) = U(z)/X(z)$ of the encoder in the absence of any quantization and the transfer function $G(z) = Y(z)/V(z)$ of the decoder for the case of each of the following predictors, and show that $G(z)$ is the inverse of $H(z)$ in each case.
    (a) $P(z) = h_1 z^{-1}$, and (b) $P(z) = h_1 z^{-1} + h_2 z^{-2}$.

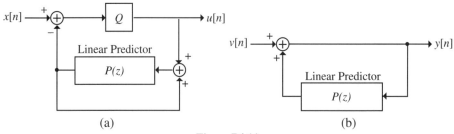

(a)   (b)

Figure P4.11

4.112 A causal stable LTI discrete-time system is characterized by an impulse response $h_1[n] = -\delta[n] + \frac{5}{3}(0.5)^n \mu[n] + \frac{1}{12}(-0.25)^n \mu[n]$. Determine the impulse response $h_2[n]$ of its inverse system which is causal and stable.

4.113 Verify the relations between the transfer parameters and the chain parameters of a two-pair given in Eqs. (4.176a) and (4.176b).

4.114 A two-pair is said to be *reciprocal* if $t_{12} = t_{21}$ [Mit73a]. Show that for a reciprocal two-pair, $AD - BC = 1$.

4.115 Consider the Γ-cascade of Figure P4.12(a), where the two two-pairs are described by the transfer matrices

$$\tau_1 = \begin{bmatrix} k_1 & (1-k_1^2)z^{-1} \\ 1 & -k_1 z^{-1} \end{bmatrix}, \qquad \tau_2 = \begin{bmatrix} k_2 & (1-k_2^2)z^{-1} \\ 1 & -k_2 z^{-1} \end{bmatrix}.$$

Determine the transfer matrix of the cascade.

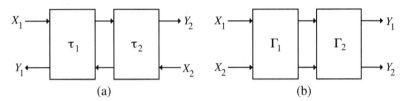

(a)   (b)

Figure P4.12

4.116 Consider the τ-cascade of Figure P4.12(b), where the two two-pairs are described by the chain matrices

$$\Gamma_1 = \begin{bmatrix} 1 & k_1 z^{-1} \\ k_1 & z^{-1} \end{bmatrix}, \qquad \Gamma_2 = \begin{bmatrix} 1 & k_2 z^{-1} \\ k_2 & z^{-1} \end{bmatrix}.$$

Determine the chain matrix of the cascade.

4.117 Determine the transfer parameters and the chain parameters of the digital two-pairs of Figure P4.13.

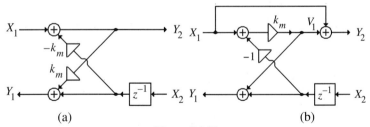

(a)   (b)

Figure P4.13

4.118 A transfer function $H(z)$ is realized in the form of Figure 4.40, where the constraining transfer function is given by $G(z)$. If the relation between $H(z)$ and $G(z)$ is of the form

$$H(z) = \frac{k_m + z^{-1}G(z)}{1 + k_m z^{-1}G(z)},$$

determine the transfer matrix and the chain matrix parameters of the two-pair of Figure 4.40.

4.119 Determine chain parameters of the cascade of three lattice two-pairs of Figure P4.14. Using these chain parameters determine the expression for the transfer function $A_3(z)$.

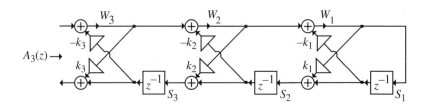

Figure P4.14

4.120 Derive the inequality of Eq. (4.187).

4.121 Determine by inspection which one of the following second-order polynomials has roots inside the unit circle:

(a) $D_a(z) = 1 + 0.92z^{-1} + 0.1995z^{-2}$,

(b) $D_b(z) = 2 + 0.4z^{-1} - 2.86z^{-2}$,

(c) $D_c(z) = 1 + 1.4562z^{-1} + 0.81z^{-2}$,

(d) $D_d(z) = 1 + 2.1843z^{-1} + 0.81z^{-2}$.

4.122 Test analytically the BIBO stability of the following causal IIR transfer functions:

(a) $H_a(z) = \dfrac{z^2 + 0.3z - 99.17}{z^3 - \frac{1}{2}z^2 - \frac{1}{4}z + \frac{1}{12}}$,

(b) $H_b(z) = \dfrac{z^2 + 11.63z + 0.001}{z^3 + \frac{13}{6}z^2 + \frac{1}{6}z - \frac{1}{3}}$,

(c) $H_c(z) = \dfrac{9.25z^3 + 13.7z^2 + 4.04z + 0.3}{18z^4 + 30z^3 + 18.5z^2 + 5z + 0.5}$,

(d) $H_d(z) = \dfrac{9 + z^{-1} - 3z^{-2} + 7z^{-4} + z^{-5}}{1 + \frac{5}{2}z^{-1} + \frac{5}{2}z^{-2} + \frac{5}{4}z^{-3} + \frac{5}{16}z^{-4} + \frac{1}{32}z^{-5}}$,

(e) $H_e(z) = \dfrac{1}{6 + 5z^{-1} + 4z^{-2} + 3z^{-3} + z^{-4} + z^{-5}}$.

4.123 Determine analytically whether all roots of the following polynomials are inside the unit circle:

(a) $D_a(z) = 5 + 4z^{-1} + 3z^{-2} + 2z^{-3} + z^{-4}$,

(b) $D_b(z) = z^3 + 0.2z^2 + 0.3z + 0.4$.

4.124 A polynomial $A(s)$ in the complex variable s is called a *strictly Hurwitz polynomial* if all zeros of $A(s)$ are in the left-half s-plane; i.e., if $s = s_k$ is a zero of $A(s)$, then $\text{Re}(s_k) < 0$. Let $D(z)$ be a polynomial in z of degree N with all roots inside the unit circle. If we replace z in $D(z)$ by the function $(1 + s)/(1 - s)$, we arrive at a rational function of s given by $B(s)/(1 - s)^N$, where $B(s)$ is a polynomial in s of degree N. Show that $B(s)$ is a strictly Hurwitz polynomial.

4.125 A zero-mean WSS white noise sequence $x[n]$ with a variance σ_x^2 is being processed by a causal LTI discrete-time system with an impulse response $h[n] = \delta[n] - \alpha\delta[n-1]$, generating the WSS sequence $y[n]$ at its output. Determine the expressions for the power spectrum $\mathcal{P}_{yy}(\omega)$, the autocorrelation $\phi_{yy}[\ell]$, and the average power of the output $y[n]$. What is the effect of α on the average output power?

4.126 A zero-mean WSS white noise sequence $x[n]$ with a variance σ_x^2 is being processed by a causal LTI discrete-time system with an impulse response $h[n] = (0.5)^n \mu[n]$ generating the WSS sequence $y[n]$ at its output. Determine the expressions for the power spectrum $\mathcal{P}_{yy}(\omega)$, and the autocorrelation $\phi_{yy}[\ell]$ of the output $y[n]$.

4.127 Consider the structure of Figure P4.15, where $A(z) = (\alpha - z^{-1})/(1 - \alpha z^{-1})$ is a causal stable allpass filter with $|\alpha| < 1$. Let the input $x[n]$ be a stationary noise with power spectrum given by

$$\mathcal{P}_{xx}(\omega) = \frac{1}{1 + d\cos\omega}.$$

(a) Calculate the power spectrum $\mathcal{P}_{yy}(\omega)$ of the output $y[n]$.

(b) Does your answer depend upon α? If not, where did you use the information $|\alpha| < 1$?

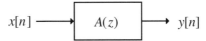

Figure P4.15

4.128 A real, white zero-mean WSS random signal $x[n]$ is processed by an LTI digital filter with a real, causal, stable impulse response $h[n]$, as indicated in Figure P4.16. Let $\mathcal{P}_{xy}(\omega)$ and $\mathcal{P}_{xu}(\omega)$ denote the respective cross-power spectrums. Justify your answers.

(a) Is $\mathcal{P}_{xy}(\omega)$ a real function of ω?

(b) Is $\mathcal{P}_{xu}(\omega)$ a real function of ω?

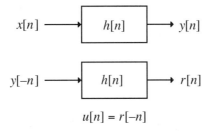

Figure P4.16

4.17 MATLAB Exercises

M 4.1 Write a MATLAB program to compute the group delay using the expression of Problem 4.5 at a prescribed set of discrete frequencies.

M 4.2 Write a MATLAB program to simulate the filter designed in Problem 4.16 and verify its filtering operation.

M 4.3 Write a MATLAB program to simulate the filter designed in Problem 4.19 and verify its filtering operation.

M 4.4 The following third-order IIR transfer function has been proposed for clutter rejection in MTI radar [Whi58]:

$$H(z) = \frac{z^{-1}(1 - z^{-1})^2}{(1 - 0.4z^{-1})(1 - 0.88z^{-1} + 0.61z^{-2})}.$$

Using MATLAB determine and plot its gain response and show that it has a highpass response.

M 4.5 Show that for each case listed below, $H(z)$ and $H(-z)$ are power-complementary.

(a) $H(z) = \dfrac{0.75z^{-1} + 2z^{-2} + 2z^{-3} + z^{-4} + 0.75z^{-5}}{3 + 3.5z^{-2} + z^{-4}}$,

(b) $H(z) = \dfrac{1 - 1.5z^{-1} + 3.75z^{-2} - 2.75z^{-3} + 2.75z^{-4} - 3.75z^{-5} + 1.5z^{-6} - z^{-7}}{6 + 6.5z^{-2} + 4.5z^{-4} + z^{-6}}.$

To verify the power-complementary property, write a MATLAB program to evaluate $H(z)H(z^{-1}) + H(-z)H(-z^{-1})$ and show that this expression is equal to unity for each of the transfer functions given above.

M 4.6 Plot the magnitude and phase responses of the causal IIR digital transfer function

$$H(z) = \frac{0.0534(1 + z^{-1})(1 - 1.0166z^{-1} + z^{-2})}{(1 - 0.683z^{-1})(1 - 1.4461z^{-1} + 0.7957z^{-2})}.$$

What type of filter does this transfer function represent? Determine the difference equation representation of the above transfer function.

M 4.7 Plot the magnitude and phase responses of the causal IIR digital transfer function

$$H(z) = \frac{(1 - z^{-1})^4}{(1 - 1.499z^{-1} + 0.8482z^{-2})(1 - 1.5548z^{-1} + 0.6493z^{-2})}.$$

What type of filter does this transfer function represent? Determine the difference equation representation of the above transfer function.

M 4.8 Design an FIR lowpass filter with a 3-dB cutoff frequency at 0.24π using a cascade of five first-order lowpass filters of Eq. (4.103). Plot its gain response.

M 4.9 Using the result of Problem 4.62 design an FIR highpass filter with a 3-dB cutoff frequency at 0.24π using a cascade of six first-order highpass filters of Eq. (4.112). Plot its gain response.

M 4.10 Design a first-order IIR lowpass and a first-order IIR highpass filter with a 3-dB cutoff frequency of 0.3π. Using MATLAB plot their magnitude responses on the same figure. Using MATLAB show that these filters are both allpass-complementary and power-complementary.

M 4.11 Design a second-order IIR bandpass and a second-order IIR notch filter with a center (notch) frequency $\omega_o = 0.4\pi$ and a 3-dB bandwidth B_w (notch width) of 0.15π. Using MATLAB plot their magnitude responses on the same figure. Using MATLAB show that these filters are both allpass-complementary and power-complementary.

M 4.12 Design a stable second-order IIR bandpass filter with a center frequency at 0.7π and a 3-dB bandwidth of 0.15π. Plot its gain response.

M 4.13 Design a stable second-order IIR notch filter with a center frequency at 0.4π and a 3-dB bandwidth of 0.1π. Plot its gain response.

M 4.14 Using MATLAB show that the transfer function pairs of Problem 4.108 are both allpass-complementary and power-complementary.

M 4.15 Develop the pole-zero plots of the transfer functions of Problem 4.109 using the function zplane of MATLAB and show that they are stable. Next, plot the magnitude response of each transfer function using MATLAB and show that it satisfies the bounded real property.

M 4.16 Develop the pole-zero plots of the transfer functions of Problem 4.122 using the function zplane of MATLAB and then test their stability.

M 4.17 Using Program 4_4 test the stability of the transfer functions of Problem 4.122.

M 4.18 Using Program 4_4 determine whether the roots of the polynomials of Problem 4.123 are inside the unit circle or not.

M 4.19 The FIR digital filter structure of Figure P4.17 is used for aperture correction in television to compensate for high-frequency losses [Dre90]. A cascade of two such circuits is used, with one correcting the vertical aperture and the other correcting the horizontal aperture. In the former case, the delay z^{-1} is a line delay, whereas in the latter case it is 70 ns for the CCIR standard and the weighting factor k provides an adjustable amount of correction. Determine the transfer function of this circuit and plot its magnitude response using MATLAB for two different values of k.

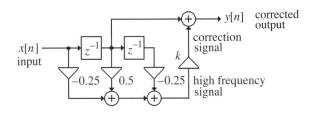

Figure P4.17

M 4.20 An improved aperture correction circuit for digital television is the FIR digital filter structure of Figure P4.18, where the delay z^{-1} is 70 ns for the CCIR standard, and the two weighting factors k_1 and k_2 provide an adjustable amount of correction with $k_1 > 0$ and $k_2 < 0$ [Dre90]. Determine the transfer function of this circuit and plot its magnitude response using MATLAB for two different values of k_1 and k_2.

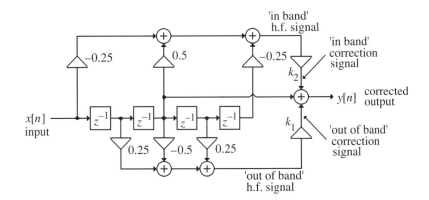

Figure P4.18

5 Digital Processing of Continuous-Time Signals

5.1 Introduction

Even though this book is concerned primarily with the processing of discrete-time signals, most signals we encounter in the real world are continuous in time, such as speech, music, and images. Increasingly, discrete-time signal processing algorithms are being used to process such signals and are implemented employing discrete-time analog or digital systems. For processing by digital systems, the discrete-time signals are represented in digital form with each discrete-time sample as a binary word. Therefore, we need the analog-to-digital and digital-to-analog interface circuits to convert the continuous-time signals into discrete-time digital form, and vice versa. As a result, it is necessary to develop the relations between the continuous-time signal and its discrete-time equivalent in the time-domain and also in the frequency-domain.[1] The latter relations are important in determining conditions under which the discrete-time processing of continuous-time signals can be done free of error under ideal situations.

The interface circuit performing the conversion of a continuous-time signal into a digital form is called the *analog-to-digital* (A/D) *converter*. Likewise, the reverse operation of converting a digital signal into a continuous-time signal is implemented by the interface circuit called the *digital-to-analog* (D/A) *converter*. In addition to these two devices, we need several additional circuits. Since the analog-to-digital conversion usually takes a finite amount of time, it is often necessary to ensure that the analog signal at the input of the A/D converter remains constant in amplitude until the conversion is complete to minimize the error in its representation. This is accomplished by a device called the *sample-and-hold* (S/H) circuit, which has dual purposes. It not only samples the input continuous-time signal at periodic intervals but also holds the analog sampled value constant at its output for sufficient time to permit accurate conversion by the A/D converter. In addition, the output of the D/A converter is a staircase-like waveform. It is therefore necessary to smooth the D/A converter output by means of an analog *reconstruction (smoothing) filter*. Finally, in most applications, the continuous-time signal to be processed has usually a larger bandwidth than the bandwidth of the available discrete-time processors. To prevent a detrimental effect called *aliasing*, an analog *anti-aliasing filter* is often placed before the S/H circuit. The complete block diagram illustrating the functional requirements for the discrete-time digital processing of a continuous-time signal is thus as indicated in Figure 5.1.

To understand the conditions under which the above system can work, we need to examine each of the interface circuits indicated in Figure 5.1. First, we assume a simpler mathematical equivalent of Figure 5.1, which enables us to derive the most fundamental condition that permits the discrete-time processing of continuous-time signals. To this end we assume that the A/D and D/A converters have infinite precision wordlengths, resulting in the simplified representation of Figure 5.2.[2] In this representation, the S/H circuit in cascade with an infinite precision A/D converter has been replaced with the ideal continuous-time–to–

---

[1]These relations also apply to discrete-time analog signal processing systems, such as the switched-capacitor networks.

[2]Effects of finite wordlength are considered in Chapter 9.

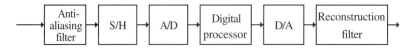

Figure 5.1: Block diagram representation of the discrete-time digital processing of a continuous-time signal.

Figure 5.2: A simplified representation of Figure 5.1.

discrete-time (CT-DT) converter (i.e., an ideal sampler), which develops a discrete-time equivalent $x[n]$ of the continuous-time signal $x_a(t)$. Likewise, the infinite-precision D/A converter in cascade with the ideal reconstruction filter has been replaced with the ideal discrete-time–to–continuous-time (DT-CT) converter (i.e., an ideal interpolator), which develops a continuous-time equivalent $y_a(t)$ of the processed discrete-time signal $y[n]$.

In this chapter, we first derive the conditions for discrete-time representation of a bandlimited continuous-time signal under ideal sampling and its exact recovery from the sampled version. If these conditions are not met, it is not possible to recover the original continuous-time signal from its sampled discrete-time equivalent, resulting in an undesirable distorted representation caused by an effect called aliasing. Since both the anti-aliasing filter and the reconstruction filter in Figure 5.1 are analog lowpass filters, we briefly review next the basic theory behind some commonly used analog filter design methods and illustrate them using MATLAB. It should be noted also that the most widely used digital filter design methods are based on the conversion of an analog prototype, and a knowledge of analog filter design is thus useful in digital signal processing. We then examine the basic properties of the various interface circuits depicted in Figure 5.1.

5.2 Sampling of Continuous-Time Signals

As indicated above, discrete-time signals in many applications are generated by sampling continuous-time signals. We also observed in Example 2.12 that identical discrete-time sequences may result from the sampling of more than one distinct continuous-time function. In fact, in general, there exists an infinite number of continuous-time signals, which when sampled lead to the same discrete-time signal. However, under certain conditions, it is possible to relate a unique continuous-time signal to a given discrete-time sequence, and it is possible to recover the original continuous-time signal from its sampled values. We develop this correspondence and the associated conditions next by considering the relation between the spectra of the original continuous-time signal and the discrete-time signal obtained by sampling.

5.2.1 Effect of Sampling in the Frequency-Domain

Let $g_a(t)$ be a continuous-time signal that is sampled uniformly at $t = nT$, generating the sequence $g[n]$ where

$$g[n] = g_a(nT), \qquad -\infty < n < \infty \tag{5.1}$$

with T being the *sampling period*. The reciprocal of T is called the *sampling frequency* F_T, i.e., $F_T = 1/T$. Now, the frequency-domain representation of $g_a(t)$ is given by its continuous-time Fourier transform (CTFT) $G_a(j\Omega)$,

$$G_a(j\Omega) = \int_{-\infty}^{\infty} g_a(t)e^{-j\Omega t}\, dt, \tag{5.2}$$

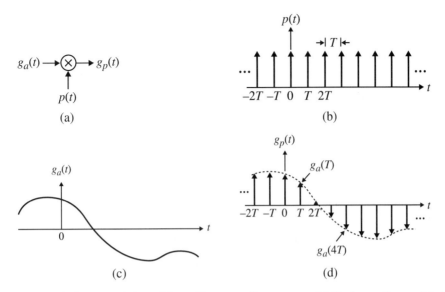

Figure 5.3: Mathematical representation of the uniform sampling process: (a) ideal sampling model, (b) impulse train, (c) continuous-time signal, and (d) its sampled version.

whereas the frequency-domain representation of $g[n]$ is given by its discrete-time Fourier transform $G(e^{j\omega})$,

$$G(e^{j\omega}) = \sum_{n=-\infty}^{\infty} g[n] e^{-j\omega n}. \tag{5.3}$$

To establish the relations between these two different types of Fourier spectra, $G_a(j\Omega)$ and $G(e^{j\omega})$, we treat the sampling operation mathematically as a multiplication of the continuous-time signal $g_a(t)$ by a periodic impulse train $p(t)$:

$$p(t) = \sum_{n=-\infty}^{\infty} \delta(t - nT), \tag{5.4}$$

consisting of a train of ideal impulses with a period T, as indicated in Figure 5.3. The multiplication operation yields an impulse train $g_p(t)$:

$$g_p(t) = g_a(t)p(t) = \sum_{n=-\infty}^{\infty} g_a(nT)\delta(t - nT), \tag{5.5}$$

which is seen to be a continuous-time signal consisting of a train of uniformly spaced impulses with the impulse at $t = nT$ weighted by the sampled value $g_a(nT)$ of $g_a(t)$ at that instant.

There are two different forms of the continuous-time Fourier transform $G_p(j\Omega)$ of $g_p(t)$. One form is given by the weighted sum of the continuous-time Fourier transforms of $\delta(t - nT)$:

$$G_p(j\Omega) = \sum_{n=-\infty}^{\infty} g_a(nT)e^{-j\Omega nT}. \tag{5.6}$$

To derive the second form, we note that the periodic impulse train $p(t)$ can be expressed as a Fourier series (Problem 5.1):

$$p(t) = \frac{1}{T} \sum_{k=-\infty}^{\infty} e^{j(2\pi/T)kt} = \frac{1}{T} \sum_{k=-\infty}^{\infty} e^{j\Omega_T kt}, \tag{5.7}$$

where $\Omega_T = 2\pi/T$ denotes the angular sampling frequency. The impulse train $g_p(t)$ therefore can be expressed as

$$g_p(t) = \left(\frac{1}{T} \sum_{k=-\infty}^{\infty} e^{j\Omega_T kt} \right) \cdot g_a(t). \tag{5.8}$$

Now from the frequency-shifting property of the continuous-time Fourier transform, the continuous-time Fourier transform of $e^{j\Omega_T kt} g_a(t)$ is given by $G_a(j(\Omega - k\Omega_T))$. Hence, an alternative form of the continuous-time Fourier transform of $g_p(t)$ is given by

$$G_p(j\Omega) = \frac{1}{T} \sum_{k=-\infty}^{\infty} G_a(j(\Omega - k\Omega_T)). \tag{5.9}$$

Therefore, $G_p(j\Omega)$ is a periodic function of frequency Ω consisting of a sum of shifted and scaled replicas of $G_a(j\Omega)$, shifted by integer multiples of Ω_T and scaled by $1/T$. The term on the right-hand side of Eq. (5.9) for $k = 0$ is the *baseband* portion of $G_p(j\Omega)$, and each of the remaining terms are the *frequency-translated* portions of $G_p(j\Omega)$. The frequency range $-\Omega_T/2 \le \Omega \le \Omega_T/2$ is called the *baseband* or *Nyquist band*.

Figure 5.4 illustrates the frequency-domain effects of time-domain sampling. Assume $g_a(t)$ is a bandlimited signal with a frequency spectrum $G_a(j\Omega)$, as sketched in Figure 5.4(a) where Ω_m is the highest frequency contained in $g_a(t)$. The spectrum $P(j\Omega)$ of the periodic impulse train $p(t)$ with a sampling period $T = 2\pi/\Omega_T$ is indicated in Figure 5.4(b) and (d). Two possible spectra of $G_p(j\Omega)$ are shown in Figure 5.4(c) and (e). It is evident from Figure 5.4(c) that if $\Omega_T > 2\Omega_m$, there is no overlap between the shifted replicas of $G_a(j\Omega)$ generating $G_p(j\Omega)$. On the other hand, as indicated in Figure 5.4(e), if $\Omega_T < 2\Omega_m$, there is an overlap of the spectra of the shifted replicas of $G_a(j\Omega)$ generating $G_p(j\Omega)$. Consequently, if $\Omega_T > 2\Omega_m$, $g_a(t)$ can be recovered exactly from $g_p(t)$ by passing it through an ideal lowpass filter $H_r(j\Omega)$ with a gain T and a cutoff frequency Ω_c greater than Ω_m and less than $\Omega_T - \Omega_m$, as illustrated in Figure 5.5. However, if $\Omega_T < 2\Omega_m$, due to the overlap of the shifted replicas of $G_a(j\Omega)$, the spectrum $G_p(j\Omega)$ cannot be separated by filtering to recover $G_a(j\Omega)$ because of the distortion caused by a part of the replicas immediately outside the baseband folded back or *aliased* into the baseband. The frequency $\Omega_T/2$ is often referred to as the *folding frequency* or *Nyquist frequency*.

The above result is more commonly known as the *sampling theorem*,[3] which can be summarized as follows. Let $g_a(t)$ be a bandlimited signal with $G_a(j\Omega) = 0$ for $|\Omega| > \Omega_m$. Then $g_a(t)$ is uniquely determined by its samples $g_a(nT)$, $-\infty \le n \le \infty$, if

$$\Omega_T \ge 2\Omega_m, \tag{5.10}$$

where

$$\Omega_T = 2\pi/T. \tag{5.11}$$

Equation (5.10) is often referred to as the *Nyquist condition*. Given $\{g_a(nT)\}$, we can recover exactly $g_a(t)$ by generating an impulse train $g_p(t)$ of the form of Eq. (5.5) and then passing $g_p(t)$ through an ideal lowpass filter $H_r(j\Omega)$ with a gain T and a cutoff frequency Ω_c greater than Ω_m and less than $\Omega_T - \Omega_m$, i.e.,

$$\Omega_m < \Omega_c < (\Omega_T - \Omega_m). \tag{5.12}$$

---

[3]Also called either the *Nyquist sampling theorem* or *Shannon sampling theorem*.

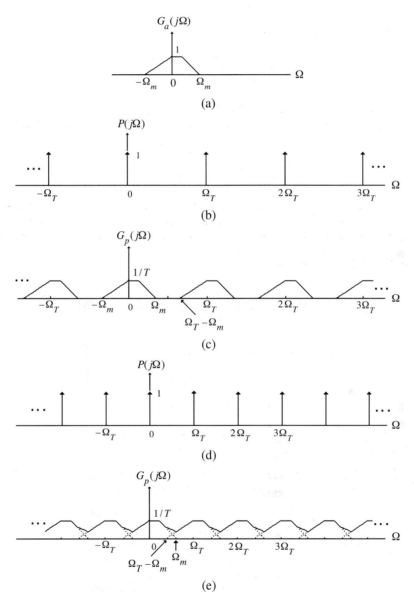

Figure 5.4: Illustration of the frequency-domain effects of time-domain sampling. (a) Spectrum of original continuous-time signal $g_a(t)$, (b) spectrum of the periodic impulse train $p(t)$, (c) spectrum of the sampled signal $g_p(t)$ with $\Omega_T > 2\omega_m$, (d) spectrum of the periodic impulse train $p(t)$ with a sampling period smaller than that shown in (b), and (e) spectrum of the sampled signal $g_p(t)$ with $\Omega_T < 2\Omega_m$. [Note: The spectrum of $g_a(t)$ is shown as not being an even function to emphasize the effect of sampling.]

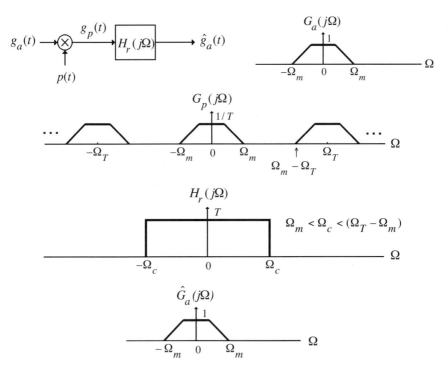

Figure 5.5: Reconstruction of the original continuous-time signal from its sampled version obtained by ideal sampling.

The highest frequency Ω_m contained in $g_a(t)$ is usually called the *Nyquist frequency* since it determines the minimum sampling frequency $\Omega_T = 2\Omega_m$ that must be used to fully recover $g_a(t)$ from its sampled version. The frequency $2\Omega_m$ is called the *Nyquist rate*.

If the sampling frequency is higher than the Nyquist rate, the sampling operation is referred to as *oversampling*. On the other hand, if the sampling frequency is lower than the Nyquist rate, it is called *undersampling*. Finally, if the sampling frequency is exactly equal to the Nyquist rate, it is called *critical sampling*.[4]

Typical sampling rates used in practice are, for example, 8 kHz in digital telephony and 44.1 kHz in compact disc (CD) music systems. In the former case, a 3.4-kHz signal bandwidth is acceptable for telephone conversation. Hence, a sampling rate of 8 kHz, which is greater than 2 times that of the acceptable bandwidth of 6.8 kHz, is quite adequate. In high-quality analog music signal processing, on the other hand, a bandwidth of about 20 kHz has been determined to preserve the fidelity. As a result, here the analog music signal is sampled at a slightly higher rate of 44.1 kHz to ensure almost negligible aliasing distortion.

EXAMPLE 5.1   Consider the three pure sinusoidal signals of Example 2.11: $g_1(t) = \cos(6\pi t)$, $g_2(t) = \cos(14\pi t)$, and $g_3(t) = \cos(26\pi t)$. The corresponding continuous-time Fourier transforms consist of two impulses each and are given by

$$G_1(j\Omega) = \pi[\delta(\Omega - 6\pi) + \delta(\Omega + 6\pi)],$$
$$G_2(j\Omega) = \pi[\delta(\Omega - 14\pi) + \delta(\Omega + 14\pi)],$$
$$G_3(j\Omega) = \pi[\delta(\Omega - 26\pi) + \delta(\Omega + 26\pi)].$$

---

[4]It should be noted that a pure sinusoid may not be recoverable from its critically sampled version.

These transforms are plotted in Figure 5.6(a) to (c), respectively.

The above signals are sampled at a rate of $T = 0.1$ sec, i.e., with a sampling frequency $\Omega_T = 20\pi$ rad/sec, generating the continuous-time impulse trains $g_{1p}(t)$, $g_{2p}(t)$, and $g_{3p}(t)$. From Eq. (5.9) their corresponding continuous-time Fourier transforms are given by

$$G_{\ell p}(j\Omega) = \frac{1}{T} \sum_{k=-\infty}^{\infty} G_\ell(j\Omega - jk\Omega_T), \qquad \ell = 1, 2, 3,$$

which have been plotted in Figure 5.6(d) to (f). These figures also indicate by dotted lines the frequency response of an ideal lowpass filter with a cutoff at $\Omega_c = \Omega_T/2 = 10\pi$ and a gain of $T = 0.1$ (not shown to scale). The continuous-time Fourier transforms of the output of the lowpass filter in each case are also indicated in Figure 5.6(d) to (f). Note that in the case of $g_1(t)$, the sampling rate satisfies the Nyquist condition and there is no aliasing. The continuous-time reconstructed output is precisely the original continuous-time signal $g_1(t)$. However, in the other two cases, the sampling rate does not satisfy the Nyquist condition, resulting in aliasing, and the outputs here are all equal to $\cos(6\pi t)$. Note that in Figure 5.6(e), the impulse appearing at $\Omega = 6\pi$ in the positive frequency passband of the lowpass filter results from the aliasing of the impulse in $G_2(j\Omega)$ at $\Omega = -14\pi$. On the other hand, the impulse appearing at $\Omega = 6\pi$ in Figure 5.6(f) has resulted from the aliasing of the impulse in $G_3(j\Omega)$ at $\Omega = 26\pi$.

We now establish the relation between the discrete-time Fourier transform $G(e^{j\omega})$ of the sequence $g[n]$ and the continuous-time Fourier transform $G_a(j\Omega)$ of the analog signal $g_a(t)$. If we compare Eq. (5.3) with Eq. (5.6) and make use of Eq. (5.1), we observe that

$$G(e^{j\omega}) = G_p(j\Omega)\big|_{\Omega=\omega/T}, \tag{5.13a}$$

or equivalently,

$$G_p(j\Omega) = G(e^{j\omega})\big|_{\omega=\Omega T}. \tag{5.13b}$$

Therefore, from the above and Eq. (5.9), we arrive at the desired result given by

$$G(e^{j\omega}) = \frac{1}{T} \sum_{k=-\infty}^{\infty} G_a(j\Omega - jk\Omega_T)\bigg|_{\Omega=\omega/T}$$

$$= \frac{1}{T} \sum_{k=-\infty}^{\infty} G_a\left(j\frac{\omega}{T} - jk\Omega_T\right)$$

$$= \frac{1}{T} \sum_{k=-\infty}^{\infty} G_a\left(j\frac{\omega}{T} - j\frac{2\pi k}{T}\right), \tag{5.14a}$$

which can be alternatively expressed as

$$G(e^{j\Omega T}) = \frac{1}{T} \sum_{k=-\infty}^{\infty} G_a(j\Omega - jk\Omega_T). \tag{5.14b}$$

As can be seen from Eq. (5.13a) or (5.13b), $G(e^{j\omega})$ is obtained from $G_p(j\Omega)$ simply by scaling the frequency axis Ω according to the relation

$$\Omega = \frac{\omega}{T}. \tag{5.15}$$

Now, the continuous-time Fourier transform $G_p(j\Omega)$ is a periodic function of Ω with a period $\Omega_T = 2\pi/T$. Because of the above normalization, the discrete-time Fourier transform $G(e^{j\omega})$ is a periodic function of ω with a period 2π.

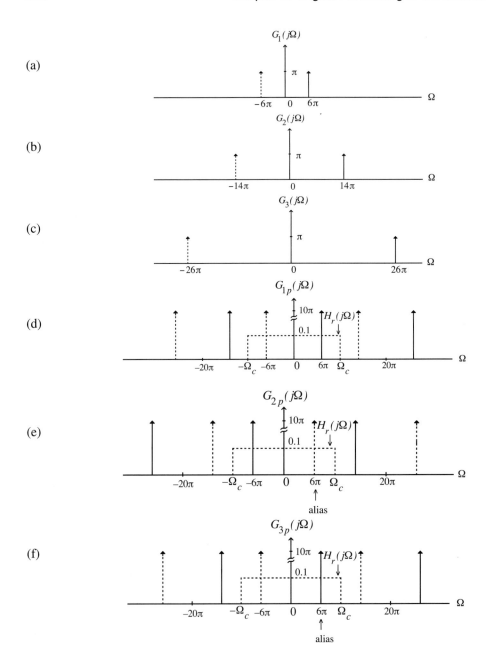

Figure 5.6: Effect of sampling on a pure cosine signal: (a) spectrum of $\cos(6\pi t)$, (b) spectrum of $\cos(14\pi t)$, (c) spectrum of $\cos(26\pi t)$ (d) spectrum of sampled version of $\cos(6\pi t)$ with $\Omega_T = 20\pi > 2\Omega_m = 12\pi$, (e) spectrum of sampled version of $\cos(14\pi t)$ with $\Omega_T = 20\pi < 2\Omega_m = 28\pi$, and (f) spectrum of sampled version of $\cos(26\pi t)$ with $\Omega_T = 20\pi < 2\Omega_m = 52\pi$.

5.2.2 Recovery of the Analog Signal

We indicated earlier that if the discrete-time sequence $g[n]$ has been obtained by uniformly sampling a bandlimited continuous-time signal $g_a(t)$ with a highest frequency Ω_m at a rate $\Omega_T = 2\pi/T$ satisfying the condition of Eq. (5.10), then the original continuous-time signal $g_a(t)$ can be fully recovered by passing the equivalent impulse train $g_p(t)$ through an ideal lowpass filter $H_r(j\Omega)$ with a cutoff at Ω_c satisfying Eq. (5.12) and with a gain of T. We next derive the expression for the output $\hat{g}_a(t)$ of the ideal lowpass filter as a function of the samples $g[n]$.

Now, the impulse response $h_r(t)$ of the above ideal lowpass filter is obtained simply by taking the inverse continuous-time Fourier transform of its frequency response $H_r(j\Omega)$:

$$H_r(j\Omega) = \begin{cases} T, & |\Omega| \leq \Omega_c, \\ 0, & |\Omega| > \Omega_c, \end{cases} \tag{5.16}$$

and is given by

$$h_r(t) = \frac{1}{2\pi} \int_{-\infty}^{\infty} H_r(j\Omega) e^{j\Omega t} \, d\Omega = \frac{T}{2\pi} \int_{-\Omega_c}^{\Omega_c} e^{j\Omega t} \, d\Omega$$

$$= \frac{\sin(\Omega_c t)}{\Omega_T t/2}, \quad -\infty \leq t \leq \infty. \tag{5.17}$$

Observe that the impulse train $g_p(t)$ is given by

$$g_p(t) = \sum_{n=-\infty}^{\infty} g[n]\delta(t - nT). \tag{5.18}$$

Therefore, the output $\hat{g}_a(t)$ of the ideal lowpass filter is given by the convolution of $g_p(t)$ with the impulse response $h_r(t)$ of the analog reconstruction filter:

$$\hat{g}_a(t) = \sum_{n=-\infty}^{\infty} g[n]h_r(t - nT). \tag{5.19}$$

Substituting $h_r(t)$ from Eq. (5.17) in Eq. (5.19) and assuming for simplicity $\Omega_c = \Omega_T/2 = \pi/T$, we arrive at

$$\hat{g}_a(t) = \sum_{n=-\infty}^{\infty} g[n] \frac{\sin[\pi(t - nT)/T]}{\pi(t - nT)/T}. \tag{5.20}$$

The above expression indicates that the reconstructed continuous-time signal $\hat{g}_a(t)$ is obtained by shifting in time the impulse response of the lowpass filter $h_r(t)$ by an amount nT and scaling it in amplitude by the factor $g[n]$ for all integer values of n in the range $-\infty < n < \infty$ and then summing up all shifted versions. The ideal bandlimited interpolation process is illustrated in Figure 5.7.

Now, it can be shown that when $\Omega_c = \Omega_T/2$ in Eq. (5.17), $h_r(0) = 1$ and $h_r(nT) = 0$ for $n \neq 0$ (Problem 5.6). As a result, from Eq. (5.20), $\hat{g}_a(rT) = g[r] = g_a(rT)$ for all integer values of r in the range $-\infty < r < \infty$, whether or not the condition of the sampling theorem has been satisfied. However, $\hat{g}_a(t) = g_a(t)$ for all values of t only if the sampling frequency Ω_T satisfies the condition of Eq. (5.10) of the sampling theorem.

It should be noted that the ideal analog lowpass filter of Eq. (5.17) has a doubly infinite length impulse response and, thus, is unstable and noncausal, and does not have a rational transfer function, making it unrealizable. An analog lowpass filter is also needed to bandlimit the continuous-time signal before it

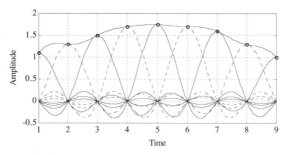

Figure 5.7: The ideal bandlimited reconstruction by interpolation.

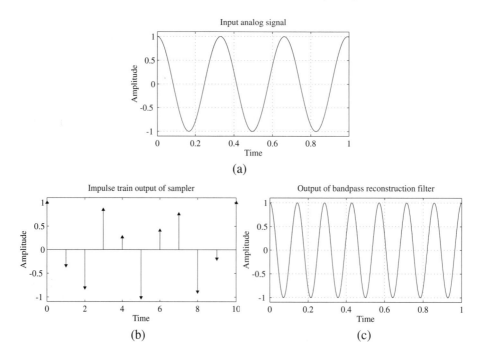

Figure 5.8: Illustration of frequency-translation effect of sampling.

is sampled to ensure that the conditions of the sampling theorem are satisfied. The magnitude response specification for both the anti-aliasing lowpass filter and the analog reconstruction filter therefore must be modified to make them realizable. We review in Section 5.4 some commonly used methods for the determination of the transfer function of realizable and stable analog lowpass filters approximating the ideal magnitude characteristic of Eq. (5.16). In Sections 5.6 and 5.10 we consider specifically the issues concerning the design of an anti-aliasing filter and the reconstruction filter, respectively.

5.2.3 Implications of the Sampling Process

Consider again the sampling of the three continuous-time signals of Example 5.1. From Figure 5.6(d) it should be apparent that from the sampled version $g_{1p}(t)$ of the continuous-time signal $g_1(t) = \cos(6\pi t)$, we can also recover any of its frequency-translated versions $\cos[(20k \pm 6)\pi t]$ outside the baseband by passing $g_{1p}(t)$ through an ideal analog bandpass filter with a passband centered at $\Omega = (20k \pm 6)\pi$. For example, to recover the signal $\cos(34\pi t)$, it will be necessary to employ a bandpass filter with a frequency response

$$H_r(j\Omega) = \begin{cases} 0.1, & (34 - \Delta)\pi \le |\Omega| \le (34 + \Delta)\pi, \\ 0, & \text{otherwise,} \end{cases} \tag{5.21}$$

where Δ is a small number. Likewise, we can recover the aliased baseband component $\cos(6\pi t)$ from the sampled version of either $g_{2p}(t)$ or $g_{3p}(t)$ by passing it through an ideal lowpass filter:

$$H_r(j\Omega) = \begin{cases} 0.1, & (6 - \Delta)\pi \le |\Omega| \le (6 + \Delta)\pi, \\ 0, & \text{otherwise.} \end{cases} \tag{5.22}$$

There is no *aliasing distortion* unless the original continuous-time signal also contains the component $\cos(6\pi t)$. Similarly, from the sampled versions of either $g_{2p}(t)$ or $g_{3p}(t)$, we can recover any one of the frequency-translated versions, including the parent continuous-time signal, $\cos(14\pi t)$ or $\cos(26\pi t)$ as the case may be, by employing suitable ideal bandpass filters.

The frequency-translation effect of sampling is further illustrated in the following two examples.

EXAMPLE 5.2   Figure 5.8(a) shows the continuous-time sinusoidal signal $g_1(t) = \cos(6\pi t)$ of frequency 3 Hz. The impulse train $g_{1p}(t)$ generated from $g_1(t)$ by uniformly sampling it at a rate of 10 Hz is depicted in Figure 5.8(b). The continuous-time output $y(t)$ by passing $g_{1p}(t)$ through an analog bandpass reconstruction filter with a passband from 5 Hz to 10 Hz is shown in Figure 5.8(c), which is seen to be a continuous-time sinusoidal signal $\cos(14\pi t)$ of frequency 7 Hz. In general, the continuous-time output of an analog bandpass reconstruction filter with a passband from $5k$ Hz to $5k + 5$ Hz, with k a positive integer, will be a sinusoidal signal of frequency of $5k + 3$ for k even and $5(k + 1) - 3$ for k odd.

EXAMPLE 5.3   The continuous-time sinusoidal signal $g_3(t) = \cos(26\pi t)$ of frequency 13 Hz shown in Figure 5.9(a) is sampled uniformly at a 10-Hz rate, resulting in an impulse train $g_{3p}(t)$ sketched in Figure 5.9(b). The output generated from $g_{3p}(t)$ by passing it through an analog lowpass reconstruction filter with a passband edge at 5 Hz is a continuous-time sinusoidal signal of frequency 3 Hz, as shown in Figure 5.9(c). Note that the original high-frequency sinusoidal signal can be fully recovered from $g_{3p}(t)$ by passing it instead through a bandpass filter with a passband from 10 Hz to 15 Hz, as indicated in Figure 5.9(d).

Generalizing the above discussions, consider the continuous-time signals $g_1(t)$, $g_2(t)$, and $g_3(t)$ with bandlimited frequency spectrums $G_1(j\Omega)$, $G_2(j\Omega)$, and $G_3(j\Omega)$, as shown in Figure 5.10(a) to (c), respectively. Each of these continuous-time signals, when sampled at a sampling frequency of Ω_T, develops a continuous-time signal $g_{\ell p}(t)$, $\ell = 1, 2, 3$, with an identical periodic frequency spectrum, as indicated in Figure 5.10(d). Therefore, by passing the sampled signal through an appropriate analog lowpass or bandpass filter of bandwidth greater than $\Omega_2 - \Omega_1$ but less than or equal to $\Omega_T/2$, we can recover either the original continuous-time signal or any one of its frequency-translated versions. Note that as long as the spectrum of the continuous-time signal being sampled at a sampling rate of Ω_T is bandlimited to the frequency range $k\Omega_T/2 \le |\Omega| \le (k + 1)\Omega_T/2$, there is no aliasing distortion due to sampling, and it or its frequency-translated version can always be recovered from the sampled signal by appropriate filtering. There will be aliasing distortion only if there are frequency components in a wider frequency range than that indicated.

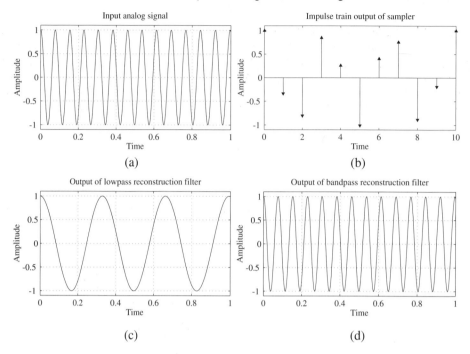

Figure 5.9: Illustration of the effect of undersampling.

5.3 Sampling of Bandpass Signals

The conditions developed in Section 5.2.1 for the unique representation of a continuous-time signal by the discrete-time signal obtained by uniform sampling assumed that the spectrum of the continuous-time signal is bandlimited in the frequency range from dc to some frequency Ω_m. Such continuous-time signals are commonly referred to as *lowpass* signals. There are applications where the continuous-time signal is bandlimited to a higher range $\Omega_L \leq |\Omega| \leq \Omega_H$, where $\Omega_L > 0$. Such a signal is usually referred to as a *bandpass* signal and is often obtained by modulating a lowpass signal. We can of course sample such a bandpass continuous-time signal with a sampling rate greater than twice the highest frequency, i.e., by ensuring

$$\Omega_T \geq 2\Omega_H,$$

to prevent aliasing. However, in this case, due to the bandpass spectrum of the continuous-time signal, the spectrum of the discrete-time signal obtained by sampling will have spectral gaps with no signal components present in these gaps. Moreover, if Ω_H is very large, the sampling rate also has to be very large which may not be practical in some situations.

   We outline next a more practical and efficient approach [Por97]. Let $\Delta\Omega = \Omega_H - \Omega_L$ define the *bandwidth* of the bandpass signal. Assume first that the highest frequency Ω_H contained in the signal is an integer multiple of the bandwidth, i.e.,

$$\Omega_H = M(\Delta\Omega).$$

We choose the sampling frequency Ω_T to satisfy the condition

$$\Omega_T = 2(\Delta\Omega) = \frac{2\Omega_H}{M}, \tag{5.23}$$

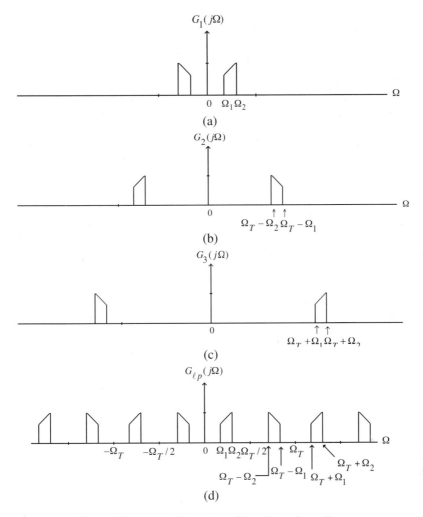

Figure 5.10: Further illustration of the effect of sampling.

which is smaller than $2\Omega_H$, the Nyquist rate. Substituting Eq. (5.23) in Eq. (5.9) we arrive at the expression for the Fourier transform $G_p(j\Omega)$ of the impulse-sampled signal $g_p(t)$:

$$G_p(j\Omega) = \frac{1}{T} \sum_{k=-\infty}^{\infty} G_a \left(j(\Omega - 2k(\Delta\Omega)) \right). \tag{5.24}$$

As before, $G_p(j\Omega)$ consists of a sum of the original Fourier transform $G_a(j\Omega)$ and replicas of $G_a(j\Omega)$ shifted by integer multiples of twice the bandwidth $\Delta\Omega$, and then scaled by $1/T$. The amount of the shift for each value of k ensures that there will be no overlap between all shifted replicas, and hence no aliasing. Figure 5.11 shows the spectrum of the original continuous-time signal $g_a(t)$ and that of the sampled version $g_p(t)$, sampled at the rate given by Eq. (5.23) for $M = 4$. As can be seen from this figure, $g_a(t)$ can be recovered from $g_p(t)$ by passing the latter through an ideal bandpass filter with a passband given by $\Omega_L \le |\Omega| \le \Omega_H$ and a gain of T.

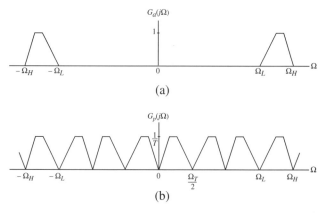

Figure 5.11: Illustration of the effect in the frequency-domain of sampling below the Nyquist rate a bandpass signal with highest frequency that is an integer multiple of its bandwidth: (a) spectrum of original bandpass signal, and (b) spectrum of sampled bandpass signal.

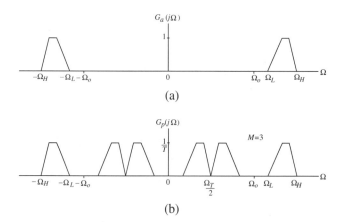

Figure 5.12: Illustration of the effect in the frequency-domain of sampling below the Nyquist rate a bandpass signal with highest frequency that is not an integer multiple of its bandwidth: (a) spectrum of original bandpass signal, and (b) spectrum of sampled bandpass signal.

Note that any of the replicas in the lower frequency bands can be retained by passing $g_p(t)$ through bandpass filters with passbands $\Omega_L - k(\Delta\Omega) \leq |\Omega| \leq \Omega_H - k(\Delta\Omega)$, $1 \leq k \leq M - 1$ providing a translation of the original bandpass signal to lower frequency ranges. If the bandpass signal has been obtained by modulating a lowpass signal, then the latter can be recovered by passing $g_p(t)$ through a lowpass filter with passband $0 \leq |\Omega| \leq \Omega_H$ which retains the replica in the baseband. This approach is often employed in digital radio receivers.

If Ω_H is not an integer multiple of the bandwidth $\Omega_H - \Omega_L$, we can artificially extend the bandwidth either to the right or to the left so that the highest frequency contained in the bandpass signal is an integer multiple of the extended bandwidth. For example, if we extend the bandwidth to the left by assuming the lowest frequency contained in the bandpass signal to be Ω_o, then Ω_o is chosen such that the extended bandwidth $\Omega_H - \Omega_o$ is an integer multiple of Ω_H. In both cases the spectrum of the sampled signal obtained by sampling $g_a(t)$ will have small spectral gaps between the replicas. This is illustrated in Figure 5.12 when the bandwidth is extended to the left and M is chosen as 3.

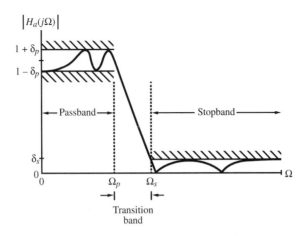

Figure 5.13: Typical magnitude specifications for an analog lowpass filter.

As in the previous case, any of the replicas in the lower frequency bands can be retained by passing $g_p(t)$ through appropriate bandpass filters.

5.4 Analog Lowpass Filter Design

There are a number of established approximation techniques for the design of analog lowpass filters [Vla69], [Dan74], [Tem73], [Tem77]. We describe four widely used design techniques here without their detailed derivations. Further details of these methods can be found in texts on analog filter design. Extensive tables for the design of analog lowpass filters are also available [Chr66], [Skw65], [Zve67]. As indicated earlier, a commonly used technique for the design of IIR digital filters is based on the conversion of a prototype analog transfer function that has been designed employing one of the methods discussed here. The digital filter design techniques are the subject of discussion in Chapter 7.

5.4.1 Filter Specifications

Both the anti-aliasing filter and the reconstruction filter of Figure 5.1 are of the lowpass type, and ideally they should have a magnitude response of the form shown in Figure 5.5. In practice, the magnitude response characteristics in the passband and in the stopband cannot be constant and are therefore specified with some acceptable tolerances. Moreover, a transition band is specified between the passband and the stopband to permit the magnitude to drop off smoothly. For example, the magnitude $|H_a(j\Omega)|$ of a lowpass filter may be given as shown in Figure 5.13. As indicated in the figure, in the *passband* defined by $0 \leq \Omega \leq \Omega_p$, we require

$$1 - \delta_p \leq |H_a(j\Omega)| \leq 1 + \delta_p, \qquad \text{for } |\Omega| \leq \Omega_p, \tag{5.25}$$

or in other words, the magnitude approximates unity within an error of $\pm\delta_p$. In the *stopband*, defined by $\Omega_s \leq \Omega \leq \infty$, we require

$$|H_a(j\Omega)| \leq \delta_s, \qquad \text{for } \Omega_s \leq |\Omega| \leq \infty, \tag{5.26}$$

implying that the magnitude approximates zero within an error of δ_s. The frequencies Ω_p and Ω_s are, respectively, called the *passband edge frequency* and the *stopband edge frequency*.

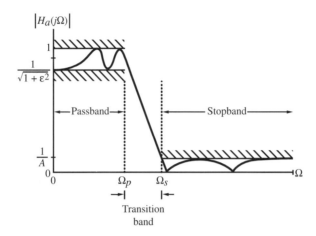

Figure 5.14: Normalized magnitude specifications for an analog lowpass filter.

The limits of the tolerances in the passband and stopband, δ_p and δ_s, are called *ripples*. Usually these ripples are specified in dB in terms of the *peak passband ripple* α_p and the *minimum stopband attenuation* α_s, defined by

$$\alpha_p = -20 \log_{10}(1 - \delta_p) \, \text{dB}, \tag{5.27}$$

$$\alpha_s = -20 \log_{10}(\delta_s) \, \text{dB}. \tag{5.28}$$

Often, the filter specifications are given in terms of the *loss function* or *attenuation function* $a(\Omega)$ in dB, which is defined as the negative of the gain in dB, i.e., $-20 \log_{10} |H_a(j\Omega)|$.

EXAMPLE 5.4   Let the desired peak passband ripple of a lowpass filter be 0.01 dB, and the minimum attenuation in the stopband be 70 dB. Determine δ_p and δ_s.
    Substituting $\alpha_p = 0.01$ in Eq. (5.27) and solving we get

$$\delta_p = 1 - 10^{-\alpha_p/20} = 0.00115.$$

Likewise, substituting $\alpha_s = 70$ in Eq. (5.28) we obtain

$$\delta_s = 10^{-\alpha_s/20} = 0.0003162.$$

The magnitude response specifications for an analog lowpass filter, in some applications, are given in a normalized form, as indicated in Figure 5.14. Here the maximum value of the magnitude in the passband is assumed to be unity and the passband ripple, denoted as $1/\sqrt{1 + \varepsilon^2}$, is given by the minimum value of the magnitude in the passband. The maximum passband gain or the minimum passband loss is therefore 0 dB. The maximum stopband ripple is denoted by $1/A$, and the minimum stopband attenuation is therefore given by $-20 \log_{10}(1/A)$.

In analog filter theory two additional parameters are defined. The first one, called the *transition ratio* or *selectivity parameter*, is defined by the ratio of the passband edge frequency Ω_p and the stopband edge frequency Ω_s, and is usually denoted by k, i.e.,

$$k = \frac{\Omega_p}{\Omega_s}. \tag{5.29}$$

Note that for a lowpass filter, $k < 1$. The second one, called the *discrimination parameter* and denoted by k_1, is defined as

$$k_1 = \frac{\varepsilon}{\sqrt{A^2 - 1}}. \tag{5.30}$$

Usually, $k_1 \ll 1$.

5.4.2 Butterworth Approximation

The magnitude-squared response of an analog lowpass Butterworth filter $H_a(s)$ of Nth order is given by

$$|H_a(j\Omega)|^2 = \frac{1}{1 + (\Omega/\Omega_c)^{2N}}. \tag{5.31}$$

It can be easily shown that the first $2N - 1$ derivatives of $|H_a(j\Omega)|^2$ at $\Omega = 0$ are equal to zero, and as a result, the Butterworth lowpass filter is said to have a *maximally flat magnitude* at $\Omega = 0$. The gain of the Butterworth filter in dB is given by

$$\mathcal{G}(\Omega) = 10\log_{10}|H_a(j\Omega)|^2 \text{ dB}.$$

At dc, i.e., at $\Omega = 0$, the gain in dB is equal to zero, and at $\Omega = \Omega_c$, the gain is

$$\mathcal{G}(\Omega_c) = 10\log_{10}\left(\tfrac{1}{2}\right) = -3.0103 \cong -3\,\text{dB}$$

and, therefore, Ω_c is often called the *3-dB cutoff frequency*. Since the derivative of the squared-magnitude response or, equivalently, of the magnitude response is always negative for positive values of Ω, the magnitude response is monotonically decreasing with increasing Ω. For $\Omega \gg \Omega_c$, the squared-magnitude function can be approximated by

$$|H_a(j\Omega)|^2 \approx \frac{1}{(\Omega/\Omega_c)^{2N}}.$$

The gain $\mathcal{G}(\Omega_2)$ in dB at $\Omega_2 = 2\Omega_1$ with $\Omega_1 \gg \Omega_c$ is given by

$$\mathcal{G}(\Omega_2) = -10\log_{10}\left(\frac{\Omega_2}{\Omega_c}\right)^{2N} = \mathcal{G}(\Omega_1) - 6N \text{ dB},$$

where $\mathcal{G}(\Omega_1)$ is the gain in dB at Ω_1. As a result, the gain roll-off per octave in the stopband decreases by 6 dB or, equivalently, by 20 dB per decade for an increase of the filter order by one. In other words, the passband and the stopband behaviors of the magnitude response improve with a corresponding decrease in the transition band as the filter order N increases. A plot of the magnitude response of the normalized Butterworth lowpass filter with $\Omega_c = 1$ for some typical values of N is shown in Figure 5.15.

The two parameters completely characterizing a Butterworth filter are therefore the 3-dB cutoff frequency Ω_c and the order N. These are determined from the specified passband edge Ω_p, the minimum passband magnitude $1/\sqrt{1 + \varepsilon^2}$, the stopband edge Ω_s, and the maximum stopband ripple $1/A$. From Eq. (5.31) we get

$$|H_a(j\Omega_p)|^2 = \frac{1}{1 + (\Omega_p/\Omega_c)^{2N}} = \frac{1}{1 + \varepsilon^2}, \tag{5.32a}$$

$$|H_a(j\Omega_s)|^2 = \frac{1}{1 + (\Omega_s/\Omega_c)^{2N}} = \frac{1}{A^2}. \tag{5.32b}$$

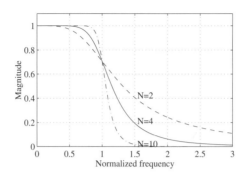

Figure 5.15: Typical Butterworth lowpass filter responses.

Solving the above we arrive at the expression for the order N as

$$N = \frac{1}{2} \frac{\log_{10}\left[\left(A^2 - 1\right)/\varepsilon^2\right]}{\log_{10}\left(\Omega_s/\Omega_p\right)} = \frac{\log_{10}(1/k_1)}{\log_{10}(1/k)}. \tag{5.33}$$

Since the order N of the filter must be an integer, the value of N computed using the above expression is rounded up to the next higher integer. This value of N can be used next in either Eq. (5.32a) or (5.32b) to solve for the 3-dB cutoff frequency Ω_c. It is a usual practice to determine Ω_c using Eq. (5.32b) which satisfies the stopband specification at Ω_s exactly while the passband specification is exceeded providing a safety margin at Ω_p [Tem77]. However, if Eq. (5.32a) is used to solve for Ω_c, then the passband specification at Ω_p is met exactly while the stopband specification at Ω_s is exceeded.

The expression for the transfer function of the Butterworth lowpass filter is given by

$$H_a(s) = \frac{C}{D_N(s)} = \frac{\Omega_c^N}{s^N + \sum_{\ell=0}^{N-1} d_\ell s^\ell} = \frac{\Omega_c^N}{\prod_{\ell=1}^{N} (s - p_\ell)}, \tag{5.34}$$

where

$$p_\ell = \Omega_c e^{j[\pi(N+2\ell-1)/2N]}, \qquad \ell = 1, 2, \ldots, N. \tag{5.35}$$

The denominator $D_N(s)$ of Eq. (5.34) is known as the *Butterworth polynomial* of order N and is easy to compute. These polynomials have been tabulated for easy design reference [Chr66], [Skw65], [Zve67]. The analog lowpass Butterworth filters can be readily designed using MATLAB (see Section 5.4.6).

EXAMPLE 5.5   Determine the lowest order of a transfer function $H_a(s)$ having a maximally flat lowpass characteristic with a 1-dB cutoff frequency at 1 kHz and a minimum attenuation of 40 dB at 5 kHz.

We first determine ε and A. From Eq. (5.32a),

$$10 \log_{10}\left(\frac{1}{1 + \varepsilon^2}\right) = -1,$$

which yields $\varepsilon^2 = 0.25895$. Likewise, from Eq. (5.32b),

$$10 \log_{10}\left(\frac{1}{A^2}\right) = -40,$$

which leads to $A^2 = 10,000$. Substituting the values in Eq. (5.30), we get $1/k_1 = 196.51334$.

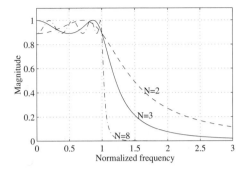

Figure 5.16: Typical Type 1 Chebyshev lowpass filter responses with 1-dB passband ripple.

The inverse transition ratio here is $1/k = 5000/1000 = 5$. Substituting these values in Eq. (5.33), we get

$$N = \frac{\log_{10}(1/k_1)}{\log_{10}(1/k)} = \frac{\log_{10}(196.51334)}{\log_{10}(5)} = 3.2811022. \tag{5.36}$$

Since the order of the transfer function must be an integer, we round the above up to the nearest integer $N = 4$.

5.4.3 Chebyshev Approximation

In this case, the approximation error, defined as the difference between the ideal brickwall characteristic and the actual response, is minimized over a prescribed band of frequencies. In fact, the magnitude error is equiripple in the band. There are two types of *Chebyshev transfer functions*. In the *Type 1* approximation, the magnitude characteristic is equiripple in the passband and monotonic in the stopband, whereas in the *Type 2* approximation, the magnitude response is monotonic in the passband and equiripple in the stopband.

Type 1 Chebyshev Approximation

The magnitude-squared response of the analog lowpass Type 1 Chebyshev filter $H_a(s)$ of Nth order is given by

$$|H_a(j\Omega)|^2 = \frac{1}{1 + \varepsilon^2 T_N^2(\Omega/\Omega_p)}, \tag{5.37}$$

where $T_N(\Omega)$ is the Chebyshev polynomial of order N:

$$T_N(\Omega) = \begin{cases} \cos(N\cos^{-1}\Omega), & |\Omega| \leq 1, \\ \cosh(N\cosh^{-1}\Omega), & |\Omega| > 1. \end{cases} \tag{5.38}$$

The above polynomial can also be derived via a recurrence relation given by

$$T_r(\Omega) = 2\Omega T_{r-1}(\Omega) - T_{r-2}(\Omega), \qquad r \geq 2, \tag{5.39}$$

with $T_0(\Omega) = 1$ and $T_1(\Omega) = \Omega$.

Typical plots of the magnitude responses of the Type 1 Chebyshev lowpass filter are shown in Figure 5.16 for three different values of filter order N with the same passband ripple ε. From these plots it is seen that the squared-magnitude response is equiripple between $\Omega = 0$ and $\Omega = 1$, and it decreases monotonically for all $\Omega > 1$.

The order N of the transfer function is determined from the attenuation specification in the stopband at a particular frequency. For example, if at $\Omega = \Omega_s$ the magnitude is equal to $1/A$, then from Eqs. (5.37) and (5.38),

$$
\begin{aligned}
|H_a(j\Omega_s)|^2 &= \frac{1}{1 + \varepsilon^2 T_N^2(\Omega_s/\Omega_p)} \\
&= \frac{1}{1 + \varepsilon^2 \left\{ \cosh[N \cosh^{-1}(\Omega_s/\Omega_p)] \right\}^2} = \frac{1}{A^2}.
\end{aligned}
\tag{5.40}
$$

Solving the above, we get

$$
N = \frac{\cosh^{-1}(\sqrt{A^2 - 1}/\varepsilon)}{\cosh^{-1}(\Omega_s/\Omega_p)} = \frac{\cosh^{-1}(1/k_1)}{\cosh^{-1}(1/k)}.
\tag{5.41}
$$

In computing N using the above expression, it is usually convenient to use the identity $\cosh^{-1}(y) = \ln\left(y + \sqrt{y^2 - 1}\right)$. As in the case of the Butterworth filter, the order N of the filter is chosen as the nearest integer greater than or equal to the number given by Eq. (5.41).

The transfer function $H_a(s)$ is again of the form of Eq. (5.34), with

$$
p_\ell = \sigma_\ell + j\,\Omega_\ell, \qquad \ell = 1, 2, \ldots, N,
\tag{5.42}
$$

where

$$
\sigma_\ell = -\Omega_p \xi \sin\left[\frac{(2\ell - 1)\pi}{2N}\right], \qquad \Omega_\ell = \Omega_p \zeta \cos\left[\frac{(2\ell - 1)\pi}{2N}\right],
\tag{5.43a}
$$

$$
\xi = \frac{\gamma^2 - 1}{2\gamma}, \qquad \zeta = \frac{\gamma^2 + 1}{2\gamma}, \qquad \gamma = \left(\frac{1 + \sqrt{1 + \varepsilon^2}}{\varepsilon}\right)^{1/N}.
\tag{5.43b}
$$

Type 2 Chebyshev Approximation

The magnitude-squared response of the the analog lowpass Type 2 Chebyshev filter, also known as the *inverse Chebyshev approximation*, exhibits a monotonic behavior in the passband with a maximally flat response at $\Omega = 0$ and an equiripple behavior in the stopband. The squared-magnitude response expression here is given by

$$
|H_a(j\Omega)|^2 = \frac{1}{1 + \varepsilon^2 \left[\frac{T_N(\Omega_s/\Omega_p)}{T_N(\Omega_s/\Omega)}\right]^2}.
\tag{5.44}
$$

Typical responses are as shown in Figure 5.17. The transfer function of a Type 2 Chebyshev lowpass filter is no longer an all-pole function and has both poles and zeros. If we write

$$
H_a(s) = C_0 \frac{\prod_{\ell=1}^{N}(s - z_\ell)}{\prod_{\ell=1}^{N}(s - p_\ell)},
\tag{5.45}
$$

the zeros are on the $j\Omega$-axis and are given by

$$
z_\ell = j\frac{\Omega_s}{\cos\left[\frac{(2\ell-1)\pi}{2N}\right]}, \qquad \ell = 1, 2, \ldots, N.
\tag{5.46}
$$

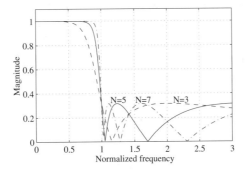

Figure 5.17: Typical Type 2 Chebyshev lowpass filter responses with 10-dB minimum stopband attenuation.

If N is odd, then for $\ell = (N+1)/2$, the zero is at $s = \infty$. The poles are located at

$$p_\ell = \sigma_\ell + j\Omega_\ell, \quad \ell = 1, 2, \ldots, N, \tag{5.47}$$

where

$$\sigma_\ell = \frac{\Omega_s \alpha_\ell}{\alpha_\ell^2 + \beta_\ell^2}, \qquad \Omega_\ell = -\frac{\Omega_s \beta_\ell}{\alpha_\ell^2 + \beta_\ell^2}, \tag{5.48a}$$

$$\alpha_\ell = -\Omega_p \zeta \sin\left[\frac{(2\ell-1)\pi}{2N}\right], \qquad \beta_\ell = \Omega_p \xi \cos\left[\frac{(2\ell-1)\pi}{2N}\right], \tag{5.48b}$$

$$\zeta = \frac{\gamma^2 - 1}{2\gamma}, \qquad \xi = \frac{\gamma^2 + 1}{2\gamma}, \qquad \gamma = \left(A + \sqrt{A^2 - 1}\right)^{1/N}. \tag{5.48c}$$

The order N of the Type 2 Chebyshev lowpass filter is determined from given ε, Ω_s, and A using Eq. (5.41).

EXAMPLE 5.6   We wish to determine the minimum order N required to design a lowpass filter with a Chebyshev or an inverse Chebyshev response with the specifications given in Example 5.5.
   From Example 5.5, we have the following parameters: $1/k_1 = 196.51334$ and $1/k = 5$. Substituting these values in Eq. (5.41), we arrive at

$$N = \frac{\cosh^{-1}(1/k_1)}{\cosh^{-1}(1/k)} = \frac{\cosh^{-1}(196.51334)}{\cosh^{-1}(5)} = 2.60591. \tag{5.49}$$

Since the order of the filter must be an integer, we choose the next highest integer value 3 for N. Note that the order of the Chebyshev lowpass filter is lower than that of a Butterworth lowpass filter meeting the same specifications as given by Eq. (5.36). This is invariably the case for $N \geq 2$.

5.4.4 Elliptic Approximation

An *elliptic filter*, also known as a *Cauer filter*, has an equiripple passband and an equiripple stopband magnitude response, as indicated in Figure 5.18 for typical elliptic lowpass filters. The transfer function of an elliptic filter meets a given set of filter specifications, passband edge frequency Ω_p, stopband edge frequency Ω_s, passband ripple ε, and minimum stopband attenuation A, with the lowest filter order N. The theory of elliptic filter approximation is mathematically quite involved, and a detailed treatment of this topic is beyond the scope of this text. Interested readers are referred to the books by Antoniou [Ant93], Parks and Burrus [Par87], and Temes and LaPatra [Tem77].

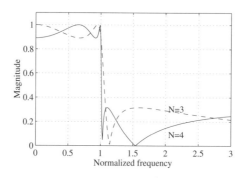

Figure 5.18: Typical elliptic lowpass filter responses with 1-dB passband ripple and 10-dB minimum stopband attenuation.

The square-magnitude response of an elliptic lowpass filter is given by

$$|H_a(j\Omega)|^2 = \frac{1}{1 + \varepsilon^2 R_N^2(\Omega/\Omega_p)}, \tag{5.50}$$

where $R_N(\Omega)$ is a rational function of order N satisfying the property $R_N(1/\Omega) = 1/R_N(\Omega)$, with the roots of its numerator lying within the interval $0 < \Omega < 1$ and the roots of its denominator lying in the interval $1 < \Omega < \infty$. For most applications, the filter order meeting a given set of specifications of passband edge frequency Ω_p, passband ripple ε, stopband edge frequency Ω_s, and the minimum stopband ripple A can be estimated using the approximate formula [Ant93]

$$N \cong \frac{2\log_{10}(4/k_1)}{\log_{10}(1/\rho)}, \tag{5.51}$$

where k_1 is the discrimination parameter defined in Eq. (5.30) and ρ is computed as follows:

$$k' = \sqrt{1 - k^2}, \tag{5.52a}$$

$$\rho_0 = \frac{1 - \sqrt{k'}}{2(1 + \sqrt{k'})}, \tag{5.52b}$$

$$\rho = \rho_0 + 2(\rho_0)^5 + 15(\rho_0)^9 + 150(\rho_0)^{13}. \tag{5.52c}$$

In Eq. (5.52a), k is the selectivity parameter defined in Eq. (5.29).

EXAMPLE 5.7   We wish to determine the minimum order N required to design a lowpass elliptic filter with the specifications given in Example 5.5.

From Example 5.5, we note that $k = 1/5 = 0.2$ and $1/k_1 = 196.5134$. Substituting these values in Eqs. (5.52a) to (5.52c), we arrive at

$$k' = 0.979796, \quad \rho_0 = 0.00255135, \quad \rho = 0.0025513525.$$

Using the above values in Eq. (5.51), we get

$$N = 2.23308. \tag{5.53}$$

Rounding the above value to the nearest higher integer, we obtain $N = 3$ as the appropriate order for the elliptic lowpass filter.

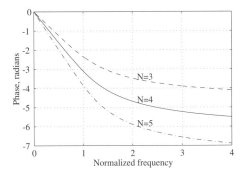

Figure 5.19: The phase responses of typical normalized Bessel lowpass filters.

5.4.5 Linear-Phase Approximation

The previous three approximation techniques are for developing analog lowpass transfer functions meeting specified magnitude or gain response specifications without any concern for their phase responses. In a number of applications it is desirable that the analog lowpass filter being designed have a linear-phase characteristic in the passband, in addition to approximating the magnitude specifications. One way to achieve this goal is to cascade an analog allpass filter with the filter designed to meet the magnitude specifications, so that the phase response of the overall cascade realization approximates linear-phase response in the passband. This approach increases the overall hardware complexity of the analog filter and may not be desirable for designing an analog anti-aliasing filter in some A/D conversion or designing an analog reconstruction filter in D/A conversion applications.

It is possible to design a lowpass filter that approximates a linear-phase characteristic in the passband. Such a filter has an all-pole transfer function of the form

$$H_a(s) = \frac{d_0}{B_N(s)} = \frac{d_0}{d_0 + d_1 s + \cdots + d_{N-1}s^{N-1} + s^N}, \tag{5.54}$$

and provides a maximally flat approximation to the linear-phase characteristic at $\Omega = 0$, i.e., has a maximally flat constant group delay at dc ($\Omega = 0$). For a normalized group delay of unity at dc, the denominator polynomial $B_N(s)$ of the transfer function, called the *Bessel polynomial*, can be derived via the recursion relation

$$B_N(s) = (2N - 1)B_{N-1}(s) + s^2 B_{N-2}(s), \tag{5.55}$$

starting with $B_1(s) = s + 1$ and $B_2(s) = s^2 + 3s + 3$. Alternatively, the coefficients of the Bessel polynomial $B_N(s)$ can be found from

$$d_\ell = \frac{(2N - \ell)!}{2^{N-\ell}\ell!(N - \ell)!}, \qquad \ell = 0, 1, \ldots, N - 1. \tag{5.56}$$

These filters are often referred to as *Bessel filters*. Figure 5.19 shows the phase responses of some typical Bessel filters. It should be noted that the Bessel filter has a poorer magnitude response than that of the lowpass filter of the same order designed using any one of the previous three techniques.

5.4.6 Analog Filter Design Using MATLAB

The *Signal Processing Toolbox* in MATLAB includes a number of M-files to directly develop analog transfer functions for each one of the above approximation techniques. We next review these functions.

Butterworth filter

The M-files for the design of analog Butterworth filters are

```
[z,p,k]     = buttap(N),
[num,den]   = butter(N,Wn,'s')
[num,den]   = butter(N,Wn,'type','s')
[N, Wn]     = buttord(Wp,Ws,Rp,Rs,'s')
```

The M-file `buttap(N)` computes the zeros, poles, and gain factor of the normalized analog Butterworth lowpass filter transfer function of order N with a 3-dB cutoff frequency of 1. The output files are the length N column vector p providing the locations of the poles, a null vector z for the zero locations, and the gain factor k. The form of the transfer function obtained is given by

$$H_a(s) = \frac{P_a(s)}{D_a(s)} = \frac{k}{(s - p(1))\,(s - p(2))\cdots(s - p(N))}. \tag{5.57}$$

To determine the numerator and denominator coefficients of the transfer function from the zeros and poles computed, we need to use the M-file `zp2tf(z,p,k)`.[5]

Alternatively, we can use the M-file `butter(N,Wn,'s')` to design an order-N lowpass transfer function with a prescribed 3-dB cutoff frequency at Wn rad/sec, a nonzero number. The output data of this M-file are the numerator and the denominator coefficient vectors, num and den, respectively, in descending powers of s. If Wn is a two-element vector [W1,W2] with W1 < W2, the M-file generates an order-2N bandpass transfer function with 3-dB bandedge frequencies at W1 and W2 with both being nonzero numbers. To design an order-N highpass or an order-2N bandstop filter, the M-file `butter(N,Wn,'type','s')` is employed where type = high for a highpass filter with a 3-dB cutoff frequency at Wn or type = stop for a bandstop filter with 3-dB stopband edges given by a two-element vector of nonzero numbers Wn = [W1,W2] with W1 < W2.

The M-file `buttord(Wp,Ws,Rp,Rs,'s')` computes the lowest order N of a Butterworth analog transfer function meeting the specifications given by the filter parameters, Wp, Ws, Rp, and Rs, where Wp is the passband edge angular frequency in rad/sec, Ws is the stopband edge angular frequency in rad/sec, Rp is the maximum passband attenuation in dB, and Rs is the minimum stopband attenuation in dB. The output data are the filter order N and the 3-dB cutoff angular frequency Wn in rad/sec. This M-file can also be used to calculate the order of any one of the four basic types of analog Butterworth filters. For lowpass design, Wp < Ws, whereas for highpass design, Wp > Ws. For the other two types, Wp and Ws are two-element vectors specifying the passband and stopband edge frequencies.

Type 1 Chebyshev Filter

The M-files for the design of analog Type 1 Chebyshev filters are as follows:

```
[z,p,k]     = cheb1ap(N,Rp)
[num,den]   = cheby1(N,Rp,Wn,'s')
[num,den]   = cheby1(N,Rp,Wn,'type','s')
[N,Wn]      = cheb1ord(Wp,Ws,Rp,Rs,'s')
```

The M-file `cheb1ap(N,Rp)` computes the zeros, poles, and gain factor of the normalized analog Type 1 Chebyshev lowpass filter transfer function of order N with a passband ripple in dB given by Rp. The normalized passband edge frequency is set to 1. The output files are the column vector p providing

---

[5] See Section 3.8.

the locations of the poles, a null vector z for the zero locations, and the gain factor k. The form of the transfer function is as in Eq. (5.57).

As in the previous case, the numerator and denominator coefficients of the transfer function can be determined using the M-file zp2tf(z,p,k). The rational form of the Type 1 Chebyshev lowpass filter transfer function can also be determined directly using the M-file cheby1(N,Rp,Wn,'s'), where Wn is the passband edge angular frequency in rad/sec and Rp is the passband ripple in dB. The output data are the vectors, num and den, containing the numerator and the denominator coefficients of the transfer function in descending powers of s. If Wn is a two-element vector [W1,W2] with W1 < W2, the M-file computes the transfer function of an order-2N bandpass filter with passband edge angular frequencies in rad/sec given by W1 and W2. The M-file cheby1(N,Rp,Wn,'type','s') is employed for the other two types of filter designs, where the rtype = high for the highpass case and type = stop for the bandstop case. Wn is a scalar representing the passband edge frequency for the highpass filter design, and is a two-element vector defining the stopband edge frequencies for a bandstop filter design.

The M-file cheb1ord(Wp,Ws,Rp,Rs,'s') determines the lowest order N of a Type 1 Chebyshev analog transfer function meeting the specifications given by the filter parameters, Wp, Ws, Rp, and Rs, where Wp is the passband edge angular frequency, Ws is the stopband angular edge frequency, Rp is the passband ripple in dB, and Rs is the minimum stopband attenuation in dB. The output data are the filter order N and the cutoff angular frequency Wn. This M-file can also be used to calculate the order of any one of the four basic types of analog Type 1 Chebyshev filters. For the lowpass design, Wp < Ws, whereas for the highpass design, Wp > Ws. For the other two types, Wp and Ws are two-element vectors specifying the passband and stopband edge frequencies. All bandedge frequencies are specified in rad/sec.

Type 2 Chebyshev Filter

The M-files for the design of analog Type 2 Chebyshev filters are

```
[z,p,k]    = cheb2ap(N,Rs)
[num,den] = cheby2(N,Rs,Wn,'s')
[num,den] = cheby2(N,Rs,Wn,'type','s')
[N,Wn]     = cheb2ord(Wp,Ws,Rp,Rs,'s')
```

The M-file cheb2ap(N,Rs) returns the zeros, poles, and gain factor of a normalized analog Type 2 Chebyshev lowpass filter of order N with a minimum stopband attenuation of Rs in dB. The normalized stopband edge angular frequency is set to 1. The output data are the length-N column vectors z and p, providing the locations of the zeros and the poles, respectively, and the gain factor k. If N is odd, z is of length N-1. The form of the transfer function obtained is given by

$$H_a(s) = \frac{P_a(s)}{D_a(s)} = k\frac{(s - z(1))\,(s - z(2)) \cdots (s - z(N))}{(s - p(1))\,(s - p(2)) \cdots (s - p(N))}. \tag{5.58}$$

The M-file cheby2(N,Rs,Wn,'s') can be employed to determine the transfer function of a Type 2 Chebyshev lowpass filter when Wn is a scalar defining the stopband edge angular frequency in rad/sec or a bandpass filter when Wn is a two-element vector defining the stopband edge angular frequencies in rad/sec. The M-file cheby2(N,Rs, Wn,'type','s') provides the transfer function of a Type 2 Chebyshev highpass filter when type = high or a bandstop filter when type = stop. In all cases, the specified minimum stopband attenuation is Rs in dB. The output data are the vectors, num and den, containing the numerator and denominator coefficients in descending powers of s.

The M-file cheb2ord(Wp,Ws,Rp,Rs,'s') determines the lowest order N of a Type 2 Chebyshev analog transfer function meeting the specifications given by the filter parameters, Wp, Ws, Rp, and Rs, as defined for the Type 1 Chebyshev filter.

Elliptic (Cauer) Filter

The M-files for the design of analog elliptic filters are

```
[z,p,k]   = ellipap(N,Rp,Rs)
[num,den] = ellip(N,Rp,Rs,Wn,'s')
[num,den] = ellip(N,Rp,Rs,Wn,'type','s')
[N,Wn]    = ellipord(Wp,Ws,Rp,Rs,'s')
```

The M-file ellipap(N,Rp,Rs) determines the zeros, poles, and gain factor of a normalized analog elliptic lowpass filter of order N with a passband ripple of Rp dB and a minimum stopband attenuation of Rs dB. The normalized passband edge angular frequency is set to 1. The output files are the length-N column vectors z and p, providing the locations of the zeros and the poles, respectively, and the gain factor k. If N is odd, z is of length N-1. The form of the transfer function obtained is as given in Eq. (5.58).

The M-file ellip(N,Rp,Rs,Wn,'s') returns the transfer function of an elliptic analog lowpass filter when Wn is a scalar defining the passband edge angular frequency in rad/sec or a bandpass filter when Wn is a two-element vector defining the passband edge frequencies in rad/sec. The M-file ellip(N,Rp,Rs,Wn,'type','s') is used to determine the transfer function of an elliptic highpass when type = high, and Wn is a scalar defining the stopband edge angular frequency in rad/sec, or a bandstop filter when type = stop, and Wn is a two-element vector defining the stopband edge angular frequencies in rad/sec. In all cases, the specified passband ripple is Rp dB and the minimum stopband attenuation is Rs in dB. The output files are the vectors, num and den, containing the numerator and denominator coefficients in descending powers of s.

The M-file ellipord(Wp,Ws,Rp,Rs,'s') determines the lowest order N of an elliptic analog transfer function meeting the specifications given by the filter parameters, Wp, Ws, Rp, and Rs, as defined for the Type 1 Chebyshev filter.

Bessel Filter

For the design of a Bessel filter, the available M-files are

```
[z,p,k]   = besselap(N)
[num,den] = besself(N,Wn)
[num,den] = besself(N,Wn,'type')
```

The M-file besselap(N) is employed to compute the zeros, poles, and gain factor of an order-N Bessel lowpass filter prototype. The output data of these M-files are the length-N column vector p, providing the locations of the poles, and the gain factor k. Since there are no zeros, the output vector z is a null vector. The form of the transfer function is as in Eq. (5.57).

The M-file besself(N,Wn) is used to design an Nth-order analog lowpass Bessel filter with a 3-dB cutoff angular frequency given by the scalar Wn in rad/sec. It generates the length-(N+1) vectors, num and den, containing the numerator and the denominator coefficients in descending powers of s. If Wn is a two-element vector, then it returns the transfer function of an order-2N analog bandpass Bessel filter. For designing the other two types of Bessel filters, the function besself(N,Wn,'type') is used. Here the type = high with Wn representing the 3-dB stopband edge frequency in rad/sec for the highpass case, or type = stop with Wn a two-element vector defining the 3-dB stopband edge frequencies in rad/sec for the bandstop case.

Limitations

The zero-pole-gain form is more accurate than the transfer function form for the design of Butterworth, Type 2 Chebyshev, elliptic, or Bessel filter design. It is recommended that the filter design function in these cases be used only for filter orders less than 15 since numerical problems may arise for filter orders equal to or greater than 15.

Analog Lowpass Filter Design Examples

We provide below several examples illustrating the use of some of the above functions in the design of analog filters. In the first three examples, we repeat Examples 5.5 through 5.7 to determine the order of the transfer function using the respective M-files. In the remaining examples, we determine the corresponding transfer functions and then compute the frequency response using the M-file `freqs(num,den,w)`, where `num` and `den` are the vectors of the numerator and denominator coefficients in descending powers of `s`, and `w` is a set of specified discrete angular frequencies. This function generates a complex vector of frequency response samples from which magnitude and/or phase response samples can be readily computed.

EXAMPLE 5.8 To determine the order of the analog Butterworth lowpass filter meeting the frequency response specifications given in Example 5.5, we use the command `[N,Wn] = buttord (Wp,Ws,Rp,Rs,'s')` where `Wp = 2π(1000)`, `Ws = 2π(5000)`, `Rp = 1`, and `Rs = 40`. The outputs generated are given by `N = 4` and `Wn = 7947.77`, where `Wn` is the 3-dB cutoff angular frequency. The computed order is the same as that determined in Example 5.5 using the formula of Eq. (5.33).

EXAMPLE 5.9 We next determine the order of analog Type 1 and Type 2 Chebyshev lowpass filters meeting the same specifications as above. To this end, we employ the commands `[N,Wn] = cheb1ord (Wp,Ws,Rp,Rs,'s')` and `[N,Wn] = cheb2ord(Wp,Ws,Rp, Rs,'s')`, respectively. For the Type 1 Chebyshev filter, the computed output data are `N = 3` and `Wn = 6283.18`, and for the Type 2 Chebyshev filter, the computed output data are `N = 3` and `Wn = 23440.97`. The computed `Wn` in the former case is the 3-dB passband edge angular frequency in rad/sec, whereas in the second case, it is the 40-dB stopband edge angular frequency in rad/sec. The order determined is identical to that derived in Example 5.6 using the formula of Eq. (5.41).

EXAMPLE 5.10 To determine the order of an analog elliptic lowpass filter meeting the above specifications, we use the command `[N,Wn] = ellipord(Wp,Ws,Rp,Rs,'s')`, resulting in the output data `N = 3` and `Wn = 6283.185` and verifying the order obtained in Example 5.7. Here `Wn` is the 1-dB passband edge angular frequency.

EXAMPLE 5.11 We consider first the design of a fourth-order maximally flat analog lowpass filter with a 3-dB cutoff frequency at $\Omega = 1$. The pertinent MATLAB program is given below.

```
% Program 5_1
% 4-th Order Analog Butterworth Lowpass Filter Design
%
format long
% Determine zeros and poles
[z,p,k] = buttap(4);
disp('Poles are at');disp(p);
% Determine transfer function coefficients
```

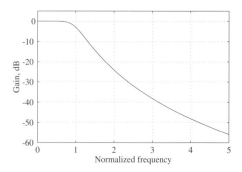

Figure 5.20: The gain response of the normalized fourth-order Butterworth lowpass filter of Example 5.11.

```
[pz, pp] = zp2tf(z, p, k);
% Print coefficients in ascending powers of s
disp('Numerator polynomial'); disp(pz)
disp('Denominator polynomial'); disp(real(pp))
omega = [0: 0.01: 5];
% Compute and plot frequency response
h = freqs(pz,pp,omega);
plot (omega,20*log10(abs(h)));grid
xlabel('Normalized frequency'); ylabel('Gain, dB');
```

The output data generated by running this program are as follows:

```
Poles are at
 -0.38268343236509 + 0.92387953251129i
 -0.92387953251129 + 0.38268343236509i
 -0.92387953251129 - 0.38268343236509i
 -0.38268343236509 - 0.92387953251129i

Numerator polynomial
       0      0      0      0      1
Denominator polynomial
  Columns 1 through 4

   1.00000000000000    2.61312592975275    3.41421356237310
   2.61312592975275

  Column 5
   1.00000000000000
```

The plot of the gain response generated by this program is sketched in Figure 5.20.

EXAMPLE 5.12   We now use MATLAB to complete the design of the Butterworth lowpass filter of Example 5.5. To this end we modify Program 5.1 of Example 5.11 as indicated below in Program 5.2. During execution, the program asks for the order of the filter and the 3-dB cutoff angular frequency (determined in Example 5.8 to be equal to 4 and 7947.77, respectively) to be typed in. It then generates the gain response plot, as shown in Figure 5.21. As can be seen from this plot, the 1-dB passband edge is at 1 kHz, as desired, and at 5 kHz the attenuation is more than 40 dB, as expected.

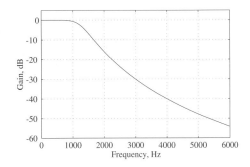

Figure 5.21: The gain response of the Butterworth lowpass filter of Example 5.12.

```
% Program 5_2
% Program to Design Butterworth Lowpass Filter
%
% Type in the filter order and passband edge frequency
N = input('Type in filter order = ');
Wn = input('3-dB cutoff frequency = ');
% Determine the transfer function
[num,den] = butter(N,Wn,'s');
%  Compute and plot the frequency response
omega = [0: 200: 12000*pi];
h = freqs(num,den,omega);
plot (omega/(2*pi),20*log10(abs(h)));
xlabel('Frequency, Hz'); ylabel('Gain, dB');
```

EXAMPLE 5.13   We next illustrate the design of a Type 1 Chebyshev lowpass filter meeting the specifications of Example 5.5. The corresponding MATLAB program is given below in Program 5_3. As the program is run, it asks for the order of the filter, the passband edge angular frequency (determined in Example 5.9 to be equal to 3 and 6283.18, respectively), and the passband ripple (1 dB) to be typed in. It then generates the gain response plot as shown in Figure 5.22.

```
% Program 5_3
% Program to Design Type 1 Chebyshev Lowpass Filter
%
% Read in the filter order, passband edge frequency
% and passband ripple in dB
N = input('Order = ');
Fp = input('Passband edge frequency in Hz = ');
Rp = input('Passband ripple in dB = ');
% Determine the coefficients of the transfer function
[num,den] = cheby1(N,Rp,Fp,'s');
% Compute and plot the frequency response
omega = [0: 200: 12000*pi];
h = freqs(num,den,omega);
plot (omega/(2*pi),20*log10(abs(h)));
xlabel('Frequency, Hz'); ylabel('Gain, dB');
```

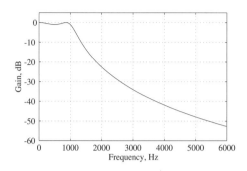

Figure 5.22: The gain response of the Type 1 Chebyshev lowpass filter of Example 5.13.

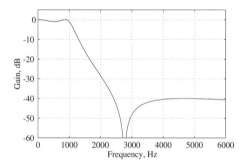

Figure 5.23: The gain response of the elliptic lowpass filter of Example 5.14.

EXAMPLE 5.14   The design of an elliptic lowpass filter meeting the specifications of Example 5.5 follows a similar format. The pertinent MATLAB program is given below in Program 5.4. As the program is run, it asks for the order of the filter (determined in Example 5.10 to be equal to 3), the passband edge angular frequency ($2000\pi = 6283.185$), the passband ripple (1 dB), and the minimum stopband attenuation (40 dB) to be typed in. It then generates the gain response plot as shown in Figure 5.23.

```
% Program 5_4
% Program to Design Elliptic Lowpass Filter
%
% Read in the filter order, passband edge frequency,
% passband ripple in dB and minimum stopband
% attenuation in dB
N = input('Order = ');
Fp = input('Passband edge frequency in Hz = ');
Rp = input('Passband ripple in dB = ');
Rs = input('Minimum stopband attenuation in dB = ');
% Determine the coefficients of the transfer function
[num,den] = ellip(N,Rp,Rs,2*pi*Fp,'s');
% Compute and plot the frequency response
omega = [0: 200: 12000*pi];
h = freqs(num,den,omega);
plot (omega/(2*pi),20*log10(abs(h)));
xlabel('Frequency, Hz'); ylabel('Gain, dB');
```

5.4.7 A Comparison of the Filter Types

In the previous four sections we have described four types of analog lowpass filter approximations, three of which have been developed primarily to meet the magnitude response specifications while the fourth has been developed primarily to provide a near linear-phase approximation. In order to determine which filter type to choose to meet a given magnitude response specification, we need to compare the performances of the four types of approximations. To this end, we compare here the frequency responses of the normalized Butterworth, Chebyshev, and elliptic analog lowpass filters of the same order. The passband ripple of the Type 1 Chebyshev and the equiripple filters are assumed to be the same, while the minimum stopband attenuation of the Type 2 Chebyshev and the equiripple filters are assumed to be the same. The filter specifications used for comparison are as follows: filter order of 6, passband edge at $\Omega = 1$, maximum passband deviation of 1 dB, and minimum stopband attenuation of 40 dB. The frequency responses computed using MATLAB are plotted in Figure 5.24.

As can be seen from Figure 5.24, the Butterworth filter has the widest transition band, with a monotonically decreasing gain response. Both types of Chebyshev filters have a transition band of equal width that is smaller than that of the Butterworth filter but greater than that of the elliptic filter. The Type 1 Chebyshev filter provides a slightly faster roll-off in the transition band than the Type 2 Chebyshev filter. The magnitude response of the Type 2 Chebyshev filter in the passband is nearly identical to that of the Butterworth filter. The elliptic filter has the narrowest transition band, with an equiripple passband and an equiripple stopband response.

The Butterworth and Chebyshev filters have a nearly linear phase response over about three-fourths of the passband, whereas the elliptic filter has a nearly linear phase response over about one-half of the passband. On the other hand, the Bessel filter may be more attractive if the linearity of the phase response over a larger portion of the passband is desired at the expense of a poorer gain response. Figure 5.25 shows the gain and phase responses of a sixth-order Bessel filter frequency scaled to have a passband edge at $\Omega = 1$ with a maximum passband deviation of 1 dB. However, the Bessel filter provides a minimum of 40 dB attenuation at approximately $\Omega = 6.4$ and, as a result, has the largest transition band compared to the other three types.

Another way of comparing the performances of the Butterworth, Chebyshev, and elliptic filters would be to compare the order of these filters required to meet the same filter specifications. For example, consider the specifications of a lowpass filter: passband edge at $\Omega = 1$, maximum passband deviation of 1 dB, stopband edge at $\Omega = 1.2$, and minimum stopband attenuation of 40 dB. These specifications are met by a Butterworth filter of order 29, a Chebyshev Type 1 or 2 filter of order 10, and an elliptic filter of order 6.

5.5 Design of Analog Highpass, Bandpass, and Bandstop Filters

All of the four types of approximations discussed in the previous section dealt with the design of analog lowpass filters meeting the prescribed specifications. Design of the other three classes of analog filters, namely, the highpass, bandpass, and bandstop filters, can be carried out by simple spectral transformations of the frequency variables [Tem77]. The design process involves the development of the specifications of a prototype analog lowpass filter from the specifications of the desired analog filter using a frequency transformation, design of the analog prototype lowpass filter, and then determination of the transfer function of the desired analog filter by applying the inverse of the frequency transformation used to determine the specifications of the prototype lowpass filter.

To eliminate the confusion between the Laplace transform variable of the prototype analog lowpass transfer function $H_{LP}(s)$ and that of the desired analog transfer function $H_D(s)$, we shall use different symbols. Thus we shall use s to denote the Laplace transform variable of the prototype analog lowpass

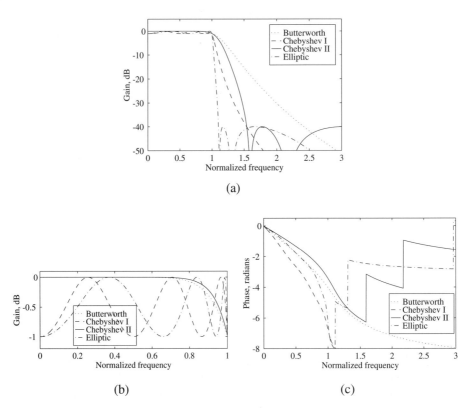

Figure 5.24: A comparison of the frequency responses of the four types of analog lowpass filters: (a) gain responses, (b) passband details, and (c) phase responses.

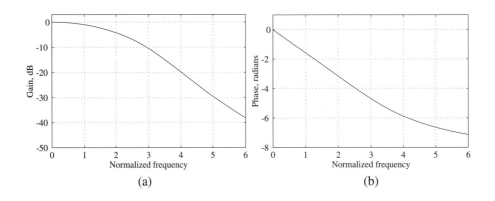

Figure 5.25: The frequency responses of a sixth-order analog Bessel filter: (a) gain response, and (b) unwrapped phase response.

filter $H_{LP}(s)$ and $\hat{s}$ to denote the Laplace transform variable of the desired analog filter $H_D(\hat{s})$. The angular frequency variables in the s- and $\hat{s}$-domains are given by Ω and $\hat{\Omega}$, respectively.

The mapping from s-domain to $\hat{s}$-domain is given by the invertible transformation

$$s = F(\hat{s}).$$

The transfer functions $H_{LP}(s)$ and $H_D(\hat{s})$ are related through

$$H_D(\hat{s}) = H_{LP}(s)\big|_{s=F(\hat{s})},$$
$$H_{LP}(s) = H_D(\hat{s})\big|_{\hat{s}=F^{-1}(s)}.$$

5.5.1 Analog Highpass Filter Design

A prototype analog lowpass transfer function $H_{LP}(s)$ with a passband edge frequency Ω_p can be transformed into an analog highpass transfer function $H_{HP}(\hat{s})$ with a passband edge frequency $\hat{\Omega}_p$ using the spectral transformation

$$s = \frac{\Omega_p \hat{\Omega}_p}{\hat{s}}. \tag{5.59}$$

On the imaginary axis the above transformation reduces to

$$\Omega = -\frac{\Omega_p \hat{\Omega}_p}{\hat{\Omega}}. \tag{5.60}$$

The above mapping implies that the passband of the lowpass filter in the positive frequency range $0 \leq \Omega \leq \Omega_p$ is mapped into the passband of the highpass filter in the negative frequency range $-\infty \leq \hat{\Omega} \leq -\hat{\Omega}_p$, and the passband of the lowpass filter in the negative frequency range $-\Omega_p \leq \Omega \leq 0$ is mapped into the passband of the highpass filter in the positive frequency range $\hat{\Omega}_p \leq \hat{\Omega} \leq \infty$. Likewise, the stopband of the lowpass filter in the positive frequency range $\Omega_s \leq \Omega \leq \infty$ is mapped into the stopband of the highpass filter in the negative frequency range $\hat{\Omega}_s \leq \hat{\Omega} \leq 0$, and the stopband of the lowpass filter in the negative frequency range $-\infty \leq \Omega \leq -\Omega_s$ is mapped into the stopband of the highpass filter in the positive frequency range $0 \leq \hat{\Omega} \leq \hat{\Omega}_s$. The mapping of Eq. (5.59) ensures that the gain value of $|H_{LP}(j\Omega)|$ of the prototype lowpass filter in its passband will appear in the passband $|\hat{\Omega}| \geq \hat{\Omega}_p$ of the desired highpass filter. Likewise, the gain value of the prototype lowpass filter in its stopband $|\Omega| \geq \Omega_s$ will appear in the stopband $0 \leq |\hat{\Omega}| \leq \hat{\Omega}_s$ of the desired highpass filter.

EXAMPLE 5.15 Design an analog Butterworth highpass filter with the following specifications: passband edge at 4 kHz, stopband edge at 1 kHz, passband ripple of 0.1 dB, and minimum stopband attenuation of 40 dB.

For the prototype analog lowpass filter we choose the normalized passband edge to be $\Omega_p = 1$. Hence, from Eq. (5.60), the normalized stopband edge of the lowpass filter is given by

$$\Omega_s = \frac{2\pi \times 4000}{2\pi \times 1000} = 4.$$

The specifications of the lowpass filter are therefore as follows: passband edge at $\Omega_p = 1$ rad/sec, stopband edge $\Omega_s = 4$ rad/sec, passband ripple of 0.1 dB, and minimum stopband attenuation of 40 dB.

We first use the function buttord to determine the order N and the 3-dB cutoff frequency Wn of the lowpass filter. Next we use the function butter to determine the transfer function $H_{LP}(s)$ of the prototype lowpass filter. The lowpass filter is then transformed into the desired highpass filter $H_{HP}(s)$ using the function lp2hp. The code fragments used to design $H_{HP}(s)$ are given below.

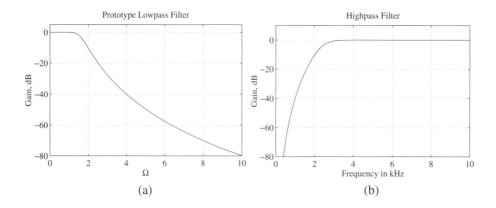

Figure 5.26: (a) Gain response of the prototype analog lowpass filter, and (b) gain response of the desired analog highpass filter of Example 5.15.

```
[N,Wn] = buttord(1,4,0.1,40,'s');
[B,A]  = butter(N,Wn,'s');
[num,den] = lp2hp(B,A,2*pi*4000);
```

The transfer function $H_{LP}(s)$ of the analog lowpass filter can be obtained by displaying the numerator coefficient vector B and denominator coefficient vector A and is given by

$$H_{LP}(s) = \frac{10.2405}{s^5 + 5.1533s^4 + 13.278s^3 + 21.1445s^2 + 20.8101s + 10.2405}.$$

The gain response of the above lowpass filter is shown in Figure 5.26(a). The transfer function $H_{HP}(s)$ of the desired analog highpass filter can be obtained by displaying the numerator coefficient vector num and denominator coefficient vector den. The gain response of the designed highpass filter is shown in Figure 5.26(b).

It should be noted that the desired highpass filter could have been designed directly using the statements

```
[N,Wn] = buttord(8000*pi, 2000*pi, 0.1, 40, 's');
[num,den] = butter(N,Wn,'high','s');
```

5.5.2 Analog Bandpass Filter Design

A prototype analog lowpass transfer function $H_{LP}(s)$ with a passband edge frequency Ω_p can be transformed into an analog bandpass transfer function $H_{BP}(\hat{s})$ with a lower passband edge frequency $\hat{\Omega}_{p1}$ and an upper passband edge frequency $\hat{\Omega}_{p2}$ using the spectral transformation

$$s = \Omega_p \frac{\hat{s}^2 + \hat{\Omega}_o^2}{\hat{s}(\hat{\Omega}_{p2} - \hat{\Omega}_{p1})}. \tag{5.61}$$

On the imaginary axis, the above transformation reduces to

$$\Omega = -\Omega_p \frac{\hat{\Omega}_o^2 - \hat{\Omega}^2}{\hat{\Omega} \, B_w}, \tag{5.62}$$

where $B_w = (\hat{\Omega}_{p2} - \hat{\Omega}_{p1})$ denotes the width of the passband of the bandpass filter. It follows from the above equation that the frequency $\Omega = 0$ is mapped onto the frequency $\hat{\Omega}_o$ which is called the *passband center frequency* of the bandpass filter. It also follows from Eq. (5.62) that the frequency Ω_p maps into the frequencies $\hat{\Omega}_{p2}$ and $-\hat{\Omega}_{p1}$. Moreover, the frequency range $-\Omega_p \le \Omega \le \Omega_p$ of the lowpass filter is mapped into the frequency ranges $-\hat{\Omega}_{p2} \le \hat{\Omega} \le -\hat{\Omega}_{p1}$ and $\hat{\Omega}_{p1} \le \hat{\Omega} \le \hat{\Omega}_{p2}$ of the bandpass filter. It can be shown that

$$\hat{\Omega}_{p1}\hat{\Omega}_{p2} = \hat{\Omega}_{s1}\hat{\Omega}_{s2} = \hat{\Omega}_o^2. \tag{5.63}$$

Thus, the two passband edge frequencies exhibit geometric symmetry with respect to the center frequency $\hat{\Omega}_o$. Likewise, the stopband edge frequencies exhibit geometric symmetry with respect to the center frequency. If the bandedge frequencies do not satisfy the condition of Eq. (5.63), one of them needs to be changed to a new value so that it is satisfied while introducing some safety margin [Tem77]. For example, if

$$\hat{\Omega}_{p1}\hat{\Omega}_{p2} > \hat{\Omega}_{s1}\hat{\Omega}_{s2},$$

then either $\hat{\Omega}_{p1}$ can be decreased to $\hat{\Omega}_{s1}\hat{\Omega}_{s2}/\hat{\Omega}_{p2}$ or $\hat{\Omega}_{s1}$ can be increased to $\hat{\Omega}_{p1}\hat{\Omega}_{p2}/\hat{\Omega}_{s2}$ to satisfy the condition of Eq. (5.63). In the former case, the new passband will be larger than the original desired passband, whereas, in the latter case, the left transition band will be smaller than the original value. On the other hand, if

$$\hat{\Omega}_{p1}\hat{\Omega}_{p2} < \hat{\Omega}_{s1}\hat{\Omega}_{s2},$$

then either $\hat{\Omega}_{p2}$ can be increased or $\hat{\Omega}_{s2}$ can be decreased to satisfy the condition of Eq. (5.63). Moreover, if the gain value of the lowpass filter at a frequency is α dB, the same gain value is obtained for the bandpass filter at the positive frequencies $\hat{\Omega}_a$ and $\hat{\Omega}_b$, with the latter frequencies exhibiting geometric symmetry with respect to $\hat{\Omega}_o$.

EXAMPLE 5.16 Design an analog elliptic bandpass filter with the following specifications: passband edges at 4 kHz and 7 kHz, stopband edges at 3 kHz and 8 kHz, passband ripple of 1 dB, and a minimum stopband attenuation of 22 dB.

The product of the passband edge frequencies is 28×10^6, whereas the product of the stopband edge frequencies is 24×10^6. Since the former product is greater than the latter, we decrease the first passband edge to $24/7 = 3.42857$ kHz. The passband center frequency is thus at $\sqrt{24} = 4.8989795$ kHz. The width of the passband then increases from 3 kHz to $25/7 = 3.571428$ kHz.

For the prototype analog lowpass filter we choose the passband edge frequency $\Omega_p = 1$. From Eq. (5.62) we get the stopband edge frequency of the lowpass filter to be

$$\Omega_s = \frac{24 - 9}{(25/7) \times 3} = 1.4.$$

The specifications of the elliptic lowpass filter are thus given by normalized passband edge at 1 rad/sec, normalized stopband edge at 1.4 rad/sec, passband ripple of 1 dB, and minimum stopband attenuation of 22 dB.

We first use the function `ellipord` to determine the filter order N and the passband edge angular frequency Wn of the prototype lowpass filter $H_{LP}(s)$. We then use the function `ellip` to design the desired lowpass filter $H_{LP}(s)$. Next, using the function `lp2bp` we transform the above lowpass transfer function to the transfer function of the desired bandpass filter $H_{BP}(s)$. The code fragments used to design $H_{BP}(s)$ are given below.

```
[N,Wn] = ellipord(1,1.4,1,22,'s');
[B,A]  = ellip(N,1,22,Wn,'s');
[num,den] = lp2bp(B,A,2*pi*4.8989795,2*pi*25/7);
```

The transfer function of the prototype analog lowpass filter $H_{LP}(s)$ can be obtained by printing the numerator coefficient vector B and the denominator coefficient vector A and is given by

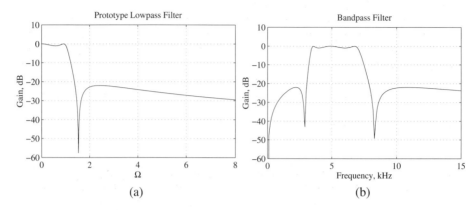

Figure 5.27: (a) Gain response of the prototype analog lowpass filter, and (b) gain response of the desired analog bandpass filter of Example 5.16.

$$H_{LP}(s) = \frac{0.275032211648s^2 + 0.638449761}{s^3 + 0.965577206s^2 + 1.243426s + 0.63844976}.$$

The gain response of the lowpass filter is plotted in Figure 5.27(a). The transfer function of the desired analog bandpass filter $H_{BP}(s)$ can be obtained by printing the numerator coefficient vector num and the denominator coefficient vector den. The gain response of $H_{BP}(s)$ is plotted in Figure 5.27(b).

As in the case of the previous example, the desired bandpass filter could have also been designed directly using the statements

```
Wp = [3.42857 7]*2*pi; Ws = [3 8]*2*pi;
[N,Wn] = ellipord(Wp,Ws,1,22,'s');
[num,den] = ellip(N,1,22,Wn,'s');
```

5.5.3 Analog Bandstop Filter Design

An analog prototype lowpass transfer function $H_{LP}(s)$ with a passband edge frequency Ω_p can be transformed into an analog bandstop transfer function $H_{BP}(\hat{s})$ with a lower stopband edge frequency $\hat{\Omega}_{s1}$ and an upper stopband edge frequency $\hat{\Omega}_{s2}$ using the spectral transformation

$$s = \Omega_s \frac{\hat{s}(\hat{\Omega}_{s2} - \hat{\Omega}_{s1})}{\hat{s}^2 + \hat{\Omega}_o^2}. \tag{5.64}$$

On the imaginary axis, the above transformation reduces to

$$\Omega = \Omega_s \frac{\hat{\Omega} B_w}{\hat{\Omega}_o^2 - \hat{\Omega}^2}, \tag{5.65}$$

where $B_w = (\hat{\Omega}_{s2} - \hat{\Omega}_{s1})$ is the width of the stopband of the bandpass filter. Here the frequency $\Omega = 0$ is mapped onto the frequency $\hat{\Omega}_o$ which is called the *stopband center frequency* of the bandpass filter. It follows from Eq. (5.65) that the frequency range $-\Omega_p \leq \Omega \leq \Omega_p$ of the lowpass filter is mapped into the frequency ranges $-\hat{\Omega}_{s2} \leq \hat{\Omega} \leq -\hat{\Omega}_{s1}$ and $\hat{\Omega}_{s1} \leq \hat{\Omega} \leq \hat{\Omega}_{s2}$ of the bandstop filter. Moreover, as in the case of the analog bandpass filter, the bandedge frequencies here also exhibit geometric symmetry with respect to the center frequency, i.e.,

$$\hat{\Omega}_{s1}\hat{\Omega}_{s2} = \hat{\Omega}_o^2, \qquad \hat{\Omega}_{p1}\hat{\Omega}_{p2} = \hat{\Omega}_o^2.$$

Since the design of the analog bandstop filter is very similar to that of the analog bandpass filter, we leave it as an exercise for the reader (Problem 5.27 and Exercise M5.10).

5.6 Anti-Aliasing Filter Design

According to the sampling theorem of Section 5.2.1, a bandlimited continuous-time signal $g_a(t)$ can be fully recovered from its uniformly sampled version if the condition of Eq. (5.10) is satisfied, i.e., if $g_a(t)$ is sampled at a sampling frequency Ω_T that is at least twice the highest frequency Ω_m contained in $g_a(t)$. If this condition is not satisfied, the original continuous-time signal $g_a(t)$ cannot be recovered from its sampled version because of distortion caused by aliasing. In practice, $g_a(t)$ is passed through an analog anti-aliasing lowpass filter prior to sampling to enforce the condition of Eq. (5.10). This analog filter is the first circuit in the interface between the continuous-time and the discrete-time domains and is studied in this section.

Ideally, the anti-aliasing filter $H_a(s)$ should have a lowpass frequency response $H_a(j\Omega)$ given by

$$H_a(j\Omega) = \begin{cases} 1, & |\Omega| < \Omega_T/2, \\ 0, & |\Omega| \geq \Omega_T/2. \end{cases} \tag{5.66}$$

Such a "brickwall" type frequency response cannot be realized using practical analog circuit components and, hence, must be approximated. A practical anti-aliasing filter therefore should have a magnitude response approximating unity in the passband with an acceptable tolerance, a stopband magnitude response exceeding a minimum attenuation level, and an acceptable transition band separating the passband and the stopband, with a transmission zero at infinity. In addition, in many applications, it is also desirable to have a linear-phase response in the passband. The passband edge frequency Ω_p, the stopband edge frequency Ω_s, and the sampling frequency Ω_T must satisfy the relation

$$\Omega_p < \Omega_s \leq \frac{\Omega_T}{2}. \tag{5.67}$$

The passband edge frequency Ω_p is determined by the highest frequency in the continuous-time signal $g_a(t)$ that must be faithfully preserved in the sampled version. Since signal components with frequencies greater than $\Omega_T/2$ appear as frequencies less than $\Omega_T/2$ due to aliasing, the attenuation level of the anti-aliasing filter at frequencies greater than $\Omega_T/2$ is determined by the amount of aliasing that can be tolerated in the passband. The maximum aliasing distortion comes from the signal components in the replicas of the input spectrum adjacent to the baseband.[6] It follows from Figure 5.28 that the frequency $\Omega_o = \Omega_T - \Omega_p$ is aliased into Ω_p, and if the acceptable amount of aliased spectrum at Ω_p is $\alpha_p = -20\log_{10}(1/A)$, then the minimum attenuation of the anti-aliasing filter at Ω_0 must also be α_p [Jac96].

EXAMPLE 5.17   Consider the design of an anti-aliasing filter with a Butterworth lowpass magnitude response. From Eq. (5.31) the difference in dB in the attenuation levels at Ω_p and Ω_0 is given by

$$10\log_{10}\left[\frac{1 + (\Omega_o/\Omega_c)^{2N}}{1 + (\Omega_p/\Omega_c)^{2N}}\right] \cong 10\log_{10}\left(\frac{\Omega_o}{\Omega_p}\right)^{2N} \tag{5.68}$$

---

[6]It is tacitly assumed here that the magnitude response of the anti-aliasing filter is monotonically decreasing in the stopband.

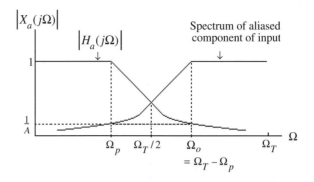

Figure 5.28: Anti-aliasing filter magnitude response and its effect in the signal band of interest.

Table 5.1: Approximate minimum stopband attenuation of a Butterworth lowpass filter.

Ω_0	$2\Omega_p$	$3\Omega_p$	$4\Omega_p$
Attenuation (dB)	$6.02N$	$9.54N$	$12.04N$
Ω_T	$3\Omega_p$	$4\Omega_p$	$5\Omega_p$

This difference has been tabulated in Table 5.1 for several values of the ratio Ω_o/Ω_p as a function of the filter order N.

This table can be used to estimate the minimum sampling frequency Ω_T as a function of the passband edge Ω_p for a specified filter order and a specified minimum stopband attenuation level at a frequency $\Omega_o = \Omega_T - \Omega_p$. For example, if the minimum stopband attenuation at $\Omega_T - \Omega_p$ is 60 dB, then from Table 5.1 we observe that for a sampling frequency $\Omega_T = 3\Omega_p$, the filter order should be $N = \lceil 60/6.02 \rceil = \lceil 9.967 \rceil = 10$.[7] If we increase the sampling frequency to $4\Omega_p$, then the filter order reduces to $N = \lceil 60/9.54 \rceil = \lceil 6.29 \rceil = 7$. For a still higher sampling frequency of $5\Omega_p$, the filter order becomes $N = \lceil 60/12.04 \rceil = \lceil 4.98 \rceil = 5$.

In practice, the sampling frequency chosen depends on the specific application. In applications requiring minimal aliasing, the sampling rate is typically chosen to be 3 to 4 times the passband edge Ω_p of the anti-aliasing analog filter. In noncritical applications, a sampling rate of twice the passband edge Ω_p of the anti-aliasing analog filter is more than adequate. For example, in pulse-code modulation (PCM) telephone systems, the voice signal is first bandlimited to 4 kHz by an anti-aliasing analog filter with a passband edge at 3.6 kHz and a stopband edge at 4 kHz. These specifications are typically met by a third-order elliptic lowpass filter. The output of the filter is then sampled at 8 kHz.

Requirements for the analog anti-aliasing filter can be relaxed by oversampling the analog signal and then decimating the high-sampling-rate digital signal to the desired low-rate digital signal. The decimation process can be implemented completely in the digital domain by first passing the high-rate digital signal through a digital anti-aliasing filter and then down-sampling its output. To understand the advantages of the oversampling approach, consider the sampling of an analog signal bandlimited to a frequency Ω_m. Figure 5.29 shows the spectra of the sampled version of this signal sampled at two different rates, Ω_T

---

[7] $\lceil x \rceil$ denotes the smallest integer greater than or equal to x.

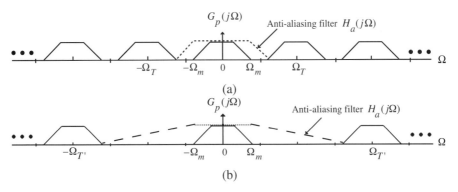

Figure 5.29: Analog anti-aliasing filter requirements for two different oversampling rates.

and $\Omega'_T = 2\Omega_T$, where Ω_T is slightly higher than the Nyquist rate of $2\Omega_m$. These figures also show the desired frequency response of the analog anti-aliasing filter in both cases. Note that the transition band of the analog anti-aliasing filter in the latter case is considerably more than 3 times that needed in the former situation. As a result, the filter specifications are met more easily with a much lower order analog filter.

For the design of the anti-aliasing filter, any one of the four approximations described in Section 5.4 can be employed. Of the four types, the Butterworth approximation provides a reasonable compromise between the desired magnitude response and a linear-phase response in the passband for a given filter order. For an improved phase response at the expense of a poorer magnitude response, the Bessel approximation can be used. On the other hand, for an improved magnitude response with a poorer phase response, either the Chebyshev or the elliptic approximation can be used, with the latter providing the smallest aliasing error for a given filter order. However, in the latter cases, it is necessary to ensure that the transfer function has a zero at infinity. Otherwise, the tails of all the shifted spectra will add up to infinity.

Once the transfer function $H_a(s)$ for the anti-aliasing filter meeting the requirements has been determined, it can be implemented in a number of ways, such as a passive RLC filter, an active-RC filter, or a switched-capacitor filter [Lar93]. A detailed discussion of these implementations is beyond the scope of this book, and we refer the reader to the texts listed at the end of this book [Dar76], [Tem77], [Tem73].

5.7 Sample-and-Hold Circuit[8]

As indicated earlier, the bandlimited output of the analog anti-aliasing filter is fed into a sample-and-hold (S/H) circuit, which is the second circuit in the interface between the continuous-time and discrete-time domains. It samples the analog signal at uniform intervals and holds the sampled value after each sampling operation for sufficient time for accurate conversion by the A/D converter.

The basic S/H circuit, shown in Figure 5.30, operates as follows: During the sampling phase the periodically operated analog switch S remains closed allowing the capacitor C to track the input analog signal $x_a(t)$ and to charge up to a voltage equal to that of the input. During the hold period, the switch remains open, permitting the charged capacitor to hold the voltage across it until the next sampling phase begins. The operation of the switch is controlled by a digital clock signal. The voltage follower at the output of the S/H circuit acts as a buffer between the capacitor C and the input stage of the A/D converter.

Typical input-output waveforms of an S/H circuit are as shown in Figure 5.31, where the dotted line represents the input continuous-time signal $x_a(t)$ and the solid line represents the output $x_d(t)$, assuming instantaneous change from the sample mode to the hold mode and vice versa.

---

[8]This section has been adapted from [Mit80] by permission of the author and the publisher.

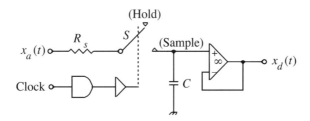

Figure 5.30: The basic sample-and-hold circuit.

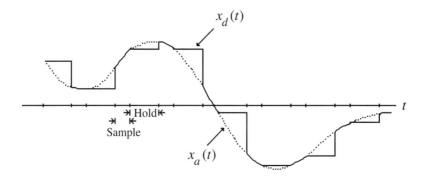

Figure 5.31: Input-output waveforms of a sample-and-hold circuit.

Practical S/H circuits are often much more complex than the basic circuit of Figure 5.30. They typically include an additional operational amplifier at the input to provide better isolation between the source and the capacitor and better tracking of the input signal. They may contain additional circuit components to minimize the effect of the hold voltage decay, which otherwise may occur due to the leakage through the input resistance of the output buffer amplifier and through the finite OFF resistance of the switch. The major parameters characterizing the performance of a practical S/H circuit are acquisition time, aperture time, and droop. The total time needed to switch from the hold mode to the sample mode and acquire the input analog signal within a specified accuracy is defined as its *acquisition time*. This parameter depends on the switching delay time, the time constant of the RC circuit, and the dynamic performance of the output operational amplifier determined primarily by its slew time and settling time. The time taken by the switch to change from the sample mode to the hold mode is defined as the *aperture time*. The hold voltage drift per second due to leakage out of the holding capacitor is called the *droop*. In addition to these parameters, nonideal properties of the operational amplifiers used in the design of the S/H circuit should also be taken into account in determining the overall performance of the circuit.

5.8 Analog-to-Digital Converter

The next step in the digital processing of an analog signal is the conversion of the output of the S/H circuit in its hold mode to a digital form by means of an analog-to-digital (A/D) converter. For digital signal processing applications, the output of the A/D converter is usually in binary code. The output is a sequence of *words*, with each word representing a sample of the sequence. The *wordlength* of the A/D converter output, given by the number of bits, limits the achievable dynamic range of the converter and its accuracy

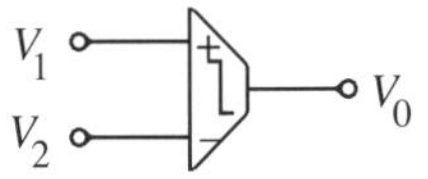

Figure 5.32: Block diagram representation of an analog comparator.

in representing the input analog signal. The accuracy of conversion of an ideal A/D converter is expressed in terms of its *resolution*, which is determined by the number of discrete levels that can be assumed by the A/D converter output. For an output coded in natural binary form with an N-bit wordlength, the number of available discrete levels is 2^N, and as a result, the resolution or accuracy is 1 part in 2^N or $100/2^N$ percent.

There is a variety of A/D converters that are used in signal processing applications [Lar93]. In all of these converters, the analog comparator is an important circuit component. The analog comparator is a device that compares two analog voltages at its input and develops a binary output indicating which input voltage level is larger. With respect to its circuit symbol shown in Figure 5.32, the input-output relation of an analog comparator is thus as follows:

$$V_0 = \begin{cases} V^+, & \text{if } V_1 > V_2, \\ V^-, & \text{if } V_1 < V_2, \end{cases} \tag{5.69}$$

where $V^+ > V^-$. Usually these circuits are designed such that their output voltage levels are compatible with the logic levels of conventional digital circuits.

We briefly review next the operations of the following types of A/D converters: (1) flash A/D converter, (2) serial-parallel A/D converter, (3) successive-approximation A/D converter, (4) counting A/D converter, and (5) oversampling A/D converter. A detailed discussion of their design and implementation is beyond the scope of this book.

5.8.1 Flash A/D Converters

In an N-bit flash converter, shown in block diagram form in Figure 5.33, the input analog voltage V_A is compared simultaneously with a set of $2^N - 1$ uniformly separated reference voltage levels by means of a set of $2^N - 1$ analog comparators, and the location of the adjacent comparators for which the outputs changes from V^- to V^+ indicates the range of the input voltage. A logic encoder circuit is then used to convert the location information into an N-bit binary code. The set of reference voltages is usually derived by a potential-divider resistor string. In a flash A/D converter, all output bits are developed simultaneously, and as a result, it is the fastest converter with a conversion time given by the comparator switching time plus the propagation delay of the encoder circuit. However, the hardware requirements of this type of converter increase very rapidly (exponentially) with an increase in the resolution. As a result, flash converters are employed for low-resolution (typically 8-bit or less) and high-speed conversion applications.

5.8.2 Serial-Parallel A/D Converter

Two $N/2$-bit flash converters in a serial-parallel configuration can be employed to reduce the hardware complexity of an N-bit flash converter at a slight increase in the conversion time [Lar93]. One such scheme, called a *subranging A/D converter*, is shown in Figure 5.34. Here, a coarse approximation of the input analog voltage V_A composed of the $N/2$ most significant bits (MSBs) is first generated by means of one of the $N/2$-bit flash converters. These MSBs are then fed into a D/A converter whose analog output is subtracted from the V_A. The difference voltage is scaled by an amplifier of gain $2^{N/2}$ and converted into digital form by the second (fine) $N/2$-bit flash converter, which provides the least significant bits (LSBs).

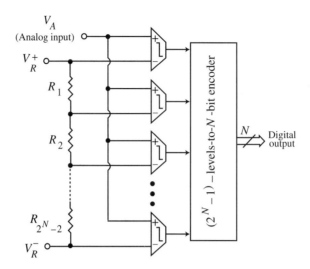

Figure 5.33: Block diagram representation of a flash unipolar A/D converter.

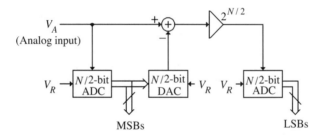

Figure 5.34: Block diagram representation of a subranging A/D converter.

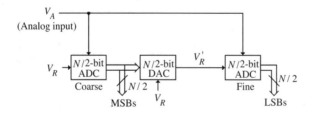

Figure 5.35: Block diagram representation of a ripple A/D converter.

The second scheme utilizing also two $N/2$-bit flash converters is called the *ripple A/D converter*, as shown in Figure 5.35. In this scheme, the coarse $N/2$-bit A/D converter has two functions. It generates the $N/2$ MSBs, and it controls a reference voltage generator that develops a reference voltage V_R' for the fine $N/2$-bit A/D converter generating the $N/2$ LSBs.

In both of the two serial-parallel A/D converters, while one of the $N/2$-bit A/D converters is operating, the other is idle, permitting the use of a single $N/2$-bit A/D converter twice in one conversion period. A generalization of the above two schemes employing N 1-bit A/D converters is called the *pipelined A/D converter* [Lar93].

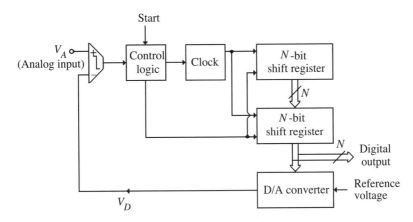

Figure 5.36: Block diagram representation of a successive-approximation A/D converter. (Adapted from [Mit80] by permission of the author and the publisher.)

5.8.3 Successive-Approximation A/D Converter

In this type of converter, essentially a trial-and-error approach is successively used to obtain the digital word representing the input analog voltage V_A [Mit80]. The basic idea behind the operation of this converter can be explained with the aid of its block diagram representation given in Figure 5.36. The conversion procedure is an iterative process. At the kth step of the iteration, the digital approximation stored in the shift register is converted into an analog voltage V_D by the D/A converter in the system. If $V_D < V_A$, then the digital number is increased by setting to ONE the $(k+1)$th bit to the right of the kth bit, which is assumed to be a ONE. If, on the other hand, $V_D > V_A$, then the digital number is decreased by setting the kth bit to a ZERO, and the $(k+1)$th bit to a ONE. The above process is followed for all k from $k = 1, 2, \ldots, N$. After the Nth bit has been examined, the conversion process is terminated, with the contents of the shift register representing the digital equivalent of the input analog voltage. In practice, to round off the approximation to the nearest discrete level, an analog equivalent to one-half of the value of the LSB is added to the analog input signal before conversion begins.

The successive-approximation A/D converter can be designed with high resolution and reasonably high speed at a moderate cost and is therefore widely used in digital signal processing applications. The performance of this type of A/D converter depends primarily on the performance of its constituent D/A converter and the analog comparator.

5.8.4 Counting A/D Converter

In the *counting A/D converter*, shown in block diagram form in Figure 5.37, an N-bit counter begins counting clock pulses when the conversion process is started [Mit80]. The analog equivalent V_D of the digital word of the counter formed by means of a D/A converter, at each clock cycle, is compared with the input analog voltage level V_A. The conversion process is terminated as soon as $V_D > V_A$. Here, the conversion time depends on the value of V_A. It is maximum when the digital word in the counter becomes all ONEs, and thus, for an N-bit binary counter, the conversion time is equal to $(2^N - 1)T$, where T is the clock period. This is a long time for $N > 10$, and hence, this technique can be used only in low-speed applications.

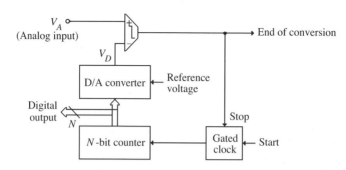

Figure 5.37: Block diagram representation of a counting A/D converter. (Adapted from [Mit80] by permission of the author and publisher.)

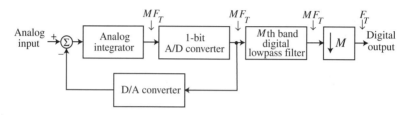

Figure 5.38: Block diagram representation of an oversampling sigma-delta A/D converter.

5.8.5 Oversampling Sigma-Delta A/D Converter

As the name implies, in this type of converter the analog signal is sampled at a rate much higher than the Nyquist rate, resulting in very closely spaced samples. As a consequence, the difference between the amplitudes of two consecutive samples is very small, permitting it to be represented in digital form using very few bits, usually by one bit. The sampling rate is then decreased by passing the digital signal first through a factor-of-M decimator to lower the sampling rate from MF_T to F_T. The decimator is designed by cascading an anti-aliasing lowpass Mth band digital filter to reduce its bandwidth to π/M and a factor-of-M down-sampler.[9] The wordlength of the down-sampler output determines the resolution of the oversampling A/D converter, and it is much higher than that of the high-rate digital signal due to the effect of digital filtering. The basic block diagram representation of the oversampling A/D converter is shown in Figure 5.38. This type of converter, often called a *sigma-delta A/D converter*, is discussed in more detail in Section 11.12.

5.8.6 Characteristics of a Practical A/D Converter [10]

A practical A/D converter is a nonideal device and exhibits a variety of errors that are usually determined in terms of the input analog values at which the transition in the digital output takes place, since these transitions can be measured accurately. In order to understand the effects of these errors on the performance of an A/D converter, consider first its operation as an ideal device. The input-output relation of an ideal 3-bit A/D converter is shown in Figure 5.39. The error introduced by an ideal A/D converter is simply the difference between the value of the analog input and the analog equivalent of the digital representation; this difference is called the *quantization error*. It follows from Figure 5.39 that the quantization error $e[n]$

---

[9]The design of a decimator is considered in Section 10.2.

[10]This section has been adapted from [Mit80] by permission of the author and the publisher.

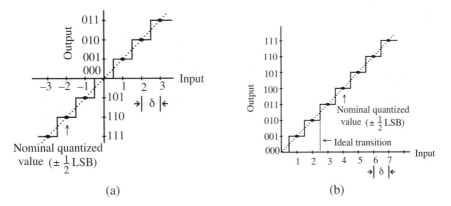

Figure 5.39: Input-output characteristics of an ideal 3-bit A/D converter: (a) bipolar converter, and (b) unipolar converter.

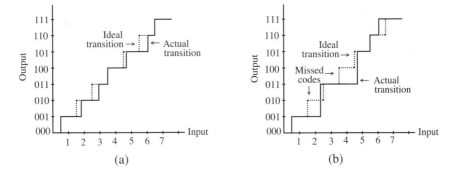

Figure 5.40: Linearity errors in a practical A/D converter.

for an ideal converter satisfies

$$-\frac{\delta}{2} < e[n] \le \frac{\delta}{2}, \tag{5.70}$$

where δ is called the quantization step and is given by

$$\delta = 2^{-N} \tag{5.71}$$

for an N-bit wordlength. Note that δ is precisely the value of the LSB.

A practical A/D converter exhibits *linearity error* if the difference between two consecutive transition values of the input is not equal for the complete range of the input as illustrated in Figure 5.40. The maximum value of this difference value over the full range is called the *differential nonlinearity (DNL) error*. Note from Figure 5.40(b) that in some cases severe nonlinearity in the input-output relation may result in missing codes at the output.

The A/D converter exhibits *gain* or *scale-factor error* if the difference between the last and first transitions is not equal to the full-scale value minus 1/2 LSB. An *offset error* occurs if all transitions are shifted by an equal amount from the ideal transition locations. These errors are illustrated in Figure 5.41.

The time needed by the A/D converter to generate the full digital equivalent of the input analog signal is called its *word-conversion time*, whereas the time required to generate a single bit is called the *bit-conversion time*. In the flash A/D converter, since all bits of the output digital word are generated at the same time, the word-conversion time is essentially equal to the bit-conversion time. On the other hand, in

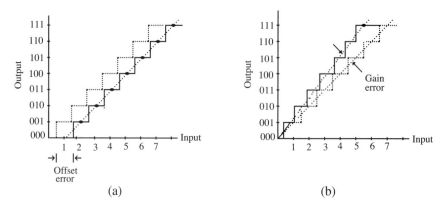

Figure 5.41: (a) Offset and (b) gain errors in a practical ideal A/D converter.

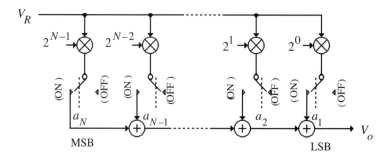

Figure 5.42: Block diagram representation of an N-bit D/A converter.

the successive approximation A/D converter, the word-conversion time is equal to the bit-conversion time multiplied by the wordlength.

It should be noted that an *overflow error* occurs if the input analog voltage V_A exceeds the dynamic range of the A/D converter. It is therefore important to ensure that V_A is properly scaled before it is fed into the A/D converter.

5.9 Digital-to-Analog Converter

The final step in the digital processing of analog signals is the conversion of the digital filter output into an analog form, which is accomplished by a digital-to-analog (D/A) converter followed by a reconstruction filter. The basic idea behind the most commonly used D/A converter can be explained by means of the simplified block diagram representation shown in Figure 5.42, where we have assumed, without any loss of generality, that the digital sample is positive and represented in a natural binary code. Here the ℓth switch S_ℓ is in its ON position if the ℓth binary bit $a_\ell = 1$, and it is in the OFF position if $a_\ell = 0$. The output V_o of the D/A converter is then given by

$$V_o = \sum_{\ell=1}^{N} 2^{\ell-1} a_\ell V_R. \tag{5.72}$$

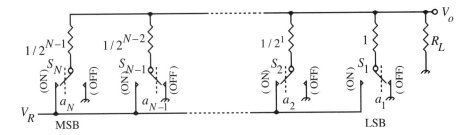

Figure 5.43: Schematic representation of an N-bit weighted resistor unipolar D/A converter.

There are a variety of D/A converters that are used in signal processing applications [Lar93]. We discuss below only the following types: (1) weighted-resistor D/A converter, (2) resistor-ladder D/A converter, and (3) oversampling D/A converter.

5.9.1 Weighted-Resistor D/A Converter

The schematic of an N-bit weighted-resistor D/A converter is shown in Figure 5.43. Here the operation of the switches are as shown in Figure 5.42. It can be shown that the output V_o of the D/A converter is given simply by

$$V_o = \sum_{\ell=1}^{N} 2^{\ell-1} a_\ell \left(\frac{R_L}{(2^N - 1)R_L + 1} \right) V_R. \tag{5.73}$$

The full-scale output voltage $V_{o,FS}$ is obtained when all a_ℓ's are ONEs. Then from Eq. (5.73),

$$V_{o,FS} = \left(\frac{(2^N - 1)R_L}{(2^N - 1)R_L + 1} \right) V_R.$$

In practice, usually $(2^N - 1)R_L \gg 1$ and, as a result, $V_{o,FS} \cong V_R$.

Usually a buffer amplifier is placed at the output to provide gain and prevent loading. For a D/A converter with a moderate to high resolution, the spread of the resistor values becomes very large, making this type of converter unsuitable for many applications.

Based on the same principle as discussed above, a weighted-capacitor D/A converter can be designed. Such circuits are more popular in IC design.

5.9.2 Resistor-Ladder D/A Converter

This type of converter is probably the most widely used in practice. From its schematic representation shown in Figure 5.44, it can be shown that the D/A converter output V_o is given by

$$V_o = \sum_{\ell=1}^{N} 2^{\ell-1} a_\ell \left(\frac{R_L}{2(R_L + R)} \right) \frac{V_R}{2^{N-1}}. \tag{5.74}$$

Because of the resistor values used and the ladder-like circuit connection, the structure is often referred to as the R–$2R$ ladder D/A converter. In practice, often $2R_L \gg R$ and, hence, the full-scale output voltage $V_{o,FS}$ is given by

$$V_{o,FS} \cong \left(\frac{2^N - 1}{2^N} \right) V_R.$$

As in the previous case, a buffer amplifier is also placed at the output to provide gain and prevent loading.

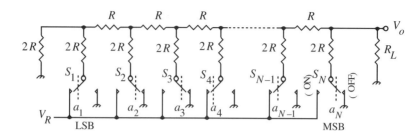

Figure 5.44: Schematic representation of an N-bit resistor-ladder ($R - 2R$ ladder) unipolar D/A converter.

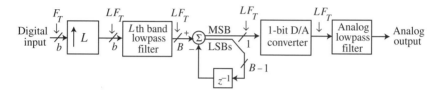

Figure 5.45: Block diagram representation of an oversampling sigma-delta D/A converter.

5.9.3   Oversampling Sigma-Delta D/A Converter

The basic scheme employed in an oversampling sigma-delta D/A converter is shown in Figure 5.45. Here, the sampling rate of the input b-bit digital signal is first increased from F_T to LF_T by a factor-of-L interpolator implemented by an up-sampler followed by an Lth band digital lowpass filter.[11] The output of the interpolator is fed into a digital sigma-delta quantizer, which creates a single bit output by extracting only the MSB. The MSB is then converted to analog form by a 1-bit D/A converter followed by an analog lowpass filter. The LSBs form the error signal that is subtracted from the interpolator output in the summer of the sigma-delta quantizer. A detailed analysis of the operation of the oversampling sigma-delta D/A converter is provided in Section 11.13.

5.9.4   Characteristics of a Practical D/A Converter[12]

A practical D/A converter is characterized by a number of parameters. The effects of these parameters on the performance of a D/A converter are best understood by first examining the input-output relation of an idealized device. Figure 5.46(a) shows the input-output relation of a 3-bit unipolar D/A converter. Here the analog outputs for all possible digital inputs are shown as vertical "bars."

The *resolution* of a D/A converter is defined in a manner identical to that of an A/D converter. For an N-bit D/A converter with the input digital word coded in natural binary form, the resolution is therefore 1 part in 2^{N-1}.

In an ideal D/A converter, the analog outputs as a function of the input discrete levels will be on a straight line going through the origin, as shown in Figure 5.46(a), with the difference between the outputs for two consecutive input digital signals being 1 LSB. In a practical D/A converter, the actual outputs evaluated for each possible input digital signal may be unevenly distributed instead of being on a straight line as indicated in Figure 5.46(b). The *integral linearity (INL) error* is defined as the maximum deviation from the straight line. A measure of the variation in the difference of the analog outputs corresponding to

---

[11]The design of an interpolator is treated in Section 10.2.

[12]This section has been adapted from [Mit80] by permission of the author and publisher.

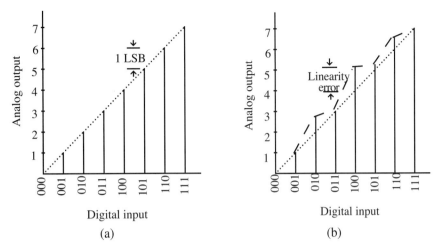

Figure 5.46: Input-output relation of a unipolar D/A converter: (a) ideal converter, and (b) linearity errors in a practical D/A converter.

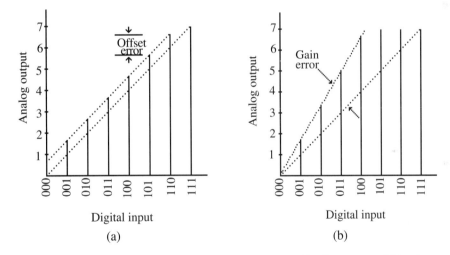

Figure 5.47: Input-output relation of a practical D/A converter: (a) offset error, and (b) gain error.

two consecutive digital inputs is called the *differential nonlinearity*. If the analog outputs are on a straight line that is uniformly shifted by an equal amount for each input discrete value, as shown in Figure 5.47(a), the D/A converter is said to have an *offset error*. If the differences between the actual analog outputs and their corresponding ideal analog outputs increase linearly for increases in the digital inputs, as indicated in Figure 5.47(b), the D/A converter is said to exhibit a *gain error*.

The *accuracy* of a practical D/A converter is defined by the maximum deviation of its measured output from that of an ideal D/A converter. The *word-conversion time* of a D/A converter is given by the time taken to decode a digital word.

The finite turn-on and turn-off times of analog switches in the D/A converter, in general, are not equal, and as a result, they give rise to a dynamic error called a *glitch*. Consider the situation when at time $t = t_n$, the N-bit word is given as the MSB being ONE and the remaining bits as ZEROs. Assume that at time $t = t_{n+1}$, the MSB changes to a ZERO with all other bits becoming ONEs. If the turn-on time of the analog

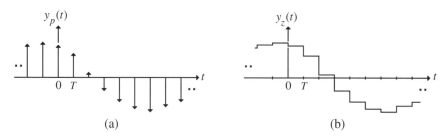

Figure 5.48: Typical output waveforms: (a) ideal D/A converter, and (b) a practical D/A converter.

switch is greater than its turn-off time, then the Nth switch is turned off first and momentarily all switches in the D/A converter are in their OFF positions, resulting in a false analog output equal to zero before the switches settle to their correct positions. The temporary state of all switches being in the OFF positions causes a narrow pulse or spike of half the height of full scale to appear at the converter output. These pulses appearing during the transition periods are called *glitches*. If such glitches are undesirable, they can be eliminated by placing an S/H circuit at the output of the D/A converter, which holds the previous D/A converter output until the glitches disappear and then acquires and holds the new output.

5.10 Reconstruction Filter Design

The output of the D/A converter is finally passed through an analog reconstruction or smoothing filter to eliminate all the replicas of the spectrum outside the baseband. As indicated earlier in Section 5.2.2, this filter ideally should have a frequency response such as given by Eq. (5.16). If the cutoff frequency Ω_c of the reconstruction filter is chosen as $\Omega_T/2$, where Ω_T is the sampling angular frequency, the corresponding frequency response is given by

$$H_r(j\Omega) = \begin{cases} T, & |\Omega| \leq \Omega_T/2, \\ 0, & |\Omega| > \Omega_T/2. \end{cases} \tag{5.75}$$

If we denote the input to the D/A converter as $y[n]$, then from Eq. (5.20) the reconstructed analog equivalent $y_a(t)$ is given by

$$y_a(t) = \sum_{n=-\infty}^{\infty} y[n] \frac{\sin[\pi(t - nT)/T]}{\pi(t - nT)/T}. \tag{5.76}$$

Since the ideal reconstruction filter of Eq. (5.75) has a doubly infinite impulse response, it is noncausal and thus unrealizable. In practice, it is necessary to use filters that approximate the ideal lowpass frequency response.

Almost always, a practical D/A converter unit contains a *zero-order hold circuit* at its output producing a staircase-like analog waveform $y_z(t)$, as shown in Figure 5.48(b). It is therefore important to analyze the effect of the zero-order hold circuit in order to determine the specifications for the smoothing lowpass filter that should follow the overall D/A converter structure.

The zero-order hold operation can be modeled by an ideal impulse-train D/A output $y_p(t)$ followed by a linear, time-invariant analog circuit with an impulse response $h_z(t)$ that is a rectangular pulse of width T and unity height, as indicated in Figure 5.49. It follows from this figure that if $Y_p(j\Omega)$ denotes the continuous-time Fourier transform of $y_p(t)$, the output of the ideal D/A converter, then the continuous-time Fourier transform $Y_z(j\Omega)$ of $y_z(t)$, the output of the zero-order hold circuit, is simply given by

$$Y_z(j\Omega) = H_z(j\Omega)Y_p(j\Omega), \tag{5.77}$$

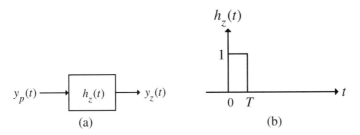

Figure 5.49: (a) Modeling of the zero-order hold operation and (b) impulse response of the zero-order hold circuit.

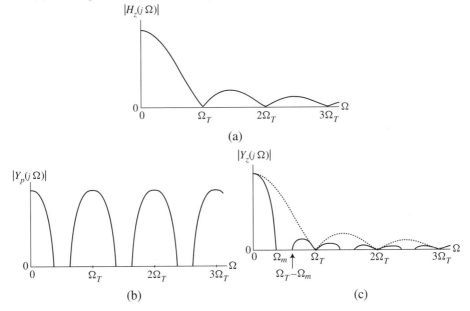

Figure 5.50: Magnitude responses of (a) the zero-order hold circuit, (b) the output of the ideal D/A converter, and (c) the output of the practical D/A converter.

where

$$H_z(j\Omega) = \int_0^T e^{-j\Omega t}\, dt = -\left.\frac{e^{-j\Omega t}}{j\Omega}\right|_0^T = \frac{1 - e^{-j\Omega T}}{j\Omega} = e^{-j\frac{\Omega T}{2}}\left[\frac{\sin(\Omega T/2)}{\Omega/2}\right]. \qquad (5.78)$$

The magnitude response of the zero-order hold circuit, as indicated in Figure 5.50(a), has a lowpass characteristic with zeros at $\pm\Omega_T, \pm 2\Omega_T, \ldots$, where $\Omega_T = 2\pi/T$ is the sampling angular frequency. Figure 5.50(b) shows the magnitude response $|Y_p(j\Omega)|$, which is a periodic function of Ω with a period Ω_T. Since the frequency response $Y_z(j\Omega)$ of the analog output $y_z(t)$ of the zero-order hold circuit is a product of $Y_p(j\Omega)$ and $H_z(j\Omega)$, the zero-order hold circuit somewhat attenuates the unwanted replicas centered at multiples of the sampling frequency Ω_T, as sketched in Figure 5.50(c). An *analog reconstruction filter*, also called a *smoothing filter*, $H_r(j\Omega)$ thus follows a practical D/A converter unit and is designed to further attenuate the residual portions of the signal spectrum centered at multiples of the sampling frequency Ω_T. Moreover, it should also compensate for the amplitude distortion, more commonly called *droop*, caused by the zero-order hold circuit in the band from dc to $\Omega_T/2$.

The general specifications for the analog reconstruction filter $H_r(j\Omega)$ can be easily determined if the effect of the droop is neglected. If Ω_c denotes the highest frequency of the signal $y_p(t)$ that should be preserved at the output of the reconstruction filter, then the lowest-frequency component present in the residual images in the output of the zero-order hold circuit is of frequency $\Omega_o = \Omega_T - \Omega_c$. The zero-order hold circuit has a gain at frequency Ω_o given by

$$20\log_{10}|H_z(j\Omega_o)| = 20\log_{10}\left[\frac{\sin(\Omega_o T/2)}{\Omega_o/2}\right]. \tag{5.79}$$

Therefore, if the system specification calls for a minimum attenuation of A_s dB of all frequency components in the residual images, then the reconstruction filter should provide at least an attenuation of $A_s + 20\log_{10}|H_z(j\Omega_o)|$ dB at Ω_o. For example, if the normalized value of Ω_c is 0.7π, then the gain of the zero-order hold circuit at 0.7π is -7.2 dB. Now, the lowest normalized frequency of the residual images is given by 1.3π. For a minimum attenuation of 50 dB of all signal components in the residual images at the output of the zero-order hold, the reconstruction filter must therefore provide at least an attenuation of 42.8 dB at frequency 1.3π.

The droop caused by the zero-order hold circuit can be compensated either before the D/A converter by means of a digital filter or after the zero-order hold circuit by the analog reconstruction filter. For the latter approach, we observe that the cascade of the zero-order hold circuit and the analog reconstruction filter must have a frequency response of an ideal reconstruction filter following an ideal D/A converter. If we denote the frequency response of the ideal reconstruction filter as $H_r(j\Omega)$ and that of the actual reconstruction filter as $\hat{H}_r(j\Omega)$, then we require

$$H_r(j\Omega) = H_z(j\Omega)\hat{H}_r(j\Omega), \tag{5.80}$$

where $H_r(j\Omega)$ is as given by Eq. (5.16). Therefore, from Eq. (5.80) the desired frequency response of the actual reconstruction filter is given by

$$\hat{H}_r(j\Omega) = \begin{cases} \frac{\Omega T/2}{\sin(\Omega T/2)}, & |\Omega| \leq \Omega_c, \\ 0, & |\Omega| > \Omega_c, \end{cases} \tag{5.81}$$

to ensure a faithful reconstruction of the original signal $g_a(t)$. The modified reconstruction filter has also a noncausal impulse response defined for $-\infty < t < \infty$ and is therefore unrealizable. As a result, an analog filter approximating the magnitude response of $\hat{H}_r(j\Omega)$ must therefore be designed.

Alternatively, the effect of the droop can be compensated by including a digital compensation filter $G(z)$ prior to the D/A converter circuit with a modest increase in the digital hardware requirements. The digital compensation filter can be either an FIR or an IIR type. The frequency response of the digital compensation filter is given by

$$G(e^{j\omega}) = \frac{\omega/2}{\sin(\omega/2)}, \qquad 0 \leq |\omega| \leq \pi. \tag{5.82}$$

Two very low order digital compensation filters are as follows [Jac96]:

$$G_{\text{FIR}}(z) = -\frac{1}{16} + \frac{9}{8}z^{-1} - \frac{1}{16}z^{-2}, \tag{5.83a}$$

$$G_{\text{IIR}}(z) = \frac{9}{8 + z^{-1}}. \tag{5.83b}$$

Figure 5.51 shows the gain responses of the uncompensated and the droop-compensated D/A converters in the baseband. Since the above digital compensation filters have a periodic frequency response of period Ω_T, the replicas of the baseband magnitude response outside the baseband need to be suppressed sufficiently to ensure minimal effect from aliasing. Even though the zero-order hold circuit in the D/A converter provides some attenuation of these unwanted replicas [see Figure 5.50(c)], it may be necessary for the analog reconstruction filter following the D/A converter to provide additional attenuation.

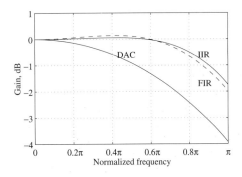

Figure 5.51: Gain responses of the uncompensated and compensated DAC in the baseband.

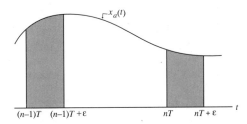

Figure 5.52: Illustration of the averaging operation caused by the S/H circuit.

5.11 Effect of Sample-and-Hold Operation

The frequency-domain analysis of the sampling of a continuous-time signal discussed in Section 5.2.1 assumed ideal sampling generating an impulse train representation of the sampled signal. As indicated in Figure 5.1, in most applications, the sampling operation is provided by an S/H circuit. In principle, the S/H circuit samples the analog signal at each sampling instant and holds a constant value equal to the sampled value for a finite and short period of time to permit the A/D converter to convert it into its digital form. However, in practice, as indicated in Figure 5.31, the S/H circuit tracks the analog signal $x_a(t)$ over a small interval ε. The overall effect, as illustrated in Figure 5.52, is to develop an average of the analog signal over this interval which is held constant at the input of the A/D converter. We now analyze in the frequency-domain the effect of the averaging operation of the S/H circuit [Por97].

From Figure 5.52 it follows that the nth sample value $x[n]$ of the impulse train output $x_p(t)$ of a practical S/H circuit is given by

$$x[n] = \frac{1}{\varepsilon} \int_{nT}^{nT+\varepsilon} x_a(t)dt. \tag{5.84}$$

To understand the effect of the above averaging operation, denote

$$g_a(t) = \int_{-\infty}^{t} x_a(\tau)\, d\tau + K, \tag{5.85}$$

where K is a constant of arbitrary value. Then, from Eq. (5.84) we get

$$x[n] = \frac{1}{\varepsilon} \{g_a(nT + \varepsilon) - g_a(nT)\}. \tag{5.86}$$

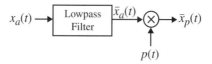

Figure 5.53: Equivalent representation of a practical S/H circuit.

The impulse train $g_p(t)$ with sample value $g_a(nT)$ is obtained by an ideal sampling of the analog signal $g_a(t)$, and it follows from Eq. (5.85) and the differentiation property of the continuous-time Fourier transform (CTFT), that the CTFT of $g_p(t)$ is simply $\frac{1}{j\Omega}X_a(j\Omega)$, where $X_a(j\Omega)$ is the CTFT of $x_a(t)$. Using the time-shifting property, the CTFT of the impulse train with sample value $g_a(nT+\varepsilon)$ is therefore given by $\frac{e^{-j\Omega\varepsilon}}{j\Omega}X_a(j\Omega)$. Hence, from Eqs. (5.14a) and (5.86), the discrete-time Fourier transform (DTFT) of the discrete-time signal $x[n]$ appearing as the input to the A/D converter can be expressed as

$$X(e^{j\omega}) = \frac{1}{T}\sum_{k=-\infty}^{\infty}\bar{X}_a\left(j\frac{\omega-2\pi k}{T}\right), \tag{5.87}$$

where

$$\bar{X}_a(j\Omega) = \frac{1-e^{-j\Omega\varepsilon}}{j\Omega\varepsilon}X_a(j\Omega) = e^{-j\Omega\varepsilon/2}\left(\frac{\sin(\Omega\varepsilon/2)}{\Omega\varepsilon/2}\right)X_a(j\Omega). \tag{5.88}$$

It follows from the above equation that the averaging operation performed by a practical S/H circuit is equivalent to passing the continuous-time signal $x_a(t)$ through an LTI discrete-time system with a frequency response $e^{-j\Omega\varepsilon/2}(\sin(\Omega\varepsilon/2))/(\Omega\varepsilon/2)$ followed by an ideal impulse train sampling as indicated in Figure 5.53. Note that the frequency response of the discrete-time system is similar in form to that of the zero-order hold circuit as given in Eq. (5.78) as shown in Figure 5.50(a). Thus the discrete-time system of Figure 5.53 acts like a narrowband lowpass filter which performs the averaging operation. If the tracking period ε is much smaller compared to the sampling period T, as is usually the case, the effect of the lowpass filter can be neglected and the practical S/H circuit can be considered as an ideal sampler.

5.12 Summary

Various issues concerned with the digital processing of continuous-time signals are studied in this chapter. A discrete-time signal is obtained by uniformly sampling a continuous-time signal. The discrete-time representation is unique if the sampling frequency is greater than twice the highest frequency contained in the continuous-time signal, and the latter can be fully recovered from its discrete-time equivalent by passing it through an ideal analog lowpass reconstruction filter with a cutoff frequency that is half the sampling frequency. If the sampling frequency is lower than twice the highest frequency contained in the continuous-time signal, in general, the latter cannot be recovered from its discrete-time version due to aliasing. In practice, the continuous-time signal is first passed through an analog lowpass anti-aliasing filter with the cutoff frequency chosen as half of the sampling frequency whose output is sampled to prevent aliasing. It is also shown that a bandpass continuous-time signal can be recovered from its discrete-time equivalent by undersampling provided the highest frequency is an integer multiple of the bandwidth of the continuous-time signal and the sampling frequency is greater than twice the bandwidth.

A brief review of the theory behind some popular analog lowpass filter design techniques is included and their design using MATLAB is illustrated. Also discussed are the procedures for designing analog highpass, bandpass, and bandstop filters, and their implementations using MATLAB. The specifications of the analog filters are usually given in terms of the locations of the passband and stopband edge frequencies, and the passband and stopband ripples. Effects of these parameters on the performances of the anti-aliasing and reconstruction filters are examined.

Other interface devices involved in the digital processing of continuous-time signals are the sample-and-hold circuit, comparator, analog-to-digital converter, and digital-to-analog converter. A brief introduction to these devices is included for completeness.

5.13 Problems

5.1 Show that the periodic impulse train $p(t)$ defined in Eq. (5.4) can be expressed as a Fourier series as given by Eq. (5.7).

5.2 A continuous-time signal $x_a(t)$ is composed of a linear combination of sinusoidal signals of frequencies 250 Hz, 450 Hz, 1.0 kHz, 2.75 kHz, and 4.05 kHz. The signal $x_a(t)$ is sampled at a 1.5-kHz rate and the sampled sequence is passed through an ideal lowpass filter with a cutoff frequency of 750 Hz, generating a continuous-time signal $y_a(t)$. What are the frequency components present in the reconstructed signal $y_a(t)$?

5.3 A continuous-time signal $x_a(t)$ is composed of a linear combination of sinusoidal signals of frequencies F_1 Hz, F_2 Hz, F_3 Hz, and F_4 Hz. The signal $x_a(t)$ is sampled at an 8-kHz rate, and the sampled sequence is then passed through an ideal lowpass filter with a cutoff frequency of 3.8 kHz, generating a continuous-time signal $y_a(t)$ composed of three sinusoidal signals of frequencies 450 Hz, 625 Hz, and 950 Hz, respectively. What are the possible values of F_1, F_2, F_3, and F_4? Is your answer unique? If not, indicate another set of possible values of these frequencies.

5.4 The continuous-time signal $x_a(t) = 3\cos(400\pi t) + 5\sin(1200\pi t) + 6\cos(4400\pi t) + 2\sin(5200\pi t)$ is sampled at a 4-kHz rate generating the sequence $x[n]$. Determine the exact expression of $x[n]$.

5.5 The left and right channels of an analog stereo audio signal are sampled at a 48.2-kHz rate with each channel then being converted into a digital bit stream using a 15-bit A/D converter. Determine the combined bit rate of the two channels after sampling and digitization.

5.6 Show that the impulse response $h_r(t)$ of an ideal lowpass filter as derived in Eq. (5.17) takes the value $h_r(nT) = \delta[n]$ for all n if the cutoff frequency $\Omega_c = \Omega_T/2$, where Ω_T is the sampling frequency.

5.7 Consider the system of Figure 5.2, where the input continuous-time signal $x_a(t)$ has a bandlimited spectrum $X_a(j\Omega)$ as sketched in Figure P5.1(a) and is being sampled at the Nyquist rate. The discrete-time processor is an ideal lowpass filter with a frequency response $H(e^{j\omega})$ as shown in Figure P5.1(b) and has a cutoff frequency $\omega_c = \Omega_m T/3$, where T is the sampling period. Sketch as accurately as possible the spectrum $Y_a(j\Omega)$ of the output continuous-time signal $y_a(t)$.

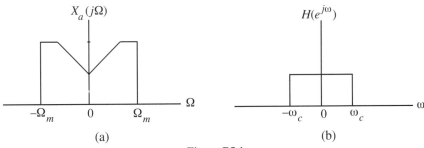

Figure P5.1

5.8 A continuous-time signal $x_a(t)$ has a bandlimited spectrum $X_a(j\Omega)$ as indicated in Figure P5.2. Determine the smallest sampling rate F_T that can be employed to sample $x_a(t)$ so that it can be fully recovered from its sampled version $x[n]$ for each of the following sets of values of the bandedges Ω_1 and Ω_2. Sketch the Fourier transform of the sampled version $x[n]$ obtained by sampling $x_a(t)$ at the smallest sampling rate F_T and the frequency response of the ideal reconstruction filter needed to fully recover $x_a(t)$ for each case.
    (a) $\Omega_1 = 200\pi$, $\Omega_2 = 160\pi$;   (b) $\Omega_1 = 160\pi$, $\Omega_2 = 120\pi$;   (c) $\Omega_1 = 150\pi$, $\Omega_2 = 110\pi$.

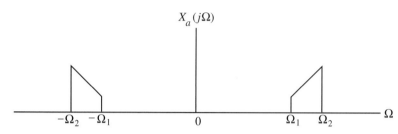

Figure P5.2

5.9 For each set of desired peak passband deviation α_p and the minimum stopband attenuation α_s of an analog lowpass filter given below, determine the corresponding passband and stopband ripples, δ_p and δ_s:
    (a) $\alpha_p = 0.15\,\text{dB}$, $\alpha_s = 43\,\text{dB}$;   (b) $\alpha_p = 0.04\,\text{dB}$, $\alpha_s = 57\,\text{dB}$;   (c) $\alpha_p = 0.23\,\text{dB}$, $\alpha_s = 39\,\text{dB}$.

5.10 Show that the analog transfer function

$$H_1(s) = \frac{a}{s+a}, \qquad a > 0, \tag{5.89}$$

has a lowpass magnitude response with a monotonically decreasing magnitude response with $|H_a(j0)| = 1$ and $|H_a(j\infty)| = 0$. Determine the 3-dB cutoff frequency Ω_c at which the gain response is 3 dB below the maximum value of 0 dB at $\Omega = 0$.

5.11 Show that the analog transfer function

$$H_2(s) = \frac{s}{s+a}, \qquad a > 0, \tag{5.90}$$

has a highpass magnitude response with a monotonically increasing magnitude response with $|H_a(j0)| = 0$ and $|H_a(j\infty)| = 1$. Determine the 3-dB cutoff frequency Ω_c at which the gain response is 3 dB below the maximum value of 0 dB at $\Omega = \infty$.

5.12 The lowpass transfer function $H_1(s)$ of Eq. (5.89) and the highpass transfer function $H_2(s)$ of Eq. (5.90) can be expressed in the form
$$H_1(s) = \tfrac{1}{2}\{A_1(s) - A_2(s)\}, \qquad H_2(s) = \tfrac{1}{2}\{A_1(s) + A_2(s)\},$$
where $A_1(s)$ and $A_2(s)$ are stable analog allpass transfer functions. Determine $A_1(s)$ and $A_2(s)$.

5.13 Show that the analog transfer function

$$H_1(s) = \frac{bs}{s^2 + bs + \Omega_o^2}, \qquad b > 0, \tag{5.91}$$

has a bandpass magnitude response with $|H_a(j0)| = |H_a(j\infty)| = 0$ and $|H_a(j\Omega_o)| = 1$. Determine the frequencies Ω_1 and Ω_2 at which the gain is 3 dB below the maximum value of 0 dB at Ω_o. Show that $\Omega_1\Omega_2 = \Omega_o^2$. The difference $\Omega_2 - \Omega_1$ is called the 3-dB bandwidth of the bandpass transfer function. Show that $b = \Omega_2 - \Omega_1$.

5.14 Show that the analog transfer function

$$H_2(s) = \frac{s^2 + \Omega_o^2}{s^2 + bs + \Omega_o^2}, \qquad b > 0, \tag{5.92}$$

has a bandstop magnitude response with $|H_a(j0)| = |H_a(j\infty)| = 1$ and $|H_a(j\Omega_o)| = 0$. Since the magnitude is exactly zero at Ω_o, it is called the notch frequency, and $H_2(s)$ is often called the notch transfer function. Determine the frequencies Ω_1 and Ω_2 at which the gain is 3 dB below the maximum value of 0 dB at $\Omega = 0$ and $\Omega = \infty$. Show that $\Omega_1\Omega_2 = \Omega_o^2$. The difference $\Omega_2 - \Omega_1$ is called the 3-dB notch bandwidth of the bandpass transfer function. Show that $b = \Omega_2 - \Omega_1$.

5.15 The bandpass transfer function $H_1(s)$ of Eq. (5.91) and the bandstop transfer function $H_2(s)$ of Eq. (5.92) can be expressed in the form

$$H_1(s) = \tfrac{1}{2}\{A_1(s) - A_2(s)\}, \qquad H_2(s) = \tfrac{1}{2}\{A_1(s) + A_2(s)\},$$

where $A_1(s)$ and $A_2(s)$ are stable analog allpass transfer functions. Determine $A_1(s)$ and $A_2(s)$.

5.16 Show that the first $2N - 1$ derivatives of the squared-magnitude response $|H_a(j\Omega)|^2$ of a Butterworth filter of order N as given by Eq. (5.31) are equal to zero at $\Omega = 0$.

5.17 Using Eq. (5.33) determine the lowest order of a lowpass Butterworth filter with a 0.5-dB cutoff frequency at 2.1 kHz and a minimum attenuation of 30 dB at 8 kHz.

5.18 Using Eq. (5.35) determine the pole locations and the coefficients of a fifth-order Butterworth polynomial with unity 3-dB cutoff frequency.

5.19 Show that the Chebyshev polynomial $T_N(\Omega)$ defined in Eq. (5.38) satisfies the recurrence relation given in Eq. (5.39) with $T_0(\Omega) = 1$, and $T_1(\Omega) = \Omega$.

5.20 Using Eq. (5.41) determine the lowest order of a lowpass Type 1 Chebyshev filter with a 0.5-dB cutoff frequency at 2.1 kHz and a minimum attenuation of 30 dB at 8 kHz.

5.21 Using Eq. (5.51) determine the lowest order of a lowpass elliptic filter with a 0.5-dB cutoff frequency at 2.1 kHz and a minimum attenuation of 30 dB at 8 kHz.

5.22 Determine the Bessel polynomials $B_N(s)$ for the following values of N: (a) $N = 5$, and (b) $N = 6$.

5.23 The transfer function of a second-order analog Butterworth lowpass filter with a passband edge at 0.2 Hz and a passband ripple of 0.5 dB is given by

$$H_{LP}(s) = \frac{4.52}{s^2 + 3s + 4.52}.$$

Determine the transfer function $H_{HP}(s)$ of an analog highpass filter with a passband edge at 2 Hz and a passband ripple of 0.5 dB by applying the spectral transformation of Eq. (5.59).

5.24 The transfer function of a second-order analog elliptic lowpass filter with a passband edge at 0.16 Hz and a passband ripple of 1 dB is given by

$$H_{LP}(s) = \frac{0.056(s^2 + 17.95)}{s^2 + 1.06s + 1.13}.$$

Determine the transfer function $H_{BP}(s)$ of an analog bandpass filter with a center frequency at 3 Hz and a bandwidth of 0.5 Hz by applying the spectral transformation of Eq. (5.61).

5.25 A Butterworth analog highpass filter is to be designed with the following specifications: $F_p = 5$ MHz, $F_s = 0.5$ MHz, $\alpha_p = 0.3$ dB, and $\alpha_s = 45$ dB. What are the bandedges and the order of the corresponding analog lowpass filter? What is the order of the highpass filter? Verify your results using the function `buttord`.

5.26 An elliptic analog bandpass filter is to be designed with the following specifications: passband edges at 20 kHz and 45 kHz, stopband edges at 10 kHz and 60 kHz, peak passband ripple of 0.5 dB, and minimum stopband attenuation of 40 dB. What are the bandedges and the order of the corresponding analog lowpass filter? What is the order of the bandpass filter? Verify your results using the function `ellipord`.

5.27 A Type 1 Chebyshev analog bandstop filter is to be designed with the following specifications: passband edges at 10 MHz and 70 MHz, stopband edges at 20 MHz and 45 MHz, peak passband ripple of 0.5 dB, and minimum stopband attenuation of 30 dB. What are the bandedges and the order of the corresponding analog lowpass filter? What is the order of the bandstop filter? Verify your results using the function `cheb1ord`.

5.28 Verify Table 5.1.

5.29 Derive Eq. (5.73).

5.30 Derive Eq. (5.74).

5.31 An alternative to the zero-order hold circuit of Figure 5.49 used for signal reconstruction at the output of a D/A converter is the *first-order hold circuit* which approximates $y_a(t)$ according to the following relation:

$$y_f(t) = y_p(nT) + \frac{y_p(nT) - y_p(nT - T)}{T}(t - nT), \quad nT \le t \le (n+1)T.$$

As indicated by the above equation, the first-order hold circuit approximates $y_a(t)$ by straight-line segments. The slope of the segment between $t = nT$ and $t = (n+1)T$ is determined from the sample values $y_p(nT)$ and $y_p(nT - T)$. Determine the impulse response $h_f(t)$ and the frequency response $H_f(j\Omega)$ of the first-order hold circuit, and compare its performance with that of the zero-order hold circuit.

5.32 A more improved signal reconstruction at the output of a D/A converter is provided by a linear interpolation circuit which approximates $y_a(t)$ by connecting successive sample points of $y_p(t)$ with straight-line segments. The input-output relation of this circuit is given by

$$y_f(t) = y_p(nT - T) + \frac{y_p(nT) - y_p(nT - T)}{T}(t - nT), \quad nT \le t \le (n+1)T.$$

Determine the impulse response $h_f(t)$ and the frequency response $H_f(j\Omega)$ of the linear interpolation circuit, and compare its performance with that of the first-order hold circuit.

5.14 MATLAB Exercises

M 5.1 Write a MATLAB program to compute the required order of a lowpass Butterworth analog filter according to Eq. (5.33). Using this program determine the lowest order of the lowpass filter of Problem 5.17.

M 5.2 Write a MATLAB program to compute the required order of a lowpass Chebyshev analog filter according to Eq. (5.36). Using this program determine the lowest order of the lowpass filter of Problem 5.20.

M 5.3 Write a MATLAB program to compute the required order of a lowpass elliptic analog filter according to Eq. (5.51). Using this program determine the lowest order of the lowpass filter of Problem 5.21.

M 5.4 Determine the transfer function of a lowpass Butterworth analog filter with specifications as given in Problem 5.17 using Program 5_2. Plot the gain response and verify that the filter designed meets the given specifications. Show all steps.

M 5.5 Determine the transfer function of a lowpass Type 1 Chebyshev analog filter with specifications as given in Problem 5.20 using Program 5_3. Plot the gain response and verify that the filter designed meets the given specifications. Show all steps.

M 5.6 Modify Program 5_3 to design lowpass Type 2 Chebyshev analog filters. Using this program, determine the transfer function of a lowpass Type 2 Chebyshev analog filter with specifications as given in Problem 5.20. Plot the gain response and verify that the filter designed meets the given specifications. Show all steps.

M 5.7 Determine the transfer function of a lowpass elliptic analog filter with specifications as given in Problem 5.21 using Program 5_4. Plot the gain response and verify that the filter designed meets the given specifications. Show all steps.

M 5.8 Design a Butterworth analog highpass filter with specifications given in Problem 5.25. Show the transfer functions of the prototype analog lowpass and the highpass filters. Plot their gain responses and verify that both filters meet their respective specifications. Show all steps.

M 5.9 Design an elliptic analog bandpass filter with specifications given in Problem 5.26. Show the transfer functions of the prototype analog lowpass and the bandpass filters. Plot their gain responses and verify that both filters meet their respective specifications. Show all steps.

M 5.10 Design a Type 1 analog bandstop filter with specifications given in Problem 5.27. Show the transfer functions of the prototype analog lowpass and the bandstop filters. Plot their gain responses and verify that both filters meet their respective specifications. Show all steps.

M 5.11 Write a MATLAB program to verify the plots of Figure 5.52.

6 Digital Filter Structures

The description of the discrete-time system of Eq. (2.64a) or (2.64b) expresses the nth output sample as a convolution sum of the input with the impulse response of the system, and in some sense, it is the most fundamental characterization of an LTI digital filter. The convolution sum, in principle, can be employed to implement a digital filter with a known impulse response, and the implementation involves addition, multiplication, and delay, which are fairly simple operations. For an LTI system with an infinite-length impulse response, the approach is not practical. However, for the infinite impulse response (IIR) LTI digital filter described by a constant coefficient difference equation of the form of Eq. (2.82), and for the finite impulse response (FIR) LTI digital filter described by Eq. (2.97), the input-output relation involves a finite sum of products, and a direct implementation based on these equations is quite practical. In this text we deal only with these two types of LTI digital filters.

Now, the actual implementation of a digital filter could be either in software or hardware form, depending on applications. In both types of implementation, the signal variables and the filter coefficients cannot be represented with infinite precision. As a result, a direct implementation of a digital filter based on either Eq. (2.82) or Eq. (2.97) may not provide satisfactory performance due to the finite precision arithmetic. It is thus of interest to develop alternative realizations based on other types of time-domain representations with equivalent input-output relations to either Eq. (2.82) or Eq. (2.97), depending on the type of digital filter being implemented, and choose the realization that provides satisfactory performance under finite precision arithmetic.

In this chapter, we consider the realization problem of causal IIR and FIR transfer functions and outline realization methods based on both the time-domain and the transform-domain representations. In Chapter 9, we develop the methods for the analysis of such structures when implemented with finite precision arithmetic and present additional realizations that have been developed to minimize the effects of finite wordlength.

A structural representation using interconnected basic building blocks is the first step in the hardware or software implementation of an LTI digital filter. The structural representation provides the relations between some pertinent internal variables with the input and the output that in turn provide the keys to the implementation. There are various forms of the structural representation of a digital filter. We review in this chapter two such representations, and then describe some popular schemes for the realization of real causal IIR and FIR digital filters. In addition, we outline a method for the realization of IIR digital filter structures that can be used for the generation of a pair of orthogonal sinusoidal sequences.

6.1 Block Diagram Representation

As indicated earlier, the input-output relations of an LTI digital filter can be expressed in various ways. In the time-domain, it is given by the convolution sum

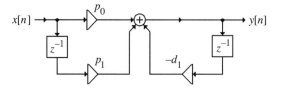

Figure 6.1: A first-order LTI digital filter.

$$y[n] = \sum_{k=-\infty}^{\infty} h[k]x[n-k], \qquad (6.1)$$

or by the linear constant coefficient difference equation

$$y[n] = -\sum_{k=1}^{N} d_k y[n-k] + \sum_{k=0}^{M} p_k x[n-k]. \qquad (6.2)$$

A digital filter can be implemented on a general-purpose digital computer in software form or with special-purpose hardware. To this end, it is necessary to describe the input-output relationship by means of a computational algorithm. To illustrate what we mean by a computational algorithm, consider a first-order causal LTI IIR digital filter described by

$$y[n] = -d_1 y[n-1] + p_0 x[n] + p_1 x[n-1]. \qquad (6.3)$$

Using Eq. (6.3) we can compute $y[n]$ for $n = 0, 1, 2, \ldots$, knowing the initial condition $y[-1]$ and the input $x[n]$ for $n = -1, 0, 1, 2, \ldots$:

$$y[0] = -d_1 y[-1] + p_0 x[0] + p_1 x[-1],$$
$$y[1] = -d_1 y[0] + p_0 x[1] + p_1 x[0],$$
$$y[2] = -d_1 y[1] + p_0 x[2] + p_1 x[1],$$
$$\vdots$$

We can continue this calculation for any value of n we desire. Each step of the calculation process requires a knowledge of the previously calculated value of the output sample (delayed value of the output), the present value of the input sample, and the previous value of the input sample (delayed value of the input). Knowing the data values, we multiply each of them appropriately with coefficients $-d_1$, p_0, and p_1, and then sum the products to compute the present value of the output. As a result, the difference equation of Eq. (6.3) can be interpreted as a valid computational algorithm.

6.1.1 Basic Building Blocks

The computational algorithm of an LTI digital filter can be conveniently represented in block diagram form using the basic building blocks representing the unit delay, the multiplier, the adder, and the pick-off node as shown earlier in Figure 2.5. Thus, a block diagram representation of the first-order digital filter described by Eq. (6.3) is as indicated in Figure 6.1. The adder shown here is a three-input adder. Note also the pick-off nodes at the input and the output.

There are several advantages in representing the digital filter in block diagram form: (1) it is easy to write down the computational algorithm by inspection, (2) it is easy to analyze the block diagram to determine the explicit relation between the output and the input, (3) it is easy to manipulate a block diagram to derive other "equivalent" block diagrams yielding different computational algorithms, (4) it is easy to determine the hardware requirements, and finally (5) it is easier to develop block diagram representations from the transfer function directly leading to a variety of "equivalent" representations.

6.1.2 Analysis of Block Diagrams

Digital filter structures represented in block diagram form can often be analyzed by writing down the expressions for the output signals of each adder as a sum of its input signals, developing a set of equations relating the filter input and output signals in terms of all internal signals. Eliminating the unwanted internal variables then results in the expression for the output signal as a function of the input signal and the filter parameters that are the multiplier coefficients.

The following two examples illustrate the analysis approach.

EXAMPLE 6.1   Consider the single-loop feedback structure of Figure 6.2. There is only one adder in this structure whose output signal $E(z)$ is given by

$$E(z) = X(z) + G_2(z)Y(z).$$

But from the figure,

$$Y(z) = G_1(z)E(z).$$

Eliminating $E(z)$ from the above two equations, we arrive at

$$[1 - G_1(z)G_2(z)]Y(z) = G_1(z)X(z),$$

which leads to the transfer function

$$H(z) = \frac{Y(z)}{X(z)} = \frac{G_1(z)}{1 - G_1(z)G_2(z)}. \tag{6.4}$$

EXAMPLE 6.2   Analyze the cascaded lattice digital filter structure of Figure 6.3. The output signals of the four adders in Figure 6.3 are given by

$$W_1 = X - \alpha S_2, \tag{6.5a}$$
$$W_2 = W_1 - \delta S_1, \tag{6.5b}$$
$$W_3 = S_1 + \varepsilon W_2, \tag{6.5c}$$
$$Y = \beta W_1 + \gamma S_2. \tag{6.5d}$$

But, from the figure, $S_2 = z^{-1}W_3$ and $S_1 = z^{-1}W_2$. Substituting these relations in Eqs. (6.5a) to (6.5d), we obtain

$$W_1 = X - \alpha z^{-1}W_3, \tag{6.6a}$$
$$W_2 = W_1 - \delta z^{-1}W_2, \tag{6.6b}$$
$$W_3 = z^{-1}W_2 + \varepsilon W_2, \tag{6.6c}$$
$$Y = \beta W_1 + \gamma z^{-1}W_3. \tag{6.6d}$$

From Eq. (6.6b), $W_2 = W_1/(1 + \delta z^{-1})$ and, from Eq. (6.6c), $W_3 = (\varepsilon + z^{-1})W_2$. Combining these two equations, we get

$$W_3 = \frac{\varepsilon + z^{-1}}{1 + \delta z^{-1}} W_1. \tag{6.7}$$

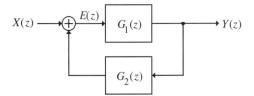

Figure 6.2: A single-loop digital filter structure.

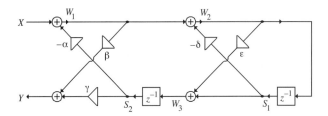

Figure 6.3: A cascaded lattice digital filter structure.

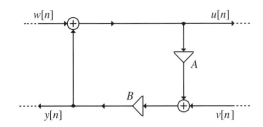

Figure 6.4: An example of a delay-free loop.

Substituting Eq. (6.7) in Eqs. (6.6a) and (6.6d), we finally obtain after some algebra the expression for the transfer function $H(z)$ of the digital filter structure as

$$H(z) = \frac{Y}{X} = \frac{\beta + (\beta\delta + \gamma\varepsilon)z^{-1} + \gamma z^{-2}}{1 + (\delta + \alpha\varepsilon)z^{-1} + \alpha z^{-2}}. \tag{6.8}$$

6.1.3 The Delay-Free Loop Problem

For physical realizability of the digital filter structure, it is necessary that the block diagram representation contains no delay-free loops, i.e., feedback loops without any delay elements. Figure 6.4 illustrates a typical delay-free loop that may appear unintentionally in a specific structure. Analysis of this structure yields

$$y[n] = B\{A(w[n] + y[n]) + v[n]\}.$$

The above expression implies that the determination of the current value of $y[n]$ requires the knowledge of the same value that is physically impossible to achieve due to the finite time required to carry out all arithmetic operations in a digital machine.

A simple graph-theoretic–based method has been proposed to detect the presence of delay-free loops in an arbitrary digital filter structure, along with the methods to locate and remove these loops without

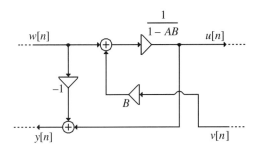

Figure 6.5: An equivalent realization of the structure of Figure 6.4 with no delay-free loop.

altering the overall input-output relations [Szc75]. The removal is achieved by replacing the portion of the structure containing the delay-free loops by an equivalent realization with no delay-free loops. For example, Figure 6.5 shows an equivalent realization of the block diagram of Figure 6.4 without the delay-free loop.

6.1.4 Canonic and Noncanonic Structures

A digital filter structure is said to be *canonic* if the number of delays in the block diagram representation is equal to the order of the difference equation (i.e., the order of the transfer function). Otherwise, it is a *noncanonic* structure. For example, the structure of Figure 6.1 is a noncanonic realization since it employs two delays to realize the first-order difference equation of Eq. (6.3).

6.2 Equivalent Structures

Our main objective in this chapter is to develop various realizations of a given transfer function. We define two digital filter structures to be *equivalent* if they have the same transfer function. We outline in this chapter and in Chapter 9 a number of methods for the generation of equivalent structures. However, a fairly simple way to generate an equivalent structure from a given realization is via the *transpose operation*, which is as follows [Jac70a]:

(i) Reverse all paths,

(ii) Replace pick-off nodes by adders, and vice versa, and

(iii) Interchange the input and the output nodes.

All other methods for developing equivalent structures are based on a specific algorithm for each structure. There are literally an infinite number of equivalent structures realizing the same transfer function, and it is impossible to develop all such realizations. Moreover, a large variety of algorithms have been advanced by various authors, and space limitations prevent us from reviewing each method in this text. We therefore restrict ourselves to a discussion of some commonly used structures.

It should be noted that under infinite precision arithmetic any given realization of a digital filter behaves identically to any other equivalent structure. However, in practice, due to the finite wordlength limitations, a specific realization behaves totally differently from its other equivalent realizations. Hence, it is important to choose a structure that has the least quantization effects from the finite wordlength implementation. One way to arrive at such a structure is to determine a large number of equivalent structures, analyze the finite wordlength effects of each one, and then select the one showing the least effects. In certain cases, it is possible to develop a structure that by construction has the least quantization effects. The analysis of quantization effects is the subject of Chapter 9, which also describes additional structures specifically developed to minimize certain quantization effects. In this chapter, we discuss some simple realizations that in many applications are adequate. We do, however, compare each of the realizations discussed here with regard to their computational complexity determined in terms of the total number of multipliers and the total number of adders required for their implementation. This latter issue is important where the cost of implementation is critical.

6.3 Basic FIR Digital Filter Structures

We first consider the realization of FIR digital filters. Recall that a causal FIR filter of order N is characterized by a transfer function $H(z)$,

$$H(z) = \sum_{k=0}^{N} h[k]z^{-k}, \tag{6.9}$$

which is a polynomial in z^{-1} of degree N. In the time-domain the input-output relation of the above FIR filter is given by

$$y[n] = \sum_{k=0}^{N} h[k]x[n-k], \tag{6.10}$$

where $y[n]$ and $x[n]$ are the output and input sequences, respectively. Since FIR filters can be designed to provide exact linear phase over the whole frequency range and are always BIBO stable independent of the filter coefficients, such filters are often preferred in many applications. We now outline several realization methods for such filters.

6.3.1 Direct Forms

An FIR filter of order N is characterized by $N + 1$ coefficients and, in general, requires $N + 1$ multipliers and N two-input adders for implementation. Structures in which the multiplier coefficients are precisely the coefficients of the transfer function are called *direct form* structures. A direct form realization of an FIR filter can be readily developed from Eq. (6.10), as indicated in Figure 6.6(a) for $N = 4$. An analysis of this structure yields

$$y[n] = h[0]x[n] + h[1]x[n-1] + h[2]x[n-2] + h[3]x[n-3] + h[4]x[n-4],$$

which is precisely of the form of Eq. (6.10).

The structure of Figure 6.6(a) is also called a *tapped delay line* or a *transversal filter*. Its transpose as sketched in Figure 6.6(b) is the second direct form structure. Both direct form structures are canonic with respect to delays.

6.3.2 Cascade Form

A higher-order FIR transfer function can also be realized as a cascade of FIR sections with each section characterized by either a first-order or a second-order transfer function. To this end, we factor the FIR transfer function $H(z)$ of Eq. (6.9) and write it in the form

$$H(z) = h[0] \prod_{k=1}^{K} \left(1 + \beta_{1k}z^{-1} + \beta_{2k}z^{-2}\right), \tag{6.11}$$

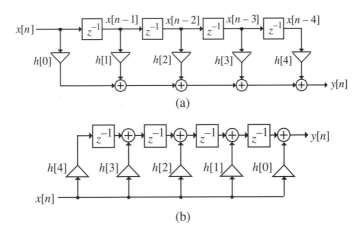

(a)

(b)

Figure 6.6: Direct form FIR structures.

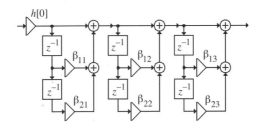

Figure 6.7: Cascade form FIR structure for a sixth-order FIR filter.

where $K = N/2$ if N is even, and $K = (N+1)/2$ if N is odd, with $\beta_{2K} = 0$. A cascade realization of Eq. (6.11) for $N = 6$ is shown in Figure 6.7 requiring three second-order sections. Each second-order stage in Figure 6.7 can of course also be realized in the transposed direct form. Note that the cascade form is canonic, and also employs N two-input adders and $N + 1$ multipliers for an Nth-order FIR transfer function.

6.3.3 Polyphase Realization

Another interesting realization of an FIR filter is based on the polyphase decomposition of its transfer function and results in a parallel structure [Bel76]. To illustrate this approach, consider for simplicity a causal FIR transfer function $H(z)$ of length 9:

$$H(z) = h[0] + h[1]z^{-1} + h[2]z^{-2} + h[3]z^{-3} + h[4]z^{-4}$$
$$+ h[5]z^{-5} + h[6]z^{-6} + h[7]z^{-7} + h[8]z^{-8}. \tag{6.12}$$

The above transfer function can be expressed as a sum of two terms, with one term containing the even-indexed coefficients and the other containing the odd-indexed coefficients, as indicated below:

$$H(z) = \left(h[0] + h[2]z^{-2} + h[4]z^{-4} + h[6]z^{-6} + h[8]z^{-8} \right)$$
$$+ \left(h[1]z^{-1} + h[3]z^{-3} + h[5]z^{-5} + h[7]z^{-7} \right)$$

$$= \left(h[0] + h[2]z^{-2} + h[4]z^{-4} + h[6]z^{-6} + h[8]z^{-8} \right)$$
$$+ z^{-1} \left(h[1] + h[3]z^{-2} + h[5]z^{-4} + h[7]z^{-6} \right). \tag{6.13}$$

By using the notations

$$E_0(z) = h[0] + h[2]z^{-1} + h[4]z^{-2} + h[6]z^{-3} + h[8]z^{-4},$$
$$E_1(z) = h[1] + h[3]z^{-1} + h[5]z^{-2} + h[7]z^{-3}, \tag{6.14}$$

we can rewrite Eq. (6.13) as

$$H(z) = E_0(z^2) + z^{-1} E_1(z^2). \tag{6.15}$$

In a similar manner, by grouping the terms of Eq. (6.12) differently, we can reexpress it in the form

$$H(z) = E_0(z^3) + z^{-1} E_1(z^3) + z^{-2} E_2(z^3), \tag{6.16}$$

where now

$$E_0(z) = h[0] + h[3]z^{-1} + h[6]z^{-2},$$
$$E_1(z) = h[1] + h[4]z^{-1} + h[7]z^{-2}, \tag{6.17}$$
$$E_2(z) = h[2] + h[5]z^{-1} + h[8]z^{-2}.$$

The decomposition of $H(z)$ in the form of Eqs. (6.15) and (6.16) is more commonly known as the *polyphase decomposition*. In the general case, an L-branch polyphase decomposition of the transfer function of Eq. (6.9) of order N is of the form

$$H(z) = \sum_{m=0}^{L-1} z^{-m} E_m(z^L), \tag{6.18}$$

where[1]

$$E_m(z) = \sum_{n=0}^{\lfloor (N+1)/L \rfloor} h[Ln + m]z^{-n}, \qquad 0 \le m \le L - 1, \tag{6.19}$$

with $h[n] = 0$ for $n > N$. A realization of $H(z)$ based on the decomposition of Eq. (6.18) is called a *polyphase realization*. Figure 6.8 shows the four-branch, the three-branch, and the two-branch polyphase realizations of a transfer function. As indicated in Eqs. (6.14) and (6.17), the expression for the transfer function $E_0(z)$ is different for each structure and so are the expressions for $E_1(z)$, etc.

The subfilters $E_m(z^L)$ in the polyphase realization of an FIR transfer function are also FIR filters and can be realized using any of the methods described earlier. However, to obtain a canonic realization of the overall structure, the delays in all subfilters must be shared. Figure 6.9 illustrates a canonic polyphase realization of a length-9 FIR transfer function obtained by delay-sharing. It should be noted that in developing this realization, we have used the transpose of the structure of Figure 6.8(b). Other canonic polyphase realizations can be similarly derived (Problems 6.11 to 6.13).

The polyphase structures are often used in multirate digital signal processing applications for computationally efficient realizations (see Section 10.4.3).

---

[1] $\lfloor x \rfloor$ is the integer part of x.

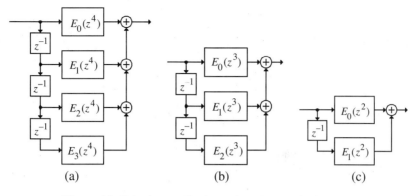

(a) (b) (c)

Figure 6.8: Polyphase realizations of an FIR transfer function.

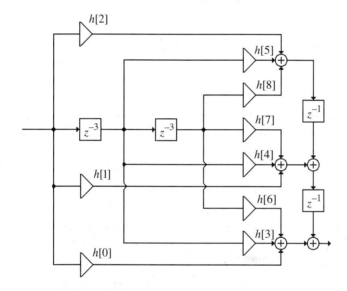

Figure 6.9: Canonic three-branch polyphase realization of a length-9 FIR filter.

6.3.4 Linear-Phase FIR Structures

We showed in Section 4.4.3 that a linear-phase FIR filter of order N is either characterized by a symmetric impulse response

$$h[n] = h[N - n], \tag{6.20}$$

or by an antisymmetric impulse response

$$h[n] = -h[N - n]. \tag{6.21}$$

The symmetry (or antisymmetry) property of a linear-phase FIR filter can be exploited to reduce the total number of multipliers into almost half of that in the direct form implementations of the transfer function. To this end consider the realization of a length-7 Type 1 FIR transfer function with a symmetric impulse response:

$$H(z) = h[0] + h[1]z^{-1} + h[2]z^{-2} + h[3]z^{-3} + h[2]z^{-4} + h[1]z^{-5} + h[0]z^{-6},$$

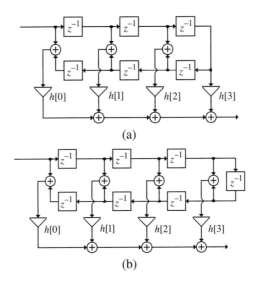

Figure 6.10: Linear-phase FIR structures: (a) Type 1, and (b) Type 2.

which can be rewritten in the form

$$H(z) = h[0]\left(1 + z^{-6}\right) + h[1]\left(z^{-1} + z^{-5}\right)$$
$$+ h[2]\left(z^{-2} + z^{-4}\right) + h[3]z^{-3}. \tag{6.22}$$

A realization of $H(z)$ based on the decomposition of Eq. (6.22) is shown in Figure 6.10(a). A similar decomposition can be applied for the realization of a Type 2 FIR transfer function. For example, for a length-8 Type 2 FIR transfer function, the pertinent decomposition is given by

$$H(z) = h[0]\left(1 + z^{-7}\right) + h[1]\left(z^{-1} + z^{-6}\right)$$
$$+ h[2]\left(z^{-2} + z^{-5}\right) + h[3]\left(z^{-3} + z^{-4}\right), \tag{6.23}$$

leading to the realization shown in Figure 6.10(b).

It should be noted that the structure of Figure 6.10(a) requires 4 multipliers, whereas a direct form realization of the original length-7 FIR filter would require 7 multipliers. Likewise, the structure of Figure 6.10(b) requires 4 multipliers, compared to 8 multipliers in the direct form realization of the original length-8 FIR filter. A similar savings occurs in the case of an FIR filter with an antisymmetric impulse response.

6.4 Basic IIR Digital Filter Structures

The causal IIR digital filters we are concerned with in this text are characterized by a real rational transfer function of the form of Eq. (4.48) or, equivalently, by the constant coefficient difference equation of Eq. (2.82). From the difference equation representation, it can be seen that the computation of the nth output sample requires the knowledge of several past samples of the output sequence, or in other words, the realization of a causal IIR filter requires some form of feedback. We outline here several simple and straightforward realizations of IIR filters.

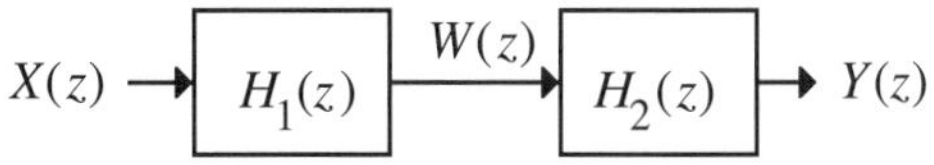

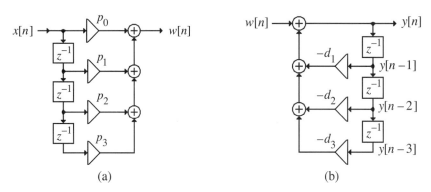

Figure 6.11: A possible IIR filter realization scheme.

(a)   (b)

Figure 6.12: (a) Realization of the transfer function $H_1(z) = W(z)/X(z)$ and (b) realization of the transfer function $H_2(z) = Y(z)/W(z)$.

6.4.1 Direct Forms

An Nth-order IIR digital filter transfer function is characterized by $2N + 1$ unique coefficients and, in general, requires $2N + 1$ multipliers and $2N$ two-input adders for implementation. As in the case of FIR filter realization, IIR filter structures in which the multiplier coefficients are precisely the coefficients of the transfer function are called *direct form* structures. We now describe the development of these structures.

Consider for simplicity a third-order IIR filter characterized by a transfer function

$$H(z) = \frac{P(z)}{D(z)} = \frac{p_0 + p_1 z^{-1} + p_2 z^{-2} + p_3 z^{-3}}{1 + d_1 z^{-1} + d_2 z^{-2} + d_3 z^{-3}}, \tag{6.24}$$

which we can implement as a cascade of two filter sections as shown in Figure 6.11 where

$$H_1(z) = \frac{W(z)}{X(z)} = P(z) = p_0 + p_1 z^{-1} + p_2 z^{-2} + p_3 z^{-3}, \tag{6.25}$$

and

$$H_2(z) = \frac{Y(z)}{W(z)} = \frac{1}{D(z)} = \frac{1}{1 + d_1 z^{-1} + d_2 z^{-2} + d_3 z^{-3}}. \tag{6.26}$$

The filter section $H_1(z)$ of Eq. (6.25) is seen to be an FIR filter and can be realized as shown in Figure 6.12(a). We next consider the realization of $H_2(z)$ given by Eq. (6.26). Note that a time-domain representation of this transfer function is given by

$$y[n] = w[n] - d_1 y[n-1] - d_2 y[n-2] - d_3 y[n-3], \tag{6.27}$$

resulting in the realization indicated in Figure 6.12(b).

A cascade of the structures of Figure 6.12(a) and (b) as indicated in Figure 6.11 leads to a realization of the original IIR transfer function $H(z)$ of Eq. (6.24). The resulting structure is sketched in Figure 6.13(a) and is commonly known as the *direct form I* structure. Note that the overall realization is noncanonic since it employs six delays to implement a third-order transfer function. The transpose of this structure is

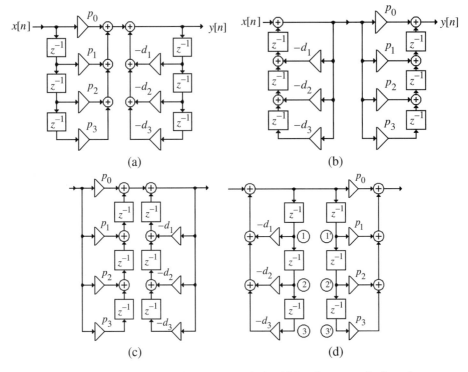

Figure 6.13: (a) Direct form I, (b) direct form I_t, (c) and (d) additional noncanonic direct form structures.

sketched in Figure 6.13(b). Various other noncanonic direct forms can be derived by simple block diagram manipulations. Two such realizations are shown in Figure 6.13(c) and (d).

To derive a canonic realization, we observe that in Figure 6.13(d) the signal variables at nodes ① and ①ʹ are the same, and hence, the two top delays can be shared. Likewise, the signal variables at nodes ② and ②ʹ are the same, which permits the sharing of the two middle delays. Following the same argument, we can share the two delays at the bottom leading to the final canonic structure shown in Figure 6.14(a), which is called the *direct form II* realization. The transpose of this is indicated in Figure 6.14(b).

The structures for direct form I and direct form II realizations of an Nth-order IIR transfer function should be evident from the third-order structures of Figures 6.13 and 6.14.

6.4.2 Cascade Realizations

By expressing the numerator and the denominator polynomials of the transfer function $H(z)$ as a product of polynomials of lower degree, a digital filter is often realized as a cascade of low-order filter sections. Consider for example, $H(z) = P(z)/D(z)$ expressed as

$$H(z) = \frac{P_1(z)P_2(z)P_3(z)}{D_1(z)D_2(z)D_3(z)}. \tag{6.28}$$

Various different cascade realizations of $H(z)$ can be obtained by different pole-zero polynomial pairings. Some examples of such realizations are shown in Figure 6.15. Additional cascade realizations are obtained by simply changing the ordering of the sections.[2] Figure 6.16 illustrates examples of different structures

---

[2] See Section 9.7.7.

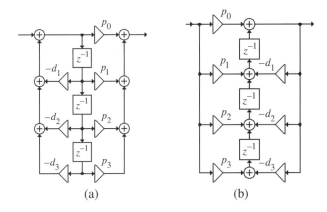

Figure 6.14: Direct form II and II$_t$ structures.

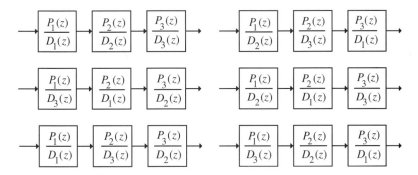

Figure 6.15: Examples of different equivalent cascade realizations obtained by different pole-zero pairings.

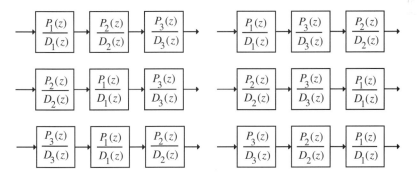

Figure 6.16: Different cascade realizations obtained by changing the ordering of the sections.

obtained by changing the ordering of the sections. There are altogether a total of 36 cascade realizations for the factored form indicated in Eq. (6.28) based on pole-zero pairings and ordering (Problem 6.19). In practice due to the finite wordlength effects, each such cascade realization behaves differently from others.

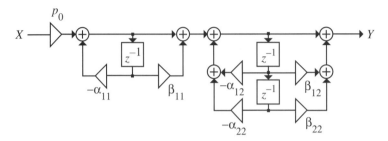

Figure 6.17: Cascade realization of a third-order IIR transfer function.

Usually, the polynomials are factored into a product of first-order and second-order polynomials. In this case, $H(z)$ is expressed as

$$H(z) = p_0 \prod_k \left(\frac{1 + \beta_{1k}z^{-1} + \beta_{2k}z^{-2}}{1 + \alpha_{1k}z^{-1} + \alpha_{2k}z^{-2}} \right). \tag{6.29}$$

In the above, for a first-order factor, $\alpha_{2k} = \beta_{2k} = 0$. A possible realization of a third-order transfer function

$$H(z) = p_0 \left(\frac{1 + \beta_{11}z^{-1}}{1 + \alpha_{11}z^{-1}} \right) \left(\frac{1 + \beta_{12}z^{-1} + \beta_{22}z^{-2}}{1 + \alpha_{12}z^{-1} + \alpha_{22}z^{-2}} \right)$$

is shown in Figure 6.17.

6.4.3 Parallel Realizations

An IIR transfer function can be realized in a parallel form by making use of the partial-fraction expansion of the transfer function. A partial-fraction expansion of the transfer function in the form of Eq. (3.131) leads to the *parallel form I*. Thus, assuming simple poles, $H(z)$ is expressed in the form

$$H(z) = \gamma_0 + \sum_k \left(\frac{\gamma_{0k} + \gamma_{1k}z^{-1}}{1 + \alpha_{1k}z^{-1} + \alpha_{2k}z^{-2}} \right). \tag{6.30}$$

In the above, for a real pole, $\alpha_{2k} = \gamma_{1k} = 0$.

A direct partial-fraction expansion of the transfer function $H(z)$ expressed as a ratio of polynomials in z leads to the second basic form of the parallel structure, called the *parallel form II* [Mit77c]. Assuming simple poles, we arrive at

$$H(z) = \delta_0 + \sum_k \left(\frac{\delta_{1k}z^{-1} + \delta_{2k}z^{-2}}{1 + \alpha_{1k}z^{-1} + \alpha_{2k}z^{-2}} \right). \tag{6.31}$$

Here, for a real pole, $\alpha_{2k} = \delta_{2k} = 0$.

The two basic parallel realizations of a third-order IIR transfer function are sketched in Figure 6.18.

EXAMPLE 6.3   We develop different realizations of the third-order IIR transfer function:

$$H(z) = \frac{0.44z^2 + 0.362z + 0.02}{z^3 + 0.4z^2 + 0.18z - 0.2} = \frac{0.44z^{-1} + 0.362z^{-2} + 0.02z^{-3}}{1 + 0.4z^{-1} + 0.18z^{-2} - 0.2z^{-3}}. \tag{6.32}$$

A direct form II realization of the above transfer function is shown in Figure 6.19(a).

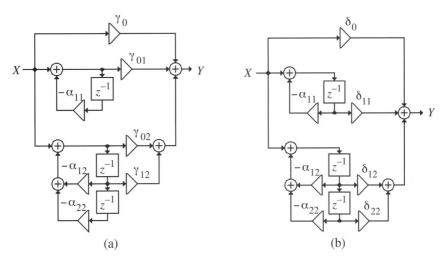

Figure 6.18: Parallel realizations of a third-order IIR transfer function: (a) parallel form I, and (b) parallel form II.

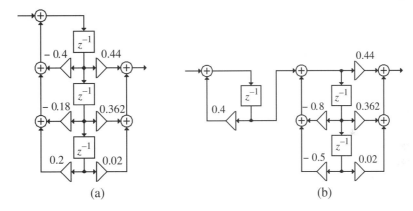

Figure 6.19: (a) Direct form II realization, and (b) cascade realization based on direct form II realization of each section.

By factoring the numerator and the denominator polynomials of $H(z)$ as given in Eq. (6.29), we obtain

$$H(z) = \frac{0.44z^2 + 0.362z + 0.02}{(z^2 + 0.8z + 0.5)(z - 0.4)}$$

$$= \left(\frac{0.44 + 0.362z^{-1} + 0.02z^{-2}}{1 + 0.8z^{-1} + 0.5z^{-2}} \right) \left(\frac{z^{-1}}{1 - 0.4z^{-1}} \right).$$

From the above, we arrive at a cascade realization of $H(z)$, as shown in Figure 6.19(b). Another cascade realization is obtained by using a different pole-zero pairing:

$$H(z) = \left(\frac{z^{-1}}{1 + 0.8z^{-1} + 0.5z^{-2}} \right) \left(\frac{0.44 + 0.362z^{-1} + 0.02z^{-2}}{1 - 0.4z^{-1}} \right), \qquad (6.33)$$

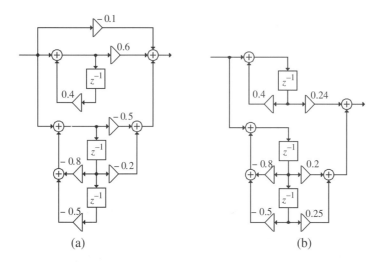

Figure 6.20: (a) Parallel form I realization, and (b) parallel form II realization.

whose realization is left as an exercise (Problem 6.24). However, it should be noted that the realization based on the above factored form will be noncanonic since it would require four delays instead of three delays employed in Figure 6.19(b). The total number of multipliers in both realizations remains the same.

Next, we make a partial-fraction expansion of $H(z)$ of the form of Eq. (3.131), resulting in

$$H(z) = -0.1 + \frac{0.6}{1 - 0.4z^{-1}} + \frac{-0.5 - 0.2z^{-1}}{1 + 0.8z^{-1} + 0.5z^{-2}},$$

which leads to the parallel form I realization indicated in Figure 6.20(a).

Finally, a direct partial-fraction expansion of $H(z)$ expressed as a ratio of polynomials in z is given by

$$H(z) = \frac{0.24}{z - 0.4} + \frac{0.2z + 0.25}{z^2 + 0.8z + 0.5} = \frac{0.24z^{-1}}{1 - 0.4z^{-1}} + \frac{0.2z^{-1} + 0.25z^{-2}}{1 + 0.8z^{-1} + 0.5z^{-2}},$$

resulting in the parallel form II realization sketched in Figure 6.20(b).

6.5 Realization of Basic Structures Using MATLAB

The basic FIR and IIR structures described in the previous two sections can be easily developed using MATLAB. In this section we describe this approach.

The cascade realization of an FIR transfer function $H(z)$ involves its factorization in the form of Eq. (6.11). Likewise, the cascade realization of an IIR transfer function $H(z)$ involves its factorization in the form of Eq. (6.29). The factorization of a polynomial can be carried out in MATLAB using the function roots. For example, r = roots(h) will return the roots of the polynomial vector h containing the coefficients of a polynomial in z^{-1} in ascending powers of z^{-1} in the output vector r. From the computed roots, the coefficients of the quadratic factors can be determined.

A much simpler approach is to use the function zp2sos which determines the second-order factors directly from the specified transfer function $H(z)$. The function sos = zp2sos(z,p,k) generates a matrix sos containing the coefficients of each second-order section of the equivalent transfer function $H(z)$ determined from its zero-pole form. sos is an $L \times 6$ matrix of the form

$$\text{sos} = \begin{bmatrix} p_{01} & p_{11} & p_{21} & d_{01} & d_{11} & d_{21} \\ p_{02} & p_{12} & p_{22} & d_{02} & d_{12} & d_{22} \\ \vdots & \vdots & \vdots & \vdots & \vdots & \vdots \\ p_{0L} & p_{1L} & p_{2L} & d_{0L} & d_{1L} & d_{2L} \end{bmatrix},$$

whose ith row contains the coefficients, $\{p_{i\ell}\}$ and $\{d_{i\ell}\}$, of the numerator and denominator polynomials of the ith second-order section with L denoting the total number of sections in the cascade. The form of the overall transfer function is thus given by

$$H(z) = \prod_{i=1}^{L} H_i(z) = \prod_{i=1}^{L} \frac{p_{0i} + p_{1i}z^{-1} + p_{2i}z^{-2}}{d_{0i} + d_{1i}z^{-1} + d_{2i}z^{-2}}.$$

We illustrate this approach in the following two examples. We first consider the factorization of an FIR transfer function.

EXAMPLE 6.4 We develop the second-order factors of the sixth-order FIR transfer function:

$$H(z) = 50.4 + 28.02z^{-1} + 13.89z^{-2} + 7.42z^{-3} + 6.09z^{-4} + 3z^{-5} + z^{-6}.$$

To this end we can use Program 6_1 given below.

```
% Program 6_1
% Factorization of a Rational Transfer Function
%
format long
num = input('Numerator coefficient vector = ');
den = input('Denominator coefficient vector = ');
[z,p,k] = tf2zp(num,den);
sos = zp2sos(z,p,k)
```

Program 6_1 has been written to develop the factorization of the numerator and denominator polynomials of a rational IIR transfer function. To apply it for the factorization of an FIR transfer function, we can treat it as an IIR transfer function with a constant numerator of unity, and a denominator which is the polynomial describing the FIR transfer function. Thus, the input data to this program are

```
num = 1
den = [50.4   28.02   13.89   7.42   6.09   3   1]
```

The output data generated by running this program are as follows:

```
sos =

  Columns 1 through 4
     0        0    0.01984126984127    1.00000000000000
     0        0    1.00000000000000    1.00000000000000
     0        0    1.00000000000000    1.00000000000000
```

Columns 5 through 6
    0.38095238095238   0.23809523809524
    0.87500000000000   0.25000000000000
   -0.70000000000000   0.33333333333333

The factorized form of $H(z)$ is therefore given by

$$H(z) = 50.4(1 + 0.38095z^{-1} + 0.2381z^{-2})(1 + 0.875z^{-1} + 0.25z^{-2})(1 - 0.7z^{-1} + 0.33333z^{-2}).$$

The function zp2sos works only for a stable transfer function. As a result, the denominator polynomial must have all its roots inside the unit circle. Hence, the above method can be used only for a minimum-phase FIR transfer function. In the case of a nonminimum-phase FIR transfer function, the first- and second-order factors can be obtained by computing the zeros of the transfer function using the function roots and then combining the complex conjugate root pairs using the function conv to form the second-order factors.

The factorization of an IIR transfer function is demonstrated next.

EXAMPLE 6.5   Consider the sixth-order IIR transfer function:

$$\frac{6 + 17.1z^{-1} + 33.03z^{-2} + 24.72z^{-3} + 19.908z^{-4} - 5.292z^{-5} + 18.144z^{-6}}{1 + 2.2z^{-1} + 2.56z^{-2} + 1.372z^{-3} + 0.118z^{-4} - 0.332z^{-5} - 0.168z^{-6}}.$$

The input data to Program 6_1 are now:

    num = [6   17.1   33.03   24.72   19.098   5.292   18.144]
    den = [1   2.2   2.56   1.372   0.118   -0.332 -0.168]

The output data generated by running this program are then

sos =

Columns 1 through 4
    6.00000000000000   -4.80000000000000   3.59999999999999   1.00000000000000
    1.00000000000000   1.89999999999999   2.40000000000000   1.00000000000000
    1.00000000000000   1.75000000000001   2.10000000000000   1.00000000000000

Columns 5 through 6
    0.20000000000000   -0.30000000000000
    0.80000000000000   0.70000000000000
    1.20000000000000   0.80000000000000

From the above we arrive at the factored form of the specified IIR transfer function as given below:

$$H(z) = \frac{6(1 - 0.8z^{-1} + 0.6z^{-2})(1 + 1.9z^{-1} + 2.4z^{-2})(1 + 1.75z^{-1} + 2.1z^{-2})}{(1 + 0.2z^{-1} - 0.3z^{-2})(1 + 0.8z^{-1} + 0.7z^{-2})(1 + 1.2z^{-1} + 0.8z^{-2})},$$

obtained by factoring out 6 from the first numerator factor.

The two parallel realizations described above can be developed readily in MATLAB using functions residue and residuez. We illustrate their use in the following example.

EXAMPLE 6.6   We repeat the parallel realizations of Example 6.3 using Program 6.2 given below.

```
% Program 6_2
% Parallel Realizations of an IIR Transfer Function
%
num = input('Numerator coefficient vector = ');
den = input('Denominator coefficient vector = ');
[r1,p1,k1] = residuez(num,den);
[r2,p2,k2] = residue(num,den);
disp('Parallel Form I')
disp('Residues are');disp(r1);
disp('Poles are at');disp(p1);
disp('Constant value');disp(k1);
disp('Parallel Form II')
disp('Residues are');disp(r2);
disp('Poles are at');disp(p2);
disp('Constant value');disp(k2);
```

The input data requested by the program are the vectors num and den, containing the numerator and denominator coefficients, respectively. For our example, these are given by

num = [0 0.44 0.362 0.02],   den = [1 0.4 0.18 -0.2].

The results generated by this program are

```
        Parallel Form I
        Residues are
          -0.2500 - 0.0000i
          -0.2500 + 0.0000i
           0.6000
          -0.1000

        Poles are at
          -0.4000 + 0.5831i
          -0.4000 - 0.5831i
           0.4000
                0

        Constant value
          -0.1000

        Parallel Form II
        Residues are
           0.1000 - 0.1458i
           0.1000 + 0.1458i
           0.2400

        Poles are at
          -0.4000 + 0.5831i
          -0.4000 - 0.5831i
           0.4000

        Constant value
```

The complex conjugate pairs in the partial-fraction expansion for parallel form I can be combined in MATLAB using the statement [b1,a1] = residuez[R1, P1, 0] where R1 is the two-element vector of the complex conjugate residues, and P1 is the two-element vector of the complex conjugate poles. For this example, we use R1 = [r1(1) r1(2)] and P1 = [p1(1) p1(2)] resulting in

```
    b1 =
        -0.5000    -0.2000           0

    a1 =
         1.0000     0.8000      0.5000
```

verifying the correctness of the results obtained in Example 6.3.

Likewise, the complex conjugate pairs in the partial-fraction expansion for parallel form II can be combined in MATLAB using the statement [b2,a2] = residue[R2, P2, 0] where now R2 is the two-element vector of the complex conjugate residues, and P2 is the two-element vector of the complex conjugate poles. For this example, we use R2 = [r2(1) r2(2)] and P2 = [p2(1) p2(2)] which yields

```
    b2 =
         0.2000     0.2500

    a2 =
         1.0000     0.8000      0.5000
```

The above results agree with those derived in Example 6.3.

6.6 Allpass Filters

We now turn our attention to the realization of a very special type of IIR transfer function, the allpass function, introduced earlier in Section 4.6. We recall from its definition in Eq. (4.126) that an IIR allpass transfer function $A(z)$ has a unity magnitude response for all frequencies, i.e., $|A(e^{j\omega})| = 1$ for all values of ω. The digital allpass filter is a versatile building block for signal processing applications. Some of its possible applications have already been pointed out earlier. For example, as indicated in Section 4.6.3, it is used often as a delay equalizer, in which case it is designed such that when cascaded with another discrete-time system, the overall cascade has a constant group delay in the frequency range of interest. Another application described in Section 4.8.4 is in the efficient implementation of a set of transfer functions satisfying certain complementary properties. For example, a pair of power-complementary first-order lowpass and highpass transfer functions can be implemented simultaneously employing only a single first-order allpass filter. Likewise, a pair of power-complementary second-order bandpass and bandstop transfer functions can be implemented simultaneously employing only a single second-order allpass filter. In fact, we shall demonstrate later in Section 6.10 that a large class of power-complementary transfer function pairs can be realized as a parallel connection of two allpass filters. It is thus of interest to develop techniques for the computationally efficient realization of allpass transfer functions, which is the subject of this section.

An Mth-order real-coefficient allpass transfer function is of the form of Eq. (4.127), with the numerator being the mirror-image polynomial of the denominator. A direct realization of an Mth-order allpass transfer function requires $2M$ multipliers. Since an Mth-order allpass transfer function is characterized by M unique coefficients, our objective here is to develop realization methods requiring only M multipliers. We outline here two different approaches to the minimum-multiplier realizations of allpass transfer functions.

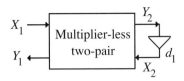

Figure 6.21: A multiplier-less two-pair constrained by a single multiplier.

6.6.1 Realization Based on the Multiplier Extraction Approach

Since an arbitrary allpass transfer function can be expressed as the product of second-order and/or first-order allpass transfer functions, we consider the realization of these lower-order transfer functions here. Even though an allpass transfer function can be realized using any of the methods discussed in this chapter, our objective here is to develop structures that remain allpass despite changes in the multiplier coefficients that may occur due to coefficient quantization [Mit74a].

Consider first the realization of a first-order allpass transfer function given by

$$A_1(z) = \frac{d_1 + z^{-1}}{1 + d_1 z^{-1}}. \tag{6.34}$$

Since the above transfer function is uniquely characterized by a single constant d_1, we attempt to realize it using a structure containing a single multiplier d_1 in the form of Figure 6.21. Substituting $G(z) = d_1$ in Eq. (4.181b), we express the input transfer function $A_1(z) = Y_1/X_1$ in terms of the transfer parameters of the two-pair as

$$A_1(z) = t_{11} + \frac{t_{12}t_{21}d_1}{1 - d_1 t_{22}} = \frac{t_{11} - d_1(t_{11}t_{22} - t_{12}t_{21})}{1 - d_1 t_{22}}. \tag{6.35}$$

A comparison of Eqs. (6.34) and (6.35) yields

$$t_{11} = z^{-1}, \qquad t_{22} = -z^{-1}, \tag{6.36}$$

$$t_{11}t_{22} - t_{12}t_{21} = -1. \tag{6.37}$$

Substituting Eq. (6.36) in Eq. (6.37), we arrive at

$$t_{12}t_{21} = 1 - z^{-2},$$

which leads to four possible solutions:

$$\text{Type 1A}: \ t_{11} = z^{-1}, \ t_{22} = -z^{-1}, \ t_{12} = 1 - z^{-2}, \ t_{21} = 1; \tag{6.38a}$$
$$\text{Type 1B}: \ t_{11} = z^{-1}, \ t_{22} = -z^{-1}, \ t_{12} = 1 + z^{-1}, \ t_{21} = 1 - z^{-1}; \tag{6.38b}$$
$$\text{Type 1A}_t: \ t_{11} = z^{-1}, \ t_{22} = -z^{-1}, \ t_{12} = 1, \ t_{21} = 1 - z^{-2}; \tag{6.38c}$$
$$\text{Type 1B}_t: \ t_{11} = z^{-1}, \ t_{22} = -z^{-1}, \ t_{12} = 1 - z^{-1}, \ t_{21} = 1 + z^{-1}. \tag{6.38d}$$

We now outline the development of the two-pair structure implementing the transfer parameters of Eq. (6.38a). From these equations we arrive at

$$Y_2 = X_1 - z^{-1}X_2,$$
$$Y_1 = z^{-1}X_1 + (1 - z^{-2})X_2 = z^{-1}Y_2 + X_2.$$

A realization of the above is sketched in Figure 6.22. By constraining the X_2, Y_2 terminal-pair with the multiplier d_1, we arrive at the single multiplier realization of the first-order allpass transfer function, as

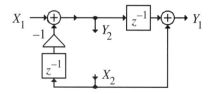

Figure 6.22: Development of the Type 1A first-order allpass structure.

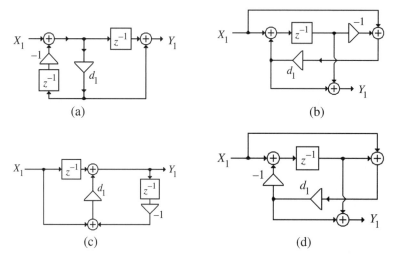

Figure 6.23: First-order single-multiplier allpass structures: (a) Type 1A, (b) Type 1B, (c) Type $1A_t$, (d) Type $1B_t$.

sketched in Figure 6.23(a). In a similar fashion, the other three single-multiplier allpass structures can be realized from their transfer parameter descriptions of Eqs. (6.38b) to (6.38d) (Problem 6.35). The final realizations are indicated in Figure 6.23(b) to (d). These structures have been called the *Type 1 allpass networks*. It should be noted that the structures of Figure 6.23(c) and (d) are, respectively, the transpose of the structures of Figure 6.23(a) and (b).

We next consider the realization of a second-order allpass transfer function. Such a function is characterized by two unique coefficients and, hence, can be realized by a structure with two multipliers with multiplier constants d_1 and d_2. Various forms of the second-order allpass transfer functions exist. Circuits realizing the transfer function of the form

$$A_2(z) = \frac{d_1 d_2 + d_1 z^{-1} + z^{-2}}{1 + d_1 z^{-1} + d_1 d_2 z^{-2}} \qquad (6.39)$$

are called the *Type 2 allpass networks* and are sketched in Figure 6.24. Additional Type 2 allpass structures can be derived by transposing the structures of Figure 6.24.[3]

Another form of the second-order allpass transfer function is given by

$$A_2(z) = \frac{d_2 + d_1 z^{-1} + z^{-2}}{1 + d_1 z^{-1} + d_2 z^{-2}}. \qquad (6.40)$$

The corresponding circuits are called the *Type 3 allpass structures* and are indicated in Figure 6.25.[4]

---

[3]The labeling of the structures here is as given in [Mit74a].
[4]The labeling of the structures here is as given in [Mit74a].

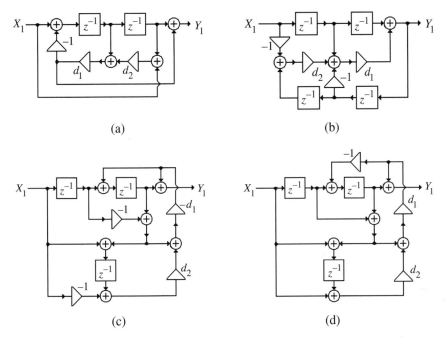

Figure 6.24: Second-order two-multiplier Type 2 allpass structures. (a) Type 2A, (b) Type 2D, (c) Type 2B, and (d) Type 2C.

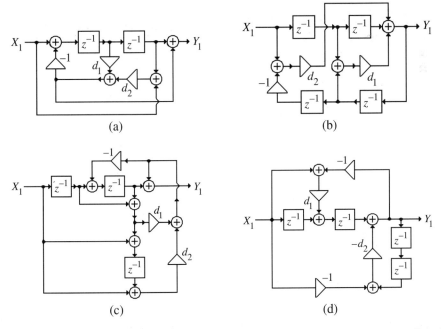

Figure 6.25: Second-order two-multiplier Type 3 allpass structures. (a) Type 3A, (b) Type 3D, (c) Type 3C, and (d) Type 3B.

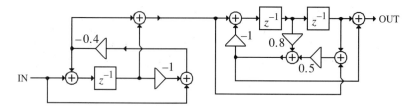

Figure 6.26: A three-multiplier realization of the allpass transfer function of Eq. (6.42).

EXAMPLE 6.7   Realize the third-order allpass transfer function

$$A_3(z) = \frac{-0.2 + 0.18z^{-1} + 0.4z^{-2} + z^{-3}}{1 + 0.4z^{-1} + 0.18z^{-2} - 0.2z^{-3}}. \qquad (6.41)$$

By factoring the numerator and the denominator, we rewrite the above as

$$A_3(z) = \frac{(-0.4 + z^{-1})(0.5 + 0.8z^{-1} + z^{-2})}{(1 - 0.4z^{-1})(1 + 0.8z^{-1} + 0.5z^{-2})}. \qquad (6.42)$$

A three-multiplier realization of the above allpass transfer function is shown in Figure 6.26, where the first-order allpass section has been implemented using the structure of Figure 6.23(b) and the second-order allpass section has been implemented using the structure of Figure 6.25(a).

6.6.2 Realization Based on the Two-Pair Extraction Approach

The stability test algorithm described in Section 4.12.2 also leads to an elegant realization of an Mth-order allpass transfer function $A_M(z)$ in the form of a cascaded two-pair [Vai87e]. This algorithm is based on the development of a series of $(m-1)$th-order allpass transfer functions $A_{m-1}(z)$ from an mth-order allpass transfer function $A_m(z)$:

$$A_m(z) = \frac{d_m + d_{m-1}z^{-1} + d_{m-2}z^{-2} + \cdots + d_1 z^{-(m-1)} + z^{-m}}{1 + d_1 z^{-1} + d_2 z^{-2} + \cdots + d_{m-1}z^{-(m-1)} + d_m z^{-m}} \qquad (6.43)$$

using the recursion

$$A_{m-1}(z) = z \left[\frac{A_m(z) - k_m}{1 - k_m A_m(z)} \right], \qquad m = M, M-1, \ldots 1, \qquad (6.44)$$

where $k_m = A_m(\infty) = d_m$. It has been shown in Section 4.12.2 that $A_M(z)$ is stable if and only if

$$k_m^2 < 1, \quad \text{for } m = M, M-1, \ldots, 1. \qquad (6.45)$$

If the allpass transfer function $A_{m-1}(z)$ is expressed in the form

$$A_{m-1}(z) = \frac{d'_{m-1} + d'_{m-2}z^{-1} + \cdots + d'_1 z^{-(m-2)} + z^{-(m-1)}}{1 + d'_1 z^{-1} + \cdots + d'_{m-2}z^{-(m-2)} + d'_{m-1}z^{-(m-1)}}, \qquad (6.46)$$

then the coefficients of $A_{m-1}(z)$ are simply related to the coefficients of $A_m(z)$ through the expression

$$d'_i = \frac{d_i - d_m d_{m-i}}{1 - d_m^2}, \qquad i = 1, 2, \ldots, m-1. \qquad (6.47)$$

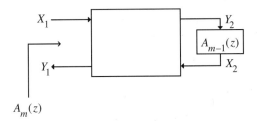

Figure 6.27: Realization of $A_m(z)$ by two-pair extraction.

To develop a realization of $A_m(z)$ using the above algorithm, we rewrite Eq. (6.44) as

$$A_m(z) = \frac{k_m + z^{-1}A_{m-1}(z)}{1 + k_m z^{-1}A_{m-1}(z)}. \tag{6.48}$$

We realize $A_m(z)$ by extracting a two-pair constrained by $A_{m-1}(z)$ in the form of Figure 6.27.
Now from Eq. (4.181b), $A_m(z)$ can be expressed in terms of the transfer parameters of the two-pair as

$$A_m(z) = \frac{t_{11} - (t_{11}t_{22} - t_{12}t_{21})A_{m-1}(z)}{1 - t_{22}A_{m-1}(z)}. \tag{6.49}$$

Comparing Eqs. (6.48) and (6.49), we readily obtain

$$t_{11} = k_m, \qquad t_{22} = -k_m z^{-1}, \tag{6.50a}$$

$$t_{11}t_{22} - t_{12}t_{21} = -z^{-1}. \tag{6.50b}$$

Substituting Eq. (6.50a) in Eq. (6.50b), we get

$$t_{12}t_{21} = (1 - k_m^2)z^{-1}. \tag{6.51}$$

As can be seen from the above, there are a number of solutions for t_{12} and t_{21} leading to different realizations for the two-pair. Some possible solutions are as indicated below:

$$t_{11} = k_m, \; t_{22} = -k_m z^{-1}, \; t_{12} = (1 - k_m^2)z^{-1}, \; t_{21} = 1, \tag{6.52a}$$

$$t_{11} = k_m, \; t_{22} = -k_m z^{-1}, \; t_{12} = (1 - k_m)z^{-1}, \; t_{21} = (1 + k_m), \tag{6.52b}$$

$$t_{11} = k_m, \; t_{22} = -k_m z^{-1}, \; t_{12} = \sqrt{1 - k_m^2}\, z^{-1}, \; t_{21} = \sqrt{1 - k_m^2}, \tag{6.52c}$$

$$t_{11} = k_m, \; t_{22} = -k_m z^{-1}, \; t_{12} = z^{-1}, \; t_{21} = (1 - k_m^2). \tag{6.52d}$$

The input-output relations of the two-pair described by Eq. (6.52a) are given by

$$Y_1 = k_m X_1 + (1 - k_m^2)z^{-1}X_2, \tag{6.53a}$$

$$Y_2 = X_1 - k_m z^{-1}X_2. \tag{6.53b}$$

A direct realization of the above equations leads to the three-multiplier two-pair shown in Figure 6.28(a). In a similar manner, direct realizations based on Eqs. (6.52b) and (6.52c) result in the four-multiplier structures of Figure 6.28(b) and (c), respectively.[5] A direct realization of Eq. (6.52d) results in a three-multiplier structure and is left as an exercise (Problem 6.42).

[5]The structure of Figure 6.28(b) is called the *Kelly-Lochbaum form* [Kel62].

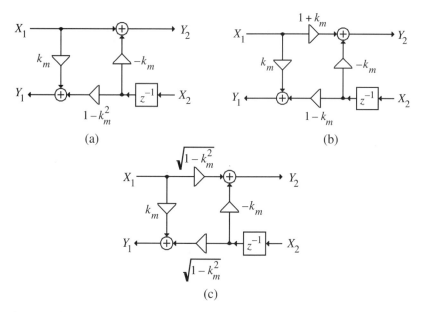

Figure 6.28: Direct realization of the two-pairs described by Eq. (6.52a) through (6.52c). (a) The two-pair described by Eq. (6.52a), (b) the two-pair described by Eq. (6.52b), and (c) the two-pair described by Eq. (6.52c).

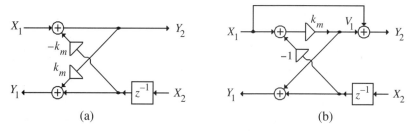

Figure 6.29: (a) A two-multiplier realization of the two-pair described by Eq. (6.52a), and (b) a one-multiplier realization of the two-pair described by Eq. (6.52b).

A two-multiplier realization can be derived by manipulating the input-output relations of Eqs. (6.53a) and (6.53b). Using Eq. (6.53b), we can rewrite Eq. (6.53a) as

$$Y_1 = k_m Y_2 + z^{-1} X_2. \tag{6.54}$$

A realization of the two-pair based on Eqs. (6.53b) and (6.54) is given in Figure 6.29(a). The two-pair of Figure 6.29(a) is often referred to as a *lattice structure*. A two-multiplier lattice realization of Eq. (6.52d) can be derived accordingly and is left as an excercise (Problem 6.43) [Lar99].

The two-pair described by Eq. (6.52b) can be realized using a single multiplier. To this end, we first write its input-output relation from Eq. (6.52b) as

$$Y_1 = k_m X_1 + (1 - k_m) z^{-1} X_2, \tag{6.55a}$$
$$Y_2 = (1 + k_m) X_1 - k_m z^{-1} X_2. \tag{6.55b}$$

Defining

$$V_1 = k_m (X_1 - z^{-1} X_2), \tag{6.56}$$

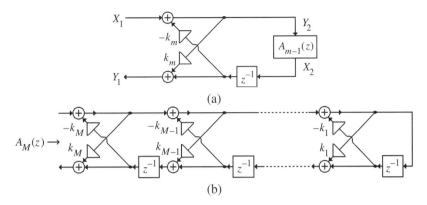

Figure 6.30: (a) Realization of $A_m(z)$ by extracting the lattice two-pair of Figure 6.29(a), and (b) cascaded lattice realization of $A_M(z)$.

we can rewrite Eqs. (6.55a) and (6.55b) as

$$Y_1 = V_1 + z^{-1}X_2, \tag{6.57a}$$
$$Y_2 = X_1 + V_1. \tag{6.57b}$$

A realization based on Eqs. (6.56), (6.57a), and (6.57b) leads to the single-multiplier two-pair of Figure 6.29(b). Note that the two-input adder with an incoming multiplier with a coefficient -1 is implemented as a subtractor.

The realization of the mth-order allpass transfer function $A_m(z)$ is therefore obtained by constraining any one of the two-pairs of Figures 6.28 and 6.29 by the $(m-1)$th-order allpass transfer function $A_{m-1}(z)$. For example, Figure 6.30(a) shows the realization of $A_m(z)$ by extracting the lattice two-pair of Figure 6.29(a).

Following the above algorithm, we can next realize $A_{m-1}(z)$ as a lattice two-pair constrained by the allpass transfer function $A_{m-2}(z)$. This process is repeated until the constraining transfer function is $A_0(z) = 1$. The complete realization of the original allpass transfer function $A_M(z)$ based on the extraction of the lattice two-pair of Figure 6.29(a) is therefore as indicated in Figure 6.30(b). It follows from our discussion in Section 4.12.2 that $A_M(z)$ is stable if the magnitudes of all multiplier coefficients in the realization are less than unity, i.e., $|k_m| < 1$ for $m = M, M-1, \ldots, 1$.

Note that the above allpass structure requires $2M$ multipliers, which is twice that needed in the realization of an Mth-order allpass transfer function. However, a realization of $A_M(z)$ with M multipliers is obtained by extracting the two-pair of Figure 6.29(b). Here, also, the stability of $A_M(z)$ is ensured if $|k_m| < 1$ for $m = M, M-1, \ldots, 1$.

EXAMPLE 6.8 Realize the third-order allpass transfer function $A_3(z)$ of Example 6.7 by means of a cascaded lattice structure. Here,

$$A_3(z) = \frac{d_3 + d_2 z^{-1} + d_1 z^{-2} + z^{-3}}{1 + d_1 z^{-1} + d_2 z^{-2} + d_3 z^{-3}}$$

$$= \frac{-0.2 + 0.18 z^{-1} + 0.4 z^{-2} + z^{-3}}{1 + 0.4 z^{-1} + 0.18 z^{-2} - 0.2 z^{-3}}. \tag{6.58}$$

Using the method outlined above, we realize $A_3(z)$ first in the form of a lattice two-pair characterized by the multiplier coefficient $k_3 = d_3 = -0.2$ and constrained by an allpass $A_2(z)$ as indicated in Figure 6.31(a). The allpass $A_2(z)$ is of the form

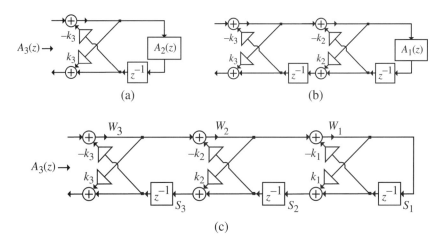

Figure 6.31: Cascaded lattice realization of the third-order allpass transfer function of Eq. (6.58): $k_3 = d_3 = -0.2$, $k_2 = d_2' = 0.2708333$, and $k_1 = d_1'' = 0.3573771$.

$$A_2(z) = \frac{d_2' + d_1' z^{-1} + z^{-2}}{1 + d_1' z^{-1} + d_2' z^{-2}}, \tag{6.59}$$

whose coefficients are obtained using Eq. (6.47) and are given by

$$d_1' = \frac{d_1 - d_3 d_2}{1 - d_3^2} = \frac{0.4 - (-0.2)(0.18)}{1 - (-0.2)^2} = 0.4541667, \tag{6.60a}$$

$$d_2' = \frac{d_2 - d_3 d_1}{1 - d_3^2} = \frac{0.18 - (-0.2)(0.4)}{1 - (-0.2)^2} = 0.2708333. \tag{6.60b}$$

Next, the allpass $A_2(z)$ is realized as a lattice two-pair characterized by the multiplier coefficient $k_2 = d_2' = 0.2708333$ and constrained by an allpass $A_1(z)$, as indicated in Figure 6.31(b). The allpass $A_1(z)$ is of the form

$$A_1(z) = \frac{d_1'' + z^{-1}}{1 + d_1'' z^{-1}}, \tag{6.61}$$

where

$$d_1'' = \frac{d_1' - d_2' d_1'}{1 - (d_2')^2} = \frac{d_1'}{1 + d_2'} = \frac{0.4541667}{1.2708333} = 0.3573771. \tag{6.62}$$

Finally, the allpass $A_1(z)$ is realized as a lattice structure characterized by the multiplier coefficient $k_1 = d_1'' = 0.3573771$, resulting in the complete realization of the allpass $A_3(z)$, as shown in Figure 6.31(c). Note from Figure 6.31(c) that all multiplier coefficients have magnitudes less than unity. Therefore, the allpass transfer function of Eq. (6.58) is stable.

The M-file `poly2rc` in MATLAB can be employed to realize an allpass transfer function in the cascaded lattice form. We illustrate its application in the following example.

EXAMPLE 6.9   We consider the development of the cascaded lattice realization of the allpass transfer function given in Eq. (6.41) of Example 6.7. To this end we can use Program 6_3 given below. Note the similarity between this program and Program 4_4 used for the stability test.

```
% Program 6_3
% Cascaded Lattice Realization of an
% Allpass Transfer Function
%
format long
den = input('Denominator coefficient vector = ');
k = poly2rc(den);
knew = fliplr(k);
disp('The lattice section multipliers are');disp(knew);
```

The program requests the input data, which for this example is the vector of denominator coefficients in descending powers of z:

$$\text{den} = \begin{bmatrix} 1 & 0.4 & 0.18 & -0.2 \end{bmatrix}$$

It then prints the lattice section multiplier coefficients as given below:

```
The lattice section multipliers are
 -0.2    0.27083333    0.35737705
```

These values are seen to be identical to those derived in Example 6.8.

6.7 Tunable IIR Digital Filters

In Section 4.5.2, we described two first-order and two second-order IIR digital transfer functions with tunable frequency response characteristics. As we show next, these transfer functions can be realized easily using allpass structures, and the resulting realizations provide independent tuning of the filter parameters such as the cutoff frequencies and bandwidth [Mit90a].

6.7.1 Tunable Lowpass and Highpass First-Order Filters

In Example 4.14, we showed that the first-order lowpass transfer function $H_{LP}(z)$ of Eq. (4.109) and the first-order highpass transfer function $H_{HP}(z)$ of Eq. (4.112) are a doubly-complementary pair and can be expressed as

$$
\begin{aligned}
H_{LP}(z) &= \frac{1}{2}\left[\frac{1-\alpha+z^{-1}-\alpha z^{-1}}{1-\alpha z^{-1}}\right] = \frac{1}{2}\left[1+\frac{-\alpha+z^{-1}}{1-\alpha z^{-1}}\right] \\
&= \tfrac{1}{2}[1+A_1(z)],
\end{aligned}
\tag{6.63}
$$

$$
\begin{aligned}
H_{HP}(z) &= \frac{1}{2}\left[\frac{1+\alpha-z^{-1}-\alpha z^{-1}}{1-\alpha z^{-1}}\right] = \frac{1}{2}\left[1-\frac{-\alpha+z^{-1}}{1-\alpha z^{-1}}\right] \\
&= \tfrac{1}{2}[1-A_1(z)],
\end{aligned}
\tag{6.64}
$$

where

$$
A_1(z) = \frac{-\alpha+z^{-1}}{1-\alpha z^{-1}},
\tag{6.65}
$$

is a first-order allpass transfer function. A combined realization of $H_{LP}(z)$ and $H_{HP}(z)$ based on the decompositions of Eqs. (6.63) and (6.64) is as indicated in Figure 6.32, in which the allpass filter given by Eq. (6.65) can be realized using any one of the four single-multiplier allpass structures of Figure 6.23.

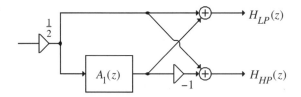

Figure 6.32: Allpass-based realization of the doubly-complementary first-order lowpass and highpass filters.

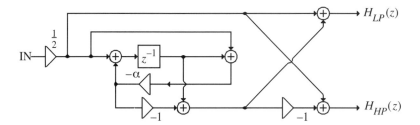

Figure 6.33: A tunable first-order lowpass/highpass filter structure.

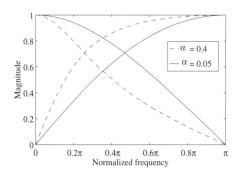

Figure 6.34: Magnitude responses of the lowpass/highpass filter structure of Figure 6.33 for two different values of the parameter α.

Figure 6.33 shows one such realization in which the 3-dB cutoff frequency of both filters can be simultaneously varied by changing the multiplier coefficient α. Figure 6.34 shows the composite magnitude responses of the two filters for two different values of α.

6.7.2 Tunable Bandpass and Bandstop Second-Order Filters

The second-order bandpass transfer function $H_{BP}(z)$ of Eq. (4.113) and the second-order bandstop transfer function $H_{BS}(z)$ of Eq. (4.118) also form a doubly-complementary pair and can be expressed as

$$H_{BP}(z) = \tfrac{1}{2}\left[1 - A_2(z)\right], \tag{6.66}$$

$$H_{BS}(z) = \tfrac{1}{2}\left[1 + A_2(z)\right], \tag{6.67}$$

where $A_2(z)$ is a second-order allpass transfer function given by

$$A_2(z) = \frac{\alpha - \beta(1+\alpha)z^{-1} + z^{-2}}{1 - \beta(1+\alpha)z^{-1} + \alpha z^{-2}}. \tag{6.68}$$

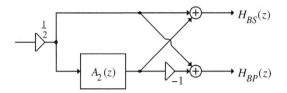

Figure 6.35: Allpass-based realization of doubly-complementary second-order bandpass/bandstop filter.

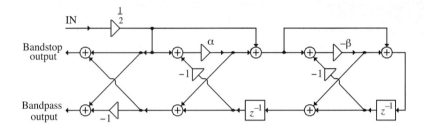

Figure 6.36: Tunable second-order bandpass/bandstop filter structure.

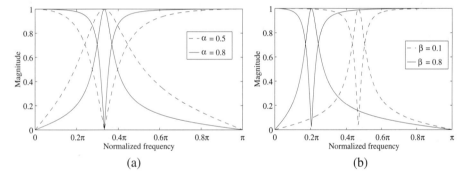

Figure 6.37: Magnitude responses of the bandpass/bandstop filter structure of Figure 6.36 for different values of the parameters α and β. (a) $\beta = 0.5$ and (b) $\alpha = 0.8$.

Therefore, the bandpass transfer function $H_{BP}(z)$ of Eq. (4.113) and the bandstop transfer function $H_{BS}(z)$ of Eq. (4.118) can be realized together, as indicated in Figure 6.35, where the allpass filter $A_2(z)$ is given by Eq. (6.68). A tunable bandpass/bandstop filter structure is then obtained by realizing the allpass $A_2(z)$ by means of a cascaded lattice structure described in Section 6.6.2, with the lattice two-pair realized using its single-multiplier equivalent of Figure 6.29(a). The final structure is indicated in Figure 6.36. Note that in the structure of Figure 6.36, the multiplier β controls the center frequency and the multiplier α controls the 3-dB bandwidth. Figure 6.37 illustrates the parametric tuning property of the bandpass/bandstop filter structure of Figure 6.36.

6.8 IIR Tapped Cascaded Lattice Structures

Several simple straightforward realizations of FIR and IIR transfer functions have been outlined in Sections 6.3 and 6.4, respectively. In most applications, these realizations work reasonably well even under finite wordlength constraints. However, in some cases, digital filters with more robust properties are needed to provide satisfactory performances. We describe in this section two such realizations.

The cascaded lattice structure of Figure 6.30(b) can also realize an all-pole IIR digital filter which finds application in the power spectrum estimation of random signals (see Section 11.4.2). It also forms the basis of an often used method for the realization of an arbitrary Mth-order transfer function $H(z)$ originally proposed by Gray and Markel [Gra73]. We first demonstrate the realization of an all-pole IIR transfer function and then outline the Gray and Markel realization method. We also provide a MATLAB program implementing both IIR structures.

6.8.1   Realization of an All-Pole IIR Transfer Function

We now show that the transfer function $W_1(z)/X_1(z)$ of the cascaded lattice structure of Figure 6.31(c) is an all-pole transfer function with the same denominator as the allpass transfer function $A_3(z)$ of Eq. (6.58). We first observe that a typical lattice two-pair here is described by the equations:

$$W_i(z) = W_{i+1}(z) - k_i z^{-1} S_i(z),$$
$$S_{i+1}(z) = k_i W_I(z) + z^{-1} S_i(z).$$

From the above we obtain the chain matrix description of the two-pair as

$$\begin{bmatrix} W_{i+1}(z) \\ S_{i+1}(z) \end{bmatrix} = \begin{bmatrix} 1 & k_i z^{-1} \\ k_i & z^{-1} \end{bmatrix} \begin{bmatrix} W_i(z) \\ S_i(z) \end{bmatrix}. \tag{6.69}$$

Thus, we can express the chain matrix of the cascaded lattice structure of Figure 6.31(b) as

$$\begin{bmatrix} X_1(z) \\ Y_1(z) \end{bmatrix} = \begin{bmatrix} 1 & k_3 z^{-1} \\ k_3 & z^{-1} \end{bmatrix} \begin{bmatrix} 1 & k_2 z^{-1} \\ k_2 & z^{-1} \end{bmatrix} \begin{bmatrix} 1 & k_1 z^{-1} \\ k_1 & z^{-1} \end{bmatrix} \begin{bmatrix} W_1(z) \\ S_1(z) \end{bmatrix}$$

$$= \begin{bmatrix} 1 + k_2(k_1 + k_3)z^{-1} + k_3 k_1 z^{-2} & k_1 z^{-1} + k_2(1 + k_1 k_3)z^{-2} + k_3 z^{-3} \\ k_3 + k_2(1 + k_1 k_3)z^{-1} + k_1 z^{-2} & k_1 k_3 z^{-1} + k_2(k_1 + k_3)z^{-2} + z^{-3} \end{bmatrix} \begin{bmatrix} W_1(z) \\ S_1(z) \end{bmatrix}. \tag{6.70}$$

From Eq. (6.70) we finally arrive at

$$X_1(z) = \left(1 + [k_1(1 + k_2) + k_2 k_3]z^{-1} + [k_2 + k_1 k_3(1 + k_2)]z^{-2} + k_3 z^{-3}\right) W_1(z),$$

where we have used the relation $S_1(z) = W_1(z)$. The transfer function $W_1(z)/X_1(z)$ is thus an all-pole transfer function with the same denominator polynomial as the third-order allpass transfer function of Eq. (6.58), i.e.,

$$\frac{W_1(z)}{X_1(z)} = \frac{1}{1 + [k_1(1 + k_2) + k_2 k_3]z^{-1} + [k_2 + k_1 k_3(1 + k_2)]z^{-2} + k_3 z^{-3}} \tag{6.71}$$

$$= \frac{1}{1 + d_1 z^{-1} + d_2 z^{-2} + d_3 z^{-3}}, \tag{6.72}$$

where we have used $k_1 = d_1''$, $k_2 = d_2'$, and $k_3 = d_3$, and Eqs. (6.60a), (6.60b), and (6.62).

It follows from the above discussion that, in the general case, the cascaded lattice structure of Figure 6.30(b) realizes an Mth-order all-pole transfer function with the same denominator as $A_M(z)$ if the output is tapped from the input of the rightmost delay. Such an all-pole IIR structure results from modeling an autoregressive process as described in Section 11.4. The multiplier coefficients $\{k_i\}$ of the cascaded lattice structure are called *reflection coefficients*.

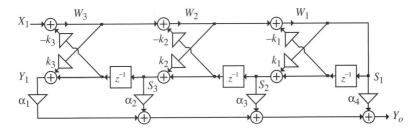

Figure 6.38: The Gray-Markel structure for a third-order transfer function.

6.8.2 Gray-Markel Method

The method of realizing an IIR transfer function $H(z) = P_M(z)/D_M(z)$ consists of two steps. In the first step, an intermediate allpass transfer function $A_M(z) = z^{-M} D_M(z^{-1})/D_M(z) = \tilde{D}_M(z)/D_M(z)$ is realized in the form of a cascaded lattice structure.[6] A set of independent variables of this structure are then summed in the second step, with appropriate weights to yield the desired numerator $P_M(z)$.

To illustrate the method of realizing the numerator, we consider for simplicity the implementation of a third-order IIR transfer function

$$H(z) = \frac{P_3(z)}{D_3(z)} = \frac{p_0 + p_1 z^{-1} + p_2 z^{-2} + p_3 z^{-3}}{1 + d_1 z^{-1} + d_2 z^{-2} + d_3 z^{-3}}. \tag{6.73}$$

In the first step, we form an intermediate allpass transfer function $A_3(z) = Y_1(z)/X_1(z) = \tilde{D}_3(z)/D_3(z)$. Realization of $A_3(z)$ has been illustrated in Example 6.8 resulting in the structure of Figure 6.31(c). Our objective is to sum the linearly independent signal variables Y_1, S_1, S_2, and S_3 with weights $\{\alpha_i\}$ as shown in Figure 6.38 to arrive at the desired numerator $P_3(z)$. To this end, we need to analyze the digital filter structure of Figure 6.31(c) and determine the transfer functions $S_1(z)/X_1(z)$, $S_2(z)/X_1(z)$, and $S_3(z)/X_1(z)$.

From Eq. (6.71) we have

$$\frac{S_1(z)}{X_1(z)} = \frac{1}{D_3(z)}. \tag{6.74}$$

Next, we observe from Figure 6.38 that $S_2(z) = (d_1'' + z^{-1})S_1(z)$ and, hence, from the above we get

$$\frac{S_2(z)}{X_1(z)} = \frac{d_1'' + z^{-1}}{D_3(z)}. \tag{6.75}$$

Finally, from Figure 6.38 we have $S_3(z) = d_2' W_2(z) + z^{-1} S_2(z)$ and $S_1(z) = W_2(z) - d_1'' z^{-1} S_1(z)$. From these equations, the relation $S_2(z) = (d_1'' + z^{-1})S_1(z)$, and Eqs. (6.60a), (6.60b), and (6.62), we arrive at $S_3(z) = (d_2' + d_1' z^{-1} + z^{-2})S_1(z)$ yielding

$$\frac{S_3(z)}{X_1(z)} = \frac{d_2' + d_1' z^{-1} + z^{-2}}{D_3(z)}. \tag{6.76}$$

It should be noted that the numerator of the transfer function $S_i(z)/X_1(z)$ is precisely the numerator of the allpass transfer function $S_i(z)/W_i(z)$.

---

[6] $\tilde{D}_M(z)$ denotes the mirror-image polynomial of $D_M(z)$ of degree M, i.e., $\tilde{D}_M(z) = z^{-M} D_M(z^{-1})$.

We now form

$$\frac{Y_o(z)}{X_1(z)} = \alpha_1 \frac{Y_1(z)}{X_1(z)} + \alpha_2 \frac{S_3(z)}{X_1(z)} + \alpha_3 \frac{S_2(z)}{X_1(z)} + \alpha_4 \frac{S_1(z)}{X_1(z)}. \tag{6.77}$$

Substituting Eqs. (6.74) to (6.76) and the expression for $Y_1(z)/X_1(z) = A_3(z)$ as given by Eq. (6.58) in Eq. (6.77), we arrive at

$$\frac{Y_o(z)}{X_1(z)} = \frac{\begin{array}{c} \alpha_1(d_3 + d_2 z^{-1} + d_1 z^{-2} + z^{-3}) \\ + \alpha_2(d_2' + d_1' z^{-1} + z^{-2}) + \alpha_3(d_1'' + z^{-1}) + \alpha_4 \end{array}}{D_3(z)}. \tag{6.78}$$

Comparing the numerators of the right-hand sides of Eqs. (6.73) and (6.78), and by equating the coefficients of like powers of z^{-1}, we thus obtain

$$\alpha_1 d_3 + \alpha_2 d_2' + \alpha_3 d_1'' + \alpha_4 = p_0,$$
$$\alpha_1 d_2 + \alpha_2 d_1' + \alpha_3 = p_1,$$
$$\alpha_1 d_1 + \alpha_2 = p_2,$$
$$\alpha_1 = p_3.$$

Solving the above we get

$$\alpha_1 = p_3,$$
$$\alpha_2 = p_2 - \alpha_1 d_1,$$
$$\alpha_3 = p_1 - \alpha_1 d_2 - \alpha_2 d_1',$$
$$\alpha_4 = p_0 - \alpha_1 d_3 - \alpha_2 d_2' - \alpha_3 d_1'', \tag{6.79}$$

which is in a form suitable for step-by-step calculation of the feedforward tap coefficients α_i.

EXAMPLE 6.10   Let us realize the transfer function of Eq. (6.32) in Example 6.3, repeated below for convenience:

$$H(z) = \frac{P_3(z)}{D_3(z)} = \frac{0.44z^{-1} + 0.362z^{-2} + 0.02z^{-3}}{1 + 0.4z^{-1} + 0.18z^{-2} - 0.2z^{-3}}, \tag{6.80}$$

using the Gray-Markel method.

We form the intermediate allpass transfer function $A_3(z)$ having the same denominator as $H(z)$ in Eq. (6.80). Its realization has been carried out in Example 6.8 and is shown in Figure 6.31(c), where $d_3 = -0.2$, $d_2' = 0.2708333$, and $d_1'' = 0.3573771$.

To realize the numerator, we compute the tap coefficients $\{\alpha_i\}$ using Eq. (6.79). Substituting the values of $d_1 = 0.4$, $d_2 = 0.18$, $d_3 = -0.2$, $d_1' = 0.4541667$, $d_1'' = 0.3573771$, $d_2' = 0.2708333$, and the values of the numerator coefficients, $p_0 = 0$, $p_1 = 0.44$, $p_2 = 0.36$, and $p_3 = 0.02$ in Eq. (6.79), we arrive at

$$\alpha_1 = 0.02, \quad \alpha_2 = 0.352, \quad \alpha_3 = 0.2765333, \quad \alpha_4 = -0.19016.$$

The final realization is thus as indicated in Figure 6.38 with the multiplier values as given above.

6.8.3 Realization Using MATLAB

Both the pole-zero and the all-pole IIR cascaded lattice structures can be developed from their specified transfer functions using the M-file `tf2latc` in the *Signal Processing Toolbox*. Various forms of this function relevant to IIR transfer functions are

```
[k,alpha] = tf2latc(num,den)
k = tf2latc(1,den)
[k,v] = tf2latc(1,den)
```

where k is the lattice parameter vector, alpha is the feedforward coefficient vector, and the vectors num and den contain, respectively, the coefficients of the numerator and the denominator polynomials in ascending powers of z^{-1}.

The function latc2tf implements the reverse process and can be used to verify the realization. Various forms of this function are

```
[num,den] = latc2tf(k,alpha)
[num,den] = latc2tf(k,'iir')
```

We illustrate the use of the above two functions in the next several examples.

EXAMPLE 6.11   Using MATLAB we determine the lattice and the feedforward parameters of the Gray-Markel structure for the IIR transfer function of Eq. (6.80). To this end, Program 6_4 can be employed.

```
% Program 6_4
% Gray-Markel Cascaded Lattice Structure
% Development
% den is the denominator coefficient vector
% num is the numerator coefficient vector
% k is the lattice parameter vector
% alpha is the vector of feedforward multipliers
%
format long
% Read in the transfer function coefficients
num = input('Numerator coefficient vector = ');
den = input('Denominator coefficient vector = ');
num = num/den(1);
den = den/den(1);
[k,alpha] = tf2latc(num,den);
disp('Lattice parameters are');disp(k');
disp('Feedforward multipliers are');disp(fliplr(alpha'));
```

The input data called by the above program are

```
num = [0   0.44   0.36    0.02]
den = [1   0.4    0.18   -0.2]
```

which results in the following output data generated by the program:

```
Lattice parameters are
    0.35737704918033    0.27083333333333   -0.2

Feedforward multipliers are
    0.02   0.352   0.27653333333333   -0.19016
```

which are seen to be identical to those derived in Example 6.7.

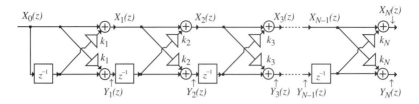

Figure 6.39: The FIR cascaded lattice structure.

Program 6_4 can also be employed to develop the cascaded lattice realization of an all-pole IIR transfer function as illustrated in the following example.

EXAMPLE 6.12   We now realize the all-pole transfer function

$$H(z) = \frac{1}{1 + 0.5\,z^{-1} + 0.4\,z^{-2} + 0.3\,z^{-3} + 0.2\,z^{-4}}.$$

The output data generated by running this program for the above transfer function are

```
Lattice parameters are
    0.325581395   0.24863884    0.20833333    0.20000000

Feedforward multipliers are
    0      0      0      0      1
```

Since the absolute values of all lattice parameters are less than 1, $H(z)$ is guaranteed stable.

Program 6_5, given below, can be used to verify the cascaded lattice structure developed using Program 6_4.

```
% Program 6_5
% Transfer Function of Gray-Markel Cascaded
% Lattice Structure from the Lattice and
% Feedforward Parameters
% k1 is the lattice parameter vector
% alpha is the vector of feedforward multipliers
% den is the denominator coefficient vector
% num is the numerator coefficient vector
%
format long
% Read in the lattice and feedforward parameters
k1 = input('Lattice parameter vector = ');
alpha = input('Feedforward parameter vector = ');
[num,den] = latc2tf(k1,fliplr(alpha1);
disp('Numerator coefficients are');disp(num)
disp('Denominator coefficients are');disp(den)
```

The following example illustrates the application of Program 6_5.

EXAMPLE 6.13   We verify the cascaded lattice structure generated in Example 6.11. The input data called by Program 6_5 are

```
k1 = [0.35737704918033     0.27083333333333    -0.2]
alpha = [0.02    0.352    0.27653333333333    -0.19016]
```

The program then generates the following output data,

```
Numerator coefficients are
  -0.00    0.44    0.36    0.02

Denominator coefficients are
  1.00    0.40    0.180   -0.20
```

which are seen to be the same as that given in Eq. (6.80).

6.9   FIR Cascaded Lattice Structures

There are two types of cascaded lattice structures for the realization of FIR transfer functions. In this section, we describe methods for their realizations.

6.9.1   Realization of Arbitrary FIR Transfer Functions

The cascaded lattice structure that can be used to realize an arbitrary FIR transfer function is shown in Figure 6.39 [Mak75]. The realization algorithm is developed first followed by its MATLAB implementation.

Realization Method

For the cascaded lattice realization, the Nth-order FIR transfer function is assumed to be of the form

$$H_N(z) = 1 + \sum_{n=1}^{N} p_n z^{-n}. \tag{6.81}$$

An arbitrary FIR transfer function $H(z)$ of the form of Eq. (6.9) can be expressed in the form $H(z) = h[0]H_N(z)$, where $p_n = h[n]/h[0]$, $0 \le n \le N$. Hence, by placing a multiplier $h[0]$ at the input of the structure realizing $H_N(z)$ we obtain a realization of $H(z)$.

The normalized transfer function $H_N(z)$ is realized in the form of Figure 6.39. To develop the realization algorithm, we first analyze it to determine the relations between the input $X(z) = X_0(z)$ and the intermediate variables $X_i(z)$ and $Y_i(z)$, i.e., between the series of transfer functions $H_i(z) = X_i(z)/X_0(z)$ and $G_i(z) = Y_i(z)/X_0(z)$, $1 \le i \le N$. Note that $X_N(z)$ and $Y_N(z)$ are the output variables.

From Figure 6.39 we obtain

$$X_1(z) = X_0(z) + k_1 z^{-1} X_0(z), \tag{6.82a}$$

$$Y_1(z) = k_1 X_0(z) + z^{-1} X_0(z). \tag{6.82b}$$

The corresponding transfer functions are given by

$$H_1(z) = \frac{X_1(z)}{X_0(z)} = 1 + k_1 z^{-1}, \tag{6.83a}$$

$$G_1(z) = \frac{Y_1(z)}{X_0(z)} = k_1 + z^{-1}, \tag{6.83b}$$

which are seen to be first-order FIR transfer functions. Moreover, it can be seen from Eqs. (6.83a) and (6.83b) that

$$\tilde{H}_1(z) = z^{-1}H_1(z^{-1}) = z^{-1} + k_1 = G_1(z),$$

indicating that the FIR transfer function $G_1(z)$ of Eq. (6.83b) is the mirror image of the FIR transfer function $H_1(z)$ of Eq. (6.83a).

Next, we observe from Figure 6.39 that

$$X_2(z) = X_1(z) + k_2 z^{-1} Y_1(z), \tag{6.84a}$$

$$Y_2(z) = k_2 X_1(z) + z^{-1} Y_1(z). \tag{6.84b}$$

The corresponding transfer functions are given by

$$H_2(z) = \frac{X_2(z)}{X_0(z)} = H_1(z) + k_2 z^{-1} G_1(z), \tag{6.85a}$$

$$G_2(z) = \frac{Y_2(z)}{X_0(z)} = k_2 H_1(z) + z^{-1} G_1(z). \tag{6.85b}$$

Substituting Eqs. (6.83a) and (6.83b) in the above we observe that $H_2(z)$ and $G_2(z)$ are second-order FIR transfer functions. Moreover, since $G_1(z) = \tilde{H}_1(z)$, it follows from the above that $G_2(z) = z^{-2}H_2(z^{-1}) = \tilde{H}_2(z)$, implying that $G_2(z)$ is the mirror image of $H_2(z)$.

Now, the input-output relation of the ith lattice section is given by

$$X_i(z) = X_{i-1}(z) + k_i z^{-1} Y_{i-1}(z), \tag{6.86a}$$

$$Y_i(z) = k_i X_{i-1}(z) + z^{-1} Y_{i-1}(z). \tag{6.86b}$$

The corresponding transfer functions, $H_i(z) = X_i(z)/X_0(z)$ and $G_i(z) = Y_i(z)/X_0(z)$, are related through

$$H_i(z) = H_{i-1}(z) + k_i z^{-1} G_{i-1}(z), \tag{6.87a}$$

$$G_i(z) = k_i H_{i-1}(z) + z^{-1} G_{i-1}(z), \tag{6.87b}$$

where $H_{i-1}(z) = X_{i-1}(z)/X_0(z)$ and $G_{i-1}(z) = Y_{i-1}(z)/X_0(z)$. It should be noted that the relations of Eqs. (6.87a) and (6.87b) hold for all $1 \le i \le N$.

If we assume that $H_{i-1}(z)$ and $G_{i-1}(z)$ are $(i-1)$th-order FIR transfer functions, then it follows from Eqs. (6.87a) and (6.87b) that $H_i(z)$ and $G_i(z)$ are ith-order FIR transfer functions. Moreover, if $G_{i-1}(z) = \tilde{H}_{i-1}(z)$, then it also follows from Eqs. (6.87a) and (6.87b) that $G_i(z) = \tilde{H}_i(z)$. We have already shown that these two observations hold for $i = 1$ and $i = 2$. Therefore, by induction, they also hold for all $1 \le i \le N$.

To develop the realization algorithm, the above process is reversed. That is, given the expressions for $H_i(z)$ and $\tilde{H}_i(z)$, we find the expressions for $H_{i-1}(z)$ and $\tilde{H}_{i-1}(z)$ for $i = N, N-1, \ldots, 2, 1$. We now develop a recursion relation to compute these intermediate transfer functions.

Solving Eqs. (6.87a) and (6.87b) for $i = N$, we obtain

$$H_{N-1}(z) = \frac{1}{(1 - k_N^2)z^{-1}} \left\{ z^{-1} H_N(z) - k_N z^{-1} \tilde{H}_N(z) \right\}, \tag{6.88}$$

$$\tilde{H}_{N-1}(z) = \frac{1}{(1 - k_N^2)z^{-1}} \left\{ -k_N H_N(z) + \tilde{H}_N(z) \right\}. \tag{6.89}$$

It can be easily verified that $\tilde{H}_{N-1}(z)$ of Eq. (6.89) is indeed the mirror image of $H_{N-1}(z)$ of Eq. (6.88). Substituting the expression for $H_N(z)$ of Eq. (6.81) in Eq. (6.88) we get

$$H_{N-1}(z) = \frac{1}{(1-k_N^2)} \left\{ (1 - k_N p_N) + (p_1 - k_N p_{N-1})z^{-1} + \cdots \right.$$
$$\left. + (p_{N-1} - k_N p_1)z^{-(N-1)} + (p_N - k_N)z^{-N} \right\}. \tag{6.90}$$

If we choose $k_N = p_N$, $H_{N-1}(z)$ reduces to an FIR transfer function of degree $N-1$ and can be written in the form

$$H_{N-1}(z) = 1 + \sum_{n=1}^{N-1} p'_n z^{-n}, \tag{6.91}$$

where

$$p'_n = \frac{p_n - k_N p_{N-n}}{1 - k_N^2}, \qquad 1 \le n \le N-1. \tag{6.92}$$

Continuing the above recursion algorithm, all the multiplier coefficients of the structure of Figure 6.39 can be computed. The following example illustrates the procedure.

EXAMPLE 6.14   Realize the fourth-order FIR transfer function

$$H_4(z) = 1 + 1.2z^{-1} + 1.12z^{-2} + 0.12z^{-3} - 0.08z^{-4}. \tag{6.93}$$

From the above the multiplier coefficient of the rightmost lattice is given by $k_4 = p_4 = -0.08$. Using Eq. (6.92), we next determine the coefficients p'_n of the third-order transfer function $H_3(z)$ as,

$$p'_3 = \frac{p_3 - k_4 p_1}{1 - k_4^2} = \frac{0.12 - (-0.08)(1.2)}{1 - (-0.08)^2} = 0.2173913,$$

$$p'_2 = \frac{p_2 - k_4 p_2}{1 - k_4^2} = \frac{1.12}{1 - 0.08} = 1.2173913,$$

$$p'_1 = \frac{p_1 - k_4 p_3}{1 - k_4^2} = \frac{1.2 - (-0.08)(0.12)}{1 - (-0.08)^2} = 1.2173913.$$

As a result,

$$H_3(z) = 1 + 1.2173913z^{-1} + 1.2173913z^{-2} + 0.2173913z^{-3}.$$

Thus, the multiplier coefficient of the third lattice two-pair is given by $k_3 = p'_3 = 0.2173913$.

We continue the above recursion process to determine the coefficients p''_n of the second-order transfer function $H_2(z)$ which are given by

$$p''_2 = \frac{p'_2 - k_3 p'_1}{1 - k_3^2} = \frac{p'_2}{1 + k_3} = \frac{1.2173913}{1.2173913} = 1.0,$$

$$p''_1 = \frac{p'_1 - k_3 p'_2}{1 - k_3^2} = \frac{p'_1}{1 + k_3} = 1.0,$$

leading to

$$H_2(z) = 1 + z^{-1} + z^{-2}.$$

Consequently, the multiplier coefficient of the second lattice two-pair is given by $k_2 = p''_2 = 1.0$.

The last recursion yields the multiplier coefficient of the first lattice two-pair:

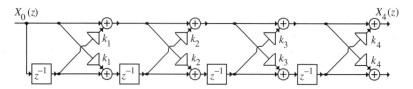

Figure 6.40: Cascaded lattice realization of the FIR transfer function of Eq. (6.93). $k_1 = 0.5$, $k_2 = 1.0$, $k_3 = 0.2173913$, and $k_4 = -0.08$.

$$k_1 = \frac{p_1''}{1 + k_2} = \frac{1.0}{2.0} = 0.5.$$

The final realization of the FIR transfer function of Eq. (6.93) is thus as shown in Figure 6.40.

Realization Using MATLAB

The function `tf2latc` in MATLAB can again be employed to compute the multiplier coefficients of the cascade lattice structure of Figure 6.39. To this end, we make use of Program 6.6 given below.

```
% Program 6_6
% FIR Cascaded Lattice Realization
%
format long
num = input('Transfer function coefficients = ');
k = tf2latc(num);
disp('Lattice coefficients are'); disp(fliplr(k)');
```

The input data called by the program is the vector `num` of transfer function coefficients entered in ascending powers of z^{-1}.

We illustrate its application in the following example.

EXAMPLE 6.15   Using Program 6.6 develop the cascaded lattice realization of the FIR transfer function of Eq. (6.93). For our example the input data are

```
num = [1   1.2   1.12   0.12   -0.08]
```

The output data is the vector of lattice coefficients as given below.

```
Lattice coefficients are
  -0.08      0.2173913      1.0       0.5
```

which are seen to be identical to those computed in Example 6.9.

The coefficients of the FIR cascaded lattice can also be computed using the M-file `poly2rc`. To this end, the statement to use is

```
k = poly2rc(num);
```

The FIR cascaded lattice structure developed using Program 6.6 can also be verified using the function `latc2tf`. To this end, the pertinent forms of this function are

```
num = latc2tf(k,'fir')
num = latc2tf(k)
```

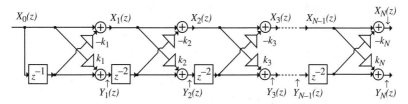

Figure 6.41: Power-symmetric FIR cascaded lattice structure.

where num is the vector of transfer function coefficients in ascending powers of z^{-1} and k is the vector of the FIR lattice coefficients.

6.9.2 Power-Symmetric FIR Cascaded Lattice Structure

Another type of a cascaded lattice structure for the realization of a real-coefficient FIR transfer function $H_N(z)$ of order N is shown in Figure 6.41 [Vai86b]. However, for realizability, the transfer function must satisfy the *power-symmetric condition* given by Eq. (4.146) and repeated below for convenience:

$$H_N(z)H_N(z^{-1}) + H_N(-z)H_N(-z^{-1}) = K_N, \qquad (6.94)$$

where K_N is a constant. We first analyze the structure of Figure 6.41 and then develop the synthesis procedure.

We define $H_i(z) = X_i(z)/X_0(z)$ and $G_i(z) = Y_i(z)/X_0(z)$. From Figure 6.41 we observe that

$$X_1(z) = X_0(z) + k_1 z^{-1} X_0(z),$$
$$Y_1(z) = -k_1 X_0(z) + z^{-1} X_0(z). \qquad (6.95)$$

Therefore,

$$H_1(z) = 1 + k_1 z^{-1},$$
$$G_1(z) = -k_1 + z^{-1}. \qquad (6.96)$$

It can be easily verified that

$$G_1(z) = z^{-1} H_1(-z^{-1}). \qquad (6.97)$$

Next, from Figure 6.41, it follows that the transfer functions $H_i(z)$ and $G_i(z)$ can be expressed in terms of $H_{i-2}(z)$ and $G_{i-2}(z)$ as

$$H_i(z) = H_{i-2}(z) + k_i z^{-2} G_{i-2}(z),$$
$$G_i(z) = -k_i H_{i-2}(z) + z^{-2} G_{i-2}(z). \qquad (6.98)$$

It can be easily shown that

$$G_i(z) = z^{-i} H_i(-z^{-1}) \qquad (6.99)$$

provided $G_{i-2}(z) = z^{-(i-2)} H_{i-2}(-z^{-1})$. However, as can be seen from Eq. (6.97), Eq. (6.99) holds for $i = 1$. Hence, the relation of Eq. (6.99) holds for all odd integer values of i. This result also implies that N must be an odd integer.

It is a simple exercise to show that both $H_i(z)$ and $G_i(z)$ satisfy the power-symmetry condition of Eq. (6.94). In addition, $H_i(z)$ and $G_i(z)$ are power-complementary, i.e.,

$$\left| H_i(e^{j\omega}) \right|^2 + \left| G_i(e^{j\omega}) \right|^2 = C_i, \qquad (6.100)$$

where C_i is a constant.

To develop the synthesis equation, we invert Eq. (6.98) to arrive at

$$(1 + k_i^2)H_{i-2}(z) = H_i(z) - k_i G_i(z),$$
$$(1 + k_i^2)z^{-2}G_{i-2}(z) = k_i H_i(z) + G_i(z). \tag{6.101}$$

Note that at the ith step, the multiplier coefficient k_i is chosen to eliminate the coefficient of z^{-i}, the highest power of z^{-1}, in $H_i(z) - k_i G_i(z)$. For this choice of k_i the coefficient of z^{-i+1} also vanishes making $H_{i-2}(z)$ a polynomial of degree $i - 2$.

We start the process by setting $i = N$, and compute $G_N(z)$ using Eq. (6.99). Next, we determine the transfer functions $H_{N-2}(z)$ and $G_{N-2}(z)$ using the above recurrence relations. This process is repeated until all coefficients of the lattice have been determined.

We illustrate the above synthesis method in the following example.

EXAMPLE 6.16   Let us realize the FIR transfer function

$$H_5(z) = 1 + 0.3z^{-1} + 0.2z^{-2} - 0.376z^{-3} - 0.06z^{-4} + 0.2z^{-5},$$

in the form of the cascaded lattice structure of Figure 6.41.

It can be easily verified that the above transfer function satisfies the power-symmetric condition of Eq. (6.94). Next, we form

$$G_5(z) = z^{-5}H_5(-z^{-1}) = -0.2 - 0.06z^{-1} + 0.376z^{-2} + 0.2z^{-3} - 0.3z^{-4} + z^{-5}.$$

To determine $H_3(z)$ we first form

$$H_5(z) - k_5 G_5(z) = (1 + 0.2k_5) + (0.3 + 0.06k_5)z^{-1} + (0.2 - 0.376k_5)z^{-2}$$
$$+ (-0.376 - 0.2k_5)z^{-3} + (-0.06 + 0.3k_5)z^{-4} + (0.2 - k_5)z^{-5}.$$

To cancel the coefficient of z^{-5} in the above expression, we choose

$$k_5 = 0.2.$$

Then from Eq. (6.101) and the above equation we get

$$H_3(z) = \frac{1}{1.04}\left(1.04 + 0.312z^{-1} + 0.1248z^{-2} - 0.416z^{-3}\right)$$
$$= 1 + 0.3z^{-1} + 0.12z^{-2} - 0.4z^{-3}.$$

We now compute

$$0.2H_5(z) + G_5(z) = (0.2 - 0.2) + (0.06 - 0.06)z^{-1} + (0.04 + 0.376)z^{-2}$$
$$+ (-0.0752 + 0.2)z^{-3} + (-0.012 - 0.3)z^{-4} + 1.04z^{-5}.$$

Therefore, from Eq. (6.101) and the above equation we get

$$G_3(z) = \frac{1}{1.04}\left(0.416 + 0.1248z^{-1} - 0.312z^{-2} + 1.04z^{-3}\right)$$
$$= 0.4 + 0.12z^{-1} - 0.3z^{-2} + z^{-3}.$$

Continuing the above process we arrive at the remaining two multiplier coefficients:

$$k_3 = -0.4, \qquad k_1 = 0.3.$$

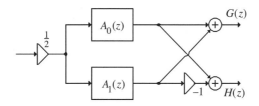

Figure 6.42: Realization of a doubly-complementary transfer function pair using a parallel allpass structure.

6.10 Parallel Allpass Realization of IIR Transfer Functions

In Section 6.6, we have demonstrated that a pair of doubly-complementary first-order lowpass and highpass transfer functions can be realized as shown in Figure 6.32. Likewise, we have shown that a pair of doubly-complementary second-order bandpass and bandstop transfer functions can be realized, as indicated in Figure 6.35. These two structures can be considered as special cases of a structure composed of a parallel connection of two stable allpass filters of the form of Figure 6.42, where one of the allpass sections is a zeroth-order transfer function. We consider in this section the realization of an Nth-order transfer function $G(z)$ in the form of Figure 6.42, along with its power-complementary transfer function $H(z)$ [Vai86a]. As we shall point out later, such structures have a number of very attractive properties from an implementation point of view.

From Figure 6.42 we observe that

$$G(z) = \tfrac{1}{2}\{A_0(z) + A_1(z)\}, \tag{6.102a}$$

$$H(z) = \tfrac{1}{2}\{A_0(z) - A_1(z)\}. \tag{6.102b}$$

It is a simple exercise to show that a necessary condition for $G(z)$ and $H(z)$ to have the sum of allpass decompositions of the form of Eqs. (6.102a) and (6.102b) is that each be a bounded real (BR) IIR transfer function (Problem 4.94).[7] The BR condition can be easily satisfied by any stable transfer function by a simple scaling. Let

$$G(z) = \frac{P(z)}{D(z)} = \frac{p_0 + p_1 z^{-1} + \cdots + p_N z^{-N}}{1 + d_1 z^{-1} + \cdots + d_N z^{-N}} \tag{6.103}$$

be an Nth-order BR transfer function with its power-complementary transfer function given by

$$H(z) = \frac{Q(z)}{D(z)} = \frac{q_0 + q_1 z^{-1} + \cdots + q_N z^{-N}}{1 + d_1 z^{-1} + \cdots + d_N z^{-N}}. \tag{6.104}$$

The power-complementary property implies that

$$\left|G(e^{j\omega})\right|^2 + \left|H(e^{j\omega})\right|^2 = 1. \tag{6.105}$$

From Eq. (6.102a) it follows that $G(z)$ must have a symmetric numerator, i.e.,

$$p_n = p_{N-n}, \tag{6.106}$$

and from Eq. (6.102b) it follows that $H(z)$ must have an antisymmetric numerator, i.e.,

$$q_n = -q_{N-n}. \tag{6.107}$$

---

[7] See Section 4.4.5 for the definition of a bounded real transfer function.

We develop next the procedure to identify the two allpass transfer functions from a specified transfer function $G(z)$. Now, the symmetric property of the numerator of $G(z)$ as given by Eq. (6.106) implies that

$$P(z^{-1}) = z^N P(z). \tag{6.108}$$

Likewise, the antisymmetric property of the numerator of $H(z)$ as given by Eq. (6.107) implies that

$$Q(z^{-1}) = -z^N Q(z). \tag{6.109}$$

By analytic continuation, we can rewrite Eq. (6.105) as

$$G(z^{-1})G(z) + H(z)H(z^{-1}) = 1. \tag{6.110}$$

Substituting $G(z) = P(z)/D(z)$ and $H(z) = Q(z)/D(z)$ in the above equation, and making use of Eqs. (6.108) and (6.109), we arrive at

$$[P(z) + Q(z)][P(z) - Q(z)] = z^{-N}D(z^{-1})D(z). \tag{6.111}$$

From the relations of Eqs. (6.108) and (6.109), we can write

$$P(z) - Q(z) = z^{-N}[P(z^{-1}) + Q(z^{-1})]. \tag{6.112}$$

If we denote the zeros of $[P(z) + Q(z)]$ as $z = \xi_k$, $1 \le k \le N$, then it follows from Eq. (6.111) that $z = 1/\xi_k$, $1 \le k \le N$, are the zeros of $[P(z) - Q(z)]$, i.e., $[P(z) - Q(z)]$ is the mirror-image polynomial of $[P(z) + Q(z)]$. From Eq. (6.112) it also follows that the zeros of $[P(z) + Q(z)]$ inside the unit circle are zeros of $D(z)$, whereas the zeros of $[P(z) + Q(z)]$ outside the unit circle are zeros of $D(z^{-1})$, since $G(z)$ and $H(z)$ have been assumed to be stable transfer functions. Let $z = \xi_k$, $1 \le k \le r$, be the r zeros of $[P(z) + Q(z)]$ inside the unit circle, and the remaining $N - r$ zeros, $z = \xi_k$, $r + 1 \le k \le N$, be outside the unit circle. Hence, it can be seen from Eq. (6.112) that the N zeros of $D(z)$ are given by

$$z = \begin{cases} \xi_k, & 1 \le k \le r, \\ \frac{1}{\xi_k}, & r + 1 \le k \le N. \end{cases}$$

All we need to do now is to identify the above zeros with the appropriate allpass transfer functions $A_0(z)$ and $A_1(z)$. To this end, we obtain from Eqs. (6.102a) and (6.102b)

$$A_0(z) = G(z) + H(z) = \frac{P(z) + Q(z)}{D(z)}, \tag{6.113a}$$

$$A_1(z) = G(z) - H(z) = \frac{P(z) - Q(z)}{D(z)}. \tag{6.113b}$$

Therefore, the two allpass transfer functions can be expressed as

$$A_0(z) = \prod_{k=r+1}^{N} \frac{z^{-1} - (\xi_k^*)^{-1}}{1 - \xi_k^{-1}z^{-1}}, \tag{6.114a}$$

$$A_1(z) = \prod_{k=1}^{r} \frac{z^{-1} - \xi_k^*}{1 - \xi_k z^{-1}}. \tag{6.114b}$$

In order to arrive at the above expressions for the two allpass transfer functions, it is necessary to determine the transfer function $H(z)$ that is power-complementary to $G(z)$. Denoting the $2N$th-degree polynomial $P^2(z) - z^{-N}D(z^{-1})D(z)$ as $U(z)$,

$$P^2(z) - z^{-N}D(z^{-1})D(z) = U(z) = \sum_{n=0}^{2N} u_n z^{-n}, \tag{6.115}$$

we can rewrite Eq. (6.111) as

$$Q^2(z) = \sum_{n=0}^{2N} u_n z^{-n}. \tag{6.116}$$

Solving the above equation for the coefficients q_k of $Q(z)$ we finally arrive at

$$q_0 = \sqrt{u_0}, \quad q_1 = \frac{u_1}{2q_0}, \tag{6.117a}$$

$$q_k = -q_{N-k} = \frac{u_k - \sum_{\ell=1}^{k-1} q_\ell q_{n-\ell}}{2q_0}, \quad k \geq 2, \tag{6.117b}$$

where we have used the antisymmetric property of the coefficients. After $Q(z)$ has been determined, we form the polynomial $[P(z) + Q(z)]$, find its zeros $z = \xi_k$, and then determine the two allpass transfer functions using Eqs. (6.114a) and (6.114b).

It can be shown that IIR digital transfer functions derived from the analog Butterworth, Chebyshev, and elliptic filters[8] via the bilinear transformation approach discussed in Section 7.2 can be decomposed in the form of Eqs. (6.102a) and (6.102b) [Vai86a]. Moreover, for lowpass-highpass filter pairs, the order N of the transfer function must be odd with the orders of $A_0(z)$ and $A_1(z)$ differing by 1. Likewise, for bandpass-bandstop filter pairs, the order N of the transfer function must be even with the orders of $A_0(z)$ and $A_1(z)$ differing by 2 [Vai86a].

In the case of odd-order digital Butterworth, Chebyshev, and elliptic lowpass or highpass transfer functions, there is a simple approach to identify the poles of the allpass transfer functions $A_0(z)$ and $A_1(z)$ from the poles λ_k, $0 \leq k \leq N - 1$, of the parent lowpass transfer function $G(z)$ or $H(z)$. Let θ_k denote the angle of the pole λ_k. If we assume that the poles are numbered such that $\theta_k < \theta_{k+1}$, then the poles of $A_0(z)$ are given by λ_{2k} and the poles of $A_1(z)$ are given by λ_{2k+1} [Gas85]. Figure 6.43 illustrates this *pole interlacing property* of the two allpass transfer functions.

EXAMPLE 6.17 Consider the transfer function of a seventh-order Butterworth lowpass digital filter with a 3-dB cutoff frequency of 0.3π:[9]

$$G(z) = \frac{0.000963(1 + 7z^{-1} + 21z^{-2} + 35z^{-3} + 35z^{-4} + 21z^{-5} + 7z^{-6} + z^{-7})}{1 - 2.7825z^{-1} + 3.9668z^{-2} - 3.4051z^{-3} + 1.8757z^{-4}} \tag{6.118}$$
$$\overline{\; - 0.6509z^{-5} + 0.1308z^{-6} - 0.01166z^{-7}}$$

Its seven poles are located at (numbered according to increasing angles):

$$\xi_0 = 0.4981133 - j0.668405, \quad \xi_1 = 0.3907071 - j0.4204395,$$
$$\xi_2 = 0.3399766 - j0.2030305, \quad \xi_3 = 0.32491969623291,$$
$$\xi_4 = 0.3399766 + j0.2030305, \quad \xi_5 = 0.3907071 + j0.4204395,$$
$$\xi_6 = 0.4981133 + j0.668405.$$

[8] See Section 5.4 for a review of these analog filter approximation techniques.
[9] This transfer function has been obtained using MATLAB. For details, see Section 7.10.1.

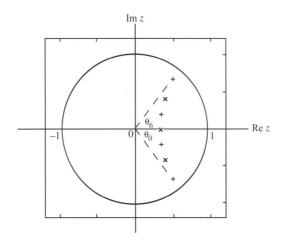

Figure 6.43: Pole interlacing property in the case of a seventh-order digital Butterworth lowpass filter. The poles marked "+" belong to the allpass function $A_0(z)$ and the poles marked "×" belong to the allpass function $A_1(z)$.

The corresponding z-plane pole-plot is indicated in Figure 6.43. Making use of the pole interlacing property, we identify the two allpass transfer functions as

$$A_0(z) = \frac{(z^{-1} - \xi_0)(z^{-1} - \xi_2)(z^{-1} - \xi_4)(z^{-1} - \xi_6)}{(1 - \xi_0 z^{-1})(1 - \xi_2 z^{-1})(1 - \xi_4 z^{-1})(1 - \xi_6 z^{-1})}$$

$$= \frac{(0.69488 - 0.99623z^{-1} + z^{-2})(0.15681 - 0.67995z^{-1} + z^{-2})}{(1 - 0.99623z^{-1} + 0.69488z^{-2})(1 - 0.67995z^{-1} + 0.15681z^{-2})},$$
(6.119a)

$$A_1(z) = \frac{(z^{-1} - \xi_1)(z^{-1} - \xi_3)(z^{-1} - \xi_5)}{(1 - \xi_1 z^{-1})(1 - \xi_3 z^{-1})(1 - \xi_5 z^{-1})}$$

$$= \frac{(-0.32492 + z^{-1})(0.32942 - 0.781414z^{-1} + z^{-2})}{(1 - 0.32492z^{-1})(1 - 0.781414z^{-1} + 0.32942z^{-2})}.$$
(6.119b)

It can be easily verified that by substituting the above two allpass transfer functions in Eq. (6.102a), we indeed arrive at the transfer function $G(z)$ of Eq. (6.118). In addition, making use of these in Eq. (6.102b), we also get its power-complementary highpass transfer function $H(z)$ given by

$$H(z) = \frac{0.108(1 - 7z^{-1} + 21z^{-2} - 35z^{-4} - 21z^{-5} + 7z^{-6} - z^{-7})}{\begin{array}{c} 1 - 2.7825z^{-1} + 3.9668z^{-2} - 3.4051z^{-3} + 1.8757z^{-4} \\ - 0.6509z^{-5} + 0.1308z^{-6} - 0.01166z^{-7} \end{array}}.$$
(6.120)

Figure 6.44 shows the magnitude response plots of $G(z)$ of Eq. (6.118) and $H(z)$ of Eq. (6.120) demonstrating their power-complementary property.

The parallel allpass structure has two other very attractive properties. As indicated in Section 6.5, an Mth-order allpass transfer function can be realized using only M multipliers. Therefore, the realization of an Nth-order IIR transfer function $G(z)$ based on the allpass decomposition of Eq. (6.102a) requires only $r + (N - r) = N$ multipliers. On the other hand, in general, a direct form realization of an Nth-order IIR transfer function uses $2N + 1$ multipliers. Moreover, with an additional adder and no additional multipliers, one can easily implement its power-complementary transfer function $H(z)$, also of Nth-order,

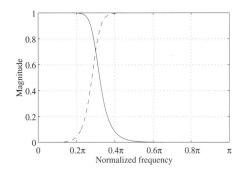

Figure 6.44: Magnitude responses of a pair of power-complementary seventh-order Butterworth lowpass and highpass transfer functions with a 3-dB cutoff frequency at 0.3π.

as indicated in Figure 6.42. As a result, the parallel allpass realization of an IIR transfer function, if it exists, is computationally very efficient. We shall show later in Section 9.9.2, that the parallel allpass realization also has very low passband sensitivity with respect to multiplier coefficients if the allpass sections are realized in a structurally lossless form.

6.11 Digital Sine-Cosine Generator

We now consider the realization of digital sinusoidal oscillators. In particular, we consider here the design of digital sine-cosine generators that produce two sinusoidal sequences that are exactly 90 degrees out of phase with each other [Mit75]. These circuits have a number of applications, such as the computation of the discrete Fourier transform and certain digital communication systems. In some applications, it is more convenient to use numerical algorithms for the computation of the sine or cosine functions. These are discussed in Section 8.8.1.

Let $s_1[n]$ and $s_2[n]$ denote the two outputs of a digital sine-cosine generator given by

$$s_1[n] = \alpha \sin(n\theta), \tag{6.121a}$$
$$s_2[n] = \beta \cos(n\theta). \tag{6.121b}$$

From the above we arrive at

$$
\begin{aligned}
s_1[n+1] &= \alpha \sin\left((n+1)\,\theta\right) \\
&= \alpha \sin(n\theta) \cos\theta + \alpha \cos(n\theta) \sin\theta, \tag{6.122a}
\end{aligned}
$$
$$
\begin{aligned}
s_2[n+1] &= \beta \cos\left((n+1)\,\theta\right) \\
&= \beta \cos(n\theta) \cos\theta - \beta \sin(n\theta) \sin\theta. \tag{6.122b}
\end{aligned}
$$

Making use of Eqs. (6.121a) and (6.121b), we can rewrite Eqs. (6.122a) and (6.122b) in a matrix form as

$$
\begin{bmatrix} s_1[n+1] \\ s_2[n+1] \end{bmatrix} = \begin{bmatrix} \cos\theta & \frac{\alpha}{\beta}\sin\theta \\ -\frac{\beta}{\alpha}\sin\theta & \cos\theta \end{bmatrix} \begin{bmatrix} s_1[n] \\ s_2[n] \end{bmatrix}. \tag{6.123}
$$

To generate $s_1[n]$ from $s_1[n+1]$ we need a delay, and similarly, to generate $s_2[n]$ from $s_2[n+1]$ we need a delay. Thus, such a structure must have at least two delays.

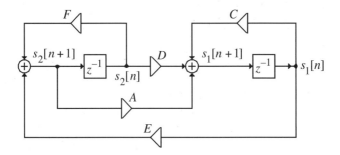

Figure 6.45: A general second-order digital filter structure with no input node.

In order to arrive at a realization of the sine-cosine generator, we need to compare Eq. (6.123) with the equivalent expression of a general second-order structure with no delay-free loops. Such a structure is characterized by the following equations

$$\begin{bmatrix} s_1[n+1] \\ s_2[n+1] \end{bmatrix} = \begin{bmatrix} 0 & A \\ 0 & 0 \end{bmatrix} \begin{bmatrix} s_1[n+1] \\ s_2[n+1] \end{bmatrix} + \begin{bmatrix} C & D \\ E & F \end{bmatrix} \begin{bmatrix} s_1[n] \\ s_2[n] \end{bmatrix} \tag{6.124}$$

and can be implemented using five multipliers, as indicated in Figure 6.45.

From Figure 6.45 we readily arrive at the time-domain description of this structure:

$$\begin{bmatrix} s_1[n+1] \\ s_2[n+1] \end{bmatrix} = \begin{bmatrix} AE+C & AF+D \\ E & F \end{bmatrix} \begin{bmatrix} s_1[n] \\ s_2[n] \end{bmatrix}. \tag{6.125}$$

Comparing Eqs. (6.123) and (6.125), we get

$$E = -\frac{\beta}{\alpha}\sin\theta, \qquad F = \cos\theta, \tag{6.126a}$$

$$AE+C = \cos\theta, \qquad AF+D = \frac{\alpha}{\beta}\sin\theta. \tag{6.126b}$$

From the above two equations, we obtain after some algebra

$$C\cos\theta + D\frac{\beta}{\alpha}\sin\theta = 1. \tag{6.126c}$$

Equations (6.126a) and (6.126c) ensure that the structure of Figure 6.45 will be a sine-cosine generator.

Expressing the multiplier constants A and D as a function of the multiplier constant C, we obtain after some simple manipulations, the five-multiplier characterization of the sine-cosine generator as

$$\begin{bmatrix} s_1[n+1] \\ s_2[n+1] \end{bmatrix} = \begin{bmatrix} 0 & \frac{\alpha(C-\cos\theta)}{\beta\sin\theta} \\ 0 & 0 \end{bmatrix} \begin{bmatrix} s_1[n+1] \\ s_2[n+1] \end{bmatrix}$$
$$+ \begin{bmatrix} C & \frac{\alpha(1-C\cdot\cos\theta)}{\beta\sin\theta} \\ -\frac{\beta}{\alpha}\sin\theta & \cos\theta \end{bmatrix} \begin{bmatrix} s_1[n] \\ s_2[n] \end{bmatrix}. \tag{6.127}$$

To reduce the total number of multipliers to 4, we can choose the following specific values for the multiplier constant C: $\cos\theta, 0, +1$, and -1. For example, if $C = \cos\theta$, then Eq. (6.127) reduces to

$$\begin{bmatrix} s_1[n+1] \\ s_2[n+1] \end{bmatrix} = \begin{bmatrix} \cos\theta & \frac{\alpha}{\beta}\sin\theta \\ -\frac{\beta}{\alpha}\sin\theta & \cos\theta \end{bmatrix} \begin{bmatrix} s_1[n] \\ s_2[n] \end{bmatrix}, \tag{6.128}$$

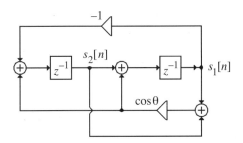

Figure 6.46: A single multiplier sine-cosine generator.

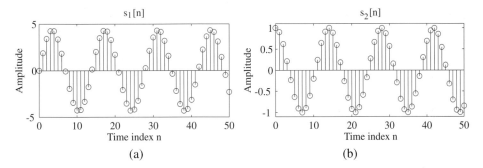

Figure 6.47: Sine-cosine sequences generated by the structure of Figure 6.46 for $\cos\theta = 0.9$.

which for $\beta = \pm\alpha$, leads to the four-multiplier sine-cosine generator described by Gold and Rader [Gol69b]. On the other hand, if we set $\alpha\sin\theta = \pm\beta$, then Eq. (6.128) leads to a three-multiplier realization (Problem 6.54). Another three-multiplier sine-cosine generator is obtained if we set $\alpha = \pm\beta\sin\theta$ in Eq. (6.128) (Problem 6.55).

A single-multiplier structure can be derived by setting $\beta\sin\theta/\alpha = 1 - \cos\theta$ or, equivalently, $\beta = \alpha\tan(\theta/2)$, in Eq. (6.128) resulting in

$$\begin{bmatrix} s_1[n+1] \\ s_2[n+1] \end{bmatrix} = \begin{bmatrix} \cos\theta & \cos\theta+1 \\ \cos\theta-1 & \cos\theta \end{bmatrix} \begin{bmatrix} s_1[n] \\ s_2[n] \end{bmatrix}, \tag{6.129}$$

leading to the realization of Figure 6.46. Another single-multiplier realization can be derived by setting $-\beta\sin\theta/\alpha = (1+\cos\theta)$ or, equivalently, $\alpha = \beta\tan(\theta/2)$, in Eq. (6.128) (Problem 6.56). Various other single-multiplier sine-cosine generators can be developed for $C = 0$ and $C = \pm1$ and are left as exercises (Problems 6.57 and 6.58).

It should be noted that to start the generation of the sinusoidal sequences at least one of the signal variables $s_1[n]$ and $s_2[n]$ should have initially a nonzero value. Moreover, the actual amplitudes and the relative phases of the cosinusoidal and the sinusoidal sequences generated by the sine-cosine generator depend on the initial values chosen for the signal variables $s_1[n]$ and $s_2[n]$.

Figure 6.47 shows the plots of the sequences $s_1[n]$ and $s_2[n]$ generated by simulating Eq. (6.129) in MATLAB for $\cos\theta = 0.9$. As can be seen from this figure, $s_1[n]$ and $s_2[n]$ are, respectively, cosinusoidal and sinusoidal sequences. Note that the maximum value of the amplitudes of the two sequences can be made equal by scaling one of the sequences appropriately.

The single-multiplier structures retain their characteristic roots on the unit circle under finite wordlength constraints. On the other hand, in other realizations of the sine-cosine generators, roots may go inside

Table 6.1: Computational complexity comparison of various realizations of an FIR filter of order N.

Structure	No. of multipliers	No. of two-input adders
Direct form	$N + 1$	N
Cascade form	$N + 1$	N
Polyphase	$N + 1$	N
Cascaded lattice	$2(N + 1)$	$2N + 1$
Linear phase	$\left\lfloor \dfrac{N + 2}{2} \right\rfloor$	N

or outside the unit circle due to the quantization of the multiplier coefficients, causing the oscillations to decay or to build up as n increases. In addition, due to product round-off errors, the sequences generated by the sine-cosine generator may not retain their sinusoidal behaviors, even in the case of a single-multiplier generator. It is therefore advisable to reset the variables $s_1[n]$ and $s_2[n]$ after some iterations at prescribed time instants, so that the accumulated errors do not become unacceptable.

6.12 Computational Complexity of Digital Filter Structures

The computational complexity of a digital filter structure is given by the total number of multipliers and the total number of two-input adders required for its implementation which roughly provides an indication of its cost of implementation. We summarize this measure here for various realizations discussed in this chapter. It should be emphasized, though, that the computational complexity measure is not the only criterion that should be used in selecting a particular structure for implementation of a given transfer function. The performances of all equivalent realizations under finite wordlength constraints should also be considered together with the costs of implementation in selecting a structure.

Tables 6.1 and 6.2 show the computational complexity measures of all realizations discussed in this chapter. As can be seen from these tables, in general, the direct form realizations for the case of both the FIR and the IIR transfer functions require the least number of multipliers and two-input adders. However, in the case of the IIR transfer function, the parallel allpass realization is the most efficient if such a realization exists. Likewise, it is possible to realize certain special types of FIR transfer functions with a lot fewer multipliers and adders than that indicated in Table 6.1.

6.13 Summary

This chapter considered the realization of a causal digital transfer function. Such realizations, called structures, are usually represented in block diagram forms that are formed by interconnecting adders, multipliers, and unit delays. A digital filter structure represented in block diagram form can be analyzed to develop its input-output relationship either in the time-domain or in the transform-domain. Often, for analysis purposes, it is convenient to represent the digital filter structure as a signal flow-graph.

Several basic FIR and IIR digital filter structures are then reviewed. These structures, in most cases, can be developed from the transfer function description of the digital filter essentially by inspection.

Table 6.2: Computational complexity comparison of various realizations of an IIR filter of order N.

Structure	No. of multipliers	No. of two-input adders
Direct form II and II$_t$	$2N + 1$	$2N$
Cascade form	$2N + 1$	$2N$
Parallel form	$2N + 1$	$2N$
Gray-Markel	$3N + 1$	$3N$
	$2N + 1$	$4N$
Parallel allpass	N	$1.5(N + 1)$

The digital allpass filter is a versatile building block and has a number of attractive digital signal processing applications. Since the numerator and the denominator polynomials of a digital transfer function exhibit mirror-image symmetry, an Mth-order digital allpass filter can be realized with only M distinct multipliers. Two different approaches to the minimum-multiplier realization of a digital allpass transfer function are described. One approach is based on a realization in the form of a cascade of first- and second-order allpass filters. The second approach results in a cascaded lattice realization. The final realizations in both cases remain allpass independent of the actual values of the multiplier coefficients and are thus less sensitive to multiplier coefficient quantization. An elegant application of the first- and second-order minimum multiplier allpass structures, considered here, is in the implementation of some simple transfer functions with parametrically tunable properties.

The Gray-Markel method to realize any arbitrary transfer function using the cascaded lattice form of allpass structure is outlined. The realization of a large class of arbitrary Nth-order transfer functions using a parallel connection of two allpass filters is described. The final structure is shown to require only N multipliers. In Section 9.9.2 we demonstrate the low passband sensitivity property of these structures to small changes in the multiplier coefficients.

The cascaded lattice realization of an FIR transfer function is considered. The realization of a digital sine-cosine generator is then described and various oscillator structures are systematically developed. The chapter concludes with a comparison of the computational complexities of FIR and IIR digital filter structures.

The digital filter realization methods outlined in this chapter assume the existence of the corresponding transfer function. The following chapter considers the development of such transfer functions meeting the given frequency response specifications. The analysis of the finite wordlength effects on the performance of the digital filter structures is treated in Chapter 9.

6.14 Problems

6.1 The digital filter structure of Figure P6.1 has a delay-free loop and is therefore unrealizable. Determine a realizable equivalent structure with identical input-output relations and without any delay-free loop. (*Hint*: Express the output variables $y[n]$ and $w[n]$ in terms of the input variables $x[n]$ and $u[n]$ only and develop the correresponding block diagram representation from these input-output relations.)

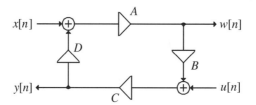

Figure P6.1

6.2 Determine by inspection whether or not the digital filter structures in Figure P6.2 have delay-free loops. Identify these loops if they exist. Develop equivalent structures without delay-free loops.

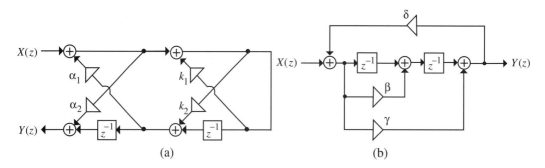

Figure P6.2

6.3 An unstable discrete-time system $G_1(z)$ can be made stable by placing it in the forward path of a single-loop feedback structure as indicated in Figure 6.2 with a multiplier in the feedback path with an appropriate value. Let $G_1(z) = \frac{1}{1-2z^{-1}}$ and $G_2(z) = K$. Determine the range of values of K for which the feedback structure is stable.

6.4 Analyze the digital filter structure of Figure P6.3 and determine its transfer function $H(z) = Y(z)/X(z)$.

(a) Is this a canonic structure?

(b) What should be the value of the multiplier coefficient K so that $H(z)$ has a unity gain at $\omega = 0$?

(c) What should be the value of the multiplier coefficient K so that $H(z)$ has a unity gain at $\omega = \pi$?

(d) Is there a difference between these two values of K? If not, why not?

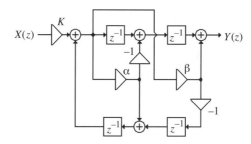

Figure P6.3

6.5 Analyze the digital filter structure of Figure P6.4, where all multiplier coefficients are real, and determine the transfer function $H(z) = Y(z)/X(z)$. What are the range of values of the multiplier coefficients for which the filter is BIBO stable?

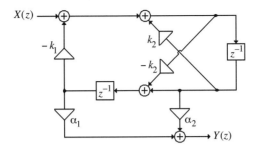

Figure P6.4

6.6 By using the block diagram analysis approach, determine the transfer function $H(z) = Y(z)/X(z)$ of the digital filter structure of Figure P6.5.

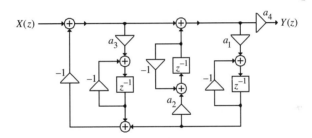

Figure P6.5

6.7 By using the block diagram analysis approach, determine the transfer function $H(z) = Y(z)/X(z)$ of the digital filter structure of Figure P6.6.

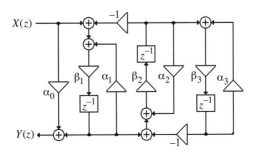

Figure P6.6

6.8 By using the block diagram analysis approach, determine the transfer function $H(z) = Y(z)/X(z)$ of the digital filter structure of Figure P6.7.

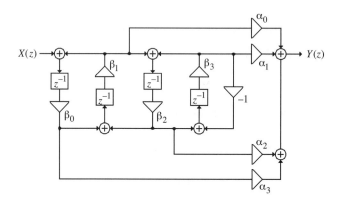

Figure P6.7

6.9 Determine the transfer function of the digital filter structure of Figure P6.8 [Kin72].

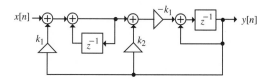

Figure P6.8

6.10 Realize the FIR transfer function

$$H(z) = (1 - 0.7z^{-1})^5 = 1 - 3.5z^{-1} + 4.9z^{-2} - 3.43z^{-3} + 1.2005z^{-4} - 0.16807z^{-5}$$

in the following forms:

(a) Two different direct forms,

(b) Cascade of five first-order sections,

(c) Cascade of one first-order section and two second-order sections, and

(d) Cascade of one second-order section and one third-order section.

Compare the computational complexity of each of the above realizations.

6.11 Consider a length-8 FIR transfer function given by

$$H(z) = h[0] + h[1]z^{-1} + h[2]z^{-2} + h[3]z^{-3} + h[4]z^{-4}$$
$$+ h[5]z^{-5} + h[6]z^{-6} + h[7]z^{-7}.$$

(a) Develop a three-branch polyphase realization of $H(z)$ in the form of Figure 6.8(b) and determine the expressions for the polyphase transfer functions $E_0(z)$, $E_1(z)$, and $E_2(z)$.

(b) From this realization develop a canonic three-branch polyphase realization.

6.12   (a) Develop a two-branch polyphase realization of $H(z)$ of Problem 6.11 in the form of Figure 6.8(c) and determine the expressions for the polyphase transfer functions $E_0(z)$ and $E_1(z)$.

(b) From this realization develop a canonic two-branch polyphase realization of $H(z)$.

6.13   (a) Develop a four-branch polyphase realization of $H(z)$ of Problem 6.11 in the form of Figure 6.8(a) and determine the expressions for the polyphase transfer functions $E_0(z)$, $E_1(z)$, $E_2(z)$, and $E_3(z)$.

(b) From this realization develop a canonic four-branch polyphase realization of $H(z)$.

6.14 Develop a minimum-multiplier realization of a length-9 Type 3 FIR transfer function.

6.15 Develop a minimum-multiplier realization of a length-10 Type 4 FIR transfer function.

6.16 The nested form of the FIR transfer function of Eq. (6.9) is given by

$$H(z) = b_0 + b_1 z^{-1} \left(1 + b_2 z^{-1} \left(1 + b_3 z^{-1} \left(1 + \cdots + b_{N-1} z^{-1} (1 + b_N z^{-1}) \cdots \right) \right) \right).$$

Express the coefficients b_i in terms of the coefficients $h[k]$. Develop the nested form realization of a 7th-order FIR transfer function $H(z)$ based on the above expansion [Mah82].

6.17 Let $H(z)$ be a Type 1 linear-phase FIR filter of order N with $G(z)$ denoting its delay-complementary filter. Develop a realization of both filters using only N delays and $(N + 2)/2$ multipliers.

6.18 Show that a Type 1 linear-phase FIR transfer function $H(z)$ of length $2M + 1$ can be expressed as

$$H(z) = z^{-M} \left[h[M] + \sum_{n=1}^{M} h[M - n] \left(z^n + z^{-n} \right) \right]. \tag{6.130}$$

By using the relation

$$z^{\ell} + z^{-\ell} = 2T_{\ell} \left(\frac{z + z^{-1}}{2} \right),$$

where $T_{\ell}(x)$ is the ℓth-order Chebyshev polynomial[10] in x, express $H(z)$ in the form

$$H(z) = z^{-M} \sum_{n=0}^{M} a[n] \left(\frac{z + z^{-1}}{2} \right)^n. \tag{6.131}$$

Determine the relation between $a[n]$ and $h[n]$. Develop a realization of $H(z)$ based on Eq. (6.131) in the form of Figure P6.9, where $F_1(z^{-1})$ and $F_2(z^{-1})$ are causal structures. Determine the form of $F_1(z^{-1})$ and $F_2(z^{-1})$. The structure of Figure P6.9 is called the *Taylor structure* for linear-phase FIR filters [Sch72].

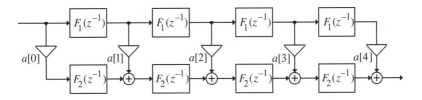

Figure P6.9   The Taylor structure shown for $M = 4$.

---

[10]For a definition of the Chebyshev polynomial and a recursive equation for generating such a polynomial, see Section 5.4.3.

6.19 Show that there are 36 distinct cascade realizations of the transfer function $H(z)$ of Eq. (6.28) obtained by different pole-zero pairings and different orderings of the individual sections.

6.20 Consider a real coefficient IIR transfer function $H(z)$ with its numerator and denominator expressed as a product of polynomials,

$$H(z) = \prod_{i=1}^{K} \frac{P_i(z)}{D_i(z)},$$

where $P_i(z)$ and $D_i(z)$ are either first-order or second-order polynomials with real coefficients. Determine the total number of distinct cascade realizations that can be obtained by different pole-zero pairings and different orderings of the individual sections.

6.21 Develop a canonic direct form realization of the transfer function

$$H(z) = \frac{3 + 4z^{-2} - 2z^{-5}}{1 + 3z^{-1} + 5z^{-2} + 4z^{-4}},$$

and then determine its transpose configuration.

6.22 Develop two different cascade canonic realizations of the following causal IIR transfer functions:

(a) $H_1(z) = \dfrac{(0.3 - 0.5z^{-1})(2 + 3.1z^{-1})}{(1 + 2.1z^{-1} - 3z^{-2})(1 + 0.67z^{-1})}$,

(b) $H_2(z) = \dfrac{4.1z\left(z - \frac{1}{3}\right)\left(z^2 + 0.4\right)}{(z + 0.3)(z - 0.5)\left(z^2 - z + 0.1\right)}$,

(c) $H_3(z) = \dfrac{(3 - 2.1z^{-1})(2.7 + 4.2z^{-1} - 5z^{-2})}{(1 + 0.52z^{-1})(1 + z^{-1} - 0.34z^{-2})}$.

6.23 Realize the transfer functions of Problem 6.22 in parallel forms I and II.

6.24 Develop a cascade realization of the transfer function of Eq. (6.32) using the factorization given in Eq. (6.33). Compare the computational complexity of this realization with the one shown in Figure 6.19(b).

6.25 Consider the cascade of three causal first-order LTI discrete-time systems shown in Figure P6.10 where

$$H_1(z) = \frac{2 + 0.1z^{-1}}{1 + 0.4z^{-1}}, \quad H_2(z) = \frac{3 + 0.2z^{-1}}{1 - 0.3z^{-1}}, \quad H_3(z) = \frac{1}{1 - 0.2z^{-1}}.$$

(a) Determine the transfer function of the overall system as a ratio of two polynomials in z^{-1},

(b) Determine the difference equation characterizing the overall system,

(c) Develop the realization of the overall system with each section realized in direct form II,

(d) Develop a parallel form I realization of the overall system,

(e) Determine the impulse response of the overall system in closed form.

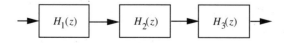

Figure P6.10

6.26 A causal LTI discrete-time system develops an output $y[n] = (0.4)^n \mu[n] - 0.3(0.4)^{n-1} \mu[n-1]$, for an input $x[n] = (0.2)^n \mu[n]$.

(a) Determine the transfer function of the system,

(b) Determine the difference equation characterizing the system,

(c) Develop a canonic direct form realization of the system with no more than three multipliers,

(d) Develop a parallel form I realization of the system,

(e) Determine the impulse response of the system in closed form,

(f) Determine the output $y[n]$ of the system for an input $x[n] = (0.3)^n \mu[n] - 0.4(0.3)^{n-1} \mu[n-1]$.

6.27 The structure shown in Figure P6.11 was developed in the course of a realization of the IIR digital transfer function

$$H(z) = \frac{3z^2 + 18.5z + 17.5}{(z + 0.5)(z + 2)}.$$

However, by a mistake in the labeling, two of the multiplier coefficients in this structure have incorrect values. Find these two multipliers and determine their correct values.

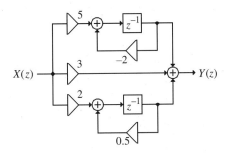

Figure P6.11

6.28 Figure P6.12 shows an incomplete realization of the causal IIR transfer function

$$H(z) = \frac{3z(5z - 2)}{(z + 0.5)(4z + 1)}.$$

Determine the values of the multiplier coefficients A and B.

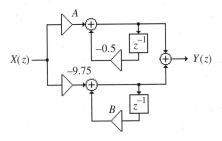

Figure P6.12

6.29 Develop a two-multiplier canonic realization of the second-order transfer function

$$H(z) = \frac{1 + \alpha + \beta}{1 + \alpha z^{-1} + \beta z^{-2}},$$

with multiplier coefficients α and β.

6.30 Develop a two-multiplier canonic realization of each of the following second-order transfer functions where the multiplier coefficients are α_1 and $-\alpha_2$, respectively [Hir73]:

(a) $H_1(z) = \dfrac{(1 - \alpha_1 + \alpha_2)(1 + z^{-1})^2}{1 - \alpha_1 z^{-1} + \alpha_2 z^{-2}},$

(b) $H_2(z) = \dfrac{(1 - \alpha_2)(1 - z^{-2})}{1 - \alpha_1 z^{-1} + \alpha_2 z^{-2}}.$

6.31 In this problem we develop an alternative cascaded lattice realization of an Nth-order IIR transfer function [Mit77b]

$$H_N(z) = \frac{p_0 + p_1 z^{-1} + \cdots + p_{N-1} z^{-N+1} + p_N z^{-N}}{1 + d_1 z^{-1} + \cdots + d_{N-1} z^{-N+1} + d_N z^{-N}}. \tag{6.132}$$

The first stage of the realization process is shown in Figure P6.13.

(a) Show that if the two-pair chain parameters are chosen as

$$A = 1, \qquad B = d_N z^{-1}, \qquad C = p_0, \qquad D = p_N z^{-1}, \tag{6.133}$$

then $H_{N-1}(z)$ is an $(N - 1)$th-order IIR transfer function of the form

$$H_{N-1}(z) = \frac{p_0' + p_1' z^{-1} + \cdots + p_{N-1}' z^{-N+1}}{1 + d_1' z^{-1} + \cdots + d_{N-1}' z^{-N+1}}, \tag{6.134}$$

with coefficients given by

$$p_k' = \frac{p_0 d_{k+1} - p_{k+1}}{p_0 d_N - p_N}, \quad k = 0, 1, \ldots, N - 1, \tag{6.135a}$$

$$d_k' = \frac{p_k d_N - p_N d_k}{p_0 d_N - p_N}, \quad k = 1, 2, \ldots, N - 1. \tag{6.135b}$$

(b) Develop a lattice realization of the two-pair.

(c) Continuing the above process, we can realize $H_N(z)$ as a cascade connection of N lattice sections constrained by a transfer function $H_0(z) = 1$. What are the total number of multipliers and two-input adders in the final realization of $H_N(z)$?

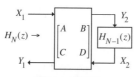

Figure P6.13

6.32 Realize the transfer function of Eq. (6.32) using the cascaded lattice realization method of Problem 6.31.

6.33 Realize the transfer functions of Problem 6.22 using the cascaded lattice realization method of Problem 6.31.

6.34 Show that the cascaded lattice realization method of Problem 6.31 results in the cascaded lattice structure described in Section 6.6.2 when $H_N(z)$ is an allpass transfer function.

6.35 Develop the structures of Types 1B, 1A$_t$, and 1B$_t$ first-order allpass transfer functions shown in Figure 6.23(b), (c), and (d), from Eqs. (6.38b), (6.38c), and (6.38d), respectively.

6.36   (a) Develop a cascade realization of the third-order allpass transfer function

$$A(z) = \left(\frac{a+z^{-1}}{1+az^{-1}}\right)\left(\frac{b+z^{-1}}{1+bz^{-1}}\right)\left(\frac{c+z^{-1}}{1+cz^{-1}}\right),$$

with each allpass section realized in Type 1A form. By sharing the delays between adjacent allpass sections, show that the total number of delays in the overall structure can be reduced from 6 to 4 [Mit74a].

(b) Repeat part (a) with each allpass section realized in Type 1A$_t$ form.

6.37 Analyze the digital filter structure of Figure P6.14 and show that it realizes a first-order allpass transfer [Sto94].

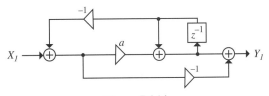

Figure P6.14

6.38 Develop formally the realizations of a second-order Type 2 allpass transfer function of Eq. (6.39) shown in Figure 6.24 using the multiplier extraction approach. Are there other Type 2 allpass structures?

6.39 Show that a cascade of two Type 2D second-order allpass structures can be realized with six delays by sharing the delays between adjacent sections. What is the minimum number of multipliers needed to implement a cascade of M Type 2D second-order allpass structures?

6.40 Develop formally the realizations of a second-order Type 3 allpass transfer function of Eq. (6.40) shown in Figure 6.2 using the multiplier extraction approach. Are there other Type 3 allpass structures?

6.41 Show that a cascade of two Type 3H second-order allpass structures can be realized with six delays by sharing the delays between adjacent sections. What is the minimum number of multipliers needed to implement a cascade of M Type 3H second-order allpass structures?

6.42 Develop a three-multiplier realization of the two-pair described by Eq. (6.52d).

6.43 Develop a lattice realization of the two-pair given by Eq. (6.52d). Determine the transfer function of an all-pole second-order cascaded lattice filter realized using this lattice structure and the transfer function of an all-pole second-order cascaded lattice filter realized using the lattice structure of Figure 6.29(a). Evaluate the approximate expressions for the gain of both second-order filters at resonance when the poles are close to the unit circle. Show that the gain of the first all-pole filter is approximately independent of the pole radius, whereas that of the second filter is not [Lar99].

6.44 Realize each of the following IIR transfer functions in the Gray-Markel form and check the BIBO stability of each transfer function:

(a) $H_1(z) = \dfrac{2 + z^{-1} + 2z^{-2}}{1 - \frac{3}{4}z^{-1} + \frac{1}{8}z^{-2}}$,

(b) $H_2(z) = \dfrac{1 + 2z^{-1} + 3z^{-2}}{1 - z^{-1} + \frac{1}{4}z^{-2}}$,

(c) $H_3(z) = \dfrac{2 + 5z^{-1} + 8z^{-2} + 3z^{-3}}{1 + 0.75z^{-1} + 0.5z^{-2} + 0.25z^{-3}}$,

(d) $H_4(z) = \dfrac{1 + 1.6z^{-1} + 0.6z^{-2}}{1 - z^{-1} - 0.25z^{-2} + 0.25z^{-3}}$,

(e) $H_5(z) = \dfrac{3z\left(z + \frac{1}{2}\right)\left(z^2 + \frac{1}{3}\right)}{\left(z - \frac{1}{2}\right)\left(z - \frac{1}{3}\right)\left(z^2 - z + \frac{1}{2}\right)}$.

6.45 Realize each of the IIR transfer functions of Problem 6.22 in the Gray-Markel form and check their BIBO stability.

6.46 Realize the IIR transfer function

$$H(z) = \frac{0.05634(1 + z^{-1})(1 - 1.0166z^{-1} + z^{-2})}{(1 - 0.683z^{-1})(1 - 1.4461z^{-1} + 0.7957z^{-2})}$$

in the following forms: (a) direct canonic form, (b) cascade form, (c) Gray-Markel form, and (d) cascaded lattice structure described in Problem 6.31. Compare their hardware requirements.

6.47 In this problem, the realization of a real-coefficient transfer function using complex arithmetic is illustrated [Reg87a]. Let $G(z)$ be an Nth-order-real coefficient transfer function with simple poles in complex-conjugate pairs and with numerator degree less than or equal to that of the denominator.

(a) Show that $G(z)$ can be expressed as a sum of two complex coefficient transfer functions of order $N/2$,

$$G(z) = H(z) + H^*(z^*), \qquad (6.136)$$

where the coefficients of $H^*(z^*)$ are complex conjugate of their corresponding coefficients of $H(z)$.

(b) Generalize the above decomposition to the case when $G(z)$ has one or more simple real poles.

(c) Consider a realization of $H(z)$ that has one real input $x[n]$ and a complex output $y[n]$. Show that the transfer function from the input to the real part of the output is simply $G(z)$.

[*Hint*: Use a partial-fraction expansion to obtain the decomposition of Eq. (6.136).]

6.48 Develop a realization of a first-order complex coefficient transfer function $H(z)$ given by

$$H(z) = \frac{A + jB}{1 + (\alpha + j\beta)z^{-1}},$$

where A, B, α, and β are real constants. Show the real and imaginary parts of all signal variables separately. Determine the transfer functions from the input to the real and imaginary parts of the output.

6.49 Develop a cascaded lattice realization of an Nth-order complex coefficient allpass transfer function $A_N(z)$.

6.50 Realize the following transfer functions in the form of a parallel connection of two allpass filters:

(a) $H_1(z) = \dfrac{2 + 2z^{-1}}{3 + z^{-1}}$,

(b) $H_2(z) = \dfrac{1 - z^{-1}}{4 + 2z^{-1}}$,

(c) $H_3(z) = \dfrac{1 - z^{-2}}{4 + 2z^{-1} + 2z^{-2}}$,

(d) $H_4(z) = \dfrac{3 + 9z^{-1} + 9z^{-2} + 3z^{-3}}{12 + 10z^{-1} + 2z^{-2}}$.

6.51 Consider a causal length-6 FIR filter described by the convolution sum

$$y[n] = \sum_{k=0}^{5} h[k]x[n - k], \qquad n \geq 0,$$

where $y[n]$ and $x[n]$ denote, respectively, the output and the input sequences.

(a) Let the output and input sequences be blocked into length-2 vectors

$$\mathbf{Y}_\ell = \begin{bmatrix} y[2\ell] \\ y[2\ell + 1] \end{bmatrix}, \quad \mathbf{X}_\ell = \begin{bmatrix} x[2\ell] \\ x[2\ell + 1] \end{bmatrix}.$$

Show that the above FIR filter can be equivalently described by a block convolution sum given by

$$\mathbf{Y}_\ell = \sum_{r=0}^{3} \mathbf{H}_r \mathbf{X}_{\ell-r},$$

where $\mathbf{H}_r$ is a 2×2 matrix composed of the impulse response coefficients. Determine the block convolution matrices $\mathbf{H}_r$. An implementation of the FIR filter based on the above block convolution sum is shown in Figure P6.15 where the block labeled "S/P" is a serial-to-parallel converter and the block marked "P/S" is a parallel-to-serial converter.

(b) Develop the block convolution sum description of the above FIR filter for an input and output block length of 3.

(c) Develop the block convolution sum description of the above FIR filter for an input and output block length of 4.

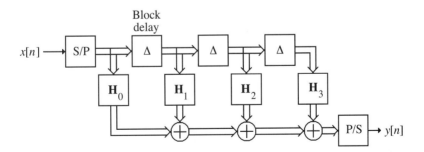

Figure P6.15

6.52 Consider a causal IIR filter described by a difference equation

$$\sum_{k=0}^{4} d_k y[n - k] = \sum_{k=0}^{4} p_k x[n - k], \qquad n \geq 0,$$

where $y[n]$ and $x[n]$ denote, respectively, the output and the input sequences.

(a) Let the output and input sequences be blocked into length-2 vectors

$$\mathbf{Y}_\ell = \begin{bmatrix} y[2\ell] \\ y[2\ell + 1] \end{bmatrix}, \mathbf{X}_\ell = \begin{bmatrix} x[2\ell] \\ x[2\ell + 1] \end{bmatrix}.$$

Show that the above IIR filter can be equivalently described by a block difference equation given by [Bur72]

$$\sum_{r=0}^{2} \mathbf{D}_r \mathbf{Y}_{\ell-r} = \sum_{r=0}^{2} \mathbf{P}_r \mathbf{X}_{\ell-r},$$

where $\mathbf{D}_r$ and $\mathbf{P}_r$ are 2×2 matrices composed of the difference equation coefficients $\{d_k\}$ and $\{p_k\}$, respectively. Determine the block difference equation matrices $\mathbf{D}_r$ and $\mathbf{P}_r$. An implementation of the IIR filter based on the above block difference equation is shown in Figure P6.16.

(b) Develop the block difference equation description of the above IIR filter for an input and output block length of 3.

(c) Develop the block difference equation description of the above IIR filter for an input and output block length of 4.

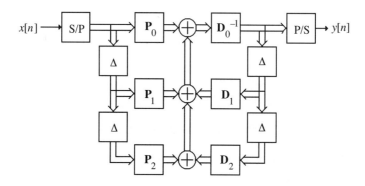

Figure P6.16

6.53 Develop a canonic realization of the block digital filter of Figure P6.16 employing only two block delays.

6.54 Develop the three-multiplier structure of a digital sine-cosine generator obtained by setting $\alpha \sin \theta = \pm \beta$ in Eq. (6.128).

6.55 Develop the three-multiplier structure of a digital sine-cosine generator obtained by setting $\alpha = \pm \beta \sin \theta$ in Eq. (6.128).

6.56 Develop a one-multiplier structure of a digital sine-cosine generator obtained by setting $-\beta \sin \theta / \alpha = 1 + \cos \theta$ in Eq. (6.128).

6.57 Develop a one-multiplier structure of a digital sine-cosine generator obtained by setting $C = 0$ in Eq. (6.127) and then choosing α and β properly.

6.58 Develop a one-multiplier structure of a digital sine-cosine generator obtained by setting $C = -1$ in Eq. (6.127) and then choosing α and β properly. Show the final structure.

6.59 Signals generated by multiple sources or multiple sensors, called a *multichannel signal*, are usually transmitted through independent channels in close proximity to each other. As a result, each component of the multichannel signal often gets corrupted by signals from adjacent channels during transmission resulting in *cross-talk*. Separation of the multichannel signal at the receiver is thus of practical interest. A model representing the cross-talk between a pair of channels for a two-channel signal is depicted in Figure P6.17(a) and the corresponding discrete-time system for channel separation is as shown in Figure P6.17(b) [Yel96]. Determine two possible sets of conditions for perfect channel separation.

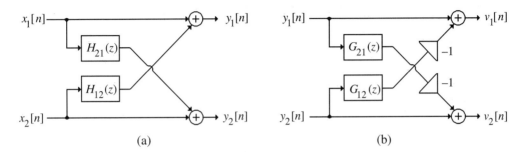

(a)   (b)

Figure P6.17

6.15 MATLAB Exercises

M 6.1 Using MATLAB develop a cascade realization of each of the following linear-phase FIR transfer functions:

(a) $H_1(z) = 4.5 + 4.05z^{-1} - 35.325z^{-2} - 71.6185z^{-3} + 63.99z^{-4} - 71.6185z^{-5} - 35.325z^{-6} + 4.05z^{-7} + 4.5z^{-8}$,

(b) $H_2(z) = 6 + 6z^{-1} - 24.66z^{-2} + 72.96z^{-3} - 88.62z^{-4} + 88.62z^{-5} - 72.96z^{-6} + 24.66z^{-7} - 6z^{-8} - 6z^{-9}$.

M 6.2 Consider the fourth-order IIR transfer function

$$G(z) = \frac{9 + 33z^{-1} + 57z^{-2} + 33z^{-3} + 12z^{-4}}{6 - 12z^{-1} + 11z^{-2} - 5z^{-3} + z^{-4}}.$$

(a) Using MATLAB express $G(z)$ in factored form,

(b) Develop two different cascade realizations of $G(z)$,

(c) Develop two different parallel form realizations of $G(z)$. Realize each second-order section in direct form II.

M 6.3 Consider the fourth-order IIR transfer function given below:

$$H(z) = \frac{12 - 2z^{-1} + 3z^{-2} + z^{-4}}{6 + 3z^{-1} + 2z^{-2} + 2z^{-3} + z^{-4}}.$$

(a) Using MATLAB express $H(z)$ in factored form,

(b) Develop two different cascade realizations of $H(z)$,

(c) Realize $H(z)$ in parallel forms I and II.

Realize each second-order section in direct form II.

M 6.4 Using Program 6_4 develop a Gray-Markel cascaded lattice realization of the IIR transfer function $G(z)$ of Problem M6.2.

M 6.5 Using Program 6_4 develop a Gray-Markel cascaded lattice realization of the IIR transfer function $H(z)$ of Problem M6.3.

M 6.6 Using Program 6_6 develop a cascaded lattice realization of each of the FIR transfer functions of Problem M6.1.

M 6.7 (a) Realize the following IIR lowpass transfer function $G(z)$ in the form of a parallel allpass structure:

$$G(z) = \frac{0.0219(1 + 5z^{-1} + 10z^{-2} + 10z^{-3} + 5z^{-4} + z^{-5})}{1 - 0.9853z^{-1} + 0.9738z^{-2} - 0.3864z^{-3} + 0.1112z^{-4} - 0.0113z^{-5}}.$$

   (b) From the allpass decomposition determine its power-complementary transfer function $H(z)$.

   (c) Plot the square of the magnitude responses of the original transfer function $G(z)$ and its power-complementary transfer function $H(z)$ derived in part (b) and verify that their sum is equal to one at all frequencies.

M 6.8 (a) Realize the following IIR highpass transfer function $G(z)$ in the form of a parallel allpass structure:

$$G(z) = \frac{0.0903(1 - 0.8971z^{-1} + 1.8122z^{-2} - 1.8122z^{-3} + 0.8971z^{-4} - z^{-5})}{1 + 1.7028z^{-1} + 2.4423z^{-2} + 1.7896z^{-3} + 0.9492z^{-4} + 0.2295z^{-5}}.$$

   (b) From the allpass decomposition determine its power-complementary transfer function $H(z)$.

   (c) Plot the square of the magnitude responses of the original transfer function $G(z)$ and its power-complementary transfer function $H(z)$ derived in part (b) and verify that their sum is equal to one at all frequencies.

M 6.9 (a) Realize the following IIR bandpass transfer function $G(z)$ in the form of a parallel allpass structure.

$$G(z) = \frac{0.1327(1 - z^{-2})^2}{1 + 0.2377z^{-1} + 0.8152z^{-2} + 0.1294z^{-3} + 0.3618z^{-4}}.$$

   (b) From the allpass decomposition determine its power-complementary transfer function $H(z)$.

   (c) Plot the square of the magnitude responses of the original transfer function $G(z)$ and its power-complementary transfer function $H(z)$ derived in part (b) and verify that their sum is equal to one at all frequencies.

M 6.10 Using MATLAB simulate the single-multiplier sine-cosine generator of Problem 6.56 with $\cos\theta = 0.9$ and plot the first 50 samples of its two output sequences. Scale the outputs so that they both have a maximum amplitude of ± 1. What is the effect of initial values of the variables $s_i[n]$?

M 6.11 Using MATLAB simulate the single-multiplier sine-cosine generator of Problem 6.57 with $\cos\theta = 0.9$ and plot the first 50 samples of its two output sequences. Scale the outputs so that they both have a maximum amplitude of ± 1. What is the effect of initial values of the variables $s_i[n]$?

M 6.12 Using MATLAB simulate the single-multiplier sine-cosine generator of Problem 6.58 with $\cos\theta = 0.9$ and plot the first 50 samples of its two output sequences. Scale the outputs so that they both have a maximum amplitude of ± 1. What is the effect of initial values of the variables $s_i[n]$ on the outputs?

7 Digital Filter Design

An important step in the development of a digital filter is the determination of a realizable transfer function $G(z)$ approximating the given frequency response specifications. If an IIR filter is desired, it is also necessary to ensure that $G(z)$ is stable. The process of deriving the transfer function $G(z)$ is called *digital filter design*. After $G(z)$ has been obtained, the next step is to realize it in the form of a suitable filter structure. In the previous chapter, we outlined a variety of basic structures for the realization of FIR and IIR transfer functions. In this chapter, we consider the digital filter design problem.

First we review some of the issues associated with the filter design problem. A widely used approach to IIR filter design based on the conversion of a prototype analog transfer function to a digital transfer function is discussed next. Typical design examples are included to illustrate this approach. We then consider the transformation of one type of filter transfer function into another type, which is achieved by replacing the complex variable z by a function of z. Four commonly used transformations are summarized. A popular approach to FIR filter design is next described. Finally, we consider the computer-aided design of both IIR and FIR digital filters. To this end, we restrict our discussion to the use of MATLAB in determining the transfer functions.

7.1 Preliminary Considerations

There are two major issues that need to be answered before one can develop the digital transfer function $G(z)$. The first and foremost issue is the development of a reasonable filter frequency response specification from the requirements of the overall system in which the digital filter is to be employed. The second issue is to determine whether an FIR or an IIR digital filter is to be designed. In this section, we examine these two issues. In addition, we review the basic approaches to the design of IIR and FIR digital filters and determination of the filter order to meet the prescribed specifications. We also discuss the scaling of the transfer function.

7.1.1 Digital Filter Specifications

As in the case of the analog filter, either the magnitude and/or the phase (delay) response is specified for the design of a digital filter for most applications. In some situations, the unit sample response or the step response may be specified. In most practical applications, the problem of interest is the development of a realizable approximation to a given magnitude response specification. As indicated in Section 4.6.3, the phase response of the designed filter can be corrected by cascading it with an allpass section. The design of allpass phase equalizers has received a fair amount of attention in the last few years.

We restrict our attention in this chapter to the magnitude approximation problem only. We pointed out in Section 4.4.1 that there are four basic types of filters, whose magnitude responses are shown in Figure 4.10. Since the impulse response corresponding to each of these is noncausal and of infinite length, these

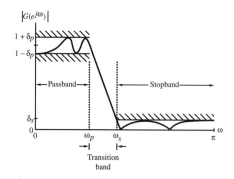

Figure 7.1: Typical magnitude specifications for a digital lowpass filter.

ideal filters are not realizable. One way of developing a realizable approximation to these filters would be to truncate the impulse response as indicated in Eq. (4.72) for a lowpass filter. The magnitude response of the FIR lowpass filter obtained by truncating the impulse response of the ideal lowpass filter does not have a sharp transition from passband to stopband but, rather, exhibits a gradual "roll-off."

Thus, as in the case of the analog filter design problem outlined in Section 5.4.1, the magnitude response specifications of a digital filter in the passband and in the stopband are given with some acceptable tolerances. In addition, a transition band is specified between the passband and the stopband to permit the magnitude to drop off smoothly. For example, the magnitude $|G(e^{j\omega})|$ of a lowpass filter may be given as shown in Figure 7.1. As indicated in the figure, in the *passband* defined by $0 \leq \omega \leq \omega_p$, we require that the magnitude approximates unity with an error of $\pm\delta_p$, i.e.,

$$1 - \delta_p \leq \left|G(e^{j\omega})\right| \leq 1 + \delta_p, \qquad \text{for } |\omega| \leq \omega_p. \tag{7.1}$$

In the *stopband*, defined by $\omega_s \leq \omega \leq \pi$, we require that the magnitude approximates zero with an error of δ_s, i.e.,

$$\left|G(e^{j\omega})\right| \leq \delta_s, \qquad \text{for } \omega_s \leq |\omega| \leq \pi. \tag{7.2}$$

The frequencies ω_p and ω_s are, respectively, called the *passband edge frequency* and the *stopband edge frequency*. The limits of the tolerances in the passband and stopband, δ_p and δ_s, are usually called the *peak ripple values*. Note that the frequency response $G(e^{j\omega})$ of a digital filter is a periodic function of ω, and the magnitude response of a real-coefficient digital filter is an even function of ω. As a result, the digital filter specifications are given only for the range $0 \leq |\omega| \leq \pi$.

Digital filter specifications are often given in terms of the loss function, $\mathcal{A}(\omega) = -20\log_{10}\left|G(e^{j\omega})\right|$, in dB. Here the *peak passband ripple* α_p and the *minimum stopband attenuation* α_s are given in dB, i.e., the loss specifications of a digital filter are given by

$$\alpha_p = -20\log_{10}(1 - \delta_p) \text{ dB}, \tag{7.3}$$

$$\alpha_s = -20\log_{10}(\delta_s) \text{ dB}. \tag{7.4}$$

EXAMPLE 7.1   The peak passband ripple α_p and the minimum stopband attenuation α_s of a digital filter are, respectively, 0.1 dB and 35 dB. Determine their corresponding peak ripple values δ_p and δ_s. From Eqs. (7.3) and (7.4) we obtain

$$\delta_p = 1 - 10^{-\alpha_p/20} = 1 - 10^{-0.005} = 0.01144690,$$

$$\delta_s = 10^{-\alpha_s/20} = 10^{-1.75} = 0.01778279.$$

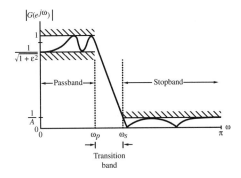

Figure 7.2: Alternate magnitude specifications for a digital lowpass filter.

As in the case of an analog lowpass filter, the specifications for a digital lowpass filter may alternatively be given in terms of its magnitude response, as in Figure 7.2. Here the maximum value of the magnitude in the passband is assumed to be unity, and the maximum passband deviation, denoted as $1/\sqrt{1+\varepsilon^2}$, is given by the minimum value of the magnitude in the passband. The maximum stopband magnitude is denoted by $1/A$.[1]

For the normalized specification, the maximum value of the gain function or the minimum value of the loss function is therefore 0 dB. The quantity $\alpha_{\max}$ given by

$$\alpha_{\max} = 20 \log_{10}\left(\sqrt{1+\varepsilon^2}\right) \text{ dB} \tag{7.5}$$

is called the *maximum passband attenuation*. For $\delta_p \ll 1$, as is typically the case, it can be shown that

$$\alpha_{\max} \cong -20 \log_{10}(1 - 2\delta_p) \cong 2\alpha_p. \tag{7.6}$$

The passband and stopband edge frequencies, in most applications, are specified in Hz, along with the sampling rate of the digital filter. Since all filter design techniques are developed in terms of normalized angular frequencies ω_p and ω_s, the specified critical frequencies need to be normalized before a specific filter design algorithm can be applied. Let F_T denote the sampling frequency in Hz, and F_p and F_s denote, respectively, the passband and stopband edge frequencies in Hz. Then the normalized angular edge frequencies in radians are given by

$$\omega_p = \frac{\Omega_p}{F_T} = \frac{2\pi F_p}{F_T} = 2\pi F_p T, \tag{7.7}$$

$$\omega_s = \frac{\Omega_s}{F_T} = \frac{2\pi F_s}{F_T} = 2\pi F_s T. \tag{7.8}$$

EXAMPLE 7.2   Let the specified passband and stopband edge frequencies of a digital highpass filter operating at a sampling rate of 25 kHz be 7 kHz and 3 kHz, respectively. Using Eqs. (7.7) and (7.8) we determine the corresponding normalized angular bandedge frequencies as

$$\omega_p = \frac{2\pi(7 \times 10^3)}{25 \times 10^3} = 0.56\pi, \qquad \omega_s = \frac{2\pi(3 \times 10^3)}{25 \times 10^3} = 0.24\pi.$$

---

[1]The minimum stopband attenuation is therefore $20 \log_{10}(A)$.

7.1.2   Selection of the Filter Type

The second issue of interest is the selection of the digital filter type, i.e., whether an IIR or an FIR digital filter is to be employed. The objective of digital filter design is to develop a causal transfer function $H(z)$ meeting the frequency response specifications. For IIR digital filter design, the IIR transfer function is a real rational function of z^{-1}:

$$H(z) = \frac{p_0 + p_1 z^{-1} + p_2 z^{-2} + \cdots + p_M z^{-M}}{d_0 + d_1 z^{-1} + d_2 z^{-2} + \cdots + d_N z^{-N}}. \tag{7.9}$$

Moreover, $H(z)$ must be a stable transfer function, and for reduced computational complexity, it must be of lowest order N. On the other hand, for FIR filter design, the FIR transfer function is a polynomial in z^{-1}:

$$H(z) = \sum_{n=0}^{N} h[n] z^{-n}. \tag{7.10}$$

For reduced computational complexity, the degree N of $H(z)$ must be as small as possible. In addition, if a linear phase is desired, then the FIR filter coefficients must satisfy the constraint:

$$h[n] = \pm h[N - n]. \tag{7.11}$$

There are several advantages in using an FIR filter, since it can be designed with exact linear phase and the filter structure is always stable with quantized filter coefficients. However, in most cases, the order N_{FIR} of an FIR filter is considerably higher than the order N_{IIR} of an equivalent IIR filter meeting the same magnitude specifications. In general, the implementation of the FIR filter requires approximately N_{FIR} multiplications per output sample, whereas the IIR filter requires $2N_{\text{IIR}} + 1$ multiplications per output sample. In the former case, if the FIR filter is designed with a linear phase, then the number of multiplications per output sample reduces to approximately $(N_{\text{FIR}} + 1)/2$. Likewise, most IIR filter designs result in transfer functions with zeros on the unit circle, and the cascade realization of an IIR filter of order N_{IIR} with all of the zeros on the unit circle requires $\lfloor (3N_{\text{IIR}} + 3)/2 \rfloor$ multiplications per output sample.[2] It has been shown that for most practical filter specifications, the ratio $N_{\text{FIR}}/N_{\text{IIR}}$ is typically of the order of tens or more and, as a result, the IIR filter usually is computationally more efficient [Rab75]. However, if the group delay of the IIR filter is equalized by cascading it with an allpass equalizer, then the savings in computation may no longer be that significant [Rab75]. In many applications, the linearity of the phase response of the digital filter is not an issue, making the IIR filter preferable because of the lower computational requirements.

7.1.3   Basic Approaches to Digital Filter Design

In the case of IIR filter design, the most common practice is to convert the digital filter specifications into analog lowpass prototype filter specifications, to determine the analog lowpass filter transfer function $H_a(s)$ meeting these specifications, and then to transform it into the desired digital filter transfer function $G(z)$. This approach has been widely used for many reasons:

   (a) Analog approximation techniques are highly advanced.

   (b) They usually yield closed-form solutions.

   (c) Extensive tables are available for analog filter design.

---

[2] $\lfloor x \rfloor$ denotes the integer part of x.

(d) Many applications require the digital simulation of analog filters.

In the sequel, we denote an analog transfer function as

$$H_a(s) = \frac{P_a(s)}{D_a(s)}, \tag{7.12}$$

where the subscript "a" specifically indicates the analog domain. The digital transfer function derived from $H_a(s)$ is denoted by

$$G(z) = \frac{P(z)}{D(z)}. \tag{7.13}$$

The basic idea behind the conversion of an analog prototype transfer function $H_a(s)$ into a digital IIR transfer function $G(z)$ is to apply a mapping from the s-domain to the z-domain so that the essential properties of the analog frequency response are preserved. This implies that the mapping function should be such that

(a) The imaginary ($j\Omega$) axis in the s-plane be mapped onto the unit circle of the z-plane.

(b) A stable analog transfer function be transformed into a stable digital transfer function.

To this end, the most widely used transformation is the bilinear transformation described in Section 7.2.

Unlike IIR digital filter design, the FIR filter design does not have any connection with the design of analog filters. The design of FIR filters is therefore based on a direct approximation of the specified magnitude response, with the often added requirement that the phase response be linear. As pointed out in Eq. (7.10), a causal FIR transfer function $H(z)$ of length $N + 1$ is a polynomial in z^{-1} of degree N. The corresponding frequency response is given by

$$H(e^{j\omega}) = \sum_{n=0}^{N} h[n]e^{-j\omega n}. \tag{7.14}$$

It has been shown in Section 3.2.1 that any finite duration sequence $x[n]$ of length $N + 1$ is completely characterized by $N + 1$ samples of its discrete-time Fourier transform $X(e^{j\omega})$. As a result, the design of an FIR filter of length $N + 1$ may be accomplished by finding either the impulse response sequence $\{h[n]\}$ or $N + 1$ samples of its frequency response $H(e^{j\omega})$. Also, to ensure a linear-phase design, the condition of Eq. (7.11) must be satisfied. Two direct approaches to the design of FIR filters are the windowed Fourier series approach and the frequency sampling approach. We describe the former approach in Section 7.6. The second approach is treated in Problem 7.59. In Section 7.7 we outline computer-based digital filter design methods.

7.1.4 Estimation of the Filter Order

For the design of an IIR digital lowpass filter $G(z)$ based on the conversion of an analog lowpass filter $H_a(s)$, the filter order of $H_a(s)$ is first estimated from its specifications using the appropriate formula given in Eq. (5.33), (5.41), or (5.51), depending on whether a Butterworth, Chebyshev, or equiripple filter approximation is desired. The order of $G(z)$ is then determined automatically from the transformation being used to convert $H_a(s)$ into $G(z)$. There are several M-files in MATLAB that can be used to directly estimate the minimum order of an IIR digital transfer function meeting the filter specifications for the class of approximations discussed in Section 5.4. These are discussed in Section 7.10.1.

For the design of FIR lowpass digital filters, several authors have advanced formulas for estimating the minimum value of the filter order N directly from the digital filter specifications: normalized passband edge angular frequency ω_p, normalized stopband edge angular frequency ω_s, peak passband ripple δ_p, and peak stopband ripple δ_s [Her73], [Kai74], and [Rab73]. A rather simple approximate formula developed by Kaiser [Kai74] is given by

$$N \cong \frac{-20 \log_{10} \left(\sqrt{\delta_p \delta_s} \right) - 13}{14.6(\omega_s - \omega_p)/2\pi}. \tag{7.15}$$

Note from the above formula that the filter order N of an FIR filter is inversely proportional to the transition band width $(\omega_s - \omega_p)$ and does not depend on the actual location of the transition band. This implies that a sharp cutoff FIR filter with a narrow transition band would be of very long length, whereas an FIR filter with a wide transition band will have a very short length. Another interesting property of Kaiser's formula is that the length depends on the product $\delta_p \delta_s$. This implies that if the values of δ_p and δ_s are interchanged, the length remains the same.

The formula of Eq. (7.15) provides a reasonably good estimate of the filter order in the case of FIR filters with a moderate passband width and may not work well in the case of very narrow passband or very wide passband FIR filters. In the case of a narrowband filter, the stopband ripple essentially controls the order and an alternative approximate formula provides a more reasonable estimate of the filter order [Par87]:

$$N \cong \frac{-20 \log_{10}(\delta_s) + 0.22}{(\omega_s - \omega_p)/2\pi}. \tag{7.16}$$

On the other hand, in the case of a very wide band filter, the passband ripple has more effect on the order, and a more reasonable estimate of the filter order can be obtained using the following formula [Par87]:

$$N \cong \frac{-20 \log_{10}(\delta_p) + 5.94}{27(\omega_s - \omega_p)/2\pi}. \tag{7.17}$$

We illustrate the application of Kaiser's formula of Eq. (7.15) in the following two examples.

EXAMPLE 7.3   Estimate the order of a linear-phase lowpass FIR filter with the following specifications: passband edge $F_p = 1.8$ kHz, stopband edge $F_s = 2$ kHz, peak passband ripple $\alpha_p = 0.1$ dB, minimum stopband attenuation $\alpha_s = 35$ dB, and sampling rate $F_T = 12$ kHz.

From Example 7.1, we get $\delta_p = 0.0114469$ and $\delta_s = 0.01778279$. Substituting these values along with the bandedge frequencies in Eq. (7.15) we get

$$N \cong \frac{-20 \log_{10}(\sqrt{0.00020355796}) - 13}{14.6(200 - 180)/12,000} = \frac{23.913119}{0.2433333} = 98.2730.$$

Since the order N must be an integer, we round up the above value yielding $N = 99$.

As the order is odd, a Type 2 FIR filter can be designed to meet the specifications. A Type 1 FIR filter can be designed by increasing the order by 1 to 100.

Kaiser's formula can also be used to estimate the length of highpass, bandpass, and bandstop FIR filters.

EXAMPLE 7.4   Estimate the order of a linear-phase bandpass FIR filter with the following specifications: passband edges $F_{p1} = 0.35$ kHz and $F_{p2} = 1$ kHz, stopband edges $F_{s1} = 0.3$ kHz and $F_{s2} = 1.1$ kHz, passband ripple $\delta_p = 0.002$, stopband ripple $\delta_s = 0.001$, and sampling rate $F_T = 10$ kHz.

Note that the widths of the transition bands are not equal. We therefore use the width of the smallest transition band to compute the order N. Substituting the appropriate values in Eq. (7.15) we get

$$N \cong \frac{-20 \log_{10}(\sqrt{0.00004}) - 13}{14.6(350 - 300)/10,000} = \frac{30.9794}{0.0730} = 424.3753.$$

The order of the FIR filter is therefore 425 which corresponds to a Type 2 filter. As before, by increasing the order by 1 we can design a Type 1 FIR filter to meet the specifications.

The formula due to Hermann et al. [Her73] gives a slightly more accurate value for the order and is given by [3]

$$N \cong \frac{D_\infty(\delta_p, \delta_s) - F(\delta_p, \delta_s) \left[(\omega_s - \omega_p)/2\pi\right]^2}{(\omega_s - \omega_p)/2\pi}, \tag{7.18}$$

where

$$D_\infty(\delta_p, \delta_s) = \left[a_1(\log_{10}\delta_p)^2 + a_2(\log_{10}\delta_p) + a_3\right]\log_{10}\delta_s$$
$$- \left[a_4(\log_{10}\delta_p)^2 + a_5(\log_{10}\delta_p) + a_6\right], \tag{7.19a}$$

and

$$F(\delta_p, \delta_s) = b_1 + b_2 \left[\log_{10}\delta_p - \log_{10}\delta_s\right], \tag{7.19b}$$

with

$$a_1 = 0.005309, \quad a_2 = 0.07114, \quad a_3 = -0.4761, \tag{7.19c}$$

$$a_4 = 0.00266, \quad a_5 = 0.5941, \quad a_6 = 0.4278, \tag{7.19d}$$

$$b_1 = 11.01217, \quad b_2 = 0.51244. \tag{7.19e}$$

The formula given in Eq. (7.18) is valid for $\delta_p \geq \delta_s$. If $\delta_p < \delta_s$, then the filter order formula to be used is obtained by interchanging δ_p and δ_s in Eq. (7.19).

For small values of δ_p and δ_s, both Eqs. (7.15) and (7.18) provide reasonably close and accurate results. On the other hand, when the values of δ_p and δ_s are large, Eq. (7.18) yields a more accurate value for the order.

EXAMPLE 7.5   Estimate the order of a linear-phase FIR transfer function with the specifications given in Example 7.3 using Eq. (7.18).

Substituting the values of the parameters a_i given in Eqs. (7.19c) and (7.19d) and the values of the specified ripples $\delta_p = 0.0114469$ and $\delta_s = 0.01778279$ in Eq. (7.19a), we get $D_\infty(\delta_p, \delta_s) = 1.7553533358$. Next, substituting the values of b_1 and b_2 from Eq. (7.19e) and the ripples in Eq. (7.19b), we get $F(\delta_p, \delta_s) = 10.9141341$. Now, $(\omega_s - \omega_p)/2\pi = (F_s - F_p)/F_T = 0.017$. Finally, using the computed values of $D_\infty(\delta_p, \delta_s)$, $F(\delta_p, \delta_s)$, and $(\omega_s - \omega_p)/2\pi$ in Eq. (7.18) we arrive at

$$N \cong 105.139.$$

We choose the next higher integer 106 as the estimated filter order.

Note that the filter order computed using Eq. (7.18) is slightly higher than that obtained using Eq. (7.15) in Example 7.3. Both formulas provide only an estimate of the required filter order. The frequency response of the FIR filter designed using this estimated order may or may not meet the given specifications. If the specifications are not met, it is recommended that the filter order be gradually increased until the specifications are met. Estimation of the FIR filter order using MATLAB is discussed in Section 7.10.2.

---

[3]It has been obtained by carrying out a least-squares approximation to N as a function of the bandedges and ripples.

7.1.5 Scaling the Digital Transfer Function

After a digital filter has been designed following any one of the techniques outlined in this chapter, the corresponding transfer function $G(z)$ has to be scaled in magnitude before it can be implemented. In magnitude scaling, the transfer function is multiplied by a scaling constant K so that the maximum magnitude of the scaled transfer function $G_t(z) = K\,G(z)$ in the passband is unity, i.e., the scaled transfer function has a maximum gain of 0 dB. For a stable transfer function $G(z)$ with real coefficients, the scaled transfer function $K\,G(z)$ is then a bounded real (BR) function.[4]

For a frequency selective transfer function $G(z)$, if G_{max} is the maximum value of $|G(e^{j\omega})|$ in the frequency range $0 \leq \omega \leq \pi$, then $K = 1/G_{max}$, which results in a maximum gain of 0 dB in the passband of the scaled transfer function. For example, in the case of a lowpass transfer function with a maximum magnitude at dc, it is usual practice to use $K = 1/G(1)$, implying a dc gain of 0 dB for the scaled transfer function. Likewise, in the case of a highpass transfer function with a maximum magnitude at $\omega = \pi$, K is selected equal to $1/G(-1)$ yielding a gain of 0 dB at $\omega = \pi$ for the scaled transfer function. For a bandpass transfer function, it is common to use K equal to $1/|G(e^{j\omega_c})|$, where ω_c is the center frequency.

EXAMPLE 7.6   Consider the fourth-order highpass transfer function

$$G(z) = \frac{p_0 + p_1 z^{-1} + p_2 z^{-2} + p_3 z^{-3} + p_4 z^{-4}}{1 + d_1 z^{-1} + d_2 z^{-2} + d_3 z^{-3} + d_4 z^{-4}}.$$

Its magnitude scaled version is then given by

$$G_t(z) = K \frac{p_0 + p_1 z^{-1} + p_2 z^{-2} + p_3 z^{-3} + p_4 z^{-4}}{1 + d_1 z^{-1} + d_2 z^{-2} + d_3 z^{-3} + d_4 z^{-4}},$$

where

$$K = \frac{1 - d_1 + d_2 - d_3 + d_4}{p_0 - p_1 + p_2 - p_3 + p_4}.$$

7.2 Bilinear Transformation Method of IIR Filter Design

A number of transformations has been proposed to convert an analog transfer function $H_a(s)$ into a digital transfer function $G(z)$ so that essential properties of the analog transfer function in the s-domain are preserved for the digital transfer function in the z-domain. Of these, the bilinear transformation is more commonly used to design IIR digital filters based on the conversion of analog prototype filters.

7.2.1 The Bilinear Transformation

The bilinear transformation from the s-plane to the z-plane is given by [Kai66]

$$s = \frac{2}{T}\left(\frac{1 - z^{-1}}{1 + z^{-1}}\right). \tag{7.20}$$

The above transformation is a one-to-one mapping, i.e., it maps a single point in the s-plane to a unique point in the z-plane, and vice versa. The relation between the digital transfer function $G(z)$ and the parent analog transfer function $H_a(s)$ is then given by

$$G(z) = H_a(s)\Big|_{s = \frac{2}{T}\left(\frac{1-z^{-1}}{1+z^{-1}}\right)}. \tag{7.21}$$

---

[4] See Section 4.4.5 for a definition of a bounded real (BR) function.

The bilinear transformation is derived by applying the trapezoidal numerical integration approach to the differential equation representation of $H_a(s)$ that leads to the difference equation representation of $G(z)$ (see Example 2.36). The parameter T represents the step size in the numerical integration. As we shall see later in this section, the digital filter design procedure consists of two steps: first, the inverse bilinear transformation is applied to the digital filter specifications to arrive at the specifications of the analog filter prototype; then the bilinear transformation of Eq. (7.20) is employed to obtain the desired digital transfer function $G(z)$ from the analog transfer function $H_a(s)$ designed to meet the analog filter specifications. As a result, the parameter T has no effect on the expression for $G(z)$, and we shall choose, for convenience, $T = 2$ to simplify the design procedure.

The corresponding inverse transformation for $T = 2$ is given by

$$z = \frac{1+s}{1-s}. \tag{7.22}$$

Let us now examine the above transformation. Note that for $s = j\Omega_o$,

$$z = \frac{1 + j\Omega_o}{1 - j\Omega_o}, \tag{7.23}$$

which has a unity magnitude. This implies that a point on the imaginary axis in the s-plane is mapped onto a point on the unit circle in the z-plane. In the general case, for $s = \sigma_o + j\Omega_o$,

$$z = \frac{1 + (\sigma_o + j\Omega_o)}{1 - (\sigma_o + j\Omega_o)} = \frac{(1 + \sigma_o) + j\Omega_o}{(1 - \sigma_o) - j\Omega_o}. \tag{7.24}$$

Therefore,

$$|z|^2 = \frac{(1 + \sigma_o)^2 + (\Omega_o)^2}{(1 - \sigma_o)^2 + (\Omega_o)^2}. \tag{7.25}$$

Thus, a point on the $j\Omega$-axis in the s-plane ($\sigma_o = 0$) is mapped onto a point on the unit circle in the z-plane as $|z| = 1$. A point in the left-half s-plane with $\sigma_o < 0$ is mapped onto a point inside the unit circle in the z-plane as $|z| < 1$. Likewise, a point in the right-half s-plane with $\sigma_o > 0$ is mapped onto a point outside the unit circle in the z-plane, as $|z| > 1$. Any point in the s-plane is mapped onto a unique point in the z-plane and vice versa. The mapping of the s-plane into the z-plane via the bilinear transformation is illustrated in Figure 7.3 and is seen to have all the desired properties. Also, there is no aliasing due to the one-to-one mapping.

The exact relation between the imaginary axis in the s-plane ($s = j\Omega$) and the unit circle in the z-plane ($z = e^{j\omega}$) is of interest. From Eq. (7.20) with $T = 2$ it follows that

$$j\Omega = \frac{1 - e^{-j\omega}}{1 + e^{-j\omega}} = j \tan\left(\frac{\omega}{2}\right),$$

or

$$\Omega = \tan\left(\frac{\omega}{2}\right), \tag{7.26}$$

which has been plotted in Figure 7.4. Note from this plot that the positive (negative) imaginary axis in the s-plane is mapped into the upper (lower) half of the unit circle in the z-plane. However, it is clear that the mapping is highly nonlinear since the complete negative imaginary axis in the s-plane from $\Omega = -\infty$ to $\Omega = 0$ is mapped into the lower half of the unit circle from $\omega = -\pi$ (i.e., $z = -1$) to $\omega = 0$ (i.e., $z = +1$), and the complete positive imaginary axis in the s-plane from $\Omega = 0$ to $\Omega = +\infty$ is mapped into the upper half of the unit circle from $\omega = 0$ (i.e., $z = +1$) to $\omega = +\pi$ (i.e., $z = -1$). This introduces a distortion in the frequency axis called *frequency warping*. The effect of warping is more evident in

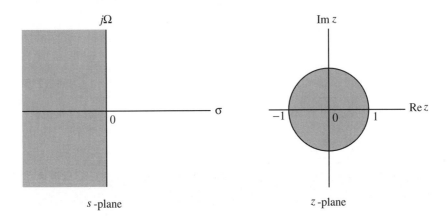

Figure 7.3: The bilinear transformation mapping.

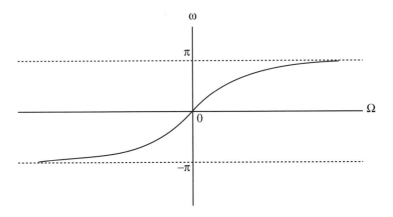

Figure 7.4: Mapping of the angular analog frequencies Ω to the angular digital frequencies ω via the bilinear transformation.

Figure 7.5, which shows the transformation of a typical analog filter magnitude response to a digital filter magnitude response derived via the bilinear transformation. Thus, to develop a digital filter meeting a specified magnitude response, we must first prewarp the critical bandedge frequencies (ω_p and ω_s) to find their analog equivalents (Ω_p and Ω_s) using the relation of Eq. (7.26), design the analog prototype $H_a(s)$ using the prewarped critical frequencies, and then transform $H_a(s)$ using the bilinear transformation to obtain the desired digital filter transfer function $G(z)$.

It should be noted that the bilinear transformation preserves the magnitude response of an analog filter only if the specification requires piecewise constant magnitude. However, the phase response of the analog filter is not preserved after transformation. Hence, the transformation can be used only to design digital filters with prescribed magnitude response with piecewise constant values.

EXAMPLE 7.7   It follows from Eqs. (5.34) and (5.35) that a first-order Butterworth lowpass transfer function with a 3-dB cutoff frequency at Ω_c is given by

$$H_a(s) = \frac{\Omega_c}{s + \Omega_c}. \tag{7.27}$$

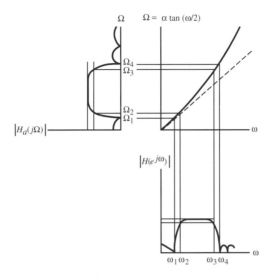

Figure 7.5: Illustration of frequency warping effect.

Applying a bilinear transformation to Eq. (7.27), we arrive at the expression for a first-order digital lowpass transfer function $G(z)$ with a maximally flat magnitude response given by

$$G(z) = H_a(s)|_{s=(1-z^{-1})/(1+z^{-1})} = \frac{\Omega_c(1+z^{-1})}{(1-z^{-1}) + \Omega_c(1+z^{-1})}. \tag{7.28}$$

Rearranging terms, Eq. (7.28) can be written as

$$G(z) = \frac{1-\alpha}{2} \cdot \frac{1+z^{-1}}{1-\alpha z^{-1}}, \tag{7.29}$$

where

$$\alpha = \frac{1-\Omega_c}{1+\Omega_c}. \tag{7.30}$$

The 3-dB cutoff frequency ω_c of the digital transfer function of Eq. (7.29) is related to the 3-dB cutoff frequency Ω_c of the analog transfer function through Eq. (7.26). Using Eq. (7.26) in Eq. (7.30), we can express α as a function of ω_c, which is thus given by

$$\alpha = \frac{1-\tan(\omega_c/2)}{1+\tan(\omega_c/2)}. \tag{7.31}$$

It should be noted that the first-order lowpass transfer function of Eq. (7.29) is identical to that of Eq. (4.109), where it was introduced without any derivation. It can be easily shown that the relation between the 3-dB bandwidth and the transfer function parameter as given by Eq. (7.31) is the same as that given in Eq. (4.244) in Problem 4.52.

7.2.2 Design of Digital IIR Notch Filters

We consider next the design of a second-order IIR notch filter as an example of the application of the bilinear transformation method [Hir74]. Now, a second-order analog notch filter has a transfer function given by

$$H_a(s) = \frac{s^2 + \Omega_o^2}{s^2 + Bs + \Omega_o^2}. \tag{7.32}$$

Its magnitude response is then

$$|H_a(j\Omega)| = \frac{\Omega_o^2 - \Omega^2}{\sqrt{(\Omega_o^2 - \Omega^2)^2 + B^2\Omega^2}}, \tag{7.33}$$

which approaches unity values, i.e., a gain of 0 dB, at $\Omega = 0$ and ∞. The magnitude has a zero value at the notch frequency $\Omega = \Omega_o$. If Ω_1 and Ω_2, $\Omega_2 > \Omega_1$, denote the frequencies at which the gain is down by -3 dB, it can be shown that the 3-dB notch bandwidth defined by $(\Omega_2 - \Omega_1)$ is equal to B.

Applying a bilinear transformation to $H_a(s)$ of Eq. (7.32), we arrive at

$$\begin{aligned} G(z) &= H_a(s)|_{s=(1-z^{-1})/(1+z^{-1})} \\ &= \frac{(1 + \Omega_o^2) - 2(1 - \Omega_o^2)z^{-1} + (1 + \Omega_o^2)z^{-2}}{(1 + \Omega_o^2 + B) - 2(1 - \Omega_o^2)z^{-1} + (1 + \Omega_o^2 - B)z^{-2}}, \end{aligned} \tag{7.34}$$

which can be rewritten as

$$G(z) = \frac{1}{2} \frac{(1+\alpha) - 2\beta(1+\alpha)z^{-1} + (1+\alpha)z^{-2}}{1 - \beta(1+\alpha)z^{-1} + \alpha z^{-2}}, \tag{7.35}$$

where

$$\alpha = \frac{1 + \Omega_o^2 - B}{1 + \Omega_o^2 + B}, \tag{7.36a}$$

$$\beta = \frac{1 - \Omega_o^2}{1 + \Omega_o^2}. \tag{7.36b}$$

It is a simple exercise to show that the notch frequency ω_o and the 3-dB notch bandwidth B_w of the digital notch filter of Eq. (7.35) are related to the constants α and β through

$$\alpha = \frac{1 - \tan(B_w/2)}{1 + \tan(B_w/2)}, \tag{7.37a}$$

$$\beta = \cos\omega_o. \tag{7.37b}$$

Equations (7.37a) and (7.37b) are the desired design formulas to determine the constants α and β for a given notch frequency ω_o and a 3-dB notch bandwidth B_w. It should be noted that Eq. (7.35) is precisely the transfer function of the second-order notch filter given in Eq. (4.118) and introduced earlier in Section 4.5.2 without any derivation.

EXAMPLE 7.8   Design a second-order digital notch filter having a notch frequency at 60 Hz and a 3-dB notch bandwidth of 6 Hz. The sampling frequency employed is 400 Hz. The normalized angular notch frequency ω_o and the normalized angular 3-dB bandwidth B_w are therefore given by

$$\omega_o = 2\pi\left(\frac{60}{400}\right) = 0.3\pi, \quad B_w = 2\pi\left(\frac{6}{400}\right) = 0.03\pi.$$

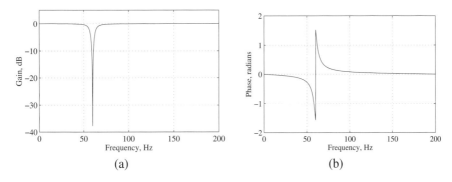

Figure 7.6: Gain and phase responses of the notch transfer function of Example 7.8.

Substituting these values in Eqs. (7.37a) and (7.37b), we arrive at

$$\alpha = 0.90993, \qquad \beta = 0.587785.$$

Then, substituting the above values in Eq. (7.35), we arrive at the desired transfer function

$$G(z) = \frac{0.954965 - 1.1226287z^{-1} + 0.954965z^{-2}}{1 - 1.1226287z^{-1} + 0.90993z^{-2}},$$

whose gain and phase responses are plotted in Figure 7.6. Observe the phase shift of π radians at 60 Hz, the notch frequency.

7.3 Design of Lowpass IIR Digital Filters

We illustrate now the development of a lowpass IIR digital transfer function meeting given specifications using the bilinear transformation method. To this end, we first obtain the specifications for a prototype lowpass analog filter from the specifications of the lowpass digital filter using the inverse transformation. We then determine the analog transfer function $H_a(s)$ meeting the specifications of the prototype analog filter. Finally, the analog transfer function $H_a(s)$ is transformed into a digital transfer function $G(z)$ using the bilinear transformation.

Specifically, consider the design of a lowpass IIR digital filter $G(z)$ with a maximally flat magnitude characteristic. The passband edge frequency ω_p is 0.25π, with a passband ripple not exceeding 0.5 dB. The minimum stopband attenuation at the stopband edge frequency ω_s of 0.55π is 15 dB. Thus, if $|G(e^{j0})| = 1$, then we require that

$$20\log_{10}\left|G(e^{j0.25\pi})\right| \geq -0.5 \text{ dB}, \tag{7.38a}$$

$$20\log_{10}\left|G(e^{j0.55\pi})\right| \leq -15 \text{ dB}. \tag{7.38b}$$

We first prewarp the digital bandedge frequencies to obtain the corresponding analog bandedge frequencies. From Eq. (7.26) the pertinent analog bandedge frequencies Ω_p and Ω_s corresponding to the two digital frequencies ω_p and ω_s are given by

$$\Omega_p = \tan\left(\frac{\omega_p}{2}\right) = \tan\left(\frac{0.25\pi}{2}\right) = 0.4142136,$$

$$\Omega_s = \tan\left(\frac{\omega_s}{2}\right) = \tan\left(\frac{0.55\pi}{2}\right) = 1.1708496.$$

From Eq. (5.29) the inverse transition ratio is

$$\frac{1}{k} = \frac{\Omega_s}{\Omega_p} = \frac{1.1708496}{0.4142135} = 2.8266809.$$

From the specified passband ripple of 0.5 dB, we obtain $\varepsilon^2 = 0.1220185$, and from the minimum stopband attenuation of 15 dB, we obtain $A^2 = 31.622777$. Therefore, from Eq. (5.30) the inverse discrimination ratio is

$$\frac{1}{k_1} = \frac{\sqrt{A^2 - 1}}{\varepsilon} = 15.841979.$$

Substituting these values in Eq. (5.33) we obtain the filter order N as

$$N = \frac{\log_{10}(1/k_1)}{\log_{10}(1/k)} = \frac{\log_{10}(15.841979)}{\log_{10}(2.8266814)} = 2.6586997.$$

The nearest higher integer 3 is thus taken as the filter order.

The filter order is used next to determine the 3-dB cutoff frequency Ω_c. To this end, either Eq. (5.32a) or Eq. (5.32b) can be used. However, it is preferable to use the latter since this ensures the smallest ripple in the passband or, in other words, the smallest amplitude distortion to the signal being filtered in its band of interest. Substituting the values of ε^2, Ω_p, and N in Eq. (5.32a), we arrive at

$$\Omega_c = 1.419915(\Omega_p) = 1.419915 \times 0.4142135 = 0.588148.$$

Using the function `buttap` of MATLAB,[5] we obtain the third-order normalized lowpass Butterworth transfer function as

$$H_{an}(s) = \frac{1}{(s+1)(s^2 + s + 1)},$$

which has a 3-dB frequency at $\Omega = 1$ and therefore has to be denormalized to bring the 3-dB frequency to $\Omega_c = 0.588148$. The denormalized transfer function is given by

$$H_a(s) = H_{an}\left(\frac{s}{0.588148}\right) = \frac{0.203451}{(s + 0.588148)(s^2 + 0.588148s + 0.345918)}.$$

Applying the bilinear transformation to the above, we finally arrive at the desired expression for the digital lowpass transfer function:

$$\begin{aligned}
G(z) &= H_a(s)|_{s=(1-z^{-1})/(1+z^{-1})} \\
&= \frac{0.0662272(1 + z^{-1})^3}{(1 - 0.2593284z^{-1})(1 - 0.6762858z^{-1} + 0.3917468z^{-2})}.
\end{aligned} \qquad (7.39)$$

The corresponding magnitude and gain responses are plotted in Figure 7.7.

The above digital lowpass filter can be designed directly in the z-domain using the M-files `buttord` and `butter`.

---

[5] See Section 5.4.6.

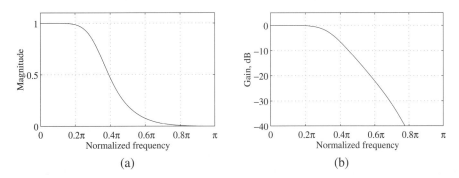

Figure 7.7: Magnitude and gain responses of the lowpass filter design based on the bilinear transformation method.

7.4 Design of Highpass, Bandpass, and Bandstop IIR Digital Filters

In the previous section we outlined the design of lowpass IIR digital filters. We now consider the design of the other three types of IIR digital filters. To this end, two approaches can be followed.

The first approach consists of the following steps:

Step 1: Prewarp the specified digital frequency specifications of the desired digital filter $G_D(z)$ using Eq. (7.26) to arrive at the frequency specifications of an analog filter $H_D(s)$ of the same type.

Step 2: Convert the frequency specifications of $H_D(s)$ into that of a prototype analog lowpass filter $H_{LP}(s)$ using an appropriate frequency transformation discussed in Section 5.5.

Step 3: Design the analog lowpass filter $H_{LP}(s)$ using the methods described in Section 5.4.

Step 4: Convert the transfer function $H_{LP}(s)$ into $H_D(s)$ using the inverse of the frequency transformation used in Step 2.

Step 5: Transform the transfer function $H_D(s)$ using the bilinear transformation of Eq. (7.20) to arrive at the desired digital IIR transfer function $G_D(z)$.

The second approach consists of the following steps:

Step 1: Prewarp the specified digital frequency specifications of the desired digital filter $G_D(z)$ using Eq. (7.26) to arrive at the frequency specifications of an analog filter $H_D(s)$ of the same type.

Step 2: Convert the frequency specifications of $H_D(s)$ into that of a prototype analog lowpass filter $H_{LP}(s)$ using an appropriate frequency transformation discussed in Section 5.5.

Step 3: Design the analog lowpass filter $H_{LP}(s)$ using the methods described in Section 5.4.

Step 4: Convert the transfer function $H_{LP}(s)$ into the transfer function $G_{LP}(z)$ of an IIR digital filter using the bilinear transformation of Eq. (7.20).

Step 5: Transform $G_{LP}(z)$ into the desired digital transfer function $G_D(z)$ using the appropriate spectral transformation discussed in Section 7.5.

We illustrate the first approach in this section with the aid of examples.

Design of Highpass IIR Digital Filter

We consider the design of a Type 1 Chebyshev IIR digital highpass filter in the following example.

EXAMPLE 7.9   The specifications of the highpass filter are: passband edge $F_p = 700$ Hz, stopband edge $F_s = 500$ Hz, passband ripple $\alpha_p = 1$ dB, minimum stopband attenuation $\alpha_s = 32$ dB, and sampling frequency $F_T = 2$ kHz.

Using Eqs. (7.7) and (7.8) we first determine the normalized angular bandedge frequencies as

$$\omega_p = \frac{2\pi F_p}{F_T} = \frac{2\pi (700)}{2000} = 0.7\pi,$$

$$\omega_s = \frac{2\pi F_s}{F_T} = \frac{2\pi (500)}{2000} = 0.5\pi.$$

We then prewarp the above digital edge frequencies using Eq. (7.26) to arrive at the following angular edge frequencies of the analog highpass filter:

$$\hat{\Omega}_p = \tan\left(\frac{\omega_p}{2}\right) = \tan\left(\frac{0.7\pi}{2}\right) = 1.9626105,$$

$$\hat{\Omega}_s = \tan\left(\frac{\omega_s}{2}\right) = \tan\left(\frac{0.5\pi}{2}\right) = 1.0.$$

For the prototype analog lowpass filter we choose the normalized passband edge to be $\Omega_p = 1$. From Eq. (5.60) the normalized stopband edge of the lowpass filter is thus $\Omega_s = 1.9626105$. The specifications of the analog lowpass filter are therefore as follows: passband edge at 1 rad/sec, stopband edge at 1.9626105 rad/sec, passband ripple of 1 dB, and a minimum stopband attenuation of 32 dB.

Using the M-file `cheb1ord` the order N and the passband edge `Wn` of the lowpass filter $H_{LP}(s)$ are determined. Then using the M-file `cheby1` the lowpass prototype filter $H_{LP}(s)$ is designed. Next, using the M-file `lp2hp` the analog highpass filter $H_{HP}(s)$ is designed by applying the lowpass-to-highpass transformation of Eq. (5.59) to $H_{LP}(s)$. Finally, using the M-file `bilinear` the desired digital IIR highpass filter $G_{HP}(z)$ is designed by applying the bilinear transformation of Eq. (7.20) to $H_{HP}(s)$. The code fragments used are

```
[N,Wn]    = cheb1ord(1,1.9626105, 1, 32,'s');
[B,A]     = cheby1(N,1,Wn,'s');
[BT,AT]   = lp2hp(B,A,1.9626105);
[num,den] = bilinear(BT,AT,0.5);
```

The transfer function $H_{LP}(s)$ of the prototype analog lowpass filter can be obtained by displaying the numerator coefficient vector `B` and the denominator coefficient vector `A`. Likewise, the transfer function $H_{HP}(s)$ of the analog highpass filter can be obtained by displaying the numerator coefficient vector `BT` and the denominator coefficient vector `AT`. Similarly, the transfer function $G_{HP}(z)$ of the desired digital IIR highpass filter can be obtained by displaying the numerator coefficient vector `num` and the denominator coefficient vector `den`. The gain response of the designed IIR digital highpass filter is shown in Figure 7.8.

The above IIR digital highpass filter can be directly designed in the z-domain using only the M-files `cheb1ord` and `cheby1` as illustrated in Example 7.22.

Design of Bandpass IIR Digital Filter

The design of a Butterworth bandpass IIR digital filter is treated in the following example.

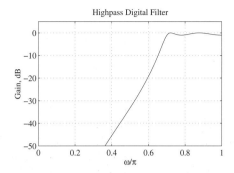

Figure 7.8: Gain response of the Type 1 Chebyshev highpass IIR digital filter of Example 7.9.

EXAMPLE 7.10 The desired specifications of the digital bandpass filter are: normalized passband edges at $\omega_{p1} = 0.45\pi$ and $\omega_{p2} = 0.65\pi$, normalized stopband edges at $\omega_{s1} = 0.3\pi$ and $\omega_{s2} = 0.75\pi$, passband ripple of 1 dB, and a minimum stopband attenuation of 40 dB.

We first prewarp the digital edge frequencies using Eq. (7.26) to arrive at the following angular edge frequencies of the analog bandpass filter:

$$\hat{\Omega}_{p1} = \tan\left(\frac{\omega_{p1}}{2}\right) = \tan\left(\frac{0.45\pi}{2}\right) = 0.8540807,$$

$$\hat{\Omega}_{p2} = \tan\left(\frac{\omega_{p2}}{2}\right) = \tan\left(\frac{0.65\pi}{2}\right) = 1.6318517,$$

$$\hat{\Omega}_{s1} = \tan\left(\frac{\omega_{s1}}{2}\right) = \tan\left(\frac{0.3\pi}{2}\right) = 0.5095254,$$

$$\hat{\Omega}_{s2} = \tan\left(\frac{\omega_{s2}}{2}\right) = \tan\left(\frac{0.75\pi}{2}\right) = 2.41421356.$$

The width of the passband of the bandpass filter is $B_w = \hat{\Omega}_{p2} - \hat{\Omega}_{p1} = 0.777771$. The product of the two passband edge frequencies is given by $\hat{\Omega}_o^2 = 1.393733$, and the product of the two stopband edge frequencies is 1.23010325. It is a usual practice not to change the specified width B_w of the passband but to adjust the value of the lower stopband edge frequency so that the two stopband edge frequencies exhibit geometric symmetry with respect to $\hat{\Omega}_o = 1.1805647$. To this end, we modify the lower stopband edge frequency and set $\hat{\Omega}_{s1} = 0.577327$.

For the prototype analog lowpass filter we choose the normalized passband edge frequency $\Omega_p = 1$. From Eq. (5.62) we get the stopband edge frequency of the lowpass filter to be

$$\Omega_s = \frac{1.393733 - 0.3332788}{0.5773031 \times 0.777771} = 2.3617627.$$

The specifications of the analog Butterworth lowpass filter are thus given by: normalized passband edge at 1 rad/sec, normalized stopband edge at 2.3617627 rad/sec, passband ripple of 1 dB, and minimum stopband attenuation of 40 dB.

We first use the M-file `buttord` to determine the filter order N and the passband edge angular frequency Wn of the prototype analog lowpass filter $H_{LP}(s)$. We then use the M-file `butter` to design the prototype analog lowpass filter $H_{LP}(s)$. Next, using the M-file `lp2bp` we apply the lowpass-to-bandpass transformation of Eq. (5.62) to $H_{LP}(s)$ to design the analog bandpass filter $H_{BP}(s)$. Finally, using the M-file `bilinear` we apply the bilinear transformation of Eq. (7.20) to $H_{BP}(s)$ to arrive at the transfer function $G_{BP}(z)$ of the desired bandpass IIR digital filter. The code fragments used are

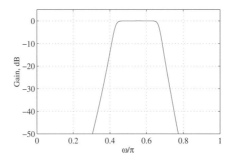

Figure 7.9: Gain response of the Butterworth bandpass IIR digital filter of Example 7.10.

```
[N,Wn] = buttord(1,2.3617627, 1, 40,'s');
[B,A] = butter(N,Wn,'s');
[BT,AT] = lp2bp(B,A,1.1805647,0.777771);
[num,den] = bilinear(BT,AT,0.5);
```

The transfer function $H_{LP}(s)$ of the prototype analog lowpass filter can be obtained by displaying the numerator coefficient vector B and the denominator coefficient vector A. Likewise, the transfer function $H_{BP}(s)$ of the analog bandpass filter can be obtained by displaying the numerator coefficient vector BT and the denominator coefficient vector AT. Similarly, the transfer function $G_{BP}(z)$ of the desired Butterworth bandpass IIR digital filter can be obtained by displaying the numerator coefficient vector num and the denominator coefficient vector den. The gain response of the designed Butterworth bandpass IIR digital filter is shown in Figure 7.9.

The above IIR digital bandpass filter can be directly designed in the z-domain using only the M-files buttord and butter as illustrated in Example 7.23.

Design of Bandstop IIR Digital Filter

We consider next the design of an elliptic bandstop IIR digital filter.

EXAMPLE 7.11   The specifications of the bandstop filter are: normalized stopband edges at $\omega_{s1} = 0.45\pi$ and $\omega_{s2} = 0.65\pi$, normalized passband edges at $\omega_{p1} = 0.3\pi$ and $\omega_{p2} = 0.75\pi$, passband ripple of 1 dB, and a minimum stopband attenuation of 40 dB.

As the bandedges of the above digital bandstop filter are the same as those of the digital bandpass filter of the previous example, the prewarped angular edge frequencies of the analog bandstop filter are obtained directly from the previous example:

$$\hat{\Omega}_{s1} = 0.8540806, \qquad \hat{\Omega}_{s2} = 1.6318517, \qquad \hat{\Omega}_{p1} = 0.5095254, \qquad \hat{\Omega}_{p2} = 2.4142136.$$

For this example, the width of the stopband of the bandstop filter is $B_w = \hat{\Omega}_{s2} - \hat{\Omega}_{s1} = 0.777771$. The product of the two stopband edge frequencies is given by $\hat{\Omega}_o^2 = 1.393733$, and the product of the two passband edge frequencies is 1.230103. In this case, we keep the width B_w of the stopband as specified and adjust the value of the lower passband edge frequency so that the two passband edge frequencies exhibit geometric symmetry with respect to $\hat{\Omega}_o = 1.1805647$. To this end, we modify the lower passband edge frequency and set $\hat{\Omega}_{p1} = 0.577303$.

The normalized stopband edge frequency of the prototype analog lowpass filter is set at $\Omega_s = 1$. From Eq. (5.65) we get the passband edge frequency of the lowpass filter to be

$$\Omega_p = \frac{0.5095254 \times 0.777771}{1.393733 - 0.3332787} = 0.4234126.$$

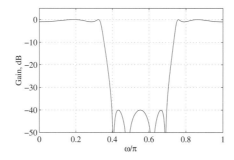

Figure 7.10: Gain response of the elliptic bandstop IIR digital filter of Example 7.11.

The specifications of the analog elliptic lowpass filter are thus given by: normalized passband edge at 0.3494297 rad/sec, normalized stopband edge at 1 rad/sec, passband ripple of 1 dB, and minimum stopband attenuation of 40 dB.

The code fragments used to design the desired digital bandstop filter are

```
[N,Wn] = ellipord(0.4234126,1, 1, 40,'s');
[B,A]  = ellip(N,1, 40,Wn,'s');
[BT,AT] = lp2bs(B,A,1.1805647,0.777771);
[num,den] = bilinear(BT,AT,0.5);
```

The gain response of the designed elliptic bandstop IIR digital filter is indicated in Figure 7.10. The transfer function $G_{BS}(z)$ of the elliptic bandstop IIR digital filter can be obtained by displaying the numerator coefficient vector num and the denominator coefficient vector den.

As in the case of the previous two examples, the above bandstop filter can also be designed directly in the z-domain using only the M-files ellipord and ellip.

7.5 Spectral Transformations of IIR Filters

Often, in practice, it may be necessary to modify the characteristics of a filter to meet the new specifications without repeating the filter design procedure. For example, after a lowpass filter with a passband edge at 2 kHz has been designed, it may be required to move the passband edge to 2.1 kHz. It is also possible to design a digital filter with highpass or bandpass or bandstop characteristics by transforming a given digital lowpass filter. We describe here the spectral transformations that can be used to transform a given lowpass digital IIR transfer function $G_L(z)$ to another digital transfer function $G_D(z)$ that could be a lowpass, highpass, bandpass, or bandstop filter [Con70]. Figure 4.10 shows the magnitude responses of these four types of ideal filters.

To eliminate the confusion between the complex variable z of the lowpass transfer function $G_L(z)$ and that of the desired transfer function $G_D(z)$, we shall use different symbols. Thus we shall use z^{-1} to denote the unit delay in the prototype lowpass digital filter $G_L(z)$ and $\hat{z}^{-1}$ to denote the unit delay in the transformed filter $G_D(\hat{z})$. The unit circles in the z- and $\hat{z}$-planes are defined by

$$z = e^{j\omega}, \qquad \hat{z} = e^{j\hat{\omega}}.$$

We denote the transformation from the z-domain to the $\hat{z}$-domain as

$$z = F(\hat{z}). \tag{7.40}$$

Then, $G_L(z)$ is transformed to $G_D(\hat{z})$ through

$$G_D(\hat{z}) = G_L\{F(\hat{z})\}. \tag{7.41}$$

To transform a rational $G_L(z)$ into a rational $G_D(\hat{z})$, $F(\hat{z})$ must be a rational function of $\hat{z}$. In addition, to guarantee the stability of $G_D(\hat{z})$, the transformation should be such that the inside of the unit circle of the z-plane is mapped into the inside of the unit circle of the $\hat{z}$-plane. Finally, to ensure that a lowpass magnitude response is mapped into one of the four basic types of magnitude responses, points on the unit circle of the z-plane should be mapped to points on the unit circle of the $\hat{z}$-plane.

Now, in the z-plane, a point on the unit circle is characterized by $|z| = 1$, a point inside the unit circle is given by $|z| < 1$, and a point outside the unit circle is defined by $|z| > 1$. Thus, from Eq. (7.40), $|F(\hat{z})| = |z|$ and, therefore,

$$\left|F(\hat{z})\right| \begin{cases} > 1, & \text{if } |z| > 1, \\ = 1, & \text{if } |z| = 1, \\ < 1, & \text{if } |z| < 1. \end{cases} \tag{7.42}$$

Thus, from the above and Eq. (4.132), it follows that $1/F(\hat{z})$ is a stable allpass function. From Eq. (4.129) we observe that the most general form of $F(\hat{z})$ with real coefficients is thus given by

$$F(\hat{z}) = \pm \prod_{\ell=1}^{L} \left(\frac{\hat{z} - \alpha_\ell}{1 - \alpha_\ell^* \hat{z}} \right), \tag{7.43}$$

where α_ℓ is either real or occurs in complex conjugate pairs with $|\alpha_\ell| < 1$ for stability.

7.5.1 Lowpass-to-Lowpass Transformation

To transform a prototype lowpass filter $G_L(z)$ with a cutoff frequency ω_c to another lowpass filter $G_D(\hat{z})$ with a cutoff frequency $\hat{\omega}_c$, we use the transformation

$$z^{-1} = F^{-1}(\hat{z}) = \frac{1 - \alpha \hat{z}}{\hat{z} - \alpha}, \tag{7.44}$$

with α real. On the unit circle, the above transformation reduces to

$$e^{-j\omega} = \frac{e^{-j\hat{\omega}} - \alpha}{1 - \alpha e^{-j\hat{\omega}}},$$

from which we arrive at

$$\tan\left(\frac{\omega}{2}\right) = \left(\frac{1+\alpha}{1-\alpha}\right) \tan\left(\frac{\hat{\omega}}{2}\right). \tag{7.45}$$

A plot of the relation between ω and $\hat{\omega}$ is given in Figure 7.11 for three values of α. Note that the mapping is nonlinear except for $\alpha = 0$, resulting in a warping of the frequency scale for nonzero values of α. However, if $G_L(z)$ is a piecewise constant lowpass magnitude response, then the transformed filter $G_D(\hat{z})$ will likewise have a similar piecewise constant lowpass magnitude response due to the monotonicity of the transformation of Eq. (7.45). The relation between the cutoff frequency ω_c of $G_L(z)$ with the cutoff frequency $\hat{\omega}_c$ of $G_D(\hat{z})$ follows from Eq. (7.45):

$$\tan\left(\frac{\omega_c}{2}\right) = \left(\frac{1+\alpha}{1-\alpha}\right) \tan\left(\frac{\hat{\omega}_c}{2}\right),$$

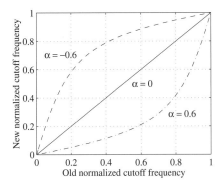

Figure 7.11: Mapping of the angular frequencies in the lowpass-to-lowpass transformation for three different values of the parameter α.

which can be solved for α yielding:

$$\alpha = \frac{\tan(\omega_c/2) - \tan(\hat{\omega}_c/2)}{\tan(\omega_c/2) + \tan(\hat{\omega}_c/2)} = \frac{\sin\left(\frac{\omega_c - \hat{\omega}_c}{2}\right)}{\sin\left(\frac{\omega_c + \hat{\omega}_c}{2}\right)}. \tag{7.46}$$

EXAMPLE 7.12 Consider the third-order lowpass digital transfer function $G(z)$ of Eq. (7.39), which has a passband from dc to 0.25π with a 0.5-dB ripple, and a stopband from 0.55π to π with an attenuation greater than 15 dB.

We wish to redesign this lowpass filter by applying the lowpass-to-lowpass transformation of Eq. (7.44) so that its passband edge moves from 0.25π to 0.35π. Here we have $\omega_c = 0.25\pi$ and $\hat{\omega}_c = 0.35\pi$. Substituting these values in Eq. (7.46), we get

$$\alpha = \frac{\sin\left(\frac{0.25\pi - 0.35\pi}{2}\right)}{\sin\left(\frac{0.25\pi + 0.35\pi}{2}\right)} = -\frac{\sin(0.05\pi)}{\sin(0.3\pi)} = -0.1933636.$$

Thus the desired lowpass transfer function $G_D(\hat{z})$ is given by

$$G_D(\hat{z}) = G(z)\Big|_{z^{-1} = \frac{\hat{z}^{-1} + 0.1933636}{1 + 0.1933636\hat{z}^{-1}}}$$

$$= \frac{0.2172235(1 + \hat{z}^{-1})^3}{(1 - 0.06944472\hat{z}^{-1})(1 - 0.1848053\hat{z}^{-1} + 0.337566\hat{z}^{-2})}. \tag{7.47}$$

A plot of the gain responses of the original lowpass filter $G(z)$ of Eq. (7.39) and the new lowpass filter $G_D(\hat{z})$ of Eq. (7.47) is shown in Figure 7.12. Note that the transformed transfer function $G_D(\hat{z})$ of Eq. (7.47) has been scaled in the plot to have a 0-dB dc gain.

It should be noted that the lowpass-to-lowpass transformation can also be used to transform a highpass filter with a cutoff at ω_c to another highpass filter with a cutoff at $\hat{\omega}_c$ (Problem 7.30), a bandpass filter with a center frequency at ω_o to another bandpass filter with a center frequency at $\hat{\omega}_o$ (Problem 7.31), and a bandstop filter with a center frequency at ω_o to another bandstop filter with a center frequency at $\hat{\omega}_o$ (Problem 7.32).

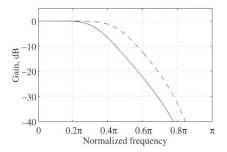

Figure 7.12: Gain responses of the prototype lowpass filter (solid line) and the transformed lowpass filter (dashed line).

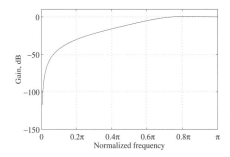

Figure 7.13: Gain response of the highpass filter of Example 7.13.

7.5.2 Other Transformations

Table 7.1 lists other useful transformations such as the lowpass-to-highpass, lowpass-to-bandpass, and lowpass-to-bandstop transformations, in addition to the lowpass-to-lowpass transformation discussed above. It should be noted that these spectral transformations can be used only to map one frequency point ω_c in the magnitude response of the lowpass prototype filter into a new position $\hat{\omega}_c$, with the same magnitude response value for the transformed lowpass and highpass filters, or into two new positions, $\hat{\omega}_{c1}$ and $\hat{\omega}_{c2}$, with the same magnitude response values for the transformed bandpass and bandstop filters. Hence, it is possible only to map either the passband edge or the stopband edge of the lowpass prototype filter onto the desired position(s), but not both.

EXAMPLE 7.13   We now consider the design of a highpass filter by applying a spectral transformation to the third-order lowpass digital transfer function $G(z)$ of Eq. (7.39). The desired passband edge $\hat{\omega}_c$ of the highpass filter is 0.55π, whereas the passband edge of the prototype lowpass filter is at $\omega_c = 0.25\pi$. Substituting these values in the lowpass-to-highpass transformation given in Table 7.1, we arrive at

$$\alpha = -\frac{\cos\left(\frac{0.55\pi + 0.25\pi}{2}\right)}{\cos\left(\frac{0.55\pi - 0.25\pi}{2}\right)} = -\frac{\cos(0.4\pi)}{\cos(0.15\pi)} = -0.3468179.$$

Thus, from Table 7.1, the desired lowpass-to-highpass transformation is given by

$$z^{-1} = -\frac{\hat{z}^{-1} - 0.3468179}{1 - 0.3468179\hat{z}^{-1}}.$$

Table 7.1: Spectral transformations of a lowpass filter with a cutoff frequency ω_c.

Filter type	Spectral transformation	Design parameters
Lowpass	$z^{-1} = \dfrac{\hat{z}^{-1} - \alpha}{1 - \alpha\hat{z}^{-1}}$	$\alpha = \dfrac{\sin\left(\frac{\omega_c - \hat{\omega}_c}{2}\right)}{\sin\left(\frac{\omega_c + \hat{\omega}_c}{2}\right)}$ $\hat{\omega}_c =$ desired cutoff frequency
Highpass	$z^{-1} = -\dfrac{\hat{z}^{-1} + \alpha}{1 + \alpha\hat{z}^{-1}}$	$\alpha = -\dfrac{\cos\left(\frac{\omega_c + \hat{\omega}_c}{2}\right)}{\cos\left(\frac{\omega_c - \hat{\omega}_c}{2}\right)}$ $\hat{\omega}_c =$ desired cutoff frequency
Bandpass	$z^{-1} = -\dfrac{\hat{z}^{-2} - \frac{2\alpha\beta}{\beta+1}\hat{z}^{-1} + \frac{\beta-1}{\beta+1}}{\frac{\beta-1}{\beta+1}\hat{z}^{-2} - \frac{2\alpha\beta}{\beta+1}\hat{z}^{-1} + 1}$	$\alpha = \dfrac{\cos\left(\frac{\hat{\omega}_{c2} + \hat{\omega}_{c1}}{2}\right)}{\cos\left(\frac{\hat{\omega}_{c2} - \hat{\omega}_{c1}}{2}\right)}$ $\beta = \cot\left(\frac{\hat{\omega}_{c2} - \hat{\omega}_{c1}}{2}\right)\tan\left(\frac{\omega_c}{2}\right)$ $\hat{\omega}_{c2}, \hat{\omega}_{c1} =$ desired upper and lower cutoff frequencies
Bandstop	$z^{-1} = \dfrac{\hat{z}^{-2} - \frac{2\alpha}{1+\beta}\hat{z}^{-1} + \frac{1-\beta}{1+\beta}}{\frac{1-\beta}{1+\beta}\hat{z}^{-2} - \frac{2\alpha}{1+\beta}\hat{z}^{-1} + 1}$	$\alpha = \dfrac{\cos\left(\frac{\hat{\omega}_{c2} + \hat{\omega}_{c1}}{2}\right)}{\cos\left(\frac{\hat{\omega}_{c2} - \hat{\omega}_{c1}}{2}\right)}$ $\beta = \tan\left(\frac{\hat{\omega}_{c2} - \hat{\omega}_{c1}}{2}\right)\tan\left(\frac{\omega_c}{2}\right)$ $\hat{\omega}_{c2}, \hat{\omega}_{c1} =$ desired upper and lower cutoff frequencies

Using the above transformation, we obtain from Eqs. (7.39) and (7.41) the desired highpass transfer function

$$G_D(\hat{z}) = G(z)\Big|_{z^{-1} = -\frac{\hat{z}^{-1} - 0.3468179}{1 - 0.3468179\hat{z}^{-1}}}$$

$$= \frac{0.2187917(1 - \hat{z}^{-1})^3}{(1 - 0.091006022\hat{z}^{-1})(1 + 0.7460066\hat{z}^{-1} + 0.3414851\hat{z}^{-2})}.$$

The gain response of the above transfer function has been plotted in Figure 7.13.

The lowpass-to-bandpass transformation given in Table 7.1 can be simplified if we consider the case when the bandwidth of the passband for the prototype lowpass filter is the same as that of the transformed bandpass filter, i.e., $\omega_c = \hat{\omega}_{c2} - \hat{\omega}_{c1}$. Applying this constraint to the respective spectral transformation in

Table 7.1, we arrive at the modified spectral transformation given by

$$z^{-1} = -\hat{z}^{-1} \frac{\hat{z}^{-1} - \alpha}{1 - \alpha \hat{z}^{-1}}. \tag{7.48}$$

The parameter α is determined from the desired location of the center frequency $\hat{\omega}_o$ of the bandpass filter derived via the transformation of Eq. (7.48), which maps the zero frequency of the lowpass filter, i.e., $\omega = 0$, to $\hat{\omega}_o$. From Eq. (7.48) we get

$$e^{-j\omega} = -e^{-j\hat{\omega}} \frac{e^{-j\hat{\omega}} - \alpha}{1 - \alpha e^{-j\hat{\omega}}}. \tag{7.49}$$

Substituting $\omega = 0$ and $\hat{\omega} = \hat{\omega}_o$ in the above equation, we arrive at

$$\alpha = \cos \hat{\omega}_o. \tag{7.50}$$

EXAMPLE 7.14   We develop the transfer function of a second-order bandpass filter by applying the transformation of Eq. (7.48) to the first-order lowpass transfer function of Eq. (7.29). The desired transfer function is thus given by

$$G_D(\hat{z}) = G(z)|_{z^{-1} = -\hat{z}^{-1}(\hat{z}^{-1} - \beta)/(1 - \beta \hat{z}^{-1})}$$
$$= \frac{1 - \alpha}{2} \left[\frac{1 - \hat{z}^{-2}}{1 - \beta(1 + \alpha)\hat{z}^{-1} + \alpha \hat{z}^{-2}} \right], \tag{7.51}$$

whose passband center frequency is given by $\beta = \cos \hat{\omega}_o$, and the 3-dB passband bandwidth $\hat{B}_w$ is given by $\alpha = [1 - \tan(\hat{B}_w)]/[1 + \tan(\hat{B}_w)]$. It should be noted that the above bandpass transfer function is precisely the same as in Eq. (4.113), where it was introduced without any derivation. The relation between the transfer function parameters β and α with the passband center frequency $\hat{\omega}_o$ and the 3-dB bandwidth $\hat{B}_w$ given above are identical to those given in Eqs. (4.115) and (4.116), respectively.

It should be noted that the lowpass-to-highpass transformation can also be used to transform a highpass filter with a cutoff at ω_c to a lowpass filter with a cutoff at $\hat{\omega}_c$ (Problem 7.33).

7.6   FIR Filter Design Based on Windowed Fourier Series

So far we have considered only the design of real-coefficient IIR digital filters that are described by a real rational transfer function that is a ratio of polynomials in z^{-1} with real coefficients. Since the transfer function of an analog filter is also a real rational transfer function in the complex frequency variable s, it is more convenient to design IIR digital filters by the conversion of a prototype analog transfer function, and a commonly used method based on this approach was discussed. In addition, we outlined a method to transform one type of IIR digital filter into another type. We now turn our attention to the design of real-coefficient FIR filters. These filters are described by a transfer function that is a polynomial in z^{-1} and therefore require different approaches for their design.

A variety of approaches have been proposed for the design of FIR digital filters. A direct and straightforward method is based on truncating the Fourier series representation of the prescribed frequency response and is discussed in this section. The second method is based on the observation that for a length-N FIR digital filter, N distinct equally spaced frequency samples of its frequency response constitute the N-point

DFT of its impulse response, and hence, the impulse response sequence can be readily computed by applying an inverse DFT to these frequency samples (Problem 7.59).

7.6.1 Least Integral-Squared Error Design of FIR Filters

Let $H_d(e^{j\omega})$ denote the desired frequency response function. Since $H_d(e^{j\omega})$ is a periodic function of ω with a period 2π, it can be expressed as a Fourier series,

$$H_d(e^{j\omega}) = \sum_{n=-\infty}^{\infty} h_d[n]e^{-j\omega n}, \qquad (7.52)$$

where the Fourier coefficients $\{h_d[n]\}$ are precisely the corresponding impulse response samples and are given by

$$h_d[n] = \frac{1}{2\pi} \int_{-\pi}^{\pi} H_d(e^{j\omega})e^{j\omega n} d\omega, \qquad -\infty \leq n \leq \infty. \qquad (7.53)$$

Thus, given a frequency response specification $H_d(e^{j\omega})$, we can compute $h_d[n]$ using Eq. (7.53) and, hence, determine the transfer function $H_d(z)$. However, for most practical applications, the desired frequency response is piecewise constant with sharp transitions between bands, in which case, the corresponding impulse response sequence $\{h_d[n]\}$ is of infinite length and noncausal.

Our objective is to find a finite-duration impulse response sequence $\{h_t[n]\}$ of length $2M + 1$ whose DTFT $H_t(e^{j\omega})$ approximates the desired DTFT $H_d(e^{j\omega})$ in some sense. One commonly used approximation criterion is to minimize the integral-squared error

$$\Phi = \frac{1}{2\pi} \int_{-\pi}^{\pi} \left| H_t(e^{j\omega}) - H_d(e^{j\omega}) \right|^2 d\omega, \qquad (7.54)$$

where

$$H_t(e^{j\omega}) = \sum_{n=-M}^{M} h_t[n]e^{-j\omega n}. \qquad (7.55)$$

Using Parseval's relation (Table 3.2) we can rewrite Eq. (7.54) as

$$\Phi = \sum_{n=-\infty}^{\infty} |h_t[n] - h_d[n]|^2$$

$$= \sum_{n=-M}^{M} |h_t[n] - h_d[n]|^2 + \sum_{n=-\infty}^{-M-1} h_d^2[n] + \sum_{n=M+1}^{\infty} h_d^2[n]. \qquad (7.56)$$

It is evident from Eq. (7.56) that the integral-squared error is minimum when $h_t[n] = h_d[n]$ for $-M \leq n \leq M$, or in other words, the best finite-length approximation to the ideal infinite-length impulse response in the mean-square error sense is simply obtained by truncation.

A causal FIR filter with an impulse response $h[n]$ can be derived from $h_t[n]$ by delaying the latter sequence by M samples, i.e., by forming

$$h[n] = h_t[n - M]. \qquad (7.57)$$

Note that the causal filter $h[n]$ has the same magnitude response as that of the noncausal filter $h_t[n]$ and its phase response has a linear phase shift of ωM radians with respect to that of the noncausal filter.

7.6.2 Impulse Responses of Ideal Filters

Four commonly used frequency selective filters are the lowpass, highpass, bandpass, and bandstop filters whose ideal frequency responses are shown in Figure 4.10. It is straightforward to develop their corresponding impulse responses. For example, the ideal *lowpass filter* of Figure 4.10(a) has a zero-phase frequency response

$$H_{LP}(e^{j\omega}) = \begin{cases} 1, & |\omega| \leq \omega_c, \\ 0, & \omega_c < |\omega| \leq \pi. \end{cases} \tag{7.58}$$

The corresponding impulse response coefficients were determined in Example 3.3 and are given by

$$h_{LP}[n] = \frac{\sin \omega_c n}{\pi n}, \qquad -\infty \leq n \leq \infty. \tag{7.59}$$

As can be seen from the above equation, the impulse response of an ideal lowpass filter is doubly infinite, not absolutely summable, and therefore unrealizable. By setting all impulse response coefficients outside the range $-M \leq n \leq M$ equal to zero, we arrive at a finite-length noncausal approximation of length $N = 2M + 1$, which when shifted to the right yields the coefficients of a causal FIR lowpass filter:

$$\hat{h}_{LP}[n] = \begin{cases} \frac{\sin(\omega_c(n - M))}{\pi(n - M)}, & 0 \leq n \leq N - 1, \\ 0, & \text{otherwise.} \end{cases} \tag{7.60}$$

It should be noted that the above expression also holds for N even, in which case M is a fraction.

Likewise, the impulse response coefficients $h_{HP}[n]$ of the ideal *highpass filter* of Figure 4.10(b) are given by

$$h_{HP}[n] = \begin{cases} 1 - \frac{\omega_c}{\pi}, & \text{for } n = 0, \\ -\frac{\sin(\omega_c n)}{\pi n}, & \text{for } |n| > 0. \end{cases} \tag{7.61}$$

Correspondingly, the impulse response coefficients $h_{BP}[n]$ of the ideal *bandpass filter* of Figure 4.10(c) with cutoffs at ω_{c1} and ω_{c2} are given by

$$h_{BP}[n] = \frac{\sin(\omega_{c2}n)}{\pi n} - \frac{\sin(\omega_{c1}n)}{\pi n}, \qquad |n| \geq 0, \tag{7.62}$$

and those of the ideal *bandstop filter* of Figure 4.10(d) with cutoffs at ω_{c1} and ω_{c2} are given by

$$h_{BS}[n] = \begin{cases} 1 - \frac{(\omega_{c2} - \omega_{c1})}{\pi}, & \text{for } n = 0, \\ \frac{\sin(\omega_{c1}n)}{\pi n} - \frac{\sin(\omega_{c2}n)}{\pi n}, & \text{for } |n| > 0. \end{cases} \tag{7.63}$$

All the above design methods are for single passband or single stopband filters with two magnitude levels. However, it is quite straightforward to generalize the method to the design of multilevel FIR filters and obtain the expression for the impulse response coefficients. The zero-phase frequency response of an ideal L-band digital filter $H_{ML}(z)$ is given by

$$H_{ML}(e^{j\omega}) = A_k, \quad \text{for } \omega_{k-1} \leq \omega \leq \omega_k, \quad k = 1, 2, \ldots, L, \tag{7.64}$$

where $\omega_0 = 0$ and $\omega_L = \pi$. Figure 7.14 shows the zero-phase frequency response of a typical multilevel filter. Its impulse response $h_{ML}[n]$ is given by

$$h_{ML}[n] = \sum_{\ell=1}^{L} (A_\ell - A_{\ell+1}) \cdot \frac{\sin(\omega_\ell n)}{\pi n}, \tag{7.65}$$

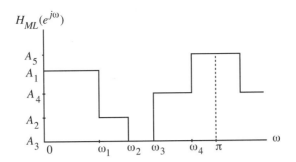

Figure 7.14: A typical zero-phase multilevel frequency response.

with $A_{L+1} = 0$.

Two other types of FIR digital filters that find applications are the discrete-time *Hilbert transformer* and the *differentiator*. The ideal Hilbert transformer, also called a *90-degree phase shifter*, is characterized by a frequency response

$$H_{HT}(e^{j\omega}) = \begin{cases} j, & -\pi < \omega < 0, \\ -j, & 0 < \omega < \pi. \end{cases} \tag{7.66}$$

It finds application in the generation of analytic signals (see Section 11.7). The impulse response $h_{HT}[n]$ of the Hilbert transformer is obtained by computing the inverse discrete-time Fourier transform of Eq. (7.66) and is given by

$$h_{HT}[n] = \begin{cases} 0, & \text{for } n \text{ even}, \\ \frac{2}{\pi n}, & \text{for } n \text{ odd}. \end{cases} \tag{7.67}$$

The ideal discrete-time differentiator is employed to perform the differentiation operation in discrete-time on the sampled version of a continuous-time signal. It is characterized by a frequency response given by

$$H_{DIF}(e^{j\omega}) = j\omega, \quad 0 \le |\omega| \le \pi. \tag{7.68}$$

The impulse response $h_{DIF}[n]$ of the ideal discrete-time differentiator is determined by an inverse discrete-time Fourier transform of Eq. (7.68) and is given by

$$h_{DIF}[n] = \begin{cases} 0, & n = 0, \\ \frac{\cos \pi n}{n}, & |n| > 0. \end{cases} \tag{7.69}$$

Like the ideal lowpass filter, all the above five ideal filters are also characterized by doubly infinite impulse responses that are not absolutely summable, making them unrealizable. They can be made realizable by truncating the impulse response sequences to finite lengths and shifting the truncated coefficients to the right appropriately.

7.6.3 Gibbs Phenomenon

The causal FIR filters obtained by simply truncating the impulse response coefficients of the ideal filters given in the previous section exhibit an oscillatory behavior in their respective magnitude responses, which is more commonly referred to as the *Gibbs phenomenon*. We illustrate here the occurrence of the Gibbs phenomenon by considering the design of lowpass filters. Figure 7.15 shows the magnitude responses of a lowpass filter with a cutoff at $\omega_c = 0.4\pi$ designed using the formula of Eq. (7.60) for two different values of filter lengths. The oscillatory behavior of the magnitude response on both sides of the cutoff frequency is

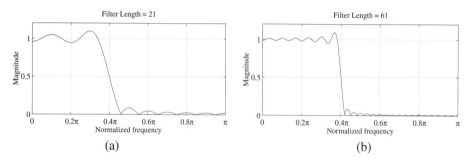

(a)   (b)

Figure 7.15: Magnitude responses of lowpass filters designed using the truncated impulse response of Eq. (7.60): (a) length $N = 21$, and (b) length $N = 61$.

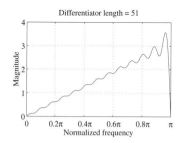

Figure 7.16: Magnitude response of a length-51 differentiator designed by truncating the impulse response of Eq. (7.69).

clearly visible in both cases. Moreover, as the length of the filter is increased, the number of ripples in both passband and stopband increases, with a corresponding decrease in the widths of the ripples. However, the heights of the largest ripples, which occur on both sides of the cutoff frequency, remain the same independent of the filter length and are approximately 11 percent of the difference between the passband and stopband magnitudes of the ideal filter [Par87].

A similar oscillatory behavior is also observed in the frequency responses of the truncated versions of the impulse responses of other types of ideal filters described in the previous section (Exercises M7.13 to M7.15). For example, Figure 7.16 shows the magnitude response of a length-51 differentiator designed by truncating the impulse response coefficients of Eq. (7.69).

The reason behind the Gibbs phenomenon can be explained by considering the truncation operation as multiplication by a finite-length window sequence $w[n]$ and by examining the windowing process in the frequency domain. Thus, the FIR filter obtained by truncation can be alternatively expressed as

$$h_t[n] = h_d[n] \cdot w[n]. \tag{7.70}$$

From the modulation theorem of Table 3.2, the Fourier transform of Eq. (7.70) is given by

$$H_t(e^{j\omega}) = \frac{1}{2\pi} \int_{-\pi}^{\pi} H_d(e^{j\varphi}) \Psi(e^{j(\omega-\varphi)}) \, d\varphi, \tag{7.71}$$

where $H_t(e^{j\omega})$ and $\Psi(e^{j\omega})$ are the Fourier transforms of $h_t[n]$ and $w[n]$, respectively. Equation (7.71) implies that $H_t(e^{j\omega})$ is obtained by a periodic continuous convolution of the desired frequency response $H_d(e^{j\omega})$ with the Fourier transform $\Psi(e^{j\omega})$ of the window. The process is illustrated in Figure 7.17 with all Fourier transforms shown as real functions for convenience.

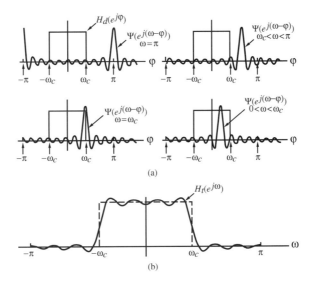

Figure 7.17: Illustration of the effect of windowing in the frequency domain.

From Eq. (7.71) it follows that if $\Psi(e^{j\omega})$ is a very narrow pulse centered at $\omega = 0$ (ideally a delta function) compared to variations in $H_d(e^{j\omega})$, then $H_t(e^{j\omega})$ will approximate $H_d(e^{j\omega})$ very closely. This implies that the length $2M + 1$ of the window function $w[n]$ should be very large. On the other hand, the length $2M + 1$ of $h_t[n]$, and hence that of $w[n]$, should be as small as possible to make the computational complexity of the filtering process small.

Now, the window used to achieve simple truncation of the ideal infinite-length impulse response is called a *rectangular window* and is given by

$$w_R[n] = \begin{cases} 1, & 0 \le |n| \le M, \\ 0, & \text{otherwise.} \end{cases} \tag{7.72}$$

The presence of the oscillatory behavior in the Fourier transform of a truncated Fourier series representation of an ideal filter is basically due to two reasons. First, the impulse response of an ideal filter is infinitely long and not absolutely summable, and as a result, the filter is unstable. Second, the rectangular window has an abrupt transition to zero. The oscillatory behavior can be easily explained by examining the Fourier transform $\Psi_R(e^{j\omega})$ of the rectangular window function of Eq. (7.72):

$$\Psi_R(e^{j\omega}) = \sum_{n=-M}^{M} e^{-j\omega n} = \frac{\sin([2M + 1]\omega/2)}{\sin(\omega/2)}. \tag{7.73}$$

A plot of the above is sketched in Figure 7.18 for $M = 4$ and 10. The frequency response $\Psi_R(e^{j\omega})$ has a narrow "main lobe" centered at $\omega = 0$. All the other ripples in the frequency response are called the "sidelobes." The main lobe is characterized by its width $4\pi/(2M + 1)$ defined by the first zero crossings on both sides of $\omega = 0$. Thus, as M increases, the width of the main lobe decreases as desired. However, the area under each lobe remains constant, while the width of each lobe decreases with an increase in M. This implies that with increasing M, ripples in $H_t(e^{j\omega})$ around the point of discontinuity occur more closely but with no decrease in amplitude.

Recall also that ideally the Fourier transform of the window function should closely resemble an impulse function centered at $\omega = 0$, with its length $2M + 1$ being as small as possible to reduce the

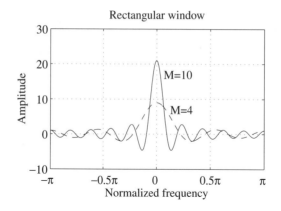

Figure 7.18: Frequency responses of the rectangular window for $M = 4$ and $M = 10$.

computational complexity of the FIR filter. An increase in the length of the rectangular window function reduces the main lobe width but unfortunately increases the computational complexity.

The rectangular window has an abrupt transition to zero outside the range $-M \leq n \leq M$, which is the reason behind the appearance of the Gibbs phenomenon in the magnitude response of the windowed ideal filter impulse response sequence. The Gibbs phenomenon can be reduced by either using a window that tapers smoothly to zero at each end or by providing a smooth transition from the passband to the stopband. Use of a tapered window causes the height of the sidelobes to diminish, with a corresponding increase in the main lobe width resulting in a wider transition at the discontinuity. We review in the next two sections a few of these windows and study their properties. Elimination of the Gibbs phenomenon by introducing a smooth transition in the filter specifications is considered in Section 7.6.6.

7.6.4 Fixed Window Functions

Many tapered windows have been proposed by various authors. A discussion of all these suggested windows is beyond the scope of this text. We restrict our discussion to three commonly used tapered windows of length $2M + 1$, which are listed below [Sar93]:[6]

$$Hann: [8] \quad w[n] = \frac{1}{2}\left[1 + \cos\left(\frac{2\pi n}{2M+1}\right)\right], \qquad -M \leq n \leq M, \tag{7.74}$$

$$Hamming: \; w[n] = 0.54 + 0.46\cos\left(\frac{2\pi n}{2M+1}\right), \quad -M \leq n \leq M, \tag{7.75}$$

$$Blackman: \; w[n] = 0.42 + 0.5\cos\left(\frac{2\pi n}{2M+1}\right)$$
$$+ 0.08\cos\left(\frac{4\pi n}{2M+1}\right), \qquad -M \leq n \leq M. \tag{7.76}$$

A plot of the magnitude of the Fourier transform of each of the above windows in the dB scale is shown in Figure 7.19 for $M = 25$. As can be seen from these plots, the magnitude spectrum of each window is

---

[6]The expressions for the window functions given here are slightly different from that given in the literature.
[8]In the literature, this window is often called the Hanning window or the von Hann window.

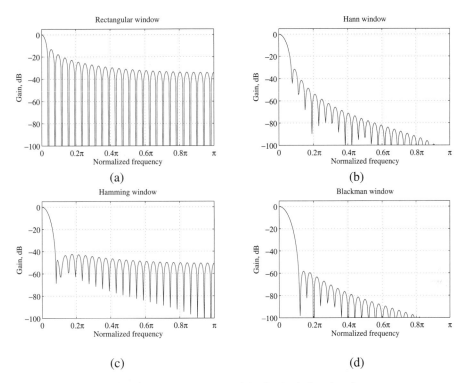

Figure 7.19: Gain response of the fixed window functions.

characterized by a large main lobe centered at $\omega = 0$ followed by a series of sidelobes with decreasing amplitudes. Two parameters that somewhat predict the performance of a window in FIR filter design are its *main lobe width* and the *relative sidelobe level*. The main lobe width Δ_{ML} is the distance between the nearest zero crossings on both sides of the main lobe, and the relative sidelobe level $A_{s\ell}$ is the difference in dB between the amplitudes of the largest sidelobe and the main lobe.

To understand the effect of the window function on FIR filter design, we show in Figure 7.20 a typical relation among $H_t(e^{j\omega})$, $\Psi(e^{j\omega})$, and $H_d(e^{j\omega})$, the frequency responses of the windowed lowpass filter, the window function, and the desired ideal lowpass filter, respectively [Sar93]. Since the corresponding impulse responses are symmetric with respect to $n = 0$, the frequency responses are of zero-phase. From this figure, we observe that for the windowed filter, $H_t(e^{j(\omega_c+\omega)}) + H_t(e^{j(\omega_c-\omega)}) \cong 1$, around the cutoff frequency ω_c. As a result, $H_t(e^{j\omega_c}) \cong 0.5$. Moreover, the passband and stopband ripples are the same. In addition, the distance between the maximum passband deviation and the minimum stopband value is approximately equal to the width Δ_{ML} of the main lobe of the window, with the center at ω_c. The width of the transition band, defined by $\Delta\omega = \omega_s - \omega_p$, is less than Δ_{ML}. Therefore, to ensure a fast transition from the passband to the stopband, the window should have a very small main lobe width. On the other hand, to reduce the passband and stopband ripple δ, the area under the sidelobes should be very small. Unfortunately, these two requirements are contradictory.

In the case of the window functions of Eqs. (7.72) and (7.74) to (7.76), the value of the ripple δ does not depend on the filter length, or the cutoff frequency ω_c, and is essentially constant. In addition, the

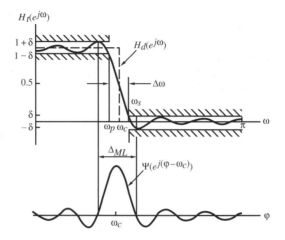

Figure 7.20: Relations among the frequency responses of an ideal lowpass filter, a typical window, and the windowed filter.

Table 7.2: Properties of some fixed window functions.[9]

Type of window	Main lobe width Δ_{ML}	Relative sidelobe level $A_{s\ell}$	Minimum stopband attenuation	Transition bandwidth $\Delta\omega$
Rectangular	$4\pi/(2M+1)$	13.3 dB	20.9 dB	$0.92\pi/M$
Hann	$8\pi/(2M+1)$	31.5 dB	43.9 dB	$3.11\pi/M$
Hamming	$8\pi/(2M+1)$	42.7 dB	54.5 dB	$3.32\pi/M$
Blackman	$12\pi/(2M+1)$	58.1 dB	75.3 dB	$5.56\pi/M$

transition bandwidth is approximately given by

$$\Delta\omega \approx \frac{c}{M}, \tag{7.77}$$

where c is a constant for most practical purposes [Sar93].

Table 7.2 summarizes the essential properties of the above window functions.

For designing an FIR filter using one of the above windows, first the cutoff frequency ω_c is determined from the specified passband and stopband edge frequencies, ω_p and ω_s, by setting $\omega_c = (\omega_p + \omega_s)/2$. Next, M is estimated using Eq. (7.77), where the value of the constant c is obtained from Table 7.2 for the window chosen. The following example illustrates the effect of each of the above windows on the frequency response of an FIR lowpass filter designed using the windowed Fourier series approach.

EXAMPLE 7.15   The impulse response $h_{LP}[n]$ of the ideal lowpass filter is given in Eq. (7.59). To create a finite-duration zero-phase FIR filter of length N, we form $h_t[n] = h_{LP}[n] \cdot w[n]$, where $M = (N-1)/2$. Figure 7.21 shows the impulse response samples $h_t[n]$ for $N = 51$ and $\omega_c = \pi/2$. The gain response of each of the filters obtained by windowing the above impulse-response samples with various fixed window functions is sketched in

---

[9]This table has been adapted from [Sar93] with the values shown in the table for $\omega_c = 0.4\pi$ and $M = 128$.

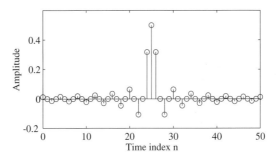

Figure 7.21: Impulse response of the truncated ideal lowpass FIR filter of length 51 with a cutoff at $\pi/2$.

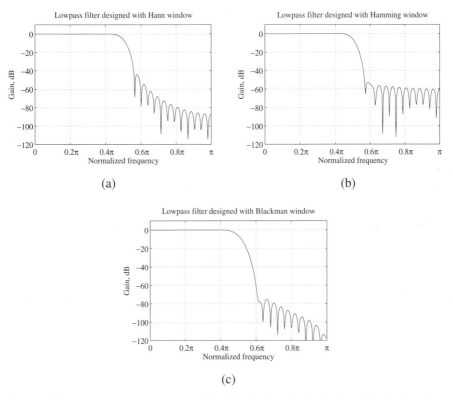

Figure 7.22: Gain responses of a lowpass FIR filter designed using the fixed window functions.

Figure 7.22.[10] It should be noted from this figure that the increase in the main lobe width of the window functions is clearly associated with an increase in the transition width. Likewise, a decrease in the sidelobe amplitude results in an increase in the stopband attenuation as expected.

---

[10]These window functions have been generated using the M-files `hanning`, `hamming`, and `blackman` (see Section 7.10.4).

7.6.5 Adjustable Window Functions

As indicated above, the ripple δ of the filter designed using any one of the fixed window functions is fixed. Several windows have been developed that provide control over δ by means of an additional parameter characterizing the window. We describe here two such windows.

The *Dolph-Chebyshev window* of length $2M + 1$ is defined by [Hel68]

$$w[n] = \frac{1}{2M+1} \left[\frac{1}{\gamma} + 2 \sum_{k=1}^{M} T_k \left(\beta \cos \frac{k\pi}{2M+1} \right) \cos \frac{2nk\pi}{2M+1} \right], \quad -M \leq n \leq M, \quad (7.78)$$

where γ is the relative sidelobe amplitude expressed as a fraction,

$$\gamma = \frac{\text{amplitude of sidelobe}}{\text{main lobe amplitude}}, \quad (7.79)$$

$$\beta = \cosh \left(\frac{1}{2M} \cosh^{-1} \frac{1}{\gamma} \right), \quad (7.80)$$

and $T_\ell(x)$ is the ℓth-order Chebyshev polynomial in x defined by

$$T_\ell(x) = \begin{cases} \cos(\ell \cos^{-1} x), & \text{for } |x| \leq 1, \\ \cosh(\ell \cosh^{-1} x), & \text{for } |x| > 1. \end{cases} \quad (7.81)$$

The above window can be designed with any specified relative sidelobe level and, as in the case of the other windows, its main lobe width can be adjusted by choosing the length appropriately. The filter order $N = 2M$ is estimated using the formula [Sar93]

$$N = \frac{2.056\alpha_s - 16.4}{2.285(\Delta\omega)} \quad (7.82)$$

where $\Delta\omega$ is the normalized transition bandwidth. In the case of a lowpass filter with normalized angular passband and stopband edge frequencies ω_p and ω_s, $\Delta\omega = \omega_s - \omega_p$.

Figure 7.23 shows the gain response of a Dolph-Chebyshev window for $M = 25$, i.e., a window length of 51, and a relative sidelobe level of 50 dB.[11] As can be seen from this plot, all sidelobes are of equal height. As a result, the stopband approximation error of filters designed using this window have essentially an equiripple behavior. Another interesting property of this window is that for a given window length it has the smallest main lobe width compared to other windows, resulting in filters with the smallest transition band.

The most widely used adjustable window is the *Kaiser window* given by [Kai74]:

$$w[n] = \frac{I_0 \left\{ \beta \sqrt{1 - (n/M)^2} \right\}}{I_0(\beta)}, \quad -M \leq n \leq M, \quad (7.83)$$

where β is an adjustable parameter and $I_0(u)$ is the modified zeroth-order Bessel function, which can be expressed in a power series form

$$I_0(u) = 1 + \sum_{r=1}^{\infty} \left[\frac{(u/2)^r}{r!} \right]^2, \quad (7.84)$$

which is seen to be positive for all real values of u. In practice, it is sufficient to keep only the first 20 terms in the summation of Eq. (7.84) to arrive at a reasonably accurate value of $I_0(u)$.

---

[11] The coefficients of the window have been computed using the M-file chebwin (see Section 7.10.4).

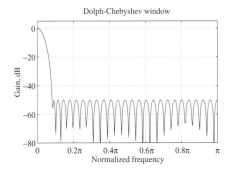

Figure 7.23: Gain response of a length-51 Dolph-Chebyshev window with a relative sidelobe level of 50 dB.

The parameter β controls the minimum attenuation $\alpha_s = -20 \log_{10}(\delta_s)$ in the stopband of the windowed filter response. Formulas for estimating β and the filter order $N = 2M$, for specified α_s and normalized transition bandwidth $\Delta\omega$, have been developed by Kaiser [Kai74]. The parameter β is computed from [12]

$$\beta = \begin{cases} 0.1102(\alpha_s - 8.7), & \text{for } \alpha_s > 50, \\ 0.5842(\alpha_s - 21)^{0.4} + 0.07886(\alpha_s - 21), & \text{for } 21 \leq \alpha_s \leq 50, \\ 0, & \text{for } \alpha_s < 21. \end{cases} \quad (7.85)$$

The filter order N is estimated using the formula

$$N = \frac{\alpha_s - 8}{2.285(\Delta\omega)}, \quad (7.86)$$

where $\Delta\omega$ is the normalized transition bandwidth. It should be noted that the Kaiser window provides no independent control over the passband ripple δ_p. However, in practice, δ_p is approximately equal to δ_s.

We illustrate the design of a linear-phase lowpass filter using the Kaiser window in the following example.

EXAMPLE 7.16   The desired specifications for the lowpass filter are $\omega_p = 0.3\pi$, $\omega_s = 0.5\pi$, and $\alpha_s = 40$ dB. The cutoff frequency is therefore given by $\omega_c = (\omega_p + \omega_s)/2 = 0.4\pi$.

For the specified minimum stopband attenuation, the peak stopband ripple δ_s is computed using Eq. (7.4) and is found to be $\delta_s = 0.01$. Using Eq. (7.85), the parameter β is next calculated and is determined to be $\beta = 3.3953$. The normalized transition bandwidth $\Delta\omega$ is given by

$$\Delta\omega = \omega_s - \omega_p = 0.5\pi - 0.3\pi = 0.2\pi.$$

Applying Eq. (7.86), we next determine the filter order N to be 22.2886, which when rounded up to the next higher integer is 23. We choose the next higher even integer value of 24 as the filter order, implying $M = 12$.

From Eqs. (7.59) and (7.83), we then arrive at the impulse response coefficients of the FIR filter obtained by windowing as

$$h_t[n] = \frac{\sin \omega_c n}{\pi n} \cdot w[n], \quad -M \leq n \leq M,$$

where $M = 12$, $\omega_c = 0.4\pi$, and $w[n]$ is the nth coefficient of a length-25 Kaiser window. It should be noted that the above filter is noncausal and can be converted into a causal filter by delaying the filter coefficients by M samples. Since M is even for our example, the delayed filter here is a Type 1 FIR linear-phase filter.

---

[12]Determined empirically.

Table 7.3: Coefficients of the truncated ideal lowpass filter, the Kaiser window, and the windowed lowpass filter of Example 7.16.

n	Truncated ideal filter	Kaiser window	Windowed filter
0	0.4	1.0	0.4
1	0.30273069145626	0.99018600306076	0.22975969337690
2	0.09354892837886	0.96116367723306	0.08991583299184
3	−0.06236595225258	0.91416906485180	−0.05701302424933
4	−0.07568267286407	0.85118713849723	−0.06442011774899
5	0	0.77484399922791	0
6	0.05045511524271	0.68826531740711	0.03472650590734
7	0.02672826525110	0.59490963898165	0.01590090263114
8	−0.02338723209472	0.49838664295215	−0.01165588409163
9	−0.02752097195057	0.40227124342555	−0.01353109463058
10	0	0.30992453405416	0
11	0.02752097195057	0.22433197338245	0.00617383394707
12	0.01559148806314	0.14796795346662	0.00230704058020

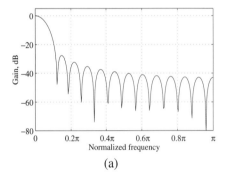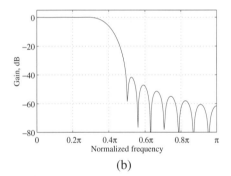

Figure 7.24: (a) Gain response of the Kaiser window of Example 7.16, and (b) gain response of the lowpass filter designed using this window.

The coefficients of the truncated ideal lowpass filter, the Kaiser window, and the windowed lowpass filter are listed in Table 7.3 for $0 \leq n \leq M$.[13] The gain responses of the Kaiser window and the windowed filter are sketched in Figure 7.24.

7.6.6 Impulse Responses of FIR Filters with a Smooth Transition

We showed earlier that the FIR filter obtained by truncating the infinite-length impulse response of a digital filter developed from a frequency response specification with sharp discontinuities exhibits oscillatory behavior, called the Gibbs phenomenon, in its frequency response. One way of reducing the heights

---

[13]The coefficients of the window have been computed using the M-file `kaiser` (see Section 7.10.4).

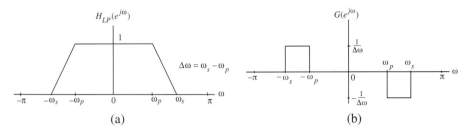

Figure 7.25: (a) Lowpass filter specification with a transition region and (b) the specification of its derivative function.

of the ripples to acceptable values is to truncate the infinite-length impulse response by tapered window functions. Another approach to eliminate the Gibbs phenomenon is by modifying the frequency response specification of the digital filter to have a transition band between the passband and the stopband and to provide a smooth transition between the bands [Orm61]. We now discuss this second approach for lowpass filter design. Similar modifications can be carried out for the design of the other types of filters.

The simplest modification to the zero-phase lowpass filter specification is to provide a transition band between the passband and the stopband responses and to connect these two with a first-order spline function (straight line), as indicated in Figure 7.25(a). An inverse discrete-time Fourier transform of the modified frequency response $H_{LP}(e^{j\omega})$ leads to the expression for its corresponding impulse response coefficients $h_{LP}[n]$. However, as illustrated below, a simpler method is to compute $h_{LP}[n]$ from the inverse discrete-time Fourier transform of the derivative of the specified frequency response $H_{LP}(e^{j\omega})$.

Let $G(e^{j\omega})$ denote $dH_{LP}(e^{j\omega})/d\omega$ with a corresponding inverse DTFT $g[n]$. Its specification will be thus as indicated in Figure 7.25(b). It follows from the differentiation-in-frequency property of the DTFT given in Table 3.2, $h_{LP}[n] = jg[n]/n$. From the inverse DTFT $g[n]$ of $G(e^{j\omega})$ given in Figure 7.25(b), we thus arrive at the impulse response of the modified lowpass filter [Par87]:

$$h_{LP}[n] = \begin{cases} \frac{\omega_c}{\pi}, & n = 0, \\ \frac{2\sin(\Delta\omega n/2)}{\Delta\omega n} \cdot \frac{\sin(\omega_c n)}{\pi n}, & |n| > 0, \end{cases} \tag{7.87}$$

where $\Delta\omega = \omega_s - \omega_p$ and $\omega_c = (\omega_p + \omega_s)/2$.

A still smoother transition between the passband and the stopband of the lowpass filter can be provided by specifying the transition function by a higher-order spline. The corresponding impulse response for a Pth-order spline as the transition function is given by [Bur92], [Par87]

$$h_{LP}[n] = \begin{cases} \frac{\omega_c}{\pi}, & n = 0, \\ \left(\frac{\sin(\Delta\omega n/2P)}{\Delta\omega n/2P}\right)^P \cdot \frac{\sin(\omega_c n)}{\pi n}, & |n| > 0. \end{cases} \tag{7.88}$$

EXAMPLE 7.17   Design a lowpass filter with a passband edge at 0.35π and a stopband edge at 0.45π using a first-order and a second-order spline as the transition functions. The magnitude response of a length-41 filter obtained by truncating the impulse response of Eq. (7.87) is shown by the solid line in Figure 7.26. The magnitude response of a length-61 FIR filter derived by truncating the impulse response of Eq. (7.88) for $P = 2$ (second-order spline) is shown in Figure 7.26 by the dashed line.

As the above example points out, the effect of P on the frequency response of the truncated filter is not that obvious. For a given filter order $2M$ and transition bandwidth $\Delta\omega$, the optimum value of P minimizing the integral-squared approximation error has been shown to be given by [Bur92]

$$P = 1.248(\Delta\omega)M. \tag{7.89}$$

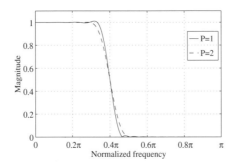

Figure 7.26: Magnitude responses of FIR lowpass filters with smooth transition between bands obtained using spline transition functions.

Various other transition functions have been investigated (see for example, Problem 7.52) [Bur92], [Par87]. These transition functions can also be employed for the design of other types of filters. So far, no specific guidelines have been advanced to select the optimum transition function for a given filter design problem. As a result, it should be selected by a trial-and-error procedure.

7.7 Computer-Aided Design of Digital Filters

The filter design algorithms described in Sections 7.2, 7.5, and 7.6 can be easily implemented on a computer. In addition, a number of filter design algorithms have been advanced that rely on some type of iterative optimization techniques that are used to minimize the error between the desired frequency response and that of the computer-generated filter. In this section, we consider the computer-aided design of FIR and IIR digital filters. First, we describe two specific design approaches based on iterative optimization techniques. A variety of software packages are presently commercially available that have made the design of digital filters rather simple to implement on a computer. In this section, we consider only the application of MATLAB.

The basic idea behind the computer-based iterative technique is as follows. Let $H(e^{j\omega})$ denote the frequency response of the digital transfer function $H(z)$ to be designed so that it approximates the desired frequency response $D(e^{j\omega})$, given as a piecewise linear function of ω, in some sense. The objective is to determine iteratively the coefficients of the transfer function so that the difference between $H(e^{j\omega})$ and $D(e^{j\omega})$ for all values of ω over closed subintervals of $0 \leq \omega \leq \pi$ is minimized. Usually, this difference is specified as a weighted error function $\mathcal{E}(\omega)$ given by

$$\mathcal{E}(\omega) = W(e^{j\omega})\left[H(e^{j\omega}) - D(e^{j\omega})\right], \tag{7.90}$$

where $W(e^{j\omega})$ is some user-specified positive weighting function.

A commonly used approximation measure, called the *Chebyshev* or *minimax criterion*, is to minimize the peak absolute value of the weighted error $\mathcal{E}(\omega)$,

$$\varepsilon = \max_{\omega \in R} |\mathcal{E}(\omega)|, \tag{7.91}$$

where R is the set of disjoint frequency bands in the range $0 \leq \omega \leq \pi$, on which the desired frequency response is defined. In filtering applications, R is composed of the passbands and stopbands of the filter to be designed. For example, for a lowpass filter design, R is the disjoint union of the frequency ranges $[0, \omega_p]$ and $[\omega_s, \pi]$, where ω_p and ω_s are, respectively, the passband edge and the stopband edge.

A second approximation measure, called the *least-p criterion*, is to minimize the integral of pth power of the weighted error function $\mathcal{E}(\omega)$:

$$\varepsilon = \int_{\omega \in R} \left| W(e^{j\omega}) \left(H(e^{j\omega}) - D(e^{j\omega}) \right) \right|^p d\omega, \tag{7.92}$$

over the specified frequency range R with p a positive integer. The *least-squares criterion* obtained from Eq. (7.92) with $p = 2$ is used often for simplicity. If the weighting function $W(e^{j\omega})$ is 1 over the frequency range $[0, \pi]$, we have shown in Section 7.6.1 that the FIR filter obtained by simply truncating the Fourier series of the desired amplitude response $D(e^{j\omega})$ has the least integral-squared error. However, the resulting FIR filter exhibits large peak errors near the bandedges due to the Gibbs phenomenon. Hence, $W(e^{j\omega}) = 1$ is usually not used.

It can be shown that as $p \to \infty$, the least pth solution approaches the minimax solution. In practice, the integral error measure of Eq. (7.92) is approximated by a finite sum given by

$$\varepsilon = \sum_{i=1}^{K} \left\{ W(e^{j\omega_i}) \left(H(e^{j\omega_i}) - D(e^{j\omega_i}) \right) \right\}^p, \tag{7.93}$$

where ω_i, $1 \le i \le K$, is a suitably chosen dense grid of digital angular frequencies. The *least-squares criterion* obtained from Eq. (7.93) with $p = 2$ is used often for simplicity.

In the case of linear-phase FIR filter design, $H(e^{j\omega})$ and $D(e^{j\omega})$ are zero-phase frequency responses. On the other hand, for IIR filter design, these functions are replaced with their magnitude functions. The design objective is thus to iteratively adjust the filter parameters so that ε defined by either Eq. (7.91) or Eq. (7.93) is a minimum.

7.7.1 Design of Equiripple Linear-Phase FIR Filters

The linear-phase FIR filter obtained by minimizing the peak absolute value of the weighted error ε given by Eq. (7.91) is usually called the *equiripple FIR filter*, since here, after ε has been minimized, the weighted error function $\mathcal{E}(\omega)$ exhibits an equiripple behavior in the frequency range of interest. We briefly outline below the basic idea behind the *Parks-McClellan algorithm*, the most widely used and highly efficient algorithm, for designing the equiripple linear-phase FIR filter [Par72].

In Section 4.4.3, we defined the four types of linear-phase FIR filters. The general form of the frequency response $H(e^{j\omega})$ of a causal linear-phase FIR filter of length $N + 1$ is given by

$$H(e^{j\omega}) = e^{-jN\omega/2} e^{j\beta} \breve{H}(\omega), \tag{7.94}$$

where the amplitude response $\breve{H}(\omega)$ is a real function of ω. The weighted error function in this case involves the amplitude response and is given by

$$\mathcal{E}(\omega) = W(\omega) \left[\breve{H}(\omega) - D(\omega) \right], \tag{7.95}$$

where $D(\omega)$ is the desired amplitude response and $W(\omega)$ is a positive weighting function. $W(\omega)$ is chosen to control the relative size of the peak errors in the specified frequency bands. The Parks-McClellan algorithm is based on iteratively adjusting the coefficients of the amplitude response until the peak absolute value of $\mathcal{E}(\omega)$ is minimized.

If the minimum value of the peak absolute value of $\mathcal{E}(\omega)$ in a band $\omega_a \leq \omega \leq \omega_b$ is ε_o, then the absolute error satisfies

$$|\check{H}(\omega) - D(\omega)| \leq \frac{\varepsilon_o}{|W(\omega)|}, \qquad \omega_a \leq \omega \leq \omega_b.$$

In typical filter design applications, the desired amplitude response is given by

$$D(\omega) = \begin{cases} 1, & \text{in the passband,} \\ 0, & \text{in the stopband,} \end{cases}$$

and the amplitude response $\check{H}(\omega)$ is required to satisfy the above desired response with a ripple of $\pm\delta_p$ in the passband, and a ripple of δ_s in the stopband. As a result, it is evident from Eq. (7.95) that the weighting function can be chosen either as

$$W(\omega) = \begin{cases} 1, & \text{in the passband,} \\ \delta_p/\delta_s, & \text{in the stopband,} \end{cases}$$

or as

$$W(\omega) = \begin{cases} \delta_s/\delta_p, & \text{in the passband,} \\ 1, & \text{in the stopband.} \end{cases}$$

By a clever manipulation, the expression for the amplitude response for each of the four types of linear-phase FIR filters can be expressed in the same form and, as a result, basically the same algorithm can be adapted to design any one of the four types of filters. To develop this general form for the amplitude response expression, we consider each of the four types of filters separately.

For the Type 1 linear-phase FIR filter, the amplitude response given by Eq. (4.81) can be rewritten using the notation $N = 2M$ in the form

$$\check{H}(\omega) = \sum_{k=0}^{M} a[k] \cos(\omega k), \tag{7.96}$$

where

$$a[0] = h[M], \qquad a[k] = 2\,h[M-k], \qquad 1 \leq k \leq M. \tag{7.97}$$

For the Type 2 linear-phase FIR filter, the amplitude response given by Eq. (4.86) can be rewritten in the form

$$\check{H}(\omega) = \sum_{k=1}^{(2M+1)/2} b[k] \cos\left(\omega(k - \tfrac{1}{2})\right), \tag{7.98}$$

where

$$b[k] = 2h\left[\tfrac{2M+1}{2} - k\right], \qquad 1 \leq k \leq \tfrac{2M+1}{2}. \tag{7.99}$$

Equation (7.98) can be expressed in the form

$$\check{H}(\omega) = \cos\left(\tfrac{\omega}{2}\right) \sum_{k=0}^{(2M-1)/2} \tilde{b}[k] \cos(\omega k), \tag{7.100}$$

where

$$b[1] = \tfrac{1}{2}\left(\tilde{b}[1] + 2\tilde{b}[0]\right), \tag{7.101}$$

$$b[k] = \tfrac{1}{2}\left(\tilde{b}[k] + \tilde{b}[k-1]\right), \qquad 2 \leq k \leq \tfrac{2M-1}{2},$$

$$b[\tfrac{2M+1}{2}] = \tfrac{1}{2}\tilde{b}\left[\tfrac{2M-1}{2}\right].$$

The amplitude response for the case of the Type 3 linear-phase FIR filter given by Eq. (4.91) can be rewritten in the form

$$\breve{H}(\omega) = \sum_{k=1}^{M} c[k] \sin(\omega k), \tag{7.102}$$

where

$$c[k] = 2 h[M - k], \qquad 1 \le k \le M. \tag{7.103}$$

Equation (7.102) can be expressed in the form

$$\breve{H}(\omega) = \sin \omega \sum_{k=0}^{M-1} \tilde{c}[k] \cos(\omega k), \tag{7.104}$$

where

$$c[1] = \tilde{c}[0] - \tfrac{1}{2}\tilde{c}[1], \tag{7.105}$$
$$c[k] = \tfrac{1}{2} \left(\tilde{c}[k-1] - \tilde{c}[k] \right), \qquad 2 \le k \le M - 1,$$
$$c[M] = \tfrac{1}{2}\tilde{c}[M - 1].$$

Likewise, the amplitude response for the case of the Type 4 linear-phase FIR filter given by Eq. (4.96) can be rewritten in the form

$$\breve{H}(\omega) = \sum_{k=1}^{(2M+1)/2} d[k] \sin \omega (k - \tfrac{1}{2}), \tag{7.106}$$

where

$$d[k] = 2 h[\tfrac{2M+1}{2} - k], \qquad 1 \le k \le \tfrac{2M+1}{2}. \tag{7.107}$$

Equation (7.106) can be expressed in the form

$$\breve{H}(\omega) = \sin(\tfrac{\omega}{2}) \sum_{k=0}^{(2M-1)/2} \tilde{d}[k] \cos(\omega k), \tag{7.108}$$

where

$$d[1] = \tilde{d}[0] - \tfrac{1}{2}\tilde{d}[1], \tag{7.109}$$
$$d[k] = \tfrac{1}{2} \left(\tilde{d}[k-1] - \tilde{d}[k] \right), \qquad 2 \le k \le \tfrac{2M-1}{2},$$
$$d[\tfrac{2M+1}{2}] = \tilde{d}[\tfrac{2M-1}{2}].$$

If we now examine Eqs.(7.96), (7.100), (7.104), and (7.108), we observe that the amplitude response for all four types of linear-phase FIR filters can be expressed in the form

$$\breve{H}(\omega) = Q(\omega) A(\omega), \tag{7.110}$$

where the first factor $Q(\omega)$ is given by

$$Q(\omega) = \begin{cases} 1, & \text{for Type 1,} \\ \cos(\omega/2), & \text{for Type 2,} \\ \sin(\omega), & \text{for Type 3,} \\ \sin(\omega/2), & \text{for Type 4,} \end{cases} \tag{7.111}$$

and the second factor $A(\omega)$ is of the form

$$A(\omega) = \sum_{k=0}^{L} \tilde{a}[k] \cos(\omega k), \tag{7.112}$$

where

$$\tilde{a}[k] = \begin{cases} a[k], & \text{for Type 1,} \\ \tilde{b}[k], & \text{for Type 2,} \\ \tilde{c}[k], & \text{for Type 3,} \\ \tilde{d}[k], & \text{for Type 4,} \end{cases} \tag{7.113}$$

with

$$L = \begin{cases} M, & \text{for Type 1,} \\ \frac{2M-1}{2}, & \text{for Type 2,} \\ M-1, & \text{for Type 3,} \\ \frac{2M-1}{2}, & \text{for Type 4.} \end{cases} \tag{7.114}$$

Substituting Eq. (7.110) in Eq. (7.95) we arrive at a modified form of the weighted approximation function given by

$$\mathcal{E}(\omega) = W(\omega)\left[Q(\omega)A(\omega) - D(\omega)\right] \tag{7.115}$$
$$= W(\omega)Q(\omega)\left[A(\omega) - \frac{D(\omega)}{Q(\omega)}\right].$$

Using the notations $\tilde{W}(\omega) = W(\omega)Q(\omega)$ and $\tilde{D}(\omega) = D(\omega)/Q(\omega)$ we can rewrite the above equation as

$$\mathcal{E}(\omega) = \tilde{W}(\omega)\left[A(\omega) - \tilde{D}(\omega)\right]. \tag{7.116}$$

The optimization problem now becomes the determination of the coefficients $\tilde{a}[k]$, $0 \le k \le L$, which minimize the peak absolute value ε of the weighted approximation error $\mathcal{E}(\omega)$ of Eq. (7.116) over the specified frequency bands $\omega \in R$. After the coefficients $\{\tilde{a}[k]\}$ have been determined, the corresponding coefficients of the original amplitude response are computed from which the filter coefficients are then obtained. For example, if the filter being designed is of Type 2, we observe from Eq. (7.113) that $\tilde{b}[k] = \tilde{a}[k]$, and from Eq. (7.114) $M = (2L + 1)/2$. Knowing $\tilde{b}[k]$ and M, we determine next $b[k]$ using Eq. (7.101). Substituting these values of $b[k]$ in Eq. (7.99) we finally arrive at the filter coefficients $h[n]$. In a similar manner, the filter coefficients for the other three types of FIR filters can be determined from $\tilde{a}[k]$.

Parks and McClellan solved the above problem applying the following theorem from the theory of Chebyshev approximation [Par72]:

Alternation Theorem. The amplitude function $A(\omega)$ of Eq. (7.112) is the best unique approximation of the desired amplitude response obtained by minimizing the peak absolute value ε of $\mathcal{E}(\omega)$ given by Eq. (7.115) if and only if there exist at least $L + 2$ extremal angular frequencies, $\omega_0, \omega_1, \ldots, \omega_{L+1}$, in a closed subset R of the frequency range $0 \le \omega \le \pi$ such that $\omega_0 < \omega_1 < \cdots < \omega_L < \omega_{L+1}$ and $\mathcal{E}(\omega_i) = -\mathcal{E}(\omega_{i+1})$ with $|\mathcal{E}(\omega_i)| = \varepsilon$ for all i in the range $0 \le i \le L + 1$.

Let us examine the behavior of the amplitude response for a Type 1 equiripple lowpass FIR filter whose approximation error $\mathcal{E}(\omega)$ satisfies the condition of the above theorem. The peaks of $\mathcal{E}(\omega)$ are at

$\omega = \omega_i$, $0 \leq i \leq L + 1$, where $\frac{d\mathcal{E}(\omega)}{d\omega} = 0$. Since in the passband and the stopband, $\tilde{W}(\omega)$ and $\tilde{D}(\omega)$ are piecewise constant, it follows from Eq. (7.116) that

$$\frac{d\mathcal{E}(\omega)}{d\omega} = \frac{dA(\omega)}{d\omega} = 0,$$

at $\omega = \omega_i$, or in other words, the amplitude response $A(\omega)$ also has peaks at $\omega = \omega_i$. Using the relation

$$\cos(\omega k) = T_k(\cos \omega),$$

where $T_k(x)$ is the kth order Chebyshev polynomial defined by

$$T_k(x) = \cos(k \cos^{-1} x),$$

the amplitude response $A(\omega)$ given by Eq. (7.112) can be expressed as a power series in $\cos \omega$,

$$A(\omega) = \sum_{k=0}^{L} \alpha[k] (\cos \omega)^k,$$

which is seen to be an Lth order polynomial in $\cos \omega$. As a result, $A(\omega)$ can have at most $L - 1$ local minima and maxima inside the specified passband and stopband. Moreover, at the bandedges, $\omega = \omega_p$ and $\omega = \omega_s$, $|\mathcal{E}(\omega)|$ is a maximum, and hence $A(\omega)$ has extrema at these angular frequencies. In addition, $A(\omega)$ may also have extrema at $\omega = 0$ and $\omega = \pi$. Therefore, there are at most, $L + 3$ extremal frequencies of $\mathcal{E}(\omega)$. Similarly, in the case of a linear-phase FIR filter with K specified bandedges and designed using the Remez algorithm, there can be at most $L + K + 1$ extremal frequencies. An equiripple linear-phase FIR filter with more than $L + 2$ extremal frequencies has been called an *extra-ripple filter*.

To arrive at the optimum solution we need to solve the set of $L + 2$ equations

$$\tilde{W}(\omega_i)[A(\omega_i) - \tilde{D}(\omega_i)] = (-1)^i \varepsilon, \qquad 0 \leq i \leq L + 1, \tag{7.117}$$

for the unknowns $\tilde{a}[i]$ and ε, provided the $L + 2$ extremal angular frequencies are known. To this end, Eq. (7.117) is rewritten in a matrix form as

$$\begin{bmatrix} 1 & \cos(\omega_0) & \cdots & \cos(L\omega_0) & -1/\tilde{W}(\omega_0) \\ 1 & \cos(\omega_1) & \cdots & \cos(L\omega_1) & 1/\tilde{W}(\omega_1) \\ \vdots & \vdots & \ddots & \vdots & \vdots \\ 1 & \cos(\omega_L) & \cdots & \cos(L\omega_L) & (-1)^{L-1}/\tilde{W}(\omega_L) \\ 1 & \cos(\omega_{L+1}) & \cdots & \cos(L\omega_{L+1}) & (-1)^L/\tilde{W}(\omega_{L+1}) \end{bmatrix} \begin{bmatrix} \tilde{a}[0] \\ \tilde{a}[1] \\ \vdots \\ \tilde{a}[L] \\ \varepsilon \end{bmatrix}$$

$$= \begin{bmatrix} \tilde{D}(\omega_0) \\ \tilde{D}(\omega_1) \\ \vdots \\ \tilde{D}(\omega_L) \\ \tilde{D}(\omega_{L+1}) \end{bmatrix}, \tag{7.118}$$

which can, in principle, be solved for the unknowns if the locations of the $L + 2$ extremal frequencies are known a priori. The *Remez exchange algorithm*, a highly efficient iterative procedure, is used to determine the locations of the extremal frequencies and consists of the following steps at each iteration stage:

Step 1: A set of initial values for the extremal frequencies are either chosen or are available from the completion of the previous iteration.

Step 2: The value ε is then computed by solving Eq. (7.118), resulting in the expression

$$\varepsilon = \frac{c_0 \tilde{D}(\omega_0) + c_1 \tilde{D}(\omega_1) + \cdots + c_{L+1} \tilde{D}(\omega_{L+1})}{\frac{c_0}{\tilde{W}(\omega_0)} - \frac{c_1}{\tilde{W}(\omega_1)} + \cdots + \frac{(-1)^{L+1} c_{L+1}}{\tilde{W}(\omega_{L+1})}}, \tag{7.119}$$

where the constant c_n is given by

$$c_n = \prod_{\substack{i=0 \\ i \neq n}}^{L+1} \frac{1}{\cos(\omega_n) - \cos(\omega_i)} \tag{7.120}$$

Step 3: The values of the amplitude response $A(\omega)$ at $\omega = \omega_i$ are then computed using

$$A(\omega_i) = \frac{(-1)^i \varepsilon}{\tilde{W}(\omega_i)} + \tilde{D}(\omega_i), \qquad 0 \leq i \leq L + 1.$$

Step 4: The polynomial $A(\omega)$ is determined by interpolating the above values at the $L + 2$ extremal frequencies using the Lagrange interpolation formula:

$$A(\omega) = \sum_{i=0}^{L+1} A(\omega_i) \, P_i(\cos \omega),$$

where

$$P_i(\cos \omega) = \prod_{\substack{\ell=0, \\ \ell \neq i}}^{L+1} \left(\frac{\cos \omega - \cos \omega_\ell}{\cos \omega_i - \cos \omega_\ell} \right), \qquad 0 \leq i \leq L + 1.$$

Step 5: The new weighted error function $\mathcal{E}(\omega)$ of Eq. (7.116) is computed at a dense set S ($S \geq L$) of frequencies. In practice, $S = 16L$ is adequate. Determine the $L + 2$ new extremal frequencies from the values of $\mathcal{E}(\omega)$ evaluated at the dense set of frequencies.

Step 6: If the peak values ε of $\mathcal{E}(\omega)$ are equal in magnitude, the algorithm has converged. Otherwise, go back to Step 2.

Figure 7.27 demonstrates how the weighted error function changes from one iteration stage to the next and how the new extremal frequencies are determined. Finally, the iteration process is stopped after the difference between the value of the peak error ε calculated at any stage and that at the previous stage is below a preset threshold value, such as 10^{-6}. In practice, the process converges after very few iterations. The basic principle of the Remez exchange algorithm is illustrated in the following example [Par87].

EXAMPLE 7.18 Let the desired function be a quadratic function $D(x) = 1.1x^2 - 0.1$ defined for the range $0 \leq x \leq 2$. We wish to approximate $D(x)$ by a linear function $a_1 x + a_0$ by minimizing the peak value of the absolute error, i.e.,

$$\max_{x \in [0,2]} \left| 1.1x^2 - 0.1 - a_0 - a_1 x \right|. \tag{7.121}$$

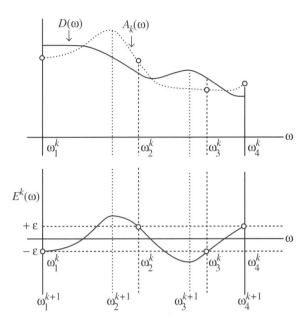

Figure 7.27: Plots of the desired response $D(\omega)$, the amplitude response $A_k(\omega)$, and the error $E^k(\omega)$ at the end of the kth iteration. The locations of the new extremal frequencies are given by ω_i^{k+1}.

Since there are three unknowns, a_0, a_1, and ε, we need three extremal points on x which we arbitrarily choose as $x_1 = 0$, $x_2 = 0.5$, and $x_3 = 1.5$. We then solve the three linear equations

$$a_0 + a_1 x_\ell - (-1)^\ell \varepsilon = D(x_\ell), \qquad \ell = 1, 2, 3, \tag{7.122}$$

which leads to

$$\begin{bmatrix} 1 & 0 & 1 \\ 1 & 0.5 & -1 \\ 1 & 1.5 & 1 \end{bmatrix} \begin{bmatrix} a_0 \\ a_1 \\ \varepsilon \end{bmatrix} = \begin{bmatrix} -0.1 \\ 0.175 \\ 2.375 \end{bmatrix}, \tag{7.123}$$

for the given extremal points. The above matrix equation, when solved, yields

$$a_0 = -0.375, \qquad a_1 = 1.65, \qquad \varepsilon = 0.275. \tag{7.124}$$

Figure 7.28(a) shows the plot of the corresponding error $\mathcal{E}_1(x) = 1.1x^2 - 1.65x + 0.275$, along with the values of error at the chosen extremal points. As expected, at the extremal points, the errors are equal in magnitude and alternate in sign.

The next set of extremal points are those points where $\mathcal{E}_1(x)$ assumes its maximum absolute values. These extremal points are given by $x_1 = 0$, $x_2 = 0.75$, and $x_3 = 2$. The new values of the unknowns are obtained by solving

$$\begin{bmatrix} 1 & 0 & 1 \\ 1 & 0.75 & -1 \\ 1 & 2 & 1 \end{bmatrix} \begin{bmatrix} a_0 \\ a_1 \\ \varepsilon \end{bmatrix} = \begin{bmatrix} -0.1 \\ 0.5188 \\ 4.3 \end{bmatrix}, \tag{7.125}$$

which yields

$$a_0 = -0.6156, \qquad a_1 = 2.2, \qquad \varepsilon = 0.5156. \tag{7.126}$$

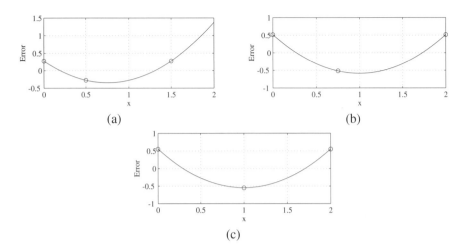

Figure 7.28: Illustration of the Remez algorithm.

The plot of the new error $\mathcal{E}_2(x) = 1.1x^2 - 2.2x + 0.5156$ and the values of the error at the new set of extremal points are shown in Figure 7.28(b).

We continue the process and determine the new set of extremal points, where $\mathcal{E}_2(x)$ takes on the maximum absolute values. These are now given by $x_1 = 0$, $x_2 = 1$, and $x_3 = 2$. Using these extremal points we determine the next set of values for the unknowns by solving

$$\begin{bmatrix} 1 & 0 & 1 \\ 1 & 1 & -1 \\ 1 & 2 & 1 \end{bmatrix} \begin{bmatrix} a_0 \\ a_1 \\ \varepsilon \end{bmatrix} = \begin{bmatrix} -0.1 \\ 1.0 \\ 4.3 \end{bmatrix}, \tag{7.127}$$

which results in

$$a_0 = -0.65, \quad a_1 = 2.2, \quad \varepsilon = 0.55. \tag{7.128}$$

Figure 7.28(c) depicts the new error $\mathcal{E}_3(x) = 1.1x^2 - 2.2x + 0.55$ and the values of the error at the new extremal points. This time the algorithm has converged as ε is also the maximum value of the absolute error.

7.8   Design of FIR Digital Filters with Least-Mean-Square Error

For the design of a linear-phase FIR filter with a minimum mean-square error criterion, the error measure of Eq. (7.93) reduces to

$$\varepsilon = \sum_{i=1}^{K} \left\{ W(\omega_i) \left[\breve{H}(\omega_i) - D(\omega_i) \right] \right\}^2, \tag{7.129}$$

where $\breve{H}(\omega)$ is the amplitude response of the designed filter, $D(\omega)$ is the desired amplitude response, and $W(\omega)$ is the weighting function. Now, as shown in Section 7.7.1, the amplitude response for all four types of linear-phase FIR filters can be expressed in the form

$$\breve{H}(\omega) = Q(\omega) \sum_{k=0}^{L} \tilde{a}[k] \cos(\omega k), \tag{7.130}$$

where $Q(\omega)$, $\tilde{a}[k]$, and L are given by Eqs. (7.111), (7.113), and (7.114), respectively. Hence, the mean-square error of Eq. (7.129) is a function of the filter parameters $\tilde{a}[k]$. To arrive at the minimum value of ε, we set

$$\frac{\partial \varepsilon}{\partial \tilde{a}[k]} = 0, \qquad 0 \le k \le L,$$

which results in a set of $(L + 1)$ linear equations that can be solved for $\tilde{a}[k]$.

Without any loss of generality, we consider here the design of a Type 1 linear-phase FIR filter. In this case, $Q(\omega) = 1$, $\tilde{a}[k] = a[k]$, and $L = M$. The expression for the mean-square error then takes the form

$$\varepsilon = \sum_{i=1}^{K} \left\{ W(\omega_i) \left[\sum_{k=0}^{M} a[k] \cos(\omega_i k) - D(\omega_i) \right] \right\}^2$$

$$= \sum_{i=1}^{K} \left\{ \sum_{k=0}^{M} W(\omega_i) a[k] \cos(\omega_i k) - W(\omega_i) D(\omega_i) \right\}^2. \tag{7.131}$$

Using the notation,

$$\mathbf{H} = \begin{bmatrix} W(\omega_1) & W(\omega_1)\cos(\omega_1) & \cdots & W(\omega_1)\cos(M\omega_1) \\ W(\omega_2) & W(\omega_2)\cos(\omega_2) & \cdots & W(\omega_2)\cos(M\omega_2) \\ \vdots & \vdots & \ddots & \vdots \\ W(\omega_K) & W(\omega_K)\cos(\omega_K) & \cdots & W(\omega_K)\cos(M\omega_K) \end{bmatrix},$$

$$\mathbf{a} = [\, a[0]\, a[1]\, \cdots\, a[M]\,]^T,$$

and

$$\mathbf{d} = [\, W(\omega_1) D(\omega_1)\, W(\omega_2) D(\omega_2)\, \cdots\, W(\omega_K) D(\omega_K)\,]^T,$$

we can express Eq. (7.131) in the form

$$\varepsilon = \mathbf{e}^T \mathbf{e},$$

where

$$\mathbf{e} = \mathbf{H}\mathbf{a} - \mathbf{d}.$$

The minimum mean-square solution is then obtained by solving the *normal equations* [Par87]:

$$\mathbf{H}^T \mathbf{H} \mathbf{a} = \mathbf{H}^T \mathbf{d}.$$

If $K \ge M$, which is typically the case, the above equation should be solved using an iterative method such as the *Levinson-Durbin algorithm* [Lev47], [Dur59], as the direct solution is often ill-conditioned. A similar formulation can be carried out for the other three types of linear-phase FIR filters. Note that the design approach outlined here can be used to design a linear-phase FIR filter meeting any arbitrarily shaped desired response.

7.9 Constrained Least-Square Design of FIR Digital Filters

For many specialized filter design problems, the amplitude response $\breve{H}(\omega)$ of the FIR filter is required to satisfy some side constraints. For example, it might be desired that the frequency response have a null at a specified frequency ω_o. FIR filters with constraints on their frequency response can be designed using the least-mean-squares approach by incorporating the constraints into the design algorithm. To illustrate this approach, assume, without any loss of generality, that the filter to be designed is a Type 1 linear-phase FIR filter of order $N = 2M$ with an amplitude response given by Eq. (7.96) which is constrained to have a null at ω_o. This can be written as a single equality constraint $\mathbf{Ga} = \mathbf{d}$, where

$$
\begin{aligned}
\mathbf{G} &= [1, \; \cos(\omega_o), \; \cos(2\omega_o), \; \cdots, \cos(M\omega_o)], \\
\mathbf{a} &= [a[0], \; a[1], \; \cdots, \; a[M]]^T \\
\mathbf{d} &= [0].
\end{aligned}
\tag{7.132}
$$

In the general case, if there are r constraints, then $\mathbf{G}$ is an $r \times (M+1)$ matrix and $\mathbf{d}$ is an $r \times 1$ column vector. Such a filter can be designed by using the constrained least-square method. This method minimizes the square error

$$
\varepsilon = \left(\frac{1}{\pi} \int_0^\pi W(\omega)[\breve{H}(\omega) - D(\omega)]^2 \, d\omega \right)^{1/2},
\tag{7.133}
$$

subject to the side constraints

$$
\mathbf{Ga} = \mathbf{d}.
\tag{7.134}
$$

In Eq. (7.133), as before, $D(\omega)$ is the desired amplitude response and $W(\omega)$ is the weighting function. The side constraints of Eq. (7.134) need not be linear, but the solution is more easily obtained if they are.

To minimize ε^2 subject to the constraints, we first form the Lagrangian:

$$
\Phi = \varepsilon^2 + \boldsymbol{\mu}^T \cdot [\mathbf{Ga} - \mathbf{d}],
\tag{7.135}
$$

where

$$
\boldsymbol{\mu} = [\mu_1, \; \mu_2, \; \ldots, \mu_r]^T
$$

is the vector of the so-called Lagrange multipliers. We can derive the necessary conditions for the minimization of ε^2 by setting the derivatives of Φ with respect to the filter parameters $a[k]$ and the Lagrange multipliers μ_i to zero resulting in the following equations:

$$
\begin{aligned}
\mathbf{Ra} + \mathbf{G}^T \boldsymbol{\mu} &= \mathbf{c}, \\
\mathbf{Ga} &= \mathbf{d},
\end{aligned}
\tag{7.136}
$$

where the coefficients of the vector $\mathbf{c} = [c[0], \; c[1], \; \ldots, \; c[M]]$ are given by

$$
\begin{aligned}
c[0] &= \frac{1}{\pi} \int_0^\pi W(\omega)D(\omega) \, d\omega, \\
c[k] &= \frac{1}{\pi} \int_0^\pi W(\omega)D(\omega) \cos(k\omega) \, d\omega, \qquad 1 \le k \le M,
\end{aligned}
$$

and the (i, k)th element $R_{i,k}$ of the matrix $\mathbf{R}$ is given by

$$
R_{i,k} = \int_0^\pi W(\omega) \cos(i\omega) \cos(k\omega) \, d\omega.
$$

The two matrix equations of Eq. (7.136) can be written as

$$
\begin{bmatrix} \mathbf{R} & \mathbf{G}^T \\ \mathbf{G} & \mathbf{0} \end{bmatrix} \begin{bmatrix} \mathbf{a} \\ \boldsymbol{\mu} \end{bmatrix} = \begin{bmatrix} \mathbf{c} \\ \mathbf{d} \end{bmatrix}.
\tag{7.137}
$$

Solving the above equation we get

$$\mu = (\mathbf{G}\,\mathbf{R}^{-1}\mathbf{G}^T)^{-1}(\mathbf{G}\,\mathbf{R}^{-1}\mathbf{c} - \mathbf{d}),$$
$$\mathbf{a} = \mathbf{R}^{-1}(\mathbf{c} - \mathbf{G}^T\mu). \tag{7.138}$$

When the integrals needed to form $\mathbf{R}$ and $\mathbf{c}$ cannot be calculated simply, then $\mathbf{R}$ and $\mathbf{c}$ can be approximated using the discrete forms

$$\mathbf{R} \cong \mathbf{H}^T\mathbf{H}, \qquad \mathbf{c} \cong \mathbf{H}^T\mathbf{d},$$

as in Section 7.8.

In the special case when the error function is not weighted, i.e., $W(\omega) = 1$, $\mathbf{R}$ becomes an identity matrix, and the c_i are simply the coefficients of the Fourier series expansion of $D(\omega)$. As a result, Eq. (7.138) reduces to

$$\mu = (\mathbf{G}\,\mathbf{G}^T)^{-1}(\mathbf{G}\,\mathbf{c} - \mathbf{d}),$$
$$\mathbf{a} = \mathbf{c} - \mathbf{G}^T\mu.$$

One useful application of the constrained least-square approach is the design of filters by a criterion that takes into account both the square error and the peak-ripple error (or Chebyshev error). The constrained least-square approach to filter design allows a compromise between the square error and the Chebyshev criteria, and produces the filter with least-square error and the best Chebyshev error filter as special cases.

The least-square error filter design is based on the assumption that the size of the peak error can be ignored, whereas filter design according to the Chebyshev norm assumes the integral-squared error is irrelevant. In practice, however, both of these criteria are often important [Ada91]. The design problem can thus be formulated as the minimization of the square error subject to constraints on the peak error; and the solution can be obtained by an iterative constrained least-square algorithm.

In all of the FIR filter design methods outlined earlier, the frequency response specifications include transition (or "don't care") bands to either reduce the oscillations near the bandedges due to the Gibbs phenomenon for least-squares approximation or to allow the use of the minimax approximation. There are applications where the spectrum of the desired signal is in a narrowband of the frequency range $0 \le \omega \le \pi$, while the spectrum of the interfering signal (or noise) occupies the whole range, and there is no transition band separating the two spectra.

The constrained least-square filter design method can be used to design both linear-phase and minimum-phase FIR filters without specifying explicitly the transition bands [Sel96]. It minimizes the weighted integral-square error of Eq. (7.133) over the whole frequency range such that the local minima and maxima of $\check{H}(\omega)$ remain within the specified lower and upper bound functions $L(\omega)$ and $U(\omega)$. As ε defined above is simply the $\mathcal{L}_2$ norm of the error function $[\check{H}(\omega) - D(\omega)]$, it has also been referred to as the $\mathcal{L}_2$ error. For lowpass filter design with a cutoff frequency ω_o, the functions $L(\omega)$ and $U(\omega)$ are defined by

$$L(\omega) = 1 - \delta_p, \ \ U(\omega) = 1 + \delta_p, \quad \text{for } 0 \le \omega \le \omega_o,$$
$$L(\omega) = -\delta_s, \ \ U(\omega) = \delta_s, \quad \text{for } \omega_o \le \omega \le \pi. \tag{7.139}$$

Because this design problem has inequality constraints, an iterative algorithm is employed to minimize the error ε of Eq. (7.133) subject to the constraints on the values of $\check{H}(\omega_i)$, where the frequency points ω_i are contained in a constraint set $S = \{\omega_1, \omega_2, \ldots, \omega_m\}$ with $\omega_i \in [0, \pi]$. Let the set S be partitioned into two sets, the set S_ℓ containing the frequency points ω_i, $1 \le i \le q$, where the equality constraint

$$\check{H}(\omega) = L(\omega)$$

is imposed, and the set S_u containing the frequency points ω_i, $q+1 \le i \le m$, where the equality constraint

$$\check{H}(\omega) = U(\omega)$$

is imposed. Then the equality constrained problem is solved on each iteration.

When the Lagrange multipliers are all nonnegative, Kuhn-Tucker conditions [Fle87] state that the solution of the equality constrained problem minimizes ε of Eq. (7.133) while satisfying the inequality constraints:

$$\breve{H}(\omega_i) \geq L(\omega_i), \quad 1 \leq i \leq q,$$
$$\breve{H}(\omega_i) \leq U(\omega_i), \quad q+1 \leq i \leq m.$$

The constrained least-square design algorithm therefore consists of the following steps:

Step 1: Initialization. Choose the constraint set to be an empty set, i.e., $S = \emptyset$.

Step 2: Minimization with Equality Constraints. Solve Eq. (7.138) for the Lagrange multipliers by minimizing the mean-square error ε of Eq. (7.133) satisfying the equality constraints $\breve{H}(\omega_i) = L(\omega_i)$ for $\omega_i \in S_\ell$ and $\breve{H}(\omega_i) = U(\omega_i)$ for $\omega_i \in S_u$.

Step 3: Kuhn-Tucker Conditions. If a Lagrange multiplier μ_j is negative, then remove the corresponding frequency ω_j from the constrained set S, and return to Step 2. Otherwise, calculate the coefficients $a[k]$ using Eq. (7.138) and proceed to Step 4.

Step 4: Multiple Exchange of Constraint Set. Set the constraint set S equal to $S_\ell \cup S_u$. Note that at the frequency points ω_i in S_ℓ, $\partial \breve{H}(\omega)/\partial\omega|_{\omega=\omega_i} = 0$ and $\breve{H}(\omega_i) \leq L(\omega_i)$. Likewise, at the frequency points ω_i in S_u, $\partial \breve{H}(\omega)/\partial\omega|_{\omega=\omega_i} = 0$ and $\breve{H}(\omega_i) \geq U(\omega_i)$.

Step 5: Convergence Check. The algorithm converges if $\breve{H}(\omega) \geq L(\omega) - \Delta$ for all ω_i in S_ℓ, and if $\breve{H}(\omega) \leq U(\omega) + \Delta$ for all ω_i in S_u. Otherwise, go back to Step 2.

In Step 5, Δ is a very small number, typically 10^{-6}, chosen a priori based on the desired numerical accuracy. For an additional discussion on the algorithm and its properties, see [Sel96], [Sel98].

7.10   Digital Filter Design Using MATLAB

The *Signal Processing Toolbox* of MATLAB includes a variety of M-files for the design of both IIR and FIR digital filters. We illustrate the use of some of these functions in this section.

7.10.1   IIR Digital Filter Design Using MATLAB

The IIR digital filter design process involves two steps. In the first step, the filter order N and the frequency scaling factor Wn are determined from the given specifications. Using these parameters and the specified ripples, the coefficients of the transfer function are then determined in the next step. We describe the MATLAB implementation of these two steps below.

Order Estimation

For IIR digital filter design using the bilinear transformation method, the MATLAB statements to use are as follows:

```
[N,Wn] = buttord(Wp,Ws,Rp,Rs)
[N,Wn] = cheb1ord(Wp,Ws,Rp,Rs)
[N,Wn] = cheb2ord(Wp,Ws,Rp,Rs)
[N,Wn] = ellipord(Wp,Ws,Rp,Rs)
```

For lowpass filters, Wp and Ws are the normalized passband and stopband edge frequencies, respectively, with Wp < Ws. These frequency points must be between 0 and 1, where the sampling frequency is equal to 2. If the sampling frequency FT, the passband edge frequency Fp, and the stopband edge frequency Fs are specified in Hz, then Wp = 2Fp/FT and Ws = 2Fs/FT. The other two parameters, Rp and Rs, are the passband ripple and the minimum stopband attenuation specified in dB, respectively. The outputs of these functions are the filter order N and the frequency scaling factor Wn. Both of these two parameters are needed in the MATLAB functions for filter design meeting the desired specifications. For the Butterworth filter Wn is the 3-dB cutoff frequency. For a Type 1 Chebyshev filter and an elliptic filter, Wn is the passband edge frequency, whereas for a Type 2 Chebyshev filter, it is the stopband edge frequency.

For highpass filters Wp > Ws. For bandpass and bandstop digital filters, Wp and Ws are vectors of length-2 specifying the transition bandedges, with the lower-frequency edge being the first element of the vectors. In the latter two cases, Wn is also a length-2 vector, and N is half of the order of the filter to be designed.

The use of the M-files for order estimation is illustrated in the next two examples.

EXAMPLE 7.19   Determine the minimum order of the transfer function of a Type 2 Chebyshev digital highpass filter operating at a sampling rate of 4 kHz with the following specifications: passband edge at 1000 Hz, stopband edge at 600 Hz, passband ripple of 1 dB, and minimum stopband attenuation of 40 dB.

The normalized passband edge Wp is $2 \times 1000/4000 = 0.5$ and the normalized stopband edge Ws is $2 \times 600/4000 = 0.3$. For the design using the bilinear transformation method, we use the statement

```
[N,Wn] = cheb2ord(0.5,0.3,1,40);
```

which yields N = 5 and Wn = 0.3224.

EXAMPLE 7.20   Determine the minimum order of an elliptic bandpass filter operating at a sampling rate of 1600 Hz with the following specifications: passband edges at 200 Hz and 280 Hz, stopband edges at 160 Hz and 300 Hz, passband ripple of 0.1 dB, and minimum stopband attenuation of 70 dB.

The normalized passband edges are $2(200)/1600 = 0.25$ and $2(280)/1600 = 0.35$. Hence, Wp = [0.25 0.35]. Likewise, the normalized stopband edges are at $2(160)/1600 = 0.2$ and $2(300)/1600 = 0.375$ implying Ws = [0.2   0.375]. For the design using the bilinear transformation method, we use the statement

```
[N,Wn] = ellipord([0.25 0.35], [0.2 0.375],0.1, 70);
```

to determine the minimum order of the filter. Execution of the above statement results in N = 8 and Wn = [0.25 0.35]. Note that the order N = 8 here is the order of the prototype lowpass filter. The order of the bandpass filter is $2N = 16$.

Filter Design

For IIR filter design based on the bilinear transformation, the *Signal Processing Toolbox* of MATLAB includes functions for each one of the four magnitude approximation techniques, i.e., Butterworth, Type 1 and 2 Chebyshev, and elliptic approximations. Specifically, the following M-files are available:

```
[b,a] = butter(N,Wn)
[b,a] = cheby1(N,Rp,Wn)
[b,a] = cheby2(N,Rs,Wn)
[b,a] = ellip(N,Rp,Rs,Wn)
```

These functions directly determine the digital lowpass filter transfer function of order N with a frequency scaling factor Wn that must be a number between 0 and 1 with a sampling frequency assumed to be equal to 2 Hz. These two parameters are those determined in the order estimation stage. The additional parameters for the Chebyshev and the elliptic filters are Rp, specifying the passband ripple in dB, and Rs, specifying the minimum stopband attenuation in dB. The output files of these functions are the length N+1 column vectors b and a providing, respectively, the numerator and denominator coefficients in ascending powers of z^{-1}. The form of the transfer function obtained is given by

$$G(z) = \frac{B(z)}{A(z)} = \frac{b(1) + b(2)z^{-1} + \cdots + b(N+1)z^{-N}}{1 + a(2)z^{-1} + \cdots + a(N+1)z^{-N}}. \tag{7.140}$$

After these coefficients have been computed, the frequency response can be computed using the M-file freqz(b,a,w), where w is a set of specified angular frequencies. [14] The function freqz(b,a,w) generates a complex vector of frequency response samples from which magnitude and/or phase response samples can be readily computed.

The following four IIR lowpass filter design functions can be employed to determine the zeros and poles of the transfer function:

```
[z,p,k]  =  butter(N,Wn)
[z,p,k]  =  cheby1(N,Rp,Wn)
[z,p,k]  =  cheby2(N,Rs,Wn)
[z,p,k]  =  ellip(N,Rp,Rs,Wn)
```

The output files of these functions are the zeros and poles of the transfer function given as the N-length vectors z and p, and the scalar gain factor k. The numerator and denominator coefficients of the transfer function can then be determined using the function zp2tf.

We illustrate the design of a digital lowpass filter using MATLAB in the next example.

EXAMPLE 7.21   Determine the transfer function and plot the gain response of an elliptic IIR lowpass filter with the following specifications: passband edge $F_p = 800$ Hz, stopband edge $F_s = 1$ kHz, passband ripple of 0.5 dB, minimum stopband attenuation of 40 dB, and sampling rate $F_T = 4$ kHz.

From Eqs. (7.7) and (7.8), we arrive at the normalized bandedges given by $\omega_p = 2\pi F_p/F_T = 0.4\pi$ and $\omega_s = 2\pi F_s/F_T = 0.5\pi$. The following MATLAB program is next used to design the above filter.

```
% Program 7_1
% Elliptic IIR Lowpass Filter Design
%
Wp = input('Normalized passband edge = ');
Ws = input('Normalized stopband edge = ');
Rp = input('Passband ripple in dB = ');
Rs = input('Minimum stopband attenuation in dB = ');
[N,Wn] = ellipord(Wp,Ws,Rp,Rs)
[b,a] = ellip(N,Rp,Rs,Wn);
[h,omega] = freqz(b,a,256);
plot (omega/pi,20*log10(abs(h)));grid;
xlabel('\omega/\pi'); ylabel('Gain, dB');
title('IIR Elliptic Lowpass Filter');
```

As the program is run, it asks for the filter specifications to be typed in. It first computes the minimum order N of the filter and the desired cutoff frequency Wn necessary to meet the given specifications. In the elliptic filter case, Wn = Wp = 0.4. The gain response plot generated is shown in Figure 7.29. The numerator and the denominator coefficients of the transfer function can be obtained by printing in the MATLAB Command Window b and a.

---

[14]The range of frequencies w in freqz(b,a,w) should be between 0 and π, whereas the range of frequencies in any of the filter functions are between 0 and 1.

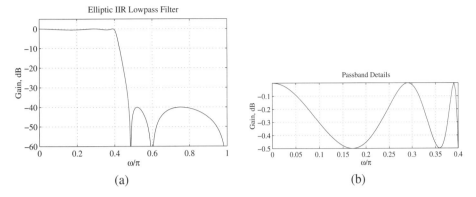

Figure 7.29: IIR elliptic lowpass filter of Example 7.21: (a) gain response, and (b) passband details.

Other types of digital filters can be designed by simple modifications of the filter function commands given in Program 7_1. Bandpass and bandstop digital filters of order 2N are designed by using Wn as a two-element vector, Wn = [w1 w2], where the frequency range w1 < ω < w2 is the passband for a bandpass filter and the stopband for a bandstop filter. For designing highpass digital filters, the string high is used as the last argument for the filter function, e.g.,

$$[b,a] = \text{cheby1}(N,Rp,Wn,'high')$$

Similarly, the string stop is included as a final argument for designing bandstop digital filters, e.g.,

$$[b,a] = \text{ellip}(N,Rp,Rs,Wn,'stop')$$

The following two examples illustrate the design of a highpass and a bandpass digital filter, respectively.

EXAMPLE 7.22   Design a Type 1 Chebyshev IIR highpass filter with normalized passband edge at 0.7 Hz, normalized stopband edge at 0.5 Hz, passband ripple of 1 dB, and minimum stopband attenuation of 32 dB. The sampling frequency is assumed to be 2 Hz.

Program 7_2 below provides the MATLAB program that can be used to design such a filter. During execution the program asks for the filter specifications. It first determines the minimum filter order N and the cutoff frequency Wn of the highpass filter neeeded to meet the specifications. In this case, Wn = Wp = 0.7.

```
% Program 7_2
% Type 1 Chebyshev IIR Highpass Filter Design
%
Wp = input('Normalized passband edge = ');
Ws = input('Normalized stopband edge = ');
Rp = input('Passband ripple in dB = ');
Rs = input('Minimum stopband attenuation in dB = ');
[N,Wn] = cheb1ord(Wp,Ws,Rp,Rs)
[b,a] = cheby1(N,Rp,Wn,'high');
[h,omega] = freqz(b,a,256);
plot (omega/pi,20*log10(abs(h)));grid;
xlabel('\omega/\pi'); ylabel('Gain, dB');
title('Type I Chebyshev Highpass Filter');
```

The gain response generated by this program is shown in Figure 7.30. The transfer function coefficients can be obtained by displaying the numerator coefficient vector b and the denominator coefficient vector a.

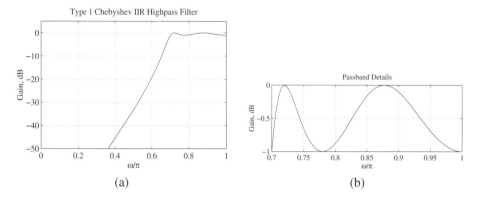

Figure 7.30: IIR Type 1 Chebyshev highpass filter of Example 7.22: (a) Gain response, and (b) passband details.

EXAMPLE 7.23   Design an IIR Butterworth bandpass filter with the following specifications: normalized passband edges at 0.45 and 0.65, normalized stopband edges at 0.3 and 0.75, passband ripple of 1 dB, and minimum stopband attenuation of 40 dB.

The MATLAB program for designing such a filter is given in Program 7_3.

```
% Program 7_3
% Design of IIR Butterworth Bandpass Filter
%
Wp = input('Passband edge frequencies = ');
Ws = input('Stopband edge frequencies = ');
Rp = input('Passband ripple in dB = ');
Rs = input('Minimum stopband attenuation = ');
[N,Wn] = buttord(Wp, Ws, Rp, Rs);
[b,a] = butter(N,Wn);
[h,omega] = freqz(b,a,256);
gain = 20*log10(abs(h));
plot (omega/pi,gain);grid;
xlabel('\omega/\pi'); ylabel('Gain, dB');
title('IIR Butterworth Bandpass Filter');
```

The input data are the vector of passband edges $Wp = [0.45 \ 0.65]$, the vector of stopband edges $Ws = [0.3 \ 0.75]$, the passband ripple $Rp = 1$, and the minimum stopband attenuation $Rs = 40$. The computed gain response is plotted in Figure 7.31. The transfer function coefficients can be obtained by displaying the numerator coefficient vector b and the denominator coefficient vector a. It should be noted that the parameter N used here is half the filter order. The filter designed above has $N = 6$, i.e., it is a 12th-order bandpass filter.

For the design of higher-order filters, the functions computing the zeros and poles of the transfer function are more accurate than the functions computing the transfer function coefficients. Moreover, numerical accuracy could be a problem in designing filters of order 15 or above [Kra94].

7.10.2   FIR Digital Filter Order Estimation Using MATLAB

As in the case of IIR digital filter design, the FIR digital filter design process also consists of two steps. In the first step, the filter order is estimated from the given specifications. In the second step, the coefficients of the transfer function are determined using the estimated order and the filter specifications. We consider

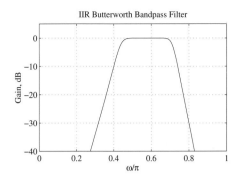

Figure 7.31: Gain response of the IIR Butterworth bandpass filter of Example 7.23.

in this section the order estimation problem. In the next two sections we treat the FIR filter design problem.

Either of the two formulas of Eqs. (7.15) and (7.18) can be used to estimate the order of an FIR filter. Program 7_4 given below can be employed to estimate the order using Kaiser's formula of Eq. (7.15).

```
% Program 7_4
% Computation of the order of a linear-phase
% FIR lowpass filter using Kaiser's formula
%
dp = input('Type in the passband ripple = ');
ds = input('Type in the stopband ripple = ');
Fp = input('Type in the passband edge in Hz = ');
Fs = input('Type in the stopband edge in Hz = ');
FT = input('Type in the sampling frequency in Hz = ');
num = -20*log10(sqrt(dp*ds))-13;
den = 14.6*(Fs - Fp)/FT;
N = ceil(num/den);
fprintf('Filter order is %d \n',N);
```

The following example illustrates the application of the above program.

EXAMPLE 7.24 We recompute the order of the linear-phase FIR filter of Example 7.3.

During execution, the program requests the relevant filter specifications which are entered as follows: `dp = 0.0114469, ds = 0.01778279, Fp = 1800, Fs = 2000,` and `FT = 12000`. It displays the filter order at the end of execution as indicated below:

```
        Filter order is 99
```

which is seen to be the same as that derived in Example 7.3.

The *Signal Processing Toolbox* of MATLAB includes the M-file `remezord` which determines the FIR filter order employing the formula of Eq. (7.18). There are two different options available with this function:

```
[N,fpts,mag,wt] = remezord(fedge,mval,dev)
[N,fpts,mag,wt] = remezord(fedge,mval,dev,FT)
```

where the input data are the vector `fedge` of bandedges, the vector `mval` of desired magnitude values in each frequency band, and a vector `dev` specifying the maximum allowable deviation between the magnitude response of the designed filter and the desired magnitude. The length of `fedge` is two times that of `mval` minus 2, while the length of `dev` is the same as that of `mval`. The output data are the estimated value N of the filter order, the normalized frequency bandedge vector `fpts`, the frequency band magnitude vector `mag`, and the weight vector `wt` meeting the filter specifications. The output data can then be directly used in filter design using the function `remez`. The sampling frequency FT can be specified using the second version of this function. A default of 2 Hz is employed if FT is not explicitly indicated. In which case, the frequency points in `fedge` must have values between 0 and 1.

EXAMPLE 7.25   Estimate the order of a linear-phase FIR transfer function with the specifications given in Example 7.3 using `remezord`. The MATLAB code used for this example is given below:

```
% Program 7_5
% Estimation of FIR Filter Order Using remezord
%
fedge = input('Type in the bandedges = ');
mval = input('Desired magnitude values in each band = ');
dev = input('Allowable deviation in each band = ');
FT = input('Type in the sampling frequency = ');
[N, fpts, mag, wt] = remezord(fedge, mval, dev, FT);
fprintf('Filter order is %d \n',N);
```

The input data called by this program are the vectors fedge, mval, and dev, and the scalar FT. For this example, these are given by fedge = [1800 2000], mval = [1 0], dev = [0.0114469 0.01778279], and FT = 12000. The output data printed by this program is given by

```
     Filter order is 106
```

which is seen to be the same as that derived in Example 7.5.

For FIR filter design using the Kaiser window, the window order should be estimated using the formula of Eq. (7.86). To this end, the M-file `kaiserord` in the *Signal Processing Toolbox* of MATLAB can be employed. The basic forms of this M-file are

```
[N,Wn,beta,ftype] = kaiserord(fedge,mval,dev)
[N,Wn,beta,ftype] = kaiserord(fedge,mval,dev,FT)
c = kaiserord(fpts,mval,dev,FT,'cell')
```

This M-file estimates the minimum filter order N using Eq. (7.86) and the parameter $\beta = $ beta using Eq. (7.85). It also determines the normalized frequency bandedges Wn. The input data required are the filter specifications given by the vector `fpts` of specified bandedges, the vector `mval` of the specified amplitudes on the frequency bands defined by `fpts`, and the vector `dev` of the specified ripples in each frequency band. The length of `fpts` is two times that of `mag` minus 2. The second form of `kaiserord` specifies the sampling frequency FT. The sampling frequency is assumed to be 2 if FT is not included in the function argument.

The function `kaiserord`, in some cases, can generate a value for N which is either greater or smaller than the required minimum value. If the FIR filter designed using N does not meet the specifications, the order should be either gradually increased or decreased by 1 until the specifications are met. The filter order N estimated using Eq. (7.86) has been found to be within ± 2 of the required value for a broad range of filter specifications.

The output data generated by this function can be directly used to design the FIR filter based on the windowed Fourier series approach using the function `firl` discussed later in this section. The third form of `kaiserord` generates a cell array c whose elements are the parameters needed to run `firl`.

We demonstrate its application in the following example.

EXAMPLE 7.26   Using MATLAB we determine the order N and the parameter β of the Kaiser window needed to design the lowpass filter of Example 7.16. Using the program statement

```
[N,Wn,beta,ftype] = kaiserord([0.3 0.5],[1 0],[0.01 0.01])
```

we get $N = 23$ and $\beta = 3.3953$, which are seen to be the same as those derived in Example 7.16.

7.10.3   Equiripple Linear-Phase FIR Filter Design Using MATLAB

For designing an equiripple linear-phase FIR filter employing the Parks-McClellan algorithm (Section 7.7.1) we can use the MATLAB M-file `remez`. There are various versions of this function:

```
b = remez(N,fpts,mag)
b = remez(N,fpts,mag,wt)
b = remez(N,fpts,mag,'ftype')
b = remez(N,fpts,mag,wt,'ftype')
```

It can design any type of multiband linear-phase filters. Its output is a vector b of length N+1 containing the impulse response coefficients of the linear-phase FIR filter with bandedges specified by the vector `fpts`. The frequency points of the vector `fpts` must be in the range between 0 and 1, with sampling frequency being equal to 2, and must be specified in an increasing order. The first element of `fpts` must be 0 and the last element must be 1. The bandedges between a passband and its adjacent stopband must be separated by at least 0.1. If same values are used for these frequency points, the program automatically separates them by 0.1. The desired magnitudes of the FIR filter frequency response at the specified bandedges are given by the vector `mag`, with the elements given in equal-valued pairs. The desired magnitudes between two specified consecutive frequency points `fpts(k)` and `fpts(k+1)` are determined according to the following rules. For k odd, the magnitude is a line segment joining the points {`mag(k)`, `fpts(k)`} and {`mag(k+1)`, `fpts(k+1)`}, whereas for k even, it is unspecified, with the frequency range [`fpts(k)`, `fpts(k+1)`] being a *transition* or *"don't care"* region. The vectors `fpts` and `mag` must be of the same length with the length being even. Figure 7.32 illustrates the relationship between the vectors `fpts` and `mag` given by

```
fpts = [0     0.2     0.4    0.7    0.8      1.0]
 mag = [0     0.5     1.0    0      0.25     0.25]
```

The desired magnitude responses in the passband(s) and the stopband(s) can be weighted by an additional vector `wt` included as the argument of the function `remez`. The function can design equiripple Types 1, 2, 3, and 4 linear-phase FIR filters. Types 1 and 2 are the default designs for order N even and odd, respectively. Types 3 (N even) and 4 (N odd) are used for two specialized filter designs, the Hilbert transformer and differentiator. To design these two types of FIR filters the flags 'hilbert' and 'differentiator' are used for 'ftype' in the last two versions of `remez`.

There are several other options available with the function `remez`:

```
            b = remez(...., {S})
        [b,err] = remez(....)
    [b,err,res] = remez(....)
```

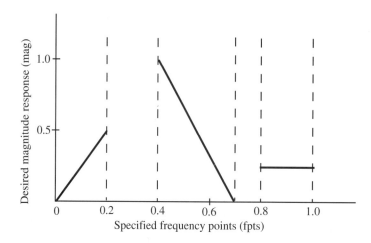

Figure 7.32: Illustration of the relationship between vectors `fpts` and `mag`.

The density of the frequency grid needed to evaluate the error function $\mathcal{E}(\omega)$ can be increased from the default value of $S = 16L$ by using the statement `b = remez(...., S)` where S is a one-by-one cell array containing an integer greater than 16. Use of a larger value usually results in a filter with exactly equiripple error at the cost of an increased computation time. The maximum ripple height can be obtained using the statement `[b,err] = remez(....)`. Optional results computed by `remez` can be obtained using the statement `[b,err,res] = remez(....)` which can provide the following information if needed: the vector of frequency grid point in `res.fgrid`; the desired response in `res.des`; the weights used on `fgrid` in `res.wt`; actual frequency response in `res.H`; error at each point of the frequency grid in `res.error`; the vector of indices into `fgrid` of extremal frequencies; and the vector of extremal frequencies in `res.fextr`.

Incorrect use of `remez` results in self-explanatory diagnostic messages. The use of the function `remez` is illustrated next for the design of frequency selective filters, differentiators, and Hilbert transformers.

Lowpass FIR Filter Design Examples

We consider first the design of a lowpass linear-phase filter using the function `remez`. Recall from our discussion in Section 4.4.4 that the transfer function of a Type 3 FIR filter has zeros at both $z = 1$ and $z = -1$, and the transfer function of a Type 4 FIR filter has a zero at $z = 1$. These filters cannot be used to design a lowpass filter. On the other hand, the transfer function of a Type 2 FIR filter has a zero at $z = -1$ and hence it can be used to design a lowpass filter along with the Type 1 FIR filter which has no restrictions.

The following example illustrates the design using MATLAB.

EXAMPLE 7.27   Determine the transfer function and plot the gain response of an equiripple linear-phase FIR lowpass filter with the same specifications as given in Example 7.22, i.e., passband edge $F_p = 800$ Hz, stopband edge $F_s = 1000$ Hz, passband ripple $\alpha_p = 0.5$ dB, minimum stopband attenuation $\alpha_s = 40$ dB, and sampling frequency $F_T = 4000$ Hz. From Eqs. (7.3) and (7.4), we thus obtain $\delta_p = 0.0559$, and $\delta_s = 0.01$.

The program for designing an equiripple linear-phase FIR filter is as follows:

```
% Program 7_6
% Design of Equiripple Linear-Phase FIR Filters
```

```
%
format long
fedge = input('Band edges in Hz = ');
mval = input('Desired magnitude values in each band = ');
dev = input('Desired ripple in each band =');
FT = input('Sampling frequency in Hz = ');
[N,fpts,mag,wt] = remezord(fedge,mval,dev,FT);
b = remez(N,fpts,mag,wt);
disp('FIR Filter Coefficients'); disp(b)
[h,w] = freqz(b,1,256);
plot(w/pi,20*log10(abs(h)));grid;
xlabel('\omega/\pi'); ylabel('Gain, dB');
```

The input data are the filter specifications which are entered as follows:

```
fedge = [800 1000]
mval = [1 0]
dev = [0.0559 0.01]
FT = 4000
```

The output data generated by this program are given below.

```
FIR Filter Coefficients
     Columns 1 through 4
     -0.00528821565206   -0.02427656045325
     -0.01763627129682    0.00748393833738

     Columns 5 through 8
      0.02235487646567   -0.00197921004415
     -0.03172161448863   -0.01082775442469

     Columns 9 through 12
      0.04175020832597    0.03549072964251
     -0.05040443020573   -0.08768797471498

     Columns 13 through 16
      0.05623651758405    0.31190360510442
      0.44170620151624    0.31190360510442

     Columns 17 through 20
      0.05623651758405   -0.08768797471498
     -0.05040443020573    0.03549072964251

     Columns 21 through 24
      0.04175020832597   -0.01082775442469
     -0.03172161448863   -0.00197921004415

     Columns 25 through 28
      0.02235487646567    0.00748393833738
     -0.01763627129682   -0.02427656045325

     Column 29
     -0.00528821565206
```

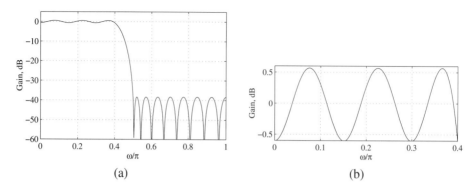

Figure 7.33: FIR equiripple lowpass filter of Example 7.27 for $N = 28$: (a) gain response, and (b) passband details.

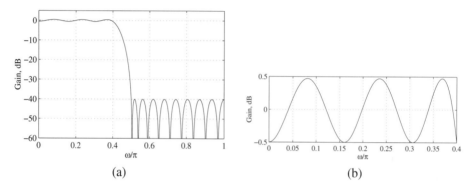

Figure 7.34: FIR equiripple lowpass filter of Example 7.27 for $N = 30$: (a) gain response, and (b) passband details.

In the above list, the first number is the filter coefficient $h[0]$, the second is the filter coefficient $h[1]$, and so on. As can be seen from above, the filter coefficients satisfy the symmetry constraint $h[k] = h[N - k]$, as expected. Since the filter length is 29 (the filter order $N = 28$), it is a Type 1 FIR filter. The computed gain response is plotted in Figure 7.33(a) with passband details in Figure 7.33(b).

The passband ripple and the minimum stopband attenuation of the designed filter are 0.6 dB and 38.7 dB, respectively, which do not meet the given specifications. We next increase the filter order by 1 and find that for a filter order of $N = 30$, the passband ripple and the minimum stopband attenuation are now 0.5 dB and 40.02 dB, respectively, which are now satisfactory. The gain response of this filter along with the passband details are shown in Figure 7.34.

Bandpass FIR Filter Design Examples

In this case, all four types of linear-phase FIR filters can be used. The following example considers the design of a bandpass linear-phase FIR filter using MATLAB and investigates the relation between the filter order and the ripple ratio on the frequency response of the filter.

EXAMPLE 7.28   Consider the design of a linear-phase FIR bandpass filter of order 26 with a passband from 0.3 to 0.5 and stopbands from 0 to 0.25 and from 0.55 to 1.0. The desired passband and stopband magnitudes are, respectively, 1 and 0. Assume equal weights in the passband and the stopbands.

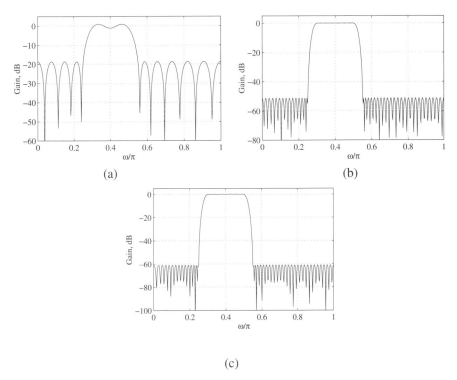

Figure 7.35: Gain response of the FIR equiripple bandpass filter of Example 7.28: (a) $N = 26$, weight ratio = 1, (b) $N = 110$, weight ratio = 1, and (c) $N = 110$, weight ratio = 1/10.

We use Program 7.6 with the following input data:

```
N = 26
fpts = [0 0.25 0.3 0.5 0.55 1]
mag = [0 0 1 1 0 0]
wt = [1 1 1]
```

The computed gain response is plotted in Figure 7.35(a). The passband ripple and the minimum stopband attenuation of the designed filter are about 1 dB and 18.7 dB, respectively.

We next repeat the previous example with the order increased to 110. The gain response obtained is shown in Figure 7.35(b). The new filter has now a passband ripple of 0.024 dB which is much smaller than that of the previous design and a minimum stopband attenuation of 51.2 dB which is larger than that obtained earlier. Hence, an increase in the filter order improves the frequency response characteristics at the expense of an increasing computational complexity.

It should be noted that the minimum stopband attenuation can be increased at the expense of a larger passband ripple while keeping the order as given earlier by decreasing the relative weight ratio $W(\omega) = \delta_p/\delta_s$. Figure 7.35(c) shows the gain response of the bandpass filter of order 110 obtained with a weight vector [1, 0.1, 1]. This weight ratio puts more weight on the stopband response relative to the passband response. As expected, this unequal weighting has increased the passband ripple to 0.076 dB while increasing the minimum stopband attenuation to 60.86 dB.

The absolute error for each of the above three bandpass filter design examples has been plotted in Figure 7.36. As can be seen from Figure 7.36(a), the absolute error, as expected, has the same peak value

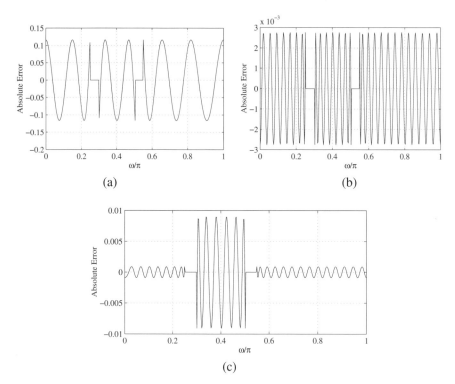

Figure 7.36: Absolute error of the FIR equiripple bandpass filter of Example 7.28: (a) $N = 26$, weight ratio = 1, (b) $N = 110$, weight ratio = 1, and (c) $N = 110$, weight ratio = 1/10.

in all three bands. Since $L = 13$, and there are four bandedges, there can be at most $L - 1 + 6 = 18$ extrema in this design. The error plot exhibits 17 extrema. In Figure 7.36(b), we again observe the error plot has the same peak value in each band. On the other hand, as can be seen from Figure 7.36(c), the peak absolute value of the error in the passband is 10 times the peak absolute error in the stopbands as expected.

The Remez algorithm can create some unusual results in the design of linear-phase FIR filters with more than two bands. We investigate this problem in the following example.

EXAMPLE 7.29   Consider now the design of a linear-phase FIR bandpass filter of order 60 with a normalized passband from 0.3π to 0.5π, and stopbands from 0 to 0.25π and from 0.6π to π. The desired passband and stopband magnitudes are, respectively, 1 and 0. Assume equal weights in the passband and the low-frequency stopband, and a weight in the high-frequency stopband that is 30 percent of that in the other two bands.

We use Program 7_6 with the following input data:

```
N = 60
fpts = [0 0.25 0.3 0.5 0.6 1]
mag = [0 0 1 1 0 0]
wt = [1 1 0.3]
```

From the computed gain response shown in Figure 7.37(a) we observe that the response in the second transition band is nonmonotonic and shows a peak with a value higher than that in the passband. This indicates the presence of extremal frequencies in the second transition band. However, this result does not contradict the alternation theorem which is concerned with extremal frequencies only in the passbands and the stopbands. As the FIR filter

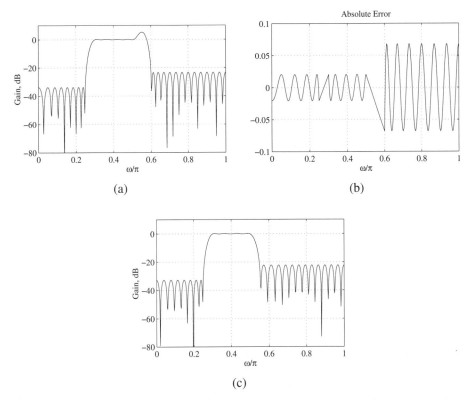

(a)   (b)

(c)

Figure 7.37: FIR equiripple bandpass filter of Example 7.29: (a) amplitude response with original bandedge specifications, (b) absolute error, and (c) amplitude response with a slightly modified stopband edge.

order is 60, $M = 30$. Thus according to the alternation theorem, there must be at least $M + 2 = 32$ extremal frequencies. We note that the plot of the absolute error given in Figure 7.37(b) also shows the presence of 32 extremal frequencies.

Even though the Remez algorithm guarantees equiripple error in the specified bands, as the response in a transition band is left unspecified, it cannot ensure that the gain response of a filter with more than two bands will decrease monotonically from its value in the passband. If the gain response of the filter designed exhibits a nonmonotonic behavior, it is recommended that either the filter order or the bandedges or the weighting function be adjusted until a satisfactory gain response is obtained. For example, the bandpass filter of the above example has an acceptable gain response as shown in Figure 7.37(c) if the second stopband edge is moved from 0.6π to 0.55π with all other parameters kept at their original values.

FIR Differentiator Design Examples

Now, from Eq. (7.69), we observe that an ideal differentiator is characterized by an antisymmetric impulse response which implies that either a Type 3 or a Type 4 FIR filter can be used for its realization. However, from Eq. (7.68), for an ideal differentiator $\breve{H}(\pi) = \pi$. Hence, a Type 3 FIR filter cannot be used for its realization as its transfer function has a zero at $z = -1$, which forces the amplitude response to go to zero at $\omega = \pi$. As a result, only a Type 4 FIR filter can be used for its design.

In most practical applications, signals of interest are in a frequency range $0 \leq \omega \leq \omega_p$. Consequently, a lowpass differentiator with a bandlimited frequency response

$$H_{DIF}(e^{j\omega}) = \begin{cases} j\omega, & 0 \leq |\omega| \leq \omega_p, \\ 0, & \omega_s \leq |\omega| \leq \pi, \end{cases} \tag{7.141}$$

can be employed. In the above equation, $0 \leq \omega \leq \omega_p$ and $\omega_s \leq \omega \leq \pi$ represent, respectively, the passband and the stopband of the differentiator. The frequency ω_p is usually called its bandwidth. Hence, both Type 3 and Type 4 FIR filters can be used to design a lowpass differentiator. For the design phase, we choose the weighting function as

$$P(\omega) = \frac{1}{\omega}, \qquad 0 \leq \omega \leq \omega_p,$$

and the desired amplitude response as

$$D(\omega) = 1, \qquad 0 \leq \omega \leq \omega_p.$$

The function `remezord` cannot be used to estimate the order of an FIR differentiator as the formula of Eq. (7.18) has been developed for conventional filters with two or more bands with constant gain levels. However, the function `remez` can be employed to design an equiripple FIR differentiator as demonstrated in the following two examples.

EXAMPLE 7.30   Consider the design of a full-band differentiator of order 11. Since the length is even, we can use a Type 4 FIR filter for its design. We use the program statement

```
        b = remez(N,fpts,mag,'differentiator');
```

where   N = 11
             fpts = [0 1]
             mag = [0 pi]

The plots of the amplitude response of the differentiator designed and the absolute error between the amplitude response of the differentiator and the desired amplitude response $[A(\omega) - \omega]$ are shown in Figure 7.38. We observe from Figure 7.38(b) that the absolute error increases as ω increases from 0 to π. This is expected as the application of the Remez algorithm results in an equiripple error of the function $[\frac{A(\omega)}{\omega} - 1]$.

The filter coefficients can be displayed by typing b in the MATLAB Command Window which yields

```
  b =
  Columns 1 through 4
  -0.00467247274580    0.00833867946220
  -0.00926393325297    0.01683341669736


  Columns 5 through 8
  -0.04553782313238    0.40574536683417
  -0.40574536683417    0.04553782313238


  Columns 9 through 12
  -0.01683341669736    0.00926393325297
  -0.00833867946220    0.00467247274580
```

It can be seen from the above that the filter coefficients satisfy the antisymmetric condition $h[n] = -h[11 - n]$.

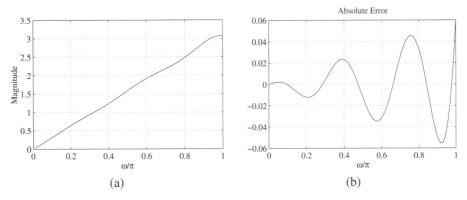

Figure 7.38: FIR equiripple differentiator of Example 7.30 of length 11: (a) magnitude response, and (b) absolute error.

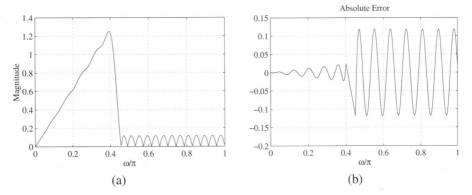

Figure 7.39: FIR equiripple lowpass differentiator of Example 7.31 of length 51: (a) magnitude response, and (b) absolute error.

EXAMPLE 7.31   We now consider the design of a lowpass differentiator with the following specifications: passband edge $\omega_p = 0.4\pi$, stopband edge $\omega_s = 0.45\pi$. Here, we can use either a Type 3 or a Type 4 FIR filter for the implementation. We consider here the design using a Type 4 filter of length 51.

   Here we use the program statement

```
b = remez(N,fpts,mag,'differentiator');
```

where   N = 50
             fpts = [0 0.4 0.45 1]
             mag = [0 0.4*pi 0 0]

Figure 7.39 shows the magnitude response of the Type 4 FIR lowpass differentiator and the resulting absolute error.

   Rabiner and Schafer [Rab74a] have investigated the relations between the filter order N, the bandwidth ω_p, and the peak absolute error ε of a lowpass differentiator through extensive designs. Their results, available in the form of design charts, can be used to estimate the filter order for a specified bandwidth and the peak absolute error in dB.

FIR Hilbert Transformer Design Examples

Like the ideal differentiator, as can be seen from Eq. (7.67), the ideal Hilbert transformer also has an antisymmetric impulse response implying either a Type 3 or a Type 4 FIR filter for its realization. However, from Eq. (7.68) we observe that the magnitude response of an ideal Hilbert transformer is unity for all ω which cannot be satisfied by a Type 3 FIR filter whose magnitude response has a zero at $\omega = 0$ or by a Type 4 FIR filter whose magnitude response has a zero at both $\omega = 0$ and $\omega = \pi$. In practice, the signals of interest are in a finite range $\omega_L \leq |\omega| \leq \omega_H$, and as a consequence, the Hilbert transformer can be designed with a bandpass amplitude response given by

$$D(\omega) = 1, \qquad \omega_L \leq |\omega| \leq \omega_H, \tag{7.142}$$

with the weighting function $P(\omega)$ set to unity in the band of interest.

As indicated by Eq. (7.67), the impulse response samples of an ideal Hilbert transformer satisfy the condition

$$h_{HT}[n] = 0, \qquad \text{for } n \text{ even.} \tag{7.143}$$

It can be shown that the above attractive property can be maintained by a Type 3 linear-phase FIR filter if the desired amplitude response $\breve{H}(\omega)$ is symmetric with respect to $\pi/2$, i.e.,

$$\breve{H}(\omega) = \breve{H}(\pi - \omega).$$

However, the condition of Eq. (7.143) cannot be met by a Type 4 linear-phase filter with an antisymmetric amplitude response (Problem 7.64).

As in the case of the FIR differentiator design, the function `remezord` cannot be used to estimate the order of an FIR Hilbert transformer. Rabiner and Schafer [Rab74b] have investigated the relations between the order N, passband ripple δ_p, and the normalized transition bandwidth ω_L of a bandpass Hilbert transformer using extensive designs. Based on their investigations, the following formula has been developed to estimate the order of the Hilbert transformer:

$$N \cong -\frac{3.833 \, \log_{10} \delta_p}{\omega_L}.$$

The design of an equiripple FIR bandpass Hilbert transformer using `remez` is demonstrated in the following example.

EXAMPLE 7.32   Design a linear-phase bandpass FIR Hilbert transformer of order 20 having a magnitude of unity in the passband, and passband edges at 0.1π and 0.9π, respectively. To this end we use the program statement

```
b = remez(N,fpts,mag,'Hilbert');
```

where

```
N = 20
fpts = [0.1 0.9]
mag = [1 1]
```

The magnitude response of the Type 3 FIR bandpass Hilbert transformer obtained and its absolute error are shown in Figure 7.40. The impulse response samples of the designed Hilbert transformer can be obtained by typing b in the MATLAB Command Window which yields

```
b =
Columns 1 through 4
   0  -0.02728070544840    0  -0.04788280428795
```

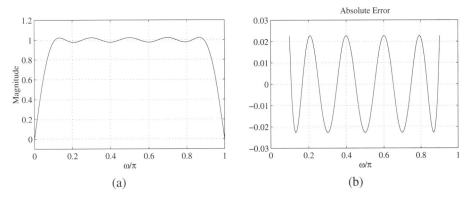

Figure 7.40: FIR equiripple bandpass Hilbert transformer of Example 7.32 of length 21: (a) magnitude response, and (b) absolute error.

```
Columns 5 through 8
0   -0.09317642235235    0   -0.19023268768181

Columns 9 through 12
0   -0.62903099358176    0    0.62903099358176

Columns 13 through 16
0    0.19023268768181    0    0.09317642235235

Columns 17 through 20
0    0.04788280428795    0    0.02728070544840

Column 21
0
```

We observe that the impulse response samples for n even are zero, as expected because of the symmetry imposed on the desired amplitude response. It can be shown that if the desired amplitude response does not satisfy the symmetry condition, the even-indexed samples of the impulse response of the Hilbert transformer designed will not have zero values. To ensure zero values for the even-indexed samples, an elegant solution has been provided by Vaidyanathan and Nguyen [Vai87b]. We describe their technique in Section 11.6.

7.10.4 Window-Based FIR Filter Design Using MATLAB

The window-based FIR filter design process involves three steps. In the first step, the order of the FIR filter to be designed is estimated using either Eq. (7.15) or (7.18). For FIR filter design using the Kaiser window, it is recommended that the order be estimated using the formula of Eq. (7.86). In the second step, the type of window to be used is selected and its coefficients are then computed. Finally, in the last step, the desired impulse response of the ideal filter is computed which is then multiplied by the window coefficients generated in the first step to yield the coefficients of the FIR filter.

FIR filter order estimation using MATLAB has been described in Section 7.10.2. We now discuss the window generation using MATLAB.

Window Generation

The *Signal Processing Toolbox* of MATLAB includes the following functions for generating the windows discussed earlier in Section 7.6:

```
w = blackman(L),    w = hamming(L),    w = hanning(L)
w = chebwin(L,Rs),  w = kaiser (L,beta)
```

The above functions generate a vector w of window coefficients of odd length L.[15] The parameter *beta* in the Kaiser window is the same as the parameter β of Eq. (7.83) and can be computed using Eq. (7.85).

The following example illustrates the generation of the coefficients of the Kaiser window.

EXAMPLE 7.33   Consider the determination of the coefficients of a Kaiser window to be used for the design of a lowpass FIR filter. The filter specifications are $\omega_p = 0.3\pi$, $\omega_s = 0.4\pi$, and $\alpha_s = 50$ dB. Substituting the value of α_s in Eq. (7.4), we arrive at the value of the peak stopband ripple $\delta_s = 0.003162$.

We use the following MATLAB program to determine the window coefficients and plot its gain response.

```
% Program 7_7
% Kaiser Window Generation Program
%
fpts = input('Type in the bandedges = ');
mag = input('Type in the desired magnitude values = ');
dev = input('Type in the ripples in each band = ');
[N,Wn,beta,ftype] = kaiserord(fpts,mag,dev)
w = kaiser(N+1,beta); w = w/sum(w);
[h,omega] = freqz(w,1,256);
plot(omega/pi,20*log10(abs(h)));grid;
xlabel('\omega/\pi'); ylabel('Gain, dB');
```

The program requests the desired bandedges and the desired magnitude values in each band followed by the specified ripples in each band which are entered as follows:

```
fpts = [0.3 0.4]
mag = [1 0]
dev = [0.003162 0.003162]
```

It then plots the gain response of the window as indicated in Figure 7.41. The window coefficients can be printed by typing w in the MATLAB Command Window.

Filter Design

The functions available in MATLAB for the design of FIR filters using the windowed Fourier series approach are `fir1` and `fir2`. The function `fir1` is used to design conventional lowpass, highpass, bandpass, bandstop, and multiband FIR filters, whereas the function `fir2` is employed to design FIR filters with arbitrarily shaped magnitude response.

The various forms of the function `fir1` are as follows:

```
b = fir1(N,Wn)
b = fir1(N,Wn,'ftype')
```

[15]It should be noted that the Hann and Blackman windows generated by MATLAB have zero-valued first and last coefficients. As a result, to keep the length the same for all windows, these two windows should be generated using a value of L that is greater by 2 than the value of L used for the Hamming and the Kaiser windows.

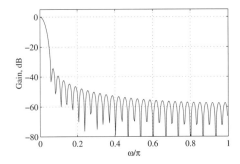

Figure 7.41: Gain response of the Kaiser window generated in Example 7.33.

```
b = fir1(N,Wn,window)
b = fir1(N,Wn,'ftype',window)
b = fir1(....,'noscale')
```

The basic form, b = fir1(N,Wn), generates the vector b of length N+1 containing the impulse response coefficients of a lowpass FIR filter of order N with a normalized cutoff frequency of Wn between 0 and 1. The sampling frequency is assumed to be equal to 2. The transfer function of the designed filter is of the form:

$$H(z) = b(1) + b(2)z^{-1} + \cdots + b(N+1)z^{-N}.$$

For the design of bandpass filters, Wn is a length-2 vector, Wn = [W1 W2], containing the passband edge frequencies where W1 < W2. For the design of a multiband filter, Wn is a multielement vector containing bandedge frequencies of increasing magnitudes, Wn = [W1 W2 W3 W4 W5 ..., Wm], where the m+1 bands are defined by (0, W1), (W1, W2),... (Wm, 1). The designed filter, obtained after windowing, is scaled to ensure that the center of the first passband has a default magnitude of 1.

The option b = fir1(N,Wn,'ftype') specifies one of four other types of filters. For designing a highpass filter with a cutoff frequency Wn, ftype is high. In the case of a bandstop filter, ftype is stop with Wn a length-2 vector, Wn = [W1 W2], containing the stopband edge frequencies where W1 < W2. Since a Type 2 FIR filter cannot be used to design a highpass or a bandstop filter due to the presence of a zero at $z = -1$, the order N must be an even integer for these two cases. If N is specified as an odd integer, fir1 automatically increases it by 1 to make the order even. ftype is DC-1 if the first band of a multiband filter should be a passband, otherwise it is DC-0.

The Hamming window for filter design is used as a default if no window is specified in the argument of fir1. The option b = fir1(N,Wn,window) is employed with the vector window of length N+1 containing the coefficients of the specified window. The option b = fir1(N,Wn,'ftype',window) is used with specified filter type and window coefficients. The default scaling is turned off in the option b = fir1(....,'noscale').

The various forms of the function fir2 are as follows:

```
b = fir2(N,f,m)
b = fir2(N,f,m,window)
b = fir2(N,f,m,npt)
b = fir2(N,f,m,npt,window)
b = fir2(N,f,m,npt,lap)
b = fir2(N,f,m,npt,lap,window)
```

As in the previous case, the basic form b = fir2(N,f,m) is used to design a lowpass FIR filter of order N whose magnitude response samples at the specified frequency points given by the vector f match the specified magnitude response samples defined by the vector m. The output vector b is of length N+1 containing the FIR filter coefficients ordered in ascending powers of z^{-1}. The vectors f and m must be of the same length. Elements of f must be in the range from 0 to 1 in increasing order, with the sampling frequency assumed to be of value 2. Moreover, the first and last elements of f must be 0 and 1, respectively. Duplicate frequency points in f are permissible for approximating magnitude response specifications with jumps. Any one of the window functions available in the MATLAB *Signal Processing Toolbox* can be used for designing FIR filters by making use of its corresponding M-file to generate the window coefficients and entering them into the function fir2 by means of the length N+1 vector window in the argument. As before, if the window coefficients are not explicitly included, the Hamming window is used as the default.

In the options b = fir2(N,f,m,npt) and b = fir2(N,f,m,npt,window), npt specifies the number of points for the grid onto which the frequency response is interpolated by fir2. The default value of npt is 512. In the remaining two options, lap specifies the size of the region inserted between duplicate frequency points in f.

In the first step of the algorithm, the desired amplitude response is interpolated onto a dense grid of size npt with grid points evenly spaced in the frequency range $(0, 1)$. The filter coefficients are then determined by applying an inverse DFT to the sample values on the grid and then multiplying the IDFT values by the coefficients of the window.

FIR Filter Design Examples

We illustrate the use of the above two functions in designing linear-phase FIR filters in the next several examples. Without any loss of generality, we restrict our attention to filter design using the Kaiser window. The first example considered is for a lowpass filter design.

EXAMPLE 7.34   We continue here with the remaining steps in the design of the lowpass FIR filter of Example 7.33 using the Kaiser window. The desired specifications for the filter are $\omega_p = 0.3\pi$, $\omega_s = 0.4\pi$, and $\delta_s = 0.003162$. The MATLAB program used for this filter design problem is indicated below.

```
% Program 7_8
% Lowpass Filter Design Using the Kaiser Window
%
fpts = input('Type in the bandedges = ');
mag = input('Type in the desired magnitude values = ');
dev = input('Type in the ripples in each band = ');
[N,Wn,beta,ftype] = kaiserord(fpts,mag,dev)
kw = kaiser(N+1,beta);
b = fir1(N,Wn, kw);
[h,omega] = freqz(b,1,512);
plot(omega/pi,20*log10(abs(h)));grid;
xlabel('\omega/\pi'); ylabel('Gain, dB');
```

The input data for this program are the same as in the previous example. The gain response of the resulting lowpass filter is shown in Figure 7.42. The filter order computed by the above program can be printed by typing N in the Command Window of MATLAB which yields N = 59. The filter designed is therefore of Type 2.

The next example considers the design of a highpass filter using the Kaiser window. It should be noted that for highpass and bandstop filter design, the function fir1 requires that the order of the filter N be even. If the order generated using kaiserord is not even, it should be increased by 1 before the window and filter coefficients are generated.

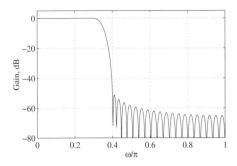

Figure 7.42: Gain response of the lowpass filter of Example 7.34.

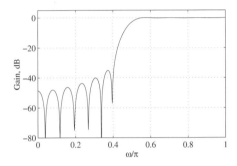

Figure 7.43: Gain response of the highpass filter of Example 7.35.

EXAMPLE 7.35   A highpass FIR filter is to be designed meeting the following specifications: passband edge $\omega_p = 0.55\pi$, stopband edge $\omega_s = 0.4\pi$, and stopband ripple $\delta_s = 0.02$. To design a highpass filter (or a bandstop filter) we modify Program 7.8 by changing the statement b = fir1(N,Wn, kw); to b = fir1(N,Wn,'ftype', kw);. With this modification and the input data

```
fpts = [0.4 0.55]
mag  = [0 1]
dev  = [0.02 0.02]
```

we arrive at the computed gain response indicated in Figure 7.43.

In the following example we consider the application of the function fir2 of MATLAB in designing an FIR filter with passbands having different gain levels.

EXAMPLE 7.36   Consider the design of an FIR filter of order 100 with three different constant magnitude levels: 0.3 in the frequency range 0 to 0.28, 1.0 in the frequency range 0.3 to 0.5, and 0.7 in the frequency range 0.52 to 1.0. To this end we use the following MATLAB program:

```
% Program 7_9
% Design of Multiband Filter Using Hamming Window
%
fpts = [0 0.28 0.3 0.5 0.52 1];
```

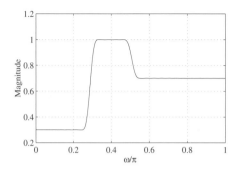

Figure 7.44: Magnitude response of the multilevel filter of Example 7.36.

```
mval = [0.3 0.3 1.0 1.0 0.7 0.7];
b = fir2(100,fpts,mval);
[h,omega] = freqz(b,1,512);
plot(omega/pi,abs(h));grid;
xlabel('\omega/\pi'); ylabel('Magnitude');
```

The filter coefficients are contained in the vector b. The resulting magnitude response is as indicated in Figure 7.44.

7.10.5 Least-Squares Error FIR Filter Design Using MATLAB

The function `firls` in the *Signal Processing Toolbox* of MATLAB can be used to design any type of multiband linear-phase FIR filter based on the least-squares method. The various options available with this function are

```
b = firls(N,fpts,mag)
b = firls(N,fpts,mag, wt)
b = firls(N,fpts,mag,'ftype')
b = firls(N,fpts,mag,wt,'ftype')
```

The specified order is given by N. The vectors `fpts`, `mag`, and `wt`, and the string `ftype` are defined the same way as in the case of the function `remez` in Section 7.10.3.

Typical filter design applications of the function `firls` are considered below.

EXAMPLE 7.37 We consider the design of linear-phase FIR lowpass filter with the same specifications as given in Example 7.27, i.e., passband edge $F_p = 800$ Hz, stopband edge $F_s = 1000$ Hz, passband ripple $\alpha_p = 0.5$ dB, minimum stopband attenuation $\alpha_s = 40$ dB, and sampling frequency $F_T = 4000$ Hz. From Eqs. (7.3) and (7.4), we thus obtain $\delta_p = 0.0559$, and $\delta_s = 0.01$. Next, using Program 7_5 we determine the filter order which is found to be 27.

Program 7_10 given below is then used for designing the lowpass linear-phase FIR filter using the least-squares approach.

```
% Program 7_10
% Design of Least-Squares Error Linear-Phase FIR Filters
%
```

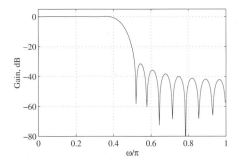

Figure 7.45: Gain response of the lowpass filter of Example 7.37.

```
N = input('Type in the order = ');
fpts = input('Type in the bandedge vector = ');
mag = input('Type in the magnitude value vector = ');
wt = input('Type in the weight vector = ');
b = firls(N,fpts,mag,wt);
[h,w] = freqz(b,1,256);
plot(w/pi,20*log10(abs(h)));grid;
xlabel('\omega/\pi'); ylabel('Gain, dB');
```

For our example, the input data are

```
N = 27
fpts = [0,4 0.5]
mag = [1 0]
wt = [1 1]
```

The computed gain response generated by this program is indicated in Figure 7.45. The passband ripple and the minimum stopband attenuation of the designed filter are, respectively, 0.57 dB and 20.43 dB, and, hence, do not meet the specifications. It can be shown that the stopband specifications are met with $N = 40$ with considerably reduced passband ripple.

For linear-phase FIR filter design using the constrained least-squares method, the functions `fircls` and `fircls1` are available. The latter is employed for the design of lowpass and highpass filters. Various options available with this function are

```
b = fircls1(N,wo,dp,ds)
b = fircls1(N,wo,dp,ds,'high')
b = fircls1(N,wo,dp,ds,wt)
b = fircls1(N,wo,dp,ds,wt,'high')
b = fircls1(N,wo,dp,ds,wp,ws,K)
b = fircls1(N,wo,dp,ds,wp,ws,K,'high')
b = fircls1(N,wo,dp,ds,...,'design_flag')
```

The basic form `b = fircls1(N,wo,dp,ds)` generates the length-`(N+1)` vector `b` of filter coefficients with a normalized cutoff frequency of `wo` in the range between 0 and 1 where the sampling frequency is assumed to be 2. The passband and the stopband ripples are given by `dp` and `ds`, respectively. The string `'high'` is included in the program statement for highpass filter design.

The parameter `wt` in the argument of the function `cls1` specifies a frequency above which (for `wt` > `wo`) or below which (for `wt` < `wo`) the filter designed is guaranteed to meet the given passband or stopband edge requirement. For the lowpass case, if 0 < `wt` < `wo` < 1, the amplitude response of the designed filter is within `dp` over the frequency range 0 < ω < `wt`; otherwise, the amplitude response is within `ds` over the frequency range `wt` < ω < 1. For the highpass case, if 0 < `wt` < `wo` < 1, the amplitude response of the designed filter is within `ds` over the frequency range 0 < ω < `wt`; otherwise, the amplitude response is within `dp` over the frequency range `wt` < ω < 1.

The parameter K in the function argument indicates the relative weight of the $\mathcal{L}_2$ norm of the passband error and the $\mathcal{L}_2$ norm of the stopband error. Thus the error that is minimized is given by

$$\int_0^{\omega_p} |A(\omega) - D(\omega)|^2 \, d\omega + K \int_{\omega_s}^{\pi} |A(\omega) - D(\omega)|^2 \, d\omega.$$

Thus by choosing a higher value of K, the stopband error can be heavily weighted relative to the passband error.

The string `'design_flag'` in the function argument is for monitoring the filter design process. It is `'trace'` for a textual display of the design table being used, `'plots'` for displaying the full-band magnitude response, passband and stopband details of the filter, and `'both'` for displaying both the textual information and plots. All plots are updated at each iteration step.

The function `fircls` is used for the constrained least-squares design of multiband linear-phase FIR filters. The two forms of this function are

```
b = fircls(N,f,amp,up,lo)
b = fircls(N,f,amp,up,lo,'design_flag')
```

The basic form `b = fircls(N,f,amp,up,lo)` generates a length-(`N+1`) vector of filter coefficients meeting the amplitude response specifications given by the vectors `f` and `amp`. The vector `f` contains the frequency points defining the transition frequencies in increasing order in the range 0 and 1 where the sampling frequency is assumed to be 2. The first and last frequency points must be a 0 and a 1, respectively. The desired piecewise constant values of the amplitude response are specified through the vector `amp`, whose length is equal to the length of `f` minus 1, i.e., the number of frequency bands.

The vectors `up` and `lo` specify, respectively, the upper and lower bounds of the amplitude response in each frequency band, and have the same length as that of the vector `amp`.

The string `'display_flag'` is used to monitor the design process the same way as in the function `fircls1`.

EXAMPLE 7.38   We consider the constrained least-squares design of a linear-phase FIR lowpass filter with the same specifications as given in Example 7.27: normalized passband edge $\omega_p = 0.4\pi$, normalized stopband edge $\omega_s = 0.5\pi$, passband ripple $\delta_p = 0.0559$, and stopband ripple $\delta_s = 0.01$. The filter order is 27. This filter was also designed using the least-squares method in Example 7.37.

We make use of the function `fircls1` for the filter design. The code fragments used for the design are as given below:

```
n = 27; wo = 0.4;
dp = 0.0559, ds = 0.01;
b = fircls1(n,wo,dp,ds);
```

The gain response of the designed lowpass filter is shown in Figure 7.46(a). Using the string `'plots'` for the string `'design_flag'` we can generate a plot of the magnitude response, passband, and stopband details showing the frequency points used for the final design. Such a plot for this example is indicated in Figure 7.46(b).

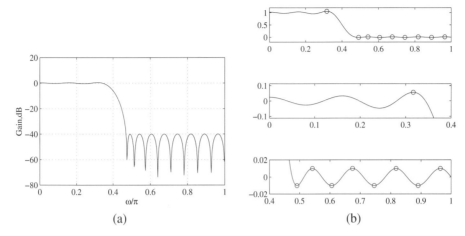

Figure 7.46: FIR lowpass filter of Example 7.38: (a) gain response, and (b) plots of magnitude response, passband, and stopband details showing frequency points selected.

EXAMPLE 7.39   As a second example, we consider the constrained least-squares design of the linear-phase FIR bandpass filter of order 60 with the same specifications as given in Example 7.29: normalized passband edges $\omega_{p1} = 0.3\pi$ and $\omega_{p2} = 0.5\pi$, normalized stopband edges $\omega_{s1} = 0.25\pi$ and $\omega_{s2} = 0.6\pi$. We assume the amplitude response in the passband to lie in the range $(0.98, 1.02)$, and the amplitude response in both the lower and the upper stopbands to lie in the range $(-0.02, 0.02)$.

Here we use the function `fircls` to design the filter. The code fragments used are as given below:

```
n = 60; f = [0 0.3 0.5 1];
amp = [0 1 0];
up = [0.02 1.02 0.02];
lo = [-0.02 0.98 -0.02];
b = fircls(n,f,amp,up,lo);
```

Note from the values given in vector `f` of the frequency points, only the passband edges are specified here. This vector does not include data on the stopband edges and, consequently, no transition bands are specified. Moreover, the negative numbers have been used in the vector `lo` to specify the lower limits of the amplitude response in the stopband as in general the amplitude response of a linear-phase FIR filter is permitted to assume negative values in the stopband. However, the lower limits can be set equal to 0 in the stopbands. In which case, the amplitude response of the desgined FIR filter will remain nonnegative over the whole frequency range. These types of FIR filters can be factored into a product of a minimum-phase FIR filter and a maximum-phase FIR filter using the spectral factorization approach (Problem 4.80). This method can be used to design minimum-phase FIR filters.

The gain response of the designed bandpass filter is shown in Figure 7.47(a). Using the string `'plots'` for the string `'design_flag'` we can generate a plot of the magnitude response, passband, and stopband details showing the frequency points used for the final design. Such a plot for this example is indicated in Figure 7.47(b).

7.11 Summary

This chapter considered the design of both infinite impulse response (IIR) and finite impulse response (FIR) causal digital filters. The digital filter design problem is concerned with the development of a suitable transfer function meeting the frequency response specifications, which, in this chapter, is restricted to magnitude (or, equivalently, gain) response specifications. These specifications are usually given in terms

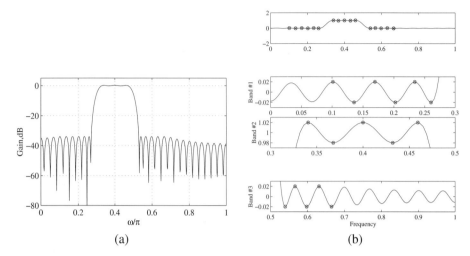

Figure 7.47: FIR bandpass filter of Example 7.39: (a) gain response, and (b) plots of magnitude response, passband, and stopband details showing frequency points selected.

of the desired passband edge and stopband edge frequencies and the allowable deviations from the desired passband and stopband magnitude (gain) levels.

IIR filter design is usually carried out by transforming a prototype analog transfer function by means of a suitable mapping of the complex frequency variable s into the complex variable z. The widely used bilinear transform method, discussed in this chapter, is based on this approach.

FIR filter design, on the other hand, is carried out directly from the digital filter specifications. The method outlined in this chapter is based on the truncated Fourier series expansion of the desired frequency response. Formulas for computing the Fourier series coefficients of some ideal frequency response specifications are included. To reduce the effect of the Gibbs phenomenon, the truncation can be carried out by applying a suitable window function. Some commonly used window functions are reviewed along with their properties. The effect of the Gibbs phenomenon can also be reduced by providing a smooth transition between the passband and the stopband.

Finally, the chapter considers the computer-aided design of digital filters. To this end it primarily discusses design algorithms that are available in the *Signal Processing Toolbox* of MATLAB as functions. For the design of IIR digital filters, MATLAB provides functions for designing digital Butterworth, Chebyshev, and elliptic filters. For the design of FIR filters, it includes functions for the windowed Fourier series approach, the Parks-McClellan algorithm, the least-squares algorithm, and the constrained least-squares algorithm. The Parks-McClellan algorithm develops a transfer function with equiripple passband and stopband magnitude responses and makes use of the Remez optimization algorithm.

7.12 Problems

7.1 Determine the peak ripple values δ_p and δ_s for each of the following sets of peak passband ripple α_p and minimum stopband attenuation α_s:

(a) $\alpha_p = 0.15$ dB, $\alpha_s = 41$ dB,   (b) $\alpha_p = 0.23$ dB, $\alpha_s = 73$ dB.

7.2 Determine the peak passband ripple α_p and minimum stopband attenuation α_s in dB for each of the following sets of peak ripple values δ_p and δ_s:

(a) $\delta_p = 0.01$, $\delta_s = 0.01$,   (b) $\delta_p = 0.035$, $\delta_s = 0.023$.

7.3 Let $H(z)$ be the transfer function of a lowpass digital filter with a passband edge at ω_p, stopband edge at ω_s, passband ripple of δ_p, and stopband ripple of δ_s, as indicated in Figure 7.1. Consider a cascade of two identical filters with a transfer function $H(z)$. What are the passband and stopband ripples of the cascade at ω_p and ω_s, respectively? Generalize the results for a cascade of M identical sections.

7.4 Let $H_{LP}(z)$ denote the transfer function of a real-coefficient lowpass filter with a passband edge at ω_p, stopband edge at ω_s, passband ripple of δ_p, and stopband ripple of δ_s as indicated in Figure 7.1. Sketch the magnitude response of the highpass transfer function $H_{LP}(-z)$ for $-\pi \leq \omega \leq \pi$ and determine its passband and stopband edges in terms of ω_p and ω_s.

7.5 Consider the transfer function $G(z) = H_{LP}(e^{j\omega_o}z)$, where $H_{LP}(z)$ is the lowpass transfer function of Problem 7.4. Sketch its magnitude response for $-\pi \leq \omega \leq \pi$, and determine its passband and stopband edge frequencies in terms of ω_p, ω_s, and ω_o.

7.6 The impulse invariance method is another approach to the design of a causal IIR digital filter $G(z)$ based on the transformation of a prototype causal analog transfer function $H_a(s)$. If $h_a(t)$ is the impulse response of $H_a(s)$, in the impulse invariance method, we require that the unit sample response $g[n]$ of $G(z)$ be given by the sampled version of $h_a(t)$ sampled at uniform intervals of T seconds, i.e.,

$$g[n] = h_a(nT), \quad n = 0, 1, 2, \ldots.$$

(a) Show that $G(z)$ and $H_a(s)$ are related through

$$G(z) = \mathcal{Z}\{g[n]\} = \mathcal{Z}\{h_a(nT)\}$$

$$= \frac{1}{T} \sum_{k=-\infty}^{\infty} H_a\left(s + j\frac{2\pi k}{T}\right)\bigg|_{s=(1/T)\ln z}. \tag{7.144}$$

(b) Show that the transformation

$$s = \frac{1}{T} \ln z, \tag{7.145}$$

has the desirable properties enumerated in Section 7.1.3.

(c) Develop the condition under which the frequency response $G(e^{j\omega})$ of $G(z)$ will be a scaled replica of the frequency response $H_a(j\Omega)$ of $H_a(s)$.

(d) Show that the normalized digital angular frequency ω is related to the analog angular frequency Ω as

$$\omega = \Omega T. \tag{7.146}$$

7.7 Let

$$H_a(s) = \frac{A}{s + \alpha} \tag{7.147}$$

be a causal first-order analog transfer function. Show that the causal first-order digital transfer function $G(z)$ obtained from $H_a(s)$ via the the impulse invariance method is given by

$$G(z) = \frac{A}{1 - e^{-\alpha T}z^{-1}}. \tag{7.148}$$

7.8 Let

$$H_a(s) = \frac{\lambda}{(s + \beta)^2 + \lambda^2} \tag{7.149}$$

be a causal second-order analog transfer function. Show that the causal second-order digital transfer function $G(z)$ obtained from $H_a(s)$ via the impulse invariance method is given by

$$G(z) = \frac{ze^{-\beta T} \sin \lambda T}{z^2 - 2ze^{-\beta T} \cos \lambda T + e^{-2\beta T}}. \tag{7.150}$$

7.9 Let

$$H_a(s) = \frac{s + \beta}{(s + \beta)^2 + \lambda^2} \tag{7.151}$$

be a causal second-order analog transfer function. Show that the causal second-order digital transfer function $G(z)$ obtained from $H_a(s)$ via the the impulse invariance method is given by

$$G(z) = \frac{z^2 - ze^{-\beta T}\cos\lambda T}{z^2 - 2ze^{-\beta T}\cos\lambda T + e^{-2\beta T}}. \tag{7.152}$$

7.10 Show that the digital transfer function $G(z)$ obtained from an arbitrary rational analog transfer function $H_a(s)$ with simple poles via the impulse invariance method is given by

$$G(z) = \sum_{\substack{\text{all poles of} \\ H_a(s)}} \text{Residues}\left[\frac{H_a(s)}{1 - e^{sT}z^{-1}}\right]. \tag{7.153}$$

7.11 Verify the relation between Eqs. (7.147) and (7.148) using the above formula.

7.12 Determine the digital transfer functions obtained by transforming the following causal analog transfer functions using the impulse invariance method. Assume $T = 0.2$ sec.

(a) $H_a(s) = \dfrac{16(s + 2)}{(s + 3)(s^2 + 2s + 5)}$, (b) $H_b(s) = \dfrac{4s^2 + 10s + 8}{(s^2 + 2s + 3)(s + 1)}$, (c) $H_c(s) = \dfrac{3s^3 + 7s^2 + 10s + 7}{(s^2 + s + 1)(s^2 + 2s + 3)}$.

7.13 The following causal IIR digital transfer functions were designed using the impulse invariance method with $T = 0.3$ sec. Determine their respective parent causal analog transfer functions.

(a) $G_a(z) = \dfrac{2z}{z - e^{-0.9}} + \dfrac{3z}{z - e^{-1.2}}$, (b) $G_b(z) = \dfrac{z^2 - ze^{-0.6}\cos(0.9)}{z^2 - 2ze^{-0.6}\cos(0.9) + e^{-1.2}}$.

7.14 The following causal IIR digital transfer functions were designed using the bilinear transformation method with $T = 2$. Determine their respective parent causal analog transfer functions.

(a) $G_a(z) = \dfrac{5z^2 + 4z - 1}{8z^2 + 4z}$, (b) $G_b(z) = \dfrac{8(z^3 + 3z^2 + 3z + 1)}{(3z + 1)(7z^2 + 6z + 3)}$.

7.15 An IIR digital lowpass filter is to be designed by transforming an analog lowpass filter with a passband edge frequency F_p at 0.5 kHz using the impulse invariance method with $T = 0.5$ ms. What is the normalized passband edge angular frequency ω_p of the digital filter if there is no aliasing? What would be the normalized passband edge angular frequency ω_p of the digital filter if it is designed using the bilinear transformation with $T = 0.5$ ms?

7.16 An IIR lowpass digital filter has a normalized passband edge frequency $\omega = 0.3\pi$. What is the passband edge frequency in Hz of the prototype analog lowpass filter if the digital filter has been designed using the impulse invariance method with $T = 0.1$ ms? What is the passband edge frequency in Hz of the prototype analog lowpass filter if the digital filter has been designed using the bilinear transformation method with $T = 0.1$ ms?

7.17 Design an IIR lowpass digital filter $G(z)$ with a maximally flat magnitude response and meeting the specifications given by Eqs. (7.38a) and (7.38b) using the impulse invariance method. How does this filter compare with that designed via the bilinear transformation method in Section 7.3?

7.18 This problem illustrates how aliasing can be suitably exploited in order to realize interesting frequency response characteristics. An ideal causal analog lowpass filter with an impulse response $h_a(t)$ has a frequency response given by

$$H_a(j\Omega) = \begin{cases} 1, & |\Omega| < \Omega_c, \\ 0, & \text{otherwise.} \end{cases}$$

Let $H_1(e^{j\omega})$ and $H_2(e^{j\omega})$ be the frequency responses of digital filters obtained by sampling $h_a(t)$ at $t = nT$, where $T = 3\pi/2\Omega_c$ and π/Ω_c, respectively. Assume the transfer functions are later normalized so that $H_1(e^{j0}) = H_2(e^{j0}) = 1$.

  (a) Sketch the frequency responses $G_1(e^{j\omega})$ and $G_2(e^{j\omega})$ of the two digital filter structures shown in Figure P7.1.

  (b) What type of filters are $G_1(z)$ and $G_2(z)$ (lowpass, highpass, etc.)?

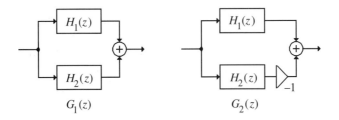

Figure P7.1

7.19 This problem illustrates the method of digital filter design by the *step-response invariance method* [Hay70], [Whi71]. Let $H_a(s)$ be a real-coefficient causal and stable analog transfer function. Its unit step response $h_{a,\mu}(t)$ is given by the inverse Laplace transform of $H_a(s)/s$. Let $G(z)$ be a causal digital transfer function with a unit step response $g_\mu[n]$ such that

$$g_\mu[n] = h_{a,\mu}(nT), \qquad n = 0, 1, 2, \ldots \tag{7.154}$$

Determine the expression for $G(z)$ if

$$H_a(s) = \sum_{k=0}^{R} \frac{A_k}{s - \alpha_k}, \tag{7.155}$$

and show that if $\omega \ll 1/T$ and

$$\frac{H_a(j\omega)}{j\omega} \approx 0, \qquad \text{for } |\omega| \geq \frac{1}{2T}, \tag{7.156}$$

then

$$G(e^{j\omega}) \cong H_a(j\omega), \qquad \text{for } |\omega| < \frac{1}{2T}. \tag{7.157}$$

7.20 An LTI continuous-time system described by a linear constant coefficient differential equation is often solved numerically by developing an equivalent linear constant coefficient difference equation by replacing the derivative operators in the differential equation by their approximate difference equation representation. A commonly used difference equation representation of the first derivative at time $t = nT$ is given by

$$\left. \frac{d\,y(t)}{dt} \right|_{t=nT} \cong \frac{1}{T}(y[n] - y[n-1]),$$

where T is the sampling period and $y[n] = y(nT)$. The corresponding mapping from the s-domain to the z-domain is obtained by replacing s with the *backward difference operator* $\frac{1}{T}(1 - z^{-1})$. Investigate the above mapping and its properties. Does a stable $H_a(s)$ result in a stable $H(z)$? How useful is this mapping for digital filter design?

7.21 Let $H_a(s)$ be a real-coefficient causal and stable analog transfer function with a magnitude response bounded above by unity. Show that the digital transfer function $G(z)$ obtained by a bilinear transformation of $H_a(s)$ is a BR function.

7.22 Show that the second-order analog bandpass transfer function

$$H_a(s) = \frac{Bs}{s^2 + Bs + \Omega_o^2} \tag{7.158}$$

has a magnitude response that goes to zero values at $\Omega = 0$ and ∞, and has a value of unity at $\Omega = \Omega_o$. If Ω_1 and Ω_2, $\Omega_2 > \Omega_1$, denote the frequencies at which the gain is down by -3 dB, it can be shown that the 3-dB bandwidth defined by $(\Omega_2 - \Omega_1)$ is equal to B. Develop the second-order digital transfer function $G(z)$ from the above $H_a(s)$ via the bilinear transformation. Show that $G(z)$ can be expressed in the form of Eq. (4.113) if the constants α and β are chosen according to Eqs. (7.36a) and (7.36b).

7.23 We have shown in Section 6.7.2 that the transfer function $G(z)$ of a second-order IIR notch filter as given in Eq. (7.35) can be expressed in the form $G(z) = \frac{1}{2}[1 + A_2(z)]$, where $A_2(z)$ is a second-order allpass transfer function given by Eq. (6.68). Consider a notch filter with a notch frequency at $\omega = \pi/2$. Show that a notch filter with multiple notch frequencies is obtained if z^{-1} is replaced with z^{-N} [Reg88]. What are the locations of the new notch frequencies?

7.24 A notch filter with N notch frequencies can be realized by replacing the allpass filter $A_2(z)$ in the above problem with a cascade of N second-order allpass filters [Jos99]. In this problem, we consider the design of a notch filter with two notch frequencies ω_1, ω_2, and corresponding 3-dB notch bandwidths B_1, B_2. We thus replace $A_2(z)$ with a fourth-order allpass transfer function $A_4(z)$,

$$A_4(z) = \left(\frac{\alpha_1 - \beta_1(1 + \alpha_1)z^{-1} + z^{-2}}{1 - \beta_1(1 + \alpha_1)z^{-1} + \alpha_1 z^{-2}}\right)\left(\frac{\alpha_2 - \beta_2(1 + \alpha_2)z^{-1} + z^{-2}}{1 - \beta_2(1 + \alpha_2)z^{-1} + \alpha_2 z^{-2}}\right),$$

obtained by cascading two second-order allpass filters. The constants α_1 and α_2 are chosen as

$$\alpha_i = \frac{1 - \tan(B_i/2)}{1 + \tan(B_i/2)}, \quad i = 1, 2.$$

The transfer function of the modified structure is now given by $H(z) = \frac{1}{2}[1 + A_4(z)] = N(z)/D(z)$.

(a) Show that $N(z)$ is a mirror-image polynomial of the form $a(1 + b_1 z^{-1} + b_2 z^{-2} + b_1 z^{-3} + z^{-4})$, and express the constants b_1 and b_2 in terms of the coefficients of $A_4(z)$.

(b) Show that $a = (1 + \alpha_1\alpha_2)/2$.

(c) By setting $N(e^{j\omega_i}) = 0$, $i = 1, 2$, solve for the constants b_1 and b_2 in terms of ω_1 and ω_2. From the equations in parts (a) and (b), determine the expressions for the coefficients β_1 and β_2.

(d) Using the design equations derived above, design a double notch filter with the following specifications: $\omega_1 = 0.3\pi$, $\omega_2 = 0.5\pi$, $B_1 = 0.1\pi$, and $B_2 = 0.15\pi$. Using MATLAB plot the magnitude response of the designed notch filter.

7.25 Let $H_{LP}(z)$ be an IIR lowpass transfer function with a zero (pole) at $z = z_k$. Let $H_D(\hat{z})$ denote the lowpass transfer function obtained by applying the lowpass-to-lowpass transformation given in Table 7.1 which moves the zero (pole) at $z = z_k$ of $H_{LP}(z)$ to a new location at $\hat{z} = \hat{z}_k$. Express $\hat{z}_k$ in terms of z_k. If $H_{LP}(z)$ has a zero at $z = -1$, show that $H_D(\hat{z})$ also has a zero at $z = -1$.

7.26 Let $H_{LP}(z)$ be an IIR lowpass transfer function with a zero (pole) at $z = z_k$. Let $H_D(\hat{z})$ denote the bandpass transfer function obtained by applying the lowpass-to-bandpass transformation given in Table 7.1 which moves the zero (pole) at $z = z_k$ of $H_{LP}(z)$ to a new location at $\hat{z} = \hat{z}_k$. Express $\hat{z}_k$ in terms of z_k. If $H_{LP}(z)$ has a zero at $z = -1$, show that $H_D(\hat{z})$ also has a zero at $z = \pm 1$.

7.27 A second-order lowpass IIR digital filter with a 3-dB cutoff frequency at $\omega_c = 0.42\pi$ has a transfer function

$$G_{LP}(z) = \frac{0.223(1 + z^{-1})^2}{1 - 0.2952z^{-1} + 0.187z^{-2}}. \qquad (7.159)$$

Design a second-order lowpass filter $H_{LP}(z)$ with a 3-dB cutoff frequency at $\hat{\omega}_c = 0.57\pi$ by transforming the above lowpass transfer function using a lowpass-to-lowpass spectral transformation. Using MATLAB plot the gain responses of the two lowpass filters on the same figure.

7.28 Design a second-order highpass filter $H_{HP}(z)$ with a 3-dB cutoff frequency at $\hat{\omega}_c = 0.61\pi$ by transforming the lowpass transfer function of Eq. (7.159) using a lowpass-to-highpass spectral transformation. Using MATLAB plot the gain responses of the highpass and the lowpass filters on the same figure.

7.29 A second-order lowpass Type 1 Chebyshev IIR digital filter $G_{LP}(z)$ with a 0.5-dB cutoff frequency at $\omega_c = 0.27\pi$ has a transfer function

$$G_{LP}(z) = \frac{0.1494(1 + z^{-1})^2}{1 - 0.7076z^{-1} + 0.3407z^{-2}}. \qquad (7.160)$$

Design a fourth-order bandpass filter $H_{BP}(z)$ with a center frequency at $\hat{\omega}_c = 0.45\pi$ by transforming the above lowpass transfer function using a lowpass-to-bandpass spectral transformation. Using MATLAB plot the gain responses of the lowpass and the bandpass filters on the same figure.

7.30 A third-order Type 1 Chebyshev highpass filter with a passband edge at $\omega_p = 0.6\pi$ has a transfer function

$$G_{HP}(z) = \frac{0.0916(1 - 3z^{-1} + 3z^{-2} - z^{-3})}{1 + 0.7601z^{-1} + 0.7021z^{-2} + 0.2088z^{-3}}.$$

Design a highpass filter $H_{HP}(z)$ with a passband edge at $\omega_p = 0.5\pi$ by transforming the above highpass transfer function using the lowpass-to-lowpass spectral transformation. Using MATLAB plot the gain responses of the two highpass filters on the same figure.

7.31 Design a second-order bandpass filter with a center frequency at $\omega_o = 0.5\pi$ by transforming the bandpass transfer function of Eq. (4.117b) using the lowpass-to-lowpass spectral transformation. Using MATLAB plot the gain responses of the two bandpass filters on the same figure.

7.32 A second-order notch filter with a notch frequency at 100 Hz and operating a sampling rate of 400 Hz is to be designed. Design this filter by transforming the notch transfer function of Example 7.8 using the lowpass-to-lowpass spectral transformation. Using MATLAB plot the gain responses of the two notch filters on the same figure.

7.33 Design a lowpass filter with a cutoff at $\omega_p = 0.5\pi$ by transforming the highpass transfer function of Problem 7.30 using the lowpass-to-highpass spectral transformation. Using MATLAB plot the gain responses of the highpass and the lowpass filters on the same figure.

7.34 Verify the expression for the impulse-response coefficients $h_{ML}[n]$ given in Eq. (7.65) for the zero-phase multiband filter with a frequency response $H_{ML}(e^{j\omega})$ defined in Eq. (7.64) and shown in Figure 7.14.

7.35 Show that the ideal Hilbert transformer with a frequency response $H_{HT}(j^\omega)$ defined in Eq. (7.66) has an impulse response $h_{HT}[n]$ as given in Eq. (7.67). Since the impulse response is doubly infinite, the ideal discrete-time Hilbert transformer is not realizable. To make it realizable, the impulse response has to be truncated to $|n| \leq M$. What type of linear-phase FIR filter is the truncated impulse response? Plot the frequency response of the truncated approximation for various values of M. Comment on your results.

7.36 Let $\mathcal{H}\{\cdot\}$ denote the ideal operation of Hilbert transformation defined by

$$\mathcal{H}\{x[n]\} = \sum_{\ell=-\infty}^{\infty} h_{HT}[n - \ell]x[\ell],$$

where $h_{HT}[n]$ is as given in Eq. (7.67). Evaluate the following quantities:

$$\text{(a) } \mathcal{H}\{\mathcal{H}\{\mathcal{H}\{\mathcal{H}\{x[n]\}\}\}\}, \quad \text{(b) } \sum_{\ell=-\infty}^{\infty} x[\ell]\mathcal{H}\{x[\ell]\}.$$

7.37 Show that the ideal differentiator with a frequency response $H_{DIF}(e^{j\omega})$ defined in Eq. (7.68) has an impulse response $h_{DIF}[n]$ as given in Eq. (7.69). Since the impulse response is doubly infinite, the ideal discrete-time differentiator is not realizable. To make it realizable, the impulse response has to be truncated to $|n| \leq M$. What type of linear-phase FIR filter is the truncated impulse response? Plot the frequency response of the truncated approximation for various values of M. Comment on your results.

7.38 Develop the expression for the impulse response $\hat{h}_{HP}[n]$ of a causal highpass FIR filter of length $N = 2M + 1$ obtained by truncating and shifting the impulse response $h_{HP}[n]$ of the ideal highpass filter given by Eq. (7.61). Show that the causal lowpass FIR filter $\hat{h}_{LP}[n]$ of Eq. (7.60) and $\hat{h}_{HP}[n]$ are a delay-complementary pair.

7.39 Determine the impulse response $h_{LLP}[n]$ of a zero-phase ideal linear passband lowpass filter characterized by a frequency response shown in Figure P7.2(a).

7.40 Determine the impulse response $h_{BLDIF}[n]$ of a zero-phase ideal bandlimited differentiator characterized by a frequency response shown in Figure P7.2(b).

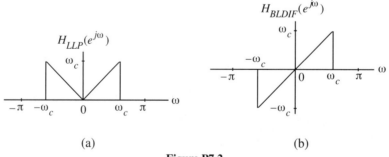

(a)   (b)

Figure P7.2

7.41 The desired frequency response of an ideal *integrator* is given by

$$H_{\text{int}}(e^{j\omega}) = \frac{1}{j\omega}. \tag{7.161}$$

Determine the transfer function $H_R(z)$ of an IIR integrator derived via the rectangular numerical integration method[16] and the transfer function $H_T(z)$ of an IIR integrator derived via the trapezoidal numerical integration method.[17] Using MATLAB plot the magnitude responses of $H_{\text{int}}(z)$, $H_R(z)$, and $H_T(z)$ for $T = 1$. Comment on your results.

---

[16] See Eq. (2.181).
[17] See Eq. (2.98).

7.42 An improved IIR digital integrator can be obtained by interpolating the rectangular and the trapezoidal integrators according to [Ala93]

$$H_N(z) = \frac{3}{4}H_R(z) + \frac{1}{4}H_T(z).$$

Using MATLAB plot the magnitude responses of $H_N(z)$, $H_R(z)$, and $H_T(z)$ for $T = 1$. Comment on your results.

7.43 Develop an IIR digital differentiator by inverting the IIR digital integrator of Problem 7.42 [Ala93]. Is this a stable transfer function? If not, develop a stable equivalent. Using MATLAB plot the magnitude responses of the ideal differentiator and the digital differentiator designed here. Comment on your results.

7.44 The magnitude response of an ideal *notch filter* $H_{\text{notch}}(z)$ is defined by

$$|H_{\text{notch}}(e^{j\omega})| = \begin{cases} 0, & |\omega| = \omega_o, \\ 1, & \text{otherwise}, \end{cases} \tag{7.162}$$

where ω_o is the *notch frequency*. Determine its impulse response $h_{\text{notch}}[n]$ [Yu90].

7.45 In this problem we consider the design of an FIR digital filter approximating a fractional delay z^{-D} :

$$z^{-D} \cong \sum_{n=0}^{N} h[n]z^{-n},$$

where the delay D is a positive real rational number.

(a) Show that the filter coefficients obtained using the Lagrange interpolation method [18] are given by [Lak96]

$$h[n] = \prod_{\substack{k=0 \\ k \neq n}}^{N} \frac{D-k}{n-k}, \qquad 0 \leq n \leq N.$$

(b) Design a length-17 FIR fractional delay filter with a delay of 100/11 samples. Plot using MATLAB the magnitude response of the designed filter along with that of the ideal fractional delay filter. Comment on your results.

7.46 A maximally flat group delay IIR allpass filter can also be designed to approximate a fractional delay z^{-D}:

$$z^{-D} \cong \frac{d_N + d_{N-1}z^{-1} + \cdots + d_1 z^{-(N-1)} + z^{-N}}{1 + d_1 z^{-1} + d_2 z^{-2} + \cdots + d_{N-1}z^{-(N-1)} + d_N z^{-N}}.$$

By expressing the desired positive delay as $D = N + \delta$, where N is a positive integer and δ a fractional number, it can be shown that the coefficient $\{d_k\}$ of the allpass filter is given by [Fet71]

$$d_k = (-1)^k C_k^N \prod_{n=0}^{N} \frac{D-N+n}{D-N+k+n},$$

where $C_k^N = N!/k!(N-k)!$ is a binomial coefficient. Design an allpass fractional delay filter of order 9 with a delay of 100/11 samples. Plot using MATLAB the magnitude response of the designed filter along with that of the ideal fractional delay filter. Comment on your results.

---

[18]See Section 10.5.2.

7.47 An ideal zero-phase comb filter with notches at a fundamental frequency ω_o and its harmonics has a frequency response given by

$$H_{\text{comb}}(e^{j\omega}) = \begin{cases} 0, & \omega = k\omega_o, \ 0 \le k \le M, \\ 1, & \text{otherwise.} \end{cases} \tag{7.163}$$

If the input to the comb filter is of the form $x[n] = s[n] + r[n]$, where $s[n]$ is the desired signal and $r[n] = \sum_{k=0}^{M} A_k \sin(k\omega_o n + \phi_k)$ is the harmonic interference with a fundamental frequency ω_o, the comb filter suppresses the interference and generates $s[n]$ as its output. Let $D = 2\pi/\omega_o$ denote the fractional sample delay.

(a) Show that $r[n - D] = r[n]$.

(b) Next, by computing the output $y[n]$ of a filter $H(z) = 1 - z^{-D}$ whose input is $x[n]$, show that $y[n]$ does not contain any harmonic interference.

(c) Even though the filter $H(z) = 1 - z^{-D}$ eliminates the harmonic interference completely, it does not have a unity magnitude at frequencies $\omega \ne k\omega_o$ thus introducing signal distortion at its output. The distortion in the passband of $H(z)$ can be eliminated by modifying the filter according to [Pei98]

$$H_c(z) = \frac{1 - z^{-D}}{1 - \rho^D z^{-D}},$$

where $0 < \rho < 1$. In practice, ρ should be close to 1. Using MATLAB plot the magnitude response of $H_c(z)$ for $\omega_o = 0.22\pi$ and $\rho = 0.99$.

(d) Develop an efficient realization of the improved comb filter $H_c(z)$.

7.48 Using the method of Problem 7.47 and the FIR fractional delay filter design method of Problem 7.45, design a comb filter of order 16 for $\omega_o = 0.22\pi$, and $\rho = 0.99$. Using MATLAB plot the magnitude response of the designed filter.

7.49 Using the method of Problem 7.47 and the allpass fractional delay filter design method of Problem 7.46, design a comb filter of order 9 for $\omega_o = 0.22\pi$ and $\rho = 0.99$. Using MATLAB plot the magnitude response of the designed filter.

7.50 By computing the inverse discrete-time Fourier transform of the frequency response $H_{LP}(e^{j\omega})$ of the zero-phase modified lowpass filter of Figure 7.25(a) with a first-order spline as the transition function, verify the expression for its impulse response $h_{LP}[n]$ as given in Eq. (7.87). Show that $h_{LP}[n]$ of Eq. (7.87) can also be derived by computing the inverse discrete-time Fourier transform of the derivative function $G(e^{j\omega})$ of Figure 7.25(b) and then using the differentiation-in-frequency property of the discrete-time Fourier transform given in Table 3.2.

7.51 Show that the impulse response $h_{LP}[n]$ of the zero-phase modified lowpass filter with a Pth-order spline as the transition function is given by Eq. (7.88).

7.52 The frequency response of a zero-phase lowpass filter with a passband edge at ω_p, a stopband edge at ω_s, and a raised cosine transition function is given by [Bur92], [Par87].

$$H_{LP}(e^{j\omega}) = \begin{cases} 1, & 0 \le |\omega| < \omega_p, \\ \frac{1}{2}\left(1 + \cos\left(\frac{\pi(\omega - \omega_p)}{\omega_s - \omega_p}\right)\right), & \omega_p \le |\omega| \le \omega_s, \\ 0, & \omega_s < |\omega| < \pi. \end{cases} \tag{7.164}$$

Show that its impulse response is of the form

$$h_{LP}[n] = \left[\frac{\cos(\Delta\omega n/2)}{1 - (\Delta\omega/\pi)^2 n^2}\right] \cdot \frac{\sin(\omega_c n)}{\pi n}, \tag{7.165}$$

where $\Delta\omega = \omega_s - \omega_p$ and $\omega_c = (\omega_p + \omega_s)/2$.

7.53 Show that the length-$(2M + 1)$ *Bartlett window* sequence given by

$$w[n] = 1 - \frac{|n|}{M + 1}, \qquad -M \leq n \leq M, \tag{7.166}$$

can be obtained by a linear convolution of two scaled length-N rectangular windows. Determine N and the scale factor. From this relation, determine the expression for the frequency response of the length-$(2M+1)$ Bartlett window. Determine the main lobe width Δ_{ML} and the relative sidelobe level $A_{s\ell}$ of the Bartlett window sequence.

7.54 The length-$(2M + 1)$ Hann, Hamming, and Blackman window sequences given in Eqs. (7.74) to (7.76) are all of the form of raised cosine windows and can be expressed as

$$w_{GC}[n] = \left[\alpha + \beta \cos\left(\frac{2\pi n}{2M + 1}\right) + \gamma \cos\left(\frac{4\pi n}{2M + 1}\right) \right] w_R[n], \tag{7.167}$$

where $w_R[n]$ is a length-$(2M+1)$ rectangular window sequence. Express the Fourier transform of the above generalized cosine window in terms of the Fourier transform of the rectangular window $\Psi_R(e^{j\omega})$. From this expression, determine the Fourier transform of the Hann, Hamming, and Blackman window sequences.

7.55 Many applications require the fitting of a set of $2L+1$ equally spaced data samples $x[n]$ by a smooth polynomial $x_a(t)$ of degree N where $N < 2L$. In the least-squares fitting approach the polynomial coefficients $\alpha_i, i = 0, 1, \ldots < N$, are determined so that the mean-square error

$$\varepsilon(\alpha_i) = \sum_{k=-L}^{L} \{x[k] - x_a(k)\}^2 \tag{7.168}$$

is a minimum [Ham89]. In smoothing a very long data sequence $x[n]$ based on the least-squares fitting approach, the central sample in a set of consecutive $2L + 1$ data samples is replaced by the polynomial coefficient minimizing the corresponding mean-square error.

(a) Develop the smoothing algorithm for $N = 1$ and $L = 5$, and show that it is a moving average FIR filter of length 5.

(b) Develop the smoothing algorithm for $N = 2$ and $L = 5$. What type of digital filter is represented by this algorithm?

(c) By comparing the frequency responses of the previous two FIR smoothing filters, select the filter that provides better smoothing.

7.56 An improved smoothing algorithm is *Spencer's* 15-point smoothing formula given by [Ham89]

$$\begin{aligned}
y[n] = \tfrac{1}{320} \{ &-3x[n - 7] - 6x[n - 6] - 5x[n - 5] + 3x[n - 4] \\
&+ 21x[n - 3] + 46x[n - 2] + 67x[n - 1] + 74x[n] \\
&+ 67x[n + 1] + 46x[n + 2] + 21x[n + 3] + 3x[n + 4] \\
&- 5x[n + 5] - 6x[n + 6] - 3x[n + 7] \}.
\end{aligned} \tag{7.169}$$

Evaluate its frequency response and, comparing it with that of the two smoothing filters of Problem 7.55, show why Spencer's formula yields the better result.

7.57 In Problem 7.3, we considered filtering by a cascade of a number of identical filters. While the cascade provides more stopband attenuation than that obtained by a single filter section, it also increases the passband ripple or in effect decreases the passband width for a given maximum passband deviation. In the case of an FIR filter $H(z)$ with a symmetric impulse response, improved passband and stopband performances can be achieved by employing the *filter sharpening approach* [Kai77] in which the overall system $G(z)$ is implemented as

$$G(z) = \sum_{\ell=1}^{L} \alpha_\ell [H(z)]^\ell, \tag{7.170}$$

where $\{\alpha_\ell\}$ are real constants. In this problem, we outline the method of selecting the weighting coefficients $\{\alpha_\ell\}$ for a given L. It follows from above that $G(z)$ is also an FIR filter with a symmetric impulse response. Let x denote a specific value of the amplitude response of $H(z)$ at a given angular frequency ω. If we denote the value of the amplitude response of $G(z)$ at this value of ω as $P(x)$, then it is related to x through

$$P(x) = \sum_{\ell=1}^{L} \alpha_\ell x^\ell. \tag{7.171}$$

$P(x)$ is called the *amplitude change function*. For a BR transfer function $H(z)$, $0 \le x \le 1$, where $x = 0$ is in the stopband and $x = 1$ is in the passband. If we further desire $G(z)$ to be a BR transfer function, then the amplitude change function must satisfy the two basic properties $P(0) = 0$ and $P(1) = 1$. Additional conditions on the amplitude change function are obtained by constraining the behavior of its slope at $x = 0$ and $x = 1$. To improve the performance of $G(z)$ in the stopband, we need to ensure

$$\left. \frac{d^k P(x)}{dx^k} \right|_{x=0} = 0, \quad k = 1, 2, \ldots, n, \tag{7.172}$$

and to improve the performance of $G(z)$ in the passband, we need to ensure

$$\left. \frac{d^k P(x)}{dx^k} \right|_{x=1} = 0, \quad k = 1, 2, \ldots, m, \tag{7.173}$$

where $m + n = L - 1$. Determine the coefficients $\{\alpha_\ell\}$ for $L = 3, 4$, and 5.

7.58 Consider a Type 3 linear-phase FIR filter with an amplitude response as given in Eq. (7.102). Show that if the amplitude response is symmetric, i.e., $\breve{H}(\omega) = \breve{H}(\pi - \omega)$, then it is possible to choose the parameters $c[k]$ of Eq. (7.102) so that the even-indexed impulse response samples $h[n]$ are zero.

7.59 In the frequency sampling approach of FIR filter design, the specified frequency response $H_d(e^{j\omega})$ is first uniformly sampled at M equally spaced points $\omega_k = 2\pi k / M$, $0 \le k \le M - 1$, providing M *frequency samples* $H[k] = H_d(e^{j\omega_k})$. These M frequency samples constitute an M-point DFT $H[k]$ whose M-point inverse-DFT thus yields the impulse response coefficients $h[n]$ of the FIR filter of length M [Gol69a]. The basic assumption here is that the specified frequency response is uniquely characterized by the M frequency samples and, hence, can be fully recovered from these samples.

  (a) Show that the transfer function $H(z)$ of the FIR filter can be expressed as

$$H(z) = \frac{1 - z^{-M}}{M} \sum_{k=0}^{M-1} \frac{H[k]}{1 - W_M^{-k} z^{-1}}.$$

  (b) Develop a realization of the FIR filter based on the above expression.

  (c) Show that the frequency response $H(e^{j\omega})$ of the FIR filter designed via the frequency sampling approach has exactly the specified frequency samples $H(e^{j\omega_k}) = H[k]$ at $\omega_k = 2\pi k / M$, $0 \le k \le M - 1$.

7.60 Let $|H_d(e^{j\omega})|$ denote the desired magnitude response of a real linear-phase FIR filter of length M.

(a) For M odd (Type 1 FIR filter), show that the DFT samples $H[k]$ needed for a frequency sampling–based design are given by

$$H[k] = \begin{cases} \left|H_d\left(e^{j2\pi k/M}\right)\right| e^{-j2\pi k(M-1)/2M}, & k = 0, 1, \ldots, \frac{M-1}{2}, \\ \left|H_d\left(e^{j2\pi k/M}\right)\right| e^{j2\pi(M-k)(M-1)/2M}, & k = \frac{M+1}{2}, \ldots, M-1. \end{cases} \qquad (7.174)$$

(b) For M even (Type 2 FIR filter), show that the DFT samples $H[k]$ needed for a frequency sampling–based design are given by

$$H[k] = \begin{cases} \left|H_d\left(e^{j2\pi k/M}\right)\right| e^{-j2\pi k(M-1)/2M}, & k = 0, 1, \ldots, \frac{M}{2}-1, \\ 0, & k = \frac{M}{2}, \\ \left|H_d\left(e^{j2\pi k/M}\right)\right| e^{j2\pi(M-k)(M-1)/2M}, & k = \frac{M}{2}+1, \ldots, M-1. \end{cases} \qquad (7.175)$$

7.61 Design a linear-phase FIR lowpass filter of length 17 with a passband edge at $\omega_p = 0.5\pi$ using the frequency sampling approach. Assume an ideal brickwall characteristic for the desired magnitude response.

(a) Using Eq. (7.174) develop the exact values for the desired frequency samples.

(b) Using MATLAB plot the magnitude response of the designed filter.

7.62 Design a linear-phase FIR lowpass filter of length 37 with a passband edge at $\omega_p = 0.3\pi$ using the frequency sampling approach. Assume an ideal brickwall characteristic for the desired magnitude response.

(a) Using Eq. (7.174) develop the exact values for the desired frequency samples.

(b) Using MATLAB plot the magnitude response of the designed filter.

7.63 By solving Eq. (7.118), derive the value of ε given by Eq. (7.119).

7.64 Show that the condition of Eq. (7.143) on the impulse response samples $h_{HT}[n]$ of an ideal Hilbert transformer cannot be met by a Type 4 linear-phase FIR filter.

7.65 The *warped discrete Fourier transform* (WDFT) can be employed to determine the N frequency samples of the z-transform $X(z)$ of a length-N sequence $x[n]$ at a warped frequency scale on the unit circle. The N-point WDFT $\check{X}[k]$ of $x[n]$ is given by the N equally spaced frequency samples on the unit circle of the modified z-transform $X(\check{z})$ obtained by applying an allpass first-order spectral transformation to $X(z)$ [Mit98a]:

$$X(\check{z}) = X(z)\Big|_{z^{-1} = \frac{-\alpha + \check{z}^{-1}}{1 - \alpha\check{z}^{-1}}} = \frac{P(\check{z})}{D(\check{z})}, \qquad (7.176)$$

where $|\alpha| < 1$. Thus, the N-point WDFT $\check{X}[k]$ of $x[n]$ is given by

$$\check{X}[k] = X(\check{z})\big|_{\check{z} = e^{j2\pi k/N}}, \quad 0 \le k \le N-1. \qquad (7.177)$$

(a) Develop the expressions for $P(\check{z})$ and $D(\check{z})$.

(b) If we denote

$$P(\check{z}) = \sum_{n=0}^{N-1} p[n]\check{z}^{-n} \text{ and } D(\check{z}) = \sum_{n=0}^{N-1} d[n]\check{z}^{-n},$$

show that $\check{X}[k] = P[k]/D[k]$, where $P[k]$ and $D[k]$ are, respectively, the N-point DFTs of the sequences $p[n]$ and $d[n]$.

(c) If we denote $\mathbf{P} = [p[0]\ p[1]\ \cdots\ p[N-1]]^T$, and $\mathbf{X} = [x[0]\ x[1]\ \cdots x[N-1]]^T$, show that $\mathbf{P} = \mathbf{Q} \cdot \mathbf{X}$, where $\mathbf{Q} = [q_{r,s}]$ is a real $N \times N$ matrix whose first row is given by $q_{0,s} = \alpha^s$, first column is given by $q_{r,0} = {}^{N-1}C_r\alpha^r$, and remaining elements $q_{r,s}$ can be derived using the recursion relation

$$q_{r,s} = q_{r-1,s-1} + \alpha q_{r,s-1} - \alpha q_{r-1,s}.$$

7.13   MATLAB Exercises

M 7.1 Design a digital filter by an impulse invariant transformation of a fourth-order analog Bessel transfer function for the following values of sampling frequencies: (a) $F_T = 1$ Hz, and (b) $F_T = 2$ Hz. Plot the gain and the group delay responses of both designs using MATLAB, and compare these responses with that of the original Bessel transfer function. Comment on your results.

M 7.2 Design a digital Butterworth lowpass filter operating at a sampling rate of 80 kHz with a 0.5-dB cutoff frequency at 4 kHz and a minimum stopband attenuation of 45 dB at 20 kHz using the bilinear transformation method. Determine the order of the analog filter prototype using the formula given in Eq. (5.33) and then design the analog prototype filter using the M-file `buttap` of MATLAB. Transform the analog filter transfer function to the desired digital transfer function using the M-file `bilinear`. Plot the gain and phase responses using MATLAB. Show all steps used in the design.

M 7.3 Modify Program 7_3 to design a digital Butterworth lowpass filter using the bilinear transformation method. The input data required by the modified program should be the desired passband and stopband edges, and maximum passband deviation and the minimum stopband attenuation in dB. Using the modified program, design the digital Butterworth lowpass filter of Exercise M7.2.

M 7.4 Using the M-file `impinvar` design the digital Butterworth lowpass filter of Exercise M7.2. Use the analog prototype filter order determined using the formula given in Eq. (5.36).

M 7.5 Design a digital Type 1 Chebyshev lowpass filter operating at a sampling rate of 80 kHz with a passband edge frequency at 4 kHz, a passband ripple of 0.5 dB, and a minimum stopband attenuation of 45 dB at 20 kHz using the impulse invariance method and the bilinear transformation method. Determine the order of the analog filter prototype using the formula given in Eq. (5.41) and then design the analog prototype filter using the M-file `cheb1ap` of MATLAB. Transform the analog filter transfer function to the desired digital transfer function using the M-file `bilinear`. Plot the gain and phase responses of both designs using MATLAB. Compare the performances of the two filters. Show all steps used in the design.

M 7.6 Modify Program 7_2 to design a digital Type 1 Chebyshev lowpass filter using the bilinear transformation method. The input data required by the modified program should be the desired passband and stopband edges, and maximum passband deviation and the minimum stopband attenuation in dB. Using the modified program, design the digital Type 1 Chebyshev lowpass filter of Exercise M7.5.

M 7.7 Using the M-file `impinvar` write a MATLAB program to design a digital Type 1 Chebyshev lowpass filter using the impulse invariance method. The input data required by the modified program should be the desired passband and stopband edges, and maximum passband deviation and the minimum stopband attenuation in dB. Using your program, design the digital Type 1 Chebyshev lowpass filter of Exercise M7.5.

M 7.8 Design a digital elliptic lowpass filter operating at a sampling rate of 80 kHz with a passband edge frequency at 4 kHz, a stopband edge frequency at 20 kHz, passband ripple of 0.5 dB, and a stopband ripple of 45 dB using the impulse invariance method and the bilinear transformation method. Determine the order of the analog filter prototype using the formula given in Eq. (5.51) and then design the analog prototype filter using the M-file `ellipap` of MATLAB. Transform the analog filter transfer function to the desired digital transfer function using the M-file `bilinear`. Plot the gain and phase responses of both designs using MATLAB. Compare the performances of the two filters. Show all steps used in the design.

M 7.9 Modify Program 7_3 to design a digital elliptic lowpass filter using the bilinear transformation method. The input data required by the modified program should be the desired passband and stopband edges, and the maximum passband deviation and the minimum stopband attenuation in dB. Using the modified program, design the digital elliptic lowpass filter of Exercise M7.8.

M 7.10 Design using the bilinear transformation method a digital elliptic highpass filter operating at a sampling rate of 1 MHz with the following specifications: passband edge at 325 kHz, stopband edge at 225 kHz, peak passband ripple of 0.5 dB, and minimum stopband attenuation of 50 dB. (a) What are the specifications of the analog highpass filter? (b) What are the specifications of the analog prototype lowpass filter? (c) Show all pertinent transfer functions. Plot the gain responses of the prototype analog lowpass filter, analog highpass filter, and desired digital highpass filter. Show all steps.

M 7.11 Design using the bilinear transformation method a digital Type 2 Chebyshev bandpass filter operating at a sampling rate of 2500 Hz with the following specifications: passband edges at 560 Hz and 780 Hz, stopband edges at 375 Hz and 1000 Hz, peak passband ripple of 1.2 dB, and minimum stopband attenuation of 25 dB. (a) What are the specifications of the analog bandpass filter? (b) What are the specifications of the analog prototype lowpass filter? (c) Show all pertinent transfer functions. Plot the gain responses of the prototype analog lowpass filter, the analog bandpass filter, and desired digital bandpass filter. Show all steps.

M 7.12 Design using the bilinear transformation method a digital Butterworth bandstop filter operating at a sampling rate of 5 kHz with the following specifications: passband edges at 500 Hz and 2125 Hz, stopband edges at 1050 Hz and 1400 Hz, peak passband ripple of 2 dB, and minimum stopband attenuation of 40 dB. (a) What are the specifications of the analog bandstop filter? (b) What are the specifications of the analog prototype lowpass filter? (c) Show all pertinent transfer functions. Plot the gain responses of the prototype analog lowpass filter, the analog bandstop filter, and desired digital bandstop filter. Show all steps.

M 7.13 Plot the magnitude response of a linear-phase FIR highpass filter by truncating the impulse response $h_{HP}[n]$ of the ideal highpass filter of Eq. (7.61) to length $N = 2M + 1$ for two different values of M and show that the truncated filter exhibits oscillatory behavior on both sides of the cutoff frequency.

M 7.14 Plot the magnitude response of a linear-phase FIR bandpass filter by truncating the impulse response $h_{BP}[n]$ of the ideal bandpass filter of Eq. (7.62) to length $N = 2M + 1$ for two different values of M and show that the truncated filter exhibits oscillatory behavior on both sides of the cutoff frequency.

M 7.15 Plot the magnitude response of a linear-phase FIR Hilbert transformer by truncating the impulse response $h_{HT}[n]$ of the ideal Hilbert transformer of Eq. (7.67) to length $N = 2M + 1$ for two different values of M and show that the truncated filter exhibits oscillatory behavior near $\omega = 0$ and $\omega = \pi$.

M 7.16 Write a MATLAB program to design a linear-phase FIR notch filter by windowing the impulse response of the ideal notch filter of Problem 7.44. Using this program, design an FIR notch filter of order 40 operating at a 400-Hz sampling rate with a notch frequency of 60 Hz.

M 7.17 Determine a quadratic approximation $a_0 x + a_1 x + a_2 x^2$ to the cubic function $D(x) = 2.2x^3 - 3x^2 + 0.5$ defined for the range $-3 \leq x \leq 3.5$ by minimizing the peak value of the absolute error $\left| D(x) - a_0 - a_1 x - a_2 x^2 \right|$, i.e.,

$$\max_{-3 \leq x \leq 3.5} \left| D(x) - a_0 - a_1 x - a_1 x^2 \right|,$$

using the Remez algorithm. Plot the error function after convergence of the algorithm.

M 7.18 Design using the windowed Fourier series approach a linear-phase FIR lowpass filter with the following specifications: passband edge at 2 rad/sec, stopband edge at 4 rad/sec, maximum passband attenuation of 0.1 dB, minimum stopband attenuation of 40 dB, and a sampling frequency of 20 rad/sec. Use each of the following windows for the design: Hamming, Hann, and Blackman. Show the impulse response coefficients and plot the gain response of the designed filters for each case. Comment on your results. Do not use the M-file `fir1`.

M 7.19 Repeat Exercise M7.18 using the Kaiser window. Do not use the M-file `fir1`.

M 7.20 Design using the windowed Fourier series approach a linear-phase FIR lowpass filter of lowest order with the following specifications: passband edge at 0.3π, stopband edge at 0.5π, and minimum stopband attenuation of 40 dB. Which window function is appropriate for this design? Show the impulse response coefficients and plot the gain response of the designed filter. Comment on your results. Do not use the M-file `fir1`.

M 7.21 Repeat Exercise M7.20 using the Dolph-Chebyshev window. Do not use the M-file `fir1`. Compare your results with that obtained in Exercise M7.20.

M 7.22 Repeat Exercise M7.20 using the M-file `fir1`. Compare your results with that obtained in Exercise M7.20.

M 7.23 Design a linear-phase highpass FIR filter of length 30 with a passband edge at $\omega_p = 0.5\pi$ using the frequency sampling approach. Show the impulse response coefficients and plot the magnitude response of the designed filter using MATLAB.

M 7.24 Design a linear-phase bandpass FIR filter of order 40 with passband edges at $\omega_{p1} = 0.4\pi$ and $\omega_{p2} = 0.6\pi$ using the frequency sampling approach. Show the impulse response coefficients and plot the magnitude response of the designed filter using MATLAB.

M 7.25 Using the frequency sampling approach redesign the length-37 linear-phase lowpass filter of Problem 7.62 by including a transition band with one frequency sample of magnitude 1/2. Plot the magnitude response of the new filter using MATLAB and compare it with that obtained in Problem 7.62.

M 7.26 Repeat Exercise M7.25 by including a transition band with two frequency samples of magnitude 2/3 and 1/3, respectively.

M 7.27 Design the linear-phase FIR lowpass filter of Exercise M7.18 using the function `fir1` of MATLAB. Use each of the following windows for the design: Hamming, Hann, Blackman, and Kaiser. Show the impulse response coefficients and plot the gain response of the designed filters for each case. Compare your results with those obtained in Exercises M7.18 and M7.19.

M 7.28 Using the M-file `fir1` design a linear-phase FIR highpass filter with the following specifications: stopband edge at 0.45π, passband edge at 0.6π, maximum passband attenuation of 0.2 dB, and minimum stopband attenuation of 45 dB. Use each of the following windows for the design: Hamming, Hann, Blackman, and Kaiser. Show the impulse response coefficients and plot the gain response of the designed filters for each case. Comment on your results.

M 7.29 Using the M-file `fir1` design a linear-phase FIR bandpass filter with the following specifications: stopband edges at 0.45π and 0.8π, passband edges at 0.55π and 0.7π, maximum passband attenuation of 0.15 dB, and minimum stopband attenuation of 40 dB. Use each of the following windows for the design: Hamming, Hann, Blackman, and Kaiser. Show the impulse response coefficients and plot the gain response of the designed filters for each case. Comment on your results.

M 7.30 Design a two-channel crossover FIR lowpass and highpass filter pair for digital audio applications. The lowpass and the highpass filters are of length 27 and have a crossover frequency of 10 kHz operating at a sampling rate of 80 kHz. Use the function `fir1` with a Hamming window to design the lowpass filter while the highpass filter is derived from the lowpass filter using the delay-complementary property. Plot the gain responses of both filters on the same figure. What is the minimum number of delays and multipliers needed to implement the crossover network?

M 7.31 Design a three-channel crossover FIR filter system for digital audio applications. All filters are of length 31 and operate at a sampling rate of 44.1 kHz. The two crossover frequencies are at 3.5 kHz and 9 kHz, respectively. Use the function `fir1` with a Hann window to design the lowpass and the highpass filters while the bandpass filter is derived from the lowpass and highpass filters using the delay-complementary property. Plot the gain responses of all filters on the same figure. What is the minimum number of delays and multipliers needed to implement the crossover network?

M 7.32 The M-file `fir2` is employed to design FIR filters with arbitrarily shaped magnitude responses. Using this function, design an FIR filter of order 80 with three different constant magnitude levels: 0.5 in the frequency range 0 to 0.4, 0.3 in the frequency range 0.42 to 0.7, and 1.0 in the frequency range 0.72 to 1.0. Plot the gain response of the designed filter.

M 7.33 Design the linear-phase FIR lowpass filter of Exercise M7.18 using the `remez` function of MATLAB and plot its magnitude response.

M 7.34 Design the linear-phase FIR highpass filter of Exercise M7.28 using the `remez` function of MATLAB and plot its magnitude response.

M 7.35 Design the linear-phase FIR bandpass filter of Exercise M7.29 using the `remez` function of MATLAB and plot its magnitude response.

M 7.36 Design a length-32 discrete-time FIR differentiator using the `remez` function of MATLAB and plot its magnitude response.

M 7.37 Design a 28th-order FIR Hilbert transformer using the `remez` function. The passband is from 0.08π to 0.92π. The two stopbands are from 0.01π to 0.07π, and from 0.94π to π. Plot its magnitude response.

M 7.38 Repeat Exercise M7.36 using the M-file `firls`.

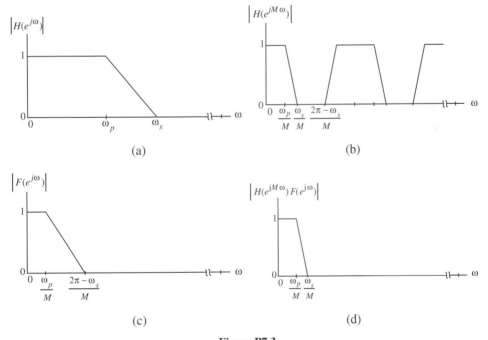

Figure P7.3

M 7.39 In this exercise, we consider the design of computationally efficient linear-phase narrowband FIR filters using the *interpolated finite impulse response* (IFIR) approach [Neu84b]. Let $H(z)$ be a lowpass FIR transfer function with a passband edge at ω_p and a stopband edge at ω_s, as shown in Figure P7.3(a) The magnitude response of the transfer function $H(z^M)$ is then as shown in Figure P7.3(b), where M is a positive integer. If we cascade $H(z^M)$ with another lowpass filter $F(z)$ with passband edge at ω_p/M and stopband edge at $(2\pi - \omega_s)/M$, as shown in Figure P7.3(c), the cascade $G(z) = H(z^M)F(z)$ is a sharp cutoff lowpass filter as indicated in Figure P7.3(d). By choosing $F(z)$ appropriately, any one of the M passbands of $H(z^M)$ can be retained and the remaining $(M-1)$ passbands attenuated resulting in either a narrowband lowpass, or highpass, or bandpass filter.

By reversing the above process, we can develop the specifications of $H(z)$ and $F(z)$ from the desired specifications of the overall narrowband filter $G(z)$ obtained by the cascade realization. Since the gain in dB of a cascade is the sum of the gains in dB of the individual sections, the passband ripples of $H(z)$ and $F(z)$ can be made one-half of the desired passband ripple δ_p of the narrowband filter $G(z)$. On the other hand, the stopband ripples of $H(z)$ and $F(z)$ can be made equal to the desired passband ripple δ_s of the narrowband filter $G(z)$ for simplicity. The computational complexity of the cascade $G(z)$ is thus equal to that of $H(z)$ together with that of $F(z)$.

Using the above approach, design a linear-phase lowpass FIR filter $G(z)$ in the form of $H(z^M)F(z)$, where $H(z)$ and $F(z)$ are linear-phase FIR filters. The specifications for $G(z)$ are as follows: $\omega_p/M = 0.15\pi$, $\omega_s/M = 0.2\pi$, $\delta_p = 0.002$, $\delta_s = 0.001$. Choose the largest possible value for M. Compare the computational complexities of $G(z)$ designed as a single stage and designed as a cascade $H(z^M)F(z)$.

M 7.40 Another approach to the design of a computationally efficient FIR filter is the *prefilter-equalizer method* [Ada83]. In this method first a computationally efficient FIR prefilter $H(z)$ with a frequency response reasonably close to the desired response is selected. Next an FIR equalizer $F(z)$ is designed so that the cascade of the prefilter and the equalizer meets the desired specifications. An attractive prefilter structure for the design of a lowpass FIR filter is the recursive running sum (RRS) FIR filter of order N which has a transfer function

$$H(z) = \frac{1 - z^{-(N+1)}}{1 - z^{-1}}.$$

The first null of the frequency response of the RRS filter is at $\omega = 2\pi/(N+1)$. Thus, if the desired stopband edge is at ω_s, the order of the RRS filter should be chosen as $N \cong 2\pi/\omega_s$. If N is a fraction, then both the integer values nearest to $2\pi/\omega_s$ are good candidates for the order of the RRS filter. The Park-McClellan algorithm can be modified to incorporate the frequency response of the RRS filter in the weighting function of the error function $W(e^{j\omega})$ of Eq. (7.95). Using the prefilter-equalizer approach, design a computationally-efficient narrowband FIR lowpass filter with the following specifications: $\omega_p = 0.042\pi$, $\omega_s = 0.14\pi$, $\alpha_p = 0.2$ dB, and $\alpha_s = 35$ dB.

8 DSP Algorithm Implementation

There are basically two types of digital signal processing (DSP) algorithms: filtering algorithms and signal analysis algorithms. These algorithms can be based on either the difference equation, both recursive and nonrecursive, or the discrete Fourier transform (DFT) and can be implemented in any one of the following forms: hardware, firmware, and software. In the hardware approach, the algorithm may be implemented using digital circuitry, such as the shift register to provide the delaying operation, the digital multiplier, and the digital adder. Alternatively, a special-purpose VLSI chip may be designed and fabricated to implement a specific filtering algorithm. In the firmware approach, the algorithm is implemented on a read-only-memory (ROM) chip. Additional control circuitry, and storage registers, are usually needed in the final hardware or firmware realization. Finally, in the software approach, the algorithm is implemented as a computer program on a general-purpose computer such as a workstation, a minicomputer, a personal computer, or a programmable DSP chip. This chapter is concerned with the implementation aspects of DSP algorithms. We first examine the two major issues concerning all the above types of approaches to implementation. We then discuss the software implementation of digital filtering and DFT algorithms on a computer using MATLAB to illustrate the main points. It is followed by a review of various schemes for the representation of numbers and signal variables on a digital machine. The number representation scheme is basic to the development of methods for the analysis of finite wordlength effects considered in the following chapter. Next, we review algorithms that are employed to implement addition and multiplication, the two key arithmetic operations in digital signal processing. We then briefly review operations developed to handle overflow. Finally, two general methods for the design and implementation of tunable digital filters are outlined, followed by a discussion of algorithms for the approximation of certain special functions. A detailed discussion of hardware, firmware, and DSP chip implementations is beyond the scope of this book. Information on programming the DSP chips can be found in the books and application notes published by the manufacturers of these chips. Discussion on selected DSP chips can also be found in the following books and chapters [Bab95], [Cha90], [Fad93], [Had91], [Ife93], [Lin87], [Mar92], [Pap90].

8.1 Basic Issues

We examine first two specific problems that may be encountered before a digital filter is actually implemented. The first problem is concerned with the computability of the equations describing the structure, and the second problem is concerned with the verification of the structure developed to realize a prescribed transfer function.

8.1.1 Matrix Representation of the Digital Filter Structure

As indicated in Chapter 6, a digital filter can be described in the time-domain by a set of equations relating the output sequence to the input sequence and, in some cases, one or more internally generated sequences. The ordering of these equations in computing the output samples is important, as discussed next.

515

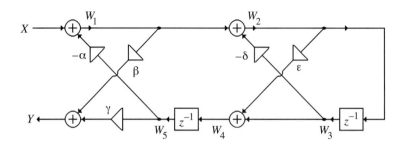

Figure 8.1: A cascaded lattice digital filter structure.

Consider the digital filter structure of Figure 8.1. We can describe this structure by the following set of equations relating the signal variables $W_k(z)$, the output $Y(z)$, and the input $X(z)$:

$$W_1(z) = X(z) - \alpha W_5(z), \tag{8.1a}$$
$$W_2(z) = W_1(z) - \delta W_3(z), \tag{8.1b}$$
$$W_3(z) = z^{-1} W_2(z), \tag{8.1c}$$
$$W_4(z) = W_3(z) + \varepsilon W_2(z), \tag{8.1d}$$
$$W_5(z) = z^{-1} W_4(z), \tag{8.1e}$$
$$Y(z) = \beta W_1(z) + \gamma W_5(z). \tag{8.1f}$$

In the time-domain, the above set of equations is equivalent to

$$w_1[n] = x[n] - \alpha w_5[n], \tag{8.2a}$$
$$w_2[n] = w_1[n] - \delta w_3[n], \tag{8.2b}$$
$$w_3[n] = w_2[n-1], \tag{8.2c}$$
$$w_4[n] = w_3[n] + \varepsilon w_2[n], \tag{8.2d}$$
$$w_5[n] = w_4[n-1], \tag{8.2e}$$
$$y[n] = \beta w_1[n] + \gamma w_5[n]. \tag{8.2f}$$

The above set of equations do not describe a valid computational algorithm since the equations cannot be implemented in the order shown with each variable on the left side computed before the variable below is computed. For example, computation of $w_1[n]$ in the first step requires the knowledge of $w_5[n]$ that is computed in the fifth step. Likewise, the computation of $w_2[n]$ in the second step requires the knowledge of $w_3[n]$ that is computed in the following step. We call the ordered set of equations of Eqs. (8.2a) to (8.2f) *noncomputable*.

Suppose we reorder the above equations and write them in the form

$$w_3[n] = w_2[n-1], \tag{8.3a}$$
$$w_5[n] = w_4[n-1], \tag{8.3b}$$
$$w_1[n] = x[n] - \alpha w_5[n], \tag{8.3c}$$
$$w_2[n] = w_1[n] - \delta w_3[n], \tag{8.3d}$$
$$y[n] = \beta w_1[n] + \gamma w_5[n], \tag{8.3e}$$
$$w_4[n] = w_3[n] + \varepsilon w_2[n]. \tag{8.3f}$$

It can be seen that the above ordered set of equations now describes a valid computational algorithm since the equations can be implemented in the sequential order shown, with each variable on the left side computed before the variable below is computed.

In most practical applications, the equations characterizing the digital filter can be put into a computable order by inspection. It is, however, instructive to examine the computability of the equations describing a digital filter in a more formal fashion, which is described next [Cro75].

To this end, we write the equations of the digital filter in matrix form. Thus, a matrix representation of Eqs. (8.2a) to (8.2f) is given by

$$
\begin{bmatrix} w_1[n] \\ w_2[n] \\ w_3[n] \\ w_4[n] \\ w_5[n] \\ y[n] \end{bmatrix} = \begin{bmatrix} x[n] \\ 0 \\ 0 \\ 0 \\ 0 \\ 0 \end{bmatrix} + \begin{bmatrix} 0 & 0 & 0 & 0 & -\alpha & 0 \\ 1 & 0 & -\delta & 0 & 0 & 0 \\ 0 & 0 & 0 & 0 & 0 & 0 \\ 0 & \varepsilon & 1 & 0 & 0 & 0 \\ 0 & 0 & 0 & 0 & 0 & 0 \\ \beta & 0 & 0 & 0 & \gamma & 0 \end{bmatrix} \begin{bmatrix} w_1[n] \\ w_2[n] \\ w_3[n] \\ w_4[n] \\ w_5[n] \\ y[n] \end{bmatrix}
$$

$$
+ \begin{bmatrix} 0 & 0 & 0 & 0 & 0 & 0 \\ 0 & 0 & 0 & 0 & 0 & 0 \\ 0 & 1 & 0 & 0 & 0 & 0 \\ 0 & 0 & 0 & 0 & 0 & 0 \\ 0 & 0 & 0 & 1 & 0 & 0 \\ 0 & 0 & 0 & 0 & 0 & 0 \end{bmatrix} \begin{bmatrix} w_1[n-1] \\ w_2[n-1] \\ w_3[n-1] \\ w_4[n-1] \\ w_5[n-1] \\ y[n-1] \end{bmatrix}, \tag{8.4}
$$

which we can write compactly as

$$ \mathbf{y}[n] = \mathbf{x}[n] + \mathbf{F}\mathbf{y}[n] + \mathbf{G}\mathbf{y}[n-1] \tag{8.5} $$

where

$$
\mathbf{y}[n] = \begin{bmatrix} w_1[n] \\ w_2[n] \\ w_3[n] \\ w_4[n] \\ w_5[n] \\ y[n] \end{bmatrix}, \quad \mathbf{x}[n] = \begin{bmatrix} x[n] \\ 0 \\ 0 \\ 0 \\ 0 \\ 0 \end{bmatrix}, \tag{8.6a}
$$

$$
\mathbf{F} = \begin{bmatrix} 0 & 0 & 0 & 0 & -\alpha & 0 \\ 1 & 0 & -\delta & 0 & 0 & 0 \\ 0 & 0 & 0 & 0 & 0 & 0 \\ 0 & \varepsilon & 1 & 0 & 0 & 0 \\ 0 & 0 & 0 & 0 & 0 & 0 \\ \beta & 0 & 0 & 0 & \gamma & 0 \end{bmatrix}, \quad \mathbf{G} = \begin{bmatrix} 0 & 0 & 0 & 0 & 0 & 0 \\ 0 & 0 & 0 & 0 & 0 & 0 \\ 0 & 1 & 0 & 0 & 0 & 0 \\ 0 & 0 & 0 & 0 & 0 & 0 \\ 0 & 0 & 0 & 1 & 0 & 0 \\ 0 & 0 & 0 & 0 & 0 & 0 \end{bmatrix}. \tag{8.6b}
$$

If we examine Eq. (8.4), we observe that, for the computation of the present value of a particular signal variable, the nonzero entries in the corresponding rows of the matrices $\mathbf{F}$ and $\mathbf{G}$ determine the variables whose present and previous values are needed. If the diagonal element in $\mathbf{F}$ is nonzero, then the computation of the present value of the corresponding variable requires the knowledge of its present value indicating the presence of a delay-free loop making the structure totally noncomputable. Any nonzero entries in the same row above the diagonal of $\mathbf{F}$ imply that the computation of the present value of the corresponding variable requires the present values of other variables that have not yet been computed, thus making the set of equations noncomputable.

It follows therefore for computability all elements of the $\mathbf{F}$ matrix on the diagonal and above diagonal must be zeros.

In the $\mathbf{F}$ matrix of this example, the diagonal elements are all zeros, indicating that there are no delay-free loops in the structure. However, there are nonzero entries above the diagonal in the first and second rows of $\mathbf{F}$, indicating that the set of equations of Eqs. (8.2a) to (8.2f) is not in proper order for computation.

On the other hand, the matrix representation of Eqs. (8.3a) to (8.3f) results in

$$
\begin{bmatrix} w_3[n] \\ w_5[n] \\ w_1[n] \\ w_2[n] \\ y[n] \\ w_4[n] \end{bmatrix} = \begin{bmatrix} 0 \\ 0 \\ x[n] \\ 0 \\ 0 \\ 0 \end{bmatrix} + \begin{bmatrix} 0 & 0 & 0 & 0 & 0 & 0 \\ 0 & 0 & 0 & 0 & 0 & 0 \\ 0 & -\alpha & 0 & 0 & 0 & 0 \\ -\delta & 0 & 1 & 0 & 0 & 0 \\ 0 & \gamma & \beta & 0 & 0 & 0 \\ 1 & 0 & 0 & \varepsilon & 0 & 0 \end{bmatrix} \begin{bmatrix} w_3[n] \\ w_5[n] \\ w_1[n] \\ w_2[n] \\ y[n] \\ w_4[n] \end{bmatrix}
$$

$$
+ \begin{bmatrix} 0 & 0 & 0 & 1 & 0 & 0 \\ 0 & 0 & 0 & 0 & 0 & 1 \\ 0 & 0 & 0 & 0 & 0 & 0 \\ 0 & 0 & 0 & 0 & 0 & 0 \\ 0 & 0 & 0 & 0 & 0 & 0 \\ 0 & 0 & 0 & 0 & 0 & 0 \end{bmatrix} \begin{bmatrix} w_3[n-1] \\ w_5[n-1] \\ w_1[n-1] \\ w_2[n-1] \\ y[n-1] \\ w_4[n-1] \end{bmatrix}, \tag{8.7}
$$

for which the $\mathbf{F}$ matrix satisfies the computability condition, and thus, the equations describing the filter are in proper order.

8.1.2 Precedence Graph

We now describe a simple algorithm for testing the computability of a digital filter structure and for developing the proper ordering sequence for a set of equations describing a computable structure. To this end, we first form a signal flow-graph description of the digital filter structure. In a signal flow-graph [Mas60], the dependent and the independent signal variables are represented by *nodes*, whereas the multiplier and the delay units are represented by *directed branches*. In the latter case, the directed branch has an attached symbol denoting the *branch-gain* or the *transmittance*, which for a multiplier branch is the multiplier coefficient value and for a delay branch is simply z^{-1}. For example, the signal flow-graph representation of the digital filter structure of Figure 8.1 is as shown in Figure 8.2.

As the output of the delay branches can always be computed at any instant since they are the delayed values of their respective input signals computed at the previous instant, we remove all delay branches from the complete signal flow-graph of the digital filter structure. Similarly, all branches coming out of

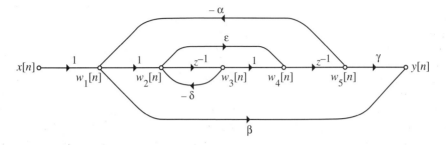

Figure 8.2: Signal flow-graph representation of the digital filter structure of Figure 8.1.

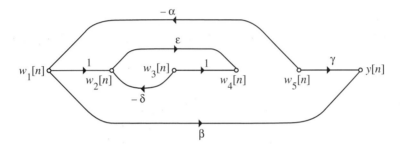

Figure 8.3: Reduced signal flow-graph obtained by removing the branches going out of the input node and the delay branches from the signal flow-graph of Figure 8.2.

the input node are also removed since the input variables are always available at each instant. For our example, the resulting reduced signal flow-graph is as shown in Figure 8.3.

We now group the remaining nodes in the reduced signal flow-graph as follows. All nodes with only outgoing branches are grouped into one set labeled $\{\mathcal{N}_1\}$. Next, we form the set $\{\mathcal{N}_2\}$ containing nodes that have branches coming in from one or more of the nodes in the set $\{\mathcal{N}_1\}$ and have outgoing branches to the other nodes. We then form a set $\{\mathcal{N}_3\}$ containing nodes that have branches coming in from one or more of the nodes in the sets $\{\mathcal{N}_1\}$ and $\{\mathcal{N}_2\}$ and have outgoing branches to the other nodes. This process is continued until we have a set $\{\mathcal{N}_f\}$ containing only nodes with only incoming branches.

Since the signal variables belonging to the set $\{\mathcal{N}_1\}$ do not depend on the present values of the other signal variables, these variables should be computed first. Next, the signal variables belonging to the set $\{\mathcal{N}_2\}$ can be computed since they depend on the present values of the signal variables contained in the set $\{\mathcal{N}_1\}$ that have already been computed. This is followed by the computation of the signal variables in the sets $\{\mathcal{N}_3\}$, $\{\mathcal{N}_4\}$, etc. Finally, in the last step, the signal variables belonging to the set $\{\mathcal{N}_f\}$ are computed. This process of sequential computation ensures the development of a valid computational algorithm. However, if there is no final set $\{\mathcal{N}_f\}$ containing only incoming branches, the signal flow-graph is noncomputable. The rearranged signal flow-graph without the delay branches and with nodes grouped as indicated above is called a *precedence graph* [Cro75].

For our example, the pertinent groupings of node variables according to their precedence relations are as follows:

$$\{\mathcal{N}_1\} = \{w_3[n], \ w_5[n]\},$$
$$\{\mathcal{N}_2\} = \{w_1[n]\},$$
$$\{\mathcal{N}_3\} = \{w_2[n]\},$$

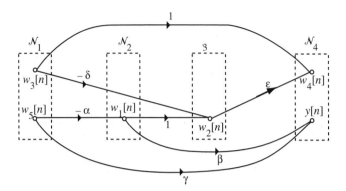

Figure 8.4: Precedence graph of Figure 8.3 redrawn with signal variables grouped according to their precedence relations.

$$\{\mathcal{N}_4\} = \{w_4[n], \ y[n]\}.$$

The precedence graph redrawn according to the above groupings is indicated in Figure 8.4. Since the final node set $\{\mathcal{N}_4\}$ has only incoming nodes, the structure of Figure 8.1 has no delay-free loops. Therefore, for our example structure, we can compute first the signal variables $w_3[n]$ and $w_5[n]$ in any order, then compute the signal variable $w_1[n]$, followed by the computation of the signal variable $w_2[n]$, and finally, compute the signal variables $w_4[n]$ and $y[n]$ in any order to arrive at a valid computational algorithm.

8.1.3 Structure Verification

An important step that needs to be considered in the hardware or software implementation of a digital transfer function is to ensure that no computational and/or other errors have taken place during the course of the realization process and that the structure obtained is indeed characterized by the prescribed transfer function $H(z)$. A simple technique to verify the structure is outlined next [Mit77a].

Without any loss of generality, consider a causal LTI digital filter structure characterized by a fourth-order transfer function

$$H(z) = \frac{P(z)}{D(z)} = \frac{p_0 + p_1 z^{-1} + p_2 z^{-2} + p_3 z^{-3} + p_4 z^{-4}}{1 + d_1 z^{-1} + d_2 z^{-2} + d_3 z^{-3} + d_4 z^{-4}}. \tag{8.8}$$

If $\{h[n]\}$ denotes its unit sample response, then

$$H(z) = \sum_{n=0}^{\infty} h[n] z^{-n}. \tag{8.9}$$

From Eqs. (8.8) and (8.9) it follows that

$$P(z) = H(z) D(z), \tag{8.10}$$

or, equivalently, in the time-domain by the convolution sum,

$$p_n = h[n] \circledast d_n, \tag{8.11}$$

which explicitly shows the relation between the numerator and the denominator coefficients of the transfer function $H(z)$ of Eq. (8.8) and its impulse response samples. Since the total number of transfer function coefficients is nine, we need only any consecutive nine equations of the set of Eq. (8.11) to have unique relations between the transfer function coefficients and the impulse response samples. Writing out Eq. (8.11) explicitly for $n = 0, 1, 2, \ldots, 8$, we obtain

$$
\begin{aligned}
p_0 &= h[0], \\
p_1 &= h[1] + h[0]d_1, \\
p_2 &= h[2] + h[1]d_1 + h[0]d_2, \\
p_3 &= h[3] + h[2]d_1 + h[1]d_2 + h[0]d_3, \\
p_4 &= h[4] + h[3]d_1 + h[2]d_2 + h[1]d_3 + h[0]d_4, \\
0 &= h[5] + h[4]d_1 + h[3]d_2 + h[2]d_3 + h[1]d_4, \\
0 &= h[6] + h[5]d_1 + h[4]d_2 + h[3]d_3 + h[2]d_4, \\
0 &= h[7] + h[6]d_1 + h[5]d_2 + h[4]d_3 + h[3]d_4, \\
0 &= h[8] + h[7]d_1 + h[6]d_2 + h[5]d_3 + h[4]d_4.
\end{aligned}
$$

In matrix form the above equations can be rewritten as

$$
\begin{bmatrix} p_0 \\ p_1 \\ p_2 \\ p_3 \\ p_4 \\ 0 \\ 0 \\ 0 \\ 0 \end{bmatrix} =
\begin{bmatrix}
h[0] & 0 & 0 & 0 & 0 \\
h[1] & h[0] & 0 & 0 & 0 \\
h[2] & h[1] & h[0] & 0 & 0 \\
h[3] & h[2] & h[1] & h[0] & 0 \\
h[4] & h[3] & h[2] & h[1] & h[0] \\
\cdots & \cdots & \cdots & \cdots & \cdots & \cdots \\
h[5] & \vdots & h[4] & h[3] & h[2] & h[1] \\
h[6] & \vdots & h[5] & h[4] & h[3] & h[2] \\
h[7] & \vdots & h[6] & h[5] & h[4] & h[3] \\
h[8] & \vdots & h[7] & h[6] & h[5] & h[4]
\end{bmatrix}
\begin{bmatrix} 1 \\ d_1 \\ d_2 \\ d_3 \\ d_4 \end{bmatrix}. \tag{8.12}
$$

In partitioned form, Eq. (8.12) can be reexpressed as

$$
\begin{bmatrix} \mathbf{p} \\ \cdots \\ \mathbf{0} \end{bmatrix} =
\begin{bmatrix} & \mathbf{H}_1 & \\ \cdots & \cdots & \cdots \\ \mathbf{h} & \vdots & \mathbf{H}_2 \end{bmatrix}
\begin{bmatrix} 1 \\ \cdots \\ \mathbf{d} \end{bmatrix} \tag{8.13}
$$

where

$$
\mathbf{p} = \begin{bmatrix} p_0 \\ p_1 \\ p_2 \\ p_3 \\ p_4 \end{bmatrix}, \quad
\mathbf{d} = \begin{bmatrix} d_1 \\ d_2 \\ d_3 \\ d_4 \end{bmatrix}, \quad
\mathbf{0} = \begin{bmatrix} 0 \\ 0 \\ 0 \\ 0 \end{bmatrix}, \tag{8.14a}
$$

$$
\mathbf{H}_1 = \begin{bmatrix}
h[0] & 0 & 0 & 0 & 0 \\
h[1] & h[0] & 0 & 0 & 0 \\
h[2] & h[1] & h[0] & 0 & 0 \\
h[3] & h[2] & h[1] & h[0] & 0 \\
h[4] & h[3] & h[2] & h[1] & h[0]
\end{bmatrix}, \quad
\mathbf{h} = \begin{bmatrix} h[5] \\ h[6] \\ h[7] \\ h[8] \end{bmatrix},
$$

$$\mathbf{H}_2 = \begin{bmatrix} h[4] & h[3] & h[2] & h[1] \\ h[5] & h[4] & h[3] & h[2] \\ h[6] & h[5] & h[4] & h[3] \\ h[7] & h[6] & h[5] & h[4] \end{bmatrix}. \tag{8.14b}$$

Equation (8.12) therefore can be written as two matrix equations:

$$\mathbf{p} = \mathbf{H}_1 \begin{bmatrix} 1 \\ \mathbf{d} \end{bmatrix}, \tag{8.15}$$

$$\mathbf{0} = [\mathbf{h} \quad \mathbf{H}_2] \begin{bmatrix} 1 \\ \mathbf{d} \end{bmatrix}. \tag{8.16}$$

Solving Eq. (8.16), we obtain the vector $\mathbf{d}$ composed of the denominator coefficients:

$$\mathbf{d} = -\mathbf{H}_2^{-1}\mathbf{h}. \tag{8.17}$$

Substituting Eq. (8.17) in Eq. (8.15), we arrive at the vector $\mathbf{p}$ containing the numerator coefficients:

$$\mathbf{p} = \mathbf{H}_1 \begin{bmatrix} 1 \\ -\mathbf{H}_2^{-1}\mathbf{h} \end{bmatrix}. \tag{8.18}$$

In the general case of an Nth-order IIR transfer function, knowing the first $2N + 1$ impulse response samples is sufficient to determine the transfer function coefficients. Here, the vector $\mathbf{p}$ is of length $N + 1$, the vector $\mathbf{d}$ is of length N, the vector $\mathbf{h}$ is of length N, the matrix $\mathbf{H}_1$ is of size $(N + 1) \times (N + 1)$, and the matrix $\mathbf{H}_2$ is of size $N \times N$.

We illustrate the above approach for the reconstruction of the transfer function of a causal IIR filter from its impulse response coefficients in the following example.

EXAMPLE 8.1   Consider the causal IIR transfer function

$$H(z) = \frac{2 + 6z^{-1} + 3z^{-2}}{1 + z^{-1} + 2z^{-2}}.$$

Since $H(z)$ is of second-order, we need to compute only the first five impulse response samples. Using long division, we determine the first five terms in the series expansion of $H(z)$:

$$H(z) = 2 + 4z^{-1} - 5z^{-2} - 3z^{-3} + 13z^{-4} + \cdots.$$

Therefore, here,

$$h[0] = 2, \quad h[1] = 4, \quad h[2] = -5, \quad h[3] = -3, \quad h[4] = 13.$$

The equation corresponding to Eq. (8.12) here is given by

$$\begin{bmatrix} p_0 \\ p_1 \\ p_2 \\ 0 \\ 0 \end{bmatrix} = \begin{bmatrix} 2 & 0 & 0 \\ 4 & 2 & 0 \\ -5 & 4 & 2 \\ -3 & -5 & 4 \\ 13 & -3 & -5 \end{bmatrix} \begin{bmatrix} 1 \\ d_1 \\ d_2 \end{bmatrix}.$$

Therefore, from Eq. (8.17),

$$\begin{bmatrix} d_1 \\ d_2 \end{bmatrix} = -\begin{bmatrix} -5 & 4 \\ -3 & -5 \end{bmatrix}^{-1} \begin{bmatrix} -3 \\ 13 \end{bmatrix} = -\frac{1}{37} \begin{bmatrix} -5 & -4 \\ 3 & -5 \end{bmatrix} \begin{bmatrix} -3 \\ 13 \end{bmatrix} = \begin{bmatrix} 1 \\ 2 \end{bmatrix}.$$

Next, from Eq. (8.18),

$$
\begin{bmatrix} p_0 \\ p_1 \\ p_2 \end{bmatrix} = \begin{bmatrix} 2 & 0 & 0 \\ 4 & 2 & 0 \\ -5 & 4 & 2 \end{bmatrix} \begin{bmatrix} 1 \\ 1 \\ 2 \end{bmatrix} = \begin{bmatrix} 2 \\ 6 \\ 3 \end{bmatrix}.
$$

This gives us a straightforward means of finding the transfer function of any discrete-time structure knowing the first $2M + 1$ samples of $\{h[n]\}$, where M is the order of the transfer function $H(z)$. This approach to structure verification is illustrated later in Examples 8.4 and 8.9. It can also be used to determine the effect of coefficient quantization by computing the transfer function realized with the multiplier coefficients quantized to the desired number of bits. This actual transfer function is then used to compute the frequency response, to determine the actual pole locations, etc. Another application of the above method is in the determination of noise transfer functions for computing the output noise power due to product round-offs in fixed-point digital filter implementations considered in Section 9.6, and in the determination of scaling transfer functions needed for dynamic range scaling discussed in Section 9.7. In most cases, we can assume that the denominator coefficients are known; therefore, the solution of Eq. (8.17) that requires a matrix inversion can be avoided. The numerator coefficients are easily found from the first $M + 1$ samples of $\{h[n]\}$ and using Eq. (8.18).

8.2 Structure Simulation and Verification Using MATLAB

As indicated earlier, we concentrate in this book only on software implementation of DSP algorithms. In this section, we consider only the implementation of digital filtering algorithms. The following section is devoted to the implementation of discrete Fourier transform algorithms.

The software implementation of a digital filtering algorithm on a computer is often carried out before the algorithm is implemented in a hardware form to verify that the algorithm chosen does indeed meet the goals of the application on hand. Moreover, such an implementation is adequate if the application under consideration does not require real-time signal processing.

For computer simulation, we basically describe the structure in the form of a set of equations. These equations must be ordered properly to ensure computability. For simplicity, the procedure is to express the output variable of each adder and the filter output variable in terms of all incoming signal variables. For example, for the structure of Figure 8.1, a valid computational algorithm involving the least number of equations is

$$
\begin{aligned}
w_1[n] &= x[n] - \alpha w_4[n - 1], \\
w_2[n] &= w_1[n] - \delta w_2[n - 1], \\
w_4[n] &= w_2[n - 1] + \varepsilon w_2[n], \\
y[n] &= \beta w_1[n] + \gamma w_4[n - 1].
\end{aligned}
$$

The above set of equations is evaluated for increasing values of n starting at $n = 0$. At the beginning, the initial conditions $w_2[-1]$ and $w_4[-1]$ can be set to any desired values, which are typically zero. After the computation of the last equation at time instant n, the computed values of $w_2[n]$ and $w_4[n]$ replace the values of $w_2[n - 1]$ and $w_4[n - 1]$ before the set of equations are evaluated for the next time instant $n + 1$.

In Chapter 6, we outlined a number of methods for the realization of both FIR and IIR digital transfer functions resulting in a variety of structures. We restrict our attention to some of these structures to demonstrate the simulation of digital filters using MATLAB. The structure being simulated can be verified by computing its transfer function using the method described in Section 8.1.3. To this end, the M-file strucver given below can be used.

```
function [p,d] = strucver(ir,N)
H = zeros(2*N+1,N+1);
H(:,1) = ir';
for n = 2:N+1;
    H(:,n) = [zeros(1,n-1) ir(1:2*(N+1)-n)]';
end
H1 = zeros(N+1,N+1);
for k = 1:N+1;
    H1(k,:) = H(k,:);
end
H3 = zeros(N,N+1);
for k = 1:N;
    H3'(k,:) = H(k+N+1,:);
end
H2 = H3(:,2:N+1);
hf = H3(:,1);
% Compute the denominator coefficients
d = -(inv(H2))*hf;
% Compute the numerator coefficients
p = H1*[1;d];
d = [1; d];
```

8.2.1   Simulation of Direct Form IIR Filters

The M-file `filter` in the *Signal Processing Toolbox* of MATLAB basically implements the IIR digital filter in the transposed direct form II structure shown in Figure 8.5 for a third-order filter.[1] As indicated in this figure, $d(1)$ has been assumed to be equal to 1. If $d(1) \neq 1$, the program automatically normalizes all filter coefficients in p and d to make $d(1) = 1$.[2] The basic forms of this function are as follows:

$$y = filter(num,den,x)$$
$$[y,sf] = filter(num,den,x,si)$$

The numerator and the denominator coefficients are contained in the vectors num and den, respectively. These vectors do not have to be of the same size. The input vector is x while the output vector generated by the filtering algorithm is y.

As indicated in Figure 8.5, the function `filter` simulates the filtering operation in the time-domain in accordance with the following representation of the digital filter:

$$
\begin{aligned}
s3(n+1) &= p(4)x(n) - d(4)y(n), \\
s2(n+1) &= p(3)x(n) - d(3)y(n) + s3(n), \\
s1(n+1) &= p(2)x(n) - d(2)y(n) + s2(n), \\
y(n) &= p(1)x(n) + s1(n).
\end{aligned}
$$

In the second form of the function `filter`, the initial conditions of the delay (state) variables, $sk(n)$, $k = 1, 2, \ldots$, can be specified through the argument si. Moreover, the function `filter` can return the final values of the delay (state) variables through the output vector sf. The size of the initial (final) condition vector si (sf) is one less than the maximum of the sizes of the filter coefficient vectors num

---

[1] See Section 6.4.1 for the development of IIR direct form structures.

[2] It should be noted that in Figure 8.5 we have used the MATLAB notations for vector elements instead of the notations used elsewhere in the text for representing filter coefficient and signal variables.

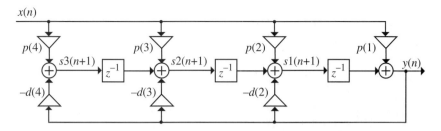

Figure 8.5: Transposed direct form II IIR structure.

and den. The final values of the state variables given as vector sf are useful if the input vector to be processed is very long and needs to be segmented into small blocks of data for processing in stages. In such a situation, after the ith block of input data has been processed, the final state vector sf is fed as the initial state vector si for the processing of the $(i + 1)$th block of input data, and so on.

For simulating a causal IIR filter realized in the direct form II structure, the function direct2 given below can be employed.

```
function [y,sf] = direct2(p,d,x,si);
% Y = DIRECT2(P,D,X) filters input data vector X with
% the filter described by vectors P and D to create the
% filtered data Y.  The filter is a "Direct Form II"
% implementation of the difference equation:
% y(n) = p(1)*x(n) + p(2)*x(n-1) + ... + p(np+1)*x(n-np)
%            - d(2)*y(n-1) - ... - d(nd+1)*y(n-nd)
% [Y,SF] = DIRECT2(P,D,X,SI) gives access to initial and
% final conditions, SI and SF, of the delays.
dlen = length(d); plen = length(p);
N = max(dlen,plen); M =length(x);
sf = zeros(1,N-1); y = zeros(1,M);
if nargin ~= 3,
    sf = si;
end
if dlen < plen,
    d = [d zeros(1,plen - dlen)];
else
    p = [pzeros(1, dlen - plen)];
end
p = p/d(1);
d = d/d(1);
for n = 1:M;
    wnew = [1 -d(2:N)]*[x(n) sf]';
    K = [wnew sf];
    y(n) = K*p';
    sf = [wnew sf(1:N-2)];
end
```

The following example illustrates the application of the M-file filter in generating the impulse response coefficients of a causal IIR digital filter.

EXAMPLE 8.2   Determine and plot the first 25 samples of the impulse response sequence, from $n = 0$ to $n = 24$, of the IIR lowpass digital filter described by the transfer function of Eq. (7.39) repeated below in a form suitable for direct form implementation:

$$H(z) = \frac{0.0662272(1 + 3z^{-1} + 3z^{-2} + z^{-3})}{1 - 0.9356142z^{-1} + 0.5671269z^{-2} - 0.10159107z^{-3}}. \qquad (8.19)$$

The MATLAB program given below can be used to compute the impulse response.

```
% Program 8_1
% Impulse Response Computation
%
n = input('Impulse response length desired = ');
num = input('Numerator coefficients = ');
den = input('Denominator coefficients = ');
x = [1 zeros(1,n-1)];
order = max(length(num),length(den))-1;
si = [zeros(1,order)];
y = filter(num,den,x,si);
stem(n-1,y)
xlabel('Time index n'); ylabel('Amplitude');
title('Impulse response samples')
```

During execution, the program first requests the length of the impulse response to be computed. The program then requests the numerator and denominator vectors to be typed in. These vectors must be entered using the square brackets. After execution, it plots the impulse response as shown in Figure 8.6 for the IIR filter of Eq. (8.19).

The digital filtering application of the M-file `filter` is considered in the next several examples.

EXAMPLE 8.3   We illustrate in this example the filtering of a signal composed of the sum of two sinusoids of normalized angular frequencies, 0.1π and 0.8π, using the lowpass IIR digital filter of Eq. (8.19). MATLAB Program 8_2 given below can be utilized for this purpose.

```
% Program 8_2
% Illustration of Filtering by a Lowpass IIR Filter
%
% Generate the input sequence
k = 1:51;
w1 = 0.8*pi;w2 = 0.1*pi;
A = 1.5;B = 2.0;
x1 = A*cos(w1*(k-1)); x2 = B*cos(w2*(k-1));
x = x1+x2;
% Generate the output sequence by filtering the input
si = [0 0 0];
num = 0.0662272*[1 3 3 1];
den = [1 -0.9356142 0.5671269 -0.10159107];
y = filter(num,den,x,si);
% Plot the input and the output sequences
subplot(2,1,1);
stem(k-1,x); axis([0 50 -4 4]);
xlabel('Time index n'); ylabel('Amplitude');
title('Input sequence');
subplot(2,1,2);
```

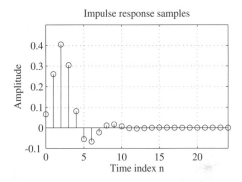

Figure 8.6: Impulse response samples of the IIR digital filter of Eq. (8.19).

```
stem(k-1,y); axis([0 50 -4 4]);
xlabel('Time index n'); ylabel('Amplitude');
title('Output sequence');
```

Figure 8.7 shows the plots generated by this program, verifying the blocking of the high-frequency sinusoidal sequence by the lowpass filter. Note also the transients generated by the zero initial conditions at the beginning of the output sequence.

8.2.2 Simulation of Cascade Form IIR Filters

In the following example, we illustrate the simulation of cascade form realizations of IIR transfer functions. We first verify the simulation using the method described in Section 8.1.3.

EXAMPLE 8.4 We repeat now Example 8.3, using instead a cascade realization of the digital filter of Eq. (8.19). From Eq. (7.39) we arrive at one possible realization using a cascade of a first-order transfer function $H_1(z)$ and a second-order transfer function $H_2(z)$ given by

$$H_1(z) = \frac{0.0662272(1 + z^{-1})}{1 - 0.2593284z^{-1}}, \tag{8.20a}$$

and

$$H_2(z) = \frac{1 + 2z^{-1} + z^{-2}}{1 - 0.6762858z^{-1} + 0.3917468z^{-2}} \tag{8.20b}$$

The MATLAB program used for simulation of the cascade realization given below computes the first 7 impulse response samples. From these impulse response samples, the transfer function of the simulated filter is determined using the function `strucver` to verify the correctness of the simulation.

```
% Program 8_3
% Illustration of Cascade Realization of IIR Filters
% and Structure Verification
%
format long
x = [1 zeros(1,6)];
b1 = 0.0662272*[1 1];
a1 = [1 -0.2593284];
```

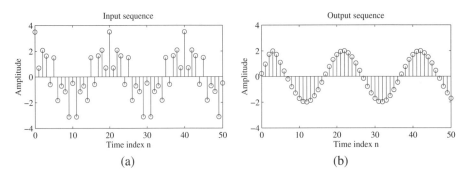

Figure 8.7: Illustration of filtering by an IIR lowpass filter: (a) input sequence, and (b) output sequence.

```
y1 = filter(b1,a1,x,0);
si = [0 0];
b2 = [1 2 1];
a2 = [1 -0.6762858 0.3917468];
y2 = filter(b2,a2,y1,si);
[p,d] = strucver(y2,3);
disp('Actual numerator coefficients are ');disp(p');
disp('Actual denominator coefficients are ');disp(d');
```

The output data generated by running this program are

```
Actual numerator coefficients are
    0.0662272    0.1986816    0.1986816    0.0662272

Actual denominator coefficients are
    1.0000000   -0.9356142    0.5671269   -0.1015910
```

which are seen to be identical to those given in Eq. (8.19).

EXAMPLE 8.5   We next modify Program 8_3 to demonstrate the lowpass filtering using the cascade realization of the IIR transfer function of Eq. (8.19).

```
% Program 8_4
% Illustration of Lowpass Filtering
% Using the Cascade Realization
%
k = 1:51;
w1 = 0.8*pi;w2 = 0.1*pi;
A = 1.5;B = 2.0;
x1 = A*cos(w1*(k-1)); x2 = B*cos(w2*(k-1));
x = x1+x2;
b1 = 0.0662272*[1 1];
a1 = [1 -0.2593284];
y1 = filter(b1,a1,x,0);
si = [0 0];
b2 = [1 2 1];
a2 = [1 -0.6762858 0.3917468];
y2 = filter(b2,a2,y1,si);
```

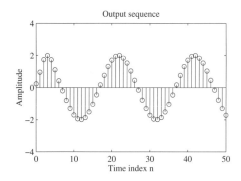

Figure 8.8: Output of the cascade realization of the lowpass filter of Eq. (8.19).

```
stem(k-1,y2);axis([0 50 -4 4]);
xlabel('Time index n'); ylabel('Amplitude');
title('Output sequence');
```

The output generated by the above program is shown in Figure 8.8.

8.2.3 Simulation of Overlap-Add Filtering Method

As indicated in Section 3.6.2, a long input sequence can be filtered using the overlap-add method in which the input sequence is segmented into a set of contiguous short input blocks, each block is then filtered separately, and the overlaps in the output blocks are added appropriately to generate the long output sequence. This method of filtering can be implemented on MATLAB using the M-file `fftfilt`. It can also be easily implemented using the second form of the M-file `filter`. Here, the final values of the internal variable vector `sf` at any stage of filtering is fed back in the following stage of filtering as the initial condition vector `si`.

EXAMPLE 8.6   We use the following program to illustrate the method by repeating the above lowpass filtering example.

```
% Program 8_5
% Illustration of Overlap-Add Filtering Approach
%
k = 1:51;
w1 = 0.8*pi; w2 = 0.1*pi;
A = 1.5;B = 2.0;
x1 = A*cos(w1*(k-1)); x2 = B*cos(w2*(k-1));
x = x1+x2;
xs = zeros(10,5);
xs(:) = x;
si = [0 0 0];
b = 0.0662272*[1 3 3 1];
a = [1 -0.9356142 0.5671269 -0.10159107];
for n = 1:5
    for m = 1:10
        [ys(m,n),sf] = filter(b,a,xs(m,n),si);
```

```
            si = sf;
        end
    end
    y = ys(:);
    stem(k-1,y);
    xlabel('Time index n'); ylabel('Amplitude');
    title('Output Sequence');
```

The output plot generated by the above program is identical to the one shown in Figure 8.8.

8.2.4 Simulation of Direct Form FIR Filters

Both `filter` and `direct2` can be used also to simulate direct form FIR filter structures by setting the denominator vector `den` equal to 1. The following example illustrates the use of the former M-file.

EXAMPLE 8.7   We consider in this example the filtering of a signal formed by the sum of two sinusoidal sequences of normalized angular frequencies, 0.2π and 0.75π, by an equiripple FIR lowpass filter. The coefficients are generated using the function `remez`. For simplicity, we consider the design of a ninth-order Type 1 FIR filter. Program 8_6 given below has been utilized to illustrate the filtering process.

```
% Program 8_6
% Illustration of Filtering by a Lowpass FIR Filter
%
% Determine the FIR filter coefficients
f = [0   0.3  0.5 1];
m = [1 1 0 0];
num = remez(9,f,m);
% Generate the input sequence
k = 1:51;
w1 = 0.75*pi;w2 = 0.2*pi;
A = 1.5;B = 2.0;
x1 = A*cos(w1*(k-1)); x2 = B*cos(w2*(k-1));
x = x1+x2;
% Filter the input sequence
si = [zeros(1,9)];
y = filter(num,1,x,si);
% Plot the input and the output sequences
subplot(2,1,1)
stem(k-1,x);
xlabel('Time index n'); ylabel('Amplitude');
title('Input Sequence');
subplot(2,1,2)
stem(k-1,y);
xlabel('Time index n'); ylabel('Amplitude');
title('Output Sequence');
```

Figure 8.9 shows the input and the output sequences generated by this program. These plots verify the lowpass filtering operation. The transients generated by the zero initial conditions at the beginning of the output sequence are again clearly visible.

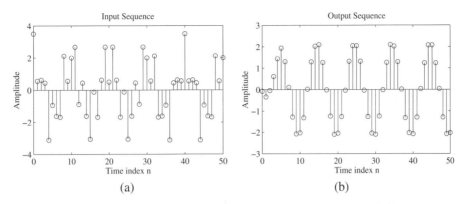

Figure 8.9: Illustration of filtering by an FIR lowpass filter: (a) input sequence, and (b) output sequence.

8.2.5 Zero-Phase Filtering

The *Signal Processing Toolbox* of MATLAB includes the M-file `filtfilt` which implements the zero-phase filtering scheme discussed in Section 4.4.2. In its basic form, `y = filtfilt(p,d,x)` implements zero-phase filtering by performing both forward and time-reversed processing operations. The resulting filter thus has double the order of the filter characterized by the coefficients p and d. If $H(z)$ denotes the transfer function of the original filter, `filtfilt` then realizes a filter with a zero-phase frequency response given by $|H(e^{j\omega})|^2$ and therefore has a passband ripple in dB and a minimum stopband attenuation in dB that are twice those of the original filter, respectively. We illustrate below one possible application of the function `filtfilt`.

EXAMPLE 8.8 We have indicated in Section 7.1.2 that IIR digital filters are, in general, computationally more efficient than equivalent FIR digital filters meeting the same magnitude specifications. On the other hand, IIR filters cannot be designed with linear phase, whereas it is much easier to design FIR filters with linear phase. One way to implement an IIR filter with linear phase is to use both forward and time-reversed filtering operations as discussed in Section 4.4.2. To this end we repeat the lowpass filtering problem of Example 8.3 and compare the results of the functions `filter` and `filtfilt` using the following MATLAB program.

```
% Program 8_7
% Illustration of Lowpass Digital Filtering
%
clf
% Generate the input sequence
k = 1:51;
w1 = 0.8*pi;w2 = 0.1*pi;
A = 1.5;B = 2.0;
x1 = A*cos(w1*(k-1)); x2 = B*cos(w2*(k-1));
x = x1+x2;
si = [0 0 0];
b = 0.0662272*[1 3 3 1];
a = [1 -0.9356142  0.56712691  -0.10159107];
y = filter(b,a,x,si);
stem(k-1,x2);axis([0 50 -2 2]);
xlabel('Time index n'); ylabel('Amplitude');
title('Desired component of the input sequence');
```

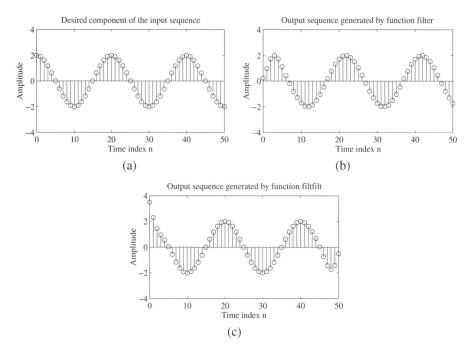

Figure 8.10: Example 8.8: (a) low-frequency component of the input, (b) the output generated by forward-only filtering, and (c) the output generated by both forward and time-reversed filtering.

```
pause
stem(k-1,y);
xlabel('Time index n'); ylabel('Amplitude');
title('Output sequence generated by function filter ');
pause
si = [0 0 0];
y2 = filtfilt(b,a,x);
stem(k-1,y2);
xlabel('Time index n'); ylabel('Amplitude');
title('Output sequence generated by function filtfilt');
```

   Figure 8.10 depicts the low-frequency component of the input, the output generated by forward-only filtering, and the output generated by both forward and time-reversed filtering. Several comments are in order here. Both outputs show the effect of the transients. In the case of forward-only filtering, as shown in Figure 8.10(b), only the first few output samples are corrupted by the transients generated by the initial conditions. On the other hand, in the case of both forward and time-reversed filtering indicated in Figure 8.10(c), effects of transients can be seen at both the beginning and the end. Next, as expected, both outputs are nearly identical to the input except for the transient part, but the output in the first case is delayed with respect to the input, whereas the output in the second case is not.

8.2.6 Simulation of Cascaded Lattice Filter Structures

The function `latcfilt` in the *Signal Processing Toolbox* can be used to simulate the IIR and the FIR cascaded lattice filter structure of Sections 6.8 and 6.9.1, respectively. The basic forms of this function are

```
[f,g] = latcfilt(k,x)
[f,g] = latcfilt(k,alpha,x)
[f,g] = latcfilt(k,1,x)
```

In the first form [f,g] = latcfilt(k,x) simulates an FIR cascaded lattice filter structure with lattice filter coefficients given by vector k and generates the forward output vector f and the backward output vector g for an input vector x. The second form [f,g] = latcfilt(k,alpha,x) simulates an IIR tapped cascaded lattice structure with lattice coefficient vector k and the feedforward multiplier vector alpha. The last form [f,g] = latcfilt(k,1,x) simulates an all-pole IIR cascaded lattice filter structure.

We illustrate in the next two examples the simulation of both IIR and FIR cascaded lattice filter structures.

EXAMPLE 8.9   Consider the tapped cascaded lattice realization of the third-order IIR transfer function of Example 6.3 repeated below for convenience.

$$H(z) = \frac{0.44z^{-1} + 0.36z^{-2} + 0.02z^{-3}}{1 + 0.4z^{-1} + 0.18z^{-2} - 0.2z^{-3}}. \tag{8.21}$$

The realization derived in this example is depicted in Figure 8.11. The multiplier coefficients as derived in Examples 6.8 and 6.10 are given by

$$d_3 = -0.2, \quad d_2' = 0.270833, \quad d_1'' = 0.357377,$$
$$\alpha_1 = 0.02, \quad \alpha_2 = 0.352, \quad \alpha_3 = 0.276533, \quad \alpha_4 = -0.19016.$$

The MATLAB program for the simulation of an IIR cascaded lattice structure of order N is given below. The input vector x consists of the first $2N + 1$ coefficients of a unit sample sequence. The output vector f contains the first $2N + 1$ impulse response coefficients which are then used to determine the numerator and the denominator coefficients of the actual transfer function implemented using the method described in Section 8.1.3.

```
% Program 8_8
% Simulation of IIR Cascaded Lattice Structure
%
k = input('Type in the lattice multiplier coefficients = ');
alpha = input('Type in the feed-forward multiplers = ');
x = [1 zeros(1,2*length(k))];
[f,g] = latcfilt(k,alpha,x);
% Verify the structure
[p,d] = strucver(f,length(k));
disp('Actual numerator coefficients are '); disp(p);
disp('Actual denominator coefficients are '); disp(d');
```

For the simulation of the structure of Figure 8.11 the input data are

```
k = [0.357377 0.270833 -0.2]
alpha = [-0.19016 0.276533 0.352 0.02]
```

The resulting transfer function coefficients generated by running this program are as follows:

```
Actual numerator coefficients are
    -0.0000    0.4400    0.3600    0.0200
Actual denominator coefficients are
     1.0000    0.4000    0.1800   -0.2000
```

which are seen to be identical to those given in Eq. (8.21), thus verifying the structure of Figure 8.11.

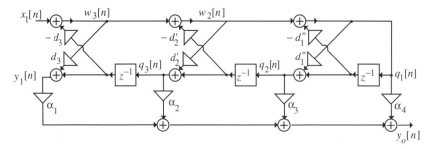

Figure 8.11: Cascaded lattice realization of the IIR transfer function of Eq. (8.21).

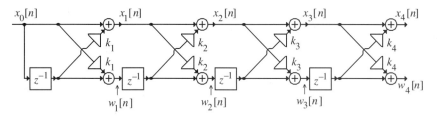

Figure 8.12: Cascaded lattice realization of the FIR transfer function of Eq. (8.22): $k_1 = 0.5$, $k_2 = 1.0$, $k_3 = 0.2173913$, and $k_4 = -0.08$.

EXAMPLE 8.10   Consider the cascaded lattice realization of the fourth-order FIR transfer function of Example 6.14 repeated below for convenience:

$$H_4(z) = 1 + 1.2z^{-1} + 1.12z^{-2} + 0.12z^{-3} - 0.08z^{-4}. \tag{8.22}$$

The realization derived in this example is shown in Figure 8.12. The multiplier coefficients as determined in Example 6.14 are given by

$$k_1 = 0.5, \quad k_2 = 1, \quad k_3 = 0.2173913, \quad k_4 = -0.08.$$

The MATLAB program for the simulation of an FIR transfer function of order N as a FIR cascaded lattice structure of the form of Figure 8.12 is given below. The input vector x0 contains the first $N + 1$ samples of a unit sample sequence, while the output vector f contains the first $N + 1$ samples of the impulse response sequence. The program also computes the mirror-image output vector g.

```
% Program 8_9
% Simulation of FIR Cascaded Lattice Structure
%
k = input('Type in the multiplier values = ');
x0 = [1 zeros(length(k))];
[f,g] = latcfilt(k,x0);
disp('Filter Coefficients');disp(f);
disp('Mirror Image Filter Coefficients');disp(g);
```

For the simulation of the structure of Figure 8.12 the input data are

```
k = [0.5   1    0.2173913    -0.08]
```

The output data displayed are the coefficients of the transfer function $H_4(z)$ and the coefficients of its mirror-image transfer function $\tilde{H}_4(z)$ as given below:

```
      Filter Coefficients
          1.0000    1.2000    1.1200    0.1200    -0.0800

      Coefficients of Mirror Image Filter
          -0.0800    0.1200    1.1200    1.2000    1.0000
```

The coefficients of $H_4(z)$ are seen to be identical to those given in Eq. (8.22).

8.3 Computation of the Discrete Fourier Transform

The discrete Fourier transform (DFT) is another widely used DSP algorithm. As indicated earlier, it can be employed to implement the linear convolution of two sequences, a key digital filtering operation. It is also used for the spectral analysis of signals, discussed in Sections 11.2 to 11.4. Because of the widespread use of the DFT, it is of interest to investigate its efficient implementation methods which are considered in this section.

In Section 3.2 we introduced the concept of the N-point DFT $X[k]$ of a sequence $x[n]$ of length N as the N samples of its Fourier transform, $X(e^{j\omega})$, evaluated uniformly on the ω-axis at $\omega_k = 2\pi k/N$, $0 \le k \le N - 1$:

$$X[k] = X(e^{j\omega})\Big|_{\omega=2\pi k/N} = \sum_{n=0}^{N-1} x[n]e^{-j2\pi kn/N}, \qquad 0 \le k \le N - 1. \tag{8.23}$$

Since a finite-length sequence is always absolutely summable, the DFT is thus the samples of its z-transform $X(z)$ evaluated on the unit circle at N equally spaced points:

$$X[k] = X(z)\big|_{z=e^{j2\pi k/N}}, \qquad 0 \le k \le N - 1. \tag{8.24}$$

As can be seen from Eq. (8.23), the computation of each sample of the DFT sequence requires N complex multiplications and $N - 1$ complex additions. Hence, the computation of the N-point DFT sequence requires N^2 complex multiplications and $(N - 1)N$ complex additions. In the case of a sequence of length N, it can be shown that the computation of its N-point DFT sequence requires $4N^2$ real multiplications and $(4N - 2)N$ real additions (Problem 8.12). As a result, the total number of computations to compute an N-point DFT increases very rapidly as N increases. For large N, the number of complex multiplications and additions is approximately equal to N^2. Hence, it is of practical interest to develop more efficient or fast algorithms for computing the DFT.

8.3.1 Goertzel's Algorithm

An elegant approach to computing the DFT is to use a recursive computation scheme. To this end, the most popular approach is the *Goertzel's algorithm* derived next [Goe58]. This algorithm makes use of the identity

$$W_N^{-kN} = 1, \tag{8.25}$$

obtained using the periodicity of W_N^{-kn}. Using the above identity, we can rewrite Eq. (8.23) as

$$X[k] = \sum_{\ell=0}^{N-1} x[\ell]W_N^{k\ell} = W_N^{-kN} \sum_{\ell=0}^{N-1} x[\ell]W_N^{k\ell} = \sum_{\ell=0}^{N-1} x[\ell]W_N^{-k(N-\ell)}. \tag{8.26}$$

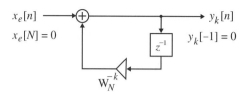

Figure 8.13: Recursive computation of the kth DFT sample.

The above expression can be expressed in the form of a convolution. To this end, we define a new sequence

$$y_k[n] = \sum_{\ell=0}^{n} x_e[\ell] W_N^{-k(n-\ell)}, \tag{8.27}$$

which is a direct convolution of the causal sequence $x_e[n]$ defined by

$$x_e[n] = \begin{cases} x[n], & 0 \le n \le N - 1, \\ 0, & n < 0, n \ge N, \end{cases} \tag{8.28a}$$

with a causal infinite-length sequence

$$h_k[n] = \begin{cases} W_N^{-kn}, & n \ge 0, \\ 0, & n < 0. \end{cases} \tag{8.28b}$$

It follows from Eqs. (8.26) and (8.27) that

$$X[k] = y_k[n]|_{n=N} .$$

By taking the z-transform of both sides of Eq. (8.27) we arrive at

$$Y_k(z) = \frac{X_e(z)}{1 - W_N^{-k} z^{-1}}, \tag{8.29}$$

where $1/(1 - W_N^{-k} z^{-1})$ is the z-transform of the causal sequence $h_k[n]$ and $X_e(z)$ is the z-transform of $x_e[n]$. The above equation implies that $y_k[n]$ is the output of an initially relaxed LTI digital filter with a transfer function

$$H_k(z) = \frac{1}{1 - W_N^{-k} z^{-1}}, \tag{8.30}$$

with an input $x[n]$ as indicated in Figure 8.13. When $n = N$, the output of the filter $y_k[N]$ is precisely $X[k]$.

From the above figure, the DFT computation algorithm is given by

$$y_k[n] = x_e[n] + W_N^{-k} y_k[n - 1] \qquad 0 \le n \le N, \tag{8.31}$$

with $y_k[-1] = 0$ and $x_e[N] = 0$. Since a complex multiplication can be implemented using four real multiplications and two real additions, computation of each new value of $y_k[n]$ thus, in general, requires four real multiplications and four real additions.[3] As a result, the computation of $X[k] = y_k[N]$ involves

---

[3]It can be shown that a simple modification of the complex multiplication algorithm can reduce the number of real multiplications to 3 while increasing the number of real additions to 5 (see Problem 8.13).

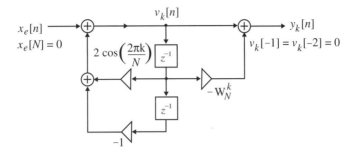

Figure 8.14: Modified approach to the recursive computation of the kth DFT sample.

$4N$ real multiplications and $4N$ real additions, resulting in $4N^2$ real multiplications and $4N^2$ real additions for the computation of all N DFT samples.

　　Hence, the above algorithm, in comparison to the direct DFT computation, requires the same number of real multiplications but $2N$ more real additions. As a result, it is computationally slightly more inefficient than the direct approach. The advantage of the recursive algorithm, however, is that the N complex coefficients W_N^{kn} required to compute $X[k]$ do not have to be either computed or stored in advance, but are computed recursively as needed.

　　The algorithm can be made more efficient by observing that $H_k(z)$ of Eq. (8.30) can be rewritten as

$$H_k(z) = \frac{1}{1 - W_N^{-k}z^{-1}} = \frac{1 - W_N^k z^{-1}}{(1 - W_N^{-k}z^{-1})(1 - W_N^k z^{-1})}$$
$$= \frac{1 - W_N^k z^{-1}}{1 - 2\cos\left(\frac{2\pi}{N}k\right)z^{-1} + z^{-2}}, \tag{8.32}$$

resulting in the new realization shown in Figure 8.14.

　　The DFT computation equations are now given by

$$v_k[n] = x_e[n] + 2\cos\left(\frac{2\pi k}{N}\right)v_k[n-1] - v_k[n-2], \quad 0 \le n \le N, \tag{8.33a}$$

$$X[k] = y_k[N] = v_k[N] - W_N^k v_k[N-1]. \tag{8.33b}$$

　　Note that the computation of each sample of the intermediate variable $v_k[n]$ involves only two real multiplications and four real additions.[4] The complex multiplication by the constant W_N^k needs to be performed only once at $n = N$. Thus, to compute one sample $X[k]$ of the N-point DFT, we need $(2N + 4)$ real multiplications and $(4N + 4)$ real additions. As a result, the modified Goertzel's algorithm for computing the N-point DFT requires $2(N + 2)N$ real multiplications and $4(N + 1)N$ real additions.

　　Further savings in computational requirements can be obtained by comparing the realization of $H_{N-k}(z)$ with that of $H_k(z)$ shown in Figure 8.14. In the case of the former, the multiplier in the feedback path is $2\cos(2\pi(N-k)/N) = 2\cos(2\pi k/N)$ which is the same as in Figure 8.14. Hence, $v_{N-k}[n] = v_k[n]$, indicating that the intermediate variables computed for determining $X[k]$ need no longer be computed again for determining $X[N-k]$. The only difference between these two structures is in the feedforward path, where the multiplier is instead $W_N^{N-k} = W_N^{-k}$, which is the complex conjugate of the coefficient W_N^k used in Figure 8.14. Thus, the computation of the two samples of the N-point DFT, $X[k]$ and $X[N-k]$,

---

[4]In general, $x[n]$ and $v_k[n]$ are complex sequences.

requires $2(N + 4)$ real multiplications and $4(N + 2)$ real additions. Or in other words, all N samples of the DFT can be determined using approximately N^2 real multiplications and approximately $2N^2$ real additions. The number of real multiplications is thus about one-fourth and the number of real additions is about one-half of those needed in the direct DFT computation.

The MATLAB M-file gfft(x,N,k) given below implements the modified Goertzel's algorithm to compute the kth DFT sample of an N-point DFT of the sequence x. The computed DFT sample is XF. The length of the input sequence must be less than or equal to N. If the length is less than N, the sequence length is increased to N by zero-padding.

```
% Function to Compute a DFT Sample
% Using Goertzel's Algorithm
% XF = gfft(x,N,k)
% x is the input sequence of length <= N
% N is the DFT length
% k is the specified bin number
% XF is the desired DFT sample
%
function XF = gfft(x,N,k)
if length(x) < N
       xe = [x zeros(1,N-length(x))];
else
       xe = x;
end
x1 = [xe 0];
d1 = 2*cos(2*pi*k/N);W = exp(-i*2*pi*k/N);
y = filter(1,[1 -d1 1],x1);
XF = y(N+1) - W*y(N);
```

Goertzel's algorithm is attractive in applications requiring the computation of a few samples of the DFT. One such example is the *dual-tone multifrequency* (DTMF) signal detection in the TOUCH-TONE® telephone dialing system discussed in Section 11.1. In such an application, the input $x[n]$ is a real sequence and the square magnitude of the DFT sample, $|X[k]|^2$, is of interest. Since $x[n]$ is a real sequence, the intermediate sequence $v_k[n]$ generated in the modified Goertzel's algorithm is also a real sequence. As a result, we obtain from Eq. (8.33b):

$$|X[k]|^2 = |y_k[N]|^2 = v_k^2[N] + v_k^2[N-1] - 2\cos(2\pi k/N)v_k[N]v_k[N-1]. \qquad (8.34)$$

The above scheme uses only real multiplications, avoiding the complex multiplication required for the computation of $y_k[N]$ as indicated in Eq. (8.33b).

We next describe a fast algorithm for the computation of the DFT when the length N is a composite number. As we shall demonstrate, in the new algorithm in the case of N that is a power of 2, the total number of computations can be made proportional to $N \log_2(N)$ and is highly preferable in applications requiring the computation of all DFT samples.

8.3.2 Fast Fourier Transform Algorithms

The basic idea behind all fast algorithms for computing the discrete Fourier transform (DFT), commonly called the *fast Fourier transform* (FFT) algorithms, is to decompose successively the N-point DFT computation into computations of smaller-size DFTs and to take advantage of the periodicity and symmetry of the complex number W_N^{kn}. Such decompositions, if properly carried out, can result in a significant savings in the computational complexity. There are various versions of the FFT algorithms. We review here the main concepts behind the two most basic FFT algorithms [Coo65].

Decimation-in-Time FFT Algorithm

Consider a sequence $x[n]$ of length N that is assumed to be a power of 2. Using a two-band polyphase decomposition[5] of $x[n]$ we can express its z-transform $X(z)$ as

$$X(z) = X_0(z^2) + z^{-1}X_1(z^2), \tag{8.35}$$

where

$$X_0(z) = \sum_{n=0}^{\frac{N}{2}-1} x_0[n]z^{-n} = \sum_{n=0}^{\frac{N}{2}-1} x[2n]z^{-n}, \tag{8.36a}$$

$$X_1(z) = \sum_{n=0}^{\frac{N}{2}-1} x_1[n]z^{-n} = \sum_{n=0}^{\frac{N}{2}-1} x[2n+1]z^{-n}. \tag{8.36b}$$

Thus, $X_0(z)$ is the z-transform of the $(N/2)$-length sequence $x_0[n] = x[2n]$ formed from the even-indexed samples of $x[n]$, while $X_1(z)$ is the z-transform of the $(N/2)$-length sequence $x_1[n] = x[2n+1]$ formed from the odd-indexed samples of $x[n]$.

Evaluating $X(z)$ on the unit circle at N equally spaced points, $z = W_N^{-k}$, we arrive at the N-point DFT of $x[n]$ given by[6]

$$X[k] = X_0[\langle k \rangle_{N/2}] + W_N^k X_1[\langle k \rangle_{N/2}], \qquad 0 \le k \le N - 1, \tag{8.37}$$

where $X_0[k]$ and $X_1[k]$ are the $(N/2)$-point DFTs of the $(N/2)$-length sequences $x_0[n]$ and $x_1[n]$, respectively, i.e.,

$$X_0[k] = \sum_{r=0}^{\frac{N}{2}-1} x_0[r]W_{N/2}^{rk}$$

$$= \sum_{r=0}^{\frac{N}{2}-1} x[2r]W_{N/2}^{rk}, \quad 0 \le k \le \frac{N}{2} - 1, \tag{8.38a}$$

$$X_1[k] = \sum_{r=0}^{\frac{N}{2}-1} x_1[r]W_{N/2}^{rk}$$

$$= \sum_{r=0}^{\frac{N}{2}-1} x[2r+1]W_{N/2}^{rk}, \quad 0 \le k \le \frac{N}{2} - 1. \tag{8.38b}$$

It is instructive at this point to examine a block diagram interpretation of the modified DFT computation scheme of Eq. (8.37) that computes an N-point DFT of the original length-N sequence $x[n]$ by forming

---

[5] See Section 6.3.3 for a discussion on the polyphase decomposition.
[6] $\langle k \rangle_{N/2} = k$ modulo $(N/2)$.

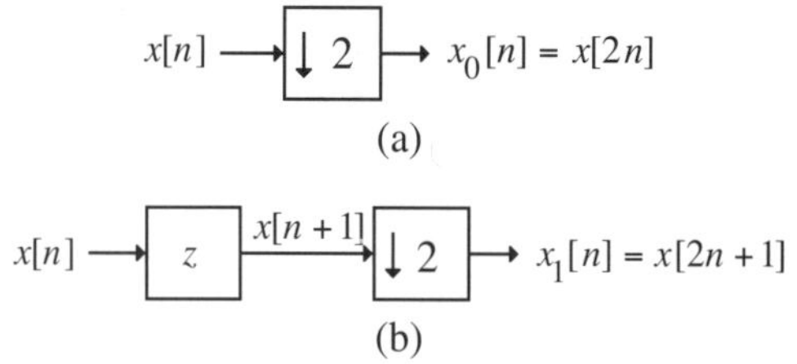

$$x[n] \longrightarrow \boxed{\downarrow 2} \longrightarrow x_0[n] = x[2n]$$

(a)

$$x[n] \longrightarrow \boxed{z} \overset{x[n+1]}{\longrightarrow} \boxed{\downarrow 2} \longrightarrow x_1[n] = x[2n+1]$$

(b)

Figure 8.15: (a) Generation of a subsequence containing even-indexed input samples, and (b) generation of a subsequence containing odd-indexed input samples.

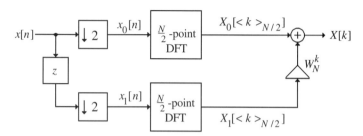

Figure 8.16: Structural interpretation of the DFT decomposition scheme of Eq. (8.37).

a weighted sum of two $(N/2)$-point DFTs of two $(N/2)$-length subsequences formed from the even-indexed samples $x_0[n] = x[2n]$ and the odd-indexed samples $x_1[n] = x[2n + 1]$. To this end we need the *down-sampler*, introduced earlier in Section 2.1.2, to develop the two subsequences $x_0[n]$ and $x_1[n]$ from $x[n]$. It follows from the definition of Eq. (2.18), if the input $x[n]$ to a factor-of-2 down-sampler is a length-N sequence defined for $0 \leq n \leq N - 1$, its output $x_0[n]$ is a length-$(N/2)$ sequence defined for $0 \leq n \leq (N/2) - 1$ and is composed of the even-indexed samples of $x[n]$, i.e., $x_0[n] = x[2n]$, as shown in Figure 8.15(a). To generate the subsequence $x_1[n]$ composed of the odd-indexed samples of $x[n]$, i.e., $x_1[n] = x[2n + 1], \ 0 \leq n \leq (N/2) - 1$, we need to pass $x[n + 1]$ through a factor-of-2 down-sampler. The sequence $x[n + 1]$ can be developed from the sequence $x[n]$ by means of an advance operation. The process is illustrated in Figure 8.15(b).

From the above discussion it follows then that a block diagram interpretation of the DFT computation scheme of Eq. (8.37) is as indicated in Figure 8.16. Figure 8.17 shows its flow-graph representation for the case of $N = 8$.

Before proceeding further, let us evaluate the computational requirements for computing an N-point DFT using two $(N/2)$-point DFTs based on the decomposition of Eq. (8.37). Now a direct computation of an N-point DFT requires N^2 complex multiplications and $N^2 - N \cong N^2$ complex additions. On the other hand, computation of an N-point DFT using the decomposition of Eq. (8.37) requires the computations of two $(N/2)$-point DFTs that need to be combined with N complex multiplications and N complex additions, resulting in a total of $N + (N^2/2)$ complex multiplications and approximately $N + (N^2/2)$ complex additions. It can easily be verified that for $N \geq 3$, $N + (N^2/2) < N^2$.

We can continue the above process by expressing each of the two $(N/2)$-point DFTs, $X_0[k]$ and $X_1[k]$, as a weighted combination of two $(N/4)$-point DFTs since by assumption $N/2$ is even. For example, we can express $X_0[k]$ as

$$X_0[k] = X_{00}[\langle k \rangle_{N/4}] + W_{N/2}^k X_{01}[\langle k \rangle_{N/4}], \quad 0 \leq k \leq \tfrac{N}{2} - 1, \tag{8.39}$$

where $X_{00}[k]$ and $X_{01}[k]$ are the $(N/4)$-point DFTs of the $(N/4)$-length sequences, $x_{00}[n]$ and $x_{01}[n]$,

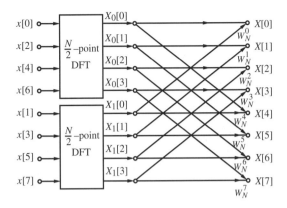

Figure 8.17: Flow-graph of the first stage in the decimation-in-time FFT algorithm for $N = 8$.

generated from the even and odd samples of $x_0[n]$, respectively. Likewise, we can express $X_1[k]$ as

$$X_1[k] = X_{10}[\langle k \rangle_{N/4}] + W_{N/2}^k X_{11}[\langle k \rangle_{N/4}], \quad 0 \leq k \leq \frac{N}{2} - 1, \qquad (8.40)$$

where $X_{10}[k]$ and $X_{11}[k]$ are the $(N/4)$-point DFTs of the $(N/4)$-length sequences, $x_{10}[n]$ and $x_{11}[n]$, generated from the even and odd samples of $x_1[n]$, respectively.

Substituting Eqs. (8.39) and (8.40) in Eq. (8.37) and making use of the identity $W_{N/2}^k = W_N^{2k}$, we then arrive at the two-stage decomposition of an N-point DFT computation in terms of four $(N/4)$-point DFTs, as indicated by the block diagram of Figure 8.18. The corresponding flow-graph representation is shown in Figure 8.19 for $N = 8$. In the case of the 8-point DFT computation illustrated in Figure 8.20, the $(N/4)$-point DFT is a 2-point DFT and no further decomposition is possible. The 2-point DFTs, $X_{00}[k]$, $X_{01}[k]$, $X_{10}[k]$, and $X_{11}[k]$, can be easily computed. For example, for the computation of $X_{00}[k]$ the pertinent expressions are

$$X_{00}[k] = \sum_{n=0}^{1} x_{00}[n] W_2^{nk} = x[0] + W_2^k x[4], \quad k = 0, 1. \qquad (8.41)$$

The corresponding flow-graph is indicated in Figure 8.20 where we have used the identity $W_2^k = W_N^{(N/2)k}$. Replacing each of the 2-point DFTs in Figure 8.19 with their respective flow-graph representations, we finally arrive at the complete flow-graph of the basic decimation-in-time DFT algorithm as shown in Figure 8.21.

If we examine the flow-graph of Figure 8.21, we note that it consists of three stages. The first stage computes the four 2-point DFTs, the second stage computes the two 4-point DFTs, and finally the last stage computes the desired 8-point DFT. Moreover, the number of complex multiplications and additions performed at each stage is equal to 8, the size of the transform. As a result, the total number of complex multiplications and additions in computing all 8 DFT samples is equal to $3 \times 8 = 24$.

It follows from the above observation that in the general case when $N = 2^\mu$, the number of stages of computation of the (2^μ)-point DFT in the fast algorithm will be $\mu = \log_2 N$. Therefore, the total number of complex multiplications and additions in computing all N DFT samples is equal to $N(\log_2 N)$. In developing this count, we have for the present considered multiplications with $W_N^0 = 1$ and $W_N^{N/2} = -1$ to be complex. In addition, we have not taken advantage of the symmetry property $W_N^{(N/2)+k} = -W_N^k$. These properties can be made use of in reducing the computational complexity further.

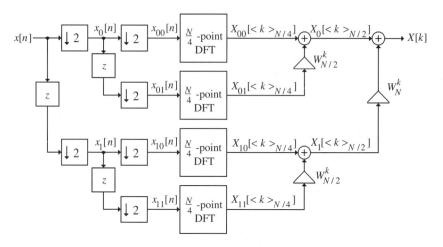

Figure 8.18: Structural interpretation of the two-stage DFT decomposition scheme.

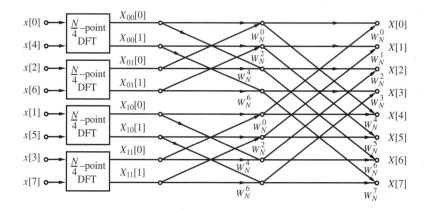

Figure 8.19: Flow-graph of the second stage in the decimation-in-time FFT algorithm for $N = 8$.

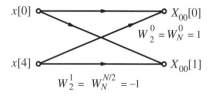

Figure 8.20: Flow-graph of the 2-point DFT.

Computational Considerations

Examination of the flow-graph of Figure 8.21 also reveals that each stage of the DFT computation process employs the same basic computational module in which two output variables are generated from a weighted combination of two input variables. To see this aspect more clearly, let us label the N-input and the N-output variables in the rth stage of the DFT computation as $\Psi_r[m]$ and $\Psi_{r+1}[m]$, respectively, with $r = 1, 2, \ldots, \mu$, and $m = 0, 1, \ldots, N-1$. According to this labeling, in the case of Figure 8.22 for the 8-point

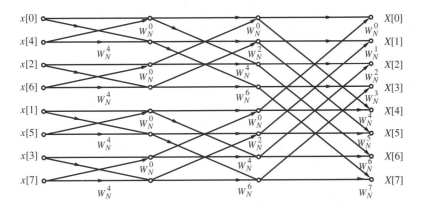

Figure 8.21: Complete flow-graph of the basic decimation-in-time FFT algorithm for $N = 8$.

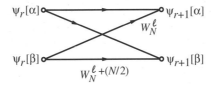

Figure 8.22: Flow-graph of the basic computational module in the decimation-in-time FFT method.

DFT computation, $\Psi_1[0] = x[0]$, $\Psi_1[1] = x[4]$, and so on. Similarly, here, $\Psi_4[0] = X[0]$, $\Psi_4[1] = X[1]$, and so on. Based on this labeling scheme, it can be easily verified that the basic computational module is represented by the flow-graph of Figure 8.22 and described by the following input-output relations:

$$\Psi_{r+1}[\alpha] = \Psi_r[\alpha] + W_N^\ell \Psi_r[\beta], \tag{8.42a}$$

$$\Psi_{r+1}[\beta] = \Psi_r[\alpha] + W_N^{\ell+(N/2)} \Psi_r[\beta]. \tag{8.42b}$$

Because of its shape, the basic computational module of Figure 8.22 is referred to in the literature as the *butterfly computation*.

Substituting $W_N^{\ell+(N/2)} = -W_N^\ell$ in Eq. (8.42b) we can rewrite it as

$$\Psi_{r+1}[\beta] = \Psi_r[\alpha] - W_N^\ell \Psi_r[\beta]. \tag{8.42c}$$

The modified butterfly computation is thus as indicated in Figure 8.23 and requires only one complex multiplication. Use of this modified butterfly computational module in the FFT computation leads to a reduction in the total number of complex multiplications by 50%, as can be seen from the new flow-graph for the $N = 8$ case illustrated in Figure 8.24. Further savings in the computational complexity arise by taking into consideration that multiplications by $W_N^0 = 1$, $W_N^{N/2} = -1$, $W_N^{N/4} = j$, and $W_N^{3N/4} = -j$ can be avoided in the DFT computation process.

Another attractive feature of the FFT algorithm described above is with regard to memory requirements. Since each stage employs the same butterfly computation to compute the two output variables $\Psi_{r+1}[\alpha]$ and $\Psi_{r+1}[\beta]$ from the input variables $\Psi_r[\alpha]$ and $\Psi_r[\beta]$, after $\Psi_{r+1}[\alpha]$ and $\Psi_{r+1}[\beta]$ have been determined, they can be stored in the same memory locations where $\Psi_r[\alpha]$ and $\Psi_r[\beta]$ were previously stored. Thus, at the end of the computation at any stage, the output variables $\Psi_{r+1}[m]$ can be stored in the same registers

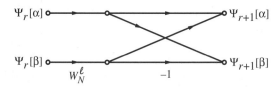

Figure 8.23: Flow-graph of the modified butterfly computational module.

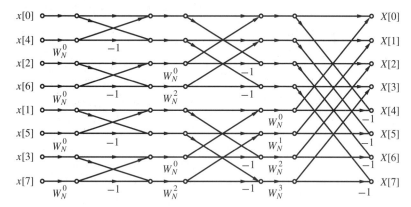

Figure 8.24: Flow-graph of the modified decimation-in-time FFT algorithm.

previously occupied by the corresponding input variables $\Psi_r[m]$. This type of memory location sharing feature is commonly known as the *in-place computation*, resulting in a significant saving in the overall memory requirements.

It should be noted, however, from Figure 8.24 that while the DFT samples $X[k]$ appear at the output in a sequential order, the input time-domain samples $x[n]$ appear in a different order. Thus, a sequentially ordered input $x[n]$ must be reordered appropriately before the FFT algorithm described by the above structure can begin. To understand the basic scheme, in the input reordering scheme consider the 8-point DFT computation illustrated in Figure 8.24. If we represent the arguments of the input samples $x[n]$ and their sequentially ordered new representations $\Psi_1[m]$ in binary forms, we arrive at the following relations between m and n:

m	n
000	000
001	100
010	010
011	110
100	001
101	101
110	011
111	111

It follows from the above that if $(b_2 b_1 b_0)$ represents the index n of $x[n]$ in binary form, then the sample $x[b_2 b_1 b_0]$ appears in the location $m = b_0 b_1 b_2$ as $\Psi[b_0 b_1 b_2]$ before the DFT computation is started, or in other words, the location of $\Psi_1[m]$ is in *bit-reversed order* from that of the original input array $x[n]$.

Various alternative forms of the FFT computation can be easily obtained by reordering the computations, such as the input in normal order and the output in bit-reversed order (Problem 8.16), and both input and output in normal order (Problem 8.17).

The FFT algorithm outlined above assumed the length of the input sequence $x[n]$ to be a power of 2. If it is not, we can extend the length by zero-padding $x[n]$ (see Section 3.2.3) so that the length of the extended sequence is a power of 2.[7] Even after zero-padding, the DFT computation based on the fast algorithm derived above may be computationally more efficient than the direct DFT computation of the original shorter sequence. Alternatively, we can develop fast algorithms that make use of polyphase decomposition with more than two subsequences. To illustrate this modification to the basic FFT algorithm, consider a sequence $x[n]$ of a length that is a power of 3. Here, the DFT samples in the first stage are computed using a three-band polyphase decomposition of $X(z)$,

$$X(z) = X_0(z^3) + z^{-1}X_1(z^3) + z^{-2}X_2(z^3), \tag{8.43}$$

resulting in

$$X[k] = X_0[\langle k \rangle_{N/3}] + W_N^k X_1[\langle k \rangle_{N/3}] + W_N^{2k} X_2[\langle k \rangle_{N/3}], \quad 0 \le k \le N - 1, \tag{8.44}$$

where $X_0[k]$, $X_1[k]$, and $X_2[k]$ are now $(N/3)$-point DFTs. This process can be repeated to compute each of the $(N/3)$-point DFTs in terms of three $(N/9)$-point DFTs, and so on, until the smallest computational module is a 3-point DFT and no further decomposition is possible.

The FFT computation schemes described above are called *decimation-in-time (DIT) FFT algorithms* since here the input sequence $x[n]$ is first decimated to form a set of subsequences before the DFT is computed. For example, the relation between the input sequence $x[n]$ and its even and odd parts, $x_0[n]$ and $x_1[n]$, respectively, generated by the first stage of the DIT algorithm shown in Figure 8.16 is as follows:

$x[n]$:	$x[0]$	$x[1]$	$x[2]$	$x[3]$	$x[4]$	$x[5]$	$x[6]$	$x[7]$
$x_0[n]$:	$x[0]$		$x[2]$		$x[4]$		$x[6]$	
$x_1[n]$:	$x[1]$		$x[3]$		$x[5]$		$x[7]$	

Likewise, the relation between the input sequence $x[n]$ and the sequences $x_{00}[n]$, $x_{01}[n]$, $x_{10}[n]$, and $x_{11}[n]$, generated by the two-stage decomposition of the DIT algorithm and illustrated in Figure 8.19, is given by

$x[n]$:	$x[0]$	$x[1]$	$x[2]$	$x[3]$	$x[4]$	$x[5]$	$x[6]$	$x[7]$
$x_{00}[n]$:	$x[0]$				$x[4]$			
$x_{01}[n]$:	$x[2]$				$x[6]$			
$x_{10}[n]$:	$x[1]$				$x[5]$			
$x_{11}[n]$:	$x[3]$				$x[7]$			

Or in other words, the subsequences $x_{00}[n]$, $x_{01}[n]$, $x_{10}[n]$, and $x_{11}[n]$ can be generated directly by a factor-of-4 decimation process, leading to the single-stage decomposition given in Figure 8.25.

If at each stage the decimation is by a factor of R, the resulting FFT algorithm is called a *radix-R* FFT algorithm. Thus, Figure 8.24 shows a radix-2 DIT FFT algorithm. Likewise, Figure 8.25 illustrates the first stage of a radix-4 DIT FFT algorithm. It also follows from the above discussion that, depending on the value of N, various combinations of decompositions of $X[k]$ can be used to develop different types of DIT FFT algorithms. If the scheme uses a mixture of decimations by different factors, it is called a *mixed radix* FFT algorithm.

For N which is a composite number expressible in the form of a product of integers,

$$N = r_1 \cdot r_2 \cdots r_\nu,$$

---

[7]It should be noted that zero-padding increases the effective length of the original sequence, and hence, the longer-length DFT samples are different frequency samples of the frequency response and are more closely spaced on the unit circle than that of the original shorter-length DFT samples.

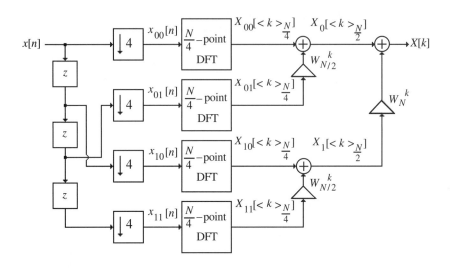

Figure 8.25: First stage in a radix-4 decimation-in-time FFT algorithm.

it can be shown that the total number of complex multiplications (additions) in a DIT FFT algorithm based on a ν-stage decomposition is given by (Problem 8.18)

$$\text{No. of multiply (add) operations} = \left(\sum_{i=1}^{\nu} r_i - \nu \right) N. \tag{8.45}$$

Modified Goertzel's Algorithm Based on DIT Decomposition

In some applications, it may be convenient to apply the Goertzel's algorithm in computing the smaller-size DFTs after one or two or more stages of decimation-in-time decompositions of the input sequence $x[n]$. For example, the four $(N/4)$-DFTs in Figure 8.18 can be computed using Goertzel's algorithm. This approach reduces the overall computational complexity of the direct Goertzel's algorithm while still permitting the computation of a few DFT samples.

It should be noted, however, that the structural interpretations of the DIT FFT algorithms given in Figures 8.16, 8.18, and 8.25 make use of the advance block with a transfer function z that is not physically realizable. If a hardware implementation of these structures is desired, we can insert an appropriate amount of delays at the input and then move them through the chain of advance blocks to make the overall structure physically realizable. Figure 8.26 shows the realizable version of Figure 8.25. Moreover, the input sequence $x[n]$ should be delayed by 3 sample periods before the down-sampling operations are carried out. The relations between the subsequences $x_{00}[n]$, $x_{01}[n]$, $x_{10}[n]$, and $x_{11}[n]$ and the original sequence $x[n]$ are now

$x[n]$:	$x[0]$	$x[1]$	$x[2]$	$x[3]$	$x[4]$	$x[5]$	$x[6]$	$x[7]$
$x_{00}[n]$:				$x[0]$				$x[4]$
$x_{01}[n]$:				$x[2]$				$x[6]$
$x_{10}[n]$:				$x[1]$				$x[5]$
$x_{11}[n]$:				$x[3]$				$x[7]$

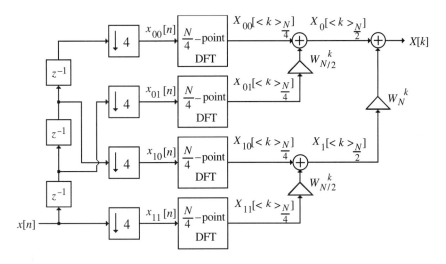

Figure 8.26: Modified structure for the first stage of a radix-4 decimation-in-time FFT algorithm.

Decimation-in-Frequency FFT Algorithm

The basic idea behind the decimation-in-time FFT algorithm is to decompose sequentially the N-point sequence $x[n]$ into sets of smaller and smaller subsequences and then form a weighted combinations of the DFTs of these subsequences. The same idea can be applied to the N-point DFT sequence $X[k]$ to decompose it sequentially into sets of smaller and smaller subsequences. This approach leads to another class of DFT computation schemes collectively known as the *decimation-in-frequency (DIF) FFT algorithm.*

To illustrate the basic difference between the above two decomposition schemes, we develop below the first stage of the DIF FFT algorithm for the case when N is a power of 2. We first express the z-transform $X(z)$ of $x[n]$ as

$$X(z) = X_0(z) + z^{-N/2}X_1(z), \tag{8.46}$$

where

$$X_0(z) = \sum_{n=0}^{(N/2)-1} x[n]z^{-n}, \quad X_1(z) = \sum_{n=0}^{(N/2)-1} x\left[\tfrac{N}{2}+n\right]z^{-n}. \tag{8.47}$$

Evaluating $X(z)$ on the unit circle at $z = W_N^{-k}$, we get from Eqs. (8.46) and (8.47),

$$X[k] = \sum_{n=0}^{(N/2)-1} x[n]W_N^{nk} + W_N^{(N/2)k} \sum_{n=0}^{(N/2)-1} x\left[\tfrac{N}{2}+n\right]W_N^{nk}. \tag{8.48}$$

The above equation can be rewritten as

$$X[k] = \sum_{n=0}^{(N/2)-1} \left(x[n] + (-1)^k x\left[\tfrac{N}{2}+n\right]\right) W_N^{nk}, \tag{8.49}$$

where we have used the identity $W_N^{(N/2)k} = (-1)^k$. Two different forms of Eq. (8.49) are obtained, depending on whether k is even or odd:

$$X[2\ell] = \sum_{n=0}^{(N/2)-1} \left(x[n] + x\left[\tfrac{N}{2}+n\right]\right) W_N^{2n\ell}$$

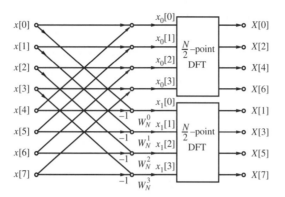

Figure 8.27: Flow-graph of the first stage of the decimation-in-frequency FFT algorithm for $N = 8$.

$$= \sum_{n=0}^{(N/2)-1} \left(x[n] + x\left[\tfrac{N}{2} + n\right]\right) W_{N/2}^{n\ell}, \qquad 0 \le \ell \le \tfrac{N}{2} - 1, \tag{8.50a}$$

$$X[2\ell + 1] = \sum_{n=0}^{(N/2)-1} \left(x[n] - x\left[\tfrac{N}{2} + n\right]\right) W_N^{n(2\ell+1)}$$

$$= \sum_{n=0}^{(N/2)-1} \left(x[n] - x\left[\tfrac{N}{2} + n\right]\right) W_N^{n} W_{N/2}^{n\ell}, \qquad 0 \le \ell \le \tfrac{N}{2} - 1. \tag{8.50b}$$

The above two expressions represent the $\frac{N}{2}$-point DFTs of the following two $\frac{N}{2}$-point sequences:

$$x_0[n] = \left(x[n] + x\left[\tfrac{N}{2} + n\right]\right),$$

$$x_1[n] = \left(x[n] - x\left[\tfrac{N}{2} + n\right]\right) W_N^{n}, \qquad 0 \le n \le \tfrac{N}{2} - 1, \tag{8.51}$$

respectively. The flow-graph of the first stage of the DFT computation scheme defined by Eqs. (8.50a) and (8.50b) is shown in Figure 8.27 for $N = 8$. As can be seen from this figure, here the input samples are in a sequential order, while the output DFT samples appear in a decimated form with the even-indexed samples appearing as the outputs of one $(N/2)$-point DFT and the odd-indexed samples appearing as the outputs of the other $(N/2)$-point DFT.

We can continue the above decomposition process by expressing the even- and odd-indexed samples of each one of the two $(N/2)$-point DFTs as a sum of two $(N/4)$-point DFTs. This process can continue until the smallest DFTs are 2-point DFTs. The complete flow-graph of the decimation-in-frequency DFT computation scheme for the $N = 8$ case is shown in Figure 8.28.

It can be seen from Figure 8.28 that in the DIF FFT algorithm, the input $x[n]$ appears in the normal order while the output $X[k]$ appears in the bit-reversed order. Just as in the case of the radix-2 DIT FFT algorithm, the total number of complex multiplications per stage in the radix-2 DIF FFT algorithm is $N/2$. Hence, the total number of complex multiplications for computing the N-point DFT samples here is also equal to $(N/2) \log_2 N$ ignoring the fact that multiplications by $W_N^{(N/4)\ell}$, $\ell = 0, 1, 2, 3, 4$, can be avoided. As before, various forms of the DIF FFT algorithm can be generated.

The DIT and DIF FFT algorithms described here are often referred to as the *Cooley-Tukey* FFT algorithms.

An examination of the flow-graph of the first stage of the radix-2 DIF FFT algorithm given in Figure 8.27 reveals that the even- and odd-indexed samples of $X[k]$ can be computed independently of each other.

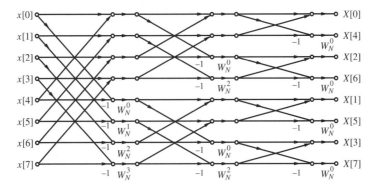

Figure 8.28: Complete flow-graph of the decimation-in-frequency FFT algorithm for $N = 8$.

It has been shown by Duhamel [Duh86] that significant reduction in the computational complexity can be obtained by using a radix-2 DIF algorithm to compute the even-indexed DFT samples and a radix-4 DIF FFT algorithm to compute the odd-indexed DFT samples. This type of computational scheme has been called a *split-radix FFT algorithm* (Problem 8.29).

Inverse DFT Computation

An FFT algorithm for computing the DFT samples can also be used to calculate efficiently the inverse DFT (IDFT). To show this, consider an N-point sequence $x[n]$ with an N-point DFT $X[k]$. The sequence $x[n]$ is related to the samples $X[k]$ through

$$x[n] = \frac{1}{N} \sum_{k=0}^{N-1} X[k] W_N^{-nk}. \tag{8.52}$$

If we multiply both sides of the above equation by N and take the complex conjugate, we arrive at

$$N x^*[n] = \sum_{k=0}^{N-1} X^*[k] W_N^{nk}. \tag{8.53}$$

The right-hand side of the above expression can be recognized as the N-point DFT of a sequence $X^*[k]$ and can be computed using any one of the FFT algorithms discussed earlier. The desired IDFT $x[n]$ is then obtained as

$$x[n] = \frac{1}{N} \left\{ \sum_{k=0}^{N-1} X^*[k] W_N^{nk} \right\}^*. \tag{8.54}$$

In summary, given an N-point DFT $X[k]$, we first form its complex conjugate sequence $X^*[k]$, and then compute the N-point DFT of $X^*[k]$, form the complex conjugate of the DFT computed, and finally, divide each sample by N. The inverse DFT computation process is illustrated in Figure 8.29. Two other approaches to inverse DFT computation are described in Problems 8.23 and 8.24.

8.3.3 DFT and IDFT Computation Using MATLAB

The following functions are included in the MATLAB package for the computation of the DFT and the IDFT:

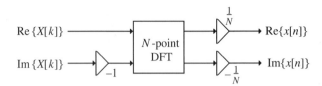

Figure 8.29: Inverse DFT computation via DFT.

```
fft(x),              fft(x,N)
ifft(X),             ifft(X,N)
```

These programs employ efficient FFT algorithms for the computation. The M-file `fft(x)` computes the DFT of a vector x of the same length as that of x. For computing the DFT of a specific length N, the M-file `fft(x,N)` is used. Here, if the length of x is greater than N, it is truncated to the first N samples, whereas if the length of x is less than N, the vector x is zero-padded at the end to make it into a length-N sequence. Likewise, the M-file `ifft(X)` computes the IDFT of a vector X of the same length as that of X, whereas the M-file `ifft(X,N)` can be used to compute the IDFT of a specified length N. In the latter case, the restrictions on N are the same as that in the M-file `fft(x,N)`. It should be noted that the M-file `ifft` in MATLAB employs the IDFT computation scheme outlined in the previous section.

MATLAB uses a high-speed radix-2 algorithm when the sequence length of x or X is a power of 2. Moreover, the radix-2 FFT program has been optimized specifically to compute the DFT of a real input sequence faster than the DFT of a complex sequence. If the sequence length is not a power of 2, it employs a mixed-radix FFT algorithm and usually takes a much longer time to compute the DFT or the IDFT of a sequence of length N that is not a power of 2 than those of a sequence of a power-of-2 length that is closest to N.

Since vectors in MATLAB are indexed from 1 to N instead of 0 to $N - 1$, the DFT and the IDFT computed in the above MATLAB functions make use of the expressions

$$X[k] = \sum_{n=1}^{N} x[n] W_N^{(n-1)(k-1)}, \qquad 1 \le k \le N, \tag{8.55a}$$

$$x[n] = \frac{1}{N} \sum_{k=1}^{N} X[k] W_N^{-(n-1)(k-1)}, \quad 1 \le n \le N. \tag{8.55b}$$

EXAMPLE 8.11   We consider here the computation of the DFT of a sinusoidal sequence of finite length. The MATLAB program used for this purpose is given below:

```
% Program 8_10
% Spectral Analysis Using the DFT
%
N = input('Type in the length of DFT = ');
T = input('Type in the sampling period = ');
freq = input('Type in the sinusoid's frequency = ');
k = 0:N-1;
f = sin(2*pi*freq*k*(1/T));
F = fft(f);
stem(k,abs(F));grid
xlabel('k');
ylabel('|X[k]|');
```

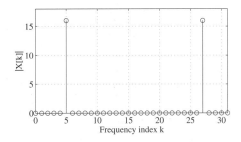

Figure 8.30: The magnitudes of a 32-point DFT of a sinusoid of frequency 10 Hz sampled at a 64-Hz rate.

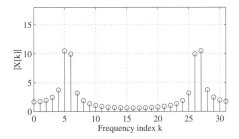

Figure 8.31: The magnitudes of a 32-point DFT of a sinusoid of frequency 11 Hz sampled at a 64-Hz rate.

During execution, the program requests the following input data: the length N of the DFT, the sampling frequency F_T in Hz, and the frequency of the sinusoid in Hz. We compute here the 32-point DFT $X[k]$ of a length-32 sinusoidal sequence $x[n]$ of frequency 10 Hz with a sampling rate of 64 Hz. Note that the sampling frequency of 64 Hz is considerably higher than 20 Hz, the Nyquist frequency, and as a result, no aliasing distortion will occur after sampling. The magnitudes of the DFT samples generated by this program are plotted in Figure 8.30.

Several comments are in order here. Since the time-domain sequence $x[n]$ is a pure sinusoid, its discrete-time Fourier transform $X(e^{j2\pi f})$ contains two impulses at $f = \pm 10$ Hz and is zero everywhere else. Its 32-point DFT is obtained by sampling $X(e^{j2\pi f})$ at $f = 64 \times 2/32 = 2k$ Hz, $0 \le k \le 31$. As a result, the impulse at $f = 10$ Hz appears at the DFT frequency bin location $k = 5$ as $X[5]$, while the impulse at $f = -10$ Hz appears at the bin location $k = 32 - 5 = 27$ as $X[27]$. Thus, the DFT has correctly identified the frequency of the sinusoid in this case. Note also that the first half of the DFT samples for $k = 0$ to $k = (N/2) - 1$ corresponds to the positive frequency axis from $f = 0$ to $f = F_T/2$ excluding the point $F_T/2$, while the second half for $k = N/2$ to $k = N - 1$ corresponds to the negative frequency axis from $f = -F_T/2$ to $f = 0$ excluding the point $f = 0$.

We next compute the 32-point DFTs $X[k]$ of a sinusoid $x[n]$ of frequency 11 Hz with a sampling frequency of 64 Hz. Since the time-domain sequence $x[n]$ is a pure sinusoid, its discrete-time Fourier transform $X(e^{j2\pi f})$ contains two impulses at $f = \pm 11$ Hz and is zero everywhere else. However, these frequency locations are between the bins $k = 5$ and $k = 6$, and $k = 26$ and $k = 27$. From the computed magnitudes of the DFT samples, shown in Figure 8.31, it can be seen that the spectrum contains frequency components at all bins, with two strong components at bins $k = 5$ and $k = 6$ and with two strong components at bins $k = 26$ and $k = 27$. This type of phenomenon is called the *leakage*. It is further discussed in detail in Section 11.2.

EXAMPLE 8.12 In this example we illustrate the application of the FFT and IFFT functions of MATLAB in performing the linear convolution of two finite-length sequences. The MATLAB program that can be employed for this purpose is as follows:

```
% Program 8_11
% Computation of Linear Convolution Using DFT
%
g = input('Type in first sequence = ');
h = input('Type in second sequence = ');
ga = [g zeros(1,length(h)-1)];
ha = [h zeros(1,length(g)-1)];
G = fft(ga);H = fft(ha);
Y = G.*H;
y = ifft(Y);
disp('New Sequence')
disp(real(y))
```

We illustrate its use in developing the convolution of the two length-4 sequences of Example 3.15:

$$g[n] = \{1 \quad 2 \quad 0 \quad 1\}, \quad h[n] = \{2 \quad 2 \quad 1 \quad 1\}.$$

The input data requested by this program are the two sequences entered as vectors with square brackets. The result of the linear convolution using Program 8_11 for the above two sequences is given by

```
New Sequence
   2.0000    6.0000    5.0000    5.0000    4.0000
   1.0000    1.0000
```

which is seen to be identical to that determined in Example 3.17.

8.4 Number Representation

The binary representation is used to represent numbers (and signal variables) in most digital computers and special-purpose digital signal processors used for implementing digital filtering algorithms. In this form, the number is represented using the symbols 0 and 1, called *bits*,[8] with the *binary point* separating the integer part from the fractional part. For example, the binary representation of the decimal number 11.625 is given by

$$1011_\triangle 101$$

where $\triangle$ denotes the binary point. The four bits, 1011, to the left of the binary point form the integer part and the three bits, 101, to the right of the binary point represent the fractional part. In general, the decimal equivalent of a binary number η consisting of B integer bits and b fractional bits,

$$a_{B-1}a_{B-2}\cdots a_1 a_{0\triangle}a_{-1}a_{-2}\cdots a_{-b}$$

is given by

$$\sum_{i=-b}^{B-1} a_i 2^i,$$

where each bit a_i is either a 0 or a 1. The leftmost bit a_{B-1} is called the *most significant bit (MSB)* and the rightmost bit a_{-b} is called the *least significant bit (LSB)*.

To avoid the confusion between a decimal number containing the digits 1 and 0, and a binary number containing the bits 1 and 0 we shall include a subscript 10 to the right of the least significant digit to indicate

---

[8]**bit** is an abbreviated form of **binary digit**.

a decimal number and a subscript 2 to the right of the least significant bit to indicate a binary number. Thus, for example, 1101_{10} represents a decimal number, whereas 1101_2 represents a binary number whose decimal equivalent is 13_{10}. If there is no ambiguity in the representation, then the subscript is dropped.

The block of bits representing a number is called a *word*, and the number of bits in the word is called the *wordlength* or *word size*. The wordlength is typically a positive integer power of 2, such as 8 or 16 or 32. The word size is often expressed in units of eight bits called a *byte*. For example, a 4-byte word is equivalent to a 32-bit-size word.

Digital circuits implementing the arithmetic operations, addition and multiplication of two binary numbers, are specifically designed to develop the results, the sum and the product, respectively, in binary form with their binary points in the assumed locations. There are two basic types of binary representations of numbers, fixed-point and floating-point, as discussed below.

8.4.1 Fixed-Point Representation

In this type of representation, the binary point is assumed to be fixed at a specific location, and the hardware implementation of the arithmetic circuits takes into account the fixed location in performing the arithmetic operations. Implementation of the addition operation carried out by a digital adder circuit is independent of the location of the binary points in the two numbers being added as long as they are in the same location for both numbers. On the other hand, it is not simple to locate the binary point in the product of two binary numbers unless they are both integers or are both fractions. In the case of the multiplication of two integers by the multiplier circuit, the result is also an integer. Likewise, multiplying two fractions results in a fraction. In digital signal processing applications, therefore, fixed-point numbers are always represented as fractions.

The range of nonnegative integers η that can be represented by B bits in a fixed-point representation is given by

$$0 \le \eta \le 2^B - 1. \tag{8.56}$$

Similarly, the range of positive fractions η that can be represented by B bits in a fixed-point representation is given by

$$0 \le \eta \le 1 - 2^{-B}. \tag{8.57}$$

In either case, the range is fixed. If η_{max} and η_{min} denote, respectively, the maximum and the minimum values of the numbers that can be represented in a B-bit fixed-point representation, then the dynamic range R of the numbers that can be represented with B bits is given by $R = \eta_{max} - \eta_{min}$, and the *resolution* of the representation is defined by

$$\delta = \frac{R}{2^B - 1}, \tag{8.58}$$

where δ is also known as the *quantization level*.

8.4.2 Floating-Point Representation

In the normalized floating-point representation, a positive number η is represented using two parameters, the *mantissa M* and the *exponent* or the *characteristic E* in the form

$$\eta = M \cdot 2^E, \tag{8.59}$$

where the mantissa M is a binary fraction restricted to lie in the range

$$\tfrac{1}{2} \le M < 1, \tag{8.60}$$

and the exponent E is either a positive or a negative binary integer.

Figure 8.32: IEEE 32-bit floating-point format.

The floating-point system provides a variable resolution for the range of numbers being represented. The resolution increases exponentially as the magnitude of the number being represented increases. Floating-point numbers are stored in a register by assigning B_E bits of the register to the exponent and the remaining B_M bits to the mantissa. If the same number of total bits is used in both floating-point and fixed-point representations, i.e., $B = B_M + B_E$, then the former provides a larger dynamic range than the latter (Problem 8.44).

The most widely followed floating-point formats for 32-bit and 64-bit representations are those given by the ANSI/IEEE Standard 754-1985 [IEEE85]. In this format, a 32-bit number is divided into fields. The exponent field is of 8-bit length, the mantissa[9] field is of 23-bit length, and 1 bit is assigned for the sign, as indicated in Figure 8.32. The exponent is coded in a biased form as $E - 127$. Thus, a floating-point number η under this scheme is represented as

$$\eta = (-1)^S \cdot 2^{E-127}(M), \tag{8.61}$$

with the mantissa in the range

$$0 \leq M < 1. \tag{8.62}$$

The following conventions are followed in interpreting the representation of Eq. (8.61):

1. If $E = 255$ and $M \neq 0$, then η is not a number (abbreviated as *NaN*).

2. If $E = 255$ and $M = 0$, then $\eta = (-1)^S \cdot \infty$.

3. If $0 < E < 255$, then $\eta = (-1)^S \cdot 2^{E-127}(1_\Delta M)$.

4. If $E = 0$ and $M \neq 0$, then $\eta = (-1)^S \cdot 2^{-126}(0_\Delta M)$.

5. If $E = 0$ and $M = 0$, then $\eta = (-1)^S \cdot 0$.

where $1_\Delta M$ is a number with one integer bit and 23 fractional bits and $0_\Delta M$ is a fraction. The range of a 32-bit floating-point number in the above format is from 1.18×10^{-38} to 3.4×10^{38} (Problem 8.45). The IEEE 32-bit standard has been adopted for number representation in almost all commercial floating-point digital signal processor chips.

8.4.3 Representation of Negative Numbers

To accommodate the representation of both positive and negative b-bit fractions, an additional bit, called the *sign bit*, is placed at the leading position of the register to indicate the sign of the number (Figure 8.33). Independent of the scheme being used to represent the negative number, the *sign bit* is 0 for a positive number and 1 for a negative number.

A fixed-point negative number is represented in one of three different forms. In the *sign-magnitude* format, if the sign bit $s = 0$, the b-bit fraction is a positive number with a magnitude $\sum_{i=1}^b a_{-i} 2^{-i}$, and if $s = 1$, the b-bit fraction is a negative number with a magnitude $\sum_{i=1}^b a_{-i} 2^{-i}$.

---

[9]In the IEEE floating-point standard, the mantissa is called the *significand*.

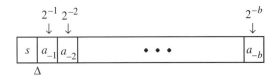

Figure 8.33: Representation of a general signed b-bit fixed-point fraction.

In the *ones'-complement* form, a positive fraction is represented as in the sign-magnitude form, while its negative is represented by complementing each bit of the binary representation of the positive fraction. In this representation, the decimal equivalent of a positive or a negative fraction is thus given by $-s(1 - 2^{-b}) + \sum_{i=1}^{b} a_{-i} 2^{-i}$.

Finally, in the *two's-complement* representation, the positive fraction is represented as in the sign-magnitude form, while its negative is represented by complementing each bit (i.e., by replacing each 0 with a 1, and vice-versa) of the binary representation of the positive fraction, and adding a 1 to the LSB, the bth bit. In this form, the decimal equivalent of a positive or a negative fraction is thus given by $-s + \sum_{i=1}^{b} a_{-i} 2^{-i}$.

EXAMPLE 8.13   We determine the 5-bit binary representations of the decimal fraction -0.625_{10}. Now, the sign-magnitude representation of 0.625_{10} is given by $0_\triangle 1010$ ($= 2^{-1} + 2^{-3} = 0.5 + 0.125$). Therefore, the sign-magnitude representation of -0.625_{10} is simply $1_\triangle 1010$. The ones'-complement representation -0.625_{10} is obtained by complementing each bit of $0_\triangle 1010$ individually, leading to $1_\triangle 0101$. Finally, the two's-complement representation of -0.625_{10} is obtained by adding a 1 to the LSB of the ones'-complement representation, resulting in $1_\triangle 0110$.

Table 8.1 illustrates the above three representations for a 4-bit number (3-bit fraction and a sign bit).

8.4.4 Offset Binary Representation

In the offset binary representation, used primarily in bipolar digital-to-analog conversion, a b-bit fraction with an additional sign bit is considered as a $(b + 1)$-bit number representing 2^{b+1} decimal numbers. About half of these numbers represent the negative fractions and the remaining half represent the positive fractions, as illustrated in Table 8.1 for $b = 3$. It should be noted that the two's-complement representation can be converted to the offset binary representation by simply complementing the sign bit.

8.4.5 Signed Digit Representation

The radix-2 signed digit (SD) format is a 3-valued representation of a radix-2 number and employs three digit values, 0, 1, and $\bar{1}$ (with the last symbol representing -1). In many instances, the SD representation of a binary number requires fewer nonzero digits and has been exploited in developing an algorithm for a faster hardware implementation of the multiplication operation (see Section 8.5.2). A simple algorithm for the conversion of a radix-2 binary number into an equivalent SD representation is as follows [Boo51]. Let $a_{-1} a_{-2} \cdots a_b$ denote a binary number with its equivalent SD representation denoted by $c_{-1} c_{-2} \cdots c_b$. The bits c_{-i} of the SD number are determined through the relation

$$c_{-i} = a_{-i-1} - a_{-i}, \qquad i = b, b-1, \ldots, 1,$$

where $a_{-b-1} = 0$.

The following example illustrates the radix-2 SD representation.

EXAMPLE 8.14   Consider the binary number $\eta = 0_\Delta 0111_2$ whose decimal equivalent is 0.4375_{10}. Its various equivalent SD representations are indicated below, along with their decimal equivalents:

$$0_\Delta 100\bar{1} = (0.5)_{10} - (0.0625)_{10}$$
$$0_\Delta 10\bar{1}1 = (0.5625)_{10} - (0.125)_{10}$$
$$0_\Delta 1\bar{1}11 = (0.6875)_{10} - (0.25)_{10}$$

As can be seen from Example 8.14, the SD representation is not unique, and a representation with the fewest number of nonzero digits is called a *minimal* SD representation. A minimal SD representation containing no adjacent nonzero digits is called a *canonic signed-digit (CSD)* representation. An algorithm to derive the CSD representation of a binary number is given in Hwang [Hwa79].

8.4.6 Hexadecimal Representation

In writing the codes for programmable DSP chips, we often prefer the hexadecimal representation of a binary number for compactness. In this format, each binary number is divided into groups of four bits beginning at the binary point. The decimal equivalent of each 4-bit group is then represented by one of 16 symbols formed by the 10 decimal digits 0 through 9 and the six letters of the alphabet A through F, with A representing the decimal 10, B representing decimal 11, etc. The conversion process is illustrated next.

EXAMPLE 8.15   Consider the 16-bit binary number 1101001011000101. We divide this number into four groups of four bits each as indicated below. The decimal equivalent of the first group of four bits is 13, the decimal equivalent of the second group of four bits is 2, the decimal equivalent of the third group of four bits is 12, and finally the decimal equivalent of the last group of four bits is 5:

$$\underbrace{1101}_{D}\,\underbrace{0010}_{2}\,\underbrace{1100}_{C}\,\underbrace{0101}_{5} \quad \leftarrow \text{binary}$$
$$ \leftarrow \text{hexadecimal}$$

Since the hexadecimal digits denoting 13 and 12 are, respectively, D and C, the hexadecimal representation of the 16-bit binary number 1101001011000101 is $D2C5$.

8.5 Arithmetic Operations

The two basic arithmetic operations in the implementation of digital filtering algorithms are addition (subtraction) and multiplication. We illustrate the corresponding implementation algorithms for binary numbers.

8.5.1 Fixed-Point Addition

The procedure for the addition of two binary numbers represented in the signed-magnitude representation is similar to that used in the addition of two decimal numbers. The serial addition of two positive binary fractions $0_\Delta a_{-1}a_{-2}\cdots a_{-b}$ and $0_\Delta d_{-1}d_{-2}\cdots d_{-b}$ is carried out in b steps. At the first step, we add the two LSBs a_{-b} and d_{-b} whose sum will consist of two bits; the first bit is a carry bit c_{-b} and the second bit is the LSB s_{-b} of the final sum. At all other steps, we add three bits, a_{-i}, d_{-i}, and c_{-i-1}, where c_{-i-1} is the carry bit generated at the previous step. Their sum again will consist of a carry bit c_i and a sum bit s_i. If the addition of the two positive binary fractions is still a binary fraction, the above process yields a correct result. This happens if at the bth step, the addition of a_{-1}, d_{-1}, and c_{-2} does not create a carry

Table 8.1: Binary number representations.

Decimal equivalent	Sign-magnitude	Ones'-complement	Two's-complement	Offset binary
7/8	$0_\triangle 111$	$0_\triangle 111$	$0_\triangle 111$	$1_\triangle 111$
6/8	$0_\triangle 110$	$0_\triangle 110$	$0_\triangle 110$	$1_\triangle 110$
5/8	$0_\triangle 101$	$0_\triangle 101$	$0_\triangle 101$	$1_\triangle 101$
4/8	$0_\triangle 100$	$0_\triangle 100$	$0_\triangle 100$	$1_\triangle 100$
3/8	$0_\triangle 011$	$0_\triangle 011$	$0_\triangle 011$	$1_\triangle 011$
2/8	$0_\triangle 010$	$0_\triangle 010$	$0_\triangle 010$	$1_\triangle 010$
1/8	$0_\triangle 001$	$0_\triangle 001$	$0_\triangle 001$	$1_\triangle 001$
0	$0_\triangle 000$	$0_\triangle 000$	$0_\triangle 000$	$1_\triangle 000$
-0	$1_\triangle 000$	$1_\triangle 111$	N/A	N/A
$-1/8$	$1_\triangle 001$	$1_\triangle 110$	$1_\triangle 111$	$0_\triangle 111$
$-2/8$	$1_\triangle 010$	$1_\triangle 101$	$1_\triangle 110$	$0_\triangle 110$
$-3/8$	$1_\triangle 011$	$1_\triangle 100$	$1_\triangle 101$	$0_\triangle 101$
$-4/8$	$1_\triangle 100$	$1_\triangle 011$	$1_\triangle 100$	$0_\triangle 100$
$-5/8$	$1_\triangle 101$	$1_\triangle 010$	$1_\triangle 011$	$0_\triangle 011$
$-6/8$	$1_\triangle 110$	$1_\triangle 001$	$1_\triangle 010$	$0_\triangle 010$
$-7/8$	$1_\triangle 111$	$1_\triangle 000$	$1_\triangle 001$	$0_\triangle 001$
$-8/8$	N/A	N/A	$1_\triangle 000$	$0_\triangle 000$

bit 1. However, if their sum is no longer a fraction, an *overflow* is said to have occurred, and the carry bit $c_i = 1$ generated is used to indicate the overflow.

We illustrate the addition process in the following example.

EXAMPLE 8.16 Consider the binary addition of the two positive binary fractions $0_\triangle 1001$ and $0_\triangle 0101$, whose decimal equivalents are, respectively, 0.5625_{10} and 0.3125_{10}. The serial addition of these two numbers is shown below and is carried out in four steps:

$$
\begin{array}{ccccccc}
0 \leftarrow & 0 \leftarrow & 0 \leftarrow & 1 \leftarrow & & \leftarrow \text{carry} \\
0_\triangle & 1 & 0 & 0 & 1 & \\
+\ 0_\triangle & 0 & 1 & 0 & 1 & \\
\hline
0_\triangle & 1 & 1 & 1 & 0 & \leftarrow \text{sum} \\
\end{array}
$$

sign bit ⏎

Note that the sum of these two numbers is the positive binary fraction $0_\triangle 1110$ whose decimal equivalent is 0.875_{10} and no overflow has occurred.

Now, consider the binary addition of the two positive binary fractions $0_\triangle 1101$ and $0_\triangle 0111$, whose decimal equivalents are, respectively, 0.8175_{10} and 0.4375_{10}. The serial addition of these two numbers is illustrated below:

$$
\begin{array}{cccccc}
1 \leftarrow & 1 \leftarrow & 1 \leftarrow & 1 \leftarrow & & \leftarrow \text{ carry} \\
0 _\Delta 1 & 1 & 0 & 1 & & \\
+ 0 _\Delta 0 & 1 & 1 & 1 & & \\
\hline
1 _\Delta 0 & 1 & 0 & 0 & & \leftarrow \text{ sum}
\end{array}
$$

overflow   sign bit

In this case, the addition of two positive fractions has resulted in a sum whose decimal equivalent, 1.25_{10}, is no longer a fraction and thus appears as a negative fraction. The appearance of a sign bit of 1 in the sum of two positive fractions indicates an overflow.

The subtraction of a positive binary number from another positive binary number, or, equivalently, the addition of a positive binary number to a negative binary number in signed-magnitude form, is slightly more complicated since here we may have to introduce a borrowing process. The following example illustrates the method.

EXAMPLE 8.17   Consider the subtraction of $0_\Delta 0111$ (called the *subtrahend*) from $0_\Delta 1101$ (called the *minuend*). The subtraction is carried out sequentially in four steps as shown below.

$$
\begin{array}{ccccc}
-1 & -1 & & & \leftarrow \text{ borrow} \\
0 _\Delta 1 & 11 & 10 & 1 & \leftarrow \text{ minuend} \\
- 0 _\Delta 0 & 1 & 1 & 1 & \leftarrow \text{ subtrahend} \\
\hline
0 _\Delta 0 & 1 & 1 & 0 & \leftarrow \text{ difference}
\end{array}
$$

In the first step, the LSB of the subtrahend is subtracted from the LSB of the minuend. Since both LSBs are 1, their difference is 0. Next, the difference of the two bits to the left of the LSB is formed. Note that here the bit in the minuend is 0, whereas the corresponding bit in the subtrahend is 1. Thus, to form the difference, we borrow a 1 from the bit to its left in the minuend, making its effective value 10. Subtracting a 1 from it results in a difference of 1. This process is continued until the difference of the MSBs is computed.

Note that in the above example, the subtrahend is smaller in value than the minuend, resulting in the correct difference. However, in the opposite case, the arithmetic operation will cause an *overflow*.

As illustrated above, the algorithms for the addition and subtraction operations for binary numbers in the signed-magnitude form are different, and therefore, separate circuits are needed in their hardware implementations. On the other hand, both operations can be carried out using essentially the same algorithm for binary numbers in either the ones'-complement or the two's-complement form. The basic difference in the arithmetic operations between the two representations is in the handling of the carry bit resulting in the last step. The following two examples illustrate the respective algorithms.

EXAMPLE 8.18   Consider the computation of the difference $0_\Delta 1101 - 0_\Delta 0111$. We can treat the subtraction operation as an addition of a positive number $0_\Delta 1101$ to a negative number $-0_\Delta 0111$. The two's-complement representation of $-0_\Delta 0111$ is simply $1_\Delta 1001$. In the addition of the positive binary fraction $0_\Delta 1101$ and the two's-complement negative binary fraction $1_\Delta 1001$, we treat the sign bits as the rest of the number as indicated below.

$$
\begin{array}{cccccc}
 & 0_\Delta 1 & 1 & 0 & 1 \\
+ & 1_\Delta 1 & 0 & 0 & 1 \\
\hline
1 & 0_\Delta 0 & 1 & 1 & 0 \\
\end{array}
$$

↑
drop

By dropping the carry bit generated at the fifth step, we arrive at the correct result given by $0_\Delta 0110$.

EXAMPLE 8.19 We repeat the problem of the previous example by representing the negative number $-0_\Delta 0111$ in its ones'-complement form, which is given by $1_\Delta 1000$. The addition of $0_\Delta 1101$ and $1_\Delta 1000$ is shown below.

$$
\begin{array}{cccccc}
 & 0_\Delta 1 & 1 & 0 & 1 \\
+ & 1_\Delta 1 & 0 & 0 & 0 \\
\hline
1 & 0_\Delta 0 & 1 & 0 & 1 \\
+ & & & & 1 \leftarrow \text{end around carry} \\
\hline
 & 0_\Delta 0 & 1 & 1 & 0 \\
\end{array}
$$

Here also, the addition operation has created an extra bit 1 to the left of the sign bit due to the carry bit 1 generated at the last step. This bit is next added to the LSB position of the intermediate sum resulting in the correct sum $0_\Delta 0110$.

8.5.2 Fixed-Point Multiplication

The multiplication of two b-bit binary fractions, $A = a_{s\Delta} a_{-1} a_{-2} \cdots a_{-b}$ (called the multiplicand) and $D = d_{s\Delta} d_{-1} d_{-2} \cdots d_{-b}$ (called the multiplier), in sign-magnitude form is carried out by forming the product $P^{(b)}$ of their respective magnitudes first, and then assigning the appropriate sign to the product from the signs of the multiplier and the multiplicand. The product of the magnitudes can be implemented serially in b steps where at the ith step the partial product $P^{(i)}$ is determined as follows:

$$P^{(i)} = \left(P^{(i-1)} + d_{-b+i-1} \cdot A \right) \cdot 2^{-1}, \qquad i = 1, 2, \ldots, b, \tag{8.63}$$

with $P^{(0)} = 0$. It follows from the above equation that if $d_{-b+i-1} = 0$, the new partial product is obtained by simply shifting the previous partial product to the right by one bit position. On the other hand, if $d_{-b+i-1} = 1$, the multiplicand A is added to the previous partial product and then shifted to the right by one bit position to arrive at the new partial product. The final product is a $2b$-bit fraction. The following example illustrates the algorithm.

EXAMPLE 8.20 Consider the multiplication of the two 3-bit binary fractions, $a_{s\Delta} 110$ and $d_{s\Delta} 101$. We first form the product of their magnitudes using the recursive algorithm of Eq. (8.63) as illustrated below. The final answer in sign-magnitude form is $p_{s\Delta} 011110$ where the sign of the product $p_s = 0$ if $a_s = d_s$; otherwise it is 1. Note that the product is a 6-bit fraction.

$$
\begin{array}{rcl}
1\ 1\ 0 & = & A \\
\times\ 1\ 0\ 1 & = & D \\
\hline
0\ 0\ 0 & = & P^{(0)} \\
+\ 1\ 1\ 0 & & \\
\hline
1\ 1\ 0 & & \\
\hline
0\ 1\ 1\ 0 & & P^{(1)} \\
\hline
0\ 0\ 1\ 1\ 0 & & P^{(2)} \\
+\ 1\ 1\ 0 & & \\
\hline
1\ 1\ 1\ 1\ 0 & & \\
\hline
0\ 1\ 1\ 1\ 1\ 0 & = & P^{(3)}
\end{array}
$$

If the multiplicand is a negative fraction in either two's-complement form or in ones'-complement form, and the multiplier is a positive fraction, the algorithm of Eq. (8.63) can be followed without any change, except that the addition of the negative multiplicand is carried out according to the method outlined in Section 8.5.1, and after shifting the sum, the sign bit of the sum is left as is. On the other hand, if the multiplier is a negative fraction, a correction step is needed at the end to arrive at the correct result [Kor93].

In Booth's multiplication algorithm, employed in most DSP chips, the serial multiplication process described above is implemented, with the multiplier recoded in the SD form resulting, in general, in a faster operation [Boo51]. For example, a multiplier $0_\Delta 011001111_2$ requires six additions, whereas its SD representation would require four add/subtract operations. Booth's algorithm generates the correct product if both the multiplicand and the multiplier are represented in the two's-complement form provided the sign bit of the multiplicand is employed in determining whether to perform an add or a subtract operation at the last step.

EXAMPLE 8.21   Let $A = 1_\Delta 101 = -0.375_{10}$ and $D = 1_\Delta 011 = -0.625_{10}$ represented in two's-complement form. The SD representation of D is given by $0_\Delta \bar{1}0\bar{1}$. The multiplication of $1_\Delta 101$ with $0_\Delta \bar{1}0\bar{1}$ using Booth's algorithm is illustrated below:

$$
\begin{array}{rcl}
\text{sign bit} \urcorner & & \\
1\ 1\ 0\ 1 & = A = & -0.375_{10} \\
\times\ \ 0\ \bar{1}\ 0\ \bar{1} & & \\
\hline
0\ 0\ 0\ 0 & = P^{(0)} & \\
+\ 0\ 0\ 1\ 1 & \text{Add} & -A \\
\hline
0\ 0\ 1\ 1 & & \\
\hline
0\ 0\ 0\ 1\ 1 & = P^{(1)} & \\
\hline
0\ 0\ 0\ 0\ 1\ 1 & = P^{(2)} & \\
+\ 0\ 0\ 1\ 1 & \text{Add} & -A \\
\hline
0\ 0\ 1\ 1\ 1\ 1 & & \\
\hline
0_\Delta 0\ 0\ 1\ 1\ 1\ 1 & = P^{(3)} &
\end{array}
$$

The final product is a positive binary fraction $0_\Delta 001111$ of value 0.234375_{10}, as expected.

8.5.3 Floating-Point Arithmetic

Addition of two floating-point binary numbers can be carried out easily, if their exponents are equal. Thus, in a floating-point adder, the mantissa of the smaller number is shifted to the right by an appropriate number of bits to make its exponent equal to that of the larger number, and then the two mantissas are added. If the sum of the two mantissas is within the range of Eq. (8.58), then no additional normalization is necessary; otherwise, a normalization is carried out to bring it to the proper range with a corresponding adjustment to the exponent.

EXAMPLE 8.22   Consider the addition of the two floating-point numbers

$$\eta_1 = (0_\Delta 1110)2^{01}$$
$$\eta_2 = (0_\Delta 1101)2^{10}$$

whose decimal equivalents are, respectively, 1.75_{10} and 3.25_{10}. We first rewrite the floating-point representation of the smaller number η_1 so that its characteristic is equal to that of the larger number η_2:

$$\eta_1 = (0_\Delta 0111)2^{10}.$$

Then the sum of these two numbers is given by

$$\eta_1 + \eta_2 = (0_\Delta 1101 + 0_\Delta 0111)2^{10} = (1_\Delta 0100)2^{10}.$$

Since the mantissa of the sum is greater than 1, its characteristic is therefore increased by one unit and the mantissa is shifted to the right by one bit, resulting in

$$\eta_1 + \eta_2 = (0_\Delta 1010)2^{11},$$

whose decimal equivalent is 5_{10} as expected.

In a floating-point multiplier, on the other hand, the multiplication is carried out by multiplying the two mantissas and adding their corresponding exponents. Since the range of the product of the mantissas is now between $\frac{1}{4}$ and 1, a normalization of the product is carried out if it is less than $\frac{1}{2}$ along with a corresponding adjustment to the new exponent.

EXAMPLE 8.23   Consider the multiplication of the two floating-point numbers

$$\eta_1 = (0_\Delta 1010)2^{01},$$
$$\eta_2 = (0_\Delta 1001)2^{10},$$

whose decimal equivalents are, respectively, 1.25_{10} and 2.25_{10}. Their product is thus given by

$$\eta_1 \eta_2 = (0_\Delta 1010)(0_\Delta 1001)2^{01+10} = (0_\Delta 0101101)2^{11}.$$

As the mantissa of the product is less than $\frac{1}{2}$, the characteristic is reduced by one unit and the mantissa is shifted to the left by one bit resulting in the correct floating-point representation

$$\eta_1 \eta_2 = (0_\Delta 1011010)2^{10},$$

whose decimal equivalent is 2.8125_{10} as expected.

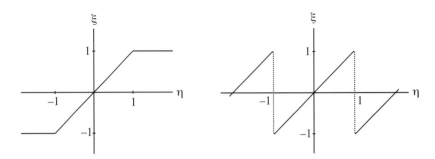

Figure 8.34: (a) Saturation overflow, and (b) two's-complement overflow.

8.6 Handling of Overflow

In Section 8.5.1, we pointed out that the results of the addition of two fixed-point fractions can result in a sum exceeding the dynamic range of the register storing the result of the addition, thus resulting in an overflow. Occurrence of overflow leads to severe output distortion and may often result in large-amplitude oscillations at the filter output (see Section 9.11.2). Therefore, the sum should be substituted with another number that is within the dynamic range. Two widely used schemes for handling the overflow are described next.

Let η denote the sum; then in either of the two schemes, if η exceeds the dynamic range $[-1, 1)$, it is substituted with a number ξ which is within the range. In the *saturation overflow* scheme shown in Figure 8.34(a), if $\eta \geq 1$, it is replaced with $1 - 2^{-b}$, where the number is assumed to be a b-bit fraction with an additional bit for the sign, and if $\eta \leq -1$, it is replaced with -1. On the other hand, in the *two's-complement overflow* scheme shown in Figure 8.34(b), whenever η is outside the range $[-1, 1)$, it is replaced with $\xi = \langle \eta + 1 \rangle_2$. Basically, here, when η is outside the range, the bits to the left of the sign bit (overflow bits) are ignored. The second scheme is usually implemented in nonrecursive digital filters employing two's-complement arithmetic. In most applications, the first scheme is usually preferred.

We shall discuss in Section 9.7 the dynamic range scaling of the digital filter structure to either eliminate completely or reduce the probability of overflow. The effects of the two overflow handling schemes on the performance of the digital filter are also considered in Section 9.7.

8.7 Tunable Digital Filters

Many applications require the use of digital filters with easily tunable characteristics. In Section 6.7, we outlined the design of tunable first-order and second-order digital filters that may provide adequate solutions in some applications. There are other applications where use of higher-order tunable digital filters may be necessary. The design of such tunable filters is the subject of this section.

8.7.1 Tunable IIR Digital Filters

The basis for the design of tunable digital filters is the spectral transformation discussed in Section 7.5, which can be used to tune a given digital filter realization with a specified cutoff frequency to another realization with a different cutoff frequency. Thus, if $G_{old}(z)$ is the transfer function of the original realization, the transfer function of the new structure is $G_{\text{new}}(z)$ where

$$G_{\text{new}}(z) = G_{\text{old}}(z)|_{z^{-1}=F^{-1}(z)}, \tag{8.64}$$

in which $F^{-1}(z)$ is a stable allpass function of the form given in Table 7.1 with the parameters of the transformation being the tuning parameters. One straightforward way to implement this transformation would be to replace each delay block in the realization of $G_{\text{old}}(z)$ with an allpass structure realizing $F^{-1}(z)$. However, such an approach leads, in general, to a structure realizing $G_{\text{new}}(z)$ with delay-free loops and cannot be implemented, as explained in Section 6.1.3.

We describe now a very simple practical modification to the above approach that does not result in a structure with delay-free loops [Mit90b]. In Section 6.10, we outlined a method of realization of a large class of stable IIR transfer functions $G(z)$ in the form [Vai86a]

$$G(z) = \tfrac{1}{2}\{A_0(z) + A_1(z)\}, \tag{8.65}$$

where $A_0(z)$ and $A_1(z)$ are stable allpass filters. The conditions for the realization of $G(z)$ as a parallel connection of two allpass sections are that $G(z)$ be a bounded real transfer function with a symmetric numerator and that it have a power complementary transfer function $H(z)$ with an antisymmetric numerator. These conditions are satisfied by all odd-order lowpass Butterworth, Chebyshev, and elliptic transfer functions.

The allpass filters $A_0(z)$ and $A_1(z)$ can be realized using any one of the approaches discussed in Section 6.6. We consider here their realizations as a cascade of first-order and second-order sections. As indicated in Section 6.6, there are a large variety of structurally lossless canonic structures realizing the first-order and second-order allpass transfer functions [Mit74a], [Szc88]. These structures use only one multiplier and one delay for the realization of a first-order allpass function, and two multipliers and two delays for the realization of a second-order allpass function.

Consider first the tuning of the cutoff frequency of a lowpass IIR filter realized by a parallel allpass structure. From Section 7.5, we note that the lowpass-to-lowpass transformation is given by

$$z^{-1} \rightarrow F^{-1}(z^{-1}) = \frac{z^{-1} - \alpha}{1 - \alpha z^{-1}}, \tag{8.66}$$

where the parameter α is related to the old and new cutoff frequencies, ω_c and $\hat{\omega}_c$, respectively, through

$$\alpha = \frac{\sin[(\omega_c - \hat{\omega}_c)/2]}{\sin[(\omega_c + \hat{\omega}_c)/2]}. \tag{8.67}$$

Substituting the transformation of Eq. (8.66) on a Type 1 first-order allpass transfer function

$$a_1(z) = \frac{d_1 + z^{-1}}{1 + d_1 z^{-1}}, \tag{8.68}$$

we obtain the new first-order allpass transfer function

$$\hat{a}_1(z) = a_1(z)|_{z^{-1}=(z^{-1}-\alpha)/(1-\alpha z^{-1})} = \frac{d_1 + \left(\frac{z^{-1}-\alpha}{1-\alpha z^{-1}}\right)}{1 + d_1\left(\frac{z^{-1}-\alpha}{1-\alpha z^{-1}}\right)}$$

$$= \frac{(d_1 - \alpha) + (1 - \alpha d_1)z^{-1}}{(1 - \alpha d_1) + (d_1 - \alpha)z^{-1}} = \frac{\left(\frac{d_1-\alpha}{1-\alpha d_1}\right) + z^{-1}}{1 + \left(\frac{d_1-\alpha}{1-\alpha d_1}\right)z^{-1}}. \tag{8.69}$$

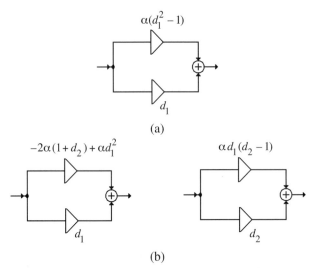

(b)

Figure 8.35: Multiplier replacement schemes in the constituent allpass sections for designing a tunable IIR filters: (a) Type 1 allpass network and (b) Type 3 allpass network.

If α is very small, we can make a Taylor series expansion of the coefficient $(d_1 - \alpha)/(1 - \alpha d_1)$ of the allpass function $\hat{a}_1(z)$ of Eq. (8.69) and arrive at an approximation

$$\hat{a}_1(z) \cong \frac{\left[d_1 + \alpha(d_1^2 - 1)\right] + z^{-1}}{1 + \left[d_1 + \alpha(d_1^2 - 1)\right] z^{-1}}, \tag{8.70}$$

which is seen to be a Type 1 first-order allpass transfer function with a coefficient that is now a linear function of α. The approximated allpass section $\hat{a}_1(z)$ can be simply implemented by replacing each multiplier d_1 in Figure 6.23 with a parallel connection of two multipliers, as indicated in Figure 8.35(a). Note that for $\alpha = 0$, $\hat{a}_1(z)$ of Eq. (8.70) reduces to $a_1(z)$ of Eq. (8.68), as expected.

Applying a similar procedure to the Type 3 second-order allpass transfer function

$$a_2(z) = \frac{d_2 + d_1 z^{-1} + z^{-2}}{1 + d_1 z^{-1} + d_2 z^{-2}}, \tag{8.71}$$

we arrive at

$$\hat{a}_2(z) = a_2(z)|_{z^{-1}=(z^{-1}-\alpha)/(1-\alpha z^{-1})} = \frac{d_2 + d_1\left(\frac{z^{-1}-\alpha}{1-\alpha z^{-1}}\right) + \left(\frac{z^{-1}-\alpha}{1-\alpha z^{-1}}\right)^2}{1 + d_1\left(\frac{z^{-1}-\alpha}{1-\alpha z^{-1}}\right) + d_2\left(\frac{z^{-1}-\alpha}{1-\alpha z^{-1}}\right)^2}. \tag{8.72}$$

Again, if we assume α to be very small, we can rewrite the above expression as

$$\begin{aligned}
\hat{a}_2(z) &\cong \frac{(d_2 - \alpha d_1) + (d_1 - 2\alpha[1 + d_2]) z^{-1} + (1 - \alpha d_1) z^{-2}}{(1 - \alpha d_1) + (d_1 - 2\alpha[1 + d_2]) z^{-1} + (d_2 - \alpha d_1) z^{-2}} \\
&= \frac{\left(\frac{d_2 - \alpha d_1}{1 - \alpha d_1}\right) + \left(\frac{d_1 - 2\alpha[1 + d_2]}{1 - \alpha d_1}\right) z^{-1} + z^{-2}}{1 + \left(\frac{d_1 - 2\alpha[1 + d_2]}{1 - \alpha d_1}\right) z^{-1} + \left(\frac{d_2 - \alpha d_1}{1 - \alpha d_1}\right) z^{-2}},
\end{aligned} \tag{8.73}$$

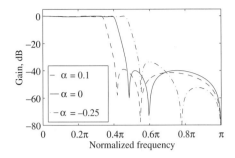

Figure 8.36: Gain responses of a fifth-order tunable elliptic lowpass filter for three values of α.

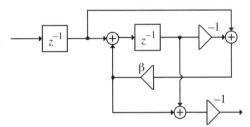

Figure 8.37: Lowpass-to-bandpass transformation.

by neglecting coefficients containing α^2. Next, an approximation based on the Taylor series expansion of the last term in the above equation results in

$$\hat{a}_2(z) \cong \frac{[d_2 + \alpha d_1(d_2 - 1)] + [d_1 - 2\alpha(1 + d_2) + \alpha d_1^2] z^{-1} + z^{-2}}{1 + [d_1 - 2\alpha(1 + d_2) + \alpha d_1^2] z^{-1} + [d_2 + \alpha d_1(d_2 - 1)] z^{-2}}, \tag{8.74}$$

which is seen to be a Type 3 second-order allpass transfer function with coefficients that are a linear function of α. The approximated allpass section $\hat{a}_2(z)$ can be simply implemented by replacing the multipliers d_1 and d_2 in Figure 6.25 with a parallel connection of two multipliers, as indicated in Figure 8.35(b). Note that for $\alpha = 0$, $\hat{a}_2(z)$ of Eq. (8.74) reduces to $a_2(z)$ of Eq. (8.71) as expected.

Note that in the case of both the first-order and the second-order allpass filters, the tuning rule is now a linear function of α. Even though this tuning algorithm is approximate and has been derived for small values of α, in practice, tuning ranges of several octaves have been observed to hold in the case of narrowband lowpass elliptic filters [Mit90b]. Figure 8.36 shows the gain responses of a fifth-order tunable elliptic lowpass filter for three values of the tuning parameter α. The prototype filter ($\alpha = 0$) has a passband edge at $\omega_p = 0.4\pi$, passband ripple of 0.5 dB, and minimum stopband attenuation of 40 dB.

By applying a lowpass-to-bandpass transformation

$$z^{-1} \rightarrow F^{-1}(z^{-1}) = -z^{-1} \frac{z^{-1} + \beta}{1 + \beta z^{-1}} \tag{8.75}$$

to a tunable lowpass IIR filter, we can design a tunable bandpass filter whose center frequency ω_0 is tuned by adjusting the parameter $\beta = \cos \omega_0$, and the bandwidth is tuned by changing α [Mit90b]. Unlike the lowpass-to-lowpass transformation of Eq. (8.66), the transformation of Eq. (8.75) can be directly implemented on the structure realizing the tunable lowpass filter by replacing each delay with the structure shown in Figure 8.37 to arrive at a tunable bandpass filter structure.

8.7.2 Tunable FIR Digital Filters

The spectral transformation approach of Section 7.5 can also be applied to an FIR filter to develop a filter structure with tunable characteristics. However, the resulting structure is no longer FIR since the replacement of the delays in the prototype FIR structure by allpass sections implementing the spectral transformation makes it an IIR filter. We outline below a straightforward method for designing tunable linear-phase lowpass FIR filters and later show how to modify the method to the design of other types of tunable FIR filters [Jar88]. The method discussed preserves the FIR structure and permits easy tuning of the cutoff frequency.

The basic idea behind the tuning procedure is the observation that, for an ideal lowpass FIR filter with a zero-phase response given by

$$H_d(e^{j\omega}) = \begin{cases} 1, & \text{for } 0 \le |\omega| \le \omega_c, \\ 0, & \text{for } \omega_c < |\omega| \le \pi. \end{cases} \tag{8.76}$$

the impulse response coefficients are given by

$$h_d[n] = \frac{\sin(\omega_c n)}{\pi n}, \quad 0 \le |n| < \infty, \tag{8.77}$$

as derived in Example 3.3. We can truncate the above expression and obtain the coefficients of a realizable approximation given by

$$h_{LP}[n] = \begin{cases} c[n]\omega_c, & \text{for } n = 0, \\ c[n]\sin(\omega_c n), & \text{for } 1 \le |n| \le N, \\ 0, & \text{otherwise}, \end{cases} \tag{8.78}$$

where ω_c is the 6-dB cutoff frequency, and

$$c[n] = \begin{cases} 1/\pi, & \text{for } n = 0, \\ 1/\pi n, & \text{for } 1 \le |n| \le N. \end{cases} \tag{8.79}$$

It follows from the above that once an FIR lowpass filter has been designed for a given cutoff frequency, it can be tuned simply by changing ω_c and recomputing the filter coefficients according to the above expression. It can be shown that Eq. (8.78) can also be used to design a tunable FIR lowpass filter by equating the coefficients of a prototype filter developed using any of the FIR filter design methods outlined in Chapter 7 with those of Eq. (8.78) and solving for $c[n]$ [Jar88]. Thus, if $h_{LP}[n]$ denotes the coefficients of the prototype lowpass filter designed for a cutoff frequency ω_c, from Eq. (8.78) the constants $c[n]$ are given by

$$c[0] = \frac{h_{LP}[0]}{\omega_c}, \tag{8.80a}$$

$$c[n] = \frac{h_{LP}[n]}{\sin(\omega_c n)}, \quad 1 \le |n| \le N. \tag{8.80b}$$

Then, the coefficients $\hat{h}_{LP}[n]$ of the transformed FIR filter with a cutoff frequency $\hat{\omega}_c$ are given by

$$\hat{h}_{LP}[0] = c[0]\hat{\omega}_c = \left(\frac{\hat{\omega}_c}{\omega_c}\right) h_{LP}[0], \tag{8.81a}$$

$$\hat{h}_{LP}[n] = c[n]\sin(\hat{\omega}_c n) = \left(\frac{\sin(\hat{\omega}_c n)}{\sin(\omega_c n)}\right) h_{LP}[n], \quad 1 \le |n| \le N. \tag{8.81b}$$

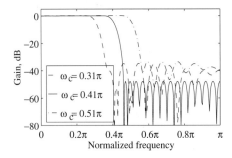

Figure 8.38: Tunable lowpass FIR filter of length 51.

This tuning procedure has been recommended for filters with equal passband and stopband ripples. It has also been recommended that the prototype filter be designed such that its coefficients have values not too close to zero.

We illustrate the design of a tunable FIR filter in the following example.

EXAMPLE 8.24 The prototype filter is a linear-phase lowpass FIR filter of length 51 with a passband edge ω_p at 0.36π and a stopband edge ω_s at 0.46π and a transition bandwidth of 0.1π. The cutoff frequency ω_c is thus at 0.41π. It has been designed using the function remez of MATLAB, assuming equal weights to the passband and the stopband. Using the method discussed above, the filter coefficients for a cutoff at 0.31π and at 0.51π have been computed. Figure 8.38 shows the plots of the gain responses of the three FIR filters.

For the design of tunable FIR filters with unequal passband and stopband ripples, the following modification is used. The impulse response coefficients of the tunable filter of odd length are now given by

$$h_{LP}[n] = \begin{cases} c[n]\omega_c + d[n], & \text{for } n = 0, \\ c[n]\sin(\omega_c n) + d[n]\cos(\omega_c n), & \text{for } 1 \leq |n| \leq N. \end{cases} \tag{8.82}$$

The constants $c[n]$ and $d[n]$ are determined from two different optimal prototype filters, substituting them in Eq. (8.82), and then solving for these constants. For example, if the passband weight W_p is greater than the stopband weight W_s, the cutoff frequencies ω_{ca} and ω_{cb} of the two filters are chosen as

$$\omega_{ca} = 0.8 - \frac{0.25}{N}, \quad \omega_{cb} = 0.8 + \frac{0.25}{N}.$$

On the other hand, if the passband weight W_p is less than the stopband weight W_s, the cutoff frequencies ω_{ca} and ω_{cb} of the two filters are chosen as

$$\omega_{ca} = 0.2 - \frac{0.25}{N}, \quad \omega_{cb} = 0.2 + \frac{0.25}{N}.$$

The two optimal filters $h_a[n]$ and $h_b[n]$ are then designed using the Parks-McClellan algorithm. As before, the prototype filters should be designed with nonzero coefficients.

The above modification is not recommended for the design of tunable wideband or very narrowband filters. Figure 8.39 shows the gain responses of a lowpass FIR filter of length 41 and a transition width of 0.1π with a variable 6-dB cutoff frequency for various values of passband and stopband weights. The cutoff frequencies of the filters vary from 0.2 to 0.9, as indicated in Figure 8.39.

The above methods can be directly applied to the design of a tunable highpass FIR filter from a highpass FIR prototype filter. Alternatively, the latter can be designed as a delay-complementary tunable Type 1 FIR lowpass filter.[10]

---

[10]See Section 4.8.1.

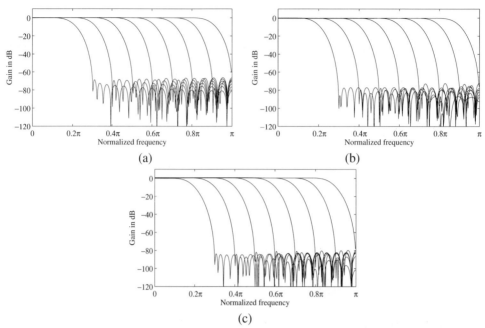

Figure 8.39: Length 41, transition bandwidth of 0.1π, and passband/stopband weights of (a) $1/1$, (b) $1/10$, and (c) $1/50$.

To develop the methods for designing a tunable bandpass FIR filter, we observe that a filter $H_{BP}(z)$ with a symmetrical bandpass magnitude response can be derived from a lowpass prototype filter $H_{LP}(z)$ by applying a frequency translation to its frequency response. This process results in

$$H_{BP}(e^{j\omega}) = H_{LP}(e^{j(\omega+\omega_0)}) + H_{LP}(e^{j(\omega-\omega_0)}), \tag{8.83}$$

where ω_0 is the desired center frequency of the bandpass filter. The relation between the impulse responses of the bandpass and the prototype lowpass filters is given by

$$h_{BP}[n] = \left(e^{-j\omega_0 n} + e^{j\omega_0 n}\right) h_{LP}[n] = 2\cos(\omega_0 n)h_{LP}[n]. \tag{8.84}$$

If δ_p and δ_s denote the passband and stopband ripples of the lowpass prototype, the corresponding ripples of the bandpass filter are $\delta_p + \delta_s$ for the passband and $2\delta_s$ for the stopband. Equation (8.84) forms the basis for designing a tunable bandpass FIR filter with adjustable center frequency. A similar approach can be followed for the design of a tunable bandstop FIR filter, in which case the prototype is a highpass FIR filter.

It should be noted that the tunable FIR filters designed using the methods suggested above have the same hardware requirements as that of their prototypes. They also have linear phase if the prototype filter has linear phase.

8.8 Function Approximation

Often there is a need to use transcendental functions and random numbers in certain DSP applications. For example, the FFT computations require the generation of the complex exponential sequences. Certain

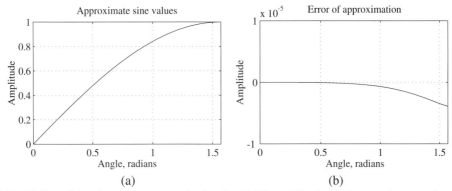

Figure 8.40: (a) Plot of the sine value computed using Eq. (8.85) and (b) plot of the error between the actual sine value and the approximation.

digital communication systems require sinusoidal sequences. Both of these sequences can of course be generated by second-order digital filter structures described in Section 6.11. We outline below an alternative approach to the generation of transcendental functions based on truncated polynomial expansions. These types of expansions are often used in DSP chip implementations [Mar92].

8.8.1 Trigonometric Function Approximation

The sine of a number x can be approximated using the expansion [Abr72]

$$\sin(x) \cong x - 0.166667x^3 + 0.008333x^5 - 0.0001984x^7 + 0.0000027x^9, \tag{8.85}$$

where the argument x is in radians, and its range is restricted to the first quadrant, i.e., from 0 to $\pi/2$. If x is outside this range, its sine can be computed by making use of the identities $\sin(-x) = -\sin(x)$, and $\sin[(\pi/2) + x] = \sin[(\pi/2) - x]$. Figure 8.40 shows the plots of the sine approximation computed using Eq. (8.85) and the error due to the approximation.

For computing the arctangent of a number x where $-1 \leq x \leq 1$, a recommended expansion is given by [Abr72]

$$\tan^{-1}(x) \cong 0.999866x - 0.3302995x^3 + 0.180141x^5$$
$$- 0.085133x^7 + 0.0208351x^9. \tag{8.86}$$

Figure 8.41 shows the plots of the arctangent approximation computed using Eq. (8.86) and the error due to the approximation. If $x \geq 1$, then its arcangent can be computed by making use of the identity $\tan^{-1}(x) = (\pi/2) - \tan^{-1}(1/x)$.

Exercises M8.12 and M8.13 list several polynomial approximations for computing these trigonometric functions. Various trigonometric functions can also be computed from Eq. (8.85) using trigonometric identities.

8.8.2 Square-Root and Logarithm Approximation

The square root of a positive number x in the range $0.5 \leq x \leq 1$ can be evaluated using the truncated polynomial approximation [Mar92]:

$$\sqrt{x} \cong 0.2075806 + 1.454895x - 1.34491x^2 + 1.106812x^3$$
$$- 0.536499x^4 + 0.1121216x^5. \tag{8.87}$$

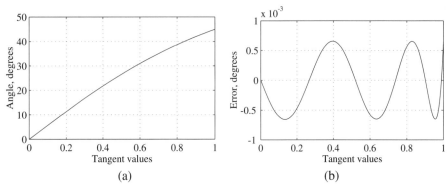

Figure 8.41: (a) Plot of the arctangent value computed using Eq. (8.86) and (b) plot of the error between the actual arctangent value and the approximation.

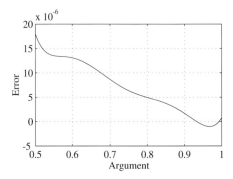

Figure 8.42: Plot of the error between the actual square-root value and the approximation given by Eq. (8.87).

Figure 8.42 shows the plot of the error due to the above approximation. If x is outside the range from 0.5 to 1, it can be multiplied by a binary constant K^2 to bring the product $x' = K^2 x$ into the desirable range, compute $\sqrt{x'}$ using Eq. (8.83), and then determine $\sqrt{x} = \sqrt{x'}/K$.

Polynomial expansions have also been advanced for the approximate computation of both the logarithm (base 10) and natural logarithm (base e) of any number x between one and two. Two such expressions are given by [Bur73], [Mar92]:

$$\log_{10}(x) \cong 0.4339142(x-1) - 0.21278385(x-1)^2 + 0.1240692(x-1)^3$$
$$- 0.05778505(x-1)^4 + 0.0136261(x-1)^5, \tag{8.88}$$
$$\log_e(x) \cong 0.999115(x-1) - 0.4899597(x-1)^2 + 0.2856751(x-1)^3$$
$$- 0.13305665(x-1)^4 + 0.031372071(x-1)^5. \tag{8.89}$$

For the calculation of the logarithm of a number x that is outside the range from 1 to 2, the number x must be scaled by an appropriate factor K with known logarithm to bring the product $x' = Kx$ to this range and then the logarithm of the product x' must be computed. From the logarithm of the product x', the logarithm of K is subtracted to get the logarithm of x.

8.8.3 Random Number Generation

Random number generation is used in a number of applications. For example, it is used as a training signal for the adaptive equalizer in high-speed modems [Mar92]. Another application is in the elimination of limit cycles in IIR digital filter structures using the random rounding method (Section 9.11.4). A variety of approaches has been proposed for the generation of uniformly distributed random numbers [Knu69]. One rather simple method is based on the linear congruence method and is given by the recursive equation

$$x[n + 1] = \langle \alpha x[n] + \beta \rangle_M, \tag{8.90}$$

where the modulus M is a positive integer. The rules for the selection of the constants α and β are given in [Knu69]. The above equation generates a periodic pseudo-random sequence having a period M with proper choice of α and β fairly independent of the *seed value* $x[0]$. The samples of the sequence generated by Eq. (8.90) within a period are uniformly distributed integers from 0 to $M - 1$. A family of such random sequences can be generated by choosing different seed values for each sequence. To ensure the randomness of each sequence generated, M should be chosen as large as possible.

The M-file `rand` can be used to generate random numbers and matrices with elements uniformly distributed in the interval $(0, 1)$. Various versions of these functions are available as indicated below:

```
rand(N),          rand(M,N),          rand(size(A)),
rand('seed',N),   rand('seed').
```

The output of `rand(N)` is an $N \times N$ matrix and that of `rand(M,N)` is an $M \times N$ matrix with random entries. `rand(size(A))` generates an output of the same size as A. If `rand` is used without any argument, its output is a scalar with a value that changes each time it is called. The output of `rand('seed')` is the current value of the seed employed by the random number generator. Finally, `rand('seed',N)` uses the value N for the seed.

A similar set of M-files for the generation of random numbers and matrices with elements that are normally (Gaussian) distributed with zero mean and unity variance are given by

```
randn(N),          randn(M,N),          randn(size(A)),
randn('seed',N),   randn('seed').
```

The function `rand` was used in Programs 2_3 and 2_4 of Example 2.14 to generate a random sequence.

8.9 Summary

Some of the common factors in the implementation of DSP algorithms, which are independent of the type of implementation being carried out, are discussed in this chapter. The implementation is usually carried out by a sequential implementation of the set of equations describing the algorithm. These equations can be derived directly from the structure realizing the algorithm. The computability condition of these equations is derived and an algorithm for testing the computability is described. A simple algebraic technique to verify the structure from the input and output samples is also outlined.

MATLAB-based software implementation of digital filtering algorithms and the discrete Fourier transform (DFT) are then considered. The basic ideas behind the fast Fourier transform (FFT) used for faster computation of the DFT samples are explained and several commonly used FFT algorithms are derived.

Various schemes for the binary representation of the numbers and the signal variables that are employed in the digital computers and special-purpose DSP chips are reviewed, followed by a discussion of the algorithms used for the implementation of the addition and the multiplication operation. In certain cases, the result of an addition of numbers may cause an overflow of the dynamic range of the register storing the sum. Two methods of preventing physically the overflow are outlined.

Methods for the design of IIR and FIR digital filters with tunable characteristics are introduced next. The chapter concludes with a discussion on the approximation of certain functions that are needed in a number of applications requiring implementation using DSP chips. In particular, the approximation of trigonometric functions, square roots, logarithms, and the generation of random numbers are considered.

8.10 Problems

8.1 Develop a set of time-domain equations describing the digital filter structure of Figure P8.1 in terms of the input $x[n]$, output $y[n]$, and the intermediate variables $w_k[n]$ in a sequential order. Does this set describe a valid computational algorithm? Justify your answer by developing a matrix representation of the digital filter structure and by examining the matrix **F**.

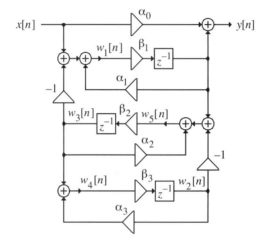

Figure P8.1

8.2 Develop a computable set of time-domain equations describing the digital filter structure of Figure P8.1. Verify the computability condition by forming an equivalent matrix representation and by examining the matrix **F**.

8.3 Develop a set of time-domain equations describing the digital filter structure of Figure P8.2 in terms of the input $x[n]$, output $y[n]$, and the intermediate variables $w_k[n]$ in a sequential order. Does this set of equations describe a valid computational algorithm? Justify your answer by developing a matrix representation of the digital filter structure and by examining the matrix **F**.

8.4 Develop the precedence graph of the digital filter structure of Figure P8.1 and investigate its realizability. If the structure is found to be realizable, then from the precedence graph, determine a valid computational algorithm describing the structure.

8.5 Develop the precedence graph of the digital filter structure of Figure P8.2 and investigate its realizability. If the structure is found to be realizable, then from the precedence graph, determine a valid computational algorithm describing the structure.

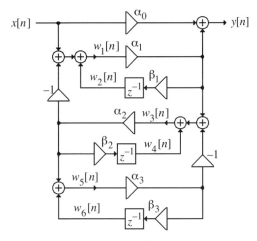

Figure P8.2

8.6 (a) Write down the time-domain equations relating the node variables $w_i[n]$, $y[n]$ and the input $x[n]$ of the digital filter structure of Figure P8.3. Check formally the computability of the set of equations if the equations are ordered sequentially with increasing values of the node indices.

(b) Develop a signal flow-graph representation of this digital filter structure and then determine its precedence graph. From the precedence graph, develop a set of computable equations describing the structure and show formally that these equations are indeed computable.

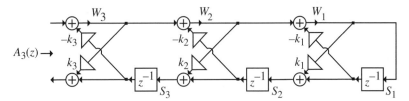

Figure P8.3

8.7 The measured impulse response samples of a causal second-order IIR digital filter with a transfer function $H(z) = P(z)/(1 - 3z^{-1} - 5z^{-2})$ are given by

$$h[0] = 2.1, \ h[1] = 1.1, \ h[2] = -3.2, \ h[3] = -15.1, \ h[4] = -29.33, \ldots$$

Determine the numerator $P(z)$ of the transfer function $H(z)$.

8.8 The first five impulse response samples of a causal second-order IIR digital filter are given by

$$h[0] = 3, \ h[1] = -2, \ h[2] = -6, \ h[3] = 12, \ h[4] = -15.$$

Determine the transfer function $H(z)$.

8.9 Determine the transfer function of a third-order causal IIR digital filter whose first ten impulse response samples are given by

$$\{h[n]\} = \{2 \quad -5 \quad 6 \quad -2 \quad -9 \quad 18 \quad -7 \quad -31 \quad 65 \quad -30\}.$$

8.10 The first four impulse response samples of a causal third-order IIR digital filter with a transfer function $G(z) = P(z)/(2 - 6z^{-1} + 8z^{-2} + 10z^{-3})$ are given by

$$g[0] = 3, \quad g[1] = 7, \quad g[2] = 13, \quad g[3] = -3.$$

Determine the numerator polynomial $P(z)$ of the transfer function.

8.11 The first 11 impulse response samples of a causal IIR transfer function

$$H(z) = \frac{p_0 + p_1 z^{-1} + p_2 z^{-2} + p_3 z^{-3} + p_4 z^{-4}}{1 + 2z^{-1} + 2z^{-2} + 3z^{-3} + 3z^{-4}},$$

are given by

$$\{h[n]\} = \{2 \quad 0 \quad -5 \quad -10 \quad -10 \quad 15 \quad 90 \quad 185 \quad 125 \quad -455 \quad -1830\}.$$

Determine its numerator coefficients $\{p_i\}$.

8.12 Show that the direct computation of the N-point DFT of a length-N sequence requires $4N^2$ real multiplications and $(4N - 2)N$ real additions.

8.13 Develop an algorithm for the complex multiplication of two complex numbers using only three real multiplications and five real additions.

8.14 A digital oscillator working at a sampling rate of 2500 Hz can generate any one of four sinusoidal signals of frequencies 150 Hz, 375 Hz, 620 Hz, and 850 Hz. We would like to detect the tone frequency of the signal being generated by computing only four samples of an N-point DFT using Goertzel's method. What is the smallest value of the DFT length N so that the four tone frequencies fall as close as possible to four DFT bin indices to make the leakage to adjacent bins as small as possible? The DFT length N should be the same in all four cases.

8.15 Let $H_k(z)$ denote the transfer function of the kth filter used in the Goertzel's algorithm to calculate the N-point DFT $X[k]$ of a length-N sequence $x[n]$. Consider the following input sequence applied to $H_k(z)$:

$$x[n] = \{1, \ 0, \ 0, \ \ldots, \ 1, \ 0, \ 0, \ldots\},$$
$$\quad\quad\quad \underset{0}{\uparrow} \quad\quad\quad\quad \underset{N/2}{\uparrow}$$

where $N/2$ is assumed to be divisible by k. What is the output sequence $y[n]$ for $k = 1$? What is the output sequence $y[n]$ for $k = N/2$? Give a qualitative sketch.

8.16 Develop the flow-graph for the FFT algorithm from the radix-2 DIT FFT algorithm of Figure 8.24 for the $N = 8$ case in which the input is in normal order and the output is in the bit-reversed order.

8.17 Develop the flow-graph for the FFT algorithm from the radix-2 DIT FFT algorithm of Figure 8.24 for the $N = 8$ case in which both the input and the output are in normal order.

8.18 Verify Eq. (8.45).

8.19 Develop the structural interpretation of the first stage of the radix-3 DIT FFT algorithm.

8.20 Develop the flow-graph for the radix-3 DIT FFT algorithm for the $N = 9$ case in which the input is in digit-reversed order and the output is in the normal order.

8.21 Develop the flow-graph for a mixed-radix DIT FFT algorithm for the $N = 15$ case in which the input is in digit-reversed order and the output is in the normal order.

8.22 Form the transposed graph of the flow-graph of Figure 8.25. Replace each complex multiplicand W_N^r in the transposed flow-graph with $\frac{1}{2} W_N^{-r}$. Show that the final flow-graph implements an inverse DFT if the input is $X[k]$.

8.23 A second approach to the inverse DFT computation using a DFT algorithm is illustrated in Figure P8.4. Let $X[k]$ be the N-point DFT of a length-N sequence $x[n]$. Define a length-N time-domain sequence $q[n]$ as

$$\text{Re}\{q[n]\} = \text{Im}\,X[k]|_{k=n}, \quad \text{Im}\{q[n]\} = \text{Re}\,X[k]|_{k=n},$$

with $Q[k]$ denoting its N-point DFT. Show that

$$\text{Re}\{x[n]\} = \frac{1}{N} \cdot \text{Im}\,Q[k]\bigg|_{k=n}, \quad \text{Im}\{x[n]\} = \frac{1}{N} \cdot \text{Re}\,Q[k]\bigg|_{k=n}.$$

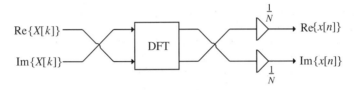

Figure P8.4

8.24 A third approach to the inverse DFT computation using a DFT algorithm is described below. Let $X[k]$ be the N-point DFT of a length-N sequence $x[n]$. Define a length-N time-domain sequence $r[n]$ as

$$r[n] = X[k]|_{k=\langle -n \rangle_N},$$

with $R[k]$ denoting its N-point DFT. Show that

$$x[n] = \frac{1}{N} \cdot R[k]\bigg|_{k=n}.$$

8.25 We wish to determine the sequence $y[n]$ generated by a linear convolution of a length-8 real sequence $x[n]$ and a length-5 real sequence $h[n]$. To this end, we can follow one of the following methods:

Method # 1. Direct computation of the linear convolution.

Method # 2. Computation of the linear convolution via a single circular convolution.

Method # 3. Computation of the linear convolution using radix-2 FFT algorithms.

Determine the least number of real multiplications needed in each of the above methods. For the radix-2 FFT algorithm, do not include in the count multiplications by ± 1, $\pm j$, and W_N^0.

8.26 Repeat Problem 8.25 for the computation of the linear convolution of a length-8 sequence $x[n]$ with a length-6 sequence $h[n]$.

8.27 An input sequence $x[n]$ of length 1024 is to be filtered using a linear-phase FIR filter $h[n]$ of length 34. This filtering process involves the linear convolution of two finite-length sequences and can be computed using the overlap-add algorithm discussed in Section 3.6.2 where the short linear convolutions are performed using the DFT-based approach of Figure 3.14 with the DFTs implemented by the Cooley-Tukey FFT algorithm.

(a) Determine the appropriate power-of-2 transform length that would result in a minimum number of multiplications and calculate the total number of multiplications that would be required.

(b) What would be the total number of multiplications if the direct convolution method is used?

8.28 We pointed out in Eq. (3.37) that the vector $\mathbf{X}$ of DFT samples can be expressed as the product of the DFT matrix $\mathbf{D}_N$ and the vector $\mathbf{x}$ of input samples, where $\mathbf{D}_N$ is given by Eq. (3.40).

(a) Show that the 8-point DIT FFT algorithm shown in Figure 8.18 is equivalent to expressing the DFT matrix as a product of four matrices as indicated below:

$$\mathbf{D}_N = \mathbf{V}_8 \mathbf{V}_4 \mathbf{V}_2 \mathbf{E}. \tag{8.91}$$

Determine the matrices given above and show that multiplication by each matrix $\mathbf{V}_k$, $k = 8, 4, 2$, requires at most eight complex multplications.

(b) Since the DFT matrix $\mathbf{D}_N$ is its own transpose, i.e., $\mathbf{D}_N = \mathbf{D}_N^T$, another FFT algorithm is readily obtained by forming the transpose of the right-hand side of Eq. (8.91), resulting in a factorization of $\mathbf{D}_N$ given by

$$\mathbf{D}_N = \mathbf{E}^T \mathbf{V}_2^T \mathbf{V}_4^T \mathbf{V}_8^T. \tag{8.92}$$

Show that the flow-graph representation of the above factorization is precisely the 8-point DIF FFT algorithm of Figure 8.27.

8.29 The basic idea behind the split-radix FFT (SRFFT) algorithm is explored in this problem [Duh86]. As in the case of the decimation-in-frequency FFT algorithm, in the SRFFT algorithm, the even-indexed and the odd-indexed samples of the DFT are computed separately. For the computation of the even-indexed samples we write, as in the DIF FFT approach,

$$X[2\ell] = \sum_{n=0}^{N-1} x[n] W_N^{2\ell n}$$

$$= \sum_{n=0}^{(N/2)-1} x[n] W_N^{2\ell n} + \sum_{n=N/2}^{N-1} x[n] W_N^{2\ell n}, \quad \ell = 0, 1, \ldots, \tfrac{N}{2} - 1. \tag{8.93}$$

Show that the above can be reexpressed as an $(N/2)$-point DFT in the form

$$X[2\ell] = \sum_{n=0}^{(N/2)-1} \left(x[n] + x\left[n + \tfrac{N}{2}\right] \right) W_{N/2}^{\ell n}, \quad \ell = 0, 1, \ldots, \tfrac{N}{2} - 1. \tag{8.94}$$

For the computation of the odd-indexed samples, we write two different expressions, depending on whether the frequency index k can be expressed as $4\ell + 1$ or $4\ell + 3$:

$$X[4\ell + 1] = \sum_{n=0}^{(N/4)-1} x[n] W_N^{(4\ell+1)n} + \sum_{n=\frac{N}{4}}^{(N/2)-1} x[n] W_N^{(4\ell+1)n}$$

$$+ \sum_{n=N/2}^{(3N/4)-1} x[n] W_N^{(4\ell+1)n} + \sum_{n=3N/4}^{N-1} x[n] W_N^{(4\ell+1)n},$$

$$\ell = 0, 1, \ldots, \tfrac{N}{4} - 1, \tag{8.95a}$$

$$X[4\ell + 3] = \sum_{n=0}^{(N/4)-1} x[n] W_N^{(4\ell+3)n} + \sum_{n=N/4}^{(N/2)-1} x[n] W_N^{(4\ell+3)n}$$

$$+ \sum_{n=N/2}^{(3N/4)-1} x[n] W_N^{(4\ell+3)n} + \sum_{n=3N/4}^{N-1} x[n] W_N^{(4\ell+3)n},$$

$$\ell = 0, 1, \ldots, \tfrac{N}{4} - 1. \tag{8.95b}$$

Show that Eqs. (8.95a) and (8.95b) can be rewritten as two $(N/4)$-point DFTs of the form

$$X[4\ell + 1] = \sum_{m=0}^{(N/4)-1} \left\{ \left(x[m] - x\left[m + \frac{N}{2}\right] \right) \right.$$
$$\left. - j \left(x\left[m + \frac{N}{4}\right] - x\left[m + \frac{3N}{4}\right] \right) \right\} W_N^m W_{N/4}^{\ell m}, \tag{8.96a}$$

$$X[4\ell + 3] = \sum_{m=0}^{(N/4)-1} \left\{ \left(x[m] - x\left[m + \frac{N}{2}\right] \right) \right.$$
$$\left. + j \left(x\left[m + \frac{N}{4}\right] - x\left[m + \frac{3N}{4}\right] \right) \right\} W_N^{3m} W_{N/4}^{\ell m}, \tag{8.96b}$$

where $\ell = 0, 1, \ldots, \frac{N}{4} - 1$. Sketch the flow-graph of a typical butterfly in the above algorithm for computing two even-numbered and two odd-numbered points, and show that the split-radix FFT algorithm requires only two complex multiplications per butterfly.

8.30 Develop the flow-graph for the computation of an 8-point DFT based on the split-radix algorithm described in Problem 8.29. What is the total number of real multiplications needed to implement this algorithm? How does this number compare with that required in a radix-2 DIF FFT algorithm? Ignore multiplications by ± 1 and $\pm j$.

8.31 Develop the flow-graph for the computation of a 16-point DFT based on the split-radix algorithm described in Problem 8.29. What is the total number of real multiplications needed to implement this algorithm? How does this number compare with that required in a radix-2 DIF FFT algorithm? Ignore multiplications by ± 1 and $\pm j$.

8.32 An alternative approach to the development of FFT algorithms is via index mapping [Coo65], which is studied in this problem. Consider a length-N sequence $x[n]$ with $X[k]$ denoting its N-point DFT where $N = N_1 N_2$. Define the index mappings

$$n = n_1 + N_1 n_2 \qquad \begin{cases} 0 \le n_1 \le N_1 - 1, \\ 0 \le n_2 \le N_2 - 1, \end{cases} \tag{8.97a}$$

$$k = N_2 k_1 + k_2 \qquad \begin{cases} 0 \le k_1 \le N_1 - 1, \\ 0 \le k_2 \le N_2 - 1. \end{cases} \tag{8.97b}$$

(a) Using the above mappings, show that $X[k]$ can be expressed as

$$X[k] = X[N_2 k_1 + k_2]$$
$$= \sum_{n_1=0}^{N_1-1} \left[\left(\sum_{n_2=0}^{N_2-1} x[n_1 + N_1 n_2] W_{N_2}^{k_2 n_2} \right) W_N^{k_2 n_1} \right] W_{N_1}^{k_1 n_1},$$
$$0 \le k_1 \le N_1 - 1, \quad 0 \le k_2 \le N_2 - 1. \tag{8.98}$$

As indicated above, the computation of the N-point DFT $X[k]$ is now carried out in three steps: (1) compute the N_2-point DFTs $H[n_1, k_2]$ of the set of N_1 sequences $x[n_1 + N_1 n_2]$ of length N_2, (2) multiply these DFTs with the twiddle factors $W_N^{k_2 n_1}$ to form $\hat{H}[n_1, k_2]$, and (3) compute the N_2-point DFTs $X[N_2 k_1 + k_2]$ of the set of N_2 sequences $\hat{H}[n_1, k_2]$ of length N_1.

(b) Show that for $N_1 = 2$ and $N_2 = N/2$, the above decomposition scheme leads to the DIT FFT algorithm and for $N_1 = N/2$ and $N_2 = 2$, the above decomposition scheme leads to the DIF FFT algorithm .

(c) If $\mathcal{R}(N)$ denotes the total number of multiplications needed to compute an N-point DFT, show that for the DFT computation algorithm based on the above decomposition scheme outlined above,

$$\mathcal{R}(N) = N_1 \mathcal{R}(N_2) + N_2 \mathcal{R}(N_1) + N$$
$$= N \left[\frac{1}{N_1} \mathcal{R}(N_1) + \frac{1}{N_2} \mathcal{R}(N_2) + 1 \right]. \tag{8.99}$$

(d) Determine the total number of multiplications needed for the FFT algorithm based on the above decomposition scheme for $N = 2^{\nu}$.

8.33 Develop the index mapping for implementing an N-point DFT $X[k]$ of a length-N sequence $x[n]$ using the Cooley-Tukey FFT algorithm for (a) $N = 12$, (b) $N = 15$, (c) $N = 21$, and (d) $N = 35$.

8.34 (a) The twiddle factors needed in the DFT computation scheme outlined in Problem 8.32 can be eliminated, resulting in a more computationally efficient FFT algorithm for the case when the factors N_1 and N_2 are relatively prime by using the following index mappings [Bur77]:

$$n = \langle An_1 + Bn_2 \rangle_N, \quad \begin{cases} 0 \le n_1 \le N_1 - 1, \\ 0 \le n_2 \le N_2 - 1, \end{cases} \tag{8.100a}$$

$$k = \langle Ck_1 + Dk_2 \rangle_N, \quad \begin{cases} 0 \le k_1 \le N_1 - 1, \\ 0 \le k_2 \le N_2 - 1. \end{cases} \tag{8.100b}$$

Show that the twiddle factors are eliminated totally if the constants A, B, C, and D in the above mappings are chosen satisfying the following conditions:

$$\langle AC \rangle_N = N_2, \quad \langle BD \rangle_N = N_1, \quad \text{and} \quad \langle AD \rangle_N = \langle BC \rangle_N = 0.$$

(b) Show that the following set of constants satisfy the above condition

$$A = N_2, \quad B = N_1, \quad C = N_2 \langle N_2^{-1} \rangle_{N_1}, \quad \text{and} \quad D = N_1 \langle N_1^{-1} \rangle_{N_2},$$

where $\langle N_1^{-1} \rangle_{N_2}$ denotes the multiplicative inverse of N_1 evaluated modulo N_2. Based on this choice of constants, develop the algorithm for the computation of $X[k]$. FFT computation schemes based on this approach are called the *prime factor algorithms* [Bur81], [Kol77].

8.35 Develop the index mapping for implementing an N-point DFT $X[k]$ of a length-N sequence $x[n]$ using the prime factor algorithm for (a) $N = 12$, (b) $N = 15$, (c) $N = 21$, and (d) $N = 35$.

8.36 Consider the computation of a 12-point DFT $X[k]$ of a length-N sequence $x[n]$. The index mapping to be used can be either

$$n = \langle An_1 + Bn_2 \rangle_{12}, \quad k = \langle Ck_1 + Dk_2 \rangle_{12},$$

or

$$n = \langle Cn_1 + Dn_2 \rangle_{12}, \quad k = \langle Ak_1 + Bk_2 \rangle_{12},$$

where A, B, C, and D are appropriately chosen constants. Use of the mapping $g[n_1, n_2] = x[\langle An_1 + Bn_2 \rangle_{12}]$ results in

$$G[k_1, k_2] = \sum_{n_1=0}^{2} \sum_{n_2=0}^{3} g[n_1, n_2] W_4^{n_2 k_2} W_3^{n_1 k_1}, \tag{8.101}$$

$$X[\langle Ck_1 + Dk_2 \rangle_{12}] = G[k_1, k_2]. \tag{8.102}$$

If, on the other hand, the index mapping used is $h[n_1, n_2] = x[\langle Cn_1 + Dn_2 \rangle_{12}]$, we obtain

$$H[k_1, k_2] = \sum_{n_1=0}^{2} \sum_{n_2=0}^{3} h[n_1, n_2] W_4^{n_2 k_2} W_3^{n_1 k_1}, \tag{8.103}$$

$$Y[\langle Ak_1 + Bk_2 \rangle_{12}] = H[k_1, k_2]. \tag{8.104}$$

What is the relation between $X[k]$ and $Y[k]$?

8.37 Develop the flow-graph for the computation of a 10-point DFT based on the prime factor algorithm.

8.38 Develop the flow-graph for the computation of a 15-point DFT based on the prime factor algorithm.

8.39 Develop a scheme to compute the 3072-point DFT of a sequence of length 3072 using 512-point FFT modules and complex multiplications and additions. Show the scheme in block diagram form. How many FFT modules and complex multiplications and additions are needed for the overall computation?

8.40 A 1024-point DFT of a length-1000 sequence $x[n]$ is to be computed. How many zero-valued samples should be appended to $x[n]$ prior to the computation of the DFT? What are the total number of complex multiplications and additions needed for the direct evaluation of all DFT samples? What are the total number of complex multiplications and additions needed if a Cooley-Tukey type FFT is used to compute the DFT samples?

8.41 Let $H(z) = \sum_{n=0}^{N-1} h[n]z^{-n}$ and $X(z) = \sum_{n=0}^{N-1} x[n]z^{-n}$ be two real polynomials of degree $(N-1)$. Their product $Y(z)$ is then a polynomial of degree $(2N-2)$ and a direct evaluation of $Y(z)$ requires N^2 multiplications and $N+1$ additions. Note also that the coefficients $y[n]$ of $Y(z)$ are the same as that obtained by a linear convolution of the two length-N sequences $h[n]$ and $x[n]$. The number of multiplications in computing the product $H(z)X(z)$ can be reduced but with an increase in the number of additions by making use of the Cook-Toom algorithm [Aga77], [Knu69] which is studied in this problem.

Let $z_k, k = 0, 1, \ldots, 2N-1$, be $2N-1$ distinct points in the z-plane. Then $\hat{Y}[z_k] = H(z_k)X(z_k)$ represent the $(2N-1)$-point NDFT of $y[n]$, and from these $2N-1$ samples we can uniquely determine $Y(z)$ using the Lagrange interpolation formula as discussed in Problem 3.110. By choosing the values of z_k appropriately, $\hat{Y}[z_k]$ can be evaluated with only $2N-1$ multiplications if we ignore multiplications (or divisions) by a power-of-2 integer that can be implemented using simple shifts in a binary representation.

(a) We first develop the Cook-Toom algorithm for $N = 2$. Here,

$$Y(z) = y[0] + y[1]z^{-1} + y[2]z^{-2} = \left(h[0] + h[1]z^{-1}\right)\left(x[0] + x[1]z^{-1}\right).$$

Using the Lagrange interpolation formula express $Y(z)$ in terms of its 3-point NDFT samples evaluated at $z_0 = -1, z_1 = \infty$, and $z_2 = +1$. Develop the expression for $y[0]$, $y[1]$, and $y[2]$ in terms of the parameters $X(z_k)$ and $H(z_k)$ and show that the computations of $\{y[n]\}$ require only a total of three multiplications. Determine the total number of additions required by this algorithm. Note that in many applications $h[n]$ represents a fixed FIR filter and, hence, the multiplications by integers needed to evaluate $y[n]$ can be included in the constants $H(z_k)$ to eliminate them from future computations.

(b) Develop the Cook-Toom algorithm for the linear convolution of two length-3 sequences.

8.42 Let $H(z) = h[0] + h[1]z^{-1}$ and $X(z) = x[0] + x[1]z^{-1}$. Show that $Y(z) = H(z)X(z) = y[0] + y[1]z^{-1} + y[2]z^{-2}$ can be written as [Jen91]

$$Y(z) = h[0]x[0] + [(h[0] + h[1])(x[0] + x[1])$$
$$- (h[0]x[0] + h[1]x[1])] z^{-1} + h[1]x[1]z^{-2}.$$

The evaluation of the product of the two first-order real polynomials thus can be carried out using three multiplications instead of four as required in the direct product. Equivalently, a linear convolution of two length-2 sequences can thus be implemented using only three multiplications.

8.43 Develop a multistage algorithm to compute the linear convolution of two length-N real sequences based on the scheme outlined in Problem 8.42 for the case when N is a power of 2 [Jen91]. What is the least number of multiplications required to compute the convolution using the multistage algorithm?

8.44 Let a 32-bit register be used to represent a floating-point number with E bits assigned for the exponent and M bits plus a sign bit for the mantissa. Determine the approximate dynamic range of this floating-point representation for the following pairs of bit assignments for the exponent and the mantissa by evaluating the values of the smallest and the largest numbers that can be represented by the floating-point representation: (a) $E = 6$ and $M = 25$, (b) $E = 7$ and $M = 24$, and (c) $E = 8$ and $M = 23$. Determine the dynamic range of a 32-bit fixed-point representation of a signed integer. Show that the floating-point representation provides a larger dynamic range than the fixed-point representation.

8.45 Show that the range of a 32-bit floating-point number in the IEEE standard is from 1.18×10^{-38} to 3.4×10^{38}.

8.46 Show that the decimal equivalent of a positive or a negative binary fraction given by $s_\Delta a_{-1} a_{-2} \cdots a_{-b}$ in two's-complement form is $-s + \sum_{i=1}^{b} a_{-i} 2^{-i}$.

8.47 Show that the decimal equivalent of a positive or a negative binary fraction given by $s_\Delta a_{-1} a_{-2} \cdots a_{-b}$ in ones'-complement form is $-s(1 - 2^{-b}) + \sum_{i=1}^{b} a_{-i} 2^{-i}$.

8.48 Determine the 9-bit sign-magnitude, ones'-complement, and two's-complement representations of the following negative decimal fractions:
(a) -0.625_{10}, (b) -0.7734375_{10}, (c) -0.36328125_{10}, (d) -0.94921875_{10}.

8.49 Determine the 7-bit offset binary representations of the following decimal numbers:
(a) 0.625_{10}, (b) -0.625_{10}, (c) 0.359375_{10}, (d) -0.359375_{10}, (e) 0.90625_{10}, (f) -0.90625_{10}.

8.50 Develop the signed-digit (SD) representation of the following binary numbers:
(a) $0_\Delta 11101101$, (b) $0_\Delta 01111101$, (c) $0_\Delta 10101111$.

8.51 Develop the hexadecimal representation of the following binary numbers:
(a) 1101011011000111, (b) 0101111110101001, (c) 1011010000101110.

8.52 Perform the following binary additions of positive binary fractions and comment on your results:
(a) $0_\Delta 10101 + 0_\Delta 01111$, (b) $0_\Delta 01011 + 0_\Delta 10001$.

8.53 Compute the following differences of positive binary fractions by performing binary additions of a positive fraction and a negative number represented in two's-complement form:
(a) $0_\Delta 10101 - 0_\Delta 01111$, (b) $0_\Delta 10001 - 0_\Delta 01011$.

8.54 Repeat Problem 8.53 by representing the negative number in ones'-complement form.

8.55 Consider the binary addition of the following three numbers represented in two's-complement form with five bits: $\eta_1 = 0.6875_{10}$, $\eta_2 = 0.8125_{10}$, and $\eta_3 = -0.5625_{10}$. The addition is carried out in two steps: first we form the sum $\eta_1 + \eta_2$ and then we form the sum $(\eta_1 + \eta_2) + \eta_3$. Since the magnitudes of η_1 and η_2 are both greater than 0.5, their sum will lead to overflow indicated by a 1 in the sign bit of the sum. Ignore this overflow by keeping all bits in the partial sum and add η_3. Show that the final sum is correct in spite of the overflow generated by the first addition.

8.56 Develop the products of the following binary fractions, where the negative numbers are in two's-complement form:
(a) $(0_\Delta 11101) \times (1_\Delta 10111)$, (b) $(1_\Delta 10101) \times (0_\Delta 10111)$.

8.57 Repeat Problem 8.56 with the negative numbers considered to be in ones'-complement form.

8.58 The Taylor structure of Figure P8.5(a) developed for the realization of a Type 1 linear-phase FIR transfer function in Problem 6.18 has been proposed for designing tunable FIR filters [Cro76b], [Opp76]. From Eq. (6.131) observe that the zero-phase frequency response, also called the amplitude response, of a Type 1 FIR transfer function of length $2M + 1$ is given by

$$\breve{H}(\omega) = \sum_{n=0}^{M} a[n]\,(\cos\omega)^n. \tag{8.105}$$

Let $\hat{\omega}$ denote the angular frequency variable of the transformed FIR filter $\hat{H}(z)$. Show that a lowpass-to-lowpass transformation can be achieved by substituting

$$\cos\omega = \alpha + \beta\cos\hat{\omega} \tag{8.106}$$

in Eq. (8.105). Show that this transformation can be implemented by replacing each block with a transfer function $(1 + z^{-2})/2$ in Figure P8.5 by a block with a transfer function

$$\alpha z^{-1} + \frac{\beta}{2}(1 + z^{-2}). \tag{8.107}$$

Let ω_c and $\hat{\omega}_c$ denote, respectively, the cutoff frequency of the prototype filter and the desired cutoff frequency of the transformed filter. Show that if $\hat{\omega}_c < \omega_c$, it is convenient to choose $\beta = 1 - \alpha$, with $0 \le \alpha < 1$. Sketch the mapping from $\cos\omega_c$ to $\cos\hat{\omega}_c$ for this case. On the other hand, if $\hat{\omega}_c > \omega_c$, show that it is convenient to choose $\beta = 1 + \alpha$, with $-1 < \alpha \le 0$. Sketch the mapping from $\cos\omega_c$ to $\cos\hat{\omega}_c$ for this second case.

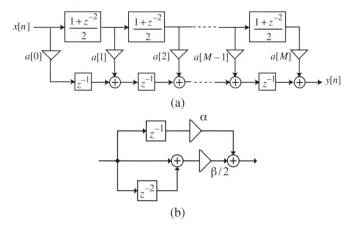

Figure P8.5

8.11 MATLAB Exercises

M 8.1 Using Program 8_1 determine the first 30 samples of the impulse response coefficients of the fifth-order elliptic lowpass filter developed in Example 7.21.

M 8.2 Using Program 8_1 determine the first 35 samples of the impulse response coefficients of the fourth-order Type 1 Chebyshev highpass filter developed in Example 7.22.

M 8.3 Using Program 8_1 determine the first 30 samples of the impulse response coefficients of the eighth-order Butterworth bandpass filter developed in Example 7.23.

M 8.4 Modify Program 8_2 to demonstrate filtering of a sum of two sinusoidal sequences by an arbitrary IIR causal digital filter. The input data to the modified program should be the angular frequencies of the sinusoidal sequences, and the numerator and denominator coefficients of the transfer function of the IIR digital filter. Apply the modified program to filter a sum of two sinusoidal sequences of angular frequencies 0.3π and 0.6π by the filter developed in Example 7.21 and verify its lowpass filtering property.

M 8.5 Apply the modified program developed in Problem M8.4 to filter a sum of two sinusoidal sequences of angular frequencies 0.3π and 0.6π by the filter developed in Example 7.22 and verify its highpass filtering property.

M 8.6 Modify Program 8_3 to demonstrate the filtering of a sum of two sinusoidal sequences by an arbitrary IIR causal digital filter implemented in a cascade form. The individual sections in the cascade are either second-order or first-order with real coefficients. The input data to the modified program should be the angular frequencies of the sinusoidal sequences, the numerator and denominator coefficients of the transfer function of the individual sections in the cascade. Apply the modified program to filter a sum of two sinusoidal sequences of angular frequencies 0.3π and 0.6π by the filter developed in Example 7.21 and verify its lowpass filtering property. The output plot generated by the modified Program 8_3 should be identical to that generated in Problem M8.4.

M 8.7 Apply the modified program developed in Problem M8.6 to filter a sum of two sinusoidal sequences of angular frequencies 0.3π and 0.6π by the filter developed in Example 7.22 and verify its highpass filtering property. The output plot generated by the modified Program 8_3 should be identical to that generated in Problem M8.5.

M 8.8 Write a MATLAB program using the function `direct2` of Section 8.2.1 to simulate the direct form II structure and demonstrate the filtering of input sequences. Apply this program to filter a sum of two sinusoidal sequences of angular frequencies 0.3π and 0.6π by the filter developed in Example 7.21 and verify its lowpass filtering property. The output plot generated by your new program should be identical to that generated in Problem M8.4.

M 8.9 Using the M-file function `gfft` described in Section 8.3.1 for computing a single DFT sample by Goertzel's algorithm, write a MATLAB program to compute the N-point DFT of an arbitrary length-N sequence and compare the DFT samples generated with those obtained using the M-file function `fft`. Verify the DFT computation of several sequences of lengths $N = 8$, 12, and 16.

M 8.10 Write a MATLAB program to verify the plots given in Figure 8.38.

M 8.11 Write a MATLAB program to verify the plots given in Figure 8.39.

M 8.12 A polynomial expansion suggested for the approximation of the sine of a number x is given by [Mar92]:

$$\sin(\pi x) \cong 3.140625x + 0.02026367x^2 - 5.325196x^3$$
$$+ 0.5446778x^4 + 1.800293x^5, \tag{8.108}$$

where x is normalized by π, and its range is restricted to the first quadrant given by $0 < x < 0.5$. (*Note*: 0.5 is the normalized value of $\pi/2$.) Using MATLAB compute and plot the values of $\sin(\pi x)$ given by Eq. (8.108) and the error due to the above approximation. Compare this approximation with that given by Eq. (8.85).

M 8.13 A polynomial expansion suggested for the approximation of the arctangent of a number x in the range $-1 \le x \le 1$ is given by [Mar92]

$$\tan^{-1}(x/\pi) \cong 0.318253x + 0.003314x^2 - 0.130908x^3$$
$$+ 0.068542x^4 - 0.009159x^5. \tag{8.109}$$

Using MATLAB compute and plot the values of $\tan^{-1}(x)$ given by Eq. (8.109) and the error due to the above approximation. Compare this approximation with that given by Eq. (8.86).

9 Analysis of Finite Wordlength Effects

So far, we have assumed that we are dealing with discrete-time systems characterized by linear difference equations with constant coefficients, where both the coefficients and the signal variables have infinite precision taking any value between $-\infty$ and ∞. However, when implemented in either software form on a general-purpose computer or in special-purpose hardware form, the system parameters along with the signal variables can take only discrete values within a specified range since the registers of the digital machine where they are stored are of finite length. The discretization process results in nonlinear difference equations characterizing the discrete-time systems. These nonlinear equations, in principle, are almost impossible to analyze and deal with exactly. Fortunately, if the quantization amounts are small compared to the values of the signal variables and filter constants, a simpler approximate theory based on a statistical model can be applied, and it is possible to derive the effects of discretization and develop results that can be verified experimentally.

To illustrate the various sources of errors arising from the discretization process in the implementation of a digital filter, consider for simplicity, the first-order IIR digital filter of Figure 9.1 defined by the linear constant coefficient difference equation

$$y[n] = \alpha y[n-1] + x[n], \tag{9.1}$$

where $y[n]$ and $x[n]$ are the output and the input signal variables, respectively. The corresponding transfer function describing the above digital filter is given by

$$H(z) = \frac{1}{1 - \alpha z^{-1}} = \frac{z}{z - \alpha}. \tag{9.2}$$

When implemented on a digital machine, the filter coefficient α can assume only certain discrete values $\hat{\alpha}$ and, in general, can only approximate the original design value of α. As a result, the actual transfer function implemented is given by

$$\hat{H}(z) = \frac{z}{z - \hat{\alpha}}, \tag{9.3}$$

which may be different from the desired transfer function $H(z)$ of Eq. (9.2). Therefore, the actual frequency response may be quite different from the desired frequency response. This coefficient quantization problem is similar to the sensitivity problem encountered in analog filter implementation.

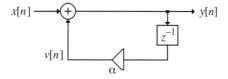

Figure 9.1: A first-order IIR digital filter.

If we assume that the input sequence $x[n]$ has been obtained by sampling an analog signal $x_a(t)$, it is discretized by the A/D converter being employed to convert the output of the sample-and-hold to digital samples. If we represent the output of the A/D converter as $\hat{x}[n]$, then the actual input to the digital filter of Figure 9.1 is given by

$$\hat{x}[n] = x[n] + e[n], \tag{9.4}$$

where $e[n]$ is the A/D conversion error generated by the input quantization process.

The quantization of arithmetic operations leads to another source of errors. In the case of our simple digital filter of Eq. (9.1), the output of the multiplier $v[n]$ generated by multiplying the signal $y[n-1]$ with α,

$$v[n] = \alpha y[n-1], \tag{9.5}$$

is quantized to fit the register containing the product. The quantized signal $\hat{v}[n]$ can be represented as

$$\hat{v}[n] = v[n] + e_\alpha[n], \tag{9.6}$$

where $e_\alpha[n]$ is the error sequence generated by the product quantization process. The properties of this type of round-off error are somewhat similar to those of the A/D conversion error.

In addition to the above sources of errors, another type of error occurs in digital filters due to the nonlinearity caused by the quantization of arithmetic operations. These errors manifest themselves in the form of oscillations, called limit cycles, at the output of the filter, usually in the absence of input or sometimes in the presence of constant input signals or sinusoidal input signals. In this chapter, we analyze the effects of the above sources of quantization errors and then describe structures that are less sensitive to these effects.

9.1 The Quantization Process and Errors

There are two basic types of the binary representations of data, fixed-point and floating-point formats, as described in Section 8.4. In each of these formats, a negative number can be represented in one of three different forms. The arithmetic operations involved in digital signal processing are the addition (subtraction) and the multiplication operations, discussed in Section 8.5. Various problems can arise in the digital implementation of the arithmetic operations involving the binary data due to the finite wordlength limitations of the registers storing the numbers and the results of the arithmetic operations. For example, in fixed-point arithmetic, as demonstrated in Example 8.20, the product of two b-bit numbers is $2b$ bits long, which has to be quantized to b bits to fit the prescribed wordlength of the registers. Moreover, in fixed-point arithmetic, the addition operation can result in a sum exceeding the register wordlength, causing an overflow as illustrated in Example 8.16. On the other hand, there is essentially no overflow in a floating-point addition. However, the results of both addition and multiplication may have to be quantized to fit the prescribed wordlength of the registers.

An analysis of the various quantization effects on the performance of a digital filter in practice depends on whether the numbers are in fixed-point or floating-point format, the type of representation for the negative numbers being used, the quantization method being employed to quantize the data, and the digital filter structure being used for implementation. Since the number of all possible combinations of the type of arithmetic, type of quantization method, and digital filter structure (of which there are literally thousands) is very large, we consider in this chapter analysis of quantization effects in some selected practical cases. However, the analysis presented can be easily extended to other cases.

We now describe the three different types of quantization that can be employed. As indicated in Section 8.4, it is a common practice in digital signal processing applications, to represent data in a digital machine either as a fixed-point fraction or as a floating-point binary number with the mantissa as a binary fraction.

Figure 9.2: A general $(b + 1)$-bit fixed-point fraction.

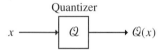

Figure 9.3: The quantization process model.

We assume the available wordlength is $(b + 1)$ bits with the *most significant bit (MSB)* representing the sign of the number. Consider first the data to be a $(b + 1)$-bit fixed-point fraction with the binary point just to the right of the sign bit, as indicated in Figure 9.2. The smallest positive number that can be represented in this format will have a *least significant bit (LSB)* of 1, with the remaining bits being all 0's. Its decimal equivalent is 2^{-b}. Numbers represented with $(b + 1)$ bits are thus quantized in steps of 2^{-b}, called the *quantization step* or the *width of quantization*.

Before quantization, the wordlength is much larger than that indicated above. Assume that the original data x is represented as a $(\beta + 1)$-bit fraction with $\beta >> b$. To convert it into a $(b + 1)$-bit fraction, to be denoted as $\mathcal{Q}(x)$, we can employ either *truncation* or *rounding*. In either case, the quantization process can be modeled as shown in Figure 9.3. Since the representation of a positive binary fraction is the same independent of the format being used to represent the negative binary fraction, the effect of quantization of a positive fraction remains unchanged. However, the effect on negative fractions is not the same for the three different types of representations.

9.2 Quantization of Fixed-Point Numbers

To truncate a fixed-point number from $(\beta + 1)$ bits to $(b + 1)$ bits, we simply discard the least significant $(\beta - b)$ bits, as indicated in Figure 9.4. Let ε_t denote the truncation error defined by

$$\varepsilon_t = \mathcal{Q}(x) - x. \tag{9.7}$$

For a positive number x, the magnitude of the number $\mathcal{Q}(x)$ obtained after truncation is less than or equal to the magnitude of x. Therefore, $\varepsilon_t \leq 0$ for a positive x. The error ε_t is equal to zero if all bits being discarded are 0's and is largest if all bits being discarded are 1's. In the latter case, the decimal equivalent of the portion being discarded is equal to $2^{-b} - 2^{-\beta}$. Hence the range of the error ε_t in the case of truncation of a positive number x is given by

$$-(2^{-b} - 2^{-\beta}) \leq \varepsilon_t \leq 0. \tag{9.8}$$

For a negative number x, each one of the three different representations needs to be examined individually. For a negative fraction in sign-magnitude form, the magnitude of the truncated number $\mathcal{Q}(x)$ is smaller than that of the unquantized negative number x. Thus, it follows from the definition of the quantization error ε_t given by Eq. (9.7) that here

$$0 \leq \varepsilon_t \leq 2^{-b} - 2^{-\beta}. \tag{9.9}$$

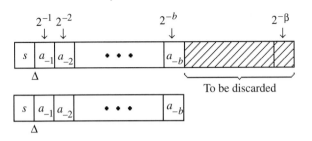

Figure 9.4: Illustration of the truncation operation.

For a negative fraction x in ones'-complement form $1_\Delta a_{-1}a_{-2}\ldots a_{-\beta}$, the numerical value is $-(1 - 2^{-\beta}) + \sum_{i=1}^{\beta} a_{-i}2^{-i}$. The numerical value of its quantized version $Q(x)$ is thus $-(1 - 2^{-b}) + \sum_{i=1}^{b} a_{-i}2^{-i}$ and hence, the error ε_t is given by

$$\varepsilon_t = Q(x) - x = -(1 - 2^{-b}) + \sum_{i=1}^{b} a_{-i}2^{-i} + (1 - 2^{-\beta}) - \sum_{i=1}^{\beta} a_{-i}2^{-i}$$

$$= (2^{-b} - 2^{-\beta}) - \sum_{i=b+1}^{\beta} a_{-i}2^{-i}. \tag{9.10}$$

The truncation error for this representation is always positive and has a range

$$0 \le \varepsilon_t \le 2^{-b} - 2^{-\beta}. \tag{9.11}$$

Now, consider a negative fraction x given in the two's-complement format $1_\Delta a_{-1}a_{-2}\ldots a_{-\beta}$. Its numerical value is given by $(-1 + \sum_{i=1}^{\beta} a_{-i}2^{-i})$. The representation of $Q(x)$, obtained after truncating x, is given by $1_\Delta a_{-1}a_{-2}\ldots a_{-b}$, with a numerical value $(-1 + \sum_{i=1}^{b} a_{-i}2^{-i})$. Therefore,

$$\varepsilon_t = Q(x) - x = \left(-1 + \sum_{i=1}^{b} a_{-i}2^{-i}\right) - \left(-1 + \sum_{i=1}^{\beta} a_{-i}2^{-i}\right) = -\sum_{i=b+1}^{\beta} a_{-i}2^{-i}. \tag{9.12}$$

The truncation error is here always negative and has a range

$$-(2^{-b} - 2^{-\beta}) \le \varepsilon_t \le 0. \tag{9.13}$$

In the case of rounding, the number is quantized to the nearest quantization level. We assume that a number exactly halfway between two quantization levels is rounded up to the nearest higher level. Therefore, if the bit $a_{-(b+1)}$ is 0, rounding is equivalent to truncation, and if this bit is 1, then 1 is added to the LSB position of the truncated number. It should be noted that the rounding error ε_r does not depend on the format being used to represent the negative fraction since the operation is solely based on the magnitude of the number. To determine the range of ε_r, we observe that the quantization step after rounding has a value 2^{-b}. The maximum rounding error ε_r therefore has a magnitude $(2^{-b})/2$. As a result, the range of ε_r is given by

$$-\tfrac{1}{2}(2^{-b} - 2^{-\beta}) < \varepsilon_r \le \tfrac{1}{2}(2^{-b} - 2^{-\beta}). \tag{9.14}$$

In practice, $\beta \gg b$. For example, the wordlength of a product is typically twice that of the numbers being multiplied. Hence, we can set $2^{-\beta} \cong 0$ in the inequalities of Eqs. (9.9), (9.11), (9.13), and (9.14),

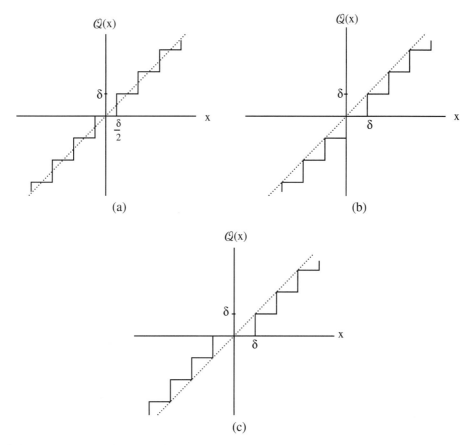

Figure 9.5: Input-output relationships of the quantizer: (a) rounding, (b) two's complement truncation, and (c) ones'-complement and sign-magnitude truncation. Here $\delta = 2^{-b}$.

leading to the simpler inequalities shown in Table 9.1, where we have set the quantization step as $\delta = 2^{-b}$. Plots of the input-output characteristics of the quantizer for the three different representations and the two different quantization methods are shown in Figure 9.5.

9.3 Quantization of Floating-Point Numbers

In the case of floating-point numbers, quantization is carried out only on the mantissa. As a result, here it is more relevant to consider the relative error caused by the quantization process. To this end, we define the relative error ε in terms of the numerical values of quantized floating-point number $\mathcal{Q}(x) = 2^E \mathcal{Q}(M)$ and the unquantized number $x = 2^E M$ as

$$\varepsilon = \frac{\mathcal{Q}(x) - x}{x} = \frac{\mathcal{Q}(M) - M}{M}. \tag{9.15}$$

It can be shown (Problem 9.1) that the range of the relative errors for the different representations of a floating-point binary number are as indicated in Table 9.2, assuming $2^{-\beta} << 2^{-b}$.

Table 9.1: Range of quantization error.

Type of quantization	Number representation	Range of error $Q(x) - x$
Truncation	Positive number Two's-complement negative number	$-\delta < \varepsilon_t \leq 0$
Truncation	Sign-magnitude negative number Ones'-complement negative number	$0 \leq \varepsilon_t < \delta$
Rounding	All positive and negative numbers	$-\frac{\delta}{2} < \varepsilon_r \leq \frac{\delta}{2}$

Note: $\delta = 2^{-b}$.

Table 9.2: Range of relative error $\varepsilon = (Q(x) - x)/x$.

Type of quantization	Number representation	Range of relative error
Truncation	Two's-complement	$-2\delta < \varepsilon_t \leq 0, \quad x > 0$ $0 \leq \varepsilon_t < 2\delta, \qquad x < 0$
Truncation	Sign-magnitude Ones'-complement	$-2\delta < \varepsilon_t \leq 0$
Rounding	All numbers	$-\delta < \varepsilon_r \leq \delta$

Note: $\delta = 2^{-b}$.

A discussion of the analysis of quantization effects of digital filters implemented using floating-point arithmetic is beyond the scope of this text. Interested readers are referred to the publications listed at the end of this book [Kan71], [Liu69], [Opp75], [San67], [Wei69b], [Wei69a]. We consider here the fixed-point implementation case.

9.4 Analysis of Coefficient Quantization Effects

The effect of multiplier coefficient quantization on digital filters is similar to that observed in analog filters. The transfer function $\hat{H}(z)$ of the digital filter implemented in hardware or software form with quantized coefficients is different from the desired transfer function $H(z)$. The main effect of the coefficient quantization is therefore on the poles and zeros that move to different locations from the original desired locations. As a result, the actual frequency response $\hat{H}(e^{j\omega})$ is different from the desired frequency response $H(e^{j\omega})$ and may not be acceptable to the user. Moreover, the poles may move outside the unit circle causing the implemented digital filter to become unstable even though the original transfer function with unquantized coefficients is stable, as illustrated earlier in Example 4.6. However, as demonstrated in Example 4.7, the same transfer function when implemented in a cascade form remained stable after coefficient quantization.

9.4.1 Analysis Using MATLAB

Before we study the effect of the coefficient quantization on the performance of a digital filter analytically, it is instructive to investigate the effect on a computer using MATLAB.

As MATLAB uses decimal numbers and arithmetic, to study the quantization effects on the digital filter implemented using binary numbers and arithmetic, we need to develop the decimal equivalents of the quantized representations of binary numbers and signals. The latter can be quantized using either truncation or rounding. We provide below two M-files, a2dT and a2dR, which develop the decimal equivalent beq of the binary representation of a vector d of decimal numbers with N bits for the magnitude part by truncation and rounding, respectively.

```
function beq = a2dT(d,n)
% BEQ = A2DT(D, N) generates the decimal
% equivalent beq of the binary representation
% of a decimal number D with N bits for the
% magnitude part obtained by truncation
%
m = 1; d1 = abs(d);
while fix(d1) > 0
    d1 = abs(d)/(2^m);
    m = m+1;
end
beq = fix(d1*2^n);
beq = sign(d).*beq.*2^(m-n-1);

function beq = a2dR(d,n)
% BEQ = A2DR(D, N) generates the decimal
% equivalent beq of the binary representation
% of a decimal number D with N bits for the
% magnitude part obtained by rounding
%
m = 1; d1 = abs(d);
while fix(d1) > 0
    d1 = abs(d)/(2^m);
    m = m+1;
end
beq = fix(d1*2^n+.5);
beq = sign(d).*beq.*2^(m-n-1);
```

To illustrate the effect of the coefficient quantization on the frequency response and the pole-zero locations of a digital filter we need to evaluate these characteristics with both infinite and finite precision for the filter coefficients. As MATLAB uses double-precision decimal numbers and arithmetic, the filter coefficients and signals generated using MATLAB can be considered to be of infinite precision for all practical purposes. To evaluate the effect of quantized binary representation, we can use either the M-file a2dT or the M-file a2dR given above to develop the decimal equivalent of the quantized binary numbers and signals.

We first illustrate the effect of coefficient quantization of an IIR digital filter implemented in direct form using Program 9_1 given below. In it present form, the program evaluates the frequency response of a fifth-order elliptic lowpass digital filter with a cutoff at 0.4π, a passband ripple of 0.4 dB, and a minimum stopband attenuation of 50 dB. With simple modifications, the program can be used to study the effect on other types of filters with different specifications. For truncating the transfer function coefficients, this program uses the M-file a2dT. In addition, it uses the M-file plotzp, which is the same as the M-file zplane in the *Signal Processing Toolbox* of MATLAB except that here the poles are shown with the symbol + and the zeros are shown with ∗ in the pole-zero plot.[1]

```
% Program 9_1
% Coefficient Quantization Effects on the
% Frequency Response of a Direct Form IIR Filter
%
clf;
[b,a] = ellip(5,0.4,50,0.4);
[h,w] = freqz(b,a,512); g = 20*log10(abs(h));
bq = a2dT(b,5); aq = a2dT(a,5);
[hq,w] = freqz(bq,aq,512); gq = 20*log10(abs(hq));
plot(w/pi,g,'b',w/pi,gq,'r:');grid
axis([0 1 -80 5]);
xlabel('\omega/\pi');ylabel('Gain, dB');
title('original - solid line, quantized - dashed line');
pause
zplane(b,a);
plotzp(bq,aq);
```

Figure 9.6(a) and (b) shows the gain response of the ideal filter with infinite precision coefficients (shown with a solid line) and the gain response obtained when the transfer function coefficients are truncated to 5-bit length (shown with a dashed line). It can be seen from this figure that the effect of the coefficient quantization is more severe around the bandedges with a higher passband ripple and a smaller transition band. The minimum stopband attenuation has also become smaller. Moreover, the transmission zeros closest to the stopband edge have moved closer to the passband edge.

Figure 9.6(c) shows the locations of the poles and zeros of the original elliptic lowpass filter transfer function with unquantized coefficients and of the transfer function of the elliptic filter implemented with quantized coefficients. As can be seen from this plot, coefficient quantization can cause substantial displacement of the poles and zeros from their desired nominal locations. In this example, the zero closest to the pole has moved the farthest from its original location and has moved closer to the new location of its nearest pole, which is now much closer to the unit circle.

It is of interest to compare the performance of the direct form realization of an IIR transfer function with that of a cascade realization when implemented with quantized coefficients. Program 9_2 given below can be used to evaluate the effect of the quantization of the transfer function coefficients of each section of a cascade form realization of the above elliptic lowpass filter. However, this program can be easily modified to study the effect on other types of filters with different specifications.

```
% Program 9_2
% Coefficient Quantization Effects on the
% Frequency Response of a Cascade Form IIR Filter
%
```

---

[1]The modification to the function zplane is with permission from The Mathworks, Inc., Natick, MA.

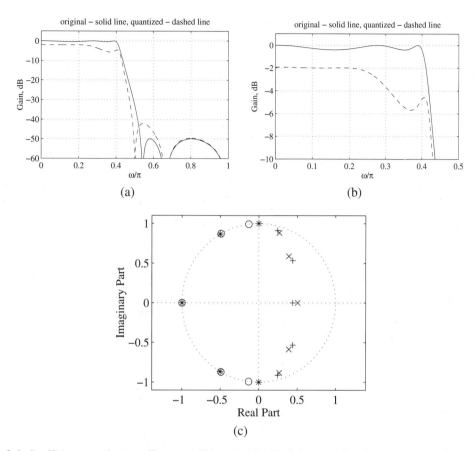

Figure 9.6: Coefficient quantization effects on a fifth-order IIR elliptic lowpass filter implemented in direct form: (a) fullband gain responses with unquantized (shown with solid line) and quantized coefficients (shown with dashed line), (b) passband details, and (c) pole-zero movements: Pole and zero locations of the filter with quantized coefficients denoted by "x" and "o", respectively, and pole and zero locations of the filter with unquantized coefficients denoted by "+" and "*", respectively.

```
clf;
[z,p,k] = ellip(5,0.4,50,0.4);
[b,a] = zp2tf(z,p,k);
[h,w] = freqz(b,a,512); g = 20*log10(abs(h));
sos = zp2sos(z,p,k);
sosq = a2dT(sos,5);
R1 = sosq(1,:); R2 = sosq(2,:); R3 = sosq(3,:);
b1 = conv(R1(1:3),R2(1:3)); bq = conv(R3(1:3),b1);
a1 = conv(R1(4:6),R2(4:6)); aq = conv(R3(4:6),a1);
[hq,w] = freqz(bq,aq,512); gq = 20*log10(abs(hq));
plot(w/pi,g,'b',w/pi,gq,'r:');grid
axis([0 1 -70 5]);
xlabel('\omega/\pi');ylabel('Gain, dB');
title('original - solid line, quantized - dashed line');
```

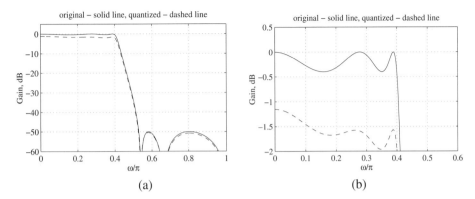

Figure 9.7: Coefficient quantization effects on a fifth-order IIR elliptic lowpass filter implemented in a cascade form: (a) fullband gain responses with unquantized (shown with solid line) and quantized coefficients (shown with dashed line), and (b) passband details.

Figure 9.7 shows the fullband gain response and the passband gain response details of the ideal cascade for realization with infinite precision coefficients (shown with solid line) and the gain response obtained when the transfer function coefficients of each section are truncated to 5-bit length (shown with dashed line). It can be seen from this figure that the effect of the coefficient quantization here is not as severe as in the previous case. A flat loss has been added to the passband response with an increase in the passband ripple. Here each complex zero-pair is realized by a second-order section, and hence, the zeros remain on the unit circle. In fact, all of the zeros remain pretty much at their original locations. The overall effect on the stopband response is minimal.

In general, a higher-order IIR transfer function should never be realized as a single direct form structure, but realized as a cascade of second-order and first-order sections to minimize the effect of coefficient quantization.

The above two programs can be easily modified to study the effect of coefficient quantization on the performance of an FIR digital filter. Program 9_3 given below can be employed to study the effect of the coefficient quantization on the frequency response of a lowpass equiripple FIR digital filter implemented in direct form.

```
% Program 9_3
% Coefficient Quantization Effects on the
% Frequency Response of a Direct Form FIR Filter
%
fpts = [0 0.5 0.55 1]; mag = [1 1 0 0];
b = remez(39,fpts,mag);
[h,w] = freqz(b,1,512); g = 20*log10(abs(h));
bq = a2dT(b,5);
[hq,w] = freqz(bq,1,512); gq = 20*log10(abs(hq));
plot(w/pi,g,'b',w/pi,gq,'r:');grid
axis([0 1 -60 5]);
xlabel('\omega/\pi');ylabel('Gain, dB');
title('original - solid line, quantized - dashed line');
```

Figure 9.8 shows the gain responses of the FIR filter generated by the above program. As can be seen from this figure, the effect of the coefficient quantization on an FIR filter implemented in direct form is

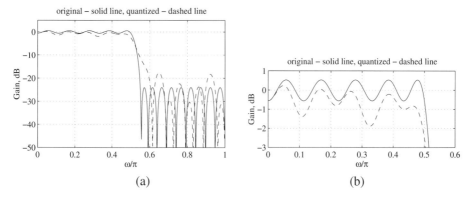

Figure 9.8: Coefficient quantization effects on a 39th-order FIR equiripple lowpass filter implemented in direct form: (a) fullband gain responses with unquantized (shown with solid line) and quantized coefficients (shown with dashed line), and (b) passband details.

to reduce the passband width, increase the passband ripple, increase the transition band, and reduce the minimum stopband attenuation.

9.4.2 Estimation of Pole-Zero Displacements

A measure of the coefficient quantization effects on the performance of a digital filter is given by the pole-zero displacements from their original positions. These displacements can be evaluated analytically as described next [Mit74c].

Consider an Nth-degree polynomial $B(z)$ with simple roots:

$$B(z) = \sum_{i=0}^{N} b_i z^i = \prod_{k=1}^{N} (z - z_k), \tag{9.16}$$

where $b_N = 1$. The roots z_k of $B(z)$ are given by

$$z_k = r_k e^{j\theta_k}. \tag{9.17}$$

Note that $B(z)$ can be either the denominator polynomial or the numerator polynomial of the digital transfer function. The effect of coefficient quantization is manifested by the change of the polynomial coefficients from b_i to $b_i + \Delta b_i$, and as a result, the polynomial $B(z)$ changes to a new polynomial $\hat{B}(z)$ given by

$$\hat{B}(z) = \sum_{i=0}^{N} (b_i + \Delta b_i) z^i = B(z) + \sum_{i=0}^{N-1} (\Delta b_i) z^i = \prod_{k=1}^{N} (z - \hat{z}_k), \tag{9.18}$$

with $\hat{z}_k$ denoting the roots of the polynomial $\hat{B}(z)$. Note that $\hat{z}_k$ are the new locations to which the roots z_k of $B(z)$ have moved. For small changes, $\hat{z}_k$ will be close to z_k and can be expressed as

$$\hat{z}_k = (r_k + \Delta r_k) e^{j(\theta_k + \Delta\theta_k)}, \tag{9.19}$$

where Δr_k and $\Delta\theta_k$ represent the changes in the radius and the angle of the kth root due to coefficient quantization. Our aim is to develop simple expressions for estimating Δr_k and $\Delta\theta_k$ knowing the changes

Δb_i in the coefficients of the polynomial $B(z)$. If we assume the change Δb_i to be very small, the changes in the radius and the angle, Δr_k and $\Delta \theta_k$, of the kth root can be also considered to be very small, and we can rewrite the expression for $\hat{z}_k$ in Eq. (9.19) as

$$\begin{aligned} \hat{z}_k &= (r_k + \Delta r_k)e^{j\Delta\theta_k}e^{j\theta_k} \cong (r_k + \Delta r_k)(1 + j\Delta\theta_k)e^{j\theta_k} \\ &\cong r_k e^{j\theta_k} + (\Delta r_k + jr_k\Delta\theta_k)e^{j\theta_k}, \end{aligned} \tag{9.20}$$

neglecting higher-order terms. The root displacement can now be expressed as

$$\hat{z}_k - z_k = \Delta z_k \cong (\Delta r_k + jr_k\Delta\theta_k)e^{j\theta_k}. \tag{9.21}$$

Now consider the rational function $1/B(z)$. Its partial-fraction expansion is given by

$$\frac{1}{B(z)} = \sum_{k=1}^{N} \frac{\rho_k}{z - z_k}, \tag{9.22}$$

where ρ_k is the residue of $1/B(z)$ at the pole $z = z_k$, i.e.,

$$\rho_k = \left.\frac{(z - z_k)}{B(z)}\right|_{z=z_k} = R_k + jX_k. \tag{9.23}$$

If we assume that $\hat{z}_k$ is very close to z_k, then we can write

$$\frac{1}{B(\hat{z}_k)} \cong \frac{\rho_k}{\hat{z}_k - z_k}, \tag{9.24}$$

or

$$\Delta z_k = \rho_k \cdot B(\hat{z}_k). \tag{9.25}$$

But

$$\hat{B}(\hat{z}_k) = 0 = B(\hat{z}_k) + \sum_{i=0}^{N-1} (\Delta b_i)(\hat{z}_k)^i. \tag{9.26}$$

Therefore, from Eqs. (9.25) and (9.26), we arrive at

$$\Delta z_k = -\rho_k \left\{ \sum_{i=0}^{N-1} (\Delta b_i)(\hat{z}_k)^i \right\} \cong -\rho_k \left\{ \sum_{i=0}^{N-1} (\Delta b_i)(z_k)^i \right\}, \tag{9.27}$$

assuming that $\hat{z}_k$ is very close to z_k. Rewriting Eq. (9.27) we obtain

$$(\Delta r_k + jr_k\Delta\theta_k)e^{j\theta_k} = -(R_k + jX_k)\left\{ \sum_{i=0}^{N-1} (\Delta b_i)(r_k e^{j\theta_k})^i \right\}. \tag{9.28}$$

Equating real and imaginary parts of the above, we arrive at, after some algebra,

$$\Delta r_k = (-R_k \mathbf{P}_k + X_k \mathbf{Q}_k) \cdot \Delta\mathbf{B} = \mathbf{S}_b^{r_k} \cdot \Delta\mathbf{B}, \tag{9.29a}$$

$$\Delta\theta_k = -\frac{1}{r_k}(X_k \mathbf{P}_k + R_k \mathbf{Q}_k) \cdot \Delta\mathbf{B} = \mathbf{S}_b^{\theta_k} \cdot \Delta\mathbf{B}, \tag{9.29b}$$

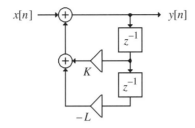

Figure 9.9: Direct form II realization of the second-order IIR transfer function of Eq. (9.31).

where

$$\mathbf{P}_k = \left[\cos\theta_k \quad r_k \quad r_k^2\cos\theta_k \cdots r_k^{N-1}\cos(N-2)\theta_k\right],\tag{9.30a}$$

$$\mathbf{Q}_k = \left[-\sin\theta_k \quad 0 \quad r_k^2\sin\theta_k \cdots r_k^{N-1}\sin(N-2)\theta_k\right],\tag{9.30b}$$

$$\Delta\mathbf{B} = \left[\Delta b_0 \ \Delta b_1 \ \Delta b_2 \cdots \Delta b_{N-1}\right]^T.\tag{9.30c}$$

It should be noted that the sensitivity vectors $\mathbf{S}_b^{r_k}$ and $\mathbf{S}_b^{\theta_k}$ depend on only $B(z)$ and are independent of $\Delta\mathbf{B}$. Hence, once these vectors have been calculated, pole-zero displacements for any sets of $\Delta\mathbf{B}$ can be rapidly calculated using Eqs. (9.29a) and (9.29b). Moreover, the elements of $\Delta\mathbf{B}$ are multiplier coefficient changes only for the direct form realization.

We illustrate in the following example the application of Eqs. (9.29a) and (9.29b) in computing the pole displacements of a second-order direct form IIR digital filter structure due to coefficient quantization.

EXAMPLE 9.1   Consider the direct form II realization of

$$H(z) = \frac{z^2}{z^2 - Kz + L} = \frac{z^2}{B(z)}\tag{9.31}$$

as indicated in Figure 9.9. Here,

$$B(z) = z^2 - 2r\cos\theta z + r^2 = (z - z_1)(z - z_2),\tag{9.32}$$

where

$$z_1 = re^{j\theta}, \qquad z_2 = re^{-j\theta}.\tag{9.33}$$

From Eqs. (9.31) and (9.32), we get

$$K = 2r\cos\theta, \qquad L = r^2.\tag{9.34}$$

The residue ρ_1 of $1/B(z)$ at $z = z_1$ is given by

$$\rho_1 = \left.\frac{z - z_1}{B(z)}\right|_{z=z_1} = -\frac{j}{2r\sin\theta}.\tag{9.35}$$

Therefore, for the IIR filter of Figure 9.9, we have

$$\Delta\mathbf{B} = [\Delta L \quad -\Delta K]^T,\tag{9.36a}$$

$$\mathbf{Q}_1 = [-\sin\theta \quad 0],\tag{9.36b}$$

$$\mathbf{P}_1 = [\cos\theta \quad r].\tag{9.36c}$$

Substituting Eqs. (9.36a) to (9.36c) in Eqs. (9.29a) and (9.29b), we arrive at

$$\Delta r = X_1 \mathbf{Q}_1 \Delta \mathbf{B} = \frac{1}{2r} \Delta L, \tag{9.37a}$$

$$\Delta \theta = -\frac{1}{r}(X_1 \mathbf{P}_1 \Delta \mathbf{B}) = \frac{\Delta L}{2r^2 \tan \theta} - \frac{\Delta K}{2r \sin \theta}. \tag{9.37b}$$

It is evident from the above expressions that the second-order direct form IIR structure is highly sensitive to coefficient quantizations for transfer functions with poles close to $\theta = 0$ or π, i.e., for narrow bandwidth lowpass and highpass filters.

We now extend the results of Eqs. (9.29a) and (9.29b) to an arbitrary structure with R multipliers given by α_k, $k = 1, 2, \ldots, R$. Due to coefficient quantization these coefficients change to $\alpha_k + \Delta \alpha_k$. Since the multiplier coefficients α_i are multilinear functions of the coefficients b_i of the polynomial $B(z)$, we can express the change Δb_i in the transfer function coefficient b_i to the change $\Delta \alpha_k$ in the multiplier coefficient α_k through

$$\Delta b_i = \sum_{k=1}^{R} \frac{\partial b_i}{\partial \alpha_k} \Delta \alpha_k, \qquad i = 0, 1, \ldots, N - 1. \tag{9.38}$$

In matrix form the above can be written as

$$\Delta \mathbf{B} = \mathbf{C} \cdot \Delta \boldsymbol{\alpha}, \tag{9.39}$$

where

$$\mathbf{C} = \begin{bmatrix} \frac{\partial b_0}{\partial \alpha_1} & \frac{\partial b_0}{\partial \alpha_2} & \cdots & \frac{\partial b_0}{\partial \alpha_R} \\ \frac{\partial b_1}{\partial \alpha_1} & \frac{\partial b_1}{\partial \alpha_2} & \cdots & \frac{\partial b_1}{\partial \alpha_R} \\ \vdots & \vdots & \ddots & \vdots \\ \frac{\partial b_{N-1}}{\partial \alpha_1} & \frac{\partial b_{N-1}}{\partial \alpha_2} & \cdots & \frac{\partial b_{N-1}}{\partial \alpha_R} \end{bmatrix}, \tag{9.40a}$$

$$\Delta \boldsymbol{\alpha} = [\Delta \alpha_1 \quad \Delta \alpha_2 \cdots \Delta \alpha_R]^T. \tag{9.40b}$$

Substituting Eq. (9.39) in Eqs. (9.29a) and (9.29b), we arrive at the desired result:

$$\Delta r_k = \mathbf{S}_b^{r_k} \cdot \mathbf{C} \cdot \Delta \boldsymbol{\alpha}, \tag{9.41a}$$

$$\Delta \theta_k = \mathbf{S}_b^{\theta_k} \cdot \mathbf{C} \cdot \Delta \boldsymbol{\alpha}, \tag{9.41b}$$

where the sensitivity vectors are as given in Eqs. (9.29a) and (9.29b). It should be noted that $\mathbf{C}$ depends on the structure but has to be computed only once.

The application of Eqs. (9.41a) and (9.41b) in computing the pole displacements of a second-order IIR digital filter structure is treated in the following example.

EXAMPLE 9.2   Consider the coupled-form structure [Gol69b] of Figure 9.10 for which the transfer function is given by

$$H(z) = \frac{\gamma z^2}{z^2 - (\alpha + \delta)z + (\alpha \delta - \beta \gamma)}. \tag{9.42}$$

If $\alpha = \delta = r \cos \theta$ and $\beta = -\gamma = r \sin \theta$, then the transfer function of Eq. (9.42) becomes

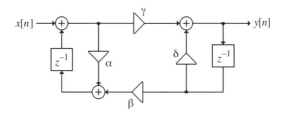

Figure 9.10: A second-order coupled form structure.

$$H(z) = \frac{\gamma z^2}{z^2 - 2r\cos\theta z + r^2}. \tag{9.43}$$

Comparing the denominator polynomial of Eq. (9.42) with that of Eq. (9.31), we observe

$$K = \alpha + \delta = 2\alpha, \tag{9.44a}$$

$$L = \alpha\delta - \beta\gamma = \alpha^2 + \beta^2, \tag{9.44b}$$

from which we arrive at

$$\begin{bmatrix} \Delta L \\ \Delta K \end{bmatrix} = \begin{bmatrix} 2r\cos\theta & 2r\sin\theta \\ 2 & 0 \end{bmatrix} \begin{bmatrix} \Delta\alpha \\ \Delta\beta \end{bmatrix}. \tag{9.45}$$

Making use of Eqs. (9.37a) and (9.37b) in the above we then get

$$\begin{bmatrix} \Delta r \\ \Delta\theta \end{bmatrix} = \begin{bmatrix} \frac{1}{2r} & 0 \\ \frac{1}{2r^2\tan\theta} & -\frac{1}{2r\sin\theta} \end{bmatrix} \begin{bmatrix} 2r\cos\theta & 2r\sin\theta \\ 2 & 0 \end{bmatrix} \begin{bmatrix} \Delta\alpha \\ \Delta\beta \end{bmatrix}$$

$$= \begin{bmatrix} \cos\theta & \sin\theta \\ -\frac{1}{r}\sin\theta & \frac{1}{r}\cos\theta \end{bmatrix} \begin{bmatrix} \Delta\alpha \\ \Delta\beta \end{bmatrix}. \tag{9.46}$$

As can be seen from the above, the coupled-form structure is less sensitive to multiplier coefficient quantization than the direct form structure. Several other low-sensitivity second-order structures are considered in Problems 9.4 and 9.5.

9.4.3 Analysis of Coefficient Quantization Effects in FIR Filters

The analysis of the displacement of the roots of a polynomial due to coefficient quantization outlined in the previous section can of course be applied to FIR transfer function to determine the sensitivity of its zeros to changes in the coefficients. A more meaningful analysis is obtained by examining the changes in the frequency response due to coefficient quantization as described next.

Consider an Nth-order FIR transfer function:

$$H(z) = \sum_{n=0}^{N} h[n]z^{-n}. \tag{9.47}$$

Quantization of the filter coefficients results in a new transfer function

$$\hat{H}(z) = \sum_{n=0}^{N} \hat{h}[n]z^{-n} = \sum_{n=0}^{N} (h[n] + e[n])\,z^{-n}, \tag{9.48}$$

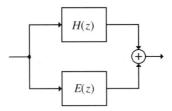

Figure 9.11: Model of the FIR filter with quantized coefficients.

which can be rewritten as

$$\hat{H}(z) = H(z) + E(z), \tag{9.49}$$

where

$$E(z) = \sum_{n=0}^{N} e[n]z^{-n}. \tag{9.50}$$

Thus, the FIR filter with quantized coefficients can be modeled as a parallel connection of two FIR filters, $H(z)$ and $E(z)$, as shown in Figure 9.11, where $H(z)$ represents the desired FIR filter with unquantized coefficients, and $E(z)$ is the FIR filter representing the error in the transfer function due to coefficient quantization.

Without any loss of generality, assume the FIR filter $H(z)$ to be a Type 1 linear-phase filter of order N with impulse response $h[n]$. Hence, $E(z)$ is also a Type 1 linear-phase FIR transfer function. The frequency response of $H(z)$ can be expressed as

$$H(e^{j\omega}) = e^{-j\omega N/2} \left(h\left[\tfrac{N}{2}\right] + \sum_{n=0}^{(N-2)/2} 2h[n] \cos\left[\left(\tfrac{N}{2} - n\right)\omega\right] \right). \tag{9.51}$$

The frequency response of the actual FIR filter with quantized coefficients $\hat{h}[n]$ can be expressed as

$$\hat{H}(e^{j\omega}) = H(e^{j\omega}) + E(e^{j\omega}), \tag{9.52}$$

where $E(e^{j\omega})$ represents the error in the desired frequency response $H(e^{j\omega})$:

$$E(e^{j\omega}) = \sum_{n=0}^{N} e[n]e^{-j\omega n}, \tag{9.53}$$

with $e[n] = \hat{h}[n] - h[n]$. The error in the frequency response is thus bounded by

$$\left| E(e^{j\omega}) \right| = \left| \sum_{n=0}^{N} e[n]e^{-j\omega n} \right| \leq \sum_{n=0}^{N} |e[n]| \left| e^{-j\omega n} \right| \leq \sum_{n=0}^{N} |e[n]|. \tag{9.54}$$

Assume each impulse response coefficient $h[n]$ is a $(b+1)$-bit signed fraction. In this case, the range of the coefficient quantization error $e[n]$ is exactly the same as that indicated in Table 9.1. Using the data given in this table, an upper bound on $|E(e^{j\omega})|$ can be derived from Eq. (9.54). For example, for rounding, $|e[n]| \leq \delta/2$, where $\delta = 2^{-b}$ is the quantization step. As a result

$$\left| E(e^{j\omega}) \right| \leq \frac{(N+1)\delta}{2}. \tag{9.55}$$

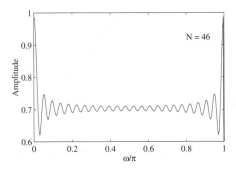

Figure 9.12: Plot of a typical $W_N(\omega)$.

The above bound is rather conservative and can be reached only if all errors in Eq. (9.54) are of same sign and have the maximum value in the range. A more realistic bound can be derived assuming $e[n]$ are statistically independent random variables [Cha73]. From Eqs. (9.51) to (9.53), we obtain

$$E(e^{j\omega}) = e^{-j\omega N/2} \left(e\left[\tfrac{N}{2}\right] + \sum_{n=0}^{(N-2)/2} 2e[n] \cos\left[\left(\tfrac{N}{2} - n\right)\omega\right] \right). \tag{9.56}$$

From the above, we observe that $E(e^{j\omega})$ is a sum of independent random variables. If we denote the variance of $e[n]$ as σ_e^2, then the variance of $E(e^{j\omega})$ is simply given by

$$\sigma_E^2(\omega) = \sigma_e^2 \left(1 + 4 \sum_{n=1}^{N/2} \cos^2(\omega n) \right) = \sigma_e^2 \left(N + \frac{\sin(N+1)\omega}{\sin\omega} \right). \tag{9.57}$$

Using the notation

$$W_N(\omega) = \left[\frac{1}{2N+1} \left(N + \frac{\sin(N+1)\omega}{\sin\omega} \right) \right]^{1/2}, \tag{9.58}$$

the standard deviation of $E(e^{j\omega})$ can be expressed as

$$\sigma_E(\omega) = \sigma_e \left(\sqrt{2N+1} \right) W_N(\omega). \tag{9.59}$$

For uniformly distributed $e[n]$, $\sigma_e = \delta/\sqrt{12} = 2^{-b-1}/\sqrt{3}$.

A plot of a typical weighting function $W_N(\omega)$ is sketched in Figure 9.12. In fact, it can be shown that $W_N(\omega)$ is in the range $(0, 1)$, and hence, the standard deviation $\sigma_E(\omega)$ is bounded by

$$\sigma_E(\omega) \le \delta \sqrt{\frac{2N+1}{12}}. \tag{9.60}$$

Based on the above bound, Chan and Rabiner [Cha73] have advanced a method to estimate the wordlength of the FIR filter coefficients to meet the prescribed filter specifications.

Figure 9.13: Model of a practical A/D conversion system.

9.5   A/D Conversion Noise Analysis

In many applications, digital signal processing techniques are employed to process continuous-time (analog) bandlimited signals that are either voltage or current waveforms. These analog signals must be converted into digital form before they can be processed digitally. According to the sampling theorem given in Section 5.2.1, an analog bandlimited signal $x_a(t)$ can be represented uniquely by its sampled version $x_a(nT)$ if the sampling frequency $\Omega_T = 1/T$ is greater than twice the highest frequency Ω_m contained in $x_a(t)$. In order to ensure that the sampling frequency chosen does satisfy this condition, an anti-aliasing filter is used to bandlimit the analog signal to half of the sampling frequency. The discrete-time sequence $x_a(nT) = x[n]$ is then converted into a digital sequence for digital signal processing. As indicated by Figure 5.1, the conversion of an analog signal into a digital sequence is implemented in practice by a cascade of two devices, a sample-and-hold (S/H) circuit followed by an analog-to-digital (A/D) converter.

The digital samples produced by the A/D converter are usually represented in a binary form. As indicated in Section 8.4, there are several different forms of binary representations of which the two's-complement representation is usually employed in digital signal processing for convenience in the implementation of the arithmetic operations. Consequently, the A/D converters used for the digital signal processing of analog signals in general are those that employ the two's-complement fixed-point representation to represent the digital equivalent of the input analog signal. Moreover, for the processing of bipolar analog signals, the A/D converter generates a bipolar output represented as a fixed-point signed binary fraction.

9.5.1   Quantization Noise Model

The digital sample generated by the A/D converter is the binary representation of the quantized version of that produced by an ideal sampler with infinite precision. Because of the finite wordlength of the output register, the digital equivalent can take a value from a finite set of discrete values within the dynamic range of the register. For example, if the output word is of length $(b + 1)$ bits including the sign bit, the total number of discrete levels available for representation is 2^{b+1}. The dynamic range of the output register depends on the binary number representation selected for the A/D converter. The operation of a practical analog-to-digital conversion system consisting of a sample-and-hold circuit followed by an A/D converter therefore can be modeled as shown in Figure 9.13. The quantizer maps the input analog sample $x[n]$ into $\hat{x}[n]$, one of a set of discrete values, and the coder determines its binary equivalent $\hat{x}_{eq}[n]$ based on the binary representation scheme adopted by the A/D converter.

The quantization process employed by the quantizer in the A/D converter can be either rounding or truncation. Assuming rounding is used, the input-output characteristic of a 3-bit A/D converter with the output in two's-complement form is as shown in Figure 9.14. This figure also shows the binary equivalents of the quantized samples.

As indicated earlier and shown explicitly in Figure 9.14, for a two's-complement binary representation, the binary equivalent $\hat{x}_{eq}[n]$ of the quantized input analog sample $\hat{x}[n]$ is a binary fraction in the range

$$-1 \leq \hat{x}_{eq}[n] < 1. \tag{9.61}$$

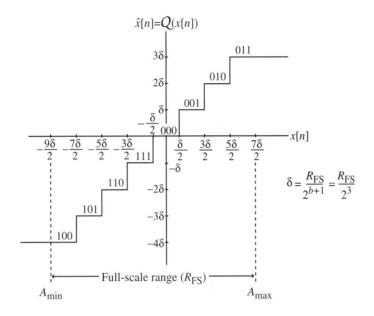

Figure 9.14: Input-output characteristic of a 3-bit bipolar A/D converter with two's-complement representation.

It is related to the quantized sample $\hat{x}[n]$ through

$$\hat{x}_{eq}[n] = \frac{2\hat{x}[n]}{R_{FS}}, \tag{9.62}$$

where R_{FS} denotes the *full-scale range* of the A/D converter. We shall assume that the input signal has been scaled to be in the range of ± 1 by dividing its amplitude by $R_{FS}/2$, as is usually the case. Then the decimal equivalent of $\hat{x}_{eq}[n]$ is equal to $\hat{x}[n]$, and we shall not differentiate between these two numbers.

For a $(b+1)$-bit bipolar A/D converter, the total number of quantization levels is 2^{b+1} and the full-scale range R_{FS}, usually given in volts or amperes, is given by

$$R_{FS} = 2^{b+1}\delta, \tag{9.63}$$

where δ is the quantization step size, also called the quantization width, in volts or amperes. If the input signal is in the range given by Eq. (9.61), $R_{FS} = 2$, and then $\delta = 2^{-b}$. For the 3-bit A/D converter depicted in Figure 9.14, the total number of levels is $2^3 = 8$ and the full-scale range is $R_{FS} = 8\delta$, with a maximum value of $A_{max} = 7\delta/2$ and a minimum value of $A_{min} = -9\delta/2$. If the input analog sample $x_a(nT)$ is within the full-scale range,

$$-\frac{9\delta}{2} < x_a(nT) \le \frac{7\delta}{2}, \tag{9.64}$$

it is quantized to one of the 8 discrete values indicated in Figure 9.14. In general, for a $(b + 1)$-bit wordlength A/D converter employing two's-complement representation, the full-scale range is given by

$$-(2^{b+1} + 1)\frac{\delta}{2} < x_a(nT) \le (2^{b+1} - 1)\frac{\delta}{2}. \tag{9.65}$$

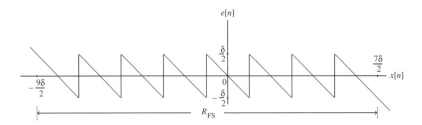

Figure 9.15: Quantization error as a function of the input for a 3-bit bipolar A/D converter.

If we denote the difference between the quantized value $Q(x[n]) = \hat{x}[n]$ and the input sample $x_a(nT) = x[n]$ as the quantization error $e[n]$,

$$e[n] = Q(x[n]) - x[n] = \hat{x}[n] - x[n], \tag{9.66}$$

it follows from Figure 9.14 that $e[n]$ is within the range

$$-\frac{\delta}{2} < e[n] \le \frac{\delta}{2}, \tag{9.67}$$

assuming that a sample exactly halfway between two levels is rounded up to the nearest higher level and assuming that the analog input is within the A/D converter full-scale range as given by Eq. (9.61). In this case, the quantization error $e[n]$, called the *granular noise*, is bounded in magnitude according to Eq. (9.67). A plot of the quantization error $e[n]$ of the above 3-bit A/D converter as a function of the input sample $x[n]$ is given in Figure 9.15.

As can be seen from this figure, when the input analog sample is outside the A/D converter full-scale range, the magnitude of the error $e[n]$ increases linearly with an increase in the magnitude of the input. In the latter situation, the A/D converter error $e[n]$ is called the *saturation error* or the *overload noise* as the A/D converter output is "clipped" to the maximum value $(1 - 2^{-b})$ if the analog input is positive or the minimum value -1 if the analog input is negative independent of the actual value of the input. A clipping of the A/D converter output causes signal distortion with highly undesirable effects and must be avoided by scaling down the analog input $x_a(nT)$ to ensure that it remains within the A/D converter full-scale range.

In order to develop the necessary mathematical model for analyzing the effect of the finite wordlength of the A/D converter output, we assume that analog input samples are within the full-scale range, and as a result, there is no saturation error at the converter output. Since the input-output characteristic of an A/D converter is nonlinear and the analog input signal, in most practical cases, is not known a priori, it is reasonable to assume for analysis purposes that the quantization error $e[n]$ is a random signal and to use a statistical model of the quantizer operation as indicated in Figure 9.16. Furthermore, for simplified analysis, we make the following assumptions:

(a) The error sequence $\{e[n]\}$ is a sample sequence of a wide-sense stationary (WSS) white noise process, with each sample $e[n]$ being uniformly distributed over the range of the quantization error as indicated in Figure 9.17, where δ is the quantization step.

(b) The error sequence is uncorrelated with its corresponding input sequence $\{x[n]\}$.

(c) The input sequence is a sample sequence of a stationary random process.

These assumptions hold in most practical situations for input signals whose samples are large and change in amplitude very rapidly in time relative to the quantization step in a somewhat random fashion, and

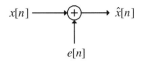

Figure 9.16: A statistical model of the A/D quantizer.

have been verified to be valid assumptions experimentally [Ben48], [Wid56], [Wid61] and by computer simulations [DeF88]. The statistical model also makes the analysis of A/D conversion noise more tractable, and results derived have been found to be useful for most applications. It should be pointed out that in the case of an A/D converter employing ones'-complement or sign-magnitude truncation, the quantization error is correlated to the input signal since here the sign of each error sample $e[n]$ is exactly opposite to the sign of the corresponding input sample $x[n]$. As a result, practical A/D converters use either rounding or two's-complement truncation. The mean and the variance of the uniformly distributed random variables were computed in Example 2.42. Using the results of this example, we observe that the mean and variance of the error sample in the case of rounding are given by

$$m_e = \frac{(\delta/2) - (\delta/2)}{2} = 0, \tag{9.68}$$

$$\sigma_e^2 = \frac{((\delta/2) - (-\delta/2))^2}{12} = \frac{\delta^2}{12}. \tag{9.69}$$

The corresponding parameters for the two's-complement truncation are as follows:

$$m_e = \frac{0 - \delta}{2} = -\frac{\delta}{2}, \tag{9.70}$$

$$\sigma_e^2 = \frac{(0 - \delta)^2}{12} = \frac{\delta^2}{12}. \tag{9.71}$$

9.5.2 Signal-to-Quantization Noise Ratio

Based on the model of Figure 9.16, we can evaluate the effect of the additive quantization noise $e[n]$ on the input signal $x[n]$ by computing the *signal-to-quantization noise ratio* (SNR$_{A/D}$) in dB defined by

$$\text{SNR}_{A/D} = 10 \log_{10} \left(\frac{\sigma_x^2}{\sigma_e^2} \right) \text{dB}, \tag{9.72}$$

where σ_x^2 is the input signal variance representing the signal power and σ_e^2 is the noise variance representing the quantization noise power. For rounding, the quantization error is uniformly distributed in the range $(-\delta/2, \delta/2)$ and for two's-complement truncation, the quantization error is uniformly distributed in the range $(-\delta, 0)$ as indicated in Figure 9.17(a) and (b), respectively. In the case of a bipolar $(b + 1)$-bit A/D converter, $\delta = 2^{-(b+1)} R_{FS}$, and hence,

$$\sigma_e^2 = \frac{2^{-2b}(R_{FS})^2}{48}. \tag{9.73}$$

Substituting Eq. (9.73) in Eq. (9.72), we arrive at

$$\text{SNR}_{A/D} = 10 \log_{10} \left(\frac{48\sigma_x^2}{2^{-2b}(R_{FS})^2} \right)$$

$$= 6.02b + 16.81 - 20 \log_{10} \left(\frac{R_{FS}}{\sigma_x} \right) \text{dB}. \tag{9.74}$$

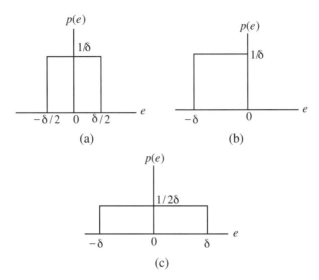

Figure 9.17: Quantization error probability density functions: (a) rounding, (b) two's-complement truncation, and (c) ones'-complement truncation.

Table 9.3: Signal-to-quantization noise ratio of an A/D converter as a function of wordlength and full-scale range.

	$b = 7$	$b = 9$	$b = 11$	$b = 13$	$b = 15$
$K = 4$	46.91	58.95	70.99	83.04	95.08
$K = 6$	43.39	55.43	67.47	79.51	91.56
$K = 8$	40.89	52.93	64.97	77.01	89.05

The above expression is used to determine the minimum A/D converter wordlength needed to meet a specified signal-to-quantization noise ratio. As can be seen from this expression, the SNR increases by approximately 6 dB for each bit added to the wordlength. For a given wordlength, the actual SNR depends on the last term in Eq. (9.74), which in turn depends on the σ_x, the rms value of the input signal amplitude, and the full-scale range R_{FS} of the converter, as illustrated by the following example.

EXAMPLE 9.3   Determine the signal-to-quantization noise ratio in the digital equivalent of an analog sample $x[n]$ with a zero-mean Gaussian distribution using a $(b+1)$-bit A/D converter having a full-scale range $R_{FS} = K\sigma_x$. From Eq. (9.74) we obtain

$$\text{SNR}_{A/D} = 6.02\, b + 16.81 - 20\log_{10}(K). \tag{9.75}$$

Table 9.3 shows the computed values of the SNR for various values of b and K.

Now, the probability of a particular input analog sample with a zero-mean Gaussian distribution staying within the full-scale range $K\sigma_x$ is given by [Par60]

$$2\Phi(K) - 1 = \sqrt{\frac{2}{\pi}} \int_0^K e^{-y^2/2}\, dy. \tag{9.76}$$

For example, for $K = 4$, the probability of an input analog sample staying within the full-scale range of $4\sigma_x$ is 0.9544. This implies that on average about 456 samples out of 10,000 samples will fall outside the range and be clipped. If we increase the full-scale range to $6\sigma_x$, the probability of an input analog sample staying within the expanded full-scale range increases to 0.9974, in which case on average about 26 samples out of 10,000 samples will now be outside the range. In most applications, a full-scale range of $8\sigma_x$ is more than adequate to ensure no clipping in conversion.

9.5.3 Effect of Input Scaling on SNR

Now, consider the effect of scaling of the input on the SNR. Let the input scaling factor be A with $A > 0$. Since the variance of the scaled input $Ax[n]$ is $A^2\sigma_x^2$, the signal-to-quantization noise expression of Eq. (9.75) changes to

$$\text{SNR}_{\text{A/D}} = 6.02\,b + 16.81 - 20\log_{10}(K) + 20\log_{10}(A). \tag{9.77}$$

For a given b, the SNR can also be increased by scaling up the input analog signal by making $A > 1$. However, this process also increases the probability of some of the input analog samples being outside the full-scale range R_{FS}, and as a result, Eq. (9.77) no longer holds. Furthermore, the output is clipped, causing severe distortion in the digital representation of the analog input. On the other hand, a scaling down of the input analog signal by making $A < 1$ decreases the SNR. It is therefore necessary to ensure that the analog sample range matches as close as possible to the full-scale range of the A/D converter to get the maximum possible SNR without any signal distortion.

It should be noted here that the above analysis assumed an ideal A/D conversion. However, as pointed out in Section 5.8.6, a practical A/D converter is a nonideal device exhibiting a variety of errors, resulting in the actual signal-to-quantization noise ratio being smaller than that predicted by Eq. (9.74). Hence, the effective wordlength of the A/D converter, in general, is less than the wordlength computed using Eq. (9.74) by 1 to 2 bits. This factor should be taken into consideration in selecting an appropriate A/D converter for a given application.

9.5.4 Propagation of Input Quantization Noise to Digital Filter Output

In most applications, the quantized signal $\hat{x}[n]$ generated by the A/D converter is processed by a linear time-invariant discrete-time system $H(z)$. It is thus of interest to determine how the input quantization noise propagates to the output of the digital filter. In determining the noise at the filter output generated by the input noise, we can assume that the digital filter is implemented using infinite precision. However, as we shall point out later, in practice, the quantization of the arithmetic operations generates errors inside the digital filter structure, which also propagate to the output and appear as noise. These noise sources are assumed to be independent of the input quantization noise, and their effects can be analyzed separately and added to that due to the input noise.

As indicated in Figure 9.16, the quantized signal $\hat{x}[n]$ can be considered as a sum of two sequences: the unquantized input $x[n]$ and the input quantization noise $e[n]$. Because of the linearity property and the assumption that $x[n]$ and $e[n]$ are uncorrelated, the output $\hat{y}[n]$ of the LTI system can thus be expressed as a sum of two sequences: $y[n]$ generated by the unquantized input $x[n]$ and $v[n]$ generated by the error sequence $e[n]$, as shown in Figure 9.18. As a result, we can compute the output noise component $v[n]$ as a linear convolution of $e[n]$ with the impulse response sequence $h[n]$ of the LTI system:

$$v[n] = \sum_{m=-\infty}^{\infty} e[m]h[n-m]. \tag{9.78}$$

From Eq. (4.204), the mean m_v of the output noise $v[n]$ is given by

$$m_v = m_e H(e^{j0}), \tag{9.79}$$

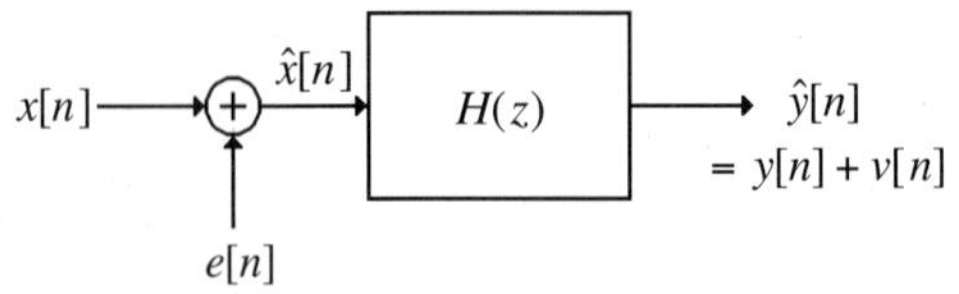

Figure 9.18: Model for the analysis of the effect of processing a quantized input by an LTI discrete-time system.

and, from Eq. (4.223), its variance σ_v^2 is given by

$$\sigma_v^2 = \frac{\sigma_e^2}{2\pi} \int_{-\pi}^{\pi} \left| H(e^{j\omega}) \right|^2 d\omega. \tag{9.80}$$

The output noise power spectrum is given by

$$P_{vv}(\omega) = \sigma_e^2 \left| H(e^{j\omega}) \right|^2. \tag{9.81}$$

The normalized output noise variance is given by

$$\sigma_{v,n}^2 = \frac{\sigma_v^2}{\sigma_e^2} = \frac{1}{2\pi} \int_{-\pi}^{\pi} \left| H(e^{j\omega}) \right|^2 d\omega, \tag{9.82a}$$

which can be alternately expressed as

$$\sigma_{v,n}^2 = \frac{1}{2\pi j} \oint_C H(z) H(z^{-1}) z^{-1} \, dz, \tag{9.82b}$$

where C is a counterclockwise contour in the ROC of $H(z)H(z^{-1})$. An equivalent expression for Eq. (9.82b) obtained using Eq. (4.225) is given by

$$\sigma_{v,n}^2 = \sum_{n=-\infty}^{\infty} |h[n]|^2. \tag{9.82c}$$

9.5.5 Algebraic Computation of Output Noise Variance

We now outline a simple algebraic approach for computing the normalized output noise variance using Eq. (9.82b) [Mit74b]. Several other algebraic approaches have been suggested for computing integrals of the form of Eq. (9.82b)[Chu95] [Dug80], [Dug82].

In general, $H(z)$ is a causal stable real rational function with all poles inside the unit circle in the z-plane. It can be expressed in a partial-fraction form as

$$H(z) = \sum_{i=1}^{R} H_i(z), \tag{9.83}$$

where $H_i(z)$ is a low-order real rational causal stable transfer function. Substituting the above in Eq. (9.82b), we arrive at

$$\sigma_{v,n}^2 = \frac{1}{2\pi j} \sum_{k=1}^{R} \sum_{\ell=1}^{R} \oint_C H_k(z) H_\ell(z^{-1}) z^{-1} \, dz. \tag{9.84}$$

Since $H_k(z)$ and $H_\ell(z)$ are stable transfer functions, it can be shown that

$$\oint_C H_k(z) H_\ell(z^{-1}) z^{-1}\, dz = \oint_C H_\ell(z) H_k(z^{-1}) z^{-1}\, dz. \tag{9.85}$$

As a result, Eq. (9.84) can be rewritten as

$$\sigma_{v,n}^2 = \frac{1}{2\pi j} \left\{ \sum_{k=1}^{R} \oint_C H_k(z) H_k(z^{-1}) z^{-1}\, dz \right.$$

$$\left. + 2 \sum_{k=1}^{R-1} \sum_{\ell=k+1}^{R} \oint_C H_k(z) H_\ell(z^{-1}) z^{-1}\, dz \right\}. \tag{9.86}$$

In most practical cases, $H(z)$ has only simple poles with $H_k(z)$ being either a first-order or a second-order transfer function. As a result, each of the above contour integrations is much simpler to perform using the Cauchy's residue theorem, and the results are tabulated for easy reference. Typical terms in the partial-fraction expansion of $H(z)$ are as follows:[2]

$$A, \qquad \frac{B_k}{z - a_k}, \qquad \frac{C_k z + D_k}{z^2 + b_k z + d_k}. \tag{9.87}$$

Let us denote a typical contour integral in Eq. (9.86) as

$$I_i = \frac{1}{2\pi j} \oint_C H_k(z) H_\ell(z^{-1}) z^{-1}\, dz. \tag{9.88}$$

The expressions obtained after performing the contour integrations for different I_i are listed in Table 9.4.

EXAMPLE 9.4 Let us determine the variance of the noise generated by the A/D quantization noise at the output of a first-order digital filter with a transfer function

$$H(z) = \frac{1}{1 - \alpha z^{-1}}. \tag{9.89}$$

We rewrite $H(z)$ as

$$H(z) = \frac{z}{z - \alpha},$$

whose partial-fraction expansion is given by

$$H(z) = 1 + \frac{\alpha}{z - \alpha}. \tag{9.90}$$

The two terms in the partial-fraction expansion are

$$H_1(z) = 1, \qquad H_2(z) = \frac{\alpha}{z - \alpha}.$$

Therefore, using the results of Table 9.4, we finally obtain the expression for the normalized output noise variance as

---

[2]It should be noted that $H(z)$ is expressed as the ratio of polynomials in z before a direct partial-fraction expansion is carried out to arrive at the terms given in Eq. (9.87)

Table 9.4: Expressions for typical contour integrals in the output noise variance calculation.

$H_k(z)$	A	$\dfrac{B_\ell}{z^{-1} - a_\ell}$	$\dfrac{C_\ell z^{-1} + D_\ell}{z^{-2} + b_\ell z^{-1} + d_\ell}$
			$H_\ell(z^{-1})$
A	I_1	0	0
$\dfrac{B_k}{z - a_k}$	0	I_2	I_4'
$\dfrac{C_k z + D_k}{z^2 + b_k z + d_k}$	0	I_4	I_3

$$I_1 = A^2$$

$$I_2 = \frac{B_k B_\ell}{1 - a_k a_\ell}$$

$$I_3 = \frac{(C_k C_\ell + D_k D_\ell)(1 - d_k d_\ell) - (D_k C_\ell - C_k D_\ell d_k)b_\ell - (C_k D_\ell - D_k C_\ell d_\ell)b_k}{(1 - d_k d_\ell)^2 + d_k b_\ell^2 + d_\ell b_k^2 - (1 + d_k d_\ell)b_k b_\ell}$$

$$I_4 = \frac{B_\ell (C_k + D_k a_\ell)}{1 + b_k a_\ell + d_k a_\ell^2}$$

$$I_4' = \frac{B_k (C_\ell + D_\ell a_k)}{1 + b_\ell a_k + d_\ell a_k^2}$$

$$\sigma_{v,n}^2 = 1^2 + \frac{\alpha^2}{1 - \alpha^2} = \frac{1}{1 - \alpha^2}. \tag{9.91}$$

If the pole is close to the unit circle, we can write $\alpha = 1 - \varepsilon$, where $\varepsilon \cong 0$. In which case, we can rewrite Eq. (9.91) as

$$\sigma_{v,n}^2 = \frac{1}{1 - (1 - \varepsilon)^2} \cong \frac{1}{2\varepsilon}, \tag{9.92}$$

indicating that as the pole gets closer to the unit circle, the output noise increases rapidly to very high values approaching infinity. Thus for high Q realizations, the wordlengths of the registers storing the signal variables should be of longer length to keep the output round-off noise below a prescribed level.

The exact value of the output noise variance can be obtained from Eq. (9.91) by multiplying it with the variance of the A/D conversion noise σ_e^2.

9.5.6 Computation of Output Noise Variance Using MATLAB

The algebraic method of calculating the output noise variance developed in the previous section can be carried out much more easily using MATLAB. To this end, it is more convenient to develop a partial-fraction expansion of the real coefficient transfer function $H(z)$ using the function `residue`, which results in terms of the form A and $B_k/(z - a_k)$ only, where the residues B_k and the poles a_k are either real or complex numbers. Thus, for the variance calculation, only the terms I_1 and I_2 of Table 9.4 are employed. Program

9_4 given below is based on the approach of Section 9.5.5. The input data called by the program are the numerator and denominator coefficients in ascending powers of z^{-1} entered as vectors inside square brackets. The program first determines the partial-fraction expansion of the transfer function and then computes the normalized output round-off noise variance using Eq. (9.86) and the tabulated integrals I_1 and I_2 in Table 9.4. The output data is the desired noise variance.

```
% Program 9_4
%   Computation of the Output Noise Variance Due
%   to Input Quantization of a Digital Filter
%   Based on a Partial-Fraction Approach
%
num = input('Type in the numerator = ');
den = input('Type in the denominator = ');
[r,p,K] = residue(num,den); % partial fraction expansion
R = size(r,1);
R2 = size(K,1);
    if R2 > 1
        disp('Cannot continue...');
        return;
    end

    if R2 == 1
        nvar = K^2;
    else
        nvar = 0;
    end
% Compute output noise variance
for k = 1:R,
    for m = 1:R,
        integral = r(k)*conj(r(m))/(1-p(k)*conj(p(m)));
        nvar = nvar + integral;
    end
end
disp('Output Noise Variance = ');disp(real(nvar))
```

We illustrate its use in the following example.

EXAMPLE 9.5 Determine using Program 9_4 the output noise variance due to the input quantization of a causal fourth-order elliptic lowpass digital filter with a transfer function given by

$$H(z) = \frac{\begin{array}{c}0.06891875 + 0.13808186z^{-1} + 0.18636107z^{-2} \\ + 0.13808186z^{-3} + 0.06891875z^{-4}\end{array}}{\begin{array}{c}1 - 1.30613249z^{-1} + 1.48301305z^{-2} \\ - 0.77709026z^{-3} + 0.2361457z^{-4}\end{array}}. \tag{9.93}$$

The input data are the numerator and the denominator coefficients entered as vectors:

```
num = [0.06891875  0.13808186  0.18636107  0.13808186
       0.06891875]
```

```
den = [1.0    -1.30613249    1.48301305    -0.77709026
       0.2361457].
```

The output generated by the program is

```
Output Noise Variance = 0.40263012511156
```

An alternative, fairly simple and straightforward computer-based method for the computation of the approximate value of the normalized output noise makes use of its equivalent expression given by Eq. (9.82c). For a causal and stable digital filter, the impulse response decays rapidly to zero values, and hence, Eq. (9.82c) can be approximated as a finite sum

$$\sigma_{v,n}^2 = S_L \cong \sum_{n=0}^{L} |h[n]|^2. \qquad (9.94)$$

To determine $\sigma_{v,n}^2$, we can iteratively compute the above partial sum for $L = 1, 2, \ldots$, and stop the computation when the difference $S_L - S_{L-1}$ becomes smaller than a specified value κ, which is typically chosen as 10^{-6}.

The following example illustrates this approach.

EXAMPLE 9.6   We determine again the normalized round-off noise variance of the fourth-order digital filter of Example 9.5 due to input quantization. To this end, we make use of the following MATLAB program.

```
% Program 9_5
% Computation of Approximate Output Round-Off Noise
%
num = input('Numerator coefficients = ');
den = input('Denominator coefficients = ');
x = 1;
order = max(length(num),length(den))-1;
si = [zeros(1,order)];
varnew = 0; k = 1;
while k > 0.000001
    [y,sf] = filter(num,den,x,si);
    si = sf;
    varold = varnew;
    varnew = varnew + abs(y)*abs(y);
    k = varnew - varold;
    x = 0;
end
disp('Output Noise Variance = ');disp(varnew)
```

During execution the program calls for the numerator and the denominator polynomial coefficients in ascending powers of z^{-1}. These data are entered as vectors inside square brackets as in Example 9.5. The program computes the impulse response coefficients of the filter and then evaluates the partial sum using Eq. (9.94) and displays the normalized round-off noise variance when the difference between the two successive partial sums falls below 10^{-6}. For the transfer function of Eq. (9.93), the output is given as indicated below:

```
Output Noise Variance = 0.40254346992755
```

The approximate value computed using the method of Eq. (9.94) is seen to be quite close to the actual value computed in the previous example.

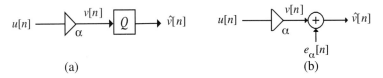

(a)   (b)

Figure 9.19: (a) Product quantization process and (b) its statistical model for the product round-off error analysis.

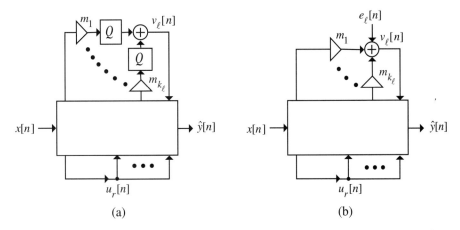

(a)   (b)

Figure 9.20: (a) Representation of a digital filter structure with product round-off before summation, and (b) its statistical model.

9.6 Analysis of Arithmetic Round-Off Errors

As illustrated earlier in Section 8.5.2, in the fixed-point implementation of a digital filter the result of only the multiplication operation is quantized. In this section we develop the tools for the analysis of product round-off errors. Figure 9.19(a) shows the representation of a practical multiplier with the quantizer at its output. Its statistical model is indicated in Figure 9.19(b), which will be used to develop the error analysis methods. Here, the output $v[n]$ of the ideal multiplier is quantized to a value $\hat{v}[n]$, where $\hat{v}[n] = v[n] + e_\alpha[n]$. For analysis purposes we again make assumptions similar to those made for the A/D conversion error analysis, i.e.,

(a) The error sequence $\{e_\alpha[n]\}$ is a sample sequence of a stationary white noise process, with each sample $e_\alpha[n]$ being uniformly distributed over the range of the quantization error.

(b) The quantization error sequence $\{e_\alpha[n]\}$ is uncorrelated with the sequence $\{v[n]\}$, the input sequence $\{x[n]\}$ to the digital filter, and all other quantization noise sources.

Recall that the assumption of $\{e_\alpha[n]\}$ being uncorrelated with $\{v[n]\}$ holds only for rounding and two's-complement truncation. The range of the error sample $e_\alpha[n]$ for these two cases are as given in Table 9.1. The mean and variance of the error sample for rounding are given by Eqs. (9.68) and (9.69), respectively, while those for the two's-complement truncation are given by Eqs. (9.70) and (9.71), respectively.

Using the above model for each multiplier, the representation of a digital filter to determine the effect of product quantizations at the output of the digital filter is as indicated in Figure 9.20(a), which explicitly shows the ℓth adder with an output $v_\ell[n]$ summing the quantized outputs of the k_ℓ multipliers at its input. This figure also shows the internal rth branch node associated with the signal variable $u_r[n]$ that needs to be scaled to prevent overflows at these nodes. These nodes are typically the inputs to the multipliers, as

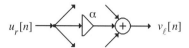

Figure 9.21: A typical multiplier branch with input as a branch node and output feeding into an adder.

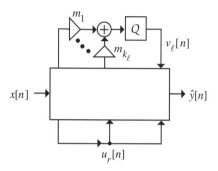

Figure 9.22: Digital filter structure with product round-off after summation.

indicated in Figure 9.21. In digital filters employing two's-complement arithmetic, these nodes are outputs of adders forming sums of products, since here the sums will still have the correct values even though some of the products and/or the partial sums overflow (Problem 8.55). If we assume the error sources are statistically independent of each other, then each error source develops a round-off noise at the output of the digital filter. An equivalent statistical model is then as shown in Figure 9.20(b).

Let the impulse response from the digital filter input to the rth branch node be denoted as $f_r[n]$ and the impulse response from the input of the ℓth adder to the digital filter output be denoted as $g_\ell[n]$, with their corresponding z-transforms denoted by $F_r(z)$ and $G_\ell(z)$, respectively. $F_r(z)$ is called the *scaling transfer function* and plays a role in the dynamic range scaling schemes employed in fixed-point digital filter structures to be discussed in Section 9.7. $G_\ell(z)$ is called the *noise transfer function*, which is used in computing the noise power at the filter output due to product round-off as described next.

If σ_0^2 denotes the variance of each individual noise source at the output of each multiplier, the variance of $e_\ell[n]$ in Figure 9.20(b) is simply $k_\ell \sigma_0^2$ since we have assumed each noise source to be statistically independent of all others. The variance of the output noise caused by $e_\ell[n]$ is then given by

$$\sigma_0^2 \left[k_\ell \left(\frac{1}{2\pi j} \oint_C G_\ell(z) G_\ell(z^{-1}) z^{-1} dz \right) \right] = \sigma_0^2 \left[k_\ell \left(\frac{1}{2\pi} \int_{-\pi}^{\pi} \left| G_\ell(e^{j\omega}) \right|^2 d\omega \right) \right]. \quad (9.95)$$

If there are L such adders in the digital filter structure, the total output noise power due to all product round-offs is given by

$$\sigma_\gamma^2 = \sigma_0^2 \sum_{\ell=1}^{L} k_\ell \left(\frac{1}{2\pi j} \oint_C G_\ell(z) G_\ell(z^{-1}) z^{-1} dz \right). \quad (9.96)$$

In many hardware implementation schemes such as those employing DSP chips, the multiplication operation is carried out as a multiply-add operation with the result stored in a double-precision register. In such cases, if the signal variable being generated is by a sum of product operations, the quantization operation can be carried out after all the multiply-add operations have been completed, reducing the number of quantization error sources to one for each such sum of product operations, as indicated in Figure 9.22. In such a case, the statistical model of Figure 9.20(b) can still be used except the variance of the noise source $e_\ell[n]$ is now σ_0^2, thus resulting in considerably lower noise at the digital filter output.

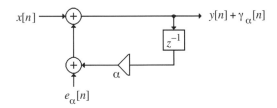

Figure 9.23: Model for the arithmetic round-off error analysis of Figure 9.1.

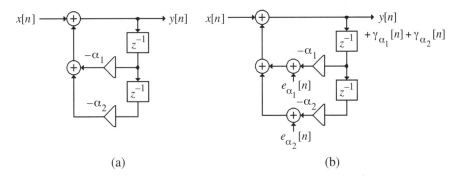

(a)   (b)

Figure 9.24: (a) A second-order digital filter structure, and (b) its model for product round-off error analysis.

We illustrate the product round-off error analysis method for two simple digital filter structures.

EXAMPLE 9.7   Consider first the filter of Figure 9.1. The model for the arithmetic round-off error analysis of this structure is as given in Figure 9.23, where the output is now given by a sum of two sequences, the desired output $y[n]$ generated by the input sequence $x[n]$ and the round-off noise $\gamma_\alpha[n]$ generated by the error sequence $e_\alpha[n]$. It follows from the figure that the noise transfer function $G_\alpha(z)$ is the same as the filter transfer function $H(z)$ as given by Eq. (9.2),

$$G_\alpha(z) = \frac{z}{z - \alpha},\tag{9.97}$$

and thus the output noise variance due to product round-off is the same as that due to input quantization given in Eq. (9.91) and is given by

$$\sigma_\gamma^2 = \frac{\sigma_0^2}{1 - \alpha^2},\tag{9.98}$$

where σ_0^2 is the variance of the error sequence $e_\alpha[n]$. The quantity $\sigma_\gamma^2/\sigma_0^2$ is called the *noise gain* or the *normalized round-off noise variance* denoted as $\sigma_{\gamma,n}^2$.

EXAMPLE 9.8   Consider next the causal second-order digital filter structure of Figure 9.24(a). Its transfer function is given by

$$H(z) = \frac{Y(z)}{X(z)} = \frac{z^2}{z^2 + \alpha_1 z + \alpha_2}.\tag{9.99}$$

A model for the analysis of the product round-off error is as indicated in Figure 9.24(b), where we have assumed that each product is rounded first before addition. It follows from this figure that the two noise transfer functions $G_{\alpha_1}(z)$ and $G_{\alpha_2}(z)$ are the same as the transfer function $H(z)$ of the digital filter given by Eq. (9.99). Thus, the total normalized variance $\sigma_{\gamma,n}^2$ of the output noise generated by the error sources $e_{\alpha_1}[n]$ and $e_{\alpha_2}[n]$ is given by

$$\sigma_{\gamma,n}^2 = 2\left[\frac{1}{2\pi j}\oint_C H(z)H(z^{-1})z^{-1}\,dz\right], \qquad (9.100)$$

where $H(z)$ is as given in Eq. (9.99).

Now a direct partial-fraction expansion of $H(z)$ is given by

$$H(z) = 1 + \frac{-\alpha_1 z - \alpha_2}{z^2 + \alpha_1 z + \alpha_2}. \qquad (9.101)$$

Applying the technique of Section 9.5.5, we arrive at the expression for the normalized output noise variance,

$$\sigma_{\gamma,n}^2 = 2\left[1 + \frac{(\alpha_1^2 + \alpha_2^2)(1 - \alpha_2^2) - 2(\alpha_1\alpha_2 - \alpha_1\alpha_2^2)\alpha_1}{(1 - \alpha_2^2)^2 + 2\alpha_2\alpha_1^2 - (1 + \alpha_2^2)\alpha_1^2}\right], \qquad (9.102)$$

which, after some algebra, simplifies to

$$\sigma_{\gamma,n}^2 = 2\left(\frac{1 + \alpha_2}{1 - \alpha_2}\right)\left(\frac{1}{1 + \alpha_2^2 + 2\alpha_2 - \alpha_1^2}\right). \qquad (9.103)$$

It is instructive to express the multiplier coefficients in terms of the pole positions $re^{\pm j\theta}$ in the above expression. In this case, $\alpha_1 = -2r\cos\theta$ and $\alpha_2 = r^2$. Substituting these values in Eq. (9.103), we arrive after some simple manipulations at

$$\sigma_{\gamma,n}^2 = 2\left(\frac{1 + r^2}{1 - r^2}\right)\left(\frac{1}{1 + r^4 - 2r^2\cos 2\theta}\right). \qquad (9.104)$$

If the pole is close to the unit circle, we can replace r in Eq. (9.104) with $1 - \varepsilon$, where ε is a very small positive number. This leads to

$$\sigma_{\gamma,n}^2 \cong \frac{1}{2\sin^2\theta}\left(\frac{1 - \varepsilon}{\varepsilon}\right)\left(\frac{1}{1 - 2\varepsilon}\right). \qquad (9.105)$$

It can be seen from the above that as the poles get closer to the unit circle, $\varepsilon \to 0$, and as a result, the total output noise variance increases rapidly.

9.7 Dynamic Range Scaling

In a digital filter implemented using fixed-point arithmetic, overflow may occur at certain internal nodes such as inputs to multipliers and/or the adder outputs that may lead to large amplitude oscillations at the filter output causing unsatisfactory operation as discussed in Section 9.11.2. The probability of overflow can be minimized significantly by properly scaling the internal signal levels with the aid of scaling multipliers inserted at appropriate points in the digital filter structure. In many cases, most of these scaling multipliers can be absorbed with existing multipliers in the structure, thus reducing the total number needed to implement the scaled filter.

To understand the basic concepts involved in scaling, consider again the digital filter structure of Figure 9.20 showing explicitly the rth node variable $u_r[n]$ that needs to be scaled. We assume that all fixed-point numbers are represented as binary fractions and the input sequence of the filter is bounded by unity, i.e.,

$$|x[n]| \leq 1, \qquad \text{for all values of } n. \qquad (9.106)$$

The objective of scaling is to ensure that

$$|u_r[n]| \leq 1, \qquad \text{for all } r \text{ and for all values of } n. \qquad (9.107)$$

We now derive three different conditions to ensure that $u_r[n]$ satisfies the above bound.

9.7.1 An Absolute Bound

The inverse z-transform of the scaling transfer function $F_r(z)$ is the impulse response $f_r[n]$ from the filter input to the rth node. Therefore, $u_r[n]$ can be expressed as the linear convolution of $f_r[n]$ and the input $x[n]$:

$$u_r[n] = \sum_{k=-\infty}^{\infty} f_r[k]x[n-k]. \tag{9.108}$$

From Eq. (9.108) it follows then that

$$|u_r[n]| = \left| \sum_{k=-\infty}^{\infty} f_r[k]x[n-k] \right| \leq \sum_{k=-\infty}^{\infty} |f_r[k]|.$$

Thus, Eq. (9.107) is satisfied if

$$\sum_{k=-\infty}^{\infty} |f_r[k]| \leq 1, \qquad \text{for all } r. \tag{9.109}$$

The above condition is both necessary and sufficient to guarantee no overflow [Jac70a].[3] If it is not satisfied in the unscaled realization, we scale the input signal with a multiplier K of value

$$K = \frac{1}{\max\limits_{r} \sum_{k=-\infty}^{\infty} |f_r[k]|}. \tag{9.110}$$

It should be noted that the above scaling rule is based on a worst case bound and does not fully utilize the dynamic range of all adder output registers and, as a result, reduces the SNR significantly. Even though it is difficult to compute analytically the value of K, an approximate value can be computed on a computer using an approach similar to that given by Eq. (9.94).

More practical and easy to use scaling rules can be derived in the frequency domain if some information about the input signals is known a priori [Jac70a]. To this end, we define the $\mathcal{L}_p$-norm ($p \geq 1$) of a Fourier transform $F(e^{j\omega})$ as

$$\|F\|_p \overset{\Delta}{=} \left(\frac{1}{2\pi} \int_{-\pi}^{\pi} \left| F(e^{j\omega}) \right|^p d\omega \right)^{1/p}. \tag{9.111}$$

From the above definition it follows that the $\mathcal{L}_2$-norm, $\|F\|_2$, is the root-mean-squared (rms) value of $F(e^{j\omega})$, and the $\mathcal{L}_1$-norm, $\|F\|_1$, is the mean absolute value of $F(e^{j\omega})$ over ω. Moreover $\lim_{p\to\infty} \|F\|_p$ exists for a continuous $F(e^{j\omega})$ and is given by the peak absolute value:

$$\|F\|_\infty = \max_{-\pi \leq \omega \leq \pi} \left| F(e^{j\omega}) \right|. \tag{9.112}$$

The bounds to be derived assume that the input $x[n]$ to the digital filter is a deterministic signal with a Fourier transform $X(e^{j\omega})$.

9.7.2 $\mathcal{L}_\infty$-Bound

Now from Eq. (9.108) we get

$$U_r(e^{j\omega}) = F_r(e^{j\omega})X(e^{j\omega}), \tag{9.113}$$

---

[3]It also follows from the BIBO stability requirements discussed in Section 2.5.3 and Eq. (2.73).

where $U_r(e^{j\omega})$ and $F_r(e^{j\omega})$ are the Fourier transforms of $u_r[n]$ and $f_r[n]$, respectively. An inverse Fourier transform of Eq. (9.113) yields

$$u_r[n] = \frac{1}{2\pi} \int_{-\pi}^{\pi} F_r(e^{j\omega}) X(e^{j\omega}) e^{j\omega n} d\omega. \tag{9.114}$$

As a result,

$$\begin{aligned} |u_r[n]| &\leq \frac{1}{2\pi} \int_{-\pi}^{\pi} \left| F_r(e^{j\omega}) \right| \cdot \left| X(e^{j\omega}) \right| d\omega \\ &\leq \left\| F_r(e^{j\omega}) \right\|_{\infty} \left[\frac{1}{2\pi} \int_{-\pi}^{\pi} \left| X(e^{j\omega}) \right| d\omega \right] \\ &\leq \| F_r \|_{\infty} \cdot \| X \|_1 . \end{aligned} \tag{9.115}$$

If $\| X \|_1 \leq 1$, then the dynamic range constraint of Eq. (9.107) is satisfied if

$$\left\| F_r(e^{j\omega}) \right\|_{\infty} \leq 1. \tag{9.116}$$

Thus, if the mean absolute value of the input spectrum is bounded by unity, then there will be no adder overflow if the peak gains from the filter input to all adder outputs are scaled satisfying the bound of Eq. (9.116). In general, this scaling rule is rarely used since, with most input signals encountered in practice, $\| X \|_1 \leq 1$ does not hold!

9.7.3 $\mathcal{L}_2$-Bound

Applying the Schwartz inequality to Eq. (9.114), we arrive at [Jac70a]

$$|u_r[n]|^2 \leq \left(\frac{1}{2\pi} \int_{-\pi}^{\pi} \left| F_r(e^{j\omega}) \right|^2 d\omega \right) \cdot \left(\frac{1}{2\pi} \int_{-\pi}^{\pi} \left| X(e^{j\omega}) \right|^2 d\omega \right) \tag{9.117}$$

or equivalently,

$$|u_r[n]| \leq \| F_r \|_2 \cdot \| X \|_2 . \tag{9.118}$$

Thus, if the input to the filter has finite energy bounded by unity, i.e., $\| X \|_2 \leq 1$, then the adder overflow can be prevented by scaling the filter such that the rms values of the transfer functions from the input to all adder outputs are bounded by unity:

$$\| F_r \|_2 \leq 1, \qquad r = 1, 2, \ldots, R. \tag{9.119}$$

9.7.4 A General Scaling Rule

A more general scaling rule obtained using Holder's inequality is given by [Jac70a]

$$|u_r[n]| \leq \| F_r \|_p \cdot \| X \|_q , \tag{9.120}$$

for all $p, q \geq 1$ satisfying $(1/p) + (1/q) = 1$. Note that for the $\mathcal{L}_{\infty}$-bound of Eq. (9.115), $p = \infty$ and $q = 1$, and for the $\mathcal{L}_2$-bound of Eq. (9.118), $p = q = 2$. Another useful scaling rule, $\mathcal{L}_1$-bound, is obtained for $p = 1$ and $q = \infty$.

After scaling, the scaling transfer functions become $\left\| \bar{F}_r \right\|_p$ and the scaling constants should be chosen such that

$$\left\| \bar{F}_r \right\|_p \leq 1, \qquad r = 1, 2, \ldots, R. \tag{9.121}$$

In many structures, all scaling multipliers can be absorbed into the existing feedforward multipliers without any increase in the total number of multipliers and, hence, noise sources. However, in some cases, the scaling process may introduce additional multipliers in the system. If all scaling multipliers are regular b-bit units, then Eq. (9.121) can be satisfied with an equality sign, providing a full utilization of the dynamic range of each adder output and, thus, yielding a maximum SNR. An attractive option from a hardware point of view, and preferred in cases where scaling introduces new multipliers, is to make as many scaling multiplier coefficients as possible in the scaled structure take a value that is a power of 2 [Jac70a]. In which case, these multipliers can be implemented simply by a shift operation. Now, the norm of the scaling transfer function satisfies

$$\frac{1}{2} < \left\| \bar{F}_r \right\|_p \leq 1 \tag{9.122}$$

with a slight decrease in the SNR.

It should be pointed out here that the output round-off noise should be always computed after the digital filter structure has been scaled. For the scaled structure, the expression for the output round-off noise of Eq. (9.95) thus changes to

$$\sigma_\gamma^2 = \sigma_0^2 \sum_{\ell=1}^{L} \bar{k}_\ell \left(\frac{1}{2\pi j} \oint_C \bar{G}_\ell(z) \bar{G}_\ell(z^{-1}) z^{-1} \, dz \right), \tag{9.123}$$

where $\bar{k}_\ell$ is the total number of multipliers feeding the ℓth adder with $\bar{k}_\ell \geq k_\ell$ and $\bar{G}_\ell(z)$ is the modified noise transfer function from the input of the ℓth adder to the filter output.

We illustrate next the application of the above method in the scaling of a cascade realization of an IIR transfer function [Jac70b].

9.7.5 Scaling of a Cascade Form IIR Digital Filter Structure

Consider the unscaled structure of Figure 9.25 consisting of a cascade of R second-order IIR sections realized in direct form II. Its transfer function is given by

$$H(z) = K \prod_{i=1}^{R} H_i(z), \tag{9.124}$$

where

$$H_i(z) = \frac{B_i(z)}{A_i(z)} = \frac{1 + b_{1i} z^{-1} + b_{2i} z^{-2}}{1 + a_{1i} z^{-1} + a_{2i} z^{-2}}. \tag{9.125}$$

The branch nodes to be scaled are marked by $(*)$ in the figure, which are seen to be the inputs to the multipliers in each second-order section. The transfer functions from the input to these branch nodes are the scaling transfer functions, which are given by

$$F_r(z) = \frac{K}{A_r(z)} \prod_{\ell=1}^{r-1} H_\ell(z), \qquad r = 1, 2, \ldots, R. \tag{9.126}$$

The scaled version of the above structure is shown in Figure 9.26 with new values for the feedforward multipliers. Note that the scaling process has introduced a new multiplier $\bar{b}_{0\ell}$ in each second-order section. If the zeros of the transfer function are on the unit circle, as is usually the case in practice, then $b_{2\ell} = \pm 1$ in which case we can choose $\bar{b}_{0\ell} = \bar{b}_{2\ell} = 2^{-\beta}$ to reduce the total number of actual multiplications in the final realization.

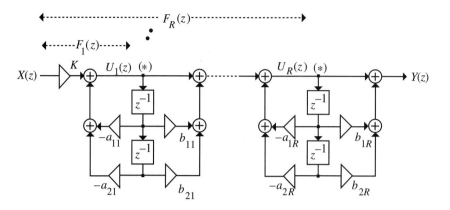

Figure 9.25: An unscaled cascade realization of second-order IIR sections.

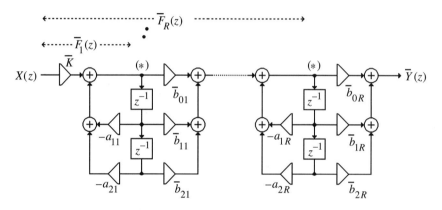

Figure 9.26: The scaled cascade realization.

It can be seen from Figure 9.26 that

$$\bar{F}_r(z) = \frac{\bar{K}}{A_r(z)} \prod_{\ell=1}^{r-1} \bar{H}_\ell(z), \tag{9.127a}$$

$$\bar{H}(z) = \bar{K} \prod_{\ell=1}^{R} \bar{H}_\ell(z), \tag{9.127b}$$

where

$$\bar{H}_\ell(z) = \frac{\bar{b}_{0\ell} + \bar{b}_{1\ell}z^{-1} + \bar{b}_{2\ell}z^{-2}}{1 + a_{1\ell}z^{-1} + a_{2\ell}z^{-2}}. \tag{9.128}$$

Denote

$$\|F_r\|_p \overset{\Delta}{=} \alpha_r, \qquad r = 1, 2, \ldots, R, \tag{9.129a}$$

$$\|H\|_p \overset{\Delta}{=} \alpha_{R+1}, \tag{9.129b}$$

and choose the scaling constants as

$$\bar{K} = \beta_0 K; \qquad \bar{b}_{\ell r} = \beta_r b_{\ell r}, \qquad \ell = 1, 2, 3; \quad r = 1, 2, \ldots, R. \tag{9.130}$$

Now, it follows from Eqs. (9.130), (9.127a), and (9.127b) that

$$\bar{F}_r(z) = \left(\prod_{i=0}^{r-1} \beta_i \right) F_r(z), \qquad r = 1, 2, \ldots, R, \tag{9.131a}$$

$$\bar{H}(z) = \left(\prod_{i=0}^{R} \beta_i \right) H(z). \tag{9.131b}$$

After scaling we require

$$\|\bar{F}_r\|_p = \left(\prod_{i=0}^{r-1} \beta_i \right) \|F_r\|_p = \alpha_r \left(\prod_{i=0}^{r-1} \beta_i \right) = 1, \quad r = 1, 2, \ldots, R, \tag{9.132a}$$

$$\|\bar{H}\|_p = \left(\prod_{i=0}^{R} \beta_i \right) \|H\|_p = \alpha_{R+1} \left(\prod_{i=0}^{R} \beta_i \right) = 1. \tag{9.132b}$$

Solving the above for the scaling constants, we arrive at

$$\beta_0 = \frac{1}{\alpha_1}, \tag{9.133a}$$

$$\beta_r = \frac{\alpha_r}{\alpha_{r+1}}, \qquad r = 1, 2, \ldots, R. \tag{9.133b}$$

9.7.6 Dynamic Range Scaling Using MATLAB

The dynamic range scaling using the $\mathcal{L}_2$-norm rule can be easily performed using MATLAB by actual simulation of the digital filter structure. If we denote the impulse response from the input of the digital filter to the output of the rth branch node as $\{f_r[n]\}$ and assume that the branch nodes have been ordered in accordance to their precedence relations with increasing i (see Section 8.1.2), then we can compute the $\mathcal{L}_2$-norm $\|F_1\|_2$ of $\{f_1[n]\}$ first, and divide the input by a multiplier $k_1 = \|F_1\|_2$. Next, we compute the $\mathcal{L}_2$-norm $\|F_2\|_2$ of $\{f_2[n]\}$ and scale the multipliers feeding into the second adder by dividing with a constant $k_2 = \|F_2\|_2$. This process is continued until the output node has been scaled to yield an $\mathcal{L}_2$-norm of unity.

We illustrate the method by means of the following example.

EXAMPLE 9.9   Consider the cascade realization of the third-order lowpass digital filter of Eq. (8.19). The transfer functions of the two sections in the cascade are given by Eqs. (8.20a) and (8.20b) repeated below for convenience:

$$H_1(z) = \frac{0.0662272(1 + z^{-1})}{1 - 0.2593284z^{-1}}, \tag{9.134a}$$

$$H_2(z) = \frac{1 + 2z^{-1} + z^{-2}}{1 - 0.6762858z^{-1} + 0.3917468z^{-2}}. \tag{9.134b}$$

Their realization in the direct form II structure is as shown in Figure 9.27. The MATLAB program simulating this structure is given by Program 9_6 below. The program is first run with all scaling constants set to unity, i.e., $k1 = k2 = k3 = 1$. In the line indicating the approximate $\mathcal{L}_2$-norm calculation, the output variable is chosen as $y1$. The program computes the square of the $\mathcal{L}_2$-norm of the impulse response at node $y1$ as 1.07210002757252, which is used to set $k1 = \sqrt{1.07210002757252}$ with other scaling constants still set to unity. A second run of the program shows the $\mathcal{L}_2$-norm of the impulse response at node $y1$ as 1.00000000000000, verifying the success of scaling the input. In the second step, in the line indicating the approximate $\mathcal{L}_2$-norm calculation, the output variable is chosen as $y2$. The program then yields the square of the $\mathcal{L}_2$-norm of the impulse response at node $y2$ as 0.02679820762398, which is used to set $k2 = \sqrt{0.02679820762398}$ with $k3$ still set to unity. This process is repeated for node $y3$, resulting in $k3 = \sqrt{11.96975400608943}$ with the final value of the $\mathcal{L}_2$-norm of the impulse response at node $y3$ as 0.99999683131540. The program listing shown below is after all adder outputs have been scaled.

```
% Program 9_6
% Illustration of Scaling & Round-Off Noise Calculation
%
format long
k1 = sqrt(1.07210002757252);
k2 = sqrt(0.02679820762398);
k3 = sqrt(11.96975400608943);
x1 = 1/k1;
si1 = 0;
si2 = [0 0];
b1 = 1;
varnew = 0; k = 1;
while k > 0.000001
    y1 = 0.2593284*si1 + x1;
    x2 = (0.0662272/k2)*(y1 + si1);
    si1 = y1;
    y2 = 0.6762858*si2(1) - 0.3917468*si2(2) + x2;
    y3 = (y2 + 2*si2(1) + si2(2))/k3;
    si2(2) = si2(1); si2(1) = y2;
    varold = varnew;
    varnew = varnew + abs(y3)*abs(y3);
    % Compute approximate L2 norm square
    k = varnew - varold;
    x1 = 0;
end
disp('L2 norm square = ');disp(varnew);
```

The above program can be easily modified to compute the product round-off noise variance at the output of the scaled structure. To this end, we set the digital filter input to zero and apply an impulse at the input of the first adder. This is equivalent to setting $x1 = 1$ in Program 9_6. The normalized output noise variance due to a single error source is computed as 1.07209663042567. Next, we apply an impulse at the input of the second adder with the digital filter input set to zero. This is achieved by replacing $x2$ in the calculation of $y2$ as $x1$. The program yields the normalized output noise variance due to a single error source at the second adder as 1.26109014071707. The total normalized output round-off noise, assuming all products to be quantized before addition, is then equal to

$$2 \times 1.07209663042567 + 4 \times 1.26109014071707 + 3 = 10.18855382371962.$$

On the other hand, if we assume quantization after addition of all products at each adder, the total normalized output round-off noise becomes

$$1.07209663042567 + 1.26109014071707 + 1 = 3.33318677114274.$$

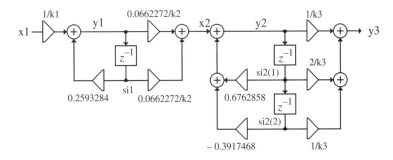

Figure 9.27: Cascade realization of the transfer functions of Eqs. (9.134a) in direct form II.

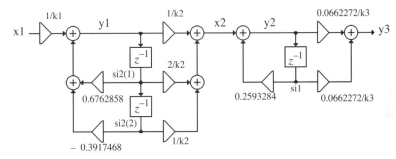

Figure 9.28: Alternate cascade realization of the transfer functions of Eqs. (9.134b) in direct form II.

EXAMPLE 9.10   We repeat the problem of Example 9.9 by interchanging the locations of the two sections $G_1(z)$ and $G_2(z)$ in the cascade connection, as indicated in Figure 9.28. By modifying Program 9.6, we arrive at the following values for the scaling constants:

$$k1 = \sqrt{1.54643736440090}, \quad k2 = \sqrt{14.44567893169408}, \quad k3 = \sqrt{0.01539326750782}.$$

The total normalized output round-off noise assuming all products to be quantized before addition in this case is equal to

$$3 \times 1.54652209477041 + 4 \times 0.76938948942268 + 2 = 9.717124200195,$$

whereas the total normalized output round-off noise assuming quantization after addition is given by

$$1.5465220947704 + 0.76938948942268 + 1 = 3.31591158419309.$$

Note from above that the second cascade connection of Figure 9.28 yields a slightly lower round-off noise than the first of Figure 9.27.

9.7.7 Optimum Ordering and Pole-Zero Pairing of the Cascade Form IIR Digital Filter

As indicated in Section 6.4.2, and illustrated in Figures 6.15 and 6.16, there are many possible cascade realizations of a higher-order IIR transfer function obtained by different pole-zero pairings and ordering. In fact, for a cascade of R second-order sections, there are $(R!)^2$ different possible realizations (Problem 9.30). Each one of these realizations will have different output noise power, as illustrated in the previous two examples, and hence, it is of interest to determine the cascade realization with the lowest output noise power.

A fairly simple heuristic set of rules for determining an optimum pole-zero pairing and ordering of sections in a cascade realization has been advanced by Jackson [Jac70b]. To understand the reasoning behind these rules, we first develop the expression for the output noise variance due to product round-off in a cascade IIR structure implemented in fixed-point arithmetic. To this end we make use of the noise model of the scaled cascade structure of Figure 9.26 as shown in Figure 9.29.

It follows from Figure 9.29, and Eqs. (9.128) and (9.130) that the scaled noise transfer functions are given by

$$\bar{G}_\ell(z) = \prod_{i=\ell}^{R} \bar{H}_i(z) = \left(\prod_{i=\ell}^{R} \beta_i \right) G_\ell(z), \qquad \ell = 1, 2, \ldots, R, \tag{9.135a}$$

$$\bar{G}_{R+1}(z) = 1, \tag{9.135b}$$

where $\bar{H}_i(z)$ is as given in Eq. (9.128) and the unscaled noise transfer function is given by

$$G_\ell(z) = \prod_{i=\ell}^{R} H_i(z). \tag{9.136}$$

Hence, the output noise power spectrum due to product round-off is given by

$$P_{\gamma\gamma}(\omega) = \sigma_0^2 \left[\sum_{\ell=1}^{R+1} k_\ell \left| \bar{G}_\ell(e^{j\omega}) \right|^2 \right], \tag{9.137}$$

and the output noise variance is thus

$$\sigma_\gamma^2 = \sigma_0^2 \left[\sum_{\ell=1}^{R+1} k_\ell \left(\frac{1}{2\pi} \int_{-\pi}^{\pi} \left| \bar{G}_\ell(e^{j\omega}) \right|^2 d\omega \right) \right]$$

$$= \sigma_0^2 \left[\sum_{\ell=1}^{R+1} k_\ell \left\| \bar{G}_\ell \right\|_2^2 \right], \tag{9.138}$$

where we have used the fact that the quantity inside the parentheses in the middle expression is the square of the $\mathcal{L}_2$-norm of $\bar{G}_\ell(e^{j\omega})$ as defined in Eq. (9.111). In Eqs. (9.137) and (9.138), k_ℓ is the total number of multipliers connected to the ℓth adder. If all products are rounded before summation, then

$$k_1 = k_{R+1} = 3, \tag{9.139a}$$

$$k_\ell = 5, \qquad \text{for } \ell = 2, 3, \ldots, R. \tag{9.139b}$$

On the other hand, if all products are rounded after summation, then

$$k_\ell = 1, \qquad \text{for } \ell = 1, 2, \ldots, R+1. \tag{9.140}$$

From Eqs. (9.133a) and (9.133b),

$$\prod_{i=\ell}^{R} \beta_i = \frac{\alpha_\ell}{\alpha_{R+1}} = \frac{\|F_\ell\|_p}{\|H\|_p}. \tag{9.141}$$

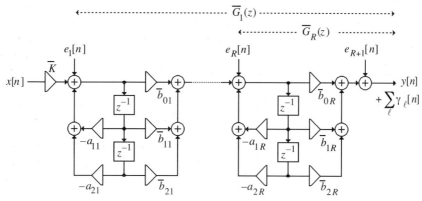

Figure 9.29: The noise model of the scaled cascade realization of Figure 9.25.

Substituting Eq. (9.141) in Eq. (9.137), we obtain for the output noise power spectrum

$$P_{\gamma\gamma}(\omega) = \frac{\sigma_0^2}{\|H\|_p^2}\left[k_{R+1}\|H\|_p^2 + \sum_{\ell=1}^{R} k_\ell \|F_\ell\|_p^2 \left|G_\ell(e^{j\omega})\right|^2\right]. \qquad (9.142)$$

Correspondingly, the output noise variance is given by

$$\sigma_\gamma^2 = \frac{\sigma_0^2}{\|H\|_p^2}\left[k_{R+1}\|H\|_p^2 + \sum_{\ell=1}^{R} k_\ell \|F_\ell\|_p^2 \|G_\ell\|_2^2\right]. \qquad (9.143)$$

Now, the scaling transfer function $F_\ell(z)$ contains a product of section transfer functions, $H_i(z)$, $i = 1, 2, \ldots, \ell-1$, whereas the noise transfer function $G_\ell(z)$ contains the product of section transfer functions, $H_i(z)$, $i = \ell, \ell+1, \ldots, R$. Thus, every term in the sum in Eqs. (9.142) and (9.143) includes the transfer function of all R sections in the cascade realization. This implies that to minimize the output noise power, the norms of $H_i(z)$ should be minimized for all values of i by appropriately pairing the poles and zeros. To this end, Jackson has proposed the following rule [Jac70b]:

Pole-Zero Pairing Rule. First, the complex pole-pair closest to the unit circle should be paired with the nearest complex zero-pair. Next, the complex pole-pair that is closest to the previous set of poles should be matched with its nearest complex zero-pair. This process should be continued until all poles and zeros have been paired.

The above type of pairing of poles and zeros is likely to lower the peak gain of the section characterized by the paired poles and zeros. Lowering of the peak gain in turn reduces the possibility of overflow and attenuates the round-off noise. The pole-zero pairing rule is illustrated in Figure 9.30 for a fifth-order elliptic lowpass IIR digital filter with passband edge at 0.3π, passband ripple of 0.5 dB, and minimum stopband attenuation of 40 dB.

Once the appropriate pole-zero pairings have been made, the next question that needs to be answered is how to order the sections in the cascade structure [Jac70b]. A section in the front part of the cascade has its transfer function $H_i(z)$ appearing more frequently in the scaling transfer function expressions in Eqs. (9.142) and (9.143), whereas a section near the output end of the cascade has its transfer function $H_i(z)$ appearing more frequently in the noise transfer function expressions. The best location for $H_i(z)$ obviously depends on the type of norms being applied to the scaling and noise transfer functions.

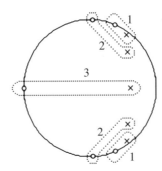

Figure 9.30: Illustration of the pole-zero pairing and ordering rules.

A careful examination of Eqs. (9.142) and (9.143) reveals that if the $\mathcal{L}_2$-scaling is used, then ordering of paired sections does not influence too much the output noise power since all norms in the expressions are $\mathcal{L}_2$-norms. This fact is evident from the results of Examples 9.9 and 9.10 where the total normalized output round-off noise power of the cascade realization of Figure 9.27 is seen to be quite close to that of the cascade realization of Figure 9.28. If, however, an $\mathcal{L}_\infty$-scaling is being employed, the sections with the poles closest to the unit circle exhibit a peaking magnitude response and should be placed closer to the output end. The ordering rule in this case is therefore to place the least-peaked section to the most-peaked section starting at the input end. On the other hand, the ordering scheme is exactly opposite if the objective is to minimize the peak noise $\| P_{\gamma\gamma}(\omega) \|_\infty$ and an $\mathcal{L}_2$-scaling is used. However, the ordering has no effect on the peak noise if an $\mathcal{L}_\infty$-scaling is used.

The MATLAB *Signal Processing Toolbox* includes the function zp2sos that can be employed to determine the optimum pole-zero pairing and ordering according to the above discussed rules. The basic form of this function is

```
sos = zp2sos(z,p,k)
```

This function generates a matrix sos containing the coefficients of each second-order section of the equivalent transfer function $H(z)$ determined from the specified zero-pole form. sos is an $L \times 6$ matrix of the form

$$
\text{sos} = \begin{bmatrix}
b_{01} & b_{11} & b_{21} & a_{01} & a_{11} & a_{21} \\
b_{02} & b_{12} & b_{22} & a_{02} & a_{12} & a_{22} \\
\vdots & \vdots & \vdots & \vdots & \vdots & \vdots \\
b_{0L} & b_{1L} & b_{2L} & a_{0L} & a_{1L} & a_{2L}
\end{bmatrix},
$$

whose ith row contains the coefficients, $\{b_{i\ell}\}$ and $\{a_{i\ell}\}$, of the numerator and denominator polynomials of the ith second-order section with L denoting the total number of sections in the cascade. The form of the overall transfer function is thus given by

$$
H(z) = \prod_{i=1}^{L} H_i(z) = \prod_{i=1}^{L} \frac{b_{0i} + b_{1i}z^{-1} + b_{2i}z^{-2}}{a_{0i} + a_{1i}z^{-1} + a_{2i}z^{-2}}.
$$

The rows are ordered so that the first row ($i = 1$) contains the coefficients of the pole pair farthest from the unit circle and its nearest zero pair. If a reverse ordering is desired with the first row containing the coefficients of the pole pair closest to the unit circle and its nearest zero pair, then the function to be used is

```
sos = zp2sos(z,p,k,'down')
```

EXAMPLE 9.11   We develop the optimum pairing and ordering of the fifth-order elliptic lowpass filter whose pole-zero plot is shown in Figure 9.30. The filter specifications are $\alpha_p = 0.5$ dB, $\alpha_s = 40$ dB, and $\omega_p = 0.3\pi$.

To this end we first determine the zeros, poles, and the gain constant of the desired lowpass filter using the command [z,p,k] = ellip(5,0.5,40,0.3) and then employ the command sos = zp2sos(z,p,k) to determine the coefficients of the second-order sections. The matrix sos obtained is given by

```
sos =
  Columns 1 through 3

    0.03000180248135      0.03000180248135    0
    1.00000000000000     -0.08610163536316    1.00000000000000
    1.00000000000000     -0.76382924065789    1.00000000000000

  Columns 4 through 6

    1.00000000000000     -0.61356860505629    0
    1.00000000000000     -1.15286286138340    0.61164840151481
    1.00000000000000     -1.09833407780228    0.89907598271183
```

The transfer functions of the sections are thus given by

$$H_1(z) = \left[\frac{0.03 + 0.03z^{-1}}{1 - 0.6135686z^{-1}} \right],$$

$$H_2(z) = \left[\frac{1 - 0.0861z^{-1} + z^{-2}}{1 - 1.152863z^{-1} + 0.6116484z^{-2}} \right],$$

$$H_3(z) = \left[\frac{1 - 0.76382924z^{-1} + z^{-2}}{1 - 1.0983341z^{-1} + 0.899076z^{-2}} \right].$$

As indicated above, the first section in the cascade is $H_1(z)$, followed by $H_2(z)$, and the last section is $H_3(z)$. This pairing is identical to that displayed in Figure 9.30 but the ordering is reversed. This can be easily verified by using the function [z,p,k] = sos2zp(sos), which computes the zeros and poles of each individual second-order section defined by the matrix sos.

Another MATLAB function that is available for the generation of the numerator and the denominator polynomials of the transfer function from the coefficients of the matrix sos is

```
[num, den] = sos2tf(sos)
```

9.8 Signal-to-Noise Ratio in Low-Order IIR Filters

The output round-off noise variances of unscaled digital filters do not provide a realistic picture of the performances of these structures in practice since introduction of scaling multipliers can increase the number of error sources and the gain for the scaling transfer functions. It is therefore important to scale the digital filter structure first before its round-off noise performance is analyzed. In most applications, the round-off noise variance by itself is not sufficient, and a more meaningful result is obtained by computing instead the expression for the signal to round-off noise ratio (SNR) for performance evaluation. We illustrate such computation for the first-order and the second-order IIR structures considered earlier in Examples 9.7 and 9.8, respectively [Vai87c]. Most conclusions derived from the detailed analysis of these simple structures outlined here are also valid in the case of more complex structures. Moreover, the methods followed here can be easily extended to the general case.

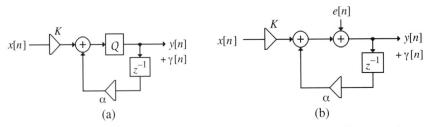

Figure 9.31: Scaled first-order section with quantizer, and its round-off noise model.

9.8.1　First-Order Section

Consider first the causal unscaled first-order IIR filter of Figure 9.1. Its round-off noise variance was computed in Example 9.7 and is given by Eq. (9.98). To determine its signal-to-noise ratio, assume the input $x[n]$ to be a wide-sense stationary (WSS) random signal with a uniform probability density function and a variance σ_x^2. The variance σ_y^2 of the output signal $y[n]$ generated by this input is then given by

$$\sigma_y^2 = \sigma_x^2 \left(\sum_{n=0}^{\infty} h^2[n] \right) = \frac{\sigma_x^2}{1 - \alpha^2}. \tag{9.144}$$

Taking the ratio of Eqs. (9.144) and (9.98), we arrive at the expression for the signal-to-noise ratio of the unscaled section as

$$\text{SNR} = \frac{\sigma_y^2}{\sigma_\gamma^2} = \frac{\sigma_x^2}{\sigma_0^2}, \tag{9.145}$$

implying that the SNR is independent of α. However, this is not a valid result since the adder is likely to overflow in an unscaled structure. It is therefore important first to scale the structure and then to compute the SNR to arrive at a more meaningful result.

The scaled structure is as indicated in Figure 9.31 along with its round-off error analysis model assuming quantization after addition of all products. With the scaling multiplier present, the output signal power now becomes

$$\sigma_y^2 = \frac{K^2 \sigma_x^2}{1 - \alpha^2}, \tag{9.146}$$

modifying the signal-to-noise ratio to

$$\text{SNR} = \frac{K^2 \sigma_x^2}{\sigma_0^2}. \tag{9.147}$$

Since the scaling multiplier coefficient K depends on the pole location and the type of scaling rule being followed, the SNR will thus reflect this dependence. The scaling parameter K can be chosen according to the rules outlined in Section 9.7. To this end we need first to determine the scaling transfer function $F(z)$ of the unscaled structure. It follows from Figure 9.31(a) that

$$F(z) = H(z) = \frac{1}{1 - \alpha z^{-1}}, \tag{9.148}$$

with a corresponding impulse response

$$f[n] = \mathcal{Z}^{-1}\{F(z)\} = \alpha^n \mu[n]. \tag{9.149}$$

Table 9.5: Signal-to-noise ratio of first-order IIR digital filters for different inputs. Adapted from [Vai87c].

Scaling rule	Input type	SNR	Typical SNR, dB $(b = 12, |\alpha| = 0.95)$
No overflow	WSS, white uniform density	$\dfrac{(1 - |\alpha|)^2}{3\sigma_0^2}$	52.24
No overflow	WSS, white Gaussian density $(\sigma_x^2 = 1/9)$	$\dfrac{(1 - |\alpha|)^2}{9\sigma_0^2}$	47.97
No overflow	Sinusoid, known frequency	$\dfrac{1 - \alpha^2}{2\sigma_0^2}$	69.91

Now, as indicated in Section 9.7, there are different scaling rules. To guarantee no overflow, we can use the scaling rule of Eq. (9.110) and arrive at

$$K = \frac{1}{\sum_{n=0}^{\infty} |f[n]|} = 1 - |\alpha|. \tag{9.150}$$

To evaluate the SNR, we need to know the type of input $x[n]$ being applied. If $x[n]$ is uniformly distributed with $|x[n]| \leq 1$, its variance is given by

$$\sigma_x^2 = \tfrac{1}{3}. \tag{9.151}$$

Substituting Eqs. (9.150) and (9.151) in Eq. (9.147), we arrive at

$$\text{SNR} = \frac{(1 - |\alpha|)^2}{3\sigma_0^2}. \tag{9.152}$$

For a $(b + 1)$-bit signed fraction with round-off or two's-complement truncation, $\sigma_0^2 = 2^{-2b}/12$. Substituting this figure in Eq. (9.152), we obtain the signal-to-noise ratio in dB as

$$\text{SNR}_{dB} = 20 \log_{10}(1 - |\alpha|) + 6.02 + 6.02\, b. \tag{9.153}$$

As a result, for a given transfer function, the SNR increases by 6 dB for each additional bit added to the register storing the product.

The above analysis can be carried out for other types of input and different scaling rules. We summarize in Table 9.5 the results for three different types of input for the scaling ensuring no overflow. Several conclusions can be made by examining this table. Independent of the type of input being applied, the SNR decreases rapidly as the pole moves closer to the unit circle. For a given transfer function, knowing the type of input being applied, the internal wordlength can be computed to achieve a desired SNR.

9.8.2 Second-Order Section

Consider next the causal unscaled second-order IIR filter of Figure 9.24(a). Its scaled version is indicated in Figure 9.32 along with the round-off noise analysis model, assuming again quantization after addition

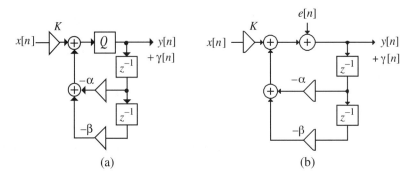

Figure 9.32: Scaled second-order section with quantizer, and its round-off noise model.

of all products. Now for a WSS input with a uniform probability density function and a variance σ_x^2, the signal power at the output is given by

$$\sigma_y^2 = \sigma_x^2 \left(K^2 \sum_{n=0}^{\infty} h^2[n] \right),$$
(9.154)

while the round-off noise power at the output is given by

$$\sigma_\gamma^2 = \sigma_0^2 \left(\sum_{n=0}^{\infty} h^2[n] \right).$$
(9.155)

Therefore the signal-to-noise ratio of the scaled structure is of the form

$$\text{SNR} = \frac{\sigma_y^2}{\sigma_\gamma^2} = \frac{K^2 \sigma_x^2}{\sigma_0^2}.$$
(9.156)

To determine the appropriate value of the scaling multiplier K, we need to compute the scaling transfer function $F(z)$. It can be seen from Figure 9.32(a) that $F(z)$ is identical to the filter transfer function $H(z)$. If the poles of the transfer function are at $z = re^{-j\theta}$, then both transfer functions take the form

$$F(z) = H(z) = \frac{1}{1 + \alpha z^{-1} + \beta z^{-2}} = \frac{1}{1 - 2r \cos \theta z^{-1} + r^2 z^{-2}}.$$
(9.157)

The corresponding impulse response is obtained by taking the inverse z-transform of the above, resulting in (Problem 3.102)

$$f[n] = h[n] = \frac{r^n \sin(n+1)\theta}{\sin \theta} \cdot \mu[n].$$
(9.158)

To eliminate completely the overflow at the output in Figure 9.32, the scaling multiplier K must be chosen as

$$K = \frac{1}{\sum_{n=0}^{\infty} |h[n]|}.$$
(9.159)

The summation in the denominator of Eq. (9.159) with $h[n]$ given by Eq. (9.158) is difficult to compute analytically. However, it is possible to establish some bounds on the summation to provide a reasonable estimate of the value of the scaling multiplier coefficient K [Opp75]. To this end, note that the amplitude of the response of the unscaled second-order section of Figure 9.24(a) to a sinusoidal input at the resonant frequency $\omega = \theta$, i.e., $x[n] = \cos(\theta n)$, is given by

$$\left| H(e^{j\theta}) \right| = \left[\frac{1}{(1-r)^2(1-2r\cos\theta+r^2)} \right]^{1/2}. \tag{9.160}$$

However, $\left| H(e^{j\theta}) \right|$ cannot be greater than $\sum_{n=0}^{\infty} |h[n]|$ since the latter is the largest possible value of output $y[n]$ for an input $x[n]$ with $|x[n]| \leq 1$. Moreover,

$$\sum_{n=0}^{\infty} |h[n]| = \frac{1}{\sin\theta} \sum_{n=0}^{\infty} r^n |\sin(n+1)\theta|$$

$$\leq \frac{1}{\sin\theta} \sum_{n=0}^{\infty} r^n = \frac{1}{(1-r)\sin\theta}. \tag{9.161}$$

A tighter upper bound on $\sum_{n=0}^{\infty} |h[n]|$ has been derived in [Unv75] and is given by

$$\sum_{n=0}^{\infty} |h[n]| \leq \frac{4}{\pi(1-r^2)\sin\theta}. \tag{9.162}$$

From Eqs. (9.159), (9.160), and (9.162), we therefore arrive at

$$(1-r)^2(1-2r\cos\theta+r^2) \geq K^2 \geq \frac{\pi^2(1-r^2)^2\sin^2\theta}{16}. \tag{9.163}$$

Following the technique carried out for the first-order section, we can derive the bounds on the SNR for the all-pole second-order section from Eqs. (9.156) and (9.163) for various types of inputs. As in the case of the first-order section, as the pole moves closer to the unit circle ($r \to 1$), the gain of the filter increases, causing the input signal to be scaled down significantly to avoid the overflow while at the same time boosting the output round-off noise. This type of interplay between round-off noise and dynamic range is a characteristic of all fixed-point digital filters [Jac70a].

In certain cases, it is possible to develop digital filter structures that have inherently the least quantization effects. In the following section we consider these structures.

9.9 Low-Sensitivity Digital Filters

In Section 9.2, we considered the effect of multiplier coefficient quantization on the performance of a digital filter. One major consequence is that the frequency response of the digital filter with quantized multiplier coefficients is different from that of the desired digital filter with unquantized coefficients, and this difference may be significant enough to make the practical digital filter unsuitable in most applications. It is thus of interest to develop digital filter structures that are inherently less sensitive to coefficient quantization. To this end, the first approach advanced is based on the conversion of an inherently low sensitivity analog network composed of inductors, capacitors, and resistors to a digital filter structure by replacing each analog network component and their interconnections to a corresponding digital filter equivalent such that the overall structure "simulates" the analog prototype [Fet71], [Fet86]. The resulting digital filter structure is called a *wave digital filter*, which also shares some additional properties of its analog prototype. An alternative approach is to determine directly the conditions for low coefficient sensitivity to be satisfied by the digital filter structure and to develop realization methods that ensure that the final structure indeed satisfies these conditions [Vai84]. In this section we examine the latter approach.

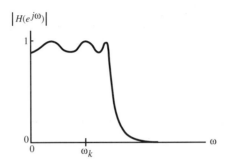

Figure 9.33: A typical magnitude response of a bounded real transfer function.

9.9.1 Requirements for Low Coefficient Sensitivity

Let the prescribed transfer function $H(z)$ be a *bounded real* (BR) function as defined in Section 4.4.5. That is, $H(z)$ is a causal stable real-coefficient function characterized by a magnitude response $\left|H(e^{j\omega})\right|$ bounded above by unity, i.e.,

$$\left|H(e^{j\omega})\right| \le 1. \tag{9.164}$$

Assume that $H(z)$ is such that at a set of frequencies ω_k the magnitude is exactly equal to 1:

$$\left|H(e^{j\omega_k})\right| = 1. \tag{9.165}$$

Since the magnitude function is bounded above by unity, the frequencies ω_k must be in the passband of the filter. A typical frequency response satisfying the above conditions is shown in Figure 9.33. Note that any causal stable transfer function can be scaled to satisfy conditions of Eqs. (9.164) and (9.165).

Let the digital filter structure $\mathcal{N}$ realizing $H(z)$ be characterized by a set of R multipliers with coefficients m_i. Moreover, let the nominal values of these multiplier coefficients, assuming infinite precision realization, be m_{i0}. Assume that the structure $\mathcal{N}$ is such that, regardless of actual values of the multiplier coefficients m_i in the immediate neighborhood of their design values m_{i0}, its transfer function remains bounded real, satisfying the condition of Eq. (9.164). Now, consider $\left|H(e^{j\omega_k})\right|$, the value of the magnitude response at $\omega = \omega_k$, which is equal to 1 when the multiplier values have infinite precision. Because of the assumption on $\mathcal{N}$ if a multiplier coefficient m_i is quantized, then $\left|H(e^{j\omega_k})\right|$ can only become less than 1 because of the BR condition of Eq. (9.164). Thus, a plot of $\left|H(e^{j\omega_k})\right|$ as a function of m_i will appear as indicated in Figure 9.34 with a zero-valued slope at $m_i = m_{i0}$,

$$\left. \frac{\partial \left|H(e^{j\omega_k})\right|}{\partial m_i} \right|_{m_i = m_{i0}} = 0, \tag{9.166}$$

implying that the first-order sensitivity of the magnitude function $|H(e^{j\omega})|$ with respect to each multiplier coefficient m_i is zero at all frequencies ω_k where $|H(e^{j\omega})|$ assumes its maximum value of unity. Since all frequencies ω_k, where the magnitude function is exactly equal to unity, are in the passband of the filter, and if these frequencies are closely spaced, we expect the sensitivity of the magnitude function to be very low at other frequencies in the passband.

A digital filter structure satisfying the above conditions for low passband sensitivity is called a *structurally bounded* system. Since the output energy of such a structure is less than the input energy for all finite energy inputs [see Eq. (4.102)], it is also called a *structurally passive* system. If Eq. (9.164) holds

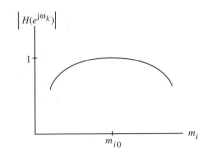

Figure 9.34: Illustration of the zero-sensitivity property.

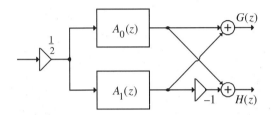

Figure 9.35: Parallel allpass realization of a power-complementary pair of transfer functions.

with an equality sign, the transfer function $H(z)$ is called a *lossless bounded real* (LBR) function, i.e., a stable allpass function. An allpass realization satisfying the LBR condition is therefore a *structurally lossless* or *LBR* system implying that the structure remains allpass under coefficient quantization.

We now outline methods for the realization of low passband sensitivity digital filters.

9.9.2 Low Passband Sensitivity IIR Digital Filter

In Section 6.10 we outlined a method for the realization of a large class of stable IIR transfer functions $G(z)$ in the form of a parallel allpass structure:

$$G(z) = \tfrac{1}{2}\left\{A_0(z) + A_1(z)\right\}, \tag{9.167}$$

where $A_0(z)$ and $A_1(z)$ are stable allpass transfer functions. Such a realization is possible if $G(z)$ is a BR function with a symmetric numerator and has a power-complementary BR transfer function $H(z)$ with an antisymmetric numerator [Vai86a]. In this case, $H(z)$ can be expressed as

$$H(z) = \frac{1}{2}\left\{A_0(z) - A_1(z)\right\}. \tag{9.168}$$

The allpass decompositions of Eqs. (9.167) and (9.168) permit the realization of the power-complementary pair $\{G(z), H(z)\}$ as indicated in Figure 9.35.

Now, on the unit circle, Eq. (9.167) becomes

$$G(e^{j\omega}) = \frac{1}{2}\left\{e^{j\theta_0(\omega)} + e^{j\theta_1(\omega)}\right\}, \tag{9.169}$$

since $A_0(z)$ and $A_1(z)$ are allpass functions with unity magnitude responses. Thus, if the allpass transfer functions are realized in structurally lossless forms, $|G(e^{j\omega})|$ will remain bounded above by unity. Moreover, at frequencies ω_k, where $|G(e^{j\omega_k})| = 1$, $\theta_0(\omega_k) = \langle \theta_1(\omega_k) \rangle_{2\pi}$, and $|e^{j\theta_0(\omega_k)} + e^{j\theta_1(\omega_k)}| = 2$. Or in

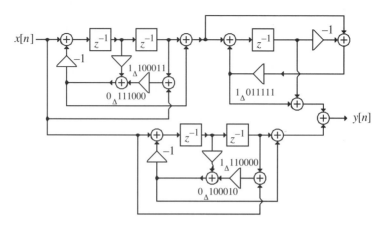

Figure 9.36: Parallel allpass realization of the lowpass filter using structurally lossless allpass sections with quantized multiplier coefficients.

other words, the realization of Figure 9.35 is structurally passive implying low passband sensitivity with respect to multiplier coefficient quantization.

Section 6.10 describes a method for determining the two allpass transfer functions $A_0(z)$ and $A_1(z)$ for a given BR transfer function $H(z)$ satisfying the conditions given above. Once these allpass transfer functions have been determined, they can then be realized in structurally lossless forms using one of the two methods discussed in Section 6.6.

The following example illustrates the low passband sensitivity characteristics of a parallel allpass structure.

EXAMPLE 9.12   Consider the realization of the fifth-order elliptic IIR lowpass filter of Example 7.21. The filter specifications are passband ripple of 0.5 dB, minimum stopband attenuation of 40 dB, and passband edge at 0.4. Using a modified version of Program 7_1, we arrive at the desired transfer function given by

$$G(z) = \frac{\begin{aligned}0.0528168 + 0.0797082z^{-1} + 0.1294975z^{-2} + 0.1294975z^{-3} \\ + 0.0797082z^{-4} + 0.0528168z^{-5}\end{aligned}}{\begin{aligned}(1 - 0.4908777z^{-1})(1 - 0.7624319z^{-1} + 0.5390008z^{-2}) \\ \times (1 - 0.5573823z^{-1} + 0.8828417z^{-2})\end{aligned}}.$$

Its parallel allpass decomposition is given by

$$G(z) = \frac{1}{2}\left[\frac{(0.4908777 - z^{-1})(0.8828417 - 0.5573823z^{-1} + z^{-2})}{(1 - 0.4908777z^{-1})(1 - 0.5573823z^{-1} + 0.8828417z^{-2})}\right.$$
$$\left. + \frac{0.5390008 - 0.7624319z^{-1} + z^{-2}}{1 - 0.7624319z^{-1} + 0.5390008z^{-2}}\right].$$

A 5-multiplier realization of the above decomposition using a signed 7-bit fractional sign-magnitude representation of each multiplier coefficient is shown in Figure 9.36. The gain responses of the filter with infinite precision multiplier coefficients and that with the quantized coefficients are shown in Figure 9.37. The low passband sensitivity properties of the parallel allpass structure is evident from this figure. Figure 9.38 depicts the passband details of the gain response of a direct form realization of the above fifth-order lowpass filter with infinite precision multiplier coefficients and that with the quantized coefficients. The poor passband sensitivity of the direct form realization is clearly seen in this plot.

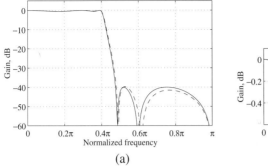
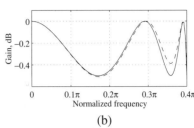

Figure 9.37: (a) Gain responses of the parallel allpass realization with infinite precision coefficients (shown with solid line), and with quantized coefficients (dashed line), and (b) passband details.

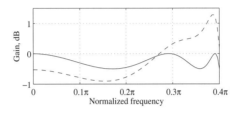

Figure 9.38: Passband responses of the direct form realization with infinite precision coefficients (shown with solid line), and with quantized coefficients (dashed line).

It should be noted that the power-complementary transfer function $H(z)$ realized in the form of Eq. (9.168) also remains BR if the allpass filters are structurally LBR. Hence, it exhibits low sensitivity in its passband [which is the stopband of $G(z)$]. However, low passband sensitivity of $H(z)$ does not imply low stopband sensitivity of $G(z)$ (Problem 9.35).

A number of other methods have been advanced for the realization of BR IIR digital transfer functions [Dep80], [Hen83], [Rao84], [Vai84], [Vai85a].

9.9.3 Low Passband Sensitivity FIR Digital Filter

In many applications linear-phase FIR filters are preferred. As indicated in Section 4.4.3, there are four types of linear-phase FIR filters of which the Type 1 filter is the most general and can realize any type of frequency response. We therefore restrict our attention here to a BR Type 1 FIR transfer function and outline a simple approach to realize it in a structurally passive form, thus ensuring low passband sensitivity to multiplier coefficients [Vai85b].

Now, from Eq. (4.80), the frequency response of a Type 1 FIR filter of order N can be expressed as

$$H(e^{j\omega}) = e^{-j\omega N/2}\breve{H}(\omega) \tag{9.170}$$

where $\breve{H}(\omega)$, a real function of ω, is its amplitude response. Since $H(z)$ is a BR function, $\breve{H}(\omega) \leq 1$. Its delay-complementary filter $G(z)$ defined by (Section 4.8.1)

$$G(z) = z^{-N/2} - H(z) \tag{9.171}$$

has a frequency response given by

$$G(e^{j\omega}) = e^{-j\omega N/2}\left[1 - \breve{H}(\omega)\right] = e^{-j\omega N/2}\breve{G}(\omega), \tag{9.172}$$

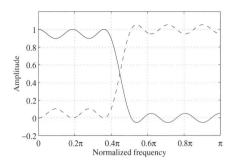

Figure 9.39: Amplitude responses of typical delay-complementary Type 1 FIR filters.

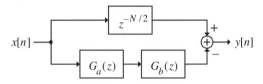

Figure 9.40: Low passband sensitivity realization of a Type 1 FIR filter $H(z)$.

where $\breve{G}(\omega) = 1 - \breve{H}(\omega)$ is its amplitude response. Amplitude responses of a typical delay-complementary FIR filter pair are depicted in Figure 9.39. It follows from this figure that at $\omega = \omega_k$, where $\left|H(e^{j\omega_k})\right| = 1$, $\breve{G}(\omega)$ has double zeros. Thus, $G(z)$ can be expressed as

$$G(z) = G_a(z) \prod_{k=1}^{L} \left(1 - 2\cos\omega_k z^{-1} + z^{-2}\right)^2 = G_a(z)G_b(z). \qquad (9.173)$$

A delay-complementary implementation of $H(z)$ based on Eq. (9.171) is sketched in Figure 9.40, where the FIR filter $G(z)$ is realized according to Eq. (9.173) as a cascade of $G_a(z)$ and L fourth-order FIR sections, with the kth section having a transfer function $(1 - 2\cos\omega_k z^{-1} + z^{-2})^2$. Now if the multiplier coefficient $2\cos\omega_k$ of the kth section is quantized, its zeros are still double and remain on the unit circle. As a result, quantization of the coefficients in $G_b(z)$ does not change the sign of the amplitude response $\breve{G}(\omega)$, and in the passband of $H(z)$, $\breve{G}(\omega) \geq 0$. Moreover, $G_a(z)$ has no zeros on the unit circle, and quantization of its coefficients also does not affect the sign of $\breve{G}(\omega)$. Hence, $\breve{H}(\omega)$ continues to remain bounded above by unity, i.e., the realization of $H(z)$ as indicated in Figure 9.39 remains structurally BR or structurally passive with regard to all multiplier coefficients, resulting in a low passband sensitivity realization.

EXAMPLE 9.13   We consider here the low passband sensitivity realization of a lowpass FIR filter using the above method. The filter specifications are length 13 with a normalized passband edge at 0.5 and a normalized stopband edge at 0.6 with equal weights to passband and stopband ripples. Using Program 7_8 we first design the lowpass filter $H(z)$ and then form its delay-complementary filter as $G(z) = z^{-6} - H(z)$. Using the function roots in MATLAB, we next determine the roots of $G(z)$ which has six zeros on the unit circle: two at $z = 1$, a pair of complex conjugate zeros at $z = -0.26463064626566 \pm j0.96434984370664$, and a pair of complex conjugate zeros at $z = -0.27683551142484 \pm j0.96091732194510$. These unit circle zeros constitute the transfer function

$$G_b(z) = (1 - z^{-1})^2(1 - 0.52926129253132z^{-1} + z^{-2})$$
$$\times(1 - 0.55367102284968z^{-1} + z^{-2}).$$

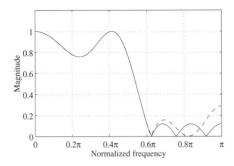

Figure 9.41: Magnitude responses of the original FIR lowpass filter (solid line) and its low sensitivity realization (dashed line).

By factoring out $G_b(z)$ from $G(z)$ using the function deconv in MATLAB, we arrive at the remaining term $G_a(z)$:

$$
\begin{aligned}
G_a(z) = {} & 0.04107997195619 + 0.05197154435126z^{-1} \\
& - 0.12094731168744z^{-2} - 0.30704562224137z^{-3} \\
& - 0.12094731168744z^{-4} + 0.05197154435126z^{-5} \\
& + 0.04107997195619z^{-6}.
\end{aligned}
$$

Next we quantize the coefficients of $G_a(z)$ and $G_b(z)$ by rounding the fractional part to two decimal digits, and from the quantized $G(z)$ determine again the delay complementary filter as indicated in Figure 9.40. Plots of the frequency responses of the original transfer function $H(z)$ and its realization as in Figure 9.40 with quantized coefficients are shown in Figure 9.41, verifying the low passband sensitivity behavior of the proposed realization. Another novel approach to the design of low passband sensitivity FIR filters is described in [Vai86b].

9.10 Reduction of Product Round-Off Errors Using Error Feedback

We have indicated earlier in Section 9.6 that in digital filters implemented using fixed-point arithmetic, the quantization of multiplication operations can be treated as a round-off noise at the output of the filter structure and can be analyzed using a statistical model of the quantization process. In many applications, this noise may decrease the output signal-to-noise ratio to an unacceptable level. It is thus of interest to investigate techniques that can reduce the output round-off noise. In this section, we describe two possible solutions that require additional hardware. In critical applications, the cost of the additional hardware may be justified.

The basic idea behind the error-feedback approach is to make use of the difference between the unquantized and quantized signal in reducing the round-off noise. This difference, called the error, is fed back to the digital filter structure in such a way as to not change the transfer function originally implemented by the structure while effectively decreasing the noise power [Cha81], [Hig84], [Mun81], [Tho77]. We illustrate the approach for a first-order and a second-order filter structure. The error-feedback approach is often used in designing high-precision oversampling A/D converters (see Section 11.12).

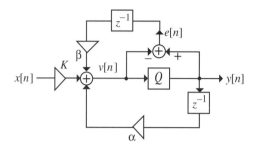

Figure 9.42: A first-order digital filter structure with error feedback.

9.10.1   First-Order Error-Feedback Structure

Consider again the scaled first-order section of Figure 9.31(a). We also assume that all multiplier coefficients are signed $(b + 1)$-bit fractions. The quantization error signal is then given by

$$e[n] = y[n] - v[n]. \tag{9.174}$$

We now modify the structure of Figure 9.31(a) as shown in Figure 9.42, where now the error signal is being fed back to the system through a delay and a multiplier with a coefficient β. In practice, the coefficient β is chosen to be a simple integer or fraction, such as ± 1, ± 2, or ± 0.5, so that the multiplication can be simply performed using a shift operation and will not introduce an additional quantization error.

Analyzing Figure 9.42, we arrive at the expression for the transfer function of the digital filter structure with $y[n]$ as the output as

$$H(z) = \left.\frac{Y(z)}{X(z)}\right|_{E(z)=0} = \frac{K}{1 - \alpha z^{-1}}. \tag{9.175}$$

The noise transfer function $G(z)$ with the error feedback, with $y[n]$ as the output, is given by

$$G(z) = \left.\frac{Y(z)}{E(z)}\right|_{X(z)=0} = \frac{1 + \beta z^{-1}}{1 - \alpha z^{-1}}, \tag{9.176}$$

and without the error feedback ($\beta = 0$) is given by

$$G(z) = \frac{1}{1 - \alpha z^{-1}}. \tag{9.177}$$

Following steps similar to that outlined in Example 9.7, we arrive at the output noise variance of the error-feedback structure as

$$\sigma_\gamma^2 = \frac{1 + 2\alpha\beta + \beta^2}{1 - \alpha^2}\sigma_0^2, \tag{9.178}$$

where σ_0^2 is the variance of the noise source $e[n]$. Note from the above that the output noise variance is a minimum when $\beta = -\alpha$. However, in practice $|\alpha| < 1$, and hence this choice for β will introduce an additional quantization noise source, making the analysis resulting in Eq. (9.178) invalid. Thus, from a practical point of view, it is more attractive to reduce the output round-off noise variance by choosing β as an integer with a value close to that of $-\alpha$, in which case the only noise source is due to the quantization of $\alpha y[n - 1]$.

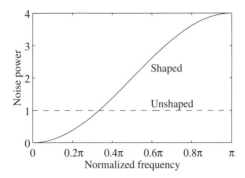

Figure 9.43: The normalized error power spectrum of the first-order section with and without error feedback.

For $|\alpha| < 0.5$, $\beta = 0$, implying no error feedback. However, in this case, the pole of $H(z)$ is far from the unit circle, and as a result, the noise variance is not high. For $|\alpha| \geq 0.5$, we choose $\beta = (-1)\mathrm{sgn}(\alpha)$.[4] Substituting this value of β in Eq. (9.178), we arrive at

$$\sigma_\gamma^2 = \frac{2}{1 + |\alpha|}\sigma_0^2. \tag{9.179}$$

Comparing Eqs. (9.179) and (9.178) with $\beta = 0$, we note that the introduction of error feedback has increased the SNR by a factor of $10 \log_{10}[2(1 - |\alpha|)]$. The above increase in SNR is quite significant if the pole is closer to the unit circle. For example, if $|\alpha| = 0.99$, the improvement is about 17 dB, which is equivalent to about 3 bits of increased accuracy compared to the case without error feedback. The additional hardware requirements for the structure of Figure 9.42 are two new adders and an additional storage register.

Comparing the two expressions of Eqs. (9.176) and (9.177), we conclude that the noise transfer function with error feedback is given by that without the error feedback multiplied by the expression $(1 + \beta z^{-1})$. Equivalently, the error-feedback circuit is *shaping* the error spectrum by modifying the input quantization noise $E(z)$ to $E_s(z) = (1 + \beta z^{-1})E(z)$. The output noise is generated by passing $E_s(z)$ through the usual noise transfer function of Eq. (9.177).

To examine the effect of noise spectrum shaping, consider the case of a narrowband lowpass first-order filter with $\alpha \to 1$. In this case, we choose $\beta = -1$, and as a result, $E_s(z)$ has a zero at $z = 1$ (i.e., $\omega = 0$). The power spectral density of the unshaped quantization noise $E(z)$ is σ_0^2, a constant. The power spectral density of the shaped noise source $E_s(z)$ is given by $4\sin^2(\omega/2)\sigma_0^2$ and is plotted in Figure 9.43 along with that of the unshaped case. The noise shaping redistributes the noise so as to move it mostly into the stopband of the lowpass filter, thus reducing the noise variance. Because of the noise redistribution caused by the error-feedback circuit, this method of round-off noise reduction has also been called the *error-spectrum shaping* approach in the literature [Hig84].

9.10.2 Second-Order Error-Feedback Structure

The error-feedback approach for round-off noise reduction has also been applied to second-order IIR digital filter structures [Hig84], [Mun81]. One proposed structure obtained by modifying Figure 9.32(a) is indicated in Figure 9.44. The transfer function $H(z)$ of this structure is given by Eq. (9.157). It should be noted that the inclusion of the error-feedback circuit to the structure of Figure 9.32(a) does not affect

---

[4]$\mathrm{sgn}(\alpha) = +1$ for $\alpha \geq 0$ and $\mathrm{sgn}(\alpha) = -1$ for $\alpha < 0$.

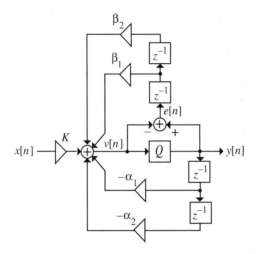

Figure 9.44: A second-order digital filter with error feedback.

either $H(z)$ or the scaling transfer function $F(z)$. Analyzing Figure 9.44, we arrive at the expression for the noise transfer function

$$G(z) = \frac{1 + \beta_1 z^{-1} + \beta_2 z^{-2}}{1 + \alpha_1 z^{-1} + \alpha_2 z^{-2}}. \tag{9.180}$$

The output round-off noise variance for $\mathcal{L}_2$-scaling is given by

$$\sigma_\gamma^2 = (\|G\|_2)^2 \, \sigma_0^2. \tag{9.181}$$

Note that a choice of $\beta_1 = \alpha_1$ and $\beta_2 = \alpha_2$ makes $\|G\|_2 = 1$, yielding $\sigma_\gamma^2 = \sigma_0^2$, an apparent optimal solution. However, this choice for the multiplier coefficients in the error-feedback path introduces additional quantization noise sources that were not taken into account in the above analysis. As in the case of the first-order section with error feedback, a more attractive solution is to make β_1 and β_2 integers with values close to α_1 and α_2, respectively. For example, for a narrowband lowpass transfer function, the poles are close to the unit circle and to the real axis, i.e., $r \approx 1$ and $\theta \approx 0$. In this case, α_1 is close to -2 and α_2 is close to 1. We therefore choose here $\beta_1 = -2$ and $\beta_2 = 1$, resulting in a noise transfer function

$$G(z) = \frac{1 - 2z^{-1} + z^{-2}}{1 + \alpha_1 z^{-1} + \alpha_2 z^{-2}}. \tag{9.182}$$

It has been shown by Vaidyanathan [Vai87c] that for a very narrowband lowpass filter with $r = 0.995$, $\theta = 0.07\pi$, and $b = 16$, the second-order error-feedback structure has an SNR that is approximately 25 dB higher than that without the error feedback. A detailed comparison of the second-order section with and without the error feedback for various types of inputs is left as an exercise (Problem 9.36).

It should be noted that here also the error-feedback circuit provides a noise shaping, as in the first-order case. In fact, for the parameters indicated above, the noise transfer function with error feedback is given by that without error-feedback multiplied by the expression $(1 - z^{-1})^2$. Or in other words, the error-feedback circuit is *shaping* the error spectrum by modifying the input quantization noise $E(z)$ to $E_s(z) = (1 - z^{-1})^2 E(z)$. The output noise is generated by passing $E_s(z)$ through the usual noise transfer function of Eq. (9.180) with $\beta_1 = \beta_2 = 0$. The power spectral density of the shaped noise source $E_s(z)$ is given by $16 \sin^4(\omega/2)\sigma_0^2$, whereas that of the unshaped case is simply σ_0^2. These power spectral densities have been plotted in Figure 9.45. Observe how the error feedback lowers the noise in the passband by pushing it into the stopband of the filter.

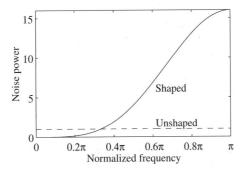

Figure 9.45: Normalized error power spectrum of the second-order section with and without error feedback.

9.11 Limit Cycles in IIR Digital Filters

So far we have treated the analysis of finite wordlength effects using a linear model of the system. However, a practical digital filter is a nonlinear system caused by the quantization of the arithmetic operations. Such nonlinearities may cause an IIR filter, which is stable under infinite precision, to exhibit an unstable behavior under finite precision arithmetic for specific input signals, such as zero or constant inputs. This type of instability usually results in an oscillatory periodic output called a *limit cycle*, and the system remains in this condition until an input of sufficiently large amplitude is applied to move the system into a more conventional operation.

In applications where the digital filter is operative at all times, oscillatory output in the absence of an input is highly undesirable. In particular, limit cycles with a frequency of oscillation in the audio frequency range can be quite annoying to the listener in audio and musical sound processing applications.

It should be noted here that limit cycles occur in IIR filters due to the presence of a feedback path. Such oscillations are absent in FIR structures which do not have any feedback path.

There are basically two types of limit cycles: (1) *granular* and (2) *overflow*. The former type of limit cycle is usually of low amplitude, whereas overflow oscillations have large amplitudes. We examine both types of limit cycles in this section.

9.11.1 Granular Limit Cycles

Two types of granular limit cycles have been observed in IIR digital filters: *inaccessible* and *accessible limit cycles* [Cla73]. The former type can appear only if the initial conditions of the digital filter at the time of starting pertain to that limit cycle, whereas in the second case, the limit cycle condition can be reached by starting the digital filter with initial conditions not pertaining to that limit cycle. We illustrate the generation of the limit cycle by analyzing the nonlinear behavior of a first-order and a second-order IIR digital filter.

EXAMPLE 9.14 Consider the causal first-order digital filter of Figure 9.1 characterized by the difference equation of Eq. (9.1), where for stability $|\alpha| < 1$. We assume that the quantization operation being performed is rounding and it is carried out on the result of the multiplication, as indicated in Figure 9.46. In this case, Eq. (9.1) becomes the nonlinear difference equation

$$\hat{y}[n] = \mathcal{Q}\left(\alpha \hat{y}[n-1]\right) + x[n], \tag{9.183}$$

where $\hat{y}[n]$ denotes the actual output of the filter. Without any loss of generality, we assume the digital filter is being implemented using a signed 5-bit fractional arithmetic with a quantization step of $\delta = 2^{-4}$. Table 9.6 shows the

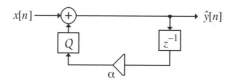

Figure 9.46: A first-order IIR filter with a quantizer.

first seven output samples for two different pole positions for an impulse input with $x[0] = 0_\wedge 1101$ and $x[n] = 0$ for $n > 0$, and an initial condition $\hat{y}[-1] = 0$. Observe that the steady-state output in the first case is a nonzero constant, i.e., periodic with a period of 1, whereas in the second case it is periodic with a period of 2. On the other hand, with infinite precision, the ideal output goes to zero exponentially as $n \to \infty$.

The above types of oscillations at the output are called zero-input limit cycles, and the amplitude ranges of the oscillations are often called *dead bands* [Bla65]. A digital filter exhibiting limit cycles at its output can be modeled by a linear system with its poles on the unit circle [Jac69]. Using this representation, it is possible to determine the range of the dead band in low-order filters. For example, in the case of the first-order IIR filter considered above, we assume that the system under the limit cycle condition has an effective pole at $z = 1$ when $\alpha > 0$ and at $z = -1$ when $\alpha < 0$. This implies that effectively

$$Q\left(\alpha \hat{y}[n-1]\right) = \begin{cases} \hat{y}[n-1], & \alpha > 0, \\ -\hat{y}[n-1], & \alpha < 0. \end{cases} \tag{9.184}$$

In addition, the quantization error due to rounding is bounded by $\pm\delta/2$, where δ is the quantization step, i.e.,

$$\left|Q\left(\alpha \hat{y}[n-1]\right) - \alpha \hat{y}[n-1]\right| \le \frac{\delta}{2}. \tag{9.185}$$

From the above two equations, we arrive at the dead band range of the first-order IIR filter as

$$\left|\hat{y}[n-1]\right| \le \frac{\delta}{2\left(1 - |\alpha|\right)}. \tag{9.186}$$

As a result, if for any value of n, the output $\hat{y}[n-1]$ of the delay unit is in the above range with the input set to zero, then the system gets trapped into a limit cycle mode. For our numerical example cited above, the dead band range is $|\hat{y}[n-1]| \le 0.1$, which definitely is satisfied by the entries in Table 9.6.

The limit cycle generation can be easily illustrated on a computer. MATLAB Program 9_7 given below can be used to study the granular limit cycle process. This program uses the function a2dR of Section 9.4.1 to develop the decimal equivalent of the binary representation of the filter coefficient with N bits for the magnitude after rounding.

```
% Program 9_7
% Granular Limit Cycles in First-Order IIR Filter
%
clf;
alpha = input('Type in the filter coefficient = ');
yi = input('Type in the initial condition = ');
x = input('Type in the value of x[0] = ');
for n = 1:21
    y(n) = a2dR(alpha*yi,5) + x;
    yi = y(n); x = 0;
end
```

Table 9.6: Limit cycle behavior of the first-order IIR digital filter.

	$\alpha = 0_{\triangle}1011,\ \hat{y}[-1] = 0$		$\alpha = 1_{\triangle}1011,\ \hat{y}[-1] = 0$	
n	$\alpha\hat{y}[n-1]$	$\hat{y}[n]$	$\alpha\hat{y}[n-1]$	$\hat{y}[n]$
0	0	$0_{\triangle}1101$	0	$0_{\triangle}1101$
1	$0_{\triangle}10001111$	$0_{\triangle}1001$	$1_{\triangle}10001111$	$1_{\triangle}1001$
2	$0_{\triangle}01100011$	$0_{\triangle}0110$	$0_{\triangle}01100011$	$0_{\triangle}0110$
3	$0_{\triangle}01000010$	$0_{\triangle}0100$	$1_{\triangle}01000010$	$1_{\triangle}0100$
4	$0_{\triangle}00101100$	$0_{\triangle}0011$	$0_{\triangle}00101100$	$0_{\triangle}0011$
5	$0_{\triangle}00100001$	$0_{\triangle}0010$	$1_{\triangle}00100001$	$1_{\triangle}0010$
6	$0_{\triangle}00010110$	$0_{\triangle}0001$	$0_{\triangle}00010110$	$0_{\triangle}0001$
7	$0_{\triangle}00001011$	$0_{\triangle}0001$	$1_{\triangle}00001011$	$1_{\triangle}0001$
8	$0_{\triangle}00001011$	$0_{\triangle}0001$	$0_{\triangle}00001011$	$0_{\triangle}0001$

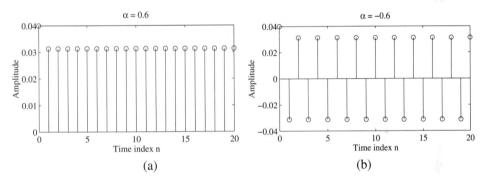

Figure 9.47: Illustration of limit cycles in a first-order IIR digital filter: (a) $\alpha = 0.6$, and (b) $\alpha = -0.6$.

```
k = 0:20;
stem(k,y)
ylabel('Amplitude'); xlabel('Time index n');
title(['\alpha = ' num2str(alpha)]);
```

Figure 9.47 shows the plots of the first 21 samples of the output response of the first-order IIR digital filter of Eq. (9.1) implemented with filter coefficients rounded to 6 bits. The input is $x[n] = 0.04\delta[n]$ and the filter coefficients are $\alpha = \pm 0.6$. The initial condition $y[-1]$ is set to 0.

EXAMPLE 9.15 In the case of a second-order IIR filter, a similar situation exists, except there are various types of limit cycles here. Consider the all-pole second-order IIR digital filter of Figure 9.24(a). It is described by the difference equation

$$y[n] = -\alpha_1 y[n-1] - \alpha_2 y[n-2] + x[n]. \tag{9.187}$$

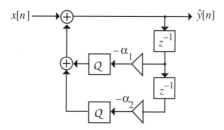

Figure 9.48: A second-order IIR section with quantizers after the multipliers.

The poles of this system are at

$$z = -\frac{\alpha_1}{2} - \frac{\sqrt{\alpha_1^2 - 4\alpha_2}}{2}, \tag{9.188}$$

which are complex for $\alpha_1^2 < 4\alpha_2$ and on the unit circle if $\alpha_2 = 1$. If we assume the products are first quantized by rounding, as indicated in Figure 9.48, the difference equation of Eq. (9.187) becomes

$$\hat{y}[n] = \mathcal{Q}\left(\alpha_1 \hat{y}[n-1]\right) - \mathcal{Q}\left(\alpha_2 \hat{y}[n-2]\right) + x[n], \tag{9.189}$$

with $\hat{y}[n]$ denoting the actual output. One way the system of Figure 9.48 could be in the limit cycle mode with zero input is if its effective pole is on the unit circle. In this case,

$$\mathcal{Q}\left(\alpha_2 \hat{y}[n-2]\right) = \hat{y}[n-2]. \tag{9.190}$$

Moreover, because of rounding, we have

$$\left|\mathcal{Q}\left(\alpha_2 \hat{y}[n-2]\right) - \alpha_2 \hat{y}[n-2]\right| \le \frac{\delta}{2}. \tag{9.191}$$

From Eqs. (9.190) and (9.191), we arrive at the dead band region governing the limit cycle mode:

$$\left|\hat{y}[n-2]\right| \le \frac{\delta}{2\left|1 + \alpha_2\right|}, \tag{9.192}$$

with α_1 determining the frequency of oscillation.

Limit cycles can also occur in second-order IIR filters if the system with quantizers has effective poles at $z = \pm 1$. In this mode the dead band is bounded by $2\delta/(1 - |\alpha_1| + \alpha_2)$ [Jac69].

It should be noted that the limit cycles occurring with the amplitude bound of Eq. (9.192) are basically inaccessible limit cycles, and it is highly unlikely, in practice, that the digital filter will start with initial conditions pertaining to these limit cycles [Cla73]. On the other hand, for the second-order IIR filter structure of Figure 9.48, with arbitrary initial conditions, accessible limit cycles are highly likely to occur and their amplitude can only be bounded from below:

$$\left|\hat{y}[n-2]\right| \ge \frac{\delta}{2\left|1 + \alpha_2\right|}. \tag{9.193}$$

Even though we have considered here the zero-input limit cycle generation in very simple IIR digital filter structures, limit cycles also occur in higher-order structures. However, their analysis is almost impossible, except for the determination of the bounds on the amplitudes of the limit cycles [Lon73]. In some structures, periodic limit cycles have been observed with nonzero constant amplitudes and also for constant amplitude sinusoidal inputs.

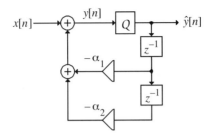

Figure 9.49: A second-order IIR section with a quantizer after the accumulator.

9.11.2 Overflow Limit Cycles

As indicated earlier, limit-cycle-like oscillations can also result from overflow in digital filters implemented with finite precision arithmetic. The amplitude of the overflow oscillations can cover the whole dynamic range of the register experiencing the overflow and are much more serious in nature than the granular type. We illustrate the generation of an overflow oscillation on a computer in the following example.

EXAMPLE 9.16   We consider the causal all-pole second-order IIR digital filter of Figure 9.24(a). We assume its implementation using sign-magnitude arithmetic with a rounding of the sum of products by a single quantizer, as indicated in Figure 9.49. In this case, the linear difference equation of Eq. (9.187) describing the ideal filter reduces to the nonlinear difference equation given by

$$\hat{y}[n] = \mathcal{Q}\left(-\alpha_1 \hat{y}[n-1] - \alpha_2 \hat{y}[n-2] + x[n]\right), \qquad (9.194)$$

where $\mathcal{Q}(\cdot)$ represents the rounding operation and $\hat{y}[n]$ denotes the actual output of the filter. All numbers are assumed to be signed 4-bit fractions.

Let the filter coefficients be given by $\alpha_1 = 1_\triangle 001 = -0.875_{10}$ and $\alpha_2 = 0_\triangle 111 = 0.875_{10}$. The initial conditions are assumed to be $\hat{y}[-1] = -0.625_{10}$ and $\hat{y}[-2] = -0.125_{10}$. We consider the zero-input case, i.e., $x[n] = 0$ for $n \geq 0$.

Program 9.8 given below can be used to illustrate the overflow limit cycle generation. It uses the function a2dR of Section 9.4.1 to perform the rounding operation on the sum of products as indicated in Eq. (9.194).

```
% Program 9_8
% Illustration of Overflow Limit Cycles
%
a = input('alpha_1 and alpha_2 values = ');
yi1 = -0.625; yi2 = -0.125;
for n = 1:41;
    y(n)  = - a(1)*yi1 - a(2)*yi2;
    y(n)  = a2dR(y(n),3);
    yi2 = yi1; yi1 = y(n);
end
k = 0:40;
stem(k,y)
xlabel('Time index n');ylabel('Amplitude');
title(['\alpha_1 = ' num2str(a(1)),'     \alpha_2 = ' num2str(a(2))]);
```

Figure 9.50 shows the output generated by the above program demonstrating the generation of overflow limit cycles with zero input. It should be noted that, as shown in the following section, the structure of Figure 9.49 does not exhibit overflow limit cycles if sign-magnitude truncation is used to quantize the sum of products. This property can be easily demonstrated by replacing the function a2dR in Program 9.8 with the function a2dT of Section 9.4.1.

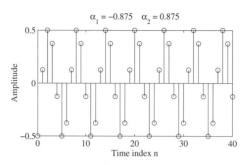

Figure 9.50: Overflow limit cycles of Figure 9.49.

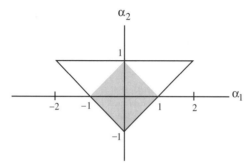

Figure 9.51: Coefficient ranges of a second-order direct form IIR filter to guarantee no overflow oscillations (shown by shaded region).

For stability we have shown in Section 4.12.1 that the filter coefficients of the second-order direct form IIR structure of Figure 9.49 must remain inside the stability triangle of Figure 9.51. However, the structure can still get into a zero-input overflow oscillation mode for a large range of values of the filter constants, satisfying this stability constraint when implemented using two's-complement arithmetic with rounding. It has been shown that overflow limit cycles under zero-input cannot occur if the filter coefficients lie in the shaded region inside the stability triangle, as indicated in Figure 9.51 [Ebe69]. This region is defined by the relation

$$|\alpha_1| + |\alpha_2| < 1. \tag{9.195}$$

As the above condition is quite restrictive, we now examine other IIR structures that do not sustain limit cycles with less severe constraint on the pole locations.

9.11.3 Limit Cycle Free Structures

A number of authors have advanced digital filter structures that are limit cycle free when implemented using specific arithmetic schemes. The most general approach to the development of such structures is based on the state-space representation [Mil78]. In Chapter 2 we considered the time-domain description of an LTI discrete-time system in terms of the convolution sum and the linear constant coefficient difference equation relating the input and the output signals. Another time-domain representation for a causal LTI discrete-time system is in terms of internal variables called the state variables, which are usually the output variables of all unit delays. For a second-order causal LTI discrete-time system, the state-space representation relating the output sequence $y[n]$ to the input sequence $x[n]$ is given by

$$\begin{bmatrix} s_1[n+1] \\ s_2[n+1] \end{bmatrix} = \begin{bmatrix} a_{11} & a_{12} \\ a_{21} & a_{22} \end{bmatrix} \begin{bmatrix} s_1[n] \\ s_2[n] \end{bmatrix} + \begin{bmatrix} b_1 \\ b_2 \end{bmatrix} x[n], \tag{9.196}$$

$$y[n] = \begin{bmatrix} c_1 & c_2 \end{bmatrix} \begin{bmatrix} s_1[n] \\ s_2[n] \end{bmatrix} + dx[n]. \tag{9.197}$$

Denoting[5]

$$\mathbf{s}[n] = [s_1[n] \quad s_2[n]]^T, \tag{9.198}$$

$$\mathbf{A} = \begin{bmatrix} a_{11} & a_{12} \\ a_{21} & a_{22} \end{bmatrix}, \qquad \mathbf{B} = \begin{bmatrix} b_1 \\ b_2 \end{bmatrix}, \qquad \mathbf{C} = [c_1 \quad c_2], \tag{9.199}$$

we can rewrite Eqs. (9.196) and (9.197) in a compact form as

$$\mathbf{s}[n+1] = \mathbf{A}\,\mathbf{s}[n] + \mathbf{B}x[n], \tag{9.200}$$

$$y[n] = \mathbf{C}\,\mathbf{s}[n] + dx[n]. \tag{9.201}$$

In the above equations $\mathbf{s}[n]$, given by Eq. (9.198), is called the *state vector* with its elements $s_i[n]$ known as the *state variables*. The matrix $\mathbf{A}$ given in Eq. (9.199) is called the *state transition matrix*.

Even though in an actual implementation, the right-hand sides of both Eqs. (9.200) and (9.201) are quantized, the quantization errors caused by the quantization of the right-hand side of Eq. (9.200) go through the feedback loop and are responsible for the generation of limit cycles. We assume that the variables $s_1[n+1]$ and $s_2[n+1]$ are quantized, and the delayed versions of these quantized signals are the state variables $s_1[n]$ and $s_2[n]$, respectively.

We define a quantizer to be *passive* if

$$|\mathcal{Q}(x)| \le |x|, \qquad \text{for all } x. \tag{9.202}$$

If x is inside the given dynamic range of the system, then for magnitude truncation (Section 9.2) it is evident that Eq. (9.202) holds, i.e., the quantizer is passive. If x is outside the dynamic range caused, for example, by overflow, it must be brought back to the range by following either the saturation arithmetic scheme or the two's-complement overflow scheme discussed in Section 8.6. As a result, magnitude truncation followed by either of the above two overflow handling schemes is again a passive quantizer.

A digital filter structure with a state transition matrix satisfying

$$\mathbf{A}^T\mathbf{A} = \mathbf{A}\mathbf{A}^T \tag{9.203}$$

has been called a *normal form structure*. It has been shown that such a structure with passive quantizers does not support zero-input limit cycles of either type [Mil78]. The matrix $\mathbf{A}$ satisfying the above condition and $\|\mathbf{A}\|_2 < 1$ is called a *normal matrix*.

EXAMPLE 9.17 Consider the digital filter structure of Figure 9.52 [Yan82]. Analysis yields

$$s_1[n+1] = cs_1[n] - cds_2[n] + cx[n], \tag{9.204a}$$

$$s_2[n+1] = cds_1[n] + cs_2[n]. \tag{9.204b}$$

---

[5]In this text, row and column vectors, in general, are indicated by boldface lowercase letters, while the matrices are represented by boldface uppercase letters.

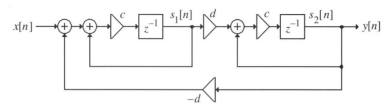

Figure 9.52: A second-order IIR section.

Figure 9.53: Register storing the signal to be quantized.

From the above, we easily identify the state transition matrix as

$$\mathbf{A} = \begin{bmatrix} c & -cd \\ cd & c \end{bmatrix}. \qquad (9.205)$$

Comparing the determinant of $(z\mathbf{I} - \mathbf{A})$ with the denominator of a second-order IIR transfer function with poles at $z = re^{\pm j\theta}$ (with $r < 1$ for stability), we obtain

$$c = r\cos\theta, \qquad d = \tan\theta. \qquad (9.206)$$

Substituting the above values in Eq. (9.205), we arrive at

$$\mathbf{A} = \begin{bmatrix} r\cos\theta & -r\sin\theta \\ r\sin\theta & r\cos\theta \end{bmatrix}. \qquad (9.207)$$

Note that $\mathbf{A}^T\mathbf{A} = \mathbf{A}\mathbf{A}^T = r^2\mathbf{I}$. Since $r < 1$, we have $\|\mathbf{A}\|_2 = r < 1$. Therefore, the filter of Figure 9.52 is a normal form structure and will not exhibit zero-input limit cycles of either type.

9.11.4 Random Rounding

A conceptually simple technique to suppress zero-input granular limit cycles in second-order IIR sections is called random rounding [But77], [Law78]. To explain the operation of this method, let the register storing the signal $v[n]$ prior to the quantization be as indicated in Figure 9.53. In the random rounding method, an uncorrelated binary random sequence $x[n]$ is generated externally by a random number generator, and the $(b+1)$th bit a_{-b-1} of $v[n]$ is replaced with $x[n]$. The modified signal $v[n]$ is then passed through a passive quantizer. In addition to the increase in hardware complexity, this approach also results in a slightly higher quantization noise that is, however, distributed over the whole frequency range.

9.12 Round-Off Errors in FFT Algorithms

Since FFT is often employed in a number of digital signal processing applications, it is of interest to analyze the effects of finite wordlengths in FFT computations. The most critical error in the computation is that due to the arithmetic round-off errors. As in the earlier sections, we assume that the DFT computations are being carried out using fixed-point arithmetic, and we thus restrict our analysis to the effect of product round-off errors in the DFT computation via the FFT algorithms and to compare them with the errors generated in the direct DFT computation [Opp89], [Pro92], [Wel69]. The model for the round-off error analysis to be employed here is the same as in the case of LTI digital filters and is shown in Figure 9.18.

Now the multiplier coefficients W_N^{kn} and the signals in the DFT computation are, in general, complex numbers resulting in complex multiplications. Since each complex multiplication usually requires four real multiplications, there are four quantization errors per multiplication. If the DFT computation requires K complex multiplications, there are $4K$ sources of quantization errors in the computational structure. We make the usual assumptions about the statistical properties of the noise sources:

(a) All $4K$ errors are uncorrelated with each other and uncorrelated with the input sequence.

(b) The quantization errors are random variables uniformly distributed with a variance $\sigma_0^2 = 2^{-2b}/12$, assuming a signed b-bit fractional fixed-point arithmetic.

9.12.1 Direct DFT Computation

Recall that the N-point DFT $X[k]$ of a length-N complex sequence $x[n]$ is given by

$$X[k] = \sum_{n=0}^{N-1} x[n] W_N^{nk}, \qquad 0 \le k \le N-1. \tag{9.208}$$

Thus, the computation of a single DFT sample requires N complex multiplications, and hence, the total number of real multiplications for the computation of a single DFT sample is $4N$. As a result, there are $4N$ quantization error sources. The variance of the error in the computation of one DFT sample is therefore[6]

$$\sigma_\gamma^2 = 4N\sigma_0^2 = \frac{2^{-2b}N}{3}, \tag{9.209}$$

indicating that the output round-off error is proportional to the DFT length.

Now, the input sequence $x[n]$ must be scaled to avoid overflow in the computation of $X[k]$. From Eq. (9.208) it follows that

$$|X[k]| \le \sum_{n=0}^{N-1} |x[n]| < N, \tag{9.210}$$

assuming that the input samples satisfy the dynamic range constraint $|x[n]| \le 1$. To prevent overflow we need to ensure that

$$|X[k]| < 1, \tag{9.211}$$

which can be guaranteed by dividing each sample $x[n]$ in the input sequence by N.

To analyze the effect of the above scaling, assume the input to be a white noise sequence with each sample uniformly distributed in the range $(-1/N, 1/N)$ [Pro92]. The input signal power is then given by

$$\sigma_x^2 = \frac{(2/N)^2}{12} = \frac{1}{3N^2}. \tag{9.212}$$

The corresponding output signal power is

$$\sigma_X^2 = N\sigma_x^2 = \frac{1}{3N}. \tag{9.213}$$

---

[6]Strictly speaking, the output noise variance will be less than the value here since multiplications with W_N^0 and $W_N^{N/2}$ do not develop any error.

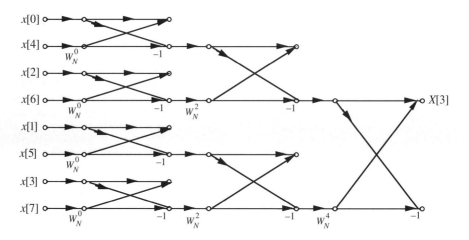

Figure 9.54: Reduced flow-graph for the computation of $X[3]$.

As a result, the signal-to-noise ratio is

$$\text{SNR} = \frac{\sigma_X^2}{\sigma_\gamma^2} = \frac{2^{2b}}{N^2}. \tag{9.214}$$

The above expression indicates that the SNR has been reduced by a factor of N^2 due to scaling and the round-off error. The wordlength size needed to compute a DFT of given length with a desired SNR can be determined using Eq. (9.214) (Problem 9.43).

9.12.2 DFT Computation via FFT Algorithm

We now consider the round-off error analysis of the DFT computation based on an FFT algorithm. Without any loss of generality, we analyze the decimation-in-time radix-2 FFT algorithm. However, the results derived can be easily extended to other types of fast DFT algorithms.

From the flow-graph of the DIT FFT algorithm given in Figure 8.24, it can be seen that the DFT samples are computed by a series of butterfly computations with a single complex multiplication per butterfly module. Some of the butterfly computations require multiplications by ± 1 or $\pm j$ that we do not treat separately here to simplify the analysis.

Consider now the computation of a single DFT sample as indicated in Figure 9.54. It follows from this figure that the computation of a single DFT sample involves $\nu = \log_2 N$ stages. The number of butterflies in a particular stage depends on the stage's location in the computational chain with $N/2^r = 2^{\nu-r}$ butterflies in the rth stage, where $r = 1, 2, \ldots, \nu$. The total number of butterflies involved per DFT sample is therefore

$$1 + 2 + 2^2 + \cdots + 2^{\nu-2} + 2^{\nu-1} = 2^\nu - 1 = N - 1. \tag{9.215}$$

It also follows from Figure 9.54 that the quantization errors introduced at the rth stage appear at the output after propagating through $(r-1)$ stages while getting multiplied by the twiddle factors at each subsequent stage. Since the magnitude of the twiddle factors is always unity, the variances of the quantization errors do not change while propagating to the output. The total number of error sources contributing to the output round-off error is $4(N-1)$. Assuming that the quantization errors introduced in each butterfly are uncorrelated with those generated at other butterflies, the variance of the output round-off error is then

$$\sigma_\gamma^2 = 4(N-1)\frac{2^{-2b}}{12} \cong \frac{2^{-2b}N}{3}. \tag{9.216}$$

It should be noted that the above expression for σ_γ^2 is identical to that derived for the direct DFT computation case given by Eq. (9.209). This is to be expected since the FFT algorithm does not alter the total number of complex multiplications to compute a single DFT sample, rather it organizes the computations more efficiently so that the number of multiplications to compute all N DFT samples is reduced.

Now, to prevent overflow at the output if we scale the input samples to satisfy the condition $|x[n]| < 1/N$, the SNR obtained in the FFT algorithm remains the same as given in Eq. (9.214). However, in the case of the FFT algorithm, the SNR can be improved by following a different scaling rule as described below.

Instead of scaling the input samples by $1/N$, we can scale the input signals at each stage by $1/2$. This scaling rule also guarantees that the output DFT samples are scaled by a factor of $(1/2)^\nu = 1/N$ as desired. However, each scaling by a factor of $1/2$ reduces the round-off noise variance by a factor of $1/4$. As a result, the round-off noise variances of the $4(2^{\nu-r})$ noise sources at the rth stage are reduced by a factor of $1/4^{r-1}$, while the noise propagates to the output. It can be shown that the total round-off noise variance at the output is now given by

$$\sigma_\gamma^2 = \frac{2}{3} \cdot 2^{-2b}\left(1 - \frac{1}{N}\right) \cong \frac{2}{3} \cdot 2^{-2b}, \tag{9.217}$$

assuming N to be large enough so that $1/N << 1$ (Problem 9.44). The SNR in this case reduces to

$$\frac{\sigma_X^2}{\sigma_\gamma^2} = \frac{2^{2b}}{2N}. \tag{9.218}$$

Hence, distribution of the scaling into each stage has increased the SNR by a factor of N. The formula given in Eq. (9.218) can be used to determine the wordlength required to achieve a specified SNR for computing DFT of a given length N (Problem 9.45).

9.13 Summary

This chapter is concerned with the effects of the finite wordlengths caused by the actual implementation of a digital signal processing algorithm that causes the results of the algorithm to be different, in general, from the desired ones obtained in the ideal case with infinite precision wordlengths. To develop the appropriate models for the analysis of these effects prior to actual implementation, we first review the quantization process used to fit the infinite precision data into a finite wordlength register and the resulting errors caused by it. The quantization of both fixed-point and floating-point numbers are considered. However, the discussion in the rest of the chapter is restricted to fixed-point implementations.

The effect of the quantization of the multiplier coefficients in the implementation of a digital filter is considered next as a coefficient sensitivity problem. Simple formulas are derived for such sensitivity analysis for the infinite impulse response (IIR) filter and the finite impulse response (FIR) filter.

In the digital processing of a continuous-time signal, the latter is sampled periodically by a sample-and-hold device and then converted into a digital form by an analog-to-digital (A/D) converter. A statistical model is developed for the analysis of the input quantization error caused by the A/D conversion and is used to derive the expression for the signal-to-quantization ratio as a function of the A/D converter wordlength. The A/D quantization error propagates to the output of the digital filter processing the digitized continuous-time signal, appearing as a noise added to the desired output, and methods for the statistical analysis of the output error are provided.

The effect of product round-off in the fixed-point implementation of a digital filter is then analyzed using a statistical model. A statistical analysis of the output noise caused by the propagation of these internally generated errors also to the output of the digital filter is provided. An overflow may occur at certain internal nodes in a digital filter implemented in fixed-point arithmetic, which can result in a large amplitude oscillation at the filter output. Methods of scaling the internal signal variables with the aid of suitably placed scaling multipliers to minimize the probability of overflow are discussed. The performance of the cascade form of digital filters is particularly examined in depth to illustrate the effect of pole-zero pairing and the ordering of the low-order sections. A detailed analysis of the output signal-to-noise ratios of scaled first-order and second-order IIR digital filter sections is then provided.

Conditions for the low passband sensitivity realizations of IIR and FIR digital filters are next derived, and a method for such low sensitivity realization for each case is outlined. Two approaches to the reduction of round-off errors in IIR digital filter structures are then described.

The discretization process can also cause the occurrence of periodic oscillation, called limit cycles, at the output of an IIR digital filter under certain conditions. These limit cycles are difficult to analyze in the general case. Their existence is demonstrated here only for the first-order and second-order IIR digital filters. Conditions for the limit cycle free operation of a state-space structure is derived.

The chapter concludes with an analysis of round-off errors in the implementation of DFT and FFT algorithms.

9.14 Problems

9.1 Derive the ranges of the relative quantization errors of Table 9.2.

9.2 Compute the pole sensitivity of the first-order lowpass transfer function $H_{LP}(z)$ of Eq. (4.109) with respect to the coefficient α.

9.3 Compute the pole sensitivities of the second-order bandpass transfer function $H_{BP}(z)$ of Eq. (4.113) with respect to the coefficients α and β.

9.4 The digital filter structure of Figure 9.52 is more commonly known as a *modified coupled-form structure* [Yan82]. Another modified coupled-form structure is shown in Figure P9.1. Determine the transfer function of both structures and then compute their respective pole sensitivities. Compare these sensitivities with those of the structure of Figure 9.10 as given in Eq. (9.46).

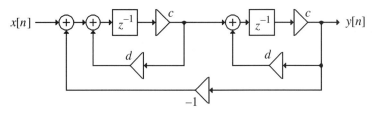

Figure P9.1

9.5 Determine the transfer function of the second-order digital filter structure shown in Figure P9.2 and compute its pole sensitivities [Aga75]. Compare these sensitivities with those of the structures of Figure 9.10 and Figure P9.1.

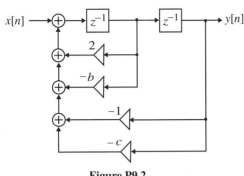

Figure P9.2

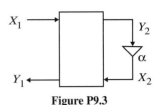

Figure P9.3

9.6 A third-order elliptic highpass transfer function

$$H(z) = \frac{0.1868(z-1)(z^2 - 0.0902z + 1)}{(z + 0.3628)(z^2 + 0.5111z + 0.7363)}$$

is realized in (1) direct form, and (2) cascade form. Compute the pole sensitivities of each structure.

9.7 Consider the digital filter structure of Figure P9.3 characterized by a transfer function $H(z) = Y_1/X_1$. Show that the sensitivity of $H(z)$ with respect to the multiplier coefficient α defined by $\partial H(z)/\partial \alpha$ is given by

$$\frac{\partial H(z)}{\partial \alpha} = F_\alpha(z) \cdot G_\alpha(z),$$

where $F_\alpha(z)$ is the scaling transfer function from the input node X_1 to the input node Y_2 of the multiplier α and $G_\alpha(z)$ is the noise transfer function from the output node X_2 of the multiplier α to the filter output node Y_1.

9.8 Show that the function $W_N(\omega)$ defined in Eq. (9.58) satisfies the following properties:

(a) $0 < W_N(\omega) \le 1$,

(b) $W_N(0) = W_N(\pi) = 1$,   for all N,

(c) $\lim_{N\to\infty} W_N(\omega) = \frac{1}{\sqrt{2}}$, $0 < \omega < \pi$.

9.9 Verify the SNR values given in Table 9.3.

9.10 An alternate approach to the algebraic calculation of output round-off noise variance is outlined in this problem [Pat80]. Let the partial-fraction expansion in z of an Nth-order real rational noise transfer function $H(z)$ with simple poles be given by

$$H(z) = \sum_{k=1}^{N} \frac{A_k}{z + a_k},$$

where A_k and a_k are, in general, complex numbers.

(a) Show that $H(z)$ can be expressed in the form

$$H(z) = \sum_{k=1}^{N} C_k \left(\frac{a_k z + 1}{z + a_k} \right) + B,$$

where C_k and B are constants. Determine the expressions for C_k and B.

(b) Show that the normalized output round-off noise variance σ_0^2 can be expressed in the form

$$\sigma_0^2 = \sum_{k=1}^{N} \frac{A_k^2}{1 - a_k^2} + 2 \sum_{k=1}^{N-1} \sum_{\ell=k+1}^{N} \frac{A_k A_\ell}{1 - a_k a_\ell}.$$

(c) Show that the above expression can be further simplified as

$$\sigma_0^2 = \sum_{k=1}^{N} \sum_{\ell=1}^{N} \frac{A_k A_\ell}{1 - a_k a_\ell}.$$

9.11 Determine the output noise variance due to the propagation of the input quantization noise for each of the following causal IIR digital filters:

(a) $H_1(z) = \dfrac{(z + 3)(z - 1)}{(z - 0.5)(z + 0.3)}$,

(b) $H_2(z) = \dfrac{3(2z + 1)(0.5z^2 - 0.3z + 1)}{(3z + 1)(4z + 1)(z^2 - 0.5z + 0.4)}$,

(c) $H_3(z) = \dfrac{(z - 1)^2}{(z^2 - 0.4z + 0.7)}$.

9.12 Determine the expression for the normalized output noise variance due to input quantization of the following digital filter realized in a parallel form

$$H(z) = C + \frac{A}{1 - \alpha z^{-1}} + \frac{B}{1 - \beta z^{-1}}.$$

Each section in the parallel structure is realized in direct form II. What is the value of the variance for $\alpha = -0.7$, $\beta = 0.9$, $A = 3$, $B = -2$, $C = 3$?

9.13 Realize the following transfer function

$$H(z) = \frac{(z - 2)(z + 3)}{(z + 0.3)(z - 0.4)},$$

in four different cascade forms with each first-order stage implemented in direct form II.

(a) Show the noise model for each unscaled structure for the computation of the product round-off noise at the output assuming quantization of products before addition assuming fixed-point implementation with either rounding or two's-complement truncation. Compute the normalized output round-off noise variance for each realization. Which cascade realization has the lowest round-off noise?

(b) Repeat part (a) assuming quantization after addition of product.

9.14 Realize the transfer function of Problem 9.13 in two different parallel forms with each first-order stage implemented in direct form II.

(a) Show the noise model for each unscaled structure for the computation of the product round-off noise at the output assuming quantization of products before addition and assuming fixed-point implementation with either rounding or two's-complement truncation. Compute the normalized output round-off noise variance for each realization. Which parallel realization has the lowest round-off noise?

(b) Repeat part (a) assuming quantization after addition of products.

9.15 Realize the following second-order transfer function

$$H(z) = \frac{2 + 2z^{-1} - 1.5z^{-2}}{1 + 0.5z^{-1} + 0.06z^{-2}},$$

in (1) direct form, (2) cascade form, and (3) parallel form. Each section in the cascade and parallel structures is realized in direct form II. Show the noise model for each unscaled structure for the computation of the product round-off noise at the output assuming quantization of products before addition and assuming fixed-point implementation with either rounding or two's-complement truncation. Compute the product round-off noise variance for each realization. Which realization has the lowest round-off noise? *Note*: There are four cascade and two parallel realizations.

9.16 Realize the transfer function of Problem 9.15 in the Gray-Markel form.

(a) Show the noise model for the unscaled structure for the computation of the product round-off noise variance at the output assuming quantization of products before addition and compute its product round-off noise variance.

(b) Repeat part (a) assuming quantization after addition of products.

9.17 One possible two-multiplier realization of a second-order Type 2 allpass transfer function

$$A_2(z) = \frac{d_2 + d_1 z^{-1} + z^{-2}}{1 + d_1 z^{-1} + d_2 z^{-2}}$$

is shown in Figure P9.4. Derive the expression for the normalized steady-state output noise variance due to product round-off assuming fixed-point implementation with either rounding or two's-complement truncation.

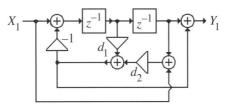

Figure P9.4

9.18 Develop the noise model for the product round-off noise analysis of the second-order coupled-form structure of Figure 9.10. Determine the normalized output round-off noise variances due to product round-off before summation and after summation.

9.19 Develop the noise model for the product round-off noise analysis of the second-order Kingsbury structure of Figure P6.8 of Problem 6.9. Determine the normalized output round-off noise variances due to product round-off before summation and after summation.

9.20 The allpass section of Figure P9.4 is employed to alter the phase response of the structure realizing the transfer function $H(z)$ of Problem 9.13 as indicated in Figure P9.5. The allpass equalizer of Figure P9.5 has a transfer function

$$A_2(z) = \frac{0.8 + 0.5z^{-1} + z^{-2}}{1 + 0.5z^{-1} + 0.8z^{-2}}.$$

Compute the normalized steady-state output noise variance due to product round-off of the phase-equalized structure of Figure P9.5 if $H(z)$ is realized in a cascade form with the lowest product round-off noise.

$$\boxed{H(z)} \rightarrow \boxed{A_2(z)} \rightarrow$$

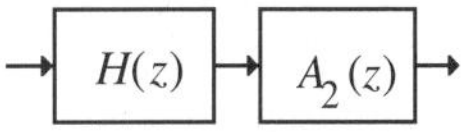

Figure P9.5

9.21 Consider the digital filter structure of Figure P9.6 that is assumed to be implemented using 9-bit signed two's-complement fixed-point arithmetic with all products quantized before additions. Draw the linear noise model of the unscaled system and compute its total output noise power.

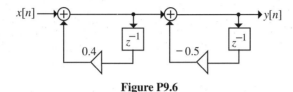

Figure P9.6

9.22 Scale the first-order digital filter structure of Figure P9.7 using the $\mathcal{L}_2$-norm scaling rule.

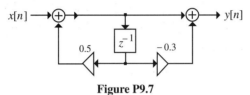

Figure P9.7

9.23 Scale the second-order digital filter structure of Figure P9.8 using the $\mathcal{L}_2$-norm scaling rule.

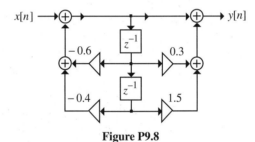

Figure P9.8

9.24 Scale the structure realized in Problem 9.12 using the $\mathcal{L}_2$-norm scaling rule and then compute its output noise variance due to product round-off assuming quantization of products before addition.

9.25 Scale the structures realized in Problem 9.13 using the $\mathcal{L}_2$-norm scaling rule and then compute the output noise variances due to product round-off assuming quantization of products before addition. What would be the output noise variances if quantization is carried out after addition?

9.26 Scale the structures realized in Problem 9.14 using the $\mathcal{L}_2$-norm scaling rule and then compute the output noise variances due to product round-off assuming quantization of products before addition. What would be the output noise variances if quantization is carried out after addition?

9.27 Scale the structures realized in Problem 9.15 using the $\mathcal{L}_2$-norm scaling rule and then compute the output noise variance due to product round-off assuming quantization of products before addition. What would be the output noise variance if quantization is carried out after addition?

9.28 (a) Calculate the output noise variance of the digital filter structure of Figure P9.9(a) due to product round-off before addition. Assume all numbers are fractions and represented in a two's-complement fixed-point representation with a wordlength of $b+1$ bits. Note that the multiplier "-1" does not generate noise. (b) Now consider the configuration of Figure P9.9(b), where the filters $A_1(z)$, $A_2(z)$, and $A_3(z)$ are implemented as in Figure P9.9(a) with the multiplier coefficients d_i replaced with d_1, d_2, and d_3, respectively. Calculate the output noise variance of this new digital filter structure due to product round-off before addition.

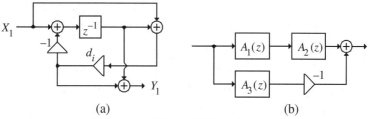

(a)   (b)

Figure P9.9

9.29 Consider the digital filter structure of Figure P9.10. Let σ_0^2 represent the total output noise variance due to product round-off in $G(z)$ and $A(z)$. If each delay in the realization of $G(z)$ is replaced with two delays, i.e., z^{-1} replaced with z^{-2}, but $A(z)$ is left unchanged, calculate the output noise variance due to product round-off in terms of σ_0.

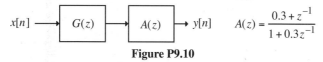

Figure P9.10

9.30 Show that there are $(R!)^2$ different possible realizations of a cascade of R second-order sections.

9.31 (a) What is the optimum pole-zero pairing and ordering of each of the following transfer functions for obtaining the smallest peak output noise due to product round-off under an $\mathcal{L}_2$-scaling rule? (b) Repeat part (a) if the objective is to minimize the output noise power due to product round-off under an $\mathcal{L}_\infty$-scaling rule.

(i) $H_1(z) = \dfrac{(z^2 + 0.8z + 0.2)(z^2 + 0.2z + 0.9)(z^2 + 0.3z + 0.5)}{(z^2 + 0.1z + 0.8)(z^2 + 0.2z + 0.4)(z^2 + 0.6z + 0.3)}$,

(ii) $H_2(z) = \dfrac{(z^2 + 0.1z + 0.7)(z^2 + 1.4z + 0.33)(z^2 + 0.5z + 0.6)}{(z^2 + 0.4z + 0.7)(z^2 - 0.2z + 0.9)(z^2 + 1.1z + 0.18)}$.

9.32 Derive the expressions for the SNR given in Table 9.5.

9.33 In Section 6.7.1, we showed that the first-order lowpass transfer function $H_{LP}(z)$ of Eq. (4.113) and the first-order highpass transfer function $H_{HP}(z)$ of Eq. (4.118) can be realized in the form of a parallel allpass structure of Figure 6.32, where $A_1(z)$ is a first-order allpass transfer function given by Eq. (6.65). The first-order allpass section can be implemented using a single multiplier α using any one of the four single-multiplier allpass structures of Figure 6.23. Show that the realization of $H_{LP}(z)$ in the form of Figure 6.32 exhibits low sensitivity in the passband at $\omega = 0$ with respect to the multiplier coefficient α, i.e.,

$$\left.\frac{\partial \left| H_{LP}(e^{j\omega}) \right|}{\partial \alpha}\right|_{\omega=0} = 0.$$

Likewise, show that the realization of $H_{HP}(z)$ in the form of Figure 6.32 exhibits low sensitivity in the passband at $\omega = \pi$ with respect to the multiplier coefficient α, i.e.,

$$\left.\frac{\partial \left| H_{HP}(e^{j\omega}) \right|}{\partial \alpha}\right|_{\omega=\pi} = 0.$$

9.34 In Section 6.7.2, we showed that the second-order bandpass transfer function $H_{BP}(z)$ of Eq. (4.107) and the second-order bandstop transfer function $H_{BS}(z)$ of Eq. (4.112) can be realized in the form of a parallel allpass structure of Figure 6.35, where $A_2(z)$ is a second-order allpass transfer function given by Eq. (6.68). The second-order allpass section can be implemented using two multipliers α and β by employing a cascaded lattice structure shown in Figure 6.36. Show that the realization of $H_{BP}(z)$ in the form of Figure 6.35 exhibits low sensitivity in the passband at the center frequency ω_0 with respect to the multiplier coefficients α and β, i.e.,

$$\frac{\partial \left| H_{BP}(e^{j\omega}) \right|}{\partial \alpha} \Bigg|_{\omega=\omega_0} = 0, \qquad \frac{\partial \left| H_{BP}(e^{j\omega}) \right|}{\partial \beta} \Bigg|_{\omega=\omega_0} = 0.$$

Likewise, show that the realization of $H_{BS}(z)$ in the form of Figure 6.35 exhibits low sensitivity in the passband at $\omega = 0$ and $\omega = \pi$ with respect to the multiplier coefficients α and β, i.e.,

$$\frac{\partial \left| H_{BS}(e^{j\omega}) \right|}{\partial \alpha} \Bigg|_{\omega=0} = 0, \qquad \frac{\partial \left| H_{BS}(e^{j\omega}) \right|}{\partial \beta} \Bigg|_{\omega=0} = 0,$$

$$\frac{\partial \left| H_{BS}(e^{j\omega}) \right|}{\partial \alpha} \Bigg|_{\omega=\pi} = 0, \qquad \frac{\partial \left| H_{BS}(e^{j\omega}) \right|}{\partial \beta} \Bigg|_{\omega=\pi} = 0.$$

9.35 In the parallel allpass realization of a bounded real (BR) transfer function $G(z)$, the passband of $G(z)$ is the stopband of its power-complementary transfer function $H(z)$. Show why low passband sensitivity of $G(z)$ does not imply low stopband sensitivity of $H(z)$.

9.36 Develop the expressions for the SNR of the second-order IIR filter structure of Figure 9.44 with and without error feedback for the following types of inputs: (a) WSS, white uniform density, (b) WSS, white Gaussian density ($\sigma_x = 1/3$), and (c) sinusoid with known frequency. The poles are at $z = (1 - \varepsilon)e^{\pm j\theta}$ with $\varepsilon \to 0$, $\theta \to 0$, and $\theta \gg \varepsilon$. Assume a $(b + 1)$-bit signed representations for the number representation.

9.37 Modify the coupled-form structure of Figure 9.10 to include error feedback and determine the normalized output round-off noise power of the modified structure. What are the appropriate values of the multiplier coefficients in the error-feedback loops to minimize the output round-off noise power? What is the expression for the output round-off noise power in this case?

9.38 Modify the Kingsbury structure of Figure P6.8 to include error feedback and determine the normalized output round-off noise power of the modified structure. What are the appropriate values of the multiplier coefficients in the error-feedback loops to minimize the output round-off noise power? What is the expression for the output round-off noise power in this case?

9.39 Show that the coupled-form structure of Figure 9.10 will not support zero-input limit cycles under magnitude quantization.

9.40 Show that the modified coupled-form structure of Figure P9.1 will not support zero-input limit cycles under magnitude quantization.

9.41 Show that the second-order structure of Figure P9.11 will not support zero-input limit cycles under magnitude quantization [Mee76].

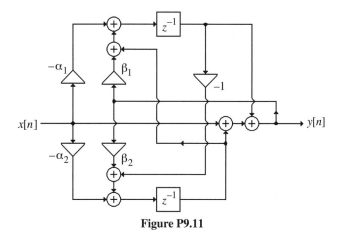

Figure P9.11

9.42 Determine the expression for the output round-off noise variance in the computation of a single sample of a length-N DFT using Goertzel's algorithm implemented in a fractional signed $(b+1)$-bit fixed-point arithmetic.

9.43 Determine the number of bits b to compute a single sample of a 512-point DFT of an input sequence of length 512 by direct computation with an SNR of 25 dB.

9.44 Derive Eq. (9.217).

9.45 Determine the number of bits needed to compute a single sample of a 512-point DFT of an input sequence of length 512 using a radix-2 decimation-in-time FFT algorithm with an SNR of 25 dB. Assume a distributed scaling to prevent overflow at the output.

9.15 MATLAB Exercises

M 9.1 Write a MATLAB program to plot the pole distribution of a second-order two-multiplier structure with multiplier coefficients represented in sign-magnitude form with b bits.

M 9.2 Using the program developed in Exercise M9.1, plot the pole distributions of the direct form, the coupled form of Figure 9.9, and the Kingsbury structure of Figure P6.9 for a 4-bit wordlength. Comment on the coefficient sensitivity of each of these structures from the pole distribution plots.

M 9.3 Modify Program 9_1 to design an elliptic lowpass IIR filter with the following specifications: passband edge at 0.5π, stopband edge at 0.55π, passband ripple of 0.01 dB, and minimum stopband attenuation of 60 dB. Quantize the transfer function coefficients to 6 bits using the M-function a2dT. Plot the magnitude responses and the pole-zero plots of the two transfer functions. Comment on your results.

M 9.4 Determine the factored form of a fifth-order elliptic lowpass transfer function with the following specifications: passband ripple of 0.5 dB, minimum stopband attenuation of 45 dB, and passband edge at 0.45π. From its pole-zero locations, determine the optimum pole-zero pairing and their ordering to minimize the output noise power under an $\mathcal{L}_\infty$-scaling rule. Verify your result using MATLAB.

M 9.5 Realize the transfer function of Exercise M9.4 as a parallel connection of two allpass filters and determine its power-complementary highpass transfer function. Plot on the same figure the gain responses of the two filters. Verify the low passband sensitivity of the parallel allpass realization by making a 1 percent change in the filter coefficients.

M 9.6 Write a MATLAB program to simulate a fourth-order transfer function of the form

$$H(z) = \frac{(1 + b_1 z^{-1} + b_2 z^{-2})(1 + b_3 z^{-1} + b_4 z^{-2})}{(1 + a_1 z^{-1} + a_2 z^{-2})(1 + a_3 z^{-1} + a_4 z^{-2})} \tag{9.219}$$

in a cascade form with each second-order section realized in direct form II. The input data to your program are the numerator coefficients $\{b_i\}$ and the denominator coefficients $\{a_i\}$ of the second-order sections. Using this program simulate two different cascade realizations of the following transfer function

$$H(z) = \frac{(1 - z^{-1})^2 (1 - 0.707 z^{-1} + z^{-2})}{(1 + 0.777 z^{-1} + 0.3434 z^{-2})(1 + 0.01877 z^{-1} + 0.801 z^{-2})}. \tag{9.220}$$

Using this program determine the impulse response of each pertinent scaling transfer functions and their approximate $\mathcal{L}_2$-norms. Based on this information, scale each realization using an $\mathcal{L}_2$-scaling rule and compute the total product round-off noise variance for each scaled structure assuming rounding before addition.

M 9.7 Write a MATLAB program to simulate a fourth-order transfer function of the form given in Eq. (9.219) in parallel forms I and II. The input data to your program are the numerator coefficients $\{b_i\}$ and the denominator coefficients $\{a_i\}$. Using this program simulate the two different parallel realizations of the transfer function of Eq. (9.220). Determine the impulse response of each pertinent scaling transfer functions and their approximate $\mathcal{L}_2$-norms. Based on this information, scale each realization using an $\mathcal{L}_2$-scaling rule and compute the total product round-off noise variance for each scaled structure assuming rounding before addition. Compare the output round-off noises of the two parallel structures with those of the cascade realizations of Exercise M9.6.

M 9.8 Write a MATLAB program to simulate a fourth-order transfer function of the form given in Eq. (9.219) using the Gray-Markel method. The input data to your program are the numerator coefficients $\{b_i\}$ and the denominator coefficients $\{a_i\}$. Using this program simulate the Gray-Markel realization of the transfer function of Eq. (9.220). Determine the impulse response of each pertinent scaling transfer function and its approximate $\mathcal{L}_2$-norm. Based on this information, scale the realization using an $\mathcal{L}_2$-scaling rule and compute the total product round-off noise variance for the scaled structure assuming rounding before addition. Compare the output round-off noise of the Gray-Markel realization with those of the cascade realizations of Exercise M9.6 and the two parallel structures of Exercise M9.7.

M 9.9 Using Program 9_7 investigate the granular limit cycles of Figure 9.46 for the following sets of values of the coefficient, the initial condition and the scale factor of the input impulse: (a) $\alpha = 0.5$, $y[-1] = 0.1$, $x[0] = 0.04$; (b) $\alpha = 0.5$, $y[-1] = 1$, $x[0] = 0.04$; and (c) $\alpha = 0.5$, $y[-1] = 10$, $x[0] = 6$. Comment on your results.

M 9.10 Modify Program 9_8 by replacing the M-function a2dR with the function a2dT and then demonstrate by running the modified program that the structure of Figure 9.49 does not exhibit overflow limit cycles if sign-magnitude truncation is used to truncate the sum of products of Eq. (9.194).

10 Multirate Digital Signal Processing

The digital signal processing structures discussed so far in this text belong to the class of single-rate systems since the sampling rates at the input and at the output and all internal nodes are the same. There are many applications where the signal of a given sampling rate needs to be converted into an equivalent signal with a different sampling rate. For example, in digital audio, three different sampling rates are presently employed: 32 kHz in broadcasting, 44.1 kHz in digital compact disk, and 48 kHz in digital audio tape (DAT) and other applications [Lag82]. Conversion of sampling rates of audio signals between these three different rates is often necessary in many situations. Another example is the pitch control of audio recordings usually performed by varying the tape recorder speed. However, such an approach in digital audio changes the sampling frequency of the digital signal and, as a result, conversion to the original sampling rate is needed [Lag82]. In the video applications, the sampling rates of NTSC (National Television Systems Committee) and PAL (Phase Alternate Line) composite video signals are, respectively, 14.3181818 MHz and 17.734475 MHz, whereas the sampling rates of the digital component video signal are 13.5 MHz and 6.75 MHz for the luminance and the color-difference signals, respectively [Lut91].[1] There are other applications where it is convenient (and often judicious) to have unequal rates of sampling at the filter input and output and at internal nodes. Examples of such sampling rate alterations are the oversampling A/D and D/A converters discussed in Sections 5.8.5 and 5.9.3, respectively, and analyzed in detail in Sections 11.12 and 11.13, respectively. Additional applications of sampling rate alterations are also given in Sections 11.8 through 11.11.

To achieve different sampling rates at different stages, multirate digital signal processing systems employ the *down-sampler* and the *up-sampler*, the two basic sampling rate alteration devices in addition to the conventional elements such as the adder, the multiplier, and the delay. Discrete-time systems with unequal sampling rates at various parts of the system are called *multirate systems* and are the subject of discussion of this chapter.

We first examine the input-output relations of the up-sampler and the down-sampler both in the time-domain and the transform-domain. As in many applications, cascade connections of the basic sampling rate alteration devices and digital filters are employed; some basic cascade equivalences are then reviewed. For sampling rate alterations, the basic sampling rate alteration devices are invariably employed together with lowpass digital filters. The frequency response specifications of these filters are developed next. A computationally more efficient approach to sampling rate alteration is based on a multistage implementation that is then illustrated by means of a specific design problem. The polyphase decomposition of a sequence is reexamined next in the framework of multirate theory, and its application in developing computationally efficient sampling rate alteration systems is illustrated.

Many multirate systems employ a bank of filters with either a common input or a summed output. These filter banks are introduced next, and the design and computationally efficient implementation of a class of such filter banks are discussed. It is followed by an introduction of certain special types of transfer

---

[1] CCIR Recommendation No. 601.

functions that are particularly attractive in the design of computationally efficient multirate systems. The rest of the chapter is devoted to a discussion of quadrature-mirror filter banks that find applications in signal compression and other areas.

10.1 The Basic Sample Rate Alteration Devices

The two basic components in sampling rate alteration are the up-sampler and the down-sampler introduced earlier in Section 2.1.2 where we examined their input-output relations in the time-domain. However, it is also instructive to analyze their operations in the frequency-domain. This will point out why these devices must be used with additional filters. In addition, the frequency-domain analysis provides the basic foundation for analyzing more complex multirate systems introduced in the latter parts of the chapter.

10.1.1 Time-Domain Characterization

We reexamine the time-domain characterizations of the two basic sampling rate alteration devices here again. An up-sampler with an up-sampling factor L, where L is a positive integer, develops an output sequence $x_u[n]$ with a sampling rate that is L times larger than that of the input sequence $x[n]$. The up-sampling operation is implemented by inserting $L-1$ equidistant zero-valued samples between two consecutive samples of the input sequence $x[n]$ according to the relation

$$x_u[n] = \begin{cases} x[n/L], & n = 0, \pm L, \pm 2L, \ldots, \\ 0, & \text{otherwise.} \end{cases} \tag{10.1}$$

The up-sampling operation is illustrated in the following example using MATLAB.

EXAMPLE 10.1   Program 10_1 given below can be used to study the up-sampling of a sinusoidal input sequence. Its input data are the length of the input sequence, the up-sampling factor, and the frequency of the sinusoid in Hz. It then plots the input sequence and its up-sampled version.

```
% Program 10_1
% Illustration of Up-Sampling by an Integer Factor
%
clf;
N = input('Input length = ');
L = input('Up-sampling Factor = ');
fo = input('Input signal frequency = ');
%  Generate the input sinusoidal sequence
n = 0:N-1;
x = sin(2*pi*fo*n);
% Generate the up-sampled sequence
y = zeros(1, L*length(x));
y([1: L: length(y)]) = x;
% Plot the input and the output sequences
subplot(2,1,1)
stem(n,x);
title('Input Sequence');
xlabel('Time index n');ylabel('Amplitude');
subplot(2,1,2)
stem(n,y(1:length(x)));
```

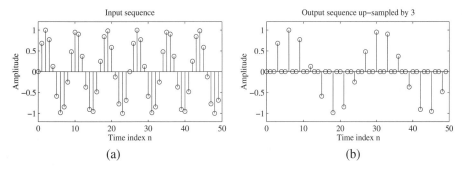

Figure 10.1: Illustration of the up-sampling process.

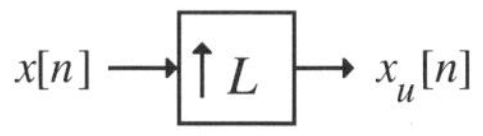

$$x[n] \longrightarrow \boxed{\uparrow L} \longrightarrow x_u[n]$$

Figure 10.2: Block diagram representation of an up-sampler.

```
title(['Output sequence up-sampled by', num2str(L)]);
xlabel('Time index n');ylabel('Amplitude');
```

Figure 10.1 shows the result obtained for a length-50 sinusoidal sequence with a frequency of 0.12 Hz and with an up-sampling factor of 3.

The block diagram representation of the up-sampler, also called a *sampling rate expander* or simply an *expander*, is shown in Figure 10.2.

In practice, the zero-valued samples inserted by the up-sampler are replaced with appropriate nonzero values using some type of filtering process in order that the new higher-rate sequence be useful. This process, called *interpolation* is discussed later in this chapter.

On the other hand, the down-sampler with a down-sampling factor M, where M is a positive integer, develops an output sequence $y[n]$ with a sampling rate that is $(1/M)$th of that of the input sequence $x[n]$. The down-sampling operation is implemented by keeping every Mth sample of the input sequence and removing $M - 1$ in-between samples, to generate the output sequence according to the relation

$$y[n] = x[nM]. \tag{10.2}$$

As a result, all input samples with indices equal to an integer multiple of M are retained at the output and all others are discarded, as illustrated in the following example.

EXAMPLE 10.2   We investigate the down-sampling of a sinusoidal input sequence using MATLAB. To this end we utilize Program 10.2 given below. Its input data are the length of the input sequence, the down-sampling factor, and the frequency of the sinusoid in Hz. It then plots the input sequence and its down-sampled version.

```
% Program 10_2
% Illustration of Down-Sampling by an Integer Factor
%
clf;
N = input('Output length = ');
M = input('Down-sampling Factor = ');
fo = input('Input signal frequency = ');
```

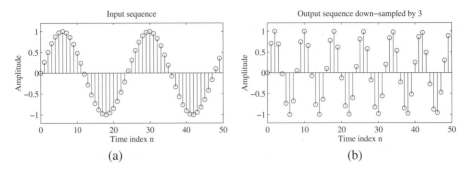

Figure 10.3: Illustration of the down-sampling process.

Figure 10.4: Block diagram representation of a down-sampler.

```
% Generate the input sinusoidal sequence
n = 0 : N-1;
m = 0 : N*M-1;
x = sin(2*pi*fo*m);
% Generate the down-sampled sequence
y = x([1 : M : length(x)]);
% Plot the input and the output sequences
subplot(2,1,1)
stem(n, x(1:N));
title('Input Sequence');
xlabel('Time index n');ylabel('Amplitude');
subplot(2,1,2)
stem(n, y);
title(['Output sequence down-sampled by ',num2str(M)]);
ylabel('Amplitude');xlabel('Time index n');
```

The result obtained for a length-50 sinusoidal sequence with a frequency 0.042 Hz and with a down-sampling factor of 3 are shown in Figure 10.3.

The block diagram representation of the *down-sampler* or *sampling rate compressor* is shown in Figure 10.4.

The sampling periods involved have not been explicitly shown in Figures 10.2 and 10.4. This is in the interest of simplicity and in view of the fact that the mathematical theory of multirate systems can be understood without bringing T or the sampling frequency F_T into the picture. It is instructive in the beginning to explicitly see the time dimensions at various stages in the sampling rate alteration process, as indicated in Figure 10.5. In the remainder of this book, however, the explicit appearance of T or the sampling frequency F_T is not shown unless their actual values are relevant.

The up-sampler and the down-sampler are linear but time-varying discrete-time systems. The time-varying property of these devices is easy to show. Consider for example, the factor-of-M down-sampler defined by Eq. (10.2). Its output $y_1[n]$ for an input $x_1[n] = x[n - n_0]$ is then given by

$$y_1[n] = x_1[Mn] = x[Mn - n_0].$$

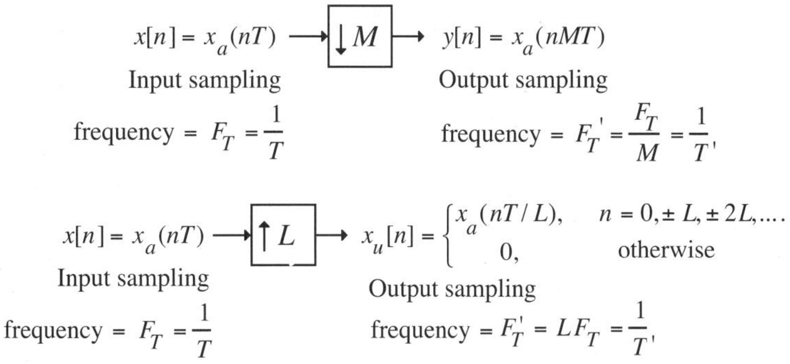

$$x[n] = x_a(nT) \longrightarrow \boxed{\downarrow M} \longrightarrow y[n] = x_a(nMT)$$

Input sampling

frequency $= F_T = \dfrac{1}{T}$

Output sampling

frequency $= F_T' = \dfrac{F_T}{M} = \dfrac{1}{T'}$

$$x[n] = x_a(nT) \longrightarrow \boxed{\uparrow L} \longrightarrow x_u[n] = \begin{cases} x_a(nT/L), & n = 0, \pm L, \pm 2L, \dots \\ 0, & \text{otherwise} \end{cases}$$

Input sampling

frequency $= F_T = \dfrac{1}{T}$

Output sampling

frequency $= F_T' = LF_T = \dfrac{1}{T'}$

Figure 10.5: The sampling rate alteration building blocks with sampling rates explicitly shown.

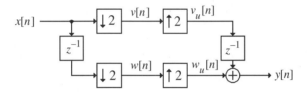

Figure 10.6: A simple multirate system.

But from Eq. (10.2),

$$y[n - n_0] = x[M(n - n_0)] = x[Mn - Mn_0] \neq y_1[n].$$

Likewise, it can be shown that the up-sampler defined by Eq. (10.1) is also a time-varying device (Problem 10.1). However, they are both linear systems (Problem 10.2).

The up-sampler and the down-sampler building blocks of Figures 10.2 and 10.4 are often used together in a number of applications involving multirate signal processing and are discussed in more detail in this and the following chapter. For example, one application of using both types of sampling rate alteration devices is to achieve a sampling rate change by a rational number rather than an integer value. Example 10.3 below illustrates another application.

EXAMPLE 10.3   Consider the multirate system of Figure 10.6. Its operation can be analyzed by writing down the relations between various signal variables and the input as depicted below:

n :	0	1	2	3	4	5	6	7	8
$x[n]$:	$x[0]$	$x[1]$	$x[2]$	$x[3]$	$x[4]$	$x[5]$	$x[6]$	$x[7]$	$x[8]$
$v[n]$:	$x[0]$	$x[2]$	$x[4]$	$x[6]$	$x[8]$	$x[10]$	$x[12]$	$x[14]$	$x[16]$
$w[n]$:	$x[-1]$	$x[1]$	$x[3]$	$x[5]$	$x[7]$	$x[9]$	$x[11]$	$x[13]$	$x[15]$
$v_u[n]$:	$x[0]$	0	$x[2]$	0	$x[4]$	0	$x[6]$	0	$x[8]$
$w_u[n]$:	$x[-1]$	0	$x[1]$	0	$x[3]$	0	$x[5]$	0	$x[7]$
$v_u[n-1]$:	0	$x[0]$	0	$x[2]$	0	$x[4]$	0	$x[6]$	0
$y[n]$:	$x[-1]$	$x[0]$	$x[1]$	$x[2]$	$x[3]$	$x[4]$	$x[5]$	$x[6]$	$x[7]$

It can be seen from the above that the output $y[n]$ of the multirate system of Figure 10.6 is given by $y[n] = v_u[n-1] + w_u[n]$, which is simply $x[n-1]$.

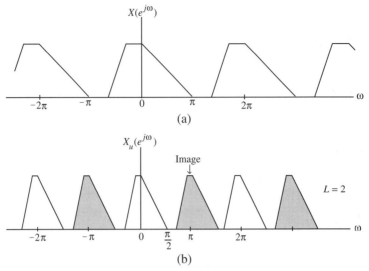

Figure 10.7: Effects of up-sampling in the frequency domain: (a) input spectrum, and (b) output spectrum for $L = 2$.

10.1.2 Frequency-Domain Characterization

We first derive the relations between the spectrums of the input and the output of a factor-of-2 up-sampler. From the input-output relation of the factor-of-L up-sampler given by Eq. (10.1), we arrive at the corresponding relation for the factor-of-2 up-sampler:

$$x_u[n] = \begin{cases} x[n/2], & n = 0, \pm 2, \pm 4, \ldots, \\ 0, & \text{otherwise.} \end{cases} \tag{10.3}$$

In terms of the z-transform, the input-output relation is then given by

$$X_u(z) = \sum_{n=-\infty}^{\infty} x_u[n]z^{-n} = \sum_{\substack{n=-\infty \\ n \text{ even}}}^{\infty} x[n/2]z^{-n}$$

$$= \sum_{m=-\infty}^{\infty} x[m]z^{-2m} = X(z^2). \tag{10.4}$$

In a similar manner, we can show that for the factor-of-L up-sampler,

$$X_u(z) = X(z^L). \tag{10.5}$$

Let us examine the implication of the above relation on the unit circle. For $z = e^{j\omega}$ the above equation becomes $X_u(e^{j\omega}) = X(e^{j\omega L})$. Figure 10.7(a) shows the DTFT $X(e^{j\omega})$ that has been assumed to be a real function for convenience. Moreover, the DTFT $X(e^{j\omega})$ shown is not an even function of ω, implying that $x[n]$ is a complex sequence. The asymmetric response has been purposely chosen to illustrate more clearly the effect of up-sampling.

As shown in Figure 10.7(b), a factor-of-2 sampling rate expansion thus leads to a 2-fold repetition of $X(e^{j\omega})$, indicating that the Fourier transform is compressed by a factor of 2. This process is called *imaging* because we get an additional "image" of the input spectrum. In the case of a factor-of-L sampling rate expansion there will be $L - 1$ additional images of the input spectrum in the baseband. Thus, a

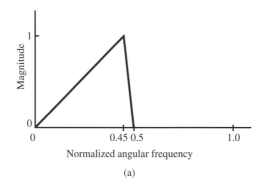

(a)

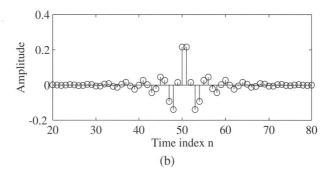

(b)

Figure 10.8: (a) Desired magnitude response and (b) corresponding time sequence.

spectrum $X(e^{j\omega})$ bandlimited to the low-frequency region does not look like a low-frequency spectrum after up-sampling, because of the insertion of zero-valued samples between the nonzero samples of $x_u[n]$. Lowpass filtering of $x_u[n]$ removes the $L - 1$ images and in effect "fills in" the zero-valued samples in $x_u[n]$ with interpolated sample values.

We next illustrate the frequency-domain properties of the up-sampler using MATLAB. The input is a causal finite-length sequence with a bandlimited frequency response generated using the M-file `fir2`. The input to `fir2` is as follows: length of the sequence is `100`, the desired magnitude response vector `mag = [0 1 0 0]`, and the vector of frequency points `freq = [0 0.45 0.5 1]`. The desired magnitude response is thus as indicated in Figure 10.8(a). A plot of the middle 61 samples of the signal generated is shown in Figure 10.8(b). To investigate the effect of up-sampling, we use Program 10_3 which follows.

The input data called by the program is the up-sampling factor L. The program determines the output of the up-sampler and then plots the input and output spectrums, as indicated in Figure 10.9 for `L = 5`. It can be seen from this figure that, as expected, the output spectrum consists of a factor-of-5 compressed version of the input spectrum followed by `L-1 = 4` images.

```
% Program 10_3
% Effect of Up-Sampling in the Frequency Domain
% Use fir2 to create a bandlimited input sequence
freq = [0 0.45 0.5 1];
mag = [0 1 0 0];
```

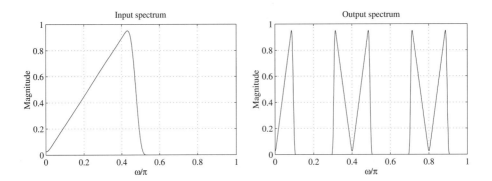

Figure 10.9: MATLAB-generated input and output spectrum of a factor-of-5 up-sampler.

```
x = fir2(99, freq, mag);
% Evaluate and plot the input spectrum
[Xz, w] = freqz(x, 1, 512);
plot(w/pi, abs(Xz)); grid
xlabel('\omega/\pi'); ylabel('Magnitude');
title('Input spectrum');
pause
% Generate the up-sampled sequence
L = input('Type in the up-sampling factor = ');
y = zeros(1, L*length(x));
y([1: L: length(y)]) = x;
% Evaluate and plot the output spectrum
[Yz, w] = freqz(y, 1, 512);
plot(w/pi, abs(Yz)); grid
xlabel('\omega/\pi'); ylabel('Magnitude');
title('Output spectrum');
```

We now derive the relations between the spectrums of the input and the output of a down-sampler. Applying the z-transform to the input-output relation given in Eq. (10.2), we arrive at

$$Y(z) = \sum_{n=-\infty}^{\infty} x[Mn]z^{-n}. \tag{10.6}$$

The expression on the right-hand side of the above equation cannot be directly expressed in terms of $X(z)$. To get around this problem, we define an intermediate sequence $x_{\text{int}}[n]$ whose sample values are the same as that of $x[n]$ at the values of n that are multiples of M and are zeros at other values of n:

$$x_{\text{int}}[n] = \begin{cases} x[n], & n = 0, \pm M, \pm 2M, \ldots, \\ 0, & \text{otherwise.} \end{cases} \tag{10.7}$$

Then,

$$Y(z) = \sum_{n=-\infty}^{\infty} x[Mn]z^{-n} = \sum_{n=-\infty}^{\infty} x_{\text{int}}[Mn]z^{-n}$$

$$= \sum_{k=-\infty}^{\infty} x_{\text{int}}[k] z^{-k/M} = X_{\text{int}}(z^{1/M}). \tag{10.8}$$

Now $x_{\text{int}}[n]$ can be formally related to $x[n]$ through $x_{\text{int}}[n] = c[n]x[n]$, where $c[n]$ is defined by

$$c[n] = \begin{cases} 1, & n = 0, \pm M, \pm 2M, \ldots, \\ 0, & \text{otherwise.} \end{cases} \tag{10.9}$$

A convenient representation of $c[n]$ is given by (Problem 10.4)

$$c[n] = \frac{1}{M} \sum_{k=0}^{M-1} W_M^{kn}, \tag{10.10}$$

where $W_M = e^{-j2\pi/M}$ is the quantity defined in Eq. (3.24). Substituting $x_{\text{int}}[n] = c[n]x[n]$ and making use of Eq. (10.10) in the z-transform of $x_{\text{int}}[n]$, we obtain

$$X_{\text{int}}(z) = \sum_{n=-\infty}^{\infty} c[n]x[n]z^{-n} = \frac{1}{M} \sum_{n=-\infty}^{\infty} \left(\sum_{k=0}^{M-1} W_M^{kn} \right) x[n]z^{-n}$$

$$= \frac{1}{M} \sum_{k=0}^{M-1} \left(\sum_{n=-\infty}^{\infty} x[n] W_M^{kn} z^{-n} \right) = \frac{1}{M} \sum_{k=0}^{M-1} X\left(z W_M^{-k} \right). \tag{10.11}$$

The desired input-output relation in the transform-domain for a factor-of-M down-sampler is then obtained by substituting Eq. (10.11) in Eq. (10.8), resulting in

$$Y(z) = \frac{1}{M} \sum_{k=0}^{M-1} X(z^{1/M} W_M^{-k}). \tag{10.12}$$

To understand the implication of the above relation, consider a factor-of-2 down-sampler with an input $x[n]$ whose spectrum is as shown in Figure 10.10(a). As before, for convenience we assume again $X(e^{j\omega})$ to be a real function with an asymmetric frequency response. From Eq. (10.12) we get

$$Y(e^{j\omega}) = \frac{1}{2} \left\{ X(e^{j\omega/2}) + X(-e^{j\omega/2}) \right\}. \tag{10.13}$$

The plot of $\frac{1}{2}X(e^{j\omega/2})$ is shown by the solid line in Figure 10.10(b). To determine the relation of the second term in Eq. (10.13) with respect to the first, we observe next that

$$X(-e^{j\omega/2}) = X(e^{j(\omega-2\pi)/2}), \tag{10.14}$$

indicating that the second term $X(e^{-j\omega/2})$ in Eq. (10.13) is obtained simply by shifting the first term $X(e^{j\omega/2})$ to the right by an amount 2π, as shown by the dotted lines in Figure 10.10(b). The plots of the two terms in Eq. (10.13) have an overlap, and hence, in general, the original "shape" of $X(e^{j\omega})$ is lost when $x[n]$ is down-sampled. This overlap causes the *aliasing* that takes place due to undersampling (i.e., down-sampling). There is no overlap, i.e., no aliasing, only if $X(e^{j\omega})$ is zero for $|\omega| \geq \pi/2$. Note that $Y(e^{j\omega})$ in Eq. (10.13) is indeed periodic with a period 2π, even though the stretched version $X(e^{j\omega})$ is periodic with a period 4π. For the general case, the situation is essentially the same, and the relations between the Fourier transform of the output and the input of the factor-of-M down-sampler is given by

$$Y(e^{j\omega}) = \frac{1}{M} \sum_{k=0}^{M-1} X(e^{j(\omega-2\pi k)/M}). \tag{10.15}$$

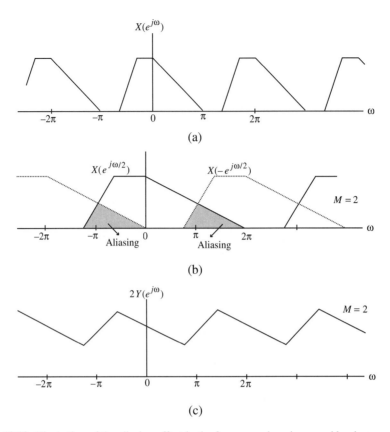

Figure 10.10: Illustration of the aliasing effect in the frequency-domain caused by down-sampling.

The above relation implies that $Y(e^{j\omega})$ is a sum of M uniformly shifted and stretched versions of $X(e^{j\omega})$ then scaled by a factor $1/M$. Aliasing due to a factor-of-M down-sampling is absent if and only if the signal $x[n]$ is bandlimited to $\pm\pi/M$, as shown in Figure 10.11 for $M = 2$.

We next illustrate the aliasing effect caused by down-sampling using MATLAB. To investigate the effect of down-sampling, we use Program 10_4 which follows. The input signal is again the signal generated using `fir2` with a triangular magnitude response, as in Figure 10.8(a). However, here the frequency vector has been selected to be `freq = [0 0.42 0.48 1]` to ensure that there are no appreciable signal components above the normalized frequency of 0.5. The input data called by the program is the down-sampling factor M. The program generates the spectrums of the original input and the down-sampled output.

```
% Program 10_4
% Effect of Down-Sampling in the Frequency Domain
% Use fir2 to create a bandlimited input sequence
freq = [0 0.42 0.48 1];
mag = [0 1 0 0];
x = fir2(101, freq, mag);
% Evaluate and plot the input spectrum
[Xz, w] = freqz(x, 1, 512);
plot(w/pi, abs(Xz)); grid
```

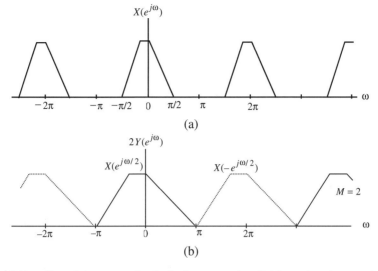

Figure 10.11: Effect of down-sampling in the frequency-domain illustrating absence of aliasing.

```
xlabel('\omega/\pi'); ylabel('Magnitude');
title('Input spectrum');
pause
% Generate the down-sampled sequence
M = input('Type in the down-sampling factor = ');
y = x([1: M: length(x)]);
% Evaluate and plot the output spectrum
[Yz, w] = freqz(y, 1, 512);
plot(w/pi, abs(Yz)); grid
xlabel('\omega/\pi'); ylabel('Magnitude');
title('Output spectrum');
```

The plots generated by the above program are shown in Figure 10.12. The input spectrum is shown in Figure 10.12(a). Since the input signal is bandlimited to $\pi/2$, the output spectrum for a down-sampling factor of M = 2 shown in Figure 10.12(b) is nearly of the same shape as the input spectrum, except it has been stretched by a factor of 2 in frequency and its magnitude is reduced by one-half as predicted by the factor 1/2 in Eq. (10.13). On the other hand, the output spectrum for a down-sampling factor of M = 3 shown in Figure 10.12(c) shows a severe distortion caused by the aliasing.

Any linear discrete-time multirate system can be analyzed in the transform-domain by using the input-output relations of the up-sampler and the down-sampler given by Eqs. (10.5) and (10.12), respectively. We illustrate the applications of these relations in the latter parts of this chapter.

10.1.3 Cascade Equivalences

As we shall observe later, a complex multirate system is formed by an interconnection of the basic sampling rate alteration devices and the components of an LTI digital filter. In many applications, these devices appear in a cascade form. An interchange of the positions of the branches in a cascade often can lead to a computationally efficient realization. We investigate certain specific cascade connections and their equivalences, which leaves input-output relations invariant.

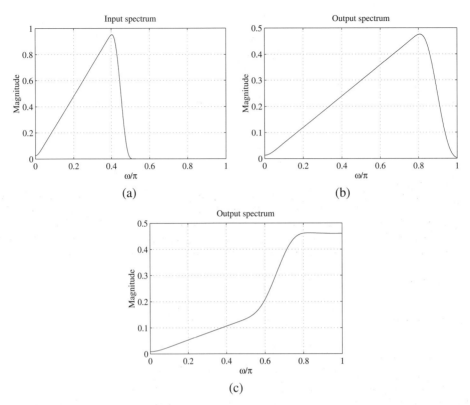

Figure 10.12: (a) Input spectrum, (b) output spectrum for a down-sampling factor of $M = 2$, and (c) output spectrum for a down-sampling factor of $M = 3$.

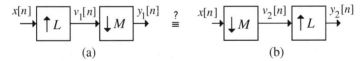

Figure 10.13: Two different cascade arrangements of a down-sampler and an up-sampler.

The basic sampling rate alteration devices can be used to change the sampling rate of a signal by an integer factor only. Therefore, to implement a fractional change in the sampling rate it follows that a cascade of a down-sampler and an up-sampler should be used. It is of interest to determine the condition under which a cascade of a factor-of-M down-sampler with a factor-of-L up-sampler (Figure 10.13) is *interchangeable*, with no change in the input-output relation. It can be shown that this interchange is possible if and only if M and L are *relatively prime*, i.e., M and L do not have a common factor that is an integer $k > 1$ (Problem 10.5) [Vai90]. We discuss fractional changes in the sampling rate in Section 10.2.2.

Two other simple cascade equivalence relations are depicted in Figure 10.14. The validity of these equivalences can be readily established using Eqs. (10.5) and (10.12) and are left as exercises (Problem 10.6). These rules enable us to move the basic sampling rate alteration devices in multirate networks to more advantageous positions. Such rules are extremely useful in the design and analysis of more complicated systems, as we shall demonstrate later.

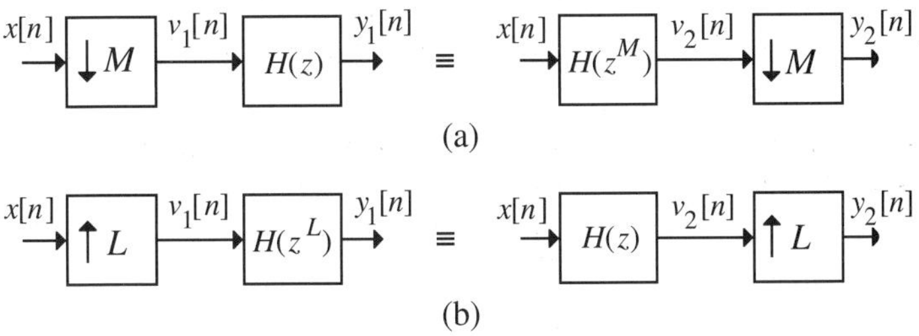

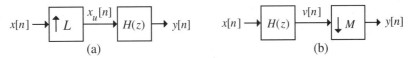

Figure 10.14: Cascade equivalences: (a) equivalence #1, and (b) equivalence #2.

Figure 10.15: Filters in sampling rate alteration systems: (a) interpolator and (b) decimator.

10.2 Filters in Sampling Rate Alteration Systems

From the sampling theorem introduced in Section 5.2.1, it is known that the sampling rate of a critically sampled discrete-time signal with a spectrum occupying the full Nyquist range cannot be reduced any further since such a reduction will introduce aliasing. Hence, the bandwidth of a critically sampled signal must first be reduced by lowpass filtering before its sampling rate is reduced by a down-sampler. Likewise, the zero-valued samples introduced by an up-sampler must be interpolated to more appropriate values for an effective sampling rate increase. As we shall show next, this interpolation can be simply achieved by digital lowpass filtering. We consider in this section some of the issues concerning these lowpass filters. We first develop the frequency response specifications of these lowpass filters. Next, we illustrate the decimation and interpolation of sequences using MATLAB. Finally, we investigate the computational complexity issues of the lowpass digital filters.

10.2.1 Filter Specifications

Since up-sampling causes periodic repetition of the basic spectrum (Figure 10.7), the unwanted images in the spectra of the up-sampled signal $x_u[n]$ must be removed by using a lowpass filter $H(z)$, called the *interpolation filter*, as indicated in Figure 10.15(a). On the other hand, as indicated in Figure 10.15(b), prior to down-sampling, the signal $v[n]$ should be bandlimited to $|\omega| < \pi/M$ by means of a lowpass filter $H(z)$, called the *decimation filter*, to avoid aliasing caused by down-sampling. The system of Figure 10.15(a) is often called an *interpolator* while the system of Figure 10.15(b) is called a *decimator*.[2]

The specifications for the lowpass filter in Figure 10.15 can now be derived. We first develop the specifications for the interpolation filter. Assume $x[n]$ has been obtained by sampling a bandlimited continuous-time signal $x_a(t)$ at the Nyquist rate. If $X_a(j\Omega)$ and $X(e^{j\omega})$ denote the Fourier transforms of

---

[2]It has been tacitly assumed here that the discrete-time signal to be interpolated or decimated has a lowpass frequency response and, as a result, the desired interpolation or the decimation filter is a lowpass filter. However, if the discrete-time signal to be interpolated or decimated has a highpass (bandpass) frequency response, then the desired interpolation or decimation filter is a highpass (bandpass) filter.

$x_a(t)$ and $x[n]$, respectively, from Eq. (5.14a), it follows that these Fourier transforms are related through

$$X(e^{j\omega}) = \frac{1}{T_0} \sum_{k=-\infty}^{\infty} X_a\left(\frac{j\omega - j2\pi k}{T_0}\right), \tag{10.16}$$

where T_0 is the sampling period. Since the sampling is being done at the Nyquist rate, there is no overlap between the shifted spectras of $X_a(j\omega/T_0)$. If we instead sample $x_a(t)$ at a much higher rate $T = T_0/L$ yielding $y[n]$, its Fourier transform $Y(e^{j\omega})$ is related to $X_a(j\Omega)$ through

$$Y(e^{j\omega}) = \frac{1}{T} \sum_{k=-\infty}^{\infty} X_a\left(\frac{j\omega - j2\pi k}{T}\right) = \frac{L}{T_0} \sum_{k=-\infty}^{\infty} X_a\left(\frac{j\omega - j2\pi k}{(T_0/L)}\right). \tag{10.17}$$

On the other hand, if we pass $x[n]$ through a factor-of-L up-sampler generating $x_u[n]$, from Eq. (10.5) the relation between the Fourier transform $X_u(e^{j\omega})$ and $X(e^{j\omega})$ is given by

$$X_u(e^{j\omega}) = X(e^{j\omega L}). \tag{10.18}$$

It follows from Eqs. (10.16) to (10.18) that if $x_u[n]$ is passed through an ideal lowpass filter with a cutoff at π/L and a gain of L, the output of the filter will be precisely $y[n]$.

In practice, a transition band is provided to ensure the realizability and stability of the lowpass interpolation filter $H(z)$. Hence, the desired lowpass filter should have a stopband edge at $\omega_s = \pi/L$ and a passband edge ω_p close to ω_s to reduce the distortion of the spectrum of the signal $x[n]$.[3] If ω_c denotes the highest frequency that needs to be preserved in the signal to be interpolated, the passband edge ω_p of the lowpass filter should be at $\omega_p = \omega_c/L$. Summarizing, the specifications for the lowpass interpolation filter are thus given by

$$\left|H(e^{j\omega})\right| = \begin{cases} L, & |\omega| \le \omega_c/L, \\ 0, & \pi/L \le |\omega| \le \pi. \end{cases} \tag{10.19}$$

It should be noted that in many interpolation applications, one requirement is to ensure that the input samples are not changed at the output. This requirement can be satisfied by a Nyquist (Lth band) interpolation filter, which is discussed in Section 10.7.

In a similar manner, we can develop the specifications for the lowpass decimation filter that are given by

$$\left|H(e^{j\omega})\right| = \begin{cases} 1, & |\omega| \le \omega_c/M, \\ 0, & \pi/M \le |\omega| \le \pi, \end{cases} \tag{10.20}$$

where ω_c denotes the highest frequency that needs be preserved in the decimated signal.

The effects of decimation and interpolation in the frequency-domain are illustrated in Figures 10.16 and 10.17, respectively, for $M = 2$ and $L = 2$. Figure 10.16(a) shows the spectrum $X(e^{j\omega})$ of the input signal $x[n]$ and the spectrum $H(e^{j\omega})$ of the decimation filter. The passband edge and the stopband edge of the decimation filters are, respectively, $\omega_c/2$ and $\pi/2$, where ω_c is the highest frequency in the input signal that is to be preserved at the output of the decimator. From Eq. (10.13), the frequency response $Y(e^{j\omega})$ of the decimator output $y[n]$ is now given by

$$Y(e^{j\omega}) = \tfrac{1}{2}\left\{V(e^{j\omega/2}) + V(-e^{j\omega/2})\right\},$$

where $V(e^{j\omega})$ is the frequency response of the filter output $v[n]$. Due to prior filtering there is no overlap in these spectra, resulting in no aliasing at the decimator output with the output spectrum $Y(e^{j\omega})$, as

[3]The meanings of ω_p, ω_s, δ_p, and δ_s in this chapter are precisely as in Section 7.1.1.

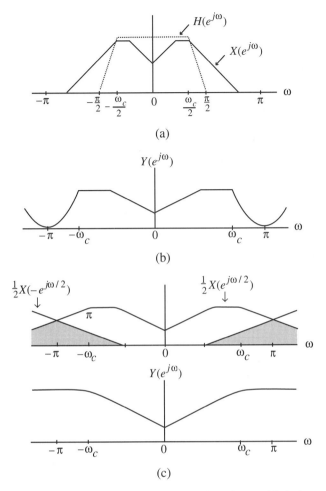

Figure 10.16: Spectrum of (a) the input $x[n]$, (b) the output $y[n]$ of the factor-of-2 decimator with $x[n]$ filtered, and (c) the output $y[n]$ of the factor-of-2 decimator with no filtering of $x[n]$ showing the effect of aliasing. The spectrum of the decimation filter $H(z)$ is shown in (a) with dotted lines.

sketched in Figure 10.16(b). However, it is not possible to recover the original input signal $x[n]$ from the decimated version $y[n]$. The filter $H(z)$ is used to preserve $X(e^{j\omega})$ in the range $-\omega_c/2 < \omega < \omega_c/2$, and one can reconstruct this portion exactly from $y[n]$. On the other hand, if there is no filtering of the input $x[n]$ prior to down-sampling, the two components $\frac{1}{2}X(e^{j\omega/2})$ and $\frac{1}{2}X(-e^{j\omega/2})$ will overlap, resulting in a severe aliasing at the decimator output, as indicated in Figure 10.16(c). It should be noted that the sequence $x[n]$ corresponding to the frequency response $X(e^{j\omega})$ of Figure 10.16(a) can be recovered from its decimated version if the decimation factor here is 4/3. We discuss fractional sampling rate alteration in Section 10.2.2.

Likewise, in the case of the factor-of-2 interpolation, the spectrum $V(e^{j\omega})$ of the up-sampler output is given by $X(e^{j2\omega})$. Figure 10.17(b) shows $V(e^{j\omega})$ for an input spectrum indicated in Figure 10.17(a). The spectrum $Y(e^{j\omega})$ of the interpolator output obtained by filtering $v[n]$ is thus as sketched in Figure 10.17(c).

The design of $H(z)$ is a standard IIR or FIR lowpass filter design problem. Any of the techniques outlined in Chapter 7 can be applied here for the design of these lowpass filters.

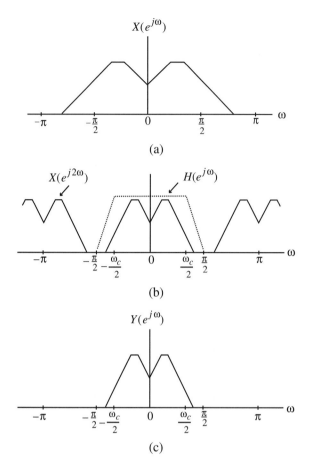

Figure 10.17: Spectrum of (a) the input $x[n]$, (b) the output $x[n]$ of the up-sampler, and (c) the output $y[n]$ of the factor-of-2 interpolator. The spectrum of the interpolation filter $H(z)$ is shown in (b) with dotted lines.

10.2.2 Filters for Fractional Sampling Rate Alteration

A fractional change in the sampling rate can be achieved by cascading a factor-of-M decimator with a factor-of-L interpolator, where M and L are positive integers. Such a cascade is equivalent to a decimator with a decimation factor of M/L or, alternatively, to an interpolator with an interpolation factor of L/M. There are two possible such cascade connections, as sketched in Figure 10.18. Of these two, the scheme shown in Figure 10.18(b) is more efficient since only one of the filters, $H_u(z)$ or $H_d(z)$, is adequate to serve as the interpolation filter and the decimation filter, depending on which one of the two stopband frequencies, π/L or π/M, is a minimum. It should be noted also that the sampling rate alteration system of Figure 10.18(a) will, in general, preserve less of the signal's frequency content than that of Figure 10.18(b). Hence, the desired configuration for the fractional sampling rate alteration is as indicated in Figure 10.18(c), where the lowpass filter $H(z)$ has a normalized stopband cutoff frequency at

$$\omega_s = \min\left(\frac{\pi}{L}, \frac{\pi}{M}\right) \tag{10.21}$$

which suppresses the imaging caused by the interpolator while at the same time ensuring the absence of aliasing that would otherwise be caused by the decimator.

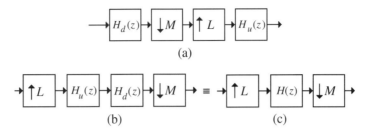

Figure 10.18: General schemes for increasing the sampling rate by L/M.

10.2.3 Computational Requirements

As indicated earlier, the lowpass decimation or interpolation filter can be designed either as an FIR or an IIR digital filter. In the case of single-rate digital signal processing, IIR filters are, in general, computationally more efficient than the FIR digital filters, and are therefore preferred in applications where computational cost needs to be minimized. This issue is not quite the same in the case of multirate digital signal processing, as elaborated next.

Consider the factor-of-M decimator structure of Figure 10.15(b). If the decimation filter $H(z)$ is an FIR filter of length N implemented in a direct form, then

$$v[n] = \sum_{m=0}^{N-1} h[m]x[n-m]. \tag{10.22}$$

Now, the down-sampler keeps only every Mth sample of $v[n]$ at its output. As a result it is sufficient to compute $v[n]$ using Eq. (10.22) only for values of n that are multiples of M and skip the computations of the in-between $M-1$ samples. This leads to a factor-of-M savings in the computational complexity. If on the other hand, $H(z)$ in Figure 10.15(b) is an IIR filter of order K with a transfer function

$$H(z) = \frac{V(z)}{X(z)} = \frac{P(z)}{D(z)}, \tag{10.23}$$

where

$$P(z) = \sum_{n=0}^{K} p_n z^{-n}, \qquad D(z) = 1 + \sum_{n=1}^{K} d_n z^{-n}, \tag{10.24}$$

its direct form implementation is given by

$$w[n] = -d_1 w[n-1] - d_2 w[n-2] - \cdots$$
$$-d_K w[n-K] + x[n], \tag{10.25a}$$
$$v[n] = p_0 w[n] + p_1 w[n-1] + \cdots + p_K w[n-K]. \tag{10.25b}$$

Since $v[n]$ is being down-sampled, it is sufficient to compute $v[n]$ using Eq. (10.25b) only for values of n that are integer multiples of M. However, the intermediate signal $w[n]$ in Eq. (10.25a) must be computed for all values of n. For example, in the computation of

$$v[M] = p_0 w[M] + p_1 w[M-1] + \cdots + p_K w[M-K],$$

$K+1$ successive values of $w[n]$ are *still* required. As a result, the savings in the computations in the case of an IIR filter is going to be less than a factor of M.

The following example provides a more detailed comparison.

EXAMPLE 10.4   In this example we compare the computational complexity of various implementations of the factor-of-M decimator of Figure 10.15(b).[4] Let the sampling frequency for the input $x[n]$ in Figure 10.15(b) be F_T. Then the number of multiplications per second, to be denoted as $\mathcal{R}_M$, are as follows for various computational schemes:

For an FIR $H(z)$ of length N,
$$\mathcal{R}_{M,\text{FIR}} = N \times F_T.$$

For an FIR $H(z)$ of length N followed by a down-sampler,

$$\mathcal{R}_{M,\text{FIR–DEC}} = N \times F_T/M.$$

For an IIR $H(z)$ of order K,
$$\mathcal{R}_{M,\text{IIR}} = (2K+1) \times F_T.$$

For an IIR $H(z)$ of order K followed by a down-sampler,

$$\mathcal{R}_{M,\text{IIR–DEC}} = K \times F_T + (K+1) \times F_T/M.$$

Thus, in the FIR case, we save computations by a factor of M. In the IIR case we save by a factor of $M(2K+1)/[(M+1)K+1]$, which is not significant for large K. For $M = 10$ and $K = 9$, the savings is only by a factor of 1.9. We shall point out later that in certain cases the IIR filter can be computationally more efficient.

For the case of the interpolation filter in Figure 10.15(a), very similar arguments hold. If $H(z)$ is an FIR filter, then the computational savings is by a factor of L (since $v[n]$ has $L-1$ zeros between its two consecutive nonzero samples). On the other hand, computational savings is significantly less with IIR filters.

10.2.4 Sampling Rate Alteration Using MATLAB

The *Signal Processing Toolbox* of MATLAB includes three specific functions for sampling rate alteration. These are `decimate`, `interp`, and `resample`.

The function `decimate` can be employed to reduce the sampling rate of an input signal vector x by an integer factor M generating the output signal vector y and is available with four options:

```
y = decimate(x,M)
y = decimate(x,M,n)
y = decimate(x,M,'fir')
y = decimate(x,M,N,'fir')
```

In the first option, the function employs an eighth-order Type 1 Chebyshev IIR lowpass filter by default and filters the input sequence in both directions to ensure zero-phase filtering. In the second option, it utilizes an order-n Type 1 Chebyshev IIR lowpass filter where n should be less than 13 to avoid numerical instability. In the third option, it employs a 30-tap FIR lowpass filter which filters the input only in the forward direction. The FIR filter is designed with a stopband edge at π/M using the function `fir1` of MATLAB. Finally, in the last option, it designs and employs a length-N FIR lowpass filter. We illustrate the application of the function `decimate` in the following example.

EXAMPLE 10.5   We consider here the decimation of a sum of two sinusoidal sequences of normalized frequencies f_1 and f_2 by an arbitrary down-sampling factor M using Program 10.5 given below. The input data to the program are the length N of the input sequence $x[n]$, the down-sampling factor M, and the two normalized

---

[4]In this chapter the "computational complexity of an implementation" is taken to be equal to the number of multiplications required per second. Also, the symmetry of FIR impulse responses, which lead to about 50 percent computational savings, is ignored.

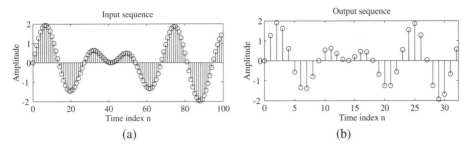

Figure 10.19: The input and output sequences of the factor-of-2 decimator of Example 10.5.

frequencies in Hz. The program uses a 30-tap FIR lowpass decimation filter designed to have a stopband edge at π/M.[5] It then plots the input and the output sequences as indicated in Figure 10.19 for $N = 100$, $M = 2$, $f_1 = 0.043$, and $f_2 = 0.031$.

```
% Program 10_5
% Illustration of Decimation Process
%
clf;
N = input('Length of input signal = ');
M = input('Down-sampling factor = ');
f1 = input('Frequency of first sinusoid = ');
f2 = input('Frequency of second sinusoid = ');
n = 0:N-1;
% Generate the input sequence
x = sin(2*pi*f1*n) + sin(2*pi*f2*n);
% Generate the decimated output sequence
y = decimate(x,M,'fir');
% Plot the input and the output sequences
subplot(2,1,1)
stem(n,x(1:N));
title('Input sequence');
xlabel('Time index n');ylabel('Amplitude');
subplot(2,1,2)
m=0:N/M-1;
stem(m,y(1:N/M));
title('Output sequence');
xlabel('Time index n'); ylabel('Amplitude');
```

The function `interp`[6] can be employed to increase the sampling rate of an input signal x by an integer factor L generating the output signal vector y. It is available with three options:

$$y = \text{interp}(x,L)$$
$$y = \text{interp}(x,L,N,alpha)$$
$$[y,h] = \text{interp}(x,L,N,alpha)$$

---

[5]The group delay of the default decimation filter is 14.5 samples. For an integer-valued group delay, a decimation filter of odd length should be used.

[6]`interp` is based on the interpolator design of Oetken et al. [Oet75].

The lowpass filter designed by the program to fill in the zero-valued samples inserted by the up-sampling operation is a symmetric FIR filter that allows the original input samples to appear as is at the output and finds the missing samples by minimizing the mean-square errors between these samples and the ideal values. The length of the FIR filter is 2NL+1, where N ≤ 10. The input signal is assumed to be bandlimited to the frequency range $0 \leq \omega \leq$ alpha, where alpha ≤ 0.5. In the first option, the default values of alpha and N are, respectively, 0.5 and 4. In the remaining two options, the bandedge alpha of the input signal and the length N of the interpolation filter are specified. In the last option, the output data contains in addition to the interpolated output vector y, the filter coefficients h.

The application of the function interp is considered in the following example.

EXAMPLE 10.6   To illustrate the interpolation process, we take the input x to be a sum of two sinusoidal sequences. Program 10.6 is then used to find its interpolated output y for an up-sampling factor of L. The input data to the program are the length N of the input vector x, the up-sampling factor L, and the two normalized frequencies in Hz. The frequencies of the two sinusoids considered here are $f_1 = 0.043$ and $f_1 = 0.031$, while the interpolation factor is 2. The input length is 50. The output plots generated by this program are shown in Figure 10.20.

```
% Program 10_6
% Illustration of Interpolation Process
%
clf;
N = input('Length of input signal = ');
L = input('Up-sampling factor = ');
f1 = input('Frequency of first sinusoid = ');
f2 = input('Frequency of second sinusoid = ');
% Generate the input sequence
n = 0:N-1;
x = sin(2*pi*f1*n) + sin(2*pi*f2*n);
% Generate the interpolated output sequence
y = interp(x,L);
% Plot the input and the output sequences
subplot(2,1,1)
stem(n,x(1:N));
title('Input sequence');
xlabel('Time index n'); ylabel('Amplitude');
subplot(2,1,2)
m=0:N*L-1;
stem(m,y(1:N*L));
title('Output sequence');
xlabel('Time index n'); ylabel('Amplitude');
```

Finally, the function resample in MATLAB can be utilized to increase the sampling rate of an input vector x by a ratio of two positive integers, L/M, generating an output vector y. There are five options available with this function:

```
       y = resample(x,L,M)
       y = resample(x,L,M,R)
       y = resample(x,L,M,R,beta)
       y = resample(x,L,M,h),
    [y,h] = resample(x,L,M)
```

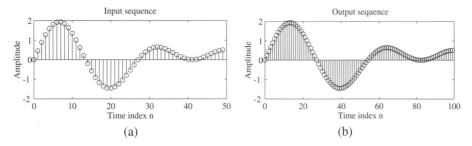

Figure 10.20: The input and output sequences of the factor-of-2 interpolator of Example 10.6.

The program employs a lowpass FIR filter designed using the function `fir1` with a Kaiser window to prevent aliasing caused by the down-sampling and eliminate the imaging caused by the up-sampling operation (see Figure 10.17). The number of samples used on both sides of the present input sample $x[n]$ can be specified by the input data R. The default value of R is 10. The design parameter β for selecting the Kaiser window can be specified through the input data `beta` (see Section 7.6.5).

Example 10.7 demonstrates the application of the function `resample`.

EXAMPLE 10.7   In this example, we consider the sampling rate increase by a ratio of two positive integers of a signal composed of two sinusoids employing Program 10_7 given below. The input data to the program are the length N of the input vector x, the up-sampling factor L, the down-sampling factor M, and the two frequencies. The up-sampling and down-sampling factors used here are 5 and 3, respectively. The frequencies of the two sinusoids considered here are $f_1 = 0.43$ and $f_2 = 0.31$. Figure 10.21 shows the output plots generated by this program.

```
% Program 10_7
% Illustration of Sampling Rate Alteration by
% a Ratio of Two Integers
%
clf;
N = input('Length of input signal = ');
L = input('Up-sampling factor = ');
M = input('Down-sampling factor = ');
f1 = input('Frequency of first sinusoid = ');
f2 = input('Frequency of second sinusoid = ');
%  Generate the input sequence
n = 0:N-1;
x = sin(2*pi*f1*n) + sin(2*pi*f2*n);
%  Generate the resampled output sequence
y = resample(x,L,M);
%  Plot the input and the output sequences
subplot(2,1,1)
stem(n,x(1:N));
title('Input sequence');
xlabel('Time index n'); ylabel('Amplitude');
subplot(2,1,2)
m=0:N*L/M-1;
stem(m,y(1:N*L/M));
title('Output sequence');
xlabel('Time index n'); ylabel('Amplitude');
```

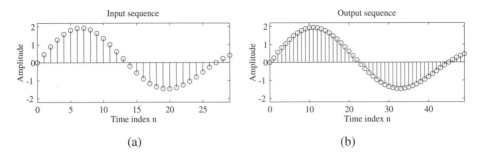

(a)   (b)

Figure 10.21: Illustration of sampling rate increase by a rational number 5/3.

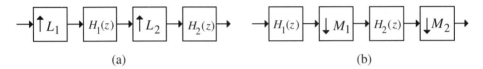

(a)   (b)

Figure 10.22: Two-stage implementation of sampling rate alteration systems: (a) interpolator and (b) decimator.

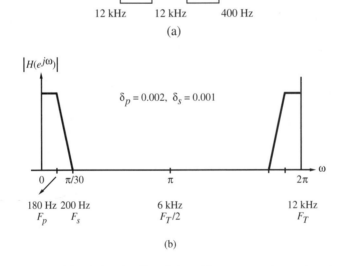

Figure 10.23: Illustration of the decimation filter design. (Frequency response plots shown not to scale.)

10.3 Multistage Design of Decimator and Interpolator

The decimator and the interpolator of Figure 10.15 are single-stage structures since here the basic scheme for the implementation involves a single lowpass filter and a single sampling rate alteration device. If the interpolation factor L can be expressed as a product of two integers L_1 and L_2, the factor-of-L interpolator of Figure 10.15(a) can also be realized in two stages, as indicated in Figure 10.22(a). Likewise, the factor-of-M decimator of Figure 10.15(b) can be implemented in two stages, as shown in Figure 10.22(b), if the decimation factor M is a product of two integers M_1 and M_2. Of course, the design can involve more than two stages, depending on the number of factors used to express L and M, respectively. It turns out that, in general, the computational efficiency is improved significantly by designing the sampling rate alteration system as a cascade of several stages. We demonstrate this feature next by means of an example which also illustrates how to choose the specifications of the individual filters in a multistage design.

EXAMPLE 10.8   Consider the design of a decimator for the reduction of the sampling rate of a signal from 12 kHz to 400 Hz [Figure 10.23(a)]. The specifications for the decimation filter $H(z)$ are assumed to be as follows: passband edge $F_p = 180$ Hz, stopband edge $F_s = 200$ Hz, passband ripple $\delta_p = 0.002$, and stopband ripple $\delta_s = 0.001$, and are shown in Figure 10.23(b). If $H(z)$ is required to be an equiripple linear-phase FIR filter, then its order N (i.e., length $N + 1$) needs to be estimated first. Recall from Section 7.1.4 that the order N can be determined from Eq. (7.15), which is repeated below for convenience:

$$N = \frac{-20 \log_{10} \sqrt{\delta_p \delta_s} - 13}{14.6 \Delta f} \tag{10.26}$$

where $\Delta f = (F_s - F_p)/F_T$ is the normalized transition bandwidth.

The order estimated using Program 7_4 is given by

$$N = 1808. \tag{10.27}$$

Thus, the number of multiplications per second in the single-stage implementation of the factor-of-30 decimator of Figure 10.23(a) is

$$\mathcal{R}_{M,H} = 1809 \times \frac{12{,}000}{30} = 723{,}600. \tag{10.28}$$

Let us now consider the implementation of $H(z)$ as a cascade realization in the form of $G(z^{15})F(z)$.[7] The parent filter $G(z)$ should now have specifications as shown in Figure 10.24(a). This corresponds to stretching the specifications in Figure 10.23(b) by 15. Figure 10.24(b) shows the magnitude response of $G(z^{15})$. The desired response of the $F(z)$ is also shown in the same figure. Note that the desired response of $F(z)$ has a wider transition band as it takes into account the spectral gaps between the passbands of $G(z^{15})$.

The implementation is shown in Figure 10.25(a). Because of the cascade connection, the overall ripple of the cascade in dB is given by the sum of the passband ripples of $F(z)$ and $G(z^{15})$ in dB (see Chapter 7). This can be compensated for by designing $F(z)$ and $G(z)$ to have a passband ripple of $\delta_p = 0.001$ (rather than 0.002) each. On the other hand, the cascade of $F(z)$ and $G(z^{15})$ has a stopband at least as good as $F(z)$ or $G(z^{15})$ individually, so the specification for δ_s for both $F(z)$ and $G(z)$ can be retained as $\delta_s = 0.001$. Therefore, the specifications for the IFIR approach are as follows:

$$G(z): \quad \delta_p = 0.001, \quad \delta_s = 0.001, \quad \Delta f = \frac{300}{12{,}000}, \tag{10.29a}$$

$$F(z): \quad \delta_p = 0.001, \quad \delta_s = 0.001, \quad \Delta f = \frac{420}{12{,}000}. \tag{10.29b}$$

The filter orders of $G(z)$ and $F(z)$ obtained using Program 7_4 are as follows:

$$\text{Order of } G(z) = 129, \tag{10.30a}$$

$$\text{Order of } F(z) = 92. \tag{10.30b}$$

It should be noted that Figure 10.25(a) has been obtained by replacing $H(z)$ in Figure 10.23(a) with $F(z)G(z^{15})$. The length of $H(z)$ for a direct implementation is 1809 from Eq. (10.27), whereas the length of $F(z)G(z^{15})$ is $92 + 15 \times 129 + 1 = 2028$. In other words, the length of the overall filter has been increased. However, the computational complexity (i.e., number of multiplications per second) of the new realization can be dramatically reduced by making use of the cascade equivalence #1 in Figure 10.14 as demonstrated next.

---

[7]The implementation in the form of $G(z^{15})F(z)$ is known as the interpolated FIR (IFIR) realization [Neu84b].

To this end we redraw Figure 10.25(a) as in Figure 10.25(b) by factoring the down-sampling factor M as $M_1 M_2$ with $M_1 = 15$ and $M_2 = 2$. Next, by invoking the equivalence of Figure 10.14(a), we replace it with Figure 10.25(c). From this figure, it can be seen that the implementation of $G(z)$ at the sampling rate 800 Hz followed by a down-sampling by a factor of 2 requires

$$\mathcal{R}_{M,G} = 130 \times \frac{800}{2} = 52{,}000 \tag{10.31}$$

multiplications per second. The implementation of $F(z)$ followed by the down-sampler of factor 15 requires about

$$\mathcal{R}_{M,F} = 93 \times \frac{12{,}000}{15} = 74{,}400 \tag{10.32}$$

multiplications per second. Thus, $F(z)$ requires more multiplications per second than the model filter $G(z)$. The total complexity of the IFIR-based implementation of Figure 10.25(c) is therefore

$$52{,}000 + 74{,}400 = 126{,}400 \tag{10.33}$$

multiplications per second, which is about 5.72 times smaller than the complexity Eq. (10.28) of a direct implementation as in Figure 10.23(a).

Let us now implement the factor-of-15 decimator in the first stage of Figure 10.25(c) also in two stages. To this end we design the decimation filter $F(z)$ as an IFIR structure of the form $R(z)S(z^5)$. Following the same procedure as described above for the design of $H(z)$ in the form of $F(z)G(z^{15})$, we arrive at the specifications of the filters $S(z)$ and $R(z)$ as given below:

$$S(z): \quad \delta_p = 0.0005, \quad \delta_s = 0.001, \quad \Delta f = \frac{420}{12{,}000} \times 5, \tag{10.34a}$$

$$R(z): \quad \delta_p = 0.0005, \quad \delta_s = 0.001, \quad \Delta f = \frac{1620}{12{,}000}. \tag{10.34b}$$

The filter lengths of $S(z)$ and $R(z)$ obtained using Program 7_4 are as follows:

$$\text{Order of } S(z) = 20, \tag{10.35a}$$

$$\text{Order of } R(z) = 26. \tag{10.35b}$$

The final realization of the decimator is as indicated in Figure 10.26. From this figure, it can be seen that the implementation of $S(z)$ at the sampling rate 2.4 kHz followed by a down-sampling by a factor of 3 requires

$$\mathcal{R}_{M,S} = 21 \times \frac{2400}{3} = 16{,}800$$

multiplications per second, whereas the implementation of $R(z)$ at the sampling rate of 12 kHz followed by a down-sampling by a factor of 5 requires

$$\mathcal{R}_{M,R} = 27 \times \frac{12{,}000}{5} = 64{,}800$$

multiplications per second. As a result, the total number of multiplications required in the three-stage realization is given by

$$\mathcal{R}_{M,G} + \mathcal{R}_{M,S} + \mathcal{R}_{M,R} = 133{,}600,$$

which is slightly higher than that required in the two-stage realization. As this example points out, a realization higher than two stages may not always lead to a computationally efficient realization.

In the general case of a k-stage implementation, as indicated in Figure 10.27(a), the decimation factor is $M = M_1 M_2 \cdots M_k$. For a given M_k there are $k - 1$ quantities, $M_1, M_2, \ldots, M_{k-1}$, which can be chosen in various ways with one combination leading to an optimum multistage realization of a decimator with

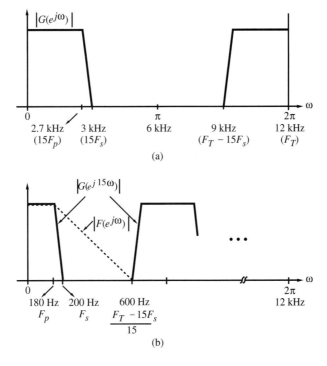

(a)

(b)

Figure 10.24: Decimation filter design based on the IFIR approach (frequency response plots not shown to scale).

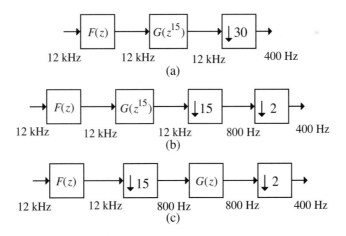

(a)

(b)

(c)

Figure 10.25: The steps in the two-stage realization of the decimator structure.

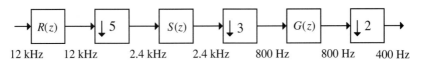

Figure 10.26: A three-stage realization of the decimator.

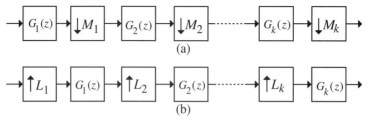

Figure 10.27: (a) A general multistage structure for decimation and (b) a general multistage structure for interpolation.

the least computational complexity. The determination of an optimum realization depends on the selection of k, and the "best" combination and ordering of $M_1, M_2, \ldots, M_k$ that minimizes the required number of multiplications per second [Cro83]. The corresponding multistage interpolator design is a very analogous problem [Figure 10.27(b)] [Cro83].

10.4 The Polyphase Decomposition

We have shown in Section 10.2.3 that a single-stage decimator or interpolator employing FIR lowpass filters can be computationally efficient since the necessary multiplications required to compute the output sample can be carried out only when needed. We demonstrated in the previous section that the computational requirements can be further decreased using a multistage design. Additional reduction in the computational complexity is possible by realizing the FIR filters using the polyphase decomposition described in Section 6.3.3. In certain cases, it is also possible to realize IIR decimation and interpolation filters in polyphase forms, resulting in reduced computational complexity realizations. We review the polyphase decomposition again here and illustrate its application in the efficient realization of the decimator and the interpolator.

10.4.1 The Decomposition

Consider an arbitrary sequence $\{x[n]\}$ with a z-transform $X(z)$:

$$X(z) = \sum_{n=-\infty}^{\infty} x[n]z^{-n}. \tag{10.36}$$

We can rewrite $X(z)$ given above as

$$X(z) = \sum_{k=0}^{M-1} z^{-k} X_k(z^M) \tag{10.37}$$

where

$$X_k(z) = \sum_{n=-\infty}^{\infty} x_k[n]z^{-n} = \sum_{n=-\infty}^{\infty} x[Mn+k]z^{-n}, \qquad 0 \le k \le M-1. \tag{10.38}$$

The subsequences $\{x_k[n]\}$ are called the polyphase components of the parent sequence $x[n]$, and the functions $X_k(z)$, given by the z-transform of $\{x_k[n]\}$, are called the *polyphase components* of $X(z)$ [Bel76]. The relation between the subsequences $\{x_k[n]\}$ and the original sequence $\{x[n]\}$ is given by

$$x_k[n] = x[Mn+k], \qquad 0 \le k \le M-1. \tag{10.39}$$

Equation (10.37) can be written in matrix form as

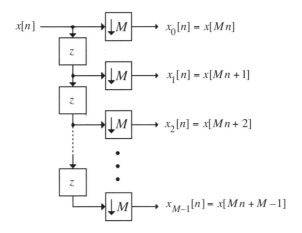

Figure 10.28: A structural interpretation of the M-band polyphase decomposition of a sequence $x[n]$.

$$X(z) = [\, 1 \quad z^{-1} \quad \cdots \quad z^{-(M-1)} \,] \begin{bmatrix} X_0(z^M) \\ X_1(z^M) \\ \vdots \\ X_{M-1}(z^M) \end{bmatrix}. \tag{10.40}$$

A multirate structural interpretation of the polyphase decomposition is given in Figure 10.28.

The polyphase decomposition of an FIR transfer function can be carried out by inspection as illustrated in Section 6.3.3, where we developed two-branch and three-branch polyphase decompositions of a length-9 FIR transfer function. Figure 6.8 illustrates the parallel realizations of an FIR transfer function based on the polyphase decomposition. In the following section, we consider several variations of the parallel realization of an FIR transfer function based on a polyphase decomposition.

The polyphase decomposition of an IIR transfer function $H(z) = P(z)/D(z)$, on the other hand, is not that straightforward. One way to arrive at an M-branch polyphase decomposition of $H(z)$ is to express it in the form $P'(z)/D'(z^M)$ by multiplying the denominator $D(z)$ and the numerator $P(z)$ with an appropriately chosen polynomial and then applying an M-branch polyphase decomposition to $P'(z)$. This approach is illustrated in the following example.

EXAMPLE 10.9 Consider an IIR transfer function

$$H(z) = \frac{1 - 2z^{-1}}{1 + 3z^{-1}}.$$

To obtain a two-band decomposition of the form of Eq. (10.38) with $M = 2$, we rewrite $H(z)$ as

$$H(z) = \frac{\left(1 - 2z^{-1}\right)\left(1 - 3z^{-1}\right)}{\left(1 + 3z^{-1}\right)\left(1 - 3z^{-1}\right)} = \frac{1 - 5z^{-1} + 6z^{-2}}{1 - 9z^{-2}} = \left(\frac{1 + 6z^{-2}}{1 - 9z^{-2}}\right) + z^{-1}\left(\frac{-5}{1 - 9z^{-2}}\right).$$

Therefore, $H(z) = E_0(z^2) + z^{-1}E_1(z^2)$, where

$$E_0(z) = \frac{1 + 6z^{-1}}{1 - 9z^{-1}}, \quad E_1(z) = \frac{-5}{1 - 9z^{-1}}.$$

Note that the above approach increases the overall order and the complexity of $H(z)$. However, when used in certain multirate structures, the approach may result in a more efficient structure. An alternative, more attractive approach is considered in the following example.

EXAMPLE 10.10   Consider the transfer function of a fifth-order Butterworth lowpass filter with a 3-dB cutoff frequency at 0.5π:

$$H(z) = \frac{0.0527864045(1 + z^{-1})^5}{1 + 0.633436854z^{-2} + 0.05572809z^{-4}}.$$

It is easy to show that $H(z)$ can be expressed as

$$H(z) = \frac{1}{2}\left[\left(\frac{0.10557281 + z^{-2}}{1 + 0.10557281z^{-2}}\right) + z^{-1}\left(\frac{0.527864045 + z^{-2}}{1 + 0.527864045z^{-2}}\right)\right].$$

Therefore, we can express $H(z) = E_0(z^2) + z^{-1}E_1(z^2)$, where

$$E_0(z) = \frac{1}{2}\left(\frac{0.10557281 + z^{-1}}{1 + 0.10557281z^{-1}}\right), \qquad E_1(z) = \frac{1}{2}\left(\frac{0.527864045 + z^{-1}}{1 + 0.527864045z^{-1}}\right).$$

Note that in the above polyphase-like decomposition, the branch transfer functions are stable allpass functions. Moreover, the decomposition has not increased the order of the overall transfer function $H(z)$.

10.4.2   FIR Filter Structures Based on the Polyphase Decomposition

We have illustrated in Section 6.3.3 that a parallel realization of an FIR filter transfer function $H(z)$ can be obtained using a polyphase decomposition. As we shall point out later in this chapter, such a realization often results in computationally efficient structures in certain multirate applications. We revisit the polyphase decomposition based FIR filter realization again here using the more commonly used notation for the polyphase components and develop several other alternate realizations.

Consider first an M-branch polyphase decomposition of $H(z)$ given by

$$H(z) = \sum_{k=0}^{M-1} z^{-k} E_k(z^M). \tag{10.41}$$

A direct realization of the above is shown in Figure 10.29(a). The transpose of this realization is indicated in Figure 10.29(b). An alternative representation of the transpose structure of Figure 10.29(b) is obtained by using the notation

$$R_\ell(z^M) = E_{M-1-\ell}(z^M), \qquad 0 \le \ell \le M - 1, \tag{10.42}$$

resulting in the structure of Figure 10.30. The corresponding polyphase decomposition is thus given by

$$H(z) = \sum_{\ell=0}^{M-1} z^{-(M-1-\ell)} R_\ell(z^M). \tag{10.43}$$

To differentiate between the two decompositions of Eqs. (10.41) and (10.43), the former is usually called the *Type I polyphase decomposition*, while the latter is called *Type II polyphase decomposition*.

10.4.3   Computationally Efficient Interpolator and Decimator Structures

Computationally efficient decimator and interpolator structures employing linear-phase lowpass filters can be derived by applying a polyphase decomposition to the lowpass filters. We demonstrate this property next.

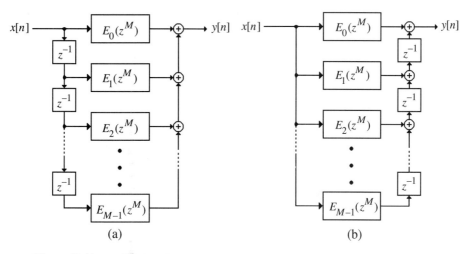

Figure 10.29: Realization of an FIR filter based on a Type I polyphase decomposition.

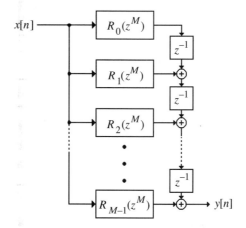

Figure 10.30: Realization of an FIR filter based on a Type II polyphase decomposition.

Consider first the use of the polyphase decomposition in the realization of the decimation filter of Figure 10.15(b). If the lowpass filter $H(z)$ is realized as in Figure 10.29(a), the overall decimator structure takes the form of Figure 10.31(a). By invoking the cascade equivalance #1 of Figure 10.14(a), this structure reduces to that indicated in Figure 10.31(b), which is computationally more efficient than the structure of Figure 10.15(b). To illustrate this point, assume that the decimation filter $H(z)$ of Figure 10.15(b) is a length-N FIR structure and the input sampling period $T = 1$. Since the decimator output $y[n]$ is obtained by down-sampling the filter output $v_1[n]$ by a factor of M, it is necessary only to compute $v[n]$ at $n = \ldots, -2M, -M, 0, M, 2M, \ldots$. The computational requirements are therefore N multiplications and $(N - 1)$ additions per output sample being computed. However, as n increases, the stored signals in the delay registers change. As a result, all computations need to be completed in one sampling period, and for the following $(M - 1)$ sampling periods the arithmetic units remain idle. Now consider the structure of Figure 10.31(b). If the lengths of the subfilter $E_k(z)$ is N_k, then $N = \sum_{k=0}^{M-1} N_k$. The computational requirements of the kth subfilter are N_k multiplications and $N_k - 1$ additions per output sample, and that for

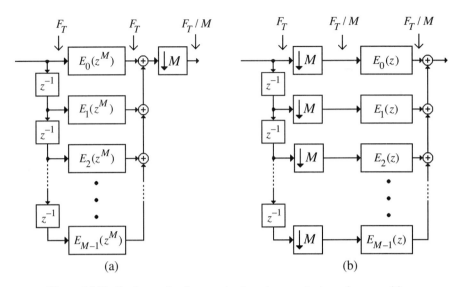

Figure 10.31: Decimator implementation based on a polyphase decomposition.

the overall structure is therefore $\sum_{k=0}^{M-1} N_k = N$ multiplications and $\sum_{k=0}^{M-1}(N_k - 1) + (M - 1) = N - 1$ additions per decimator output sample. However, in the latter structure, the arithmetic units are operative at all instants of the output sampling period, which is M times that of the input sampling period.

Similar savings are also obtained in the case of the interpolator structure employing polyphase decomposition in the realization of computationally efficient interpolators. Figure 10.32(a) shows the interpolator structure derived from Figure 10.15(a) by making use of the L-band Type I polyphase decomposition of the interpolation filter $H(z)$ and the cascade equivalence #2 of Figure 10.14(b). An alternative realization obtained using the Type II polyphase decomposition of the interpolation filter $H(z)$ and the cascade equivalence #2 is shown in Figure 10.32(b).

More efficient interpolator and decimator structures can be realized by exploiting the symmetry of the filter coefficients of $H(z)$ in the case of linear-phase filters. Consider for example the realization of a factor-of-3 ($M = 3$) decimator using a length-12 linear-phase FIR lowpass filter with a symmetric impulse response:

$$H(z) = h[0] + h[1]z^{-1} + h[2]z^{-2} + h[3]z^{-3} + h[4]z^{-4} + h[5]z^{-5} + h[5]z^{-6}$$
$$+ h[4]z^{-7} + h[3]z^{-8} + h[2]z^{-9} + h[1]z^{-10} + h[0]z^{-11}. \tag{10.44}$$

A conventional polyphase decomposition of the above $H(z)$ yields the following subfilters:

$$\begin{aligned} E_0(z) &= h[0] + h[3]z^{-1} + h[5]z^{-2} + h[2]z^{-3}, \\ E_1(z) &= h[1] + h[4]z^{-1} + h[4]z^{-2} + h[1]z^{-3}, \\ E_2(z) &= h[2] + h[5]z^{-1} + h[3]z^{-2} + h[0]z^{-3}. \end{aligned} \tag{10.45}$$

Note that the subfilter $E_1(z)$ still has a symmetric impulse response, whereas the impulse response of $E_2(z)$ is the mirror image of that of $E_0(z)$. These relations can be made use of in developing a computationally efficient realization using only six multipliers and 11 two-input adders, as depicted in Figure 10.33.

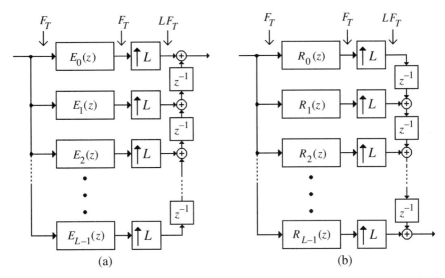

Figure 10.32: Computationally efficient interpolator structures: (a) Type I polyphase decomposition, and (b) Type II polyphase decomposition.

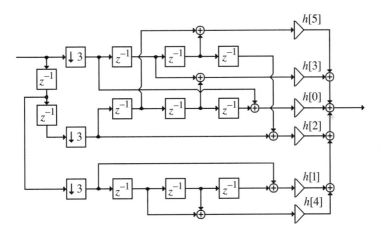

Figure 10.33: A computationally efficient realization of a factor-of-3 decimator exploiting the linear-phase symmetry of the length-12 decimation filter.

10.4.4 A Useful Identity

The cascade multirate digital filter structure of Figure 10.34(a) appears in a number of applications. If we express the transfer function $H(z)$ in its L-term Type I polyphase form $\sum_{k=0}^{L-1} z^{-k} E_k(z^L)$, it can be easily shown that the structure of Figure 10.34(a) is equivalent to the time-invariant digital filter of Figure 10.34(b), where $E_0(z)$ is the zeroth polyphase term (Problem 10.24) [Vai93]. This equivalence can be exploited in simplifying complex multirate networks containing cascade structures of the form of Figure 10.34(a) for analysis purposes.

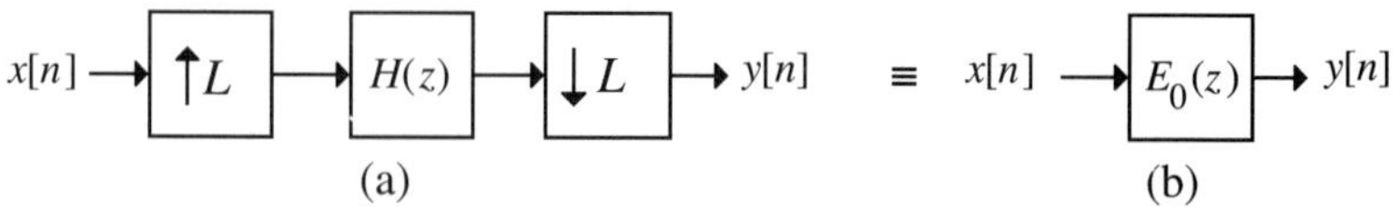

Figure 10.34: (a) A cascade multirate structure and (b) its equivalent form.

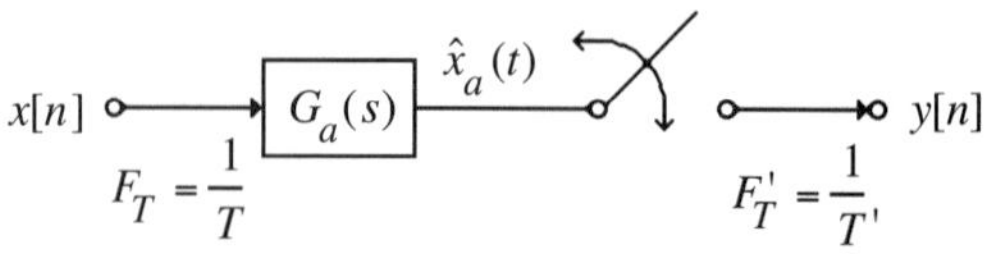

Figure 10.35: Sampling rate alteration based on conversion of input digital signal to analog form followed by a resampling at the desired output rate.

10.5 Arbitrary-Rate Sampling Rate Converter

There are many applications requiring the estimation of a discrete-time signal value at an arbitrary time instant between a consecutive pair of known samples. Applications include conversion between arbitrary sampling rates, timing adjustment in digital receivers, time-delay estimation, echo cancellation in modems, and beam steering and direction finding in antenna arrays.

The above estimation problem can be solved by using some type of interpolation which basically forms an approximating continuous-time signal from a set of known consecutive samples of the given discrete-time signal and then evaluates the value of the continuous-time signal at the desired time instant. This interpolation process can be directly implemented by designing a digital interpolation filter. An all-digital design of a sampling rate converter with an arbitrary conversion factor is not simple. In particular, the design is quite difficult and expensive when the conversion factor is a ratio of two very large integers or an irrational number.

10.5.1 Ideal Sampling Rate Converter

In principle, a sampling rate conversion by an arbitrary conversion factor can be implemented simply by passing the input digital signal through an ideal analog reconstruction lowpass filter whose output is resampled at the desired output rate, as indicated in Figure 10.35 [Ram84]. If the impulse response of the analog lowpass filter is denoted by $g_a(t)$, the output of the filter is then given by

$$\hat{x}_a(t) = \sum_{\ell=-\infty}^{\infty} x[\ell]g_a(t - \ell T). \tag{10.46}$$

If the analog filter is chosen to bandlimit its output to the frequency range $F_g < F_T'/2$, its output $\hat{x}_a(t)$ can then be resampled at the rate F_T' and, hence, the output $y[n]$ of the resampler at new time instants $t = nT'$ is given by

$$y[n] = \hat{x}_a(nT') = \sum_{\ell=-\infty}^{\infty} x[\ell]g_a(nT' - \ell T). \tag{10.47}$$

Since the impulse response $g_a(t)$ of an ideal lowpass analog filter is of infinite duration and the samples $g_a(nT' - \ell T)$ have to be computed at each output sampling instant, implementation of the ideal bandlimited interpolation algorithm of Eq. (10.47) in exact form is not practical. Thus, approximations to this ideal interpolation algorithm are usually employed in practice.

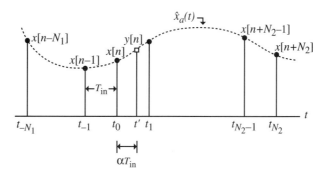

Figure 10.36: Interpolation by an arbitrary factor.

The basic interpolation problem based on a finite weighted sum of input signal samples can then be stated as follows: Given $N_2 + N_1 + 1$ input signal samples, $x[k]$, $k = -N_1, \ldots, N_2$, obtained by sampling an analog signal $x_a(t)$ at $t = t_k = t_0 + kT_{\text{in}}$, determine the sample value $x_a(t_0 + \alpha T_{\text{in}}) = y[\alpha]$ at time $t' = t_0 + \alpha T_{\text{in}}$, where $-N_1 \le \alpha \le N_2$. Figure 10.36 illustrates the interpolation process by an arbitrary factor.

We next describe a commonly employed interpolation algorithm based on a finite weighted sum of input samples.

10.5.2 Lagrange Interpolation Algorithm

In this approach, a polynomial approximation $\hat{x}_a(t)$ to $x_a(t)$ is defined as

$$\hat{x}_a(t) = \sum_{k=-N_1}^{N_2} P_k(t)x[n+k], \tag{10.48}$$

where $P_k(t)$ are the Lagrange polynomials given by

$$P_k(t) = \prod_{\substack{\ell=-N_1 \\ \ell \neq k}}^{N_2} \left(\frac{t - t_\ell}{t_k - t_\ell} \right), \qquad -N_1 \le k \le N_2. \tag{10.49}$$

Since

$$P_k(t_r) = \begin{cases} 1, & k = r \\ 0, & k \neq r \end{cases}, \qquad -N_1 \le r \le N_2, \tag{10.50}$$

it follows from Eqs. (10.48) to (10.50) that

$$\hat{x}_a(t_k) = x_a(t_k), \qquad -N_1 \le k \le N_2. \tag{10.51}$$

From Eq. (10.48), the value of $x_a(t)$ at an arbitrary value $t' = t_0 + \alpha T_{\text{in}}$ is given by

$$\hat{x}_a(t') = \hat{x}_a(t_0 + \alpha T_{\text{in}}) = y[n] = \sum_{k=-N_1}^{N_2} P_k(\alpha)x[n+k], \tag{10.52}$$

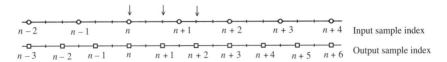

Figure 10.37: The input and output sample locations of an up-sampler with a conversion factor of 3/2.

where

$$P_k(\alpha) = P_k(t_0 + \alpha T_{in}) = \prod_{\substack{\ell=-N_1 \\ \ell \neq k}}^{N_2} \left(\frac{\alpha - \ell}{k - \ell} \right), \quad -N_1 \leq k \leq N_2. \tag{10.53}$$

We illustrate in the following example the application of the Lagrange interpolation algorithm in designing a fractional-rate sampling-rate converter.

EXAMPLE 10.11   Consider the design of a fractional-rate interpolator with an interpolation factor of 3/2. For simplicity, we make use of a third-order polynomial approximation, with $N_1 = 2$ and $N_2 = 1$, in which case Eq. (10.52) reduces to

$$y[n] = P_{-2}(\alpha)x[n-2] + P_{-1}(\alpha)x[n-1] + P_0(\alpha)x[n] + P_1(\alpha)x[n+1]. \tag{10.54}$$

Here, the Lagrange polynomials $P_k(\alpha)$ are given by

$$P_{-2}(\alpha) = \frac{(\alpha+1)\alpha(\alpha-1)}{-6} = \frac{1}{6}(-\alpha^3 + \alpha), \tag{10.55a}$$

$$P_{-1}(\alpha) = \frac{(\alpha+2)\alpha(\alpha-1)}{2} = \frac{1}{2}(\alpha^3 + \alpha^2 - 2\alpha), \tag{10.55b}$$

$$P_0(\alpha) = \frac{(\alpha+2)(\alpha+1)(\alpha-1)}{-2} = -\frac{1}{2}(\alpha^3 + 2\alpha^2 - \alpha - 2), \tag{10.55c}$$

$$P_1(\alpha) = \frac{(\alpha+2)(\alpha+1)\alpha}{6} = \frac{1}{6}(\alpha^3 + 3\alpha^2 + 2\alpha), \tag{10.55d}$$

obtained from Eq. (10.49) with $N_1 = 2$ and $N_2 = 1$.

Figure 10.37 shows the locations of the samples of the input and the output sequences for an interpolator with a conversion factor of 3/2, where input sample locations are marked with a circle and output sample locations are marked with a square. The locations of the output samples $y[0]$, $y[1]$, and $y[2]$ in the input sample domain are marked with an arrow.

Consider the computations of $y[n]$, $y[n+1]$, and $y[n+2]$ using four input samples, $x[n-2]$ through $x[n+1]$. To this end we require a third-order Lagrange polynomial, i.e., $N_1 = 2$ and $N_2 = 1$ in Eq. (10.53). From Figure 10.37 it follows that, for the computation of $y[n]$, the value of α in Eq. (10.54), to be labeled α_0, is 0. Substituting this value of α in Eqs. (10.55a) to (10.55d), we arrive at the desired values of the filter coefficients as

$$P_{-2}(\alpha_0) = 0, \quad P_{-1}(\alpha_0) = 0, \tag{10.56a}$$

$$P_0(\alpha_0) = 1, \quad P_1(\alpha_0) = 0. \tag{10.56b}$$

Next, we observe from Figure 10.37 that for the computation of $y[n+1]$ the value of α, to be labeled α_1, is equal to 2/3. Using this value of α in Eqs. (10.55a) to (10.55d), we obtain the new set of values for the interpolator filter coefficients as

$$P_{-2}(\alpha_1) = 0.0617, \quad P_{-1}(\alpha_1) = -0.2963, \tag{10.57a}$$

$$P_0(\alpha_1) = 0.7407, \quad P_1(\alpha_1) = 0.4938. \tag{10.57b}$$

Finally, for the computation of $y[n+2]$, the value of α, to be labeled α_2, is 4/3, which leads to the following filter coefficient values,

$$P_{-2}(\alpha_2) = -0.1728, \quad P_{-1}(\alpha_1) = 0.7407, \tag{10.58a}$$

$$P_0(\alpha_2) = -1.2963, \quad P_1(\alpha_1) = 1.7284. \tag{10.58b}$$

Combining Eqs. (10.54) to (10.58b) into matrix form, we arrive at the input-output relation of the interpolation filter as

$$\begin{bmatrix} y[n] \\ y[n+1] \\ y[n+2] \end{bmatrix} = \begin{bmatrix} P_{-2}(\alpha_0) & P_{-1}(\alpha_0) & P_0(\alpha_0) & P_1(\alpha_0) \\ P_{-2}(\alpha_1) & P_{-1}(\alpha_1) & P_0(\alpha_1) & P_1(\alpha_1) \\ P_{-2}(\alpha_2) & P_{-1}(\alpha_2) & P_0(\alpha_2) & P_1(\alpha_2) \end{bmatrix} \begin{bmatrix} x[n-2] \\ x[n-1] \\ x[n] \\ x[n+1] \end{bmatrix}$$

$$= \mathbf{H} \begin{bmatrix} x[n-2] \\ x[n-1] \\ x[n] \\ x[n+1] \end{bmatrix}, \tag{10.59}$$

where $\mathbf{H}$ is the block coefficient matrix, which for the above factor of 3/2 interpolator design is given by

$$\mathbf{H} = \begin{bmatrix} 0 & 0 & 1 & 0 \\ 0.0617 & -0.2963 & 0.7407 & 0.4938 \\ -0.1728 & 0.7407 & -1.2963 & 1.7284 \end{bmatrix}. \tag{10.60}$$

It should be evident from an examination of Figure 10.37 that the filter coefficients to compute $y[n+3]$, $y[n+4]$, and $y[n+5]$ are again given by the coefficients in Eqs. (10.56a) and (10.56b), Eqs. (10.57a) and (10.57b), and Eqs. (10.58a) and (10.58b), respectively. Or in other words, the desired interpolation filter is a time-varying filter with a period of three samples. A realization of the above factor-of-3/2 interpolator based on an implementation of Eq. (10.59) is indicated in Figure 10.38(a). Note that, in practice, the overall system delay will be three sample periods, and as a result, the output sample $y[n]$ actually appears at the time index $n+3$.

An alternative realization of the above fractional rate interpolator in the form of a time-varying FIR filter is indicated in Figure 10.38(b). The filter coefficients of the fifth-order time-varying FIR filter have a period of 3 and are assigned the values as indicated in Figure 10.38(c).

Another realization of the above fractional-rate interpolator is obtained by substituting the Lagrange polynomials of Eqs. (10.55a) to (10.55d) in Eq. (10.54) which yields

$$y[n] = \alpha^3 \left(-\frac{1}{6}x[n-2] + \frac{1}{2}x[n-1] - \frac{1}{2}x[n] + \frac{1}{6}x[n+1] \right)$$

$$+ \alpha^2 \left(\frac{1}{2}x[n-1] - x[n] + \frac{1}{2}x[n+1] \right)$$

$$+ \alpha \left(\frac{1}{6}x[n-2] - x[n-1] + \frac{1}{2}x[n] + \frac{1}{3}x[n+1] \right) + x[n]. \tag{10.61}$$

A digital filter realization of the above equation leads to the *Farrow structure* of Figure 10.39 [Far88] where the transfer functions of the three FIR digital filters are given by

$$H_0(z) = -\frac{1}{6}z^{-2} + \frac{1}{2}z^{-1} - \frac{1}{2} + \frac{1}{6}z,$$

$$H_1(z) = \frac{1}{2}z^{-1} - 1 + \frac{1}{2}z,$$

$$H_2(z) = \frac{1}{6}z^{-2} - z^{-1} + \frac{1}{2} + \frac{1}{3}z.$$

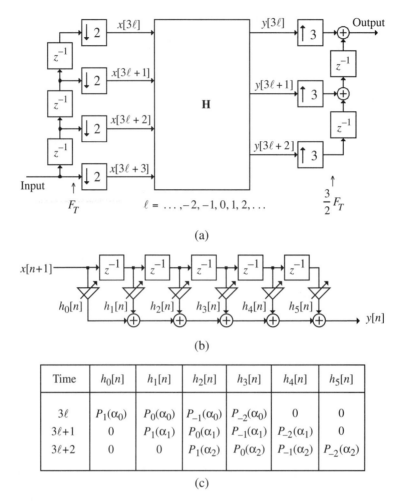

Figure 10.38: Implementation of a fractional-rate interpolator with a conversion factor of 3/2. (a) Block digital filter implementation, (b) implementation using a time-varying FIR interpolation filter, and (c) coefficients of the time-varying filter as a function of sample index.

Note that in this realization, only the value of the multiplier coefficient α is changed periodically with the remaining digital filter structure kept unchanged.

Figure 10.40 shows the plots of the input and the output of the above fractional-rate interpolator for a sinusoidal input of frequency 0.05 Hz sampled at a 1-Hz rate along with that of the error sequence defined by the sample-wise difference between the output sequence and the actual sine values. Note the beginning transient samples in Figure 10.40(c).

10.5.3 Practical Considerations

A direct design of a fractional sampling rate converter in most applications is impractical since the length of the time-varying filter needed is usually very long and the corresponding filter coefficient calculations in real time are thus nearly impossible. As a result, a fractional sampling rate converter is almost always

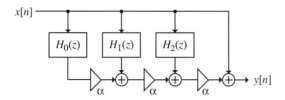

Figure 10.39: Fractional-rate interpolator implementation using the Farrow structure.

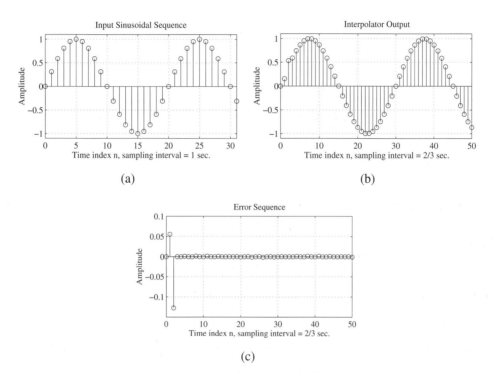

(a)

(b)

(c)

Figure 10.40: Plots of the interpolator input and output sequence, and the error sequence for a sinusoidal input of frequency 0.05 Hz.

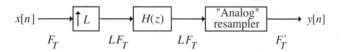

Figure 10.41: Hybrid form of a fractional-rate sampling converter.

realized in a hybrid form consisting of a digital sampling rate converter with an integer-valued conversion factor followed by an "analog" fractional rate converter as indicated in Figure 10.41 [Ram84]. The digital sampling rate converter, of course, if need be, can be implemented in a multistage form.[8]

---

[8]See Section 10.3.

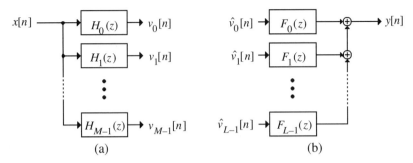

Figure 10.42: (a) Analysis filter bank, and (b) synthesis filter bank.

10.6 Digital Filter Banks

So far, we have mostly concentrated on the design, realization, and applications of single-input, single-output digital filters. There are applications, as in the case of a spectrum analyzer, where it is desirable to separate a signal into a set of subband signals occupying, usually nonoverlapping, portions of the original frequency band. In other applications, it may be necessary to combine many such subband signals into a single composite signal occupying the whole Nyquist range. To this end, digital filter banks play an important role, and are the subject of discussion in this section.

10.6.1 Definitions

The digital filter bank is a set of digital bandpass filters with either a common input or a summed output, as shown in Figure 10.42. The structure of Figure 10.42(a) is called an M-band *analysis filter bank* with the subfilters $H_k(z)$ known as the *analysis filters*. It is used to decompose the input signal $x[n]$ into a set of M subband signals $v_k[n]$ with each subband signal occupying a portion of the original frequency band. (The signal is being "analyzed" by being separated into a set of narrow spectral bands.)

The *dual* of the above operation, whereby a set of subband signals $\hat{v}_k[n]$ (typically belonging to contiguous frequency bands) is combined into one signal $y[n]$ is called a *synthesis filter bank*. Figure 10.42(b) shows an L-band synthesis bank where each filter $F_k(z)$ is called a *synthesis filter*.

10.6.2 Uniform DFT Filter Banks

We now outline a simple technique for the design of a class of filter banks with equal passband widths. Let $H_0(z)$ represent a causal lowpass digital filter with an impulse response $h_0[n]$:

$$H_0(z) = \sum_{n=0}^{\infty} h_0[n]z^{-n}, \tag{10.62}$$

which we assume to be an IIR filter without any loss of generality. Let us now assume that $H_0(z)$ has its passband edge ω_p and stopband edge ω_s around π/M, where M is some arbitrary integer, as indicated in Figure 10.43(a). Now, consider the transfer function $H_k(z)$ whose impulse response $h_k[n]$ is defined to be

$$h_k[n] = h_0[n]W_M^{-kn}, \qquad 0 \le k \le M - 1, \tag{10.63}$$

where $W_M = e^{-j2\pi/M}$, as defined in Eq. (3.24). Thus,

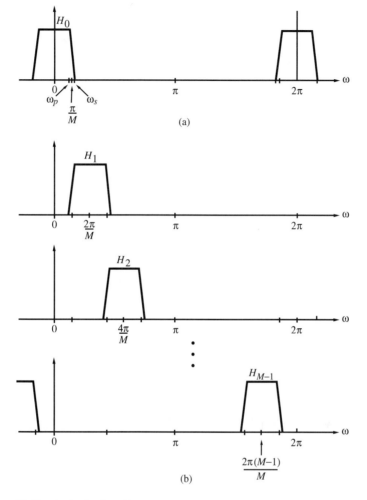

Figure 10.43: The bank of M filters $H_k(z)$ with uniformly shifted frequency responses.

$$H_k(z) = \sum_{n=0}^{\infty} h_k[n]z^{-n} = \sum_{n=0}^{\infty} h_0[n] \left(z W_M^k\right)^{-n}, \qquad 0 \le k \le M-1, \tag{10.64}$$

i.e.,

$$H_k(z) = H_0\left(z W_M^k\right), \qquad 0 \le k \le M-1, \tag{10.65}$$

with a corresponding frequency response

$$H_k(e^{j\omega}) = H_0(e^{j(\omega - 2\pi k/M)}), \qquad 0 \le k \le M-1. \tag{10.66}$$

In other words, the frequency response of $H_k(z)$ is obtained by shifting the response of $H_0(z)$ to the right, by an amount $2\pi k/M$. The responses of $H_1(z)$, $H_2(z)$, ..., $H_{M-1}(z)$ are shown in Figure 10.43(b). Note that the corresponding impulse responses $h_k[n]$ are, in general, complex and hence $|H_k(e^{j\omega})|$ does not necessarily exhibit symmetry with respect to zero frequency. Figure 10.43(b) therefore represents

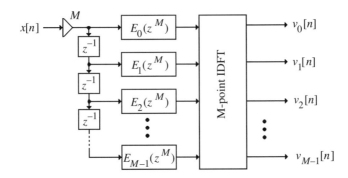

Figure 10.44: Polyphase implementation of a uniform DFT analysis filter bank where $H_k(z) = V_k(z)/X(z)$.

the responses of $M - 1$ filters $H_1(z), H_2(z), \ldots, H_{M-1}(z)$, which are uniformly shifted versions of the response of the basic *prototype* filter $H_0(z)$ of Figure 10.43(a).

The M filters $H_k(z)$ defined by Eq. (10.65) could be used as the analysis filters in the analysis filter bank of Figure 10.42(a) or as the synthesis filters $F_k(z)$ in the synthesis filter bank of Figure 10.42(b).

Since the set of magnitude responses $|H_k(e^{j\omega})|$, $k = 0, 1, \ldots, M - 1$, are uniformly shifted versions of a basic prototype $|H_0(e^{j\omega})|$, i.e.,

$$\left|H_k(e^{j\omega})\right| = \left|H_0\left(e^{j(\omega - 2\pi k/M)}\right)\right|, \tag{10.67}$$

the filter bank obtained is called a *uniform filter bank*.

10.6.3 Polyphase Implementations of Uniform Filter Banks

Let Figure 10.42(a) represent a uniform filter bank with the M analysis filters $H_k(z)$ related through Eq. (10.65). The impulse response sequences $h_k[n]$ of the analysis filters are accordingly related as in Eq. (10.63). Instead of realizing each analysis filter as a separate filter, it is possible to develop a computationally more efficient realization of the above uniform filter bank, which is described next.

Let the lowpass prototype transfer function $H_0(z)$ be represented in its M-band polyphase form:

$$H_0(z) = \sum_{\ell=0}^{M-1} z^{-\ell} E_\ell(z^M), \tag{10.68}$$

where $E_\ell(z)$ is the ℓth polyphase component of $H_0(z)$:

$$E_\ell(z) = \sum_{n=0}^{\infty} e_\ell[n]z^{-n} = \sum_{n=0}^{\infty} h_0[\ell + nM]z^{-n}, \qquad 0 \le \ell \le M - 1. \tag{10.69}$$

Substituting z with zW_M^k in Eq. (10.68), we arrive at the M-band polyphase decomposition of $H_k(z)$:

$$H_k(z) = \sum_{\ell=0}^{M-1} z^{-\ell} W_M^{-k\ell} E_\ell(z^M W_M^{kM}) = \sum_{\ell=0}^{M-1} z^{-\ell} W_M^{-k\ell} E_\ell(z^M),$$
$$k = 0, 1, \ldots, M - 1, \tag{10.70}$$

where we have used the identity $W_M^{kM} = 1$.

Note that Eq. (10.70) can be written in matrix form as

$$H_k(z) = \begin{bmatrix} 1 & W_M^{-k} & W_M^{-2k} & \cdots & W_M^{-(M-1)k} \end{bmatrix} \begin{bmatrix} E_0(z^M) \\ z^{-1}E_1(z^M) \\ z^{-2}E_2(z^M) \\ \vdots \\ z^{-(M-1)}E_{M-1}(z^M) \end{bmatrix}, \tag{10.71}$$

for $k = 0, 1, \ldots, M - 1$. All these M equations can be combined into a matrix equation as

$$\begin{bmatrix} H_0(z) \\ H_1(z) \\ H_2(z) \\ \vdots \\ H_{M-1}(z) \end{bmatrix}$$

$$= \begin{bmatrix} 1 & 1 & 1 & \cdots & 1 \\ 1 & W_M^{-1} & W_M^{-2} & \cdots & W_M^{-(M-1)} \\ 1 & W_M^{-2} & W_M^{-4} & \cdots & W_M^{-2(M-1)} \\ \vdots & \vdots & \vdots & \ddots & \vdots \\ 1 & W_M^{-(M-1)} & W_M^{-2(M-1)} & \cdots & W_M^{-(M-1)^2} \end{bmatrix} \begin{bmatrix} E_0(z^M) \\ z^{-1}E_1(z^M) \\ z^{-2}E_2(z^M) \\ \vdots \\ z^{-(M-1)}E_{M-1}(z^M) \end{bmatrix}, \tag{10.72}$$

which is equivalent to

$$\begin{bmatrix} H_0(z) \\ H_1(z) \\ H_2(z) \\ \vdots \\ H_{M-1}(z) \end{bmatrix} = M\mathbf{D}^{-1} \begin{bmatrix} E_0(z^M) \\ z^{-1}E_1(z^M) \\ z^{-2}E_2(z^M) \\ \vdots \\ z^{-(M-1)}E_{M-1}(z^M) \end{bmatrix}, \tag{10.73}$$

where $\mathbf{D}$ denotes the DFT matrix:

$$\mathbf{D} = \begin{bmatrix} 1 & 1 & 1 & \cdots & 1 \\ 1 & W_M^1 & W_M^2 & \cdots & W_M^{(M-1)} \\ 1 & W_M^2 & W_M^4 & \cdots & W_M^{2(M-1)} \\ \vdots & \vdots & \vdots & \ddots & \vdots \\ 1 & W_M^{(M-1)} & W_M^{2(M-1)} & \cdots & W_M^{(M-1)^2} \end{bmatrix}. \tag{10.74}$$

An efficient implementation of the M-band analysis filter bank based on Eq. (10.73) is thus as shown in Figure 10.44, where the prototype lowpass filter $H_0(z)$ has been implemented in a polyphase form. The structure of Figure 10.44 is more commonly known as the *uniform DFT analysis filter bank*.[9]

The computational complexity of Figure 10.44 is much smaller than that of a direct implementation, as in Figure 10.42(a). For example, an M-band uniform DFT analysis filter bank based on an N-tap prototype lowpass FIR filter requires a total of $(M/2)\log_2 M + N$ multipliers, whereas a direct implementation requires NM multiplications.

Following a development similar to that outlined above, we can derive the structure for a *uniform DFT synthesis filter bank*. Efficient realizations based on the Types I and II polyphase decompositions of the prototype lowpass filter $H_0(z)$ are indicated in Figure 10.45.

---

[9]An interesting application of the uniform DFT analysis filter with a FIR filter is in the detection of a code and the simultaneous measurement of the Doppler offset for synchronisation purposes in spread spectrum communication systems [Spa2000].

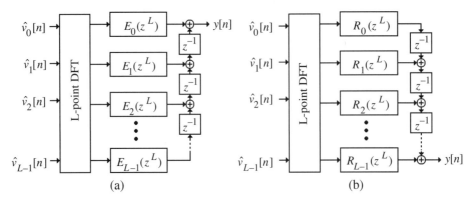

Figure 10.45: Uniform DFT synthesis filter bank.

It follows from Eq. (10.73) that the polyphase components $E_i(z^M)$ can be expressed in terms of the prototype transfer function $H_0(z)$ and its modulated versions $H_i(z)$ according to

$$
\begin{bmatrix}
E_0(z^M) \\
z^{-1} E_1(z^M) \\
z^{-2} E_2(z^M) \\
\vdots \\
z^{-(M-1)} E_{M-1}(z^M)
\end{bmatrix}
= \frac{1}{M} \mathbf{D}
\begin{bmatrix}
H_0(z) \\
H_1(z) \\
H_2(z) \\
\vdots \\
H_{M-1}(z)
\end{bmatrix},
\tag{10.75}
$$

which can be used to determine the polyphase components of an IIR transfer function (Problems 10.16 and 10.17).

10.7 Nyquist Filters

In this section, we introduce a special type of lowpass filter with a transfer function that, by design, has certain zero-valued coefficients. Due to the presence of these zero-valued coefficients, these filters are, by nature, computationally more efficient than other lowpass filters of the same order. In addition, when used as interpolator filters, they preserve the nonzero samples of the up-sampler output at the interpolator output. These filters, called Lth-band filters or Nyquist filters and discussed in this section, are often used both in single-rate and multirate signal processing. For example, they are usually preferred in decimator and interpolator design. Another application is in the design of a quadrature-mirror filter bank, discussed in Section 10.8. A third application, described in Section 11.7, is in the design of a Hilbert transformer, which is employed for the generation of analytic signals.

10.7.1 Lth-Band Filters

Consider the factor-of-L interpolator of Figure 10.15(a). The relation between the output and the input of the interpolator is given by

$$Y(z) = H(z)X(z^L). \tag{10.76}$$

If the interpolation filter $H(z)$ is realized in the L-band polyphase form, then we have

$$H(z) = E_0(z^L) + z^{-1}E_1(z^L) + z^{-2}E_2(z^L) + \cdots + z^{-(L-1)}E_{L-1}(z^L).$$

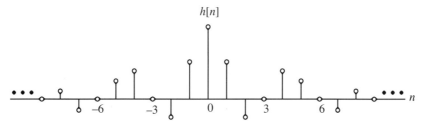

Figure 10.46: The impulse response of a typical third-band filter.

Assume that the kth polyphase component of $H(z)$ is a constant, i.e., $E_k(z) = \alpha$:

$$H(z) = E_0(z^L) + z^{-1}E_1(z^L) + \cdots + z^{-(k-1)}E_{k-1}(z^L) + \alpha z^{-k}$$
$$+ z^{-(k+1)}E_{k+1}(z^L) + \cdots + z^{-(L-1)}E_{L-1}(z^L). \qquad (10.77)$$

Then we can express $Y(z)$ as

$$Y(z) = \alpha z^{-k}X(z^L) + \sum_{\substack{\ell=0 \\ \ell \neq k}}^{L-1} z^{-\ell}E_\ell(z^L)X(z^L). \qquad (10.78)$$

As a result, $y[Ln + k] = \alpha x[n]$, i.e., the input samples appear at the output without any distortion at all values of n, whereas the in-between $(L - 1)$ samples are determined by interpolation.

A filter with the above property is called a *Nyquist filter* or an *Lth-band filter* and its impulse response has many zero-valued samples, making it computationally very attractive. For example, the impulse response of the Lth-band filter obtained for $k = 0$ satisfies the following condition:

$$h[Ln] = \begin{cases} \alpha, & n = 0, \\ 0, & \text{otherwise.} \end{cases} \qquad (10.79)$$

Figure 10.46 shows a typical impulse response of a third-band filter ($L = 3$). If $H(z)$ satisfies Eq. (10.77) with $k = 0$, i.e., $E_0(z) = \alpha$, then it can be shown that

$$\sum_{k=0}^{L-1} H(zW_L^k) = L\alpha = 1 \qquad (\text{assuming } \alpha = 1/L). \qquad (10.80)$$

Since the frequency response of $H(zW_L^k)$ is the shifted version $H(e^{j(\omega - 2\pi k/L)})$ of $H(e^{j\omega})$, the sum of all of these L uniformly shifted versions of $H(e^{j\omega})$ add up to a constant (see Figure 10.47). Lth-band filters can be either FIR or IIR filters.

10.7.2 Half-Band Filters

An Lth-band filter for $L = 2$ is called a *half-band filter*. From Eq. (10.77) the transfer function of a half-band filter is thus given by

$$H(z) = \alpha + z^{-1}E_1(z^2), \qquad (10.81)$$

with its impulse response satisfying Eq. (10.79) with $L = 2$. The condition on the frequency response given by Eq. (10.81) reduces to

$$H(z) + H(-z) = 1 \qquad (\text{assuming } \alpha = \tfrac{1}{2}). \qquad (10.82)$$

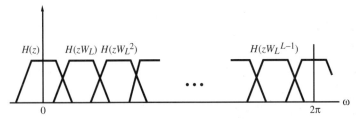

Figure 10.47: Frequency responses of $H(zW_L^k)$ for $k = 0, 1, \ldots, L - 1$.

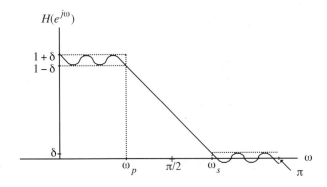

Figure 10.48: Frequency response of a zero-phase half-band filter.

If $H(z)$ has real coefficients, then $H(-e^{j\omega}) = H(e^{j(\pi - \omega)})$, and Eq. (10.82) leads to

$$H(e^{j\omega}) + H(e^{j(\pi - \omega)}) = 1. \tag{10.83}$$

The above equality implies that $H(e^{j(\pi/2 - \theta)})$ and $H(e^{j(\pi/2 + \theta)})$ add up to unity for all θ. In other words, $H(e^{j\omega})$ exhibits a symmetry with respect to the half-band frequency $\pi/2$, thus justifying the name "half-band filter." Figure 10.48 illustrates this symmetry for a half-band lowpass filter for which the passband and stopband ripples are equal, i.e., $\delta_p = \delta_s$, and the passband and stopband bandedges are symmetric with respect to $\pi/2$, i.e., $\omega_p + \omega_s = \pi$.

An important attractive property of the half-band filter is that about 50 percent of the coefficients of $h[n]$ are zero. This reduces the number of multiplications required in its implementation, making the filter computationally quite efficient. For example, if $N = 101$, an arbitrary Type 1 FIR transfer function requires about 50 multipliers, whereas a Type 1 half-band filter requires only about 25 multipliers.

An FIR half-band filter can be designed with linear phase. However, there is a constraint on its length. Consider a zero-phase half-band FIR filter for which $h[n] = ah^*[-n]$, with $|a| = 1$. Let the highest nonzero coefficient be $h[R]$. Then R is odd as a result of the condition of Eq. (10.79). Therefore, $R = 2K + 1$ for some integer K. Thus the length of the impulse response $h[n]$ is restricted to be of the form $2R + 1 = 4K + 3$ [unless $H(z)$ is a constant].

10.7.3 Design of Linear-Phase Lth-Band FIR Filters

A lowpass linear-phase Nyquist Lth-band FIR filter with a cutoff at $\omega_c = \pi/L$ and a good frequency response can be readily designed via the windowed Fourier series approach described in Section 7.6. In this approach the impulse response coefficients of the lowpass filter are chosen as

$$h[n] = h_{LP}[n] \cdot w[n], \tag{10.84}$$

where $h_{LP}[n]$ is the impulse response of an ideal lowpass filter with a cutoff at π/L, and $w[n]$ is a suitable window function. If

$$h_{LP}[n] = 0 \quad \text{for } n = \pm L, \pm 2L, \ldots, \tag{10.85}$$

then Eq. (10.79) is indeed satisfied.

Now the impulse response $h_{LP}[n]$ of an ideal Lth-band filter is obtained from Eq. (7.59) by substituting $\omega_c = \pi/L$ and is given by

$$h_{LP}[n] = \frac{\sin(\pi n/L)}{\pi n}, \quad -\infty \le n \le \infty. \tag{10.86}$$

It can be seen from the above that the impulse response coefficients do indeed satisfy the condition of Eq. (10.85). Hence, an Lth-band filter can be designed by applying a suitable window function to Eq. (10.86).

Likewise, an Lth-band filter with a frequency response as in Figure 7.25(a) can also be designed from Eq. (7.87) by replacing ω_c with π/L, resulting in an impulse response

$$h_{LP}[n] = \begin{cases} \frac{1}{L}, & n = 0, \\ \frac{2\sin(\Delta\omega n/2)}{\Delta\omega n} \cdot \frac{\sin(\pi n/L)}{\pi n}, & |n| > 0, \end{cases} \tag{10.87}$$

which again is seen to satisfy the condition of Eq. (10.85). Other candidates for Lth-band filter design are the lowpass filters of Eq. (7.88) and the raised cosine filter of Eq. (7.165) in Problem 7.52.

We illustrate the lowpass half-band filter design in the following example.

EXAMPLE 10.12   Design a length-23 half-band ($L = 2$) linear-phase lowpass filter using the windowed Fourier series approach with a Hamming window. To this end we use MATLAB Program 10_8 given below.

```
% Program 10_8
% Design of Lth Band FIR Filter Using the
% Windowed Fourier Series Approach
%
N = input('Type in the filter length = ');
L = input('Type in the value of L = ');
K = (N-1)/2;
n = -K:K;
% Generate the truncated impulse response of the ideal
% lowpass filter
b = sinc(n/L)/L;
% Generate the window sequence
win = hamming(N);
% Generate the coefficients of the windowed filter
c = b.*win';
% Plot the gain response of the windowed filter
[h,w] = freqz(c,1,256);
g = 20*log10(abs(h));
plot(w/pi,g);grid
xlabel('\omega/\pi');ylabel('Gain, dB');
```

The input data requested by the program are the desired filter length N and the value of L. The program determines the impulse response coefficients of the Lth-band FIR filter using the expression of Eq. (10.86), and computes and plots the gain response of the designed filter as shown in Figure 10.49. The filter coefficients are elements of the vector c and are given by

```
h[-11] = h[11]  =  -0.00231498033043;   h[-10] = h[10] = 0;
h[-9]  = h[9]   =   0.00541209478301;   h[-8]  = h[8]  = 0;
h[-7]  = h[7]   =  -0.01586588771510;   h[-6]  = h[6]  = 0;
h[-5]  = h[5]   =   0.03854508794537;   h[-4]  = h[4]  = 0;
h[-3]  = h[3]   =  -0.08925790518530;   h[-2]  = h[2]  = 0;
h[-1]  = h[1]   =   0.31237874418292;    h[0]   = 0.5.
```

As expected, the impulse response coefficients are equal to zero for $n = \pm 2, \pm 4, \ldots, \pm 10$.

An elegant method for the design of half-band linear-phase FIR filters is considered in Section 11.7.3. Also described here is a method for the design of half-band IIR filters. Several other design approaches have been advanced in the literature [Ren87], [Vai87a].

10.7.4 Relation between Lth-Band Filters and Power-Complementary Filters

Recall from Section 4.8.3, a set of filters is said to be power-complementary if their square magnitude responses add up to a constant. Consider an Lth-band transfer function $H(z)$ represented in the L-band polyphase decomposition form given by

$$H(z) = \sum_{\ell=0}^{L-1} z^{-\ell} E_\ell(z^L). \tag{10.88}$$

Define a new transfer function $G(z) = \tilde{H}(z)H(z)$.[10] Then the set of transfer functions $\{E_0(z), E_1(z), \ldots, E_{L-1}(z)\}$ is power-complementary if and only if $G(z)$ is an Lth-band filter [Vai90].

To prove the above property, define

$$H_r(z) = H(zW_L^{-r}) = \sum_{\ell=0}^{L-1} z^{-\ell} W_L^{\ell r} E_\ell(z^L) \tag{10.89}$$

for $0 \leq r \leq L - 1$. We can write the above set of equations in matrix form as

$$\underbrace{\begin{bmatrix} H_0(z) \\ H_1(z) \\ \vdots \\ H_{L-1}(z) \end{bmatrix}}_{\mathbf{H}(z)} = \mathbf{D} \underbrace{\begin{bmatrix} 1 & 0 & \cdots & 0 \\ 0 & z^{-1} & \cdots & 0 \\ \vdots & \vdots & \ddots & \vdots \\ 0 & 0 & \cdots & z^{-(L-1)} \end{bmatrix}}_{\mathbf{\Lambda}(z)} \underbrace{\begin{bmatrix} E_0(z^L) \\ E_1(z^L) \\ \vdots \\ E_{L-1}(z^L) \end{bmatrix}}_{\mathbf{E}(z)} \tag{10.90}$$

where $\mathbf{D}$ is the $L \times L$ DFT matrix satisfying $\mathbf{D}^\dagger \mathbf{D} = L\mathbf{I}$.[11] If the set of transfer functions $\{E_0(z), E_1(z), \ldots, E_{L-1}(z)\}$ is power-complementary, then $\tilde{\mathbf{E}}(z)\mathbf{E}(z) = \gamma$ so that[12]

$$\tilde{\mathbf{H}}(z)\mathbf{H}(z) = \tilde{\mathbf{E}}(z)\tilde{\mathbf{\Lambda}}(z)\mathbf{D}^\dagger \mathbf{D}\mathbf{\Lambda}(z)\mathbf{E}(z) = L\gamma. \tag{10.91}$$

---

[10] $\tilde{H}(z) = H(1/z^*)$.

[11] $\mathbf{D}^\dagger$ denotes the conjugate transpose of $\mathbf{D}$.

[12] $\tilde{\mathbf{H}}(z)$ is the conjugate transpose of $\mathbf{H}(1/z^*)$, i.e., $\tilde{\mathbf{H}}(z) = \mathbf{H}^\dagger(1/z^*)$.

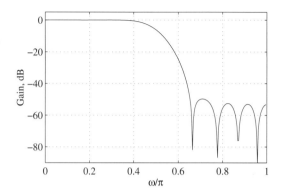

Figure 10.49: Gain response of a length-23 linear-phase half-band FIR filter of Example 10.12.

This in turn implies that the set $\{H_0(z), H_1(z), \ldots, H_{L-1}(z)\}$ is power-complementary. In other words, $G(z) = \tilde{H}(z)H(z)$ satisfies

$$\sum_{k=0}^{L-1} G(zW_L^{-k}) = L\gamma \tag{10.92}$$

so that it is an Lth-band filter.

Conversely, we can prove that the set $\{E_0(z), E_1(z), \ldots, E_{L-1}(z)\}$ is power-complementary, assuming that $G(z)$ is an Lth-band filter, simply by inverting the above matrix equation and carrying out an argument similar to that outlined above.

10.8 Two-Channel Quadrature-Mirror Filter Bank

In many applications, a discrete-time signal $x[n]$ is first split into a number of subband signals $\{v_k[n]\}$ by means of an analysis filter bank; the subband signals are then processed and finally combined by a synthesis filter bank resulting in an output signal $y[n]$. If the subband signals are bandlimited to frequency ranges much smaller than that of the original input signal, they can be down-sampled before processing. Because of the lower sampling rate, the processing of the down-sampled signals can be carried out more efficiently. After processing, these signals are up-sampled before being combined by the synthesis bank into a higher-rate signal. The combined structure employed is called a *quadrature-mirror filter (QMF) bank*. If the down-sampling and the up-sampling factors are equal to or greater than the number of bands of the filter bank, then the output $y[n]$ can be made to retain some or all of the characteristics of the input $x[n]$ by properly choosing the filters in the structure. In case of equality, the filter bank is said to be a *critically sampled filter bank*. The most common application of this scheme is in the efficient coding of a signal $x[n]$ (see Section 11.8). Another possible application is in the design of an analog voice privacy system to provide secure telephone conversation [Cox83]. In this section, we study a two-channel QMF bank.

10.8.1 The Filter Bank Structure

Figure 10.50 shows the basic two-channel QMF bank–based subband codec (coder/decoder). Here, the input signal $x[n]$ is first passed through a two-band analysis filter bank containing the filters $H_0(z)$ and $H_1(z)$, which typically have lowpass and highpass frequency responses, respectively, with a cutoff frequency at $\pi/2$, as indicated in Figure 10.51. The subband signals $\{v_k[n]\}$ are then down-sampled by a

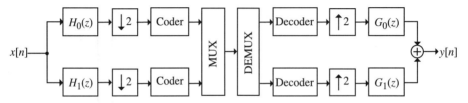

Figure 10.50: The two-channel filter bank based coder/decoder.

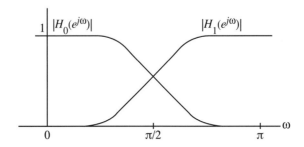

Figure 10.51: Typical frequency responses of the analysis filters.

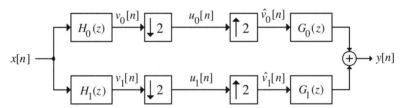

Figure 10.52: The two-channel quadrature-mirror filter (QMF) bank.

factor of 2. Each down-sampled subband signal is encoded by exploiting the special spectral properties of the signal, such as energy levels and perceptual importance (see Section 11.8). The coded subband signals are combined into one sequence by multiplexing and either stored for later retrieval or transmitted. At the receiving end, the coded subband signals are first recovered by demultiplexing and decoders are used to produce approximations of the original down-sampled signals. The decoded signals are then up-sampled by a factor of 2 and passed through a two-band synthesis filter bank composed of the filters $G_0(z)$ and $G_1(z)$ whose outputs are then added yielding $y[n]$. It follows from the figure that the sampling rates of the input signal $x[n]$ and output signal $y[n]$ are the same. The analysis and the synthesis filters in the QMF bank are chosen so as to ensure that the reconstructed output $y[n]$ is a reasonable replica of the input $x[n]$. Moreover, they are also designed to provide good frequency selectivity in order to ensure that the sum of the power of the subband signals is reasonably close to the input signal power.

   In practice, various errors are generated in this scheme. In addition to the coding error and errors caused by transmission through the channel, the QMF bank itself introduces several errors due to the sampling rate alterations and imperfect filters. We ignore the coding and channel errors, and investigate only the errors generated by the down-samplers and up-samplers in the filter bank and their effects on the performance of the system. To this end, we consider the QMF bank structure without the coders and the decoders as indicated in Figure 10.52 [Cro76a], [Est77].

10.8.2 Analysis of the Two-Channel QMF Bank

It is convenient to analyze the filter bank in the z-domain. To this end, we make use of the input-output relations of the up-sampler and the down-sampler derived earlier in Section 10.1 and given by Eqs. (10.5) and (10.12). The expressions for the z-transforms of various intermediate signals in Figure 10.52 are then given by

$$V_k(z) = H_k(z)X(z), \tag{10.93a}$$

$$U_k(z) = \tfrac{1}{2}\left\{V_k(z^{1/2}) + V_k(-z^{1/2})\right\}, \tag{10.93b}$$

$$\hat{V}_k(z) = U_k(z^2), \tag{10.93c}$$

for $k = 0, 1$. From Eqs. (10.93a) to (10.93c), we obtain after some algebra

$$\hat{V}_k(z) = \tfrac{1}{2}\left\{V_k(z) + V_k(-z)\right\} = \tfrac{1}{2}\left\{H_k(z)X(z) + H_k(-z)X(-z)\right\}. \tag{10.94}$$

The reconstructed output of the filter bank is given by

$$Y(z) = G_0(z)\hat{V}_0(z) + G_1(z)\hat{V}_1(z). \tag{10.95}$$

Substituting Eq. (10.94) in Eq. (10.95), we obtain after some rearrangement the expression for the output of the filter bank as

$$Y(z) = \tfrac{1}{2}\left\{H_0(z)G_0(z) + H_1(z)G_1(z)\right\}X(z)$$
$$+ \tfrac{1}{2}\left\{H_0(-z)G_0(z) + H_1(-z)G_1(z)\right\}X(-z). \tag{10.96}$$

The second term in the above equation is precisely due to the aliasing caused by sampling rate alteration. The above equation can be compactly expressed as

$$Y(z) = T(z)X(z) + A(z)X(-z), \tag{10.97}$$

where

$$T(z) = \tfrac{1}{2}\left\{H_0(z)G_0(z) + H_1(z)G_1(z)\right\} \tag{10.98}$$

is called the *distortion transfer function*, and

$$A(z) = \tfrac{1}{2}\left\{H_0(-z)G_0(z) + H_1(-z)G_1(z)\right\}. \tag{10.99}$$

10.8.3 Alias-Free Filter Bank

As noted in Section 10.1, the up-sampler and the down-sampler are linear time-varying components and, as a result, in general, the QMF structure of Figure 10.52 is a linear time-varying (LTV) system. It can be shown also that it has a period of 2 (Problem 10.35). However, it is possible to choose the analysis and synthesis filters such that the aliasing effect is canceled, resulting in a linear time-invariant (LTI) operation. To this end, we need to ensure that

$$A(z) = 0,$$

i.e.,

$$H_0(-z)G_0(z) + H_1(-z)G_1(z) = 0. \tag{10.100}$$

For aliasing cancellation we can choose

$$\frac{G_0(z)}{G_1(z)} = -\frac{H_1(-z)}{H_0(-z)}, \tag{10.101}$$

which yields

$$G_0(z) = C(z)H_1(-z), \qquad G_1(z) = -C(z)H_0(-z), \qquad (10.102)$$

where $C(z)$ is an arbitrary rational function.

If the above relations hold, then Eq. (10.96) reduces to

$$Y(z) = T(z)X(z), \qquad (10.103)$$

with

$$T(z) = \tfrac{1}{2}\left\{H_0(z)H_1(-z) - H_1(z)H_0(-z)\right\}. \qquad (10.104)$$

On the unit circle, we have

$$Y(e^{j\omega}) = T(e^{j\omega})X(e^{j\omega}) = \left|T(e^{j\omega})\right| e^{j\phi(\omega)}X(e^{j\omega}). \qquad (10.105)$$

If $T(z)$ is an allpass function, i.e., $\left|T(e^{j\omega})\right| = d \neq 0$, then

$$\left|Y(e^{j\omega})\right| = d\left|X(e^{j\omega})\right| \qquad (10.106)$$

indicating that the output of the QMF bank has the same magnitude response as that of the input (scaled by d) but exhibits phase distortion, and the filter bank is said to be *magnitude preserving*. If $T(z)$ has linear phase, i.e.,

$$\phi(\omega) = \alpha\omega + \beta, \qquad (10.107)$$

then

$$\arg\left\{Y(e^{j\omega})\right\} = \arg\left\{X(e^{j\omega})\right\} + \alpha\omega + \beta, \qquad (10.108)$$

and the filter bank is said to be *phase preserving* but exhibits magnitude distortion. If an alias-free QMF bank has no amplitude and phase distortion, then it is called a *perfect reconstruction (PR)* QMF bank. In such a case, $T(z) = d\,z^{-\ell}$, resulting in

$$Y(z) = d\,z^{-\ell}X(z), \qquad (10.109)$$

which in the time-domain is equivalent to

$$y[n] = d\,x[n - \ell] \qquad (10.110)$$

for all possible inputs, indicating that the reconstructed output $y[n]$ is a scaled, delayed replica of the input.

EXAMPLE 10.13   The multirate system of Figure 10.6 can be considered as a two-channel QMF bank. Comparing it with the filter bank structure of Figure 10.52 we conclude that the analysis and synthesis filters of Figure 10.6 are given by

$$H_0(z) = 1, \qquad H_1(z) = z^{-1}, \qquad G_0(z) = z^{-1}, \qquad G_1(z) = 1.$$

Substituting these values in Eqs. (10.98) and (10.99) we get

$$T(z) = \frac{1}{2}\left(z^{-1} + z^{-1}\right) = z^{-1},$$

$$A(z) = \frac{1}{2}\left(z^{-1} - z^{-1}\right) = 0,$$

confirming that the structure is an alias-free perfect reconstruction filter bank. However, the filters in the bank do not provide any frequency selectivity.

10.8.4 An Alias-Free Realization

A very simple alias-free two-band QMF bank is obtained when

$$H_1(z) = H_0(-z). \tag{10.111}$$

The above condition, in the case of a real coefficient filter, implies

$$\left| H_1(e^{j\omega}) \right| = \left| H_0(e^{j(\pi-\omega)}) \right| \tag{10.112}$$

indicating that if $H_0(z)$ is a lowpass filter, then $H_1(z)$ is a highpass filter, and vice versa. In fact, Eq. (10.112) indicates that $\left| H_1(e^{j\omega}) \right|$ is a mirror image of $\left| H_0(e^{j\omega}) \right|$ with respect to $\pi/2$, the *quadrature frequency*. This has given rise to the name quadrature-mirror filter bank.

Substituting Eq. (10.111) in Eq. (10.102), we arrive at, with $C(z) = 1$,

$$G_0(z) = H_0(z), \quad G_1(z) = -H_1(z) = -H_0(-z). \tag{10.113}$$

Equations (10.111) and (10.113) imply that the two analysis filters and the two synthesis filters in the QMF bank are essentially determined from one transfer function $H_0(z)$. Moreover, Eq. (10.113) indicates that if $H_0(z)$ is a lowpass filter, then $G_0(z)$ is also a lowpass filter, and $G_1(z)$ is a highpass filter. The distortion transfer function $T(z)$ of Eq. (10.104) in this case reduces to

$$T(z) = \tfrac{1}{2}\left\{ H_0^2(z) - H_1^2(z) \right\} = \tfrac{1}{2}\left\{ H_0^2(z) - H_0^2(-z) \right\}. \tag{10.114}$$

A computationally efficient realization of the above alias-free two-channel QMF bank is obtained by realizing the analysis and the synthesis filters in polyphase form. Let the two-band Type I polyphase representation of $H_0(z)$ be given by

$$H_0(z) = E_0(z^2) + z^{-1}E_1(z^2). \tag{10.115a}$$

From Eq. (10.111) it follows then that

$$H_1(z) = E_0(z^2) - z^{-1}E_1(z^2). \tag{10.115b}$$

In matrix form Eqs. (10.115a) and (10.115b) can be expressed as

$$\begin{bmatrix} H_0(z) \\ H_1(z) \end{bmatrix} = \begin{bmatrix} 1 & 1 \\ 1 & -1 \end{bmatrix} \begin{bmatrix} E_0(z^2) \\ z^{-1}E_1(z^2) \end{bmatrix}. \tag{10.116}$$

Likewise the synthesis filters, in matrix form, can be expressed as

$$[\, G_0(z) \quad G_1(z) \,] = [\, z^{-1}E_1(z^2) \quad E_0(z^2) \,] \begin{bmatrix} 1 & 1 \\ 1 & -1 \end{bmatrix}. \tag{10.117}$$

Using Eqs. (10.116) and (10.117), we can redraw the two-channel QMF bank as shown in Figure 10.53(a), which can be further simplified using the cascade equivalences of Figure 10.14, resulting in the computationally efficient realization of Figure 10.53(b).

The expression for the distortion transfer function in this case, obtained by substituting Eqs. (10.115a) and (10.115b) in Eq. (10.104), is given by

$$T(z) = 2z^{-1}E_0(z^2)E_1(z^2). \tag{10.118}$$

The following example illustrates the development of a very simple perfect reconstruction QMF bank.

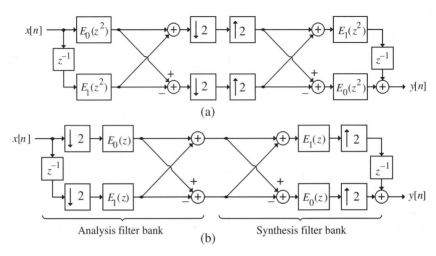

Figure 10.53: Polyphase realization of the two-channel QMF bank. (a) Direct polyphase realization and (b) computationally efficient realization.

EXAMPLE 10.14   Consider a two-channel QMF bank with the analysis filter given by

$$H_0(z) = 1 + z^{-1}.$$

Its polyphase components are thus

$$E_0(z^2) = 1, \qquad E_1(z^2) = 1.$$

From Eq. (10.115b), the second analysis filter is therefore given by

$$H_1(z) = 1 - z^{-1},$$

and the corresponding synthesis filters for an alias-free realization from Eq. (10.117) are given by

$$G_0(z) = z^{-1}E_1(z^2) + E_0(z^2) = 1 + z^{-1},$$
$$G_1(z) = z^{-1}E_1(z^2) - E_0(z^2) = -1 + z^{-1}.$$

The distortion transfer function for this realization is given by

$$T(z) = 2z^{-1},$$

resulting in the realization of a perfect reconstruction QMF bank.

10.8.5 Alias-Free FIR QMF Bank

Let the prototype analysis filter be a linear-phase FIR filter of order N with a real coefficient transfer function $H_0(z)$ given by

$$H_0(z) = \sum_{n=0}^{N} h_0[n]z^{-n}. \tag{10.119}$$

Note that $H_0(z)$ can be either a Type 1 or Type 2 linear-phase function since it has to be a lowpass filter. As a result, its impulse-response coefficients must satisfy the condition $h_0[n] = h_0[N - n]$, in which case we can write

$$H_0(e^{j\omega}) = e^{-j\omega N/2} \breve{H}_0(\omega) \tag{10.120}$$

where $\breve{H}_0(\omega)$ is the amplitude function, a real function of ω. By making use of Eq. (10.120) in Eq. (10.114) along with the property that $|H_0(e^{j\omega})|$ is an even function of ω, we can express the frequency response of the distortion transfer function as

$$T(e^{j\omega}) = \frac{e^{-jN\omega}}{2} \left\{ \left| H_0(e^{j\omega}) \right|^2 - (-1)^N \left| H_0(e^{j(\pi-\omega)}) \right|^2 \right\}. \tag{10.121}$$

From the above it can be seen that if N is even, then $T(e^{j\omega}) = 0$ at $\omega = \pi/2$, implying severe amplitude distortion at the output of the filter bank. As a result, N must be chosen to be odd, in which case Eq. (10.121) reduces to

$$\begin{aligned} T(e^{j\omega}) &= \frac{e^{-jN\omega}}{2} \left\{ \left| H_0(e^{j\omega}) \right|^2 + \left| H_0(e^{j(\pi-\omega)}) \right|^2 \right\} \\ &= \frac{e^{-jN\omega}}{2} \left\{ \left| H_0(e^{j\omega}) \right|^2 + \left| H_1(e^{j\omega}) \right|^2 \right\}. \end{aligned} \tag{10.122}$$

It follows from the above expression, the FIR two-channel filter bank with linear-phase analysis and synthesis filters will be of perfect reconstruction type if

$$\left| H_0(e^{j\omega}) \right|^2 + \left| H_1(e^{j\omega}) \right|^2 = 1, \tag{10.123}$$

i.e., the two analysis filters are power-complementary. Except for the two trivial filter banks of Examples 10.13 and 10.14, it can be shown that it is not possible to realize a perfect reconstruction two-channel filter bank with linear-phase power-complementary analysis filters [Vai85c].

As can be seen from Eq. (10.121), the QMF bank has no phase distortion, but will always exhibit amplitude distortion unless $|T(e^{j\omega})|$ is a constant for all values of ω. If $H_0(z)$ is a very good lowpass filter with $|H_0(e^{j\omega})| \cong 1$ in the passband and $|H_0(e^{j\omega})| \cong 0$ in the stopband, then $H_1(z)$ is a very good highpass filter with its passband coinciding with the stopband of $H_0(z)$, and vice versa. As a result, $|T(e^{j\omega})| \cong 1/2$ in the passbands of $H_0(z)$ and $H_1(z)$. The amplitude distortion thus occurs primarily in the transition band of these filters, with the degree of distortion being determined by the amount of overlap between their squared-magnitude responses. This distortion can be minimized by controlling the overlap, which in turn can be controlled by appropriately choosing the passband edge of $H_0(z)$.

One way to minimize the amplitude distortion is to employ a computer-aided optimization method to iteratively adjust the filter coefficients $h_0[n]$ of $H_0(z)$ such that the constraint

$$\left| H_0(e^{j\omega}) \right|^2 + \left| H_1(e^{j\omega}) \right|^2 \cong 1 \tag{10.124}$$

is satisfied for all values of ω [Joh80]. To this end, the objective function ϕ to be minimized can be chosen as a linear combination of two functions: (1) stopband attenuation of $H_0(z)$ and (2) the sum of the squared-magnitude responses of $H_0(z)$ and $H_1(z)$ as indicated in Eq. (10.124). One such objective function is given by

$$\phi = \alpha\phi_1 + (1-\alpha)\phi_2, \tag{10.125}$$

where

$$\phi_1 = \int_{\omega_s}^{\pi} \left| H_0(e^{j\omega}) \right|^2 d\omega, \quad \phi_2 = \int_0^{\pi} \left(1 - \left| H_0(e^{j\omega}) \right|^2 - \left| H_1(e^{j\omega}) \right|^2 \right)^2 d\omega, \tag{10.126}$$

and $0 < \alpha < 1$ and $\omega_s = (\pi/2) + \varepsilon$ for some small $\varepsilon > 0$. Note that since $|T(e^{j\omega})|$ is symmetric with respect to $\pi/2$, the second integral in Eq. (10.126) can be replaced with

$$2 \int_0^{\pi/2} \left(1 - \left|H_0(e^{j\omega})\right|^2 - \left|H_1(e^{j\omega})\right|^2 \right)^2 d\omega.$$

After ϕ has been made very small by the minimization procedure, both ϕ_1 and ϕ_2 will also be very small. This in turn will make $H_0(z)$ have a magnitude response satisfying $|H_0(e^{j\omega})| \cong 1$ in its passband and $|H_0(e^{j\omega})| \cong 0$ in its stopband, as desired. Moreover, since the power-complementary condition of Eq. (10.124) will be satisfied approximately, the magnitude response of the power-complementary highpass filter $H_1(z)$ to the lowpass filter will satisfy $|H_1(e^{j\omega})| \cong 0$ in the passband of $H_0(z)$ and $|H_1(e^{j\omega})| \cong 1$ in the stopband of $H_0(z)$.

Using the above approach, Johnston has designed a large class of linear-phase FIR lowpass filters $H_0(z)$ meeting a variety of specifications and has tabulated their impulse response coefficients [Joh80], [Cro83], [Ans93]. The following example examines the performance of one such filter.

EXAMPLE 10.15   The impulse response coefficients of the length-12 linear-phase lowpass filter 12B of Johnston [Joh80] are given by

$$
\begin{array}{llll}
h_0[0] & = -0.006443977 & = h_0[11], & h_0[1] & = 0.02745539 & = h_0[10], \\
h_0[2] & = -0.00758164 & = h_0[9], & h_0[3] & = -0.0913825 & = h_0[8], \\
h_0[4] & = 0.09808522 & = h_0[7], & h_0[5] & = 0.4807962 & = h_0[6].
\end{array}
$$

We use MATLAB Program 10_9 to verify the performances of the above filter $H_0(z)$. The input data requested by the program are the first half of the filter coefficients. It determines the remaining half by using the function fliplr. The program computes the gain response of the above filter and that of its complementary highpass filter $H_1(z) = H_0(-z)$, as indicated in Figure 10.54(a). It then computes the amplitude distortion function $|H_0(e^{j\omega})|^2 + |H_1(e^{j\omega})|^2$ in dB as shown in Figure 10.54(b). From Figure 10.54(a), we note that the stopband edge frequency ω_s of filter 12B is about 0.71π, which corresponds to a transition bandwidth $(\omega_s - 0.5\pi)/2\pi = 0.105$. The minimum stopband attenuation is approximately 34 dB. We also observe from Figure 10.54(b) that the amplitude distortion function is indeed very close to 0 dB in both the passbands and the stopbands of the two filters, with a peak value of ± 0.02 dB.

```
% Program 10_9
% Frequency Responses of Johnston's QMF Filters
% Type in prototype lowpass filter coefficients
B1 = input('Filter coefficients = ');
B1 = [B1 fliplr(B1)];
% Generate the complementary highpass filter
L = length(B1);
for k = 1:L
     B2(k) = ((-1)^k)*B1(k);
end
%  Compute the gain responses of the two filters
[H1z, w] = freqz(B1, 1, 256);
h1 = abs(H1z); g1 = 20*log10(h1);
[H2z, w] = freqz(B2, 1, 256);
h2 = abs(H2z); g2 = 20*log10(h2);
%  Plot the gain responses of the two filters
plot(w/pi, g1,'-',w/pi, g2,'--');grid
xlabel('\omega/\pi'); ylabel('Gain, dB')
pause
```

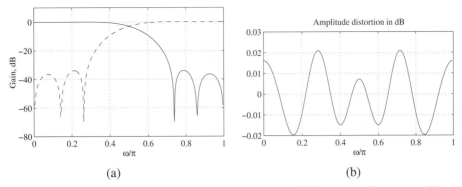

Figure 10.54: Johnston's 12B filter: (a) gain responses, and (b) reconstruction error in dB.

```
% Compute the sum of the squared-magnitude responses
for i = 1:256,
    sum(i) = (h1(i)*h1(i)) + (h2(i)*h2(i));
end
d =10*log10(sum);
% Plot the amplitude distortion
plot(w/pi,d);grid;
xlabel('\omega/\pi'); ylabel('Amplitude distortion, dB')
```

10.8.6 Alias-Free IIR QMF Bank

We now consider the design of an alias-free QMF bank employing IIR analysis and synthesis filters. Under the alias-free conditions of Eqs. (10.102) and (10.111), the distortion transfer function $T(z)$ of the two-channel QMF bank is given by $2z^{-1}E_0(z^2)E_1(z^2)$, as indicated in Eq. (10.118). If $T(z)$ is an allpass function, then its magnitude response is a constant and, as a result, the corresponding QMF bank has no magnitude distortion [Vai87f]. Let the polyphase components $E_0(z)$ and $E_1(z)$ of $H_0(z)$ be expressed as

$$E_0(z) = \tfrac{1}{2}\mathcal{A}_0(z), \qquad E_1(z) = \tfrac{1}{2}\mathcal{A}_1(z), \tag{10.127}$$

with $\mathcal{A}_0(z)$ and $\mathcal{A}_1(z)$ being stable allpass functions. Thus,

$$H_0(z) = \tfrac{1}{2}[\mathcal{A}_0(z^2) + z^{-1}\mathcal{A}_1(z^2)], \tag{10.128a}$$

$$H_1(z) = \tfrac{1}{2}[\mathcal{A}_0(z^2) - z^{-1}\mathcal{A}_1(z^2)]. \tag{10.128b}$$

Substituting Eq. (10.127) in Eq. (10.116), we obtain the expressions for the analysis filters as

$$\begin{bmatrix} H_0(z) \\ H_1(z) \end{bmatrix} = \frac{1}{2}\begin{bmatrix} 1 & 1 \\ 1 & -1 \end{bmatrix}\begin{bmatrix} \mathcal{A}_0(z^2) \\ z^{-1}\mathcal{A}_1(z^2) \end{bmatrix}. \tag{10.129}$$

The corresponding synthesis filters are obtained from Eq. (10.117) and are given by

$$[G_0(z) \quad G_1(z)] = \tfrac{1}{2}\begin{bmatrix} z^{-1}\mathcal{A}_1(z^2) & \mathcal{A}_0(z^2) \end{bmatrix}\begin{bmatrix} 1 & 1 \\ 1 & -1 \end{bmatrix}, \tag{10.130}$$

which yields

$$G_0(z) = \tfrac{1}{2}[\mathcal{A}_0(z^2) + z^{-1}\mathcal{A}_1(z^2)] = H_0(z), \qquad (10.131a)$$

$$G_1(z) = \tfrac{1}{2}[-\mathcal{A}_0(z^2) + z^{-1}\mathcal{A}_1(z^2)] = -H_1(z). \qquad (10.131b)$$

The realization of the *magnitude-preserving* two-channel QMF bank shown in Figure 10.55 is obtained by making use of Eq. (10.127) in Figure 10.53(b).

From Eq. (10.128a), we observe that the lowpass transfer function $H_0(z)$ has a polyphase-like decomposition, except here the polyphase components are stable allpass transfer functions. The existence of this type of decomposition has been illustrated before in Example 10.10. It has been shown that a bounded real (BR) lowpass transfer function $H_0(z) = P_0(z)/D(z)$ of odd order, with no common factors between its numerator and denominator, can be expressed as in Eq. (10.128a) if it satisfies the power-symmetry condition given by

$$H_0(z)H_0(z^{-1}) + H_0(-z)H_0(-z^{-1}) = 1, \qquad (10.132)$$

and the numerator $P_0(z)$ of $H_0(z)$ is a symmetric polynomial [Vai87f]. It can be easily verified that the transfer function $H(z)$ of Example 10.10 satisfies these conditions. It has also been shown that any odd-order elliptic lowpass half-band filter $H_0(z)$ with a frequency response specification given by

$$1 - \delta_p \leq \left| H_0(e^{j\omega}) \right| \leq 1, \quad \text{for } 0 \leq \omega \leq \omega_p, \qquad (10.133a)$$

$$\left| H_0(e^{j\omega}) \right| \leq \delta_s, \quad \text{for } \omega_s \leq \omega \leq \pi, \qquad (10.133b)$$

and satisfying the conditions $\omega_p + \omega_s = \pi$ and $\delta_s^2 = 4\delta_p(1 - \delta_p)$ can always be expressed as in Eq. (10.128a) [Vai87f]. The poles of the elliptic filter satisfying these two conditions lie on the imaginary axis. Using the pole-interlacing property outlined in Section 6.10, we can readily identify the expressions for the two allpass transfer functions $\mathcal{A}_0(z)$ and $\mathcal{A}_1(z)$.

We illustrate the design of an IIR half-band filter using the above approach.

EXAMPLE 10.16   The frequency response specifications of a real half-band elliptic filter $G(z)$ of odd order are given by $\omega_p = 0.4\pi$, $\omega_s = 0.6\pi$, and $\delta_s = 0.0155$. The corresponding passband ripple δ_p is found to be 0.00012013. In dB, the passband and stopband ripples are, respectively, Rp = 0.0010435178 and Rs = 36.193366. Using the function `ellipord` in MATLAB, we determine the minimum order of the elliptic lowpass filter to be 5. Next, we use the function `ellip` in MATLAB to determine the transfer function of the fifth-order half-band elliptic filter whose gain response is shown in Figure 10.56. We use the function `tf2zp` to determine its zero-pole locations and plot them using the function `zplane`, as indicated in Figure 10.57. The poles are at $z = 0$, $z = \pm j0.4866252630055$, and $z = \pm j0.84551995705794$.[13]

Using the pole-interlacing property of the parallel allpass filter structure, we arrive at the transfer functions of the two allpass sections of the half-band filter $G(z)$:

$$\mathcal{A}_0(z^2) = \frac{z^{-2} + 0.2368041466}{1 + 0.2368041466 z^{-2}}, \quad \mathcal{A}_1(z^2) = \frac{z^{-2} + 0.7149039978}{1 + 0.7149039978 z^{-2}}. \qquad (10.134)$$

10.9   Perfect Reconstruction Two-Channel FIR Filter Banks

A perfect reconstruction two-channel FIR filter bank with linear-phase FIR filters can be designed if the power-complementary requirement of Eq. (10.123) between the two analysis filters $H_0(z)$ and $H_1(z)$ is

---

[13]Because of numerical accuracy problems, the real part of the poles are not exactly zero but are quite small in value and have been ignored.

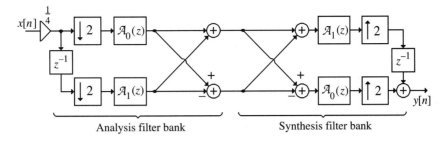

Figure 10.55: Magnitude-preserving two-channel QMF bank.

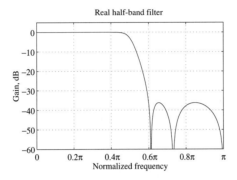

Figure 10.56: Gain response of the half-band filter of Example 10.16.

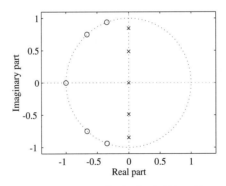

Figure 10.57: Pole-zero plot of the fifth-order elliptic IIR half-band lowpass filter of Example 10.16.

not imposed. To develop the pertinent design equations, we observe from Eq. (10.96) that $Y(z)$ can be expressed in matrix form as

$$Y(z) = \frac{1}{2}[G_0(z) \quad G_1(z)] \begin{bmatrix} H_0(z) & H_0(-z) \\ H_1(z) & H_1(-z) \end{bmatrix} \begin{bmatrix} X(z) \\ X(-z) \end{bmatrix}. \tag{10.135}$$

From the above we obtain

$$Y(-z) = \frac{1}{2}[G_0(-z) \quad G_1(-z)] \begin{bmatrix} H_0(z) & H_0(-z) \\ H_1(z) & H_1(-z) \end{bmatrix} \begin{bmatrix} X(z) \\ X(-z) \end{bmatrix}. \tag{10.136}$$

Combining the above two equations we arrive at

$$
\begin{bmatrix} Y(z) \\ Y(-z) \end{bmatrix} = \frac{1}{2} \begin{bmatrix} G_0(z) & G_1(z) \\ G_0(-z) & G_1(-z) \end{bmatrix} \begin{bmatrix} H_0(z) & H_0(-z) \\ H_1(z) & H_1(-z) \end{bmatrix} \begin{bmatrix} X(z) \\ X(-z) \end{bmatrix}
$$
$$
= \frac{1}{2} \mathbf{G}^{(m)}(z) [\mathbf{H}^{(m)}(z)]^T \begin{bmatrix} X(z) \\ X(-z) \end{bmatrix}, \tag{10.137}
$$

where

$$
\mathbf{G}^{(m)}(z) = \begin{bmatrix} G_0(z) & G_1(z) \\ G_0(-z) & G_1(-z) \end{bmatrix}, \qquad \mathbf{H}^{(m)}(z) = \begin{bmatrix} H_0(z) & H_1(z) \\ H_0(-z) & H_1(-z) \end{bmatrix} \tag{10.138}
$$

are called the *modulation matrices*.

It follows from Eq. (10.137) that for perfect reconstruction we must have $Y(z) = z^{-\ell} X(z)$ and, correspondingly, $Y(-z) = (-z)^{-\ell} X(-z)$. Substituting these relations in the above equation we conclude that the perfect reconstruction condition is satisfied if

$$
\frac{1}{2} \mathbf{G}^{(m)}(z) [\mathbf{H}^{(m)}(z)]^T = \begin{bmatrix} z^{-\ell} & 0 \\ 0 & (-z)^{-\ell} \end{bmatrix}. \tag{10.139}
$$

Thus knowing the analysis filters $H_0(z)$ and $H_1(z)$, the synthesis filters $G_0(z)$ and $G_1(z)$ are determined from

$$
\mathbf{G}^{(m)}(z) = 2 \begin{bmatrix} z^{-\ell} & 0 \\ 0 & (-z)^{-\ell} \end{bmatrix} \left([\mathbf{H}^{(m)}(z)]^T \right)^{-1},
$$

which yields after some algebra

$$
G_0(z) = \frac{2z^{-\ell}}{\det[\mathbf{H}^{(m)}(z)]} \cdot H_1(-z), \tag{10.140a}
$$

$$
G_1(z) = -\frac{2z^{-\ell}}{\det[\mathbf{H}^{(m)}(z)]} \cdot H_0(-z), \tag{10.140b}
$$

where

$$
\det[\mathbf{H}^{(m)}(z)] = H_0(z)H_1(-z) - H_0(-z)H_1(z), \tag{10.141}
$$

and ℓ is an odd positive integer.

For FIR analysis filters $H_0(z)$ and $H_1(z)$, the synthesis filters $G_0(z)$ and $G_1(z)$ will also be FIR filters if

$$
\det[\mathbf{H}^{(m)}(z)] = c\, z^{-k}, \tag{10.142}
$$

where c is a real number and k is a positive integer. In which case, the two synthesis filters are given by

$$
G_0(z) = \frac{2}{c} z^{-(\ell-k)} H_1(-z), \tag{10.143a}
$$

$$
G_1(z) = -\frac{2}{c} z^{-(\ell-k)} H_0(-z). \tag{10.143b}
$$

Orthogonal Filter Banks

Smith and Barnwell [Smi84] and Mintzer [Min85] independently showed how to choose the FIR analysis filters to satisfy the condition of Eq. (10.142) on the determinant of $\mathbf{H}^{(m)}(z)$. Let $H_0(z)$ be an FIR filter of odd order N satisfying the power-symmetric condition of Eq. (10.132). If we then choose

$$
H_1(z) = z^{-N} H_0(-z^{-1}), \tag{10.144}
$$

Eq. (10.141) reduces to

$$\det[\mathbf{H}^{(m)}(z)] = -z^{-N} \left(H_0(z) H_0(z^{-1}) + H_0(-z) H_0(-z^{-1}) \right) = -z^{-N}. \tag{10.145}$$

Comparing the above equation with Eq. (10.142) we observe that $c = -1$ and $k = N$. Using Eqs. (10.144) and (10.145) in Eqs. (10.143a) and (10.143b) with $\ell = k = N$ we get

$$G_0(z) = 2 z^{-N} H_0(z^{-1}), \qquad G_1(z) = 2 z^{-N} H_1(z^{-1}). \tag{10.146}$$

It should be noted that if $H_0(z)$ is a causal FIR filter, the other three filters are also causal FIR filters. Moreover, from Eq. (10.144) it follows that $|G_i(e^{j\omega})| = |H_i(e^{j\omega})|$, for $i = 1, 2$. In addition, $|H_1(e^{j\omega})| = |H_0(-e^{j\omega})|$, which for a real-coefficient transfer function implies that if $H_0(z)$ is a lowpass filter, then $H_1(z)$ is a highpass filter. A perfect reconstruction power-symmetric filter bank is also called an *orthogonal filter bank*.

The filter bank design problem thus reduces to the design of a power-symmetric lowpass filter $H_0(z)$. To this end, we can design an even order $F(z) = H_0(z) H_0(z^{-1})$ whose *spectral factorization* yields $H_0(z)$. Now, the power-symmetric condition of Eq. (10.132) implies that $F(z)$ be a zero-phase half-band lowpass filter with a non-negative frequency response $F(e^{j\omega})$. Such a half-band filter can be obtained by adding a constant term K to a zero-phase even-order half-band filter $Q(z)$ such that $F(e^{j\omega}) = Q(e^{j\omega}) + K \geq 0$ for all ω. The half-band filter $Q(z)$ can be designed using any of the methods outlined in Sections 10.7.3 and 11.7.3.

We summarize below the steps for the design of a real coefficient $H_0(z)$[Smi84] :

Step 1: Design a zero-phase real-coefficient FIR half-band lowpass filter $Q(z) = \sum_{n=-N}^{N} q[n] z^{-n}$, of order $2N$ with N an odd positive integer.

Step 2: Let δ denote the peak stopband ripple of $Q(e^{j\omega})$. Define $F(z) = Q(z) + \delta$ which guarantees that $F(e^{j\omega}) \geq 0$ for all ω. Note that if $q[n]$ denotes the impulse response of $Q(z)$, then the impulse response $f[n]$ of $F(z)$ is given by

$$f[n] = \begin{cases} q[n] + \delta, & \text{for } n = 0, \\ q[n], & \text{for } n \neq 0. \end{cases} \tag{10.147}$$

Step 3: Determine the spectral factor $H_0(z)$ of $F(z)$.

EXAMPLE 10.17 Consider the FIR filter

$$F(z) = z^N (1 + z^{-1})^{N+1} R(z), \tag{10.148}$$

where $R(z)$ is a polynomial in z^{-1} of degree $N - 1$ with N odd. $F(z)$ can be made a half-band filter by choosing $R(z)$ appropriately [Dau88], [Her71]. This class of half-band filters has been called the *binomial* or *maxflat* filter. The filter $F(z)$ has a frequency response $F(e^{j\omega})$ that is maximally flat at $\omega = 0$ and at $\omega = \pi$. For $N = 3$, $R(z) = \frac{1}{16}(-1 + 4z^{-1} - z^{-2})$, resulting in

$$F(z) = \frac{1}{16}(-z^3 + 9z + 16 + 9z^{-1} - z^{-3}), \tag{10.149}$$

which is seen to be a symmetric polynomial with four zeros located at $z = -1$, a zero at $z = 2 - \sqrt{3}$, and a zero at $z = 2 + \sqrt{3}$.

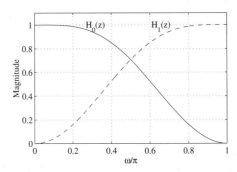

Figure 10.58: Magnitude responses of third-order power-symmetric maximally flat analysis filters of Example 10.17.

The minimum-phase spectral factor is therefore the lowpass analysis filter

$$H_0(z) = -0.3415(1 + z^{-1})^2 \left(1 - (2 - \sqrt{3})z^{-1}\right)$$

$$= -0.3415(1 + 1.7320508z^{-1} + 0.4641016z^{-2} - 0.2679492z^{-3}).$$

From Eq. (10.144) the corresponding highpass analysis filter is given by

$$H_1(z) = z^{-3} H_0(-z^{-1}) = -0.3415 \left(0.2679492 + 0.4641016z^{-1} - 1.7320508z^{-2} + z^{-3}\right).$$

The two synthesis filters are then obtained using Eq. (10.146) and are given by

$$G_0(z) = -0.683025 \left(-0.2679491 + 0.4641016z^{-1} + 1.7320508z^{-2} + z^{-3}\right),$$

$$G_1(z) = -0.683025 \left(1 - 1.7320508z^{-1} + 0.4641016z^{-2} + 0.2679491z^{-3}\right)$$

Figure 10.58 shows the magnitude responses of the two analysis filters.

Several comments are in order here. First, as shown in Section 10.7.2, the order of the half-band filter $F(z)$ is of the form $4K + 2$ where K is a positive integer. This implies that the order of $H_0(z)$ is $N = 2K + 1$, which is odd as required. Second, the zeros of $F(z)$ appear with mirror-image symmetry in the z-plane with the zeros on the unit circle being of even multiplicity. Any appropriate half of these zeros can be grouped to form the spectral factor $H_0(z)$. For example, a minimum-phase $H_0(z)$ can be formed by grouping all the zeros inside the unit circle with half of the zeros on the unit circle. Likewise, a maximum-phase $H_0(z)$ can be formed by grouping all the zeros outside the unit circle with half of the zeros on the unit circle. However, it is not possible to form a spectral factor with a linear phase. Third, the stopband edge frequency is the same for $F(z)$ and $H_0(z)$. If the desired minimum stopband attenuation of $H_0(z)$ is α_s dB, the minimum stopband attenuation of $F(z)$ is approximately $2\alpha_s + 6.02$ dB.

We illustrate the power-symmetric filter bank design in the following example.

EXAMPLE 10.18   Design a lowpass real-coefficient power-symmetric filter $H_0(z)$ with the following specifications: stopband edge at $\omega_s = 0.63\pi$ and a minimum stopband attenuation of $\alpha_s = 12$ dB. The specifications of the corresponding zero-phase half-band filter $F(z)$ are therefore as follows: stopband edge at $\omega_s = 0.63\pi$ and a minimum stopband attenuation of $\alpha_s = 40$ dB. The desired stopband ripple is thus $\delta_s = 0.01$ which is also the passband ripple. The passband edge is at $\omega_p = \pi - 0.63\pi = 0.37\pi$.

Using the function `remezord` we then estimate the order of $F(z)$ and using the function `remez` we next design $Q(z)$. To this end the code fragments used are

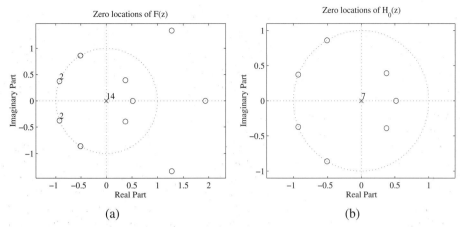

Figure 10.59: Zero locations of the zero-phase half-band filter $F(z)$ and its minimum-phase spectral factor $H_0(z)$.

```
[N,fpts,mag,wt] = remezord([0.37 0.63],[1 0],[0.01 0.01]);
[q,err] = remez(N,fpts,mag,wt);
```

The order of $F(z)$ is found to be 14 implying the order of $H_0(z)$ to be 7 which is odd as desired. Note the parameter `err` provides the maximum value of the ripple. To determine the coefficients of the half-band filter $F(z)$ we add `err` to the central coefficient $q[7]$. Next, using the function `roots`, we determine the roots of $F(z)$ which should theoretically exhibit a mirror-image symmetry with respect to the unit circle with double roots on the unit circle. However, the algorithm is numerically quite sensitive and it is found that a slightly larger value than `err` should be added to ensure double zeros of $F(z)$ on the unit circle. Choosing the roots inside the unit circle along with one set of unit circle roots we get the minimum-phase spectral factor $H_0(z)$. Figure 10.59 shows the zero locations of $F(z)$ and $H_0(z)$. The coefficients of $H_0(z)$ are as follows:

$$h_0[0] = 0.32308145852633, \quad h_0[1] = 0.51935003909222,$$
$$h_0[2] = 0.30133845772088, \quad h_0[3] = -0.07814458219037,$$
$$h_0[4] = -0.13767160620111, \quad h_0[5] = 0.03209964123472,$$
$$h_0[6] = 0.07900476042178, \quad h_0[7] = -0.04899606569875.$$

The gain responses of the two analysis filters are shown in Figure 10.60.

In realizing the analysis filter bank, if the two filters $H_0(z)$ and $H_1(z)$ are implemented independently, the overall structure would require $2(N+1)$ multipliers and $2N$ two-input adders. However, a computationally efficient realization requiring $N+1$ multipliers and $2N$ two-input adders can be developed by exploiting the relation of Eq. (10.144) (Problem 10.42).

Paraunitary Filter Banks

A p-input, q-output LTI discrete-time system with a transfer matrix $\mathbf{T}_{pq}(z)$ is called a *paraunitary system* if $\mathbf{T}_{pq}(z)$ is a *paraunitary matrix*, i.e.,

$$\tilde{\mathbf{T}}_{pq}(z)\mathbf{T}_{pq}(z) = c\mathbf{I}_p, \tag{10.150}$$

where $\tilde{\mathbf{T}}_{pq}(z)$ is the paraconjugate of $\mathbf{T}_{pq}(z)$ given by the transpose of $\mathbf{T}_{pq}(z^{-1})$ with each coefficient replaced by its conjugate, $\mathbf{I}_p$ is an $p \times p$ identity matrix, and c is a real constant. A causal, stable paraunitary system is also a lossless system.

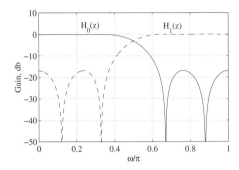

Figure 10.60: Gain responses of seventh-order power-symmetric analysis filters of Example 10.18.

Figure 10.61: The paraunitary lattice structure.

It can be easily shown that the modulation matrix $\mathbf{H}^{(m)}(z)$ defined in Eq. (10.138) of a power-symmetric filter bank is a paraunitary matrix. Hence, a power-symmetric filter bank has also been referred to as a *paraunitary filter bank*.

Since the cascade of a paraunitary system with a transfer matrix $\mathbf{T}_{pq}^{(1)}(z)$ and a paraunitary system with a transfer matrix $\mathbf{T}_{qr}^{(2)}(z)$ is also paraunitary, it is easier to design a paraunitary filter bank without resorting to spectral factorization by cascading simpler paraunitary blocks. To this end, the cascaded FIR lattice structure introduced in Section 6.9.2 can be employed. The overall structure is lossless as each lattice stage shown in Figure 10.61 is lossless, i.e., paraunitary, causal, and stable [Vai86b], [Vai88a]. The synthesis procedure outlined in [Vai86b], [Vai88a] realizes both the power-symmetric transfer function $H_0(z)$ and its conjuagate quadratic transfer function $H_1(z)$. Three important properties of the QMF lattice structure are structurally induced. First, the QMF lattice filter bank guarantees perfect reconstruction independent of the lattice parameters. Second, it exhibits very small coefficient sensitivity to lattice parameters as each stage remains lossless under coefficient quantization. Third, its computational complexity is about one-half that of any other realization as it requires $(N-1)/2$ total numbers of multipliers for an order-N filter.

EXAMPLE 10.19   Following the method outlined in Section 6.9.2, we develop the FIR cascaded lattice realization of the analysis filter bank designed in Example 10.18. The input multiplier here is $h[0] = 0.32308146$, which is used to normalize the analysis transfer function so that its constant coefficient is unity. The lattice coefficients obtained from the normalized analysis transfer function using Eq. (6.101) are given by

$$k_7 = -0.15165236, \quad k_5 = 0.23538743, \quad k_3 = -0.483934447, \quad k_1 = 1.6100196.$$

It should be noted that because of the numerical accuracy problem, the coefficients of the spectral factor obtained in Example 10.18 are not very accurate. As a result, the coefficient of $z^{-(i-1)}$ of the transfer function $H_{i-2}(z)$ generated from the transfer function $H_i(z)$ using Eq. (6.101) is not exactly zero, and has been set to zero at each iteration. Two interesting properties of the cascaded lattice filter bank can be seen from the above coefficient values. First, the signs of the lattice multiplier coefficients alternate between stages. Second, the values of the multiplier coefficients $\{k_i\}$ decrease with increasing i.

The QMF lattice structure can be used directly to design the power-symmetric analysis filter $H_0(z)$ using an iterative computer-aided optimization technique. The goal here is to determine the lattice parameters k_i by minimizing the energy in the stopband of $H_0(z)$. To this end, the objective function is given by

$$\phi = \int_{\omega_s}^{\pi} \left| H_0(e^{j\omega}) \right|^2 d\omega. \tag{10.151}$$

It should be noted that the power-symmetric property ensures good passband response.

Biorthogonal Filter Banks

In the design of an orthogonal two-channel filter bank, the zero-phase even-order half-band filter $F(z)$ is expressed in the form $F(z) = H_0(z)H_0(z^{-1})$ by spectral factorization and the analysis filter $H_0(z)$ is chosen as a spectral factor of $F(z)$. Hence, the two spectral factors $H_0(z)$ and $H_0(z^{-1})$ of $F(z)$ have the same magnitude response. As a result, it is not possible to design perfect reconstruction filter banks with linear-phase analysis and synthesis filters. However, it is possible to maintain the perfect reconstruction condition with linear-phase filters by choosing a different factorization scheme. To this end, we factorize the causal half-band filter $z^{-N}F(z)$ of order $2N$ as

$$z^{-N}F(z) = H_0(z)H_1(-z), \tag{10.152}$$

where $H_0(z)$ and $H_1(z)$ are linear-phase filters. The determinant of the modulation matrix $\mathbf{H}^{(m)}(z)$ is now given by

$$\det \mathbf{H}^{(m)}(z) = H_0(z)H_1(-z) - H_0(-z)H_1(z) = z^{-N}[F(z) + F(-z)] = z^{-N},$$

which satisfies the condition for perfect reconstruction given by Eq. (10.142). The filter bank designed using the factorization scheme of Eq. (10.152) is called a *biorthogonal filter bank*.

The two synthesis filters are given by

$$G_0(z) = H_1(-z), \qquad G_1(z) = -H_0(-z). \tag{10.153}$$

EXAMPLE 10.20 The half-band filter $F(z)$ of Eq. (10.149) can be factored several different ways to yield linear-phase analysis filters $H_0(z)$ and $H_1(z)$. For example, we can choose

$$H_0(z) = \frac{1}{8}\left(-1 + 2z^{-1} + 6z^{-2} + 2z^{-3} - z^{-4}\right), \qquad H_1(-z) = \frac{1}{2}\left(1 + 2z^{-1} + z^{-2}\right),$$

as factors of $F(z)$. Thus,

$$H_1(z) = \frac{1}{2}\left(1 - 2z^{-1} + z^{-2}\right).$$

Since the length of $H_0(z)$ is 5, and the length of $H_1(z)$ is 3, the above choices for the analysis filters are known as the 5/3 filter-pair of Daubechies [Dau88]. A plot of the gain responses of the analysis filters is shown in Figure 10.62(a).

Another choice yields the 4/4 filter pair of Daubechies [Dau88]:

$$H_0(z) = \frac{1}{8}\left(1 + 3z^{-1} + 3z^{-2} + z^{-3}\right), \qquad H_1(z) = \frac{1}{2}\left(-1 - 3z^{-1} + 3z^{-2} + z^{-3}\right).$$

A plot of the gain responses of the analysis filters is shown in Figure 10.62(b).

Several other choices for the analysis filters are possible with some not resulting in linear-phase filters.

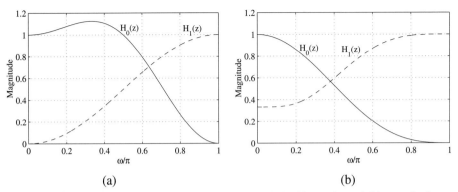

(a)   (b)

Figure 10.62: (a) Magnitude responses of Daubechies (5/3) analysis filter pair, and (b) magnitude responses of Daubechies (4/4) analysis filter pair.

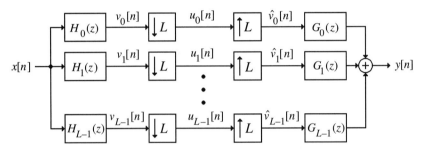

Figure 10.63: The basic L-channel QMF filter bank structure.

10.10 L-Channel QMF Banks

We now generalize the discussion of the previous section to the case of a QMF bank with more than two channels. The basic structure of the L-channel QMF bank is shown in Figure 10.63.

10.10.1   Analysis of the L-Channel Filter Bank

We analyze the operation of the L-channel QMF bank of Figure 10.63 in the z-domain. The expressions for the z-transforms of various intermediate signals in Figure 10.63 are given by

$$V_k(z) = H_k(z)X(z), \tag{10.154a}$$

$$U_k(z) = \frac{1}{L} \sum_{\ell=0}^{L-1} H_k(z^{1/L} W_L^\ell) X(z^{1/L} W_L^\ell), \tag{10.154b}$$

$$\hat{V}_k(z) = U_k(z^L), \tag{10.154c}$$

where $0 \le k \le L - 1$.

Define the vector of down-sampled subband signals $U_k(z)$ as

$$\mathbf{u}(z) = [\, U_0(z) \quad U_1(z) \quad \cdots \quad U_{L-1}(z)\,]^T, \tag{10.155}$$

the modulation vector of the input signals as

$$\mathbf{x}^{(m)}(z) = [\, X(z) \quad X(zW_L) \quad \cdots \quad X(zW_L^{L-1}) \,]^T,$$ (10.156)

and the *analysis filter bank modulation matrix* as

$$\mathbf{H}^{(m)}(z) = \begin{bmatrix} H_0(z) & H_1(z) & \cdots & H_{L-1}(z) \\ H_0(zW_L^1) & H_1(zW_L^1) & \cdots & H_{L-1}(zW_L^1) \\ \vdots & \vdots & \ddots & \vdots \\ H_0(zW_L^{L-1}) & H_1(zW_L^{L-1}) & \cdots & H_{L-1}(zW_L^{L-1}) \end{bmatrix}.$$ (10.157)

Then, Eq. (10.154b) can be compactly expressed in the form

$$\mathbf{u}(z) = \frac{1}{L}[\mathbf{H}^{(m)}(z^{1/L})]^T \mathbf{x}^{(m)}(z^{1/L}).$$ (10.158)

The output of the QMF bank is given by

$$Y(z) = \sum_{k=0}^{L-1} G_k(z)\hat{V}_k(z),$$ (10.159)

which can be expressed in a matrix form as

$$Y(z) = \mathbf{g}^T(z)\mathbf{u}(z^L),$$ (10.160)

where

$$\mathbf{g}(z) = [\, G_0(z) \quad G_1(z) \quad \cdots \quad G_{L-1}(z) \,]^T.$$ (10.161)

10.10.2 Alias-Free *L*-Channel Filter Bank

We now develop the condition for alias-free operation of the *L*-channel filter bank of Figure 10.63 [Fli94]. From Eq. (10.160), the modulated versions of the output signal are given by

$$Y(zW_L^k) = \mathbf{g}^T(zW_L^k)\mathbf{u}(z^L W_L^{kL}) = \mathbf{g}^T(zW_L^k)\mathbf{u}(z^L), \qquad 0 \le k \le L-1,$$ (10.162)

which can be expressed in a matrix form as

$$\mathbf{y}^{(m)}(z) = [\, Y(z) \quad Y(zW_L) \quad \cdots \quad Y(zW_L^{L-1}) \,]^T.$$ (10.163)

Using Eqs. (10.160) and (10.161) in the above equation, we therefore obtain

$$\mathbf{y}^{(m)}(z) = \mathbf{G}^{(m)}(z)\mathbf{u}(z^L),$$ (10.164)

where

$$\mathbf{G}^{(m)}(z) = \begin{bmatrix} G_0(z) & G_1(z) & \cdots & G_{L-1}(z) \\ G_0(zW_L^1) & G_1(zW_L^1) & \cdots & G_{L-1}(zW_L^1) \\ \vdots & \vdots & \ddots & \vdots \\ G_0(zW_L^{L-1}) & G_1(zW_L^{L-1}) & \cdots & G_{L-1}(zW_L^{L-1}) \end{bmatrix},$$ (10.165)

is the *synthesis filter bank modulation matrix*.

Combining Eqs. (10.158) and (10.164), we arrive at the input-output relationship of the L-channel filter bank as

$$\mathbf{y}^{(m)}(z) = \frac{1}{L}\mathbf{G}^{(m)}(z)[\mathbf{H}^{(m)}(z)]^T\mathbf{x}^{(m)}(z)$$

$$= \mathbf{T}(z)\mathbf{x}^{(m)}(z), \tag{10.166}$$

where $\mathbf{T}(z) = \frac{1}{L}\mathbf{G}^{(m)}(z)[\mathbf{H}^{(m)}(z)]^T$ is called the *transfer matrix* relating the input signal $X(z)$ and its frequency-modulated versions $X(zW_L^k)$, $1 \le k \le L-1$, with the output signal $Y(z)$ and its frequency-modulated versions $Y(zW_L^k)$, $1 \le k \le L-1$.

The filter bank is alias-free if the transfer matrix $\mathbf{T}(z)$ is a diagonal matrix of the form

$$\mathbf{T}(z) = \text{diag}\,[\,T(z)\quad T(zW_L)\quad \cdots \quad T(zW_L^{L-1})\,]. \tag{10.167}$$

The first element $T(z)$ of the above diagonal matrix is called the *distortion transfer function* of the L-channel filter bank. Substituting Eqs. (10.154a) to (10.154c) in Eq. (10.159) we arrive at

$$Y(z) = \sum_{\ell=0}^{L-1} a_\ell(z)X(zW_L^\ell), \tag{10.168}$$

where

$$a_\ell(z) = \frac{1}{L}\sum_{k=0}^{L-1} H_k(zW_L^\ell)G_k(z), \qquad 0 \le \ell \le L-1. \tag{10.169}$$

On the unit circle the term $X(zW_L^\ell)$ becomes

$$X(e^{j\omega}W_L^\ell) = X(e^{j(\omega-2\pi\ell/L)}). \tag{10.170}$$

Thus, from Eq. (10.168), the output spectrum $Y(e^{j\omega})$ is a weighted sum of $X(e^{j\omega})$ and its uniformly shifted versions $X(e^{j(\omega-2\pi\ell/L)})$ for $\ell = 1, 2, \ldots, L-1$, which are caused by the sampling rate alteration operations. The term $X(zW_L^\ell)$ is called the ℓth *aliasing term*, with $a_\ell(z)$ representing its *gain* at the output. In general, the QMF bank of Figure 10.61 is a linear, time-varying system with a period of L.

It follows from Eq. (10.168) that the aliasing effect can be completely eliminated at the output if and only if

$$a_\ell(z) = 0, \qquad 1 \le \ell \le L-1, \tag{10.171}$$

for all possible inputs $x[n]$. If Eq. (10.171) holds, then the L-channel QMF bank of Figure 10.63 becomes a linear time-invariant system with an input-output relation given by

$$Y(z) = T(z)X(z), \tag{10.172}$$

where $T(z)$ is the distortion transfer function given by

$$T(z) = a_0(z) = \frac{1}{L}\sum_{k=0}^{L-1} H_k(z)G_k(z). \tag{10.173}$$

If $T(z)$ has a constant magnitude, then the system of Figure 10.63 is a *magnitude-preserving QMF bank*. If $T(z)$ has a linear phase, then the QMF bank has no phase distortion. Finally, if $T(z)$ is a pure delay, it is a *perfect reconstruction QMF bank*.

Using Eqs. (10.157) and (10.161), we can express Eq. (10.169) in a matrix form as

$$L \cdot \mathbf{A}(z) = \mathbf{H}^{(m)}(z)\mathbf{g}(z), \tag{10.174}$$

where

$$\mathbf{A}(z) = [a_0(z) \quad a_1(z) \quad \cdots \quad a_{L-1}(z)]^T. \tag{10.175}$$

The aliasing cancellation condition can now be rewritten as

$$\mathbf{H}^{(m)}(z)\mathbf{g}(z) = \mathbf{t}(z), \tag{10.176}$$

where

$$\mathbf{t}(z) = [L a_0(z) \quad 0 \quad \cdots \quad 0]^T = [L \cdot T(z) \quad 0 \quad \cdots \quad 0]^T. \tag{10.177}$$

From Eq. (10.176), it follows that by knowing the set of analysis filters $\{H_k(z)\}$, we can determine the desired set of synthesis filters $\{G_k(z)\}$ as

$$\mathbf{g}(z) = [\mathbf{H}^{(m)}(z)]^{-1}\mathbf{t}(z), \tag{10.178}$$

provided of course $[\det \mathbf{H}^{(m)}(z)] \neq 0$. Moreover, a perfect reconstruction QMF bank results if we set $T(z) = z^{-n_0}$ in the expression for $\mathbf{t}(z)$ in Eq. (10.177). In practice, the above approach is difficult to carry out for a number of reasons. A more practical solution to the design of perfect reconstruction QMF bank is obtained via the polyphase representation outlined next [Vai87d].

10.10.3 Polyphase Representation

Consider the Type I polyphase representation of the kth analysis filter $H_k(z)$:

$$H_k(z) = \sum_{\ell=0}^{L-1} z^{-\ell} E_{k\ell}(z^L), \qquad 0 \leq k \leq L-1. \tag{10.179}$$

A matrix representation of the above set of equations is given by

$$\mathbf{h}(z) = \mathbf{E}(z^L)\mathbf{e}(z), \tag{10.180}$$

where

$$\mathbf{h}(z) = \begin{bmatrix} H_0(z) & H_1(z) & \cdots & H_{L-1}(z) \end{bmatrix}^T, \tag{10.181a}$$

$$\mathbf{e}(z) = \begin{bmatrix} 1 & z^{-1} & \cdots & z^{-(L-1)} \end{bmatrix}^T, \tag{10.181b}$$

and

$$\mathbf{E}(z) = \begin{bmatrix} E_{00}(z) & E_{01}(z) & \cdots & E_{0,L-1}(z) \\ E_{10}(z) & E_{11}(z) & \cdots & E_{1,L-1}(z) \\ \vdots & \vdots & \ddots & \vdots \\ E_{L-1,0}(z) & E_{L-1,1}(z) & \cdots & E_{L-1,L-1}(z) \end{bmatrix}. \tag{10.181c}$$

The matrix $\mathbf{E}(z)$ defined above is called the *Type I polyphase component matrix*. Figure 10.64(a) shows the Type I polyphase representation of the analysis filter bank.

Likewise, we can represent the L synthesis filters in a Type II polyphase form:

$$G_k(z) = \sum_{\ell=0}^{L-1} z^{-(L-1-\ell)} R_{\ell k}(z^L), \qquad 0 \leq k \leq L-1. \tag{10.182}$$

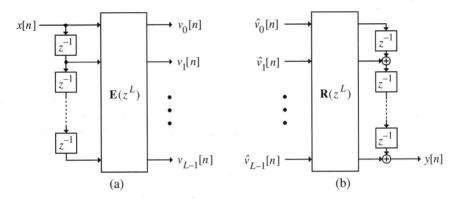

Figure 10.64: (a) Type I polyphase representation of the analysis filter bank and (b) Type II polyphase representation of the synthesis filter bank.

In matrix form, the above set of L equations can be rewritten as

$$\mathbf{g}^T(z) = z^{-(L-1)} \tilde{\mathbf{e}}(z) \mathbf{R}(z^L), \tag{10.183}$$

where

$$\mathbf{g}(z) = \begin{bmatrix} G_0(z) & G_1(z) & \cdots & G_{L-1}(z) \end{bmatrix}^T, \tag{10.184a}$$

$$\tilde{\mathbf{e}}(z) = \begin{bmatrix} 1 & z & \cdots & z^{L-1} \end{bmatrix} = \mathbf{e}^T(z^{-1}), \tag{10.184b}$$

and

$$\mathbf{R}(z) = \begin{bmatrix} R_{00}(z) & R_{01}(z) & \cdots & R_{0,L-1}(z) \\ R_{10}(z) & R_{11}(z) & \cdots & R_{1,L-1}(z) \\ \vdots & \vdots & \ddots & \vdots \\ R_{L-1,0}(z) & R_{L-1,1}(z) & \cdots & R_{L-1,L-1}(z) \end{bmatrix}. \tag{10.184c}$$

The matrix $\mathbf{R}(z)$ defined above is called the *Type II polyphase component matrix*. Figure 10.64(b) shows the Type II polyphase representation of the synthesis filter bank.

Making use of the polyphase representations of Figure 10.64 in Figure 10.62 and the cascade equivalences of Figure 10.14, we arrive at an equivalent realization of the L-channel QMF bank shown in Figure 10.65.

The relation between the modulation matrix $\mathbf{H}^{(m)}(z)$ of Eq. (10.157) and the Type I polyphase component matrix $\mathbf{E}(z)$ of Eq. (10.181c) can be easily established. From Eqs. (10.157), (10.180), and (10.181a), we observe that

$$[\mathbf{H}^{(m)}(z)]^T = \begin{bmatrix} \mathbf{h}(z) & \mathbf{h}(zW_L^1) & \cdots & \mathbf{h}(zW_L^{L-1}) \end{bmatrix}$$
$$= \mathbf{E}(z^L) \begin{bmatrix} \mathbf{e}(z) & \mathbf{e}(zW_L^1) & \cdots & \mathbf{e}(zW_L^{L-1}) \end{bmatrix}. \tag{10.185}$$

Now, from Eq. (10.181b), it follows that

$$\mathbf{e}(zW_L^k) = \mathbf{\Delta}(z) \begin{bmatrix} 1 \\ W_L^{-k} \\ \vdots \\ W_L^{-k(L-1)} \end{bmatrix}, \tag{10.186}$$

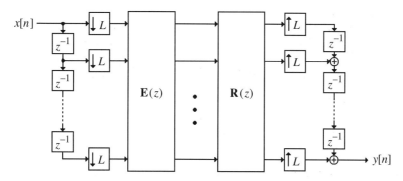

Figure 10.65: *L*-channel QMF bank structure based on the polyphase representations of the analysis and synthesis filter banks.

where

$$\mathbf{\Delta}(z) = \text{diag}\begin{bmatrix} 1 & z^{-1} & \cdots & z^{-(L-1)} \end{bmatrix}. \tag{10.187}$$

Making use of Eq. (10.187) in Eq. (10.185), we arrive at the desired result after some algebra:

$$\mathbf{H}(z) = \mathbf{D}^\dagger \mathbf{\Delta}(z) \mathbf{E}^T(z^L), \tag{10.188}$$

where **D** is the $L \times L$ DFT matrix.

10.10.4 Condition for Perfect Reconstruction

If the polyphase component matrices of Figure 10.65 satisfy the relation

$$\mathbf{R}(z)\mathbf{E}(z) = c\mathbf{I}, \tag{10.189}$$

where **I** is an $L \times L$ identity matrix and c is a constant, the structure of Figure 10.65 reduces to the one shown in Figure 10.66. Comparing Figure 10.66 with Figure 10.63, we note that the former can be considered as a special case of an L-channel QMF bank if we set

$$H_k(z) = z^{-k}, \qquad G_k(z) = z^{-(L-1-k)}, \qquad 0 \le k \le L-1. \tag{10.190}$$

Substituting the above in Eq. (10.169), we arrive at

$$a_\ell(z) = \frac{1}{L} \sum_{k=0}^{L-1} z^{-k} W_L^{-\ell k} z^{-(L-1-k)} = z^{-(L-1)} \left(\frac{1}{L} \sum_{k=0}^{L-1} W_L^{-\ell k} \right). \tag{10.191}$$

From Eqs. (10.9) and (10.10), it follows that $a_0(z) = 1$ and $a_\ell(z) = 0$ for $\ell \ne 0$. Hence, from Eq. (10.173), we note that $T(z) = z^{-(L-1)}$, or in other words, the structure of Figure 10.65 is a perfect reconstruction L-channel QMF bank if the condition of Eq. (10.189) is satisfied.

The analysis and synthesis filters of a perfect reconstruction filter bank of the form of Figure 10.65 can be easily determined as illustrated in the following example.

EXAMPLE 10.21   The three-channel analysis/synthesis filter bank of Figure 10.67 is by construction a perfect reconstruction filter bank. The output of the filter bank is simply $y[n] = dx[n-2]$. Moreover, here, $\mathbf{E}(z^3) = \mathbf{P}$ and $\mathbf{R}(z^3) = d\mathbf{P}^{-1}$. Consider

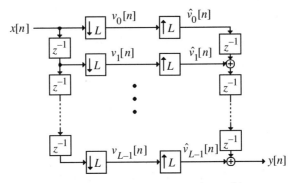

Figure 10.66: A simple perfect reconstruction multirate system.

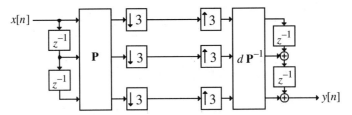

Figure 10.67: A three-channel analysis/synthesis filter bank.

$$\mathbf{P} = \begin{bmatrix} 1 & 1 & 1 \\ 1 & -1 & 1 \\ 1 & 0 & -1 \end{bmatrix}.$$

From Eqs. (10.180) and (10.181a) to (10.181c), we thus get

$$\begin{bmatrix} H_0(z) \\ H_1(z) \\ H_2(z) \end{bmatrix} = \begin{bmatrix} 1 & 1 & 1 \\ 1 & -1 & 1 \\ 1 & 0 & -1 \end{bmatrix} \begin{bmatrix} 1 \\ z^{-1} \\ z^{-2} \end{bmatrix},$$

resulting in the analysis filters

$$H_0(z) = 1 + z^{-1} + z^{-2}, \qquad H_1(z) = 1 - z^{-1} + z^{-2}, \qquad H_2(z) = 1 - z^{-2}.$$

Let $d = 4$. Then,

$$d\mathbf{P}^{-1} = \begin{bmatrix} 1 & 1 & 2 \\ 2 & -2 & 0 \\ 1 & 1 & -2 \end{bmatrix}.$$

Using Eqs. (10.183) and (10.184a) to (10.184c), we arrive at

$$\begin{bmatrix} G_0(z) \\ G_1(z) \\ G_2(z) \end{bmatrix} = \begin{bmatrix} z^{-2} & z^{-1} & 1 \end{bmatrix} \begin{bmatrix} 1 & 1 & 2 \\ 2 & -2 & 0 \\ 1 & 1 & -2 \end{bmatrix},$$

leading to the synthesis filters

$$G_0(z) = 1 + 2z^{-1} + z^{-2}, \quad G_1(z) = 1 - 2z^{-1} + z^{-2}, \quad G_2(z) = -2 + 2z^{-2}.$$

Now, for a given L-channel analysis filter bank, the polyphase matrix $\mathbf{E}(z)$ is known. From Eq. (10.189) it therefore follows that a perfect reconstruction L-channel QMF bank can be simply designed by constructing a synthesis filter bank with a polyphase matrix $\mathbf{R}(z) = [\mathbf{E}(z)]^{-1}$. In general, it is not easy to

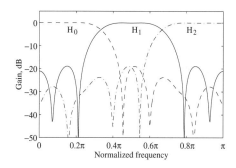

Figure 10.68: Gain responses of the three-channel FIR QMF bank of Example 10.22.

compute the inverse of a rational $L \times L$ matrix. An alternative elegant approach is to design the analysis filter bank with an invertible polyphase matrix. For example, $\mathbf{E}(z)$ can be chosen to be a paraunitary matrix satisfying the condition

$$\tilde{\mathbf{E}}(z)\mathbf{E}(z) = c\mathbf{I}, \text{ for all } z, \tag{10.192}$$

where $\tilde{\mathbf{E}}(z)$ is the paraconjugate of $\mathbf{E}(z)$ given by the transpose of $\mathbf{E}(z^{-1})$, with each coefficient replaced by its conjugate and choosing $\mathbf{R}(z) = \tilde{\mathbf{E}}(z)$.

For the design of a perfect reconstruction FIR L-channel QMF bank, the matrix $\mathbf{E}(z)$ can be expressed in a product form [Vai89]

$$\mathbf{E}(z) = \mathbf{E}_R(z)\mathbf{E}_{R-1}(z)\cdots\mathbf{E}_1(z)\mathbf{E}_0, \tag{10.193}$$

where $\mathbf{E}_0$ is a constant unitary matrix, and

$$\mathbf{E}_\ell(z) = \mathbf{I} - \mathbf{v}_\ell[\mathbf{v}_\ell^*]^T + z^{-1}\mathbf{v}_\ell[\mathbf{v}_\ell^*]^T, \tag{10.194}$$

in which $\mathbf{v}_\ell$ is a column vector of order L with unit norm, i.e., $[\mathbf{v}_\ell^*]^T\mathbf{v}_\ell = 1$. With this constraint on $\mathbf{E}(z)$, one can set up an appropriate objective function that can then be minimized to arrive at a set of L analysis filters meeting the desired passband and stopband specifications. To this end, a suitable objective function is given by

$$\phi = \sum_{k=0}^{L-1} \int_{k\text{th stopband}} \left| H_k(e^{j\omega}) \right|^2 d\omega. \tag{10.195}$$

The optimization parameters are the elements of $\mathbf{v}_\ell$ and $\mathbf{E}_0$.

EXAMPLE 10.22   We consider the design of a three-channel FIR perfect reconstruction QMF bank with a passband of width $\pi/3$ [Vai93]. The passband of the lowpass analysis filter $H_0(z)$ is from 0 to $\pi/3$, the passband of the bandpass analysis filter $H_1(z)$ is from $\pi/3$ to $2\pi/3$, and the passband of the highpass analysis filter $H_2(z)$ is from $2\pi/3$ to π. The objective function to be minimized here is thus of the form

$$\phi = \int_{\frac{\pi}{3}+\varepsilon}^{\pi} \left| H_0(e^{j\omega}) \right|^2 d\omega + \int_0^{\frac{\pi}{3}-\varepsilon} \left| H_1(e^{j\omega}) \right|^2 d\omega + \int_{\frac{2\pi}{3}+\varepsilon}^{\pi} \left| H_1(e^{j\omega}) \right|^2 d\omega + \int_0^{\frac{2\pi}{3}-\varepsilon} \left| H_2(e^{j\omega}) \right|^2 d\omega.$$

The impulse response coefficients of the analysis filters $\{h_k[n]\}$, $k = 0, 1, 2$, of length 15 obtained by minimizing ϕ are given in [Vai93]. The gain responses of these filters are plotted in Figure 10.68. It should be noted that the coefficients of the corresponding synthesis filters $\{g_k[n]\}$, $k = 0, 1, 2$, are given by $g_k[n] = h_k[14 - n]$.

10.11   Cosine-Modulated L-Channel Filter Banks

The cosine-modulated filter banks were originally developed to provide nearly perfect reconstruction with aliasing cancellation between adjacent channels and assume no aliasing between nonadjacent channels due to infinite stopband attenuation of the analysis filter in all nonadjacent bands [Rot83], [Chu85]. Even though, the latter assumption do not hold in practice, these filter banks, called *pseudo-QMF banks,* provide quite satisfactory performance if the stopband attenuation is sufficiently high.

10.11.1   Derivation of the Filter Bank

The pseudo-QMF banks are derived from a modified form of the uniform DFT filter banks. Let

$$P_0(z) = \sum_{n=0}^{N} p_0[n]z^{-n} \tag{10.196}$$

denote the prototype lowpass filter with real coefficients and a cutoff frequency at $\pi/2L$. We generate a set of filters $Q_k(z)$ from $P_0(z)$ by complex modulation at frequencies $(2k+1)\pi/2L = (k+0.5)\pi/L$ as follows:

$$Q_k(z) = P_0(zW_{2L}^{k+0.5}), \qquad 0 \le k \le 2L - 1, \tag{10.197}$$

where $W_{2L} = e^{-j\pi/L}$. Because of the complex modulation, these filters have complex-valued impulse responses. Note from the above, the response of $Q_0(z)$ is a right-shifted version of the response of $P_0(z)$ shifted by $\pi/2L$. Because of this shift, $|Q_k(e^{j\omega})| = |Q_{2L-1-k}(e^{-j\omega})|$, and the impulse response of $Q_{2L-1-k}(z)$ is complex conjugate of the impulse response of $Q_k(z)$, for $0 \le k \le L - 1$. The pair $Q_k(z)$ and $Q_{2L-1-k}(z)$ are combined to generate a filter with a real impulse response.

Define the intermediate transfer functions

$$U_k(z) = c_k Q_k(z), \qquad V_k(z) = c_k^* Q_{2L-1-k}(z). \tag{10.198}$$

The L analysis filters are then formed according to

$$H_k(z) = \sum_{n=0}^{N} h_k[n]z^{-n} = a_k U_k(z) + a_k^* V_k(z), \qquad 0 \le k \le L - 1. \tag{10.199}$$

Likewise, the L synthesis filters are formed according to

$$G_k(z) = \sum_{n=0}^{N} g_k[n]z^{-n} = b_k U_k(z) + b_k^* V_k(z), \qquad 0 \le k \le L - 1. \tag{10.200}$$

In the above equations, $a_k, b_k,$ and c_k are unit-magnitude constants. They are chosen to provide alias cancellation between adjacent channels and to ensure that the distortion transfer function $T(z)$ has a linear phase.

Consider the second channel. The output of the filter $G_2(z)$ has the components $H_2(zW_L^\ell)X(zW_L^\ell)$, $0 \le \ell \le L - 1$. However, as can be seen from Figure 10.69, the responses of $U_2(zW_L)$ and $U_2(zW_L^{-1})$ do not overlap with that of $U_2(z)$. On the other hand, responses of $U_2(zW_L^{-2})$ and $U_2(zW_L^{-3})$ overlap with that of $V_2(z)$. Likewise, the responses of $V_2(zW_L^2)$ and $V_2(zW_L^3)$ overlap with that of $U_2(z)$. In general, "significant" alias components of $X(zW_L^\ell)$ at the output of $G_k(z)$ correspond to values of

$$\ell = \langle -k+1 \rangle_L, \langle -k \rangle_L, k, k+1.$$

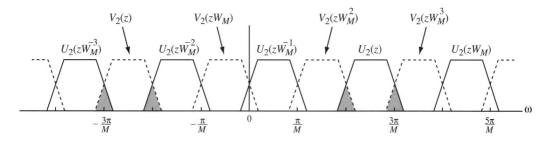

Figure 10.69: Illustration of alias components overlapping with $|G_k(e^{j\omega})|$.

Similarly, significant alias components of $X(zW_L^\ell)$ at the output of $G_{k-1}(z)$ correspond to values of

$$\ell = \langle -k \rangle_L, \langle -k+1 \rangle_L, k-1, k.$$

An additional requirement in the design of the filter bank is to have no phase distortion. To this end, the distortion transfer function $T(z)$ as given in Eq. (10.173) should have linear phase. This is achieved if the synthesis filters are related to the analysis filters according to

$$G_k(z) = z^{-N} H_k(z^{-1}). \tag{10.201}$$

It can be shown that the constants, a_k, b_k, and c_k, can be chosen to cancel the common aliasing components $X(zW_L^{\pm k})$ at the outputs of $G_k(z)$ and $G_{k-1}(z)$, and to make the distortion function $T(z)$ have linear phase resulting in closed-form expressions for the analysis and the synthesis filters which are related to the prototype filter $p_0[n]$ by cosine modulation [Chu85], [Rot83], [Vai93]:

$$h_k[n] = 2p_0[n] \cos\left(\left(k+\tfrac{1}{2}\right)\left(n-\tfrac{N}{2}\right)\frac{\pi}{L} + (-1)^k\frac{\pi}{4}\right),$$

$$g_k[n] = 2p_0[n] \cos\left(\left(k+\tfrac{1}{2}\right)\left(n-\tfrac{N}{2}\right)\frac{\pi}{L} - (-1)^k\frac{\pi}{4}\right). \tag{10.202}$$

It should be noted that if the prototype filter $P_0(z)$ has linear phase, the distortion function $T(z)$ has linear phase. However, in general, the analysis and the synthesis filters do not have linear phase.

10.11.2 Prototype Lowpass Filter Design

Hence, the design of the cosine-modulated filter bank reduces to the design of the Lth-band prototype filter $p_0[n]$ such that the magnitude response $|T(e^{j\omega})|$ of the distortion transfer function $T(z)$ is approximately flat for all values of ω. To this end, the prototype lowpass filter $P_0(z)$ should satisfy as much as possible the following two conditions:

$$\left|P_0(e^{j\omega})\right|^2 + \left|P_0(e^{j(\omega-\pi/L)})\right|^2 = 1, \qquad 0 < \omega < \frac{\pi}{L}, \tag{10.203}$$

and

$$\left|P_0(e^{j\omega})\right| = 0, \qquad \omega > \frac{\pi}{L}. \tag{10.204}$$

The QMF bank does not exhibit any amplitude distortion if Eq. (10.203) is satisfied exactly, whereas there is no aliasing between nonadjacent channels if Eq. (10.204) holds. As noted earlier, aliasing between adjacent channels is canceled structurally.

The design of the Lth-band FIR prototype filter satisfying both the conditions of Eqs. (10.203) and (10.204) is not possible. A relatively straightforward design approach makes use of the popular Parks-McClellan method [14] to design the prototype lowpass filter [Cre95]. The two conditions of Eqs. (10.203) and (10.204) are satisfied approximately by adjusting iteratively the passband edge to minimize the objective function

$$\phi = \max_{0<\omega<\pi/L} \left\{ \left| P_0(e^{j\omega}) \right|^2 + \left| P_0(e^{j(\omega-\pi/L)}) \right|^2 - 1 \right\}. \tag{10.205}$$

The filter length, stopband edge at π/L, and the relative error weighting are kept fixed during the optimization procedure.

It has been found through extensive design examples, the objective function of Eq. (10.205) is convex with respect to the passband edge. As a result, the algorithm converges to the same global minima for any initial starting value of the passband edge. The filter length is determined by fixing the initial value of the passband edge at $\omega/2L$ such that the desired stopband attenuation is obtained. At each step of the optimization process, the passband edge moves closer to zero which in turn increases slightly the stopband attenuation. It is also recommended that the passband error be weighted more heavily than the stopband attenuation.

The following MATLAB M-files can be used to design the pseudo-QMF bank based on the above method.[15]

```
function [hopt,passedge] = opt_filter(N,nbands)
%  This function creates an "optimal" lowpass prototype filter for
%  a pseudo-QMF filter bank with nbands bands of length N.
%
stopedge = 1/nbands;
delta = 0.001;
passedge = 1/(4*nbands);
toll = 0.000001;
step = 0.1 * passedge;
tcost = 0;
mpc = 1;
way = -1;
pcost = 10;
flag = 0;
s_time = cputime;
s_flops = flops;
%
while flag == 0
    hopt = remez(N,[0,passedge,stopedge,1],[1,1,0,0],[5,1]);
    H = fft(hopt,4096);
    HH = ovlp_ripple(H,nbands);
    [tcost] = max(abs(HH - ones(max(size(HH)),1)));
if tcost > pcost
    step = step/2;
    way = -way;
end
if abs(pcost - tcost) < toll
```

---
[14] See Section 7.7.1
[15] These MATLAB functions were written by C. D. Creusere.

```
        flag = 1;
    end
        pcost = tcost;
        passedge = passedge + way*step;
    end
final_time = cputime - s_time;
total_flops = flops - s_flops;
save hopt.mat hopt-ascii;
save pass passedge -ascii;


function H = comp_flat(Hin,Q)
% Calculates cost function (overlapped ripple in passband)
% being minimized.
N = max(size(Hin));
M = floor(2048/Q); H = zeros(M,1);
for k=1:M
    H(k) = abs(Hin(M - k +2))^2 + abs(Hin(k))^2;
end


function [H,G] = make_bank(h,nbands)
%   [H,G] = make_bank(h,nbands)
%   This function creates the filters for a pseudo-QMF filter banks
%   with number of bands = nbands
flen = max(size(h));
t = sqrt(2)/2;
for k = 1:nbands
    a(2*k-1) = t + i*t;
    a(2*k) = t - i*t;
end
for k=1:nbands
for l=1:flen
    m1 = cos(pi*(2*k-1)*(2*l-1)/(4*nbands));
 m2 = sin(pi*(2*k-1)*(2*l-1)/(4*nbands));
    H(k,l) = 2.0*(real(a(k))*m1 - imag(a(k))*m2)*h(l);
end
%   Form synthesis filters
for l=1:flen
    m1 = cos(pi*(2*k-1)*(2*l-1)/(4*nbands));
    m2 = sin(pi*(2*k-1)*(2*l-1)/(4*nbands));
    G(k,l) = 2.0*(real(a(k))*m1 + imag(a(k))*m2)*h(l);
end
end
```

We illustrate their use in the following example.

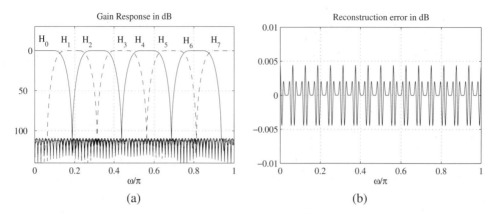

Figure 10.70: (a) Gain responses of the 8 analysis filters and (b) reconstruction error.

EXAMPLE 10.23   We consider the design of an eighth-band pseudo-QMF analysis bank using a prototype eighth-band lowpass FIR filter of length 128. Using the function `opt_filter` we designed the prototype lowpass eighth-band filter. Next, using the function `make_bank`, the coefficients of the remaining analysis filters and that of the eight synthesis filters are determined. Figure 10.70(a) shows the gain responses of the eight analysis filters $H_k(z)$ and Figure 10.70(b) shows the reconstruction error in dB.

10.12 Multilevel Filter Banks

In Section 10.10, we analyzed the general L-channel QMF bank and developed the conditions to be satisfied by the analysis and synthesis filters for perfect reconstruction. It is also possible to develop a multiband analysis/synthesis filter bank by iterating a two-channel QMF bank. Moreover, if the two-band QMF bank is of the perfect reconstruction type, the generated multiband structure also exhibits the perfect reconstruction property (Problems 10.64 and 10.65). In this section we consider this approach.

10.12.1 Filter Banks with Equal Passband Widths

By inserting a two-channel maximally decimated QMF bank in each channel of another two-channel maximally decimated QMF bank between the down-sampler and the up-sampler, we can generate a four-channel maximally decimated QMF bank, as shown in Figure 10.63. Since the analysis and the synthesis filter banks are formed like a tree, the overall system is often called a *tree-structured filter bank*. It should be noted that in the four-channel tree-structured filter bank of Figure 10.71, the 2 two-channel QMF banks in the second level do not have to be identical. However, if they are different QMF banks with different analysis and synthesis filters, to compensate for the unequal gains and unequal delays of the 2 two-channel systems, additional delays of appropriate values need to be inserted at the middle to ensure perfect reconstruction of the overall four-channel system.

An equivalent representation of the four-channel QMF system of Figure 10.71 is shown in Figure 10.72. The analysis and synthesis filters in the equivalent representation are related to those of the parent two-level tree-structured filter bank as follows:

$$H_0(z) = H_L(z)H_{10}(z^2), \qquad H_1(z) = H_L(z)H_{11}(z^2), \qquad (10.206a)$$
$$H_2(z) = H_H(z)H_{10}(z^2), \qquad H_3(z) = H_H(z)H_{11}(z^2), \qquad (10.206b)$$

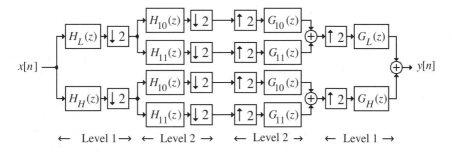

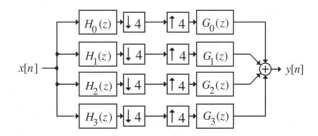

Figure 10.71: A two-level four-channel maximally decimated QMF structure.

Figure 10.72: An equivalent representation of the four-channel QMF structure of Figure 10.71.

$$G_0(z) = G_L(z)G_{10}(z^2), \qquad G_1(z) = G_L(z)G_{11}(z^2). \qquad (10.206c)$$

$$G_2(z) = G_H(z)G_{10}(z^2), \qquad G_3(z) = G_H(z)G_{11}(z^2). \qquad (10.206d)$$

EXAMPLE 10.24   We illustrate the design of a four-channel QMF bank by iterating the two-channel QMF bank based on the filter 12B of Johnston [Joh80].

From the filter's impulse response given in Example 10.15, we compute the impulse response of each of the four analysis filters using Eqs. (10.206a) and (10.206b) and then determine the gain responses of each using Program 10_10 given below.

```
% Program 10_10
% Frequency Responses of Tree-Structured QMF
% Filters
%
clf;
% Type in prototype lowpass filter coefficients
B1 = input('Filter coefficients = ');
B1 = [B1 fliplr(B1)];
% Generate the complementary highpass filter
L = length(B1);
for k = 1:L
B2(k) = ((-1)^k)*B1(k);
end
% Determine the coefficients of the four filters
B10 = zeros(1, 2*length(B1));
B10([1: 2: length(B10)]) = B1;
B11 = zeros(1, 2*length(B2));
```

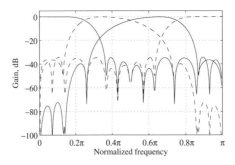

Figure 10.73: Gain responses of the four analysis filters of Example 10.24.

```
B11([1: 2: length(B11)]) = B2;
C0 = conv(B1, B10);C1 = conv(B1, B11);
C2 = conv(B2, B10);C3 = conv(B2, B11);
% Determine the frequency responses
[H0z, w] = freqz(C0, 1, 256);
h0 = abs(H0z);
M0 = 20*log10(h0);
[H1z, w] = freqz(C1, 1, 256);
h1 = abs(H1z);
M1 = 20*log10(h1);
[H2z, w] = freqz(C2, 1, 256);
h2 = abs(H2z);
M2 = 20*log10(h2);
[H3z, w] = freqz(C3, 1, 256);
h3 = abs(H3z);
M3 = 20*log10(h3);
plot(w/pi, M0,'-',w/pi, M1,'--',w/pi, M2,'--',w/pi,
M3,'-');grid
xlabel('\omega/\pi'); ylabel('Gain, dB')
```
Figure 10.73 shows the gain responses of the four analysis filters of the final four-channel QMF bank.

From Eqs. (10.206a) to (10.206d) it can be seen that each analysis filter $H_k(z)$ is a cascade of two filters, one with a single passband and a single stopband and the other with two passbands and two stopbands. The passband of the cascade is the frequency range where the passbands of the two filters overlap. On the other hand, the stopband of the cascade is formed from three different frequency ranges. In two of the frequency ranges, the passband of one coincides with the stopband of the other, while in the third range, the two stopbands overlap. As a result, the gain responses of the cascade in the three regions of the stopband are not equal, resulting in an uneven stopband attenuation characteristic. This type of behavior of the gain response can also be seen in Figure 10.73 and should be taken into account in the design of the tree-structured filter bank.

By continuing the process described above, QMF banks with more than four channels can be easily constructed. It should be noted that the number of channels resulting from this approach is restricted to a power of 2, i.e., $L = 2^\nu$. In addition, as illustrated by Figure 10.73, the filters in the analysis (synthesis) branch have passbands of equal width given by π/L. However, by a simple modification to the approach we can design QMF banks with analysis (synthesis) filters having passbands of unequal width as described next.

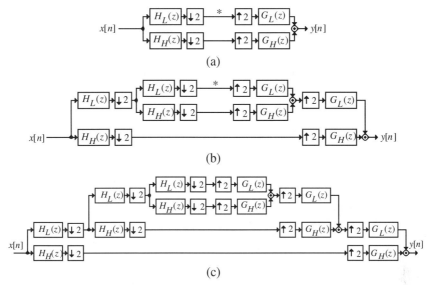

Figure 10.74: (a) A two-channel QMF bank, (b) a three-channel QMF bank derived from the two-channel QMF bank, and (c) a four-channel QMF bank derived from the three-channel QMF bank.

10.12.2 Filter Banks with Unequal Passband Widths

Consider the two-channel maximally decimated QMF bank of Figure 10.74(a). By inserting another two-channel maximally decimated QMF bank in the top subband channel between the down-sampler and the up-sampler at the position marked by a $*$, we arrive at a three-channel maximally decimated QMF bank, as shown in Figure 10.74(b). The equivalent representation of the generated three-channel filter bank is indicated in Figure 10.75(a), where the analysis and synthesis filters are given by

$$H_0(z) = H_L(z)H_L(z^2), \quad H_1(z) = H_L(z)H_H(z^2), \quad H_2(z) = H_H(z),$$
$$G_0(z) = G_L(z)G_L(z^2), \quad G_1(z) = G_L(z)G_H(z^2), \quad G_2(z) = G_H(z). \tag{10.207}$$

Typical magnitude responses of the analysis filters of the two-channel QMF bank of Figure 10.74(a) and that of the derived three-channel filter of Figure 10.74(b) are sketched in Figure 10.76(a) and (b), respectively.

We can continue this process and generate a four-channel QMF bank from the three-channel QMF bank of Figure 10.74(b) by inserting a two-channel QMF bank in the top subband channel at the position marked by a $*$, resulting in the structure of Figure 10.74(c). Its equivalent representation is indicated in Figure 10.75(b), where

$$H_0(z) = H_L(z)H_L(z^2)H_L(z^4), \quad H_1(z) = H_L(z)H_L(z^2)H_H(z^4),$$
$$H_2(z) = H_L(z)H_H(z^2), \qquad\qquad H_3(z) = H_H(z),$$
$$G_0(z) = G_L(z)G_L(z^2)G_L(z^4), \quad G_1(z) = G_L(z)G_L(z^2)G_H(z^4),$$
$$G_2(z) = G_L(z)G_H(z^2), \qquad\qquad G_3(z) = G_H(z). \tag{10.208}$$

Figure 10.76(c) shows typical magnitude responses of the analysis (synthesis) filters of the four-channel QMF bank of Figure 10.74(c) derived from a parent two-channel QMF bank with magnitude responses as indicated in Figure 10.76(a).

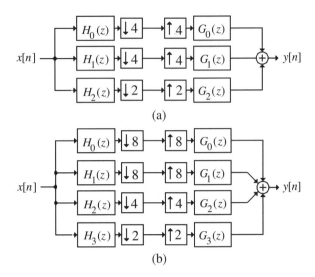

Figure 10.75: Maximally decimated QMF banks with unequal passband width analysis (synthesis) filters.

Because of the unequal passband widths of the analysis and synthesis filters, these structures belong to the class of nonuniform QMF banks. The tree-structured filter banks of Figure 10.74 are also referred to as *octave band QMF banks* [Fli94]. Various other types of nonuniform filter banks can be generated by iterating branches of a parent uniform two-channel QMF in different forms. Nonuniform filter banks are often used in speech and image coding applications. The former application is discussed in Section 11.8.

10.13 Summary

The basic theory of multirate digital signal processing is introduced in this chapter along with the design of some useful multirate systems. The two basic sampling rate alteration devices are the up-sampler and the down-sampler. We first discuss the input-output relations of these devices, both in the time-domain and in the frequency-domain. We then describe several cascade equivalences that are used to develop computationally efficient implementation of a multirate system by permitting the movement of the up-sampler and down-sampler from one part of the system to another part.

The sampling rate converter is implemented using either the up-sampler or the down-sampler or both, and a lowpass digital filter. The design issues for the sampling rate conversion are discussed next. We demonstrate that a computationally efficient sampling rate converter can often be designed as a cascade of such converters. We also outline another approach to the implementation of a computationally efficient sampling rate converter that is based on the use of the polyphase decomposition of the lowpass digital filter.

The concept of analysis and synthesis filter banks is then introduced and a method for designing such filter banks from a prototype lowpass transfer function is described. A very special type of digital lowpass filter, called a Nyquist filter, which is particularly attractive for computationally efficient sampling rate converter implementation, is considered next.

The remaining part of the chapter is devoted to the analysis and design of the so-called quadrature-mirror filter (QMF) bank that is formed by a combination of an analysis filter bank with down-sampled outputs followed by a synthesis filter bank with up-sampled inputs. Conditions satisfied by the analysis and synthesis filters for an alias-free operation of the QMF bank are derived. Several types of QMF banks

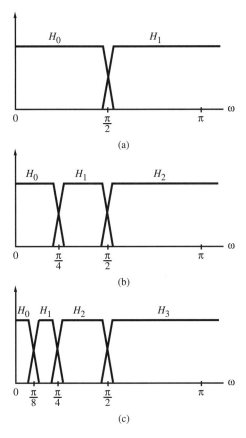

Figure 10.76: Magnitude responses of the analysis filters of a (a) two-channel QMF bank, (b) three-channel QMF bank derived from a two-channel QMF bank, and (c) four-channel QMF bank derived from a three-channel QMF bank.

are defined and their design equations are developed.

Several other applications of multirate discrete-time systems are outlined in Chapter 11. These include subband coding of speech and audio signals (Section 11.8), transmultiplexers for signal conversion between frequency-division multiplex (FDM) to time-division multiplex (TDM) communication systems (Section 11.9), discrete multitone transmission (Section 11.10), digital audio sampling rate conversion (Section 11.11), oversampling analog-to-digital conversion (Section 11.12), and oversampling digital-to-analog conversion (Section 11.13).

For additional details on multirate digital signal processing, we refer the reader to the texts by Akansu and Haddad [Aka92], Crochiere and Rabiner [Cro83], Fliege [Fli94], Vaidyanathan [Vai93], and Vetterli and Kovacevic [Vet95].

10.14 Problems

10.1 Show that the up-sampler defined by Eq. (10.1) is a time-varying system.

10.2 Show that the up-sampler defined by Eq. (10.1) and the down-sampler defined by Eq. (10.2) are linear systems.

10.3 Express the output $y[n]$ of Figure 10.6 as a function of the input $x[n]$. By simplifying the expression derived show that $y[n] = x[n - 1]$.

10.4 Prove the identity of Eq. (10.10).

10.5 Show that the two possible cascade configurations of a factor-of-L up-sampler and a factor-of-M down-sampler shown in Figure 10.13 are equivalent if and only if L and M are mutually prime.

10.6 Verify the cascade equivalences of Figure 10.14.

10.7 Develop an expression for the output $y[n]$ as a function of the input $x[n]$ for the multirate structure of Figure P10.1. [Hint: Replace the factor-of-10 down-sampler with a cascade of two down-samplers, and then make use of the identity of Figure 10.13.]

Figure P10.1

10.8 Consider the multirate structure of Figure P10.2(a) where $H_0(z)$, $H_1(z)$, and $H_2(z)$ are, respectively, ideal zero-phase real-coefficient lowpass, bandpass, and highpass filters with frequency responses as indicated in Figure P10.2(b). If the input is a real sequence with a discrete-time Fourier transform as shown in Figure P10.2(c), sketch the discrete-time Fourier transforms of the outputs $y_0[n]$, $y_1[n]$, and $y_2[n]$.

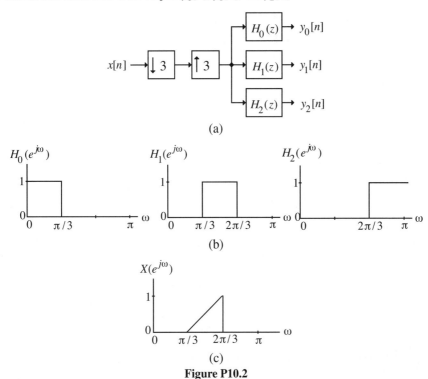

Figure P10.2

10.9 Show that the transpose of a factor-of-M decimator is a factor-of-M interpolator if the transpose of a factor-of-M down-sampler is a factor-of-M up-sampler.

10.10 Show that the transpose of an M-channel analysis filter bank is an M-channel synthesis bank.

10.11 Develop an alternate two-stage design of the decimator of Example 10.8 by designing the decimation filter in the form $H(z) = G(z^6)F(z)$. Compare its computational requirements with that of the design in Example 10.8.

10.12 Repeat Problem 10.11 for a filter of the form $H(z) = G(z^5)F(z)$. Compare its computational requirements with that of the designs in Example 10.8 and Problem 10.11.

10.13 Determine the computational complexity of a single-stage decimator designed to reduce the sampling rate from 60 kHz to 3 kHz. The decimation filter is to be designed as an equiripple FIR filter with a passband edge at 1.25 kHz, a passband ripple of 0.02, and a stopband ripple of 0.01. Use the total multiplications per second as a measure of the computational complexity.

10.14 The decimator of Problem 10.13 is to be designed as a two-stage structure. Develop an optimum design with the smallest computational complexity.

10.15 (a) Determine the computational complexity of a single-stage interpolator to be designed to increase the sampling rate from 600 Hz to 9 kHz. The interpolator is to be designed as an equiripple FIR filter with a passband edge at 200 Hz, a passband ripple of 0.002, and a stopband ripple of 0.004. Use Kaiser's formula given in Eq. (10.26) to estimate the order of the FIR filter. The measure of computational complexity is given by the total number of multiplications per second.

 (b) Develop a two-stage design of the above interpolator and compare its computational complexity with that of the single-stage design.

10.16 Using the method of Eq. (10.75) develop a two-band polyphase decomposition of each of the following IIR transfer functions:

(a) $H_1(z) = \dfrac{p_0 + p_1 z^{-1}}{1 + d_1 z^{-1}}, \ |d_1| < 1,$

(b) $H_2(z) = \dfrac{2 + 3.1 z^{-1} + 1.5 z^{-2}}{1 + 0.9 z^{-1} + 0.8 z^{-2}},$

(c) $H_3(z) = \dfrac{2 + 3.1 z^{-1} + 1.5 z^{-2} + 4 z^{-3}}{(1 - 0.5 z^{-1})(1 + 0.9 z^{-1} + 0.8 z^{-2})}.$

10.17 Using the method of Eq. (10.75) develop a three-band polyphase decomposition of each of the following IIR transfer functions:

(a) $H_1(z) = \dfrac{p_0 + p_1 z^{-1}}{1 + d_1 z^{-1}}, \ |d_1| < 1,$

(b) $H_2(z) = \dfrac{2 + 3.1 z^{-1} + 1.5 z^{-2}}{1 + 0.9 z^{-1} + 0.8 z^{-2}}.$

10.18 Develop a computationally efficient realization of a factor-of-4 interpolator employing a length-16 linear-phase FIR filter.

10.19 Develop a computationally efficient realization of a factor-of-3 decimator employing a length-15 linear-phase FIR filter.

10.20 Show that the *running-sum filter*, also called the *boxcar filter*, $H(z) = \sum_{i=0}^{N-1} z^i$, can be expressed in the form

$$H(z) = (1 + z^{-1})(1 + z^{-2}) \cdots (1 + z^{-2^{K-1}}),$$

where $N = 2^K$. Develop a computationally efficient realization of a factor-of-16 decimator using a length-17 boxcar filter.

10.21 The multirate system of Figure P10.3 is usually employed in exchanging discrete-time signals between two discrete-time systems with incommensurate sampling rates [Cro83]. Samples at the input of the digital sample-and-hold circuit may often be repeated or totally dropped resulting in an error in the overall sampling rate conversion process. Let $\mathcal{E}$ denote the ratio of the energy in the sample-to-sample difference signal to that in the original signal $y[n]$ at the output of the factor-of-L interpolator, and let $\mathcal{C}$ denote the sample-to-sample correlation of the signal $y[n]$. Express $\mathcal{E}$ as a function of $\mathcal{C}$, and show that as L becomes large, $\mathcal{E}$ becomes small, i.e., the error in the overall sampling rate conversion process becomes small.

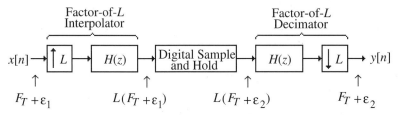

Figure P10.3

10.22 The multirate system of Figure P10.4 implements a fixed delay of L/M samples where L and M are relatively prime integers [Cro83]. Let $H(z)$ be a Type 1 length-N linear-phase FIR lowpass filter with a cutoff at π/M and a passband magnitude approximately equal to M. Develop the relation between the DTFTs $Y(e^{j\omega})$ and $X(e^{j\omega})$ of the output $y[n]$ and the input $x[n]$, respectively, assuming $N = 2KM + 1$, where K is a positive integer.

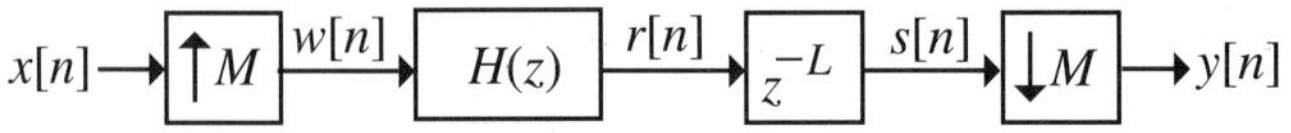

$$x[n] \longrightarrow \uparrow M \xrightarrow{w[n]} H(z) \xrightarrow{r[n]} z^{-L} \xrightarrow{s[n]} \downarrow M \longrightarrow y[n]$$

Figure P10.4

10.23 Let an ideal lowpass filter $H(z)$ with a cutoff at π/M be expressed as

$$H(z) = \sum_{k=0}^{M-1} z^{-k} H_k(z^M).$$

Show that each polyphase subfilter $H_k(z)$ is an allpass filter.

10.24 Prove the identity shown in Figure 10.34.

10.25 A generalization of the polyphase decomposition is considered in this problem. Let $H(z)$ be a causal FIR transfer function of degree $N - 1$ with N even:

$$H(z) = \sum_{n=0}^{N-1} h[n]z^{-n}.$$

(a) Show that $H(z)$ can be expressed in the form

$$H(z) = (1 + z^{-1})H_0(z^2) + (1 - z^{-1})H_1(z^2). \tag{10.209}$$

(b) Express $H_0(z)$ and $H_1(z)$ in terms of the coefficients of the polyphase components $E_0(z)$ and $E_1(z)$. The decomposition of Eq. (10.209) is an example of a generalized polyphase decomposition, called the *structural subband decomposition* [Mit93].

(c) Show that the decomposition of Eq. (10.209) can be expressed in the form

$$H(z) = [1 \ z^{-1}] \begin{bmatrix} 1 & 1 \\ 1 & -1 \end{bmatrix} \begin{bmatrix} H_0(z^2) \\ H_1(z^2) \end{bmatrix}. \tag{10.210}$$

The 2×2 matrix in Eq. (10.210) is called a Hadamard matrix of order 2 and denoted by $\mathbf{R}_2$.

(d) Show that if $N = 2^L$, then $H(z)$ can be expressed in the form

$$H(z) = [1 \ z^{-1} \dots z^{-(L-1)}] \mathbf{R}_L \begin{bmatrix} H_0(z^L) \\ H_1(z^L) \\ \vdots \\ H_{L-1}(z^L) \end{bmatrix} \tag{10.211}$$

where $\mathbf{R}_L$ is an $L \times L$ Hadamard matrix. Express the structural subband components $\{H_i(z)\}$ in terms of the polyphase components $\{E_i(z)\}$ of $H(z)$.

10.26 Develop a computationally efficient realization of a factor-of-4 interpolator employing a length-16 linear-phase FIR filter $H(z)$ and using a four-band structural subband decomposition of $H(z)$ in the form of Eq. (10.211).

10.27 Design a fractional-rate interpolator with an interpolation factor of 4/5 using the Lagrange interpolation algorithm. Use a fourth-order polynomial approximation. Develop realizations of the interpolator based on a block filtering approach and the Farrow structure.

10.28 Let $h[n]$ denote the impulse response of a lowpass half-band filter with a zero at $z = -1$. Show that

$$h[0] = \sum_{\substack{n=-\infty \\ n \neq 0}}^{\infty} h[n].$$

10.29 Show that the following FIR linear-phase transfer functions are lowpass half-band filters [Goo77]. Plot their magnitude responses using MATLAB.

(a) $H_1(z) = 1 + 2z^{-1} + z^{-2}$,

(b) $H_2(z) = -1 + 9z^{-2} + 16z^{-3} + 9z^{-4} - z^{-6}$,

(c) $H_3(z) = -3 + 19z^{-2} + 32z^{-3} + 19z^{-4} - 3z^{-6}$,

(d) $H_4(z) = 3 - 25z^{-2} + 150z^{-4} + 256z^{-5} + 150z^{-6} - 25z^{-8} + 3z^{-10}$,

(e) $H_5(z) = 9 - 44z^{-2} + 208z^{-4} + 346z^{-5} + 208z^{-6} - 44z^{-8} + 9z^{-10}$.

10.30 Consider the two-channel analysis filter bank structure of Figure P10.5 where $H_0(z)$ is a length-6 FIR filter with a transfer function given by

$$H_0(z) = h[0] + h[1]z^{-1} + h[2]z^{-2} + h[3]z^{-3} + h[4]z^{-4} + h[5]z^{-5}. \tag{10.212}$$

If $H_1(z) = H_0(-z)$, develop a realization of this filter bank with five delays and six multipliers.

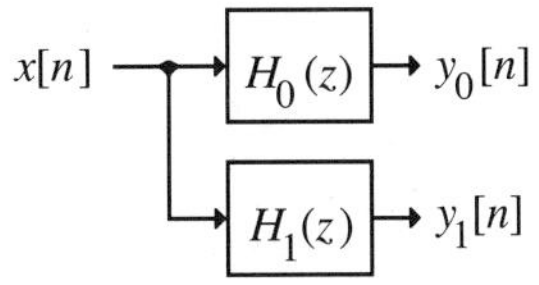

Figure P10.5

10.31 Consider the two-channel analysis filter bank structure of Figure P10.5 where $H_0(z)$ is an FIR filter of the form of Eq. (10.212). If $H_1(z)$ has a transfer function which is the mirror image of $H_0(z)$, i.e., $H_1(z) = z^{-5}H_0(z^{-1})$, develop a realization of this filter bank with only six multipliers.

10.32 The four-channel analysis filter bank of Figure P10.6(a), where **D** is a 4×4 DFT matrix, is characterized by the set of four transfer functions: $H_i(z) = Y_i(z)/X(z)$, $i = 0, 1, 2, 3$. Let the transfer functions of the four subfilters be given by

$$E_0(z) = 1 + 0.3z^{-1} - 0.8z^{-2}, \quad E_1(z) = 2 - 1.5z^{-1} + 3.1z^{-2},$$
$$E_2(z) = 4 - 0.9z^{-1} + 2.3z^{-2}, \quad E_3(z) = 1 + 3.7z^{-1} + 1.7z^{-2}.$$

(a) Determine the expressions for the four transfer functions, $H_0(z)$, $H_1(z)$, $H_2(z)$, and $H_3(z)$.

(b) Assume that the analysis filter $H_2(z)$ has a magnitude response as indicated in Figure P10.6(b). Sketch the magnitude responses of the other three analysis filters.

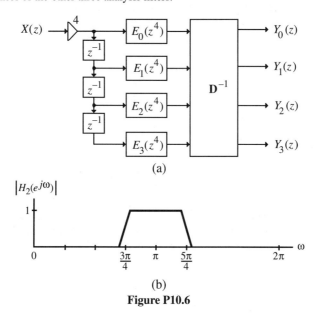

(a)

(b)

Figure P10.6

10.33 Consider the analysis-synthesis filter bank of Figure P10.7. Develop the input-output relation of this structure in the z-domain. Let $H_0(z) = (1 + z^{-1})/2$ and $H_1(z) = (1 - z^{-1})/2$. Determine the synthesis filters $G_0(z)$ and $G_1(z)$ so that the structure of Figure P10.7 is a perfect reconstruction filter bank.

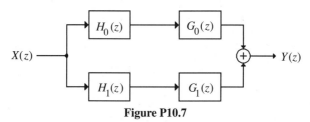

Figure P10.7

10.34 Let the analysis filters $H_0(z)$ and $H_1(z)$ of the structure of Figure P10.7 be power-complementary FIR filters of order N each.

(a) Show that this structure becomes a perfect reconstruction filter bank if the synthesis filters $G_0(z)$ and $G_1(z)$ are chosen as

$$G_0(z) = z^{-N} H_0(z^{-1}), \qquad G_1(z) = z^{-N} H_1(z^{-1}). \tag{10.213}$$

(b) Show that the synthesis filters are causal FIR filters if the analysis filters are causal.

(c) Show that the analysis and synthesis filters satisfying the perfect reconstruction condition cannot all be of linear phase.

10.35 Show that the two-channel QMF bank, in general, is a linear, time-varying system with a period of 2.

10.36 Consider the two-channel QMF structure of Figure 10.55 where $A_i(z)$ are stable allpass transfer functions. Let $\mathbf{E}(z)$ and $\mathbf{R}(z)$ denote the polyphase matrices of the analysis and synthesis filter banks. (a) Determine the expressions for $\mathbf{E}(z)$ and $\mathbf{R}(z)$ in terms of $A_i(z)$. (b) Is $\mathbf{E}(z)$ lossless? (c) Express $\mathbf{R}(z)$ in terms of $\mathbf{E}(z)$. (d) What is the product $\mathbf{R}(z)\mathbf{E}(z)$?

10.37 (a) Decompose the third-order transfer function

$$G(z) = \frac{(1 + z^{-1})^3}{6 + 2z^{-2}},$$

in the form

$$G(z) = \tfrac{1}{2}\{A_0(z) + A_1(z)\},$$

where $A_0(z)$ and $A_1(z)$ are stable allpass transfer functions.

(b) Realize $G(z)$ as a parallel connection of allpass filters with $A_0(z)$ and $A_1(z)$ realized with the fewest number of multipliers.

(c) Determine the transfer function $H(z)$ which is power-complementary to $G(z)$.

(d) Sketch the magnitude responses of $G(z)$ and $H(z)$.

10.38 Let $G_a(s)$ denote the Nth-order analog lowpass Butterworth transfer function with a 3-dB cutoff frequency at 1 rad/sec with N odd. Show that the corresponding digital Butterworth transfer function $H_0(z)$ obtained by a bilinear transformation is a half-band lowpass filter expressible in the form

$$H_0(z) = \tfrac{1}{2}\{A_0(z^2) + z^{-1}A_1(z^2)\}$$

where $A_0(z)$ and $A_1(z)$ are stable allpass transfer functions.

10.39 Using the method of Problem 10.38, develop the transfer function of a third-order lowpass half-band filter $H_0(z)$ and then determine its power-complementary transfer function $H_1(z)$. Develop a realization of a magnitude-preserving two-channel QMF bank whose analysis filters are $H_0(z)$ and $H_1(z)$, and using no more than one multiplier for the analysis stage.

10.40 Using the method of Problem 10.38, develop the transfer function of a fifth-order lowpass half-band filter $H_0(z)$ and then determine its power-complementary transfer function $H_1(z)$. Develop a realization of a magnitude-preserving two-channel QMF bank whose analysis filters are $H_0(z)$ and $H_1(z)$, and using no more than two multipliers for the analysis stage.

10.41 Let the order of the half-band lowpass IIR filter $H_0(z)$ of the QMF bank of Figure 10.52 be N where N is odd. (a) What is the total number of multipliers needed to implement the QMF bank of Figure 10.52? How many multiplications per second are needed to implement this structure? (b) If the QMF bank is of the magnitude-preserving type, the analysis and synthesis filters can be realized as a sum of IIR allpass filters resulting in the structure of Figure 10.55. What is the total number of multipliers needed in the implementation of Figure 10.55? How many multiplications per second are needed to implement this structure?

10.42 Consider a two-channel orthogonal filter bank structure of Figure 10.52 where the analysis filter $H_0(z)$ is a power-symmetric FIR transfer function of odd order N. If the second analysis filter $H_1(z)$ is chosen according to Eq. (10.144), show that both analysis filters can be realized using only $N + 1$ multipliers and $2N$ two-input adders.

10.43 Consider the QMF bank structure of Figure 10.65 with $L = 4$. Let the Type I polyphase component matrix be given by

$$\mathbf{E}(z) = \begin{bmatrix} 1 & 2 & 3 & 2 \\ 2 & 13 & 9 & 7 \\ 3 & 9 & 11 & 10 \\ 2 & 7 & 10 & 15 \end{bmatrix}.$$

Determine the Type II polyphase component matrix $\mathbf{R}(z)$ such that the four-channel QMF structure is a perfect reconstruction system with an input-output relation $y[n] = 3x[n - 3]$.

10.44 Design a three-channel perfect reconstruction QMF bank whose analysis filters are given by

$$\begin{bmatrix} H_0(z) \\ H_1(z) \\ H_2(z) \end{bmatrix} = \begin{bmatrix} 1 & 1 & 2 \\ 2 & 3 & 1 \\ 1 & 2 & 1 \end{bmatrix} \begin{bmatrix} 1 \\ z^{-1} \\ z^{-2} \end{bmatrix}.$$

Develop a computationally efficient realization of the filter bank.

10.45 Design a four-channel perfect reconstruction QMF bank whose analysis filters are given by

$$\begin{bmatrix} H_0(z) \\ H_1(z) \\ H_2(z) \\ H_3(z) \end{bmatrix} = \begin{bmatrix} 1 & 2 & 3 & 4 \\ 3 & 2 & 1 & 5 \\ 2 & 1 & 4 & 3 \\ 4 & 2 & 3 & 1 \end{bmatrix} \begin{bmatrix} 1 \\ z^{-1} \\ z^{-2} \\ z^{-3} \end{bmatrix}.$$

Develop a computationally efficient realization of the filter bank.

10.46 Design a three-channel perfect reconstruction QMF bank whose synthesis filters are given by

$$\begin{bmatrix} G_0(z) \\ G_1(z) \\ G_2(z) \end{bmatrix} = \begin{bmatrix} 4 & 2 & 3 \\ 5 & 4 & 1 \\ 2 & 1 & 3 \end{bmatrix} \begin{bmatrix} 1 \\ z^{-1} \\ z^{-2} \end{bmatrix}.$$

Develop a computationally efficient realization of the filter bank.

10.47 Show that, in general, the L-channel QMF bank is a linear, time-varying system with a period of L.

10.48 Consider an alias-free L-channel maximally decimated QMF bank with $H_k(z)$ and $G_k(z)$, $0 \leq k \leq L - 1$, denoting, respectively, the analysis and the synthesis filters. Let $T(z)$ denote its distortion transfer function. Show that if the analysis and the synthesis filters of each branch are interchanged, i.e., $G_k(z)$ are now the analysis filters, and $H_k(z)$ are the synthesis filters, the resulting system is still alias-free and has the same distortion transfer function.

10.49 Consider the multirate system of Figure 10.65 where $\mathbf{P}(z) = \mathbf{R}(z)\mathbf{E}(z)$. Show that this structure is time-invariant with no aliasing if and only if $\mathbf{P}(z)$ is a pseudo-circulant matrix of the form [Vai88b]:

$$\mathbf{P} = \begin{bmatrix} P_0(z) & P_1(z) & \cdots & P_{L-2}(z) & P_{L-1}(z) \\ z^{-1}P_{L-1}(z) & P_0(z) & \cdots & P_{L-3}(z) & P_{L-2}(z) \\ \vdots & \vdots & \ddots & \vdots & \vdots \\ z^{-1}P_2(z) & z^{-1}P_3(z) & \cdots & P_0(z) & P_1(z) \\ z^{-1}P_1(z) & z^{-1}P_2(z) & \cdots & z^{-1}P_{L-1}(z) & P_0(z) \end{bmatrix}. \tag{10.214}$$

10.50 Consider a third-order causal IIR filter described by the difference equation

$$b_0 y[n] + b_1 y[n-1] + b_2 y[n-2] + b_3 y[n-3] = a_0 x[n] + a_1 x[n-1] + a_2 x[n-2] + a_3 x[n-3],$$

where $y[n]$ and $x[n]$ are, respectively, the output and input sequences. By iterating the above difference equation for $n, n+1$, and $n+2$, we arrive at a set of three equations which can be written as a block difference equation of the form

$$\mathbf{B}_0 \mathbf{Y}_k + \mathbf{B}_1 \mathbf{Y}_{k-1} = \mathbf{A}_0 \mathbf{X}_k + \mathbf{A}_1 \mathbf{X}_{k-1},$$

where

$$\mathbf{Y}_k = \begin{bmatrix} y[3k] \\ y[3k+1] \\ y[3k+2] \end{bmatrix}, \quad \mathbf{X}_k = \begin{bmatrix} x[3k] \\ x[3k+1] \\ x[3k+2] \end{bmatrix}.$$

Determine the matrices $\mathbf{B}_0$, $\mathbf{B}_1$, $\mathbf{A}_0$, and $\mathbf{A}_1$ in terms of the coefficients $\{b_i\}$ and $\{a_i\}$. Develop a multirate structure to generate $\mathbf{X}_k$ from the input sequence $x[n]$ and a multirate structure to generate the output sequence $y[n]$ from $\mathbf{Y}_k$. Determine the expression for the block transfer function $\mathbf{H}(z)$ where $\mathbf{Y}_k = \mathbf{H}(z)\mathbf{X}_k$. Show that $\mathbf{H}(z)$ is pseudo-circulant as defined in Eq. (10.214).

10.51 The multirate system of Figure P10.8 is called an *N-path filter* and has been proposed for high-speed implementation of narrowband digital filters [Mit87]. Consider the *N*-path filter for $N = 3$. Show that the three-path filter is time-invariant and determine its transfer function. What is the transfer function for the general *N*-path filter?

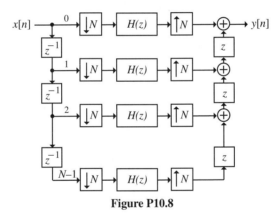

Figure P10.8

10.52 What is the transfer function of the cascade of a three-path filter and a four-path filter designed using the same filter $H(z)$ in each single path of both structures? Let $H(z)$ of the single path of an *N*-path filter of Figure P10.8 be an ideal lowpass filter with a magnitude response as shown in Figure P10.9. Sketch as accurately as possible the magnitude response of the cascade of a three-path filter and a four-path filter designed using $H(z)$.

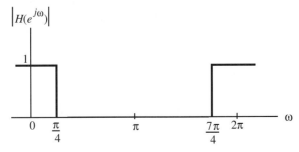

Figure P10.9

10.53 The structure of Figure P10.10 has been proposed for the computationally efficient implementation of FIR digital filters [Vet88].

(a) Show that the structure is alias-free and determine the overall transfer function $T(z) = Y(z)/X(z)$ in terms of $H_0(z)$ and $H_1(z)$.

(b) Determine the expression for $T(z)$ if

$$H_0(z^2) = \tfrac{1}{2}\{H(z) + H(-z)\}, \qquad H_1(z^2) = \tfrac{1}{2}\{H(z) - H(-z)\}z.$$

(c) If $H(z)$ is a length-$2K$ FIR filter, what are the lengths of the filters $H_0(z)$ and $H_1(z)$?

(d) Determine the computational efficiency of this structure.

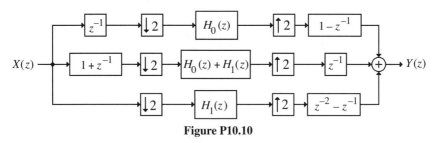

Figure P10.10

10.54 (a) Show that the structure of Figure P10.11 is time-invariant with no aliasing if and only if the following matrix is pseudo-circulant as defined by Eq. (10.214) [Lin96]:

$$\mathbf{Q}(z) = \begin{bmatrix} 0 & 1 & 0 \\ z^{-1} & 0 & 1 \end{bmatrix} \mathbf{P}(z) \begin{bmatrix} 1 & 0 \\ 0 & 1 \\ z^{-1} & 0 \end{bmatrix}.$$

(b) Develop an equivalent realization of Figure P10.11 based on a critical down-sampling and critical up-sampling.

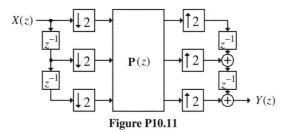

Figure P10.11

10.55 Analyze the structure of Figure P10.12 and determine its input-output relations. Comment on your results.

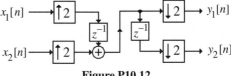

Figure P10.12

10.56 An efficient implementation of two separate single-input, single-output LTI discrete-time systems with an identical transfer function $H(z)$ by a single two-input, two-output multirate discrete-time system is obtained using the *pipelining/interleaving* (PI) technique as shown in Figure P10.13 [Jia97]. Show that the system of Figure P10.13 is time-invariant and determine the transfer functions from each input to each output.

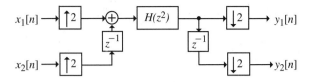

Figure P10.13

10.57 Show that the multirate system of Figure P10.14 is time-invariant and determine its transfer function [Jia97].

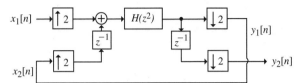

Figure P10.14

10.58 Problem 7.57 describes the filter sharpening approach [Kai77] which is used to improve the magnitude response of a filter $H(z)$ in both the passband and the stopband by employing multiple copies of the filter. For example, the *thricing method* of filter sharpening implements the transfer function

$$G(z) = z^{-2}[3H^2(z) - 2H^3(z)], \qquad (10.215)$$

where $H(z)$ is the prototype zero-phase FIR filter. Show that the multirate structure of Figure P10.15 implements the above equation using the PI technique for an appropriate value of the constant C [Jia97].

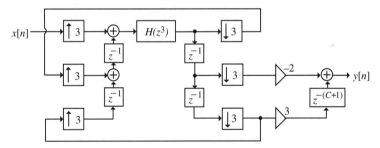

Figure P10.15

10.59 Show that each one of the following FIR transfer functions is a power-symmetric function.

(a) $H_0(z) = \frac{1}{2} - z^{-1} + \frac{21}{2}z^{-2} - \frac{27}{2}z^{-3} - 5z^{-4} - \frac{5}{2}z^{-5}$,

(b) $H_0(z) = 1 + 3z^{-1} + 14z^{-2} + 22z^{-3} - 12z^{-4} + 4z^{-5}$.

Using $H_0(z)$ as one of the analysis filter, determine the remaining three filters of the corresponding two-channel orthogonal filter bank. In each case, show that the filter bank is alias-free and satisfies the perfect reconstruction condition.

10.60 The analysis filters of a biorthogonal two-channel filter bank are given by $H_0(z) = 1 + az^{-1} + z^{-2}$, and $H_1(z) = 1 + az^{-1} + bz^{-2} + az^{-3} + z^{-4}$ [Vet89]. Determine the two synthesis filters $G_0(z)$ and $G_1(z)$ using Eq. (10.153). Show that the two-channel filter bank is alias-free and satisfies the perfect reconstruction property with $a \neq 0$ and $b \neq 2$.

10.61 Design a two-channel perfect reconstruction filter bank such that the lowpass analysis filter $H_0(z)$ is of length 4 and has two zeros at $z = -1$. Is it possible to design the analysis filters so that they have linear phase in addition to having two zeros at $z = -1$?

10.62 We have demonstrated in Example 10.3 that the multirate structure of Figure P10.6 is a perfect reconstruction system with the output $y[n]$ being a replica of the input $x[n]$ but delayed by one sample. Figure P10.16(a) shows the structure obtained from Figure 10.6 using the *lifting scheme* [Swe96]. Show that it is also a perfect reconstruction system.

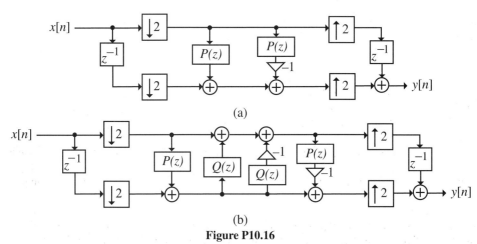

(a)

(b)

Figure P10.16

10.63 The lifting scheme can be repeatedly applied to develop perfect reconstruction systems with more desired features. Figure P10.16(b) shows a structure derived from Figure P10.16(a) by applying the lifting scheme a second time. Show that this structure is also a perfect reconstruction system.

10.64 Show that the four-channel QMF bank of Figure 10.71 is a perfect reconstruction type if the two parent two-channel QMF banks are of the perfect reconstruction type.

10.65 Show that the three-channel and the four-channel QMF banks of Figure 10.74(b) and 10.74(c), respectively, are of perfect reconstruction type if the two parent two-channel QMF banks of Figure 10.74(a) are of the perfect reconstruction type.

10.15 MATLAB Exercises

M 10.1   (a) Modify Program 10_1 to study the operation of a factor-of-4 up-sampler on the following input sequences: (i) sum of two sinusoidal sequences of normalized frequencies 0.2 and 0.35 rad/sec, (ii) ramp sequence, and (iii) square wave sequence with various duty cycles. Choose the input length to be 50. Plot the input and the output sequences.

(b) Repeat part (a) for a factor-of-5 up-sampler.

M 10.2   (a) Modify Program 10_2 to study the operation of a factor-of-4 down-sampler on the following input sequences: (i) sum of two sinusoidal sequences of normalized frequencies 0.2 and 0.35 rad/sec, (ii) ramp sequence, and (iii) square wave sequence with various duty cycles. Choose the input length to be 50. Plot the input and the output sequences.

(b) Repeat part (a) for a factor-of-5 down-sampler.

M 10.3   (a) Use a vector of frequency points `freq = [0 0.95 0.98 1]` in Program 10_3 and run it with an up-sampling factor of $L = 5$. Comment on your results.

(b) Repeat part (a) for $L = 6$. Comment on your results.

M 10.4   (a) Use a vector of frequency points `freq = [0 0.95 0.98 1]` and a magnitude response vector `mag = [1 0 0 0]` in Program 10_3, and run it with an up-sampling factor of $L = 5$. Comment on your results.

(b) Repeat part (a) for $L = 8$. Comment on your results.

M 10.5   (a) Use a vector of frequency points `freq = [0 0.22 0.25 1]` and a magnitude response vector `mag = [1 0 0 0]` in Program 10_4, and run it with a down-sampling factor of $M = 4$. Comment on your results.

(b) Repeat part (a) for $M = 5$. Comment on your results.

M 10.6 Run Program 10_5 for the following input data: (a) $N = 120$, $M = 3$, $f_1 = 0.045$, $f_2 = 0.029$; (b) $N = 120$, $M = 4$, $f_1 = 0.045$, $f_2 = 0.029$. Comment on your results.

M 10.7 Run Program 10_6 for the following input data: (a) $N = 40$, $L = 3$, $f_1 = 0.045$, $f_2 = 0.029$; (b) $N = 30$, $L = 4$, $f_1 = 0.045$, $f_2 = 0.029$. Comment on your results.

M 10.8 Run Program 10_7 for the following input data: (a) $N = 30$, $L = 3$, $M = 2$, $f_1 = 0.045$, $f_2 = 0.029$; (b) $N = 40$, $L = 3$, $M = 5$, $f_1 = 0.045$, $f_2 = 0.029$. Comment on your results.

M 10.9 Design a length-61 linear-phase FIR lowpass filter with a cutoff frequency at $\pi/4$ using the windowed Fourier series approach. Express the transfer function in a four-band polyphase decomposition. Using MATLAB compute and plot the frequency responses of each polyphase component. Show that all polyphase components have constant magnitude responses.

M 10.10 Design a fifth-order IIR half-band Butterworth lowpass filter and realize it using only two multipliers.

M 10.11 Design a seventh-order IIR half-band Butterworth lowpass filter and realize it using only three multipliers.

M 10.12 Design using MATLAB a real-coefficient elliptic half-band IIR filter $H_0(z)$ of odd order with the following specifications: $\omega_s = 0.55\pi$, and $\delta_s = 0.001$. Note that the half-band filter constraint is satisfied if $\omega_p + \omega_s = \pi$ and $(1 - \delta_p)^2 + \delta_s^2 = 1$. Express $H_0(z)$ in the form

$$H_0(z) = \tfrac{1}{2}\{\mathcal{A}_0(z^2) + z^{-1}\mathcal{A}_1(z^2)\}$$

where $\mathcal{A}_0(z)$ and $\mathcal{A}_1(z)$ are stable allpass transfer functions. Plot the magnitude responses of $H_0(z)$ and its power-complementary transfer function $H_1(z)$ in the same figure.

M 10.13 Design a linear-phase sixth-band lowpass FIR filter of order 42 with $\omega_p = 3\pi/24$ and $\omega_s = 5\pi/24$ using the windowed Fourier series approach by modifying Program 10_8. Use the Hann window.

M 10.14 Design using MATLAB a real-coefficient power-symmetric FIR lowpass filter $H_0(z)$ with a stopband edge at 0.65π and a minimum stopband attenuation of 25 dB. Design next a perfect reconstruction two-channel QMF bank based on $H_0(z)$. Show the transfer functions of all four filters. Plot the magnitude responses of the two analysis filters in the same figure.

M 10.15 Write a MATLAB program to design a two-channel QMF paraunitary lattice filter bank. Using this program design a lattice structure with filters of order 23 and a stopband edge at $\omega_s = 0.55\pi$. Plot the sum of the magnitude squares of the two analysis filters. What is the minimum stopband attenuation in dB of your filters? Plot the amplitude distortion in dB. Quantize the lattice coefficients to 6 decimal digitals and plot the gain responses of the two analysis filters along with those of the original filters on the same figure. Comment on your results.

M 10.16 Design and realize a four-channel uniform DFT analysis filter bank using a prototype linear-phase FIR filter of length 21. Design the prototype filter using the function `remez` of MATLAB. Assume a transition band of width 0.1π. Plot the magnitude responses of each filter on the same figure.

M 10.17 Design a three-channel QMF bank in the form of Figure 10.74(b) by iterating the two-channel QMF bank based on Filter 16A of Johnston [Ans93], [Cro83], [Joh80]. Plot the gain responses of the three analysis filters, $H_0(z)$, $H_1(z)$, and $H_2(z)$ on the same figure. Comment on your results.

11 Applications of Digital Signal Processing

As mentioned in Chapter 1, digital signal processing techniques are increasingly replacing conventional analog signal processing methods in many fields such as speech analysis and processing, radar and sonar signal processing, biomedical signal analysis and processing, telecommunications, and geophysical signal processing. Some typical applications of DSP chips are summarized in Table 11.1. A complete overview of these and other applications is beyond the scope of this book. Moreover, an understanding of many applications requires a knowledge of the field where they are being used. In this chapter we include a few simple applications to provide a glimpse of the potential of DSP.

We first describe several applications of the discrete Fourier transform (DFT) introduced in Section 3.2. The first application considered is the detection of the frequencies of a pair of sinusoidal signals, called tones, employed in telephone signaling. Next we discuss the use of the DFT in the determination of the spectral contents of a finite-length sequence. The effect of the DFT length and the windowing of the sequence are examined in detail here. In the following section, we introduce the concept of the short-time Fourier transform (STFT) and discuss its application in the spectral analysis of nonstationary signals. We then consider the spectral analysis of random signals using both nonparametric and parametric methods. Application of digital filtering methods to musical sound processing is considered next, and a variety of practical digital filter structures useful for the generation of certain audio effects, such as artificial reverberation, flanging, phasing, filtering, and equalization, are introduced. The digital stereo generation for FM stereo transmission is treated in the following section. Generation of discrete-time analytic signals by means of a discrete-time Hilbert transformer is then considered, and several methods of designing these circuits are outlined along with an application. The basic scheme of the subband coding of speech and audio signals is reviewed next. The theory and design of transmultiplexers are discussed in the following section. One method of digital data transmission employing digital signal processing methods is introduced then.

A method for audio sampling rate conversion is then outlined. The basic concepts behind the design of the oversampling A/D and D/A converters are reviewed in the following two sections. Finally, we review the sparse antenna array design for ultrasound scanners.

11.1 Dual-Tone Multifrequency Signal Detection

Dual-tone multifrequency (DTMF) signaling, increasingly being employed worldwide with push-button telephone sets, offers a high dialing speed over the dial-pulse signaling used in conventional rotary telephone sets. In recent years, DTMF signaling has also found applications requiring interactive control such as in voice main, electronic mail (e-mail), telephone banking, and ATM machines.

A DTMF signal consists of a sum of two tones with frequencies taken from two mutually exclusive groups of preassigned frequencies. Each pair of such tones represents a unique number or a symbol. Decoding of a DTMF signal thus involves identifying the two tones in that signal and determining their corresponding number or symbol. The frequencies allocated to the various digits and symbols of a push-

Table 11.1: Typical applications of DSP chips.

General-purpose DSP	Graphics/imaging	Instrumentation
Digital filtering	3-D rotation	Spectrum analysis
Convolution	Robot vision	Function generation
Correlation	Image transmission/	Pattern matching
Hilbert transforms	compression	Seismic processing
Fast Fourier transforms	Pattern recognition	Transient analysis
Adaptive filtering	Image enhancement	Digital filtering
Windowing	Homomorphic processing	Phase-locked loops
Waveform generation	Workstations	
	Animation/digital map	

Voice/speech	Control	Military
Voice main	Disk control	Secure communications
Speech vocoding	Servo control	Radar processing
Speech recognition	Robot control	Sonar processing
Speaker verification	Laser printer control	Image processing
Speech enhancement	Engine control	Navigation
Speech synthesis	Motor control	Missile guidance
Text to speech		Radio frequency modems

Telecommunications		Automotive
Echo cancellation	FAX	Engine control
ADPCM transcoders	Cellular telephone	Vibration analysis
Digital PBXs	Speaker phones	Antiskid brakes
Line repeaters	Digital speech	Adaptive ride control
Channel multiplexing	Interpolation (DSI)	Global positioning
1200- to 19,200-bps modems	X.25 packet switching	Navigation
Adaptive equalizers	Video conferencing	Voice commands
DTMF encoding/decoding	Spread spectrum	Digital radio
Data encryption	communications	Cellular telephones

Consumer	Industrial	Medical
Radar detectors	Robotics	Hearing aids
Power tools	Numeric control	Patient monitoring
Digital audio/TV	Security access	Ultrasound equipment
Music synthesizer	Power line monitors	Diagnostic tools
Educational toys		Prosthetics
		Fetal monitors

Reprinted from K. S. Lin, ed., *Digital Signal Processing Applications with the TMS320 Family*, vol. 1, Prentice Hall and Texas Instruments, 1987. Reprinted by Permission of Texas Instruments.

button keypad are internationally accepted standards and are shown in Figure 1.38 [ITU84]. The four keys in the last column of the keypad, as shown in this figure, are not yet available on standard handsets and are reserved for future use. Since the signaling frequencies are all located in the frequency band used for speech transmission, this is an *in-band system*. Interfacing with the analog input and output devices is provided by *codec* (coder/decoder) chips, or A/D and D/A converters (Sections 5.8 and 5.9).

Although a number of chips with analog circuitry are available for the generation and decoding of DTMF signals in a single channel, these functions can also be implemented digitally on DSP chips. Such a digital implementation surpasses analog equivalents in performance, since it provides better precision, stability, versatility, and reprogrammability to meet other tone standards, and the scope for multichannel operation by time-sharing leading to a lower chip count.

The digital implementation of a DTMF signal involves adding two finite-length digital sinusoidal sequences with the latter simply generated by using look-up tables or by computing a polynomial expansion. The digital tone detection can be easily performed by computing the DFT of the DTMF signal and then measuring the energy present at the eight DTMF frequencies. The minimum duration of a DTMF signal is 40 ms. Thus, with a sampling rate of 8 kHz, there are at most $0.04 \times 8000 = 320$ samples available for decoding each DTMF digit. The actual number of samples used for the DFT computation is less than this number and is chosen so as to minimize the difference between the actual location of the sinusoid and the nearest integer value DFT index k.

The DTMF decoder computes the DFT samples closest in frequency to the eight DTMF fundamental tones and their respective second harmonics. In addition, a practical DTMF decoder also computes the DFT samples closest in frequency to the second harmonics corresponding to each of the fundamental tone frequencies. This latter computation is employed to distinguish between human voices and the pure sinusoids generated by the DTMF signal. In general, the spectrum of a human voice contains components at all frequencies including the above second harmonic frequencies. On the other hand, the DTMF signal generated by the handset has negligible second harmonics. The DFT computation scheme employed is a slightly modified version of Goertzel's algorithm, as described in Section 8.3.1 for the computation of the squared magnitudes of the DFT samples that are needed for the energy computation.

The DFT length N determines the frequency spacing between the locations of the DFT samples and the time it takes to compute the DFT sample. A large N makes the spacing smaller, providing higher resolution in the frequency-domain but increases the computation time. The frequency f_k in Hz corresponding to the DFT index (bin number) k is given by

$$f_k = \frac{k F_T}{N}, \qquad k = 0, 1, \ldots, N - 1, \tag{11.1}$$

where F_T is the sampling frequency. If the input signal contains a sinusoid of frequency f_{in} different from that given above, its DFT will contain not only large-valued samples at values of k closest to $N f_{in}/F_T$ but also nonzero values at other values of k due to a phenomenon called leakage (see Example 8.11). To minimize the leakage it is desirable to choose N appropriately so that the tone frequencies fall as close as possible to a DFT bin, thus providing a very strong DFT sample at this index value relative to all other values. For an 8-kHz sampling frequency, the best value of the DFT length N to detect the eight fundamental DTMF tones has been found to be 205 and that for detecting the eight second harmonics is 201 [Mar92]. Table 11.2 shows the DFT index values closest to each of the tone frequencies and their second harmonics for these two values of N, respectively. Figure 11.1 shows 16 selected DFT samples computed using a 205-point DFT of a length-205 sinusoidal sequence for each of the fundamental tone frequencies.

The following MATLAB program demonstrates the DFT-based DTMF detection algorithm. It employs the function gfft(x,N,k) of Section 8.3.1 to calculate a single DFT sample using Goertzel's method. The input data is the telephone handset key symbol. The program generates a length-205 sequence consisting of two sinusoids of frequencies according to the convention shown in Figure 1.39. It then

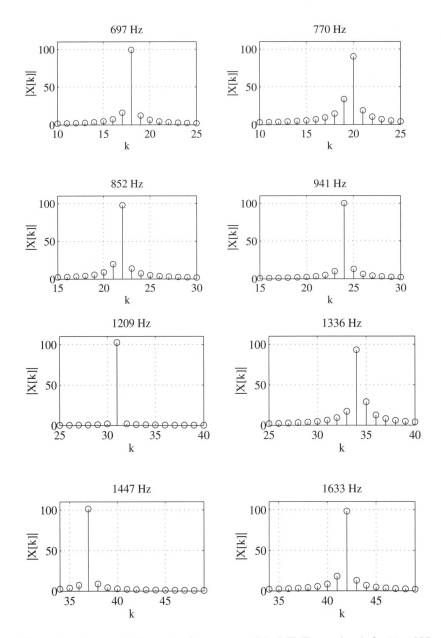

Figure 11.1: Selected DFT samples for each one of the DTMF tone signals for $N = 205$.

computes the eight DFT samples corresponding to the bin numbers of the fundamental tone frequencies given in Table 11.2 and displays these DFT samples and the decoded symbol. As only pure tones are generated, the program does not employ the test involving the second harmonic detection to distinguish between human voice and touch-tone digit. The outputs generated by this program for the input symbol # are displayed in Figure 11.2.

Table 11.2: DFT index values for DTMF tones for $N = 205$ and their second harmonics for $N = 201$ [Mar92].

Basic tone in Hz	Exact k value	Nearest integer k value	Absolute error in k
697	17.861	18	0.139
770	19.731	20	0.269
852	21.833	22	0.167
941	24.113	24	0.113
1209	30.981	31	0.019
1336	34.235	34	0.235
1477	37.848	38	0.152
1633	41.846	42	0.154
Second harmonic in Hz	Exact k value	Nearest integer k value	Absolute error in k
1394	35.024	35	0.024
1540	38.692	39	0.308
1704	42.813	43	0.187
1882	47.285	47	0.285
2418	60.752	61	0.248
2672	67.134	67	0.134
2954	74.219	74	0.219
3266	82.058	82	0.058

```
% Program 11_1
% Dual-Tone Multifrequency Tone Detection
% Using the DFT
%
clf;
d = input('Type in the telephone digit = ', 's');
symbol = abs(d);
tm = [49 50 51 65;52 53 54 66;55 56 57 67;42 48 35 68];
for p = 1:4;
for q = 1:4;
    if tm(p,q) == abs(d);break,end
end
    if tm(p,q) == abs(d);break,end
end
f1 = [697 770 852 941];
f2 = [1209 1336 1477 1633];
n = 0:204;
x = sin(2*pi*n*f1(p)/8000) + sin(2*pi*n*f2(q)/8000);
k = [18 20 22 24 31 34 38 42];
val = zeros(1,8);
for m = 1:8;
```

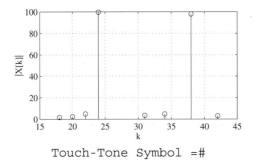

Touch-Tone Symbol =#

Figure 11.2: A typical output of Program 11_1.

```
    Fx(m) = gfft(x,205,k(m));
end
val = abs(Fx);
stem(k,val);grid; xlabel('k');ylabel('|X[k]|');
limit = 80;
for s = 5:8;
    if val(s) > limit,break,end
end
for r = 1:4;
    if val(r) > limit,break,end
end
disp(['Touch-Tone Symbol = ',setstr(tm(r,s-4))])
```

11.2 Spectral Analysis of Sinusoidal Signals

An important application of digital signal processing methods is in determining in the discrete-time domain the frequency contents of a continuous-time signal, more commonly known as *spectral analysis*. More specifically, it involves the determination of either the energy spectrum or the power spectrum of the signal. Applications of digital spectral analysis can be found in many fields and are widespread. The spectral analysis methods are based on the following observation. If the continuous-time signal $g_a(t)$ is reasonably bandlimited, the spectral characteristics of its discrete-time equivalent $g[n]$ should provide a good estimate of the spectral properties of $g_a(t)$. However, in most cases, $g_a(t)$ is defined for $-\infty < t < \infty$, and as a result, $g[n]$ is of infinite extent and defined for $-\infty < n < \infty$. Since it is difficult to evaluate the spectral parameters of an infinite-length signal, a more practical approach is as follows. First, the continuous-time signal $g_a(t)$ is passed through an analog anti-aliasing filter before it is sampled to eliminate the effect of aliasing. The output of the filter is then sampled to generate a discrete-time sequence equivalent $g[n]$. It is assumed that the anti-aliasing filter has been designed appropriately and hence, the effect of aliasing can be ignored. Moreover, it is further assumed that the A/D converter wordlength is large enough so that the A/D conversion noise can be neglected.

   This and the following two sections provide a review of some spectral analysis methods. In this section, we consider the Fourier analysis of a stationary signal composed of sinusoidal components. In Section 11.3 we discuss the Fourier analysis of nonstationary signals with time-varying parameters. Section 11.4 considers the spectral analysis of random signals. For a detailed exposition of spectral analysis and a concise review of the history of this area, see Kumaresan [Kum93].

For the spectral analysis of sinusoidal signals we assume that the parameters characterizing the sinusoidal components, such as amplitudes, frequencies, and phase, do not change with time. For such a signal $g[n]$, the Fourier analysis can be carried out by computing its DTFT $G(e^{j\omega})$:

$$G(e^{j\omega}) = \sum_{n=-\infty}^{\infty} g[n]e^{-j\omega n}. \tag{11.2}$$

In practice, the infinite-length sequence $g[n]$ is first windowed by multiplying it with a length-N window $w[n]$ to make it into a finite-length sequence $\gamma[n] = g[n] \cdot w[n]$ of length N. The spectral characteristics of the windowed finite-length sequence $\gamma[n]$ obtained from its DTFT $\Gamma(e^{j\omega})$ then is assumed to provide a reasonable estimate of the DTFT $G(e^{j\omega})$ of the discrete-time signal $g[n]$. The DTFT $\Gamma(e^{j\omega})$ of the windowed finite-length segment $\gamma[n]$ is next evaluated at a set of $R(R \geq N)$ discrete angular frequencies equally spaced in the range $0 \leq \omega \leq 2\pi$ by computing its R-point discrete Fourier transform (DFT) $\Gamma[k]$. To provide sufficient resolution, the DFT length R is chosen to be greater than the window N by zero-padding the windowed sequence with $R - N$ zero-valued samples. The DFT is usually computed using an FFT algorithm.

We examine the above approach in more detail to understand its limitations so that we can properly make use of the results obtained. In particular, we analyze here the effects of windowing and the evaluation of the frequency samples of the DTFT via the DFT.

Before we can interpret the spectral content of $\Gamma(e^{j\omega})$, i.e., $G(e^{j\omega})$, from $\Gamma[k]$, we need to reexamine the relations between these transforms and their corresponding frequencies. Now, the relation between the R-point DFT $\Gamma[k]$ of $\gamma[n]$ and its DTFT $\Gamma(e^{j\omega})$ is given by

$$\Gamma[k] = \Gamma(e^{j\omega})\Big|_{\omega=2\pi k/R}, \qquad 0 \leq k \leq R-1. \tag{11.3}$$

The normalized discrete-time angular frequency ω_k corresponding to the DFT bin number k (DFT frequency) is given by

$$\omega_k = \frac{2\pi k}{R}. \tag{11.4}$$

Likewise, the continuous-time angular frequency Ω_k corresponding to the DFT bin number k (DFT frequency) is given by

$$\Omega_k = \frac{2\pi k}{RT}. \tag{11.5}$$

To interpret the results of the DFT-based spectral analysis correctly, we first consider the frequency-domain analysis of a sinusoidal sequence. Now an infinite-length sinusoidal sequence $g[n]$ of normalized angular frequency ω_o is given by

$$g[n] = \cos(\omega_o n + \phi). \tag{11.6}$$

By expressing the above sequence as

$$g[n] = \tfrac{1}{2}\left(e^{j(\omega_o n + \phi)} + e^{-j(\omega_o n + \phi)}\right), \tag{11.7}$$

and making use of Table 3.1, we arrive at the expression for its DTFT as

$$G(e^{j\omega}) = \pi \sum_{\ell=-\infty}^{\infty} \left(e^{j\phi}\delta(\omega - \omega_o + 2\pi\ell) + e^{-j\phi}\delta(\omega + \omega_o + 2\pi\ell)\right). \tag{11.8}$$

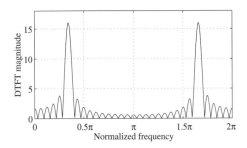

Figure 11.3: DTFT of a sinusoidal sequence windowed by a rectangular window.

Thus, the DTFT is a periodic function of ω with a period 2π containing two impulses in each period. In the frequency range, $-\pi \le \omega \le \pi$, there is an impulse at $\omega = \omega_o$ of complex amplitude $\pi e^{j\phi}$ and an impulse at $\omega = -\omega_o$ of complex amplitude $\pi e^{-j\phi}$.

To analyze $g[n]$ in the spectral domain using the DFT, we employ a finite-length version of the sequence given by

$$\gamma[n] = \cos(\omega_o n + \phi), \qquad 0 \le n \le N - 1. \tag{11.9}$$

The computation of the DFT of a finite-length sinusoid has been considered in Example 8.11. In this example, using MATLAB Program 8_10, we computed the DFT of a length-32 sinusoid of frequency 10 Hz sampled at 64 Hz, as shown in Figure 8.30. As can be seen from this figure, there are only two nonzero DFT samples, one at bin $k = 5$ and the other at bin $k = 27$. From Eq. (11.5), bin $k = 5$ corresponds to frequency 10 Hz, while bin $k = 27$ corresponds to frequency 54 Hz, or equivalently, -10 Hz. Thus, the DFT has correctly identified the frequency of the sinusoid.

Next, using the same program, we computed the 32-point DFT of a length-32 sinusoid of frequency 11 Hz sampled at 64 Hz, as shown in Figure 8.31. This figure shows two strong peaks at bin locations $k = 5$ and $k = 6$ with nonzero DFT samples at other bin locations in the positive half of the frequency range. Note that the bin locations 5 and 6 correspond to frequencies 10 Hz and 12 Hz, respectively, according to Eq. (11.5). Thus the frequency of the sinusoid being analyzed is exactly halfway between these two bin locations.

The phenomenon of the spread of energy from a single frequency to many DFT frequency locations as demonstrated by this figure is called *leakage*. To understand the cause of this effect, we recall that the DFT $\Gamma[k]$ of a length-N sequence $\gamma[n]$ is given by the samples of its discrete-time Fourier transform (DTFT) $\Gamma(e^{j\omega})$ evaluated at $\omega = 2\pi k/N$, $k = 0, 1, \ldots, N - 1$. Figure 11.3 shows the DTFT of the length-32 sinusoidal sequence of frequency 11 Hz sampled at 64 Hz. It can be seen that the DFT samples shown in Figure 8.31 are indeed obtained by the frequency samples of the plot of Figure 11.3.

To understand the shape of the DTFT shown in Figure 11.3 we observe that the sequence of Eq. (11.9) is a windowed version of the infinite-length sequence $g[n]$ of Eq. (11.6) obtained using a rectangular window $w[n]$:

$$w[n] = \begin{cases} 1, & 0 \le n \le N - 1, \\ 0, & \text{otherwise.} \end{cases} \tag{11.10}$$

Hence, the DTFT $\Gamma(e^{j\omega})$ of $\gamma[n]$ is given by the frequency-domain convolution of the DTFT $G(e^{j\omega})$ of $g[n]$ with the DTFT $\Psi_R(e^{j\omega})$ of the rectangular window $w[n]$:

$$\Gamma(e^{j\omega}) = \frac{1}{2\pi} \int_{-\pi}^{\pi} G(e^{j\varphi}) \Psi_R(e^{j(\omega - \varphi)}) \, d\varphi, \tag{11.11}$$

where

$$\Psi_R(e^{j\omega}) = e^{-j\omega(N-1)/2} \frac{\sin(\omega N/2)}{\sin(\omega/2)}. \tag{11.12}$$

Substituting $G(e^{j\omega})$ from Eq. (11.8) into Eq. (11.11), we arrive at

$$\Gamma(e^{j\omega}) = \tfrac{1}{2}e^{j\phi}\Psi_R(e^{j(\omega-\omega_o)}) + \frac{1}{2}e^{-j\phi}\Psi_R(e^{j(\omega+\omega_o)}). \tag{11.13}$$

As indicated by the above equation, the DTFT $\Gamma(e^{j\omega})$ of the windowed sequence $\gamma[n]$ is a sum of the frequency shifted and amplitude scaled DTFT $\Psi_R(e^{j\omega})$ of the window $w[n]$ with the amount of frequency shifts given by $\pm\omega_o$. Now, for the length-32 sinusoid of frequency 11 Hz sampled at 64 Hz, the normalized frequency of the sinusoid is $11/64 = 0.172$. Hence, its DTFT is obtained by frequency shifting the DTFT $\Psi_R(e^{j\omega})$ of a length-32 rectangular window to the right and to the left by the amount $0.172 \times 2\pi = 0.344\pi$, adding both shifted versions, and then amplitude scaling by a factor 1/2. In the normalized angular frequency range 0 to 2π, which is one period of the DTFT, there are two peaks, one at 0.344π and the other at $2\pi(1-0.172) = 1.656\pi$, as verified by Figure 11.3. A 32-point DFT of this DTFT is precisely the DFT shown in Figure 8.31. The two peaks of the DFT at bin locations $k = 5$ and $k = 6$ are frequency samples of the main lobe located on both sides of the peak at the normalized frequency 0.172. Likewise, the two peaks of the DFT at bin locations $k = 26$ and $k = 27$ are frequency samples of the main lobe located on both sides of the peak at the normalized frequency 0.828. All other DFT samples are given by the samples of the sidelobes of the DTFT of the window causing the leakage of the frequency components at $\pm\omega_o$ to other bin locations with the amount of leakage determined by the relative amplitude of the main lobe and the sidelobes. Since the relative sidelobe level $A_{s\ell}$, defined by the ratio in dB of the amplitude of the main lobe to that of the largest sidelobe, of the rectangular window is very high, there is a considerable amount of leakage to the bin locations adjacent to the bins showing the peaks in Figure 8.31.

The above problem gets more complicated if the signal being analyzed has more than one sinusoid, as is typically the case. We illustrate the DFT-based spectral analysis approach by means of several examples. Through these examples we examine the effects of the length R of the DFT, the type of window being used, and its length N on the results of spectral analysis.

EXAMPLE 11.1 In this example we examine the effect of the DFT length in spectral analysis. The signal to be analyzed in the spectral domain is given by

$$x[n] = \tfrac{1}{2}\sin(2\pi f_1 n) + \sin(2\pi f_2 n), \qquad 0 \le n \le N-1. \tag{11.14}$$

Let the normalized frequencies of the two length-16 sinusoidal sequences be $f_1 = 0.22$ and $f_2 = 0.34$. We compute the DFT of their sum $x[n]$ for various values of the DFT length R. To this end, we use Program 11_2 given below whose input data are the length N of the signal, length R of the DFT, and the two frequencies f_1 and f_2. The program generates the two sinusoidal sequences, forms their sum, then computes the DFT of the sum and plots the DFT samples. In this example, we fix $N = 16$ and vary the DFT length R from 16 to 128. Note that when R > N, the M-file fft(x,R) automatically zero-pads the sequence x with R-N zero-valued samples.

```
% Program 11_2
% Spectral Analysis of a Sum of Two Sinusoids
% Using the DFT
%
clf;
N = input('Signal length = ');
R = input('DFT length = ');
fr = input('Type in the sinusoid frequencies = ');
```

```
    n = 0:N-1;
    x = 0.5*sin(2*pi*n*fr(1)) + sin(2*pi*n*fr(2));
    Fx = fft(x,R);
    k = 0:R-1;
    stem(k,abs(Fx));grid
    xlabel('k'); ylabel('Magnitude');
    title(['N = ',num2str(N),', R = ',num2str(R)]);
```

Figure 11.4(a) shows the magnitude $|X[k]|$ of the DFT samples of the signal $x[n]$ of Eq. (11.14) for $R = 16$. From the plot of the magnitude $|X(e^{j\omega})|$ of the DTFT given in Figure 11.4(b), it is evident that the DFT samples given in Figure 11.4(a) are indeed the frequency samples of the frequency response, as expected. As is customary, the horizontal axis in Figure 11.4(a) has been labeled in terms of the DFT frequency sample (bin) number k, where k is related to the normalized angular frequency ω through Eq. (11.4). Thus, $\omega = 2\pi \times 8/16 = \pi$ corresponds to $k = 8$ and $\omega = 2\pi \times 15/16 = 1.875\pi$ corresponds to $k = 15$.

From the plot of Figure 11.4(a) it is difficult to determine whether there is one or more sinusoids in the signal being examined, and the exact locations of the sinusoids. To increase the accuracy of the locations of the sinusoids, we increase the size of the DFT to 32 and recompute the DFT as indicated in Figure 11.4(c). In this plot there appears to be some concentrations around $k = 7$ and around $k = 11$ in the normalized frequency range from 0 to 0.5. Figure 11.4(d) shows the DFT plot obtained for $R = 64$. In this plot there are two clear peaks occurring at $k = 13$ and $k = 22$ that correspond to the normalized frequencies of 0.2031 and 0.3438, respectively. To improve further the accuracy of the peak location, we compute next a 128-point DFT as indicated in Figure 11.4(e) in which the peak occurs around $k = 27$ and $k = 45$ corresponding to the normalized frequencies of 0.2109 and 0.3516, respectively. However, this plot also shows a number of minor peaks, and it is not clear by examining this DFT plot whether additional sinusoids of lesser strengths are present in the original signal or not.

As this example points out, in general, an increase in the DFT length improves the sampling accuracy of the DTFT by reducing the spectral separation of adjacent DFT samples.

EXAMPLE 11.2   In this example, we compute the DFT of a sum of two finite-length sinusoidal sequences as given by Eq. (11.14) with one of the sinusoids at a fixed frequency, while the frequency of the other sinusoid is varied. Specifically, we keep $f_2 = 0.34$ and vary f_1 from 0.28 to 0.31. The length of the signal being analyzed is 16, while the DFT length is 128.

Figure 11.5 shows the plots of the DFTs computed, along with the frequencies of the sinusoids. As can be seen from these plots, the two sinusoids are clearly resolved in Figure 11.5(a) and (b), while they cannot be resolved in Figure 11.5(c) and (d). The reduced resolution occurs when the difference between the two frequencies becomes less than 0.04.

As indicated by Eq. (11.11), the DTFT of a length-N sinusoid of normalized angular frequency ω_1 is obtained by frequency translating the DTFT $\Psi_R(e^{j\omega})$ of a length-N rectangular window to the frequencies $\pm\omega_1$ and scaling their amplitudes appropriately. In the case of a sum of two length-N sinusoids of normalized angular frequencies ω_1 and ω_2, the DTFT is obtained by summing the DTFTs of the individual sinusoids. As the difference between the two frequencies becomes smaller, the main lobes of the DTFTs of the individual sinusoids get closer and eventually overlap. If there is a significant overlap, it will be difficult to resolve the peaks. It follows therefore that the frequency resolution is essentially determined by the width Δ_{ML} of the main lobe of the DTFT of the window.

Now from Table 7.2, the main lobe width Δ_{ML} of a length-N rectangular window is given by $4\pi/N$. In terms of normalized frequency, the main lobe width of a length-16 rectangular window is 0.125. Hence, two closely spaced sinusoids windowed with a rectangular window of length 16 can be clearly resolved if the difference in their frequencies is about half of the main lobe width, i.e., 0.0625.

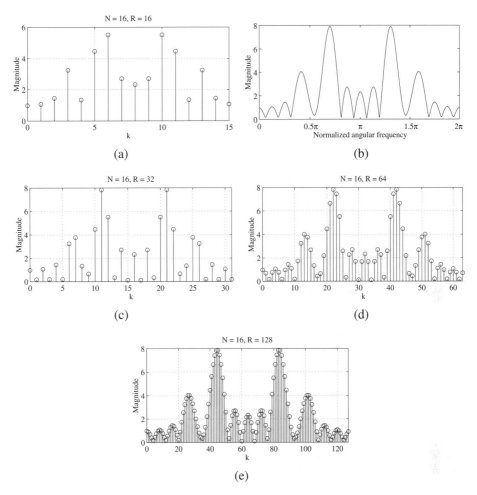

Figure 11.4: DFT-based spectral analysis of a sum of two finite-length sinusoidal sequences of normalized frequencies 0.22 and 0.34, respectively, of length 16 each for various values of DFT lengths.

Even though the rectangular window has the smallest main lobe width, it has the largest relative sidelobe amplitude and, as a consequence, causes considerable leakage. As seen from Examples 11.1 and 11.2, the large amount of leakage results in minor peaks that may be falsely identified as sinusoids. We now study the effect of windowing the signal with a Hamming window.[1]

EXAMPLE 11.3 We compute the DFT of a sum of two sinusoids windowed by a Hamming window. The signal being analyzed is given by $x[n] \cdot w[n]$, where $x[n]$ is given by

$$x[n] = 0.85 \sin(2\pi f_1 n) + \sin(2\pi f_2 n),$$

and $w[n]$ is a Hamming window of length N. The two normalized frequencies are $f_1 = 0.22$ and $f_2 = 0.26$.

---

[1]For a review of some commonly used windows, see Sections 7.6.4 and 7.6.5.

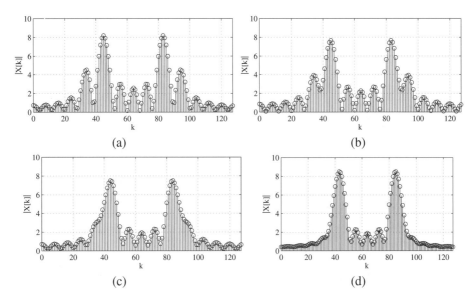

Figure 11.5: Illustration of the frequency resolution property. (a) $f_1 = 0.28$, $f_2 = 0.34$; (b) $f_1 = 0.29$, $f_2 = 0.34$; (c) $f_1 = 0.3$, $f_2 = 0.34$; and (d) $f_1 = 0.31$, $f_2 = 0.34$.

Figure 11.6(a) shows the 16-point DFT of the windowed signal with a window length of 16. As can be seen from this plot, the leakage has been reduced considerably, but it is difficult to resolve the two sinusoids. We next increase the DFT length to 64 while keeping the window length fixed at 16. The resulting plot is shown in Figure 11.6(b), indicating a substantial reduction in the leakage but with no change in the resolution. From Table 7.2, the main lobe width Δ_{ML} of a length-N Hamming window is $8\pi/N$. Thus, for $N = 16$, the normalized main lobe width is 0.25. Hence, with such a window, we can resolve two frequencies if their difference is of the order of half the main lobe width, i.e., 0.125 or better. In our example, the difference is 0.04 which is considerably smaller than this value.

In order to increase the resolution, we increase the window length to 32, which reduces the main lobe width by half. Figure 11.6(c) shows its 32-point DFT. There now appears to be two peaks. Increasing the DFT size to 64 clearly separates the two peaks, as indicated in Figure 11.6(d). This separation becomes more visible with an increase in the DFT size to 256, as shown in Figure 11.6(e). Finally, Figure 11.6(f) shows the result obtained with a window length of 64 and a DFT length of 256.

It is clear from the above examples that performance of the DFT-based spectral analysis depends on several factors, the type of window being used and its length, and the size of the DFT. To improve the frequency resolution, one must use a window with a very small main lobe width, and to reduce the leakage, the window must have a very small relative sidelobe level. The main lobe width can be reduced by increasing the length of the window. Furthermore, an increase in the accuracy of locating the peaks is achieved by increasing the size of the DFT. To this end, it is preferable to use a DFT length that is a power of 2 so that very efficient FFT algorithms can be employed to compute the DFT. Of course, an increase in the DFT size also increases the computational complexity of the spectral analysis procedure.

11.3 Spectral Analysis of Nonstationary Signals

The discrete Fourier transform can be employed for the spectral analysis of a finite-length signal composed of sinusoidal components as long as the frequency, amplitude, and phase of each sinusoidal component

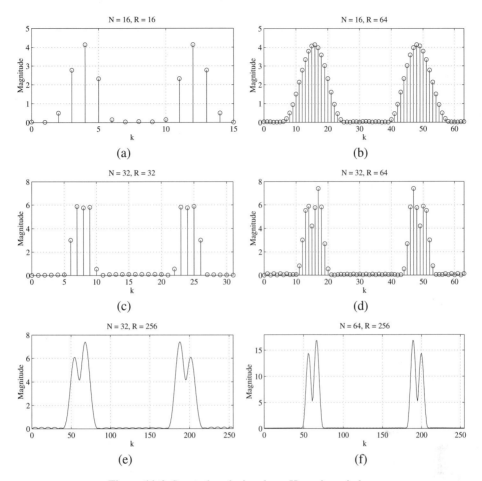

Figure 11.6: Spectral analysis using a Hamming window.

are time-invariant and independent of the signal length. There are practical situations where the signal to be analyzed is instead nonstationary, for which these signal parameters are time-varying. An example of such a time-varying signal is the chirp signal given by

$$x[n] = A \cos(\omega_o n^2) \tag{11.15}$$

and shown in Figure 11.7 for $\omega_o = 10\pi \times 10^{-5}$. Note from Eq. (11.15) that the instantaneous frequency of $x[n]$ is given by $2\omega_o n$, which is not a constant but increases linearly with time. Speech, radar, and sonar signals are other examples of such nonstationary signals. A description of such signals in the frequency-domain using a simple DFT of the complete signal will provide misleading results. To get around the time-varying nature of the signal parameters, an alternative approach would be to segment the sequence into a set of subsequences of short length, with each subsequence centered at uniform intervals of time and its DFT computed separately. If the subsequence length is reasonably small, it can be safely assumed to be stationary for practical purposes. As a result, the frequency-domain description of a long sequence is given by a set of short-length DFTs, i.e., a time-dependent DFT.

To represent a nonstationary signal $x[n]$ in terms of a set of short-length subsequences, we can multiply it with a window $w[n]$ that is stationary with respect to time and move the signal through the window. For

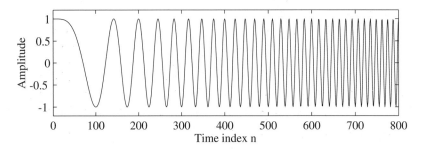

Figure 11.7: First 800 samples of a casual chirp signal $\cos(\omega_o n^2)$ with $\omega_o = 10\pi \times 10^{-5}$.

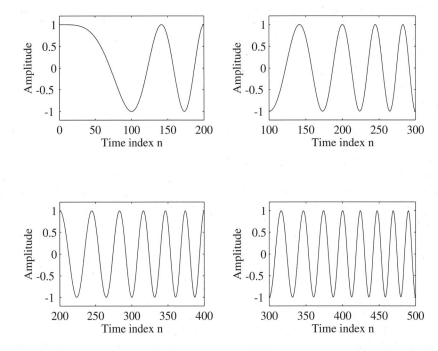

Figure 11.8: Examples of subsequences of the chirp signal of Figure 11.7 generated by a length-200 rectangular window.

example, Figure 11.8 shows four segments of the chirp signal of Figure 11.7 as seen through a stationary rectangular window of length 200. As illustrated in this figure, the segments could be overlapping in time. A discrete-time Fourier transform of the short sequence obtained by windowing is called the short-term Fourier transform which is thus a function of the location of the window relative to the original long sequence and the frequency. In this section we review the basic concepts associated with this type of transform, study some of its properties, and point out one of its important applications. A detailed exposition of this subject can be found in [All77], [Naw88], [Opp89], and [Rab78].

11.3.1 Short-Time Fourier Transform

The *short-time Fourier transform* (STFT), also known as the *time-dependent Fourier transform*, of a sequence $x[n]$ is defined by

$$X_{\text{STFT}}(e^{j\omega}, n) = \sum_{m=-\infty}^{\infty} x[n-m]w[m]e^{-j\omega m}, \qquad (11.16)$$

where $w[n]$ is a suitably chosen window sequence. It should be noted that the function of the window is to extract a finite-length portion of the sequence $x[n]$ such that the spectral characteristics of the section extracted are approximately stationary over the duration of the window for practical purposes.

Note that if $w[n] = 1$, the definition of STFT given in Eq. (11.16) reduces to the conventional discrete-time Fourier transform (DTFT) of $x[n]$. However, even though the DTFT of $x[n]$ exists under certain well-defined conditions, the windowed sequence in Eq. (11.16) being finite in length ensures the existence of the STFT for *any* sequence $x[n]$. It should be noted also that, unlike the conventional DTFT, the STFT is a function of two variables: the integer variable time index n and the continuous frequency variable ω. It also follows from the definition of Eq. (11.16), that $X_{\text{STFT}}(e^{j\omega}, n)$ is a periodic function of ω with a period 2π.

In most applications, the magnitude of the STFT is of interest. The display of the magnitude of the STFT is usually referred to as the *spectrogram*. However, since the STFT is a function of two variables, the display of its magnitude would normally require three dimensions. Often, it is plotted in two dimensions, with the magnitude represented by the darkness of the plot. Here, the white areas represent zero-valued magnitudes while the gray areas represent nonzero magnitudes, with the largest magnitudes being shown in black. In the STFT magnitude display, the vertical axis represents the frequency variable (ω) and the horizontal axis represents the time index (n). The STFT can also be visualized as a mesh plot in a three-dimensional coordinate frame in which the STFT magnitude is a point in the z-direction above the x-y plane.

Figure 11.9 shows the STFT of the chirp sequence of Eq. (11.15), with $\omega_o = 10\pi \times 10^{-5}$ for a length of 20,000 samples computed using a Hamming window of length 200 in the above two forms. In Figure 11.9, the STFT for a given value of the time index n is essentially the DFT of a segment of a sinusoidal sequence. Recall from our discussion in Section 11.2 that the shape of the DFT of such a segment is similar to that shown in Figure 11.3, with large nonzero-valued DFT samples around the frequency of the sinusoid and smaller nonzero-valued DFT samples at other frequency points. In the spectrogram plot, the large-valued DFT samples show up as narrow nearly black very short vertical lines while the other DFT samples show up as gray points. As the instantaneous frequency of the chirp signal increases linearly, the short black line moves up in the vertical direction and eventually, because of aliasing, the black line starts moving down in the vertical direction. As a result, the spectrogram of the chirp signal essentially appears as a thick line in the form of a triangular shape.

11.3.2 Sampling in the Time and Frequency Dimensions

In practice, the STFT is computed at a finite set of discrete values of ω. Moreover, due to the finite length of the windowed sequence, the STFT is accurately represented by its frequency samples as long as the number of frequency samples is greater than the window length. Moreover, the portion of the sequence $x[n]$ inside the window can be fully recovered from the frequency samples of the STFT.

To be more precise, let the window be of length R defined in the range $0 \le m \le R - 1$. We sample $X_{\text{STFT}}(e^{j\omega}, n)$ at N equally spaced frequencies $\omega_k = 2\pi k/N$, with $N \ge R$ as indicated below:

$$X_{\text{STFT}}[k, n] = X_{\text{STFT}}(e^{j\omega}, n)\Big|_{\omega=2\pi k/N} = X_{\text{STFT}}(e^{j2\pi k/N}, n)$$

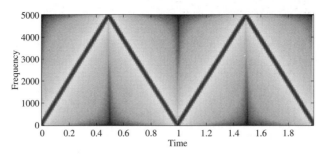

Figure 11.9: Spectrogram of a chirp signal.

$$= \sum_{m=0}^{R-1} x[n-m]w[m]e^{-j2\pi km/N}, \quad 0 \le k \le N-1. \tag{11.17}$$

It follows from Eq. (11.17), assuming $w[m] \ne 0$, that $X_{\text{STFT}}[k, n]$ is simply the DFT of $x[n-m]w[m]$. Note that $X_{\text{STFT}}[k, n]$ is a two-dimensional sequence and is periodic in k with a period N. Applying the IDFT, we thus arrive at

$$x[n-m]w[m] = \frac{1}{N} \sum_{k=0}^{N-1} X_{\text{STFT}}[k, n]e^{j2\pi km/N}, \quad 0 \le m \le R-1, \tag{11.18}$$

or in other words,

$$x[n-m] = \frac{1}{Nw[m]} \sum_{k=0}^{N-1} X_{\text{STFT}}[k, n]e^{j2\pi km/N}, \quad 0 \le m \le R-1, \tag{11.19}$$

verifying that the sequence values inside the window can be fully recovered from $X_{\text{STFT}}[k, n]$ as long as the number N of frequency samples is greater than or equal to the window length R. It should be evident by now that $x[n]$ for $-\infty < n < \infty$ can be fully recovered if $X_{\text{STFT}}(e^{j\omega}, n)$ or $X_{\text{STFT}}[k, n]$ is sampled also in the time dimension. More precisely, if we set $n = n_o$ in Eq. (11.19), we recover the signal in the interval $n_o \le n \le n_o + R - 1$ from $X_{\text{STFT}}[k, n_o]$. Likewise, by setting $n = n_o + R$ in Eq. (11.19), we recover the signal in the interval $n_o + R \le n \le n_o + 2R - 1$ from $X_{\text{STFT}}[k, n_o + R]$, and so on.

The sampled STFT for a window defined in the region $0 \le m \le R - 1$ is given by

$$X_{\text{STFT}}[k, \ell L] = X_{\text{STFT}}(e^{j2\pi k/N}, \ell L) = \sum_{m=0}^{R-1} x[\ell L - m]w[m]e^{-j2\pi km/N}, \tag{11.20}$$

where ℓ and k are integers such that $-\infty < \ell < \infty$ and $0 \le k \le N - 1$. Figure 11.10 shows lines in the (ω, n)-plane corresponding to $X_{\text{STFT}}(e^{j\omega}, n)$ and the grid of sampling points in the (ω, n)-plane for the case $N = 9$ and $L = 4$. As we have shown, it is possible to uniquely reconstruct the original signal from such a 2-D discrete representation provided $N \ge R \ge L$.

11.3.3 Window Selection

As in the case of the DFT-based spectral analysis of deterministic signals discussed in Section 11.2, in the STFT analysis of nonstationary signals, the window also plays an important role. Both the length and shape of the window are critical issues that need to be examined carefully.

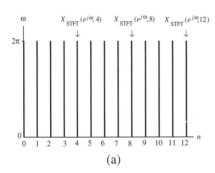 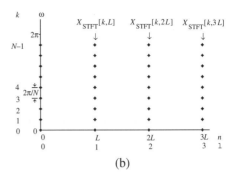

(a) (b)

Figure 11.10: Sampling grid in the (ω, n)-plane for the sampled STFT $X_{\mathrm{STFT}}[k, \ell L]$ for $N = 9$ and $L = 4$.

The function of the window $w[n]$ is to extract a portion of the signal for analysis and ensure that the extracted section of $x[n]$ is approximately stationary. To this end, the window length R should be small, in particular for signals with widely varying spectral parameters. A decrease in the window length increases the time-resolution property of the STFT, whereas the frequency-resolution property of the STFT increases with an increase in the window length. A shorter window thus provides a *wideband spectrogram* while a longer window results in a *narrowband spectrogram*.

The two frequency-domain parameters characterizing the DTFT of a window are its main lobe width Δ_{ML} and the relative sidelobe amplitude $A_{s\ell}$. The former parameter determines the ability of the window to resolve two signal components in the vicinity of each other, while the latter controls the degree of leakage of one component into a nearby signal component. It thus follows that in order to obtain a reasonably good estimate of the frequency spectrum of a time-varying signal, the window should be chosen to have a very small relative sidelobe amplitude with a length chosen based on the acceptable accuracy of the frequency and time resolutions.

11.3.4 STFT Computation Using MATLAB

The *Signal Processing Toolbox* of MATLAB includes the function `specgram` for the computation of the STFT of a signal. There are a number of versions of this function, as given below:

```
      B = specgram(x),
      B = specgram(x,nfft),
  [B,f] = specgram(x,nfft,Fs),
[B,f,t] = specgram(x,nfft,Fs),
      B = specgram(x,nfft,Fs,window),
      B = specgram(x,nfft,Fs,window,noverlap)
```

`B = specgram(x)` computes the STFT of the signal x specified as a vector using a DFT of length `nfft = min(256, length(x))` and a Hann window of length `nfft`. The sampling frequency used is Fs = 2, and the number of samples by which the consecutive segments overlap is given by `noverlap = length(window/2)`. For a real signal x, the DFT is computed at positive frequencies only, whereas, for a complex x, the DFT is computed at both positive and negative frequencies. The column indices of B refer to the time position of the window with time increasing across the columns, and the row indices of B correspond to the frequency with the first row corresponding to $\omega = 0$, as indicated in Figure 11.10(b).

The parameters `nfft`, `Fs`, and `noverlap` can be specified along with the type of window, depending on the version of `specgram` being used. If a scalar integer is used for `window`, the function uses a Hann window of length given by this integer. The specified window length must be less than or equal to `nfft`.

The vector of frequencies `f` at which the DFT is computed is returned when `[B,f] = specgram (x,nfft,Fs)` is employed, whereas both frequency and time vectors, `f` and `t`, respectively, are returned in the version `[B,f,t] = specgram(x,nfft,Fs)`. In the latter case, `t` is a column vector of scaled times and of length equal to the number of columns of `B`. The first element of `t` is always 0. `specgram` with no output arguments generates and plots the scaled logarithm of the spectrogram.

We illustrate the application of `specgram` in the following section.

11.3.5   Analysis of Speech Signals Using STFT

The short-term Fourier transform is often used in the analysis of speech, since speech signals are generally nonstationary. As indicated in Section 1.3, the speech signal, generated by the excitation of the vocal tract, is composed of two types of basic waveforms: voiced and unvoiced sounds. A typical speech signal is shown in Figure 1.17. As can be seen from this figure, a speech segment over a small time interval can be considered as a stationary signal, and as a result, the DFT of the speech segment can provide a reasonable representation of the frequency-domain characteristic of the speech in this time interval. However, in the STFT analysis, the size of the window is critical since a shorter window developing a wideband spectrogram provides a better time resolution, whereas a longer window developing a narrowband spectrogram results in an improved frequency resolution. In order to provide a reasonably good estimate of the changes in the vocal tract and the excitation, a wideband spectrogram is preferable. To this end, the window size is selected to be approximately close to one pitch period, which is adequate for resolving the formants though not adequate to resolve the harmonics of the pitch frequencies. On the other hand, to resolve the harmonics of the pitch frequencies, a narrowband spectrogram with a window size of several pitch periods is desirable.

The following example illustrates the STFT analysis of a speech signal.

EXAMPLE 11.4   The `mtlb.mat` file in the *Signal Processing Toolbox* of MATLAB contains a speech signal of duration 4001 samples sampled at 7418 Hz. We compute its STFT using a Hamming window of length 256 with an overlap of 50 samples between consecutive windowed signals using Program 11_3 given below.

```
% Program 11_3
% Spectrogram of a Speech Signal
%
load mtlb
n = 1:4001;
plot(n-1,mtlb);
xlabel('Time index n');ylabel('Amplitude');
pause
nfft = input('Type in the window length = ');
ovlap = input('Type in the desired overlap = ');
specgram(mtlb,nfft,7418,hamming(nfft),ovlap)
```

The program first loads the `mtlb.mat` file and plots the signal, as indicated in Figure 11.11(a). It then requests the desired sizes of the window and the overlap between consecutive segments, computes the STFT using the function `specgram`, and displays a plot of the spectrogram. Figure 11.11(b) and (c) show, respectively, a narrowband spectrogram and a wideband spectrogram of the speech signal of Figure 11.11(a). The frequency and time resolution tradeoff between the two spectrograms of Figure 11.11 should be evident.

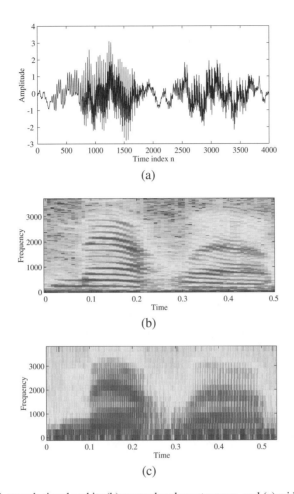

Figure 11.11: (a) A speech signal and its (b) narrowband spectrogram, and (c) wideband spectrogram.

11.4 Spectral Analysis of Random Signals

As discussed in Section 11.2, in the case of a deterministic signal composed of sinusoidal components, a Fourier analysis of the signal can be carried out by taking the discrete Fourier transform (DFT) of a finite-length segment of the signal obtained by appropriate windowing provided the parameters characterizing the components are time-invariant and independent of the window length. On the other hand, the Fourier analysis of nonstationary signals with time-varying parameters is best carried out using the short-time Fourier transform (STFT) described in Section 11.3.

Neither the DFT nor the STFT is applicable for the spectral analysis of naturally occurring random signals as here the spectral parameters are also random. These type of signals are usually classified as noise-like random signals like the unvoiced speech signal generated when letter such as "f" or "s" is spoken, and signal-plus-noise random signals such as seismic signals and nuclear magnetic resonance signals [Rob82]. Spectral analysis of a noise-like random signal is usually carried out by estimating the power density spectrum using Fourier-analysis-based nonparametric methods, whereas a signal-plus-noise random signal is best analyzed using parametric-model-based methods in which the autocovariance sequence is first estimated from the model and then the Fourier transform of the estimate is evaluated. In this section we consider both of these approaches.

11.4.1 Nonparametric Spectral Analysis

Consider a wide-sense stationary (WSS) random signal $g[n]$ with zero mean. According to the Wiener-Khintchine theorem of Eq. (3.145), the power spectrum of $g[n]$ is given by

$$\mathcal{P}_{gg}(\omega) = \sum_{\ell=-\infty}^{\infty} \phi_{gg}[\ell]e^{-j\omega\ell}, \tag{11.21}$$

where $\phi_{gg}[\ell]$ is its autocorrelation sequence, which from Eq. (2.160) is given by

$$\phi_{gg}[\ell] = E(g[n+\ell]g^*[n]). \tag{11.22}$$

In Eq. (11.22), $E(\cdot)$ denotes the expection operator as defined in Eq. (2.122).

Periodogram Analysis

Assume that the infinite-length random discrete-time signal $g[n]$ is windowed by a length-N window sequence $w[n], 0 \leq n \leq N-1$, resulting in the length-N sequence $\gamma[n] = g[n] \cdot w[n]$. The DTFT $\Gamma(e^{j\omega})$ of $\gamma[n]$ is given by

$$\Gamma(e^{j\omega}) = \sum_{n=0}^{N-1} \gamma[n]e^{-j\omega n} = \sum_{n=0}^{N-1} g[n] \cdot w[n]e^{-j\omega n}. \tag{11.23}$$

The estimate $\hat{\mathcal{P}}_{gg}(\omega)$ of the power spectrum $\mathcal{P}_{gg}(\omega)$ is then obtained using

$$\hat{\mathcal{P}}_{gg}(\omega) = \frac{1}{CN}|\Gamma(e^{j\omega})|^2, \tag{11.24}$$

where the constant C is a normalization factor given by

$$C = \frac{1}{N}\sum_{n=0}^{N-1} |w[n]|^2, \tag{11.25}$$

and included in Eq. (11.24) to eliminate any bias in the estimate occurring due to the use of the window $w[n]$. The quantity $\hat{\mathcal{P}}_{gg}(e^{j\omega})$ defined in Eq. (11.24) is called the *periodogram* when $w[n]$ is a rectangular window and called a *modified periodogram* for other types of windows.

In practice, the periodogram $\hat{\mathcal{P}}_{gg}(\omega)$ is evaluated at a discrete set of equally spaced R frequencies, $\omega_k = 2\pi k/R, 0 \leq k \leq R-1$, by replacing the DTFT $\Gamma(e^{j\omega})$ with an R-point DFT $\Gamma[k]$ of the length-N sequence $\gamma[n]$:

$$\hat{\mathcal{P}}_{gg}[k] = \frac{1}{CN}|\Gamma[k]|^2. \tag{11.26}$$

As in the case of the Fourier analysis of sinusoidal signals discussed earlier, R is usually chosen to be greater than N to provide a finer grid of the samples of the periodogram.

It can be shown that the mean value of the periodogram $\hat{\mathcal{P}}_{gg}(\omega)$ is given by

$$E\left(\hat{\mathcal{P}}_{gg}(\omega)\right) = \frac{1}{2\pi CN}\int_{-\pi}^{\pi} \mathcal{P}_{gg}(v)|\Psi(e^{j(\omega-v)})|^2 \, dv, \tag{11.27}$$

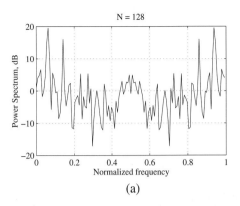
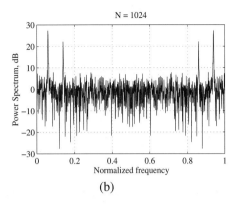

Figure 11.12: Power spectrum estimate of a signal containing two sinusoidal components corrupted with a white noise sequence of zero mean and unit variance Gaussian distribution. (a) Periodogram with a rectangular window of length $N = 128$, and (b) periodogram with a rectangular window of length $N = 1024$.

where $\mathcal{P}_{gg}(\omega)$ is the desired power spectrum and $\Psi(e^{j\omega})$ is the DTFT of the window sequence $w[n]$. The mean value being nonzero for any finite-length window sequence, the power spectrum estimate given by the periodogram is said to be *biased*. By increasing the window length N, the bias can be reduced.

We illustrate the power spectrum computation of a white noise sequence in the following example.

EXAMPLE 11.5   Let the random signal $g[n]$ be composed of two sinusoidal components of angular frequencies 0.06π and 0.14π radians corrupted with a Gaussian distributed random signal of zero mean and unity variance, and windowed by a rectangular window of two different lengths: $N = 128$ and 1024.

The random signal is generated using the M-file `randn`. Figure 11.12(a) and (b) show the plots of the estimated power spectrum for the two cases. Ideally the power spectrum should show four peaks at ω equal to $0.06, 0.14, 0.86$, and 0.94, respectively, and a flat spectral density at all other frequencies. However, Figure 11.12(a) shows four large peaks and several other smaller peaks. Moreover, the spectrum shows large amplitude variations throughout the whole frequency range. As N is increased to a much larger value, the peaks get sharper due to increased resolution of the DFT, while the spectrum shows more rapid amplitude variations.

To understand the cause behind the rapid amplitude variations of the computed power spectrum encountered in the previous example we assume $w[n]$ to be a rectangular window and rewrite the expression for the periodogram given in Eq. (11.24) using Eq. (11.23) as

$$
\begin{aligned}
\hat{\mathcal{P}}_{gg}(\omega) &= \frac{1}{N} \sum_{n=0}^{N-1} \sum_{m=0}^{N-1} g[m]g^*[n]e^{-j\omega(m-n)} \\
&= \sum_{k=-N+1}^{N-1} \left(\frac{1}{N} \sum_{n=0}^{N-1-|k|} g[n+k]g^*[n] \right) e^{-j\omega k} \\
&= \sum_{k=-N+1}^{N-1} \hat{\phi}_{gg}[k]e^{-j\omega k}.
\end{aligned}
\tag{11.28}
$$

Now $\hat{\phi}_{gg}[k]$ is the periodic correlation of $g[n]$ and is an estimate of the true correlation $\phi_{gg}[k]$. Hence, $\hat{\mathcal{P}}_{gg}(\omega)$ is actually the DTFT of $\hat{\phi}_{gg}[k]$. A few samples of $g[n]$ are used in the computation of $\hat{\phi}_{gg}[k]$ when k is near N yielding a poor estimate of the true correlation. This in turn results in rapid amplitude variations in the periodogram estimate. A smoother power spectrum estimate can be obtained by the periodogram averaging method discussed next.

Periodogram Averaging

The power spectrum estimation method, originally proposed by Bartlett [Bar48] and later modified by Welch [Wel67], is based on the computation of the modified periodogram of R overlapping portions of length-N input samples and then averaging these R periodograms. Let the overlap between adjacent segments be K samples. Consider the windowed rth segment of the input data,

$$\gamma^{(r)}[n] = g[n + rK]w[n], \qquad 0 \le n \le N - 1, \qquad 0 \le r \le R - 1, \qquad (11.29)$$

with a DTFT given by $\Gamma^{(r)}(e^{j\omega})$. Its periodogram is given by

$$\hat{\mathcal{P}}_{gg}^{(r)}(\omega) = \frac{1}{CN}|\Gamma^{(r)}(e^{j\omega})|^2. \qquad (11.30)$$

The Welch estimate is then given by the average of all R periodograms $\hat{\mathcal{P}}_{gg}^{(r)}(\omega), 0 \le r \le R - 1$:

$$\hat{\mathcal{P}}_{gg}^{W}(\omega) = \frac{1}{R}\sum_{r=1}^{R-1} \hat{\mathcal{P}}_{gg}^{(r)}(\omega). \qquad (11.31)$$

It can be shown that the variance of the above estimate is reduced approximately by a factor R if the R periodogram estimates are assumed to be independent of each other. For a fixed-length input sequence, R can be increased by decreasing the window length N which in turn decreases the DFT resolution. On the other hand, an increase in the resolution is obtained by increasing N. Thus, there is a trade-off between resolution and the bias.

It should be noted that if the data sequence is segmented by a rectangular window into contiguous segments with no overlap, the periodiogram estimate given by Eq. (11.31) reduces to Barlett estimate [Bar48].

Periodogram Estimate Computation Using MATLAB

The *Signal Processing Toolbox* of MATLAB includes the M-file psd for modified periodogram estimate computation using the Welch and Bartlett methods. Some forms of this function are

```
        Pxx  = psd(x)
        Pxx  = psd(x,nfft)
  [Pxx,f]  = psd(x,nfft,FT)
        Pxx  = psd(x,nfft,FT,window)
        Pxx  = psd(x,nfft,FT,window,noverlap)
        Pxx  = psd(x,nfft,FT,window,noverlap,'dflag')
```

where Pxx is the power spectrum estimate of the real-valued sequence x evaluated at positive frequencies. nfft is the desired FFT length. FT is the sampling frequency whose default value is two. window is the vector of the desired window sequence which needs to be generated prior to power spectrum estimation. The size of window is used by psd in sectioning the data vector x. The default window is the Hann window of length 256. If a scalar number is provided for window, then a Hann window of that length is used. The length of the window must be less than the size of the FFT given by nfft. The parameter noverlap is the number of samples by which the data segments overlap. The string 'dflag' is used

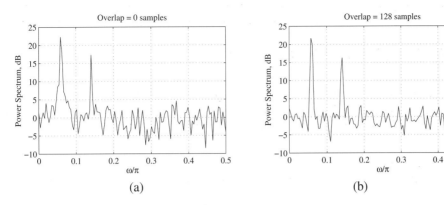

Figure 11.13: Power spectrum estimates: (a) Bartlett's method, and (b) Welch's method.

to indicate the desired detrending option, where it is `'linear'` to remove the best straight-line fit from the segments of the data segment prior to windowing, `'mean'` to remove the mean from the segments of the data segment prior to windowing, and `'none'` if no detrending is desired. Note that if `noverlap` is set to 0 and a rectangular window is used, `psd` evaluates the Bartlett periodogram.

EXAMPLE 11.6   We consider here the evaluation of the Bartlett and Welch estimates of the power spectrum of the random signal considered in the previous example. To this end the following MATLAB program can be used.

```
% Program 11_4
% Power Spectrum Estimation Using Welch's Method
%
n = 0:1000;
g = 2*sin(0.12*pi*n) + sin(0.28*pi*n) + randn(size(n));
nfft = input('Type in the fft size = ');
window = hamming(256);
noverlap =input('Type in the amount of overlap = ');
[Pxx, f] = psd(g,nfft,2,window,noverlap);
plot(f/2,10*log10(Pxx));grid
xlabel('\omega/\pi');ylabel('Power Spectrum, dB');
title(['Overlap = ',num2str(noverlap),' samples']);
```

The above program was run first with no overlap and with a rectangular window generated using the function `boxcar`. The power spectrum computed by the above program is then the Bartlett estimate as indicated in Figure 11.13(a). It was then run with an overlap of 128 samples and a Hamming window. The corresponding power spectrum is then the Welch estimate as shown in Figure 11.13(b). It should be noted from Figure 11.13 that the Welch periodogram estimate is much smoother than the Bartlett periodogram estimate as expected. Compared to the power spectrums of Figure 11.12, there is a decrease in the variance in the smoothed power spectrums of Figure 11.13 but the latter are still biased. Because of the overlap between adjacent data segments, Welch's estimate has a smaller variance than the others. It should be noted that both periodograms of Figure 11.13 show clearly two distinct peaks at 0.06 and 0.14.

11.4.2 Parametric Model–Based Spectral Analysis

In the model-based method, a causal LTI discrete-time system with a transfer function

$$H(z) = \sum_{n=0}^{\infty} h[n]z^{-n}$$

$$= \frac{P(z)}{D(z)} = \frac{\sum_{k=0}^{L} p_k z^{-k}}{1 + \sum_{k=1}^{M} d_k z^{-k}} \tag{11.32}$$

is first developed whose output, when excited by a white noise sequence $e[n]$ with zero mean and variance σ_e^2, matches the specified data sequence $g[n]$ [Kum93]. An advantage of the model-based approach is that it can extrapolate a short-length data sequence to create a longer data sequence for improved power spectrum estimation. On the other hand, in nonparametric methods, spectral leakages limit the frequency resolution if the data length is short.

The model of Eq. (11.32) is called an *autoregressive moving-average* (ARMA) process of order (L, M) if $P(z) \neq 1$, an *all-pole* or *autoregressive* (AR) process of order M if $P(z) = 1$, and an all-zero or *moving-average* (MA) process of order L if $D(z) = 1$. For an ARMA or an AR model, for stability, the denominator $D(z)$ must have all its zeros inside the unit circle. In the time-domain, the input-output relation of the model is given by

$$g[n] = -\sum_{k=1}^{M} d_k g[n-k] + \sum_{k=0}^{L} p_k e[n-k]. \tag{11.33}$$

As indicated in Section 4.13.1, the output $g[n]$ of the model is a WSS random signal. From Eq. (4.214) it follows that the power spectrum $\mathcal{P}_{gg}(\omega)$ of $g[n]$ can be expressed as

$$\mathcal{P}_{gg}(\omega) = \sigma_e^2 |H(e^{j\omega})|^2 = \sigma_e^2 \frac{|P(e^{j\omega})|^2}{|D(e^{j\omega})|^2}, \tag{11.34}$$

where $H(e^{j\omega}) = P(e^{j\omega})/D(e^{j\omega})$ is the frequency response of the model, and

$$P(e^{j\omega}) = \sum_{k=0}^{L} p_k e^{-j\omega k}, \qquad D(e^{j\omega}) = 1 + \sum_{k=1}^{M} d_k e^{-j\omega k}.$$

In the case of an AR or an MA model, the power spectrum is thus given by

$$\mathcal{P}_{gg}(\omega) = \begin{cases} \sigma_e^2 |P(e^{j\omega})|^2, & \text{for an MA model,} \\ \dfrac{\sigma_e^2}{|D(e^{j\omega})|^2}, & \text{for an AR model.} \end{cases} \tag{11.35}$$

The spectral analysis is thus carried out by first determining the model and then computing the power spectrum using either Eq. (11.34) for an ARMA model or using Eq. (11.35) for an MA or an AR model. To determine the model we need to decide the type of the model (i.e., pole-zero IIR structure, all-pole IIR structure, or all-zero FIR structure) to be used, determine an appropriate order of its transfer function $H(z)$ (i.e., both L and M for an ARMA model or M for an AR model or L for an MA model), and then from the specified length-N data $g[n]$ estimate the coefficients of $H(z)$. We restrict our discussion here to the development of the AR model, as it is simpler and often used. Applications of the AR model include spectral analysis, system identification, speech analysis and compression, and filter design. For a discussion on the development of the MA model and the ARMA model, see [Kum93].

Relation between Model Parameters and the Autocorrelation Sequence

The model filter coefficients $\{p_k\}$ and $\{d_k\}$ are related to the autocorrelation sequence $\phi_{gg}[\ell]$ of the random signal $g[n]$. To establish this relation, we obtain from Eq. (11.33),

$$\phi_{gg}[\ell] = -\sum_{k=1}^{M} d_k \phi_{gg}[\ell - k] + \sum_{k=0}^{L} p_k \phi_{eg}[\ell - k], \qquad -\infty < \ell < \infty, \tag{11.36}$$

by multiplying both sides of the equation with $g^*[n - \ell]$ and taking the expected values. In the above expression, the cross-correlation $\phi_{eg}[\ell]$ between $g[n]$ and $e[n]$ can be written as

$$\phi_{eg}[\ell] = E(g^*[n]e[n + \ell])$$
$$= \sum_{k=0}^{\infty} h^*[k] \, E(e^*[n - k]e[n + \ell]) = \sigma_e^2 h^*[-\ell], \tag{11.37}$$

where $h[n]$ is the causal impulse response of the LTI model as defined in Eq. (11.32) and σ_e^2 is the variance of the white noise sequence $e[n]$ applied to the input of the model.

For an AR model, $L = 0$, and hence Eq. (11.36) reduces to

$$\phi_{eg}[\ell] = \begin{cases} -\sum_{k=1}^{M} d_k \phi_{gg}[\ell - k], & \text{for } \ell > 0, \\ -\sum_{k=1}^{M} d_k \phi_{gg}[\ell - k] + \sigma_e^2, & \text{for } \ell = 0, \\ \phi_{gg}^*[-\ell], & \text{for } \ell < 0. \end{cases} \tag{11.38}$$

From Eq. (11.38) we obtain for $1 \leq \ell \leq M$, a set of M equations,

$$\sum_{k=1}^{M} d_k \phi_{gg}[\ell - k] = -\phi_{eg}[\ell], \qquad 1 \leq \ell \leq M,$$

which can be written in matrix form as

$$\begin{bmatrix} \phi_{gg}[0] & \phi_{gg}[-1] & \cdots & \phi_{gg}[-M + 1] \\ \phi_{gg}[1] & \phi_{gg}[0] & \cdots & \phi_{gg}[-M + 2] \\ \vdots & \vdots & \ddots & \vdots \\ \phi_{gg}[M - 1] & \phi_{gg}[M - 2] & \cdots & \phi_{gg}[0] \end{bmatrix} \begin{bmatrix} d_1 \\ d_2 \\ \vdots \\ d_M \end{bmatrix} = - \begin{bmatrix} \phi_{gg}[1] \\ \phi_{gg}[2] \\ \vdots \\ \phi_{gg}[M] \end{bmatrix}. \tag{11.39}$$

For $\ell = 0$ we also get from Eq. (11.38)

$$\phi_{gg}[0] + \sum_{k=1}^{M} d_k \phi_{gg}[-k] = \sigma_e^2.$$

Combining the above with Eq. (11.39) we arrive at

$$\begin{bmatrix} \phi_{gg}[0] & \phi_{gg}[-1] & \cdots & \phi_{gg}[-M] \\ \phi_{gg}[1] & \phi_{gg}[0] & \cdots & \phi_{gg}[-M + 1] \\ \vdots & \vdots & \ddots & \vdots \\ \phi_{gg}[M] & \phi_{gg}[M - 1] & \cdots & \phi_{gg}[0] \end{bmatrix} \begin{bmatrix} 1 \\ d_1 \\ d_2 \\ \vdots \\ d_M \end{bmatrix} = \begin{bmatrix} \sigma_e^2 \\ 0 \\ 0 \\ \vdots \\ 0 \end{bmatrix}. \tag{11.40}$$

The above matrix equation is more commonly known as the *Yule-Walker equation*. It can be seen from Eq. (11.40) that knowing the $M + 1$ autocorrelation samples $\phi_{xx}[\ell]$ for $0 \leq \ell \leq M$, we can determine the model parameters d_k for $1 \leq k \leq M$ by solving the matrix equation. The $(M + 1) \times (M + 1)$ matrix in Eq. (11.40) is a *Toeplitz matrix*.[2] Because of the structure of the Toeplitz matrix, the matrix equation of Eq. (11.40) can be solved using the fast *Levinson-Durbin algorithm* [Lev47], [Dur59]. The causal all-pole LTI system $H(z) = 1/D(z)$ resulting from the application of the Levinson-Durbin recursions is guaranteed to be BIBO stable. Moreover, the recursion automatically leads to a realization in the form of a cascaded FIR lattice structure as shown in Figure 6.39.

Power Spectrum Estimation Using an AR Model

The AR model parameters can be determined using the *Yule-Walker method*, which makes use of the estimates of the autocorrelation sequence samples as their actual values are not known a priori. The autocorrelation at lag ℓ is determined from the specified data samples $g[n]$ for $0 \leq n \leq N - 1$ using

$$\hat{\phi}_{gg}[\ell] = \frac{1}{N} \sum_{n=0}^{N-1-|\ell|} g^*[n]\, g[n + \ell], \qquad 0 \leq \ell \leq N - 1. \tag{11.41}$$

The above estimates are used in Eq. (11.39) in place of the true autocorrelation samples with the AR model parameters d_k replaced with their estimates $\hat{d}_k$. The resulting equation is next solved using the Levinson-Durbin algorithm to determine the estimates of the AR model parameters $\hat{d}_k$. The power spectrum estimate is then evaluated using

$$\hat{P}_{gg}(\omega) = \frac{\hat{\mathcal{E}}_M}{\left| 1 + \sum_{k=1}^{M} \hat{d}_k\, e^{-j\omega k} \right|^2}, \tag{11.42}$$

where $\hat{\mathcal{E}}_M$ is the prediction error for the Mth-order AR model:

$$\hat{\mathcal{E}}_M = \hat{\phi}_{gg}[0] \prod_{i=1}^{M} \left(1 - |\hat{K}_i|^2 \right). \tag{11.43}$$

The Yule-Walker method is related to the linear prediction problem. Here the problem is to predict the N-th sample $g[N]$ from the previous M data samples $g[n]$, $0 \leq n \leq M - 1$, with the assumption that data samples outside this range are zeros. The predicted value $\hat{g}[n]$ of the data sample $g[n]$ can be found by a linear combination of the previous M data samples as

$$\hat{g}[n] = -\sum_{k=1}^{M} \hat{d}_k g[n - k] = g[n] - e[n], \tag{11.44}$$

where $e[n]$ is the prediction error. For the specified data sequence, Eq. (11.44) leads to $N + M$ prediction equations given by

$$g[n] + \sum_{k=1}^{M} g[n - k]\hat{d}_k = e[n], \qquad 0 \leq n \leq N + M - 1. \tag{11.45}$$

The optimum linear predictor coefficients $\hat{d}_k$ are obtained by minimizing the error $\frac{1}{N} \sum_{n=0}^{N+M-1} |e[n]|^2$. It can be shown that the solution of the minimization problem is given by Eq. (11.39). Thus, the best all-pole linear predictor filter is also the AR model resulting from the solution of Eq. (11.39).

---

[2]A Toeplitz matrix has the same element values along each diagonal.

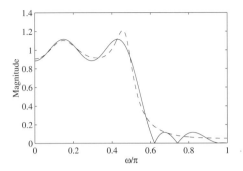

Figure 11.14: Magnitude response of the FIR filter (shown with solid line) and the all-pole IIR model (shown with dashed line).

It should be noted that the AR model is guaranteed stable. But the all-pole filter developed may not model an AR process exactly of the same order due to the windowing of the data sequence to a finite length with samples outside the window range assumed to be zeros.

The function `lpc` in MATLAB finds the AR model using the above method. Its basic form is

```
[d,K] = lpc(g,M)
```

where d and K are the denominator coefficients and the gain of the AR model of order M, respectively, and g is the specified data sequence. We illustrate its application in the following example.

EXAMPLE 11.7 We consider the approximation of an FIR digital filter of order 13 with an all-pole IIR digital filter of order 7. The coefficients of the FIR filter are obtained using the function `remez`, and the all-pole IIR filter is designed using the function `lpc`. The MATLAB program used for the design is given below.

```
% Program 11_5
% Development of an AR Model of an FIR Filter
%
b = remez(13, [0 0.5 0.6 1], [1 1 0 0]);
[h,w] = freqz(b,1,512);
[d,p0] = lpc(b,7);
[h1,w] = freqz(p0,d,512);
plot(w/pi,abs(h),w/pi,abs(h1),'--');
xlabel('\omega/\pi');ylabel('Magnitude');
```

The magnitude response plots generated by running this program are shown in Figure 11.14.

Several comments are in order here. It can be seen from this figure that the AR model has reasonably matched the passband response with peaks occurring very close to the peaks of the magnitude response of the FIR system. However, there are no nulls in the stopband response of the AR model even though the stopband response of the FIR system has nulls. Since the nulls are generated by the zeros of the transfer function, an all-pole model cannot produce nulls.

In order to apply the above method to power spectrum estimation, it is necessary to estimate first the model order M. A number of formulae has been advanced for order estimation [Kum93]. Unfortunately, none of the these formulae yields a really good estimate of the true model order in many applications.

11.5 Musical Sound Processing

Recall from our discussion in Section 1.4.1 that almost all musical programs are produced in basically two stages. First, sound from each individual instrument is recorded in an acoustically inert studio on a single track of a multitrack tape recorder. Then, the signals from each track are manipulated by the sound engineer to add special audio effects and are combined in a mix-down system to finally generate the stereo recording on a two-track tape recorder [Ble78], [Ear86]. The audio effects are artificially generated using various signal processing circuits and devices, and they are increasingly being performed using digital signal processing techniques [Ble78], [Orf96].

Some of the special audio effects that can be implemented digitally are reviewed in this section.

11.5.1 Time-Domain Operations

As indicated in Section 1.4.1, the sound reaching the listener in a closed space, such as a concert hall, consists of several components: direct sound, early reflections, and reverberation. The early reflections are composed of several closely spaced echoes that are basically delayed and attenuated copies of the direct sound, whereas the reverberation is composed of densely packed echoes. The sound recorded in an inert studio is different from that recorded inside a closed space, and, as a result, the former does not sound "natural" to a listener. However, digital filtering can be employed to convert the sound recorded in an inert studio into a natural sounding one by artificially creating the echoes and adding it to the original signal.

Echoes are simply generated by delay units. For example, the direct sound and a single echo appearing R sampling periods later can be simply generated by the FIR filter of Figure 11.15(a), which is characterized by the difference equation

$$y[n] = x[n] + \alpha x[n - R], \qquad |\alpha| < 1, \tag{11.46}$$

or, equivalently, by the transfer function

$$H(z) = 1 + \alpha z^{-R}. \tag{11.47}$$

Its impulse response is sketched in Figure 11.15(b). The magnitude response of a single echo FIR filter for $\alpha = 0.8$ and $R = 8$ is shown in Figure 11.15(c). The magnitude response exhibits R peaks and R dips in the range $0 \le \omega < 2\pi$, with the peaks occurring at $\omega = 2\pi k/R$ and the dips occurring at $\omega = (2k + 1)\pi/R, k = 0, 1, \ldots, R - 1$. Because of the *comb*-like shape of the magnitude response, such a filter is also known as a *comb filter*. The maximum and minimum values of the magnitude response are given by $1 + \alpha = 1.8$ and $1 - \alpha = 0.2$, respectively.

To generate multiple echoes spaced R sampling periods apart with exponentially decaying amplitudes, one can use an FIR filter with a transfer function of the form

$$H(z) = 1 + \alpha z^{-R} + \alpha^2 z^{-2R} + \cdots + \alpha^{N-1} z^{-(N-1)R} = \frac{1 - \alpha^N z^{-NR}}{1 - \alpha z^{-R}}. \tag{11.48}$$

An IIR realization of this filter is sketched in Figure 11.16(a). The impulse response of a multiple echo filter with $\alpha = 0.8$ for $N = 6$ and $R = 4$ is shown in Figure 11.16(b).

An infinite number of echoes spaced R sampling periods apart with exponentially decaying amplitudes can be created by an IIR filter with a transfer function of the form

$$H(z) = \frac{z^{-R}}{1 - \alpha z^{-R}}, \qquad |\alpha| < 1. \tag{11.49}$$

Figure 11.17(a) shows one possible realization of the above IIR filter whose first 61 impulse response samples for $R = 4$ are indicated in Figure 11.17(b). The magnitude response of this IIR filter for $R = 7$ is

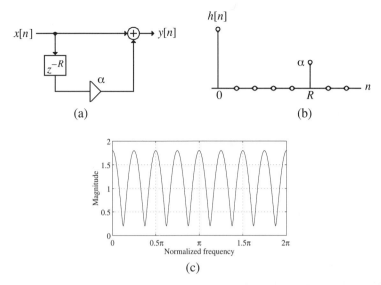

(a)   (b)

(c)

Figure 11.15: Single echo filter. (a) Filter structure, (b) typical impulse response, and (c) magnitude response for $R = 8$ and $\alpha = 0.8$.

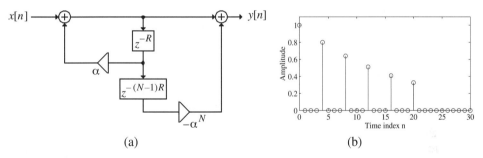

(a)   (b)

Figure 11.16: Multiple echo filter generating $N - 1$ echoes. (a) The filter structure, and (b) impulse response with $\alpha = 0.8$ for $N = 6$ and $R = 4$.

sketched in Figure 11.17(c). The magnitude response exhibits R peaks and R dips in the range $0 \le \omega < 2\pi$, with the peaks occurring at $\omega = 2\pi k/R$ and the dips occurring at $\omega = (2k+1)\pi/R, k = 0, 1, \ldots, R-1$. The maximum and minimum values of the magnitude response are given by $1/(1-\alpha) = 5$, and $1/(1+\alpha) = 0.5556$, respectively.

The IIR comb filter of Figure 11.17(a) by itself does not provide natural sounding reverberations for two reasons [Sch62]. First, as can be seen from Figure 11.17(c), its magnitude response is not constant for all frequencies, resulting in a "coloration" of many musical sounds that are often unpleasant for listening purposes. Second, the output echo density, given by the number of echoes per second, generated by a unit impulse at the input, is much lower than that observed in a real room, thus causing a "fluttering" of the composite sound. It has been observed that approximately 1000 echoes per second are necessary to create a reverberation that sounds free of flutter [Sch62]. To develop a more realistic reverberation, a reverberator with an allpass structure, as indicated in Figure 11.18(a), has been proposed [Sch62].[3] Its transfer function

---

[3]The structures shown here are the canonic single multiplier realization of a first-order allpass transfer function [Mit74a]. See also Section 6.6.1.

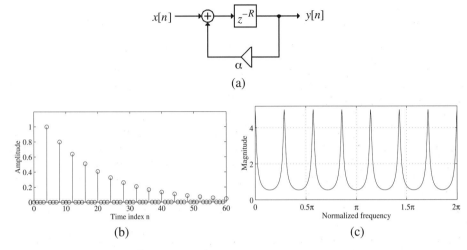

Figure 11.17: IIR filter generating an infinite number of echoes. (a) The filter structure, (b) impulse response with $\alpha = 0.8$ for $R = 4$, and (c) magnitude response with $\alpha = 0.8$ for $R = 7$.

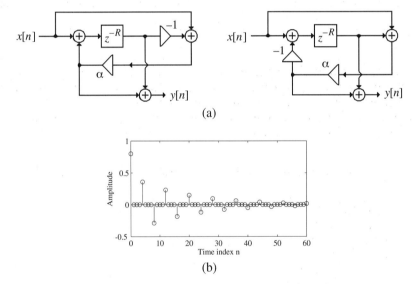

Figure 11.18: Allpass reverberator. (a) Block diagram representation, and (b) impulse response with $\alpha = 0.8$ for $R = 4$.

is given by

$$H(z) = \frac{\alpha + z^{-R}}{1 + \alpha z^{-R}}, \qquad |\alpha| < 1. \tag{11.50}$$

In the steady state, the spectral balance of the sound signal remains unchanged due to the unity magnitude response of the allpass reverberator.

The IIR comb filter of Figure 11.17(a) and the allpass reverberator of Figure 11.18(a) are basic reverberator units that are suitably interconnected to develop a natural sounding reverberation. Figure 11.19 shows one such interconnection composed of a parallel connection of four IIR echo generators in cascade with

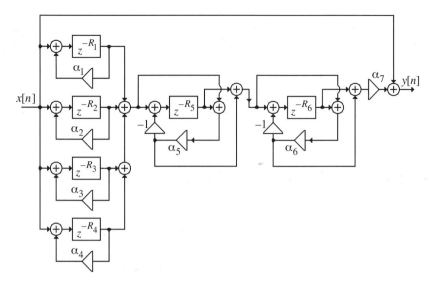

Figure 11.19: A proposed natural sounding reverberator scheme.

two allpass reverberators [Sch62]. By choosing different values for the delays in each section (obtained by adjusting R_i) and the multiplier constants α_i it is possible to arrive at a pleasant sounding reverberation, duplicating that occurring in a specific closed space, such as a concert hall.

An interesting modification of the basic IIR comb filter of Figure 11.17(a) is obtained by replacing the multiplier α with a lowpass FIR or IIR filter $G(z)$, as indicated in Figure 11.20(a). It has a transfer function given by

$$H(z) = \frac{z^{-R}}{1 - z^{-R}G(z)},\tag{11.51}$$

obtained by replacing α in Eq. (11.49) with $G(z)$. This structure has been referred to as the *teeth filter* and has been introduced to provide a natural tonal character to the artificial reverberation generated by it [Egg84]. This type of reverberator should be carefully designed to avoid the stability problem. To provide a reverberation with a higher echo density, the teeth filter has been used as a basic unit in a more complex structure such as that indicated in Figure 11.18(b).

Additional details concerning these and other such composite reverberator structures can be found in [Sch62], [Moo79].

There are a number of special sound effects that are often used in the mix-down process. One such effect is called *flanging*. Originally, it was created by feeding the same musical piece to two tape recorders and then combining their delayed outputs while varying the difference Δt between their delay times. One way of varying Δt is to slow down one of the tape recorders by placing the operator's thumb on the flange of the feed reel, which led to the name flanging [Ear86]. The FIR comb filter of Figure 11.15(a) can be modified to create the flanging effect. In this case, the unit generating the delay of R samples, or equivalently, a delay of RT seconds, where T is the sampling period, is made a time-varying delay $\beta(n)$, as indicated in Figure 11.21. The corresponding input-output relation is then given by

$$y[n] = x[n] + \alpha x[n - \beta(n)].\tag{11.52}$$

Periodically varying the delay $\beta(n)$ between 0 and R with a low frequency ω_o such as

$$\beta(n) = \frac{R}{2}(1 - \cos(\omega_o n))\tag{11.53}$$

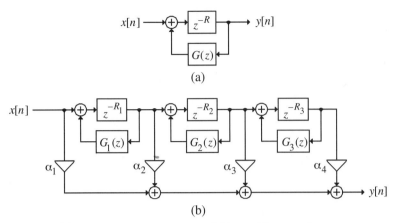

Figure 11.20: (a) Lowpass reverberator, and (b) a multitap reverberator structure.

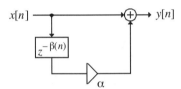

Figure 11.21: Generation of a flanging effect.

generates a flanging effect on the sound. It should be noted that, as the value of $\beta(n)$ at an instant n in general has a noninteger value, in an actual implementation, the output sample value $y[n]$ should be computed using some type of interpolation method such as that outlined in Section 10.5.

The *chorus* effect is achieved when several musicians are playing the same musical piece at the same time but with small changes in the amplitudes and small timing differences between their sounds. Such an effect can also be created synthetically by a *chorus generator* from the music of a single musician. A simple modification of the digital filter of Figure 11.21 leads to a structure that can be employed to simulate this sound effect. For example, the structure of Figure 11.22 can effectively create a chorus of four musicians from the music of a single musician. To achieve this effect, the delays $\beta_i(n)$ are randomly varied with very slow variations.

The *phasing* effect is produced by processing the signal through a narrowband notch filter with variable notch characteristics and adding a scaled portion of the notch filter output to the original signal as indicated in Figure 11.23 [Orf96]. The phase of the signal at the notch filter output can dramatically alter the phase of the combined signal, particularly around the notch frequency when it is varied slowly. The tunable notch filter can be implemented using the technique described in Section 6.7.2. The notch filter in Figure 11.23 can be replaced with a cascade of tunable notch filters to provide an effect similar to flanging. However, in flanging the swept notch frequencies are always equally spaced, whereas in phasing the locations of the notch frequencies and their corresponding 3-dB bandwidths are varied independently.

11.5.2 Frequency-Domain Operations

The frequency responses of individually recorded instruments or musical sounds of performers are frequently modified by the sound engineer during the mix-down process. These effects are achieved by

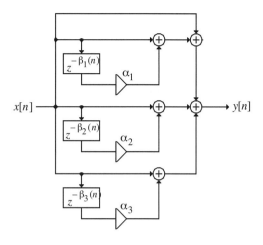

Figure 11.22: Generation of a chorus effect.

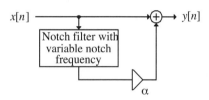

Figure 11.23: Generation of the phasing effect.

passing the original signals through an equalizer, briefly reviewed in Section 1.4.1. The purpose of the equalizer is to provide "presence" by peaking the midfrequency components in the range of 1.5 to 3 kHz, and to modify the bass-treble relationships by providing "boost" or "cut" to components outside this range. It is usually formed from a cascade of first-order and second-order filters with adjustable frequency responses. The simple digital filters described in Section 4.5.2 can be employed for implementing these functions. We review these filters here to point out their specific properties that are suitable for musical sound processing. In addition, we describe some new structures with more flexible frequency responses.

First-Order Filters and Shelving Filters

The transfer functions of a first-order lowpass filter and a highpass filter with tunable cutoff frequencies are given by Eqs. (4.109) and (4.112), respectively, and repeated below for convenience:

$$H_{LP}(z) = \frac{1 - \alpha}{2} \cdot \frac{1 + z^{-1}}{1 - \alpha z^{-1}}, \tag{11.54a}$$

$$H_{HP}(z) = \frac{1 + \alpha}{2} \cdot \frac{1 - z^{-1}}{1 - \alpha z^{-1}}. \tag{11.54b}$$

The 3-dB cutoff frequency ω_c for both transfer functions is related to the constant α through

$$\omega_c = \cos^{-1}\left(\frac{2\alpha}{1 + \alpha^2}\right). \tag{11.55}$$

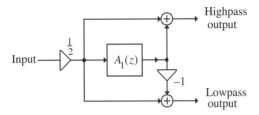

Figure 11.24: A parametrically tunable first-order lowpass/highpass filter.

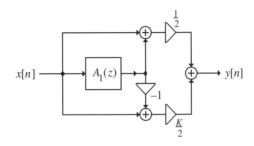

Figure 11.25: Low-frequency shelving filter.

The two transfer functions of Eqs. (11.54a) and (11.54b) can be alternatively expressed as

$$H_{LP}(z) = \frac{1}{2}\{1 - A_1(z)\},\tag{11.56a}$$

$$H_{HP}(z) = \frac{1}{2}\{1 + A_1(z)\},\tag{11.56b}$$

where $A_1(z)$ is a first-order allpass transfer function given by

$$A_1(z) = \frac{\alpha - z^{-1}}{1 - \alpha z^{-1}}.\tag{11.57}$$

A composite realization of the above two transfer functions is sketched in Figure 11.24, where the first-order allpass transfer function $A_1(z)$ can be realized using any one of the single multiplier structures of Figure 6.23. Note that in this structure the 3-dB cutoff frequency ω_c of both filters is independently controlled by the multiplier constant α of the allpass section.

Combining the outputs of the structure of Figure 11.24 as indicated in Figure 11.25, we arrive at a low-frequency shelving filter that is characterized by a transfer function $G_1(z)$ given by

$$G_1(z) = \frac{K}{2}[1 - A_1(z)] + \frac{1}{2}[1 + A_1(z)],\tag{11.58}$$

where K is a positive constant [Reg87b]. Figures 11.26 and 11.27 show the gain responses of $G_1(z)$ obtained by varying the multiplier constants K and α. Note that the parameter K controls the amount of boost or cut at low frequencies, while the parameter α controls the boost or cut bandwidth.

A high-frequency shelving filter is obtained by combining the outputs of the structure of Figure 11.24 as indicated in Figure 11.28, which is characterized by a transfer function

$$G_1(z) = \frac{1}{2}\{1 - A_1(z)\} + \frac{K}{2}\{1 + A_1(z)\}.\tag{11.59}$$

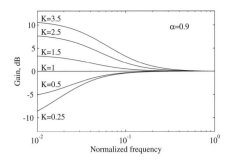

Figure 11.26: Gain responses of the low-frequency shelving filter of Figure 11.25 for various values of the parameter K with $\alpha = 0.9$.

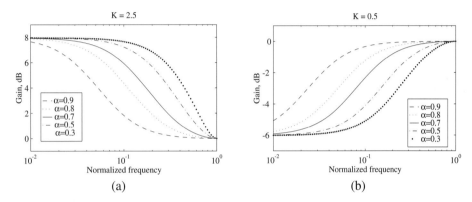

Figure 11.27: Gain responses of the low-frequency shelving filter of Figure 11.25 for various values of the parameter α with two fixed values of the parameter K.

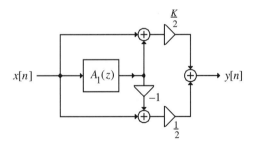

Figure 11.28: High-frequency shelving filter.

Figures 11.29 and 11.30 show the gain responses of the above transfer function obtained by varying the multiplier constants K and α. Note that, as in the case of the low-frequency shelving filter, here the parameter K controls the amount of boost or cut at high frequencies, while the parameter α controls the boost or cut bandwidth.

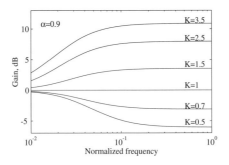

Figure 11.29: Gain responses of the high-frequency shelving filter of Figure 11.28 for various values of the parameter K with $\alpha = 0.9$.

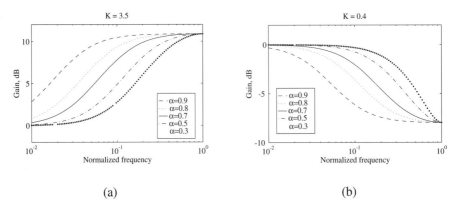

(a)   (b)

Figure 11.30: Gain responses of the high-frequency shelving filter of Figure 11.28 for various values of the parameter α with two fixed values of the parameter K.

Second-Order Filters and Equalizers

The transfer functions of a second-order bandpass and a bandstop filter with tunable cutoff frequencies and 3-dB bandwidths are given by Eqs. (4.113) and (4.118), respectively, and are repeated here for convenience:

$$H_{BP}(z) = \frac{1-\alpha}{2} \cdot \frac{1 - z^{-2}}{1 - \beta(1+\alpha)z^{-1} + \alpha z^{-2}}, \tag{11.60a}$$

$$H_{BS}(z) = \frac{1+\alpha}{2} \cdot \frac{1 - 2\beta z^{-1} + z^{-2}}{1 - \beta(1+\alpha)z^{-1} + \alpha z^{-2}}. \tag{11.60b}$$

The center frequency ω_o of the bandpass filter and the notch frequency ω_o of the bandstop filter are related to the constant β through

$$\omega_o = \cos^{-1}(\beta), \tag{11.61}$$

while the 3-dB bandwidth B_w of both transfer functions is related to the constant α through

$$B_w = \cos^{-1}\left(\frac{2\alpha}{1+\alpha^2}\right). \tag{11.62}$$

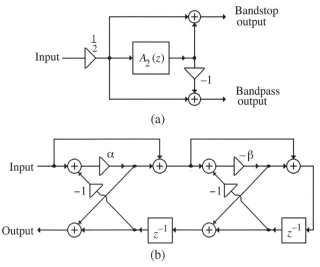

Figure 11.31: A parametrically tunable second-order bandpass/bandstop filter: (a) overall structure, and (b) allpass section.

A composite realization of both filters is as indicated in Figure 11.31(a) based on a sum of allpass decompositions of Eqs. (11.60a) and (11.60b):

$$H_{BP}(z) = \frac{1}{2}[1 - A_2(z)], \tag{11.63a}$$

$$H_{BS}(z) = \frac{1}{2}[1 + A_2(z)], \tag{11.63b}$$

where $A_2(z)$ is a second-order allpass transfer function given by

$$A_2(z) = \frac{\alpha - \beta(1+\alpha)z^{-1} + z^{-2}}{1 - \beta(1+\alpha)z^{-1} + \alpha z^{-2}}, \tag{11.64}$$

and is realized using the cascaded lattice structure of Figure 11.31(b) for independent tuning of the center (notch) frequency ω_o and the 3-dB bandwidth B_w.

As in the first-order case, a weighted combination of the outputs of the structure of Figure 11.31(a) results in a second-order equalizer (Figure 11.32) with a transfer function $G_2(z)$ given by

$$G_2(z) = \frac{K}{2}\{1 - A_2(z)\} + \frac{1}{2}\{1 + A_2(z)\}, \tag{11.65}$$

where K is a positive constant [Reg87b]. It follows from the above discussion that the peak or the dip of the magnitude response occurs at the frequency ω_o, which is controlled independently by the parameter β according to Eq. (11.61), and the 3-dB bandwidth B_w of the magnitude response is determined solely by the parameter α as indicated by Eq. (11.62). Moreover, the height of the peak or the dip of the magnitude response is given by $K = G_2(e^{j\omega_o})$. Figures 11.33 to 11.35 show the gain responses of $G_2(z)$ obtained by varying the parameters K, α, and β.

Higher-Order Equalizers

A graphic equalizer with tunable gain response can be built using a cascade of first-order and second-order equalizers with external control of the maximum gain values of each section in the cascade. Figure 11.36(a)

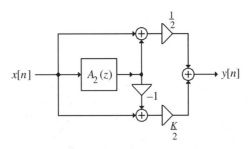

Figure 11.32: A parametrically tunable second-order equalizer.

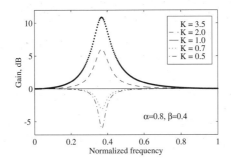

Figure 11.33: Gain responses of the second-order equalizer of Figure 11.32 for various values of the parameter K with $\alpha = 0.8$ and $\beta = 0.4$.

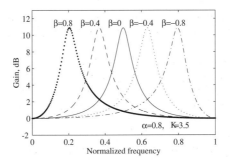

Figure 11.34: Gain responses of the second-order equalizer of Figure 11.32 for various values of the parameter β with $\alpha = 0.8$ and $K = 3.5$.

shows the block diagram of a cascade of one first-order and three second-order equalizers with nominal frequency response parameters as indicated. Figure 11.36(b) shows its gain response for some typical values of the parameter K (maximum gain values) of the individual sections.

11.6 Digital FM Stereo Generation

Frequency-division multiplexing of stereo used in FM (frequency modulation) radio for the transmission of left and right channel audio signals has been described earlier in Section 1.4.3. We now consider the

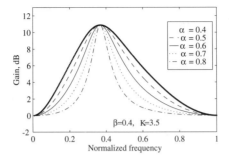

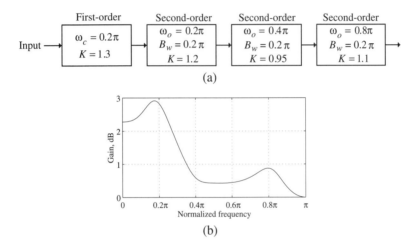

Figure 11.35: Gain responses of the second-order equalizer of Figure 11.32 for various values of the parameter α with $\beta = 0.4$ and $K = 3.5$.

Figure 11.36: (a) Block diagram of a typical graphic equalizer, and (b) its gain response for the section parameter values shown.

digital stereo generation for FM transmission. A simplified block diagram representation of a digital FM stereo generator is shown in Figure 11.37. As indicated here, the analog outputs of the left and right microphones, $s_L(t)$ and $s_R(t)$, are first converted into digital signals, $s_L[n]$ and $s_R[n]$, by means of individual A/D converters. In practice, high-frequency components of the modulating signal have much smaller amplitudes, while low-frequency components have much larger amplitudes. However, smaller amplitude components produce a correspondingly smaller frequency deviation. Consequently, the resulting FM signal does not fully utilize the bandwidth allotted for its transmission, lowering the signal-to-noise ratio considerably at the high-frequency end. The output SNR at the FM receiver is increased by emphasizing the higher frequencies of $s_L[n]$ and $s_R[n]$ by means of digital preemphasis filters, as indicated in Figure 11.37.

The sum of the preemphasized left input discrete-time signal $x_L[n]$ and the preemphasized right input discrete-time signal $x_R[n]$ is transmitted in its baseband form for monophonic reception. The difference signal $x_L[n] - x_R[n]$ is transmitted by DSB-SC (double-sideband, suppressed carrier) modulation[4] using a 38-kHz subcarrier. The transmitted multiplexed signal $y[n]$ includes the sum signal, the DSB-SC modulated

---

[4]See Section 1.2.4.

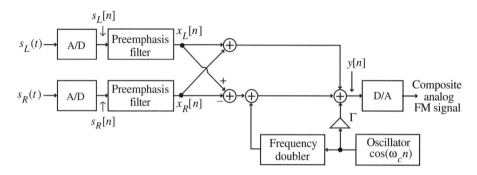

Figure 11.37: Block diagram representation of the FM stereo transmitter.

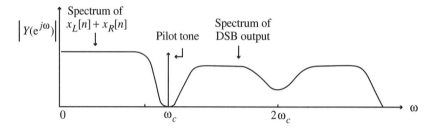

Figure 11.38: Spectrum of a composite stereo discrete-time broadcast signal.

difference signal, and a 19-kHz pilot carrier:

$$y[n] = (x_L[n] + x_R[n]) + (x_L[n] - x_R[n])\cos(2\omega_c n) + \Gamma\cos(\omega_c n), \qquad (11.66)$$

where $\omega_c = 2\pi F_c/F_T$ is the normalized angular frequency of the pilot carrier $F_c = 19\,\text{kHz}$ and F_T is the sampling frequency in Hz, which is typically 32 kHz. The 19-kHz pilot carrier is included to provide a reference for coherent demodulation at the stereo receiver. Figure 11.38 shows the power spectrum of a typical composite baseband signal $y[n]$. The composite signal $y[n]$ frequency-modulates the main carrier to generate the transmitted signal. The value of the gain constant Γ for the pilot signal is chosen such that the pilot is allotted about 10 percent of the peak frequency deviation. The original signal-power distribution is restored at the receiver output through a deemphasis network.

In the analog FM stereo generation, a first-order analog preemphasis network with a transfer function

$$G_a(s) = \frac{s + \Omega_1}{s + \Omega_1 + \Omega_2} \qquad (11.67a)$$

is employed [Pan65]. Typically, $\Omega_1 \ll \Omega_2$, in which case, the asymptotic magnitude response of the preemphasis network is of the form shown in Figure 11.39. In the frequency range of interest, the transfer function of the analog preemphasis network can be approximated as

$$G_a(s) \cong 1 + \frac{s}{\Omega_1}, \qquad (11.67b)$$

where $1/\Omega_1$ is typically 75 msec. Equation (11.67a) can be rewritten as

$$G_a(s) = \frac{s + K\lambda}{s + \lambda}, \qquad (11.68)$$

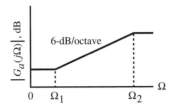

Figure 11.39: Asymptotic magnitude response of the preemphasis network.

where K is the dc gain

$$K = \frac{\Omega_1}{\Omega_1 + \Omega_2},$$
(11.69)

and

$$\lambda = \Omega_1 + \Omega_2.$$
(11.70)

Now, $G_a(s)$ can be decomposed as [Reg87b]

$$H_a(s) = \frac{1}{2}[1 + A_a(s)] + \frac{K}{2}[1 - A_a(s)],$$
(11.71)

where $A_a(s)$ is a stable analog allpass transfer function:

$$A_a(s) = \frac{s - \lambda}{s + \lambda}.$$
(11.72)

The transfer function of the corresponding digital preemphasis network is obtained by applying the bilinear transformation of Eq. (7.20), repeated below for convenience:

$$s = \frac{2}{T}\left(\frac{1 - z^{-1}}{1 + z^{-1}}\right),$$
(11.73)

where $T = 1/F_T$. Applying the above transformation to Eq. (11.71), we arrive at

$$G(z) = \frac{1}{2}[1 + A_1(z)] + \frac{K}{2}[1 - A_1(z)],$$
(11.74)

where $A_1(z)$ is a stable first-order digital allpass transfer function

$$A_1(z) = \frac{\alpha - z^{-1}}{1 - \alpha z^{-1}},$$
(11.75)

with the parameter α given by

$$\alpha = \frac{(2/T) - \lambda}{(2/T) + \lambda}.$$
(11.76)

The expression for $G(z)$ given by Eq. (11.74) is precisely the same as that of the first-order equalizer discussed in the previous section and given in Eq. (11.58) with a realization of $G(z)$ as indicated in Figure 11.25. Now, from Eq. (11.73), the pole of the analog transfer function at $s = \lambda$ is mapped onto the corresponding digital angular frequency ω_c according to

$$\lambda = \frac{2}{T}\tan(\omega_c/2).$$
(11.77)

The 19-kHz tone signal in Figure 11.37 can be generated using a look-up table method, using the trigonometric function approximation method of Section 8.8.1, or using the sine generator described in Section 6.11.

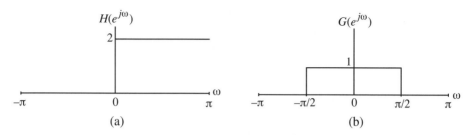

Figure 11.40: (a) Frequency response of the discrete-time filter generating an analytic signal, and (b) half-band lowpass filter.

11.7 Discrete-Time Analytic Signal Generation

As discussed in Section 1.2.3, an analytic continuous-time signal has a zero-valued spectrum for all negative frequencies. Such a signal finds applications in single-sideband analog communication systems and analog frequency-division multiplex systems. A discrete-time signal with a similar property finds applications in digital communication systems and is the subject of this section. We illustrate here the generation of an analytic signal $y[n]$ from a discrete-time real signal $x[n]$ and describe some of its applications.

Now, the Fourier transform $X(e^{j\omega})$ of a real signal $x[n]$, if it exists, is nonzero for both positive and negative frequencies. On the other hand, a signal $y[n]$ with a single-sided spectrum $Y(e^{j\omega})$ that is zero for negative frequencies must be a complex signal. Consider the complex analytic signal

$$y[n] = x[n] + j\hat{x}[n], \tag{11.78}$$

where $x[n]$ and $\hat{x}[n]$ are real. Its DTFT $Y(e^{j\omega})$ is given by

$$Y(e^{j\omega}) = X(e^{j\omega}) + j\hat{X}(e^{j\omega}), \tag{11.79}$$

where $\hat{X}(e^{j\omega})$ is the DTFT of $\hat{x}[n]$. Now, $x[n]$ and $\hat{x}[n]$ being real, their corresponding DTFTs are conjugate symmetric, i.e., $X(e^{j\omega}) = X^*(e^{-j\omega})$ and $\hat{X}(e^{j\omega}) = \hat{X}^*(e^{-j\omega})$. Hence, from Eq. (11.79) we obtain

$$X(e^{j\omega}) = \tfrac{1}{2}\left[Y(e^{j\omega}) + Y^*(e^{-j\omega})\right], \tag{11.80a}$$

$$j\hat{X}(e^{j\omega}) = \tfrac{1}{2}\left[Y(e^{j\omega}) - Y^*(e^{-j\omega})\right]. \tag{11.80b}$$

Since by assumption, $Y(e^{j\omega}) = 0$ for $-\pi \le \omega < 0$, we obtain from Eq. (11.80a)

$$Y(e^{j\omega}) = \begin{cases} 2X(e^{j\omega}), & 0 \le \omega < \pi, \\ 0, & -\pi \le \omega < 0. \end{cases} \tag{11.81}$$

Thus, the analytic signal $y[n]$ can be generated by passing $x[n]$ through a linear discrete-time system with a frequency response $H(e^{j\omega})$ given by

$$H(e^{j\omega}) = \begin{cases} 2, & 0 \le \omega < \pi, \\ 0, & -\pi \le \omega < 0, \end{cases} \tag{11.82}$$

as indicated in Figure 11.40(a).

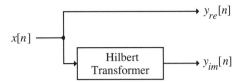

Figure 11.41: Generation of an analytic signal using a Hilbert transformer.

11.7.1 The Discrete-Time Hilbert Transformer

We now relate the imaginary part $\hat{x}[n]$ of the analytic signal $y[n]$ to its real part $x[n]$. From Eq. (11.80b),

$$\hat{X}(e^{j\omega}) = \frac{1}{2j}\left[Y(e^{j\omega}) - Y^*(e^{-j\omega})\right]. \tag{11.83}$$

For $0 \le \omega < \pi$, $Y(e^{-j\omega}) = 0$, and for $-\pi \le \omega < 0$, $Y(e^{j\omega}) = 0$. Using this property and Eq. (11.81) in Eq. (11.83), it can be easily shown that

$$\hat{X}(e^{j\omega}) = \begin{cases} -jX(e^{j\omega}), & 0 \le \omega < \pi, \\ jX(e^{j\omega}), & -\pi \le \omega < 0. \end{cases} \tag{11.84}$$

Equation (11.84) that the imaginary part $\hat{x}[n]$ of the analytic signal $y[n]$ can be generated by passing its real part $x[n]$ through a linear discrete-time system with a frequency response $H_{HT}(e^{j\omega})$ given by

$$H_{HT}(e^{j\omega}) = \begin{cases} -j, & 0 \le \omega < \pi, \\ j, & -\pi \le \omega < 0. \end{cases} \tag{11.85}$$

The linear system defined by Eq. (11.85) is usually referred to as the ideal *Hilbert transformer*. Its output $\hat{x}[n]$ is called the *Hilbert transform* of its input $x[n]$. The basic scheme for the generation of an analytic signal $y[n] = y_{re}[n] + jy_{im}[n]$ from a real signal $x[n]$ is thus as indicated in Figure 11.41. Observe that $\left|H_{HT}(e^{j\omega})\right| = 1$ for all frequencies and has a -90-degree phase-shift for $0 \le \omega < \pi$ and a $+90$-degree phase-shift for $-\pi \le \omega < 0$. As a result, an ideal Hilbert transformer is also called a 90-*degree phase-shifter*.

The impulse response $h_{HT}[n]$ of the ideal Hilbert transformer is obtained by taking the inverse DTFT of $H_{HT}(e^{j\omega})$ and can be shown to be (Problem 7.35)

$$h_{HT}[n] = \begin{cases} 0, & \text{for } n \text{ even}, \\ \frac{2}{\pi n}, & \text{for } n \text{ odd}. \end{cases} \tag{11.86}$$

Since the ideal Hilbert transformer has a two-sided infinite-length impulse response defined for $-\pi < n < \pi$, it is an unrealizable system. Moreover, its transfer function $H_{HT}(z)$ exists only on the unit circle. We describe later two approaches for developing a realizable approximation.

11.7.2 Relation with Half-Band Filters

Consider the filter with a frequency response $G(e^{j\omega})$ obtained by shifting the frequency response $H(e^{j\omega})$ of Eq. (11.82) by $\pi/2$ radians and scaling by a factor $\frac{1}{2}$ (see Figure 11.40):

$$G(e^{j\omega}) = \tfrac{1}{2}H(e^{j(\omega+\pi/2)}) = \begin{cases} 1, & 0 < |\omega| < \frac{\pi}{2}, \\ 0, & \frac{\pi}{2} < |\omega| < \pi. \end{cases} \tag{11.87}$$

From our discussion in Section 10.7.2, we observe that $G(e^{j\omega})$ is a half-band lowpass filter. Because of the relation between $H(e^{j\omega})$ of Eq. (11.82) and the real coefficient half-band lowpass filter $G(e^{j\omega})$ of Eq. (11.87), the filter $H(e^{j\omega})$ has been referred to as a *complex half-band filter* [Reg93].

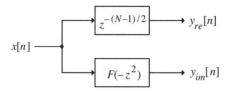

Figure 11.42: FIR realization of a complex half-band filter.

11.7.3 Design of the Hilbert Transformer

It also follows from the above relation that a complex half-band filter can be designed simply by shifting the frequency response of a half-band lowpass filter by $\pi/2$ radians and then scaling by a factor 2. Equivalently, the relation between the transfer functions of a complex half-band filter $H(z)$ and a real half-band lowpass filter $G(z)$ is given by

$$H(z) = j2G(-jz). \tag{11.88}$$

Design of the real half-band filter was briefly touched upon in Section 10.7.3. We outline below two other novel approaches.

FIR Complex Half-Band Filter

The real half-band filter design problem can be transformed into the design of a single passband FIR filter with no stopband, which can be easily designed using the popular Parks-McClellan algorithm[5] [Vai87b]. An inverse transformation on the realization of this wideband filter then yields the implementation of the desired real half-band filter.

Let the specifications of the real half-band lowpass filter $G(z)$ of length N be as follows: passband edge at ω_p, stopband edge at ω_s, passband ripple of δ_p, and stopband ripple of δ_s. As discussed in Section 10.7.2, the passband and stopband ripples of a linear-phase lowpass half-band FIR transfer function $G(z)$ are equal, i.e., $\delta_p = \delta_s = \delta$, and the length N is odd. Moreover, the passband and stopband edge frequencies are related through $\omega_p + \omega_s = \pi$. It has also been shown that $(N-1)/2$ must be odd.

Now, consider a wideband linear-phase filter $F(z)$ of degree $(N-1)/2$ with a passband from 0 to $2\omega_p$, a transition band from $2\omega_p$ to π, and a passband ripple of 2δ. Since $(N-1)/2$ is odd, $F(z)$ has a zero at $z = -1$. The wideband filter $F(z)$ can be designed using the Parks-McClellan method. Define

$$G(z) = \tfrac{1}{2}\left[z^{-(N-1)/2} + F(z^2)\right]. \tag{11.89}$$

It follows from Eq. (11.89) that $G(z)$ is indeed the desired half-band lowpass filter and has an impulse response

$$g[n] = \begin{cases} \tfrac{1}{2}f\left[\tfrac{n}{2}\right], & n \text{ even,} \\ 0, & n \text{ odd, } n \neq \tfrac{N-1}{2} \\ \tfrac{1}{2}, & n = \tfrac{N-1}{2}, \end{cases} \tag{11.90}$$

where $f[n]$ is the impulse response of $F(z)$.

Substituting Eq. (11.89) in Eq. (11.88), we obtain

$$H(z) = j\left[(-jz)^{-(N-1)/2} + F(-z^2)\right] = z^{-(N-1)/2} + jF(-z^2). \tag{11.91}$$

---

[5] See Section 7.7.1.

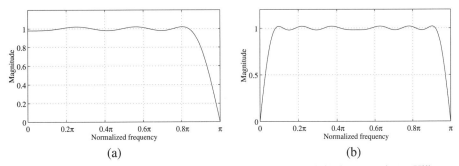

Figure 11.43: Magnitude responses of (a) the wideband FIR filter $F(z)$ and (b) the approximate Hilbert transformer $F(-z^2)$.

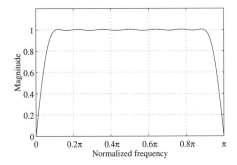

Figure 11.44: The magnitude response of the Hilbert transformer designed directly using MATLAB.

An FIR implementation of the complex half-band filter based on the above decomposition is indicated in Figure 11.42. The linear-phase FIR filter $F(-z^2)$ is thus an approximation to a Hilbert transformer.

We illustrate the above approach by means of an example.

EXAMPLE 11.8   Using MATLAB we design a wideband FIR filter $F(z)$ of degree 13 with a passband from 0 to 0.85π and an extremely small stopband from 0.9π to π. We use the function remez with a magnitude vector m = [1 1 0 0]. The weight vector used to weigh the passband and the stopband is wt = [2 0.05]. The magnitude responses of the wideband filter $F(z)$ and the Hilbert transformer $F(-z^2)$ are shown in Figure 11.43(a) and (b).

The FIR Hilbert transformer can be designed directly using MATLAB. To this end, the version of the function remez employed is either b = remez(N, f, m,'Hilbert') or b = remez(N, f, m, wt,'Hilbert'). The following example illustrates this approach.

EXAMPLE 11.9   We design a 26th-order FIR Hilbert transformer with a passband from 0.1π to 0.9π. It should be noted that for the design of a Hilbert transformer, the first frequency point in the vector f containing the specified bandedges cannot be a 0. The magnitude response of the designed Hilbert transformer obtained using the statement b = remez(26, [0.1 0.9], [1 1], 'Hilbert') is indicated in Figure 11.44. It should be noted that due to numerical round-off problems, unlike the design of Example 11.8, the odd impulse response coefficients of the Hilbert transformer here are not exactly zero.

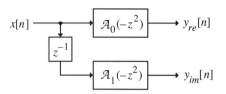

Figure 11.45: IIR realization of a complex half-band filter.

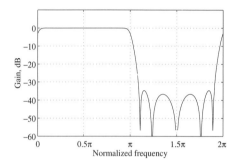

Figure 11.46: Gain responses of the complex half-band filter (normalized to 0 dB maximum gain).

IIR Complex Half-Band Filter

We indicated in Section 10.8.6 that a large class of stable IIR real coefficient half-band filters of odd order can be expressed as [Vai87f]

$$G(z) = \tfrac{1}{2}[\mathcal{A}_0(z^2) + z^{-1}\mathcal{A}_1(z^2)], \qquad (11.92)$$

where $\mathcal{A}_0(z)$ and $\mathcal{A}_1(z)$ are stable allpass transfer functions. Substituting Eq. (11.92) in Eq. (11.88), we therefore arrive at

$$H(z) = \mathcal{A}_0(-z^2) + jz^{-1}\mathcal{A}_1(-z^2). \qquad (11.93)$$

A realization of the complex half-band filter based on the above decomposition is thus as shown in Figure 11.45.

We illustrate the above approach to Hilbert transformer design in the following example.

EXAMPLE 11.10   In Example 10.16, we designed a real half-band elliptic filter with the following frequency response specifications: $\omega_p = 0.4\pi$, $\omega_s = 0.6\pi$, and $\delta_s = 0.0155$. The transfer function of the real half-band filter $G(z)$ can be expressed as in Eq. (11.92), where the transfer functions of the two allpass sections are given in Eq. (10.134) and are repeated below for convenience:

$$\mathcal{A}_0(z^2) = \frac{z^{-2} + 0.2368041466}{1 + 0.2368041466z^{-2}}, \quad \mathcal{A}_1(z^2) = \frac{z^{-2} + 0.7149039978}{1 + 0.7149039978z^{-2}}. \qquad (11.94)$$

The gain response of the complex half-band filter $H(z)$ obtained using Eq. (11.93) is sketched in Figure 11.46. Figure 11.47 shows the phase difference between the two allpass functions $\mathcal{A}_0(-z^2)$ and $z^{-1}\mathcal{A}_1(-z^2)$ of the complex half-band filter. Note that, as expected, the phase difference is 90 degrees for most of the positive frequency range and 270 degrees for most of the negative frequency range. In plotting the gain response of the complex half-band filter and the phase difference between its constituent two allpass sections, the M-file `freqz(num,den,n,'whole')` has been used to compute the pertinent frequency response values over the whole normalized frequency range from 0 to 2π.

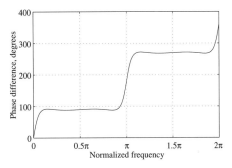

Figure 11.47: Phase difference between the two allpass sections of the complex half-band filter.

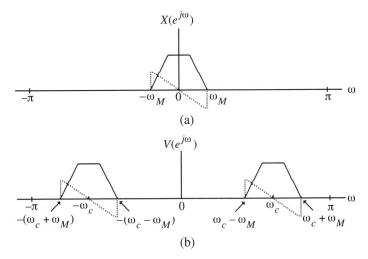

Figure 11.48: Spectra of a real signal and its modulated version. (Solid lines represent the real parts while dashed lines represent the imaginary parts.)

11.7.4 Single-Sideband Modulation

For efficient transmission over long distances, a real low-frequency bandlimited signal $x[n]$, such as speech or music, is modulated by a very high frequency sinusoidal carrier signal $\cos \omega_c n$, with the carrier frequency ω_c being less than half of the sampling frequency. The spectrum $V(e^{j\omega})$ of the resulting signal $v[n] = x[n] \cos \omega_c n$ is given by

$$V(e^{j\omega}) = \tfrac{1}{2} \left[X(e^{j(\omega-\omega_c)}) + X(e^{j(\omega+\omega_c)}) \right]. \tag{11.95}$$

As indicated in Figure 11.48, if $X(e^{j\omega})$ is bandlimited to ω_M, the spectrum $V(e^{j\omega})$ of the modulated signal $v[n]$ has a bandwidth of $2\omega_M$ centered at $\pm\omega_c$. By choosing widely separated carrier frequencies, one can modulate a number of low-frequency signals to high-frequency signals, combine them by frequency-division multiplexing, and transmit over a common channel. The carrier frequencies are chosen appropriately to ensure that there is no overlap in the spectra of the modulated signals when combined by frequency-division multiplexing. At the receiving end, each of the modulated signals is then separated by a bank of bandpass filters of center frequencies corresponding to the different carrier frequencies.

It is evident from Figure 11.48 that, for a real low-frequency signal $x[n]$, the spectrum of its modulated version $v[n]$ is symmetric with respect to the carrier frequency ω_c. Thus, the portion of the spectrum in the frequency range from ω_c to $(\omega_c + \omega_M)$, called the *upper sideband*, has the same information content as the portion in the frequency range from $(\omega_c - \omega_M)$ to ω_c, called the *lower sideband*. Hence, for a more efficient utilization of the channel bandwidth, it is sufficient to transmit either the upper or the lower sideband signal. A conceptually simple way of eliminating one of the sidebands is to pass the modulated signal $v[n]$ through a sideband filter whose passband covers the frequency range of one of the sidebands.

An alternative, often preferred, approach for single-sideband signal generation is by modulating the analytic signal whose real and imaginary parts are, respectively, the real signal and its Hilbert transform. To illustrate this approach, let $y[n] = x[n] + j\hat{x}[n]$ where $\hat{x}[n]$ is the Hilbert transform of $x[n]$. Consider

$$
\begin{aligned}
s[n] = y[n]e^{j\omega_c n} &= (y_{\text{re}}[n] + j y_{\text{im}}[n]) (\cos \omega_c n + j \sin \omega_c n) \\
&= \left(x[n] \cos \omega_c n - \hat{x}[n] \sin \omega_c n\right) \\
&\quad + j \left(x[n] \sin \omega_c n + \hat{x}[n] \cos \omega_c n\right).
\end{aligned}
\tag{11.96}
$$

From Eq. (11.96), the real and imaginary parts of $s[n]$ are thus given by

$$
s_{\text{re}}[n] = x[n] \cos \omega_c n - \hat{x}[n] \sin \omega_c n,
\tag{11.97a}
$$
$$
s_{\text{im}}[n] = x[n] \sin \omega_c n + \hat{x}[n] \cos \omega_c n.
\tag{11.97b}
$$

Figure 11.49 shows the spectra of $x[n]$, $\hat{x}[n]$, $y[n]$, $s[n]$, $s_{\text{re}}[n]$, and $s_{\text{im}}[n]$. It therefore follows from these plots that a single-sideband signal can be generated using either one of the modulation schemes described by Eqs. (11.97a) and (11.97b), respectively. A block diagram representation of the scheme of Eq. (11.97a) is sketched in Figure 11.50.

11.8   Subband Coding of Speech and Audio Signals

In many applications, the digital signal obtained by the analog-to-digital conversion of an analog signal has to be transmitted over a channel with a limited bandwidth or stored for future use in a storage medium with limited capacity. Signal compression methods are being increasingly employed to increase the efficiency of transmission and/or storage. For example, the speech signal in a telephone communication system has a bandwidth of about 3.4 kHz. For digital transmission, it is sampled at a rate of 8 kHz and then converted by means of an 8-bit A/D converter, resulting in a 64,000 bits/sec or, equivalently, 64 kbits/sec digital signal. Up to 24 such 64 kbits/sec digital signals can be transmitted over the T-1 carrier channel in the United States by time-division multiplexing. By compressing the speech signal to a rate of 32 kbits/sec, the number of telephone signals sharing the channel can be doubled to a total of 48. Likewise, the storage capacity of a compact disc can be increased by compressing the digitized music.

One of the most popular schemes for signal compression is the subband coding that employs the quadrature-mirror filter bank discussed in Sections 10.9 to 10.12, and efficiently encodes the signal by exploiting the nonuniform distribution of signal energy in the frequency band. In this section, we review the principles of subband coding of speech and audio signals and describe one specific coding system.

A simplified representation of the subband coding scheme is shown in Figure 11.51. As indicated here, in this scheme, the input signal $x[n]$ at the transmitting end is decomposed into a set of narrowband signals occupying contiguous frequency bands by means of an analysis filter bank. These narrowband signals are then down-sampled, producing the subband signals. Each subband signal is next compressed by an encoder, and all compressed subband signals are multiplexed and sent over the channel to the receiver. At the receiving end, the composite signal is first demultiplexed into a set of subband signals. Each subband signal is then decompressed by a decoder, up-sampled, and passed through a synthesis filter bank. The

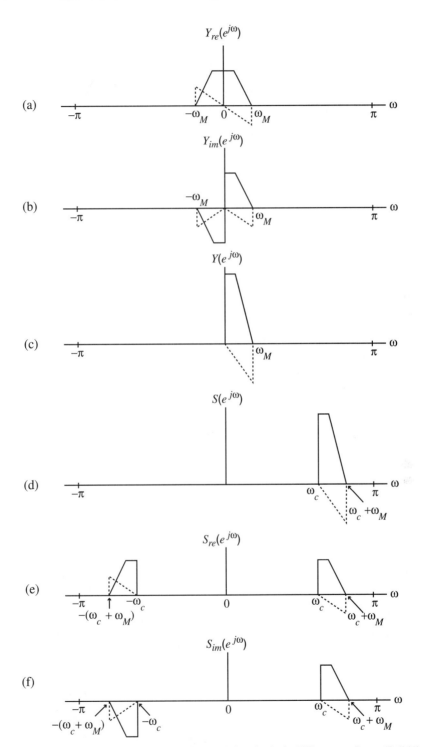

Figure 11.49: Illustration of the generation of single-sideband signals via the Hilbert transform. (Solid lines represent the real parts while dashed lines represent the imaginary parts.)

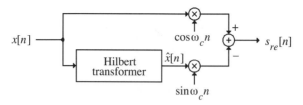

Figure 11.50: Schematic of the single-sideband generation scheme of Eq. (11.97a).

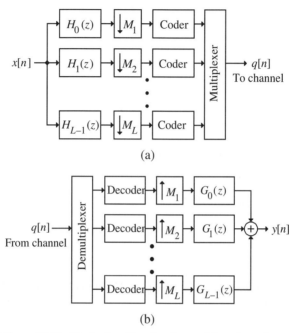

Figure 11.51: The basic subband coding and decoding scheme.

outputs of the synthesis filter bank are finally combined to produce a signal $y[n]$ that is an acceptable replica of the original input signal $x[n]$. The subband coding offers a better compression ratio than the direct compression of the original signal $x[n]$ since it allocates a different number of bits to each subband signal by taking advantage of its spectral characteristics resulting in a lower average bit rate per sample.

Subband coding of 7-kHz wideband audio at 56 kbits/sec based on a five-band QMF bank has been investigated by Richardson and Jayant [Ric86]. The five bands occupy the frequency ranges 0–875 Hz, 875–1750 Hz, 1750–3500 Hz, 3500–5250 Hz, and 5250–7000 Hz. The gain responses of the five filters used are shown in Figure 11.52 along with the sum of the squares of their magnitude responses.

The five-band partition has been obtained using a three-stage cascade of two-channel QMF banks as indicated in Figure 11.53, which also shows the sampling rate of the input and the subband signals along with the order of the filters used. The encoding delay caused by the subband filtering operations is around 10 msec. The subband signals have been encoded using an adaptive differential PCM (ADPCM) scheme [Jay74]. The bit allocations used are 5 bits/sample for each of the two low-frequency channels, 3 bits/sample for the highest-frequency channel, and 4 bits/sample for the two intermediate-frequency channels. The total bit rate of the subband ADPCM coder is thus 56 kbits/sec, which is equivalent to an average bit rate of 4 bits/sample of the input signal at the 14-kHz sampling rate.

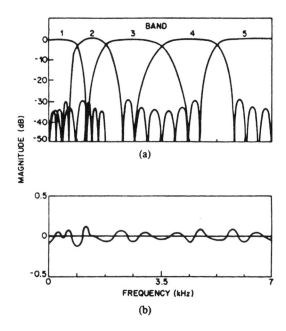

Figure 11.52: (a) Gain responses of the analysis filters and (b) sum of magnitude-squared responses in dB. (Reproduced with permission from [Ric86] ©1986 IEEE.)

Further details of the above subband coder along with a comparison of its performance with that of two other coding schemes can be found in [Ric86].

11.9 Transmultiplexers

In the United States and most other countries, the telephone service employs two types of multiplexing schemes to transmit multiple low-frequency voice signals over a wideband channel. In the *frequency-division multiplex* (FDM) telephone system, multiple analog voice signals are first modulated by single-sideband (SSB) modulators onto several subcarriers, combined, and transmitted simultaneously over a common wideband channel. To avoid *cross-talk*, the subcarriers are chosen to ensure that the spectra of the modulated signals do not overlap. At the receiving end, the modulated subcarrier signals are separated by analog bandpass filters, and demodulated to reconstruct the individual voice signals. On the other hand, in the *time-division multiplex* (TDM) telephone system, the voice signals are first converted into digital signals by sampling and A/D conversion. The samples of the digital signals are time-interleaved by a digital multiplexer, and the combined signal is transmitted. At the receiving end, the digital voice signals are separated by a digital demultiplexer and then passed through a D/A converter and an analog reconstruction filter to recover the original analog voice signals.

The TDM system is usually employed for short-haul communication, while the FDM scheme is preferred for long-haul transmission. Until the telephone service becomes all digital, it is necessary to translate signals between the two formats. This is achieved by the transmultiplexer system discussed next.

The *transmultiplexer* is a multi-input, multi-output, multirate structure, as shown in Figure 11.54. It is exactly the opposite to that of the QMF bank of Figure 10.63 and consists of an *L*-channel synthesis filter

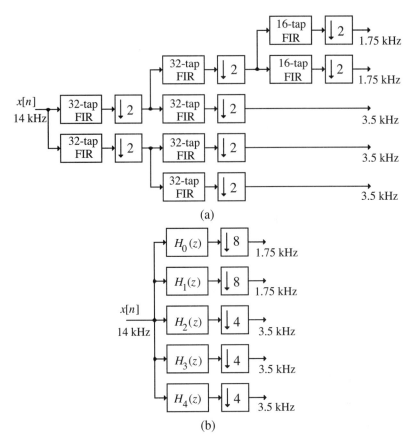

Figure 11.53: (a) Three-stage realization of the five-band analysis bank and (b) its equivalent representation.

bank at the input end followed by an L-channel analysis filter bank at the output end. To determine the input-output relation of the transmultiplexer, consider one typical path from the kth input to the ℓth output as indicated in Figure 11.55(a) [Vai93]. A polyphase representation of the structure of Figure 11.54 is shown in Figure 11.55(a). Invoking the identity of Section 10.4.4, we note that the structure of Figure 11.55(a) is equivalent to that shown in Figure 11.55(b), consisting of an LTI branch with a transfer function $F_{k\ell}(z)$ that is the zeroth polyphase component of $H_k(z)G_\ell(z)$. The input-output relation of the transmultiplexer is therefore given by

$$Y_k(z) = \sum_{\ell=0}^{L-1} F_{k\ell}(z)X_\ell(z), \quad 0 \le k \le L - 1. \tag{11.98}$$

Denoting

$$\mathbf{Y}(z) = \begin{bmatrix} Y_0(z) & Y_1(z) & \cdots & Y_{L-1}(z) \end{bmatrix}^T, \tag{11.99a}$$

$$\mathbf{X}(z) = \begin{bmatrix} X_0(z) & X_1(z) & \cdots & X_{L-1}(z) \end{bmatrix}^T, \tag{11.99b}$$

we can rewrite Eq. (11.98) as

$$\mathbf{Y}(z) = \mathbf{F}(z)\mathbf{X}(z), \tag{11.100}$$

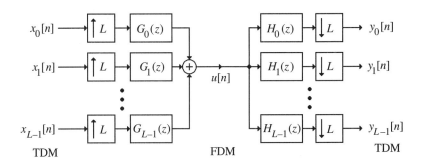

Figure 11.54: The basic L-channel transmultiplexer structure.

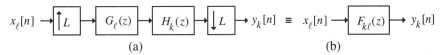

Figure 11.55: The k, ℓ-path of the L-channel transmultiplexer structure.

where $\mathbf{F}(z)$ is an $L \times L$ matrix whose (k, ℓ)th element is given by $F_{k\ell}(z)$. The objective of the transmultiplexer design is to ensure that $y_k[n]$ is a reasonable replica of $x_k[n]$. If $y_k[n]$ contains contributions from $x_r[n]$ with $r \neq n$, then there is cross-talk between these two channels. It follows from Eq. (11.100) that cross-talk is totally absent if $F(z)$ is a diagonal matrix, in which case Eq. (11.100) reduces to

$$Y_k(z) = F_{kk}(z)X_k(z), \quad 0 \leq k \leq L - 1. \tag{11.101}$$

As in the case of the QMF bank, we can define three types of transmultiplexer systems. It is a phase-preserving system, if $F_{kk}(z)$ is a linear-phase transfer function for all values of k. Likewise, it is a magnitude-preserving system, if $F_{kk}(z)$ is an allpass function. Finally, for a perfect reconstruction transmultiplexer,

$$F_{kk}(z) = \alpha_k z^{-n_k}, \quad 0 \leq k \leq L - 1, \tag{11.102}$$

where n_k is an integer and α_k is a nonzero constant. For a perfect reconstruction system, $y_k[n] = \alpha_k x_k[n - n_k]$.

The perfect reconstruction condition can also be derived in terms of the polyphase components of the synthesis and analysis filter banks of the transmultiplexer of Figure 11.54, as shown in Figure 11.56(a) [Kol91]. Using the cascade equivalences of Figure 10.14, we arrive at the equivalent representation indicated in Figure 11.56(b). Note that the structure in the center part of this figure is a special case of the system of Figure 11.54, where $G_\ell(z) = z^{-(L-1-\ell)}$ and $H_k(z) = z^{-k}$, with $\ell, k = 0, 1, \ldots, L - 1$. Here the zeroth polyphase component of $H_{\ell+1}(z)G_\ell(z)$ is z^{-1} for $\ell = 0, 1, \ldots, L - 2$, the zeroth polyphase component of $H_0(z)G_{L-1}(z)$ is 1, and the zeroth polyphase component of $H_k(z)G_\ell(z)$ is 0 for all other cases. As a result, a simplified equivalent representation of Figure 11.56(b) is as shown in Figure 11.57.

The transfer matrix characterizing the transmultiplexer is thus given by

$$\mathbf{M}(z) = \mathbf{E}(z) \begin{bmatrix} \mathbf{0} & 1 \\ z^{-1}\mathbf{I}_{L-1} & \mathbf{0} \end{bmatrix} \mathbf{R}(z), \tag{11.103}$$

where $\mathbf{I}_{L-1}$ is an $(L-1) \times (L-1)$ identity matrix. Now, for a perfect reconstruction system it is sufficient to ensure that

$$\mathbf{M}(z) = dz^{-n_o}\mathbf{I}_L, \tag{11.104}$$

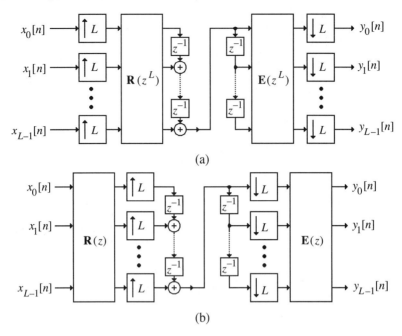

(a)

(b)

Figure 11.56: (a) Polyphase representation of the L-channel transmultiplexer, and (b) its computationally efficient realization.

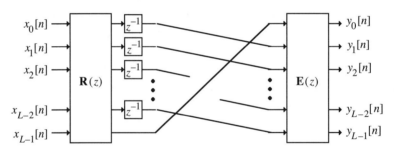

Figure 11.57: Simplified equivalent circuit of Figure 11.56.

where n_o is a positive integer. From Eqs. (11.103) and (11.104) we arrive at the condition for perfect reconstruction in terms of the polyphase components as

$$\mathbf{R}(z)\mathbf{E}(z) = dz^{-m_o}\begin{bmatrix} \mathbf{0} & \mathbf{I}_{L-1} \\ z^{-1} & \mathbf{0} \end{bmatrix}, \tag{11.105}$$

where m_o is a suitable positive integer.

It is possible to develop a perfect reconstruction transmultiplexer from a perfect reconstruction QMF bank with analysis filters $H_\ell(z)$ and synthesis filters $G_\ell(z)$, with a distortion transfer function given by $T(z) = dz^{-K}$, where d is a nonzero constant and K is a positive integer. It can be shown that a perfect reconstruction transmultiplexer can then be designed using the analysis filters $H_\ell(z)$ and synthesis filters $z^{-R}G_\ell(z)$, where R is a positive integer less than L such that $R + K$ is a multiple of L [Kol91]. We illustrate this approach in the following example.

EXAMPLE 11.11 Consider the perfect reconstruction analysis/synthesis filter bank of Example 10.21 with an input-output relation $y[n] = 2x[n - 2]$. In this case, the analysis and synthesis filters are given by

$$H_0(z) = 1 + z^{-1} + z^{-2}, \quad H_1(z) = 1 - z^{-1} + z^{-2}, \quad H_2(z) = 1 - z^{-2},$$
$$G_0(z) = 1 + 2z^{-1} + z^{-2}, \quad G_1(z) = 1 - 2z^{-1} + z^{-2}, \quad G_2(z) = -2 + 2z^{-2}.$$

Here, $d = 2$ and $K = 2$. We thus choose $R = 1$ so that $R + K = 3$. The synthesis filters of the transmultiplexer are thus given by $z^{-1}G_\ell(z)$.

We now examine the products $z^{-1}G_\ell(z)H_k(z)$, for $\ell, k = 0, 1, 2$, and determine their zeroth polyphase components. Thus,

$$z^{-1}G_0(z)H_0(z) = z^{-1} + 3z^{-2} + 4z^{-3} + 3z^{-4} + z^{-5},$$

whose zeroth polyphase component is given by $4z^{-1}$, and hence, $y_0[n] = 4x_0[n - 1]$. Likewise,

$$z^{-1}G_1(z)H_1(z) = z^{-1} - 3z^{-2} + 4z^{-3} - 3z^{-4} + z^{-5},$$

with a zeroth polyphase component $4z^{-1}$ resulting in $y_1[n] = 4x_1[n - 1]$. Similarly,

$$z^{-1}G_2(z)H_2(z) = -2z^{-1} + 4z^{-3} - 2z^{-5},$$

whose zeroth polyphase component is again $4z^{-1}$, implying $y_2[n] = 4x_2[n - 1]$. It can be shown that the zeroth polyphase components for all other products $z^{-1}G_\ell(z)H_k(z)$, with $\ell \neq k$, is 0, indicating a total absence of cross-talk between channels.

In a typical TDM-to-FDM format translation, 12 digitized speech signals are interpolated by a factor of 12, modulated by single-sideband modulation, digitally summed, and then converted into an FDM analog signal by D/A conversion. At the receiving end, the analog signal is converted into a digital signal by A/D conversion and passed through a bank of 12 single-sideband demodulators whose outputs are then decimated, resulting in the low-frequency speech signals. The speech signals have a bandwidth of 4 kHz and are sampled at an 8-kHz rate. The FDM analog signal occupies the band 60 kHz to 108 kHz, as illustrated in Figure 11.58. The interpolation and the single-sideband modulation can be performed by up-sampling and appropriate filtering. Likewise, the single-sideband demodulation and the decimation can be implemented by appropriate filtering and down-sampling.

11.10 Discrete Multitone Transmission of Digital Data

Binary data are normally transmitted serially as a pulse train, as indicated in Figure 11.59(a). However, in order to faithfully extract the information transmitted, the receiver requires complex equalization procedures to compensate for channel imperfection and to make full use of the channel bandwidth. For example, the pulse train of Figure 11.59(a) arriving at the receiver may appear as indicated in Figure 11.59(b). To alleviate the problems encountered with the transmission of data as a pulse train, frequency-division multiplexing with overlapping subchannels has been proposed. In such a system, each binary digit a_r, $r = 0, 1, 2, \ldots, N - 1$, modulates a subcarrier sinusoidal signal $\cos(2\pi r t / T)$, as indicated in Figure 11.59(c) for the transmission of the data of Figure 11.59(a), and then the modulated subcarriers are summed and transmitted as one composite analog signal. At the receiver, the analog signal is passed through a bank of coherent demodulators whose outputs are tested to determine the digits transmitted. This is the basic idea behind the multicarrier modulation/demodulation scheme for digital data transmission.

A widely used form of the multicarrier modulation is the discrete multitone transmission (DMT) scheme in which the modulation and demodulation processes are implemented via the discrete Fourier transform (DFT) efficiently realized using fast Fourier transform (FFT) methods. This approach leads to an all-digital system, eliminating the arrays of sinusoidal generators and the coherent demodulators [Cio91], [Pel80].

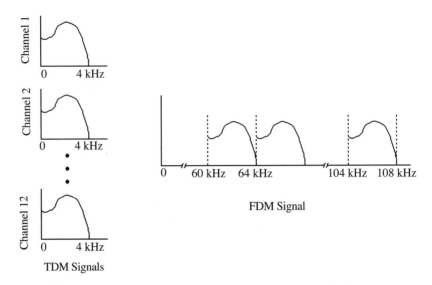

Figure 11.58: Spectrums of TDM signals and the FDM signal.

We outline here the basic idea behind the DMT scheme. Let $\{a_k[n]\}$ and $\{b_k[n]\}$, $0 \leq k \leq M - 1$, be two $M - 1$ real-valued data sequences operating at a sampling rate of F_T that are to be transmitted. Define a new set of complex sequences $\{\alpha_k[n]\}$ of length $N = 2M$ according to

$$\alpha_k[n] = \begin{cases} 0, & k = 0, \\ a_k[n] + jb_k[n], & 1 \leq k \leq \frac{N}{2} - 1, \\ 0, & k = \frac{N}{2}, \\ a_{N-k}[n] - jb_{N-k}[n], & \frac{N}{2} + 1 \leq k \leq N - 1. \end{cases} \tag{11.106}$$

We apply an inverse DFT, and the above set of N sequences is transformed into another new set of N signals $\{u_\ell[n]\}$ given by

$$u_\ell[n] = \frac{1}{N} \sum_{k=0}^{N-1} \alpha_k[n] W_N^{-\ell k}, \qquad 0 \leq \ell \leq N - 1, \tag{11.107}$$

where $W_N = e^{-j2\pi/N}$. Note that the method of generation of the complex sequence set $\{\alpha_k[n]\}$ ensures that its IDFT $\{u_\ell[n]\}$ will be a real sequence. Each of these N signals is then up-sampled by a factor of N and time-interleaved, generating a composite signal $\{x[n]\}$ operating at a rate of NF_T that is assumed to be equal to $2F_c$. The composite signal is converted into an analog signal $x_a(t)$ by passing it through a D/A converter followed by an analog reconstruction filter. The analog signal $x_a(t)$ is then transmitted over the channel.

At the receiver, the received analog signal $y_a(t)$ is passed through an analog anti-aliasing filter and then converted into a digital signal $\{y[n]\}$ by an S/H circuit followed by an A/D converter operating at a rate of $NF_T = 2F_c$. The received digital signal is then deinterleaved by a delay chain containing $N - 1$ unit delays whose outputs are next down-sampled by a factor of N, generating the set of signals $\{v_\ell[n]\}$. Applying the DFT to these N signals, we finally arrive at N signals $\{\beta_k[n]\}$

$$\beta_k[n] = \sum_{\ell=0}^{N-1} v_\ell[n] W_N^{\ell k}, \qquad 0 \leq k \leq N - 1. \tag{11.108}$$

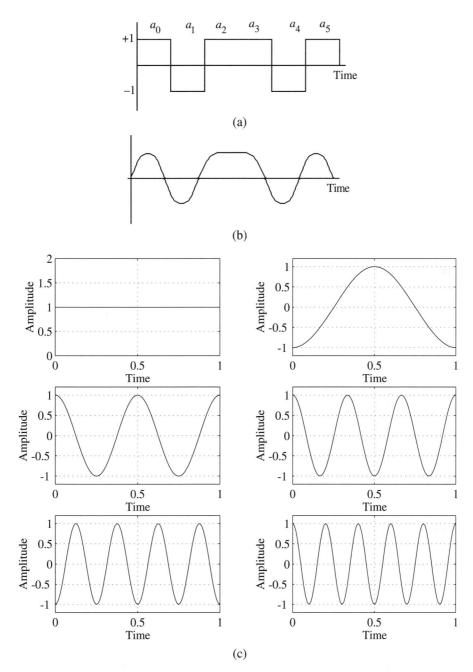

Figure 11.59: (a) Serial binary data stream, (b) baseband serially transmitted signal at the receiver, and (c) signals generated by modulating a set of subcarriers by the digits of the pulse train in (a).

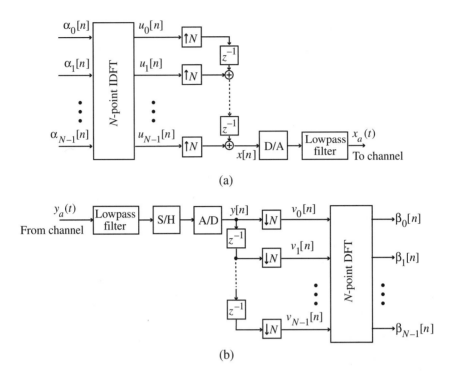

Figure 11.60: The DMT scheme: (a) transmitter, and (b) receiver.

Figure 11.60 shows schematically the overall DMT scheme. If we assume the frequency response of the channel to have a flat passband, and assume the analog reconstruction and anti-aliasing filters to be ideal lowpass filters, then neglecting the nonideal effects of the D/A and the A/D converters, we can assume $y[n] = x[n]$. Hence, the interleaving circuit of the DMT structure at the transmitting end connected to the deinterleaving circuit at the receiving end is identical to the same circuit in the transmultiplexer structure of Figure 11.56(b) (with $L = N$). From the equivalent representation given in Figure 11.57, it follows then that

$$v_k[n] = u_{k-1}[n-1], \qquad 0 \le k \le N-2,$$
$$v_0[n] = u_{N-1}[n], \tag{11.109}$$

or in other words,

$$\beta_k[n] = \alpha_{k-1}[n-1], \qquad 0 \le k \le N-2,$$
$$\beta_0[n] = \alpha_{N-1}[n]. \tag{11.110}$$

Transmission channels, in general, have a bandpass frequency response $H_{\text{ch}}(f)$ with a magnitude response dropping to zero at some frequency F_c. In some cases, in the passband of the channel, the magnitude response, instead of being flat, drops very rapidly outside its passband, as indicated in Figure 11.61. For reliable digital data transmission over such a channel and its recovery at the receiving end, the channel's frequency response needs to be compensated by essentially a highpass equalizer at the receiver. However, such an equalization also amplifies high-frequency noise that is invariably added to the data signal as it passes through the channel.

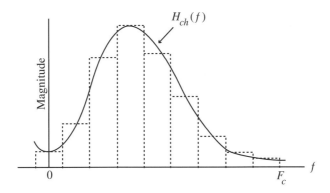

Figure 11.61: Frequency response of a typical bandlimited channel.

For a large value of the DFT length N, the channel can be assumed to be composed of a series of contiguous narrow-bandwidth bandpass subchannels. If the bandwidth is reasonably narrow, the corresponding bandpass subchannel can be considered to have an approximately flat magnitude response, as indicated by the dotted lines in Figure 11.61, and the channel can be approximately characterized by a single complex number given by the value of its frequency response at $\omega = 2\pi k/N$. The values can be determined by first transmitting a known training signal of unmodulated carriers and generating the respective channel frequency response samples. The real data samples are then divided by these complex numbers at the receiver to compensate for channel distortion.

Further details on the performance of the above DMT scheme under nonideal conditions can be found in [Bin90], [Cio91], [She95].

11.11 Digital Audio Sampling Rate Conversion

As indicated in Chapter 10, fractional sampling rate conversion of digital signals is often required in professional audio and video applications. A classical approach to such conversion is to transform the input digital signal into an analog signal by means of a D/A converter followed by an analog lowpass filter, resample the analog signal at the desired sampling rate, and convert the resampled signal into a digital form by means of an A/D converter. To reduce the effect of aliasing to a minimum, the analog reconstruction filter must have a very sharp cutoff with sufficiently large stopband attenuation and very small passband ripple, in addition to exhibiting nearly linear phase response in the passband. It is difficult to build an economical analog filter with such stringent requirements. Moreover, to achieve the desired accuracy in the output digital signal, both the D/A converter and the A/D converter must have large enough resolutions. An alternative approach is to use an all-digital conversion technique, the basics of which have been discussed in Section 10.2.2.

The complexity of the design of the fractional sampling rate converter depends on the ratio of the sampling rates between the input and the output digital signals. For example, in digital audio applications, the three different sampling frequencies employed are 44.1 kHz, 32 kHz, and 48 kHz. As a consequence, there are three different values for the sampling rate conversion factor – 2:3, 147:160, and 320:441. Likewise, in digital video applications, the sampling rates of composite video signals are 14.3181818 MHz and 17.734475 MHz, whereas the sampling rates of the digital component video signal are 13.5 MHz and 6.75 MHz for the luminance and the color-difference signals, respectively, for the NTSC and PAL systems. Here, the sampling rates for the component and the NTSC composite video signals are related by a ratio 35:33, whereas the sampling rates for the component and the PAL composite video signals are related by

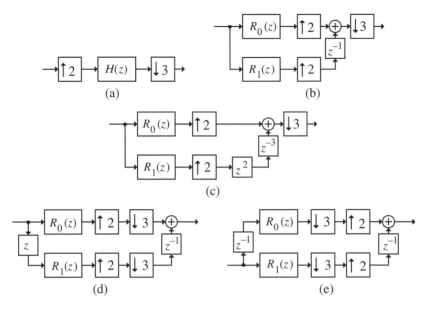

Figure 11.62: Various steps in the development of the 48–32-kHz sampling rate converter.

a ratio 709,379:540,000. There are applications, such as in the pitch control of audio signals, where the ratio is irrational, in which case there is no periodic relation between the sampling instants of the input and the output digital signals.

The design of a fractional sampling rate converter for conversion between two digital signals related by a ratio of two low-valued integers can be carried out using the method discussed in Section 10.2.2. However, for a sampling rate conversion factor that is a ratio of two very large integers or an irrational number, the design is somewhat more complex.

In this section we discuss the design of fractional sampling rate converters for both of the above two cases as encountered in audio applications [Cuc91], [Lag81], [Ram84]. A similar procedure is followed in video applications [Lut91].

11.11.1 Conversion between 32-kHz and 48-kHz Sampling Rates

The conversion of a digital audio signal of 48-kHz sampling rate to one of 32-kHz sampling rate requires the design of a fractional sampling rate decimator with a decimation factor of 2/3. Its basic form is as indicated in Figure 11.62(a), where the lowpass filter $H(z)$ has a stopband edge at $\pi/3$. Note that in this structure, the filter $H(z)$ operates at the 96-kHz rate. We now outline the development of a computationally efficient realization in which all filters operate at the 16-kHz rate [Hsi87], [Vai90]. Replacing the lowpass filter with its Type II polyphase realization and then making use of the cascade equivalence of Figure 10.14(b), we arrive at the equivalent realization indicated in Figure 11.62(b) in which now the filters $R_0(z)$ and $R_1(z)$ operate at the input sampling rate of 48 kHz. Simple block diagram manipulations and use of the cascade equivalences of Figure 10.13 lead finally to an equivalent realization depicted in Figure 11.62(e).

The computational efficiency of the sampling rate converter can be improved further by realizing the filters $R_0(z)$ and $R_1(z)$ in Type I polyphase forms and then applying the cascade equivalence of Figure 10.14(a). The final realization is as shown in Figure 11.63, in which all filters now operate at the 16-kHz rate.

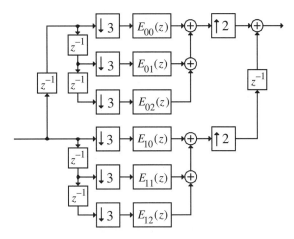

Figure 11.63: Computationally efficient realization of the 48–32-kHz sampling rate converter.

The transpose of the structure of Figure 11.63 yields the realization of a fractional rate interpolator with an interpolation factor of 3/2 and can be used as the 32-48-kHz sampling rate converter (Problem 11.23).

11.12   Oversampling A/D Converter

For the digital processing of an analog continuous-time signal, the signal is first passed through a sample-and-hold circuit whose output is then converted into a digital form by means of an A/D converter. However, according to the sampling theorem, discussed in Section 5.2.1, a bandlimited continuous-time signal with a lowpass spectrum can be fully recovered from its uniformly sampled version if it is sampled at a sampling frequency that is at least twice the highest frequency contained in the analog signal. If this condition is not satisfied, the original continuous-time signal cannot be recovered from its sampled version because of aliasing. To prevent aliasing, the analog signal is thus passed through an analog anti-aliasing lowpass filter prior to sampling, which enforces the condition of the sampling theorem. The passband cutoff frequency of the lowpass filter is chosen equal to the frequency of the highest signal frequency component that needs to be preserved at the output. The anti-aliasing filter also cuts off all out-of-band signal components and any high-frequency noise that may be present in the original analog signal, which otherwise would alias into the baseband after sampling. The filtered signal is then sampled at a rate that is at least twice that of the cutoff frequency.

Let the signal band of interest be the frequency range $0 \leq f \leq F_m$. Then, the Nyquist rate is given by $F_N = 2F_m$. Now, if the sampling rate F_T is the same as the Nyquist rate, we need to design an anti-aliasing lowpass filter with a very sharp cutoff in its frequency response, satisfying the requirements as given by Eq. (5.66).[6] This requires the design of a very high-order anti-aliasing filter structure built with high-precision analog components, and it is usually difficult to implement such a filter in VLSI technology. Moreover, such a filter also introduces undesirable phase distortion in its output. An alternative approach mentioned in Section 5.8.5 is to sample the analog signal at a rate much higher than the Nyquist rate, use a fast low-resolution A/D converter, and then decimate the digital output of the converter to the Nyquist rate. This approach relaxes the sharp cutoff requirements of the analog anti-aliasing filter, resulting in a simpler filter structure that can be built using low-precision analog components while requiring fast, more complex digital signal processing hardware at later stages. The overall structure is not only amenable to

---

[6]Recall that $F_T = 1/T$, where T is the sampling period.

VLSI fabrication but also can be designed to provide linear-phase response in the signal band of interest.

The oversampling approach is an elegant application of multirate digital signal processing and is increasingly being employed in the design of high-resolution A/D converters for many practical systems [Can92], [Fre94]. In this section, we analyze the quantization noise performance of the conventional A/D converter and show analytically how the oversampling approach decreases the quantization noise power in the signal band of interest [Fre94]. We then show that further improvement in the noise performance of an oversampling A/D converter can be obtained by employing a sigma-delta ($\Sigma\Delta$) quantization scheme.[7] For simplicity, we restrict our discussion to the case of a basic first-order sigma-delta quantizer.

To illustrate the above property, consider a b-bit A/D converter operating at F_T Hz. Now, for a full-scale peak-to-peak input analog voltage of R_{FS}, the smallest voltage step represented by b bits is

$$\Delta V = \frac{R_{FS}}{2^b - 1} \simeq \frac{R_{FS}}{2^b}. \tag{11.111}$$

From Eq. (9.69), the rms quantization noise power σ_e^2 of the error voltage, assuming a uniform distribution of the error between $-\Delta V/2$ and $\Delta V/2$, is given by

$$\sigma_e^2 = \frac{(\Delta V)^2}{12}. \tag{11.112}$$

The rms noise voltage, given by σ_e, therefore has a flat spectrum in the frequency range from 0 to $F_T/2$.

The noise power per unit bandwidth, called the *noise density*, is then given by

$$P_{e,n} = \frac{(\Delta V)^2/12}{F_T/2} = \frac{(\Delta V)^2}{6F_T}. \tag{11.113}$$

A plot of the noise densities for two different sampling rates is shown in Figure 11.64, where the shaded portion indicates the signal band of interest. As can be seen from this figure, the total amount of noise in the signal band of interest for the high sampling rate case is smaller than that for the low sampling rate case. The total noise in the signal band of interest, called the *in-band noise power*, is given by

$$P_{\text{total}} = \frac{(R_{FS}/2^b)^2}{12} \cdot \frac{F_m}{F_T/2}. \tag{11.114}$$

It is interesting to compute the needed wordlength β of the A/D converter operating at the Nyquist rate in order that its total noise in the signal band of interest be equal to that of a b-bit A/D converter operating at a higher rate. Substituting $F_T = 2F_m$ and replacing b with β in Eq. (11.114), we arrive at

$$P_{\text{total}} = \frac{(R_{FS}/2^\beta)^2}{12} = \frac{(R_{FS}/2^b)^2}{12} \cdot \frac{F_m}{F_T/2}, \tag{11.115}$$

which leads to the desired relation

$$\beta = b + \frac{1}{2} \log_2 M, \tag{11.116}$$

where $M = F_T/2F_m$ denotes the *oversampling ratio* (OSR). Thus, $\beta - b$ denotes the increase in the resolution of a b-bit converter whose oversampled output is filtered by an ideal brick-wall lowpass filter. A plot of the increase in resolution as a function of the oversampling ratio is shown in Figure 11.65. For example, for an OSR of $M = 1000$, an 8-bit oversampling A/D converter has an effective resolution equal to that of a 13-bit A/D converter operating at the Nyquist rate. Note that Eq. (11.116) implies that the increase in the resolution is $1/2$-bit per doubling of the OSR.

---

[7]Also known as delta-sigma A/D converter.

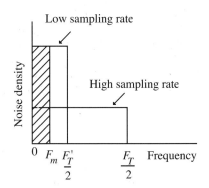

Figure 11.64: A/D converter noise density.

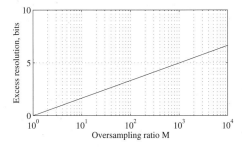

Figure 11.65: Excess resolution as a function of the oversampling ratio M.

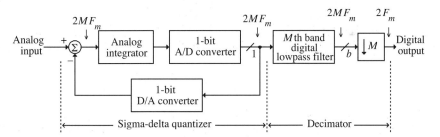

Figure 11.66: Oversampling sigma-delta A/D converter structure.

We now illustrate the improvement in the noise performance obtained by employing a sigma-delta ($\Sigma\Delta$) quantization scheme. The sigma-delta A/D converter was briefly introduced in Section 5.8.5 and is shown in block diagram form in Figure 11.66 for convenience. This figure also indicates the sampling rates at various stages of the structure. It should be noted here that the 1-bit output samples of the quantizer after decimation become b-bit samples at the output of the sigma-delta A/D converter due to the filtering operations involving b-bit multiplier coefficients of the Mth-band digital lowpass filter.

Since the oversampling ratio M is typically very large in practice, the sigma-delta A/D converter is most useful in low-frequency applications such as digital telephony, digital audio, and digital spectrum analyzers. For example, Figure 11.67 shows the block diagram of a typical compact disc encoding system used to convert the input analog audio signal into a digital bit stream that is then applied to generate the master disc [Hee82]. Here the oversampling sigma-delta A/D converter employed has a typical input sampling rate of 3175.2 kHz and an output sampling rate of 44.1 kHz [Kam86].

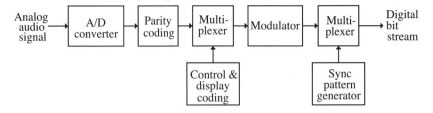

Figure 11.67: Compact disc encoding system for one channel of a stereo audio.

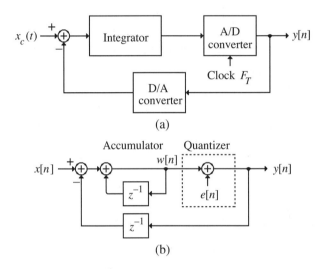

Figure 11.68: Sigma-delta quantization scheme.

To understand the operation of the sigma-delta A/D converter of Figure 11.66, we need to study the operation of the sigma-delta quantizer shown in Figure 11.68(a). To this end it is convenient to use the discrete-time equivalent circuit of Figure 11.68(b), where the integrator has been replaced with an accumulator.[8] Here, the input $x[n]$ is a discrete-time sequence of analog samples developing an output sequence of binary-valued samples $y[n]$. From this diagram, we observe that, at each discrete instant of time, the circuit forms the difference (Δ) between the input and the delayed output, which is accumulated by a summer (Σ) whose output is then quantized by a one-bit A/D converter, i.e., a comparator.

Even though the input-output relation of the sigma-delta quantizer is basically nonlinear, the low-frequency content of the input $x_c(t)$ can be recovered from the output $y[n]$ by passing it through a digital lowpass filter. This property can be easily shown for a constant input analog signal $x_a(t)$ with a magnitude less than $+1$. In this case, the output $w[n]$ of the accumulator is a bounded sequence with sample values equal to either -1 or $+1$. This can happen only if the input to the accumulator has an average value of zero. Or in other words, the average value of $w[n]$ must be equal to the average value of the input $x[n]$ [Sch91]. The following two examples illustrate the operation of a sigma-delta quantizer.

---

[8]In practice, the integrator is implemented as a discrete-time switched-capacitor integrator.

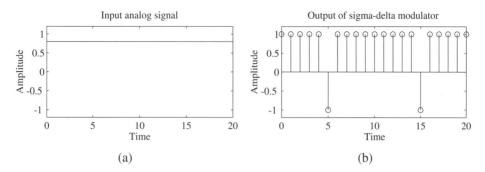

Figure 11.69: Input and output waveforms of the sigma-delta quantizer of Figure 11.68(a) for a constant input.

EXAMPLE 11.12   We first consider the operation for the case of a constant input signal using MATLAB. To this end we can use Program 11_6 given below.

```
% Program 11_6
% Sigma-Delta Quantizer Operation
N = input('Type in the length of input sequence = ');
n = 1:1:N;
m = n-1;
A = input('Type in the input amplitude = ');
x = A*ones(1,N);
plot(m,x);
axis([0 N-1 -1.2 1.2]);
xlabel('Time'); ylabel('Amplitude');
title('Input analog signal');
pause
y = zeros(1,N+1);
v0 = 0;
    for k = 2:1:N+1;
        v1 = x(k-1) - y(k-1) + v0;
        y(k) = sign(v1);
        v0 = v1;
    end
yn = y(2:N+1);
axis([0 N-1 -1.2 1.2]);
stem(m, yn);
xlabel('Time'); ylabel('Amplitude');
title('Output of sigma-delta modulator');
ave = sum(yn)/N;
disp('Average value of output is = ');disp(ave)'
```

The plots generated by this program are the input and output waveforms of the sigma-delta quantizer of Figure 11.68(a) and are shown in Figure 11.69. The program also prints the average value of the output as indicated below:

```
Average value of output is =
    0.8095
```

which is very close to the amplitude 0.8 of the constant input. It can be easily verified that the average value of the output gets closer to the amplitude of the constant input as the length of the input increases.

EXAMPLE 11.13   We now verify the operation of the sigma-delta A/D converter for a sinusoidal input of frequency 0.01 Hz using MATLAB. To this end we make use of the Program 11_7 given below. Because of the short length of the input sequence, the filtering operation is performed here in the DFT domain [Sch91].

```
% Program 11_7
% Sigma-Delta A/D Converter Operation
%
wo = 2*pi*0.01;
N = input('Type in the length of input sequence = ');
n = 1:1:N;
m = n-1;
A = input('Type in the amplitude of the input = ');
x = A*cos(wo*m);
axis([0 N-1 -1.2 1.2]);
plot(m,x);
xlabel('Time'); ylabel('Amplitude');
title('Input analog signal');
pause
y = zeros(1,N+1);
v0 = 0;
for k = 2:1:N+1;
v1 = x(k-1) - y(k-1) + v0;
    if v1 >= 0;
       y(k) = 1;
    else
       y(k) = -1;
    end
v0 = v1;
end
yn = y(2:N+1);
axis([0 N-1 -1.2 1.2]);
stairs(m, yn);
xlabel('Time'); ylabel('Amplitude');
title('Output of sigma-delta quantizer');
Y = fft(yn);
pause
H = [1 1 0.5 zeros(1,N-5) 0.5 1];
YF = Y.*H;
out = ifft(YF);
axis([0 N-1 -1.2 1.2]);
plot(m,out);
xlabel('Time'); ylabel('Amplitude');
title('Lowpass filtered output');
```

Figure 11.70 shows the input and output waveforms of the sigma-delta quantizer of Figure 11.68(a) for a sinusoidal input. Figure 11.71 depicts the lowpass filtered version of the output signal shown in Figure 11.70(b). As can be seen from these figures, the filtered output is nearly an exact replica of the input.

It follows from Fig. 11.68(b) that the output $y[n]$ of the quantizer is given by

$$y[n] = w[n] + e[n], \tag{11.117}$$

where

$$w[n] = x[n] - y[n-1] + w[n-1]. \tag{11.118}$$

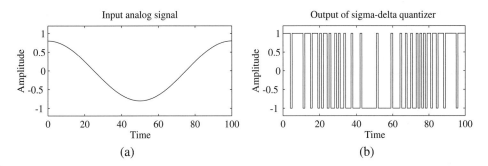

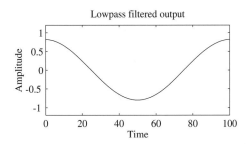

Figure 11.70: Input and output waveforms of the sigma-delta quantizer of Figure 11.68(a) with a sine wave input.

Figure 11.71: The lowpass filtered version of the waveform of Figure 11.70(b).

From the above equations, we obtain after some algebra

$$y[n] = x[n] + (e[n] - e[n-1]), \tag{11.119}$$

where the quantity inside the parentheses represents the noise due to sigma-delta modulation. The noise transfer function is simply $G(z) = (1 - z^{-1})$. The power spectral density of the modulation noise is therefore given by

$$P_y(f) = \left| G(e^{j2\pi fT}) \right|^2 P_e(f) = 4\sin^2\left(\frac{2\pi fT}{2}\right) P_e(f), \tag{11.120}$$

where we have assumed the power spectral density $P_e(\omega)$ of the quantization noise to be the one-sided power spectral density defined for positive frequencies only. For a random signal input $x[n]$, $P_e(f)$ is constant for all frequencies and is given by

$$P_e(f) = \frac{(\Delta V)^2/12}{F_T/2}. \tag{11.121}$$

Substituting the above in Eq. (11.120), we arrive at the power spectral density of the output noise, given by

$$P_y(f) = \frac{2}{3}\frac{(\Delta V)^2}{F_T}\sin^2(\pi fT). \tag{11.122}$$

The noise-shaping provided by the sigma-delta quantizer is similar to that encountered in the first-order error-feedback structures of Section 9.10.1 and shown in Figure 9.42. For a very large OSR, as is usually the case, the frequencies in the signal band of interest are much smaller than F_T, the sampling frequency. Thus, we can approximate $P_y(f)$ of Eq. (11.122) as

$$P_y(f) \cong \tfrac{2}{3} \frac{(\Delta V)^2}{F_T} (\pi f T)^2 = \tfrac{2}{3} \pi^2 (\Delta V)^2 T^3 f^2, \qquad f << F_T. \tag{11.123}$$

From the above, the in-band noise power of the sigma-delta A/D converter is thus given by

$$P_{\text{total,sd}} = \int_0^{F_m} P_y(f)\, df = \tfrac{2}{3}\pi^2 (\Delta V)^2 T^3 \int_0^{F_m} f^2\, df = \frac{2}{9}\pi^2 (\Delta V)^2 T^3 (F_m)^3. \tag{11.124}$$

It is instructive to compare the noise performance of the sigma-delta A/D converter with that of a direct oversampling A/D converter operating at a sampling rate of F_T with a signal band of interest from dc to F_m. From Eq. (11.115), the in-band noise power of the latter is given by

$$P_{\text{total,os}} = \tfrac{1}{6} (\Delta V)^2 T F_m. \tag{11.125}$$

The improvement in the noise performance is therefore given by

$$10 \log_{10} \left(\frac{P_{\text{total,os}}}{P_{\text{total,sd}}} \right) = 10 \log_{10} \left(\frac{3M^2}{\pi^2} \right) = -5.1718 + 20 \log_{10}(M)\ \text{dB}, \tag{11.126}$$

where we have used $M = F_T/2F_m$ to denote the OSR. For example, for an OSR of $M = 1000$, the improvement in the noise performance using the sigma-delta modulation scheme is about 55 dB. In this case, the increase in the resolution is about 1.5 bits per doubling of the OSR.

The improved noise performance of the sigma-delta A/D converter results from the shape of $|G(e^{j2\pi fT})|$, which decreases the noise power spectral density in-band ($0 \le f \le F_m$) while increasing it outside the signal band of interest ($f > F_m$). Since this type of converter also employs oversampling, it requires a less stringent analog anti-aliasing filter.

The A/D converter of Figure 11.66 employs a single-loop feedback and is often referred to as a first-order sigma-delta converter. Multiple feedback loop modulation schemes have been advanced to reduce the in-band noise further. However, the use of more than two feedback loops may result in unstable operation of the system, and care must be taken in the design to ensure stable operation [Can92].

As indicated in Figure 11.66, the quantizer output is passed through an Mth-band lowpass digital filter whose output is then down-sampled by a factor of M to reduce the sampling rate to the desired Nyquist rate. The function of the digital lowpass filter is to eliminate the out-of-band quantization noise and the out-of-band signals that would be aliased into the passband by the down-sampling operation. As a result, the filter must exhibit a very sharp cutoff frequency response with a passband edge at F_m. This necessitates the use of a very high order digital filter. In practice, it is preferable to use a filter with a transfer function having simple integer-valued coefficients to reduce the cost of hardware implementation and to permit all multiplication operations to be carried out at the down-sampled rate. In addition, most applications require the use of linear-phase digital filters, which can be easily implemented using FIR filters.

The simplest lowpass FIR filter is the moving-average filter of Eq. (2.56), repeated below for convenience:[9]

$$H(z) = 1 + z^{-1} + z^{-2} + \cdots + z^{-(N-1)}. \tag{11.127}$$

A more convenient form of the above transfer function for realization purposes is given by

$$H(z) = \frac{1 - z^{-N}}{1 - z^{-1}}, \tag{11.128}$$

---

[9]For simplicity, we have ignored the scale factor of $1/N$ which is needed to provide a dc gain of 0 dB.

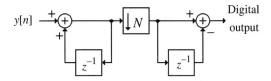

Figure 11.72: A very simple factor-of-N decimator structure.

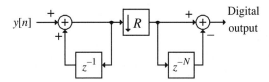

Figure 11.73: A two-stage CIC decimator structure.

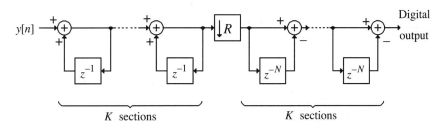

Figure 11.74: A CIC decimator structure with cascaded sections.

also known as a *recursive running-sum filter* or a *boxcar filter*. A realization of a factor-of-N decimator based on the decimation filter of Eq. (11.128) is sketched in Figure 11.72.[10]

Since the decimator based on a running-sum filter does not provide sufficient out-of-band attenuation, often a multistage decimator formed by a cascade of the running-sum decimators, more commonly known as *cascaded integrator comb* (CIC) filters, is used in practice [Hog81]. The structure of a two-stage CIC decimator is shown in Figure 11.73. It can be easily shown that the structure is equivalent to a factor-of-R decimator with a length-RN running sum decimation filter. Further flexibility in the design of a CIC decimator is obtained by including K feedback paths before and K feedforward paths after the down-sampler, as indicated in Figure 11.74. The corresponding transfer function is then given by

$$H(z) = \left(\frac{1 - z^{-RN}}{1 - z^{-1}} \right)^K.$$

(11.129)

The parameters N and K can be adjusted for a given down-sampling factor R to yield the desired out-of-band attenuation.

In some applications, it may be preferable to use a multistage decimation process in which all but the last stage employs the CIC decimators in various forms, followed by an FIR lowpass filter providing a much sharper cutoff before the final down-sampling. For example, in digital telephone applications, the following IIR transfer function provides a very good frequency response and can be used to design a

---

[10]The integrator overload caused by the input adder overflow can be easily handled with binary arithmetic.

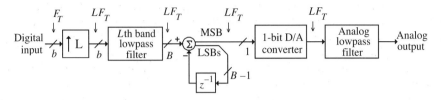

Figure 11.75: Block diagram representation of an oversampling sigma-delta D/A converter.

factor-of-4 decimator [Can92]:

$$H(z) = (1 - \tfrac{3}{2}z^{-1} + z^{-2})(1 - \tfrac{11}{8}z^{-1} + \tfrac{5}{8}z^{-2})(1 - \tfrac{5}{4}z^{-1} + z^{-2})$$

$$\times (1 - \tfrac{101}{64}z^{-1} + \tfrac{7}{8}z^{-2}) \cdot \frac{(1 + z^{-1})(1 + z^{-2})(1 - z^{-5})}{(1 - z^{-1})}. \tag{11.130}$$

Further details on first- and higher-order sigma-delta converters can be found in Candy and Temes [Can92].

11.13   Oversampling D/A Converter

As indicated earlier in Section 5.1, the digital-to-analog conversion process consists of two steps: the conversion of input digital samples into a staircase continuous-time waveform by means of a D/A converter with a zero-order hold at its output, followed by an analog lowpass reconstruction filter. If the sampling rate F_T of the input digital signal is the same as the Nyquist rate, the analog lowpass reconstruction filter must have a very sharp cutoff in its frequency response, satisfying the requirements of Eq. (5.75). As in the case of the anti-aliasing filter, this involves the design of a very high order analog reconstruction filter requiring high-precision analog circuit components. To get around the above problem, here also an oversampling approach is often used, in which case a wide transition band can be tolerated in the frequency response of the reconstruction filter allowing its implementation using low-precision analog circuit components while requiring a more complex digital interpolation filter at the front end.

Further improvement in the performance of an oversampling D/A converter is obtained by employing a digital sigma-delta 1-bit quantizer at the output of the digital interpolator, as indicated in Figure 5.45 and repeated in Figure 11.75 for convenience [Can86], [Lar93]. The quantizer extracts the MSB from its input and subtracts the remaining LSBs, the quantization noise, from its input. The MSB output is then fed into a 1-bit D/A converter and passed through an analog lowpass reconstruction filter to remove all frequency components beyond the signal band of interest. Since the signal band occupies a very small portion of the baseband of the high-sample-rate signal, the reconstruction filter in this case can have a very wide transition band, permitting its realization with a low-order filter that, for example, can be implemented using a Bessel filter to provide an approximately linear phase in the signal band.[11]

The spectrum of the quantized 1-bit output of the digital sigma-delta quantizer is nearly the same as that of its input. Moreover, it also shapes the quantization noise spectrum by moving the noise power out of the signal band of interest. To verify this result analytically, consider the sigma-delta quantizer shown separately in Figure 11.76. It follows from this figure that the input-output relation of the quantizer is given by

$$y[n] - e[n] = x[n] - e[n-1],$$

---

[11]See Section 5.4.5.

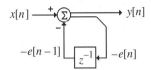

Figure 11.76: The sigma-delta quantizer.

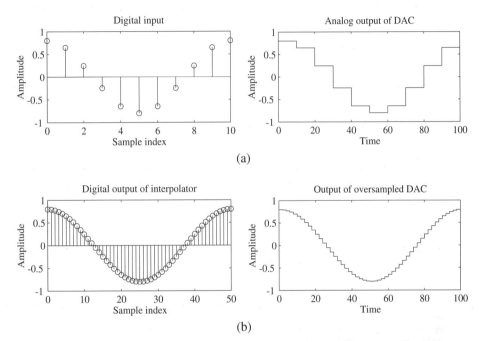

Figure 11.77: Input and output signals of (a) lower-rate D/A converter and (b) oversampling D/A converter.

or equivalently, by

$$y[n] = x[n] + e[n] - e[n-1], \qquad (11.131)$$

where $y[n]$ is the MSB of the nth sample of the adder output, and $e[n]$ is the nth sample of the quantization noise composed of all bits except the MSB. From Eq. (11.131) it can be seen that the transfer function of the quantizer with no quantization noise is simply unity and the noise transfer function is given by $G(z) = 1 - z^{-1}$, which is the same as that for the first-order sigma-delta modulator employed in the oversampling A/D converter discussed in the previous section.

The following example illustrates by computer simulation the operation of a sigma-delta D/A converter for a discrete-time sinusoidal input sequence.

EXAMPLE 11.14 Let the input to the D/A converter be a sinusoidal sequence of frequency 100 Hz operating at a sampling rate F_T of 1 kHz. Figure 11.77(a) shows the digital input sequence and the analog output generated by a D/A converter operating at F_T from this input. Figure 11.77(b) depicts the interpolated sinusoidal sequence operating at a higher sampling rate of 5 kHz obtained by passing the low-sampling-rate sinusoidal signal through a factor-of-5 digital interpolator and the corresponding analog output generated by a D/A converter operating at $5F_T$ rate. If we compare the two D/A converter outputs, we can see that the staircase

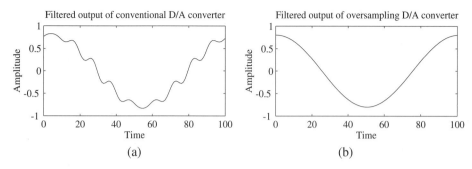

Figure 11.78: Lowpass filtered output signals of (a) conventional D/A converter, and (b) oversampling D/A converter.

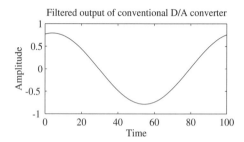

Figure 11.79: Filtered output signals of the conventional D/A converter employing a sharp cutoff lowpass filter.

waveform of the oversampling D/A converter output is much smoother with smaller jumps than that of the lower-rate D/A converter output. Thus, the oversampling D/A converter output has considerably smaller high-frequency components in contrast to the lower-rate D/A converter. This fact can be easily verified by examining their spectra.

The high-frequency components in the baseband outside the signal band of interest can be removed by passing the D/A converter output through an analog lowpass filter, which also eliminates any leftover replicas of the baseband not completely removed by the zero-order hold in the D/A converter. Since the signal band of interest occupies a small portion of the baseband, the replicas of the signal band immediately outside the baseband are widely separated from the signal band inside the baseband. Hence, the lowpass filter can be designed with a very wide transition band. Moreover, due to reduced high-frequency components in the D/A converter output caused by oversampling, the stopband attenuation also does not have to be very large. On the other hand, the replicas of the signal band in the spectrum of the output of the low-rate D/A converter are closely spaced, and the high-frequency components are relatively large in amplitudes. In this case, the lowpass filter must have a sharp cutoff with much larger stopband attenuation to effectively remove the undesired components in the D/A converter output.

Figure 11.78 shows the filtered outputs of the conventional lower-rate and oversampled D/A converters when the same lowpass filter with a wide transition band is used in both cases. As can be seen from this figure, the analog output in the case of the low-rate D/A converter still contains some high-frequency components, while that in the case of the oversampled D/A converter is very close to a perfect sinusoidal signal. A much better output response is obtained in the case of a conventional D/A converter if a sharp cutoff lowpass filter is employed, as indicated in Figure 11.79.

EXAMPLE 11.15   In this example, we verify using MATLAB the operation of the sigma-delta D/A converter for a sinusoidal input sequence of frequency 100 Hz operating at a sampling rate F_T of 5 kHz. The signal is clearly over-sampled since the sampling rate is much higher than the Nyquist rate of 200 Hz. Program 11_8 given below first generates the input digital signal, then generates a two-valued digital signal by quantizing the output of the sigma-delta quantizer, and finally, develops the output of the D/A converter by lowpass filtering the quantized output. As in the case of the sigma-delta converter of Example 11.14, the filtering operation here has also been performed in the DFT domain due to the short length of the input sequence [Sch91].

```
% Program 11_8
% Sigma-Delta D/A Converter Operation
%
clf;
% Generate the input sinusoidal sequence
N = input('Type in length of the input sequence = ');
A = input('Type in amplitude of the input = ');;
w0 = 2*pi*0.02;
n = 1:N;
m = n-1;
x = A*cos(w0*m);
axis([0 N -1 1]);
stem(m,x);
xlabel('Time index'); ylabel('Amplitude');
title ('Input digital signal');
pause
% Generation of quantized output
x = (x)/(A);
y = zeros(1,N+1);
a = zeros(1,N+1);
e = 0;
for k = 2:N+1
    a(k) = x(k-1) - e;
        if a(k) >= 0,
            y(k) = 1;
        else
            y(k) = -1;
        end
    e = y(k) - a(k);
end
yn = y(2:N+1);
axis([0 N -1.2 1.2]);
stem(m, yn); % Plot the quantized output
xlabel('Time'); ylabel('Amplitude');
title ('Digital output of sigma-delta quantizer');
pause
Y = fft(yn);
H = [1 1 0.5 zeros(1,N-5) 0.5 1];% Lowpass filter
YF = Y.*H; % Filtering in the DFT domain
out = ifft(YF);
plot(m,out);
xlabel('Time'); ylabel('Amplitude');
title ('Lowpass filtered analog output');
```

Figure 11.80 shows the digital input signal, the quantized digital output of the sigma-delta quantizer, and the filtered output of the D/A converter generated by this program. As can be seen from these plots, the lowpass filtered output is nearly a scaled replica of the desired sinusoidal analog signal.

One of the most common applications of the oversampling sigma-delta D/A converter is in the compact disc (CD) player. Figure 11.81 depicts the block diagram of the basic components in the signal processing part of a CD player, where typically a factor-of-4 oversampling D/A converter is employed for each audio

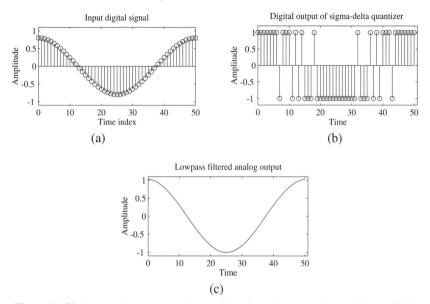

Figure 11.80: Input and output waveforms of the sigma-delta quantizer of Figure 11.77.

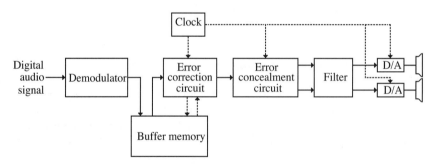

Figure 11.81: Signal processing part of a CD player.

channel [Goe82]. Here, the 44.1-kHz input digital audio signal is interpolated first by a factor of 4 to the 176.4-kHz rate and then converted into an analog audio signal.

11.14 Sparse Antenna Array Design

Linear-phased antenna arrays are used in radar, sonar, ultrasound imaging, and seismic signal processing. Sparse arrays with certain elements removed are economical and as a result are of practical interest. There is a mathematical similarity between the far-field radiation pattern for a linear antenna array of equally spaced elements and the frequency response of an FIR filter. This similarity can be exploited to design sparse arrays with specific beam patterns. In this section, we point out this similarity and outline a few simple designs of sparse arrays. We restrict our attention here on the design of sparse arrays for ultrasound scanners.

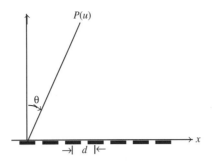

Figure 11.82: Uniform linear antenna array.

Consider a linear array of $N+1$ isotropic, equispaced elements with interelement spacing d and located at $x_n = n \cdot d$ for $0 \le n \le N$ as shown in Figure 11.82. The far-field radiation pattern at an angle θ away from the broadside (i.e., the normal to the array), is given by

$$P(u) = \sum_{n=0}^{N} w[n]e^{j[2\pi(u/\lambda)d]n}, \qquad (11.132)$$

where $w[n]$ is the complex excitation or weight of the nth element, λ is the wavelength, and

$$u = \sin\theta.$$

The function $P(u)$ thus can be considered as the discrete-time Fourier transform of $w[n]$ with the frequency variable given by $2\pi(u/\lambda)d$. The array element weighting as a function of the element position is called the *aperture function*. For a uniformly excited array, $w[n] =$ a constant, and the grating lobes in the radiation pattern are avoided if $d \le \lambda/2$. Typically $d = \lambda/2$, in which case the range of u is between $-\pi$ and π. From Eq. (11.132) it can be seen that the expression for $P(u)$ is identical to the frequency response of an FIR filter of length $N+1$. Hence, the FIR filter design methods can be readily applied to design antenna arrays with specific radiation patterns. An often used element weight is $w[n] = 1$ whose radiation pattern is this same as the frequency response of a running-sum or boxcar FIR filter.

Sparse arrays with fewer elements are obtained by removing some of the elements which increases the interelement spacing between some consecutive pairs of elements to more than $\lambda/2$. This usually results in an increase of sidelobe levels and can possibly cause the appearance of grating lobes in the radiation pattern. However, these unwanted lobes can be reduced significantly by selecting array element locations appropriately. In the case of ultrasound scanners, a two-way radiation pattern is generated by a transmit array and a receive array. The design of such arrays is simplified by treating the problem as the design of an "effective aperture function" which is given by the convolution of the transmit aperture function and the receive aperture function [Loc96].

The effective aperture function of a single-element transmit array and a 16-element nonsparse receive array is also a 16-element nonsparse array. This array system thus requires 17 elements. However, the total number of elements can be reduced by using either a sparse transmit array or a sparse receive array or both. Consider for example a nonsparse transmit array and a sparse receive array with aperture functions given by [Loc96]

$$w_T[n] = \{1 \ \ 1\}, \qquad w_R[n] = \{1 \ \ 0 \ \ 1 \ \ 0 \ \ 1 \ \ 0 \ \ 1 \ \ 0 \ \ 1 \ \ 0 \ \ 1 \ \ 0 \ \ 1 \ \ 0 \ \ 1\}, \qquad (11.133)$$

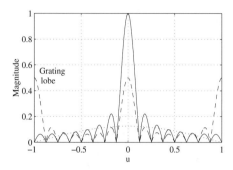

Figure 11.83: Radiation patterns of transmit array (dotted line), receive array (dashed line), and two-way radiation pattern (solid line). The radiation patterns have been scaled by a factor of 16 to make the value of the two-way radiation pattern at $u = 0$ unity.

where 0 in $w_R[n]$ indicates the absence of an element. It is easy to show that the effective aperture function here is given by

$$w_{\text{eff}}[n] = w_T[n] \circledast w_R[n] = \{1 \quad 1 \quad 1 \quad 1 \quad 1 \quad 1 \quad 1 \quad 1 \quad 1 \quad 1 \quad 1 \quad 1 \quad 1 \quad 1 \quad 1 \quad 1\}. \qquad (11.134)$$

Figure 11.83 shows the radiation patterns of the individual arrays and the two-way radiation pattern of the composite array. Note that the grating lobes in the radiation pattern of the receive array are being suppressed by the radiation pattern of the transmit array. Thus, a nonsparse transmit array with two elements and a sparse receive array with eight elements has the same radiation pattern of a single-element transmit array and a 16-element nonsparse receive array.

More economic sparse array designs with eight elements are as follows [Loc96]:

$$w_T[n] = \{1 \quad 1 \quad 0 \quad 0 \quad 0 \quad 0 \quad 0 \quad 1 \quad 1\}, \qquad w_R[n] = \{1 \quad 0 \quad 1 \quad 0 \quad 1 \quad 0 \quad 1\},$$
$$w_T[n] = \{1 \quad 1 \quad 1 \quad 1\}, \qquad w_R[n] = \{1 \quad 0 \quad 0 \quad 0 \quad 1 \quad 0 \quad 0 \quad 0 \quad 1 \quad 0 \quad 0 \quad 0 \quad 1\},$$

with the same effective aperture function $w_{\text{eff}}[n]$ as given in Eq. (11.134).

The basic idea behind the sparse transmit and receive array designs given above is that one sparse array essentially "fills in" the missing elements in the second sparse array by some type of interpolation similar to the concept of interpolated FIR filter design of Neuvo et al. [Neu84b]. The shape of the effective aperture function can be made smoother to reduce the grating lobes by controlling the shape of the transmit and receive aperture functions. For example, the transmit and the receive aperture functions given by

$$w_T[n] = \{1 \quad 1 \quad 0 \quad 0 \quad 1 \quad 1 \quad 0 \quad 0 \quad 1 \quad 1\},$$
$$w_R[n] = \{1 \quad 0 \quad 1 \quad 0 \quad 1 \quad 0 \quad 1 \quad 0 \quad 1\}$$

result in a triangular-shaped effective aperture function given by

$$w_{\text{eff}}[n] = \{1 \quad 1 \quad 1 \quad 1 \quad 2 \quad 2 \quad 2 \quad 2 \quad 3 \quad 3 \quad 2 \quad 2 \quad 2 \quad 2 \quad 1 \quad 1 \quad 1 \quad 1\}.$$

The corresponding scaled radiation patterns are shown in Figure 11.84(a). Additional smoothing can be obtained by apodizing the individual aperture functions implemented by reducing the weights applied to the outer elements. For example, the transmit and the receive aperture functions given by

$$w_T[n] = \{1 \quad 1 \quad 0 \quad 0 \quad 1 \quad 1 \quad 0 \quad 0 \quad 1 \quad 1\},$$
$$w_R[n] = \{0.5 \quad 0 \quad 1 \quad 0 \quad 1 \quad 0 \quad 1 \quad 0 \quad 0.5\}$$

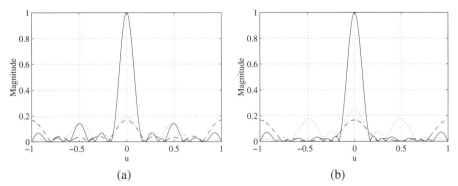

Figure 11.84: Illustration of effective aperture smoothing by shaping transmit and receive aperture functions. The radiation patterns have been scaled to make the value of the two-way radiation pattern at $u = 0$ unity.

result in a triangular-shaped effective aperture function given by

$$w_{\text{eff}}[n] = \{0.5 \quad 0.5 \quad 1 \quad 1 \quad 1.5 \quad 1.5 \quad 2 \quad 2 \quad 2 \quad 2 \quad 2 \quad 2 \quad 1.5 \quad 1.5 \quad 1 \quad 1 \quad 0.5 \quad 0.5\}.$$

The corresponding scaled radiation patterns are shown in Figure 11.84(b).

11.15 Summary

The discrete Fourier transform (DFT) is a widely used digital signal processing algorithm. One of the reasons for its widespread use is the availability of fast Fourier transform (FFT) algorithms for its computation. Earlier in Section 3.6 we discussed the implementation of high-speed convolution using the DFT. It also plays a major role in the implementation of many functions in the *Signal Processing Toolbox* of MATLAB. Three other applications are considered in this chapter. The first application discussed is in the efficient and robust detection of dual-tone multifrequency (DTMF) tones employed in dial-pulse telephone signaling, ATM machines, voice mains, etc. The next application treated here is in the spectral analysis of stationary signals and is the basis of most commercial spectrum analyzers. For the spectral analysis of nonstationary signals, such as speech and radar signals, the DFTs of small windowed segments of these signals are computed, and a three-dimensional display of the resulting spectrums, called spectrograms, are employed. This type of signal analysis is more popularly called the short-term Fourier transform, or the time-dependent Fourier transform which is covered next. This is followed by a brief discussion of spectral analysis of random signals. Here both nonparametric and parametric spectral analysis methods are reviewed.

Digital signal processing methods are increasingly being employed in digital audio applications for spectral shaping and for the generation of special audio effects. A number of these applications are outlined. These are followed by a review of digital stereo generation for FM stereo transmission and a discussion on the use of digital filtering for preemphasis of audio signals.

The discrete-time analytic signal has a zero-valued spectrum for all negative frequencies and, as a result, is a complex signal. Such a signal can be generated from a real signal by passing the latter through a discrete-time Hilbert transformer. Several methods for designing the Hilbert transformer are described. One application of the analytic signal considered in this chapter is in the design of a digital single-sideband communication system.

The next four applications treated involve multirate digital signal processing. The first two applications are concerned with the subband coding of speech and audio signals for signal compression, and the design of transmultiplexers for interconnecting frequency-division multiplex (FDM) and time-division multiplex (TDM) communication systems. Each of these applications requires the use of the so-called quadrature-mirror filter (QMF) banks.

A method for the efficient digital data transmission based on multirate techniques is discussed next. This method makes use of the discrete Fourier transform (DFT) computation and, as a result, can be implemented using fast Fourier transform (FFT) algorithms. A novel method of designing a computationally efficient fractional sampling rate converter for digital audio applications is then discussed.

The next two applications considered in this chapter are the oversampling sigma-delta analog-to-digital (A/D) and digital-to-analog (D/A) converter design. Such converters are being increasingly employed in many systems because of their improved signal-to-noise ratios.

Finally, the chapter shows the similarity between the far-field pattern of a linear antenna array of equally spaced elements and the frequency response of an FIR filter. This similarity is then made use of in the design of sparse antenna arrays.

This chapter has touched upon a variety of practical applications of digital signal processing. There are numerous other applications with many requiring a knowledge of other subjects, and as a result, they are beyond the scope of this book. Additional applications can be found in various other books; see, for example, [Bab95], [Fre94], [Lin87], [Mar92], [Opp78], [Pap90], [She95].

11.16   Problems

11.1 A bandlimited continuous-time signal is sampled at a rate of 7500 Hz to ensure no aliasing. The sampled signal is windowed generating a length-1250 sequence whose R-point DFT is then computed. (a) What is the frequency resolution of the DFT samples in Hz if $R = 1250$? (b) What should be the value of R if a frequency resolution of 4.5 Hz is desired?

11.2 A speech signal is sampled at the 12-kHz rate. We wish to analyze, using the DFT, the spectrum of the sampled speech of length 256.

(a) If we take a 256-point DFT of this segment, what would be the resolution of the DFT samples?

(b) Describe a technique to achieve a 16-Hz resolution of the DFT samples.

(c) Describe a method to achieve a 128-Hz resolution using a minimum length DFT.

11.3 A bandlimited continuous-time real signal $g_a(t)$ is sampled at a rate of F_T Hz where $F_T = 2F_m$ with F_m denoting the highest frequency contained in $g_a(t)$. An R-point DFT $G[k]$ of the sequence $g[n]$ obtained by sampling $g_a(t)$ is next computed.

(a) If $F_m = 5$ kHz and $R = 1000$, determine the continuous-time frequencies corresponding to the DFT sample indices $k = 200, 350$, and 824.

(b) If $F_m = 7$ kHz and $R = 1010$, determine the continuous-time frequencies corresponding to the DFT sample indices $k = 195, 339$, and 917.

(c) If $F_m = 5$ kHz and $R = 517$, determine the continuous-time frequencies corresponding to the DFT sample indices $k = 97, 187$, and 301.

11.4 A continuous-time real sinusoidal signal with a continuous-time Fourier transform $G_a(j\Omega)$ is sampled at a rate of F_T Hz and passed through a rectangular window generating a length-N sequence $\gamma[n], 0 \le n \le N - 1$. Let $\Gamma[k]$ denote its N-point DFT. If $\Gamma[k_0]$ is known, for what values of Ω can the values of $G_a(j\Omega)$ be determined from this value of the DFT? What are the values of $G_a(j\Omega)$ at these angular frequencies?

11.5 A continuous-time real sinusoidal signal $g_a(t) = \cos(200\pi t)$ is sampled at a rate of F_T Hz and passed through a rectangular window generating a length-512 sequence $\gamma[n], 0 \le n \le 511$. Let $\Gamma[k]$ denote its 512-point DFT. What should be the value of F_T for which $\Gamma[k] = 0$ for all values k except $k = 64$ and $k = 448$?

11.6 A bandlimited continuous-time real signal $g_a(t)$ is sampled at a rate of F_T Hz, where $F_T = 2F_m$ with F_m denoting the highest frequency contained in $g_a(t)$. An R-point DFT $G[k]$ of the sequence $g[n]$ obtained by sampling $g_a(t)$ is next computed.

(a) If $F_m = 6$ kHz, what are the values of R and the sampling rate F_T to provide a DFT resolution of 2.6 Hz with no aliasing?

(b) If R has to be a power of 2 to make use of efficient FFT algorithms for the computation of the DFT, what should then be the value of R to achieve the resolution of 3 Hz if $g_a(t)$ is sampled at a rate to produce no aliasing?

11.7 Consider the length-64 sequence

$$x[n] = A \cos(2\pi f_1 n/64) + B \cos(2\pi f_2 n/64), \quad 0 \le n \le 63.$$

It is known that its 64-point DFT $X[k]$ has zero-valued samples for all values of k except $k = 15, 27, 37$, and 49. If $|X[15]| = 32$ and $|X[27]| = 16$, determine the exact expression for $x[n]$ without evaluating the IDFT. Is your answer unique? If not, how many other sequences have the same DFT magnitudes as $X[k]$? Determine the exact expressions for these sequences.

11.8 A bandlimited continuous-time real signal $x_a(t)$ is sampled at a rate of 100 samples/sec, generating a discrete-time signal $x[n]$ of length 500 whose 500-point DFT $X[k]$ is then computed.

(a) Determine the total number of complex multiplications and additions needed for the direct evaluation of $X[k]$.

(b) What is the digital frequency resolution of the DFT samples?

(c) What is the analog frequency resolution of the DFT samples?

(d) What should be the stopband edge frequency in Hz of the analog anti-aliasing filter prior to sampling?

(e) Determine the digital frequency in radians and the analog frequency in rad/sec of the DFT samples $X[31]$ and $X[390]$.

(f) If a Cooley-Tukey type FFT algorithm is used to compute the DFT samples, what should be the minimum length of the DFT and how many zero-valued samples should be appended to $x[n]$?

(g) What are the samples of the modified DFT that are closest to the DFT samples $X[31]$ and $X[390]$?

(h) What are the total number of complex multiplications and additions needed to compute the modified DFT?

11.9 A spectral analysis of a signal composed of two sinusoidal sequences of normalized frequencies f_1 and f_2, with $f_1 < f_2$, is to be carried out using the DFT-based approach. To this end, the signal is windowed by a length-N window generating a length-N sequence $x[n]$. If $N = 60$ and the frequency of one of the sinusoidal sequence is $f_1 = 0.25$, determine the minimum value of the frequency f_2 of the second sinusoidal sequence so that both sinusoids can be resolved using a length-R DFT, $R > N$, for each of the following windows:
   (a) Rectangular window, (b) Hamming window, (c) Hann window, and (d) Blackman window.

11.10 Repeat Problem 11.9 for $N = 110$.

11.11 A bandlimited continuous-time signal $g_a(t)$ is sampled at a sampling rate of F_T Hz generating the sequence $g[n]$ which is then windowed by a length-N window resulting in a length-N sequence $\gamma[n]$. Let the highest frequency contained in $g_a(t)$ be F_m Hz. To analyze the spectral contents of $g_a(t)$ an N-point DFT of $\gamma[n]$ is carried out.

(a) What is the minimum value of F_T?

(b) What is the maximum value of F_T if the desired frequency resolution of the DFT samples is less than ΔF?

(c) If $F_m = 4$ kHz and $\Delta F = 10$ Hz, and $N = 2^\ell$ with ℓ a positive integer, determine the minimum and maximum values of the sampling rate F_T.

11.12 Let $X_{STFT}(e^{j\omega}, n)$ and $Y_{STFT}(e^{j\omega}, n)$ denote the short-term Fourier transforms of two sequences $x[n]$ and $y[n]$, obtained by applying the same window sequence $w[n]$. Prove the following properties of the STFT:

(a) Linearity property: If $g[n] = \alpha x[n] + \beta y[n]$, then $G_{\text{STFT}}(e^{j\omega}, n) = \alpha X_{\text{STFT}}(e^{j\omega}, n) + \beta Y_{\text{STFT}}(e^{j\omega}, n)$ where $G_{\text{STFT}}(e^{j\omega}, n)$ is the STFT of $g[n]$ obtained using the window $w[n]$.

(b) Shifting property: If $y[n] = x[n - n_0]$, then $Y_{\text{STFT}}(e^{j\omega}, n) = X_{\text{STFT}}(e^{j\omega}, n - n_0)$, where n_0 is any positive or negative integer.

(c) Modulation property: If $y[n] = e^{j\omega_0}x[n]$, then $Y_{\text{STFT}}(e^{j\omega}, n) = X_{\text{STFT}}(e^{j(\omega-\omega_0)}, n)$.

11.13 An alternate definition of the short-term Fourier transform is given by

$$\bar{X}_{\text{STFT}}(e^{j\omega}, n) = \sum_{m=-\infty}^{\infty} x[m]w[n - m]e^{-j\omega m}. \tag{11.135}$$

Express $\bar{X}_{\text{STFT}}(e^{j\omega}, n)$ in terms of the short-term Fourier transform $X_{\text{STFT}}(e^{j\omega}, n)$ defined in Eq. (11.16). What are the main differences between the two definitions?

11.14 Show that the inverse STFT can be computed using the following expression:

$$x[n] = \frac{1}{2\pi w[0]} \int_0^{2\pi} X_{\text{STFT}}(e^{j\omega}, n) \, d\omega, \tag{11.136}$$

provided $w[0] \neq 0$.

11.15 Show that the inverse STFT can also be computed using the following expression [Naw88]:

$$x[n] = \frac{1}{2\pi W(0)} \int_0^{2\pi} \sum_{\ell=-\infty}^{\infty} \bar{X}_{\text{STFT}}(e^{j\omega}, \ell)e^{j\omega n} \, d\omega, \tag{11.137}$$

where $W(0) = \sum_{n=-\infty}^{\infty} w[n]$.

11.16 Show that the sampled short-term discrete Fourier transform $X_{\text{STFT}}[k, n]$ defined by Eq. (11.17) can be interpreted as a bank of N linear time-invariant filters as indicated in Figure P11.1 whose outputs $y_k[n]$ are precisely $X_{\text{STFT}}[k, n]$, $k = 0, 1, \ldots, N - 1$. Determine the expressions for the impulse responses $h_k[n]$ of these N filters.

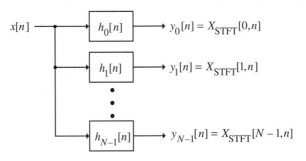

Figure P11.1

11.17 Let $X_{\text{STFT}}(e^{j\omega}, n)$ denote the STFT of a real sequence $x[n]$. Denote the inverse DTFT of $\left|X_{\text{STFT}}(e^{j\omega}, n)\right|^2$ as $r[k, n]$. What is the relation between $r[k, n]$ and $x[n]$?

11.18 A nonstationary signal $x_a(t)$ with time-varying signal parameters is sampled at a sampling rate of F_T and the spectral analysis of the sampled signal $x[n]$ is carried out using the short-time discrete Fourier transform method with an N-point DFT. (a) If the window used is of duration τ seconds, what is the length of the window in samples? (b) If the window is advanced by K samples between two consecutive DFT computations, what is the number of DFT computations per second?

11.19 The short-term autocorrelation function of a deterministic sequence $x[n]$ is defined by [Rab78]

$$\varphi_{ST}[k, n] = \sum_{m=-\infty}^{\infty} x[m]w[n-m]x[m+k]w[n-k-m], \tag{11.138}$$

where $w[n]$ is an appropriately chosen window sequence.

(a) Show that $\varphi_{ST}[k, n]$ is an even function of k, i.e., $\varphi_{ST}[k, n] = \varphi_{ST}[-k, n]$.

(b) Show that $\varphi_{ST}[k, n]$ can be computed using the digital filter structure of Figure P11.2, where $h_k[n]$ is the impulse response of an LTI discrete-time system. Determine the expression for $h_k[n]$.

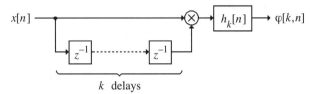

Figure P11.2

11.20 The structure of Figure P11.3 has been proposed as an allpass reverberator [Sch62]. Develop an equivalent realization with the smallest number of multipliers.

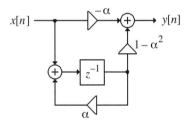

Figure P11.3

11.21 The structure of Figure P11.4 has been proposed as an allpass reverberator with a variable ratio of direct-to-reverberated sound, where the box labeled "allpass reverberator" is typically a cascade of the allpass reverberators of the form of Figure P11.3 [Sch62]. Develop an equivalent realization with the smallest number of multipliers.

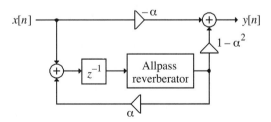

Figure P11.4

11.22 A generalization of the second-order equalizer of Eq. (11.65) is given by

$$G_2(z) = \frac{K_1}{2}\{1 - A_2(z)\} + \frac{K_2}{2}\{1 + A_2(z)\}, \tag{11.139}$$

where $A_2(z)$ is the second-order allpass transfer function of Eq. (11.64). Develop the pertinent design equations for this equalizer and plot the gain responses for various values of the filter parameters K_1, K_2, α, and β.

11.23 Develop a computationally efficient structure for realizing a fractional-rate interpolator with an interpolation factor of 3/2 by taking the transpose of Figure 11.63.

11.24 Develop a computationally efficient structure for realizing a fractional-rate interpolator with an interpolation factor of 3/4 using the method outlined in Section 11.11.1.

11.17 MATLAB Exercises

M 11.1 Verify the dual-tone multifrequency (DTMF) tone detection Program 11_1 by running it for various telephone key symbols as inputs.

M 11.2 Using Program 11_2 analyze the spectral content of a length-16 sequence $x[n]$ composed of a sum of two sinusoidal sequences of the form of Eq. (11.14) with frequencies $f_1 = 0.18$ and $f_2 = 0.3$ for DFT lengths $R = 16$, 32, 64, and 128, respectively. Comment on your results.

M 11.3 Using Program 11_2 analyze the spectral content of a length-16 sequence $x[n]$ composed of a sum of two sinusoidal sequences of the form of Eq. (11.14) with frequencies $f_1 = 0.18$ and $f_2 = 0.3$, 0.27, 0.24, and 0.21, respectively. Use a DFT length of $R = 128$. Comment on your results.

M 11.4 Modify Program 11_2 to investigate the effect of a tapered window on the spectral analysis of a sequence $x[n]$ composed of a sum of two sinusoidal sequences with closely spaced frequencies. Let $x[n]$ be of the form $x[n] = \sin(2\pi f_1 n) + 0.9\sin(2\pi f_2 n)$, where $f_1 = 0.18$ and $f_2 = 0.21$. Use a length-N Hamming window, where $N = 16$, 32, 64, and 128, respectively. Comment on your results.

M 11.5 Repeat Exercise M11.4 using a Hann window.

M 11.6 Repeat Exercise M11.4 using a Blackman window.

M 11.7 Modify Program 11_3 to perform a DFT-based spectral analysis of a noise-corrupted sinusoidal sequence of frequency 0.11. Are you able to detect the frequency of the sinusoid? Justify your answer.

M 11.8 Repeat Exercise M11.7 for a frequency of 0.16.

M 11.9 A signal composed of two sinusoidal components of angular frequencies 0.1π and 0.2π is corrupted with a Gaussian distributed random signal of zero mean and unity variance, and is windowed by a rectangular window. Evaluate and plot the power spectrum of the noise-corrupted signal for two different window lengths: $N = 64$ and $N = 1024$. Comment on your results.

M 11.10 Evaluate and plot the Bartlett and Welch estimates of the power spectrum of the noise-corrupted signal of Exercise M11.9 and windowed by a Hann window of length 1024. Evaluate the Welch estimate for overlaps of 64 and 128 samples, respectively.

M 11.11 Design using the function remez a linear-phase lowpass FIR filter of order 11 with passband edge at 0.3π and stopband edge at 0.5π. Use equal weights in the passband and the stopband. Next, using the function lpc, develop an equivalent all-pole IIR model of the FIR lowpass filter for the following values of the order: 4, 5, and 6. Plot the magnitude response of the FIR filter and the IIR equivalent in the same figure for each value of the order. Comment on your results.

M 11.12 Design using the function remez a wideband FIR filter $F(z)$ of degree 17 with a passband from 0 to 0.9π and a stopband from 0.95π to π. Use an appropriate weight vector to weigh the passband and the stopband of the wideband filter. Plot the magnitude response of $F(z)$ and its corresponding Hilbert transformer $F(-z^2)$. How many multipliers are needed to implement the Hilbert transformer?

M 11.13 Design using the function remez a Hilbert transformer of degree 34 with a passband from 0.05π to 0.95π. Compare its computational complexity with that designed in Exercise M11.13.

M 11.14 Design a real-coefficient half-band elliptic filter $G(z)$ with the following specifications: $\omega_p = 0.36\pi$, $\omega_s = 0.64\pi$, and $\delta_s = 0.014$. Plot its pole-zero locations using the function z-plane. Express $G(z)$ in the form of Eq. (11.92) and determine the transfer functions of the allpass sections $\mathcal{A}_0(z^2)$ and $\mathcal{A}_1(z^2)$. Next, determine the expression for the complex half-band filter $H(z)$ according to Eq. (11.93). Plot the gain responses of the real half-band filter $G(z)$ and the complex half-band filter $H(z)$, along with the phase difference between $\mathcal{A}_0(-z^2)$ and $z^{-1}\mathcal{A}_1(-z^2)$.

M 11.15 Verify the operation of the sigma-delta quantizer on a constant-amplitude discrete-time input sequence using Program 11_6. Choose different values for the length N of the sequence and its amplitude A.

M 11.16 Verify the operation of the sigma-delta A/D converter on a discrete-time sinusoidal input sequence using Program 11_7. Choose different values for the length N of the sequence, its amplitude A, and angular frequency ω_0.

M 11.17 Verify the operation of the sigma-delta D/A converter on a discrete-time sinusoidal input sequence using Program 11_8. Choose different values for the length N of the sequence, its amplitude A, and angular frequency ω_o.

Bibliography

[Abr83]   B. Abraham and J. Ledolter. *Statistical Methods for Forecasting*. Wiley, New York NY, 1983.

[Abr72]   M. Abramowitz and I.A. Stegun, editors. *Handbook of Mathematical Functions*. Dover Publications, New York NY, 1972.

[Aca83]   A. Acampora. Wideband picture detail restoration in a ditigal NTSC comb-filter system. *RCA Engineer*, 28(5):44–47, September/October 1983.

[Ada83]   J.W. Adams and A.N. Willson, Jr. A new approach to FIR digital filters with fewer multipliers and reduced sensitivity. *IEEE Trans. on Circuits and Systems*, CAS-31:277–283, May 1983.

[Ada91]   J.W. Adams. FIR digital filters with least squares stopbands subject to peak-gain constraints. *IEEE Trans. on Circuits and Systems*, 39:376–388, April 1991.

[Aga75]   R.C. Agarwal and C.S. Burrus. New recursive digital filter structures having very low sensitivity and low roundoff noise. *IEEE Trans. on Circuits and Systems*, CAS-22:921–927, March 1975.

[Aga77]   R.C. Agarwal and J.W. Cooley. New algorithms for digital convolution. *IEEE Trans. on Acoustics, Speech, and Signal Processing*, ASSP-25:392–410, October 1977.

[Aka92]   A.N. Akansu and R.A. Haddad. *Multiresolution Signal Decomposition*. Academic Press, New York NY, 1992.

[Ala93]   M.A. Al-Alaoui. Novel digital integrator and differentiator. *Electronic Letters*, 29:376–378, 18 February 1993.

[All77]   J.B. Allen and L.R. Rabiner. A unified approach to short-term Fourier analysis and synthesis. *Proc. IEEE*, 65:1558–1564, November 1977.

[All80]   H.G. Alles. Music synthesis using real time digital techniques. *Proc. IEEE*, 68:436–449, April 1980.

[Ans93]   R. Ansari and B. Liu. Multirate signal processing. In S.K. Mitra and J.F. Kaiser, editors, *Handbook for Digital Signal Processing*, chapter 14, pages 981–1084. Wiley-Interscience, New York NY, 1993.

[Ant93]   A. Antoniou. *Digital Filters: Analysis, Design, and Applications*. McGraw-Hill, New York NY, 2nd edition, 1993.

[Bab95]   J. Babst, editor. *Digital Signal Processing Applications Using the ADSP-2100 Family*, Prentice Hall, Englewood Cliffs NJ, 1995.

[Bag98]   S. Bagchi and S.K. Mitra. *Nonuniform Discrete Fourier Transform and Its Signal Processing Applications*, Kluwer Academic Publishers, Norwell MA, 1998.

[Bar48]   M.S. Bartlett. Smoothing periodograms from the time series with continuous spectra. *Nature (London)*, 161:686-687, 1948.

[Bel76]   M. Bellanger, G. Bonnerot, and M. Coudreuse. Digital filtering by polyphase network: Application to sample rate alteration and filter banks. *IEEE Trans. Acoustics, Speech, and Signal Processing*, ASSP-24:109–114, April 1976.

[Bel84]   M. Bellanger. *Digital Processing of Signals*. Wiley, New York NY, 1984.

[Ben48]   W.R. Bennett. Spectra of quantized signals. *Bell System Technical Journal*, 27:446–472, 1948.

[Ber76]   P.A. Bernhardt, D.A. Antoniadis and A.D. Da Rossa. Lunar perturbations in columnar electron content and their interpretations in terms of dynamo electrostatic fields. *Journal of Geophysics Research*, 43:5957–5963, December 1976.

[Bin90]   J.A.C. Bingham. Multicarrier modulation for data transmission: An idea whose time has come. *IEEE Communications Magazine*, pages 5–14, May 1990.

[Bla65]   R.B. Blackman. *Linear Data Smoothing and Prediction in Theory and Practice*. Addison-Wesley, Reading MA, 1965.

[Ble78]   B. Blesser and J.M. Kates. Digital processing in audio signals. In A.V. Oppenheim, editor, *Applications of Digital Signal Processing*, chapter 2. Prentice Hall, Englewood Cliffs NJ, 1978.

[Bol93]   B.A. Bolt. *Earthquakes*. W.H. Freeman, New York NY, 1993.

[Bon76]   G. Bongiovanni, P. Corsini, and G. Forsini. One-dimensional and two-dimensional generalized discrete Fourier transform. *IEEE Trans. on Acoustics, Speech and Signal Processing*, ASSP-24:97–99, February 1976.

[Boo51]   A.D. Booth. A signed binary multiplication technique. *Quart. J. Mech. Appl. Math.*, 4(Part 2):236–240, 1951.

[Box70]   G.E.P. Box and G.M. Jenkins. *Time Series Analysis: Forecasting and Control*. Holden-Day, San Francisco CA, 1970.

[Bra83]   R.N. Bracewell. The discrete Hartley transform. *Journal of the Optical Society of America*, 73:1832–1835, December 1983.

[Bur73]   R.S. Burrington. *Handbook of Mathematical Tables and Formulas*. McGraw-Hill, New York NY, 5th edition, 1973.

[Bur72]   C.S. Burrus. Block realization of digital filters. *IEEE Trans. on Audio and Electroacoustics*, AU-20:230–235, October 1972.

[Bur77]   C.S. Burrus. Index mappings for multidimensional formulation of the DFT and convolution. *IEEE Trans. on Acoustics, Speech, and Signal Processing*, ASSP-25:239–242, June 1977.

[Bur81]   C.S. Burrus and P.W. Eschenbacher. An in-place, in-order prime factor FFT algorithm. *IEEE Trans. on Acoustics, Speech, and Signal Processing*, ASSP-29:806–817, April 1981.

[Bur92]   C.S. Burrus, A.W. Soewito, and R.A. Gopinath. Least squared error FIR filter design with transition bands. *IEEE Trans. on Signal Processing*, 40:1327–1340, 1992.

[But77]   M. Buttner. Elimination of limit cycles in digital filters with very low increase in the quantization noise. *IEEE Trans. on Circuits and Systems*, CAS-24:300–304, 1977.

[Cad73]   J.A. Cadzow. *Discrete-Time Systems*. Prentice Hall, Englewood Cliffs NJ, 1973.

[Cad87]   J.A. Cadzow. *Foundations of Digital Signal Processing and Data Analysis*. Macmillan, New York NY, 1987.

[Can86]   J.C. Candy and A-N. Huynh. Double interpolation for digital-to-analog conversion. *IEEE Trans. on Communications*, COM-34:77–81, January 1986.

[Can92]   J.C. Candy and G.C. Temes. Oversampling methods for A/D and D/A conversion. In J.C. Candy and G.C. Temes, editors, *Oversampling Delta-Sigma Data Converters*, pages 1–25. IEEE Press, New York NY, 1992.

[Cha73]   D.S.K. Chan and L.R. Rabiner. Analysis of quantization errors in the direct form for finite impulse response digital filters. *IEEE Trans. on Audio and Electroacoustics*, AU-21:354–366, August 1973.

[Cha81]   T.L. Chang and S.A. White. An error cancellation digital filter structure and its distributed arithmetic implementation. *IEEE Trans. on Circuits and Systems*, CAS-28:339–342, April 1981.

[Cha90]   R. Chassaing and D.W. Horning. *Digital Signal Processing with the TMS320C25*. Wiley, New York NY, 1990.

[Chr66]   E. Christian and E. Eisenmann. *Filter Design Tables and Graphs*. Wiley, New York NY, 1966.

[Chu85]   P.L. Chu. Quadrature mirror filter design for an arbitrary number of equal bandwidth channels. *IEEE Trans. on Acoustics, Speech, and Signal Processing*, ASSP-33:203–218, February 1985.

[Chu90]   R.V. Churchill and J.W. Brown. *Introduction to Complex Variables and Applications*. McGraw-Hill, New York NY, 5th edition, 1990.

[Chu95]   J. Chun and N.K. Bose. Fast evaluation of an integral occuring in digital filtering applications. *IEEE Trans. on Signal Processing*, 43:1982–1986, August 1995.

[Cio91]   J.M. Cioffi. *A multicarrier primer*. ANSI T1E1.4 Committee Contribution, Boca Raton FL, November 1991.

[Cio93]   J.M. Cioffi and Y-S Byun. Adaptive filtering. In S.K. Mitra and J.F. Kaiser, editors, *Handbook for Digital Signal Processing*, chapter 15, pages 1085–1142. Wiley Interscience, New York NY, 1993.

[Cla73]   T.A.C.M. Classen, W.F.G. Mecklenbräuker, and J.B.H. Peek. Some remarks on the classifications of limit cycles in digital filters. *Philips Research Reports*, 28:297–305, August 1973.

[Coh86]   A. Cohen. *Biomedical Signal Processing*, volume II. CRC Press, Boca Raton FL, 1986.

[Con70]   A.C. Constantinides. Spectral transformations for digital filters. *Proc. IEE*, 117:1585–1590, August 1970.

[Coo65]   J.W. Cooley and J.W. Tukey. An algorithm for the machine calculation of complex Fourier series. *Math. Computation*, 19:297–301, 1965.

[Cou83]   L.W. Couch II. *Digital and Analog Communication Systems*. Macmillan, New York NY, 1983.

[Cox83]   R.V. Cox and J.M. Tribolet. Analog voice privacy systems using TFSP scrambling: Full duplex and half duplex. *Bell System Technical Journal*, 62:47–61, January 1983.

[Cre95]   C.D. Creusere and S.K. Mitra. A simple method for designing high quality prototype filters for M-band pseudo QMF banks. *IEEE Trans. Signal Processing*, 43:1005–1007, April 1995.

[Cro75]   R.E. Crochiere and A.V. Oppenheim. Analysis of linear digital networks. *Proc. IEEE*, 62:581–595, April 1975.

[Cro76a]   A. Croisier, D. Esteban, and C. Galand. Perfect channel splitting by use of interpolation/decimation/tree decomposition techniques. In *Proc. International Symposium on Information Science and Systems*, Patras, Greece, 1976.

[Cro76b]   R.E. Crochiere and L.R. Rabiner. On the properties of frequency transformations for variable cutoff linear phase filters. *IEEE Trans. on Circuits and Systems*, CAS-23:684–686, 1976.

[Cro83]   R.E. Crochiere and L.R. Rabiner. *Multirate Digital Signal Processing*. Prentice Hall, Englewood Cliffs NJ, 1983.

[Cuc91]   S. Cucchi, F. Desinan, G. Parladori, and G. Sicuranza. DSP implementation of arbitrary sampling frequency conversion for high quality sound application. In *Proc. IEEE International Conference on Acoustics, Speech, and Signal Processing*, pages 3609–3612, Toronto Canada, May 1991.

[Dan74]   R.W. Daniels. *Approximation Methods for Electronic Filter Design*. McGraw-Hill, New York NY, 1974.

[Dar76]   G. Daryanani. *Principles of Active Network Synthesis and Design*. Wiley, New York NY, 1976.

[Dau88]   I. Daubechies. Orthonormal bases of compactly supported wavelets. *Comm. Pure Appl. Math.*, 41:909–996, 1988.

[DeF88]   D.J. DeFatta, J.G. Lucas, and W.S. Hodgkiss. *Digital Signal Processing: A System Design Approach*. Wiley, New York NY, 1988.

[Dep80]   E. Deprettere and P. DeWilde. Orthogonal cascade realization of a real multiport digital filter. *International Journal on Circuit Theory Appl*, 8:245–277, 1980.

[Dre90]   J.O. Drewery. Digital filtering of television signals. In C.P. Sandbank, editor, *Digital Television*, chapter 5. Wiley, New York NY, 1990.

[Dug80]   J.P. Dugre, A.A.L. Beex, and L.L. Scharf. Generating covariance sequences and the calculation of quantization and roundoff error variances in digital filters. *IEEE Trans. on Acoustics, Speech, and Signal Processing*, ASSP-28:102–104, 1980.

[Dug82]   J.P. Dugre and E.I. Jury. A note on the evaluation of complex integrals using filtering interpretations. *IEEE Trans. on Acoustics, Speech, and Signal Processing*, ASSP-30:804–807, 1982.

[Duh86]   P. Duhamel. Implementation of "split-radix" FFT algorithms for complex, real, and real-symmetric data. *IEEE Trans. on Acoustics, Speech, and Signal Processing*, ASSP-34:285–295, April 1986.

[Dur59]   J. Durbin. Efficient estimation of parameters in moving average model. *Biometrika*, 46:306–316, 1959.

[Dut80]   D.L. Duttweiler. Bell's echo-killer chip. *IEEE Spectrum*, 17:34–37, October 1980.

[Dut83]   S.C. Dutta Roy. Comments on "On the construction of a digital transfer function from its real part on the unit circle." *Proceedings of the IEEE (Letters)*, 71:1009–1010, August 1983.

[Ear76]   J. Eargle. *Sound Recording*. Van Nostrand Reinhold, New York NY, 1976.

[Ear86]   J.M. Eargle. *Handbook of Recording Engineering*. Van Nostrand Reinhold, New York NY, 1986.

[Ebe69]   P.M. Ebert, J.E. Mazo, and M.G. Taylor. Overflow oscillations in digital filters. *Bell System Technical Journal*, 48:2999–3020, November 1969.

[Egg84]   L.D.J. Eggermont and P.J. Berkhout. Digital audio circuits: Computer simulations and listening tests. *Philips Technical Review*, 41(3):99–103, 1983/84.

[Est77]   D. Esteban and C. Galand. Application of quadrature mirror filters to split-band voice coding schemes. In *Proc. IEEE International Conference on Acoustics, Speech, and Signal Processing*, pages 191–195, May 1977.

[Fad93]   J. Fadavi-Ardekani and K. Mondal. Software considerations. In S.K. Mitra and J.F. Kaiser, editors, *Handbook for Digital Signal Processing*, chapter 12, pages 783–905. Wiley-Interscience, New York NY, 1993.

[Far88]   C.W. Farrow. A continuously variable digital delay element. In *Proc. IEEE International Symposium on Circuits and Systems*, Helsinki, Finland, pages 2641–2645, June 1988.

[Fet71]   A. Fettweis. Digital filter structures related to classical filter networks. *Archiv Elektrotechnik und Übertragungstechnik*, 25:79–81, 1971.

[Fet72]   A. Fettweis. A simple design method of maximally flat delay digital filters. *IEEE Trans. on Audio and Electroacoustics*, AU-20:112–114, June 1972.

[Fet86]   A. Fettweis. Wave digital filters: Theory and practice. *Proc. IEEE*, 74:270–316, February 1986.

[Fla79]   J.L. Flanagan, M.R. Schroeder, B.S. Atal, R.E. Chochiere, N.S. Jayant, and J.M. Tribolet. Speech coding. *IEEE Trans. on Communications*, COM-27:710–737, April 1979.

[Fle87]   R. Fletcher. *Practical Methods of Optimization*. Wiley, New York NY, 1987.

[Fli94]   N.J. Fliege. *Multirate Digital Signal Processing*. Wiley, New York NY, 1994.

[Fre78]   S.L. Freeny, J.F. Kaiser, and H.S. McDonald. Some applications of digital signal processing in telecommunications. In A.V. Oppenheim, editor, *Applications of Digital Signal Processing*, chapter 1, pages 1–28. Prentice Hall, Englewood Cliffs NJ, 1978.

[Fre94]   M.E. Frerking. *Digital Signal Processing in Communication Systems*. Van Nostrand Reinhold, New York NY, 1994.

[Gab87]   R.A. Gabel and R.A. Roberts. *Signals and Linear Systems*. Wiley, New York NY, 1987.

[Gar80]   P. Garde. Allpass crossover systems. *Journal of the Audio Engineering Society*, 28:575–584, September 1980.

[Gas85]   L. Gaszi. Explicit formulas for lattice wave digital filters. *IEEE Trans. on Circuits and Systems*, CAS-32:68–88, January 1985.

[Goe58]   G. Goertzel. An algorithm for evaluation of finite trigonometric series. *American Mathematical Monthly*, 65:34–35, January 1958.

[Goe82]   D. Goedhart, R.J. Van de Plassche, and E.F. Stikvoort. Digital-to-analog conversion in playing a compact disc. *Philips Technical Review*, 40(6):174–179, 1982.

[Gol68]   B. Gold and K.L. Jordan. A note on digital filter synthesis. *Proc. IEEE*, 56:1717–1718, October 1968.

[Gol69a]   B. Gold and K.L. Jordan. A direct search procedure for designing finite-duration impulse response filters. *IEEE Trans. Audio and Electroacoustics*, AU-17:33–36, March 1969.

[Gol69b]   B. Gold and C.M. Radar. *Digital Processing of Signals*. McGraw-Hill, New York, NY, 1969.

[Gon87]   R.C. Gonzalez and P. Wintz. *Digital Image Processing*. Addison-Wesley, Reading MA, 1987.

[Goo77]   D.J. Goodman and M.J. Carey. Nine digital filters for decimation and interpolation. *IEEE Trans. on Acoustics, Speech, and Signal Processing*, ASSP-25:121–126, April 1977.

[Gra73]   A.H. Gray Jr. and J.D. Markel. Digital lattice and ladder filter synthesis. *IEEE Trans. on Audio and Electroacoustics*, AU-21:491–500, December 1973.

[Had91]   R.A. Haddad and T.W. Parsons. *Digital Signal Processing: Theory, Applications, and Hardware*. Computer Science Press, New York NY, 1991.

[Ham89]   R.W. Hamming. *Digital Filters*. Prentice Hall, Englewood Cliffs NJ, 3rd edition, 1989.

[Hay70]   S.S. Haykin and R. Carnegie. New method of synthetising linear digital filters based on convolution integral. *IEE Proc.*, 117:1063–1072, June 1970.

[Hay99]   S.S. Haykin and B. Van Veen. *Signal and Systems*. Wiley, New York NY, 1999.

[Hee82]   J.P.J. Heemskerk and K.A.S. Immink. Compact disc: System aspects and modulation. *Philips Technical Review*, 40(6):157–165, 1982.

[Hel68]   H.D. Helms. Nonrecursive digital filters: Design method for achieving specifications on frequency response. *IEEE Trans. on Audio and Electroacoustics*, AU-16:336–342, September 1968.

[Hen83]   D. Henrot and C.T. Mullis. A modular and orthogonal digital filter structure for parallel processing. In *Proc. IEEE International Conference Acoustics, Speech, and Signal Processing*, pages 623–626, 1983.

[Her70]   O. Herrmann and H.W. Schüssler. Design of nonrecursive digital filters with minimum phase. *Electronics Letters*, 6:329–330, 1970.

[Her71]   O. Herrmann. On the approximation problem in nonrecursive digital filter design. *IEEE Trans. Circuit Theory*, CT-18:411–413, 1971.

[Her73]   O. Herrmann, L.R. Rabiner, and D.S.K. Chan. Practical design rules for optimum finite impulse response lowpass digital filters. *Bell System Tech. J.*, 52:769–799, 1973.

[Hig84]   W.E. Higgins and D.C. Munson Jr. Optimal and suboptimal error spectrum shaping for cascade form digital filters. *IEEE Trans. on Circuits and Systems*, CAS-31:429–437, May 1984.

[Hir73]   K. Hirano, T. Saito, S. Nishimura, and S.K. Mitra. Time-sharing realization of Butterworth digital filters. In *Monograph of the Circuits and Systems Group*. Institution of Electronic and Communication Engineers (Japan), No. CST 73–59, December 1973. (In Japanese).

[Hir74]   K. Hirano, S. Nishimura, and S.K. Mitra. Design of digital notch filters. *IEEE Trans. on Circuits and Systems*, CAS-21:540–546, July 1974.

[Hog81]   E.B. Hogenauer. An economical class of digital filters for decimation and interpolation. *IEEE Trans. Acoustics, Speech, and Signal Processing*, ASSP-29:155–162, April 1981.

[Hsi87]   C-C. Hsiao. Polyphase filter matrix for rational sampling rate conversions. In *Proc. IEEE International Conference on Acoustics, Speech, and Signal Processing*, pages 2173–2176, Dallas TX, April 1987.

[Hub89]   D.M. Huber and R.A. Runstein. *Modern Recording Techniques*. Howard W Sams, Indianapolis IN, 3rd edition, 1989.

[Hwa79]   K. Hwang. *Computer Arithmetic: Principles, Architecture and Designs*. Wiley, New York NY, 1979.

[IEEE85]   Institute of Electrical and Electronic Engineers. IEEE Standard for Binary Floating-Point Arithmetic, 1985.

[Ife93]   E.C. Ifeachor and B.W. Jervis. *Digital Signal Processing: A Practical Approach*. Addison-Wesley, Reading MA, 1993.

[ITU84]   International Telecommunication Union. *CCITT Red Book*, volume VI. Fascicle VI.1, October 1984.

[Jac69]   L.B. Jackson. An analysis of limit cycles due to multiplicative rounding in recursive digital filters. In *Proc. 7th Allerton Conference on Circuit and System Theory*, pages 69–78, Monticello IL, 1969.

[Jac70a]   L.B. Jackson. On the interaction of roundoff noise and dynamic range in digital filters. *Bell System Technical Journal*, 49:159–184, February 1970.

[Jac70b]   L.B. Jackson. Roundoff-noise analysis for fixed-point digital filters realized in cascade or parallel form. *IEEE Trans. on Audio and Electroacoustics*, AU-18:107–122, June 1970.

[Jac91]   L.B. Jackson. *Signals, Systems, and Transforms*. Addison-Wesley, Reading MA, 1991.

[Jac96]   L.B. Jackson. *Digital Filters and Signal Processing*. Kluwer, Boston MA, 3rd edition, 1996.

[Jai89]   A.K. Jain. *Fundamentals of Digital Image Processing*. Prentice Hall, Englewood Cliffs NJ, 1989.

[Jar88]   P. Jarske, Y. Neuvo, and S.K. Mitra. A simple approach to the design of FIR digital filters with variable characteristics. *Signal Processing*, 14:313–326, 1988.

[Jay74]   N.S. Jayant. Digital coding of speech waveforms. *Proc. IEEE*, 62:611–632, May 1974.

[Jay84]   N.S. Jayant and P. Knoll. *Digital Coding of Waveforms*. Prentice Hall, Englewood Cliffs NJ, 1984.

[Jen91]   Y-C. Jenq. Digital convolution algorithm for pipelining multiprocessor system. *IEEE Trans. on Computers*, C-30:966–973, December 1991.

[Jia97]   Z. Jiang and A.N. Willson, Jr. Efficient digital filtering architectures using pipelining/interleaving. *IEEE Trans. on Circuits and Systems*, Part II. 44:110–119, February 1997.

[Joh80]   J.D. Johnston. A filter family designed for use in quadrature mirror filter banks. In *Proc. IEEE International Conference on Acoustics, Speech, and Signal Processing*, pages 291–294, April 1980.

[Joh89]   J.R. Johnson. *Introduction to Digital Signal Processing*. Prentice Hall, Englewood Cliffs NJ, 1989.

[Jos99]   Y.V. Joshi and S.C. Dutta Roy. Design of IIR multiple notch filters based on all-pass filters. *IEEE Trans. on Circuits and Systems*, Part II. 46:134–138, February 1999.

[Jur81]   R.K. Jurgen. Detroit bets on electronics to stymie Japan. *IEEE Spectrum*, 18:29–32, July 1981.

[Kai66]   J.F. Kaiser. Digital filters. In F. Kuo and J.F. Kaiser, editors, *System Analysis by Digital Computers*, chapter 7. Wiley, New York NY, 1966.

[Kai74]   J.F. Kaiser. Nonrecursive digital filter design using the I_0-sinh window function. In *Proc. 1974 IEEE International Symposium on Circuits and Systems*, pages 20–23, San Francisco CA, April 1974.

[Kai77]   J.F. Kaiser and R.W. Hamming. Sharpening the response of a symmetric nonrecursive filter by multiple use of the same filter. *IEEE Trans. on Acoustics, Speech, and Signal Processing*, ASSP-25:415–422, October 1977.

[Kai80]   J.F. Kaiser. On a simple algorithm to calculate the 'energy' of a signal. *Proc. IEEE International Conference on Acoustics, Speech, and Signal Processing*, pages 381–384, Albuquerque NM, April 1980.

[Kam86]   J.J. Van der Kam. A digital "decimating" filter for analog-to-digital conversion of hi-fi audio signals. *Philips Technical Review*, 42:230–238, 1986.

[Kan71]   E.P.F. Kan and J.K. Aggarwal. Error analysis in digital filters employing floating point arithmetic. *IEEE Trans. on Circuit Theory*, CT-18:678–686, November 1971.

[Kar83]   K. Karplus and A. Strong. Digital synthesis of plucked-string and drum timbres. *Computer Music Journal*, 7:43–55, Summer 1983.

[Kel62]   J.L. Kelly Jr. and C. Lochbaum. Speech synthesis. In *Proc. Stockholm Speech Communication Seminar*, Stockholm, Sweden, September 1962. Royal Institute of Technology.

[Kin72]   N. Kingsbury. Second-order recursive digital filter element for poles near the unit circle and the real-z axis. *Electronics Letters*, 8: 155-156, March 1972.

[Knu69]   D.E. Knuth. *The Art of Computer Programming: Volume 2 – Seminumerical Algorithms*. Addison-Wesley, Reading MA, 2nd edition, 1969.

[Kol77]   D.P. Kolba and T.W. Parks. A prime factor FFT algorithm using high speed convolution. *IEEE Trans. on Acoustics, Speech, and Signal Processing*, ASSP-25:281–294, August 1977.

[Kol91]   R.D. Kolipillai, T.Q. Nguyen, and P.P. Vaidyanathan. Some results in the theory of cross-talk free transmultiplexers. *IEEE Trans. on Signal Processing*, 39:2174–2183, October 1991.

[Kor93]   I. Koren. *Computer Arithmetic Algorithms*. Prentice Hall, Englewood Cliffs NJ, 1993.

[Kra94]   T.P. Krauss, L. Shure, and J. Little. *Signal Processing Toolbox for use with* MATLAB. The Mathworks Inc, South Natick MA, 1994.

[Kum93]   R. Kumaresan. Spectral analysis. In S.K. Mitra and J.F. Kaiser, editors, *Handbook for Digital Signal Processing*, chapter 16, pages 1143–1242. Wiley-Interscience, New York NY, 1993.

[Lag81]   R. Lagadec and H.O. Kunz. A universal digital sampling frequency converter for digital audio. In *Proc. IEEE International Conference on Acoustics, Speech, and Signal Processing*, pages 595–598, Atlanta GA, April 1981.

[Lag82]   R. Lagadec, D. Pelloni, and D. Weiss. A 2-channel, 16-bit digital sampling rate converter for professional digital audio. In *Proc. IEEE International Conference on Acoustics, Speech, and Signal Processing*, pages 93–96, April 1982.

[Lak96]   T.I. Laakso, V. Valimiki, M. Karjalainen, and U.K. Laine. Splitting the unit delay. *IEEE Signal Processing Magazine*, 13:30–60, January 1996.

[Lar93]   L.E. Larson and G.C. Temes. Signal conditioning and interface circuits. In S.K. Mitra and J.F. Kaiser, editors, *Handbook for Digital Signal Processing*, chapter 10, pages 677–720. Wiley-Interscience, New York NY, 1993.

[Lar99]   J. Laroche. A modified lattice structure with pleasant scaling properties. *IEEE Trans. on Signal Processing*, 47:3423–3425, December 1999.

[Lat98]   B.P. Lathi. *Signals Processing and Linear Systems*. Berkley-Cambridge, Carmichael CA, 1998.

[Law78]   V.B. Lawrence and K.V. Mina. Control of limit cycle oscillations in second-order recursive digital filters by constrained random quantization. *IEEE Trans. on Acoustics, Speech, and Signal Processing*, ASSP-26:127–134, February 1978.

[Ler83]   E.L. Lerner. Electronically synthesized music. *IEEE Spectrum*, 17:46–51, June 1983.

[Lev47]   N. Levinson. The Wiener RMS criterion in filter design and prediction. *J. Math. Phys.*, 25:261–278, 1947.

[Lim90]   J.S. Lim. *Two-Dimensional Signal and Image Processing*. Prentice Hall, Englewood Cliffs NJ, 1990.

[Lin87]   K.S. Lin, editor. *Digital Signal Processing Applications with the TMS320 Family*. Prentice Hall, Englewood Cliffs NJ, 1987.

[Lin96]   I-S. Lin and S.K. Mitra. Overlapped block digital filtering. *IEEE Trans. on Circuits and Systems, II: Analog and Digital Signal Processing*, 43:586–596, August 1996.

[Liu69]   B. Liu and T. Kaneko. Error analysis of digital filters realized in floating-point arithmetic. *Proc. IEEE*, 57:1735–1747, October 1969.

[Loc96]   G.R. Lockwood, P-C. Li, M. O'Donnell, and F.S. Foster. Optimizing the radiation pattern of sparse periodic linear arrays. *IEEE Trans. on Ultrasonics Ferroelectrics, and Frequency Control*, 43:7-14, January 1996.

[Lon73]   J.L. Long and T.N. Trick. An absolute bound on limit cycles due to roundoff errors in digital filters. *IEEE Trans. on Audio and Electroacoustics*, AU-21:27–30, February 1973.

[Lut91]   A. Luthra and G. Rajan. Sampling rate conversion of video signals. *SMPTE J.*, pages 869–879, November 1991.

[Lüt91]   H. Lütkepohl. *Introduction to Multiple Time Series Analysis*. Springer-Verlag, New York NY 1991.

[Mad98]   U. Madhow. Blind adaptive interference suppression for direct-sequence CDMA. *Proc. IEEE*, 86:2049–2068, October 1998.

[Mah82]   A. Mahanta, R.C. Agarwal, and S.C. Dutta Roy. FIR filter structures having low sensitivity and roundoff noise. *IEEE Trans. on Acoustics, Speech, and Signal Processing*, ASSP-30:913–920, December 1982.

[Mak75]   J. Makhoul. Linear prediction: A tutorial review. *Proc. IEEE*, 62:561–580, April 1975.

[Mar87]   S.L. Marple, Jr. *Digital Spectral Analysis with Applications*. Prentice Hall, Englewood Cliffs NJ, 1987

[Mar92]   A. Mar, editor. *Digital Signal Processing Applications Using the ADSP-2100 Family*. Prentice Hall, Englewood Cliffs NJ, 1992.

[Mas60]   S.J. Mason and H.J. Zimmerman. *Electronic Circuits, Signals and Systems*, pages 122–123. Wiley, New York, NY, 1960.

[Mee76]   K. Meerkötter. Realization of limit cycle-free second-order digital filters. In *Proc. 1976 IEEE International Symposium on Circuits and Systems*, pages 295–298, 1976.

[Mes82]   D.G. Messerschmitt. Echo cancellation in speech and data transmission. *IEEE J. on Selected Areas in Communications*, SAC-2:283–297, March 1982.

[Mik92]   N. Mikami, M. Kobayashi, and Y. Tokoyama. A new DSP-oriented algorithm for the calculation of the square-root using a nonlinear digital filter. *IEEE Trans. on Acoustics, Speech and Signal Processing*, 40:1663–1669, July 1992.

[Mil78]   W.L. Mills, C.T. Mullis, and R.A. Roberts. Digital filter realizations without overflow oscillations. *IEEE Trans. on Acoustics, Speech, and Signal Processing*, ASSP-26:334–338, August 1978.

[Min85]   F. Mintzer. Filters for distortion-free two-band multirate filter banks. *IEEE Trans. on Acoustics, Speech, and Signal Processing*, ASSP-33:626–630, June 1985.

[Mit73a]   S.K. Mitra. On reciprocal digital two-pairs. *Proc. IEEE (Letters)*, 61:1647–1648, November 1973.

[Mit73b]   S.K. Mitra and R.J. Sherwood. Digital ladder networks. *IEEE Trans. on Audio and Electroacoustics*, AU-21:30–36, February 1973.

[Mit74a]   S.K. Mitra and K. Hirano. Digital allpass networks. *IEEE Trans. on Circuits and Systems*, CAS-21:688–700, 1974.

[Mit74b]   S.K. Mitra, K. Hirano, and H. Sakaguchi. A simple method of computing the input quantization and the multiplication round-off errors in digital filters. *IEEE Trans. on Acoustics, Speech, and Signal Processing*, ASSP-22:326–329, October 1974.

[Mit74c]   S.K. Mitra and R.J. Sherwood. Estimation of pole-zero displacements of a digital filter due to coefficient quantization. *IEEE Trans. on Circuits and Systems*. CAS-21:116–124, January 1974.

[Mit75]   S.K. Mitra, K. Hirano, and K. Furuno. Digital sine-cosine generator. In *Proc. Second Florence International Conference on Digital Signal Processing*, pages 142–149, Florence, Italy, September 1975.

[Mit77a]   S.K. Mitra and C.S. Burrus. A simple efficient method for the analysis of structures of digital and analog systems. *Archiv für Elektrotechnik und Übertrangungstechnik*, 31:33–36, 1977.

[Mit77b]   S.K. Mitra, P.S. Kamat, and D.C. Huey. Cascaded lattice realization of digital filters. *International Journal on Circuit Theory and Applications*, 5:3–11, 1977.

[Mit77c]   S.K. Mitra, K. Mondal, and J. Szczupak. An alternate parallel realization of digital transfer functions. *Proc. IEEE (Letters)*, 65:577–578, April 1977.

[Mit80]   S.K. Mitra. *An Introduction to Digital and Analog Integrated Circuits, and Applications*. Harper and Row, New York NY, 1980.

[Mit87]   S.K. Mitra, K. Hirano, and K. mensa-Ababio. Theory and applications of all-digital N-path filters. *IEEE Trans. on Circuits and Systems*. CAS-34:1045–1052, September 1987.

[Mit90a]   S.K. Mitra, K. Hirano, S. Nishimura, and K. Sugahara. Design of digital bandpass/bandstop digital filters with tunable characteristics. *Frequenz*, 44:117–121, March/April 1990.

[Mit90b]   S.K. Mitra, Y. Neuvo, and H. Roivainen. Design and implementation of recursive digital filters with variable characteristics. *International Journal on Circuit Theory and Applications*, 18:107–119, 1990.

[Mit93]   S.K. Mitra, A. Mahalanobis, and T. Saramäki. A generalized structural subband decomposition of FIR filters and its application in efficient FIR filter design and implementation. *IEEE Trans. on Circuits and Systems II: Analog and Digital Signal Processing*, 40:363–374, June 1993.

[Mit98a]   S.K. Mitra and A. Makur. Warped discrete Fourier transform. *Proc. IEEE Workshop on Digital Signal Processing*, Bryce UT, August 1998.

[Mit98b] S.K. Mitra and H. Babic. Partial-fraction expansion of rational z-transforms. *Electronics Letters*, 34:1726, 3 September 1998.

[Moo77]   J.A. Moorer. Signal processing aspects of computer music: A survey. *Proc. IEEE*, 65:1108–1137, August 1977.

[Moo79]   J.A. Moorer. About this reverberation business. *Computer Music Journal*, 3(2):13–28, 1979.

[Mun81]   D.C. Munson, Jr., and B. Liu. Narrowband recursive filters with error spectrum shaping. *IEEE Trans. on Circuits and Systems*, CAS-28:160–163, February 1981.

[Naw88]   S.H. Nawab and T.F. Quatieri. Short-time Fourier transform. In J.S. Lim and A.V. Oppenheim, editors, *Advanced Topics in Signal Processing*, chapter 6. Prentice Hall, Englewood Cliffs NJ, 1988.

[Neu84a] Y. Neuvo and S.K. Mitra. Complementary IIR digital filters. In *Proc. IEEE International Symposium on Circuits and Systems*, pages 234–237, Montreal, Canada, May 1984.

[Neu84b] Y. Neuvo, C-Y. Dong, and S.K. Mitra. Interpolated finite impulse response filters. *IEEE Trans. on Acoustics, Speech, and Signal Processing*, ASSP-32:563–570, June 1984.

[Oet75]   G. Oetken, T.W. Parks, and H.W. Schüssler. New results in the design of digital interpolators. *IEEE Trans. on Acoustics, Speech, and Signal Processing*, ASSP-23:301–309, June 1975.

[Opp75]   A.V. Oppenheim and R.W. Schafer. *Digital Signal Processing*. Prentice Hall, Englewood Cliffs NJ, 1975.

[Opp76]   A.V. Oppenheim, W.F.G. Mecklenbräuker, and R.M. Mersereau. Variable cutoff linear phase digital filters. *IEEE Trans. on Circuits and Systems*, CAS-23:199–203, 1976.

[Opp78]   A.V. Oppenheim, editor. *Applications of Digital Signal Processing*. Prentice Hall, Englewood Cliffs NJ, 1978.

[Opp83]   A.V. Oppenheim and A.S. Willsky. *Signals and Systems*. Prentice Hall, Englewood Cliffs NJ, 1983.

[Opp89]   A.V. Oppenheim and R.W. Schafer. *Discrete-Time Signal Processing*. Prentice Hall, Englewood Cliffs NJ, 1989.

[Orf96]   S.J. Orfanidis. *Introduction to Signal Processing*. Prentice Hall, Englewood Cliffs NJ, 1996.

[Orm61]   J.F.A. Ormsby. Design of numerical filters with applications to missile data processing. *Journal of ACM*, 8:440-466, July 1961.

[Pan65]   P.F. Panter. *Modulation, Noise, and Spectral Analysis*. McGraw-Hill, New York NY, 1965.

[Pap65]   A. Papoulis. *Probability, Random Variables, and Stochastic Processes*. McGraw-Hill, New York NY, 1965.

[Pap90]   P. Papamichalis. *Digital Signal Processing Applications with the TMS320 Family, Theory, Algorithms, and Implementations*, volume 3. Prentice Hall, Englewood Cliffs NJ, 1990.

[Par60]   E. Parzen. *Modern Probability Theory and Its Applications*. Wiley, New York NY, 1960.

[Par72]   T.W. Parks and J.H. McClellan. Chebyshev approximation for nonrecursive digital filters with linear phase. *IEEE Trans. on Circuit Theory*, CT-19:189–194, 1972.

[Par87]   T.W. Parks and C.S. Burrus. *Digital Filter Design*. Wiley, New York NY, 1987.

[Pat80]   R.K. Patney and S.C. Dutta Roy. A different look at round-off noise in digital filters. *IEEE Trans. on Circuits and Systems*, CAS-27:59–62, January 1980.

[Pee87]   P.Z. Peebles. *Probability, Random Variables, and Random Signal Principles*. McGraw-Hill, New York NY, 2nd edition, 1987.

[Pei98]   S-C. Pei and C-C. Tseng. A comb filter design using fractional-sample delay. *IEEE Trans. on Circuit and Systems*, 45:649–653, June 1998.

[Pel80]   A. Peled and A. Ruiz. Frequency domain data transmission using reduced computational complexity algorithm. In *Proc. IEEE International Conference on Acoustics, Speech and Signal Processing*, pages 964–967, Denver CO, April 1980.

[Por97]   B. Porat. *A Course in Digital Signal Principles*. Wiley, New York NY, 1997.

[Pou87]   K. Poulton, J.J. Corcoran, and T. Hornak. A 1-GHz 6-bit ADC system. *IEEE J. on Solid-State Circuits*, SC-22:962–970, December 1987.

[Pri80]   D.H. Pritchard. A CCD comb filter for color TV receiver picture enhancement. *RCA Review*, 41:3–28, 1980.

[Pro92]   J.G. Proakis and D.G. Manolakis. *Digital Signal Processing: Principles, Algorithms and Applications*. Prentice Hall, Englewood Cliffs NJ, 2nd edition, 1992.

[Rab69]   L.R. Rabiner, R.W. Schafer, and C.M. Rader. The chirp-z transform algorithm. *IEEE Trans. on Audio and Electroacoustics*, AU-17:86–92, June 1969.

[Rab73]   L.R. Rabiner. Approximate design relationships for lowpass FIR digital filters. *IEEE Trans. on Audio and Electroacoustics*, AU-21:456–460, 1973.

[Rab74a] L.R. Rabiner and R.W. Schafer. On the behavior of minimax relative error FIR digital differentiators. *Bell System Technical Journal*, 53:333-362, February 1974.

[Rab74b] L.R. Rabiner and R.W. Schafer. On the behavior of minimax relative error FIR digital Hilbert transformers. *Bell System Technical Journal*, 53:363-394, February 1974.

[Rab75]   L.R. Rabiner and B. Gold. *Theory and Application of Digital Signal Processing*. Prentice Hall, Englewood Cliffs NJ, 1975.

[Rab78]   L.R. Rabiner and R.W. Schafer. *Digital Processing of Speech Signals*. Prentice Hall, Englewood Cliffs NJ, 1978.

[Ram84]   T. Ramstad. Digital methods for conversion between arbitrary sampling frequencies. *IEEE Trans. Acoustics, Speech and Signal Processing*, ASSP-32:577–591, June 1984.

[Ram89]   V. Ramachandran. Determination of discrete transfer function from its real (or imaginary) parts on the unit circle. *IEEE Trans. Acoustics, Speech and Signal Processing*, ASSP-37:440–442, March 1989.

[Rao84]   S.K. Rao and T. Kailath. Orthogonal digital filters for VLSI implementation. *IEEE Trans. on Circuits and Systems*, CAS-31:933–945, 1984.

[Reg87a]   P.A. Regalia, S.K. Mitra, and J. Fadavi. Implementation of real digital filters using complex arithmetic. *IEEE Trans. on Circuits and Systems*, CAS-34:345–353, April 1987.

[Reg87b]   P.A. Regalia and S.K. Mitra. Tunable digital frequency response equalization filters. *IEEE Trans. Acoustics, Speech, and Signal Processing*, ASSP-35:118–120, February 1987.

[Reg87c]   P.A. Regalia and S.K. Mitra. A class of magnitude complementary loudspeaker crossovers. *IEEE Trans. on Acoustics, Speech, and Signal Processing*, ASSP-35:1509–1515, November 1987.

[Reg88]   P.A. Regalia, S.K. Mitra, and P.P. Vaidyanathan. The digital allpass network: A versatile signal processing building block. *Proc. IEEE*, 76:19–37, January 1988.

[Reg93]   P.A. Regalia. Special filter designs. In S.K. Mitra and J.F. Kaiser, editors, *Handbook for Digital Signal Processing*, chapter 13, pages 967–980. Wiley-Interscience, New York NY, 1993.

[Ren87]   M. Renfors and T. Saramäki. Recursive *n*-th band digital filters, Parts I and II. *IEEE Trans. on Circuits and Systems*, CAS-34:24–51, January 1987.

[Ric86]   E.B. Richardson and N.S. Jayant. Subband coding with adaptive prediction for 56 kbits/s audio. *IEEE Trans. Acoustics, Speech, and Signal Processing*, ASSP-34:691–696, August 1986.

[Rob80]   E.A. Robinson and S. Treitel. *Geophysical Signal Analysis*. Prentice Hall, Englewood Cliffs NJ, 1980.

[Rob82]   E.A. Robinson. A historical perspective of spectrum estimation. *Proc. IEEE*, 70:885-907, 1982.

[Ros75]   J.P. Rossi. Digital television image enhancement. *SMPTE J.*, 84:545–551, July 1975.

[Rot83]   J.H. Rothweiler. Polyphase quadrature filters, a new subband coding technique. In *Proc. IEEE International Conference on Acoustics, Speech and Signal Processing*, pages 1980–1983, Boston MA, April 1983.

[San67]   I.W. Sandberg. Floating-point-roundoff accumulation in digital filter realization. *Bell System Technical Journal*, 46:1775–1791, October 1967.

[Sar93]   T. Saramäki. Finite impulse response filter design. In S.K. Mitra and J.F. Kaiser, editors, *Handbook for Digital Signal Processing*, chapter 4, pages 155–278. Wiley-Interscience, New York NY, 1993.

[Sch62]   M.R. Schroeder. Natural sounding artificial reverberation. *Journal of the Audio Engineering Society*, 10:219–223, 1962.

[Sch72]   H.W. Schüssler. On structures for nonrecursive digital filters. *Archiv für Electrotechnik und Überstragunstechnik*, 26:255–258, June 1972.

[Sch75]   M. Schwartz and L. Shaw. *Signal Processing: Discrete Spectral Analysis, Detection, and Estimation*. McGraw-Hill, New York NY, 1975.

[Sch91]   R. Schreier. *Noise-Shaped Coding*. PhD thesis, University of Toronto, Toronto Canada, 1991.

[Sel96]   I.W. Selsnick, M. Lang, and C. S. Burrus. Constrained least square design of FIR filters without specified transition bands. *IEEE Trans. on Signal Processing*, 44:1879–1892, August 1996.

[Sel98]   I.W. Selsnick, M. Lang, and C. S. Burrus. A modified algorithm for constrained least square design of multiband FIR filters without specified transition bands. *IEEE Trans. on Signal Processing*, 46:497–501, February 1998.

[Sha81]   A.F. Shackil. Microprocessors and the MD. *IEEE Spectrum*, 18:45–49, April 1981.

[She95]   K. Shenoi. *Digital Signal Processing in Telecommunication*. Prentice Hall, Englewood Cliffs NJ, 1995.

[Sko62]   M.I. Skolnik. *Introduction to Radar Systems*. McGraw-Hill, New York NY, 1962.

[Skw65]   J.K. Skwirzynski. *Design Theory and Data for Electrical Filters*. Van Nostrand Reinhold, New York NY, 1965.

[Slu84]   R.J. Sluyter. Digitization of speech. *Philips Technical Review*, 41(7/8):201–223, 1983–84.

[Smi84]   M.J.T. Smith and T.P. Barnwell III. A procedure for designing exact reconstruction filter banks for tree-structured subband coders. In *Proc. IEEE Conf. on Acoustics, Speech, and Signal Processing*, pages 27.1.1–27.1.4, San Diego, CA, March 1984.

[Spa2000] S.M. Spangenberg, I. Scott, S. McLaughlin, G.J.R. Povey, D.G.M. Cruickshank, and P.M. Grant. An FFT-based approach for fast acquisition in spread spectrum communication systems. *Wireless Personal Communications*, 13:27–55, May 2000.

[Sta94]   H. Stark and J.W. Woods. *Probability, Random Processes, and Estimation Theory for Engineers*. Prentice Hall, Englewood Cliffs NJ, 2nd edition, 1994.

[Ste93]   K. Steiglitz. Mathematical foundations of signal processing. In S.K. Mitra and J.F. Kaiser, editors, *Handbook for Digital Signal Processing*, chapter 2, pages 57–99. Wiley-Interscience, New York NY, 1993.

[Ste96]   K. Steiglitz. *A Digital Signal Processing Primer*. Addison Wesley, Menlo Park CA, 1996.

[Sto66]   T.G. Stockham Jr. High speed convolution and correlation. *1966 Spring Joint Computer Conference, AFIPS Proc*, 28:229–233, 1966.

[Sto94]   G. Stoyanov and H. Clausert. A comprative study of first-order digital allpass filter sections. *Frequenz*, 48:221–226, September–October 1994.

[Stu88]   R.D. Strum and D.E. Kirk. *First Principles of Discrete Systems and Signal Processing*. Addison-Wesley, Reading MA, 1988.

[Swe96]   W. Sweldens. The lifting scheme: A custom-design construction of biorthogonal wavelets. *Applied and Computational Harmonic Analysis*, 3:186-200, 1996.

[Szc75]   J. Szczupak and S.K. Mitra. Detection, location, and removal of delay-free loops in a digital filter configuration. *IEEE Trans. Acoustics, Speech, and Signal Processing*, ASSP-23:558–562, December 1975.

[Szc88]   J. Szczupak, S.K. Mitra, and J. Fadavi-Ardekani. A computer-based synthesis method of structurally LBR digital allpass networks. *IEEE Trans. on Circuits and Systems*, CAS-35:755–760, 1988.

[Tem73]   G.C. Temes and S.K. Mitra, editors. *Modern Filter Theory and Design*. Wiley, New York NY, 1973.

[Tem77]   G.C. Temes and J.W. LaPatra. *Introduction to Circuit Synthesis and Design*. McGraw-Hill, New York NY, 1977.

[The92]   C.W. Therrien. *Discrete Random Signals and Statistical Signal Processing*. Prentice Hall, Englewood Cliffs NJ, 1992.

[Tho77]   T. Thong and B. Liu. Error spectrum shaping in narrowband recursive digital filters. *IEEE Trans. on Acoustics, Speech, and Signal Processing*, ASSP-25:200–203, 1977.

[Thu00]   S. Thurnhofer. Two-dimensional Teager filters. In S.K. Mitra and G. Sicuranza, editors, *Nonlinear Image Processing*, chapter 6. Academic Press, New York NY, 2000.

[Tom81]   W.J. Tompkins and J.G. Webster, editors. *Design of Microcomputer-Based Medical Instrumentation*. Prentice Hall, Englewood Cliffs NJ, 1981.

[Tri79]   J.M. Tribolet. *Seismic Applications of Homomorphic Signal Processing*. Prentice Hall, Englewood Cliffs NJ, 1979.

[Unv75]   Z. Unver and K. Abdullah. A tighter practical bound on quantization errors in second-order digital filters with complex conjugate poles. *IEEE Trans. on Circuits and Systems*, CAS-22:632–633, July 1975.

[Urk58]   H. Urkowitz. An extension to the theory of airborne moving target indicators. *IRE Trans*, ANE-5:210–214, December 1958.

[Vai84]   P.P. Vaidyanathan and S.K. Mitra. Low passband sensitivity digital filters: A generalized viewpoint and synthesis procedures. *Proc. IEEE*, 72:404–423, 1984.

[Vai85a]   P.P. Vaidyanathan. The doubly terminated lossless digital two-pair in digital filtering. *IEEE Trans. on Circuits and Systems*, CAS-32:197–200, 1985.

[Vai85b]   P.P. Vaidyanathan and S.K. Mitra. Very low-sensitivity FIR filter implementation using "structural passivity" concept. *IEEE Trans. on Circuits and Systems*, CAS-32:360–364, April 1985.

[Vai85c]   P.P. Vaidyanathan. On power-complementary FIR filter. *IEEE Trans. on Circuits and Systems*, CAS-32:1308–1310, December 1985.

[Vai86a]   P.P. Vaidyanathan, S.K. Mitra, and Y. Neuvo. A new approach to the realization of low sensitivity IIR digital filters. *IEEE Trans. on Acoustics, Speech, and Signal Processing*, ASSP-34:350–361, April 1986.

[Vai86b]   P.P. Vaidyanathan. Passive cascaded lattice structures for low-sensitivity FIR filter design, with application to filter banks. *IEEE Trans. on Circuits and Systems*, CAS-33:1045–1064, November 1986.

[Vai87a]   P.P. Vaidyanathan and T.Q. Nguyen. Eigenfilters: A new approach to least-squares FIR filter design and applications including Nyquist filters. *IEEE Trans. on Circuits and Systems*, CAS-34:11–23, January 1987.

[Vai87b]   P.P. Vaidyanathan and T.Q. Nguyen. A 'TRICK' for the design of FIR half-band filters. *IEEE Trans. on Circuits and Systems*, CAS-34:11–23, January 1987.

[Vai87c]   P.P. Vaidyanathan. Low-noise and low-sensitivity digital filters. In D.F. Elliot, editor, *Handbook of Digital Signal Processing*, chapter 5. Academic Press, New York NY, 1987.

[Vai87d]   P.P. Vaidyanathan. Quadrature mirror filter banks, *M*-band extension and perfect reconstruction techniques. *IEEE ASSP Magazine*, 4:4–20, 1987.

[Vai87e]   P.P. Vaidyanathan and S.K. Mitra. A unified structural interpretation and tutorial review of stability test procedures for linear systems. *Proc. IEEE*, 75:478–497, April 1987.

[Vai87f]   P.P. Vaidyanathan, P.A. Regalia, and S.K. Mitra. Design of doubly-complementary IIR digital filters using a single complex allpass filter, with multirate applications. *IEEE Trans. on Circuits and Systems*, CAS-34:378–389, April 1987.

[Vai88a]   P.P. Vaidyanathan and P.Q. Hoang. Lattice structures for optimal design and robust implementation of two-channel perfect reconstruction QMF banks. *IEEE Trans. on Acoustics, Speech, and Signal Processing*, ASSP-36:81–84, January 1988.

[Vai88b]   P.P. Vaidyanathan and S.K. Mitra. Polyphase networks, block digital filtering, LPTV systems, and alias-free QMF banks: A unified approach based on pseudo-circulants. *IEEE Trans. on Acoustics, Speech, and Signal Processing*, ASSP-36:381–391, March 1988.

[Vai89]   P.P. Vaidyanathan, T.Q. Nguyen, Z. Doganata, and T. Saramäki. Improved technique for design of perfect reconstruction FIR QMF filter banks with lossless polyphase matrices. *IEEE Trans. on Acoustics, Speech, and Signal Processing*, ASSP-37:1042–1046, July 1989.

[Vai90]   P.P. Vaidyanathan. Multirate digital filters, filter banks, polyphase networks, and applications: A tutorial. *Proc. IEEE*, 78:56–93, January 1990.

[Vai93]   P.P. Vaidyanathan. *Multirate Systems and Filter Banks*. Prentice Hall, Englewood Cliffs NJ, 1993.

[Vet88]   M. Vetterli. Running FIR and IIR filtering using multirate filter banks. *IEEE Trans. on Acoustics, Speech and Signal Processing*, ASSP-36:381–391, May 1988.

[Vet89]   M. Vetterli and D. LeGall. Perfect reconstruction filter banks: Some properties and factorization. *IEEE Trans. on Acoustics, Speech and Signal Processing*, 37:1057–1071, July 1989.

[Vet95]   M. Vetterli and J. Kovacevic. *Wavelets and Subband Coding*. Prentice Hall, Englewood Cliffs NJ, 1995.

[Vit95]   A.J. Viterbi. *Principles of Spread Spectrum Communication*. Addison-Wesley, Reading MA, 1995.

[Vla69]   J. Vlach. *Computerized Approximation and Synthesis of Linear Networks*. Wiley, New York NY, 1969.

[Wei69a]   C.J. Weinstein and A.V. Oppenheim. A comparison of roundoff noise in fixed point and floating point digital filter realizations. *Proc. IEEE*, 57:1181–1183, June 1969.

[Wei69b]   C.J. Weinstein. Roundoff noise in floating point fast Fourier transform computation. *IEEE Trans. on Audio and Electroacoustics*, AU-17:209–215, September 1969.

[Wel67]   P.D. Welch. The use of fast Fourier transform for the estimation of power spectra: A method based on time averaging over short modified periodograms. *IEEE Trans. on Audio and Electroacoustics*, AU-15:70–73, 1967.

[Wel69]   P.D. Welch. A fixed-point fast Fourier transform error analysis. *IEEE Trans. on Audio and Electroacoustics*, AU-17:153–157, June 1969.

[Whi58]   W.D. White. Synthesis of comb filters. *Proc. National Electronics Conference*, pages 279–285, 1958.

[Whi71]   S.A. White. New method of synthesizing linear digital filters based on convolution integral. *IEE Proc. (Corr.)*, 118:348, February 1971.

[Wid56]   B. Widrow. A study of rough amplitude quantization by means of Nyquist sampling theory. *IRE Trans. on Circuit Theory*, CT-3:266–276, December 1956.

[Wid61]   B. Widrow. Statistical analysis of amplitude-quantized sampled-data systems. *AIEE Trans. (Appl. Industry)*, 81:555–568, January 1961.

[Wil95]   A. Williams and F.J. Taylor. *Electronic Filter Designer's Handbook*, McGraw-Hill, New York NY, 3rd edition, 1995.

[Wor89]   J.M. Worham. *Sound Recording Handbook*. Howard W. Sams, Indianapolis IN, 1989.

[Yan82]   G-T. Yan and S.K. Mitra. Modified coupled form digital-filter structures. *Proc. IEEE (Letters)*, 70:762–763, July 1982.

[Yel96]   D. Yellin and E. Weinstein. Multichannel signal separation: Methods and analysis. *IEEE Trans. on Acoustics, Speech and Signal Processing*, ASSP-44:106–118, January 1996.

[Yu90]   T-H. Yu, S.K. Mitra, and H. Babic. Design of linear-phase FIR notch filters. *Sadhana*, 15:133–155, November 1990.

[Zie83]   R.E. Ziemar, W.H. Tranter, and D.R. Fannin. *Signals and Systems: Continuous and Discrete*. Macmillan, NY, 1983.

[Zve67]   A.I. Zverev. *Handbook of Filter Synthesis*. Wiley, New York NY, 1967.

Index